Nobel Laureates in Physiology/Medicine

Nobel Laureate	Country*	Year of Award	Accomplishment
George Hoyt Whipple	United States	1934	Discoveries concerning liver therapy against anemias
George Richards Minot	United States		
William Parry Murphy	United States		
Hans Spemann	Germany	1935	Discovery of the organizer effect in embryonic development
Henry Dale	England	1936	Chemical transmission of nerve impulses
Otto Loewi	Germany		
Albert Szent-Györgyi von Nagyrapolt	Hungary	1937	Discoveries of biological combustion
Corneille Heymans	Belgium	1938	Discovery of the role played by the sinus and aortic mechanisms in the regulation of respiration
Gerhard Domagk	Germany	1939	Discovery of the chemotherapeutic effects of prontosil
No Award		1940–1942	
Henrik Dam	Denmark	1943	Discovery of vitamin K
Edward A. Doisy	United States		
Joseph Erlanger	United States	1944	Functions of nerve fibers
Herbert Spencer Gasser	United States		
Sir Alexander Fleming	England	1945	Discovery and development of penicillin
Ernst Boris Chain	England (Germany)		
Sir Howard W. Florey	England (Australia)		
Hermann Joseph Muller	United States	1946	Discovery of the production of mutations by means of X-ray irradiation
Bernardo Alberto Houssay	United States	1947	Discovery of how glycogen is catalytically converted
Carl F. Cori	United States (Hungary)		
Gerty T. Cori	United States (Hungary)		
Paul Muller	Swtz.	1948	Discovery of the insect-killing properties of DDT
Walter Rudolf Hess	Switzerland	1949	Research on brain control of body
Egas Moniz	Portugal		
Edward Calvin Kendall	United States	1950	Discoveries concerning the suprarenal cortex hormones
Philip Showalter Hench	United States		
Tadeus Reichstein	Swtz. (Poland)		
Max Theiler	United States (South Africa)	1951	Development of a vaccine for yellow fever
Selman A. Waksman	United States	1952	Discovery and development of streptomycin
Fritz A. Lipmann	United States (Ger.)	1953	Studies on the metabolism of carbohydrates in cells
Hans Adolph Krebs	England (Germany)		
John F. Enders	United States	1954	Cultivation of polio viruses in cell culture
Thomas H. Weller	United States		
Frederick C. Robbins	United States		
Hugo Theorell	Sweden	1955	Studies on oxidation enzymes
Dickinson W. Richards, Jr.	United States	1956	Techniques in treating heart disease
Andre F. Cournand	United States (France)		
Werner Forssmann	Germany		
Daniel Bovet	Italy (Swtz.)	1957	Investigations on antihistamines
Joshua Lederberg	United States	1958	Studies on the biochemistry of microbial genetics
George W. Beadle	United States		
Edward L. Tatum	United States		
Severo Ochoa	United States (Spain)	1959	Discoveries of the mechanisms of synthesis of DNA and RNA
Arthur Kornberg	United States		
Sir R. Macfarlane Burnet	Australia	1960	Studies on immunologic tolerance
Peter Brian Medawar	England (Brazil)		
George von Bekesy	United States (Hungary)	1961	Studies on the function of cochlea
James D. Watson	United States	1962	Determination of the structure of DNA
Francis H. C. Cricks	England		
Maurice H. F. Wilkins	England		

Nobel Laureates in Physiology/Medicine

Nobel Laureate	Country*	Year of Award	Accomplishment
Alan Lloyd Hodgkin	Australia	1963	Research on nerve cells
Andrew Fielding Huxley	England		
John C. Eccles			
Konrad Emil Bloch	Germany	1964	Regulation of cholesterol and fatty acid metabolism
Feodor Lynen	Germany		
Francois Jacob	France	1965	Studies on the regulation of gene activity in the cell
Jacques Monod	France		
Andre Lwoff	France		
Charles Brenton Huggins	United States (Can.)	1966	Treatment of prostate cancer
Francis Peyton Rous	United States		Discovery of tumor-producing viruses
Haldan Keffer Hartline	United States	1967	Research on the function of the eye
George Wald	United States		
Raynar Arthur Granit	Sweden (Finland)		
Robert Holley	United States	1968	Discovery of the genetic codes for amino acids
Har Gobind Khorana	United States (India)		
Marshall W. Nirenberg	United States		
Max Delbruck	United States (Ger.)	1969	Studies on the mechanism of the viral infection of cells
Alfred D. Hershey	United States		
Salvadore E. Luria	United States (Italy)		
Julius Axelrod	United States	1970	Studies on the mechanism of neural transmission
Ulf S. Von Euler	Sweden		
Sir Bernard Katz	England (Germany)		
Earl W. Sutherland, Jr.	United States	1971	Research on how hormones work; discovery of cyclic AMP
Gerald M. Edelman	United States	1972	Elucidation of the nature and structure of antibody molecules
Rodney R. Porter	England		
Karl von Frisch	Austria	1973	Studies of individual and social behavior patterns
Konrad Lorenz	Austria		
Nikolaas Tinbergen	Netherlands		
George E. Palade	United States (Romania)	1974	Studies on cellular physiology
Christian de Duve	United States (England)		
Albert Claude	United States (Belgium)		
David Baltimore	United States	1975	Studies on the transformation of cells by tumor viruses
Howard Temin	United States		
Renato Dulbecco	United States (Italy)		
Baruch Blumberg	United States	1976	Discovery of the Australia antigen
D. Carleton Gajdusek	United States		Description of slow virus diseases
Rosalyn Yalow	United States	1977	Development of the radio-immunoassay procedure
Roger C. L. Guillemin	United States (France)		Synthesis of peptide hormones
Andrew V. Schally	United States (Poland)		
Daniel Nathans	United States	1978	Studies on restriction enzymes and their use in genetic engineering
Hamilton Smith	United States		
Werner Arber	Swtz.		
Allen M. Cormack	United States (South Africa)	1979	Developing CAT scan technique
Godfrey N. Hoursfield	England		
Baruj Benacerraf	United States	1980	Discovery of the histo-compatibility antigens used in tissue typing
George D. Snell	United States		
Jean Dausset	France		
Roger W. Sperry	United States	1981	Organization and function of the brain
David H. Habel	United States		
Tursten N. Wiesel	United States		
John R. Vane	England	1982	Function of prostaglandins
Sune K. Bergstrom	Sweden		
Bengt I. Samuelsson	Sweden		
Barbara McClintock	United States	1983	Jumping-genes
Ce'sar Milstein	Argentina	1984	Fusion of tumor cells with antibody cells creating hybrids which have rapid production of pure antibodies
Georges Köhler	West Germany		
Neils Jerne	Swtz.		
Michael S. Brown	United States	1985	Regulation of cholesterol metabolism
Joseph L. Goldstein	United States		

Concepts

of Human Anatomy
and Physiology

Concepts

of Human Anatomy
and Physiology

Kent M. Van De Graaff
Brigham Young University

Stuart Ira Fox
Pierce College

Wm. C. Brown Publishers
Dubuque, Iowa

Book Team

Edward G. Jaffe
Executive Editor

Nicholas J. Murray
Senior Developmental Editor

Lynne M. Meyers
Associate Editor

Barbara J. Grantham
Designer

Mary K. Sailer
Design Layout Assistant

Renée Pins
Art Production

Ann Helling
Production Editor

Faye M. Schilling
Senior Photo Editor

Vicki Krug
Permissions Editor

wcb group

Wm. C. Brown
Chairman of the Board

Mark C. Falb
President and Chief Executive Officer

wcb

Wm. C. Brown Publishers, College Division

Lawrence E. Cremer
President

James L. Romig
Vice-President, Product Development

David A. Corona
Vice-President, Production and Design

E. F. Jogerst
Vice-President, Cost Analyst

Bob McLaughlin
National Sales Manager

Craig Marty
Director of Marketing Research

Catherine M. Faduska
Director of Marketing Services

Marilyn A. Phelps
Manager of Design

Eugenia M. Collins
Production Editorial Manager

Mary M. Heller
Photo Research Manager

Cover Photo

Magnetic resonance scan of brain by Howard J. Sochurek
Photography by Bob Coyle

The credits section for this book begins on page 1056, and
is considered an extension of the copyright page.

To our friend, Edward G. Jaffe, for his vision of
the value of this text and his persuasive
encouragements to complete it

Contents

Brief Contents

Contents

Expanded Contents

Unit **2** **Support and
Movement of the
Human Body
190**

Unit **4 Regulation and
Maintenance of
the Human Body
614**

Preface

This text is designed to provide students studying anatomy and physiology with an organizational framework of unifying concepts. Too many introductory anatomy and physiology texts simply confront students with an array of terms and facts and the overwhelming task of memorizing and making sense of that information. Little effort is made to organize the knowledge into meaningful concepts and objectives that would facilitate understanding, learning, and application. Consequently, many students who have completed an introductory course in human anatomy and physiology view the subject as composed of trillions of trivial facts and a collection of isolated processes. From such a perspective, many of the dynamic and integrated interactions of the systems are neither recognized nor understood and appreciated. It is like having the "forest" obscured because of the attention paid to the individual "trees." Careful study, much thought, and the input of numerous reviewers have gone into the preparation of this text to insure that the principal concepts of the entire discipline of human anatomy and physiology are identified, stated succinctly, and explained clearly.

Anatomy and physiology are dynamic, applied sciences, which are changing rapidly as new discoveries are being made. There still remain gaps in our knowledge of the structure, function, and numerous dysfunctions of the human body. In this text, many of the bodily processes and conditions that are not yet fully understood are pointed out to the students. Students must understand that learning and knowledge are continuous and that scientific information is obtained through careful research and the interpretation of evidence. In many explanations of concepts, the supporting evidences, or data, are presented so that the students better understand the conclusions.

Concepts of Human Anatomy and Physiology is written for students who are taking a basic introductory course in anatomy and physiology and for course instructors. Both share a common desire for a quality textbook, but their concerns arise from different perspectives. *Students* desire a text that is readable, interesting, accurate, and explains factual information in such a way that it can be readily understood and applied. They desire a text that is well organized for easy reference in studying and reviewing. It also should be aesthetically pleasing with useful tables and quality illustrations to support and clarify the narrative.

Instructors desire an anatomy and physiology text that is based on the most current and scientifically accepted research and conclusions. There should be carefully planned organization (pedagogy) that utilizes sound learning aids effectively. Sufficient explanation should be given to each topic so that understanding becomes easy and meaningful. Instructors need a book that has a breadth of information to satisfy students with diverse backgrounds and interests. There should be enough information to challenge the ambitious students, but not so much detail that the others are discouraged. An anatomy and physiology text should be a source of informative visual pleasure, enticing students to learn while providing an intellectual adventure. The text must be organized and written to support the teaching innovations of instructors and not to limit their variations in approach and emphasis.

With the needs of students and instructors in mind, the fivefold objectives of *Concepts of Human Anatomy and Physiology* were formulated:

1. Provide a balanced human anatomy and physiology text in which the appropriate amount of developmental, chemical, cellular and histological, gross anatomical, physiological, and clinical information is presented to give students a complete and correlated perspective of the subject matter.
2. Identify and state clearly the basic concepts and accompanying learning objectives of the disciplines of human anatomy and physiology.
3. Provide an accurate and current text that is visually appealing, interesting, and readable.
4. Develop in the student a sufficient anatomical and physiological vocabulary so that he or she understands and is conversant with medical terminology.
5. Encourage proper care of the body so that a more healthy and productive life can be enjoyed.

Organization

The thirty chapters in this text are grouped into five units. Unit 1 provides a foundation for the study of anatomy and physiology. Homeostasis is introduced in this unit and is emphasized as the major organizing concept in understanding the structure and function of the body. The chapters in this unit are both comprehensive and contemporary so that those students who may have a weak background in biology and chemistry will be able to understand the physiological concepts that are presented later.

Unit 2 is concerned with the support, protection, and movement of the human body. The body's external support and protection are provided by the integumentary system, and the internal support and protection of certain bodily organs are provided by the skeletal system. Movement is possible at the joints of the skeleton as the associated skeletal muscles are contracted.

Unit 3 is devoted to the nervous system, endocrine system, and sensory organs. The concepts identified and discussed in these chapters are concerned with the integration and coordination of bodily functions and the perception of environmental stimuli.

Unit 4 discusses the structure and function of the circulatory, immune, respiratory, digestive, and urinary systems as they regulate and maintain bodily processes. The processes of metabolism and homeostasis are stressed in each of the chapters within this unit.

Unit 5 describes the male and female reproductive systems and the continuance of the human species through sexual reproduction. An overview is presented for the entire sequence of human life, including prenatal development and postnatal growth, development, and aging.

Pedagogical Aids

Each of the thirty chapters in this text contains a complete system of pedagogical aids, which organizes, stimulates an interest in, and greatly facilitates the study of anatomy and physiology. A student who anticipates and utilizes these aids will find the study of this material effective and more enjoyable.

Chapter Introductions. The chapter introductions contain an overview of the *contents of each chapter* in outline form, which will help the students locate sections dealing with particular topics. All of the *concepts* for each chapter are stated in the chapter introductions so that the students gain a perspective of the salient points that will be discussed. A list of the *key terms* with their *phonetic pronunciations* is also found at the beginning of each chapter.

Understanding Terminology. Where each technical term first appears in the narrative, it is set off in *boldface* or *italic type* and followed by a bracketed *phonetic pronunciation.* Many of the terms are further elucidated in footnotes explaining their *derivations.* In addition, the roots of each term can be identified by referring to the *glossary of prefixes and suffixes* on the inside of the front cover. Knowing the pronunciation, derivation, and roots of a scientific term will help the student retain its meaning.

Chapter Sections. Each chapter is divided into several *principal sections,* and each section is prefaced by a *concept statement* and a list of *learning objectives.* A concept is a carefully worded phrase that expresses the essence, or synthesis, of the voluminous information presented in a chapter section; it is the formulation of a unifying thought pertinent to the detail that follows. The learning objectives for a chapter section indicate the level of competency the student should obtain in order to understand the concept thoroughly and be able to make application in practical situations. *Review questions* at the end of each chapter section test for understanding of the concept and mastery of the learning objectives.

Clinical Information. Short paragraphs set off in boxes of colored type occur throughout each chapter. These *boxed clinical asides* contain information on the practical application of learning the preceding facts.

Selected developmental disorders, clinical procedures, and diseases or dysfunctions of the specific organ systems are described in the *clinical considerations sections* at the end of each chapter. Pathological changes are illustrated in many of the discussions.

Chapter Summaries. A summary, in outline form, at the end of each chapter presents, for the last time, the concepts and major ideas as well as the facts that support and elucidate them. Reading these *comprehensive summaries* is an excellent way for students to review the chapters.

Illustrations and Tables. Concepts of Human Anatomy and Physiology is designed to have the most *effective illustration program* of any anatomy and physiology text. Because anatomy and physiology are visual sciences, great care has been taken to construct completely labeled, informative figures that are rendered clearly and accurately. Each figure is integrated with the narrative of the text. The figures in each chapter include many original *full-color art* and two-color anatomical drawings. *Color graphics* are used to depict physiological processes where complex ideas need to be clarified. Microscopic anatomy is illustrated by *photomicrographs* at the light microscope level and electron micrographs from scanning and transmission electron microscopy. Carefully selected *photographs* are used throughout the text to provide a balanced perspective of the gross anatomy. *Color-coding* is used in certain places within the text as a pedagogical aid in learning. The bones of the skull, in chapter 9 for example, are color-coded so that each bone can be identified readily in the many renderings which are displayed. Two sets of *reference plates* are positioned in the text where they will be of most value for review purposes. One set, located at the end of Unit 1 (see page 182), illustrates the major internal organs of the body and their relative positions. The other set, found at the end of Unit 2 (see page 394), depicts important regions of the body as seen in human cadaver dissections. These photographs show the complexity and structural relationships of the body that can be appreciated only when seen in a human specimen.

Numerous *tables* have been specifically constructed for each chapter to summarize information and clarify complex data. These tables are placed near their related textual material or at the end of the chapter sections.

Appendixes, Glossary, and Index. The *Appendixes*, following chapter 30, contain valuable reference information. *Appendix A* contains the answers to the objective questions that appear at the end of each chapter (see page 1029). *Appendix B* lists the important scientific journals that publish current research relative to anatomy and physiology (see page 1030). *Appendix C* lists the medical and pharmacological abbreviations commonly used in clinical practice (see page 1031). *Appendix D* lists the clinical laboratory tests commonly performed on human blood and urine (see page 1032).

The *Glossary* provides pronunciations and definitions of the more important scientific and clinical terms used in the text. If a term has a commonly used synonym or eponym, it is also indicated in the glossary.

The comprehensive *Index* is complete and has numerous cross-references.

Supplementary Materials

Supplementary materials accompany the text and are designed to aid the students in their learning activities and help the instructors plan course work and presentations.

A *Student Study Guide to Accompany Concepts of Human Anatomy and Physiology* by Kent M. Van De Graaff contains chapter concepts and objectives, focus questions, mastery quizzes, study activities, and answer keys that correspond to the chapters of the text.

A *Laboratory Guide to Human Anatomy and Physiology: Concepts and Clinical Applications* by Stuart I. Fox and Kent M. Van De Graaff uses a concept approach to focus on learning activities, which are presented as laboratory exercises and are correlated with the lecture material presented in the text.

A *set of fifty color slides* that depicts important clinical aspects of human anatomy and physiology is also available. These slides correlate with the clinical information presented in the text and have a descriptive narrative explaining the conditions or diseases visualized on the slides.

A *set of sixty-four selected acetate transparencies* is available to instructors who adopt this text. These transparencies, which are taken from select illustrations in the text, are ideal supplements to classroom lectures or may be used as quiz material.

An *Instructor's Resource Manual and Test Item File* by Kerry Openshaw contains chapter overviews, possible student assignments and discussions of chapter concepts, lists of supplemental films, and directories of suppliers of audiovisual and laboratory materials. It also contains over fifty test items for each chapter of the text, designed to evaluate student comprehension of factual data, concepts, clinical situations, and applications of this knowledge.

Acknowledgments

Concepts of Human Anatomy and Physiology could not have been written without the support and encouragement of colleagues at our respective academic institutions. We wish to thank professors Richard W. Heninger, Robert J. Lyon, R. Ward Rhees, Alice Logrip, Chen-Hau Poon, Richard A. Heckmann, Wilford M. Hess, Lester Schneider, and James Sandoval for their professional counsel and example.

Sincere gratitude is expressed for the expert secretarial and typing help rendered by Kathlyn J. Loveridge and F. Ruth Beck. Kathy not only meticulously typed the entire manuscript, but also took a personal interest in helping us bring this work to completion. Others assisting in the preparation of the manuscript were Pamela Barker and Diane M. Zarate. For their efforts, we are truly indebted.

We wish to acknowledge friends who have assisted in specific ways. Douglas W. Hacking, M.D., and J. Phillip Freestone, M.D., provided professional input and reviewed the clinical portions of the text. Brent C. Chandler, M.D., provided many of the X rays used in the text. The photographs of surface anatomy were taken by Dr. Sheril D. Burton. The subjects for the photo sessions were Larry A. Porter, Richard N. Stapler, Jr., Rochelle Grames, Jamie Smith, Ryan L. Van De Graaff, Deon Dow, and Christina R. Iba. Several cadavers were dissected in preparing this text in order to obtain accurate descriptions of structures. A special thanks is extended to Edward Scott, Bruce E. Thomas, and David W. Hodges for their assistance in these dissections.

Quality illustrations for this text were a top priority. The talented artists assembled for this project were Ruth Krabach, William Loechel, Steven Moon, Diane Nelson, Michael P. Schenk, Thomas Sims, and Thomas Waldrop. It has been a pleasure working with these professionals, and we appreciate their tremendous contribution.

The editorial staff and production people at Wm. C. Brown Publishers have been superb to work with. Lawrence E. Cremer and James L. Romig were committed to this text from its conception. Our senior editor, Edward G. Jaffe, and his assistants, Nick Murray and Lynne Meyers, worked diligently throughout the writing and production of this text providing continual guidance and insight and urging all to achieve their very best. We are indebted to Elizabeth Blake for her dedicated and conscientious efforts rendered in copyediting the manuscript. Barbara Grantham designed the text from cover to cover with talented deftness. Ann Helling orchestrated its production, coordinating the efforts of many. Others at Wm. C. Brown Publishers who contributed many hours include Marilyn A. Phelps, Faye M. Schilling, Marcia Stout, and Vicki Krug. Wm. C. Brown Publishers assembled a review panel of competent anatomists and physiologists to review the manuscript. These professionals aided immeasurably, and we are especially grateful for their frank criticisms, comments, and encouragements: Donald S. Kisiel, Suffolk County Community College; Werner Wieland, Mary Washington College; Donna Van Wynsberghe, University of Wisconsin, Milwaukee; David E. Grosland, Iowa Central Community College; William C. Kleinelp, Jr., Middlesex County College; Roberta M. O'-Dell-Smith, University of New Orleans; and Michael P. Schenk, University of Texas Health Science Center, Dallas.

Finally, but most importantly, we wish to extend appreciation to our families for their patience and encouragement. Our wives, Karen Van De Graaff and Ellen Fox, have been supportive on a daily basis, for which we are grateful.

Kent Marshall Van De Graaff

Stuart Fox

Concepts

of Human Anatomy and Physiology

Unit 1 Orientation and Organization

The chapters of Unit I introduce the sciences of anatomy and physiology by explaining the basic organization and principal processes of the human body. The fundamental terminology for describing the structure and function of the body is presented. The various levels of body composition and organization, from the chemical and cellular to the tissue (histological) and organ systems, are defined and described. The theme of homeostasis is explained clearly to set the stage for understanding the body systems, which are discussed in the chapters that follow.

The chapters and principal concepts of Unit I follow:

Chapter 1—Introduction to Anatomy and Physiology

Anatomy and physiology are integrated and dynamic sciences, which have a precise, descriptive vocabulary of Greek and Latin derivation. The cells of the human body are highly organized into tissues and organ systems, functioning together to maintain homeostasis.

Chapter 2—Chemical Composition of the Body

The body is composed of atoms chemically bonded together to form molecules. The principal molecules in the body are proteins, nucleic acids, carbohydrates, and lipids.

of the Human Body

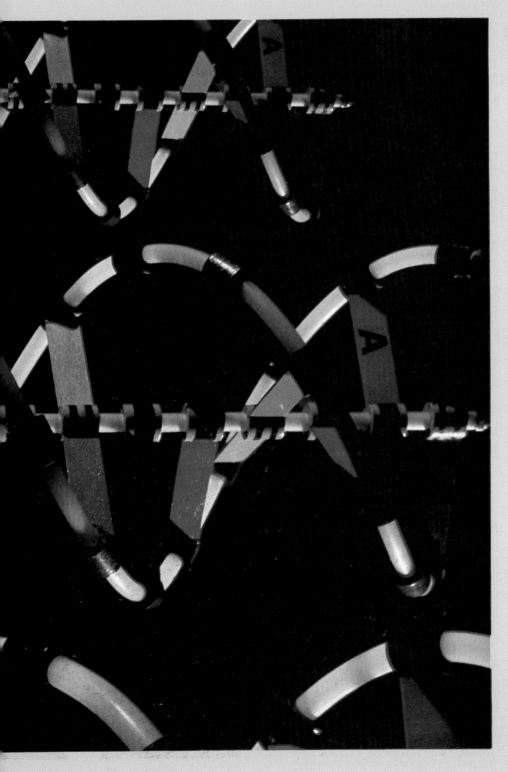

Chapter 3—Cell Structure and Genetic Regulation
Cellular structure and function depend upon the organelles contained within a cell and the controlling influence of the DNA within a cell's nucleus.

Chapter 4—Enzymes, Energy, and Metabolism
Cellular metabolism and the release of energy is dependent upon the appropriate extracellular environment, the availability of the proper nutrients to the cell, the removal of wastes, a constant optimal temperature, and the activity of enzymes.

Chapter 5—Membrane Transport and the Membrane Potential
The cell membrane is dynamic in selectively permitting the passage of materials through passive and active transport. The distribution of molecules and ions results in a potential difference, or voltage, across the membrane.

Chapter 6—Histology
Tissues are classified on the basis of their cellular composition and histological appearance into epithelial, connective, muscle, and nervous tissue; each type occurs in specific organs and performs specialized functions.

Chapter 7—Homeostasis and Control Mechanisms
Homeostasis in the body's organ systems is maintained by numerous and precise negative feedback mechanisms involving the actions of the endocrine and nervous systems.

1

Introduction to

Concepts

Anatomy and physiology are integrated
    sciences.

Anatomy and physiology are dynamic
    sciences that have long, exciting
    heritages and currently provide the
    foundation for medical, biochemical,
    developmental, cytogenetic, and
    biomechanical research.

Humans are biological organisms
    belonging to the phylum Chordata
    within the kingdom Animalia and to
    the family Hominidae within the
    class Mammalia and the order
    Primates.

The maintenance of the life of an
    organism depends upon the presence
    of certain vital factors for its
    existence.

Structural and functional levels of
    organization exist in the human body,
    with each of its parts contributing to
    the total organism.

As sciences, anatomy and physiology
    have descriptive terminologies, which
    must be mastered by students who
    would understand and be conversant
    in these subjects.

The human body is divided into regions
    and specific localized areas, which
    can be identified on the surface. Each
    region contains internal organs, the
    location of which are anatomically,
    physiologically, and clinically
    important.

For functional and protective purposes,
    the viscera are compartmentalized
    and supported in specific body
    cavities by connective and epithelial
    membranes.

All of the descriptive planes of reference
    and terms of direction used in
    anatomy are standardized because of
    their reference to the body in
    anatomical position.

Measurements are the human way of
    expressing relative lengths, masses,
    temperatures, and volumes of
    substances, which are important
    because they bring uniformity and
    consistency to scientific
    communication.

Anatomy and Physiology

Selected Terms

anatomy   (an-nat′o-me)
physiology   (fiz′′e-ol′o-je)
cadaver   (kah-dav′er)
necropsy   (nek′rop-se)
vivisection   (viv′′i-sek′shun)
phylogeny   (fi-loj′ĕny)
placenta   (plah-sen′tah)
metabolism   (mĕ-tab′o-lizm)
synthesis   (sin′thĕ-sis)
organelle   (or′′gah-nel′)
organism   (or′gah-nizm)
cranium   (kra′ne-um)
thoracic   (tho-ras′ik)
pelvic   (pel′′vik)
coelom   (see′lom)
mucous   (mu′kus)
serous   (se′rus)
pericardial   (per′′i-kar′de-al)
peritoneal   (per′′i-to-ne′al)
pleural   (ploo′ral)
parietal   (pah-ri′ĕ-tal)
visceral   (vis′er-al)

Definitions of the Sciences

Anatomy and physiology are integrated sciences.

Objective 1. Define anatomy and physiology.
Objective 2. Explain the concept of complementarity of organization between structure and function and the importance of knowing this concept in studying human anatomy and physiology.

Human anatomy and physiology are sciences concerned with the structure and function of the human body. The term **anatomy** *(an-nat′o-me)* is derived from a Greek word meaning "to cut up"; in the past, the word *anatomize* was more commonly used than the word *dissect*. The dissection of a human **cadaver** *(kah-dav′er)* has been the basis for understanding the structure of the human body for a long time. **Physiology** *(fiz′′e-ol′o-je)* is also derived from a Greek word meaning "the study of nature"—the "nature" of an organism is its function. Physiology is a dynamic science that attempts to explain physical and chemical processes occurring in the body. Much of the knowledge of physiology is gained through experimentation.

Anatomy and physiology are both subdivisions of the science of *biology*, the study of living organisms. Frequently, anatomy and physiology are studied as separate disciplines, in which case, learning the anatomy of the body precedes learning its physiology. Anatomical structure and function are complementary, however, and the simultaneous study of both facilitates learning about the body. The anatomy of every structure of your body is adapted for performing a function or perhaps several functions. Natural selection eliminates organisms with inappropriate structures and functions and determines which favorable structures will be passed from one generation to the next.

---

1. *Explain the statement that anatomy is a science of observation, and physiology is a science of experimentation and observation.*
2. *Use an example of a specific disease (such as lung cancer, which alters the anatomy of the lung) to show complementarity of organization between structure and function. Include in your answer how diagnosis (detection) of the disease is frequently determined by altered body functions.*

---

anatomy: Gk. *ana,* up; *tome,* a cutting
cadaver: L. *cadere,* to fall
physiology: Gk. *physis,* nature; *logos,* study
biology: Gk. *bios,* life; *logos,* study

Historical Perspective

Anatomy and physiology are dynamic sciences that have long, exciting heritages and currently provide the foundation for medical, biochemical, developmental, cytogenetic, and biomechanical research.

Objective 3. Discuss key historical events in the science of anatomy and physiology.

Objective 4. Relate how the scientific method is currently used in physiological experiments.

Objective 5. Explain why an understanding of anatomy and physiology is important and personally applicable.

Objective 6. List ways a person may remain informed of anatomical and physiological research, and discuss why this is important.

The sciences of human anatomy and physiology have had rich, long, and frequently troubled heritages. The histories of the sciences of human anatomy and physiology parallel that of medicine. In fact, an interest in the structure and function of the body has often resulted from the desire of the medical profession to explain a body dysfunction. Various religions, on the other hand, have stifled the study of human anatomy and physiology through their restrictions on human dissections and religious explanations of diseases and debilitations.

People have always had an innate interest in their own structure, function, and physical capabilities. The Greeks esteemed physical athletic competition and expressed the beauty of the body in their art. Many of the great masters of the Renaissance portrayed human figures in their art. Indeed, many of these artists were excellent anatomists because of their attention to detail. Such an artistic genius was Michelangelo, who was preoccupied with the detail and beauty of the human form and captured it in sculpture with the *David* (fig. 1.1) and in paintings such as those executed in the Sistine Chapel.

Even Shakespeare was awed by the structure of the human body as he wrote, "What a piece of work is a man! How noble in reason! How infinite in faculty! In form and moving how express and admirable! In action how like an angel! In apprehension how like a god! The beauty of the world! The paragon of animals" (*Hamlet* 2.2.315–319).

As a science, anatomy preceded that of physiology. It was necessary to describe the structures of the body before their functions could be determined. In addition, technology in the cognate sciences, like physics and chemistry, had to make sufficient progress so that physiological experiments could be conducted. Much current physiological research is ongoing as new technology is continuously being developed.

Figure 1.1.   Michelangelo completed the 17-foot-tall *David* in 1504, sculptured from a single block of white, unflawed Carrara marble. This masterpiece captures the physical beauty of the human body in an expression of art.

Figure 1.2. A fourteenth-century painting of the famous Greek physician Hippocrates. Hippocrates is referred to as the "father of modern medicine," and his creed is immortalized as the Hippocratic oath.

Figure 1.3. This Roman copy of a Greek sculpture is believed to be of Aristotle, the famous Grecian philosopher.

Historical Synopsis of Human Anatomy and Physiology

Grecian Period. Anatomy and physiology were first widely accepted as sciences in ancient Greece. The writings of several Greek philosophers had a tremendous impact on future scientific thinking. *Hippocrates* (460–337 B.C.) was the famous Grecian physician (fig. 1.2) who is regarded as the "father of modern medicine" because of the sound principles of medical practice that he established. Perhaps the greatest contribution of Hippocrates was that he attributed diseases to natural causes rather than to the displeasure of the gods. His application of logic and reason to medicine was the beginning of observational medicine.

Hippocrates probably had only a limited exposure to human dissections but was well disciplined in the popular humoral theory of body organization. Four body humors were recognized, and each was associated with a particular body organ: sanguine with the liver; choler, or yellow

bile, with the gallbladder; phlegm with the lungs; and melancholic, or black bile, with the spleen. A healthy person was thought to have a balance of the four humors. The concept of humors has long since been discarded, but it dominated medical thought for over two thousand years.

Aristotle (384–332 B.C.), a pupil of Plato, made careful investigations of all kinds of animals—he even included references to humans—and established a type of scientific method in obtaining data (fig. 1.3). He wrote the first known account of embryology, in which he described the development of the heart in a chick embryo. His best-known zoological works are *History of Animals, Parts of Animals,* and *Generation of Animals.*

In spite of his tremendous accomplishments, Aristotle perpetuated some erroneous views of anatomy. For example, the doctrine of the humors formed the boundaries of Aristotle's thought. Plato had described the brain as the seat of feeling and thought, but Aristotle disagreed.

physician: Gk. *physikos*, natural
sanguine: L. *sanguis*, bloody
choler: Gk. *chole*, bile

phlegm: Gk. *phlegm*, inflammation
melancholic: Gk. *melan*, black; *chole*, bile
humor: L. *humor*, fluid

He stated that the heart was the seat of intelligence and that the function of the brain, which was bathed in fluid, was to cool the blood that was pumped from the heart.

The Grecian scientist *Erasistratus* (about 300 B.C.) was more interested in body functions than structure and is, therefore, frequently referred to as the "father of physiology." Erasistratus authored a book on the causes of diseases, in which he included observations on the heart, vessels, brain, and cranial nerves. He noted the toxic effects of snake venom on various visceral organs and described changes in the liver resulting from various diseases. Although some of the writings of Erasistratus were very scientific, other concepts of his were primitive and mystical. He thought that the cranial nerves carried animal spirits and that muscles contracted because of distention by spirits.

Both *Erasistratus* and another Grecian philosopher, *Herophilus,* were greatly criticized later in history for the performance of *vivisection (viv"i-sek'shun).* Celsus (about A.D. 30) and Tertullian (about A.D. 200) were particularly critical of the practice of vivisection. Herophilus was described as a butcher of men who had dissected as many as six hundred living persons, some of them as public demonstrations.

Roman Era.   In many respects, the Roman Empire stifled scientific advancements and set the stage for the Dark Ages that were to follow. The interest and emphasis of science shifted from the theoretical to the practical during this time. Few dissections of cadavers were performed other than at *autopsies* in attempts to determine the cause of death in criminal cases. Medicine was not preventive but was almost solely limited to the treatment of soldiers injured in battle. Later in Roman history, laws were established that indicated the influence of the church on medical practice. According to Roman law, for example, no deceased pregnant women could be buried without prior removal of the fetus from the womb so that it could be baptized.

Claudius Galen (A.D. 130–201) was perhaps the best physician since Hippocrates. He was certainly the most influential writer of all times on medical subjects. For nearly fifteen hundred years, the writings of Galen were the unquestionable authority on anatomy and medical treatment. Galen probably dissected no more than two or three human cadavers during his career, of necessity limiting his anatomical descriptions to animal dissections. He compiled nearly five hundred medical papers (of which eighty-three have been preserved) from earlier works as

---

visceral: L. *viscus,* internal organ of torso
vivisection: L. *vivus,* alive; *sectio,* a cutting
autopsy: Gk. *autos,* self; *opsia,* viewing with one's eyes

---

well as his personal studies. Galen believed in the humors of the body and perpetuated this concept; he also gave authoritative explanations of nearly all body functions.

Galen's works contain many errors, primarily because of his desire to present definitive answers and his interpretation of data from nonhuman animals. He did, however, provide some astute and accurate anatomical details that are still regarded as classics. He proved to be an experimentalist, demonstrating that the heart of a pig would continue to beat when the spinal nerve was transected so that nerve impulses could not reach the heart. He showed that the squealing of a pig stopped when the particular nerve that innervated its vocal cords was cut. He also proved that arteries contained blood rather than pneuma.

Middle Ages.   The Middle Ages (Dark Ages) came with the fall of the Roman Empire in A.D. 476 and lasted nearly one thousand years. Dissections of cadavers were totally prohibited during this period, and molesting a corpse was a criminal act frequently punishable by burning at the stake. If mysterious deaths occurred, examinations by inspection and palpation were acceptable. During the plague epidemic in the sixth century, however, a few *necropsies (nek'rop-se)* and dissections were performed in hopes of determining the cause of this dreaded disease.

Renaissance.   The period of time known as the Renaissance was characterized by a rebirth of science. It lasted from the fourteenth through the sixteenth century and was a transitional period from the Middle Ages to the modern age of science. The development of movable type about 1450 revolutionized the production of printed books and helped to usher in the Renaissance.

With the increased interest in anatomy and physiology during the Renaissance, obtaining cadavers for dissection became a serious problem. Medical students regularly practiced grave robbing until finally an official decree was issued that permitted the bodies of executed criminals to be used as specimens.

The major advancement in anatomy during the Renaissance came from the artistic and scientific ability of *Andreas Vesalius* (1514–64). Vesalius apparently had enormous energy and ambitions. He completed the masterpiece of his life, *De Humani Corporis Fabrica,* by the time he was twenty-eight years old. The various body systems and individual organs were beautifully illustrated and described in the *Fabrica* (fig. 1.4). The magnitude of this work and the impact it had on science earned Vesalius the title of "father of modern anatomy." His book was especially important in that it boldly challenged hundreds of Galen's erroneous teachings. Vesalius wrote of his surprise at finding numerous anatomical errors that were

---

epidemic: Gk. *epi,* upon; *demos,* people
necropsy: Gk. *nekros,* corpse; *opsy,* view

Figure 1.4. Plates from *De Humani Corporis Fabrica,* which Vesalius completed at the age of twenty-eight. This book, published in 1543, revolutionized the sciences of anatomy and physiology.

Figure 1.5. The English physician, William Harvey, demonstrated with experiments in 1628 that blood circulates and does not flow back and forth through the same vessels.

taught as fact, and he refused to accept Galen's explanations on faith. Bitter controversies ensued between Vesalius and the traditional Galenic anatomists, including Vesalius's former teacher, Sylvius. Vesalius became so incensed by the relentless attack that he destroyed much of his later unpublished work and ceased his dissections.

Seventeenth and Eighteenth Centuries. The two most outstanding contributions to anatomy and physiology of the seventeenth and eighteenth centuries were the explanation of blood flow and the development and use of the microscope.

In 1628, the English physician *William Harvey* (1578–1657) published his outstanding work *On the Movement of the Heart and Blood in Animals.* This important research established brilliant proof of the continuous circulation of blood within contained vessels, and the technique of investigation presented in this publication is still regarded as a classic example of the scientific method for conducting research (fig. 1.5). Harvey is frequently regarded as "the father of modern physiology." Like Vesalius, Harvey was severely criticized for his departure from Galenic philosophy. The controversy over the circulation of the blood raged for twenty years until other anatomists finally repeated Harvey's experiments and added information.

Antony van Leeuwenhoek (1632–1723) was a Dutch lens grinder who improved the microscope to where he achieved a magnification of 270 times. His many contributions included developing techniques for tissue examination and describing blood cells, spermatozoa, and the striped appearance of skeletal muscle.

The development of the microscope added an entirely new dimension to anatomy and physiology and eventually led to explanations of basic body functions. In addition, the improved microscope was invaluable for understanding the etiologies (causes) of diseases and discovering cures for many of them.

Nineteenth and Twentieth Centuries. The major contribution in the nineteenth century was the formulation of the cell theory and the implications it had for a clearer understanding of the structure and functioning of the body. Once the microscope was invented, it was merely a matter of time before cells were discovered and described.

Johannes Müller (1801–58) is noted for applying the sciences of physics, chemistry, and psychology to the study of the human body. With this increased dimensional breadth to the subject, anatomy and physiology became comparative sciences.

The study of anatomy and physiology during the twentieth century became specialized, and the research more detailed and complex. In response to the increased technology and depths of understanding, new disciplines and specialities have appeared in the sciences of anatomy and physiology to categorize and use the new knowledge (fig. 1.6).

The Scientific Method

The Nobel laureates for anatomy and physiology research are listed on the inside of the front cover of this text. In addition to having an intense interest in biology and humanity, these distinguished researchers all employed the processes of scientific inquiry in conducting their investigations and making discoveries. Although there may be considerable creativity within the boundaries of these investigative processes, scientific experiments progress in well-defined, orderly ways. Such a disciplined approach for the acquisition of truth is referred to as the *scientific method*. Simply stated, the scientific method depends on a number of processes of gaining information in orderly ways. Actually, the scientific method may be used in searching for answers to questions encountered in everyday life as well as in methodical research.

All of the information in this text has been gained by the application of the scientific method. Although many different techniques are involved in the application of the scientific method, all share three attributes: *(1)* confidence that the form and function of the natural world, including ourselves, are ultimately explainable in terms we can understand; *(2)* descriptions and explanations of the natural world that are honestly based on observations and that could be modified or refuted by other observations; and *(3)* humility—the willingness to accept the fact that we could be wrong. If the study of the universe and its parts should yield conclusions that refute part or all of an idea, the idea must be accordingly modified. In short, the scientific method is based on a confidence in our rational ability, honesty, and humility. Practicing scientists may not always display these attributes, but the validity of the large body of scientific knowledge that has been accumulated—as shown by the technological applications and the predictive value of scientific hypotheses—are ample testimony to the fact that the scientific method works.

The actual techniques used in the scientific method vary. In some fields, such as anatomy, scientists carefully and extensively observe a structure or process and attempt to organize their observations in a logical way. These methods help to answer the question, "How is it put together?" In other fields, such as physiology, scientists perform experiments to answer the question, "How does it work?" This question is never answered by a single experiment. Instead, many thousands of experiments must be performed, because each experiment is designed to answer only a very small question: "Is it true (or false) that . . .?" Because such questions can only be asked after certain knowledge is obtained, the knowledge gained from experiments always leads to more questions requiring further experimentation. In this way, a large body of highly specialized information is gradually accumulated, and a scientific explanation of how "it" works can gradually be formulated. This explanation will almost always be different from preconceived notions. People who follow the scientific method will then appropriately modify their concepts, realizing that their new ideas will probably have to be changed again in the future as additional experiments are performed.

New concepts that emerge from the scientific method are not always welcome; people often prefer comfortable, older ideas, particularly when these are part of a large, cherished belief structure. In these cases, the knowledge gained by the scientific method may be rejected, and the scientists involved vilified, persecuted, or even murdered. Because the open communication of knowledge is a scientific tradition, however, those ideas that are closest to the truth have survived and eventually supplanted their competitors in the minds of most educated people.

Figure 1.6. Some of the subdivisions or specialities of human anatomy and physiology.

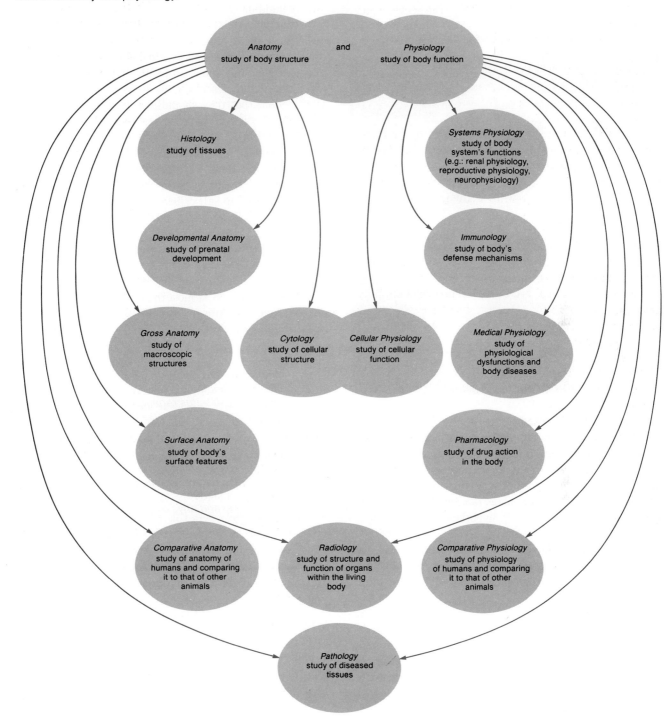

Figure 1.7.   A positron emission tomography (PET) scan of the brain following an injection of radioactive glucose. The left side of the brain, concerned with cognitive activity, and the frontal lobes, concerned with planning and organization of behavior, are seen as active areas in this PET scan.

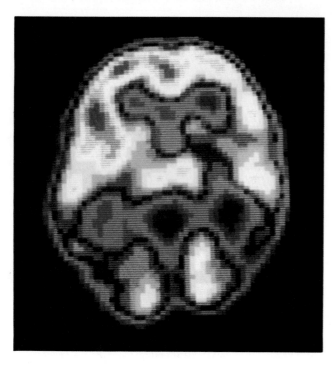

Anatomy and Physiology Today

As depicted in figure 1.6, anatomy and physiology form the bases for numerous scientific and medical disciplines. Anatomy and physiology are dynamic, applied sciences, which are changing constantly as discoveries are made. Innovative techniques and ideas for conducting research provide data for formulating hypotheses in attempts to understand physiological processes or etiologies of diseases. A positron emission tomography (PET) scan, for example, is used to quantify the relative metabolic activity within different parts of an organ such as the brain (fig. 1.7).

With such a proliferation of scientific knowledge, the question is, "How does one keep up in the dynamic subject of anatomy and physiology?" The answer is different for the scientist and the student. Being on the forefront of discovery, the scientist must specialize and stay current in a narrow realm of inquiry. The scientist collaborates with professional colleagues, conducts research, presents findings at scientific meetings, and publishes new discoveries in scientific journals (see appendix B).

In order for a student of anatomy and physiology to keep up, it is first necessary to establish a solid and broad understanding of the profession. This text as used in an introductory course is designed to provide a foundation for the student. A student must learn sufficient facts so that concepts can be understood and applied in a cognitive fashion. Not only should students of anatomy and physiology be able to describe a body structure and know how it functions, but they should also be able to predict the dysfunctions if certain processes were not to occur.

An objective of this text is to enable students to become educated and conversant in anatomy and physiology. An excellent way for students to keep up with anatomy and physiology during and after the formal course is completed is to subscribe to and read scientific magazines such as *Science, Scientific American, Discover,* or *Science 86.* These publications and others have many articles on scientific discoveries written in an understandable dialogue. There are many specialities of anatomy and physiology that are undergoing a rapid proliferation of exciting knowledge. It is most important and interesting to become and stay informed so that you may be an educated contributor to society.

1. *Define* vivisection, necropsy, *and* autopsy.
2. *What effect did religion have on the early interest in anatomy?*
3. *Why is Erasistratus regarded as the "father of physiology" and Vesalius the "father of anatomy"?*
4. *Explain the importance of experimentation and careful observation in gaining new knowledge. Why must a scientific experiment be performed many times before it is reported as fact?*
5. *How can a student stay current in the sciences of anatomy and physiology? Why is it important to do so?*

Classification and Characteristics of Humans

Humans are biological organisms belonging to the phylum Chordata within the kingdom Animalia and to the family Hominidae within the class Mammalia and the order Primates.

Objective 7.   List the taxonomic classification of humans.

Objective 8.   List the characteristics that identify humans as chordates and as mammals.

Objective 9.   Describe the characteristics that humans have but other primates do not.

Figure 1.8.   A schematic diagram of the front part of the embryo of a chordate.

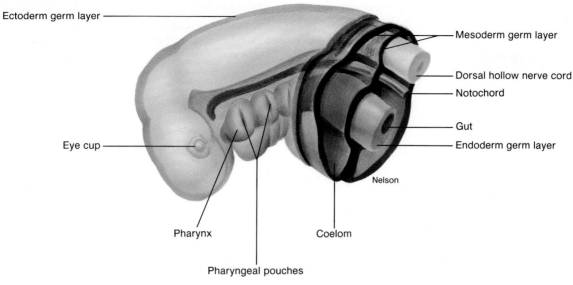

Figure 1.9.   A midsagittal section through vertebrate to show the intervertebral disc and nucleus pulposus.

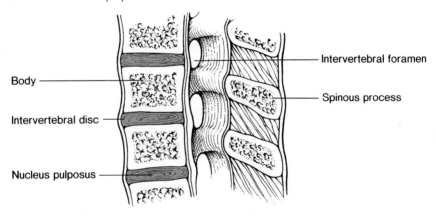

The classification, or taxonomic, scheme has been established by biologists to organize the structural and evolutionary relationships of living organisms. Each category of classification is referred to as a *taxon*. The highest taxon is the kingdom and the most specific taxon is the species. Humans are species belonging to the **animal kingdom.** *Phylogeny (fi-loj'ĕny)* is the origin and evolutionary development of animal species.

Phylum Chordata

Human beings belong to the phylum Chordata *(fi'lum kor-dah'tah)* along with fish, amphibians, reptiles, birds, and other mammals. All chordates have three structures in common: a **notochord** *(no'to-kord),* a **dorsal hollow nerve cord,** and **pharyngeal pouches** *(fah-rin'je-al)* (fig. 1.8). These chordate characteristics are well expressed during

the embryonic stage of development and, to a certain extent, are present in an adult. The notochord is a flexible rod of tissue that extends the length of the back of an embryo. A portion of the notochord persists in the adult as the **nucleus pulposus,** located within each **intervertebral disc** (fig. 1.9). The dorsal hollow nerve cord is positioned above the notochord and develops into the **brain** and **spinal cord,** which are highly functional as the **central nervous system** in the adult. Pharyngeal pouches form gill openings in fish and some amphibians. In other chordates, such as humans, embryonic pharyngeal pouches develop, but only one of the pouches persists, becoming the **auditory (eustachian) canal** *(u-sta'ke-an),* a connection between the middle ear and **pharynx** *(far'ingks)* (throat area).

Figure 1.10. Characteristics of mammals—the class to which humans belong.

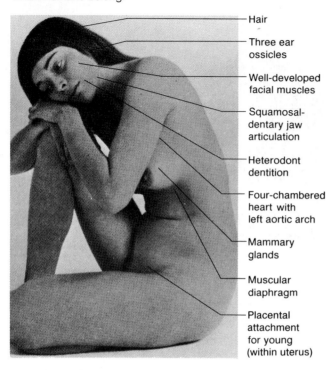

- Hair
- Three ear ossicles
- Well-developed facial muscles
- Squamosal-dentary jaw articulation
- Heterodont dentition
- Four-chambered heart with left aortic arch
- Mammary glands
- Muscular diaphragm
- Placental attachment for young (within uterus)

Table 1.1 Classification scheme of human beings

Taxon	Designated grouping	Characteristics
Kingdom	Animalia	Eucaryotic cells that lack walls, plastids, and photosynthetic pigments
Phylum	Chordata	Dorsal hollow nerve cord; notochord; pharyngeal pouches
Subphylum	Vertebrata	Vertebral column
Class	Mammalia	Mammary glands; hair
Order	Primates	Well-developed brain; prehensile hands
Family	Hominidae	Large cerebrum; bipedal locomotion
Genus	*Homo*	Flattened face; prominent chin and nose with inferiorly positioned nostrils
Species	*sapiens*	Largest cerebrum

Source: Van De Graaff, Kent M., *Human Anatomy.* © 1984 Wm. C. Brown Publishers, Dubuque, Iowa. All Rights Reserved. Reprinted by permission.

Class Mammalia

Mammals are chordate animals with hair and mammary glands. Hair is a thermoregulatory protective covering for most mammals, and mammary glands serve for suckling the young (fig. 1.10). Other characteristics of mammals include three ear ossicles (bones), heterodont dentition (differently shaped teeth), squamosal-dentary jaw articulation (a joint between lower jaw and skull), an attached placenta *(plah-sen'tah),* well-developed facial muscles, a muscular diaphragm, and a four-chambered heart with a left aortic arch.

Order Primates

There are several subdivisions of closely related groupings of mammals, called orders. Humans, along with lemurs, monkeys, and great apes, belong to the order called Primates. Members of this order have prehensile hands, digits modified for grasping, and relatively large, well-developed brains.

Family Hominidae

Humans are the sole living members of the family Hominidae. *Homo sapiens* is included within this family to which all the varieties or ethnic groups of humans belong.

heterodont: Gk. *heteros,* other; *odontos,* tooth
placenta: L. *placenta,* flat cake
Primates: L. *primas,* first
prehensile: L. *prehensus,* to grasp

Each "racial group" has distinguishing features that have been established in isolated populations over thousands of years. Our classification pedigree is presented in table 1.1.

Characteristics of Humans

We, as human beings, have a few anatomical characteristics that are so specialized that they are diagnostic in separating us from other animals and even from other closely related mammals. We also have characteristics that are equally well developed in other animals, but when these function with the human brain, they provide remarkable capabilities. Our anatomical characteristics include the following:

1. **Size and development of the brain.** The average human brain weighs between 1,350 and 1,400 g. This gives humans a large brain-to-body-weight ratio. But more important is the development of portions of the brain. Certain extremely specialized regions and structures within the brain account for emotion, thought, reasoning, memory, and precise, coordinated movement.
2. **Style of locomotion.** Because humans stand and walk on two appendages, our style of locomotion is said to be bipedal. Upright posture imposes certain other diagnostic structural features such as the sigmoid (S-shaped) curvature of the spine, the anatomy of the hip and thighs, and arched feet. Some of these features may cause clinical problems in older individuals.
3. **Opposable thumb.** The human thumb joint is structurally adapted for tremendous versatility in grasping objects. Most primates have opposable thumbs.
4. **Vocal structures.** Humans, like no other animals, have developed articulated speech. The anatomical structure of the vocal organs and the well-developed brain have made this possible.

5. **Stereoscopic vision.** Although this characteristic is well developed in several other animals, it is also keen in humans. Our eyes are directed forward so that when we focus upon an object, we view it from two angles. Stereoscopic vision gives us depth perception, or a three-dimensional image.

We also differ from other animals in the number and arrangement of vertebrae (vertebral formula), the kinds and number of teeth (tooth formula), well-developed facial muscles, and the structural specializations of various body organs.

---

1. What is a chordate? Why are humans considered members of the phylum Chordata?
2. Why are humans designated as mammals and primates? What characteristics distinguish humans from other primates?
3. Which of the characteristics of humans are adaptive for social organization?

Maintenance of Life

The maintenance of the life of an organism depends upon the presence of certain vital factors for its existence.

Objective 10. List and describe the principal requisites for life.

Objective 11. Discuss the physical needs of an organism to sustain life.

Requisites for Life

Humans, being living organisms, have several traits that are requisite for life. Some of the requisites for life are dynamic chemical processes that occur continuously. Such chemical interactions and reactions are referred to as *metabolism (mě-tab'o-lizm)*. Other requisites for life are organizational; that is, they relate to the structural complexity of an organism. The requisites for life include the following:

1. **Development.** It is during the developmental process, called **morphogenesis** *(mor''fo-jen'ĭ-sis)* that undifferentiated (unspecialized) tissues undergo cellular differentiation, resulting in organs that are structurally and functionally distinct.
2. **Organization.** Each body structure is genetically determined and adapted to carry out specific functions.
3. **Adaptation.** Just as an organism must have a tolerance range in which it can be functional, structures in the body have tolerance ranges for each of the physical needs. Narrow levels of tolerance result in a structure or organism being highly specialized and less adaptable than those that have wide levels of tolerance.

4. **Responsiveness.** Responsiveness is the ability of an organism (or a cell) to monitor internal or external conditions and respond to changes that might be threatening. A white blood cell, for example, will respond to toxins from bacteria and move toward the pathogens in an attempt to dispose of them. Pulling a hand away from a hot object demonstrates responsiveness, as does drinking water to satisfy thirst.
5. **Movement.** It is the movement of cells or within cells that results in the pumping of blood, respiratory movements, and even changes in an organism's position from one place to another as muscle fibers are stimulated to contract. Even nerve impulses result from the movements of ions across cell membranes.
6. **Reproduction.** Reproduction refers to both cellular replication, important for body growth and repair, and sexual reproduction when parents propagate (participate in having offspring) a new individual.
7. **Growth.** Growth refers to processes of both the cells and the organism. After cellular replication, it is important for the new (daughter) cells to grow and mature in order to be functional. Likewise, there is a genetically determined growth process of an organism when body structures reflect metabolic needs.
8. **Respiration.** Respiration refers to all of the processes that permit the exchange of oxygen and carbon dioxide between the body cells and the external environment and to the utilization of oxygen and production of carbon dioxide by the tissue cells.
9. **Digestion and absorption.** Digestion and absorption is the process by which complex food substances are chemically altered into simpler forms that pass through membranes lining the digestive tract into the body fluids.
10. **Circulation.** A complex, multicellular organism must have the means for all of its living cells to receive continuous sustenance; this is provided by the circulation of body fluids, which transports nutrients and oxygen to cells and removes metabolic wastes.
11. **Synthesis.** Synthesis *(sin'thě-sis)* is the process by which simple molecules are united to form more complex molecules, such as DNA, RNA, proteins, and fats.
12. **Secretion.** Secretions are specific chemical compounds synthesized (produced) within specialized cells. Various cellular secretions are necessary for intercellular communication (such as the secretion of hormones) or for other purposes. Digestive enzymes, for example, are secreted into the digestive tract to break down complex food molecules so that absorption can occur.

Table 1.2 Physical needs for life

Physical needs	Examples of physiological effects
Water	Comprises about 60% of body weight and is the medium in which most chemical reactions in the body take place; essential for body homeostasis
Oxygen	Comprises 20% of normal air and is used in the body in releasing energy from food substances
Food	Nutrient-containing substances that can be broken down into chemicals used for energy, building new compounds, or regulating body metabolic processes
Heat	A form of energy expressed as temperature; the body can tolerate a wide range in environmental temperature but maintains a narrow range in body temperature; many physiological mechanisms insure a consistent internal body temperature
Pressure	External, such as atmospheric pressure, important for respiration; internal, such as blood pressure, important for filtration of blood in kidneys
Protection	Clothing and shelter create artificial environments for our bodies

13. **Excretion.** Excretion is the removal of wastes that are produced as the result of cellular metabolism. These wastes are frequently toxic to the cell and must be eliminated from the cellular environment as well as from the body of a multicellular organism.

14. **Survival instinct.** An innate survival instinct may be essential for an organism if it is to avoid death long enough to propagate. A survival instinct is, therefore, not only important for an individual, but it is also important for the perpetuation of a species population.

Physical Conditions Required for Life

We live in an environment conducive to life. Not only are our bodies adapted to take advantage of environmental parameters, but we have become dependent upon the consistency of these factors for our survival. Life is fragile and is strictly dependent on the presence of particular environmental conditions. These are summarized in table 1.2.

1. Which of the requisites for life and the physical needs of life are necessary for metabolism?
2. Distinguish between secretion and excretion.
3. Why is maintaining a constant body temperature important for physiological mechanisms?

Body Organization

Structural and functional levels of organization exist in the human body, with each of its parts contributing to the total organism.

Objective 12. Identify the structures of a cell, tissue, organ, and system, and explain the relationships among these structures as they constitute an organism.

Objective 13. Explain the general function of each system.

Cellular Level

The **cell** is the basic structural and functional component of life. Humans are multicellular organisms *(or'gah-nizm)* composed of between 60 and 100 trillion cells. It is at the cellular level (fig. 1.11) that the vital functions of life are carried on, such as metabolism, growth, irritability and adaptability, repair, and reproduction. All cells are composed of chemical substances called **protoplasm** *(pro'to-plaz''m)*. Certain protoplasmic compounds are arranged into small functional structures called **organelles** *(or''gah-nelz)*. Each organelle carries out a specific function within the cell. The nucleus, mitochondria, and endoplasmic reticulum are all organelles.

Our bodies contain many distinct kinds of cells, each specialized to perform specific functions. Examples of specialized cells are bone cells, muscle cells, fat cells, blood cells, and nerve cells. Each of these cell types has a unique structure directly related to its function.

Tissue Level

Tissues are layers or aggregations of similar cells that perform specific functions. An example of a tissue is muscle, which functions to contract and produce movement within the body. The outer layer of skin is a tissue because it is composed of similar cells that are bound together and function to protect and contain body contents.

Organ Level

An **organ** is an aggregate of two or more tissues integrated to perform a particular function. Organs occur throughout the body and vary greatly in size and function. Examples of organs are the heart, spleen, pancreas, ovary, skin, and even any bone within the body. Each organ usually has one or more primary tissues and several secondary tissues. In the stomach, for example, the inside epithelial lining is considered the primary tissue because the basic functions of secretion and absorption occur within this layer. The secondary tissues of the stomach are the supporting connective tissue and the vascular, nervous, and muscle tissues.

cell: L. *cella,* small room
protoplasm: Gk. *protos,* first; *plassein,* to mold
tissue: Fr. *tissu,* woven; from L. *texo,* to weave
organ: Gk. *organon,* instrument

Figure 1.11. The levels of structural organization and complexity within the human body.

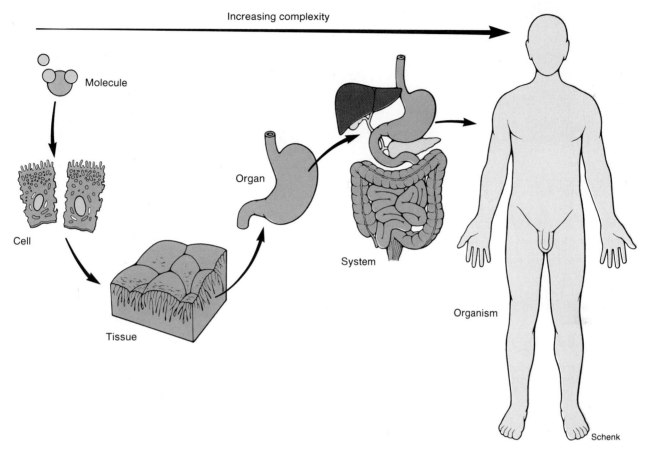

Increasing complexity

Molecule

Cell

Tissue

Organ

System

Organism

Schenk

System Level

The **systems** of the body constitute the next level of structural organization. A body system consists of various organs that have similar or related functions. Examples of systems are the circulatory system, nervous system, digestive system, and endocrine system. Certain organs may serve several systems. The pancreas, for example, functions with both the endocrine and digestive systems. All of the systems of the body are interrelated and function together, constituting the total organism.

A *systemic (sis-tem'ik) approach* to studying anatomy and physiology emphasizes the functional relationships of various organs within a system. For example, the functional role of the digestive system can be better understood if all of the organs within that system are studied together. Another approach to anatomy, the *regional approach,* has merit in professional schools because the structural relationships of portions of several systems can be observed simultaneously. Dissections of cadavers are usually conducted on a regional basis. Trauma or injury usually affects a region of the body, whereas a disease that affects a region may also involve an entire system.

This text uses a systemic approach to anatomy and physiology. In the chapters that follow, you will become acquainted system by system with the structural and functional aspects of the entire body. An overview of the structure and function of each of the body systems is presented in figure 1.12.

1. Construct a diagram to illustrate the levels of structural organization that characterize the body. Which of these levels are microscopic?
2. Why is the skin considered an organ?
3. Which body systems control the functioning of the others; which are supportive of the organism; and which serve a transportive role?

Anatomical and Physiological Terminology

As sciences, anatomy and physiology have descriptive terminologies, which must be mastered by students who would understand and be conversant in these subjects.

Objective 14. Explain how anatomical and physiological terms are derived.

Objective 15. Define what is meant by prefixes and suffixes.

system: Gk. *systema,* being together

Figure 1.12. Various systems of the human body.

Integumentary system
Function: external support
and protection of body

Skeletal system
Function: internal support and
flexible framework for body
movement; production of
blood cells

Muscular system
Function: body movement;
production of body heat

Lymphatic system
Function: body immunity;
absorption of fats; drainage
of tissue fluid

Urinary system
Function: filtration of blood;
maintenance of volume and
chemical composition
of the blood

Endocrine system
Function: secretion of
hormones for
chemical regulation

Nervous system
Function: regulation of
all body activities:
learning and memory

Respiratory system
Function: gaseous exchange
between external environment
and blood

Circulatory system
Function: transport of life-
sustaining materials to body
cells; removal of metabolic
wastes from cells

Digestive system
Function: breakdown and
absorption of food materials

Male reproductive system
Function: production of male
sex cells (sperm); transfer of
sperm to reproductive system
of female

Female reproductive system
Function: production of
female sex cells (ova);
receptacle of sperm from
male; site for fertilization
of ovum, implantation, and
development of embryo
and fetus; delivery of fetus

Moon

Anatomy and physiology are descriptive sciences. Analyzing anatomical and physiological terminology can be a rewarding experience in itself as one learns something of the character of antiquity. Not only is an understanding of the roots of words of academic interest, but a familiarity with technical terms reinforces the learning process. The majority of scientific terms are of Greek or Latin derivation. Some of the more recent terms are of German and French derivation. Unfortunately, more recent anatomical and medical terms have been coined in honor of various anatomists or physicians. Such terms have no descriptive basis, cannot be associated with anything, and must simply be memorized.

Many Greek and Latin terms were coined more than two thousand years ago. It is exciting to decipher the meanings of these terms and gain a glimpse into our medical heritage. Many terms referred to common plants or animals. Thus, the term *vermis* means "worm," *cochlea (kok'le-ah),* "snail shell," *cancer,* "crab," *uvula,* "grape," and even *muscle* comes from the Latin *musculus,* "mouse." Other terms reveal the warlike environment of the Greek and Latin era. *Thyroid,* for example, means "shield," *xiphos (zi'fōs),* "sword," and *thorax,* "breastplate." *Sella* means "saddle" and *stapes (sta'pēz),* "stirrup." Various tools or instruments were referred to in early anatomy. The malleus and anvil resemble miniatures of a blacksmith's implements, and tympanum refers to ă drum.

You will encounter many new terms throughout your study of anatomy and physiology. You can learn these terms more easily by understanding the prefixes and suffixes of the new words. Use the glossary of prefixes and suffixes (on the inside of the back cover) as an aid in learning new terms. Pronouncing these terms as you learn them will also aid your retention.

Learning the material presented in the remainder of this chapter establishes a basic foundation for anatomy and physiology, as well as for all medical and paramedical fields. Anatomy is a very precise science because there is a universally accepted and used reference language for describing body parts and locations. That language will now be introduced.

---

1. *Explain the statement that "anatomy and physiology are descriptive sciences."*
2. *Refer to the glossary of prefixes and suffixes listed on the inside of the back cover to decipher the terms:* blastocoel, hypodermic, dermatitis, *and* orchiectomy.

Body Regions

The human body is divided into regions and specific localized areas, which can be identified on the surface. Each region contains internal organs, the location of which are anatomically, physiologically, and clinically important.

Table 1.3 Body regions and localized areas

Major region	Localized area
I. Head (caput)	Cranium Face (facies)
II. Neck (cervix)	Anterior neck Posterior neck (nuchal region)
III. Trunk (truncus)	Back (dorsum) Thorax (pectus) Abdomen (venter) Pelvis
IV. Upper extremity	Shoulder (omos) Axilla Brachium (upper arm) Elbow Antebrachium (forearm) Manus (hand) Wrist Palm Dorsum of hand
V. Lower extremity	Buttock Thigh (femur) Knee Leg (crus) Ankle Pes (foot) Heel Sole (plantar surface) Dorsum of foot

Source: Van De Graaff, Kent M., *Human Anatomy.* © 1984 Wm. C. Brown Publishers, Dubuque, Iowa. All Rights Reserved. Reprinted by permission.

Objective 16. List the regions of the body and the principal localized areas comprising each region.

Objective 17. Explain why it is important to be able to describe the body areas and regions where each major internal organ is located.

The body is divided into several regions that can be identified on the surface. Learning the terminology used in reference to these regions now will help you learn the names of underlying structures later. The major body regions are the **head, neck, trunk, upper extremity,** and **lower extremity.** The trunk is frequently divided into the thorax and abdomen. Table 1.3 and figures 1.13 and 1.14 present a basic outline of the major body regions and some specific localized areas that can be identified on the surface of the body.

Head

The head, or **caput** *(kap'ut),* is divided into a **facial region,** which includes the eyes, nose, and mouth, and a **cranium** *(kra'ne-um),* or **cranial region,** which covers and supports the brain. Further regionalization of the head is important for identifying specific glands, muscles, nerves, or vessels located under the skin. The identifying names for detailed surface regions are based on associated organs—such as the orbital (eye), nasal (nose), oral (mouth), mental (chin), and auricular (ear) regions—or underlying bones—such as the frontal, zygomatic, temporal, parietal, and occipital regions.

Figure 1.13. Surface regions of the body in anterior view. The generalized body regions are shaded and shown on the figure's right side, and the more specific body regions are shown on its left side.

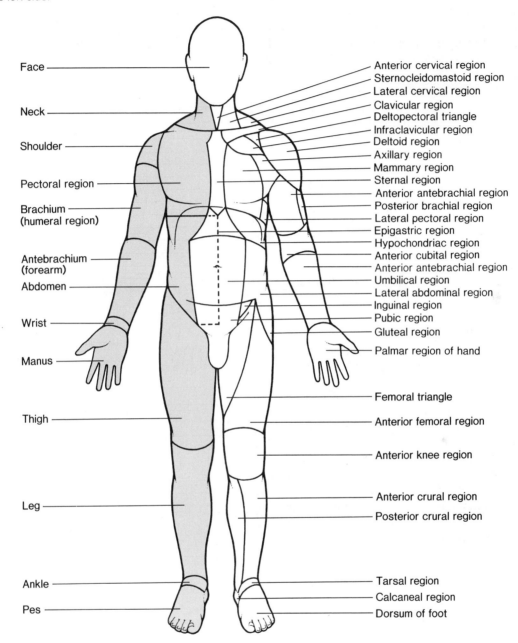

Face — Anterior cervical region
Sternocleidomastoid region
Lateral cervical region
Neck — Clavicular region
Deltopectoral triangle
Infraclavicular region
Shoulder — Deltoid region
Axillary region
Mammary region
Pectoral region — Sternal region
Anterior antebrachial region
Posterior brachial region
Brachium (humeral region) — Lateral pectoral region
Epigastric region
Hypochondriac region
Antebrachium (forearm) — Anterior cubital region
Anterior antebrachial region
Abdomen — Umbilical region
Lateral abdominal region
Inguinal region
Wrist — Pubic region
Gluteal region
Manus — Palmar region of hand

Thigh — Femoral triangle

Anterior femoral region

Anterior knee region

Leg — Anterior crural region

Posterior crural region

Ankle — Tarsal region
Calcaneal region
Pes — Dorsum of foot

Neck

The neck, referred to as the **cervix** *(ser'viks)*, or **cervical region,** supports the head and permits it to move. As with the head, detailed subdivisions of the neck can be identified and these are fully discussed in chapter 12.

Thorax

The thorax, or **thoracic** *(tho-ras'ik)* **region,** is commonly referred to as the chest. The **mammary region** of the thorax surrounds the nipple and in sexually mature females is enlarged as the breast. Between the mammary regions is the

sternal region. The armpit is called the **axillary fossa,** or simply **axilla,** and the surrounding area the **axillary region.** Paired **scapular regions** can be identified from the back of the thorax (fig. 1.14). **Suprascapular** and **interscapular regions** are located above and medial (towards the center) respectively to the scapular regions. The **vertebral region,** following the vertebral column, extends the length of the back. On either side of the thorax are the **lateral pectoral regions.**

---

axillary: L. *axilla,* armpit

---

thorax: L. *thorax,* chest
mammary: L. *mamma,* breast

Figure 1.14.   Surface regions of the body in posterior view. The generalized body regions are shaded and shown on the figure's left side, and the more specific body regions are shown on its right side.

Posterior neck region

Shoulder

Posterior thorax

Brachium (humeral region)

Abdomen

Antebrachium (forearm)

Wrist

Buttock

Manus

Thigh

Posterior region of knee

Leg

Ankle

Heel

Sole of foot

Suprascapular region
Acromial region
Scapular region
Deltoid region
Interscapular region
Vertebral region
Posterior brachial region
Lateral pectoral region
Infrascapular region
Hypochondriac region
Posterior cubital region
Lumbar region
Lateral abdominal region
Posterior antebrachial region
Sacral region
Gluteal region
Carpal region
Dorsum of hand

Anal region
Posterior femoral region

Popliteal fossa

Posterior crural region

Anterior crural region

Calcaneal region

Plantar surface

The heart and lungs are contained within the thoracic cavity. Easily identified surface landmarks greatly facilitate a physical examination of them. A physician must know, for example, where the valves of the heart can best be detected and where to listen for respiratory sounds. The axilla becomes important in examining for infected lymph nodes. When fitting a patient for crutches, a physician will instruct the patient to avoid supporting the weight of the body on the axillary area because of the possibility of damaging the underlying nerves and vessels.

Abdomen

The abdomen is located below the thorax. The **navel,** or **umbilicus,** is an obvious landmark on the front and center of the abdomen. The abdomen has been divided into nine regions in order to describe the location of internal organs. Figure 1.15 diagrams the subdivisions of the abdomen, and table 1.4 identifies the internal organs located within these regions. Subdividing the abdomen into four quadrants (fig. 1.16) is a common clinical practice so that various pains and conditions can be more easily located.

Figure 1.15.   The abdomen is subdivided into nine regions. The vertical planes are positioned just medial to the nipples; the upper horizontal plane is positioned at the level of the rib cage; and the lower horizontal plane is even with the upper border of the hipbones.

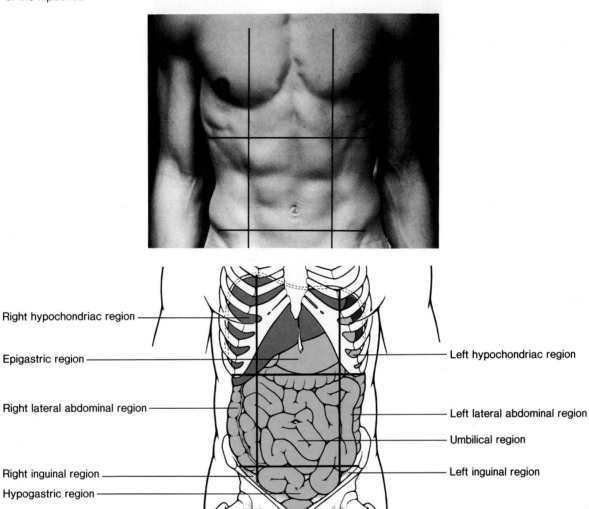

Right hypochondriac region

Epigastric region

Right lateral abdominal region

Right inguinal region

Hypogastric region

Left hypochondriac region

Left lateral abdominal region

Umbilical region

Left inguinal region

Table **1.4** Regions of the abdomen and pelvis		
Region	**Location**	**Internal organs**
Right hypochondriac	Right, upper one-third of abdomen	Gallbladder; portions of liver and right kidney
Epigastric	Upper, median abdomen	Portions of liver, stomach, pancreas, and duodenum
Left hypochondriac	Left, upper one-third of abdomen	Spleen; splenic flexure of colon; portions of right kidney and small intestine
Right lateral	Right, lateral one-third of abdomen	Cecum; ascending colon; hepatic flexure; portions of right kidney and small intestine
Umbilical	Center of abdomen	Jejunum; ileum; portions of duodenum, colon, kidneys, and major abdominal vessels
Left lateral	Left, lateral one-third of abdomen	Descending colon; portions of left kidney and small intestine
Right inguinal	Right, lower one-third of abdomen	Appendix; portions of cecum and small intestine
Pubic (hypogastric)	Lower, center one-third of abdomen	Urinary bladder; portions of small intestine and sigmoid colon
Left inguinal	Left, lower one-third of abdomen	Portions of small intestine and descending and sigmoid colon

Source: Van De Graaff, Kent M., *Human Anatomy.* © 1984 Wm. C. Brown Publishers, Dubuque, Iowa. All Rights Reserved. Reprinted by permission.

Figure 1.16.   A clinical subdivision of the abdomen into four quadrants.

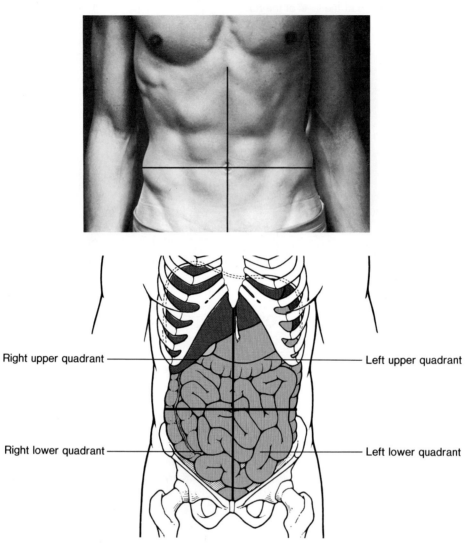

Right upper quadrant ——— Left upper quadrant

Right lower quadrant ——— Left lower quadrant

The **pelvic** *(pel''vik)* **region** forms the lower portion of the abdomen. Within the pelvic region is the **pubic area,** which is covered with pubic hair in sexually mature persons. The **perineum** *(per''i-ne'um)* (fig. 1.17) is the region containing the external sex organs and the anal opening. The center of the back side of the abdomen, commonly called the small of the back, is the **lumbar region.** The **sacral region** is located farther down, at the point where the vertebral column terminates. The large hip muscles form the **buttock,** or **gluteal region.** This region is a common injection site of hypodermic needles.

Upper Extremity

The upper extremity is anatomically divided into the **shoulder, brachium** *(bra'ke-um)* **(upper arm), antebrachium (forearm),** and **manus (hand).** The shoulder is the region between the pectoral girdle and the brachium that contains the shoulder joint. The shoulder is referred to as the **omos,** or **deltoid region.** Between the arm and forearm is a flexible joint called the **elbow.** The surface area of the elbow is known as the cubital region. The front surface of the elbow is also known as the **cubital fossa,** an important site for intravenous injections or the withdrawal of blood. The wrist is the flexible junction between the forearm and the hand. The front of the hand is referred to as the **palm,** or **palmar surface,** and the back of the hand is called the **dorsum of the hand.**

cubital: L. *cubitis,* elbow

Figure 1.17.   A superficial view of the male perineum.

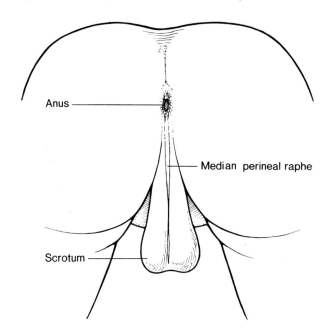

Anus

Median perineal raphe

Scrotum

Lower Extremity

The lower extremity consists of the **thigh, knee, leg,** and **foot.** The thigh is commonly called the **upper leg,** or **femoral** *(fem′o-ral)* **region.** The knee joint has two surfaces: the front surface is the **patellar region,** or kneecap; the back of the knee is called the **popliteal** *(pop″li-te′al)* **fossa.** The leg has anterior and posterior crural regions (see fig. 1.13). The **shin** is a prominent bony ridge extending longitudinally along the anterior crural region, and the **calf** is the thickened muscular mass of the posterior crural region. The **ankle** is the junction between the leg and the foot. The **heel** is the back of the foot, and the **sole** of the foot is referred to as the **plantar surface.** The **dorsum of the foot** is the top surface.

1. *Using yourself as a model, identify the various body regions that are depicted in figures 1.13 and 1.14. Which of these regions have surface landmarks that help distinguish their boundaries?*
2. *In which region of the body are intravenous injections made?*
3. *What is the distinction between the pelvic, pubic, and perineal regions?*
4. *Identify the joint between the following regions: the brachium and antebrachium, the pectoral girdle and brachium, the leg and foot, the antebrachium and hand, the thigh and leg.*
5. *Explain how a knowledge of the body regions has practical medical application.*

popliteal: L. *poples,* ham (hamstring muscle) of the knee

Body Cavities and Membranes

For functional and protective purposes, the viscera are compartmentalized and supported in specific body cavities by connective and epithelial membranes.

   Objective 18.   Identify the various body cavities and the organs found in each.
   Objective 19.   Discuss the types and functions of the various body membranes.

Body Cavities

Body cavities are confined spaces within the body. They contain organs that are protected, compartmentalized, and supported by associated membranes. There are two principal body cavities: the **dorsal body cavity** and the larger **ventral body cavity.** The dorsal body cavity contains the brain and the spinal cord.

During development, the ventral cavity forms from a cavity called the **coelom** *(see′lom).* The coelom is a body cavity within the trunk, which is lined with a membrane that secretes a lubricating fluid. As development progresses, the coelom is partitioned by the muscular diaphragm into an upper **thoracic cavity,** or chest cavity, and a lower **abdominopelvic cavity** (figs. 1.18, 1.19). Organs within the coelom are collectively called **viscera,** or **visceral** *(vis′er-al)* **organs** (fig. 1.20). Within the thoracic cavity are two **pleural** *(ploo′ral)* **cavities** for the right and left lungs and a **pericardial** *(per″ĭ-kar′de-al)* **cavity** containing the heart. The area between the two lungs is known as the **mediastinum** *(me″de-ah-sti′num).*

The abdominopelvic cavity consists of an upper **abdominal cavity** and a lower **pelvic cavity.** The abdominal cavity contains the stomach, small intestine, large intestine, liver, gallbladder, pancreas, spleen, and kidneys. The pelvic cavity is occupied by the terminal portion of the large intestine, the urinary bladder, and certain reproductive organs (the uterus, uterine tubes, and ovaries in the female; the seminal vesicles and prostate gland in the male).

B ody cavities serve to confine organs and systems that have related functions. The major portion of the nervous system occupies the dorsal cavity; the principal organs of the respiratory and circulatory systems, the thoracic cavity; the primary organs of digestion, the abdominal cavity; and the reproductive organs, the pelvic cavity. Not only do these cavities house and support various body organs, they also effectively compartmentalize them so that infections and diseases cannot spread from one compartment to another. For example, pleurisy of one lung membrane does not usually spread to the other, and an injury to the thoracic cavity will usually cause only one lung to collapse rather than both.

coelom: Gk. *koiloma,* cavity

Figure 1.18.   A median view of body cavities.

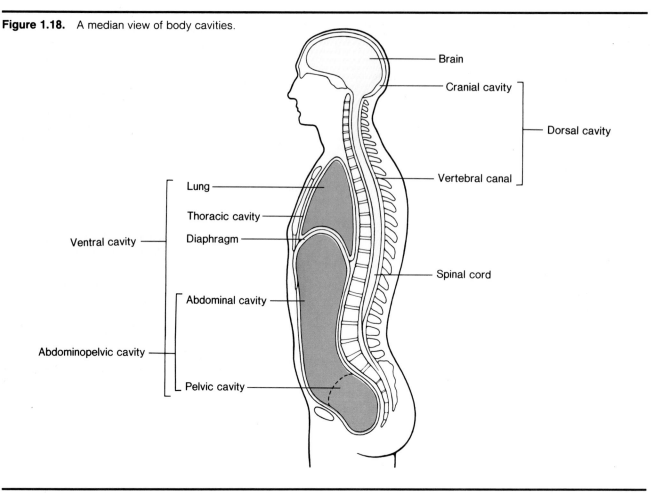

Figure 1.19.   An anterior view of body cavities.

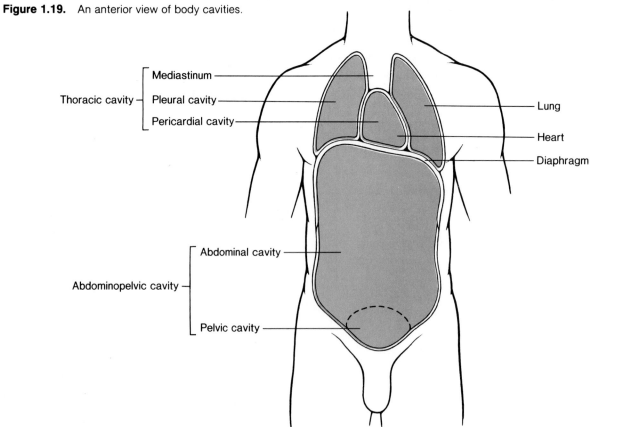

Figure 1.20. Visceral organs of the abdominal cavity and supporting serous membranes.

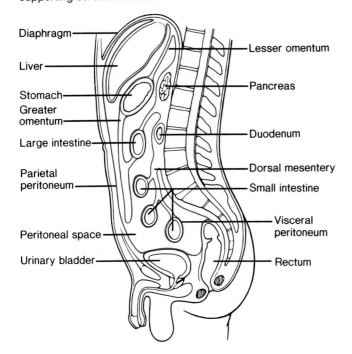

- Diaphragm
- Liver
- Stomach
- Greater omentum
- Large intestine
- Parietal peritoneum
- Peritoneal space
- Urinary bladder
- Lesser omentum
- Pancreas
- Duodenum
- Dorsal mesentery
- Small intestine
- Visceral peritoneum
- Rectum

In addition to the large ventral and dorsal cavities, there are several smaller cavities within the head. The **oral,** or **buccal** *(buk'al),* **cavity** functions primarily with digestion and secondarily with respiration. It contains the teeth and tongue. The **nasal cavity,** which is part of the respiratory system, has two chambers created by a nasal septum. There are two **orbital cavities,** each of which houses an eyeball and its associated muscles, vessels, and nerves. Likewise there are two **middle ear cavities,** which contain ear ossicles (bones) and function with hearing. Figure 1.21 shows the location of the cavities within the head.

Body Membranes

Body membranes are composed of thin layers of connective and epithelial tissue, which cover, separate, and support visceral organs and line body cavities. There are two basic types of body membranes: **mucous** *(mu'kus)* **membranes** and **serous** *(se'rus)* **membranes.**

Mucous membranes secrete a thick, viscid substance called mucus. Generally, mucus lubricates or protects the associated organs where it is secreted. Mucous membranes line various cavities and tubes that enter or exit

Figure 1.21. Cavities within the head.

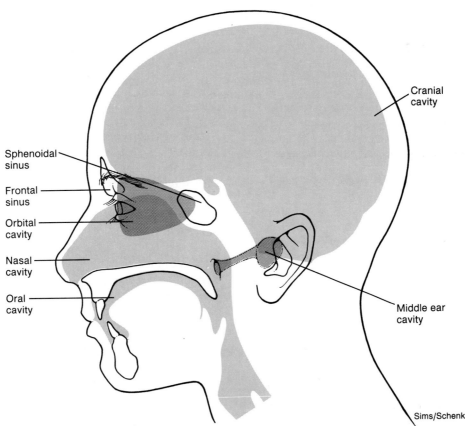

- Sphenoidal sinus
- Frontal sinus
- Orbital cavity
- Nasal cavity
- Oral cavity
- Cranial cavity
- Middle ear cavity

Sims/Schenk

from the body, such as the oral and nasal cavities and the tubes of the respiratory, reproductive, urinary, and digestive systems.

Serous membranes line the thoracic and abdominopelvic cavities and cover visceral organs, secreting a watery lubricant called serous fluid. **Pleurae** are serous membranes associated with the lungs. Each pleura (pleura of right lung and pleura of left lung) has two parts. The **visceral pleura** adheres to the outer surface of the lung, while the **parietal** *(pah-ri''ē-tal)* **pleura** lines the thoracic walls and the thoracic surface of the diaphragm. The moistened space between the two pleurae is known as the **pleural cavity.**

Pericardial membranes are the serous membranes of the heart. A thin **visceral pericardium** covers the surface of the heart, and a thicker **parietal pericardium** is the durable covering that surrounds the heart. The space between these two membranes is called the **pericardial cavity.**

Serous membranes of the abdominal cavity are called **peritoneal** *(per''i-to-ne'al)* **membranes.** The **parietal peritoneum** lines the abdominal wall, and the **visceral peritoneum** covers the visceral organs. The **peritoneal cavity** is the potential space within the abdominopelvic cavity between the parietal and visceral peritoneal membranes. Certain organs, such as the kidneys, adrenal glands, and a portion of the pancreas, which are within the abdominal cavity, are positioned behind the parietal peritoneum and are, therefore, said to be **retroperitoneal. Mesenteries** *(mes'en-ter''ez)* are double folds of peritoneum that connect the parietal to the visceral peritoneum (see fig. 1.20).

1. Describe the divisions and boundaries of the ventral body cavity, and list the major organs contained within each division of the ventral body cavity.
2. Distinguish between mucous and serous membranes. List the specific serous membranes of the thoracic and abdominopelvic cavities.
3. Explain the importance of separate and distinct body cavities.

Planes of Reference and Descriptive Terminology

All of the descriptive planes of reference and terms of direction used in anatomy are standardized because of their reference to the body in anatomical position.

Objective 20. Identify the planes of reference used to locate structures within the body.

peritoneum: Gk. *peritonaion,* stretched over

Figure 1.22. Planes of reference through the body.

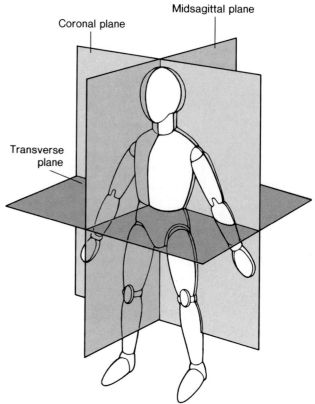

Objective 21. Describe the anatomical position.
Objective 22. Define and be able to use properly the descriptive and directional terms that have been designated to refer to body structures.

Planes of Reference

In order to visualize and study the structural arrangements of various organs, the body may be sectioned and diagrammed according to planes of reference. Three fundamental planes, **midsagittal** *(mid-saj'i-tal),* **coronal,** and **transverse,** are frequently used to depict structural arrangement (fig. 1.22).

A *midsagittal* plane passes lengthwise through the midplane of the body, dividing it into right and left halves. *Sagittal* planes also extend vertically and divide the body into unequal right and left portions. *Coronal,* or *frontal,* planes also pass lengthwise and divide the body into front and back portions. *Transverse* planes, also called *horizontal,* or *cross-sectional,* planes, divide the body into superior (upper) and inferior (lower) portions.

coronal: L. *corona,* crown

Figure 1.23. In the anatomical position, the body is erect, the feet are parallel, the eyes directed forward, the arms to the side with the palms directed forward: (*a*) anterior view; (*b*) lateral view; and (*c*) posterior view.

(a)   (b)   (c)

The value of the computerized tomographic X-ray (CT) scan is that it displays an image along a transverse plane similar to that which could otherwise be obtained only in an actual section through the body. Prior to the development of this X-ray technique, conventional X-ray images were on a vertical plane, and the dimensions of body irregularities were difficult, if not impossible, to ascertain.

Descriptive Terminology

Anatomical Position. All terms of direction that describe the relationship of one body part to another are made in reference to the **anatomical position.** In the anatomical position, the body is erect; the feet are parallel to one another and flat on the floor, the eyes directed forward, and the arms at the sides of the body with the palms of the hands turned forward (fig. 1.23).

Directional Terms. Directional terms are used to locate the position of structures, surfaces, and regions of the body. These terms are always relative to the specimen positioned in the anatomical position. A summary of directional terms is presented in table 1.5.

Clinical Procedures. Certain clinical procedures are important in determining anatomical structure and function in a living individual. The more common of these are as follows:

1. **Palpation.** Applying the fingers with firm pressure to the surface of the body to detect surface landmarks, lumps, tender spots, or pulsations.
2. **Percussion.** Tapping sharply on various locations on the thorax or abdomen to detect resonating vibrations in determining fluid concentrations and organ densities.

Table 1.5 Directional terms for the human body

Term	Definition	Example
Superior (cranial, cephalic)	Toward the head; toward the top	The thorax is superior to the abdomen.
Inferior (caudal)	Away from the head; toward the bottom	The legs are inferior to the trunk.
Anterior (ventral)	Toward the front	The navel is on the anterior side of the body.
Posterior (dorsal)	Toward the back	The kidneys are posterior to the intestine.
Medial	Toward the midline of the body	The heart is medial to the lungs.
Lateral	Toward the side of the body	The ears are on the lateral sides of the head.
Internal (deep)	Away from the surface of the body	The brain is internal to the cranium.
External (superficial)	Toward the surface of the body	The skin is external to the muscles.
Proximal	Toward the main mass of the body	The knee is proximal to the foot.
Distal	Away from the main mass of the body	The hand is distal to the elbow.
Visceral	Related to internal organs	The lungs are covered by a thin membrane called the visceral pleura.
Parietal	Related to the body walls	The parietal pleura is the inside lining of the thoracic cavity.

Source: Van De Graaff, Kent M., *Human Anatomy.* © 1984 Wm. C. Brown Publishers, Dubuque, Iowa. All Rights Reserved. Reprinted by permission.

3. **Auscultation.** Listening to the sounds that various organs make as they perform their functions (breathing sounds, heart beats, digestive sounds, etc.).

4. **Reflex response.** Reflex responses are used to determine the condition of parts of the nervous system and some associated organs. One test of a reflex mechanism involves tapping a predetermined tendon with a reflex hammer and observing the response.

---

1. *Explain why a transverse plane through the body is important in studying regional anatomy, whereas a transverse plane through an organ is more important in studying specific systems.*
2. *What does it mean that directional terms are relative and must be used in reference to a body structure or a body in anatomical position?*
3. *Write a list of statements, similar to the examples in table 1.5, that correctly expresses the directional terms used to describe the relative positions of various body structures.*

Units of Measurements

Measurements are the human way of expressing the relative lengths, masses, temperatures, and volumes of substances, which are important because they bring uniformity and consistency to scientific communication.

Objective 23. Define the frequently used metric units of length, mass, and volume, and their equivalents.

---

metric: Fr. *metron,* measure

Units of measurements are used frequently in anatomy and physiology to describe body structures and how they function. The speed of an impulse, the pressure of blood, the weight of an organ, and the temperature of the body are examples of measurements important in understanding physiological events as well as in determining a deviation from what is considered normal. Such data are also of clinical importance in determining the amounts of medication to be administered.

The length, mass, and volume units of measurement presented in this text are given in metric form followed in parentheses by the apothecary (U.S.) equivalent. The temperatures presented are in Celsius followed in parentheses by the Fahrenheit equivalent. Tables 1.6, 1.7, 1.8, and 1.9 are presented to help you become acquainted with the metric units of measurement and their apothecary equivalents and with common Celsius and Fahrenheit temperature conversions. Reference to these tables will probably be necessary as you continue your study of anatomy and physiology.

---

1. *Describe a situation that involves measurements of the human body.*
2. *Referring to tables 1.6, 1.7, 1.8, and 1.9, solve the following problems:*
   (a) *A body cell measures 100 μm in diameter. Express this measurement in nanometers.*
   (b) *A person weighs 150 lb. What is the weight in kilograms?*
   (c) *An experiment is conducted at 150°F. Express this temperature in degrees Celsius.*
   (d) *If a patient excreted 1,350 ml of urine in a day, how many liters would this be?*

---

apothecary: Gr. *apotheke,* storehouse
Celsius: from Anders Celsius, Swedish astronomer, 1701–44
Fahrenheit: from Gabriel D. Fahrenheit, German physicist, 1686–1736

Table 1.6 Units of length

Metric unit	Equivalent in meters	Apothecaries' equivalent (U.S. equivalent)
1 kilometer (km) kilo = 1,000	1,000 m	3,280.84 ft; 0.62 mi 1 mi = 1.61 km
1 hectometer (hm) hecto = 100	100 m	328 ft
1 dekameter (dam) deka = 10	10 m	3.28 ft
1 meter (m) Standard unit of length	—	39.37 in.; 3.28 ft; 1.09 yd
1 decimeter (dm) deci = $1/10$	0.1 m	3.94 in.
1 centimeter (cm) centi = $1/100$	0.01 m	0.394 in. 1 in. = 2.54 cm
1 millimeter (mm) milli = $1/1,000$	0.001 m	0.0394 in.
1 micrometer (μm) micro = $1/1,000,000$	0.000,000.1 m	3.94×10^{-5} in.
1 nanometer (nm) nano = $1/1,000,000,000$	0.000,000,001 m	3.94×10^{-8} in.
1 angstrom (Å)	0.000,000,000,1 m	3.94×10^{-9} in.

Table 1.7 Units of mass

Metric unit	Equivalent in grams	Apothecaries' equivalent
1 kilogram (kg)	1,000 g	2.205 lb
1 hectogram (hg)	100 g	
1 dekagram (dag)	10 g	
1 gram (g)	—	1 lb = 453.6 g 1 oz = 28.35 g
1 decigram (dg)	0.1 g	
1 centigram (cg)	0.01 g	
1 milligram (mg)	0.001 g	
1 microgram (μg)	0.000,001 g	

Table 1.8 Units of volume

Metric unit	Metric equivalent	Apothecaries' equivalent
1 liter (l)	1,000 ml	33.81 fl oz or 1.057 qt 1 qt = 946 ml
1 milliliter (ml)	0.001 liter	0.0338 fl oz; 1 fl oz = 30 ml 1 teaspoon = 5 ml
1 cubic centimeter (cm³)	0.999972 ml	0.0338 fl oz

Table 1.9 Units of temperature in °Celsius and °Fahrenheit

°C	°F	°C	°F
35.0	95.0	37.8	100.0
35.1	95.2	37.9	100.2
35.2	95.4	38.0	100.4
35.3	95.6	38.1	100.6
35.4	95.8	38.2	100.8
35.5	96.0	38.3	101.0
35.7	96.2	38.4	101.2
35.8	96.4	38.6	101.4
35.9	96.6	38.7	101.6
36.0	96.8	38.8	101.8
36.1	97.0	38.9	102.0
36.2	97.2	39.0	102.2
36.3	97.4	39.1	102.4
36.4	97.6	39.2	102.6
36.6	97.8	39.3	102.8
36.7	98.0	39.4	103.0
36.8	98.2	39.6	103.2
36.9	98.4	39.7	103.4
37.0	98.6	39.8	103.6
37.1	98.8	39.9	103.8
37.2	99.0	40.0	104.0
37.3	99.2	40.1	104.2
37.4	99.4	40.2	104.4
37.6	99.6	40.3	104.6
37.7	99.8	40.4	104.8
		40.6	105.0

To convert °C to °F
Multiply °C by 9/5 and add 32.
_____ °C × 9/5 + 32 = _____ °F

To convert °F to °C
Subtract 32 from °F and multiply by 5/9.
_____ °F − 32 × 5/9 = _____ °C

Chapter Summary

I. Definitions of the Sciences
Anatomy and physiology are integrated sciences.
   A. Dissections and descriptions of human cadavers are an important aspect of anatomy.
   B. Physiology is a dynamic science that explains physical and chemical processes of the body.

II. Historical Perspective
Anatomy and physiology are dynamic sciences that have long, exciting heritages and currently provide the foundation for medical, biochemical, developmental, cytogenetic, and biomechanical research.
   A. The history of anatomy and physiology parallels that of medicine.
      1. Hippocrates, the famous Grecian physician, is regarded as the "father of modern medicine," even though he erroneously practiced humoral medicine.
      2. Aristotle established a type of scientific method in obtaining data. Some basic anatomy is contained in the classic books by Aristotle.
      3. Erasistratus is frequently referred to as "the father of physiology."
      4. There was a de-emphasis of theoretical data during the Roman era. The writings of the Roman physician Galen, however, became the unquestionable authority on anatomy and medical treatment for nearly fifteen hundred years.
      5. There were no advancements in anatomy and physiology for nearly one thousand years during the period known as the Middle (Dark) Ages.
      6. The development of movable type (A.D. 1450) and the monumental work of "the father of modern anatomy," Vesalius (1514–64), accelerated an interest and research in anatomy and physiology during the Renaissance.
      7. Two major contributions of the seventeenth and eighteenth centuries were an explanation of blood flow by Harvey and the development and utilization of the microscope.
      8. Cellular biology became established as a science separate from anatomy and physiology during the nineteenth century.
      9. During the twentieth century many specialities have developed within anatomy and physiology and medicine.
   B. The scientific method is a disciplined approach for the acquisition of truth. It includes observation, gathering data, and formulating a hypothesis, which may become accepted as a theory or scientific principle.
   C. It is important for a student of anatomy and physiology to keep up with the rapid proliferation of knowledge by first establishing a strong foundation in the subject and then reading science magazines and journals that present new discoveries.

III. Classification and Characteristics of Humans
Humans are biological organisms belonging to the phylum Chordata within the kingdom Animalia and to the family Hominidae within the class Mammalia and the order Primates.
   A. Humans belong to the phylum Chordata because of the presence of a notochord, a dorsal hollow nerve cord, and pharyngeal pouches during the embryonic stage of development.
   B. Humans are mammals and as such have mammalian characteristics.
   C. Humans are also classified within the order Primates.
   D. Humans are the sole members of the family Hominidae.
   E. Some of the characteristics of humans involve the size and development of the brain, bipedal locomotion, an opposable thumb, vocal structures, and stereoscopic vision.

IV. Maintenance of Life
The maintenance of the life of an organism depends upon the presence of certain vital factors for its existence.
   A. Requisites for life include development, organization, adaptation, responsiveness, movement, reproduction, growth, respiration, digestion and absorption, circulation, synthesis, secretion, excretion, and survival instinct.
   B. Some physical needs include water, oxygen, food, heat, pressure, and protection.

V. Body Organization
Structural and functional levels of organization exist in the human body, with each of its parts contributing to the total organism.
   A. The cell is the structural and functional component of life.
   B. Tissues are aggregations of similar cells that perform specific functions.
   C. An organ is an aggregate of two or more tissues that perform specific functions.
   D. Body systems consist of various organs that have similar or interrelated functions.

VI. Anatomical and Physiological Terminology
As sciences, anatomy and physiology have descriptive terminologies, which must be mastered by students who would understand and be conversant in these subjects.
   A. The majority of anatomical and physiological terms are of Greek or Latin derivation.
   B. Learning the basic anatomical and physiological terminology establishes a foundation for anatomy and physiology, as well as for all of the medical and paramedical fields.

VII. Body Regions
The human body is divided into regions and specialized localized areas, which can be identified on the surface. Each region contains internal organs, the location of which are anatomically, physiologically, and clinically important.
   A. The head is divided into the facial and cranial regions.
   B. The neck supports the head and permits it to move.
   C. The thorax consists of a sternal region and mammary regions anteriorly, lateral pectoral regions and axillae laterally, and a vertebral region and scapular regions posteriorly.
   D. The abdomen is divided into four quadrants and nine regions. The pelvic region forms the lower portion of the abdomen.
   E. The upper extremity is divided into the shoulder, arm, forearm, and hand.
   F. The lower extremity is divided into the thigh, knee, leg, and foot.

VIII. Body Cavities and Membranes
For functional and protective purposes, the viscera are compartmentalized and supported in specific body cavities by connective and epithelial membranes.
   A. Body cavities are confined spaces within the body.
      1. The dorsal cavity is composed of the cranial and spinal cavities.
      2. The ventral cavity is composed of thoracic and abdominopelvic cavities.
      3. Other body cavities include the oral, orbital, nasal, and middle ear cavities.
   B. Body membranes are composed of thin layers of connective and epithelial tissue.

1. The body has two principal types of membranes: mucous membranes that secrete protective mucus; and serous membranes that line the ventral cavities and cover visceral organs.
2. There are three categories of serous membranes: pleural membranes, pericardial membranes, and peritoneal membranes.

IX. Planes of Reference and Descriptive Terminology
All of the descriptive planes of reference and terms of direction used in anatomy are standardized because of their reference to the body in anatomical position.
   A. In the anatomical position, the subject stands erect and faces forward with arms at the sides and palms turned forward.
   B. Directional terms are used to describe the location of one body part with respect to another part.

X. Units of Measurements
Measurements are the human way of expressing relative lengths, masses, temperatures, and volumes of substances, which are important because they bring uniformity and consistency to scientific communication.

Review Activities

Objective Questions

1. Which of the following men would be more likely to disagree with the concept of body humors?
   (a) Galen
   (b) Hippocrates
   (c) Vesalius
   (d) Aristotle
2. The most important contribution of William Harvey was his research on the
   (a) continuous circulation of blood
   (b) microscopic structure of spermatozoa
   (c) detailed structure of kidney
   (d) striped appearance of skeletal muscle
3. The taxonomic scheme from specific to general is
   (a) species, class, order, phylum
   (b) genus, family, kingdom, phylum
   (c) species, family, class, kingdom
   (d) genus, phylum, class, kingdom
4. Which of the following is not a principal chordate characteristic?
   (a) dorsal hollow nerve cord
   (b) distinct head, thorax, and abdomen
   (c) notochord
   (d) pharyngeal pouches
5. The cubital fossa is located in the
   (a) thorax
   (b) upper extremity
   (c) abdomen
   (d) lower extremity
6. Which of the following is not a fundamental plane?
   (a) coronal
   (b) transverse
   (c) vertical
   (d) midsagittal
7. The abdominal region superior to the umbilical region that contains most of the stomach is the
   (a) hypochondriac region
   (b) epigastric region
   (c) diaphragmatic region
   (d) inguinal region
8. Regarding serous membranes, which of the following word pairs is incorrect?
   (a) visceral pleura—lung
   (b) parietal peritoneum—body wall
   (c) mesentery—heart
   (d) parietal pleura—body wall
   (e) visceral peritoneum—intestines
9. In the anatomical position the
   (a) arms are extended away from the body
   (b) palms of the hands face posteriorly
   (c) body is erect and palms face anteriorly
   (d) body is in a fetal position
10. Listening to sounds that functioning visceral organs make is
   (a) percussion
   (b) palpation
   (c) audiotation
   (d) auscultation

Essay Questions

1. What is meant by the humoral theory of body organization? Which great anatomists were influenced by this theory? When did the humoral theory cease to be an influence upon the anatomical and physiological investigation of the body?
2. Discuss the impact that Galen had on the advancement of anatomy and physiology and medicine. What ideological circumstances permitted the philosophies of Galen to survive for such a long period?
3. What role did the development of the microscope play in the advancement of the sciences of anatomy and physiology and medicine? What specialities of anatomical and physiological study have arisen since the introduction of the microscope?
4. Describe the scientific method, and relate how it is of immense importance in anatomical and physiological research.
5. What is a serous membrane? Diagram the serous membranes of the thoracic cavity showing the subcompartments of the pericardial cavity, pleural cavities, and mediastinum.

Selected Readings

Bronowski, J. 1971. The identity of man. Rev. ed. New York: Doubleday and Co.
Chewning, E. B. 1979. Anatomy illustrated. New York: Simon and Schuster.
Clemente, C. D. 1981. Anatomy: a regional atlas of the human body. 2d ed. Baltimore: Urban and Schwarzenberg.
Corner, G. W. 1977. Anatomical texts of the earlier Middle Ages. Carnegie Institution of Washington, Publication No. 364. Washington, D.C.: National Publishing Co.
Fox, S. I. 1984. Human physiology. Dubuque: Wm. C. Brown Publishers.
Geller, S. A. March 1983. Autopsy. Scientific American.
Hall-Craggs, E. C. B. 1985. Anatomy as a basis for clinical medicine. Baltimore-Munich: Urban & Schwarzenberg.
Kennedy, D. 1967. Introduction to From cell to organism: readings from Scientific American. San Francisco: W. H. Freeman and Co.
Kevorkian, J. 1959. The story of dissection. New York: Philosophical Library.
Last, R. J. 1977. Anatomy, regional and applied. 6th ed. Boston: Little, Brown and Co.
Lyons, A. S., and R. J. Petrucelli. 1978. Medicine, an illustrated history. New York: Harry N. Abrams.
Morris, D. 1968. The naked ape. New York: McGraw-Hill.
Royce, J. 1973. Surface anatomy. Philadelphia: F. A. Davis Co.
Singer, C. 1957. A short history of anatomy and physiology from the Greeks to Harvey. 2d ed. New York: Dover Publications.
Sobotta, J., and F. Figge. 1974. Atlas of human anatomy. 9th ed. New York: Hafner Press.
Van De Graaff, K. M. 1984. Human anatomy. Dubuque: Wm. C. Brown Publishers.
Woodburne, R. T. 1978. Essentials of human anatomy. 6th ed. New York: Oxford Univ. Press.
Youngson, A. J. 1979. The scientific revolution in Victorian medicine. London: Croom Helm.

2 Chemical Composition of

Concepts

The anatomic structures and physiological processes of the body are based, to a large degree, on the properties and interactions of atoms, ions, and molecules.

Carbohydrates and lipids serve as principal sources of energy in the body. Although they are composed of similar atoms, they have distinct chemical structures and biological functions.

Proteins and nucleic acids are large molecules formed of amino acids and nucleotides, respectively. These molecules provide the basis for cell structure, function, and genetic regulation.

the Body

Atoms, Ions, and Chemical Bonds

The anatomical structures and physiological processes of the body are based, to a large degree, on the properties and interactions of atoms, ions, and molecules.

Objective 1. Describe the structure of an atom, using the terms *proton, neutron, electron, atomic weight* and *number,* and *orbital.*

Objective 2. Explain how different types of chemical bonds are formed.

Objective 3. Define the terms *acid, base, pH,* and *ion.*

Objective 4. Discuss the structure and chemical properties of water, and explain why some molecules are hydrophilic and others are hydrophobic.

Atoms are much too small to be seen individually, even with the most powerful electron microscope. Through the efforts of many generations of scientists, however, the structure of an atom is now well understood. At the center of an atom is its *nucleus.* The nucleus contains two types of particles—*protons,* which have a positive charge, and *neutrons,* which are noncharged. The sum of the protons and neutrons in a nucleus is the **atomic mass unit,** which is approximately equal to the *atomic weight* (table 2.1).

Orbiting the positively charged nucleus are negatively charged particles called *electrons.* The number of electrons in an atom is equal to the number of its protons; each atom, therefore, has a net charge of zero. The number of protons (or electrons) in an atom is called its *atomic number.*

The exact location of an electron at a given time cannot be predicted. One can only predict that, with a given probability, the electron will be within a given volume of space around the nucleus. The *orbital* of an electron describes the outer boundary of this volume of space. This orbital can be drawn as a circle around the nucleus, much like the orbit of a satellite (like the moon) around a planet.

Table **2.1** Atoms commonly present in organic molecules							
Atom	**Symbol**	**Atomic number**	**Atomic weight**	**Orbital 1**	**Orbital 2**	**Orbital 3**	**Number of chemical bonds**
Hydrogen	H	1	1.01	1	0	0	1
Carbon	C	6	12.01	2	4	0	4
Nitrogen	N	7	14.01	2	5	0	3
Oxygen	O	8	16.00	2	6	0	2
Phosphorous	P	15	30.97	2	8	5	5
Sulfur	S	16	32.06	2	8	6	2

Source: Fox, Stuart Ira, *Human Physiology.* © 1984 Wm. C. Brown Publishers, Dubuque, Iowa. All Rights Reserved. Reprinted by permission.

Figure 2.1. Diagrams of the hydrogen and carbon atoms. The electron orbitals on the left are represented by dots indicating probable positions of the electrons. The orbitals on the right are represented by concentric circles.

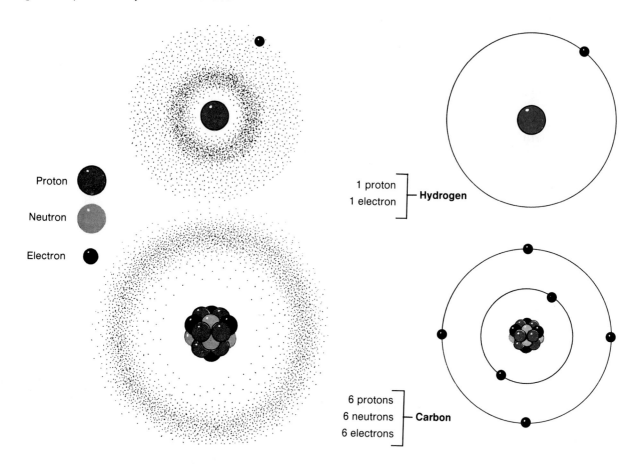

Proton

Neutron

Electron

1 proton
1 electron **Hydrogen**

6 protons
6 neutrons **Carbon**
6 electrons

The first orbital can only contain two electrons. If an atom has more than two electrons (as do all atoms except hydrogen and helium), the additional electrons must occupy orbitals that are farther removed from the nucleus than the first. The next outer orbital can contain a maximum of eight electrons. The orbitals are filled from the innermost to the outermost. Carbon—with six electrons—thus has two electrons in its first orbital and four electrons in the second orbital (fig. 2.1). It is always the electrons in the outermost orbital that participate in chemical reactions.

Chemical Bonds

Molecules are formed when two or more atoms are joined together by sharing or some other interaction of the electrons in their outer orbitals. Such sharing of electrons produces *chemical bonds* (fig. 2.2). The number of bonds that

each atom can have is determined by the number of electrons in its outer orbital. Hydrogen, for example, must obtain only one more electron—and can thus form one chemical bond—to complete the first orbital of two electrons. Carbon, in contrast, must obtain four more electrons—and can thus form four chemical bonds—to complete the second orbital of eight electrons (fig. 2.3).

Covalent Bonds. Covalent *(ko-vāl'ent)* bonds that are formed between identical atoms—as in oxygen gas (O_2) and hydrogen gas (H_2)—are the strongest because their electrons are equally shared. Since the electrons are equally distributed between the two atoms, these molecules are said to be *nonpolar*. When covalent bonds are formed between two different atoms, however, the electrons may be pulled more toward one atom than the other. The side of the molecule toward which the electrons are pulled are electrically negative in comparison to the other end. Such a molecule is said to be *polar* (has a positive and negative "pole"). Atoms of oxygen, nitrogen, and phosphorous strongly attract electrons—are *electronegative*—when they bond with other atoms.

Figure 2.2. The hydrogen molecule, showing the covalent bonds between hydrogen atoms formed by the equal sharing of electrons.

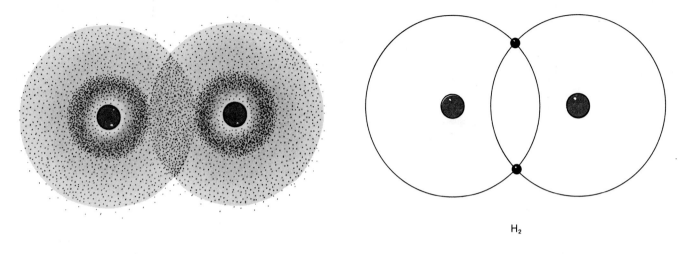

H_2

Figure 2.3. The molecules methane and ammonia represented in three different ways. Notice that a bond between two atoms consists of a pair of shared electrons (the electrons from the outer orbital of each atom).

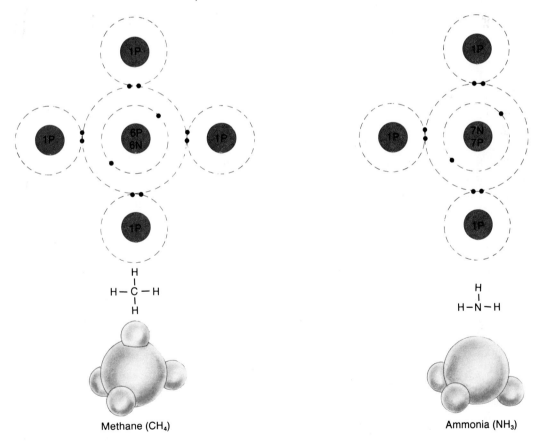

Methane (CH_4)

Ammonia (NH_3)

Figure 2.4. A model of a water molecule showing its polar nature. Notice that the oxygen side of the molecule is negative whereas the hydrogen side is positive. Polar covalent bonds are weaker than nonpolar covalent bonds. As a result, some water molecules ionize to form OH^- (hydroxyl ion) and H^+ (hydrogen ion).

Water (H_2O)

Water is the most abundant molecule in the body and serves as the solvent of body fluids. Water is a good solvent because it is polar; the oxygen atom pulls electrons from the two hydrogens toward its side of the water molecule, so that the oxygen side is negatively charged compared to the hydrogen side of the molecule (fig. 2.4). The significance of the polar nature of water in its function as a solvent is discussed in the next section.

Ionic Bonds. In covalent bonds, atoms are joined together by the sharing of electrons—equally in nonpolar covalent bonds, unequally in polar covalent bonds. In *ionic bonds* the electrons are not shared at all. Instead, they are transferred from one atom to another. Common table salt, sodium chloride (NaCl), serves as an example. Sodium, with a total of eleven electrons, has two in its first orbital, eight in its second orbital, and only one in its third orbital.

Figure 2.5. The ionization of sodium and chlorine to produce sodium and chlorine ions. The positive sodium and the negative chloride ions attract each other to produce the ionic compound sodium chloride (NaCl).

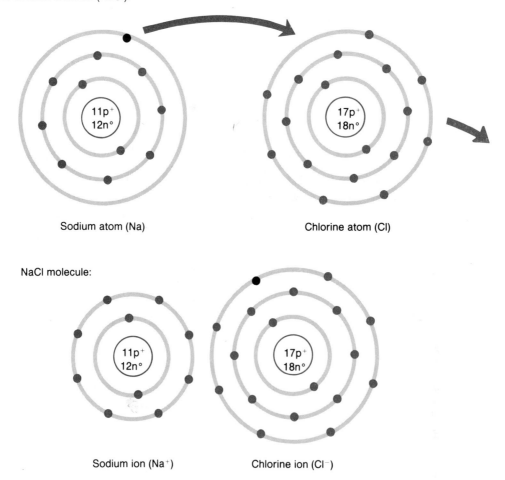

Chloride, conversely, is only one electron short of completing its outer orbital of eight electrons. The lone electron in sodium's outer orbital is attracted to chloride's outer orbital. This creates a *chloride* **ion** *(i'on)*—represented by the symbol Cl⁻—with one extra electron from sodium's outer orbital, and a *sodium ion* (Na⁺), which lost an electron to chloride's outer orbital. The attraction between these two ions produces the *ionic compound* NaCl (fig. 2.5).

Ionic bonds are weaker than polar covalent bonds and, therefore, break more easily when dissolved in water. Sodium chloride almost completely ionizes (separates into ions) when dissolved. Each of the ions released (Na⁺ and Cl⁻) attracts polar water molecules; the negative ends of water molecules are attracted to the Na⁺, and the positive ends are attracted to the Cl⁻ (fig. 2.6). The water molecules that surround these ions in turn attract other molecules of water to form *hydration spheres* around each ion.

The formation of hydration spheres makes an ion or a molecule soluble in water. Glucose, amino acids, and many other organic molecules are water-soluble because hydration spheres can form around atoms of oxygen, nitrogen, and phosphorous, which are joined by polar covalent bonds to other atoms. Such molecules are said to

be **hydrophilic** *(hi''dro-fil'ik)*. In contrast, molecules composed primarily of nonpolar covalent bonds, such as the hydrocarbon chains of fat molecules, have few charges and, thus, cannot form hydration spheres. They are insoluble in water and, in fact, actually avoid water—they are **hydrophobic.**

Hydrogen Bonds. Hydrogen bonds are weaker than ionic bonds. When hydrogen forms a polar covalent bond with an atom of oxygen or nitrogen, the hydrogen gains a slight positive charge as its electron is pulled toward the electronegative atom. Since the hydrogen has a slight positive charge, it will have a weak attraction for a second electronegative atom. This weak attraction is called a **hydrogen bond.** Hydrogen bonds are often shown with dotted lines (fig. 2.7) to distinguish them from stronger covalent bonds, which are shown with solid lines.

Water molecules are bonded to each other by weak hydrogen bonds (fig. 2.7). These bonds are responsible for many of the physical properties of water, including its *surface tension* and its ability to be pulled as a column through narrow channels in a process called *capillary action.*

---

hydrophilic: Gk. *hydor*, water; *philos*, fond
hydrophobic: Gk. *hydor*, water; *phobos*, fear

Figure 2.6.   The negatively charged oxygen-ends of water molecules are attracted to the positively charged Na⁺, whereas the positively charged hydrogen-ends of water molecules are attracted to the negatively charged Cl⁻. Other water molecules are attracted to this first concentric layer of water, forming hydration spheres around the sodium and chloride ions.

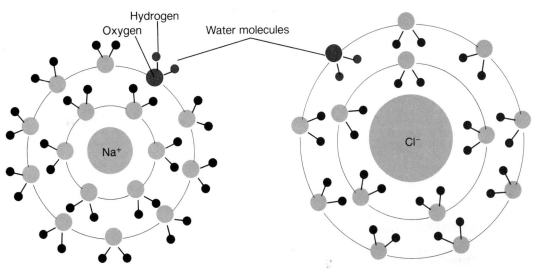

Figure 2.7. The oxygen atoms of water molecules are weakly joined together by the attraction of the electronegative oxygen for the positively charged hydrogen. These weak bonds are called *hydrogen bonds*.

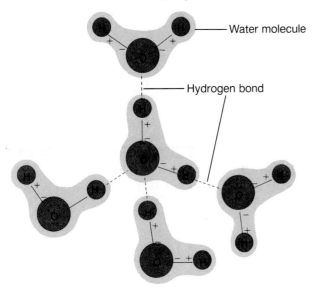

Water molecule

Hydrogen bond

Acids, Bases, and the pH Scale

The bonds joining hydrogen atoms to oxygen atoms in water molecules are, as previously discussed, polar covalent bonds. Although these bonds are stronger than ionic bonds, they are sufficiently weak to allow some water molecules to break apart, or dissociate, as a hydrogen atom completely loses its electron to oxygen. When this occurs, the hydrogen nucleus—consisting of a single proton—is released as a hydrogen ion (H^+).

The dissociation of hydrogen from water results in the production of two ions—H^+ (a hydrogen ion) and OH^- (a hydroxyl ion). Very few water molecules dissociate in this way, but those that do produce equal amounts of H^+ and OH^-. The concentration of each of these ions in pure water is equal to only 10^{-7} molar. (The term *molar* is a unit of concentration; described in chapter 5; for hydrogen, one molar equals one gram per liter.) A solution with this H^+ concentration—that is, a solution produced by the dissociation of water molecules, where the H^+ concentration is equal to the OH^- concentration—is said to be **neutral.**

A solution that contains a higher H^+ concentration than that of water is called *acidic (ă-sid'ik),* and a solution with a lower H^+ concentration than water is called *basic.* An **acid** is defined as a molecule that can release H^+ (protons) into a solution. A **base** is defined as a molecule that lowers the H^+ concentration of a solution. Most bases release OH^- into a solution, which combines with free H^+ to form water and thus lowers the H^+ concentration. Examples of common acids and bases are shown in table 2.2.

Table **2.2** Common acids and bases			
Acid	**Symbol**	**Base**	**Symbol**
Hydrochloric acid	HCl	Sodium hydroxide	NaOH
Phosphoric acid	H_3PO_4	Potassium hydroxide	KOH
Nitric acid	HNO_3	Calcium hydroxide	$Ca(OH)_2$
Sulfuric acid	H_2SO_4	Ammonium hydroxide	NH_4OH
Carbonic acid	H_2CO_3		

Source: Fox, Stuart Ira, *Laboratory Guide to Human Physiology, A: Concepts and Clinical Applications,* 3d ed. © 1976, 1980, 1984 Wm. C. Brown Publishers, Dubuque, Iowa. All Rights Reserved. Reprinted by permission.

Table **2.3** The pH scale			
	H^+ Concentration (molar)	**pH**	**OH^- Concentration (molar)**
Acids	1.0	0	10^{-14}
	0.1	1	10^{-13}
	0.01	2	10^{-12}
	0.001	3	10^{-11}
	0.0001	4	10^{-10}
	10^{-5}	5	10^{-9}
	10^{-6}	6	10^{-8}
Neutral	10^{-7}	7	10^{-7}
Bases	10^{-8}	8	10^{-6}
	10^{-9}	9	10^{-5}
	10^{-10}	10	0.0001
	10^{-11}	11	0.001
	10^{-12}	12	0.01
	10^{-13}	13	0.1
	10^{-14}	14	1.0

Source: Fox, Stuart Ira, *Human Physiology.* © 1984 Wm. C. Brown Publishers, Dubuque, Iowa. All Rights Reserved. Reprinted by permission.

pH. The H^+ concentration of a solution is usually indicated in pH units on a pH scale that runs from 0 to 14. The pH number is equal to one over the logarithm of the H^+ concentration:

$$pH = \log \frac{1}{[H^+]}$$

where $[H^+]$ = molar H^+ concentration.

Pure water has a H^+ concentration of 10^{-7} molar and, thus, has a pH of 7 (neutral). Because of the logarithmic relationship, a solution with ten times the hydrogen ion concentration (10^{-6} M) has a pH of 6, whereas a solution with one-tenth the H^+ concentration (10^{-8} M) has a pH of 8. The pH number is easier to write than the molar H^+ concentration, but it is admittedly confusing because it is *inversely related* to the H^+ concentration: a solution with a higher H^+ concentration has a lower pH number; one with a lower H^+ concentration has a higher pH number. A strong acid with a high H^+ concentration of 10^{-2} molar, for example, has a pH of 2, whereas a solution with only

Figure 2.8. Two carbon atoms joined by a single covalent bond (*above*) or by a double covalent bond (*below*). In both cases each carbon atom shares four pairs of electrons (has four bonds) to complete the eight electrons required to fill its outer orbital.

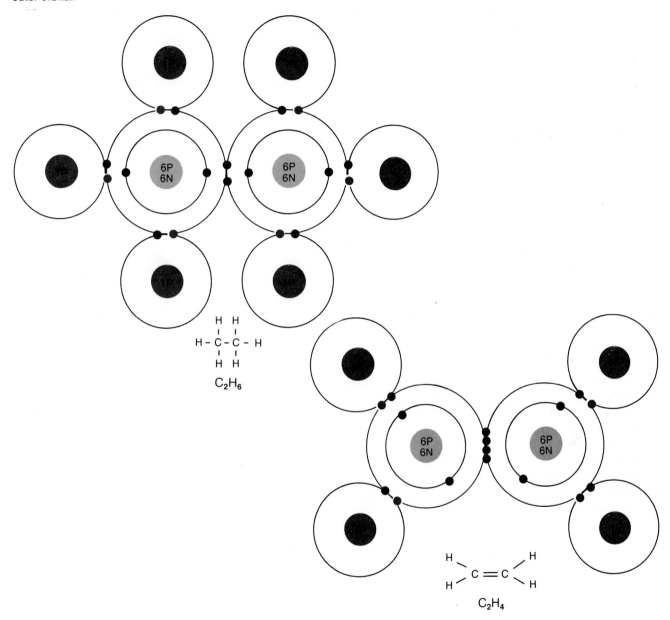

10⁻¹⁰ molar H⁺ has a pH of 10. **Acidic solutions,** therefore, have a pH of less than 7 (that of pure water), whereas **basic solutions** have a pH of between 7 and 14 (table 2.3).

Organic Molecules

Organic molecules are those that contain the atom *carbon.* Since the carbon atom has four electrons in its outer orbital, it must share four additional electrons by covalent bonding with other atoms to fill its outer orbital with eight electrons. The unique bonding requirements of carbon enable it to join with other carbon atoms to form chains and rings, while still allowing the carbon atoms to bond with hydrogen and other atoms.

Most organic molecules in the body contain hydrocarbon chains and rings as well as other atoms bonded to carbon. Two adjacent carbon atoms in a chain or ring may share one or two pairs of electrons. If the two carbon atoms share one pair of electrons, they are said to have a *single covalent bond;* this leaves each carbon atom free to bond to as many as three other atoms. If the two carbon atoms share two pairs of electrons, they have a *double covalent bond,* and each carbon atom can only bond to a maximum of two additional atoms (fig. 2.8).

Figure 2.9. Hydrocarbons that are (a) linear, (b) cyclic, and (c) aromatic rings.

Figure 2.10. Various functional groups of organic molecules.

The ends of some hydrocarbons are joined together to form rings. In the shorthand structural formulas of these molecules, the carbon atoms are not shown but are understood to be located at the corners of the ring. Some of these cyclic molecules have a double bond between two adjacent carbon atoms. Benzene and related molecules are shown as a six-sided ring with alternating double bonds. Such compounds are called *aromatic*. Since all of the carbons in an aromatic ring are equivalent, double bonds can be shown between any two adjacent carbons in the ring (fig. 2.9).

The hydrocarbon chain or ring of many organic molecules provides a relatively inactive molecular "backbone," to which more reactive groups of atoms are attached. Known as *functional groups* of the molecule, these reactive groups usually contain atoms of oxygen, nitrogen, phosphorous, or sulfur and are, in large part, responsible for the unique chemical properties of the molecule (fig. 2.10).

Classes of organic molecules can be named according to their functional groups. *Ketones (ke'tōnz)*, for example,

have a carbonyl group within the carbon chain. An organic molecule is an *alcohol* if it has a hydroxyl group at one end of the chain. All *organic acids* (such as acetic acid, citric acids, and others) have a *carboxyl (kar-bok'sil)* group (fig. 2.11).

1. Describe the structure of an atom, and define the terms atomic weight and atomic number.
2. List the types of chemical bonds in order of decreasing strength, and describe how these bonds are formed.
3. Write the definition of an acid and the equation for the definition of pH. Describe the relationship between pH and the H^+ concentration.
4. Describe the structure of water, and explain why ions and molecules with many polar covalent bonds are water-soluble.
5. Describe the structure of a carbon atom, and explain how this structure allows carbon atoms to bond together to form chains and rings.

Figure 2.11. Categories of organic molecules based on functional groups.

Ketone

Organic acid

Aldehyde

Alcohol

Figure 2.12. The structural formulas of three hexose sugars—(a) glucose, (b) galactose, and (c) fructose. All three have the same ratio of atoms—$C_6H_{12}O_6$.

(a)

Glucose

(b)

Galactose

(c)

Fructose

Carbohydrates and Lipids

Carbohydrates and lipids are large and diverse groups of organic compounds that serve as a principal source of food energy in the body. Although carbohydrates and lipids are composed of similar atoms, they have distinct chemical structures and biological functions.

Objective 5. List the different subcategories of carbohydrates, and give examples of each.

Objective 6. Describe dehydration synthesis and hydrolysis reactions, using the different subcategories of carbohydrates as examples. Describe where these reactions occur in the body.

Objective 7. Describe the different categories of lipids, explain why these molecules are all classified as lipids, and state the function of each category of lipids.

Objective 8. Explain the functions of triglycerides, steroids, phospholipids, and prostaglandins.

Carbohydrates and lipids are similar in many ways. Both groups of molecules consist primarily of the atoms carbon, hydrogen, and oxygen, and both serve as major sources of energy in the body (comprising most of the calories consumed in food). Carbohydrates and lipids differ, however, in some important aspects of their chemical structures and physical properties. Such differences significantly affect the functions of these molecules in the body.

Carbohydrates

Carbohydrates are organic molecules that contain carbon, hydrogen, and oxygen in the ratio described by their name—*carbo* (carbon) and *hydrate* (water, H_2O). The general formula of a carbohydrate molecule is thus CH_2O; the molecule contains twice the number of hydrogen atoms as it contains carbon or oxygen atoms.

Monosaccharides, Disaccharides, and Polysaccharides. Carbohydrates include simple sugars, or monosaccharides *(mon"o-sak'ah-rīdz),* and longer molecules that contain a number of monosaccharides joined together. The suffix *-ose* denotes a sugar molecule; the term *hexose,* for example, refers to a six-carbon monosaccharide with the formula $C_6H_{12}O_6$. This formula is adequate for some purposes, but it does not distinguish between related hexose sugars, which are *structural isomers* of each other. The structural isomers glucose, fructose, and galactose, for example, are monosaccharides that have the same ratio of atoms arranged in slightly different ways (fig. 2.12).

---

monosaccharide: Gk. *monos,* single; *sakcharon,* sugar

Figure 2.13. Glycogen is a polysaccharide composed of glucose subunits joined together to form a large, highly branched molecule.

Glycogen

Two monosaccharides can be joined covalently to form a **disaccharide** (dī-sak'ah-rīd), or double sugar. Common disaccharides include table sugar, or *sucrose* (composed of glucose and fructose), milk sugar, or *lactose* (composed of glucose and galactose), and malt sugar, or *maltose* (composed of two glucose molecules). When many monosaccharides are joined together, the resulting molecule is called a **polysaccharide.** *Starch,* for example, is a polysaccharide found in many plants, which is formed by the bonding together of thousands of glucose subunits. Animal starch (**glycogen**) (gli'co-gen), found in the liver and muscles, likewise consists of repeating glucose molecules but differs from plant starch in that it is more highly branched (fig. 2.13).

Dehydration Synthesis and Hydrolysis. In the formation of disaccharides and polysaccharides, the separate subunits (monosaccharides) are bonded together covalently by a type of reaction called **dehydration synthesis,** or **condensation.** In this reaction, which requires the participation of specific enzymes (chapter 4), a hydrogen atom is removed from one monosaccharide and a hydroxyl group (OH) is removed from another. As a covalent bond is formed between the two monosaccharides water (H_2O) is produced. Dehydration synthesis reactions are illustrated in figure 2.14.

When a person eats disaccharides and polysaccharides or when the stored glycogen in the liver and muscles is to be used by tissue cells, the covalent bonds that join monosaccharides into disaccharides and polysaccharides must be broken. These *digestion* reactions occur by means

Figure 2.14. Dehydration synthesis of two disaccharides, (a) maltose and (b) sucrose. Notice that as the disaccharides are formed a molecule of water is produced.

(a)

Glucose   +   Glucose   =   **Maltose**   +   Water

(b)

Glucose   Fructose

Sucrose

of **hydrolysis** *(hi-drol'ĭ-sis)*. Hydrolysis is the reverse of dehydration synthesis. A water molecule is split, as implied by the word *hydrolysis*, and the resulting hydrogen atom is added to one of the free glucose molecules as the hydroxyl group is added to the other (fig. 2.15).

When a potato is eaten, the starch within it is hydrolyzed into separate glucose molecules within the intestine. This glucose is absorbed into the blood and carried to the tissues. Some tissue cells may use this glucose for energy. Liver and muscles, however, can store excess glucose in the form of glycogen by dehydration synthesis reactions in these cells. During fasting or prolonged exercise, the liver can add glucose to the blood through hydrolysis of its stored glycogen.

Dehydration synthesis, or condensation, and hydrolysis reactions do not occur spontaneously; they require the action of specific enzymes. Similar reactions, in the presence of other enzymes, build and break down lipids, proteins, and nucleic acids. In general, therefore, hydrolysis reactions digest molecules into their subunits, and dehydration synthesis reactions build larger molecules by the bonding together of their subunits.

hydrolysis: Gk. *hydor*, water; *lysis*, dissolution, or break

Lipids

The category of molecules known as lipids includes several types of molecules that differ greatly in chemical structure. These diverse molecules are all in the lipid category by virtue of a common physical property—they are all *insoluble in polar solvents* such as water. This is because lipids consist primarily of hydrocarbon chains and rings, which are nonpolar and, thus, hydrophobic. Although lipids are insoluble in water, they can be dissolved in nonpolar solvents such as ether, benzene, and related compounds.

Triglycerides. Triglycerides are a subcategory of lipids that includes fat and oil. These molecules are formed by the condensation of one molecule of *glycerol (glis'er-ol)* (a three-carbon alcohol) with three molecules of *fatty acids*. Each fatty acid molecule consists of a nonpolar hydrocarbon chain with a carboxylic acid group (abbreviated COOH) on one end. If the carbon atoms within the hydrocarbon chain are joined by single covalent bonds, so that each carbon atom can also bond to two hydrogen atoms, the fatty acid is said to be *saturated*. If there are

Figure 2.15. The hydrolysis of starch (a) into disaccharides (maltose) and (b) into monosaccharides (glucose). Notice that as the covalent bond between the subunits breaks, a molecule of water is split. In this way the hydrogen and hydroxyl from water is added to the ends of the released subunits.

Starch

Maltose

Figure 2.16. Structural formulas for (a) saturated and (b) unsaturated fatty acids.

Palmitic acid,

a saturated fatty acid

Linolenic acid,

an unsaturated fatty acid

Figure 2.17. Dehydration synthesis of a triglyceride molecule from a glycerol and three fatty acids. A molecule of water is produced as an ester bond forms between each fatty acid and the glycerol. Sawtooth lines represent carbon chains, which are symbolized by an R.

a number of double covalent bonds within the hydrocarbon chain, so that each carbon atom can only bond to one hydrogen atom, the fatty acid is said to be *unsaturated*. Triglycerides that contain saturated fatty acids are called **saturated fats;** those that contain unsaturated fatty acids are **unsaturated fats** (fig. 2.16).

Within the adipose cells of the body, triglycerides are formed as the carboxylic acid ends of fatty acid molecules condense with the hydroxyl groups of a glycerol molecule (fig. 2.17). Since the hydrogen atoms from the carboxyl ends of fatty acid molecules form water molecules during dehydration synthesis, fatty acids that are combined with glycerol can no longer release H^+ and function as acids. For this reason, triglycerides are described as *neutral fats*.

Ketone Bodies. Hydrolysis of triglycerides within adipose tissue releases *free fatty acids* into the blood. Free fatty acids can be used as an immediate source of energy by many organs; they can also be converted by the liver into derivatives called *ketone bodies*. These include four-carbon-long acidic molecules (acetoacetic acid and β-hydroxybutyric acid) and acetone (the active ingredient in nail polish remover). A rapid breakdown of fat, such as occurs during dieting and in uncontrolled diabetes mellitus, results in elevated levels of ketone bodies in the blood, a condition called **ketosis.** If there are sufficient amounts of ketone bodies in the blood to lower the blood pH, the condition is called **ketoacidosis.**

Phospholipids. The class of lipids known as *phospholipids* contains a number of different categories, which have in common the fact that they are lipids that contain a phosphate group. The most common type of phospholipid molecule has this structure: the three-carbon alcohol molecule glycerol is attached to two fatty acid molecules; the third carbon atom of the glycerol molecule is attached to a phosphate group, and the phosphate group in turn is bonded to other molecules. If the phosphate group is attached to a nitrogen-containing choline molecule, the phospholipid molecule thus formed is known as *lecithin* (fig. 2.18). Figure 2.18 shows a simple way of illustrating the structure of a phospholipid—the parts of the molecule capable of ionizing and thus becoming charged are shown as a circle, whereas the nonpolar parts of the molecule are represented by lines.

Figure 2.18.   The structure of lecithin, a typical phospholipid (*above*), and its more simplified representation (*below*).

$H_2-C-O-\overset{\displaystyle O}{\overset{\|}{C}}$ /\/\^R_1 /\/\/\/\

$H_2-C-O-\overset{\displaystyle O}{\overset{\|}{C}}$ /\/\ R_2 /\/\/\/\

$H_2-C-O-\overset{\displaystyle O}{\overset{\|}{P}}-O$

$\overset{(-)}{O}$ CH_2

CH_2

$(+)N-CH_3$ / $-CH_3$ \ $-CH_3$

Polar end (hydrophilic)   Nonpolar hydrocarbon end (hydrophobic)

Lecithin

Since the nonpolar ends of phospholipids are hydrophobic, they tend to group together when mixed in water; this allows the hydrophilic parts (which are polar) to face the surrounding water molecules (fig. 2.19). Such aggregates of molecules are called **micelles** *(mi-selz')*. The dual nature of phospholipid molecules (part polar, part nonpolar) allows them to form the major component of the cell membrane as well as to alter the structure of water and decrease its surface tension. This latter function of phospholipids, which makes them *surfactants* (surface-active-agents), prevents collapse of the lungs.

Steroids.   The structure of steroid molecules is quite different than that of triglycerides or phospholipids, and yet steroids are still included in the lipid category of molecules because they are nonpolar and insoluble in water. All steroid molecules have the same basic structure; three six-carbon rings are joined to one five-carbon ring (fig. 2.20). However, different kinds of steroids have different

Figure 2.19.   The formation of a micelle structure by phospholipids such as lecithin. The straight lines represent the hydrophobic fatty acid parts of the molecule, and the circles represent the polar phosphate part of the molecule. The detailed structure of lecithin is shown in one part of the micelle.

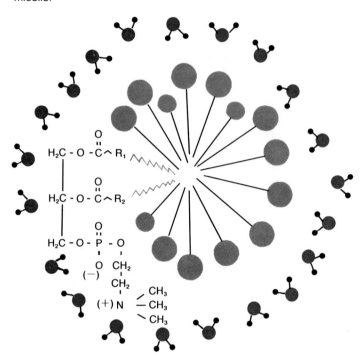

functional groups attached to this basic structure, and they vary in the number and position of the double covalent bonds between the carbon atoms in the rings.

Cholesterol is an important molecule in the body because it serves as the precursor (parent molecule) for the steroid hormones produced by the gonads and adrenal cortex. The testes and ovaries (collectively called the *gonads*) secrete **sex steroids,** which include estradiol and progesterone from the ovaries and testosterone from the testes. The adrenal cortex secretes the **corticosteroids,** including hydrocortisone and aldosterone, among others.

Prostaglandins.   Prostaglandins are a type of fatty acid (with a cyclic hydrocarbon group), which have a variety of regulatory functions. Although their name is derived from the fact that they were originally found in the semen as a secretion of the prostate gland, it has since been shown that they are produced by and active in almost all tissues. Prostaglandins are implicated in the regulation of blood vessel diameter, ovulation, uterine contraction during labor, inflammation reactions, blood clotting, and many other functions. Some of the different types of prostaglandins are shown in figure 2.21.

Figure 2.20. Cholesterol and some steroid hormones derived from cholesterol.

C_{27}
Cholesterol

C_{21}
Cortisol
(hydrocortisone)

C_{19}
Testosterone

C_{18}
Estradiol

---

1. *Define the category* carbohydrates, *and give examples of monosaccharides, disaccharides, and polysaccharides.*
2. *Describe dehydration synthesis and hydrolysis and where these reactions occur in the body, and give examples of these reactions in the metabolism of carbohydrates.*
3. *Define the different categories of lipids, and explain the structure and biological functions of these molecules.*

Figure 2.21. Structural formulas of some prostaglandins.

Prostaglandin E_1

Prostaglandin F_1

Prostaglandin E_2

Prostaglandin F_2

Proteins and Nucleic Acids

Proteins and nucleic acids are large molecules formed of amino acids and nucleotides, respectively. Proteins provide much of the structure of body cells and are important for many physiological activities, and nucleic acids provide the basis for the genetic control of cellular activity.

Objective 9. Describe the structure of amino acids.
Objective 10. Describe the primary, secondary, tertiary, and quaternary structure of proteins.
Objective 11. List some of the functions of different proteins, and explain how the great diversity of protein structure contributes to these functions.
Objective 12. Describe the structure of DNA and RNA, and explain how the law of complementary base pairing contributes to the structure of these molecules.

Figure 2.22. Representative amino acids, showing different types of functional (*R*) groups.

Functional group

Amino group   Carboxylic acid group

Nonpolar amino acids

Valine

Tryosine

Polar amino acids

Basic   Sulfur-containing   Acidic

Arginine   Cysteine   Aspartic acid

Proteins *(pro'te-in)* and nucleic acids are giant molecules, or **macromolecules.** The DNA contains the genetic code; RNA and protein provide the means by which this code can be expressed within the cells of the body. In order to understand these genetic control mechanisms (chapter 3), a knowledge of the structure of proteins and nucleic acids is required.

Proteins

Proteins consist of long chains of subunits called **amino acids** *(ah-me'no as'idz).* As the name implies, each amino acid contains an *amino group* (NH_2) on one end of the molecule and a *carboxylic acid group* (COOH) on the other end. There are approximately twenty different amino acids in the body, with different structures and chemical properties. These differences are due to differences in the

Figure 2.23. The formation of peptide bonds by a
dehydration synthesis reaction between amino acids.

functional groups of these amino acids, which is abbreviated as *R* in the general formula for an amino acid (fig. 2.22). The *R* symbol actually stands for the word *residue*, but it can be thought of as indicating the "rest of the molecule."

When amino acids are joined together by dehydration synthesis, the hydrogen from the amino end of one amino acid combines with the hydroxyl group of the carboxylic acid end of another amino acid. As a covalent bond is formed between the two amino acids water is produced (fig. 2.23). The bond between adjacent amino acids is called a **peptide bond**, and the compound formed is called a *peptide*. When many amino acids are joined in this way, a chain of amino acids, or **polypeptide,** is produced.

The lengths of polypeptide chains vary greatly. A hormone called *thyrotrophin-releasing hormone,* for example, is only three amino acids long, whereas *myosin (mi'o-sin)* (a muscle protein) contains about forty-five hundred amino acids. When the length of a polypeptide chain becomes very long (greater than about a hundred amino acids), the molecule is called a **protein.**

Protein Structure. The structure of a protein can be described at four different levels. At the first level, the sequence of amino acids in the protein, called the **primary structure** of the protein, is described. Each type of protein has a different primary structure. All of the billions of *copies* of a given type of protein in the body, however, have the same structure, because the structure of a given protein is coded by the genes. The primary structure of a protein is illustrated in figure 2.24.

Weak interactions (such as hydrogen bonds) between functional (*R*) groups of amino acids in nearby positions in the polypeptide chain cause this chain to twist into a *helix*. The extent and location of the helical structure is different for each protein because of differences in amino acid composition. A description of the helical structure of a protein is termed its **secondary structure** (fig. 2.24).

Most polypeptide chains bend and fold on themselves to produce complex three-dimensional shapes, called the **tertiary structure** of the proteins. Each type of protein has its own characteristic tertiary structure. This is because the folding and bending of the polypeptide chain is produced by chemical interactions between particular amino acids which are located in different regions of the chain.

Most of the tertiary structure of proteins is formed and stabilized by weak chemical interactions (such as hydrogen bonds) between widely spaced amino acids. The tertiary structure of some proteins, however, is made more stable by covalent bonds between sulfur atoms (called *disulfide bonds* and abbreviated S-S) in the functional (*R*) groups of amino acids known as cysteines (fig. 2.25). These strong, covalent bonds are the exception. Since most of the tertiary structure is stabilized by weak bonds, this structure can easily be disrupted by high temperature or by changes in pH. Irreversible changes in the tertiary structure of proteins produced by this means are referred to as *denaturation (de''na-chu-ra'shun)* of the proteins.

Denatured proteins retain their primary structure (the peptide bonds are not broken) but have altered chemical properties. Cooking a pot roast, for example, alters the texture of the meat proteins—it doesn't result in an amino acid soup. Denaturation is most dramatically demonstrated by frying an egg. Egg-albumin proteins are soluble in their native state, in which they form the clear, viscous fluid of a raw egg. When denatured by cooking, these proteins change shape, cross-bond with each other, and by this means form an insoluble white precipitate—the egg white.

Figure 2.24.   A polypeptide chain, showing (a) its primary structure and (b) secondary structure.

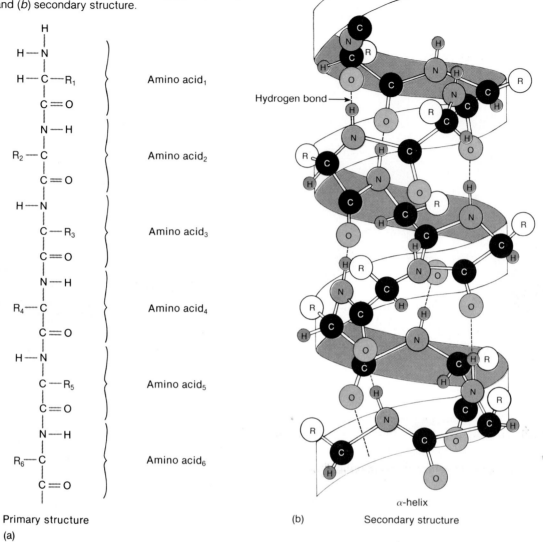

Primary structure

(a)

(b)   Secondary structure

α-helix

Hydrogen bond →

Table **2.4** Composition of selected proteins in the body			
Protein	**Number of polypeptide chains**	**Nonprotein component**	**Function**
Hemoglobin	4	Heme pigment	Carries oxygen in the blood
Myoglobin	2	Heme pigment	Stores oxygen in muscle
Insulin	2	None	Hormone-regulating metabolism
Luteinizing hormone	1	Carbohydrate	Hormone that stimulates gonads
Fibrinogen	1	Carbohydrate	Involved in blood clotting
Mucin	1	Carbohydrate	Forms mucus
Blood group proteins	1	Carbohydrate	Produces blood types
Lipoproteins	1	Lipids	Transports lipids in blood

Figure 2.25.   The tertiary structure of a protein.
(a) interactions between functional (R) groups of amino acids
result in (b) the formation of complex three-dimensional
shapes of proteins.

(a)

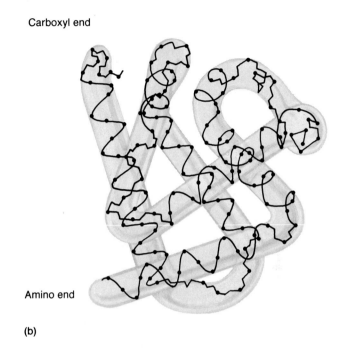

Carboxyl end

Nonhelical segment

Helical segment

Amino end

(b)

Some proteins (such as hemoglobin and insulin) are composed of a number of polypeptide chains covalently bonded together. This is the **quaternary structure** of these proteins. Insulin, for example, is composed of two polypeptide chains, one that is twenty-one amino acids long, the other that is thirty amino acids long. Hemoglobin (the protein in red blood cells that carries oxygen) is composed of four separate polypeptide chains. The composition of various body proteins is shown in table 2.4.

Many proteins in the body are normally found combined, or *conjugated,* with other types of molecules. **Glycoproteins** are proteins conjugated with carbohydrates. Examples of such molecules include certain hormones and some proteins found in the cell membrane. **Lipoproteins** are proteins conjugated with lipids. These are found in cell membranes and in the plasma (the fluid portion of the blood). Proteins conjugated with pigment molecules are chromoproteins. These include hemoglobin, which transports oxygen in red blood cells, and the cytochromes, which are needed for oxygen utilization and energy production within cells.

Functions of Proteins.   Because of their tremendous structural diversity, proteins can serve a wider variety of functions than any other type of molecule in the body. Many proteins, for example, contribute significantly to the

Figure 2.26.   A photomicrograph of collagen fibers.

structure of different tissues and in this way play a passive role in the functions of these tissues. Examples of such *structural proteins* include collagen (fig. 2.26) and keratin. Collagen is a fibrous protein that provides tensile strength to connective tissues, such as tendons and ligaments. Keratin is found in the outer layer of dead cells in the epidermis, where it serves to prevent water loss through the skin.

chromoprotein: Gk. *chroma,* color; *proteios,* of the first quality

Figure 2.27. The general structure of a nucleotide and the formation of sugar-phosphate bonds between nucleotides to form a polymer.

Phosphate   Deoxyribose sugar   Base

Nucleotide

G) Guanine

T) Thymine

C) Cytosine

A) Adenine

Many proteins serve a more active role in the body where specialized structure and function are required. *Enzymes* and *antibodies,* for example, are proteins—no other type of molecule could provide the vast array of different structures needed for these functions. Proteins in cell membranes serve as *receptors* for specific regulator molecules (such as hormones) and as *carriers* that transport specific molecules across the membrane. Proteins provide the diversity of shape and chemical properties for the specificity required by these functions.

Nucleic Acids

Nucleic acids include the macromolecules of **DNA (deoxyribonucleic acid)** and **RNA (ribonucleic acid),** which are critically important in genetic regulation, and the subunits from which these molecules are formed. These subunits are known as *nucleotides.*

Nucleotides are used as subunits in the formation of long polynucleotide chains. The nucleotides themselves, however, are composed of subunits. Each nucleotide is composed of three parts—a five-carbon sugar, a phosphate group bonded to one end of the sugar, and a *nitrogenous (ni-trahj'ĕ-nus) base* bonded to the other end of the sugar (fig. 2.27). The nucleotide bases are cyclic nitrogen-containing molecules with either one ring of carbons (the *pyrimidines) (pi-rim'i-dēnz)* or two rings (the *purines).*

Figure 2.28. The four nitrogenous bases in deoxyribonucleic acid (DNA). Notice that hydrogen bonds can form between guanine and cytosine and between thymine and adenine.

Phosphate

Deoxyribose

Guanine

Cytosine

Thymine

Adenine

Deoxyribonucleic Acid. The structure of DNA serves as the basis for the genetic code. One might, therefore, expect DNA to have an extremely complex structure. Actually, although DNA is the largest molecule in the cell, it has a simpler structure than that of most proteins. This simplicity of structure deceived some of the early scientists into believing that the protein content of chromosomes, rather than their DNA content, provided the basis for the genetic code.

Sugar molecules in the nucleotides of DNA are a type of pentose (five-carbon) sugar called **deoxyribose** *(de-ok″se-ri′bōs)* (hence the name for this nucleic acid). Each deoxyribose sugar can be covalently bonded to one of four possible bases. These bases include the two purines (adenine and guanine) and the two pyrimidines (cytosine and thymine). There are thus four different types of nucleotides that can be used to produce the long DNA chains.

When nucleotides combine to form a chain, the phosphate group of one condenses with the deoxyribose sugar of another nucleotide. This forms a sugar-phosphate chain as water is removed in dehydration synthesis. Since the nitrogenous bases are attached to the sugar molecules, the sugar-phosphate chain looks like a "backbone" from which the bases project. Each of these bases can form hydrogen bonds with other bases, which are in turn joined to a different chain of nucleotides. Such hydrogen bonding between bases thus produces a *double-stranded* DNA molecule; the two strands are like a staircase, with the paired bases as steps (fig. 2.28).

Actually, the two chains of DNA twist about each other to form a **double helix**—the molecule is like a spiral staircase (fig. 2.29). It has been shown that the number of purine bases in DNA is equal to the number of pyrimidine bases. The reason for this is explained by the law of **complementary base pairing;** adenine can only pair with thymine (through two hydrogen bonds), whereas guanine can only pair with cytosine (through three hydrogen bonds). Knowing this rule, we could predict the base sequence of one DNA strand if we knew the sequence of bases in the complementary strand.

Although we can predict which base is opposite a given base in DNA, we cannot predict which bases are above or below that position within a single polynucleotide chain. Although there are only four bases, the number of possible base sequences along a stretch of several thousand nucleotides (the length of a gene) is almost infinite. Despite the almost infinite possible variety of sequences, almost all of the billions of copies of a particular gene in a person are identical.

Ribonucleic Acid. The genetic information contained in DNA functions to direct the activities of the cell through its production of another type of nucleic acid—*RNA (ribonucleic acid)*. Like DNA, RNA consists of long chains of nucleotides joined together by sugar-phosphate bonds.

Figure 2.29. The double helix structure of DNA.

Nucleotides in RNA, however, differ from those in DNA (fig. 2.30) in three ways: (1) a **ribonucleotide** *(ri″bo-nu′kle-o-tīd)* contains the sugar *ribose* (instead of deoxyribose); (2) the base *uracil (yūr′ah-sil)* is found in place of thymine; and (3) RNA is composed of a single polynucleotide strand (it is not double stranded like DNA).

Figure 2.30. Differences between the nucleotides and sugars in DNA and RNA.

DNA nucleotides contain

Deoxyribose

RNA nucleotides contain

Ribose

Thymine

Uracil

There are three types of RNA molecules that function in the cytoplasm of cells. These are *messenger RNA (mRNA), transfer RNA (tRNA),* and *ribosomal RNA (rRNA).* All three types are made within the cell nucleus by using information contained in DNA as a guide. The synthesis of RNA and its role in gene expression are discussed in chapter 3.

1. Write the general formula for an amino acid, and describe how amino acids differ from each other.
2. Describe the structure of proteins, list some of the functions of different proteins, and explain how this structure grants specificity to the actions of proteins.
3. Define the law of complementary base pairing, and explain how this law allows us to predict the sequence of bases on one DNA strand if we know the sequence of bases on the other strand.
4. Compare the structure of DNA with that of RNA.

Chapter Summary

I. Atoms, Ions, and Chemical Bonds
The anatomic structures and physiological processes of the body are based, to a large degree, on the properties and interactions of atoms, ions, and molecules.
A. Covalent bonds are formed by atoms that share electrons; these are the strongest type of chemical bonds.
 1. Electrons are equally shared in nonpolar covalent bonds and unequally shared in polar covalent bonds.
 2. Atoms of oxygen, nitrogen, and phosphorous strongly attract electrons and become electrically negative compared to the other atoms sharing electrons with them.
B. Ionic bonds are weak bonds that join atoms together in an ionic compound.
 1. If one atom in this compound takes the electron from another atom, it gains a net negative charge and the other becomes positively charged.
 2. Ionic bonds easily break when the ionic compound is dissolved in water; dissociation of the ionic compound yields charged atoms called ions.
C. When hydrogen is bonded to an electronegative atom, it gains a slight positive charge and is weakly attracted to another electronegative atom; this weak attraction is a hydrogen bond.

D. Acids donate hydrogen ions to solution, whereas bases lower the hydrogen ion concentration of a solution.
 1. The pH scale is a negative function of the logarithm of the hydrogen ion concentration.
 2. In a neutral solution the concentration of H^+ is equal to the concentration of OH^-, and the pH is 7.
 3. Acids raise the H^+ concentration and thus lower the pH below 7; bases lower the H^+ concentration and thus raise the pH above 7.
E. Organic molecules contain atoms of carbon joined together by covalent bonds; atoms of nitrogen, oxygen, phosphorous, or sulfur may be present as specific functional groups in the organic molecule.

II. Carbohydrates and Lipids
Carbohydrates and lipids serve as principal sources of energy in the body. Although they are composed of similar atoms, they have distinct chemical structures and biological functions.
A. Carbohydrates contain carbon, hydrogen, and oxygen, usually in a ratio of 1:2:1.
 1. Carbohydrates consist of simple sugars (monosaccharides), disaccharides, and polysaccharides (such as glycogen).
 2. Covalent bonds between monosaccharides are formed by dehydration synthesis, or condensation; bonds are broken by hydrolysis reactions.

B. Lipids are organic molecules that are insoluble in polar solvents such as water.
 1. Triglycerides (fat and oil) consist of three fatty acid molecules joined to a molecule of glycerol.
 2. Ketone bodies are smaller derivatives of fatty acids.
 3. Phospholipids (such as lecithin) are phosphate-containing lipids that have a polar group, which is hydrophilic; the rest of the molecule is hydrophobic.
 4. Steroids (including the hormones of the adrenal cortex and gonads) are lipids with a characteristic five-ring structure.
 5. Prostaglandins are a family of cyclic fatty acids, which serve a variety of regulatory functions.

III. Proteins and Nucleic Acids
Proteins and nucleic acids are large molecules formed of amino acids and nucleotides, respectively. These molecules provide the basis for cell structure, function, and genetic regulation.
A. Proteins are composed of long chains of amino acids bonded together by covalent peptide bonds.
 1. Each amino acid contains an amino group, a carboxyl group, and a functional group that is different for each of the more than twenty different amino acids.
 2. The polypeptide chain may be twisted into a helix (secondary structure) and bent and folded to form the tertiary structure of the protein.

3. Proteins that are composed of two or more polypeptide chains are said to have a quaternary structure.
4. Proteins may be combined with carbohydrates, lipids, or other molecules.
5. Because of their great variety of possible structures, proteins serve a wider variety of specific functions than any other type of molecule.

B. Nucleic acids include the macromolecules DNA and RNA, and their nucleotide subunits.

1. Each nucleotide consists of a pentose sugar (ribose or deoxyribose), a phosphate, and a nitrogenous base.
2. There are four nitrogenous bases, and thus four possible nucleotides used to make either DNA or RNA.
3. DNA bases include adenine, guanine, cytosine, and thymine; in RNA, uracil is substituted for thymine.
4. DNA consists of two polynucleotide strands, in the form of a double helix, with each strand joined to the other by hydrogen bonding between complementary bases.
5. Complementary base pairing in DNA occurs because adenine can only form hydrogen bonds to thymine (and vice versa), and guanine can only pair with cytosine (and vice versa).
6. RNA consists of only one polynucleotide strand.
7. Messenger RNA (mRNA), transfer RNA (tRNA), and ribosomal RNA (rRNA) are made in the nucleus according to the instructions given in the base sequence of DNA.

Review Activities

Objective Questions

1. Which of the following statements about atoms is *true*?
   (a) They have more protons than electrons.
   (b) They have more electrons than protons.
   (c) They are electrically neutral.
   (d) They have as many neutrons as they have electrons.
2. The bond between oxygen and hydrogen in a water molecule is a(n)
   (a) hydrogen bond
   (b) polar covalent bond
   (c) nonpolar covalent bond
   (d) ionic bond
3. Which of the following is a nonpolar covalent bond?
   (a) The bond between two carbons.
   (b) The bond between sodium and chloride.
   (c) The bond between two water molecules.
   (d) The bond between nitrogen and hydrogen.
4. Solution A has a pH of 2, and solution B has a pH of 10. Which of the following statements about these solutions is *true*?
   (a) Solution A has a higher H^+ concentration than solution B.
   (b) Solution B is basic.
   (c) Solution A is acidic.
   (d) All of these are true.

5. Glucose is a
   (a) disaccharide
   (b) polysaccharide
   (c) monosaccharide
   (d) phospholipid
6. Digestion reactions occur by means of
   (a) dehydration synthesis
   (b) hydrolysis
7. Carbohydrates are stored in the liver and muscles in the form of
   (a) glucose
   (b) triglycerides
   (c) glycogen
   (d) cholesterol
8. Lecithin is a
   (a) carbohydrate
   (b) protein
   (c) steroid
   (d) phospholipid
9. Which of the following lipids have regulatory roles in the body?
   (a) steroids
   (b) prostaglandins
   (c) triglycerides
   (d) both a and b
   (e) both b and c
10. The tertiary structure of a protein is *directly* determined by
   (a) the genes
   (b) the primary structure of the protein
   (c) enzymes that "mold" the shape of the protein
   (d) the position of peptide bonds

11. If four bases in one DNA strand are A (adenine), G (guanine), C (cytosine), and T (thymine), the complementary bases in the opposite strand will be
   (a) T,C,G,A
   (b) C,G,A,T
   (c) A,G,C,T
   (d) U,C,G,A
12. Which of the following statements about RNA is *true*?
   (a) It is made in the nucleus.
   (b) It contains the base uracil.
   (c) It is double stranded.
   (d) Both a and b are true.
   (e) Both b and c are true.

Essay Questions

1. Compare and contrast nonpolar covalent bonds, polar covalent bonds, and ionic bonds.
2. Give the definition of an acid and base, and explain how these influence the pH of a solution.
3. Using dehydration synthesis and hydrolysis reactions, explain the relationships between starch in an ingested potato, liver glycogen, and blood glucose.
4. "All fats are lipids, but not all lipids are fats." Explain why this statement is true.
5. Explain the relationship between the primary structure of a protein and its secondary and tertiary structures.

Selected Readings

Jackson, R. W., and A. M. Gotto. 1974. Phospholipids in biology and medicine. *New Engl. J. of Med.* 290:24.

Kuehl, F. A., Jr., and R. W. Egan. 1980. Prostaglandins, arachidonic acid, and inflammation. *Science* 210:978.
Prockop, D. J., and N. A. Guzman. 1977. Collagen disease and the biosynthesis of collagen. *Hospital Practice* 12:61

Sharon, N. May 1974. Glycoproteins. *Scientific American*.
Sharon, N. November 1980. Carbohydrates. *Scientific American*.

3

Cell Structure and

Concepts

The cell is the basic unit of structure and function in the body. The structure and function of cells depend upon the specific membranes and organelles characteristic of each type of cell.

Lysosomes are organelles that contain digestive enzymes. Mitochondria contain enzymes and other molecules needed for obtaining energy from the combustion of food molecules.

During cell division the DNA replicates itself, and duplicated copies are distributed to the daughter cells. When a cell is not dividing, DNA serves as the information source for the formation of RNA and hence of proteins.

Translation of the messenger RNA code into the amino acid sequence of proteins requires transfer RNA, specific enzymes, and ribosomes. The secretion of proteins requires the endoplasmic reticulum, Golgi apparatus, and secretory vesicles.

Genetic Regulation

Cell Structure and Movements

The cell is the basic unit of structure and function in the body. The structure and function of cells depend upon the specific membranes and organelles characteristic of each type of cell.

Objective 1. List the parts of a generalized cell.
Objective 2. Describe the structure of the cell membrane.
Objective 3. Describe the structure of the cytoskeleton.
Objective 4. Differentiate between cilia and flagella.
Objective 5. Explain the processes of endocytosis and exocytosis.

As the basic functional unit of the body, each cell is a highly organized molecular factory. Cells come in a great variety of shapes and sizes (although most are too small to be clearly seen with the unaided eye). This great variation, which is also apparent in the subcellular structures within different cells, reflects the variation of functions of different cells in the body. All cells, however, share certain characteristics—such as the fact that they are surrounded by a cell membrane—and most cells possess the structures listed in table 3.1. Thus, although no single cell can be considered "typical," the general structure of cells can be shown with a single illustration (fig. 3.1).

For descriptive purposes, a cell can be divided into three principal parts:

1. **Cell (plasma) membrane.** The outer, differentially permeable cell membrane gives form to the cell and separates the cell's internal structures from the extracellular environment.
2. **Cytoplasm** and **organelles** *(or″gah-nelz′).* The cytoplasm is the content of a cell between the nucleus and the cell membrane. Organelles are minute structures within the cytoplasm of a cell that are concerned with specific functions.
3. **Nucleus.** The nucleus is a large, generally spheroid body within a cell that contains the DNA, or genetic material, of a cell.

Table 3.1 Structure and function of cellular components

Component	Structure	Function
Cell (plasma) membrane	Membrane composed of phospholipid and protein molecules	Gives form to cell and controls passage of materials in and out of cell
Cytoplasm	Fluid, jellylike substance in which organelles are suspended	Serves as matrix substance in which chemical reactions occur
Endoplasmic reticulum	System of interconnected membrane-forming canals and tubules	Supporting framework within cytoplasm; transports materials and provides attachment for ribosomes
Ribosomes	Granular particles composed of protein and RNA	Synthesize proteins
Golgi apparatus	Cluster of flattened, membranous sacs	Synthesizes carbohydrates and packages molecules for secretion; secretes lipids and glycoproteins
Mitochondria	Membranous sacs with folded inner partitions	Release energy from food molecules and transform energy into usable ATP
Lysosomes	Membranous sacs	Digest foreign molecules and worn and damaged cells
Peroxisomes	Spherical membranous vesicles	Contain certain enzymes; form hydrogen peroxide

Source: Van De Graaff, Kent M., *Human Anatomy.* © 1984 Wm. C. Brown Publishers, Dubuque, Iowa. All Rights Reserved. Reprinted by permission.

Figure 3.1. A generalized cell and the principal organelles.

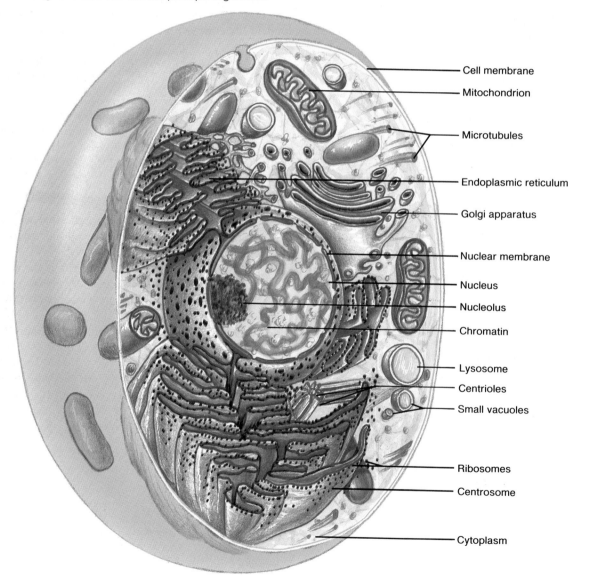

Component	Structure	Function
Centrosome	Nonmembranous mass of two rodlike centrioles	Helps organize spindle fibers and distribute chromosomes during mitosis
Vacuoles	Membranous sacs	Store and excrete various substances within the cytoplasm
Fibrils and microtubules	Thin, hollow tubes	Support cytoplasm and transport materials within the cytoplasm
Cilia and flagella	Minute cytoplasmic extensions from cell	Move particles along surface of cell or move cell
Nuclear membrane	Membrane surrounding nucleus, composed of protein and lipid molecules	Supports nucleus and controls passage of materials between nucleus and cytoplasm
Nucleolus	Dense, nonmembranous mass composed of protein and RNA molecules	Forms ribosomes
Chromatin	Fibrous strands composed of protein and DNA molecules	Controls cellular activity for carrying on life processes

Cell Membrane

Because both the intracellular and extracellular environments (or "compartments") are aqueous, a barrier must be present to prevent the loss of cellular molecules such as enzymes, nucleotides, and others that are water-soluble. Since this barrier cannot itself be composed of water-soluble molecules, it makes sense that the cell membrane is composed of lipids.

The cell membrane and indeed all of the membranes surrounding organelles within the cell are composed primarily of phospholipids. Phospholipids, as described in chapter 2, are polar on the end that contains the phosphate group and nonpolar (and hydrophobic) throughout the rest of the molecule. Since there is an aqueous environment on each side of the membrane, the hydrophobic parts of the molecules "huddle together" in the center of the membrane, leaving the polar ends exposed to water on both surfaces. This results in the formation of a double layer of phospholipids in the cell membrane.

The hydrophobic core of the membrane restricts the passage of water and water-soluble molecules and ions. Certain of these polar compounds, however, do pass through the membrane. The specialized functions and selective transport properties of the membrane are believed to be due to its protein content. Proteins are found partially submerged on each side of the membrane; some proteins span the membrane completely from one side to the other. Since the membrane is not solid—phospholipids and proteins are free to move laterally—the proteins within the phospholipid "sea" are not uniformly distributed, but rather present a mosaic pattern. This structure is known as the **fluid-mosaic model** of membrane structure (fig. 3.2).

aqueous: L. *aqua*, water

The proteins found in the cell membrane serve a variety of functions, including (1) structural support; (2) transport of molecules across the membrane; (3) enzymatic control of chemical reactions at the cell surface; (4) receptors for hormones and other regulatory molecules that arrive at the outer surface of the membrane; and (5) cellular "markers" (antigens), which identify the blood and tissue type.

The cell membranes of all higher organisms contain cholesterol. The ratio of cholesterol to phospholipids, as well as the ratio of different types of phospholipids, determines the flexibility of the cell membrane. When there is an inherited defect in these ratios, the flexibility of the cell may be reduced. This could result, for example, in the inability of red blood cells to flex at the middle when passing through narrow blood channels (thereby causing the occlusion of these small vessels).

In addition to lipids and proteins, the cell membrane also contains carbohydrates, which are primarily attached to the outer surface of the membrane as glycoproteins and glycolipids. These surface carbohydrates have many negative charges and, as a result, affect the interaction of regulatory molecules with the membrane. The negative charges at the surface also affect interactions between cells—they help keep red blood cells apart, for example. Stripping the carbohydrates from the outer red blood cell surface results in their more rapid destruction by the liver, spleen, and bone marrow.

Figure 3.2. The fluid-mosaic model of the cell membrane. The membrane consists of a double layer of phospholipids, with the phosphates (*open circles*) oriented outward and the hydrophobic hydrocarbons (*wavy lines*) oriented toward the center. Proteins may completely or partially span the membrane. Carbohydrates (*solid circles*) are attached to the outer surface.

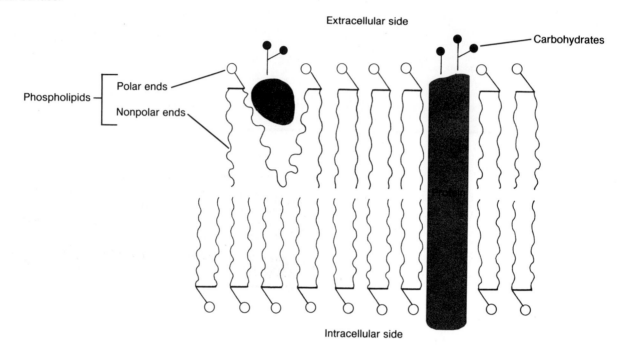

Figure 3.3. An immunofluorescence photograph of microtubules forming the cytoskeleton of a cell. Microtubules are visualized with the aid of antibodies against tubulin, the major protein component of the microtubules.

Cytoskeleton

The jellylike matrix within a cell (exclusive of that within the nucleus) is known as **cytoplasm** *(si'to-plazm)*. When viewed in a microscope without special techniques, the cytoplasm appears to be uniform and unstructured. According to recent evidence, however, the cytoplasm is not a homogenous solution; it is, rather, a highly organized structure in which protein fibers—in the form of *microtubules* and *microfilaments*—are arranged in a complex latticework. These can be seen by fluorescence microscopy with the aid of antibodies against the proteins that compose these structures (fig. 3.3). The interconnected microfilaments and microtubules are believed to provide structural organization for cytoplasmic enzymes and support for various organelles.

The latticework of microfilaments and microtubules is thus said to function as a **cytoskeleton** (fig. 3.4). The structure of this "skeleton" is not rigid; it has been shown to be capable of quite rapid reorganization. Contractile proteins—including actin and myosin, which are responsible for muscle contraction—may be able to shorten the length of some microfilaments. The cytoskeleton may thus represent the cellular "musculature." Microtubules, for example, form the *spindle apparatus* that pulls chromosomes away from each other in cell division; they also form the central parts of cilia and flagella.

Figure 3.4. A diagram of a proposed structure of the cytoskeleton. (From *The Ground Substance of the Living Cell*, by Keith Porter and Jonathon Tucker. Copyright © 1981 by Scientific American, Inc. All rights reserved.)

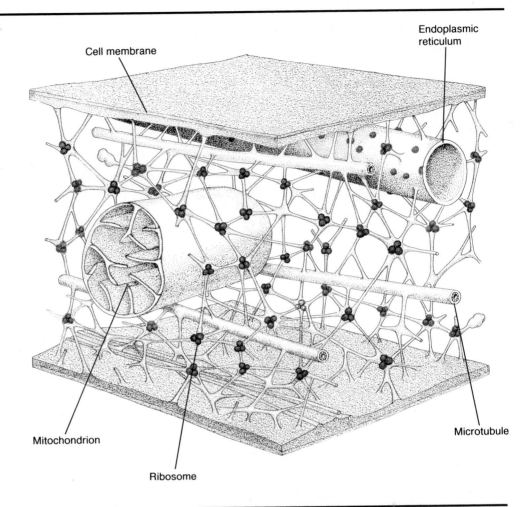

Endoplasmic reticulum

Cell membrane

Mitochondrion

Ribosome

Microtubule

Cellular Movements

Some body cells—including certain white blood cells and macrophages in connective tissues and microglial cells in the brain—are able to move like an amoeba (a single-celled animal). This "amoeboid" movement is performed by the extension of parts of the cytoplasm to form *pseudopods (soo'do-pods)*, which attach to a substrate and pull the cell along. These cells—as well as liver cells, which are not capable of amoeboid movement—use pseudopods to surround and engulf particles of organic matter (such as bacteria). This process is a type of cellular "eating" called **phagocytosis** *(fag''o-si-to'sis)*.

Phagocytic cells surround their victim with pseudopods, which join together and fuse (fig. 3.5). After the inner membrane of the pseudopods forms a continuous membranous barrier around the ingested particle it pinches off from the cell membrane, so that the ingested particle is contained in a *food vacuole* within the cell. The ingested particle will subsequently be digested by enzymes contained in a different organelle (the lysosome).

---

pseudopod: Gk. *pseudes*, false; *pons*, foot
phagocytosis: Gk. *phagein*, to eat; *kytos*, hollow body
vacuole: L. *vacuus*, empty

Figure 3.5. Scanning electron micrographs of phagocytosis showing the formation of pseudopods and the entrapment of the prey within a food vacuole.

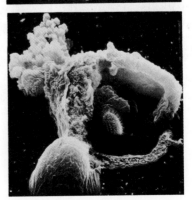

Figure 3.6.   Stages (1–4) of endocytosis, where specific bonding of extracellular particles to membrane receptor proteins is believed to occur.

Outside of cell

Cell membrane

Inside of cell

(1)

(2)

Extracellular environment

Cytoplasm

(3)

(4)

Pinocytosis *(pin″o-si-to′sis)* is a related process performed by many cells. Instead of forming pseudopods, the cell membrane invaginates to produce a deep, narrow furrow. The membrane near the surface of this furrow then fuses, and a small vacuole, containing extracellular fluid, is pinched off and enters the cell.

Endocytosis and Exocytosis. The processes by which part of the extracellular environment are brought into a cell by invagination of the cell membrane are called **endocytosis.** There are three types of endocytosis: phagocytosis, pinocytosis (as previously described), and *receptor-mediated endocytosis.* In the latter case, the interaction

of very specific molecules in the extracellular fluid with specific membrane receptor proteins causes the membrane to invaginate, fuse, and pinch off to form a *vesicle*—a small vacuole (fig. 3.6). Vesicles formed in this way contain extracellular fluid and molecules that could not have passed by other means into the cell.

Proteins and other molecules produced within the cell that are destined for export (secretion) are packaged inside of the cell within vesicles by an organelle known as the Golgi apparatus. In the process of **exocytosis,** these secretory vesicles fuse with the cell membrane and release their contents into the extracellular environment. This process adds new membrane material which replaces that which was lost from the cell membrane during endocytosis.

pinocytosis: Gk. *pinein,* to drink; *kytos,* hollow body
endocytosis: Gk. *endo,* within; *kytos,* hollow body

exocytosis: Gk. *exo,* outside; *kytos,* hollow body

Figure 3.7. Electron micrographs of cilia, showing (a) longitudinal and (b) cross sections. Notice the characteristic "9 + 2" arrangement of microtubules in the cross sections.

(a)

(b)

Endocytosis and exocytosis account for only part of the two-way traffic between the intracellular and extracellular compartments. Most of this traffic is due to membrane transport processes, the movement of molecules and ions through the cell membrane (chapter 5).

Cilia and Flagella. **Cilia** *(sil′e-ah)* are tiny hairlike structures that protrude from the cell and, like the coordinated action of oarsmen in a boat, stroke in unison. Cilia in the human body are found on the apical surface (the surface facing the lumen, or cavity) of epithelial cells in the respiratory and female genital tracts. In the respiratory system, the cilia transport strands of mucus, which are then conveyed by ciliary action to a region (the pharynx) where the mucus can either be swallowed or expectorated. In the female genital tract, ciliary movements in the epithelial lining draw the egg (ovum) into the uterine tube and move it towards the uterus.

cilia: L. *cili,* small hair

Sperm are the only cells in the human body that have **flagella** *(fla-jel′ah).* The flagellum is a single, whiplike structure that propels the sperm through its environment. Both cilia and flagella are composed of microtubules arranged in a characteristic way. Nine pairs of microtubules in the periphery of a cilium or a flagellum surround a single pair of microtubules in the center (fig. 3.7).

1. Draw the fluid-mosaic model of the cell membrane, and describe the structure of the membrane.
2. Describe the structure and possible functions of the cytoskeleton.
3. List the parts of a generalized cell, and distinguish between cilia and flagella.
4. Draw a figure showing phagocytosis; explain the events that occur during exocytosis.

Lysosomes and Mitochondria

Lysosomes and mitochondria are cellular organelles involved in enzymatic activities. Lysosomes contain digestive enzymes capable of breaking down molecules of protein, carbohydrate, and nucleic acid. Mitochondria contain enzymes and other molecules needed for obtaining energy from the combustion of food molecules.

Objective 6. Describe the structure and contents of lysosomes.

Objective 7. Describe the function of lysosomes and the significance of autophagy.

Objective 8. Describe the structure and function of mitochondria.

Lysosomes

After a phagocytic cell has engulfed the proteins, polysaccharides, and lipids present in a particle of "food" (such as a bacterium), these molecules are still kept isolated from the cytoplasm by the membranes surrounding the food vacuole. The large molecules of proteins, polysaccharides, and lipids must first be digested into their smaller subunits (amino acids, monosaccharides, and so on) before they can cross the vacuole membrane and enter the cytoplasm.

The digestive enzymes of a cell are isolated from the cytoplasm and concentrated within membrane-bound organelles called **lysosomes** (fig. 3.8). A *primary lysosome (li′so-sōm)* may fuse with a food vacuole (or with another cellular organelle) to form a *secondary lysosome.* In this way the products of phagocytosis and worn-out organelles can be digested. A lysosome that contains partially digested remnants of other organelles is called a *secondary*

flagellum: L. *flagrum,* whip

Figure 3.8.   Electron micrograph showing primary lysosomes (Lys₁) and secondary lysosomes (Lys₂). Mitochondria (Mi), Golgi apparatus (GA), and the nuclear envelope (NE) are also seen.

lysosome. A lysosome that contains undigested wastes is called a *residual body.* Residual bodies may eliminate their wastes by exocytosis, or they may accumulate within the cell as the cell ages.

Partly digested membranes of various organelles and other cellular debris are often observed within secondary lysosomes. This is a result of *autophagy,* a process that destroys worn-out organelles so that they can be continuously replaced. Lysosomes are thus aptly referred to as the "digestive system" of the cell.

---

autophagy: Gk. *autos,* self; *phagein,* to eat

Lysosomes have also been called "suicide bags," because a break in their membranes would release their digestive enzymes and thus destroy the cell. This happens normally as part of *programmed cell death,* in which the destruction of tissues is part of embryological development. It also occurs in white blood cells during an inflammation reaction.

Most, if not all, molecules in the cell have a limited life span. They are continuously destroyed and must be continuously replaced. Glycogen and some complex lipids in the brain, for example, are digested normally at a particular rate by lysosomes. If a person, because of some genetic defect, does not have the proper amount of these lysosomal enzymes, the resulting abnormal accumulation of glycogen and lipids could destroy the tissues.

Figure 3.9.   A mitochondrion (*a*). The outer membrane (CM) and the infoldings of the inner membrane—the cristae (*Cr*)—are clearly seen. The fluid in the center is the matrix (*Ma*). The structure of a mitochondrion is illustrated in (*b*).

(a)

(b)

Mitochondria

All cells in the body, with the exception of mature red blood cells, have a hundred to a few thousand organelles called **mitochondria** *(mi''to-kon'dre-ah)*. Mitochondria serve as sites for the production of most of the cellular energy (chapter 4). For this reason, mitochondria are sometimes called the "powerhouses" of the cell.

Mitochondria vary in size and shape, but all have the same basic structure (fig. 3.9). Each is surrounded by an *outer membrane* that is separated by a narrow space from an *inner membrane.* The inner membrane has many folds, called *cristae,* that extend into the central area (or *matrix*) of the mitochondrion. The cristae and the matrix provide different compartments in the mitochondrion and have different roles in the generation of cellular energy.

Mitochondria are able to migrate through the cytoplasm of a cell, and it is believed that they are able to reproduce themselves. Indeed, mitochondria contain their own DNA! This is a more primitive form of DNA than that found within the cell nucleus. For these and other reasons, many scientists believe that mitochondria evolved from separate organisms, related to bacteria, which entered animal cells and remained in a state of symbiosis.

> 1. *Describe the roles of lysosomes, and explain the importance of autophagy.*
> 2. *Describe the structure of a mitochondrion and the function it performs.*

mitochondrion: Gk. *mitos*, a thread; *chondros*, grain
symbiosis: Gk. *syn*, with; *bios*, life

Nuclear Structure and DNA Function

The nucleus of a cell contains DNA in its chromosomes and serves as the cell's control center. During cell division DNA replicates itself, and duplicated copies are distributed to the daughter cells. When a cell is not dividing, DNA serves as the information source for the formation of RNA and hence of proteins.

Objective 9.   Describe the structure of the nucleus and the chromosomes.

Objective 10.   Describe DNA synthesis, and explain why the term *semiconservative* is used.

Objective 11.   Describe the stages of the cell cycle and the different stages of mitosis.

Objective 12.   List the different types of RNA, and explain the meaning of the terms *genetic transcription* and *genetic translation.*

Most cells in the body have a single nucleus, although some—such as skeletal muscle cells—are multinucleate. The nucleus is surrounded by a *nuclear envelope* composed of an inner and an outer membrane; these two membranes fuse together to form thin sacs with openings called *nuclear pores* (figs. 3.10 and 3.11). These pores allow RNA to exit the nucleus (where it is formed) and enter the cytoplasm but prevent DNA from leaving the nucleus.

Figure 3.10.   An electron micrograph of a freeze-fractured nuclear envelope, showing the nuclear pores.

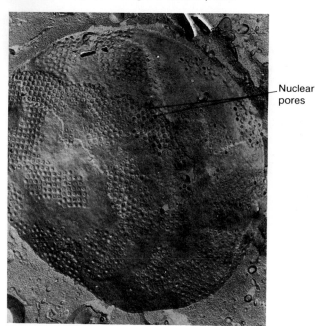

Nuclear pores

Figure 3.11.   The nucleus of a liver cell showing the nuclear envelope, heterochromatin, and nucleolus.

Nucleolus

Heterochromatin

Nuclear envelope

Many granulated threads in the nuclear fluid can be seen with an electron microscope. These threads are called *chromatin (kro'mah-tin)* and consist of a combination of DNA and protein. There are two forms of chromatin. Thin, extended chromatin—or *euchromatin (u-kro'mah-tin)* — appears to be the active form of DNA in a nondividing cell. Regions of condensed, "blotchy"-appearing chromatin, known as *heterochromatin,* are believed to contain inactive DNA.

One or more dark areas within each nucleus can be seen. These regions, which are not surrounded by membranes, are called *nucleoli (nu-kle'o-li).* The DNA within nucleoli contains genes that code for the production of ribosomal RNA (rRNA), an essential component of ribosomes.

DNA Replication

When a cell is going to divide, each DNA molecule replicates itself, and each of the identical DNA copies thus produced is distributed to the two daughter cells. Replication of DNA requires the action of a specific enzyme

chromatin: Gk. *chroma,* color

known as *DNA polymerase (pol''ah-mer-āz'').* This enzyme moves along the DNA molecule, breaking the weak hydrogen bonds between complementary bases as it travels. As a result, the bases of each of the two DNA strands become free to bond to new complementary bases (which are part of nucleotides) that are available within the surrounding environment.

Because of the rules of complementary base pairing, the bases of each original strand will bond to the appropriate free nucleotides: adenine bases pair with thymine-containing nucleotides; guanine bases pair with cytosine-containing nucleotides, and so on. In this way, two new molecules of DNA, each containing two complementary strands, are formed. The DNA polymerase enzyme links the phosphate groups and deoxyribose sugar groups together to form a second polynucleotide chain in each DNA that is complementary to the first DNA strands. Thus two new double-helix DNA molecules are produced that contain the same base sequence as the parent molecule (fig. 3.12).

When DNA replicates, therefore, each copy is composed of one new strand and one strand from the original DNA molecule. Replication is said to be **semiconservative** (half of the original DNA is "conserved" in each of the new DNA molecules). Through this mechanism, the sequence of bases in DNA—which is the basis of the genetic code—is preserved from one cell generation to the next.

Figure 3.12. The replication of DNA. Each new double helix is composed of one old and one new strand. The base sequences of each of the new molecules is identical to that of the parent DNA because of complementary base pairing.

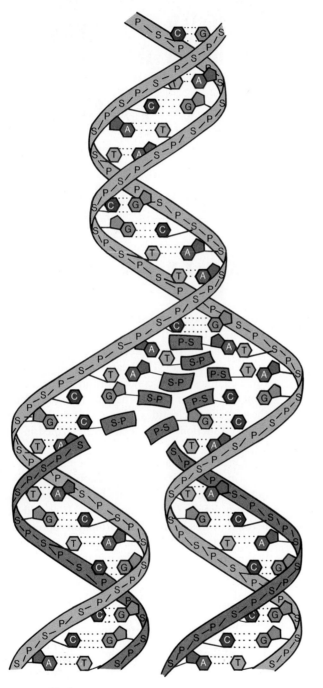

Region of parental DNA helix. (Both backbones are tan).

Region of replication. Parental DNA is unzipped and new nucleotides are pairing with those in parental strands.

Region of completed replication. Each double helix is composed of an old parental strand (tan) and a new daughter strand (red). Notice that each double helix is exactly like the other one.

Figure 3.13. The life cycle of a cell.

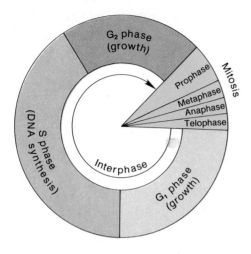

Figure 3.14. A photograph of homologous pairs of chromosomes from a human male cell at the metaphase of mitosis (homologous chromosomes have been paired and numbered according to convention).

The Cell Cycle

Unlike the life of an organism, which can be pictured as a linear progression from birth to death, the life of a cell follows a cyclical pattern. Each cell is produced as a part of its "parent" cell; when the daughter cell divides, it in turn becomes two new cells. In a sense, then, each cell is potentially immortal as long as its progeny can continue to divide. Some cells in the body divide frequently; the epidermis of the skin, for example, is renewed approximately every two weeks, and the stomach lining is renewed about every two or three days. Other cells, however, such as nerve and striated muscle cells in the adult, do not divide at all. All cells in the body, of course, live only as long as the person lives (some cells live longer than others, but eventually all cells die when vital functions cease).

The Cell Cycle.   The nondividing cell is in a part of its life cycle known as **interphase** (fig. 3.13). The chromosomes are in their extended form (as euchromatin), and their genes actively direct the synthesis of RNA. Through their direction of RNA synthesis, genes control the metabolism of the cell. During this time the cell may be growing, so this part of interphase is known as the G_1 *phase* (*G* stands for *growth*). Although sometimes described as "resting," cells in the G_1 phase perform the physiological functions characteristic of the tissue in which they are found.

If a cell is going to divide, it replicates its DNA in a part of interphase known as the *S phase* (*S* stands for *synthesis*). The mechanisms that cause transformation of a cell from the G_1 to the S phase are not known. Some experiments suggest that two nucleotides—cyclic adenosine

monophosphate (cAMP) and cyclic guanosine monophosphate (cGMP)—may regulate this transformation. An increase in the intracellular concentration of cAMP and a decrease in cGMP can stimulate cell division. Reverse changes in the concentration of the nucleotides inhibit cell division. Besides helping to regulate cell division, these cyclic nucleotides have a wide variety of other regulatory functions in the cell (as described in later chapters).

Once DNA has replicated, the chromatin condenses to form short, thick, rodlike structures. This is the more familiar form of chromosomes because they are easily seen in the ordinary (light) microscope. It should be remembered that this form of the chromatin represents a "packaged" state of DNA—not the extended, threadlike form that is active in directing the metabolism of the cell during the G_1 phase.

The matched pairs of chromosomes are called **homologous** *(ho-mol'ah-gus)* **chromosomes.** One member of each homologous pair is derived from a chromosome inherited from the father, and the other member is a copy of one of the chromosomes inherited from the mother. Homologous chromosomes do not have identical DNA base sequences; one member of the pair may code for blue eyes, for example, and the other for brown eyes. There are twenty-two homologous pairs of *autosomal (aw"to-so'mal) chromosomes* and one pair of *sex chromosomes,* described as X and Y. Females have two X chromosomes, whereas males have one X and one Y chromosome (fig. 3.14).

Figure 3.15. (a) an electron micrograph of a centriole magnified about 120,000 times. (b) the diagram shows the arrangement of the bundles of three microtubules in a centriole.

(a)

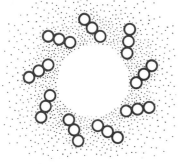

(b)

Centrioles. **Centrioles** *(sen'tre-ōlz)* are small, rodlike structures located near the nucleus. There are two centrioles, positioned at right angles to each other, within a spherical nonmembranous mass called a **centrosome.** Each centriole is composed of nine evenly spaced bundles of microtubules, with three microtubules per bundle (fig. 3.15).

Centrosomes are found only in those cells that are capable of division. During the process of cell division, mitosis *(mi-to'sis),* the centrioles take up positions on opposite sides of the nucleus. The centrioles are then involved in the production of, and are attached to, **spindle fibers,** which are also composed of microtubules. The spindle fibers and centrioles help to pull the duplicated chromosomes to opposite poles of the cell during cell division. Cells that lack centrioles, such as mature muscle and nerve cells, cannot divide.

mitosis: Gk. *mitos,* thread

Mitosis. Following the S phase of the cell cycle, each chromosome consists of two strands called **chromatids** *(kro'mah-tidz),* which are joined together by a *centromere.* The two chromatids within a chromosome contain identical DNA base sequences because each is produced by the semiconservative replication of DNA. Each chromatid, therefore, contains a complete double helix DNA molecule that is a copy of the single DNA molecule existing prior to replication. Each chromatid will become a separate chromosome once cell division has been completed.

Following a second resting phase (G_2), which is usually shorter than G_1, the cell proceeds through the various stages of cell division or **mitosis** (the *M phase* of the cell cycle). In mitosis, the chromosomes line up single file along the equator of the cell. Spindle fibers from the centrioles form and attach to the centromere of each chromosome (fig. 3.16).

When the spindle fibers shorten, the centromeres split apart and the two chromatids in each chromosome are pulled to opposite poles. Each pole therefore gets one copy of each of the forty-six chromosomes. Division of the cytoplasm (cytokinesis) results in the production of two daughter cells, which are genetically identical to each other and to the original parent cell.

Hypertrophy and Hyperplasia. The growth of an individual from a fertilized egg into an adult involves an increase in cell number and an increase in cell size. Growth due to an increase in cell number results from mitotic cell division and is termed **hyperplasia** *(hi"per-pla'ze-ah).* Growth of a tissue or organ due to an increase in cell size is termed **hypertrophy** *(hi-per'tro-fe).*

Most growth is due to hyperplasia. A callous on the palm of the hand, for example, involves thickening of the skin by hyperplasia due to frequent abrasion. An increase in skeletal muscle size as a result of exercise, in contrast, is generally believed to be produced by hypertrophy.

Meiosis. A special type of cell division occurs in the gonads (ovaries and testes) of sexually mature people. This type of cell division, known as meiosis *(mi-o'sis),* is used to produce gametes (ova and sperm). There are two stages of meiosis. In the *first meiotic division,* homologous chromosomes pair side by side along the equator of the cell (in contrast to single file in mitosis). Since each chromosome at this stage contains two identical chromatids, the paired chromosomes are known as a *tetrad.* Exchanges of genetic

meiosis: Gk. *meioun,* lessen
tetrad: Gk. *tetras,* group of four

Figure 3.16. The stages of mitosis.

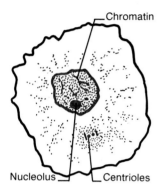

(a) Interphase

The chromosomes are in an extended form and seen as chromatin in the electron microscope.
The nucleus is visible.

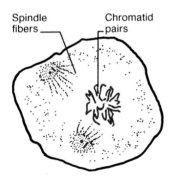

(b) Prophase

The chromosomes are seen and observed to consist of two chromatids joined together by a centromere.
The centrioles move apart towards opposite poles of the cell.
Spindle fibers are produced and extend from each centriole.
The nuclear membrane starts to disappear.
The nucleolus is no longer visible.

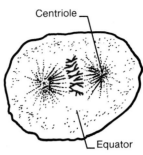

(c) Metaphase

The chromosomes are lined up at the equator of the cell.
The spindle fibers from each centriole are attached
to the centromeres of the chromosomes.
The nuclear membrane has disappeared.

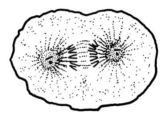

(d) Anaphase

The centromeres split, and the sister chromatids separate
as each is pulled to an opposite pole.

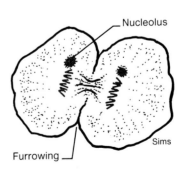

(e) Telophase

The chromosomes become longer, thinner, and less distinct.
New nuclear membranes form.
The nucleolus reappears.
Cell division (cytokinesis) is nearly complete.

Figure 3.17.   (*a*) genetic variation results from the crossing-over of tetrads, which occurs during the first meiotic prophase. (*b*) a diagram of the pairing of homologous chromosomes of a single tetrad: (*1*) before crossing-over; (*2*) chromatids crossing-over; and (*3*) results of crossing-over.

information can occur between paired homologous chromosomes, as illustrated in figure 3.17. This process, known as *crossing-over,* results in **genetic recombination.** Genetic recombination means that the gametes produced during meiosis are genetically unlike each other and unlike the parent that produced them.

The stages of the first meiotic division are labeled prophase I, metaphase I, anaphase I, and telophase I. At the end of the first meiotic division, the two daughter cells contain only one of each pair of homologous chromosomes; they have twenty-three chromosomes, although each chromosome contains duplicate chromatids. A *second meiotic division* follows, in which the identical chromatids are pulled apart and distributed to different cells (the gametes). The stages of this division are labeled prophase II, metaphase II, anaphase II, and telophase II. The details of meiosis and the production of sperm and ova will be treated in chapters 27 and 28.

Genetic Transcription—RNA Synthesis

When a cell completes mitotic division, it again enters the G_1 phase and its chromosomes become extended into the form known as euchromatin. This is the "working" form of DNA; the more familiar short, stubby form of chromosomes seen during cell division are inactive packages of DNA. The genes do not become active until the chromosomes unravel. Active DNA directs the metabolism of the cell indirectly through its regulation of RNA and protein synthesis.

One gene codes for one polypeptide chain. Each gene is a stretch of DNA that is several thousand nucleotide pairs long. The DNA in a human cell contains three to four billion base pairs—enough to code for at least three million proteins. Since the average human cell contains less than this amount (30,000 to 150,000 different proteins), it follows that only a fraction of the DNA in each cell is used to code for proteins. The remainder of the DNA may be inactive, may be redundant, and may serve to regulate those regions that do code for proteins.

In order for the genetic code to be translated into the synthesis of specific proteins, the DNA code must first be

Figure 3.18. The synthesis of ribosomal RNA on DNA in the nucleolus.

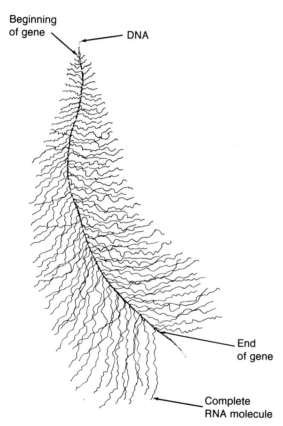

Beginning
of gene

DNA

End
of gene

Complete
RNA molecule

transcribed into an RNA code (fig. 3.18). This is accomplished by DNA-directed RNA synthesis, or **genetic transcription.**

In RNA synthesis, the enzyme *RNA polymerase* breaks the weak hydrogen bonds between paired DNA bases. This does not occur throughout the length of DNA, but only in the regions that are to be transcribed (there are base sequences that code for "start" and "stop"). Double-stranded DNA, therefore, separates in these regions so that the freed bases can pair with the complementary RNA nucleotide bases.

This pairing of bases, like that which occurs in DNA replication, follows the law of complementary base pairing: guanine bonds with cytosine (and vice versa), and adenine bonds with uracil (because uracil in RNA is equivalent to thymine in DNA). Unlike DNA replication, however, only *one* of the two freed strands of DNA serves as a guide for RNA synthesis.

Types of RNA. There are four types of RNA produced within the nucleus by genetic transcription: (1) **precursor messenger RNA (pre-mRNA)**, which is altered within the nucleus to form mRNA; (2) **messenger RNA (mRNA)**, which contains the code for the synthesis of specific proteins; (3) **transfer RNA (tRNA)**, which is needed for decoding the genetic message contained in mRNA; and (4) **ribosomal RNA (rRNA)**, which forms part of the structure of ribosomes. The genes that code for rRNA synthesis are located in the part of the nucleus called the nucleolus. The genes that code for pre-mRNA and tRNA synthesis are located elsewhere in the nucleus.

In bacteria, where the molecular biology of the gene is best understood, a gene that codes for one type of protein produces an mRNA molecule that begins to direct protein synthesis as soon as it is transcribed. This is not the case in higher organisms, including humans. In higher cells a pre-mRNA is produced that must be modified within the nucleus before it can enter the cytoplasm as mRNA and direct protein synthesis.

Precursor mRNA is much larger than the mRNA that it forms. This large size of pre-mRNA is, surprisingly, not due to excess bases at the ends of the molecule that must be trimmed. Rather, the excess bases are *within* the pre-mRNA. The genetic code for a particular protein, in other words, is split up by stretches of base pairs that do not contribute to the code. As a result, pre-mRNA must be cut and spliced to make mRNA. After cutting and splicing, the mRNA is further modified by the addition of characteristic bases at its ends; about two hundred adenine-containing nucleotides are added to one end, and a modified guanine-containing nucleotide is added to the other end of mRNA. These *post-transcriptional modifications* produce the form of mRNA that enters the cytoplasm.

1. Describe the structure of the nucieus and of the chromosomes.
2. Draw a simplified illustration of the semiconservative replication of DNA, using stick figures and two colors.
3. Describe the cell cycle, using the proper symbols to describe the different stages of the cycle.
4. List the phases of mitosis, and briefly describe the events that occur in each phase.
5. Describe how RNA is produced, and explain how pre-mRNA is converted into mRNA.
6. Define the terms genetic transcription and genetic translation, and list the different types of RNA.

Protein Synthesis and Secretion

After the genetic code is transcribed into the base sequences of messenger RNA, it is translated into the structure of proteins. Translation of the language of nucleotide bases into the language of amino acids requires transfer RNA, specific enzymes, and ribosomes. When the cell produces proteins for secretion, additional organelles—the endoplasmic reticulum, Golgi apparatus, and secretory vesicles—are required.

> *Objective 13.* Describe the functions of mRNA, tRNA, and rRNA.
>
> *Objective 14.* Explain how codons and anticodons function in protein synthesis.
>
> *Objective 15.* Describe the functions of the rough endoplasmic reticulum and the Golgi apparatus, and explain how secretory proteins are modified and packaged by these organelles.

When mRNA enters the cytoplasm it attaches to ribosomes, which are seen in the electron microscope as numerous small particles. **Ribosomes** *(ri'bo-sōm)* are composed of rRNA and protein, arranged to form two subunits of unequal size. The mRNA passes through a number of ribosomes to form a "string of pearls" structure called a *polyribosome* (or *polysome,* for short), shown in figure 3.19. Association of mRNA with ribosomes is needed for **genetic translation**—the production of specific proteins according to the code contained in the mRNA base sequence.

Each mRNA molecule contains several hundred or more nucleotides, arranged in the sequence determined by complementary base pairing with DNA during genetic transcription (RNA synthesis). Every three bases, or *base triplet,* is a code word—called a **codon**—for a specific amino acid. Sample codons and their amino acid "translations" are shown in table 3.2 and in figure 3.20. As mRNA moves through the ribosome, the sequence of codons is translated into a sequence of specific amino acids within a growing polypeptide chain.

Transfer RNA

Translation of the codons is accomplished by tRNA and particular enzymes. Each tRNA molecule, like mRNA and rRNA, is single stranded. Although tRNA is single stranded, it bends on itself to form three loop regions (fig. 3.21). One of these loops contains the anticodon—three nucleotides that are complementary to a specific codon in mRNA.

Enzymes in the cell cytoplasm called *aminoacyl-tRNA synthetase* enzymes join specific amino acids to the ends of tRNA, so that a tRNA with a given anticodon is always bonded to one specific amino acid. These enzymes

Figure 3.19. An electron micrograph of polyribosomes. An RNA strand *(arrow)* joins the ribosomes together.

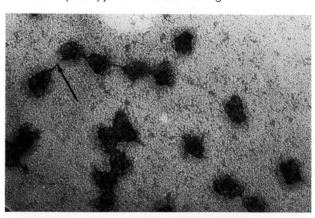

Table 3.2 Selected DNA base triplets and mRNA codons

DNA triplet	RNA codon	Amino acid
TAC	AUG	"Start"
ATC	UAG	"Stop"
AAA	UUU	Phenylalanine
AGG	UCC	Serine
ACA	UGU	Cysteine
GGG	CCC	Proline
GAA	CUU	Leucine
GCT	CGA	Arginine
TTT	AAA	Lysine
TGC	ACG	Tyrosine
CCG	GGC	Glycine
CTC	GAG	Aspartic acid

Source: Fox, Stuart Ira, *Human Physiology.* © 1984 Wm. C. Brown Publishers, Dubuque, Iowa. All Rights Reserved. Reprinted by permission.

don't actually "read" the anticodon. Rather, each tRNA with a different anticodon has a different "recognition loop" (fig. 3.21) that is detected by these enzymes.

Formation of a Polypeptide

The anticodons of tRNA bond to the codons of mRNA as the mRNA moves through the ribosome. Since each tRNA molecule carries a specific amino acid, the joining together of these amino acids by peptide bonds creates a polypeptide whose amino acid sequence has been determined by the sequence of codons in mRNA.

When the first and second tRNA bring the first and second amino acids together and a peptide bond forms between them, the first amino acid detaches from its tRNA so that a dipeptide is linked by the second amino acid to the second tRNA. When the third tRNA bonds to the third codon, the third amino acid forms a peptide bond with the second amino acid (which detaches from the second tRNA); a tripeptide is then attached by the third amino acid to the third tRNA. The polypeptide chain thus grows as new amino acids are added to its growing tip (fig. 3.22).

Figure 3.20. The genetic code is first transcribed into base triplets (codons) in mRNA and then translated into a specific sequence of amino acids in a protein.

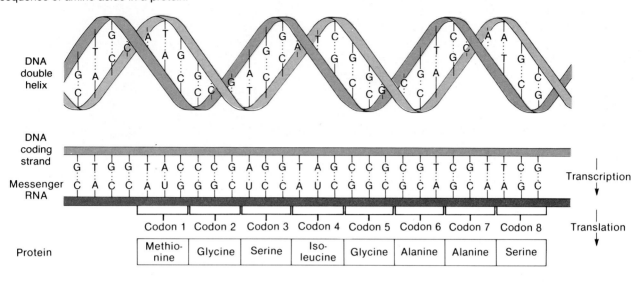

	Codon 1	Codon 2	Codon 3	Codon 4	Codon 5	Codon 6	Codon 7	Codon 8
Protein	Methio-nine	Glycine	Serine	Iso-leucine	Glycine	Alanine	Alanine	Serine

Figure 3.21. The structure of transfer RNA (tRNA).

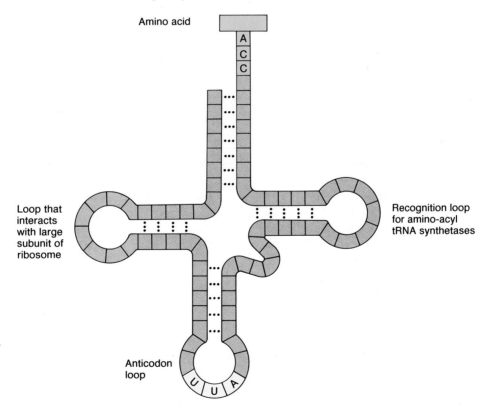

Figure 3.22.   The translation of messenger RNA (mRNA). As the anticodon of each new aminoacyl-tRNA bonds to a codon of the mRNA, new amino acids are joined to the growing tip of the polypeptide chain.

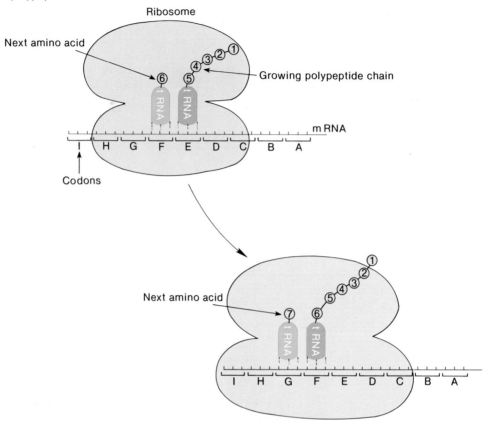

As the polypeptide chain grows in length, interactions between its amino acids cause the chain to twist into a helix (secondary structure) and to fold and bend on itself (tertiary structure). At the end of this process, the new protein detaches from the tRNA as the last amino acid is added. Many proteins are further modified after they are formed; these modifications occur in organelles called the endoplasmic reticulum and Golgi apparatus.

Endoplasmic Reticulum

Most cells contain a system of membranes known as the endoplasmic reticulum *(en''do-plas'mic rĕ-tik'u-lum),* of which there are two types: (1) a **rough,** or **granular, endoplasmic reticulum;** and (2) a **smooth endoplasmic reticulum** (fig. 3.23). A rough endoplasmic reticulum contains ribosomes on its surface, and a smooth endoplasmic reticulum does not contain ribosomes. The smooth endoplasmic reticulum is used for a variety of purposes in different cells; it serves as a site for enzyme reactions in steroid hormone production and inactivation, for example, and as a site for the storage of Ca^{++} in striated muscle cells. The rough endoplasmic reticulum is found in cells, such as those of exocrine and endocrine glands, that are active secretors of proteins.

Proteins that are to be used within the cell are produced in polyribosomes that are free in the cytoplasm. If the protein is a secretory product of the cell, however, it is made by mRNA-ribosome complexes located in the rough endoplasmic reticulum. The membranes of this system enclose fluid-filled spaces (cisternae), which the newly formed proteins may enter.

When proteins that are destined for secretion are produced, the first thirty or so amino acids are primarily hydophobic. This *leader sequence* is attracted to the lipid component of the membranes of the endoplasmic reticulum. As the polypeptide chain elongates, it is "injected" into the cisterna within the endoplasmic reticulum. The leader sequence is, in a sense, an "address" that directs secretory proteins into the endoplasmic reticulum. Once the proteins are in the cisterna, the leader sequence is removed so the protein cannot reenter the cytoplasm (fig. 3.24).

Figure 3.23. (*a*) an electron micrograph of endoplasmic reticulum magnified about 100,000 times. (*b*) rough endoplasmic reticulum has ribosomes attached to its surface, whereas (*c*) smooth endoplasmic reticulum lacks ribosomes.

Figure 3.24. A protein destined for secretion begins with a *leader sequence* that enables it to be inserted into the endoplasmic reticulum. Once it has been inserted, the leader sequence is removed and carbohydrate is added to the protein.

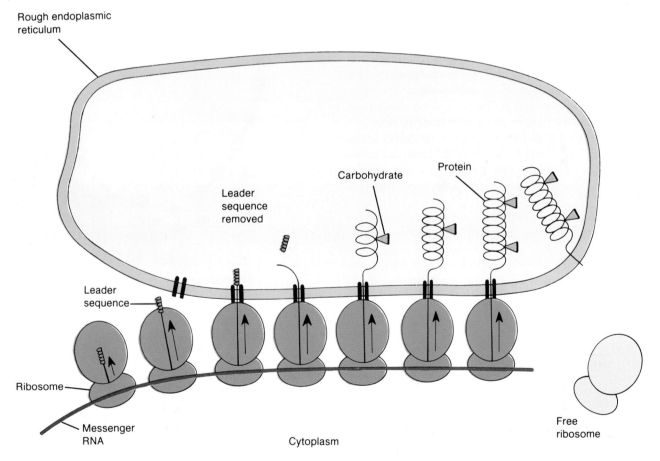

The processing of the hormone insulin can serve as an example of the post-translational changes that occur within the endoplasmic reticulum. Insulin enters the cisterna as a single polypeptide composed of 109 amino acids. The first twenty-three amino acids serve as a leader sequence and are quickly removed. The remaining chain folds within the cisterna so that the first and last amino acids in the polypeptide are brought close together. The central region is then enzymatically removed, producing two chains—one that is twenty-one amino acids long, the other that is thirty amino acids long—which are subsequently joined together by disulfide bonds. This is the form of insulin that is normally secreted from the cell.

Golgi Apparatus

Secretory proteins do not remain trapped within the endoplasmic reticulum; they are transported to another organelle within the cell—the **Golgi** *(gol'je)* **apparatus.** This organelle serves two interrelated functions: (1) further modifications of proteins (such as addition of carbohydrates) occur in the Golgi apparatus; and (2) different types of proteins are separated according to their function and destination. For example, proteins that are to be secreted are separated from those that will be incorporated into the cell membrane and from those that will be introduced into lysosomes.

The Golgi apparatus consists of several flattened sacs that enclose cisternae (spaces). These cisternae, in turn, communicate with those of the endoplasmic reticulum. This communication may be due to a direct connection between the two organelles, or it may occur by means of vesicles that pass from the endoplasmic reticulum to the Golgi apparatus. Proteins that enter the cisternae of the Golgi apparatus, in turn, are "packaged" within vesicles that bud off from the sacs of the Golgi apparatus (fig. 3.25).

The Golgi apparatus and the rough endoplasmic reticulum contain enzymes that modify the structure of proteins. These changes, combined with those that occur during genetic transcription and translation, combine to provide numerous possible sites for the regulation of genetic expression. A summary of the processes involved in genetic expression and possible sites for its regulation is provided in table 3.3.

Golgi: from Camillo Golgi, Italian histologist, 1843–1926

Figure 3.25. (*a*) an electron micrograph of a Golgi apparatus. Notice the formation of vesicles at the ends of some of the flattened sacs. (*b*) an illustration of the secretion of these vesicles by exocytosis. Portions of the Golgi apparatus function in packaging glycoproteins, which are carried to the cell membrane and released as a secretion.

Golgi apparatus

(a)

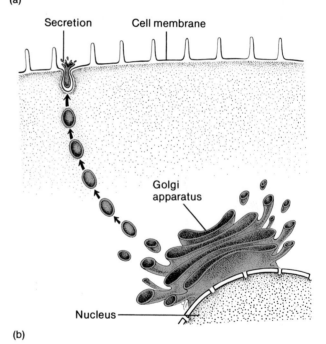

(b)

1. Explain how mRNA, rRNA, and tRNA function during the process of protein synthesis.
2. Describe the rough endoplasmic reticulum, and explain how the processing of secretory proteins differs from the processing of proteins that remain within the cell.
3. Describe the structure of the Golgi apparatus, and explain its functions.

Table 3.3 Stages in genetic expression and possible sites for its regulation

Stage of gene expression	Possible sites for regulation
RNA synthesis	Association of histone and nonhistone proteins with chromatin
Nuclear processing of RNA	Cutting and splicing of nuclear RNA
	Capping of messenger RNA with 7-methyl guanosine and addition of about two hundred adenylate nucleotides
Translation of mRNA	Availability of specific tRNA molecules
Post-translational changes in protein structure	Cleavage of parent protein into biologically active fragments
	Chemical modification of amino acids
	Association of polypeptide chains together to form quaternary structure
	"Addressing" of protein for secretion by leader sequence and insertion into rough endoplasmic reticulum
	Addition of carbohydrate to secreted protein

Source: Fox, Stuart Ira, *Human Physiology.* © 1984 Wm. C. Brown Publishers, Dubuque, Iowa. All Rights Reserved. Reprinted by permission.

Clinical Considerations

Functions of Cellular Organelles

Lysosomes. Lysosomes are important organelles in the function of phagocytic cells. Organs such as the liver, spleen, and lymph nodes have phagocytes that are immobile, or "fixed," within the organs. These are sometimes referred to as the *reticuloendothelial system.* Among other functions, the reticuloendothelial system helps to remove old red blood cells and bacteria from blood and to remove and inactivate drugs from the blood. Lysosomes also participate in *autophagy,* and by this process help to prevent the accumulation of damaged organelles and stored molecules. Genetic diseases that result in the defective production of lysosomal enzymes may produce abnormal accumulations of glycogen or lipids within particular cells. Examples of such diseases include Tay-Sachs disease and Gaucher's disease.

Endoplasmic Reticulum. The smooth endoplasmic reticulum in liver cells and other cells contains enzymes used for the inactivation of steroid hormones and many drugs. This inactivation is generally achieved by reactions that convert these compounds to more water-soluble and less active forms, which can be more easily excreted by the kidneys. When people take certain drugs, such as alcohol

and phenobarbital, for a long period of time, a larger dose of these compounds is required to produce a given effect. This phenomenon, called *tolerance,* is accompanied by an increase in the smooth endoplasmic reticulum and thus an increase in the enzymes charged with inactivation of these drugs.

Cell Growth and Reproduction

Hypertrophy and Atrophy. Hypertrophy refers to growth of an organ due to an increase in cell size (rather than to an increase in cell numbers); striated muscles—skeletal muscle and cardiac (heart) muscle—can only grow by hypertrophy. When this occurs in response to an increased workload, as occurs in skeletal muscles during weight training, it is called *compensatory hypertrophy.* The heart muscle may also demonstrate compensatory hypertrophy when its workload increases, as may occur in response to hypertension (high blood pressure). The opposite of hypertrophy is *atrophy (at'ro-fe),* where the cells become smaller than normal. This may result from the disuse of skeletal muscles (as during prolonged bed rest), various diseases, and advanced age.

Mitosis. Certain types of cells can be removed from the body and grown in nutrient solutions (outside the body, or *in vitro*). Under these artificial conditions the potential longevity of different cell lines can be studied. For unknown reasons, normal connective tissue cells (called fibroblasts) stop dividing *in vitro* after about forty to seventy population doublings. Cells that become transformed into cancer, however, apparently do not age and continue dividing indefinitely in culture. It is ironic that these potentially immortal cells may commit suicide by killing their host.

The Cell Cycle. Mature nerve and muscle cells do not replicate at all; neurons are thus particularly susceptible to damage by alcohol and other drugs. Epithelial cells, in contrast, have very rapid cell cycles that help to replace the continuous loss of cells. *Cancers* have rapid rates of cell division but not necessarily more rapid than normal tissue (table 3.4). The fast growth of some cancers is thus not due simply to a rapid rate of cell division but is rather due to the fact that the rate of cell division is much greater than the rate of cell death. The observation that many normal tissues are at least as rapid in their rates of cell divisions as cancer makes the chemotherapy of cancer much more difficult than it would be otherwise.

Table 3.4 Times for cell cycles, measured in hours, in selected normal tissues and tumors

Tissue	Cell-cycle time
	hr
Colon, epithelium of crypt cells	39
Rectum, epithelium of crypt cells	48
Bone-marrow precursor cells	18
Bronchus, epithelial cells	220
Carcinoma of stomach	72
Acute myeloblastic leukemia	80–84
Chronic myeloid leukemia	120
Carcinoma of bronchus	196–260

Reprinted by permission of *The New England Journal of Medicine*, vol. 304, p. 453, 1981.

Medical Genetics. Inherited diseases that result from defective proteins are due to changes, or *mutations,* in the DNA base sequence. In sickle-cell anemia, Tay-Sachs disease, phenylketonuria (PKU), and many other metabolic disorders, the defective protein is produced because of a single base change in the DNA, which results in a change in one mRNA codon, which in turn causes a single amino acid substitution in the protein. Inherited diseases that are produced by larger abnormalities in the chromosomes also occur. Turner's syndrome, for example, occurs when a person inherits only one sex chromosome (an X) and thus has a total of only forty-five chromosomes; Down's syndrome is produced when a person inherits an extra chromosome (chromosome number 21) and thus has a total of forty-seven chromosomes.

———————

mutation: L. *mutare,* to change

Chapter Summary

I. Cell Structure and Movements
The cell is the basic unit of structure and function in the body. The structure and function of cells depend upon the specific membranes and organelles characteristic of each type of cell.
  A. The cell, or plasma, membrane is described by a fluid-mosaic model.
    1. The membrane is composed predominately of a double layer of phospholipids with the polar ends of the molecules facing the inside and outside of the cell.
    2. The membrane also contains proteins in a mosaic pattern.
  B. Microfilaments and microtubules produce a cytoskeleton.
  C. Cellular movements are produced by contractile proteins and by structures that include cilia and flagella.
    1. Invagination of the cell membrane, followed by fusion and pinching off of vacuoles, or vesicles, occurs during endocytosis.
    2. There are three types of endocytosis: phagocytosis, pinocytosis, and receptor-mediated endocytosis.
    3. Products of the cell are secreted by a reverse process called exocytosis.
II. Lysosomes and Mitochondria
Lysosomes are organelles that contain digestive enzymes. Mitochondria contain enzymes and other molecules needed for obtaining energy from the combustion of food molecules.
  A. Lysosomes are membrane-enclosed organelles that contain digestive enzymes.
    1. Particles in a vacuole produced by phagocytosis can be digested when the vacuole fuses with a lysosome.

2. Lysosomes are also involved in autophagy—the destruction of molecules and organelles within the cell.
  B. Mitochondria are the sites where oxygen is utilized in the combustion of food molecules.
III. Nuclear Structure and DNA Function
During cell division the DNA replicates itself, and duplicated copies are distributed to the daughter cells. When a cell is not dividing, DNA serves as the information source for the formation of RNA and hence of proteins.
  A. The replication of DNA is semiconservative; each new DNA is a double helix consisting of one old and one new polynucleotide strand.
  B. In mitosis the DNA replicates and the cell divides, so that each daughter cell receives the same DNA as the parent cell.
    1. DNA replication occurs during the S phase, and mitosis during the M phase, of the cell cycle.
    2. During metaphase of mitosis the chromosomes—each consisting of a duplicate pair of chromatids—line up at the equator.
    3. Spindle fibers pull the chromatids to opposite poles during anaphase.
  C. During the G_1 phase of the cell cycle, DNA directs the synthesis of RNA.
    1. RNA consists of a single polynucleotide strand formed by complementary base pairing with parts of one strand of DNA.
    2. Messenger RNA is formed as a larger pre-mRNA, which is cut and spliced to form the mRNA that enters the cytoplasm.
    3. Transfer and ribosomal RNA are made by similar mechanisms.

IV. Protein Synthesis and Secretion
Translation of the messenger RNA code into the amino acid sequence of proteins requires transfer RNA, specific enzymes, and ribosomes. Secretion of proteins requires the endoplasmic reticulum, Golgi apparatus, and secretory vesicles.
  A. Each amino acid is bonded to a specific transfer RNA, which has a unique triplet of bases in its anticodon loop.
  B. The anticodons of tRNA bond to specific base triplets called codons in mRNA.
    1. As the mRNA moves through ribosomes, complementary base pairing between codons and anticodons occurs.
    2. As each tRNA bonds to its codon, the amino acid it carries is added to the end of a growing polypeptide chain.
    3. By this means, amino acids are placed together in the sequence determined by the sequence of codons in mRNA.
  C. Proteins destined for secretion are produced in ribosomes located in a rough endoplasmic reticulum and enter the cisternae of these organelles.
  D. Secretory proteins move from the rough endoplasmic reticulum to the Golgi apparatus.
    1. The Golgi apparatus modifies the structure of the proteins it contains and produces membranes that bud off as vesicles.
    2. Secretory vesicles from the Golgi apparatus fuse with the cell membrane and release their products by exocytosis.

Review Activities

Objective Questions

1. According to the fluid-mosaic model of the cell membrane
   (a) protein and phospholipids form a regular, repeating structure
   (b) the membrane is a rigid structure
   (c) phospholipids form a double layer, with the polar parts facing each other
   (d) proteins are free to move within a double layer of phospholipids

2. After the DNA molecule has replicated itself, the duplicate strands are called
   (a) homologous chromosomes
   (b) chromatids
   (c) centromeres
   (d) spindle fibers

3. Nerve and skeletal muscle cells in the adult, which do not divide, remain in the
   (a) G_1 phase
   (b) S phase
   (c) G_2 phase
   (d) M phase

4. The phase of mitosis in which the chromosomes line up at the equator of the cell is called
   (a) interphase
   (b) prophase
   (c) metaphase
   (d) anaphase
   (e) telophase

5. The phase of mitosis in which the chromatids separate is called
   (a) interphase
   (b) prophase
   (c) metaphase
   (d) anaphase
   (e) telophase

6. The RNA nucleotide base that pairs with adenine in DNA is
   (a) thymine
   (b) uracil
   (c) guanine
   (d) cytosine

7. Which of the following statements about RNA is *true?*
   (a) It is made in the nucleus.
   (b) It is double stranded.
   (c) It contains the sugar deoxyribose.
   (d) It is a complementary copy of the entire DNA molecule.

8. Which of the following statements about mRNA is *false?*
   (a) It is produced as a larger pre-mRNA.
   (b) It forms associations with ribosomes.
   (c) Its base triplets are called anticodons.
   (d) It codes for the synthesis of specific proteins.

9. The organelle that combines proteins with carbohydrates and packages them within vesicles for secretion is the
   (a) Golgi apparatus
   (b) rough endoplasmic reticulum
   (c) smooth endoplasmic reticulum
   (d) ribosomes

10. The organelles that contain digestive enzymes are the
    (a) mitochondria
    (b) lysosomes
    (c) endoplasmic reticulum
    (d) Golgi apparatus

Essay Questions

1. The cell membrane is an extremely dynamic structure. Using examples, explain why this statement is true.
2. Explain how one DNA molecule serves as a template for the formation of another DNA and why DNA synthesis is said to be semiconservative.
3. What is the genetic code, and how does it affect the structure and function of the body?
4. Why may tRNA be considered the "interpreter" of the genetic code?
5. Compare the processing of cellular proteins with that of proteins that are secreted by a cell.

Selected Readings

Allison, A. November 1967. Lysosomes and disease. *Scientific American.*

Berlin, R. D., T. E. Oliver, and H. H. Yin. 1975. The cell surface. *New England Journal of Medicine* 292:515.

Brown, D. D. 1981. Gene expression in eukaryotes. *Science* 211:667.

Capaldi, R. A. March 1974. A dynamic model of cell membranes. *Scientific American.*

Chambon, P. May 1981. Split genes. *Scientific American.*

Crick, F. October 1962. The genetic code. *Scientific American.*

Crick, F. 1979. Split genes and RNA splicing. *Science* 204:264.

Danielli, J. F. 1973. The bilayer hypothesis of membrane structure. *Hospital Practice* 8:63.

Dautry-Varsat, A., and H. F. Lodish. May 1984. How receptors bring proteins and particles into cells. *Scientific American.*

Davidson, E. H., and R. J. Britten. 1979. Regulation of gene expression: possible role of repetitive sequences. *Science* 204:1052.

DeDuve, C. May 1963. The lysosome. *Scientific American.*

DeDuve, C. May 1983. Microbodies in the living cell. *Scientific American.*

Dustin, P. August 1980. Microtubules. *Scientific American.*

Fox, C. F. February 1972. The structure of cell membranes. *Scientific American.*

Grivell, L. A. March 1983. Mitochondria DNA. *Scientific American.*

Hayflick, H. January 1980. The cell biology of human aging. *Scientific American.*

Lake, J. A. August 1981. The ribosome. *Scientific American.*

Lazarides, E., and J. P. Revel. May 1979. The molecular basis of cell movement. *Scientific American.*

Lodish, H. F., and J. E. Rothman. January 1979. The assembly of cell membranes. *Scientific American.*

McKusick, V. A. 1981. The anatomy of the human genome. *Hospital Practice* 16:82.

Mazia, D. January 1974. The cell cycle. *Scientific American.*

Miller, O. L., Jr. March 1973. The visualization of genes in action. *Scientific American.*

Nomura, M. October 1969. Ribosomes. *Scientific American.*

Palade, G. 1975. Intracellular aspects of the process of protein synthesis. *Science* 189:347.

Porter, K. R., and J. B. Tucker. March 1981. The ground substance of the living cell. *Scientific American.*

Racker, E. February 1968. The membrane of the mitochondrion. *Scientific American.*

Rich, A., and S. H. Kim. January 1978. The three-dimensional structure of transfer RNA. *Scientific American.*

Rothman, J. E. 1981. The Golgi Apparatus: two organelles in tandem. *Science* 213:1212.

Singer, S. J., and G. L. Nicolson. 1972. The fluid mosaic model of the structure of cell membranes. *Science* 175:720.

Singer, S. J. 1973. Biological membranes. *Hospital Practice* 8:81.

Sloboda, R. D. 1980. The role of microtubules in cell structure and cell division. *American Scientist* 68:290.

Stein, G., J. S. Stein, and L. J. Kleinsmith. February 1975. Chromosomal proteins and gene regulation. *Scientific American.*

4 Enzymes, Energy, and

Concepts

Enzymes are proteins that function as biological catalysts. The great diversity of protein structure allows different enzymes to be specialized in their catalytic actions.

The rate of enzyme-catalyzed reactions depends on numerous factors. Variations in these factors control the rate of progress along particular metabolic pathways and thus help to regulate cellular metabolism.

The coupling of energy-releasing and energy-requiring reactions is central to life processes. The transformation of energy in living systems is based on reactions that produce and destroy ATP and on oxidation-reduction reactions.

In cellular respiration, chemical reactions that transform food molecules into simpler molecules liberate energy, which is used to produce ATP. Anaerobic respiration occurs in the absence of oxygen and provides energy in some organs under special circumstances.

Aerobic respiration is equivalent to combustion in terms of its final products and the total amount of energy liberated. In aerobic respiration, however, the energy is released in small, enzymatically controlled oxidative reactions, and some of the energy released is captured in ATP.

Energy can be derived from the aerobic respiration of lipids and protein. Pyruvic acid and the Krebs cycle acids can also serve as common intermediates in the interconversion of glucose, lipids, and amino acids.

Metabolism

Enzymes as Catalysts

Enzymes are proteins that function as biological catalysts. The catalytic action of enzymes results from the complex structure of these proteins, and the great diversity in the structure of proteins allows different enzymes to be specialized in their catalytic actions.

> *Objective 1.* Define what is meant by the "energy of activation" of a reaction, and explain how this is affected by the action of a catalyst.
>
> *Objective 2.* Explain how the structure of enzyme proteins allows these molecules to function as catalysts.

The ability of yeast cells to make alcohol from glucose (a process called *fermentation*) had been known since antiquity, yet no chemist by the mid-nineteenth century could duplicate the trick in the absence of living yeast. Also, yeast and other living cells could perform a vast array of chemical reactions at body temperature that could not be duplicated in the chemical laboratory without adding a substantial amount of heat energy. These observations led many mid-nineteenth-century scientists to believe that chemical reactions in living cells were aided by a "vital force" that operated beyond the laws of the physical world. This "vitalist" concept was squashed along with the yeast cells when a pioneering biochemist, Eduard Buchner, demonstrated that juice obtained from yeast could ferment glucose to alcohol. The yeast juice was not alive—evidently some chemicals in the cells were responsible for fermentation. Buchner didn't know what these chemicals were, so he simply named them **enzymes.**

In more recent times, biochemists have demonstrated that *all enzymes are proteins* and that enzymes act as *biological catalysts*. A catalyst is a chemical that (1) increases the rate of a reaction, (2) is not itself changed at the end of the reaction, and (3) does not change the nature of the reaction or its final result. The same reaction would have occurred to the same degree in the absence of the catalyst, but it would have progressed at a much slower rate.

---

fermentation: L. *fermentum*, leaven
enzyme: Gk. *en*, in; *zyme*, yeast

Figure 4.1. A comparison of a noncatalyzed reaction with a catalyzed reaction. Upper figures compare the proportion of reactant molecules that have sufficient activation energy to participate in the reaction (shown as the shaded portion of the curves). This proportion is increased in the enzyme-catalyzed reaction, because enzymes lower the activation energy required for the reaction (shown as a barrier on top of an energy ''hill'' in the bottom figures). Reactants that can overcome this barrier are able to participate in the reaction, as shown by arrows pointing to the bottom of the energy hill.

Noncatalyzed reaction

Catalyzed reaction

In order for a given reaction to occur, the reactants must have sufficient energy. The amount of energy required for a reaction to proceed is called the **energy of activation.** In a large population of molecules, only a small fraction will possess sufficient energy for a reaction. Adding heat will raise the energy level of all the reactant molecules, thus increasing the fraction of the population that has the activation energy. Heat would make reactions go faster, but it would produce undesirable side effects in cells. Catalysts make the reaction go faster at lower temperatures by *lowering the activation energy* required so that a larger fraction of the population of reactant molecules has sufficient energy to participate in the reaction (fig. 4.1).

Since a small fraction of the reactants have the activation energy required for the reaction even in the absence of a catalyst, the reaction could theoretically occur spontaneously at a slow rate. This rate, however, would be much too slow for the needs of a cell. So, from a biological standpoint, the presence or absence of a specific enzyme catalyst acts as a switch—the reaction will occur if the enzyme is present and will not occur if the enzyme is absent.

Mechanism of Enzyme Action

The ability of enzymes to lower the activation energy of a reaction is a result of their structure. Enzymes are proteins—they are thus very large molecules with complex, highly ordered, three-dimensional shapes produced by chemical interactions between their amino acids. Each type of enzyme protein has characteristic ridges, grooves,

Figure 4.2. A lock-and-key model of enzyme action.

Substrate

Product

Enzyme

Enzyme-substrate complex

Enzyme resumes
original conformation

Table 4.1 Some examples of the names of enzymes, the reactions they catalyze, and their turnover numbers (molecules of substrate converted to product per minute under optimum conditions)

Enzyme	Reaction catalyzed	Turnover number
Catalase	$2\ H_2O_2 \rightarrow 2\ H_2O + O_2$	600×10^6
Carbonic anhydrase	$H_2CO_3 \rightarrow H_2O + CO_2$	3.6×10^6
Amylase	starch $+\ H_2O \rightarrow$ maltose	1.0×10^5
Lactate dehydrogenase	lactic acid $\rightarrow$ pyruvic acid $+\ H_2$	6.0×10^4
Ribonuclease	RNA $+\ H_2O \rightarrow$ ribonucleotides	600

Source: Fox, Stuart Ira, *Human Physiology.* © 1984 Wm. C. Brown Publishers, Dubuque, Iowa. All Rights Reserved. Reprinted by permission.

and pockets that are lined with specific amino acids. The particular pockets that are active in catalyzing a reaction are called the *active sites* of the enzyme.

The reactant molecules, which are the *substrates* of the enzyme, have shapes that allow them to fit into the active sites. The fit may not be perfect at first, but a perfect fit may be induced as the substrate gradually slips into the active site. This induced fit, together with temporary bonds that form between the substrate and the amino acids lining the active sites of the enzyme, weaken the existing bonds within the substrate molecules, which allows them to be more easily broken. New bonds are more easily formed as substrates are brought close together in the proper orientation. The *enzyme-substrate complex,* formed temporarily in the course of the reaction, then dissociates to yield *products* and the free unaltered enzyme. This model of how enzymes work is known as the **lock-and-key** model of enzyme activity (fig. 4.2).

By this mechanism, enzymes lower the activation energy needed and greatly increase the rate of reactions.

Hydrogen peroxide (H_2O_2), for example, will slowly decompose by itself if left uncapped for a few days. The addition of *catalase (kat'ah-lās),* an enzyme found in many tissues such as blood and the liver, will make hydrogen peroxide break down into water and oxygen gas a trillion times faster. The ability of enzymes to increase the rate of reactions can be measured by their **turnover number,** which is the number of substrate molecules that the enzyme can convert into products per minute (see table 4.1 for examples).

Naming of Enzymes

Although an international committee has established a uniform naming system for enzymes, the names that are in common use do not follow a completely consistent pattern. With the exception of the digestive enzymes that were discovered first (including pepsin and trypsin), all enzyme names end with the suffix *-ase.* Classes of enzymes are named according to their job category. *Hydrolases (hi'dro-lās)*, for example, promote hydrolysis reactions. Some other enzyme categories include *phosphatases*, which catalyze the removal of phosphate groups, *synthetases (sin'the-tās)*, which catalyze dehydration synthesis reactions, and *dehydrogenases (de-hi'dro-jen-ās)*, which remove hydrogen atoms from their substrates. Enzymes called *isomerases (i-som'er-ās)* rearrange atoms within their substrate molecules to form structural isomers (examples of such structural isomers include glucose and fructose).

The names of many enzymes specify both the substrate of the enzyme and the activity ("job category") of the enzyme. Lactic acid dehydrogenase, for example, removes hydrogens from lactic acid. Since enzymes are very specific as to their substrates and activity, the concentration of a specific enzyme in a sample of fluid can be measured relatively easily. This is usually done by measuring the rate of conversion of the enzyme's substrates into products under specified conditions.

Enzymes that do exactly the same job (that catalyze the same reaction) in different organs have the same name, since the name describes the activity of the enzyme. Different organs, however, may make slightly different "models" of the enzyme that differ in one or a few amino acids. These different models of the same enzyme are called **isoenzymes** *(i''so-en'zīmz)*. The differences in structure do not affect the active sites (otherwise they would not catalyze the same reaction), but they do alter the structure of the enzymes at other locations, so that the different isoenzymatic forms can be separated by standard biochemical procedures. These techniques are useful in the diagnosis of diseases (see Clinical Considerations at the end of this chapter).

---

1. Define the "energy of activation" of a reaction; explain how it is affected by a catalyst and how the structure of an enzyme makes it function as a catalyst.
2. Draw a labeled picture of the lock-and-key model of enzyme action, and write a formula for this reaction.
3. Explain how enzymes are named.

Control of Enzyme Activity

The rate of an enzyme-catalyzed reaction depends on numerous factors, including the concentrations of enzymes, substrates, cofactors, and coenzymes, and on the concentration of products formed. Variations in these factors control the rate of progress along particular metabolic pathways and thus help to regulate cellular metabolism.

---

isoenzymes: Gk. *isos*, equal; *en*, in; *zyme*, leaven

Figure 4.3.   The effect of temperature on enzyme activity, as measured by the rate of the enzyme-catalyzed reaction under standardized conditions.

Objective 3.   Describe the effects that changes in pH and temperature have on the rate of enzyme reactions.

Objective 4.   Describe the effects that changes in the concentrations of enzymes, substrates, cofactors, and coenzymes have on the rate of enzyme reactions.

Objective 5.   Describe a reversible reaction, and explain how the law of mass action can determine the relative rates of these reactions.

Objective 6.   Explain how end-product inhibition can affect the direction of a branched metabolic pathway.

The activity of an enzyme, as measured by the rate at which its substrates are converted to products, is influenced by a variety of factors, including (1) the temperature and pH of the solution; (2) the concentration of cofactors and coenzymes *(ko-en'zīm)*, which are needed by many enzymes as "helpers" for their catalytic activity; (3) the concentration of enzyme and substrate molecules in the solution; and (4) the stimulatory and inhibitory effects of some products of enzyme action on the activity of the enzymes that helped form these molecules.

Effects of Temperature and pH

An increase in temperature, as previously described, will increase the rate of non-enzyme-catalyzed reactions because a higher proportion of reactant molecules will have the activation energy required. A similar relationship between temperature and reaction rate occurs in enzyme-catalyzed reactions. At a temperature of 0°C the reaction rate is unmeasurably slow. As the temperature is raised above 0°C the reaction rate increases but only up to a

Figure 4.4. The effect of pH on the activity of three digestive enzymes.

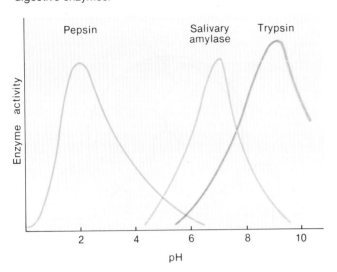

Table **4.2** The pH optima of selected enzymes

Enzyme	Reaction catalyzed	pH optimum
Pepsin (stomach)	Digestion of protein	2.0
Acid phosphatase (prostate)	Removal of phosphate group	5.5
Salivary amylase (saliva)	Digestion of starch	6.8
Lipase (pancreatic juice)	Digestion of fat	7.0
Alkaline phosphatase (bone)	Removal of phosphate group	9.0
Trypsin (pancreatic juice)	Digestion of protein	9.5
Monoamine oxidase (nerve endings)	Removal of amine group from norepinephrine	9.8

Source: Fox, Stuart Ira, *Human Physiology.* © 1984 Wm. C. Brown Publishers, Dubuque, Iowa. All Rights Reserved. Reprinted by permission.

point. At a few degrees above body temperature (which is 37°C) the reaction rate reaches a plateau; further increases in temperature actually *decrease* the rate of the reaction (fig. 4.3). This decrease is due to the fact that the tertiary structure of enzymes becomes altered at higher temperatures.

A similar relationship is observed when the rate of an enzymatic reaction is measured at different pH values. Each enzyme characteristically has its peak activity in a very narrow pH range, which is the **pH optimum** for the enzyme. If the pH is changed from this optimum, the reaction rate decreases (fig. 4.4). This decreased enzyme activity is due to changes in the conformation of the enzyme and in the charges of the R groups of the amino acids lining the active sites.

The pH optimum of an enzyme usually reflects the pH of the body fluid in which the enzyme is found. The acidic pH optimum of the protein-digesting enzyme *pepsin,* for example, allows it to be active in the strong hydrochloric acid of gastric juice. Similarly, the neutral pH optimum of *salivary amylase (am′ĭ-lās)* and the alkaline pH optimum of *trypsin (trip′sin)* in pancreatic juice allow these enzymes to digest starch and protein, respectively, in other parts of the digestive tract.

Although the pH of other body fluids shows less variation than the fluids of the digestive tract, significant differences exist between the pH optima of different enzymes found throughout the body (see table 4.2). Some of these differences can be exploited for diagnostic purposes. Disease of the prostate, for example, may be associated with elevated blood levels of a prostatic phosphatase with an acidic pH optimum (descriptively called acid phosphatase). Bone disease, on the other hand, may be associated with elevated blood levels of alkaline phosphatase, which has a higher pH optimum than the similar enzyme released from the diseased prostate.

Cofactors and Coenzymes

Many enzymes are completely inactive when they are isolated in a pure state. Evidently some of the ions and smaller organic molecules that were removed in the purification procedure are needed for enzyme activity. These ions and smaller organic molecules are called *cofactors* and *coenzymes (ko-en′zim).*

Cofactors are metal ions such as Ca^{++}, Mg^{++}, Mn^{++}, Cu^{++}, and Zn^{++}. Some enzymes with a cofactor requirement do not have a properly shaped active site in the absence of the cofactor. In these enzymes, the attachment of cofactors causes a conformational change in the protein that allows it to combine with its substrate. The cofactors of other enzymes participate in the temporary bonds between the enzyme and its substrate when the enzyme-substrate complex is formed (fig. 4.5).

Coenzymes are organic molecules that are derived from water-soluble vitamins, such as niacin and riboflavin. Coenzymes participate in enzyme-catalyzed reactions by transporting hydrogen atoms and small molecules from one enzyme to another. Examples of the actions of cofactors and coenzymes in specific reactions will be given in the context of their roles in cellular metabolism later in this chapter.

Substrate Concentration and Reversible Reactions

The rate at which an enzymatic reaction converts substrates into products depends on the enzyme concentration and on the concentration of substrates. When the enzyme concentration is at a given level, the rate of product formation will increase as the substrate concentration increases. Eventually, however, a point will be reached where additional increases in substrate concentration do not result in comparable increases in reaction rate. When the

Figure 4.5. The roles of cofactors in enzyme function. In (a) the cofactor changes the conformation of the active site, allowing a better fit between the enzyme and its substrates. In (b) the cofactor participates in the temporary bonding between the active site and the substrates.

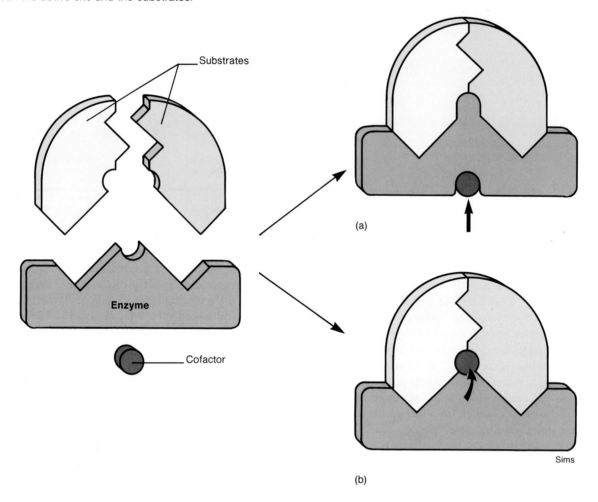

Figure 4.6. The effect of substrate concentration on the reaction rate of an enzyme-catalyzed reaction. When the reaction rate is maximum, the enzyme is said to be *saturated*.

relationship between substrate concentration and reaction rate reaches a plateau, the enzyme is said to be *saturated*. If one thinks of enzymes as workers and substrates as jobs, there is 100% employment when the enzyme is saturated; further availability of jobs (substrate) cannot further increase employment (conversion of substrate to product). This is illustrated in figure 4.6.

Some enzymatic reactions within a cell are reversible, with both the forward and backward reactions catalyzed by the same enzyme. The enzyme *carbonic anhydrase,* for example, is named because it can catalyze the following reaction:

$$H_2CO_3 \rightarrow H_2O + CO_2$$

The same enzyme, however, can also catalyze the reverse reaction:

$$H_2O + CO_2 \rightarrow H_2CO_3$$

Figure 4.7. A metabolic pathway, where the product of one enzyme becomes the substrate of the next in a multi-enzyme system.

Figure 4.8. A branched metabolic pathway.

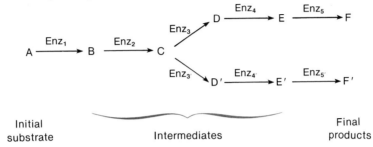

The two reactions can be more conveniently illustrated by a single equation:

$$H_2O + CO_2 \rightleftharpoons H_2CO_3$$

The direction of the reversible reaction depends, in part, on the relative concentrations of the molecules to the left and right of the arrows. If the concentration of CO_2 is very high (as it is in the tissues), the reaction will be driven to the right. If the concentration of CO_2 is low and that of H_2CO_3 is high (as it is in the lungs), the reaction will be driven to the left. The ability of reversible reactions to be driven from the side of the equation where the concentration is higher to the side where the concentration is lower is known as the **law of mass action.**

Although some enzymatic reactions are not directly reversible, the net effects of the reactions can be reversed by the action of different enzymes. The enzymes that convert glucose to pyruvic acid, for example, are different from those that reverse the pathway and produce glucose from pyruvic acid. Likewise, the formation and breakdown of glycogen (a polymer of glucose) are catalyzed by different enzymes.

Metabolic Pathways

The many thousands of different types of enzymatic reactions within a cell do not occur independently of each other. They are, rather, all linked together by intricate webs of interrelationships, the total pattern of which constitutes cellular metabolism. A part of this web that begins with an *initial substrate*, progresses through a number of *intermediates*, and ends with a *final product* is known as a **metabolic pathway.**

The enzymes in a metabolic pathway cooperate in a manner analogous to workers on an assembly line where each contributes a small part to the final product. In this process, the product of one enzyme in the line becomes the substrate of the next enzyme, and so on (fig. 4.7).

Few metabolic pathways are completely linear. Most are branched so that one intermediate at the branch point can serve as a substrate for two different enzymes. Two different products can thus be formed that serve as intermediates of two divergent pathways (fig. 4.8).

End-Product Inhibition. The activities of enzymes at the branch points of metabolic pathways are often regulated by a process called **end-product inhibition.** In this process, one of the final products of a divergent pathway inhibits the branch point enzyme that began the path toward its production. This inhibition prevents that final product from accumulating excessively and results in a shift toward the final product of the alternate divergent pathway (fig. 4.9).

The mechanism by which a final product inhibits an earlier enzymatic step in its pathway is known as **allosteric** *(al"o-ster'ik)* **inhibition.** The allosteric inhibitor combines with a part of the enzyme located away from the active site, causing the active site to change shape so that it can no longer combine properly with its substrate.

---

allosteric: Gk. *allos*, other; *stereos*, position

Figure 4.9. End-product inhibition in a branched metabolic pathway. Inhibition is shown by a dotted arrow and a negative sign.

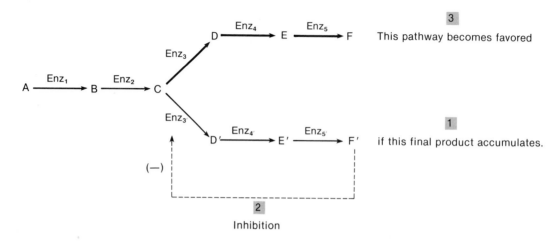

1. Draw graphs to represent the effects of changes in temperatures, pH, enzyme and substrate concentration, cofactors, and coenzymes on the rate of enzymatic reactions. Explain the mechanisms responsible for the appearance of these graphs.
2. Draw a flow chart of a metabolic pathway (using arrows and letters such as A, B, C, etc.) with one branch point.
3. Describe a reversible reaction, and explain how the law of mass action affects this reaction.
4. Define end-product inhibition, and use your diagram of a branched metabolic pathway to explain how this process will affect the concentration of different intermediates.

Bioenergetics

Life requires the constant expenditure of energy to maintain its complex structures and processes. Central to life processes are chemical reactions that are coupled, so that the energy released by one reaction is incorporated into the products of another reaction. The transformation of energy in living systems is largely based on reactions that produce and destroy molecules of ATP and on oxidation-reduction reactions.

Objective 7. Describe the first and second laws of thermodynamics, and use these laws to explain why some molecules have more chemical bond energy than other molecules.

Objective 8. Describe how the coupling of energy-releasing and energy-requiring reactions is used to transform energy in living cells, and—in this context—explain the function of ATP in cells.

Objective 9. Explain how oxidation-reduction reactions involve the transfer of hydrogen from one molecule to another.

Bioenergetics refers to the flow of energy in living systems. Organisms maintain their highly ordered structure and life-sustaining activities through the constant expenditure of energy obtained ultimately from the environment. The energy flow in living systems obeys the first and second laws of a branch of physics known as *thermodynamics.*

According to the **first law of thermodynamics,** energy can be transformed but it cannot be created nor destroyed. This is sometimes called the law of *conservation of energy.* As a result of energy transformations, according to the **second law of thermodynamics,** the universe and its parts (including living systems) have increased amounts of *entropy (en'tro-pe).* Entropy is related to the degree of disorganization of a system. Only energy that is in an organized state—called *free energy*—can be used to do work. Thus, since entropy increases in every energy transformation, the amount of free energy available to do work decreases. As a result of the increased entropy described by the second law, systems tend to go from states of higher to states of lower free energy.

Matter is a form of energy (from Einstein's $E = mc^2$). Atoms that are organized into complex organic molecules, such as glucose, have more free energy (less entropy) than six separate molecules each of carbon dioxide and water. Therefore, in order to convert carbon dioxide and water to glucose, energy must be added. Plants perform this feat using energy from the sun in the process of *photosynthesis* (fig. 4.10).

bioenergetics: Gk. *bios,* life; *energeia,* work
thermodynamics: Gk. *therme,* heat; *dynamis,* force
entropy: Gk. *entropia,* turning toward
photosynthesis: Gk. *phos,* light; *synthesis,* a putting together

Figure 4.10. A simplified diagram of photosynthesis. Some of the sun's radiant energy is captured by the plants and used to produce glucose from carbon dioxide and water. As the product of this endergonic reaction, glucose has a higher free energy content than the initial reactants.

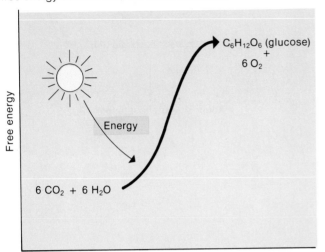

Figure 4.11. Since glucose contains more energy than carbon dioxide and water, the combustion of glucose is an exergonic reaction. The same amount of energy is released if the glucose is broken down stepwise within the cell.

Endergonic and Exergonic Reactions

Chemical reactions that require the input of energy are known as **endergonic** *(end"er-gon'ik)* **reactions.** Since energy is added to make these reactions "go," the products of endergonic reactions must contain more free energy than the reactants. A portion of the energy added, in other words, is contained within the product molecules. This follows from the fact that energy cannot be created nor destroyed (first law of thermodynamics) and from the fact that a more organized state of matter contains more free energy (less entropy) than a less organized state (as described by the second law).

The fact that glucose contains more free energy than carbon dioxide and water can be easily proven by the combustion of glucose to CO_2 and H_2O. This reaction releases energy in the form of heat. Reactions that convert molecules with more free energy to molecules with less—and, therefore, release energy as they proceed—are called **exergonic reactions.**

As illustrated in figure 4.11, the amount of energy released by an exergonic reaction is the same whether the energy is released in a single combustion reaction or in the many small, enzymatically controlled steps that occur in tissue cells. The energy that the body obtains from the consumption of particular foods can, therefore, be measured as the amount of heat energy released when these foods are combusted.

Heat is measured in units called *calories.* One calorie is defined as the amount of heat required to raise the temperature of one cubic centimeter of water one degree Celsius. The caloric value of food is usually indicated in kilocalories (one kilocalorie equals a thousand calories), which is commonly expressed with a capital letter—Calories.

Coupled Reactions: ATP

In order to remain alive a cell must maintain its highly organized, low entropy state at the expense of free energy in its environment. Accordingly, the cell contains many enzymes that catalyze exergonic reactions, using substrates that come ultimately from the environment. The energy released by these exergonic reactions is used to drive the energy-requiring processes (endergonic reactions) in the cell. Since the cell cannot use heat energy to drive energy-requiring processes, chemical bond energy that is released in the exergonic reactions must be directly transferred to chemical bond energy in the products of endergonic reactions. Energy-liberating reactions are thus *coupled* to energy-requiring reactions. This relationship is like two meshed gears; the turning of one (the energy-releasing, exergonic gear) causes turning of the other (the energy-requiring, endergonic gear—fig. 4.12).

---

endergonic: Gk. *endon*, within; *ergon*, work

---

calorie: L. *calor*, heat

Figure 4.12. A model of the coupling of exergonic and endergonic reactions. The drive shaft (representing the energy of activation) turns the exergonic gear, which turns the endergonic gear. The reactants of the exergonic reaction (represented by the larger gear) have more free energy than the products of the endergonic reaction because the coupling is not 100% efficient—some energy is lost as heat.

Exergonic reactions   Endergonic reactions

The energy released by most exergonic reactions in the cell is used, either directly or indirectly, to drive *one* endergonic reaction (fig. 4.13): the formation of **adenosine triphosphate** *(ah-den'o-sēn tri-fos'fāt)* (**ATP**) from adenosine diphosphate and inorganic phosphate (abbreviated P_i).

The formation of ATP requires the input of a fairly large amount of energy. Since this energy must be conserved (first law of thermodynamics), the bond that is produced by joining P_i to ADP must contain a part of this energy. This bond is accordingly called a *high-energy bond.* When enzymes break this bond converting ATP to ADP and P_i, a large amount of energy is released. Energy released from the breakdown of ATP is used to power the energy-requiring processes in all cells. As the **universal energy donor,** ATP serves to couple more efficiently the energy released by the breakdown of food molecules to the energy required by the diverse endergonic processes in the cell (fig. 4.14).

Coupled Reactions: Oxidation-Reduction

When an atom or a molecule gains electrons, it is said to become **reduced;** when it loses electrons, it is said to become **oxidized.** Reduction and oxidation are always coupled reactions: an atom or a molecule cannot become oxidized unless it donates electrons to another, which therefore becomes reduced. The atom or molecule that donates electrons to another is a *reducing agent,* and the one that accepts electrons from another is an *oxidizing agent.* It should be noted that an atom or a molecule may function as an oxidizing agent in one reaction and as a reducing agent in another reaction; it may gain electrons

Figure 4.13. The formation and structure of adenosine triphosphate (ATP).

Adenosine diphosphate (ADP)

+

Inorganic phosphate (P_i)

↓

Adenosine triphosphate (ATP)

from one atom or molecule and pass them on to another in a series of coupled oxidation-reduction reactions—like a bucket brigade.

Notice that the term *oxidation* does not imply that oxygen participates in the reaction. This term is derived from the fact that oxygen has a great tendency to accept electrons, that is, to act as a strong oxidizing agent. This property of oxygen is exploited by cells; oxygen acts as the final electron acceptor in a chain of oxidation-reduction reactions that provides energy for ATP production.

Oxidation-reduction reactions in cells often involve the transfer of hydrogen atoms rather than of free electrons. Since a hydrogen atom contains one electron (and one proton in the nucleus) a molecule that loses hydrogen becomes oxidized, and one that gains hydrogen becomes reduced. In many oxidation-reduction reactions, pairs of electrons—either as free electrons or as a pair of hydrogen atoms—are transferred from the reducing agent to the oxidizing agent.

Two molecules that serve important roles in the transfer of hydrogens are **nicotinamide** *(nik"o-tin'ah-mīd)* **adenine dinucleotide (NAD),** which is derived from the vitamin niacin (vitamin B_3), and **flavin adenine dinucleotide (FAD),** which is derived from the vitamin riboflavin (vitamin B_2). These molecules are *hydrogen carriers* because they accept hydrogens (becoming reduced) in one cellular location and donate hydrogens (becoming oxidized in the process) at a different cellular location (fig. 4.15).

Figure 4.14. A model of ATP as the universal energy donor of the cell. Exergonic reactions are shown as gears with arrows going down (reactions produce decrease in free energy); endergonic reactions are shown as gears with arrows going up (reactions produce increase in free energy).

Figure 4.15. Structures of (a) the oxidized form of NAD (nicotinamide adenine dinucleotide) and (b) the reduced form of FAD (flavin adenine dinucleotide). Note the two additional hydrogen atoms (shown in color) that reduce FAD.

(a)   (b)

Figure 4.16. NAD_{ox} becomes reduced by the addition of two electrons from hydrogen atoms removed from an organic molecule (X). In this reaction, NAD acts as an oxidizing agent. In another cellular location, NAD_{red} can donate these two electrons to a different organic molecule (Y). (If Y could also bind to the protons [H^+], it would receive two complete hydrogen atoms [H_2].) The molecule Y is thus reduced by this reaction in which NAD serves as a reducing agent.

NAD is oxidizing agent   NAD is reducing agent

Each FAD can accept two electrons and can bind two protons. The reduced form of FAD can, therefore, combine with the equivalent of two hydrogen atoms and can be written as $FADH_2$. Each NAD can also accept two electrons but can only bind one proton. The reduced form of NAD may, therefore, be shown as $NADH + H^+$ (the H^+ represents a free proton). In order to represent these two hydrogen carriers in a way that is both consistent and correct, the oxidized forms will subsequently be shown as NAD_{ox} and FAD_{ox}, and the reduced forms will be shown as NAD_{red} and FAD_{red} (fig. 4.16).

> 1. Describe the first and second laws of thermodynamics, and use these laws to explain why the chemical bonds in glucose represent a source of potential energy and how cells can obtain this energy.
> 2. Define the terms exergonic reaction and endergonic reaction. Use these terms to explain the function of ATP in cells.
> 3. Using the symbols X-H_2 and Y, draw a coupled oxidation-reduction reaction. Identify the molecule that is reduced and the one that is oxidized, and tell which one is the reducing agent and which is the oxidizing agent.
> 4. Describe the functions of NAD, FAD, and oxygen (in terms of oxidation-reduction reactions), and explain the meaning of the symbols NAD_{red} and FAD_{red}.

Glycolysis and Anaerobic Respiration

In cellular respiration, chemical reactions that transform food molecules into simpler molecules liberate energy, some of which is used to produce ATP. The complete combustion of a molecule requires the presence of oxygen; some energy is liberated, however, during incomplete combustion in the absence of oxygen. This anaerobic respiration can serve as an important source of energy in some organs under special circumstances.

Objective 10. Define glycolysis, and indicate the initial substrate and final products of this metabolic pathway.

Objective 11. Describe the pathway of anaerobic respiration, and explain how lactic acid is produced.

Objective 12. Explain the role of anaerobic respiration in exercise.

Objective 13. Define *gluconeogenesis,* and explain the significance of this process following exercise.

All of the reactions in the body that involve energy transformation are collectively termed **metabolism** *(me-tab'o-lizm).* Metabolism may be divided into two categories: *anabolism, (ah-nab'o-lizm)* and *catabolism (ca-tab'o-lizm).* Catabolic reactions release energy, usually by the breakdown of larger organic molecules into smaller molecules. Anabolic reactions require the input of energy and include the synthesis of large, energy-storage molecules such as glycogen, fat, and protein.

The catabolic reactions that break down glucose, fatty acids, and amino acids serve as the primary sources of energy for the cellular synthesis of ATP. These metabolic pathways are known collectively as *cellular respiration.* When oxygen serves as the final electron acceptor, these processes are called **aerobic** *(a-er-o'bik)* **cell respiration.** The final products of aerobic respiration are carbon dioxide, water, and energy (a part of which is trapped in the chemical bonds of ATP). The overall equation for aerobic respiration, therefore, is identical to the equation that describes combustion (fuel $+ O_2 \rightarrow CO_2 + H_2O +$ energy).

Notice that the term *respiration* refers to chemical reactions that liberate energy for the production of ATP. The oxygen used in aerobic respiration by tissue cells is obtained from the blood; the blood, in turn, becomes oxygenated in the lungs by the process of breathing.

metabolism: Gk. *metabole,* change
anabolism: Gk. *anabole,* a raising up
catabolism: Gk. *katabole,* a casting down

Figure 4.17. The energy expenditure and gain in glycolysis. Notice that there is a "net profit" of 2 ATP and 2 NAD$_{red}$ per glucose molecule in glycolysis. Molecules listed by number are (1) fructose–1, 6-diphosphate; (2) 1, 3-diphosphoglyceric acid; and (3) 3-phosphoglyceric acid (see fig. 4.18).

Breathing (also called *ventilation* or *external respiration*) is thus needed for, but is different from, aerobic respiration.

Unlike combustion, the conversion of glucose to carbon dioxide and water within the cells occurs in small, enzymatically catalyzed steps. Oxygen is only used at the last step (this will be described in a later section). Since a small amount of the chemical bond energy of glucose is released at early steps in the metabolic pathway, some cells in the body can obtain energy for ATP production in the temporary absence of oxygen. This process is called **anaerobic** *(an″a-er-o'bik)* **respiration.**

Glycolysis

Glycolysis *(gli-kol'ĭ-sis)* is the metabolic pathway by which glucose—a six-carbon (hexose) sugar—is converted into two molecules of *pyruvic (pi-roo'vik) acid.* Each pyruvic acid molecule contains three carbons, three oxygens, and four hydrogens. The number of carbon and oxygen atoms in glucose—$C_6H_{12}O_6$—can thus be accounted for in the two pyruvic acid molecules. Since the two pyruvic acids together account for only eight hydrogens, however, it is clear that four hydrogen atoms are removed from the intermediates in glycolysis. These hydrogen atoms are used to reduce two molecules of NAD$_{ox}$ to two molecules of NAD$_{red}$.

anaerobic: Gk. *an,* private; *aer,* air; *bios,* life
glycolysis: Gk. *glyco,* sugar; *lysis,* breaking

Glycolysis is exergonic, and a portion of the energy that is released is used to drive the endergonic reaction ADP + P_i → ATP. At the end of the glycolytic pathway there is a net gain of two ATP per glucose molecule, as indicated in the overall equation for glycolysis:

$$\text{Glucose} + 2\text{NAD}_{ox} + 2\text{ADP} + 2P_i \rightarrow$$
$$2 \text{ pyruvic acid} + 2\text{NAD}_{red} + 2\text{ATP}$$

Although the overall equation for glycolysis is exergonic, glucose must be "activated" at the beginning of the pathway before energy can be obtained. This activation requires the addition of two phosphate groups derived from two molecules of ATP. Energy from the reaction ATP → ADP + P_i is therefore consumed at the beginning of glycolysis. This is shown as an "up-staircase" in figure 4.17. At later steps in glycolysis, however, four molecules of ATP are produced (and two molecules of NAD are reduced) as energy is liberated (the "down-staircase" in figure 4.17). The two molecules of ATP used in the beginning, therefore, represent an energy investment; the net gain of two ATP and two NAD$_{red}$ by the end of the pathway represent an energy "profit."

The overall equation for glycolysis obscures the fact that this is a metabolic pathway consisting of nine separate steps. The individual steps in this pathway are shown in figure 4.18, and the enzymes that catalyze these steps are listed in table 4.3.

Figure 4.18. In glycolysis, one glucose molecule is converted into two pyruvic acid molecules in nine separate steps. Since two pyruvic acids are produced from one glucose, the products are multiplied by two in the pathway shown. In addition to two pyruvic acids, these products include two molecules of NAD_{red} and four molecules of ATP. Since two ATP molecules were used at the beginning of glycolysis, however, the net gain is 2 ATP per glucose.

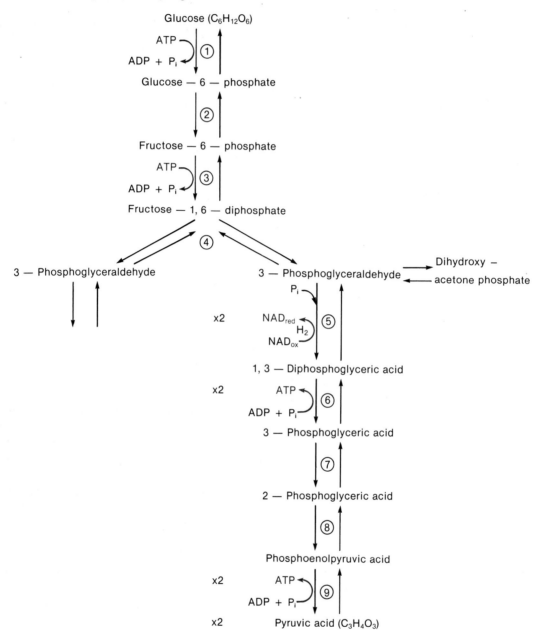

Table 4.3 Enzymes, cofactors, and coenzymes required for glycolysis

Step	Enzyme	Coenzyme or cofactor	Comments
1	Hexokinase	Mg^{++}	Liver enzyme catalyzes the phosphorylation of glucose, fructose, or mannose
2	Hexose phosphate isomerase	—	Interconverts glucose and fructose
3	Phosphofructokinase	Mg^{++}, ATP	Allosteric enzyme that is inhibited by high ATP
4	Aldolase	—	Splits hexose sugar into two three-carbon compounds
5	Phosphoglyceraldehyde dehydrogenase	NAD	Adds inorganic phosphate and oxidizes aldehyde to acid as NAD is reduced by removal of two hydrogens
6	Phosphoglycerate kinase	Mg^{++}	Two molecules of ATP formed at this step
7	Phosphoglyceromutase	—	Phosphate group transferred to different carbon
8	Enolase	Mg^{++}, Mn^{++}	Catalyzes molecular rearrangement
9	Pyruvate kinase	Mg^{++}, K^+	Two molecules of ATP formed at this step

Source: Fox, Stuart Ira, *Human Physiology.* © 1984 Wm. C. Brown Publishers, Dubuque, Iowa. All Rights Reserved. Reprinted by permission.

Anaerobic Respiration

In order for glycolysis to continue, there must be adequate amounts of NAD_{ox} available to accept hydrogen atoms. The NAD_{red} that is produced in glycolysis, therefore, must become oxidized by donating its electrons to another molecule. (In aerobic respiration this other molecule is located in the mitochondria and ultimately passes its electrons to oxygen.)

When oxygen is *not* available in sufficient amounts, the NAD_{red} produced in glycolysis are oxidized in the cytoplasm by donating their electrons to pyruvic acid. This results in the formation of NAD_{ox} and the addition of two hydrogen atoms to pyruvic acid, which is thus reduced. This addition of two hydrogen atoms to pyruvic acid produces *lactic acid* (fig. 4.19). This metabolic pathway, by which glucose is converted through pyruvic acid to lactic acid, is called **anaerobic respiration.**

Anaerobic respiration yields a net gain of two ATP (produced by glycolysis) per glucose molecule. A cell can survive anaerobically as long as it can produce sufficient energy for its needs in this way and as long as lactic acid concentrations do not become excessive. Some tissues are better adapted to anaerobic respiration than others—skeletal muscles survive better than the cardiac muscles, which in turn can survive under anaerobic conditions longer than can the brain.

Except for red blood cells, which can only respire anaerobically (thus sparing the oxygen they carry), anaerobic respiration provides only a temporary sustenance for tissues that have energy requirements in excess of their aerobic ability. Anaerobic respiration can only occur for a limited period of time (longer for skeletal muscles; shorter for the heart, and shortest for the brain) when the *ratio of oxygen supply to oxygen need* (related to the concentration of NAD_{red}) falls below a critical level. Anaerobic respiration is, in a sense, an emergency procedure that provides some ATP until the emergency (oxygen deficiency) has passed.

Figure 4.19. The addition of two hydrogen atoms (colored boxes) from reduced NAD to pyruvic acid produces lactic acid and oxidized NAD. This reaction is catalyzed by lactate dehydrogenase (LDH).

Pyruvic acid   Lactic acid

Physical Exercise. Skeletal muscles respire anaerobically for the first forty-five to ninety seconds of moderate-to-heavy exercise, because the energy requirements (needed for ATP) increase faster than does the rate of oxygen supply by the cardiopulmonary system. If exercise is moderate and the person is in good physical condition, aerobic respiration contributes the major portion of the skeletal muscle energy requirements following the first two minutes of exercise (fig. 4.20).

The maximum rate of oxygen consumption (by aerobic respiration) in the body is called the **maximal oxygen uptake.** The maximal oxygen uptake in a given person is determined primarily by the person's age, size, and sex. It is about 15 to 20% higher for males than for females and highest at age twenty for both sexes. Some world-class athletes have maximal oxygen uptakes that are twice the average for their age and sex—this appears to be due largely to genetic factors, but training can increase this value by about 20%.

In moderate to severe exercise, lactic acid concentrations in the blood increase as a result of anaerobic respiration when the maximal oxygen uptake is exceeded. The rate of this increase depends on the severity of the exercise

Figure 4.20.   The relative contributions of anaerobic and aerobic respiration to the total energy in a well-trained person performing at maximal effort.

and the physical condition of the person. Increased lactic acid concentrations, together with the depletion of stored glycogen and ATP, contribute to muscle fatigue. The amount of exercise that can be performed before muscle fatigue occurs, therefore, depends on the person's maximal oxygen uptake.

When a person stops exercising, the rate of oxygen uptake does not immediately go back to pre-exercise levels; it returns slowly (the person continues to breathe heavily for some time afterwards). This extra oxygen is used to repay the **oxygen debt** incurred during exercise. The oxygen debt includes oxygen that was withdrawn from savings deposits—hemoglobin in blood and myoglobin in muscle (chapter 11)—extra oxygen required for metabolism by tissues warmed during exercise, and oxygen needed for the metabolism of lactic acid produced during anaerobic respiration. Under aerobic conditions lactic acid can be converted to pyruvic acid, which can serve as a substrate for aerobic respiration (as in the heart) or a source of new glucose (in the liver).

Gluconeogenesis and the Cori Cycle.   Some of the lactic acid produced by exercising skeletal muscles is delivered by the blood to the liver. Lactic acid dehydrogenase within liver cells can convert lactic acid to pyruvic acid, as NAD_{ox} is reduced to NAD_{red}. Unlike most other organs, the liver contains the enzymes needed to convert pyruvic acid to glucose-6-phosphate. This is essentially the reverse of glycolysis. Glucose-6-phosphate in liver cells can be used as an intermediate for glycogen synthesis, or it can be converted to free glucose, which is secreted into the blood.

The conversion of noncarbohydrate molecules (lactic acid, amino acids, and glycerol) through pyruvic acid into glucose is called **gluconeogenesis** *(gloo″ko-ne″o-jen′ĕ-sis).*

In starvation and in prolonged exercise, when glycogen stores are depleted, the formation of new glucose in this way becomes the only means for maintaining constant blood-sugar levels. Under these conditions, gluconeogenesis in the liver is the only way that adequate blood glucose levels can be maintained to prevent brain death.

During exercise some of the lactic acid produced by skeletal muscles may be transformed through gluconeogenesis in the liver to blood glucose. This new glucose can serve as an energy source during exercise and can be used after exercise in order to replenish the depleted muscle glycogen. This two-way traffic between skeletal muscles and the liver is called the **Cori cycle** (fig. 4.21). Recent evidence also suggests that some muscle fibers may be able to convert lactic acid to glucose. Through the Cori cycle, gluconeogenesis in the liver (together perhaps with gluconeogenesis in some muscle fibers) allows depleted skeletal muscle glycogen to be restored within forty-eight hours after exercise.

1. Define the term glycolysis *in terms of its initial substrates and products. Explain why there is a net gain of two ATP in this process.*
2. Define the term anaerobic respiration, *in terms of its initial substrates and final products. Explain the significance of lactic acid formation at the end of this process.*
3. *Explain when and why anaerobic respiration occurs during exercise. What are the benefits and costs of anaerobic respiration during exercise?*
4. Define the term gluconeogenesis, *and explain how this process functions to restore the glycogen stores of skeletal muscles following exercise.*

Cori: from Carl F. Cori, U.S. biochemist, 1896-

Figure 4.21. The Cori cycle. The direction of reversible reactions that occurs in the Cori cycle is shown by heavy arrows; the sequence of steps is indicated by numbers 1 through 9.

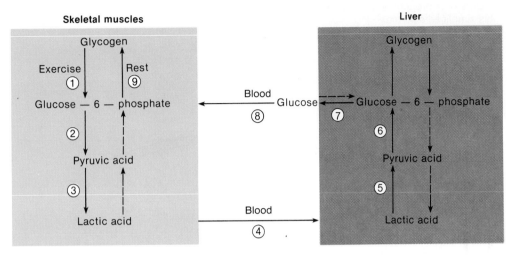

Figure 4.22. The formation of acetyl coenzyme A in aerobic respiration.

Pyruvic acid   Coenzyme A   Acetyl coenzyme A

Aerobic Respiration

Aerobic respiration is equivalent to combustion in terms of its final products (CO_2 and H_2O) and in terms of the total amount of energy liberated. In aerobic respiration, however, the energy is released in small, enzymatically controlled oxidation reactions, and a portion (38%–40%) of the energy released in this process is captured in the high energy bonds of ATP.

Objective 14. Describe the fate of pyruvic acid in aerobic respiration.

Objective 15. Describe the functional significance of the Krebs cycle, and list its products.

Objective 16. Describe the electron transport system and oxidative phosphorylation; explain the role of oxygen in aerobic respiration.

Objective 17. Explain how the metabolism of glycogen and fat are regulated by end-product inhibition.

The aerobic respiration of glucose begins with glycolysis. Glycolysis in both anaerobic and aerobic respiration results in the production of two molecules of pyruvic acid, two molecules of ATP, and two molecules of NAD_{red} per glucose. In aerobic respiration, however, the electrons in NAD_{red} are *not* donated to pyruvic acid and lactic acid is not formed, as happens in anaerobic respiration.

The enzymes that catalyze glycolysis are located in the cell cytoplasm. In aerobic respiration, pyruvic acid leaves the cell cytoplasm and enters the interior (the matrix) of mitochondria. Once pyruvic acid is inside a mitochondrion, carbon dioxide is enzymatically removed from each three-carbon-long pyruvic acid to form a two-carbon-long organic acid—acetic acid. The enzyme that catalyzes this reaction combines the acetic acid with a coenzyme (derived from the vitamin pantothenic acid) called *coenzyme A.* The combination thus produced is called **acetyl** *(as'ĕ-til)* **coenzyme A,** abbreviated **acetyl CoA** (fig. 4.22).

Figure 4.23. A simplified diagram of the Krebs cycle showing how the original four-carbon-long oxaloacetic acid is regenerated at the end of the cyclic pathway. Only the numbers of carbon atoms in the Krebs cycle intermediates are shown; the numbers of hydrogens and oxygens are not accounted for in this simplified scheme.

Since glycolysis converts one glucose molecule to two molecules of pyruvic acid, two molecules of acetyl CoA and two molecules of CO_2 are derived from each glucose. The acetyl CoA molecules serve as substrates for mitochondrial enzymes in the aerobic pathway; the carbon dioxide is a waste product in this process, which is carried by the blood to the lungs for elimination. It should be noted that the oxygen in CO_2 is derived from pyruvic acid, not from oxygen gas.

The Krebs Cycle

Once acetyl CoA is formed, the acetic acid subunit (two carbons long) is combined with oxaloacetic acid (four carbons long) to form a molecule of citric acid (six carbons long). Coenzyme A acts only as a transporter of acetic acid from one enzyme to another (similar to the transport of hydrogen by NAD). The formation of citric acid begins a cyclic metabolic pathway known as the **citric acid cycle,** or **TCA cycle** (for tricarboxylic acid; citric acid has three carboxylic acid groups). Most commonly, however, this cyclic pathway is named after its principal discoverer, Sir Hans A. Krebs, and is called the **Krebs cycle.** A simplified illustration of this pathway is shown in figure 4.23.

Through a series of reactions involving the elimination of two carbons and four oxygens (as two CO_2 molecules) and the removal of hydrogens, citric acid is eventually converted to oxaloacetic acid, which completes the cyclic metabolic pathway. In this process the following occur: (1) one guanosine triphosphate (GTP) is produced (step 5 of figure 4.24), which is converted to ATP; (2) three molecules of NAD are reduced (steps 4, 5 and 8 of figure 4.24); (3) one molecule of FAD is reduced (step 6). The production of reduced NAD and FAD by each "turn" of the Krebs cycle is far more significant, in terms of energy production, than the single GTP (converted to ATP) produced directly by the cycle.

Electron Transport and Oxidative Phosphorylation

Built into the foldings, or cristae, of the inner mitochondrial membrane are a series of molecules that serve in **electron transport** during aerobic respiration. This electron transport chain of molecules consists of a flavoprotein

Figure 4.24. The Krebs cycle.

(derived from the vitamin riboflavin), coenzyme Q (derived from vitamin E), and a group of iron-containing pigments called *cytochromes (si'to-krōmz)*. The last of this series of molecules is an *ATPase enzyme* that produces ATP from ADP and P_i, using energy derived from electron transport. This subunit protrudes from the cristae (under some conditions) like a lollipop (fig. 4.25).

In aerobic respiration, NAD_{red} and FAD_{red} become oxidized by transferring their pairs of electrons to the electron transport system of the cristae. In this way the oxidized forms of NAD and FAD are regenerated and able to continue to "shuttle" electrons from the Krebs cycle to the electron transport chain. The first molecule of the electron transport chain, in turn, becomes reduced when it accepts the electron pair from NAD_{red}. When the cytochromes receive a pair of electrons, two ferric ions

(Fe^{+++}) become reduced to two ferrous ions (Fe^{++}). Notice that the gain of an electron is indicated by the reduction of the number of positive charges.

The electron transport chain thus acts as an oxidizing agent for NAD and FAD. Each element in the chain, however, also functions as a reducing agent; one reduced cytochrome transfers its electron pair to the next cytochrome in the chain. In this way, the iron ions in each cytochrome alternately become reduced (to ferrous ions) and oxidized (to ferric ions)—like a "ferrous wheel."

The passage of electrons along the electron transport chain is an exergonic process; the reduced form of the last cytochrome contains less free energy than the first. Energy released by electron transport is used (by a process

Figure 4.25. The ATPase at the end of the electron transport system is attached to the cristae by a stalk. This unit produces ATP, using energy from electron transport (a process called oxidative phosphorylation).

whose description is beyond the scope of this book) to "drive" the reaction in which ADP is "phosphorylated"—by the addition of P_i—to ATP. The production of ATP in this way is appropriately termed **oxidative phosphorylation** *(fos″fŏr-ĭ-la′shun)* (fig. 4.26).

Function of Oxygen. If the last cytochrome remained in a reduced state, it would be unable to accept more electrons. Electron transport would then progress only to the next-to-last cytochrome. This process would continue until all of the elements of the electron transport chain remained in the reduced state. If this occurred, NAD_{red} and FAD_{red} could not become oxidized by donating their electrons to the cytochrome chain, and through inhibition of Krebs cycle enzymes, no more NAD_{red} and FAD_{red} could be produced in the mitochondria (the Krebs cycle would stop). Respiration would then become anaerobic.

Oxygen, from the air we breathe, allows electron transport to continue by functioning as the **final electron acceptor** of the electron transport chain. This oxidizes the last cytochrome so that electron transport and oxidative phosphorylation can continue. At the very last step of aerobic respiration, therefore, oxygen becomes reduced by the two electrons that were passed to the chain from NAD_{red} and FAD_{red}. This reduced oxygen binds two protons, and a molecule of water (H_2O) is formed.

ATP Balance Sheet

Each time the Krebs cycle turns, three molecules of NAD are reduced by electrons from three pairs of hydrogens removed from Krebs cycle intermediates. Each NAD_{red} donates a pair of electrons to the electron transport chain. Transport of this pair of electrons to oxygen generates energy for the production of three molecules of ATP through oxidative phosphorylation. Electrons from FAD_{red} enter

the electron transport chain "down the line" from where the first ATP is produced. Each pair of electrons from FAD_{red}, therefore, produces only two molecules of ATP from oxidative phosphorylation.

The three NAD_{red} produced per turn of the Krebs cycle result in the production of nine ATP molecules. The single FAD_{red} per turn of the Krebs cycle results in the production of two ATP. Together with the single ATP made directly by the Krebs cycle (from GTP), each turn of the Krebs cycle, therefore, yields a total of twelve ATP molecules. Since one molecule of glucose produces two pyruvic acids and thus two turns of the Krebs cycle, a total of twenty-four ATP molecules are produced by a single molecule of glucose through the Krebs cycle and oxidative phosphorylation.

The conversion of pyruvic acid to acetyl CoA also involves the reduction of one NAD. Since two pyruvic acids are produced per glucose, and since each NAD_{red} yields three ATP by oxidative phosphorylation, a total of twenty-four plus six, or thirty, ATP molecules are made in the mitochondrion from the steps that occur after pyruvic acid formation.

Two molecules of NAD_{red} are produced in the cytoplasm during glycolysis (conversion of glucose to pyruvic acid). These NAD_{red} cannot directly enter the mitochondria; instead, they donate their electrons to other molecules that "shuttle" these electrons into the mitochondria. Depending on which shuttle is used, either two or three ATP molecules can be produced from each pair of these cytoplasmic electrons through oxidative phosphorylation. A total of four or six ATP molecules are thus produced. Added to the thirty ATP previously mentioned, this brings the total to thirty-four or thirty-six (depending on which shuttle is used for the cytoplasmic electrons). Together with the two molecules of ATP produced directly by glycolysis, a grand total of thirty-six to thirty-eight ATP are produced by the aerobic respiration of glucose (table 4.4).

Figure 4.26. Electron transport and oxidative phosphorylation. Each element in the electron transport chain alternately becomes reduced and then oxidized as it transports electrons to the next member of the chain. This process provides energy for the formation of ATP. At the end of the electron transport chain the electrons are donated to oxygen, which becomes reduced (by the addition of two hydrogen atoms) to water.

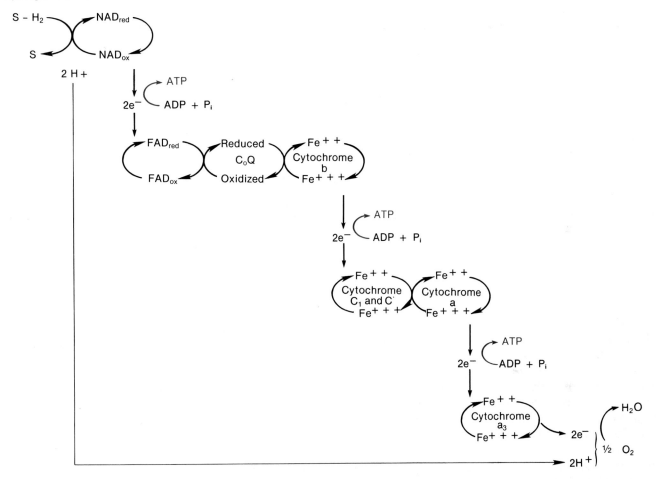

Table 4.4 Maximum ATP yield per glucose molecule respired anaerobically and aerobically

Phase of respiration	High-energy products	ATP from oxidative phosphorylation	ATP subtotal
Glycolysis (glucose to pyruvic acid)	2 ATP	—	2 (total if anaerobic)
	2 NAD_red	6	8 (if aerobic)
Pyruvic acid to acetyl CoA	1 NAD_red (X2)	6	14
Krebs cycle	1 ATP (X2)	—	16
	3 NAD_red (X2)	18	34
	1 FAD_red (X2)	4	38
		Total (aerobic)	38 ATP

Figure 4.27. Blood glucose that enters tissue cells is rapidly converted to glucose–6-phosphate. This intermediate can be metabolized for energy in glycolysis or can be converted to glycogen (*1*), a process called glycogenesis. Glycogen represents a storage form of carbohydrates, which can be used as a source for new glucose–6-phosphate (*2*), in a process called glycogenolysis. The liver contains an enzyme that can remove the phosphate from glucose–6-phosphate; liver glycogen thus serves as a source for new blood glucose.

Glycogenesis and Glycogenolysis

Many tissues, particularly the liver, skeletal muscles, and heart, store carbohydrates in the form of glycogen. The formation of glycogen (a polymer of glucose) from glucose is called **glycogenesis.** In this process, glucose is converted to glucose-6-phosphate by utilization of the terminal phosphate group of ATP. Glucose-6-phosphate is then converted into its isomer, glucose-1-phosphate. Finally, the enzyme *glycogen synthetase* removes these phosphate groups as it polymerizes glucose to form glycogen.

The reverse reactions are similar. The enzyme *glycogen phosphorylase* catalyzes the breakdown of glycogen to glucose-1-phosphate (the phosphates are derived from inorganic phosphate, not from ATP). Glucose-1-phosphate is then converted to glucose-6-phosphate. The conversion of glycogen to glucose-6-phosphate is called **glycogenolysis** *(gli″ko-jĕ-nol′i-sis).* In most tissues, glucose-6-phosphate can then be used to resynthesize glycogen, or it can be respired for energy (through glycolysis). Glucose-6-phosphate *cannot* be released from the cells into the blood because organic molecules with phosphate groups cannot cross cell membranes.

Unlike skeletal muscles, the liver contains an enzyme—known as *glucose-6-phosphatase*—which can remove the phosphate groups and produce free glucose (fig. 4.27). This free glucose can then be transported through the cell membrane; the liver, therefore, can secrete glucose into the blood, whereas skeletal muscles cannot. Liver glycogen can thus supply blood glucose for use by other organs, including exercising skeletal muscles that have depleted much of their own stored glycogen.

Control of Carbohydrate Metabolism

When oxygen is not available in sufficient amounts, the activity of the Krebs cycle enzymes decreases, and glycolysis (and anaerobic respiration) increases. This switching from aerobic to anaerobic respiration is accomplished largely by the allosteric inhibition of enzymes. Key glycolytic enzymes are inhibited by high ATP and stimulated by high ADP levels. When the Krebs cycle is inhibited (by lack of sufficient NAD_{ox} during oxygen deficiency), the decreased ATP concentrations that result stimulate glycolysis. This increased rate of glycolysis under anaerobic conditions is called the *Pasteur effect.*

Cells do not store extra energy in the form of extra ATP. When cellular ATP concentrations rise, because

---

glycogenesis: Gk. *glyco,* sugar; *genesis,* production

Pasteur: from Louis Pasteur, French chemist and bacteriologist, 1822–1895

Figure 4.28. The conversion of glucose into glycogen and fat due to the allosteric inhibition of respiratory enzymes when the cell has adequate amounts of ATP. Favored pathways are indicated by heavy arrows.

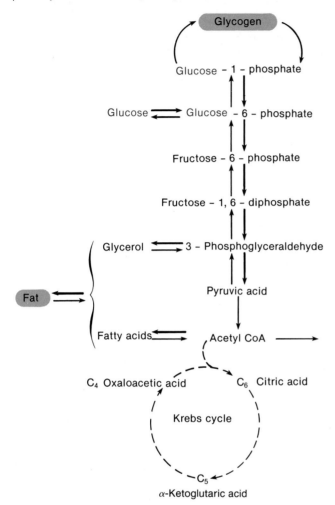

1. Describe the fate of pyruvic acid in aerobic respiration, and explain how this differs from its fate in anaerobic respiration.
2. Draw a simplified Krebs cycle, using C_2 for acetic acid, C_4 for oxaloacetic acid, C_5 for alpha-ketoglutaric acid, and C_6 for citric acid. List the high-energy products that are produced at each turn of the Krebs cycle.
3. Using a diagram, show how electrons from NAD_{red} and FAD_{red} are passed by the cytochromes. Represent the oxidized and reduced forms of the cytochromes with Fe^{+++} and Fe^{++}, respectively.
4. Explain the function of oxygen in the body and how a lack of sufficient oxygen in an organ can make respiration become anaerobic.
5. Explain how the end-product inhibition by ATP controls the rate of cell respiration and the rate of the formation of glycogen and fat.

more energy (from food) is available than can be immediately used, high ATP concentrations inhibit glycolysis. Under conditions of high cellular ATP concentrations, when glycolysis is inhibited, glucose is instead converted into glycogen.

High ATP concentrations also inhibit Krebs cycle enzymes, preventing acetyl CoA from joining oxaloacetic acid and thereby forming citric acid. The inhibition of Krebs cycle enzymes results in the increased availability of acetyl CoA, phosphoglyceraldehyde, and dihydroxyacetone phosphate (fig. 4.18) for alternate pathways. These intermediates are instead channeled into pathways leading to triglyceride (fat) production. When food intake occurs at a faster rate than energy consumption, the excess energy (calories) is stored in the form of glycogen and fat. These pathways are summarized in figure 4.28.

Metabolism of Lipids and Proteins

Energy can be derived by the cellular respiration of lipids and proteins, using the same aerobic pathway previously described for the metabolism of pyruvic acid. Indeed, some organs preferentially use molecules other than glucose as an energy source. Pyruvic acid and the Krebs cycle acids also serve as common intermediates in the interconversion of glucose, lipids, and amino acids.

Objective 18. Describe the metabolic pathway by which glucose can be converted into fat.
Objective 19. Describe the metabolic pathways by which fat can be utilized for energy.
Objective 20. Define oxidative deamination and transamination, and explain the significance of these two processes in amino acid metabolism.
Objective 21. Explain how skeletal muscles obtain energy at rest and at different stages of exercise, and explain the roles of the liver during skeletal muscle exercise.

It is common experience that the ingestion of excessive calories in the form of carbohydrates (cakes, ice cream, candy, and so on) increases fat production. The rise in blood glucose that follows carbohydrate-rich meals stimulates insulin secretion, and this hormone, in turn, promotes the entry of blood glucose into adipose cells. Increased availability of glucose within adipose cells, under conditions of high-insulin secretion, promotes the conversion of glucose to fat. The hormonal control of carbohydrate and fat metabolism is discussed in chapter 25.

Figure 4.29. Divergent metabolic pathways for acetyl coenzyme A.

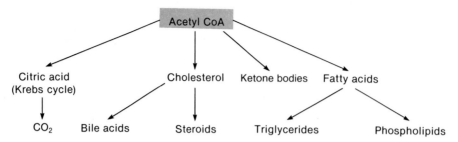

Lipid Metabolism

Phosphoglyceraldehyde and dihydroxyacetone phosphate produced by glycolysis from glucose can be converted to glycerol. Not all of the glycolytic intermediates are converted to glycerol, however; some are used to form pyruvic acid, which in turn is converted to acetyl CoA. The two-carbon acetic acid subunits of acetyl CoA can then be used to produce a variety of lipids, including steroids such as cholesterol as well as ketone bodies and fatty acids (fig. 4.29). In the formation of fatty acids, a number of acetic acid subunits are joined together to form the fatty acid chain; three of these chains can then condense with glycerol to form a triglyceride molecule.

Formation of fat, or *lipogenesis (lip"o-jen'ĭ-sis)* occurs in a number of tissues—primarily in adipose tissue and in the liver—when blood glucose is elevated. Fat represents the major form of energy storage in the body. In a nonobese 70-kilogram man, 80%–85% of the body's energy is stored as fat, which amounts to about 140,000 Calories. Stored glycogen, in contrast, accounts for less than 2,000 Calories, most of which (about 350 g) is stored in skeletal muscles and is available only for the muscle's use. The liver contains about 80–90 grams of glycogen, which can be converted to glucose and used by other organs. Protein accounts for about 15%–20% of the stored calories in the body, but this is usually not used extensively as an energy source because that would involve the loss of muscle mass.

Breakdown of Fat (Lipolysis). When fat stored in adipose tissue is going to be used as an energy source, lipase enzymes hydrolyze triglycerides into *glycerol* and *free fatty acids,* a process called lipolysis *(lĭ-pol'ĭ-sis).* These molecules (primarily the free fatty acids) serve as *blood-borne energy carriers* that can be used by the liver, skeletal muscles, and other organs for aerobic respiration.

A few organs can utilize glycerol for energy, by virtue of an enzyme that converts glycerol to phosphoglyceraldehyde. Free fatty acids, however, serve as the major energy source derived from triglycerides. Most fatty acids consist of a long hydrocarbon chain with a carboxylic acid group (COOH) at one end. In a process known as *β-oxidation* (the β is pronounced *beta*), enzymes remove two-carbon acetic acid molecules from the acid end of a fatty acid. This results in the formation of acetyl CoA, as the third carbon from the end becomes oxidized to produce a new carboxylic acid group. The fatty acid chain is thus decreased in length by two carbons. The process of β-oxidation continues until the entire fatty acid molecule is converted to acetyl CoA.

A sixteen-carbon-long fatty acid, for example, yields eight acetyl CoA molecules. Each of these can enter a Krebs cycle and produce twelve ATP per turn of the cycle, so that eight times twelve, or ninety-six, ATP are produced. In addition, each time an acetyl CoA is formed and the end-carbon of the fatty acid chain is oxidized, one NAD_{red} and one FAD_{red} are produced. Oxidative phosphorylation produces three ATP per NAD_{red} and two ATP per FAD_{red}. For a sixteen-carbon-long fatty acid, these five ATP molecules would be formed seven times (producing five times seven, or thirty-five, ATP). Not counting the single ATP used to start β-oxidation (fig. 4.30), this fatty acid could yield a grand total of 35 + 96, or 131, ATP molecules!

Ketone Bodies. There is a continuous turnover of triglycerides in adipose tissue, even when a person is not losing weight. New triglycerides are produced, while others are hydrolyzed into glycerol and fatty acids. This turnover insures that the blood normally contains a sufficient amount of fatty acids for aerobic respiration by skeletal muscles, the liver, and other organs. When the rate of lipolysis exceeds the rate of fatty acid utilization—as it may do in starvation, dieting, and in diabetes mellitus—the blood concentrations of fatty acids increase.

If the liver cells contain sufficient amounts of ATP, so that further production of ATP is not needed, some of the acetyl CoA derived from fatty acids is channeled into an alternate pathway. This pathway involves the conversion of two molecules of acetyl CoA into four-carbon-long acidic derivatives, *acetoacetic acid* and *β-hydroxybutyric (hi-drok"se-byu-ti'rik) acid.* Together with *acetone (as'ĕ-tōn),* which is a three-carbon-long derivative of acetoacetic acid, these products are known as **ketone bodies.**

Figure 4.30. β-oxidation of a fatty acid. After the attachment of coenzyme A to the carboxylic acid group (step *1*), a pair of hydrogens is removed from the fatty acid and used to reduce one molecule of FAD (step *2*). When this electron pair is donated to the cytochrome chain, two ATP are produced. The addition of a hydroxyl group from water (step *3*), followed by the oxidation of the β-carbon (step *4*), results in the production of three ATP from the electron pair donated by reduced NAD. The bond between the α and β carbons in the fatty acid is broken (step *5*), releasing acetyl coenzyme A and a fatty acid chain that is two carbons shorter than the original. With the addition of a new coenzyme A to the shorter fatty acid the process begins again (step *2*), as acetyl CoA enters the Krebs cycle and generates twelve ATP.

Ketone bodies, which can be used for energy by many organs, are normally found in the blood. Under conditions of fasting or of diabetes mellitus, however, the increased liberation of free fatty acids from adipose tissue results in an elevated production of ketone bodies in the liver. The secretion of abnormally high amounts of ketone bodies into the blood produces **ketosis,** which is one of the signs of fasting or an uncontrolled diabetic state.

Amino Acid Metabolism

Nitrogen is ingested primarily as amino acids and excreted mainly as urea in the urine. In childhood, the amount of nitrogen excreted may be less than the amount ingested because amino acids are incorporated into proteins during growth. Growing children are thus said to be in a state of *positive nitrogen balance*. People who are starving or suffering from prolonged wasting diseases, in contrast, are in a state of *negative nitrogen balance;* they excrete more nitrogen than they ingest because they are breaking down their tissue proteins.

Figure 4.31. The formation of the amino acids aspartic acid and alanine using glutamic acid as the amine donor in transamination. (GOT = glutamate oxaloacetate transaminase; GPT = glutamate pyruvate transaminase). The shaded areas show the parts of the molecules that are changed by transamination reactions.

| Glutamic acid | Oxaloacetic acid | α−Ketoglutaric acid | Aspartic acid |

| Glutamic acid | Pyruvic acid | α−Ketoglutaric acid | Alanine |

Table 4.5 The essential, semiessential, and nonessential amino acids. The semiessential amino acid can be produced by transamination, but not in sufficient quantities to classify it as a nonessential amino acid. The body cannot produce the essential amino acids, and so must obtain them in the diet.

Essential amino acids	Semiessential amino acid	Nonessential amino acids
Lysine	Arginine	Aspartic acid
Tryptophan		Glutamic acid
Phenylalanine		Proline
Threonine		Glycine
Valine		Serine
Methionine		Alanine
Leucine		Cysteine
Isoleucine		
Histidine (children)		

Source: Fox, Stuart Ira, *Human Physiology*. © 1984 Wm. C. Brown Publishers, Dubuque, Iowa. All Rights Reserved. Reprinted by permission.

Healthy adults maintain a state of nitrogen balance, in which the amount of nitrogen excreted is equal to the amount ingested. This does not imply that the amino acids ingested are unnecessary; on the contrary, they are needed to replace the approximately 400 grams of protein that is "turned over" each day. When more amino acids are ingested than are needed to replace proteins, the excess amino acids are not stored as additional protein (one cannot build muscles simply by eating large amounts of protein). Rather, the amine groups can be removed, and the "carbon skeletons" of the organic acids that are left can be used for energy or converted to carbohydrate and fat.

Transamination. An adequate amount of all twenty amino acids is required to build proteins for growth and for replacement of the proteins that are turned over. Fortunately, only eight (in adults) or nine (in children) amino acids must be obtained in the diet; these are the *essential amino acids* (table 4.5). The remaining amino acids are

Figure 4.32. Oxidative deamination. Glutamic acid is converted to α-ketoglutaric acid as it donates its amine group to a metabolic pathway that results in the formation of urea.

"nonessential" only in the sense that the body can produce them if it is given a sufficient amount of the essential amino acids and of carbohydrates.

Pyruvic acid and the Krebs cycle acids are collectively termed *keto acids* because they have a ketone group (fig. 2.11); these should not be confused with the ketone bodies (derived from acetyl CoA). Keto acids can be converted to amino acids by the addition of an amine (NH_2) group. This amine group is usually obtained by "cannibalizing" another amino acid; in this process, a new amino acid is formed as the one that was cannibalized is converted to a new keto acid. This type of reaction, in which the amine group is moved across from one amino acid to form another, is called **transamination.**

Each transamination reaction is catalyzed by a specific enzyme (a transaminase) that requires vitamin B_6 (pyridoxine) as a coenzyme. The amine group from glutamic acid, for example, may be transferred to either pyruvic acid or oxaloacetic acid. The former reaction is catalyzed by the enzyme *glutamate pyruvate transaminase* (GPT), and the latter reaction is catalyzed by *glutamate oxaloacetate (oks″ah-lo-as′e-tāt) transaminase* (GOT). The addition of an amine group to pyruvic acid produces the amino acid alanine; the addition of an amine group to oxaloacetic acid produces the amino acid known as aspartic acid (fig. 4.31).

Oxidative Deamination. As shown in figure 4.32, glutamic acid can be formed through transamination by the combination of an amine group with α-ketoglutaric acid. Glutamic acid is also produced in the liver from the ammonia that is generated by intestinal bacteria and carried to the liver in the hepatic portal vein. Since free ammonia is very toxic, its removal from the blood and incorporation into glutamic acid is an important function of the healthy liver.

If there are more amino acids than are needed for protein synthesis, the amine group from glutamic acid may be removed and excreted as *urea* in the urine (fig. 4.32). The metabolic pathway that removes amine groups from amino acids—leaving a keto acid and ammonia (which is converted to urea)—is known as **oxidative deamination** *(de-am″ĭ-na′shun).*

A number of amino acids can be converted into glutamic acid by transamination. Since glutamic acid can donate amine groups to urea (through deamination), it serves as a "funnel" through which other amino acids can be used to produce keto acids (pyruvic acid and Krebs cycle acids). These keto acids may then be used in the Krebs cycle as a source of energy (fig. 4.33).

Depending on the amino acid that is deaminated, the keto acid left over may be either pyruvic acid or one of the Krebs cycle acids. These can be respired for energy, converted via pyruvic acid to glucose (amino acids are very important in the production of glucose through *gluconeogenesis*), or converted to fat. The possible interrelationships between amino acids, carbohydrates, and fat are illustrated in figure 4.34.

Uses of Different Energy Sources

The blood serves as a common trough from which all the cells in the body are fed. If all cells used the same energy source, such as glucose, this source would quickly be depleted and cellular starvation would occur. This is normally prevented by the fact that the blood contains different energy sources: glucose and ketone bodies from the liver, fatty acids from adipose tissue, and lactic acid and amino acids from muscles. Some organs preferentially use one energy source more than the others, so that each energy source is "spared" for organs with strict energy needs.

Figure 4.33. Pathways by which amino acids can be catabolized for energy. These pathways are indirect for some amino acids, which must be transaminated into other amino acids before being converted into keto acids by deamination.

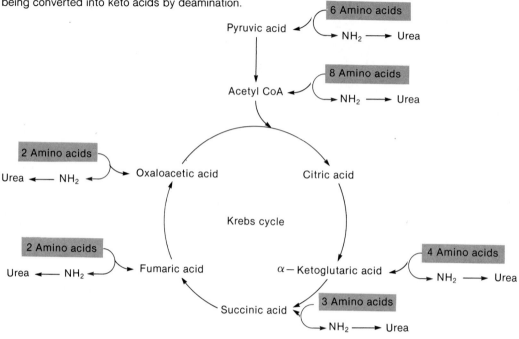

Figure 4.34. Simplified metabolic pathways showing how glycogen, fat, and protein can be interconverted.

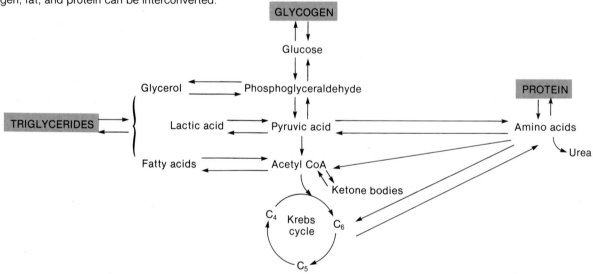

Under normal conditions, for example, the brain uses blood glucose as its major energy source. This glucose is supplied primarily by the liver, which secretes about 150 milligrams of glucose into the blood per minute. About 75% of this glucose is supplied by glycogenolysis and 25% is produced in the liver by gluconeogenesis, using amino acids, lactic acids, and glycerol as initial substrates. Many organs spare glucose by using fatty acids, ketone bodies, and lactic acid as energy sources (table 4.6).

Muscle Metabolism During Exercise. Skeletal muscles at rest use fatty acids as their major energy source. Metabolism changes at the beginning of exercise, when glucose derived from stored muscle glycogen becomes the prime source of energy. At this time the energy contribution from blood glucose is still low.

Table 4.6 Relative importance of different molecules in blood to the energy yield in various tissues

Organ	Glucose	Fatty acids	Ketone bodies	Lactic acid
Brain	+ + +	+	−	−
Skeletal muscles (resting)	+	+ + +	+	−
Liver	+	+ + +	+ +	+
Heart	+	+	+	+ + +

Source: Fox, Stuart Ira, *Human Physiology.* © 1984 Wm. C. Brown Publishers, Dubuque, Iowa. All Rights Reserved. Reprinted by permission.

Table 4.7 Relative energy yield from different sources for skeletal muscles during different phases of exercise

Phase	Free fatty acids	Glucose from muscle glycogen	Blood glucose	Liver glucose output (%)	
				From glycogenolysis	From gluconeogenesis
Rest	+ + +	−	+	75 %	25 %
5–10 min	−	+ + +	+	Decreases	Increases
10–40 min	+	+ +	+ +	Decreases	Increases
40–90 min	+ +	+	+ + +	Decreases	Increases
By 4 hours	+ + +	−	+ +	55 %	45 %

Source: Fox, Stuart Ira, *Human Physiology.* © 1984 Wm. C. Brown Publishers, Dubuque, Iowa. All Rights Reserved. Reprinted by permission.

After exercise has continued for ten to forty minutes, however, the metabolic pattern changes as muscle glycogen becomes depleted. The contribution of blood glucose to the energy requirements of exercising muscles increases to about 40%. This increased demand for blood glucose is met by the liver, which increases its rate of glycogenolysis two to five times over resting levels.

By forty minutes of exercise, the liver has added about 18 grams of glucose to the blood—roughly 25% of its stored glycogen. Liver glycogenolysis continues beyond this time, but much of the burden is removed by the increased use of fatty acids as an energy source by the exercising muscles. Since some consumption of blood glucose continues, however, the blood glucose concentration is slightly decreased by ninety minutes of exercise. True hypoglycemia (low blood glucose), however, is rare, which is fortunate because the glucose requirements of the brain do not change with exercise.

As glycogen stored in the liver becomes depleted during prolonged exercise, the contribution of gluconeogenesis to liver glycogen output increases from 25% (at rest) to 45%. By four hours of exercise, the liver contains only about 25% of its original glycogen stores, and the muscles rely increasingly on free fatty acids as an energy source (table 4.7).

During and immediately following exercise, the lactic acid produced by skeletal muscles can be used by the heart as an energy source and by the liver as a substrate for gluconeogenesis (this is part of the Cori cycle). A recovery period of about forty-eight hours is needed before the muscle glycogen that was depleted during prolonged exercise is completely restored.

1. Construct a flow chart to show the metabolic pathway by which glucose can be converted to fat. Indicate only the major intermediates involved (not all of the steps of glycolysis).
2. Define the terms *lipolysis* and *β-oxidation, and* explain in general terms how fat can be used for energy.
3. Describe transamination and deamination, and explain their functional significance.
4. List five blood-borne energy carriers, and explain in general terms how these are used as sources of energy.
5. Define the terms *glycogenolysis and* gluconeogenesis, *and explain how these processes help to maintain constant blood glucose concentrations during exercise.*

Clinical Considerations

Clinical Enzyme Measurements

Assays of Enzymes in Plasma.   When tissues become damaged due to diseases, some of the dead cells disintegrate and release their enzymes into the blood. Most of these enzymes are not normally active in the blood, because of the absence of their specific substrates, but their enzymatic activity can be measured in a test tube by the addition of the appropriate substrates to samples of plasma. Such measurements are clinically useful, because abnormally high plasma concentrations of particular enzymes are characteristic of certain diseases (table 4.8).

Table 4.8 Examples of the diagnostic value of some enzymes found in plasma	
Enzyme	**Diseases associated with abnormal plasma enzyme concentrations**
Alkaline phosphatase	Obstructive jaundice, Paget's disease (osteitis deformans), carcinoma of bone
Acid phosphatase	Benign hypertrophy of prostate, cancer of prostate
Amylase	Pancreatitus, perforated peptic ulcer
Aldolase	Muscular dystrophy
Creatine kinase (or creatine phosphokinase-CPK)	Muscular dystrophy, myocardial infarction
Lactate dehydrogenase (LDH)	Myocardial infarction, liver disease, renal disease, pernicious anemia
Transaminases (GOT and GPT)	Myocardial infarction, hepatitis, muscular dystrophy

Source: Fox, Stuart Ira, *Human Physiology.* © 1984 Wm. C. Brown Publishers, Dubuque, Iowa. All Rights Reserved. Reprinted by permission.

Table 4.9 Some clinical uses of isoenzyme measurements		
Enzyme	**Isoenzyme form number**	**Disease associated with abnormal elevation of isoenzyme**
Creatine phosphokinase (CPK)	1 2	Muscular dystrophy Myocardial infarction
Lactic acid dehydrogenase (LDH)	1 5	Myocardial infarction Liver disease (such as hepatitis)

Source: Fox, Stuart Ira, *Human Physiology.* © 1984 Wm. C. Brown Publishers, Dubuque, Iowa. All Rights Reserved. Reprinted by permission.

Figure 4.35. The effects of an inborn error of metabolism on a branched metabolic pathway.

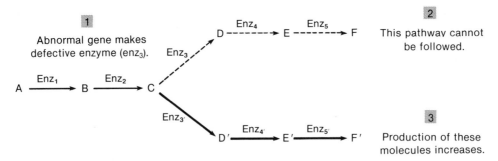

Figure 4.36. Metabolic pathways for the degradation of the amino acid phenylalanine. Defective enzyme$_1$ produces phenylketonuria (PKU), defective enzyme$_2$ produces alcaptonuria (not a clinically significant condition), and defective enzyme$_3$ produces albinism.

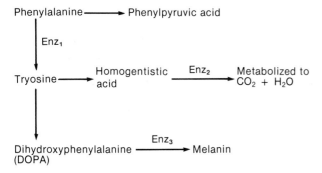

Isoenzymes.

Isoenzymes. Enzymes that do exactly the same job (catalyze the same reaction) in different organs have the same name, since the name describes the activity of the enzyme. Different organs, however, make slightly different "models" of the enzyme that differ in one or a few amino acids. These different models of the same enzyme are called *isoenzymes.* The differences in structure do not affect the active sites (which are the same because isoenzymes catalyze the same reactions), but they change the chemistry sufficiently so that isoenzymes can be separated from each other by standard biochemical procedures. Abnormal elevations in specific isoenzyme forms—which are identified by numbers (table 4.9)—occur in certain disease states.

Metabolic Disturbances

Inborn Errors of Metabolism. Each enzyme in a metabolic pathway is coded by a different gene. An inherited defect in one of these genes may result in a disease known as an *inborn error of metabolism.* In these diseases there is an increased amount of intermediates formed prior to

Table 4.10 Selected examples of inborn errors in the metabolism of amino acids, carbohydrates, and lipids

Metabolic defect	Disease	Abnormality	Clinical result
Amino acid metabolism	Phenylketonuria (PKU)	Increase in phenylalanine	Mental retardation, epilepsy
	Albinism	Lack of melanin	Carcinoma of skin
	Maple-syrup disease	Increase in leucine, isoleucine, and valine	Degeneration of brain, early death
	Homocystinuria	Accumulation of homocystine	Mental retardation, eye problems
Carbohydrate metabolism	Lactose intolerance	Lactose not utilized	Diarrhea
	Glucose-6-phosphatase deficiency (Gierke's disease)	Accumulation of glycogen in liver	Liver enlargement, hypoglycemia
	Glycogen phosphorylase deficiency (McArdle syndrome)	Accumulation of glycogen in muscle	Muscle fatigue and pain
Lipid metabolism	Gaucher's disease	Lipid accumulation (glucocerebroside)	Liver and spleen enlargement, brain degeneration
	Tay-Sachs disease	Lipid accumulation (ganglioside G_{M2})	Brain degeneration death by age 5
	Hypercholesterolemia	High blood cholesterol	Atherosclerosis of coronary and large arteries

Source: Fox, Stuart Ira, *Human Physiology.* © 1984 Wm. C. Brown Publishers, Dubuque, Iowa. All Rights Reserved. Reprinted by permission.

the defective step and a decrease in the intermediates formed after the defective enzymatic step. If the defective enzyme is active at a step after a branch point in a pathway, the intermediates and final products of the divergent pathway will increase as a result of the block in the alternate pathway (fig. 4.35).

Phenylketonuria (PKU). The branched metabolic pathway that begins with phenylalanine as the initial substrate is subject to a number of inborn errors of metabolism (fig. 4.36). When the enzyme that converts this amino acid to tyrosine is defective, the final products of a divergent pathway accumulate and can be detected in the blood and urine. This disease, *phenylketonuria (fen''el-ke''to-nu're-ah) (PKU),* can result in severe mental retardation and a shortened life span. Although no inborn error of metabolism is common, PKU occurs with enough frequency and is so easy to detect that all newborn babies are routinely tested for this defect. If this disease is detected early, brain damage can be prevented by placing the child on an artificial diet low in the amino acid phenylalanine.

Albinism and Other Defects. One of the conversion products of phenylalanine is a molecule called *DOPA,* which is an acronym for dihydroxyphenylalanine (fig. 4.36). DOPA is a precursor of the pigment molecule melanin, which gives skin, eyes, and hair their characteristic colors. An inherited defect in the enzyme that catalyzes the formation of melanin from DOPA results in an *albino (al-bi'no).* Besides PKU and albinism, there are a large number of other inherited defects of amino acid metabolism, as well as inborn errors in the metabolism of carbohydrates and lipids (table 4.10).

Myocardial Ischemia. Ischemia *(is-ke'me-ah)* refers to inadequate blood flow to an organ, such that the rate of oxygen delivery is insufficient to maintain aerobic respiration. Inadequate blood flow to the heart, or *myocardial ischemia,* may occur if the coronary blood flow is occluded by atherosclerosis, a blood clot, or by an artery spasm. People with myocardial ischemia often experience *angina (an-ji'nah) pectoris,* severe pain in the chest and left (and sometimes the right) arm area. This pain is associated with increased blood levels of lactic acid, which are produced by anaerobic respiration by the heart muscle. The degree of ischemia and angina can be decreased by vasodilator drugs such as nitroglycerin and amyl nitrite, which improve blood flow to the heart and also decrease the work of the heart by dilating peripheral blood vessels.

Endocrine Disorders and Metabolism. Since metabolism is regulated largely by hormones (chemicals secreted by endocrine glands into the blood), endocrine diseases can produce metabolic disorders. *Diabetes mellitus*—characterized by high blood glucose and the presence of glucose in the urine—for example, results from the inadequate secretion or action of the hormone insulin. This disease may also be associated with the excessive production of ketone bodies, which can alter blood pH and produce *ketoacidosis.* Abnormally low blood glucose, *hypoglycemia,* may be produced by excessive insulin secretion. Other metabolic disorders can result from diseases of the pituitary, thyroid, and adrenal glands, as described in chapter 25.

Chapter Summary

I. Enzymes as Catalysts
Enzymes are proteins that function as biological catalysts. The great diversity of protein structure allows different enzymes to be specialized in their catalytic actions.
  A. Substrates attach to specific active sites in an enzyme; dissociation of the enzyme-substrate complex yields free enzyme and products.
    1. This is called the lock-and-key model of enzyme action.
    2. Enzymes lower the activation energy required for a reaction.
  B. Enzyme names end in -ase and describe the activities of the enzymes.
II. Control of Enzyme Activity
The rate of enzyme-catalyzed reactions depends on numerous factors. Variations in these factors control the rate of progress along particular metabolic pathways and thus help to regulate cellular metabolism.
  A. Each enzyme has a characteristic pH optimum and is maximally active at a particular temperature.
  B. Many enzymes require inorganic ions as cofactors or vitamin-derivatives as coenzymes.
  C. The rate of enzymatic reactions increases with substrate concentrations up to a maximum.
    1. In a reversible reaction, the direction of the reaction is determined by the relative concentration of substrates on both sides of the double arrows.
    2. This is described by the law of mass action.
  D. In multi-enzyme metabolic pathways, enzymes near the beginning of the pathway may be inhibited by the accumulation of the final product; this is called end-product inhibition.
III. Bioenergetics
The coupling of energy-releasing and energy-requiring reactions is central to life processes. The transformation of energy in living systems is based on reactions that produce and destroy ATP and on oxidation-reduction reactions.
  A. Exergonic reactions release energy, and endergonic reactions require the input of energy.
  B. The production of ATP from ADP and P_i requires energy that is obtained from the intracellular combustion of food molecules.
    1. Since much energy is required to bond the last phosphate, much energy is released when this bond is broken.
    2. ATP is the universal energy donor of the cell, providing energy for the cell's needs when it is converted into ADP and P_i.

C. An atom is oxidized when it loses electrons and is reduced when it gains electrons.
    1. Coupled oxidation-reduction reactions often involve the transfer of pairs of electrons in the form of pairs of hydrogen atoms.
    2. When the oxidized forms of NAD and FAD (NAD_{ox} and FAD_{ox}) accept two electrons from hydrogen, they become reduced to NAD_{red} and FAD_{red}.
IV. Glycolysis and Anaerobic Respiration
In cellular respiration, chemical reactions that transform food molecules into simpler molecules liberate energy, which is used to produce ATP. Anaerobic respiration occurs in the absence of oxygen and provides energy in some organs under special circumstances.
  A. Glycolysis refers to the conversion of glucose to two molecules of pyruvic acid.
    1. In the process, two molecules of ATP are consumed and four molecules of ATP are formed; there is thus a net gain of two ATP.
    2. In the steps of glycolysis two pairs of hydrogens are released; electrons from these reduce two pairs of NAD.
  B. When respiration is anaerobic, reduced NAD is oxidized by pyruvic acid, which accepts two hydrogen atoms and is thereby reduced to lactic acid.
    1. Skeletal muscles use anaerobic respiration and thus produce lactic acid during the early stages of exercise and after the maximal oxygen uptake is reached during heavy exercise.
    2. Lactic acid can be converted to glucose in the liver by a process called gluconeogenesis.
V. Aerobic Respiration
Aerobic respiration is equivalent to combustion in terms of its final products and the total amount of energy liberated. In aerobic respiration, however, the energy is released in small, enzymatically controlled oxidative reactions, and some of the energy released is captured in ATP.
  A. The Krebs cycle begins when coenzyme A donates acetic acid to an enzyme reaction that adds it to oxaloacetic acid to form citric acid.
    1. Acetyl CoA is formed from pyruvic acid by the removal of carbon dioxide and two hydrogens.
    2. The formation of citric acid begins a cyclic pathway that ultimately forms a new molecule of oxaloacetic acid.
    3. As the Krebs cycle progresses, one ATP is formed, and three molecules of NAD and one of FAD are reduced by hydrogens from the Krebs cycle.

B. Reduced NAD and FAD donate their electrons to an electron transport chain of molecules located in the cristae.
    1. The electrons from NAD and FAD are passed from one cytochrome to the next in the electron transport chain in a series of coupled oxidation-reduction reactions.
    2. As each cytochrome ion gains an electron it becomes reduced, and as it passes the electron to the next cytochrome it becomes oxidized.
    3. The last cytochrome becomes oxidized by donating its electron to oxygen, which functions as the final electron acceptor.
    4. When one oxygen atom accepts two electrons and two protons, it becomes reduced to form water.
    5. The energy provided by electron transport is used to form ATP from ADP and P_i, which is called oxidative phosphorylation.
  C. Thirty-six to thirty-eight ATP are produced by the aerobic respiration of one glucose molecule; of these, two are produced in the cytoplasm by glycolysis and the remainder are produced in the mitochondria.
  D. The formation of glycogen from glucose is called glycogenesis, and the breakdown of glycogen is called glycogenolysis.
    1. Glycogenolysis yields glucose-6-phosphate, which can enter the pathway of glycolysis.
    2. The liver, but not skeletal muscles, contains an enzyme that can produce free glucose from glucose-6-phosphate; the liver can thus secrete glucose derived from glycogen.
  E. Carbohydrate metabolism is influenced by the availability of oxygen and by a negative feedback effect of ATP on glycolysis and the Krebs cycle.
VI. Metabolism of Lipids and Proteins
Energy can be derived from the aerobic respiration of lipids and protein. Pyruvic acid and the Krebs cycle acids can also serve as common intermediates in the interconversion of glucose, lipids, and amino acids.
  A. In lipolysis, triglycerides yield glycerol and fatty acids.
    1. Glycerol can be converted to phosphoglyceraldehyde and used for energy.
    2. In the process of the β-oxidation of fatty acid, a number of acetyl CoA molecules are produced.
    3. Processes that operate in the reverse direction can convert glucose to triglycerides.

B. Amino acids can serve as sources of energy.
  1. Through transamination, a particular amino acid and a particular keto acid (pyruvic acid or one of the Krebs cycle acids) can serve as substrates to form a new amino acid and a new keto acid.

  2. In oxidative deamination, amino acids are converted into keto acids as their amino group is incorporated into urea.
C. Each organ uses certain blood-borne energy carriers as their preferred energy source.
  1. The brain has an almost absolute requirement for blood glucose as its energy source.

  2. During exercise, the needs of skeletal muscles for blood glucose can be met by glycogenolysis and by gluconeogenesis in the liver.

Review Activities

Objective Questions

1. Which of the following statements about lactate dehydrogenase is *true*?
   (a) It is a protein.
   (b) It oxidizes lactic acid.
   (c) It reduces another molecule (pyruvic acid).
   (d) All of the above are true.
2. In a metabolic pathway
   (a) the product of one enzyme becomes the substrate of the next
   (b) the substrate of one enzyme becomes the product of the next
3. In an inborn error of metabolism
   (a) a genetic change results in the production of a defective enzyme
   (b) intermediates produced before the defective step accumulate
   (c) alternate pathways are taken by intermediates at branch points located before the defective step
   (d) all of the above
4. Which of the following represents an *endergonic* reaction?
   (a) $ADP + P_i \rightarrow ATP$
   (b) $ATP \rightarrow ADP + P_i$
   (c) $glucose + O_2 \rightarrow CO_2 + H_2O$
   (d) $CO_2 + H_2O \rightarrow glucose$
   (e) both (a) and (d)
   (f) both (b) and (c)
5. Which of the following statements about ATP is *true*?
   (a) The bond joining ADP and the third phosphate is a high-energy bond.
   (b) The formation of ATP is coupled to energy-liberating reactions.
   (c) The conversion of ATP to ADP and P_i provides energy for biosynthesis, cell movement, and other cellular processes that require energy.
   (d) ATP is the "universal energy donor" of cells.
   (e) All of the above are true.

6. When oxygen is combined with two hydrogens to make water
   (a) oxygen is reduced
   (b) the molecule that donated the hydrogens becomes oxidized
   (c) oxygen acts as a reducing agent
   (d) both (a) and (b)
   (e) both (a) and (c)
7. The oxygen in the air we breathe
   (a) functions as the final electron acceptor of the electron transport chain
   (b) combines with hydrogen to form water
   (c) combines with carbon to form CO_2
   (d) both (a) and (b)
   (e) both (a) and (c)
8. In terms of the number of ATP molecules directly produced, the major energy-yielding process in the cell is
   (a) glycolysis
   (b) the Krebs cycle
   (c) oxidative phosphorylation
   (d) gluconeogenesis
9. Ketone bodies are derived from
   (a) fatty acids
   (b) glycerol
   (c) glucose
   (d) amino acids
10. Conversion of glycogen to glucose-6-phosphate occurs in the
   (a) liver
   (b) skeletal muscles
   (c) both of these organs
11. Conversion of glucose-6-phosphate to free glucose, which can be secreted into the blood, occurs in
   (a) the liver
   (b) the skeletal muscles
   (c) both of these organs
12. Formation of glucose from pyruvic acid, which is derived from lactic acid, amino acids, or glycerol is called
   (a) glycogenesis
   (b) glycogenolysis
   (c) glycolysis
   (d) gluconeogenesis

13. Which of the following organs has an almost absolute requirement for blood glucose as its energy source?
   (a) the liver
   (b) the brain
   (c) the skeletal muscles
   (d) the heart
14. When amino acids are used as an energy source
   (a) oxidative deamination occurs
   (b) pyruvic acid or one of the Krebs cycle acids (keto acids) are formed
   (c) urea is produced
   (d) all of the above

Essay Questions

1. Explain the relationship between the chemical structure and the function of an enzyme, and describe how various conditions may alter both the structure and function of an enzyme.
2. Explain how the rate of enzymatic reactions may be regulated by the relative concentrations of substrate and products.
3. Using the first and second laws of thermodynamics, explain how ATP is formed and how it serves as the universal energy donor.
4. Explain the advantages and disadvantages of anaerobic respiration, and explain why anaerobic respiration can only occur for a limited period of time.
5. Compare and contrast the uses of glycogen and fat as energy sources.

Selected Readings

Allfrey, V. G., and A. E. Mirsky. September 1961. How cells make molecules. *Scientific American.*
Berry, H. K. March 1984. The spectrum of metabolic disorders. *Diagnostic Medicine,* p. 39.

Dyson, R. D. 1978. *Essentials of Cell Biology.* 2d ed. Boston: Allyn & Bacon.
Hinkle, P., and R. E. McCarty. March 1978. How cells make ATP. *Scientific American.*
Lehninger, A. L. September 1961. How cells transform energy. *Scientific American.*

Sheeler, P., and D. E. Bianchi. 1980. *Cell Biology: Structure, biochemistry and function.* New York: John Wiley & Sons.

5

Membrane Transport and

Outline

Simple Diffusion and Osmosis
Simple Diffusion
Osmosis
Regulation of Blood Osmolality
Carrier-Mediated Transport
Facilitated Diffusion
Active Transport
The Membrane Potential
Clinical Considerations
Simple Diffusion and Osmosis
Carrier-Mediated Transport
Membrane Potential
Chapter Summary
Review Activities
Selected Readings

Concepts

The cell (plasma) membrane separates the intracellular compartment from the extracellular compartment. This membrane is selectively permeable, allowing the passage of some molecules and ions but not others.

In order to sustain metabolism, cells must be able to take up glucose and other organic molecules that are too large to pass through the membrane pores. Protein carrier molecules within the cell membrane facilitate the transport of certain molecules into the cell.

There is a higher concentration of negatively charged molecules in the cell than in the extracellular environment. This difference in the distribution of charges results in a potential difference, or voltage, across the membrane.

the Membrane Potential

Simple Diffusion and Osmosis

The cell (plasma) membrane separates the intracellular environment from the extracellular environment. Proteins, nucleotides, and macromolecules needed for the structure and function of the cell cannot penetrate, or "permeate," the membrane. Certain molecules and many ions can penetrate the membrane to varying degrees. The cell membrane is selectively permeable; it provides two-way traffic in the nutrients and wastes needed to sustain metabolism, and it prevents the passage of other substances between the intracellular and extracellular compartments.

Objective 1. Define simple diffusion, and explain how various factors influence the rate of diffusion through a cell membrane.

Objective 2. Define the terms, *osmosis, osmolality,* and *osmotic pressure,* and explain the process of osmosis.

Objective 3. Explain the mechanisms that help to maintain a constant plasma osmolality.

There are two major categories of membrane transport processes: **passive transport** and **active transport.** Passive transport occurs when there is a movement of molecules or ions from a region of higher to a region of lower concentration and when such movement does not require the expenditure of metabolic energy. Passive transport includes *simple diffusion, osmosis,* and *facilitated diffusion.* Active transport, in contrast to these examples of passive transport, requires the expenditure of metabolic energy by the hydrolysis of ATP. Such an expenditure of metabolic energy is needed to move molecules from lower to higher concentrations.

Simple Diffusion

Molecules in a gas and molecules and ions dissolved in a solution are in a constant state of random motion as a result of their thermal (heat) energy. This random motion tends to make the gas or solution evenly mixed. Whenever a *concentration difference* (also called a *concentration gradient*) exists between two parts of a solution, therefore, random molecular motion tends to abolish the gradient and to make the molecules uniformly distributed. In terms of the second law of thermodynamics, the concentration difference represents an unstable state of high organization (low entropy), which changes to produce a uniformly distributed solution with maximum disorganization (high entropy).

---

diffusion: L. *dif-fero,* to carry; *fundo,* to pour in different directions
osmosis: Gk. *osmos,* a thrust
entropy: Gk. *entropia,* a turning toward

Figure 5.1. Diffusion occurs when there is a concentration difference (or concentration gradient) between two regions of a solution (*a*) provided that the membrane separating these regions is permeable to the diffusing substance. Diffusion tends to equalize the concentration of these solutions (*b*) and thus to abolish the concentration differences.

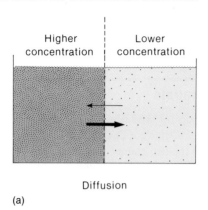

Higher concentration | Lower concentration

Diffusion

(a)

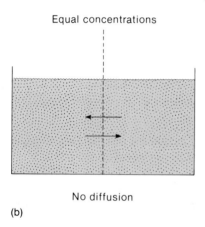

Equal concentrations

No diffusion

(b)

Figure 5.2. Gas exchange between the intracellular and extracellular compartments occurs by diffusion. The regions of higher concentration are represented by the larger symbols.

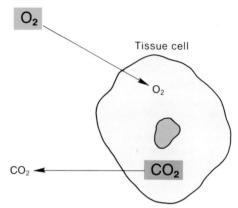

Extracellular environment

Tissue cell

As a result of random motion, molecules in the part of the solution with a higher concentration will enter the area of lower concentration. Molecules will also move in the opposite direction but not as frequently. As a result, there will be a *net movement* from the region of higher to the region of lower concentration until the concentration difference is abolished. This net movement is called **diffusion** *(dĭ-fu'zhun);* see fig. 5.1.

The oxygen concentration is relatively high, for example, in the extracellular fluid because oxygen is carried from the lungs to the body tissues by the blood. Since oxygen is converted to water in aerobic cell respiration, the oxygen concentration within the cells is lower than in the extracellular compartment. The concentration gradient for carbon dioxide is in the opposite direction because cells produce CO_2. *Gas exchange* thus occurs by diffusion between the tissue cells and their extracellular environment (fig. 5.2).

Molecules that are nonpolar and therefore lipid soluble, such as oxygen, carbon dioxide, and steroid hormones, can easily diffuse through the phospholipid layers

of the cell membrane. Inorganic ions and water molecules, which are polar, however, cannot pass through the hydrophobic lipid barrier. Despite this, the membrane does allow the diffusion of some ions under special circumstances and is freely permeable *(per'me-ah-b'l)* to water.

Tiny passages, or *pores,* must therefore be present in the membrane. Although these pores are too small to be seen, even with an electron microscope, they must be present in large numbers to account for the rapid rate with which water and some inorganic ions can penetrate the membrane. Membrane pores are believed to be part of the *integral proteins* that span the thickness of the membrane (fig. 5.3).

The *rate of diffusion,* measured by the number of diffusing molecules passing through the membrane per unit time, depends on (1) the magnitude of the concentration difference across the membrane (the "steepness" of the concentration gradient), (2) the permeability of the membrane to the diffusing substances, and (3) the surface area of the membrane through which the substances are diffusing.

The permeability of the membrane to various molecules and ions is regulated by molecular changes within the membrane structure. Changes in the permeability of the membrane to certain ions, for example, allows the diffusion of these ions and the production of electrical currents in nerve and muscle cells (chapter 13).

The rapid passage of the products of digestion across epithelial membranes in the intestine is aided by structural adaptations that increase the surface area of these cells. The surface area of the apical membrane of intestinal epithelial cells (the part of the membrane facing the lumen of the intestine) is increased by many tiny folds that form fingerlike projections. These membrane folds

pore: Gk. *poros,* a hole

Figure 5.3. Water may penetrate the membrane through pores within integral proteins that span the thickness of the double phospholipid layers.

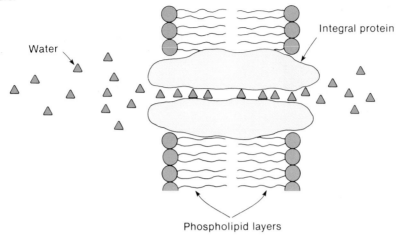

Figure 5.4. Microvilli (*MV*) in the small intestine, as seen with the transmission (*a*) and scanning (*b*) electron microscope. (From: *Tissues and Organs: A Text Atlas of Scanning Electron Microscopy* by R. G. Kessel and R. Kardon. W. H. Freeman and Company. © 1979.)

(a)

(b)

are called **microvilli** *(mi''kro-vil'i)* (fig. 5.4). Similar microvilli are also found in the kidney tubule epithelium, which must reabsorb various molecules that were filtered out of the blood.

The term *simple diffusion* refers to movement of molecules and ions by passive transport either through the phospholipid layers of the membrane (in the case of nonpolar solutes) or through pores in the membrane (in the case of water and the small, inorganic ions). Osmosis—the diffusion of water through a membrane—is thus a special case of simple diffusion.

microvillus: Gk. *mikros,* small; L. *villus,* shaggy hair

Osmosis

Osmosis *(oz-mo'sis)* refers to the simple diffusion of water (the solvent) across a membrane. Like the diffusion of solute molecules, the diffusion of water occurs when the water is more concentrated on one side of the membrane than on the other side, that is, when one solution is more dilute than the other (fig. 5.5). The more dilute solution has a higher concentration of water molecules because the less dilute solution contains more solute molecules, which reduces the water concentration. The principles of osmosis apply to the diffusion of any molecule, but the terminology is backwards because the term *concentration* is usually used to refer to the density of solute rather than solvent molecules.

Figure 5.5. A model of osmosis, or the net movement of water from the solution of lesser solute concentration to the solution of greater solute concentration.

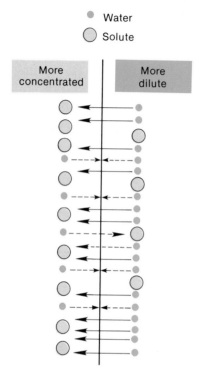

Suppose that a cylinder is divided into two equal compartments by a membrane partition that can freely move and that one compartment initially contains 180 g/L (grams per liter) of glucose and the other compartment contains 360 g/L of glucose. If the membrane is permeable to glucose, glucose will diffuse from the 360 g/L compartment to the 180 g/L compartment until both compartments contain 270 g/L of glucose. If the membrane is not permeable to glucose but is permeable to water, the same result (270 g/L solutions on both sides of the membrane) will be achieved by the diffusion of water (osmosis). As water diffused from the 180 g/L compartment to the 360 g/L compartment the former solution would become more concentrated as the latter became more dilute. This would be accompanied by volume changes, as illustrated in figure 5.6.

Osmotic Pressure.   Osmosis, and the movement of the membrane partition, could be prevented by an opposing force. If one compartment contained 180 g/L of glucose and the other compartment contained pure water, the osmosis of water into the glucose solution could be prevented by pushing against the membrane with a certain force (equal to 22.4 atmospheres pressure). This is illustrated in figure 5.7.

Figure 5.6. A movable semipermeable membrane (permeable to water but not glucose) separates two solutions of different glucose concentration (*a*). As a result, water moves by osmosis into the solution of greater concentration until (*b*) the volume changes equalize the concentrations on both sides of the membrane.

Figure 5.7. If a semipermeable membrane separates pure water from a 180 g/L glucose solution, water tends to move by osmosis into the glucose solution, thus creating a hydrostatic pressure that pushes the membrane to the left and expands the volume of the glucose solution. The amount of pressure that must be applied to just counteract this volume change is equal to the osmotic pressure of the glucose solution.

The force that would have to be exerted to prevent osmosis in this situation is the **osmotic** *(oz-mot'ik)* **pressure** of the solution. This backwards measurement indicates how strongly the solution "draws" water into it by osmosis. The greater the solute concentration of a solution, the greater its osmotic pressure. Pure water, thus, has an osmotic pressure of zero, and a 360 g/L glucose solution has twice the osmotic pressure of a 180 g/L glucose solution.

Molarity and Molality.   Glucose is a monosaccharide with a molecular weight of 180 (the sum of its atomic weights). Sucrose is a disaccharide of glucose and fructose, which have molecular weights of 180 each. When glucose and fructose join together by dehydration synthesis to form a sucrose, a molecule of water (molecular weight = 18) is split off. Therefore, sucrose has a molecular weight of 342 (the sum of 180 + 180 − 18); each sucrose molecule weighs 342/180 times as much as each

Figure 5.8.   Diagram illustrating the difference between a one molar (1.0 M) and a one molal (1.0 m) glucose solution.

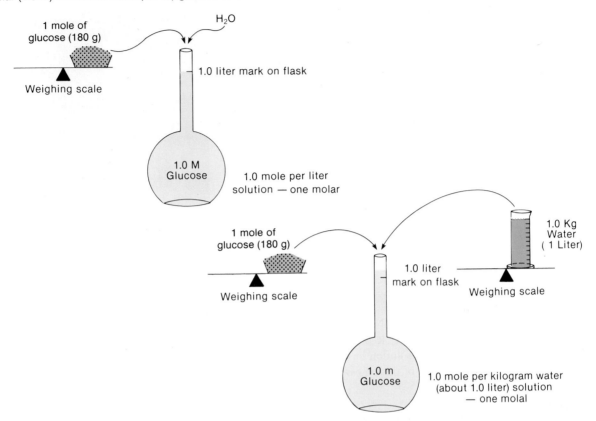

glucose molecule; and 342 grams of sucrose must contain the same number of molecules as 180 grams of glucose.

Notice that the molecular weight in grams of any compound must contain the same number of molecules as the gram molecular weight of any other compound. This unit of weight is called a *mole,* and it always contains 6.02×10^{23} molecules (**Avogadro's number**). One mole of solute dissolved in water to make one liter of solution is described as a one-**molar** solution (abbreviated 1.0 M). Although this unit of measurement is commonly used in chemistry, it is not completely desirable in discussions of osmosis because the exact ratio of solute to water is not specified. More water, for example, is needed to make a 1.0 M NaCl solution (where a mole of NaCl weighs 58.5 grams) than is needed to make a 1.0 M glucose solution, because 180 grams of glucose takes up more volume than 58.5 grams of salt.

Since the ratio of solute to water molecules is of critical importance in osmosis, a more desirable measurement of concentration is **molality.** In a one-molal solution (abbreviated 1.0 m), one mole of solute (180 grams of glucose, for example) is dissolved in one kilogram of water (equal to one liter at 4°C). A 1.0 m NaCl solution and a 1.0 m glucose solution both contain a mole of solute dissolved in exactly the same amount of water (fig. 5.8).

Figure 5.9.   The osmolality (Osm) of a solution is equal to the sum of the molalities of each solute in the solution. If a semipermeable membrane separates two solutions with equal osmolalities, no osmosis will occur.

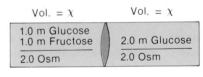

No osmosis

Osmolality.   If 180 grams of glucose and 180 grams of fructose were dissolved in the same kilogram of water, the osmotic pressure of the solution would be the same as that of a 360 g/L glucose solution. Osmotic pressure depends on the ratio of solute to solvent, *not* on the chemical nature of the solute molecules. The expression for the total molality of a solution is **osmolality (Osm).** Thus, the solution of 1.0 m glucose plus 1.0 m fructose has a total molality, or *osmolality (oz"mo-lal'i-te),* of 2.0 osmol/L (abbreviated 2.0 Osm). This is the same as the 360 g/L glucose solution, which is 2.0 m and 2.0 Osm (fig. 5.9).

Unlike glucose, fructose, and sucrose, electrolytes such as NaCl ionize when they dissolve in water. One molecule

Avogadro: from Amadeo Avogadro, Italian physicist, 1766–1856

Figure 5.10. If a semipermeable membrane (permeable to water but not to glucose, Na^+, or Cl^-) separates a 1.0 m glucose solution from a 1.0 m NaCl solution (a), water will move by osmosis into the NaCl solution. This is because NaCl can ionize to yield one molal Na^+ plus one molal Cl^-. After osmosis (b), the total concentration, or osmolality, of the two solutions are equal.

(a)

Vol. = χ Vol. = χ

1.0 m H_2O
Glucose   1.0 m NaCl
               H_2O

1.0 Osm   2.0 Osm

(b)

Vol. = 2/3 χ Vol. = 3/2 χ

1.5 m
Glucose   0.75 m NaCl

1.5 Osm   1.5 Osm

of NaCl dissolved in water yields two ions (Na^+ and Cl^-); one mole of NaCl ionizes to form one mole of Na^+ and one mole of Cl^-. Thus a 1.0 m NaCl solution has a total concentration of 2.0 Osm. The effect of this on osmosis is illustrated in figure 5.10.

Measurement of Osmolality. Plasma and other biological fluids contain many organic molecules and electrolytes. The osmolality of such complex solutions can only be estimated by calculations. Fortunately, however, there is a relatively simple method for measuring osmolality. This method is based on the fact that the freezing point of a solution, like its osmotic pressure, is affected by the total concentration of the solution and not by the chemical nature of the solute.

One mole of solute depresses the freezing point of water by $-1.86°C$. Accordingly, a 1.0-m glucose solution freezes at a temperature of $-1.86°C$, and a 1.0-m NaCl solution freezes at a temperature of $2 \times -1.86 = -3.72°C$, because of ionization. The *freezing point depression* is, thus, a measure of the osmolality. Since plasma freezes at about $-0.56°C$, its osmolality is equal to 0.56/1.86, or 0.3 Osm, which is more commonly indicated as 300 milliosmolal (or 300 mOsm).

Tonicity. A 0.3 m glucose solution has the same osmolality and osmotic pressure as plasma. The same is true of a 0.15-m NaCl solution, which ionizes to produce a total concentration of 300 mOsm. Both of these solutions are used clinically as intravenous infusions, labeled *5.0% dextrose* (5 g of glucose per 100 ml, which is 0.3 m) and

Figure 5.11. A scanning electron micrograph of normal and crenated red blood cells.

normal saline (0.9 g of NaCl per 100 ml, which is 0.15 m). Since these two solutions have the same osmolality and osmotic pressure as plasma, they are said to be *isosmotic.* If these solutions are separated from plasma by a membrane that is permeable to water but not to the solutes, osmosis will not occur. In this case the solutions are said to be **isotonic** to plasma.

Red blood cells placed in isotonic solutions neither gain nor lose water. If red blood cells are placed in a solution that has a lower osmotic pressure than plasma, water will move from the more dilute solution into the cells, causing them to swell or burst *(hemolysis [he-mol'i-sis]).* Solutions with a lower osmotic pressure than plasma are called **hypotonic.** When red blood cells are placed in a **hypertonic** solution (such as sea water), which has a higher osmolality and osmotic pressure than plasma, the osmosis of water out of the cells will occur and causes the cells to shrink. In this process, called *crenation (kre-na'shun),* the cell surface becomes scalloped in appearance (fig. 5.11).

Regulation of Blood Osmolality

The osmolality of the blood plasma is normally maintained within very narrow limits by a variety of regulatory mechanisms. When a person becomes dehydrated, for example, the blood becomes more concentrated as the total blood volume is reduced. The increased blood osmolality and osmotic pressure stimulates *osmoreceptors,* which are neurons located in a part of the brain called the hypothalamus.

As a result of increased osmoreceptor stimulation, the person becomes thirsty and, if water is available, drinks. Along with increased water intake, a person who is dehydrated excretes a lower volume of urine. This occurs as a result of the following sequence of events: (1) increased

isotonic: Gk. *isos,* equal; *tonos,* tension
hypotonic: Gk. *hypo,* under; *tonus,* tension
hypertonic: Gk. *hyper,* over; *tonus,* tension
crenation: L. *crena,* a notch

Figure 5.12. An increase in plasma osmolality (increased concentration and osmotic pressure) due to dehydration stimulates thirst and increased ADH secretion. These effects cause the person to drink more and urinate less. The blood volume, as a result, is increased while the plasma osmolality is decreased. These effects help bring the blood volume back to the normal range and complete the negative feedback (indicated by a negative sign).

plasma osmolality stimulates osmoreceptors in the hypothalamus of the brain; (2) the osmoreceptors stimulate the posterior pituitary gland, by means of a tract of nerve fibers, to secrete **antidiuretic** *(an''ti-di''u-ret'ik)* **hormone (ADH);** (3) ADH acts on the kidneys to promote water retention; so that (4) a lower volume of urine is excreted.

A person who is dehydrated, therefore, drinks more and urinates less. This represents a negative feedback loop (fig. 5.12), which acts to maintain homeostasis of the plasma concentration (osmolality) and, in the process, helps to maintain a proper blood volume.

---

1. *Define simple diffusion, and describe three factors that influence the diffusion rate.*
2. *Define the term* osmosis, *and describe the conditions required for it to occur; define the terms* osmolality *and* osmotic pressure.
3. *Define the terms* isotonic, hypotonic, *and* hypertonic, *and explain why hospitals use 5% dextrose and normal saline as intravenous infusions.*
4. *Explain how changes in the osmolality of plasma are detected and corrected by the body.*

Carrier-Mediated Transport

In order to sustain metabolism, cells must be able to take up glucose, amino acids, and other organic molecules from the extracellular environment. Molecules such as these are too large to pass through membrane pores or too polar to pass through the lipid barrier. Protein carrier molecules within the cell membrane facilitate the transport of certain molecules into the cell.

Objective 4. Define facilitated diffusion, and compare it to simple diffusion.
Objective 5. Describe the characteristics of carrier-mediated transport.
Objective 6. Explain how active transport differs from facilitated diffusion.

The transport of glucose, amino acids, and some other molecules is mediated by protein **carriers** within the membrane. Although such carriers cannot be directly observed, their presence has been inferred by the observation that this transport has characteristics in common with enzyme activity. These characteristics include (1) *specificity*, (2) *competition*, and (3) *saturation*. Because of their similarity to enzymes, the transport carriers are sometimes called *permeases (per'me-āz''ez)*.

Like enzyme proteins, carrier proteins interact with only specific molecules. Glucose carriers, for example, can only interact with glucose and not with closely related monosaccharides. As a further example of specificity, particular carriers for amino acids transport some types of amino acids but not others. Two amino acids that are transported by the same carrier compete with each other, so that the rate of transport of each is lower than it would be if each amino acid were present alone (fig. 5.13).

As the concentration of a transported molecule is increased, its rate of transport will also be increased—but only up to a maximum. Beyond this rate, called the *transport maximum* (or T_m), further increases in concentration do not further increase the transport rate. This indicates that the carriers have become saturated (fig. 5.13).

Figure 5.13. Carrier-mediated transport displays the characteristics of saturation (illustrated by the *transport maximum*) and competition. Molecules *X* and *Y* compete for the same carrier, so that when they are present together the rate of transport of each is less than when either is present separately.

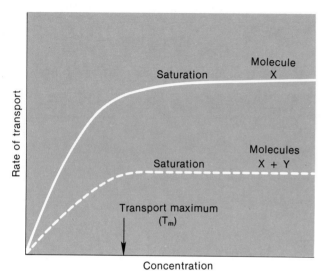

As an example of saturation, imagine a bus stop that is serviced once per hour by a bus that can hold a maximum of forty people (its "transport maximum"). If ten people wait at the bus stop, ten will be transported per hour. If twenty people wait at the bus stop, twenty will be transported per hour. This linear relationship will hold up to a maximum of forty people; if eighty people are at the bus stop, the transport rate will still be forty per hour.

The kidneys transport a number of molecules from the blood filtrate (which will become urine) back into the blood. Glucose, for example, is normally completely reabsorbed so that urine normally is free of glucose. If the glucose concentration of the blood and filtrate is too high (a condition called *hyperglycemia*), however, the transport maximum will be exceeded. In this case, glucose will be found in the urine (a condition called **glycosuria** *[(gli''ko-su're-ah)]*). This may result from eating too many sweets or from the inadequate action of the hormone **insulin** (in the disease **diabetes mellitus** *[di''ah-be'tēz mel-li'tus]*).

Facilitated Diffusion

Except for the small intestine and the kidneys (which transport glucose by active transport), the transport of glucose across cell membranes occurs by **facilitated** *(fah-sil'i-ta''tid)* **diffusion.** Facilitated diffusion, like simple diffusion, involves the passive transport of molecules across a cell membrane from the side of higher to the side of lower concentration. Both simple and facilitated diffusion are examples of passive transport because metabolic cellular energy is not required.

Unlike simple diffusion, however, the facilitated diffusion of glucose displays the properties of carrier-mediated transport: specificity, competition, and saturation. The diffusion of glucose through a cell membrane must therefore be mediated by carrier proteins, which may be conceptualized as specific revolving doors built into the membrane (fig. 5.14). The direction of the facilitated diffusion of glucose depends on the concentration gradient; the revolving doors can revolve both ways, but more glucose is moved in the direction of lower concentration.

Active Transport

There are some aspects of cell transport that cannot be explained by simple or facilitated diffusion. The epithelial lining of the intestine and of the kidney tubules, for example, moves glucose from the side of lower to the side of higher concentration (from the lumen to the blood). Similarly, all cells extrude Ca^{++} into the extracellular environment and, by this means, maintain an intracellular Ca^{++} concentration that is one thousand to ten thousand times lower than the extracellular Ca^{++} concentration.

Movement of molecules and ions against their concentration gradients, from lower to higher concentrations, requires the expenditure of cellular energy, which is obtained from ATP. This type of transport is thus termed **active transport.** If a cell is poisoned with cyanide (which inhibits oxidative phosphorylation), active transport is inhibited. This contrasts with passive transport, which can continue even when metabolic poisons kill the cell by preventing the formation of ATP.

Active transport, like facilitated diffusion, is carrier-mediated. These carriers appear to be integral proteins that span the thickness of the membrane. According to one theory of active transport, the following events may occur: (1) the molecule or ion to be transported bonds to a specific "recognition site" on one side of the carrier protein; (2) this bonding stimulates the breakdown of ATP, which in turn results in phosphorylation of the transport protein; (3) as a result of phosphorylation, the carrier protein undergoes a conformational change, like the change in enzyme proteins as a result of allosteric effects; and (4) the carrier protein undergoes a hingelike motion, which releases the transported molecule or ion on the other side of the membrane. This model of active transport is illustrated in figure 5.15.

The Sodium-Potassium Pump.
Active transport carriers are often referred to as "pumps." Although some of these carriers transport only one molecule or ion at a time, other carriers exchange one molecule or ion for another.

Figure 5.14. A model of facilitated diffusion.

Figure 5.15. A model of active transport, showing the hingelike motion of the integral protein subunits.

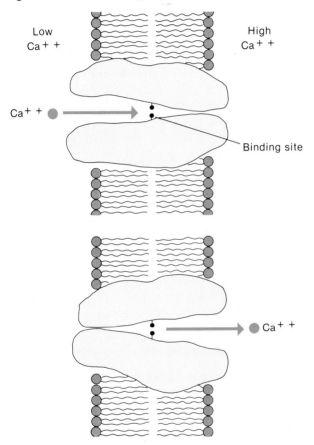

Figure 5.16. The Na$^+$/K$^+$ pump actively exchanges intracellular Na$^+$ for K$^+$. The carrier itself is an ATPase that breaks down ATP for energy. *Dotted lines* indicate the direction of passive transport (diffusion); *solid arrows* indicate the direction of active transport.

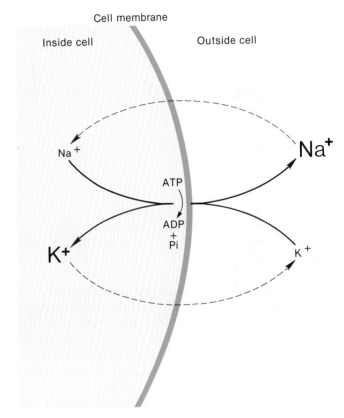

The most important of the latter type of carriers is the **Na$^+$/K$^+$ pump.** This carrier protein, which is also an ATPase enzyme that converts ATP to ADP and P$_i$, actively extrudes three Na$^+$ ions from the cell as it transports two K$^+$ into the cell. This transport is energy dependent because Na$^+$ is more highly concentrated outside the cell and K$^+$ is more concentrated within the cell. Both ions, in other words, are moved against their concentration gradients (fig. 5.16).

All cells have numerous Na$^+$/K$^+$ pumps that are constantly active. This represents an enormous expenditure of energy used to maintain a steep gradient of Na$^+$ and K$^+$ across the cell membrane. This steep gradient serves three known functions: (1) the steep Na$^+$ gradient is used to provide energy for the "co-transport" of other molecules; (2) the activity of the Na$^+$/K$^+$ pumps can be adjusted (primarily by thyroid hormones) to regulate the resting calorie expenditure and basal metabolic rate of the body; and (3) the Na$^+$ and K$^+$ gradients across the cell membranes of nerve and muscle cells are used to produce electrical impulses.

Figure 5.17.   Co-transport of glucose. Glucose accumulates inside the cell against its concentration gradient, using energy derived from the passive transport (diffusion) of Na^+ into the cell. Metabolic energy is indirectly required because the steep electrochemical gradient for Na^+ is maintained by the Na^+/K^+ pump. *Dotted arrows* indicate direction of diffusion; *solid arrows* show direction of active transport.

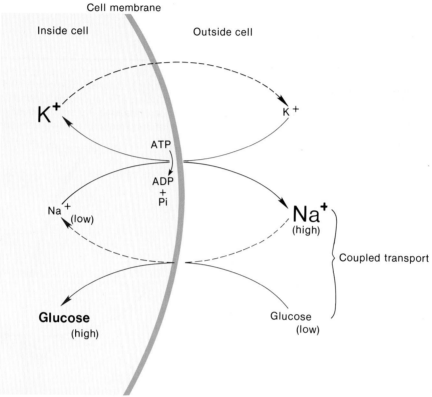

Coupled Transport (Co-transport).   In co-transport, the energy needed for the "uphill" movement of a molecule or ion is obtained from the "downhill" transport of Na^+ into the cell. The active extrusion of Ca^{++} from some cells, for example, is coupled to the passive diffusion of Na^+ into the cell. Cellular energy, obtained from ATP, is not used to move Ca^{++} directly out of the cell, but energy is constantly required to maintain the steep Na^+ gradient. The inward diffusion of Na^+ due to this concentration gradient, in turn, is used to power the active extrusion of Ca^{++}.

A similar mechanism is used to actively transport glucose across the membranes of epithelial cells that line the digestive tract and kidney tubules. In this case, the inward diffusion of Na^+ is coupled to the cellular uptake of glucose (as well as amino acids) against their concentration gradients (fig. 5.17). Active transport in the kidney tubules accounts for as much as 6% of the body's energy expenditure at rest.

1. *List the three characteristics of facilitated diffusion that distinguish it from simple diffusion.*
2. *Draw a figure that illustrates two of the characteristics of carrier-mediated transport, and explain how this differs from simple diffusion.*
3. *Describe active transport, including co-transport in your description. Explain how active transport differs from facilitated diffusion.*
4. *Explain the functional significance of the Na^+/K^+ pump.*

The Membrane Potential

As a result of the permeability properties of the cell membrane and of active transport processes across the cell membrane, there is a higher concentration of negatively charged molecules in the cell than in the extracellular environment. This difference in the distribution of charges results in a potential difference, or voltage, across the membrane.

Figure 5.18. Proteins, organic phosphates, and other organic anions that cannot leave the cell create a fixed negative charge on the inside of the membrane. This attracts positively charged inorganic ions (cations), which therefore accumulate within the cell at a higher concentration than in the extracellular fluid. The amount of cations that accumulate within the cell is limited by the fact that a concentration gradient builds up, which favors the diffusion of the cations out of the cell.

Figure 5.19. If K^+ was the only ion able to diffuse through the cell membrane, it would distribute itself between the intracellular and extracellular compartment until an equilibrium would be established. At equilibrium the K^+ concentration within the cell would be higher than outside the cell due to the attraction of K^+ for the fixed anions. Not enough K^+ would accumulate within the cell to neutralize these anions, however, so the inside of the cell would be 90 millivolts negative compared to the outside of the cell. This membrane voltage is the equilibrium potential (E_K) for potassium.

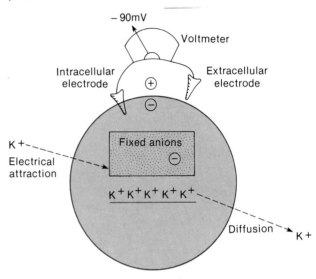

Objective 7. Describe the distribution of charges across the cell membrane; define the term *membrane potential,* and describe how it is measured.

Objective 8. Describe the potassium equilibrium potential.

Objective 9. Explain how the permeability properties of the membrane and the activity of Na^+/K^+ pumps contribute to the membrane potential.

Cellular proteins and the phosphate groups of ATP and other nucleotides are negatively charged within the cell cytoplasm. These negative ions (anions *[an'i-onz]*) are "fixed" within the cell by the fact that they cannot penetrate the cell membrane. Since these negatively charged organic molecules cannot leave the cell, they attract positively charged inorganic ions (cations *[cat'i-onz]*) from the extracellular fluid that are small enough to diffuse through the membrane pores. The distribution of small, inorganic cations (mainly K^+, Na^+ and Ca^{++}) between the intracellular and extracellular compartments, in other words, is influenced by the negatively charged fixed ions within the cell.

Since the cell membrane is much more permeable to K^+ than to any other cation, K^+ accumulates within the cell more than the others as a result of its electrical attraction for the fixed anions (fig. 5.18). Instead of being evenly distributed between the intracellular and extracellular compartments, therefore, K^+ becomes more highly concentrated within the cell. In the human body the intracellular K^+ concentration is 155 mEq/L compared to

an extracellular concentration of 4 mEq/L (mEq = milliequivalents, which is the millimolar concentration multiplied by the valence of the ion—in this case, by one).

Even without the action of the Na^+/K^+ pumps, therefore, K^+ would be more highly concentrated within the cell than in the extracellular fluid. The magnitude of the intracellular K^+ concentration depends on the amount of fixed anions and on the K^+ concentrations that are normally found in the extracellular compartment (including plasma and tissue fluid). An increase in the extracellular K^+ concentration above normal, for example, would cause a corresponding increase in the intracellular K^+ concentration.

If (1) the extracellular K^+ concentration is normal and (2) K^+ is the only inorganic cation able to diffuse through the membrane, an *equilibrium* will be established in which the K^+ concentrations remain stable. There is no net movement of K^+ at this equilibrium, because if a certain amount of K^+ were to move into the cell (by an electrical attraction for the fixed anions), an identical amount of K^+ would diffuse out of the cell (down its concentration gradient). At this equilibrium, the intracellular K^+ concentration would be such that only part of the fixed anions would be neutralized (fig. 5.19).

Figure 5.20. Because some Na^+ leaks into the cell by diffusion, the actual resting membrane potential is less than the K^+ equilibrium potential. As a result, some K^+ diffuses out of the cell (*dotted lines*).

Figure 5.21. The concentrations of Na^+ and K^+ both inside and outside the cell do not change as a result of diffusion (*dotted arrows*) because of active transport (*solid arrows*) by the Na^+/K^+ pump. Since the pump transports three Na^+ for every two K^+, the pump itself helps to create a charge separation (a potential difference, or voltage) across the membrane.

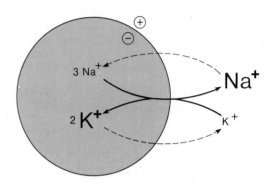

At equilibrium, therefore, the inside of the cell membrane would have a higher concentration of negative charges than the outside of the membrane. The membrane, in other words, separates charges because the fixed anions cannot leave the cell and K^+ cannot accumulate sufficiently to neutralize these negative charges. The magnitude of the charge separation—or the **potential** *(po-ten'shal)* **difference** between the two sides of the membrane—under these conditions is equal to 90 millivolts (mV). This is shown with a negative sign (as -90 mV) to indicate that the inside of the cell is the negative pole.

The potential difference of -90 mV, which would be developed if K^+ were the only diffusible cation, is sometimes called the K^+ **equilibrium** *(e''kwĭ-lib're-um)* **potential** (abbreviated E_K). It should be noted, however, that this description is only a "first approximation" to the truth. Potassium *is* the most diffusible cation, but it is not the only cation that can cross the membrane. The cell membrane is slightly permeable to Na^+, and the concentration of Na^+ in the extracellular fluid (145 mEq/L) is much higher than it is in the cell (12 mEq/L). This concentration gradient, together with sodium's electrical attraction for the fixed anions, causes Na^+ to move into the cell at the rate permitted by the permeability of the cell membrane to Na^+ (fig. 5.20).

If the cell membrane were very permeable to Na^+, so much Na^+ would enter the cell that the membrane potential would actually reverse polarity and become positively charged on the inside. This in fact happens during the production of nerve impulses. In a normal resting cell, however, the membrane permeability to Na^+ is so low that only a "trickle" of Na^+ can enter the cell. This low rate

of Na^+ entry is sufficient, however, to reduce the **resting membrane potential** from E_K (which is -90 mV) to a range of -85 mV to -65 mV, depending on the cell.

The Na^+/K^+ Pump. Since the membrane potential is reduced below E_K as a result of Na^+ entry, some K^+ leaks out of the cell. The cell is *not* at equilibrium with respect to K^+ and Na^+ concentrations. Despite this, the concentrations of K^+ and Na^+ are maintained constant; this is due to the constant expenditure of energy in active transport by the Na^+/K^+ pump. The Na^+/K^+ pump acts to counter the leaks and thus maintains the membrane potential.

Actually, the Na^+/K^+ pump does more than simply work against the ion leaks, since it transports *three* Na^+ ions out of the cell for every *two* K^+ ions that it moves in, its action helps generate a potential difference across the membrane (fig. 5.21). As a result of all of these activities, a real cell has (1) a relatively constant intracellular concentration of Na^+ and K^+; and (2) a constant membrane potential (in the absence of stimulation) in nerves and muscles of -65 mV to -85 mV, due to the action of Na^+/K^+ pumps.

1. Define the term membrane potential, *and describe how it is measured.*
2. *Describe how an equilibrium potential is produced when potassium is the only diffusible cation.*
3. *Explain why the resting membrane potential is different from the potassium equilibrium potential.*
4. *Explain the role of the Na^+/K^+ pump in the generation and maintenance of the resting membrane potential.*

Table 5.1 Inherited defects of transport carriers in the kidney and intestine

Disease	Defect	Clinical significance
Cystinuria	Excessive urinary excretion of cystine, lysine, arginine, and ornithine	Calculi (stones) in urinary tract
Phosphaturia	Excessive urinary excretion of phosphate	Rickets: treated with large doses of vitamin D
Renal glycosuria	Kidney tubules have lower than normal T_m for glucose	None known
Glucose malabsorption	Dietary glucose not absorbed from intestine	Sometimes fatal; must use fructose as only dietary carbohydrate
Hartnup disease	Delayed intestinal absorption of tryptophan and related molecules	Cerebellum dysfunction; photosensitive dermatitis

Source: Fox, Stuart Ira, *Human Physiology.* © 1984 Wm. C. Brown Publishers, Dubuque, Iowa. All Rights Reserved. Reprinted by permission.

Clinical Considerations

Simple Diffusion and Osmosis

Dialysis. The normally functioning kidneys remove waste products from the blood. After the blood is filtered through capillary pores—which are large enough to permit the passage of wastes and other molecules—the molecules needed by the body are reabsorbed back into the blood. The wastes generally remain in the filtrate and are excreted in the urine. When the kidneys do not function properly, waste molecules can be removed from the blood artificially by a process called *dialysis (di-al'i-sis).* Dialysis refers to the process of removing particular molecules from a solution by having them pass, by means of simple diffusion, through an artificial porous membrane. Since the pores in this dialysis membrane are large enough to permit the passage of some molecules but too small to permit the passage of others (the plasma proteins), small waste molecules can be removed from the blood plasma by this technique.

Tonicity of Intravenous Fluids. Fluids delivered intravenously must be isotonic to blood, in order to maintain the correct osmotic pressure and prevent cells from either expanding or shrinking due to the gain or loss of water. Common fluids used for this purpose are normal saline (approximately 0.9 g NaCl per 100 ml water) and 5% dextrose (5 g glucose per 100 ml water). These solutions have about the same osmolality as normal plasma (approximately 300 mOsm). Another isotonic solution frequently used in hospitals is *Ringer's Lactate,* which contains glucose and lactic acid in addition to a number of different salts. Isotonic solutions are also used in *blood-gas machines,* which take the place of the heart and lungs during open-heart surgery.

Ringer: from Sidney Ringer, English physiologist, 1835–1910

Carrier-Mediated Transport

Inherited Defects in Transport Carriers. Since membrane transport carriers are proteins that are coded by specific genes, inherited defects in these carriers can result when there is an alteration in the genetic code. Defective carrier proteins in the cell membranes of epithelial cells that line the intestine may produce diseases that result from an inadequate absorption of ingested molecules, and defects in transport carriers within the epithelial cells of kidney tubules may result in the abnormal excretion of particular molecules in the urine (table 5.1).

Membrane Potential

Hyperkalemia. Although changes in the extracellular concentration of many ions can affect the membrane potential, this potential is particularly sensitive to changes in plasma potassium. Since the maintenance of a particular membrane potential is critical for the generation of electrical events in nerves and muscles (including the heart), the body has a variety of mechanisms that serve to maintain plasma K^+ concentrations within very narrow limits. These mechanisms act primarily through the kidneys, which can excrete K^+ in the urine or reabsorb it into the blood. The excretion of K^+ is stimulated by hormones of the adrenal cortex (particularly by aldosterone); if the adrenal glands of an experimental animal are removed, the animal may die as a result of an accumulation of K^+ in the blood. An abnormal increase in the blood concentration of K^+ is called **hyperkalemia** *(hi"per-kah-le'me-ah).*

When hyperkalemia occurs, the diffusion gradient that favors the extrusion of K^+ from the cell is reduced. As a result, more K^+ can enter the cell and neutralize more of the fixed negative charges. This reduces the membrane potential (brings it closer to zero) and thus alters the function of many organs, particularly the heart. For these reasons, the blood electrolyte concentrations are monitored very carefully in patients with heart or kidney disease.

---

hyperkalemia: Gk. *hyper,* over; L. *kalium,* potash; Gk. *haima,* blood

Chapter Summary

I. Simple Diffusion and Osmosis
The cell (plasma) membrane separates the intracellular compartment from the extracellular compartment. This membrane is selectively permeable, allowing the passage of some molecules and ions but not others.
A. Simple diffusion is the net movement of molecules or ions from higher to lower concentration.
   1. This is a type of passive transport; metabolic energy is not needed.
   2. Nonpolar molecules diffuse across the lipid layers of the membrane; polar substances diffuse through the membrane pores.
   3. The rate of diffusion through a membrane depends on the temperature, concentration gradient, and surface area of the membrane.
B. Osmosis refers to the simple diffusion of water (the solvent) through the membrane.
   1. Osmosis occurs in response to a concentration gradient across the membrane when the membrane is more permeable to water than it is to the solute.
   2. The osmotic pressure of a solution is directly proportional to the total solute concentration, or osmolality.

   3. Solutions with the same osmolality and osmotic pressure are isosmotic; when isosmotic solutions are separated by a membrane and no osmosis occurs, the solutions are said to be isotonic.
   4. Water will move by osmosis from a hypotonic solution into a cell; water will move out of a cell placed in a hypertonic solution.
C. Osmoreceptors in the hypothalamus of the brain help to maintain a relatively constant blood osmolality.

II. Carrier-Mediated Transport
In order to sustain metabolism, cells must be able to take up glucose and other organic molecules that are too large to pass through the membrane pores. Protein carrier molecules within the cell membrane facilitate the transport of certain molecules into the cell.
A. Facilitated diffusion is a type of passive transport.
   1. Specific molecules are moved from the side of higher to the side of lower concentration by membrane carriers.
   2. Carrier-mediated transport exhibits properties of specificity, saturation, and competition.
   3. Facilitated diffusion does not require the expenditure of metabolic energy.
B. Active transport is mediated by carriers and requires metabolic energy, which is provided by the breakdown of ATP.
   1. In active transport, molecules and ions are moved from the side of lower to the side of higher concentrations.

   2. The Na^+/K^+ pump moves Na^+ out of cells and K^+ into cells against their concentration gradients.
   3. In coupled transport, or co-transport, the diffusion of Na^+ into the cell provides energy for moving a molecule (such as glucose) against its concentration gradient.

III. The Membrane Potential
There is a higher concentration of negatively charged molecules in the cell than in the extracellular environment. This difference in the distribution of charges results in a potential difference, or voltage, across the membrane.
A. Proteins and organic phosphates are negatively charged and fixed within the cell; these attract inorganic cations.
   1. As a result of the fixed anions, the permeability properties of the membrane, and the actions of the Na^+/K^+ pump, an unequal distribution of cations across the membrane is produced.
   2. The concentration of potassium is highest inside the cell, and the concentration of sodium is highest outside the cell.
B. An equilibrium state is not attained, but the constant actions of the Na^+/K^+ pumps maintain constant concentrations of diffusible cations.
   1. In this state, the inside of the cell is negatively charged compared to the outside.
   2. As a result, there is a membrane potential or voltage across the cell membrane.

Review Activities

Objective Questions

1. The movement of water across a cell membrane occurs by
   - (a) active transport
   - (b) facilitated diffusion
   - (c) simple diffusion (osmosis)
   - (d) all of these
2. Which of the following statements about the facilitated diffusion of glucose is *true*?
   - (a) There is a net movement from the region of low to the region of high concentration.
   - (b) Carrier proteins in the cell membrane are required for this transport.
   - (c) This transport requires energy obtained from ATP.
   - (d) This is an example of co-transport.
3. If a poison such as cyanide stops the production of ATP, which of the following transport processes would cease?
   - (a) the movement of Na^+ out of a cell
   - (b) osmosis
   - (c) the movement of K^+ out of a cell
   - (d) all of these
4. Red blood cells crenate in
   - (a) a hypotonic solution
   - (b) an isotonic solution
   - (c) a hypertonic solution

5. Plasma has an osmolality of about 300 mOsm. Isotonic saline has an osmolality of
   - (a) 150 mOsm
   - (b) 300 mOsm
   - (c) 600 mOsm
   - (d) none of these
6. A 0.5 m NaCl solution and a 1.0 m glucose solution
   - (a) have the same osmolality
   - (b) have the same osmotic pressure
   - (c) are isotonic to each other
   - (d) all of these
7. The diffusible ion that is most important in the establishment of the membrane potential is
   - (a) K^+
   - (b) Na^+
   - (c) Ca^{++}
   - (d) Cl^-
8. An increase in blood osmolality
   - (a) can occur as a result of dehydration
   - (b) causes a decrease in blood osmotic pressure
   - (c) is accompanied by a decrease in ADH secretion
   - (d) all of these
9. In hyperkalemia, the membrane potential
   - (a) increases
   - (b) decreases
   - (c) is not changed

10. Which of the following statements about the Na^+/K^+ pump is *true*?
    - (a) Na^+ is actively transported into the cell.
    - (b) K^+ is actively transported out of the cell.
    - (c) An equal number of Na^+ and K^+ ions are transported with each cycle of the pump.
    - (d) The pumps are constantly active in all cells.

Essay Questions

1. Describe the conditions required to produce osmosis, and explain why osmosis occurs under these conditions.
2. Explain how simple diffusion can be distinguished from facilitated diffusion and how active transport can be distinguished from passive transport.
3. Compare the theoretical membrane potential that occurs at K^+ equilibrium with the true resting membrane potential. Explain the reasons for the differences between these.
4. Explain how the Na^+/K^+ pump contributes to the resting membrane potential.

Selected Readings

Dyson, R. D. 1978. *Essentials of cell biology*. 2d ed. Boston: Allyn & Bacon.

Holter, N. September 1961. How things get into cells. *Scientific American*.

Johnson, L. G. 1983. *Biology*. Dubuque: Wm. C. Brown Company Publishers.

Keeton, W. T. 1980. *Biological Science*. 3rd ed. New York: W. W. Norton & Company.

Mader, S. S. 1982. *Inquiry into Life*. 3rd ed. Dubuque: Wm. C. Brown Company Publishers.

Sheeler, P., and D. E. Bianchi. 1980. *Cell biology: structure, biochemistry and function*. New York: John Wiley & Sons.

6

Histology

Concepts

Histology is an integral part of anatomy and physiology because it imparts an understanding of the structure and function of organs at the tissue level of study. Tissues are classified on the basis of their cellular composition and histological appearance into four principal kinds.

The various types of tissues are established during embryonic development, and as differentiation continues, organs form, each composed of a specific arrangement of tissues.

Epithelia are classified according to the physical features of the tightly packed cells that comprise these tissues. Epithelia line all body surfaces, cavities, and lumina and are adapted for protection, absorption, and secretion.

Connective tissues are classified according to the characteristics of the matrix that binds the cells. Connective tissues provide structural and metabolic support for other tissues and organs of the body.

Muscle tissues are responsible for the movement of materials through the body, the movement of one part of the body with respect to another, and for locomotion. Fibers in the three kinds of muscle tissue are adapted to contract in response to stimuli.

Nervous tissue is composed of neurons, which respond to stimuli and conduct impulses to and from all body organs, and neuroglia, which functionally support and physically bind neurons.

Definition and Classification of Tissues

Histology is an integral part of anatomy and physiology because it imparts an understanding of the structure and function of organs at the tissue level of study. Tissues are classified on the basis of their cellular composition and histological appearance into four principal kinds.

Objective 1. Define tissue, and discuss the importance of histology.
Objective 2. Explain the functional relationship between cells and tissues.
Objective 3. Classify the tissues of the body into four major types, and give the distinguishing characteristics of each type.

Although cells are the structural and functional units of the body, the cells of a multicellular organism are so specialized that they do not function independently. You will recall that tissues are aggregations of similar cells that perform specific functions. The study of tissues is referred to as **histology** *(his-tol′o-je).*

Although histology is actually microscopic anatomy, it is an essential part of anatomy and physiology because it imparts an understanding of the structure and function of organs at the tissue level. Many diseases profoundly alter the tissues within an affected organ; therefore, by knowing the normal tissue structure, a medical person can recognize the abnormal. Histology in medical schools is usually followed by a course in *pathology,* which is primarily concerned with identifying diseased tissues.

Although histologists utilize many different techniques for preparing, staining, and sectioning tissues, basically only two kinds of microscopes are used to view the prepared tissues. *Light microscopy (mi-kros′ko-pe)* is used for the general observation of tissue structure (fig. 6.1), and *electron microscopy* permits observation of the fine details of tissue and cellular structure. Most of the histological photomicrographs in this text will be at the light microscopic level to present an overview of tissue structure. A few electron micrographs from an electron microscope are used to depict the fine structural detail needed to understand a particular function.

---

histology: Gk. *histos,* web (tissue); *logos,* study
pathology: Gk. *pathos,* suffering, disease; *logos,* study

Figure 6.1. The appearance of skin at various magnifications: (a) = 10×, (b) = 25×, (c) = 50× through a compound light microscope; and (d) a hair emerging from a follicle as seen through a scanning electron microscope (SEM) at 280×.

(a)

(b)

(c)

(d)

Tissue cells are separated and bound together by a nonliving intercellular **matrix** *(ma'triks)* that the cells secrete. Matrix varies in composition from one tissue to another and may take the form of a liquid, semisolid, or solid. Blood tissue, for example, has a liquid matrix, whereas bone cells are separated by a solid matrix.

The tissues of the body are classified into four principal types, determined by structure and function: (1) *epithelial (ep''ĭ-the'le-al) tissues* cover body and organ surfaces, line body and lumen cavities, and form various glands; (2) *connective tissues* bind, support, and protect body parts; (3) *muscle tissues* contract to produce movement; and (4) *nervous tissues* initiate and transmit nerve impulses from one body part to another.

---

matrix: L. *matris,* mother

1. Define the term tissue, *and explain why histology is important to the study of anatomy and physiology and medicine.*
2. Cells are the functional units of the body. Explain how the matrix permits specific kinds of cells to even be more effective and functional as tissues.
3. List and give the distinguishing characteristics of the four principal types of tissues in the body, and explain their basic functions.

Development of Tissues

The various types of tissues are established during embryonic development, and as differentiation continues, organs form, each of which is composed of a specific arrangement of tissues.

Objective 4. Describe the location of the three primary germ layers during early embryonic development.

Figure 6.2. The early stages of embryonic development. (a) the zygote immediately following conception. (b) the morula at about the third day as it enters the uterine cavity. (c) the early blastocyst at the time of implantation between the fifth and seventh day. (d) a cross section of an implanted blastocyst at two weeks. (e) a cross section of a blastocyst at three weeks showing the three primary germ layers.

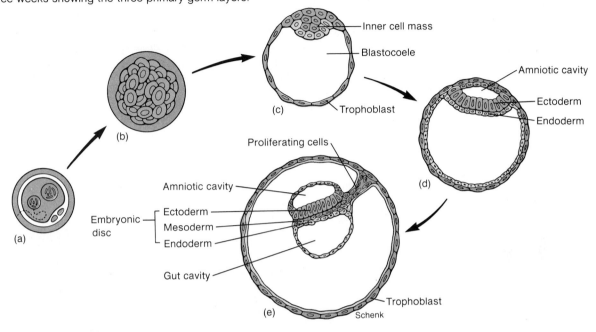

Objective 5. List the tissues and body organs derived from each of the three primary germ layers.

Human prenatal development is initiated with the fertilization of an ovulated ovum (egg) from a female by a sperm cell from a male. The chromosomes within the nucleus of a **zygote** *(zi'gōt)* (fertilized egg) contain all the genetic information necessary for the differentiation and development of all body structures.

Within thirty hours after fertilization, the zygote undergoes a mitotic division as it moves through the uterine tube toward the uterus (chapter 28). After several more cellular divisions, the embryonic mass consists of sixteen or more cells and is called a **morula** *(mor'u-lah)* (fig. 6.2). Three or four days after conception, the morula enters the uterine cavity where it remains unattached for about three days. During this time, the center of the morula fills with fluid passing in from the uterine cavity. As the fluid-filled space develops inside the morula, two distinct groups of cells form. The single layer of cells forming the outer wall is composed of **trophoblast cells,** and the small, inner aggregation of cells becomes the **inner cell mass,** or **embryoblast.** After further development the trophoblast becomes a portion of the placenta, and the inner cell mass

becomes the embryo. With the establishment of these two groups of cells, the morula becomes known as a **blastocyst** *(blas'to-sist).* Implantation of the blastocyst occurs between the fifth and seventh day (chapter 29).

As the blastocyst completes implantation during the second week of development, the inner cell mass undergoes marked differentiation. A slitlike space called the **amniotic** *(am'ne-ot-ic)* **cavity** forms between the inner cell mass and the trophoblast (see fig. 6.2). The inner cell mass now consists of two layers: an upper **ectoderm,** which is closer to the amniotic cavity, and a lower **endoderm,** which borders the blastocyst cavity. A short time later, a third layer called the **mesoderm** forms between the endoderm and ectoderm. These three layers constitute the **primary germ layers.**

The primary germ layers are important because all the cells and tissues of the body are derived from them. Ectodermal cells form the nervous system; the outer layer of skin (epidermis), including hair, nails, and skin glands; and portions of the sensory organs. Mesodermal cells form the skeleton, muscles, blood, reproductive organs, dermis of the skin, and connective tissue. Endodermal cells produce the lining of the digestive tract, the digestive organs, the respiratory tract and lungs, and the urinary bladder and urethra.

zygote: Gk. *zygotos,* yolked
morula: Gk. *morus,* mulberry
trophoblast: Gk. *trophe,* nourishment; *blastos,* germ
embryoblast: Gk. *embryon,* to be full, swell; *blastos,* germ

ectoderm: Gk. *ecto,* outside; *derm,* skin
endoderm: Gk. *endo,* within; *derm,* skin
mesoderm: Gk. *meso,* middle; *derm,* skin

Figure 6.3. The body systems and the primary germ layers from which they develop.

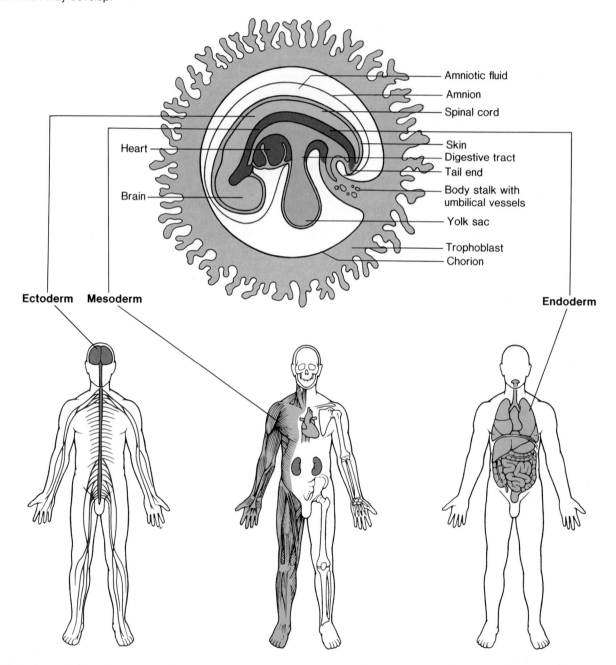

Table 6.1 Derivatives of the germ layers		
Ectoderm	**Mesoderm**	**Endoderm**
Epidermis of skin and epidermal derivatives: hair, nails, glands of the skin; linings of oral, nasal, anal, and vaginal cavities	Muscle: smooth, cardiac, and skeletal	Epithelium of pharynx, auditory canal, tonsils, thyroid, parathyroid, thymus, larynx, trachea, lungs, digestive tract, urinary bladder and urethra, and vagina
Nervous tissue; sense organs	Connective tissue: embryonic, connective tissue proper, cartilage, bone, blood	
Lens of eye; enamel of teeth	Dermis of skin; dentin of teeth	Liver and pancreas
Posterior pituitary gland	Epithelium of blood vessels, lymphatic vessels, body cavities, joint cavities	Anterior pituitary gland
Adrenal medulla	Internal reproductive organs	
	Kidneys and ureters	
	Adrenal cortex	

Figure 6.4. Types of cellular bonding in epithelia.
(a) glycoprotein deposits. (b) desmosomes. (c) tight and
intermediate junctions.

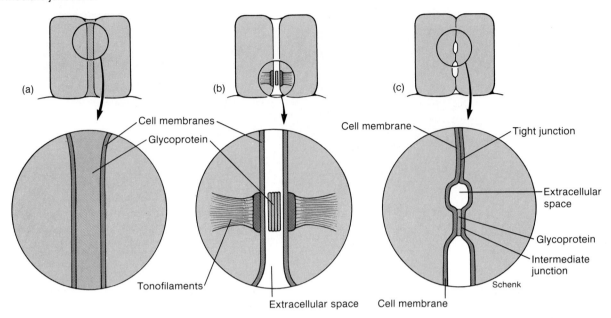

Epithelial Tissue

Epithelia are classified according to the physical features
of the tightly packed cells that comprise these tissues. Epithelia line all body surfaces, cavities, and lumina and are
adapted for protection, absorption, and secretion.

Objective 6. Differentiate between the kinds of epithelia.
Objective 7. Describe how epithelial cells are held
together.
Objective 8. Define a gland, and distinguish between
the various types of glands in the body.

Characteristics of Epithelia

Epithelia *(ep"ĭ-the'le-ah)* are located throughout the body
and form such structures as the outer layer of the skin,
the inner lining of body cavities and lumina and the covering of visceral organs, and the secretory portion of glands.

Figure 6.3 illustrates the organs and body systems that
derive from each of the three primary germ layers. Table
6.1 lists the derivatives of the primary germ layers.

> 1. *What is a morula? How does it differ from a
>   blastocyst?*
> 2. *At what developmental age (weeks following
>   conception) are the germ layers present and
>   where are they located?*
> 3. *Referring to chapter 1, list the ten body
>   systems and indicate the principal germ layers
>   that derive the organs of each.*

There is always one side of epithelia exposed to a body
cavity, lumen, or skin surface. Some epithelia are derived
from ectoderm, such as the outer layer of the skin and
integumentary glands; some from mesoderm, such as the
inside lining of blood vessels; and others from endoderm,
such as the inside lining of the digestive tract.

Epithelia may be one layer or several layers thick. The
upper surface of epithelia may be exposed to gases, as in
the case of epithelium in the integumentary and respiratory systems; to liquids, as in the circulatory and urinary
systems; or to semisolids, as in the digestive system. The
lower surface of epithelia is bound to underlying supportive tissue by a **basement membrane** consisting of glycoprotein from the epithelial cells and a meshwork of
collagen and reticular fibers from the underlying connective tissue. With few exceptions, epithelia are avascular
and must be nourished by diffusion from underlying connective tissues. Cells comprising epithelia are tightly
packed together, and there is little intercellular matrix between them.

The cells of epithelia are tightly bonded in one of three
ways (fig. 6.4): (1) some epithelial cells have a bonding
matrix of *glycoprotein deposits.* Glycoprotein is a combination of polysaccharides and protein secreted by the
cells to bind them to other like cells in a process similar
to that which binds epithelia to the basement membrane.
(2) Certain epithelial cells are bonded with *desmosomes.*
Desmosomes are characterized by V-shaped **tonofila-
ments** *(ton"o-fil'ah-ment),* reinforcing the cell membrane
where glycoprotein is secreted to bond adjacent cells.

epithelium: Gk. *epi,* upon; *thelium,* to cover

Table 6.2 Summary of epithelial tissues

Type	Structure and function	Location
Simple epithelia	Single layer of cells; diffusion and filtration	Covering visceral organs, linings of lumina and body cavities
Simple squamous epithelium	Single layer of flattened, tightly bound cells; diffusion and filtration	Capillary walls, air sacs of lungs, covering visceral organs, linings of body cavities
Simple cuboidal epithelium	Single layer of cube-shaped cells; excretion, secretion, or absorption	Surface of ovaries; linings of kidney tubules, salivary ducts, and pancreatic ducts
Simple columnar epithelium	Single, nonciliated layer of tall, columnar-shaped cells; protection, secretion, and absorption	Lining of digestive tract
Simple ciliated columnar epithelium	Single, ciliated layer of columnar-shaped cells; transportive role through ciliary motion	Lining the lumen of the uterine tubes
Pseudostratified ciliated columnar epithelium	Single layer of ciliated, irregularly shaped cells, many goblet cells; protection, secretion, ciliary movement	Lining of respiratory passageways
Stratified epithelia	Two or more layers of cells; protection, strengthening, or distension	Epidermal layer of skin; linings of body openings, ducts, urinary bladder
Stratified squamous epithelium (keratinized)	Numerous layers, contains keratin, outer layers flattened and dead; protection	Epidermis of skin
Stratified squamous epithelium (nonkeratinized)	Numerous layers, lacks keratin, outer layers moistened and alive; protection and pliability	Linings of oral and nasal cavities, vagina, and anal canal
Stratified cuboidal epithelium	Usually two layers of cube-shaped cells; strengthen luminal walls	Larger ducts of sweat glands, salivary glands, and pancreas
Transitional epithelium	Numerous layers of rounded, nonkeratinized cells; distension	Luminal walls of ureters and urinary bladder

(3) *Tight junctions* and *intermediate junctions* occur in the same type of epithelium. A tight junction exists where the cell membranes of adjacent cells form a serrated pattern. In an intermediate junction, the cell membranes of adjacent cells are bonded with glycoprotein.

Some of the functions of epithelial tissues are quite specific, but certain generalities can be made. Epithelia that cover or line surfaces provide *protection* from pathogens, physical injury, toxins, and desiccation. Epithelia lining the lumen of the digestive tract function in *absorption* and *secretion.* Other glandular epithelia elsewhere in the body are also secretory. The epithelium of the kidneys provides *filtration,* whereas the epithelium within the air sacs of the lungs allows *diffusion.* Highly specialized *neuroepithelium* in the taste buds and in the nasal region functions as *chemoreceptors (ke"mo-re-sep'tor).*

Epithelial tissue is exposed and, therefore, subject to trauma and destruction. For this reason, epithelial tissues have remarkable regenerative abilities. The mitotic replacement of the outer layer of skin and the lining of the digestive tract, for example, is a continuous process.

Epithelial tissues are histologically classified by the number of layers of cells and the shape of the cells along the exposed surface. Epithelial tissues that are composed of a single layer of cells are called *simple,* and those that are layered are said to be *stratified. Squamous* cells are flattened; *cuboidal* cells are cube shaped; and *columnar* cells are taller than they are wide.

In describing and discussing the epithelial tissues, the specific types will be presented under three major headings; simple epithelia, stratified epithelia, and glandular epithelia. A summary of simple and stratified epithelia is presented in table 6.2, and a summary of glandular epithelia is presented in table 6.3.

Simple Epithelia

Simple epithelial tissues are a single layer thick and located where diffusion, filtration, and secretion occur. The cells that constitute simple epithelia range in size from thin, flattened cells to tall, columnar cells, depending on function. These cells may also exhibit surface specializations, such as cilia and microvilli, which facilitate specific surface functions.

Simple Squamous Epithelium. Simple squamous *(skwa'mus)* epithelium is composed of flattened, irregularly shaped cells that are tightly bound together in a mosaiclike pattern (fig. 6.5). Each cell contains an oval, centrally located nucleus. Simple squamous epithelium is

---

squamous: L. *squamosus,* scaly

Figure 6.5. Simple squamous epithelium consists of a single layer of flattened cells and is highly adapted to diffusion, osmosis, and filtration.

Figure 6.6. Simple cuboidal epithelium lines the lumina of ducts, as seen in the photomicrographs of kidney tubules. (*a*) A diagrammatic drawing of this type of epithelium that consists of a single layer of tightly packed, cube-shaped cells, each with a large, centrally located nucleus; (*b*) longitudinal view of a tubule; (*c*) cross-sectional view of several tubules.

(a)

(b)   └─── Lumen

(c)   Lumen───

adapted for diffusion and filtration and occurs in such places as the lining of alveoli within the lungs where gaseous exchange occurs, portions of the kidney where blood is filtered, the inside lining of the walls of blood vessels, the lining of body cavities, and in the covering of the viscera. The simple squamous epithelium lining the lumina of blood and lymphatic vessels is sometimes referred to as **endothelium.** That which covers visceral organs and lines body cavities is called **mesothelium.**

Simple Cuboidal Epithelium. Simple cuboidal epithelium is composed of a single layer of tightly fitted, hexagonal cells (fig. 6.6). This type of epithelium is found lining small ducts and tubules that may have excretory, secretory, or absorptive functions. It occurs on the surface of the ovaries, forms a portion of the tubules within the kidney, and lines the ducts of the salivary glands and pancreas.

---

endothelium: Gk. *endon,* within; *thelium,* to cover
mesothelium: Gk. *meso,* middle; *thelium,* to cover

Figure 6.7. (*a*) simple columnar epithelium lines the lumen of the digestive tract. (*b*) a photomicrograph of this epithelium. (*c*) a diagrammatic drawing of the tall, columnar-shaped cells and the associated goblet cells.

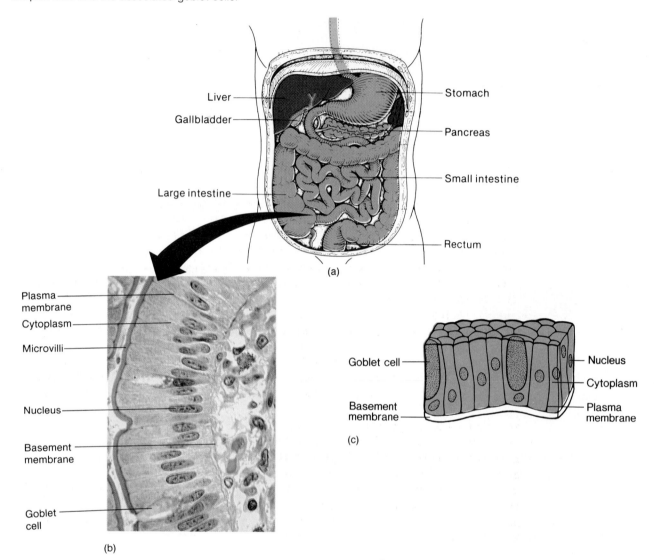

Simple Columnar Epithelium. Simple columnar epithelium is composed of tall, columnar-shaped cells (fig. 6.7). The height of the cells varies depending on the site and function of the tissue. Each cell contains a single nucleus usually located near the basement membrane. Specialized unicellular glands, called **goblet cells,** are dispersed throughout this tissue and secrete a lubricative and protective mucus along the free surfaces of the cells. This type of epithelium is found lining the lumen of the stomach and intestine. In the digestive system, simple columnar epithelium forms a highly absorptive surface and also secretes certain digestive substances. Within the stomach, simple columnar epithelium has a tremendous rate of mitotic activity—replacing itself every two or three days.

Simple Ciliated Columnar Epithelium. Simple ciliated columnar epithelium differs from the simple columnar type by the presence of cilia along the free surface (fig. 6.8). Cilia produce wavelike movements that transport materials through tubes or passageways. This type of epithelium occurs in the uterine tubes of the female where the currents generated by the cilia propel the ovum toward the uterus.

Not only do cilia of the columnar epithelium move the ovum, but recent evidence indicates that sperm introduced during sexual intercourse may be moved along the return currents, or eddies, produced by ciliary movement. This gives fertilization a much greater chance of occurring.

Figure 6.8. (*a*) simple ciliated columnar epithelium occurs within the uterine tubes of the female reproductive system, where it moves the ovum toward the uterine cavity. (*b*) a photomicrograph showing the cilia and (*c*) a diagram.

(a)

Connective tissue layer
Basement membrane
Nucleus
Cytoplasm
Cilia
Plasma membrane

(b)   (c)

Pseudostratified Ciliated Columnar Epithelium. As the name implies, this type of epithelium appears stratified but is actually simple since each cell is in contact with the basement membrane, though not all cells are exposed to the surface (fig. 6.9). The epithelium has a stratified appearance because the nuclei of these cells are located at different levels. Numerous goblet cells and a ciliated exposed surface are characteristic of this epithelium. The lumen of the trachea and the bronchial tubes are lined with this tissue; hence, it is frequently called respiratory epithelium. Its function is to remove foreign dust and bacteria entrapped in mucus from the lower respiratory system.

Coughing, sneezing, or simply "clearing the throat" are protective reflex mechanisms for clearing the respiratory passages of obstruction or of inhaled particles that have been trapped in the mucus along the ciliated lining. The material that is coughed up consists of the mucous-entrapped particles. For the health and proper functioning of the air sacs within the lungs, foreign material must not be permitted to enter this level.

Figure 6.9. (*a*) pseudostratified ciliated columnar epithelium lines most of the respiratory tract. The action of the cilia and the mucus secreted by goblet cells trap foreign material and move it away from the alveoli of the lungs. (*b*) this type of tissue appears stratified, but (*c*) all the cells are in contact with the basement membrane.

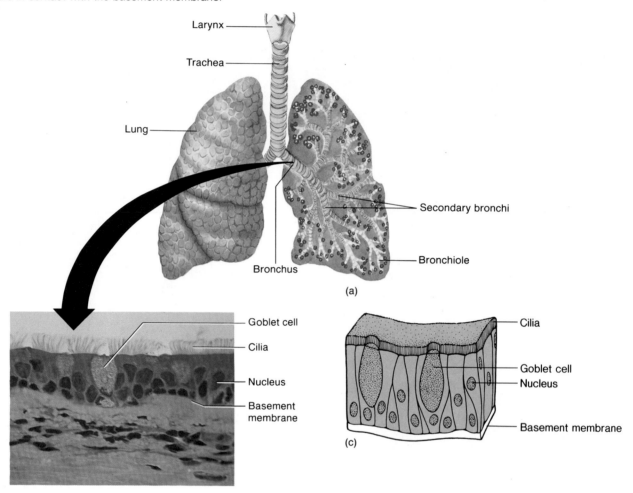

Stratified Epithelia

Stratified epithelia are tissues consisting of two or more layers of cells. In contrast to simple epithelia, stratified epithelia are poorly suited for absorption and secretion because of their thickness. Stratified epithelia have a primarily protective function, which is enhanced by a characteristically rapid mitotic activity. Stratified epithelia are classified according to the shape of the surface layer of cells, since the layer in contact with the basement membrane is cuboidal in shape.

Stratified Squamous Epithelium. Stratified squamous epithelium is composed of a variable number of cell layers that tend to flatten near the surface (fig. 6.10). Only at

the deepest layer, called the **stratum germinativum,** or basal layer, does mitosis occur. The mitotic rate approximates the rate at which cells are sloughed off at the surface. As the newly produced cells grow in size, they are pushed toward the surface where they will replace the cells that are sloughed off. Movement away from the source of nutrition, provided by the supportive basement membrane, causes the cells to become progressively dehydrated and flattened.

There are two types of stratified squamous epithelia: *keratinized* and *nonkeratinized*. Stratified squamous epithelium that is keratinized forms the outer layer, or *epidermis*, of the skin. **Keratin** *(ker'ah-tin)* is a protein that

---

keratin: Gk. *keras*, horn

Figure 6.10. Stratified squamous epithelium is a multilayered, protective tissue that forms the outer layer of skin and the lining of body openings. In the moistened areas such as in the vagina (a), it is nonkeratinized, whereas in the epidermis of the skin it is keratinized. The lower layers are cube shaped as can be seen in (b) the photomicrograph (100×) and (c) the diagram, whereas the outer layers are flattened and scalelike.

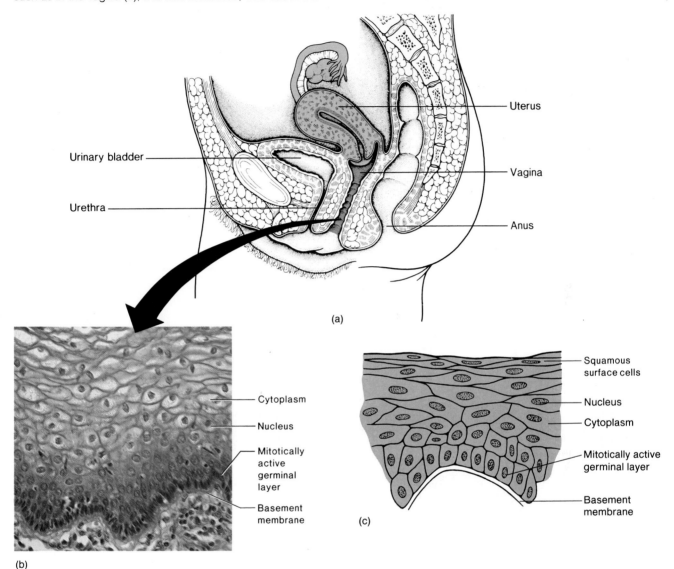

strengthens the tissue. This type of epithelium is especially durable and can generally withstand physical abrasion, desiccation, and bacterial invasion. The outer layers of the stratified squamous epithelium of the skin are dead but are kept moist by local glandular secretions.

Nonkeratinized stratified squamous epithelium lines the oral cavity and pharynx, nasal cavity, vagina, and anal canal. This type of epithelium is well adapted to withstand moderate abrasion but not fluid loss. The cells on the free surface of this tissue remain alive and are always moistened.

Stratified squamous epithelium is the first line of defense against the entry of living organisms into the body. Stratification as well as rapid mitotic activity and keratinization within the epidermis of the skin are important protective features. An acidic pH along the surfaces of this tissue also help prevent disease. The pH of the skin is between 4 and 6.8. The pH in the oral cavity ranges between 5.8 and 7.1, which tends to retard the growth of microorganisms. The pH of the anal region is about 6, and the pH along the surface of the vagina is 4 or lower.

Figure 6.11.   Stratified cuboidal epithelium consists of two or more layers of cube-shaped cells surrounding a lumen. (*a*) a photomicrograph of this tissue and (*b*) a diagram.

(a)   (b)

Figure 6.12.   (*a*) transitional epithelium lines the lumina of the urinary bladder and the ureters. (*b, c*) the cells of this tissue are stratified in a unique way to permit distension (400×).

(a)   (b)

(c)

Figure 6.13. The goblet cell is a unicellular gland that secretes mucus to lubricate and protect surface linings. (a) a photomicrograph through the ileum of the small intestine shows the numerous goblet cells (*arrow*). (b) columnar epithelia always contain mucus-secreting goblet cells.

(a)

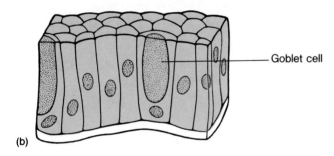

Goblet cell

(b)

Stratified Cuboidal Epithelium. Stratified cuboidal epithelium usually consists of only two or three layers of cuboidal cells forming the lining around a lumen (fig. 6.11). This type of epithelium is confined to the linings of the larger ducts of sweat glands, salivary glands, and the pancreas. The stratification of this tissue probably provides a more robust lining than would be afforded by simple epithelium.

Transitional Epithelium. Transitional epithelium is similar to nonkeratinized stratified squamous epithelium except that the surface cells are large and round rather than flat and some may have two nuclei (fig. 6.12). Transitional epithelium is located only within the urinary system, particularly in the luminal surface of the urinary bladder and the walls of the ureters. This tissue is specialized to permit distension (stretching) of the urinary bladder and to withstand the toxicity of urine.

Glandular Epithelia

During the prenatal development of epithelial tissue, certain epithelial cells invade the underlying connective tissue, forming specialized secretory accumulations called *exocrine (ek'so-krin) glands.* Exocrine glands retain a connection to the surface in the form of a duct through which

secretions flow. Exocrine glands within the integumentary system include sebaceous (oil) glands, sweat glands, and mammary glands. Within the digestive system, they include the salivary and pancreatic glands.

> Exocrine glands are derived from epithelial tissue and secrete a substance through ducts to the surface of the skin or the lumen of a body cavity. Exocrine glands should not be confused with endocrine glands, which are ductless and secrete hormones into the blood.

Exocrine glands are classified according to the structure of the gland and its means of discharging the secretory product. Structurally, there are two types, unicellular and multicellular.

Unicellular Glands. Unicellular glands are single-celled glands interspersed with the various columnar epithelia. A mucous-secreting goblet cell is a good example of a unicellular gland. Goblet cells are found in the epithelial linings of the respiratory, digestive, urinary, and reproductive systems where the mucous secretion lubricates and protects the surface linings (fig. 6.13).

Multicellular Glands. Multicellular glands, as their name implies, are composed of numerous secretory cells as well as cells forming the walls of the ducts. Multicellular glands are divided into *simple* and *compound* glands. The ducts of the simple gland either do not branch or have few branches, whereas those of the compound type have multiple branches (fig. 6.14). Multicellular glands are further classified according to the shape of the secretory portion. They are identified as *tubular* if the secretory portion resembles the ductule portion and as *acinar (as'i-nar)* if the secretory portion is flasklike. Multicellular glands with a secretory portion that resembles both a tube and a flask are termed *tubuloacinar.*

Multicellular glands are also classified according to the means by which they discharge the secretory product (fig. 6.15). In **merocrine *(mer'o-krīn)* glands,** a watery substance is secreted through the plasma membrane. Salivary glands, pancreatic glands, and certain sweat glands are of this type. In **apocrine glands** the secretion accumulates on the surfaces of the secretory cells and then a portion of the cell along with the secretion is pinched off to be discharged. Mammary glands and certain sweat glands are apocrine glands. In a **holocrine *(hol'o-krīn)* gland,** the entire secretory cell is discharged along with the secretory product. An example of a holocrine gland is a sebaceous, or oil-secreting, gland of the skin.

Table 6.3 summarizes the exocrine glands and their secretions.

merocrine: Gk. *meros*, part; *krinein*, to separate
apocrine: Gk. *apo*, off; *krinein*, to separate
holocrine: Gk. *holos*, whole; *krinein*, to separate

exocrine: Gk. *exo*, outside; *krinein*, to separate

Figure 6.14. Structural classification of multicellular exocrine glands. The secretory portions of the simple glands either do not branch or have a few branches, whereas those of the compound type have multiple branches.

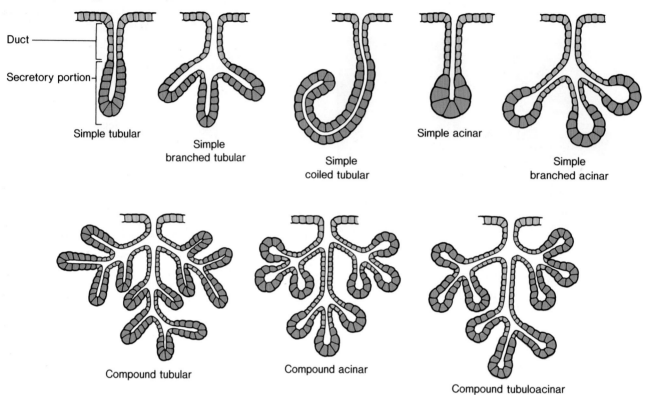

Duct

Secretory portion

Simple tubular

Simple branched tubular

Simple coiled tubular

Simple acinar

Simple branched acinar

Compound tubular

Compound acinar

Compound tubuloacinar

Figure 6.15. Secretory classification of multicellular exocrine glands: (a) merocrine gland; (b) apocrine gland; (c) holocrine gland.

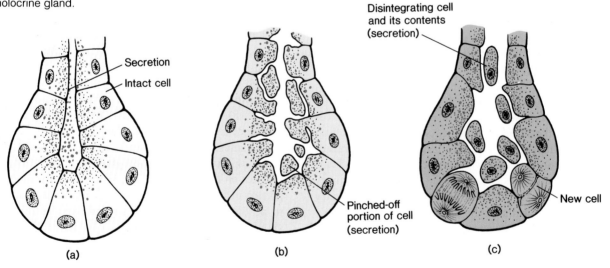

Secretion

Intact cell

Disintegrating cell and its contents (secretion)

Pinched-off portion of cell (secretion)

New cell

(a)

(b)

(c)

Table 6.3 Summary of glandular epithelia

Structural classification of exocrine glands

Type	Function	Example
I. Unicellular	Lubricate and protect	Goblet cells of digestive, respiratory, urinary, and reproductive systems
II. Multicellular	Protect, cool body, lubricate, aid in digestion, maintain body homeostasis	Sweat glands, digestive glands, mammary glands, sebaceous glands
A. Simple		
1. Tubular	Aid in digestion	Intestinal glands
2. Branched tubular	Protect, aid in digestion	Uterine glands, gastric glands
3. Coiled tubular	Regulate temperature	Certain sweat glands
4. Acinar	Additive to spermatozoa	Seminal vesicle of male reproductive system
5. Branched acinar	Skin conditioner	Sebaceous skin glands
B. Compound		
1. Tubular	Lubricate urethra of male, assist body digestion	Bulbourethral gland of male reproductive system, liver
2. Acinar	Nourishment to infant, aid in digestion	Mammary gland, salivary gland (sublingual and submandibular)
3. Tubuloacinar	Aid in digestion	Salivary gland (parotid), pancreas

Secretory classification of exocrine glands

Type	Description of secretion	Example
Merocrine glands	Water secretion for regulating temperature or aiding digestion	Salivary and pancreatic glands, certain sweat glands
Apocrine glands	Portion of secretory cell and secretion are discharged; provides nourishment to infant, assists in regulating temperature	Mammary glands, certain sweat glands
Holocrine glands	Entire secretory cell with enclosed secretion is discharged; skin conditioner	Sebaceous glands of the skin

Source: Van De Graaff, Kent M., *Human Anatomy.* © 1984 Wm. C. Brown Publishers, Dubuque, Iowa. All Rights Reserved. Reprinted by permission.

1. *List the functions of simple squamous epithelium.*
2. *What are the three types of columnarlike epithelia? What do they have in common? How are they different?*
3. *What are the two types of stratified squamous epithelia and how do they differ?*
4. *Describe the three ways that epithelial cells are bonded together.*
5. *Distinguish between unicellular and multicellular glands. Explain how multicellular glands are classified according to their mechanism of secretion.*
6. *In what ways are mammary glands and certain sweat glands similar?*

Connective Tissue

Connective tissues are classified according to the characteristics of the matrix that binds the cells. Connective tissues provide structural and metabolic support for other tissues and organs of the body.

Objective 9. Describe the general characteristics, locations, and functions of connective tissue.

Objective 10. Explain the functional relationship between embryonic and adult connective tissue.

Objective 11. List the various ground substances, fiber types, and cells that constitute connective tissue, and explain their functions.

Characteristics and Classification of Connective Tissue

Connective tissue is found throughout the body and, as the name indicates, supports or binds other tissues and provides metabolic needs to all body organs. Certain types of connective tissue store nutritional substances, whereas other types manufacture protective and regulatory materials.

Although connective tissue varies tremendously in structure and function, all connective tissues have similarities. With the exception of cartilage, connective tissues are highly vascular and well nourished. They are able to replicate and, by so doing, are responsible for the repair of body organs. Unlike epithelial tissues, which are composed of tightly fitted cells, connective tissues contain considerably more matrix than cells. Connective tissues do not occur on free surfaces of body cavities or on the surface of the body as do epithelial tissues. Furthermore, connective tissues are embryonically derived from mesoderm, whereas epithelial tissues derive from ectoderm, mesoderm, and endoderm.

The classification of connective tissues is not exact, and several schemes have been devised. In general, however, they are named according to the kind and arrangement of the matrix. Following are the basic kinds of connective tissues.

A. Embryonic connective tissue
B. Connective tissue proper
   1. Loose connective (areolar)
   2. Dense fibrous
   3. Elastic
   4. Reticular
   5. Adipose
C. Cartilage
   1. Hyaline
   2. Fibrocartilage
   3. Elastic
D. Bone
E. Vascular

Embryonic Connective Tissue

The embryonic period of development lasts six weeks (from the beginning of the third to the end of the eighth week) and is characterized by tremendous tissue differentiation and organ formation. At the beginning of the embryonic period, all connective tissue appears the same and is referred to as **mesenchyme** *(mes'en-kīm)*. Mesenchyme is undifferentiated embryonic connective tissue, which is derived from mesoderm and consists of irregularly shaped cells lying in large amounts of a homogeneous, jellylike matrix (fig. 6.16). In certain areas of development mesenchyme migrates to predisposed sites where it interacts with other tissues to form organs. Before the end of the embryonic period, once mesenchyme is in the appropriate position, it differentiates and all other kinds of connective tissues are formed. In summary, mesenchymal tissue is undifferentiated embryonic connective tissue, which can migrate and give rise to all other types of connective tissue. The causes of tissue differentiation are not fully understood.

Some mesenchymal-like tissue persists past the embryonic period in certain sites within the body. Good examples are the undifferentiated cells that surround blood vessels and form fibroblasts if the vessels are traumatized. Fibroblasts assist in healing wounds (chapter 8).

Another kind of prenatal connective tissue exists only in the fetus (the fetal period is from nine weeks to birth) and is called *mucous connective tissue*, or *Wharton's jelly*. It provides a turgid consistency to the umbilical cord.

mesenchyme: Gk. *mesos*, middle; *enchyma*, infusion
Wharton's jelly: from Thomas Wharton, English anatomist, 1614–73

Figure 6.16. Mesenchyme is undifferentiated embryonic mesodermal connective tissue, which can migrate and give rise to all other kinds of connective tissue. (*a*) it is found within an early developing embryo and (*b*) consists of irregularly shaped cells lying in a homogeneous, jellylike matrix.

Connective Tissue Proper

Connective tissue proper has a loose, flexible matrix frequently called **ground substance.** The most common cell within connective tissue proper is called a **fibroblast** *(fi'bro-blast)*. Fibroblasts are large, star-shaped cells that produce collagenous, elastic, and reticular *(rĕ-tik'u-lar)* fibers. **Collagenous** *(kol-laj'ĕ-nus),* or **white fibers,** are composed of a protein called *collagen (kol'ah-jen);* they are flexible, yet have tremendous strength. **Elastic,** or **yellow fibers,** are composed of a protein called *elastin* that gives elasticity to certain tissues. Collagenous and elastic fibers may be either sparse and irregularly arranged as in loose connective tissue or tightly packed as in dense connective tissue. Tissues with loosely arranged fibers generally form packing material that cushions and protects various organs, whereas those with tightly arranged fibers form the binding and supportive connective tissues of the body. Dense connective tissue can also vary in the orientation of the fibers. If the fibers are interwoven to provide tensile strength in any direction, it is referred to as *irregularly arranged.* Dense, irregularly arranged connective tissue forms the periosteum, or covering, around bone, the perichondrium surrounding cartilage, and the fibrous capsules around certain organs (the testes, parts of the

collagen: Gk. *kolla*, glue
elastin: Gk. *elasticus*, to drive

Figure 6.17. Loose connective tissue consists of collagenous and elastic fibers as well as fibroblasts and mast cells. This type of connective tissue is an important packing and binding material. It surrounds muscle, nerves, and vessels and binds the skin to the underlying muscles. (*a*) a photomicrograph and (*b*) a diagram.

—Elastic fiber

—Collagen fiber

—Fibroblast

(a)

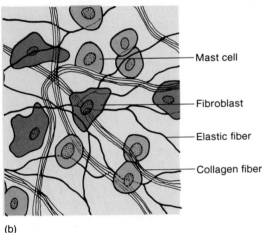

—Mast cell

—Fibroblast

—Elastic fiber

—Collagen fiber

(b)

liver, and lymph nodes). *Regularly arranged* fibers are parallel to the direction of force. Dense, regularly arranged connective tissues are found where strong, flexible bindings are required, such as in tendons, ligaments, and aponeuroses.

Tensile strength in tissues that contain elastic fibers is extremely important for several physical functions of the body. Consider, for example, that elastic fibers are found in the walls of arteries and in the walls of the lower respiratory passageways. As these walls are expanded by blood moving through vessels or by inspired air, the elastic fibers must first stretch and then recoil. This maintains the pressures of the fluid or air moving through the lumina, which provides adequate flow rates and, thus, adequate rates of diffusion through capillary and lung surfaces.

Reticular fibers reinforce by forming thin, short threads that branch and join to form a delicate lattice, or reticulum. Reticular fibers are common in lymphatic glands where they form a meshlike **stroma.**

Five basic types of connective tissue proper are generally recognized. These tissues are distinguished by the consistency of the ground substance and the type and arrangement of the reinforcement fibers.

Loose Connective (Areolar) Tissue. Loose connective tissue is distributed throughout the body as a binding and packing material. It binds the skin to the underlying muscles and is highly vascular, providing nutrients to the skin. Loose connective tissue surrounding muscle fibers and muscle groups is known as **fascia** (*fash'e-ah*). It also surrounds blood vessels and nerves where it provides both protection and nourishment. Specialized cells called **mast cells** are dispersed throughout the loose connective tissue surrounding blood vessels. Mast cells produce *heparin* (*hep'ah-rin*), an anticoagulant that prevents blood from clotting within the vessels.

The cells of loose connective tissue are predominantly fibroblasts, with collagenous and elastic fibers dispersed throughout the ground substance (fig. 6.17). The irregular arrangement of this tissue provides flexibility and yet strength in any direction. It is this tissue layer, for example, that permits the skin to move when you rub a part of your body.

Much of the fluid of the body is found within loose connective tissue and is called **tissue fluid.** Sometimes excessive tissue fluid accumulates, causing a swelled condition called **edema** (*ē-de'mah*). Edema is a symptom of a variety of dysfunctions or disease processes.

Dense Fibrous Connective Tissue. Dense fibrous connective tissue is characterized by large amounts of densely packed collagenous fibers. Because this tissue is silvery white in appearance, it is sometimes called dense, white fibrous connective tissue (fig. 6.18). The blood supply to dense fibrous connective tissue is poor, so if this tissue is injured it heals relatively slowly.

Dense fibrous connective tissue occurs where strong, flexible support is necessary. **Tendons,** which attach muscles to bones and transfer the forces of muscle contractions, and **ligaments,** which connect bone to bone across articulations, are composed of this type of tissue. The sclera of the eye and deep connective tissues of the skin are also made of dense fibrous connective tissue.

reticular: L. *rete,* net or netlike
stroma: Gk., a couch or bed
fascia: L. *fascia,* band or girdle
heparin: Gk. *hepatos,* the liver
edema: Gk. *oidema,* a swelling
tendon: L. *tendere,* to stretch
ligament: L. *ligare,* bind

Figure 6.18. Dense fibrous connective tissue forms tendons (*a*). In this tissue type, the collagenous, or white, fibers, are arranged in bundles, and fibroblasts are positioned between the bundles. The structure of this flexible, strong tissue is seen in (*b*) a photomicrograph and (*c*) a diagram.

(a)

(b)

(c)

Figure 6.19. (*a*) the wall of an artery is composed of several tissue types. (*b*) the cellular structure. (*c*) elastic connective tissue is found within the inner coat. This type of connective tissue allows stretching as the blood flows through. The outer coat consists of loose fibrous connective tissue, and the middle coat contains smooth muscle.

(a)

(b)

(c)

Figure 6.20. Reticular tissue forms the stroma, or framework, of organs such as the spleen (a), a liver, thymus, and lymph nodes. It consists of a network of woven reticular fibers, as can be seen in (b) the photomicrograph (250×) and (c) the diagram. Reticular tissue is phagocytic.

Elastic Connective Tissue. Elastic connective tissue has a predominance of elastic fibers that are irregularly arranged and yellowish in color (fig. 6.19). They can be stretched to one and a half times their original lengths and will snap back to their former size. Elastic connective tissue is found in the walls of large arteries, in portions of the larynx, and in the trachea and bronchial tubes of the lungs. It is also present between the vertebrae making up the vertebral column.

Reticular Connective Tissue. Reticular connective tissue is characterized by a network of reticular fibers woven through a jellylike matrix (fig. 6.20). Certain specialized cells within reticular tissue are *phagocytic (fag''o-sit'ik)* and therefore ingest foreign materials. The liver, spleen, lymph nodes, and bone marrow contain reticular connective tissue.

Adipose Tissue. Adipose tissue is a specialized type of loose fibrous connective tissue that contains large quantities of **adipose** *(ad'ĭ-pōs)* **cells.** Adipose cells form from fibroblasts and are, for the most part, formed prenatally and during the first year of life. Adipose cells store droplets of fat within their cytoplasm, causing them to swell and forcing their nuclei to one side (fig. 6.21).

adipose: L. *adiposus*, fat

Adipose tissue is found throughout the body but is concentrated around the kidney, in the hypodermis of the skin, on the surface of the heart, surrounding joints, and in the breasts of sexually mature females. Fat functions not only as a food reserve but supports and protects various organs. Fat is a good insulator against cold because it is a poor conductor of heat.

Excessive fat can be both unsightly and unhealthy, placing a strain on the heart and perhaps causing early death. For these reasons, good exercise programs and sensible diets are extremely important. Adipose tissue can also retain environmental pollutants that are ingested or absorbed through the skin.

Feeding an infant an excessive amount during its first year, when adipose cells are forming, may cause a greater amount of adipose tissue to develop. A person with large amounts of adipose tissue is more susceptible to developing obesity later in life than a person with a lesser amount of adipose tissue. Dieting eliminates the fat stored within the tissue but not the tissue itself.

Table 6.4 summarizes the characteristics, functions, and locations of connective tissue proper.

Figure 6.21. Adipose connective tissue is abundant in (a) the hypodermis of the skin and around various internal organs. It is protective and serves to store fat. The fat globules (b, c) are stored in the adipose cells (250×).

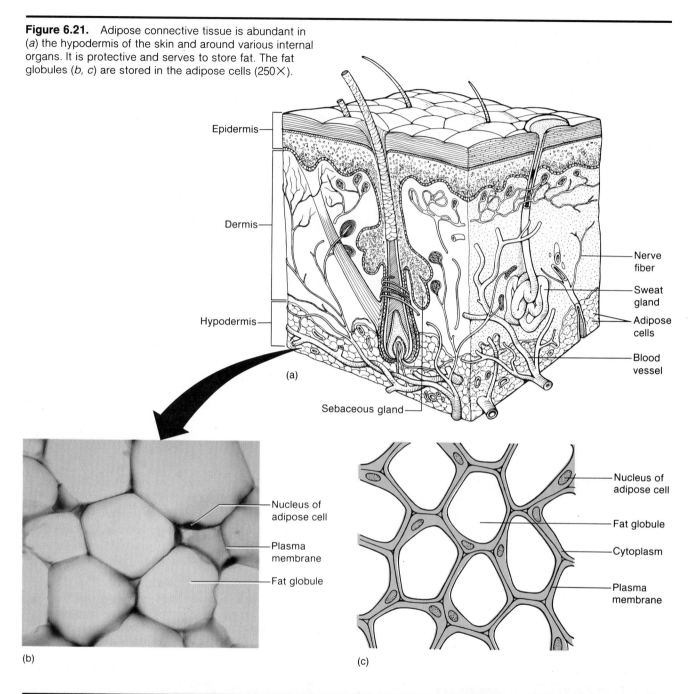

Epidermis

Dermis

Hypodermis

Nerve fiber

Sweat gland

Adipose cells

Blood vessel

Sebaceous gland

(a)

Nucleus of adipose cell

Plasma membrane

Fat globule

(b)

Nucleus of adipose cell

Fat globule

Cytoplasm

Plasma membrane

(c)

Table 6.4 Summary of connective tissue proper		
Type	**Structure and function**	**Location**
Loose connective (areolar) tissue	Predominantly fibroblast cells with lesser amounts of collagen and elastin cells; binds organs, holds tissue fluids, diffusion	Surrounding nerves and vessels, between muscles, beneath the skin
Dense fibrous connective tissue	Densely packed collagen fibers; provides strong, flexible support	Tendons, ligaments, sclera of eye, deep skin layers
Elastic connective tissue	Predominantly irregularly arranged elastic fibers; supports, provides framework	Large arteries, lower respiratory tract, between vertebrae
Reticular connective tissue	Reticular fibers forming supportive network; stores, phagocytic	Lymph nodes, liver, spleen, thymus, bone marrow
Adipose connective tissue	Adipose cells; protects, stores fat, insulates	Hypodermis of skin, surface of heart, omentum, around kidneys, back of eyeball, surrounding joints

Figure 6.22. Hyaline cartilage is the most abundant cartilage within the body. It occurs in places such as the larynx and trachea (a), articular ends of bones, ends of ribs, and embryonic skeleton. (b) the homogeneous matrix of hyaline cartilage can be seen in the photomicrograph (100×). Its structure is diagrammed in (c).

Perichondrium

Chondrocyte within lacuna

(b)

Perichondrium

Lacuna

Chondrocytes

(c)

Larynx — Thyroid cartilage

Cricoid cartilage

Trachea

Right bronchus — Left bronchus

(a)

Cartilage

Cartilage tissue consists of cartilage cells, called **chondrocytes** *(kon'dro-sīt),* and a semisolid ground substance that imparts marked elastic properties to the tissue. Cartilage is a type of supportive and protective connective tissue commonly called "gristle." Cartilage is frequently associated with bone. It forms a precursor to one type of bone and persists at the articular surfaces on the bones of all movable joints.

The chondrocytes within cartilage may occur singly but are frequently clustered. Chondrocytes occupy spaces, called **lacunae** *(lah-ku'ne),* within the matrix. Cartilage tissue is surrounded by a dense fibrous connective tissue called a **perichondrium** *(per''ĭ-kon'dre-um).* Because cartilage is avascular it must receive nutrients through diffusion from the surrounding tissue. For this reason,

cartilaginous tissue has a slow rate of mitotic activity and, if damaged, heals with difficulty.

There are three kinds of cartilage, each of which is distinguished by the types of fibers embedded within the matrix: hyaline cartilage, fibrocartilage, and elastic cartilage.

Hyaline Cartilage. Hyaline *(hi'ah-lĭn)* cartilage has a homogeneous, bluish-staining matrix, in which the collagenous fibers are so fine that they can be observed only with an electron microscope. When viewed through a microscope, hyaline cartilage has a clear, glassy appearance (fig. 6.22).

hyaline: Gk. *hyalos,* glass

lacuna: L. *lacuna,* hole or pit

Figure 6.23. Fibrocartilage is located at the symphysis pubis, within the knee joint, and between the vertebrae as the intervertebral discs (a). It consists of chondrocytes within a matrix that is reinforced with collagenous fibers (b, c) to withstand stress and compression as it provides durable support (250×).

Hyaline cartilage is the most abundant cartilage within the body. It covers the articular surfaces of bones, supports the tubular trachea and bronchi of the respiratory system, reinforces the nose, and forms the flexible bridge, called **costal cartilage,** between the ventral ends of each of the first ten ribs and the sternum. Most of the bones of the body form first as hyaline cartilage and later become bone through ossification.

Fibrocartilage. Fibrocartilage has its matrix reinforced with numerous collagenous fibers (fig. 6.23). It is a durable tissue adapted to withstand tension and compression. It is found at the symphysis pubis where the two pelvic bones articulate and between the vertebrae as intervertebral discs. It also forms the cartilaginous wedges within the knee joint.

By the end of the day the intervertebral discs of the vertebral column are somewhat compacted. So a person is actually slightly shorter in the evening than in the morning following a recuperative rest. Aging brings a gradual compression of the fibrocartilaginous discs that is irreversible.

Figure 6.24. Elastic cartilage gives support to the external ear, auditory canal, and parts of the larynx. (*a*) the matrix of elastic tissue is laced with elastic fibers to provide flexibility while maintaining shape. (*b*) a photomicrograph of elastic cartilage (100✕).

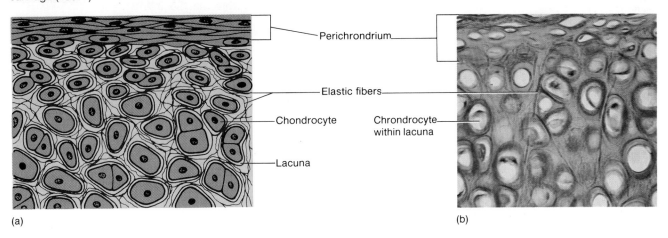

Perichrondrium

Elastic fibers

Chondrocyte

Lacuna

Chrondrocyte within lacuna

(a)   (b)

Table **6.5** Summary of cartilage tissue		
Type	**Structure and function**	**Location**
Hyaline cartilage	Homogeneous matrix with extremely fine collagenous fibers; flexible support, protection, precursor to bone	Articular surfaces of bones, nose, walls of respiratory passages, fetal skeleton
Fibrocartilage	Abundant collagenous fibers within matrix; support, withstand compression	Symphysis pubis, intervertebral discs, knee joint
Elastic cartilage	Abundant elastic fibers within matrix; support, flexibility	Framework of external ear, auditory canal, portions of larynx

Source: Van De Graaff, Kent M., *Human Anatomy.* © 1984 Wm. C. Brown Publishers, Dubuque, Iowa. All Rights Reserved. Reprinted by permission.

Elastic Cartilage.   Elastic cartilage is similar to hyaline except for the presence of abundant elastic fibers, which make it very flexible while maintaining its strength (fig. 6.24). The numerous elastic fibers also give it a yellowish appearance. This tissue is found in the external ear, portions of the larynx, and the auditory canal.

Table 6.5 summarizes the structure, function, and location of cartilage within the body.

Bone

Bone, or osseous connective tissue, is the most rigid of the connective tissues. Unlike cartilage, bone has a rich vascular supply and is the site of considerable metabolic activity. The hardness of bone is largely due to the inorganic calcium phosphate and calcium carbonate salts deposited within the intercellular matrix. Numerous collagenous fibers, also embedded within the matrix, give some flexibility to bone.

In healthy osseous tissue, a balance exists between the organic living and the inorganic nonliving materials. As a person ages, the proportion of organic material decreases and the bones become brittle. The bones of elderly persons not only fracture easily but do not heal readily.

When bone is placed in a weak acid the calcium salts dissolve away and the bone becomes pliable. It retains its basic shape but can be easily bent and twisted. In calcium deficiency diseases, such as rickets, the osseous tissue becomes pliable and bends under the weight of the body.

There are two kinds of bone tissue, based on porosity, and most bones have both types (fig. 6.25). **Compact,** or **dense,** bone tissue is the hard, outer layer, whereas **spongy,** or **cancellous,** bone tissue is the porous, highly vascular, inner portion. Compact bone tissue is covered by the periosteum, serves for attachment of muscles, provides protection, and gives durable strength to the bone. Spongy bone tissue makes the bone lighter and provides a space for bone marrow where blood cells are produced.

Figure 6.25. (*a*) osseous tissue within compact bone is arranged into osteons. (*b*) a photomicrograph of osseous tissue and (*c*) a diagram of the structure.

In compact bone tissue, the bone cells, called **osteocytes,** are arranged in concentric layers around a **Haversian** *(ha-ver′shan)* **canal,** which contains a vascular and nerve supply (fig. 6.26). Each osteocyte occupies a space called a **lacuna.** Radiating from each lacuna are numerous minute canals, called **canaliculi,** that traverse the dense matrix of the bone tissue to adjacent lacunae. Nutrients diffuse through the canaliculi to reach each osteocyte. The inorganic matrix is deposited in concentric layers called **lamellae.**

A Haversian canal, with its surrounding osteocytes, lacunae, canaliculi, and concentric lamellae, constitutes a **Haversian system,** or **osteon** (fig. 6.25). Metabolic activity within osseous tissue occurs at the osteon level. Areas between the osteons contain **interstitial** *(in″ter-stish′al)* **lamellae.** These areas possess osteocytes within lacunae and associated canaliculi but are arranged irregularly. Transverse channels, called **Volkmann's canals,** penetrate compact bone and connect various osteons with blood vessels and nerves.

Electron microscopy is a relatively new technique for examining tissues. Many students of histology do not understand or appreciate the added dimension provided by the electron microscope. The structure of the Haversian canal, for example, shown in figure 6.26, can be more accurately described using a scanning electron micrograph than when using a conventional photomicrograph. Electron microscopy is an extension of light microscopy and should be used to demonstrate the continuity of structural organization within tissues.

Haversian canal: from Clopton Havers, English anatomist, 1650–1702
osteon: Gk. *osteon,* bone

Volkmann's canals: from Alfred W. Volkmann, German physiologist, 1800–1877

Figure 6.26. A scanning electron micrograph of a Haversian canal of osseous tissue. The lacunae (*La*) provide a space for the osteocytes, which are connected to one another by canaliculi (*Ca*). Note the divisions between concentric lamellae (*arrows*). (From: *Tissues and Organs: A Text Atlas of Scanning Electron Microscopy* by R. G. Kessel and R. Kardon. W. H. Freeman and Company. © 1979.)

Figure 6.27. Vascular connective tissue, or blood, is composed of a fluid matrix, called plasma, and formed elements of which there are three types: erythrocytes (red blood cells), thrombocytes (platelets), and leukocytes (white blood cells).

Vascular Connective Tissue

Blood is a highly specialized, viscous connective tissue. **Formed elements** (red blood cells, white blood cells, and platelets) are suspended in the liquid plasma matrix. Blood plays a vital role in maintaining internal body homeostasis. The three types of formed elements found within the blood are erythrocytes, leukocytes, and thrombocytes (fig. 6.27).

Erythrocytes. Erythrocytes (*ĕ-rith'ro-sīt*), or **red blood cells** (RBC) are minute, nonnucleated, biconcave discs that transport respiratory gases. The red color results from the pigment *hemoglobin*. In an infant, erythrocytes are produced in the spleen and bone marrow, but in a mature person, the bone marrow is the only production site. Erythrocytes live between 90 and 120 days and are disposed of in the liver (chapter 19).

erythrocyte: Gk. *erythros*, red; *kytos*, hollow (cell)
hemoglobin: Gk. *haima*, blood; *globus*, globe

Leukocytes. Leukocytes, or **white blood cells** (WBC), are ameboid in movement and slightly larger than erythrocytes. Leukocytes primarily defend the body against invasions by microorganisms. They are produced in bone marrow and lymphatic tissue and have a life span that ranges from 3 to 300 days. There are five kinds of leukocytes: **neutrophils, eosinophils** (*e''o-sin'o-fil*), **basophils** (*ba'so-fil*), **lymphocytes** (*lim'fo-sīt*), and **monocytes.** The first three can be further classified as **granular leukocytes** and the remaining two as **nongranulocytes.**

Thrombocytes. Thrombocytes are the **platelets** of the blood; along with the *fibrinogen* of the plasma, they play a role in blood clotting. Platelets arise from large, multinucleated cells in the bone marrow, which are called **megakaryocytes** (*meg''ah-kar'e-o-sīt*). As these large cells fragment, numerous platelets form. Platelets, like leukocytes, have ameboid movement.

leukocyte: Gk. *leukos*, white; *kytos*, hollow (cell)
thrombocyte: Gk. *thrombos*, a clot; *kytos*, hollow (cell)
megakaryocyte: Gk. *megas*, great; *karyon*, nut or kernel; *kytos*, hollow (cell)

Figure 6.28.   The three types of muscle tissue: (*a*) smooth muscle fibers teased apart; (*b*) cardiac muscle; and (*c*) skeletal muscle.

An injury to a portion of the body may stimulate tissue repair activity, usually involving connective tissue. A minor scrape or cut results in platelet and plasma activity of the exposed blood and the formation of a scab. A severe open wound brings about connective tissue repair, in which collagenous fibers form from surrounding fibroblasts to strengthen the traumatized area. The healed area is known as a *scar*.

1. *List the kinds of connective tissues, and describe the structure, function, and location of each.*
2. *Which of the previously discussed connective tissues function to protect body organs? Which type is phagocytic? Which types bind and support various structures? Which types are associated in some way with the skin?*
3. *What is the developmental significance of mesenchyme, and how does it functionally differ from adult connective tissue?*
4. *Define or describe ground substance, reticular fibers, fibroblasts, collagenous fibers, elastic fibers, and mast cells.*

Muscle Tissue

Muscle tissues are responsible for the movement of materials through the body, the movement of one part of the body with respect to another, and for locomotion. Fibers in the three kinds of muscle tissue are adapted to contract in response to stimuli.

> *Objective 12.* Describe the structure, location, and function of the three types of muscle tissue.

Muscle tissues are unique in possessing the property of *contractility*. The muscle cells, or fibers, are elongated in the direction of contraction. Movement is accomplished through the shortening of the fibers in response to a stimulus. Muscle tissue is derived from mesoderm. There are three types of muscle tissue in the body: smooth, cardiac, and skeletal (fig. 6.28).

Smooth Muscle. Smooth muscle tissue is common throughout the body, occurring in many of the systems. For example, in the wall of the alimentary canal it provides the motive power for mechanically digesting food and the peristaltic movements through the digestive tract. Smooth muscle is also found in the walls of arteries, in the walls of respiratory passages, and in the urinary and reproductive ducts. The contraction of smooth muscle is under involuntary nervous control.

Table 6.6 Summary of muscle tissue		
Type	**Structure and function**	**Location**
Smooth	Elongated, spindle-shaped fiber with single nucleus; involuntary movements of internal organs	Walls of hollow internal organs
Cardiac	Branched, striated fiber with single nucleus and intercalated discs; involuntary rhythmic contraction	Heart muscle
Skeletal	Multinucleated, striated, cylindrical fiber that occurs in fasciculi; voluntary movement of skeletal parts	Associated with skeleton; spans joints of skeleton via tendons

Source: Van De Graaff, Kent M., *Human Anatomy.* © 1984 Wm. C. Brown Publishers, Dubuque, Iowa. All Rights Reserved. Reprinted by permission.

Smooth muscle fibers are long, spindle-shaped cells that contain a single nucleus and lack striations. These cells are usually grouped together in flattened sheets, forming the muscular portion of a wall around a lumen.

Cardiac Muscle. Cardiac muscle tissue makes up most of the wall of the heart. This tissue is characterized by branching (bifurcating) fibers, a centrally positioned nucleus, and transversely positioned **intercalated** *(in-ter' kah-lāt-ed)* **discs.** Like skeletal muscle, cardiac muscle is striated, but unlike skeletal muscle it experiences rhythmical involuntary contractions. Cardiac muscle is further discussed in chapters 16 and 19.

Skeletal Muscle. Skeletal muscle tissue attaches to the skeleton and is responsible for voluntary body movements. Each fiber is elongated, multinucleated, and has distinct transverse striations. Fibers of this muscle tissue are grouped into parallel bundles (fasciculi), which can be seen without a microscope in fresh muscle. Both cardiac and skeletal muscle fibers are striated and are so specialized at contraction that they cannot replicate once tissue formation is completed shortly after birth. Table 6.6 summarizes the three types of muscle tissue.

1. *Identify the general characteristics of muscle tissue. What is meant by a voluntary or involuntary muscle?*
2. *Distinguish between smooth, cardiac, and skeletal muscle tissue on the bases of structure, location, and function.*

Figure 6.29. The neuron. (*a*) a photomicrograph of nerve tissue. (*b*) a simplified diagram of a neuron and its principal parts. The arrows indicate the direction of the nerve impulse.

(a)

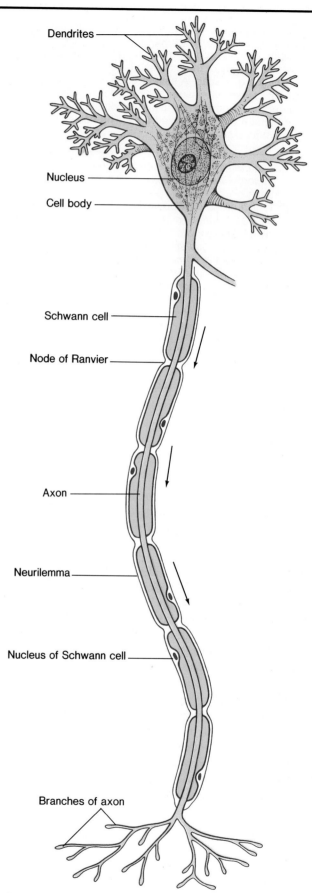

Dendrites

Nucleus

Cell body

Schwann cell

Node of Ranvier

Axon

Neurilemma

Nucleus of Schwann cell

Branches of axon

(b)

Figure 6.30. Various types of neuroglial cells that connect, support, and nourish neurons. (*a*) an astrocyte in contact with a capillary of a blood vessel serving the brain; (*b*) a microglia with processes extending to two nerve cell bodies; (*c*) oligodendrocytes near a nerve cell body.

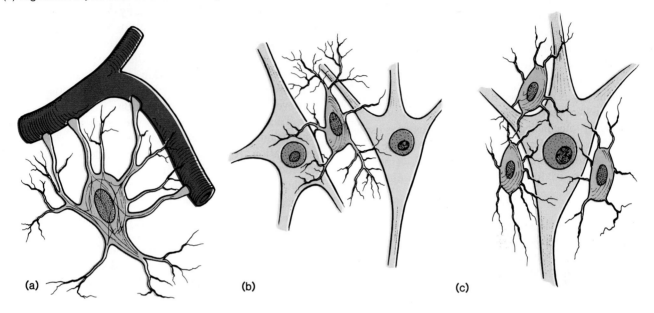

(a)   (b)   (c)

Nervous Tissue

Nervous tissue is composed of neurons, which respond to stimuli and conduct impulses to and from all body organs, and neuroglia, which functionally support and physically bind neurons.

Objective 13. Give the basic characteristics and functions of nervous tissue.
Objective 14. Distinguish between neurons and neuroglia.

Neurons. Although there are several kinds of neurons *(nu'ron)* in nervous tissue, they all have three principal components; (1) a cell body, or perikaryon, (2) dendrites, and (3) an axon (fig. 6.29). **Dendrites** function to receive a stimulus and conduct the impulse to the cell body. The **cell body,** or **perikaryon** *(per''i-kar'e-on),* contains the nucleus and specialized organelles and microtubules. The **axon** is a cytoplasmic extension that conducts an impulse away from the cell body. The term **nerve fiber** refers to an axon and its associated myelin sheath.

Neurons derive from ectoderm and are the basic structural and functional units of the nervous system. They are specialized to respond to physical and chemical stimuli, conduct nerve impulses, and perform other functions such as storing memory, thinking, and regulating the activity of other organs or glands. Of all the body's cells, neurons are probably the most specialized. As with muscle cells, the number of neurons is established shortly after birth and thereafter they lack the ability to undergo mitosis, although under certain circumstances a severed portion can regenerate.

Neuroglia. In addition to neurons, nervous tissue is composed of neuroglia *(nu'rog'le-ah)* (fig. 6.30). Neuroglial cells are about five times as abundant as neurons and have limited mitotic abilities. They do not transmit impulses but support and bind neurons together. Certain neuroglial cells are phagocytic, and others assist in providing sustenance to the neurons.

---

1. Compare the structure, function, and location of neurons and neuroglia, which comprise nervous tissue.
2. List the structures of a neuron in the sequence that a nerve impulse would pass through the cell.

---

neuroglia: Gk. *neuron,* nerve; *glia,* glue

---

perikaryon: Gk. *peri,* around; *karyon,* nut or kernel
neuron: Gk. *neuron,* sinew or nerve

Clinical Considerations

As discussed in the beginning of this chapter, understanding histology is extremely important in determining organ and system structure and function. Histology has immense clinical importance as well. Many diseases are diagnosed through the microscopic examination of tissue sections. Even in performing an autopsy, an examination of various tissues is vital for establishing the cause of death.

Several sciences are concerned with specific aspects of tissues. Histopathology is the study of diseased tissues. Histochemistry is concerned with the biochemical physiology of tissues as they function to maintain body homeostasis. Histotechnology studies the various ways tissues can be better stained and observed. In all of these disciplines, a thorough understanding of normal, or healthy, tissues is imperative for recognizing altered, or abnormal, tissues.

Changes in Tissue Composition

Most diseases alter tissue structure *locally* where the disease is prevalent, although some diseases, called *general* conditions, cause changes remote from the location of the disease. **Atrophy** (the wasting of body tissue), for example, may be confined to a particular organ where the disease interferes with the metabolism of that organ, or it may involve an entire limb if nourishment or nerve impulses are decreased or prohibited. *Muscle atrophy,* for example, can be caused by a disease of the nervous system like polio, or it can be the result of a diminished blood supply to a muscle. *Senile atrophy* is the natural aging of tissues and organs within the body. *Disuse atrophy* is a local atrophy that results from the inactivity of a tissue or organ. Muscular dystrophy causes a disuse atrophy that decreases muscle size and strength due to the loss of sarcoplasm within the muscle.

Necrosis *(ně-kro'sis)* is cellular or tissue death within the living body. It can be recognized by changes in the dead tissues. Necrosis can be caused by a number of factors, such as severe injury, physical agents (trauma, heat, radiant energy, chemical poisons), or interference with the nutrition of tissues. When thin sections are histologically examined, the necrotic tissue usually appears opaque and of a whitish or yellow color. **Gangrene** is a massive necrosis of tissue accompanied by an invasion of microorganisms that live on decaying tissues.

Somatic death is the death of the body as a whole. Following somatic death, tissues undergo irreversible changes such as **rigor mortis** (muscular rigidity), clotting of the blood, and cooling of the body. Postmortem changes occur under varying conditions at predictable rates of time, which help in estimating the approximate time of death.

Tissue Analysis

In diagnosing a disease, it is frequently important to examine tissues from a living person histologically. When this is necessary, a **biopsy,** the removal of a section of living tissue, is taken. There are several techniques for obtaining a biopsy. *Surgical removal* is usually done on large masses or tumors. *Curettage* involves cutting and scraping tissue, as may be done in examining for uterine cancer. In a *percutaneous (per''ku-ta'ne-us) needle biopsy,* a biopsy needle is inserted through a small skin incision and tissue samples are aspirated. Both normal and diseased tissues are removed for purposes of comparison.

There are several steps to preparing tissues for examination. **Fixation** is fundamental to all histological preparation. It is the rapid killing, hardening, and preservation of tissue to maintain its existing structure. **Embedding** the tissue in a supporting medium such as paraffin wax usually follows fixation. The next step is **sectioning** the tissue into extremely thin slices, followed by **mounting** the specimen on a slide. Some tissues are fixed by rapid freezing and then sectioned while frozen, making embedding unnecessary. Frozen sections enable the pathologist to make a quick diagnosis during a surgical operation. These are done frequently, for example, in cases of suspected breast cancer. **Staining** is the next step. The H and E (hematoxylin and eosin) stains are routinely used on all tissue specimens. They give a differential blue and red color to the basic and acidic structures within the tissue. Other dyes may be needed to stain for specific structures.

Examination is first done with the unaided eye and then with the use of a microscope. Practically all histological conditions can be diagnosed with low magnification (100×). Higher magnification is used to clarify specific details. Further examination may be performed with an electron microscope, which makes subcellular structures visible. Histological observation is the foundation of subsequent diagnosis, prognosis, treatment, and reevaluation.

Tissue Transplantation

In the last two decades, medical science has made tremendous advancements in tissue transplants. Tissue transplants are necessary for replacing nonfunctional, damaged, or lost body parts. The most successful transplant is one where tissue is taken from one place on a person's body and moved to another place, such as a skin graft from the thigh to replace burned tissue of the hand. This type of transplant is termed an **autotransplant. Isotransplants** are transplants between genetically closely related persons. Identical twins have the best acceptance success in these types of transplants. **Homotransplants** (between individuals of the same species) and **heterotransplants** (between two different species) have low acceptance percentages because of a *tissue rejection reaction*. When this occurs, the recipient's immune mechanisms are triggered, and the donor's tissue is identified as foreign and is destroyed. The reaction can be minimized by "matching" recipient and donor tissue. Immunosuppressive drugs may also lessen the rejection rate. These drugs act by interfering with the recipient's immune mechanisms (chapter 22). Unfortunately, immunosuppressive drugs may lower the recipient's resistance to infections as well. New techniques involving blood transfusions from donor to recipient before transplant are proving successful. In any event, tissue transplants are an important aspect of medical research, and significant breakthroughs are on the horizon.

Chapter Summary

I. Definition and Classification of Tissues
   Histology is an integral part of anatomy and physiology because it imparts an understanding of the structure and function of organs at the tissue level of study. Tissues are classified on the basis of their cellular composition and histological appearance into four principal kinds.
   A. Tissues are aggregations of similar cells that perform specific functions. The study of tissues is histology.
   B. Cells are separated and bound together by an intercellular matrix; its composition varies from solids to liquids.
   C. The four principal types of tissues are epithelial tissue, connective tissue, muscle tissue, and nervous tissue.

II. Development of Tissues
   The various types of tissues are established during embryonic development, and as differentiation continues, organs form, each of which is composed of a specific arrangement of tissues.
   A. Prenatal development following fertilization results in a blastocyst, which implants into the uterine wall between the fifth and seventh day.
   B. The three primary germ layers, from which all body tissues and organs eventually derive, form shortly after implantation.

III. Epithelial Tissue
   Epithelia are classified according to the physical features of the tightly packed cells that comprise these tissues. Epithelia line all body surfaces, cavities, and lumina and are adapted for protection, absorption, and secretion.
   A. Epithelia derive from all three germ layers and may be one or several layers thick.
      1. The lower surface is bound to a supportive basement membrane of glycoprotein.
      2. Epithelial cells are bonded by glycoprotein deposits, desmosomes, or tight and intermediate junctions.

   B. Simple epithelia vary in shape and surface characteristics; they are located where diffusion, filtration, and secretion occur.
      1. Simple squamous epithelium is found in the lungs, kidneys, inner lining of blood vessels (endothelium), covering visceral organs, and lining body cavities (mesothelium).
      2. Simple cuboidal epithelium is found lining the small ducts of such organs as the kidneys, salivary glands, and pancreas.
      3. Simple columnar epithelium lines the lumen of the digestive tract, where it has secretory and absorptive functions.
      4. Simple ciliated columnar epithelium occurs in the uterine tubes of the female for the movement of ova.
      5. Pseudostratified ciliated columnar epithelium lines the lumina of the lower respiratory tract, where it traps foreign material and moves it away from the lungs.
   C. Stratified epithelia consist of two or more layers of cells and are adaptive for protection.
      1. Stratified squamous epithelium forms the outer layer of skin, where it is keratinized, and lines the oral cavity, nasal cavity, vagina, and anal canal, where it is nonkeratinized.
      2. Stratified cuboidal epithelium surrounds the larger ducts of sweat glands, salivary glands, and the pancreas.
      3. Transitional epithelium lines the urinary bladder and is adapted for distension.
   D. Glandular epithelia derive from developing epithelial tissue and function as secretory exocrine glands.
      1. Unicellular glands (goblet cells) secrete a lubricating mucus.
      2. Multicellular glands are classified structurally as tubular, acinar, and tubuloacinar and functionally as merocrine, apocrine, and holocrine.

IV. Connective Tissue
   Connective tissues are classified according to the characteristics of the matrix that binds the cells. Connective tissues provide structural and metabolic support for other tissues and the organs of the body.
   A. Connective tissues derive from mesoderm and, with the exception of cartilage, are highly vascular.
   B. All types of connective tissue begin embryonically as mesenchyme.
   C. Connective tissue proper contains fibroblasts, collagenous fibers, and elastic fibers within a flexible ground substance.
      1. Loose connective tissue is the common binding and packing material of the body; surrounding muscle, it is known as fascia, and surrounding blood vessels, it contains mast cells that produce heparin.
      2. Dense fibrous connective tissue is composed of strong, collagenous fibers and forms tendons, ligaments, and the sclera of the eye.
      3. Elastic connective tissue has a predominance of elastic fibers and occurs in large arteries, portions of the respiratory tract, and between vertebrae.
      4. Reticular connective tissue contains specialized phagocytic cells and is found in the liver, spleen, lymph nodes, and bone marrow.
      5. Adipose tissue stores fat and provides protection and insulation.
   D. Cartilage provides a flexible framework for many organs and consists of a semisolid matrix of chondrocytes and various fibers.
      1. Hyaline cartilage is found at the ends of bone, in the nose, and throughout the respiratory tract; it forms the precursor to bone.

2. Fibrocartilage contains numerous collagenous fibers and is located at the symphysis pubis, intervertebral discs, and within the knee joint.
3. Elastic cartilage has abundant elastic fibers and forms the framework for the external ear, auditory canal, and portions of the larynx.

E. Bone tissue consists of osteocytes, collagenous fibers, and a durable matrix of mineral salts.
   1. The two types of bone tissue are compact and spongy.
   2. Osteocytes are located in lacunae arranged in concentric rings around Haversian canals. Haversian systems are the functional structures of bone tissue.

F. Vascular tissue consists of formed, cellular elements suspended in a fluid plasma matrix.
   1. Erythrocytes, leukocytes, and thrombocytes are the three types of formed elements of blood.
   2. Formed elements are produced in the bone marrow and lymphatic tissue.

V. Muscle Tissue
Muscle tissues are responsible for the movement of materials through the body, the movement of one part of the body with respect to another, and for locomotion. Fibers in the three kinds of muscle tissue are adapted to contract in response to stimuli.
   A. The three types of muscle tissue are smooth, cardiac, and skeletal; each of which is derived from mesoderm.
   B. Cardiac and skeletal muscle cannot replicate once the tissue is formed shortly after birth.

VI. Nervous Tissue
Nervous tissue is composed of neurons, which respond to stimuli and conduct impulses to and from all body organs, and neuroglia, which functionally support and physically bind neurons.
   A. Neurons are the functional units of the nervous system.
   B. Neuroglia support and bind neurons; some are phagocytic, and others provide sustenance to neurons.

Review Activities

Objective Questions

1. Which of the following is *not* a principal type of body tissue?
   (a) nervous
   (b) integumentary
   (c) connective
   (d) muscular
   (e) epithelial
2. The correct sequence of the prenatal stages of development is
   (a) ovum, zygote, blastocyst, morula
   (b) ovum, morula, blastocyst, zygote
   (c) ovum, zygote, morula, blastocyst
   (d) zygote, ovum, blastocyst, morula
3. Connective tissue, muscle, and the dermis of the skin derive from embryonic
   (a) mesoderm
   (b) endoderm
   (c) ectoderm
4. Which statement is *false* regarding epithelia?
   (a) They are derived from both ectoderm and endoderm.
   (b) They are strengthened by elastic and collagenous fibers.
   (c) One side is exposed to the lumen, cavity, or external environment.
   (d) They have very little extracellular matrix-binding cells.

5. A gastric ulcer of the stomach would involve
   (a) simple cuboidal epithelium
   (b) transitional epithelium
   (c) simple ciliated columnar epithelium
   (d) simple columnar epithelium
6. Which structural and secretory designation describes mammary glands?
   (a) acinar, apocrine
   (b) tubular, holocrine
   (c) tubular, merocrine
   (d) acinar, holocrine
7. Dense fibrous connective tissue is found in
   (a) blood vessels
   (b) the spleen
   (c) tendons
   (d) the wall of the uterus
8. The phagocytic connective tissue found in the lymph nodes, liver, spleen, and bone marrow is
   (a) reticular
   (b) areolar
   (c) mesenchyme
   (d) elastic
9. Cartilage is slow in healing following an injury because
   (a) it is located in body areas that are under constant physical strain
   (b) it is avascular
   (c) its chondrocytes cannot reproduce
   (d) it has a semisolid matrix
10. Cardiac muscle tissue has
    (a) striations
    (b) intercalated discs
    (c) rhythmical involuntary contractions
    (d) all of the above

Essay Questions

1. Define a tissue. What are the differences between cells, tissues, glands, and organs?
2. What physiological functions are epithelial tissues adapted to perform?
3. Identify the epithelial tissue found in each of the following structures or organs, and give the function of the tissue in each case.
   a. in the alveoli of the lungs
   b. lining the lumen of the digestive tract
   c. in the outer layer of skin
   d. in the urinary bladder
   e. in the uterine tube
   f. lining the lumina of the lower respiratory tract
4. Why are both keratinized and nonkeratinized epithelia found within the body?
5. Distinguish between the following: reticular fibers, collagen, elastin, fibroblasts, and mast cells.
6. What is the relationship between adipose cells and fat? Discuss the function of fat, and explain the potential danger of excessive fat.
7. Discuss the mitotic abilities of each of the four principal types of tissues.
8. Distinguish between the following terms: atrophy, necrosis, gangrene, and somatic death.

Selected Readings

Bevelander, G., and J. A. Ramaley. 1979. *Essentials of histology.* 8th ed. St. Louis: C. V. Mosby Co.

Bloom, W. B., and D. W. Fawcett. 1975. *A textbook of histology.* 10th ed. Philadelphia: W. B. Saunders Co.

Di Fiore, M. S. H. 1974. *An atlas of histology.* 4th ed. Philadelphia: Lea and Febiger.

Kessel, R. G., and R. H. Kardon. 1979. *Tissues and organs: a text-atlas of scanning electron microscopy.* San Francisco: W. H. Freeman and Co.

Porter, K. R., and M. A. Bonneville. 1973. *An introduction to the fine structure of cells and tissues.* 4th ed. Philadelphia: Lea and Febiger.

Weiss, L., and R. O. Greep. 1977. *Histology.* 4th ed. New York: McGraw-Hill.

Wheater, P. R., H. G. Burkett, and V. G. Daniels. 1979. *Functional histology: a text and colour atlas.* London: Churchill Livingstone.

Windel, W. F. 1976. *Textbook of histology.* 5th ed. New York: McGraw-Hill.

7

Homeostasis and

Concepts

The regulatory mechanisms of the body
maintain homeostasis by means of
negative feedback mechanisms. In
these mechanisms, the actions of
effector organs are regulated by
sensory information from the internal
environment.

Hormones are secreted into the blood by
endocrine glands in response to
specific stimuli. Hormonal
stimulation of effector organs
produces responses that help to
complete negative feedback loops.

Nerve fibers conduct electrochemical
nerve impulses that stimulate the
release of chemical
neurotransmitters. These events
provide pathways for negative
feedback regulation.

Control Mechanisms

Homeostasis and Feedback Control

The numerous and varied regulatory mechanisms in the body can be understood in terms of a single, shared function: that of maintaining homeostasis, or a relative constancy of the internal environment. In order to share in this common function, the actions of effector organs are regulated by sensory information from the internal environment. The activity of effector organs is thus controlled by the very changes these organs help to produce—that is, by feedback control mechanisms.

Objective 1. Define homeostasis, and describe the importance of this concept in physiology and medicine.

Objective 2. Describe negative feedback and positive feedback loops.

Objective 3. Explain how feedback control mechanisms help to maintain homeostasis.

Over a century ago the French physiologist Claude Bernard observed that the *melieu interior* (internal environment) remains remarkably constant despite changing conditions in the external environment. In a book entitled *The Wisdom of the Body* (published in 1932), Walter Cannon coined the term **homeostasis** *(ho''me-o-sta'sis)* to describe this internal constancy. Cannon further suggested that mechanisms of physiological regulation exist for one purpose—the maintenance of internal constancy.

The concept of homeostasis has been of inestimable value in the study of anatomy and physiology, because it allows diverse regulatory mechanisms to be understood in terms of their "why" as well as their "how." The concept of homeostasis also provides a major foundation for medical diagnostic procedures. When a particular measurement of the internal environment, such as blood measurements (table 7.1), deviates significantly from the normal range of values, it can be concluded that homeostasis is not maintained and the person is sick. A number of such measurements, combined with clinical observations, may allow the particular defective mechanism to be identified.

Table 7.1 Approximate normal ranges for measurements of some blood values

Measurement	Normal range	Measurement	Normal range
Arterial pH	7.35–7.43	Urea	12–35 mg/100 ml
Bicarbonate	21.3–28.5 mEq/L	Amino acids	3.3–5.1 mg/100 ml
Sodium	136–151 mEq/L	Protein	6.5–8.0 g/100 ml
Calcium	4.6–5.2 mEq/L	Total lipids	350–850 mg/100 ml
Oxygen content	17.2–22.0 ml/100 ml	Glucose	75–110 mg/100 ml

Source: Fox, Stuart Ira, *Human Physiology.* © 1984 Wm. C. Brown Publishers, Dubuque, Iowa. All Rights Reserved. Reprinted by permission.

---

homeostasis: Gk. *homoios,* like; *stasis,* a standing

Figure 7.1.   A rise in some factor of the internal environment (↑X) is detected by a sensor. The sensor activates an effector (generally, nerves or hormones), which causes a decrease in this factor (↓X). In this way the factor returns to its initial level and homeostasis is maintained. Completion of the negative feedback loop is shown by a dotted arrow and a negative sign.

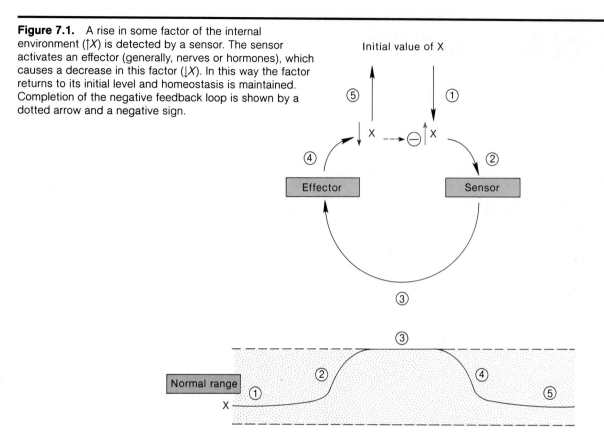

Negative Feedback Loops

In order for internal constancy to be maintained, the body must have *sensors* that are able to detect deviations from the normal range of particular internal conditions and *effectors* that are able to reduce and ultimately to reverse these deviations. A detectable change in the internal environment, thus, results in changes in the reverse direction. Internal constancy is maintained by such **negative feedback loops** (fig. 7.1).

An increase in a value (indicated as *X* in figure 7.1) above its normal range is detected by a sensor; the sensor in turn activates an effector that causes *X* to decrease. This negative feedback loop prevents *X* from rising too far above its normal range. It is important to realize, however, that even when *X* is in its normal range there is some activity of the effector. Even under normal, resting conditions, to be more specific, there is always some spontaneous nerve activity and some hormone secretion. If an increase in the activity of an effector causes a decrease in *X* (as shown in figure 7.1), a decrease in the activity of the same effector will cause an increase in *X*. Changes from the normal range in either direction can be compensated by reverse changes in effector activity (fig. 7.2).

Homeostasis is best conceived as a state of **dynamic constancy,** rather than simply a state of absolute constancy. The values of particular measurements of the internal environment fluctuate above and below the average, or *set point,* of these measurements. Values are, therefore, at their set point only in passing (fig. 7.3), and regulatory mechanisms are constantly active to greater or lesser degrees in ongoing negative feedback loops.

Antagonistic Effectors.   Most factors in the internal environment are controlled by several effectors, which often have antagonistic actions. Control by antagonistic effectors is sometimes described as "push-pull." This affords a finer degree of control than could be achieved by simply switching one effector on and off. Normal body temperature, for example, is maintained about a set point of 37°C by the antagonistic effects of sweating, shivering, and other mechanisms (fig. 7.4).

The blood concentrations of glucose, calcium, and other substances are regulated by negative feedback loops that involve hormones which promote opposite effects. While insulin, for example, lowers blood glucose, other hormones raise the blood glucose concentration. The heart rate, similarly, is controlled by nerve fibers that produce opposite effects: stimulation of one group of nerve fibers increases heart rate, and stimulation of another group slows the heart rate.

---

antagonistic: Gk. *anti,* against; *agonizomai,* to fight

Figure 7.2. A negative feedback loop in which a decrease in some factor of the internal environment is compensated for by the actions of an effector. Compare this diagram with that shown in figure 7.1.

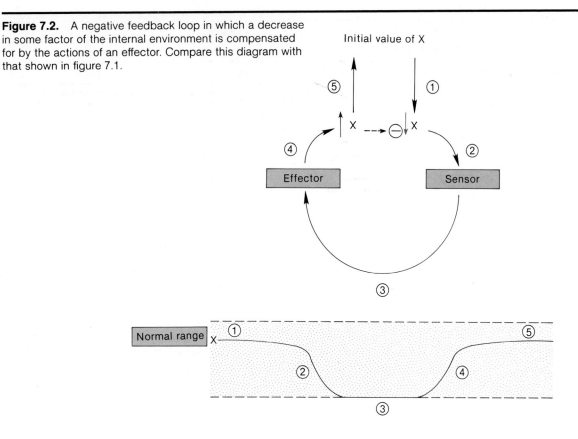

Figure 7.3. Negative feedback loops (indicated by negative signs) maintain a state of dynamic constancy within the internal environment.

Figure 7.4. A simplified scheme by which body temperature is maintained within the normal range (with a set point of 37°C) by two antagonistic mechanisms—shivering and sweating. Shivering is induced when the body temperature falls too low and gradually subsides as the temperature rises. Sweating occurs when the body temperature is too high and diminishes as the temperature falls. Most aspects of the internal environment are regulated by the antagonistic actions of different effector mechanisms.

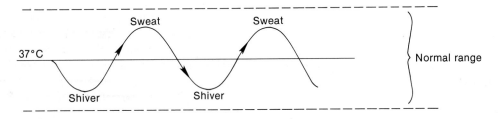

Positive Feedback

Constancy of the internal environment is maintained by effectors which act to compensate for the change that served as the stimulus for their activation; in short, by negative feedback loops. A thermostat, for example, maintains a constant temperature by increasing heat production when it is cold and decreasing heat production when it is warm. The opposite occurs during **positive feedback**—in this case, the action of effectors *amplifies* those changes that stimulated the effectors. A thermostat that works by positive feedback, for example, would increase heat production in response to a rise in temperature.

It is clear that homeostasis must ultimately be maintained by negative rather than by positive feedback mechanisms. The effectiveness of some negative feedback loops, however, is increased by positive feedback mechanisms that amplify the actions of a negative feedback response. Blood clotting, for example, occurs as a result of a sequential activation of clotting factors; the activation of one clotting factor results in activation of many in a positive feedback, avalanchelike, manner. In this way, a single change is amplified to produce a blood clot. Formation of the clot, however, can prevent further loss of blood and thus represents the completion of a negative feedback loop.

Neural and Endocrine Regulation

Homeostasis is maintained by two general categories of regulatory mechanisms: (1) those that are **intrinsic,** or "built-in," to the organ; these are usually due to the effects of chemicals that act within the organs that produce them; and (2), those that are **extrinsic;** that is, regulation of an organ by the nervous and endocrine systems.

The endocrine system functions closely with the nervous system in regulating and integrating body processes and maintaining homeostasis. The nervous system regulates body activities through the action of electrochemical nerve impulses, which are transmitted by the axons of nerve cells (neurons). Nerve impulses generally produce rapid but brief responses. In contrast, the glands of the endocrine system secrete chemical regulators (hormones) into the blood, which produce slower but more long lasting responses in their target organs. Some comparisons of the nervous and endocrine systems are presented in table 7.2.

Common Aspects of Neural and Endocrine Regulation.

It may be thought that neural regulation is electrical and therefore distinct from the chemical nature of endocrine control systems. This idea is incorrect. Electrical nerve

Table 7.2 Comparison of the basic functions of the endocrine and nervous systems

Endocrine system	Nervous system
Secretes hormones that are transported to target tissues via the blood	Transmits neurochemical impulses via nerve fibers
Causes changes in the metabolic activities in specific cells	Causes muscles to contract or glands to secrete
Exerts effects relatively slowly (seconds or even days)	Exerts effects relatively rapidly (milliseconds)
Has generally prolonged effects	Has generally brief effects

Source: Van De Graaff, Kent M., *Human Anatomy.* © 1984 Wm. C. Brown Publishers, Dubuque, Iowa. All Rights Reserved. Reprinted by permission.

impulses are in fact chemical events produced by the diffusion of ions through the neuron cell membrane. Interestingly, the action of some hormones (such as insulin) is accompanied by ion diffusion and electrical changes in the target cells. Also, most nerve fibers stimulate the cells they innervate through the release of a chemical **neurotransmitter** *(nu"ro-trans'mit-er).* Neurotransmitters differ from hormones in that they do not travel in the blood but instead diffuse only a very short distance from the nerve ending to the target cell. In other respects, however, the actions of neurotransmitters and hormones are very similar.

Indeed, many polypeptide hormones, including those secreted by the pituitary gland and by the digestive tract, have been discovered in the brain. In certain locations in the brain, some of these compounds are produced and secreted as hormones (the brain is an endocrine gland). In other brain locations, some of these compounds apparently serve as neurotransmitters. The discovery of some of these polypeptides in unicellular organisms, which of course lack a nervous and endocrine system, suggests that these regulatory molecules appeared early in evolution and became incorporated into the function of nerve and endocrine tissue as these systems evolved. This fascinating theory would help to explain, for example, why insulin, a polypeptide hormone produced in the pancreas of vertebrates, is found in neurons of invertebrates (which lack a separate endocrine system).

Regardless of whether a particular chemical is acting as a neurotransmitter or as a hormone, in order to function in physiological regulation, (1) target cells must have specific **receptor proteins** that combine with the chemical; (2) the combination of the regulator molecule with its receptor proteins must cause a specific sequence of changes in the target cells; and (3) there must be a mechanism to rapidly turn off the action of the regulator; without an "off" switch," physiological control is impossible. This last process involves rapid removal and/or chemical inactivation of the regulator molecules.

Figure 7.5. Comparison of (*a*) an endocrine gland such as a pancreatic islet in the pancreas with (*b*) an exocrine gland such as a sweat gland in the skin. An endocrine gland is a ductless gland that releases hormones into the body fluid such as the blood, whereas exocrine glands secrete their products into ducts that lead to body surfaces.

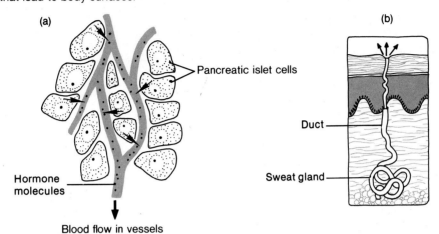

(a)

Pancreatic islet cells

Hormone molecules

Blood flow in vessels

(b)

Duct

Sweat gland

1. *Define homeostasis, and describe how this concept can be used to understand physiological control mechanisms, and explain its importance to medicine.*
2. *Describe the meaning of the term* negative feedback, *illustrate this concept by drawing a negative feedback loop, and explain how it contributes to homeostasis.*
3. *Explain how the actions of antagonistic effectors provides a greater control of the internal environment than could be afforded by a single negative feedback loop.*
4. *Define the term* positive feedback, *and explain how this control system can contribute to homeostasis.*

Hormones and the Regulation of Endocrine Gland Secretion

Hormones are specific chemicals secreted into the blood by endocrine glands, which change the metabolism of particular target cells. Some endocrine glands are devoted exclusively to the task of secreting hormones, and others secrete hormones in addition to their performance of other functions. The secretion of hormones by endocrine glands is regulated by negative feedback control mechanisms.

Objective 4. Describe how endocrine glands differ from exocrine glands.
Objective 5. Describe the different chemical classes of hormones.

Objective 6. Explain how the secretion of hormones by the endocrine system is controlled by the nervous system and by negative feedback control mechanisms.

Endocrine glands are secretory structures, derived from epithelial tissue, which secrete their products, called hormones, into the blood. In contrast, exocrine glands secrete their products via ducts to the external surface of epithelial membranes (fig. 7.5).

The hormones secreted by endocrine glands include (1) **biogenic amines** *(am'ēnz)* such as *catecholamines (kat''ē-kol'ah-mēn)*—epinephrine and norepinephrine—and other derivatives of amino acids; (2) **polypeptides** such as insulin, antidiuretic hormone (vasopressin), growth hormone, and others which are composed of chains of amino acids that are too short to be called proteins; (3) **glycoproteins** such as FSH, LH, and others that consist of large proteins combined with carbohydrates; and (4) **steroids** (fig. 7.6). Some examples of polypeptides and glycoproteins are presented in table 7.3.

Steroid hormones, which are derived from cholesterol (fig. 7.7), are lipids and thus are not water-soluble. The gonads—testes and ovaries—secrete *sex steroids;* the adrenal cortex secretes *corticosteroids (kor''ti-ko-ste'roids)* such as hydrocortisone (cortisol) and others.

---

endocrine: Gk. *endon*, within; *krinein*, to separate
hormone: Gk. *hormon*, to set in motion
exocrine: Gk. *exo*, outside; *krinein*, to separate

Figure 7.6.   Structural formulas of three basic kinds of
hormones: (a) a polypeptide hormone, (b) steroid hormone,
and (c) an amine hormone.

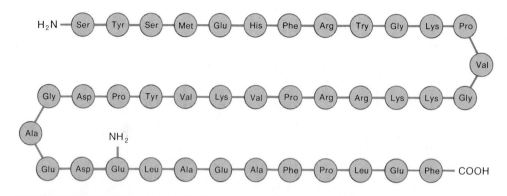

(a) ACTH (adrenocorticotrophic hormone)

(b) Cortisol

(c) Norepinephrine

Table 7.3 Some examples of polypeptide and glycoprotein hormones			
Hormone	**Structure**	**Gland**	**Primary effects**
Vasopressin	8 amino acids	Posterior pituitary	Water retention and vasoconstriction
Oxytocin	8 amino acids	Posterior pituitary	Uterine and mammary contraction
Insulin	21 and 30 amino acids (double chain)	β cells in islets of Langerhans	Cellular glucose uptake, lipogenesis, and glycogenesis
Glucagon	29 amino acids	α cells in islets of Langerhans	Hydrolysis of stored glycogen and fat
ACTH	39 amino acids	Anterior pituitary	Stimulation of adrenal cortex
Parathyroid hormone	84 amino acids	Parathyroid	Increases blood Ca^{++} concentration
FSH, LH, TSH	Glycoproteins	Anterior pituitary	Stimulates growth, development, and secretion of target glands

Source: Fox, Stuart Ira, *Human Physiology.* © 1984 Wm. C. Brown Publishers, Dubuque, Iowa. All Rights Reserved. Reprinted by permission.

Figure 7.7. Simplified biosynthetic pathways for steroid hormones. Notice that progesterone (a hormone secreted by the ovaries) is a common precursor in the formation of all other steroid hormones and that testosterone (the major androgen secreted by the testes) is a precursor in the formation of estradiol–17β, the major estrogen secreted by the ovaries.

Cholesterol

Pregnenolone

Progesterone

Secreted by ovaries

CH$_2$OH

Cortisol (hydrocortisone)

Secreted by adrenal cortex

Androstenedione

Testosterone

Secreted by testes

Estradiol – 17 β

Secreted by ovaries

Figure 7.8. The thyroid hormones: thyroxine (T_4) and triiodothyronine (T_3) are secreted in a ratio of 9:1.

Thyroxine, or tetraiodothyronine (T_4)

Triiodothyronine (T_3)

The major thyroid hormones are composed of two derivatives of the amino acid *tyrosine* that are bonded together. These hormones are unique because they contain iodine (fig. 7.8). When the hormone contains four iodine atoms it is called *tetraiodothyronine (tet″rah″i-o-do-thi′ro-nēn), T_4,* or *thyroxine (thi-rok′sin).* When it contains three atoms of iodine it is called *triiodothyronine (T_3).*

Although the thyroid hormones are not steroids, they are like steroids in that they are relatively small and nonpolar molecules. Steroid and thyroid hormones are active when taken orally (as a pill); sex steroids are contained in the contraceptive pill, and thyroid hormones are ingested by people whose own thyroid gland is deficient (who are hypothyroid). Other types of hormones cannot be taken orally because they would be digested into inactive fragments.

Control of Hormone Secretion

Hormones are secreted in response to specific chemical stimuli. The presence of proteins in the stomach, for example, stimulates the secretion of gastrin, and a rise in plasma glucose concentration stimulates insulin secretion from the islets of Langerhans in the pancreas. Hormones are also secreted in response to neurotransmitters released from nerve endings and to stimulation by other hormones.

iodine: Gk. *ion,* violet (color); *eidos,* form
thyroid: Gk. *thyreos,* oblong shield
insulin: L. *insula,* island

Figure 7.9. The regulation of hormone secretion by the hypothalamus, anterior pituitary, and target glands (thyroid, adrenal cortex, and gonads). Hormones secreted by the target glands exert negative feedback inhibition on the release of hormones from the hypothalamus and anterior pituitary.

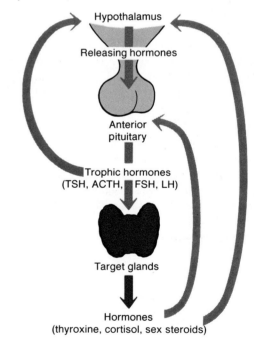

Hypothalamus

Releasing hormones

Anterior pituitary

Trophic hormones
(TSH, ACTH, FSH, LH)

Target glands

Hormones
(thyroxine, cortisol, sex steroids)

Figure 7.10. The negative feedback control of insulin secretion by changes in the blood glucose concentration.

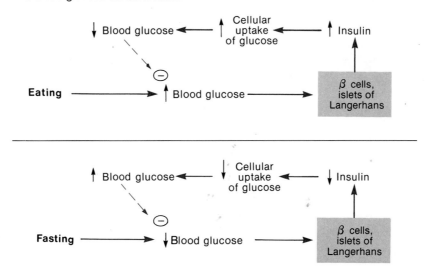

The nervous system helps to regulate hormone secretion in a variety of ways. Autonomic nerves, for example, stimulate the secretion of epinephrine from the adrenal medulla and also participate in the regulation of insulin and glucagon secretion from the islets of Langerhans in the pancreas. Hormonal secretion by the adrenal cortex, thyroid, and gonads is stimulated by hormones secreted by the anterior pituitary gland. Secretions of the anterior pituitary, in turn, are regulated by hormones released by neurons in the hypothalamus of the brain (chapters 16 and 18). This control is not one-way; many neural centers in the brain are affected by the actions of hormones (fig. 7.9).

The secretion of hormones is controlled by inhibitory as well as by stimulatory influences. The effect of a given hormone's action can inhibit its own secretion. The secretion of insulin—which acts to lower the plasma glucose concentration—is stimulated by a rise in glucose concentration, for example, and is inhibited by a fall in blood glucose. The lowering of blood glucose levels by insulin thus has an inhibitory feedback effect on further insulin secretion. This closed loop control system is called **negative feedback inhibition** (see fig. 7.10).

1. Distinguish between endocrine and exocrine glands.
2. List the different chemical classes of hormones, and give examples of each.
3. Explain how the secretion of hormones is controlled by the nervous system.
4. Explain how negative feedback mechanisms regulate the secretion of hormones by endocrine glands.

Neural Control of Homeostasis

Nerve fibers conduct electrochemical nerve impulses that stimulate the release of chemical neurotransmitters. These events provide pathways for the flow of information from sensors and for the appropriate stimulation of effectors in negative feedback regulation.

Objective 7. Distinguish between a neuron and a nerve.

Objective 8. Describe, in a general way, the nature of a nerve impulse.

Objective 9. Explain, in a general way, how nerve fibers stimulate their target organs.

Objective 10. Describe a negative feedback loop involving neural regulation.

The functional unit of the nervous system is the nerve cell, or **neuron.** A typical neuron consists of three functional parts: (1) **dendrites,** which are thin, extensively branched threads of cytoplasm; (2) a **cell body,** which contains the nucleus; and (3) an **axon** *(ak'son),* which is a thin, but often very long, cytoplasmic extension that conducts nerve impulses away from the cell body. Axons are also called **nerve fibers.**

The functional connections between the endings of the axon and another cell is called a **synapse** *(sin'aps).* Synapses may occur between an axon and the dendrites or cell body of another neuron or between an axon and an effector cell in a muscle or gland.

islets of Langerhans: from Paul Langerhans, German anatomist, 1847–88

Figure 7.11.   A schematic diagram of the flow of information from sensory neurons into the CNS and from motor neurons to effector organs.

The brain and spinal cord are together known as the **central nervous system (CNS).** Axons and some entire neurons that are located outside the brain and spinal cord are part of the **peripheral nervous system (PNS).** A **nerve** is a collection of axons, packaged together in connective tissue, which is located in the PNS. You might think of a nerve as a cable containing many wires (axons) that lead from or to the central junction box (CNS). Those axons that carry nerve impulses from sensory organs into the CNS are **sensory,** and those that carry impulses from the CNS to effector organs are **motor.** By means of these neural connections, effectors can respond appropriately to sensory information (fig. 7.11).

Electrical events in neurons are produced by the movements of ions—usually of Na^+ and K^+, but sometimes of Ca^{++} or Cl^-—across a limited region of the plasma membrane of a neuron. As these charges diffuse down their concentration gradients, either entering or leaving the neuron, they produce electrical currents across the membrane. Such electrical events are conducted to the axon endings where they stimulate the release of neurotransmitter chemicals at the synapse.

The neurotransmitters diffuse across a tiny gap to the membrane of the next cell, where receptor proteins for those transmitters are located. The bonding of the neurotransmitters to their receptor proteins in a synapse is analogous to the bonding of hormones to receptor proteins on the outer surface of target cell membranes. After the neurotransmitters have bonded to their receptor proteins, intermediate processes are stimulated that can result in the production of new electrical events. If the synapse is with an effector cell, the effector is activated, which can lead to muscle contraction or gland secretion.

Specificity of regulation is insured by the proper "wiring" (the location of synapses made by particular neurons) and by the fact that receptor proteins for neurotransmitters and hormones can combine with only specific molecules. Skeletal muscles, for example, can only be stimulated by one chemical—acetylcholine—which is released by specific motor neurons. Contractions of the heart, in contrast, are stimulated by epinephrine and norepinephrine. Both of these chemicals are released into the

Figure 7.12.   A schematic diagram showing how the homeostasis of blood pressure can be maintained by a negative feedback loop involving the nervous system.

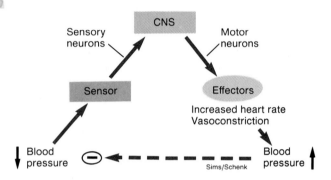

blood as hormones from the adrenal medulla gland, and norepinephrine is released as a neurotransmitter from the endings of some nerve fibers that innervate the heart. Through these mechanisms and the specific actions of sensors and effectors, the nervous system can mediate the negative feedback loops that help maintain homeostasis (fig. 7.12).

1. *Define the terms* neuron, nerve, *and* central nervous system.
2. *Distinguish between sensory and motor neurons.*
3. *Describe, in a general way, the nature of a nerve impulse and the chemical nature of synaptic transmission.*
4. *Draw a negative feedback loop that shows how homeostasis of blood pressure can be maintained by the actions of the nervous system.*

Clinical Considerations

Homeostasis

Sampling the Internal Environment.   Medical personnel obtain numerous measurements of conditions in the internal environment in order to compare these measurements to normal values. These normal values are obtained from large populations of people who are believed to be normal for the factors being tested. This circular definition of normal values can lead to some misinterpretation (for example, an endurance-trained athlete in fine health may have a lower cardiac rate than the general population); however, proper statistical analysis and frequent reevaluation of these measurements help to produce very useful reference values. When various measurements, obtained during physical examination and by laboratory tests of body tissues (usually of blood), differ from the established normal ranges, it is probable that homeostasis is not maintained. More importantly, the particular disease process that caused deviations from a homeostatic state can often be determined through such measurements of the internal environment.

Endocrine Regulation

Measurements of Hormone Levels in Blood.
There are many endocrine disorders that are characterized by either abnormally high or low levels of hormone secretion. This may result from damage to the particular gland involved (as by a tumor) or to abnormal stimulation (or a lack of stimulation) of the gland. Since most hormones are normally present in the blood at extremely low concentrations (in the nanogram or picogram range per volume of blood sample), it has been difficult in the past to accurately measure the levels of many hormones. Today, however, even amounts as small as picograms can be accurately and easily measured using a technique known as **radioimmunoassay (RIA)**. This technique, which is now a relatively routine clinical procedure, utilizes the specificity of antibodies and the sensitivity of radioactivity measurements to measure levels of hormones that would previously have been undetectable.

Clinical Use of T_4 and T_3.
People who have had their thyroid surgically removed must receive hormone replacement therapy. Since thyroxine (T_4) is the major hormone secreted by the thyroid, most people take this hormone (in the form of pills) as replacement therapy. Thyroxine, however, becomes largely bound to proteins in the plasma, and only the relatively small amounts that are free are able to stimulate target cells (by first being converted to triiodothyronine, T_3). Under some circumstances people who need thyroid replacement therapy may take pills of T_3. These have the advantage of being more immediately active than T_4. Also, since T_3 is less protein-bound than T_4, the levels in blood decrease more rapidly when hormone replacement stops. This is desirable during radiotherapy for thyroid cancer.

Clinical Uses of Hormone Receptors.
Since receptor proteins bond with high specificity to particular hormones, the receptors can be used to measure extremely low concentrations of hormones in a technique analogous to radioimmunoassay, where antibodies are used. Some pregnancy tests, for example, utilize receptors for a hormone (human chorionic gonadotrophin, HCG) secreted by the developing embryo. In other applications the technique is reversed, and radioactively labeled hormones are used to detect the presence of receptor proteins in a tissue. This technique may be used, for example, on a mammary tumor to determine if it is able to respond to estrogen (a female sex hormone); this information is useful in determining the proper course of treatment.

Hormone Receptors and Disease.
There are some rare diseases that are known to be associated with hormone receptor proteins. In a disease called testicular feminization syndrome, for example, a male embryo lacks receptors for male sex hormones, even though it has normal and active testes. The baby, as a result, is born with female genitalia. Other diseases associated with hormone receptors include those in which antibodies are produced that bond to the particular receptors. These antibodies may mimic the normal hormone and cause stimulation of the target tissue (as in Graves' disease, where the thyroid is stimulated), or the antibodies may prevent the normal action of a hormone on its receptors. The latter situation is seen in a rare form of diabetes mellitus, where the action of insulin is blocked, and with some forms of allergy, where the effectiveness of epinephrine is decreased.

Neural Regulations

There is a tremendous number of health problems associated with congenital or acquired defects in the structure and function of the nervous system. Diseases that attack neurons or the insulating layers around nerve fibers (called myelin sheaths) can have widespread adverse effects on body function. Defects in the neural regulation of skeletal muscles, including those involved in speech, produce the most obvious symptoms of these disorders.

Certain drugs interfere with either the release of neurotransmitter molecules from nerve endings or the combination of these neurotransmitters with receptor proteins in the innervated cells. These drugs can be deadly when nerve stimulation of the muscles required for respiration is blocked. Some of these drugs, however, are put to beneficial uses by modern medicine. Curare, a deadly poison when applied to darts by South American Indians, is commonly used in a controlled fashion to promote muscle relaxation in a hospital setting. Other drugs that act to enhance the effect of a neurotransmitter are used to strengthen neuromuscular transmission when this is therapeutically appropriate.

Graves' disease: from Robert James Graves, Irish physician, 1796–1853

Chapter Summary

I. Homeostasis and Feedback Control
The regulatory mechanisms of the body maintain homeostasis by means of negative feedback mechanisms. In these mechanisms, the actions of effector organs are regulated by sensory information from the internal environment.
  A. Homeostasis is a state of dynamic constancy of the internal environment.
    1. Deviations from a normal range of values in the internal environment are detected by sensors.
    2. Sensors activate effectors, which respond to cause opposing changes that help to correct the initial deviation from the homeostatic range.
    3. Homeostasis is maintained more precisely when effectors with antagonistic actions are involved.
  B. Positive feedback mechanisms amplify the events that produce them; these mechanisms make negative feedback loops more efficient.

C. Homeostasis is maintained in part through regulation by the nervous and endocrine systems.
    1. Neural regulation is quicker but generally of shorter duration than endocrine regulation.
    2. Neurotransmitters released by nerve endings and hormones from endocrine glands combine with receptor proteins in the target cells.
II. Hormones and the Regulation of Endocrine Gland Secretion
Hormones are secreted into the blood by endocrine glands in response to specific stimuli. Hormonal stimulation of effector organs produces responses that help to complete negative feedback loops.
  A. Endocrine glands secrete hormones that are of different chemical types: amines, polypeptides, glycoproteins, and steroids.
  B. The secretion of hormones may be controlled by molecules in the blood, including other hormones, and by neurotransmitters released by nerve endings.
    1. The hypothalamus secretes hormones that regulate the anterior pituitary gland, which in turn secretes hormones that

regulate other endocrine glands.
    2. Hormonal secretion is inhibited by negative feedback loops.
III. Neural Control of Homeostasis
Nerve fibers conduct electrochemical nerve impulses that stimulate the release of chemical neurotransmitters. These events produce pathways for negative feedback regulation.
  A. The axons of sensory neurons conduct impulses into the CNS, and the axons of motor neurons conduct impulses out of the CNS and stimulate effector organs.
    1. Nerve impulses are events in which ions diffuse across the neuron membrane.
    2. When nerve impulses reach the end of an axon they stimulate the release of chemical neuro-transmitters at the synapse.
  B. Through the conduction of impulses by sensory and motor neurons, effector organs such as muscles and glands can respond to deviations in the internal environment and help maintain homeostasis.

Review Activities

Objective Questions

1. Which of the following statements about homeostasis is *true*?
   (a) The internal environment is maintained absolutely constant.
   (b) Negative feedback mechanisms act to correct deviations from a normal state.
   (c) Homeostasis is maintained by switching regulator mechanisms on and off.
   (d) All of the above are true.
2. In a negative feedback loop the effector organ produces changes that are
   (a) similar in direction to that of the initial stimulus
   (b) opposite in direction to that of the initial stimulus
3. A hormone called *parathyroid hormone* acts to help raise the blood calcium concentration. According to the principles of negative feedback, an effective stimulus for parathyroid hormone secretion would be
   (a) a fall in blood calcium
   (b) a rise in blood calcium

4. The act of breathing raises the blood oxygen level, lowers the blood carbon dioxide concentration, and raises the blood pH. According to the principles of negative feedback, sensors that regulate breathing should respond to
   (a) a rise in blood oxygen
   (b) a rise in blood pH
   (c) a rise in blood carbon dioxide concentration
   (d) all of these
5. Which of the following statements about hormones is *true*?
   (a) Hormones are enzymes.
   (b) Hormones may be amines, polypeptides, glycoproteins, or steroids.
   (c) Hormones are secreted by exocrine glands.
   (d) All of the above are true.
6. Which of the following statements about neural and endocrine regulation is *true*?
   (a) Hormonal stimulation generally lasts longer than neural stimulation.
   (b) Regulation by nerves is generally faster than regulation by hormones.
   (c) Neural and endocrine regulation are both achieved by the interaction of a chemical with a receptor protein.
   (d) All of the above are true.

Essay Questions

1. Explain the role of antagonistic negative feedback processes in the maintenance of homeostasis.
2. Compare and contrast, in a general way, regulation by the nervous and endocrine systems.
3. Explain, using examples, how the secretion of a hormone is controlled by the effects of that hormone's actions.
4. Describe the marriage of neural and endocrine regulation in the control of the anterior pituitary gland.

Selected Readings

Adolph, E. F. 1968. *Origins of physiological regulations*. New York: Academic Press.

Bernard, C. 1974. *Lectures on the phenomena of life common to animals and plants*. Springfield, IL: Charles C. Thomas.

Cannon, W. B. 1939. *The wisdom of the body*. New York: Norton and Company.

Jones, R. W. 1973. *Principles of biological regulation: an introduction to feedback systems*. New York: Academic Press.

Katz, B. November 1952. The nerve impulse. *Scientific American*.

Keynes, R. D. March 1979. Ion channels in the nerve cell membrane. *Scientific American*.

Jacobs, S., and P. Cuatrecasas. 1977. Cell receptors in disease. *New Engl. J. of Med.* 297: 1383.

Soderberg, U. 1964. Neurophysiological aspects of homeostasis. *Annual Review of Physiology* 26: 271.

Reference Figures

The Human Organism

Figure 1. Anterior view of the human torso with the superficial muscles exposed.

Figure 2. The torso with the deep muscles exposed.

Figure 3. The torso with the anterior abdominal wall removed to expose the abdominal viscera.

Figure 4. The torso with the anterior thoracic wall removed to expose the thoracic viscera.

Figure 5. The torso as viewed with the thoracic viscera sectioned in a coronal plane, and the abdominal viscera as viewed with most of the small intestine removed.

Figure 6. The torso as viewed with the heart, liver, stomach, and portions of the small and large intestines removed.

Figure 7. The torso with the anterior thoracic and abdominal walls removed, along with the viscera, to expose the posterior walls and body cavities.

Figure 1. Anterior view of the human torso with the superficial muscles exposed.

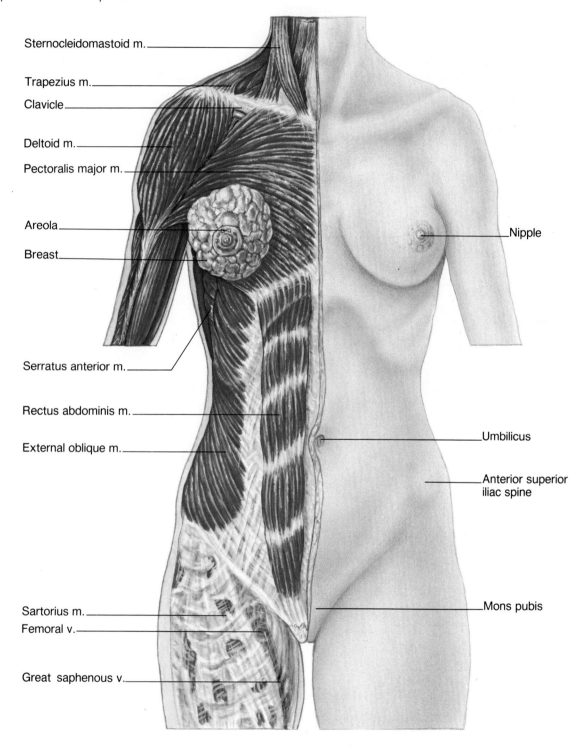

Sternocleidomastoid m.

Trapezius m.

Clavicle

Deltoid m.

Pectoralis major m.

Areola

Breast

Serratus anterior m.

Rectus abdominis m.

External oblique m.

Sartorius m.

Femoral v.

Great saphenous v.

Nipple

Umbilicus

Anterior superior iliac spine

Mons pubis

Figure 2. The torso with the deep muscles exposed.

Levator scapulae m.

Subscapularis m.

Coracobrachialis m.

Pectoralis major m.
(cut head)

Long head biceps
brachii m.

Short head biceps
brachii m.

Serratus anterior m.

Ext. intercostal m.

Latissimus dorsi m.

Rectus abdominis m.

Transversus abdominis m.

Internal oblique m.

Anterior superior
iliac spine

Femoral n.

Femoral v.

Sartorius m.

Rectus femoris m.

Sternocleidomastoid m.

Trapezius m.

External intercostal m.

Deltoid m.

Teres major m.

Pectoralis minor m.

Pectoralis major m.

External oblique m.

Linea alba

Gluteus medius m.

Tensor fasciae latae m.

Inguinal canal

Penis

Great saphenous v.

Figure 3. The torso with the anterior abdominal wall removed to expose the abdominal viscera.

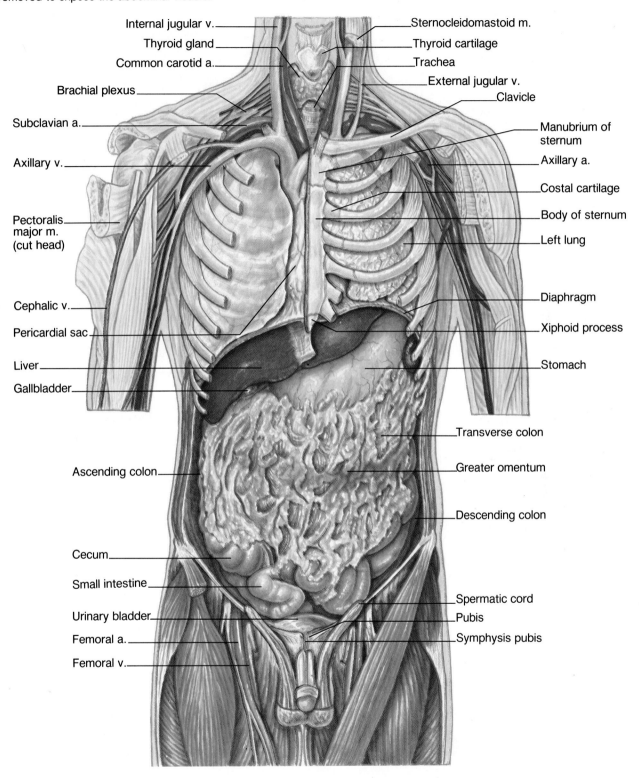

Internal jugular v.

Thyroid gland

Common carotid a.

Brachial plexus

Subclavian a.

Axillary v.

Pectoralis major m. (cut head)

Cephalic v.

Pericardial sac

Liver

Gallbladder

Ascending colon

Cecum

Small intestine

Urinary bladder

Femoral a.

Femoral v.

Sternocleidomastoid m.

Thyroid cartilage

Trachea

External jugular v.

Clavicle

Manubrium of sternum

Axillary a.

Costal cartilage

Body of sternum

Left lung

Diaphragm

Xiphoid process

Stomach

Transverse colon

Greater omentum

Descending colon

Spermatic cord

Pubis

Symphysis pubis

Figure 4.   The torso with the anterior thoracic wall removed
to expose the thoracic viscera.

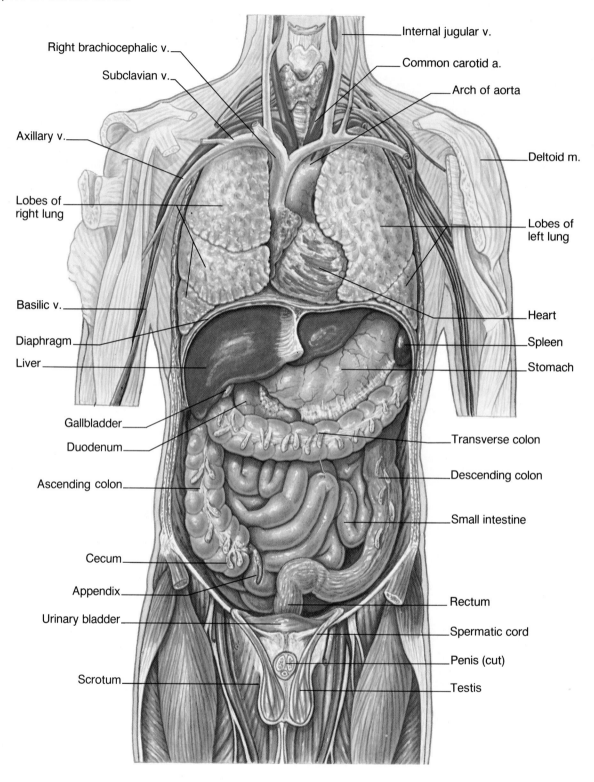

Internal jugular v.

Right brachiocephalic v.

Common carotid a.

Subclavian v.

Arch of aorta

Axillary v.

Deltoid m.

Lobes of
right lung

Lobes of
left lung

Basilic v.

Heart

Diaphragm

Spleen

Liver

Stomach

Gallbladder

Transverse colon

Duodenum

Descending colon

Ascending colon

Small intestine

Cecum

Appendix

Rectum

Urinary bladder

Spermatic cord

Penis (cut)

Scrotum

Testis

Figure 5.   The torso as viewed with the thoracic viscera sectioned in a coronal plane, and the abdominal viscera as viewed with most of the small intestine removed.

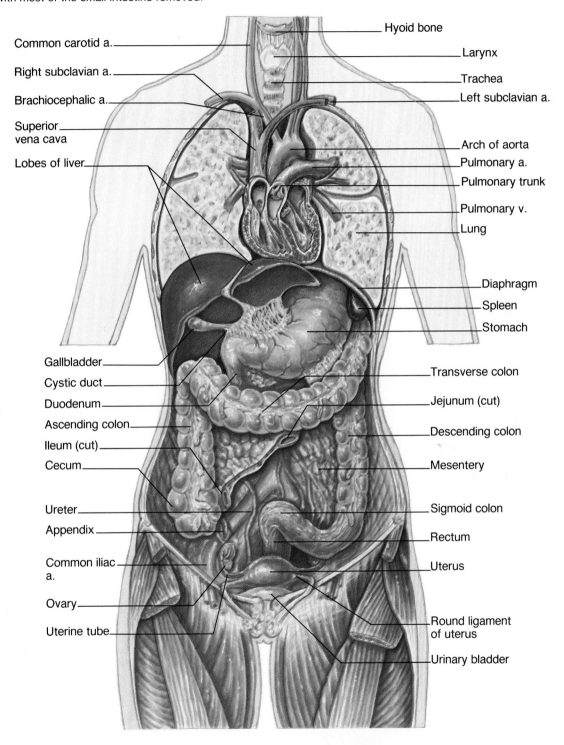

Common carotid a.

Right subclavian a.

Brachiocephalic a.

Superior vena cava

Lobes of liver

Gallbladder

Cystic duct

Duodenum

Ascending colon

Ileum (cut)

Cecum

Ureter

Appendix

Common iliac a.

Ovary

Uterine tube

Hyoid bone

Larynx

Trachea

Left subclavian a.

Arch of aorta

Pulmonary a.

Pulmonary trunk

Pulmonary v.

Lung

Diaphragm

Spleen

Stomach

Transverse colon

Jejunum (cut)

Descending colon

Mesentery

Sigmoid colon

Rectum

Uterus

Round ligament of uterus

Urinary bladder

Figure 6.   The torso as viewed with the heart, liver, stomach, and portions of the small and large intestines removed.

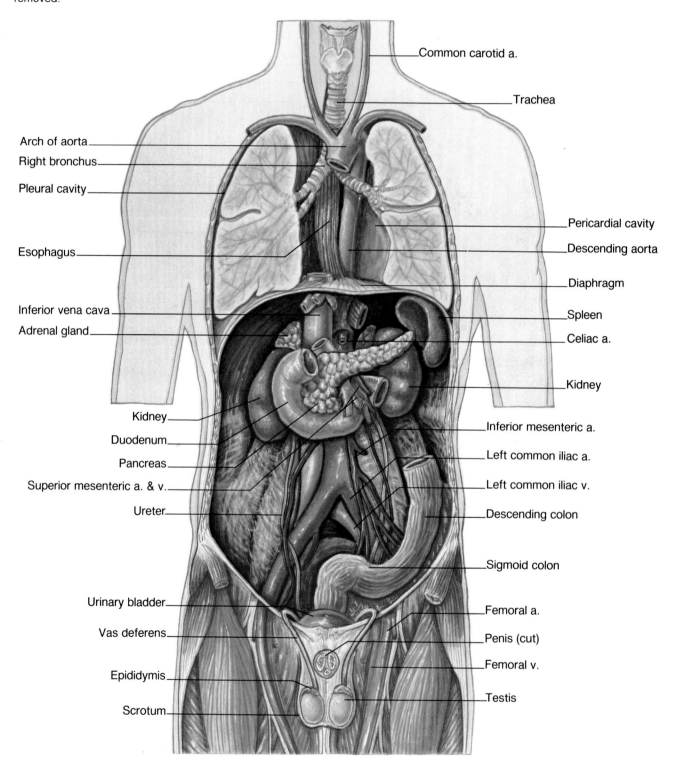

Common carotid a.

Trachea

Arch of aorta

Right bronchus

Pleural cavity

Pericardial cavity

Descending aorta

Esophagus

Diaphragm

Inferior vena cava

Spleen

Adrenal gland

Celiac a.

Kidney

Kidney

Inferior mesenteric a.

Duodenum

Left common iliac a.

Pancreas

Left common iliac v.

Superior mesenteric a. & v.

Ureter

Descending colon

Sigmoid colon

Urinary bladder

Femoral a.

Vas deferens

Penis (cut)

Femoral v.

Epididymis

Testis

Scrotum

Figure 7.   The torso with the anterior thoracic and
abdominal walls removed, along with the viscera, to expose
the posterior walls and body cavities.

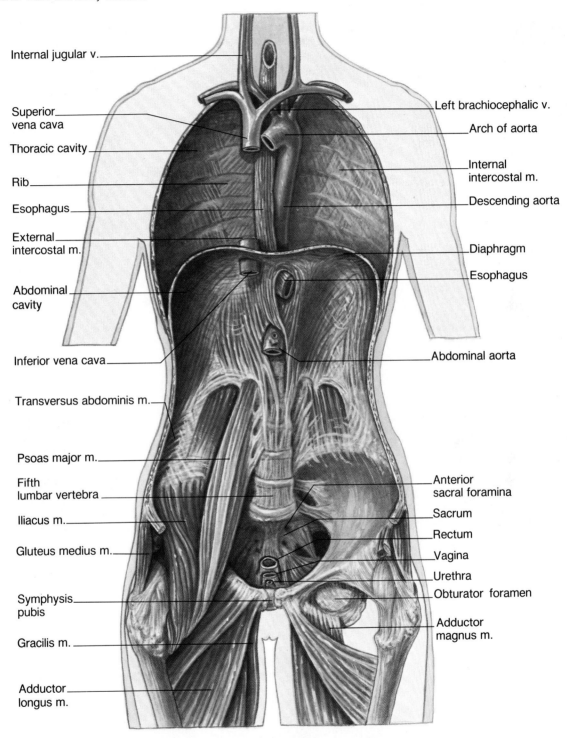

Internal jugular v.

Superior vena cava

Thoracic cavity

Rib

Esophagus

External intercostal m.

Abdominal cavity

Inferior vena cava

Transversus abdominis m.

Psoas major m.

Fifth lumbar vertebra

Iliacus m.

Gluteus medius m.

Symphysis pubis

Gracilis m.

Adductor longus m.

Left brachiocephalic v.

Arch of aorta

Internal intercostal m.

Descending aorta

Diaphragm

Esophagus

Abdominal aorta

Anterior sacral foramina

Sacrum

Rectum

Vagina

Urethra

Obturator foramen

Adductor magnus m.

Unit 2

Support and Movement

The chapters of Unit II are concerned principally with bodily support and movement. They describe how the integumentary and skeletal systems support and protect the body. The structural and functional relationships among the bones, articulations, and muscles are discussed with respect to bodily movement. The structures that can be seen or palpated on the body's surface are depicted and discussed. These chapters present the basic anatomical terminology of the integumentary, skeletal, and muscular systems and discuss how the specific organs within each of these systems function to maintain homeostasis.

The chapters and principal concepts of Unit II follow:

Chapter 8—Integumentary System
The layered integument supports hair and glands, protects the body from pathogens and injury, and is important in maintaining homeostasis.

Chapter 9—Skeletal System
The bones of the skeletal system have diagnostic features and perform the mechanical functions of support, protection, and movement, and the metabolic functions of hemopoiesis and mineral storage.

of the Human Body

Chapter 10—Articulations
The articulations between the bones of the skeleton are classified as synarthroses (immovable joints), amphiarthroses (slightly movable), or diarthroses (freely movable).

Chapter 11—Muscular System
The many skeletal muscles of the body are adapted to contract in response to electrochemical stimuli in order to carry out the functions of motion, heat production, posture, and bodily support.

Chapter 12—Surface Anatomy and Body Topography
Surface anatomy is concerned with identifying body structures through observation and palpation; it has tremendous application in physical fitness and in medical diagnosis and treatment.

8

The Integumentary

Concepts

The skin is the largest body organ, and together with the epidermal structures (hair, glands, and nails), it constitutes the integumentary system.

The integumentary system derives from ectoderm and mesoderm during embryonic development.

The outer epidermis consists of four or five layers, the dermis consists of two layers, and the hypodermis is a single layer.

The integument protects the body from pathogens and injury and is important for body homeostasis.

Hair, nails, and integumentary glands form from the epidermal layer and are ectodermal in origin. Integumentary glands are important in body defense and maintaining homeostasis.

System

The Integument as an Organ

The integument is the largest organ of the body, and together with its epidermal structures (hair, glands, and nails), it constitutes the integumentary system. It has adaptive modifications in certain body areas that accommodate protective or metabolic functions. The integument is a dynamic interface between the continually changing external environment and our internal environment, which must maintain homeostasis.

Objective 1. Explain why the integument is considered an organ and a component of the integumentary system.

Objective 2. Describe some common clinical conditions of the integument that result from nutritional deficiencies or body dysfunctions.

We are more aware of and concerned with our integumentary system than perhaps any other system of our body. One of our first actions in the morning is to examine ourselves in a mirror and assess how to make our appearance acceptable. Periodically, we examine our skin for wrinkles or gray hairs as signs of aging. We recognize other persons to a large degree by features of the skin.

Unfortunately, the appearance of the skin frequently determines first impressions and social acceptance. For example, social rejection during teenage years, imagined or real, can be directly associated with skin problems such as acne. One's image of oneself and consequent personality may be closely associated with physical appearance.

Even clothing styles are somewhat determined by how much skin we, or the designers, want to expose. But our skin is much more than a showpiece; it protects and regulates structures within the body.

The skin, or *integument (in-teg′u-ment),* and its associated structures (hair, glands, and nails) constitute the integumentary system. Included in this system are the millions of sensory receptors and the vascular network of the skin. This system accounts for approximately 7% of a person's body weight. The skin is a dynamic interface between the body and the external environment. It protects the body from the environment and at the same time allows communication with the environment.

The skin is considered an organ since it consists of several kinds of tissues that are structurally arranged to function together. It is the largest organ of the body, covering over 7,600 sq cm (3,000 sq in.) in the average adult. The skin is of variable thickness, averaging between 1.0 and 2.0 mm. It is thickest on the parts of the body exposed to wear and abrasion, such as the soles of the feet and palms of the hand, where it is about 6 mm. It is thinnest on the eyelids, external genitalia, and tympanum, where it is approximately 0.5 mm. Even the texture varies from the rough, callous skin covering the elbow and knuckle joints to the soft, sensitive areas of the eyelids, nipples, and genitalia.

---

integument: L. *integumentum,* a covering

Table 8.1 Variations of the skin and associated structures indicating nutritional deficiencies or body dysfunctions

Condition	Deficiency	Comments
General dermatitis	Zinc	Redness and itching
Scrotal or vulval dermatitis	Riboflavin	Inflammation in genital region
Hyperpigmentation	Vitamin B$_{12}$, folic acid, or starvation	Dark pigmentation occurs on backs of hands and feet
Dry, stiff, brittle hair	Protein, calories, and other nutrients	Usually in young children or infants
Follicular hyperkeratosis	Vitamin A, unsaturated fatty acids	Rough skin due to keratotic plugs from hair follicles
Pellagrous dermatitis	Niacin and tryptophan	Lesions develop on areas exposed to sun
Thickened skin at pressure points	Niacin	Noted at belt area at the hips
Spoon nails	Iron	Nails are thin and concave or spoon-shaped
Dry skin	Water or thyroid hormone	Dehydration, hypothyroidism, rough skin
Oily skin (acne)		Hyperactivity of sebaceous glands

Source: Van De Graaff, Kent M., *Human Anatomy.* © 1984 Wm. C. Brown Publishers, Dubuque, Iowa. All Rights Reserved. Reprinted by permission.

The general appearance and condition of the skin are clinically important because they provide clues to certain body conditions or dysfunctions. Pale skin may indicate shock, whereas red, flushed, overly warm skin may indicate fever and infection. A rash may indicate allergies or local infections. Abnormal textures of the skin may be the result of glandular or nutritional problems (table 8.1). Even chewed fingernails may be a clue to emotional problems.

1. Explain why the skin is considered an organ and why the skin plus the integumentary derivatives is considered a system.
2. Which vitamins and minerals are important for healthy skin?
3. Describe the appearance of the skin that may accompany each of the following conditions: allergy; shock; infection; dry, stiff hair; hyperpigmentation; and general dermatitis.

Development of the Integumentary System

All of the structures of the integumentary system are derived from ectodermal and mesodermal germ layers. Although integumentary formation is an ongoing process through prenatal development, the integument as a body covering is established early during the embryonic period.

Objective 3. Describe the development of the ectodermal and mesodermal components of the integument.

In chapter 6, you were introduced to the embryonic germ layers as they are involved in forming the various body tissues. Just as the germ layers contribute to tissue formation, they also contribute to the formation of body systems.

The skin is composed of three structural and functional layers (see figs. 8.6 and 8.7). The **epidermis** *(ep''i-der'mis)* is the thinnest, outermost layer and is derived from the embryonic *ectoderm.* The **dermis** *(der'mis),* corium, underlies the epidermis, and the **hypodermis** (subcutaneous *[sub''ku-ta'ne-us]* layer) affixes the dermis of the skin to underlying tissues or organs. Both the dermis and hypodermis derive from embryonic *mesoderm.*

Development of the Layers of the Integument

By the fourth week following conception, the embryo is surrounded by a single-cell layer of ectoderm (fig. 8.1). Just beneath the ectoderm is a thicker layer of undifferentiated mesoderm called *mesenchyme (mez'en-kim'').*

The ectodermal layer differentiates by six weeks into an outer flattened **periderm** and an inner, cuboidal **germinal (basal) layer** in contact with the mesenchyme. The periderm eventually sloughs off forming the **vernix caseosa** *(ka''se-o'sah),* a cheeselike protective coat that covers the skin of the fetus. The germinal layer gives rise to the entire epidermis as well as all the glands, nails, hair, and hair follicles of the integument.

By eleven weeks, the mesenchymal cells below the germinal cells have differentiated into the distinct collagenous and elastic connective tissue fibers of the dermis. The tensile properties of these fibers cause a buckling of the epidermis and the formation of **dermal papillae** (fig. 8.1(c)). During the early fetal period (about ten weeks), specialized *neural crest cells* called **melanoblasts** *(mel'ah-no-blastz'')* migrate into the developing dermis and differentiate into **melanocytes** *(mel'ah-no-sītz'').* The melanocytes soon migrate to the germinal layer of the epidermis (fig. 8.1(d)), where they produce *melanin (mel'ah-nin),* which colors the epidermis.

---

epidermis: Gk. *epi,* on; *derma,* skin
hypodermis: Gk. *hypo,* under; *derma,* skin
mesenchyme: Gk. *mesos,* middle; *enchyma,* infusion
periderm: Gk. *peri,* around; *derm,* skin
vernix caseosa: L. *vernix,* varnish; *caseus,* cheese
melanin: Gk. *melas,* black

Figure 8.1.   The development of the integument. (a) at four weeks only the surface ectoderm and mesenchyme of the mesoderm are apparent. (b) at seven weeks the surface ectoderm has differentiated into periderm and germinal layers. (c) by the eleventh week, the epidermis is stratified (layered) and the dermis is forming. (d) the integument at birth has distinct layers in both the epidermis and dermis. Note the development of melanocytes.

Figure 8.2.   The development of hair and the integumentary glands. (a) 12 weeks. (b) 14 weeks. (c) 16 weeks. (d) at birth.

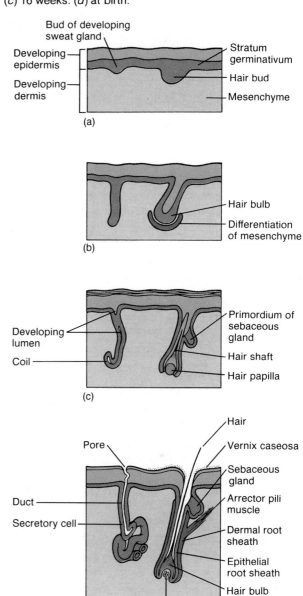

Development of the Associated Structures of the Integument

Hair, glands, and nails develop from the germinal layer of the epidermis and are therefore ectodermal in origin. Before hair can form, a hair follicle must be present. Each **hair follicle** *(fol'li-k'l)* begins to develop at about twelve weeks as a mass of germinal cells, called a **hair bud,** proliferates into the underlying mesenchyme (fig. 8.2). As the hair bud becomes club-shaped, it is known as a **hair bulb.** The **hair follicle,** which physically supports and provides

nourishment to the hair, is derived from specialized mesenchyme, called the **hair papilla** *(pah-pil'ah),* which is localized around the hair bulb, and from the epithelial cells of the hair bulb, called the **germinal matrix.** The wall of the hair follicle, known as the **epithelial root sheath,** is supported by the specialized, mesenchymal **dermal root sheath.** Continuous mitotic activity in the epithelial cells of the hair bulb results in the growth of the hair.

Figure 8.3.   The development of mammary glands.
(*a*) 12 weeks. (*b*) 14 weeks. (*c*) 16 weeks.
(*d*) about 28 weeks.

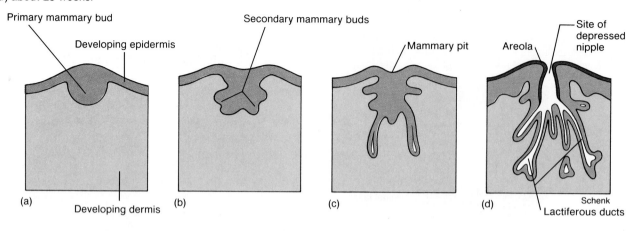

Figure 8.4.   The mammary ridge and accessory nipples.
(*a*) mammary glands are positioned along the mammary
ridges in the pectoral region. (*b*) ventral view of an embryo at
about 28 days showing location of mammary ridges.
(*c*) occasionally in humans, additional nipples (polythelia)
develop elsewhere along the mammary ridge.

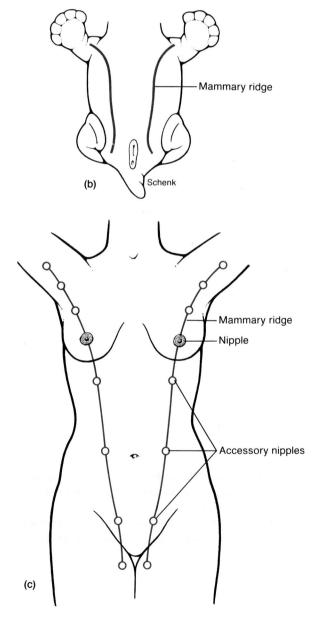

Figure 8.5. The development of nails. (*a*) 10 weeks. (*b*) about 14 weeks. (*c*) sagittal view of distal digit and nail at birth.

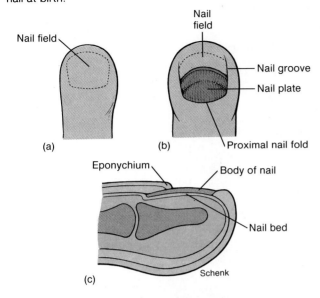

Sebaceous *(sĕ-ba'shus)* **glands** and **sweat glands** are the two principal types of integumentary glands. Both develop from the germinal layer of the epidermis. Sebaceous glands develop as proliferations from the sides of the developing hair follicle (fig. 8.2). In a mature sebaceous gland, the oily secretions empty onto the shaft of a hair in a hair follicle. Sweat glands (fig. 8.2) become coiled as the secretory portion of the developing gland proliferates into the dermal mesenchyme.

Mammary glands are modified sweat glands, which develop similar to sweat glands as downward proliferations of the germinal epithelium *(ep''i-the'le-um)* (fig. 8.3). The site of development is specific, however, along a **mammary ridge** (fig. 8.4). In humans, generally only one pair of pectoral (chest) mammary glands develop. As the mammary bud extends into the dermis, it branches (rather than coils as sweat glands) to form the secretory and ductal portions of the mammary glands (see fig. 8.3).

The mammary glands of newborns (males and females) are frequently swollen, and small amounts of secretion, called "witch's milk" may be discharged. Stimulation of the infant's mammary glands is caused by maternal hormones passing into the fetal circulation via the placenta.

Nails begin developing at about ten weeks along the dorsal aspect of each digit (fig. 8.5). The development of fingernails precedes that of toenails by about two weeks. The appearance of a thickened area of epidermis, called the **nail field,** is the first indication of nail development. The proximal and lateral borders of the nail field become thickened as the **nail folds.** Continued mitotic activity at the proximal nail fold causes a toughened **nail plate (nail)** to grow over the nail bed.

Layers of the Integument

The integument consists of three principal layers. The outer epidermis is stratified into four or five structural and functional layers, and the thick and deeper dermis consists of two layers. The hypodermis, a single layer, binds the skin to underlying tissues or organs.

Objective 4. Describe the histological characteristics of each layer of the integument.

Objective 5. Summarize the process of cellular replacement within the integument, and explain the transitional events that occur within each of the epidermal layers.

Epidermis

The epidermis is the superficial protective layer of the skin and is composed of stratified squamous epithelium. Over most of the body, this layer is between thirty and fifty cell layers thick or 0.007–0.12 mm. All but the deepest layers of the epidermis are dead cells that contain a protein called **keratin** *(ker'ah-tin),* which toughens and waterproofs the skin. The epidermis is composed of either four or five layers depending on its location within the body (figs. 8.6, 8.7). The epidermis of the palms and soles has five layers because these areas are exposed to greater friction. The epidermis of all other areas of the body has only four layers. The names and characteristics of the epidermal layer are as follows:

1. **Stratum** *(stra'tum)* **germinativum** (stratum basale). The stratum germinativum is composed of a single layer of columnarlike cells in contact with the dermis. These cells are constantly dividing mitotically and moving outward to renew the epidermis. As the cells push away from the dermis and move away from the vascular nutrient and oxygen supply, their nuclei degenerate and the cells die. In approximately six to eight weeks, the cells are sloughed off from the top layer of the epidermis.

stratum: L. *stratum,* something spread out
germinativum: L. *germinare,* sprout or growth

Figure 8.6. A diagram of the skin from the back of the hand. Note the structures in the hypodermis and dermis.

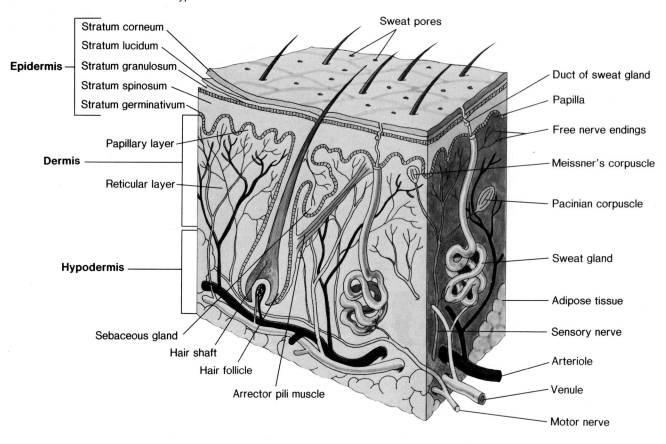

Figure 8.7. A photomicrograph of the skin from the sole of the foot (160×).

Table **8.2** Layers of the epidermis

Layer	Location	Characteristics
Stratum germinativum (basal cell layer)	Deepest layer	Consists of a single layer of columnar cells that undergo mitosis; contains pigment-producing melanocytes
Stratum spinosum	Above the stratum germinativum	Composed of several layers of cells with centrally located, large, oval nuclei and spinelike processes
Stratum granulosum	Above the stratum spinosum	Composed of one or more layers of granular cells that contain fibers of keratin and shriveled nuclei
Stratum lucidum	Above the stratum granulosum	A thin, clear layer found only in the epidermis of the palms and soles
Stratum corneum (horny layer)	Outermost layer	Consists of many layers of keratinized, dead cells that are flattened and nonnucleated; cornified

Source: Van De Graaff, Kent M., *Human Anatomy*. © 1984 Wm. C. Brown Publishers, Dubuque, Iowa. All Rights Reserved. Reprinted by permission.

2. **Stratum spinosum** (stratum malpighii). The stratum spinosum contains several stratified layers of polygonal (many-sided) cells, which are tightly attached by spinelike projections.

3. **Stratum granulosum.** The cells of the stratum granulosum are flattened and contain dark-staining granules (the source of this layer's name). The process of *keratinization (ker''ah-tin''i-za'shun),* which is associated with cellular death, is initiated in this layer.

4. **Stratum lucidum.** The nuclei, organelles, and cell membranes are no longer visible in the cells of the stratum lucidum, and so histologically this layer appears clear. This layer exists only in the thickened skin of the soles and palms.

5. **Stratum corneum** (kor'ne-um). The stratum corneum is composed of twenty-five to thirty layers of flattened, scalelike cells, which are continuously shed as flakelike residues of cells. This surface layer is cornified and is the real protective layer of the skin. *Cornification* is brought on by keratinization, and the hardening, flattening process takes place as the cells migrate to the surface.

Friction at the surface of the skin stimulates additional mitotic activity of the stratum germinativum, resulting in the formation of a *callus* for additional protection. Table 8.2 describes the specific characteristics of each epidermal layer.

Coloration of the Skin. Normal skin color is caused by the expression of a combination of three pigments: **melanin** *(mel'ah-nin),* **carotene,** and **hemoglobin** *(he''mo-glo'bin).* Melanin is a brown-black pigment formed in the pigment cells, called **melanocytes.** Melanocytes are found throughout the strata germinativum and spinosum. All races have virtually the same number of melanocytes, but the amount of melanin produced and the degree of granular aggregation of the melanin determine whether an individual's skin color is black, brown, red, tan, or white. Melanin is a protective device that guards against the damaging effect of the ultraviolet rays in sunlight. A gradual exposure to the sunlight promotes the increased production of melanin within the melanocytes and hence tanning of the skin. The skin of a genetically determined albino has the normal complement of melanocytes in the epidermis but lacks the enzyme tyrosinase, which converts the amino acid tyrosine to melanin.

Carotene is a yellowish pigment found in the epidermal cells and fatty parts of the dermis. Carotene is abundant in the skin of Oriental people and, together with melanin, accounts for their yellow-tan skin.

Hemoglobin is not a pigment of the skin but the oxygen-binding pigment found in red blood cells. Oxygenated blood flowing through the dermis gives the skin its pinkish tones.

Tattooing colors the skin permanently because pigmented dyes are injected below the mitotic germinative layer into the dermis. Because of frequent nonsterile conditions, those who administer the tatoo must take care not to introduce infections along with the dye.

Certain physical conditions or diseases cause symptomatic discoloration of the skin. Cyanosis is a bluish discoloration of the skin that appears in people with certain cardiovascular or respiratory diseases. People also become cyanotic during an interruption of breathing. In jaundice, the skin appears yellowish because of an excess of bile pigment in the bloodstream. Jaundice is usually symptomatic of liver dysfunction and sometimes of liver immaturity, as in a jaundiced newborn.

---

spinosum: L. *spina,* thorn
granulosum: L. *granum,* grain
lucidum: L. *lucidus,* light
corneum: L. *corneus,* horny

---

carotene: L. *carota,* carrot (referring to orange coloration)
hemoglobin: Gk. *haima,* blood; *globus,* globe
cyanosis: Gk. *kyanosis,* dark-blue color
jaundice: L. *galbus,* yellow

Figure 8.8. Basic dermatoglyphic patterns in the digits: (*a*) arch, (*b*) whorl, (*c*) loop, and (*d*) combination. Individual variations are commonly understood, but it is neither known why basic differences exist between men and women nor why the prints of mentally handicapped persons consistently deviate from normal patterns.

(a) (b)

(c) (d)

Surface Patterns. The exposed surface of the skin has recognizable patterns that are either congenital or acquired. Congenital patterns called fingerprints, or **friction ridges,** are present on the finger and toe pads as well as on the palms and soles. The designs formed by these lines have basic similarities but are not identical in any two individuals (fig. 8.8). They are formed by the pull of elastic fibers within the dermis and are well established prenatally. As the name implies, friction ridges function to prevent slippage when grasping objects. Because they are precise and easy to reproduce, fingerprints are customarily used for identifying individuals.

Acquired lines include the deep **flexion** *(flek'shun)* **creases** on the palms and the shallow **flexion lines** that can be seen on the knuckles and on the surface of other joints. Furrows on the forehead and face are acquired from continual contraction of facial muscles, such as from smiling or squinting in bright light or against the wind. Facial lines become more strongly delineated as a person ages.

The science known as dermatoglyphics is concerned with the classification and identification of fingerprints. As seen in figure 8.8, there are four basic dermatoglyphic patterns. Every individual's prints are unique, including those of identical twins. Fingerprints, however, are not exclusive to humans. All other primates have fingerprints, and even dogs have a characteristic "nose print" that is used for identification in the military canine corps and in certain dog kennels.

Dermis

The **dermis,** or *corium (kor'e-um),* is deeper and thicker than the epidermis (fig. 8.6). Vessels within the dermis nourish the living portion of the epidermis, and numerous collagenous, elastic, and reticular fibers give support to the skin. The fibers within the dermis radiate in definite directions and are responsible for producing lines of tension on the surface of the skin (fig. 8.9). Elastic fibers are more superficial and provide skin tone. There are considerably more elastic fibers in the dermis of a young person

corium: L. *corium,* hide or leather

Figure 8.9. Lines of tension within the skin. These lines are caused by the pull of elastic and collagen fibers within the dermis and are of special interest to surgeons. Incisions made parallel to the lines of tension heal more rapidly and create less scar tissue than transverse incisions.

Figure 8.10. Stretch marks (linea albicantes) on the abdomen of a pregnant woman. Stretch marks generally fade with time but frequently leave permanent integumentary markings.

The deeper and thicker layer of the dermis is called the **reticular layer.** Fibers within this layer are more dense and regularly arranged to form a tough, flexible meshwork. It is quite distensible, as is evident in pregnant women or obese persons, but it can be stretched too far, causing "tearing" of the dermis. The repair of a strained dermal area leaves a white streak called a stretch mark, or **linea albicans.** Lineae albicantes are frequently found on the buttocks, thighs, abdomen, and breasts (fig. 8.10).

It is the strong, resilient reticular layer of domestic mammals that is used in making leather and suede. In the tanning process, the hide of an animal is treated with various chemicals that cause the epidermis with its hair and the papillary layer of the dermis to separate from the underlying reticular layer. The reticular layer is then softened and treated with protective chemicals to make it usable.

Innervation of the Skin. The dermis of the skin has extensive innervation. Specialized integumentary *effectors* consist of muscles or glands within the dermis that respond to efferent or motor impulses transmitted from the central nervous system to the skin by autonomic nerve fibers.

Several types of **afferent** *(af'er-ent),* or **sensory, receptors** respond to various tactile (touch), pressure, temperature, tickle, or pain sensations. Some are exposed nerve endings, some form a network around hair follicles, and some extend into the papillae of the dermis. Certain areas of the body, such as the palms, soles, lips, and external genitalia, have a greater concentration of sensory receptors and are therefore more sensitive to touch. Chapter 17 contains a detailed structural and functional account of the various sensory receptors.

than in an elderly one. The decreasing amount of elastic fiber is apparently directly associated with aging. The dermis is highly vascular and glandular and contains many nerve endings and hair follicles.

Layers of the Dermis. The dermis is composed of two layers. The upper layer, called the **papillary layer,** is in contact with the epidermis and accounts for about one-fifth of the entire dermis. Numerous projections, called papillae, extend from the upper portion of the dermis into the epidermis. Papillae form the base for the friction ridges on the fingers and toes.

papilla: L. *papula,* swelling or pimple

Vascular Supply of the Skin.   Blood vessels within the dermis supply nutrients to the mitotically active stratum germinativum of the epidermis and to the cellular structures of the dermis such as glands and hair follicles. Dermal blood vessels play an important role in regulating body temperature and even blood pressure. Autonomic vasoconstriction or vasodilation responses can either shunt the blood away from the superficial dermal arterioles or permit it to flow freely throughout dermal vessels. Fever or shock can be detected by the color and temperature of the skin. Blushing is the result of involuntary vasodilation of dermal blood vessels.

A healthy circulating blood flow in debilitated bed patients is important for the prevention of bedsores, or *decubitus ulcers.* When a person lies in one position for an extended period, the dermal blood flow is restricted where the body presses against the bed. As a consequence, cells die and open wounds develop (fig. 8.11). Changing the position of the patient frequently and periodically massaging the skin to stimulate blood flow are good preventive measures against decubitus ulcers.

Hypodermis

The deepest layer of the skin is called the hypodermis, or subcutaneous layer. It binds the dermis to underlying organs. This layer is composed primarily of loose fibrous connective tissue and adipose cells interlaced with blood vessels (see fig. 8.6). Collagenous and elastic fibers reinforce the hypodermis, particularly on the palms and soles where the skin is firmly attached to underlying structures. The amount of adipose in the hypodermis varies with the sex, age, region of the body, and nutritional state of the individual. Females generally have about an 8% thicker hypodermis than do males. This layer also functions to store lipids, insulate and cushion the body, and regulate temperature.

---

1. List the layers of the epidermis and dermis, and explain how they differ in structure and function.
2. Describe the sequence of cellular replacement within the epidermis and the processes of keratinization and cornification.
3. How do both the dermis and hypodermis function in thermoregulation?
4. What two basic types of innervation are found within the dermis?

---

decubitus: L. *decumbere,* lie down
ulcer: L. *ulcus,* sore

Figure 8.11.   A decubitus ulcer on the medial surface of the ankle of the right leg. The most frequent sites for decubitus ulcers are in the skin overlying a bony projection, such as on the hip, ankle, heel, shoulder, or elbow.

Physiology of the Integument

The integument is a highly dynamic organ that not only protects the body from pathogens and external injury, but is also extremely important for body homeostasis.

Objective 6.   Discuss the role of the integument in the protection of the body from disease and external injury; the regulation of body fluids and temperature; absorption; synthesis; sensory reception; and communication.

Protection of the Body from Disease and External Injury

The integument is a physical barrier to most microorganisms, water, and excessive ultraviolet (UV) light. Oily secretions onto the surface of the skin form an acidic (pH 4.0–6.8) protective film, which waterproofs the body and retards the growth of most pathogens. The protein keratin in the epidermis also waterproofs the skin. Cornification of the outer layers of the epidermis toughens the mostly dead cells to withstand abrasion and the penetration of microorganisms. Upon exposure to UV light, the melanocytes in the lower epidermal layers are stimulated to synthesize melanin, which in turn absorbs and disperses sunlight. Surface friction causes the epidermis to thicken, as a protective response, by increasing mitosis in the cells of the stratum germinativum, which results in the formation of a protective *callus*.

Regardless of skin pigmentation, everyone is susceptible to skin cancer if his or her exposure to sunlight is sufficiently intense and continuous. There are an estimated eight hundred thousand new cases of skin cancer yearly in the United States, and approximately ninety-three hundred of these are diagnosed as the potentially life-threatening *melanoma* (cancer of melanocytes). Melanomas are usually termed malignant as they may spread rapidly. Sunscreens are advised for persons who must be in direct sunlight for long periods of time.

Regulation of Body Fluids and Temperature

Fluid Loss. Fishes submerged in fresh water are not faced with the problem of water loss from their bodies. The skin of fishes is thin and covered with mucus to protect it against pathogens and reduce friction during swimming. The skin of most amphibians is also thin and noncornified, requiring these animals to inhabit a moistened environment such as ponds or swamps. Animals that are terrestrial (land-dwelling), such as reptiles, birds, and mammals, face a serious problem of dessication (dehydration). The epidermis of the skin of terrestrial animals is adapted for continuous exposure to the air by being thickened, keratinized, and cornified. In addition, the outer layers are dead and scalelike, and a protein-polysaccharide basement membrane adheres the stratum germinativum to the dermis. Human skin is virtually waterproof, protecting the body from dessication on dry land and even from water absorption when immersed in water.

Temperature Regulation. The integument plays a crucial role in the regulation of body temperature. Body heat comes from cellular metabolism, particularly in muscle cells as they maintain tone or a degree of tension. A normal body temperature of 37°C is maintained by the antagonistic effects of sweating and shivering (see fig. 7.4) involving feedback mechanisms. Excess heat is actually lost from the body in three ways, all involving the skin: (1) through radiation from dilated blood vessels; (2) through excretion and the evaporation of sweat; and (3) through convection and the conduction of heat directly through the skin. Sweat excretion increases approximately 100–150 mL/day for each 1°C elevation in body temperature. Up to 10 L of sweat may be excreted to cool the body of a person doing hard physical work out-of-doors in the summertime.

---

amphibian: G. *amphi*, two-sided

A serious danger of continued exposure to heat and excessive water and salt loss is **heat exhaustion,** characterized by nausea, weakness, dizziness, headache, and a decreased blood pressure. **Heat stroke** is similar to heat exhaustion, with the exception that sweating is inhibited (for reasons that are not clear) and body temperature rises. Convulsions, brain damage, and death may follow.

Excessive heat loss triggers a shivering response in muscles which increases cellular metabolism. Not only do skeletal muscles contract, but tiny smooth muscles, called **arrector pili** *(ah-rek'tor pi'li),* attached to hair follicles contract involuntarily, causing goose bumps. Arrector pili muscles are extremely important in the skin of mammals that have dense fur. When arrector pili muscles are stimulated in these mammals, the individual hairs are erected, creating an insulating dead air space around the skin of these animals.

When the body's heat-producing mechanisms cannot keep pace with heat loss, *hypothermia* results. A lengthy exposure to temperatures below 20°C and dampness may lead to this condition. This is why it is so important that a hiker, for example, dress appropriately for the weather conditions, especially on cool, rainy spring or fall days. The initial symptoms of hypothermia are numbness, paleness, delirium, and uncontrolled shivering. If the core temperature falls below 32°C (90°F), the heart loses its ability to pump blood and will go into fibrillation (erratic contractions). If the victim is not warmed, extreme drowsiness, coma, and death follow.

Cutaneous Absorption

Because of the effective protective barriers of the integument already described, cutaneous (through the skin) absorption is limited. Some gases, such as oxygen and carbon dioxide may pass through the skin and enter the bloodstream. Small amounts of UV light, necessary for synthesis of vitamin D, are absorbed readily. The skin is no barrier to steroid hormones, such as cortisol, and fat-soluble vitamins (A, D, E, and K). Of clinical consideration is the fact that certain toxins and pesticides enter the body through cutaneous absorption.

Synthesis

The integumentary system synthesizes melanin and keratin, which remain in the skin, and vitamin D, which is used elsewhere in the body. The integumentary cells contain a compound called *dehydrocholesterol (de-hi"dro-ko-les'tah-rol)* from which they synthesize vitamin D in the presence of UV light. Only small amounts of UV light

Figure 8.12. (*a*) A case of rickets in a child who lives in a village in Nepal, where the people reside in windowless huts. During the five- to six-month rainy season, the children are kept indoors. (*b*) An X-ray of rickets in a ten-month-old child. Rickets develop from improper diets and also from lack of UV light in the sunlight necessary to synthesize vitamin D.

(a)

(b)

are necessary for vitamin D synthesis, but these amounts are very important to a growing child (fig. 8.12). Synthesized vitamin D enters the blood and helps regulate the metabolism of calcium and phosphorus, which are important for development of strong and healthy bones. *Rickets* is a disease caused by vitamin D deficiency.

Sensory Reception
Highly specialized sensory receptors that respond to the precise stimuli of heat, cold, pressure, touch, vibration, and pain are located abundantly throughout the dermis and hypodermis of the integument. These receptors, referred to as **cutaneous receptors,** are abundant in the skin in parts of the face, palms and fingers of the hands, soles of the feet, and the genitalia. They are less abundant along the back and on the back of the neck and are sparse in the skin over joints, especially the elbow. Generally speaking, the thinner the skin, the greater the sensitivity.

Communication
Humans are highly social animals, and the integument plays an important role in communication. Various emotions such as anger or embarrassment may be reflected in changes of skin color. Contracting specific facial muscles causes facial expressions, which convey an array of emotions, including love, surprise, happiness, sadness, or despair. Secretions from certain integumentary glands frequently elicit subconscious responses from those that detect the odors.

---

1. List five modifications of the integument that are structurally or physiologically protective.
2. Explain how the integument participates in regulating body fluids and temperature.
3. What are the four substances synthesized in the integument?

Epidermal Derivatives

Hair, nails, and integumentary glands form from the epidermal layer and are therefore of ectodermal derivation. Hair and nails are structural features of the integument and have a limited functional role, whereas integumentary glands are extremely important in body defense and maintaining homeostasis.

Objective 7. Describe the structure of hair, and list the three principal types.

Figure 8.13. The difference between male and female in the expression of hair on the body.

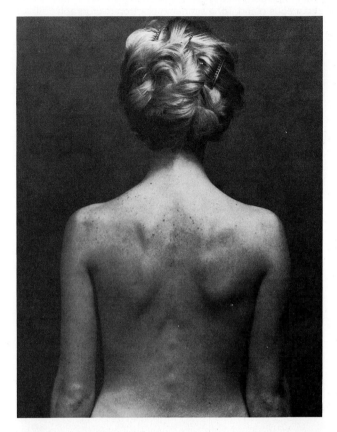

Objective 8. Discuss the structure and function of nails.

Objective 9. Compare the structure and function of the two principal kinds of integumentary glands.

Hair

Hair, or *pili (pi'li),* is characteristic of all mammals but its distribution, function, density, and texture varies among different species of mammals. Humans are relatively hairless, with only the scalp, face, pubis, and axilla being densely haired. Men and women have about the same density of hair on their bodies, but it is generally more obvious on men due to male hormones (fig. 8.13). Certain structures and regions of the body are hairless, such as the palms of the hands, soles of the feet, lips, nipples, penis, and labia minora.

The primary function of hair is protection even though its effectiveness is limited. Hair on the scalp and eyebrows protect against sunlight. The eyelashes and the hair in the nostrils protect against airborne particles. Hair on the scalp may also protect against mechanical injury. Some secondary functions of hair are to distinguish individuals and to serve as an ornamental sexual attractant.

Each hair consists of a diagonally positioned **shaft, root,** and **bulb** (fig. 8.14). The shaft is the visible, but dead, portion of the hair projecting above the surface of the skin. The bulb is the enlarged base of the root within the **hair follicle.** Each hair develops from stratum germinativum cells within the bulb of the hair, where nutrients are received from dermal blood vessels. As the cells divide, they are pushed away from the nutrient supply toward the surface, and cellular death and keratinization occur. In a healthy person, hair grows at the rate of approximately one millimeter every three days. As the hair becomes longer, however, it goes through a resting period during which it is anchored in its follicle.

The life span of a hair varies from three to four months for an eyelash to three to four years for a scalp hair. Each hair lost is replaced by a new hair that grows from the base of the follicle and pushes the old hair out. Between 10 and 100 hairs are lost each day through replacement. Baldness results when hair is lost and not replaced. This condition may be disease-related, but it is generally inherited and most frequently occurs in males because of

---

pili: L. *pilus,* hair

Figure 8.14. The structure of hair and the hair follicle. (*a*) a photomicrograph (63×) of skin showing a shaft of hair and a sebaceous gland. (*b*) a scanning electron micrograph (280×) of a hair as it extends from a follicle. (*c*) a diagram of hair, a hair follicle, a sebaceous gland, and an arrector pili muscle.

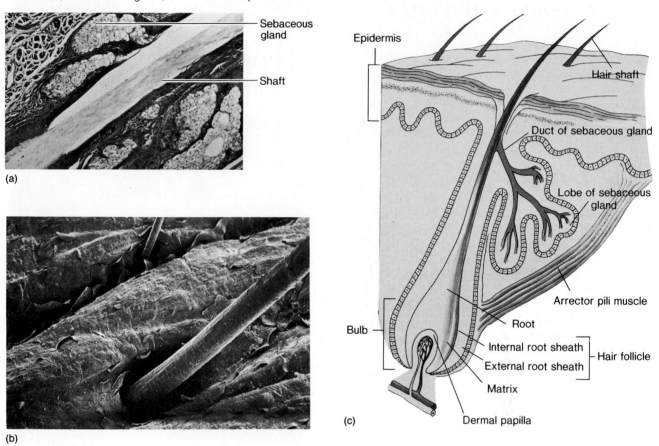

(a)

(b)

(c)

Sebaceous gland

Shaft

Epidermis

Hair shaft

Duct of sebaceous gland

Lobe of sebaceous gland

Arrector pili muscle

Bulb

Root

Internal root sheath

External root sheath

Hair follicle

Matrix

Dermal papilla

Figure 8.15. A fingertip showing the associated structures of the nail. (*a*) a diagram; (*b*) a photomicrograph of nail of neonatal human (10×).

Free edge of nail

Hyponychium

Body of nail

Nail groove

Epidermis

Dermis

Nail bed

Lunule

Eponychium

Nail root

Matrix

the action of the male sex hormone *testosterone (tes-tos'tĕ-rōn)*. No treatment is available for genetic baldness, other than grafting "plugs" of skin containing healthy follicles from hairy parts of the body to hairless regions.

Three layers can be observed in hair that is cut in cross section. An inner **medulla** *(mĕ-dul'ah)* is composed of loosely arranged cells separated by many air cells. The thick median layer, called the **cortex,** consists of hardened, tightly packed cells. A **cuticle** layer covers the cortex and forms the toughened outer portion of the hair. Cells of the cuticle have serrated edges that give a hair a scaly appearance when observed under a dissecting scope.

> People exposed to heavy metals such as lead, mercury, arsenic, or cadmium will have concentrations in the hair ten times as great as that found in their blood or urine. Because of this, hair samples may be extremely important in certain diagnostic tests.
>
> Even certain metabolic diseases or nutritional deficiencies may be detected in hair samples. For example, the hair of children with cystic fibrosis will be deficient of calcium and have excessive sodium. There is a deficiency of zinc in malnourished individuals.

Hair color is determined by the type and amount of pigment produced in the stratum germinativum at the base of the hair follicle. Varying amounts of melanin produce hair ranging in color from blond to brunette to black; the

medulla: L., marrow
cortex: L., bark
cuticle: L. *cuticula,* small skin

Matrix Nail root Eponychium Body of nail Free edge Hyponychium

(b)

Developing bone

more abundant the melanin, the darker the hair. A pigment with an iron base (trichosiderin) causes red hair. Gray and white hair is caused by air spaces within the layers of the shaft of the hair and generally accompanies aging. The texture of hair is determined by the cross-sectional shape; straight hair is round in cross section, wavy hair is oval, and kinky hair is flat.

Sebaceous glands and specialized smooth muscles, called **arrector pili,** are attached to the hair follicle (fig. 8.14). Arrector pili muscles are involuntary, responding to thermal or psychological stimuli. When these muscles contract, the hair is pulled into a more vertical position, causing goose bumps.

Humans have three distinct kinds of hair.

1. *Lanugo.* Lanugo is a fine, silky, fetal hair that appears during the last trimester of development. Lanugo is usually not evident on the baby at birth unless the baby is born prematurely.
2. *Angora.* Angora hair grows continuously in length, as on the scalp of males and females, and on the face of males.
3. *Definitive.* Definitive hair grows to a certain length and then ceases growing. It is the most common type of hair. Eyelashes, eyebrows, pubic, and axillary hair are examples.

> Anthropologists have referred to humans as the "naked apes" because of our relative hairlessness. The clothing that we wear over the exposed surface areas of our body functions to insulate and protect us as hair does in other mammals. The nakedness of our skin does lead to some problems. Skin cancer occurs frequently in humans, particularly in regions of the skin exposed to the sun. Acne, another problem unique to humans, is partly related to the fact that hair is not present to dissipate the oily secretion from the sebaceous glands.

Nails

Nails are found on the distal dorsum of each of the fingers and toes. Both fingernails and toenails serve to protect the digits, and fingernails also aid in grasping and picking up small objects. Nails form from a hardened, transparent stratum corneum of the epidermis. The hardness of the nail is due to a dense, parallel arrangement of keratin fibrils between the cells.

Each nail consists of a **body, free edge,** and **root** (fig. 8.15). The platelike body of the nail rests on a **nail bed,** which is actually the stratum spinosum of the epidermis. The body and nail bed appear pinkish because of the underlying vascular tissue. The sides of the body of the nail are protected by a **nail fold,** and the furrow between the two is the **nail groove.** The free edge is the distal exposed border, which is attached to the undersurface by the **hyponychium** *(hi''po-ni'kē-um)* (quick). The root of the nail is attached proximally.

An **eponychium** (cuticle) covers above the root of the nail. The eponychium frequently splits, causing a hangnail. The growth areas of the nail are the **matrix** and the **lunule** *(loon'ūl)*, which is the white half-moon-shaped area at the base of the nail. The nail grows by the transformation of the superficial cells of the matrix into nail cells. These harder, transparent cells are then pushed forward over the strata germinativum and spinosum of the nail bed. Fingernails grow at the rate of approximately one millimeter per week. The growth rate of toenails is somewhat slower.

The condition of nails can be an indication of the general health and personality of a person. Your nails should appear pinkish, showing the rich vascular capillaries beneath the translucent nail. A yellowish hue may indicate certain glandular dysfunctions or nutritional deficiencies. Split nails may also be caused by nutritional deficiencies. A prominent bluish tint may indicate improper oxygenation of the blood. Spoon nails (concave body) may be the result of iron-deficiency anemia, and "clubbing" at the base of the nail may be caused by lung cancer. Dirty or ragged nails may indicate poor personal hygiene, and chewed nails may suggest emotional problems.

Glands

Although they originate in the epidermal layer, all the glands of the skin are located in the dermis, where they receive physical support and nutritive sustenance. Glands of the skin are referred to as *exocrine* since they either secrete or excrete substances through ducts. The glands of the skin are of three basic types: sebaceous, sudoriferous *(soo'dor-if'er-us)*, and ceruminous *(sĕ-roo'mĭ-nus)*.

Sebaceous. Sebaceous, or oil, glands are associated with a hair follicle since they develop from the follicular epithelium of the hair. They are simple, branched glands that are connected to hair follicles, where they secrete **sebum** onto the shaft of the hair (fig. 8.14). Sebum, which consists mainly of lipids, is dispersed along the shaft of the hair to the surface of the skin, where it lubricates and waterproofs the stratum corneum layer. Sebum also prevents the hair from splitting and becoming brittle. If the drainage pathway for sebaceous glands becomes blocked for some reason, the glands may become infected, resulting in acne. Sex hormones regulate the production and secretion of sebum, and hyperactivity of sebaceous glands can result in serious acne problems, particularly during teenage years.

hyponychium: Gk. *hypo*, under; *onyx*, nail
lunule: L. *lunula*, small moon
sebum: L. *sebum*, tallow or grease

Figure 8.16. Types of skin glands.

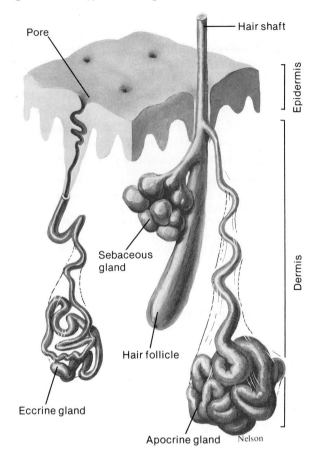

Pore, Hair shaft, Epidermis, Sebaceous gland, Dermis, Hair follicle, Eccrine gland, Apocrine gland, Nelson

Sudoriferous. Sudoriferous, or **sweat glands,** secrete perspiration, or sweat, onto the surface of the skin. Sweat glands are most numerous on the palms of the hand, soles of the feet, axillary and pubic regions, and the forehead. They are coiled and tubular shaped (fig. 8.16) and are of two types:

1. **Eccrine** *(ek'rin)* **sweat glands** are widely distributed over the body, especially on the forehead, back, palms, and soles. These glands are formed totally before birth and function in evaporative cooling in response to thermal or psychological stimuli (fig. 8.17).
2. **Apocrine** *(ap'o-krin)* **sweat glands** are much larger, localized glands found in the axillary and pubic regions and are associated with hair follicles. Apocrine glands are not functional until puberty, and their odoriferous secretion is thought to act as a sexual attractant.

Perspiration is composed of water, salts, urea, uric acid, and traces of other elements. Perspiration is therefore valuable not only for evaporative cooling, but also for the excretion of certain wastes.

sudoriferous: L. *sudorifer*, sweat; *ferre*, to bear

Figure 8.17.   A photomicrograph of an eccrine sweat gland (100×). The coiled structure of the ductule portion of the gland accounts for its irregular appearance.

Sweat gland

Mammary glands, found within the breasts, are specialized sudoriferous, or sweat, glands that secrete milk during lactation periods (see fig. 8.18). The breasts of the human female reach their greatest development during the childbearing years under the stimulus of pituitary and ovarian hormones.

Ceruminous Glands.   These highly specialized glands are found only in the external auditory meatus (ear canal). They secrete **cerumen** *(sĕ-roo'men),* or earwax, which is a water and insect repellent and also keeps the tympanum (eardrum) from drying out. Excessive amounts of cerumen may interfere with hearing.

Good routine hygiene is very important for health and social reasons. Washing away the dried residue of perspiration and sebum eliminates dirt. Excessive bathing, however, can wash off the natural sebum and dry the skin, causing it to itch or crack. The commercial lotions we use for dry skin are, for the most part, refined and perfumed lanolin, which is sebum from sheep.

cerumen: L. *cera,* wax

Figure 8.18.   Mammary gland within the breast of a human female.

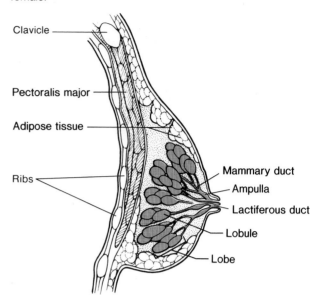

Clavicle

Pectoralis major

Adipose tissue

Ribs

Mammary duct
Ampulla
Lactiferous duct
Lobule
Lobe

1. *Diagram and label a hair. Indicate which portion is alive and discuss what causes the cells in a hair to die. What are the three principal types of hair?*
2. *Describe the structure and function of nails, and explain how nails grow and harden.*
3. *List the types of integumentary glands, and discuss the structure and function of each.*
4. *Are skin glands mesodermal or ectodermal in derivation? Are they epidermal or dermal in functional position?*

Clinical Considerations

The skin is a buffer against the external environment and is therefore subject to a variety of disease-causing microorganisms and physical assaults. A few of the many diseases and disorders of the integumentary system are briefly discussed here.

Inflammatory (Dermatitis) Conditions

Inflammatory skin disorders are caused by immunologic hypersensitivity or infectious agents. Some persons are allergic to certain foreign proteins and, because of this inherited predisposition, experience hypersensitive reactions such as asthma, hay fever, hives, drug and food allergies,

Figure 8.19. Common lesions of the skin and their usual site of occurrence.

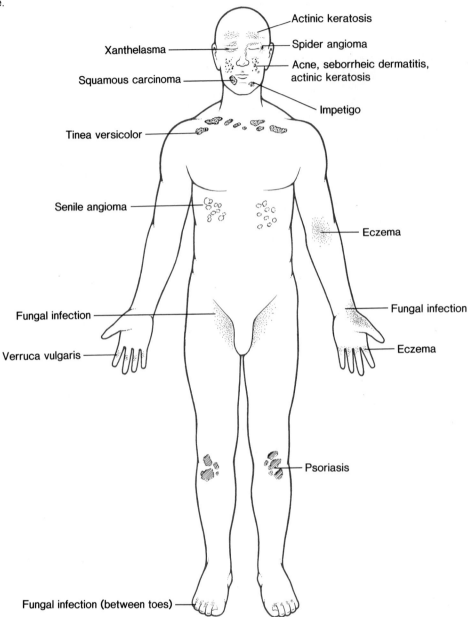

- Actinic keratosis
- Spider angioma
- Acne, seborrheic dermatitis, actinic keratosis
- Impetigo
- Xanthelasma
- Squamous carcinoma
- Tinea versicolor
- Senile angioma
- Eczema
- Fungal infection
- Fungal infection
- Verruca vulgaris
- Eczema
- Psoriasis
- Fungal infection (between toes)

and eczema. These conditions are characterized by redness, itching, and swollen vascular lesions that become dry, scaly, and crusted. **Lesions** occur commonly with skin disorders and are defined as more or less circumscribed pathologic changes in the tissue. Figure 8.19 identifies the common lesions of the skin and their usual sites.

There are a number of infectious diseases of the skin, which is not surprising considering the highly social and communal animals that we are. Effective prevention and treatment are now available for most of these diseases,

but too frequently persons do not avail themselves of safeguards or treatments. Infectious diseases of the skin include childhood viral infections (measles and chicken pox), bacteria such as staphylococcus (impetigo), venereal diseases, leprosy, fungi (ringworm, athlete's foot, candida), and mites (scabies).

Neoplasms

Both benign and malignant neoplastic conditions or diseases are common in the skin. Pigmented moles (nevi), for example, are a type of benign neoplastic growth of melanocytes. Dermal cysts and benign viral infections are also

neoplasm: Gk. *neo*, new; *plasma*, something formed
benign: L. *benignus*, good-natured
malignant: L. *malignus*, acting from malice

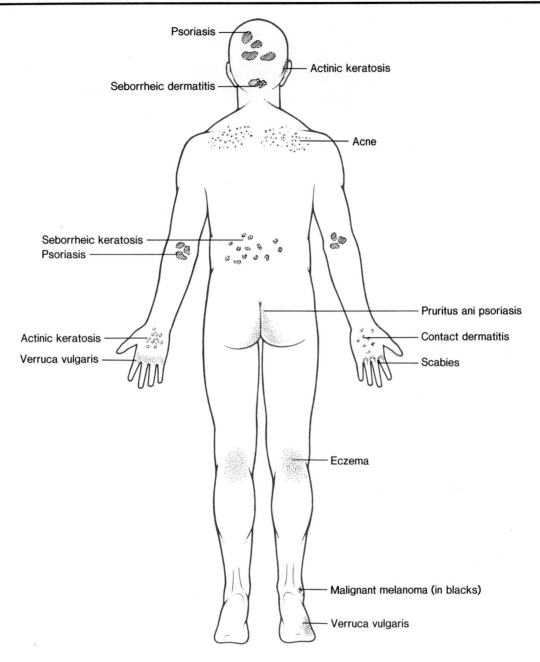

Psoriasis

Actinic keratosis

Seborrheic dermatitis

Acne

Seborrheic keratosis
Psoriasis

Pruritus ani psoriasis

Contact dermatitis

Actinic keratosis
Verruca vulgaris

Scabies

Eczema

Malignant melanoma (in blacks)

Verruca vulgaris

common. Warts are virally caused abnormal growths of tissue that occur frequently on the hands and feet. A different type of wart, called a venereal wart, occurs in the anogenital region of sexual partners. Both types of warts are easy to treat by excision or various drugs. Aging spots on elderly persons, which appear as pigmented patches on the surface of the skin, are a benign growth of melanin pigmented germinativum cells. Usually no treatment is required, unless for cosmetic purposes.

Skin cancer is the most common malignancy in the United States, but except for malignant melanomas, which arise from melanocytes within the epidermis, they are generally not life threatening. Excessive exposure to ultraviolet light from the sun is a known cause of skin cancer. The preferred treatment for this disease is complete surgical excision of the cancerous portion.

Burns

A burn is an epithelial injury caused by contact with thermal, radioactive, chemical, or electrical agents. Burns generally occur on the skin, but burns can involve the linings of the respiratory and digestive tracts. The extent and

Figure 8.20.   Skin cancer.

Neoplasms
(melanoma)

Figure 8.21.   The extent of burns, as estimated by the
"rule of nines." (*a*) anterior. (*b*) posterior.

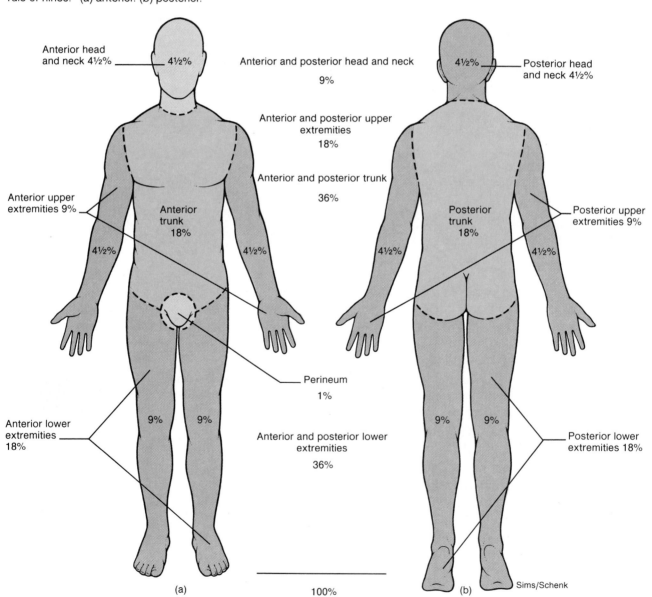

Anterior head
and neck 4½% _____ 4½%

Anterior and posterior head and neck
9%

Anterior and posterior upper
extremities
18%

Anterior and posterior trunk
36%

Anterior upper
extremities 9%

Anterior
trunk
18%

4½%   4½%

Posterior
trunk
18%

Posterior upper
extremities 9%

4½%   4½%

Posterior head
and neck 4½%

4½%

Perineum
1%

Anterior lower
extremities
18%

9%   9%

Anterior and posterior lower
extremities
36%

9%   9%

Posterior lower
extremities 18%

(a)   100%   (b)

Sims/Schenk

Figure 8.22. In a skin graft, a segment of skin is taken from a donor site (e.g., thigh) and transplanted to the recipient site.

(a)

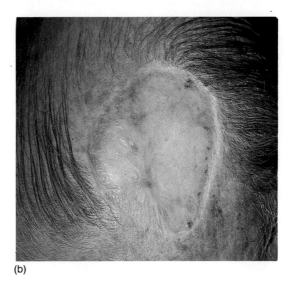

(b)

location of a burn is frequently less important than the degree to which it disrupts body homeostasis. Burns that have a **local effect** (local tissue destruction) are not as serious as those that have a **systemic effect.** Systemic effects directly or indirectly involve the entire body and are a threat to life. Possible systemic effects include body dehydration, shock, reduced circulation and urine production, and bacterial infections.

Burns are classified into three types according to their severity: first-degree, second-degree, and third-degree. In **first-degree** burns, the epidermal layers of the skin are damaged and symptoms are restricted to local effects such as redness, pain, and edema (swelling). A shedding of the surface layers (desquamation) generally follows in a few days. A sunburn is an example. **Second-degree** burns involve both the epidermis and dermis. Blisters appear and recovery is usually complete although slow. **Third-degree** burns destroy the entire thickness of the skin and frequently some of the underlying muscle. The skin appears charred and is insensitive to touch. As a result, ulcerating wounds develop and the body attempts to heal itself by forming scar tissue. Skin grafts are frequently used to assist recovery.

A procedure for estimating the extent of damaged integument suffered in burned patients is to apply the "*rule of nines*" (fig. 8.21), in which the surface area of the body is divided into regions accounting for about 9% of the area (or a multiple of 9%). Collectively, these areas account for 100% of the integumentary (body) surface. An estimation of the percent of surface area damaged is important in treating with intravenous fluid, which replaces the fluids lost from tissue damage.

Skin Grafts

If extensive areas of the stratum germinativum of the epidermis are destroyed in second-degree or third-degree burns, new skin cannot grow back. In order for this type of wound to heal, a skin graft must be performed.

A **skin graft** is a segment of skin that has been excised from a *donor site* and transplanted to the *recipient site, or graft bed.* An *autotransplant* is the most successful graft. This type of tissue transplant involves taking a thin sheet of healthy epidermis from a donor site of the burn patient and moving it to the recipient site (fig. 8.22). A *heterotransplant* (between two different species) can be a temporary treatment to prevent infection and fluid loss.

There are three types of grafts that may be used in the healing of severe skin wounds. For relatively small wounds, a *pinch graft* is used consisting of a conical piece of skin including the entire thickness of the epidermis and dermis. The epidermal area is broader than the tapered dermal area in this type of graft. A *full thickness graft,* used in extensive burns, consists of a broad sheet of skin that includes both epidermal and dermal layers. A *split thickness graft* is also used in extensive burns but includes only the upper portion of the dermis. A successful skin transplant is referred to as a *graft take.* Blood flow is usually well established by the third or fourth day in a graft take.

Recently, laboratory cultured skin from the patient's healthy epidermal cells has become available. Progress is even being made to produce synthetic skin. Both of these advancements are desirable because they minimize the problem of tissue rejection from the person's own immunity system.

Figure 8.23.   The process of wound healing. (*a*) healthy skin. (*b*) a penetrating wound into the dermis ruptures blood vessels. (*c*) blood cells, platelets, and fibrinogen flow out the wound. (*d*) a clot blocks the flow of blood. (*e*) a protective scab is formed from the clot and dried tissue fluid. (*f*) fibroblasts promote tissue granulation and regeneration. (*g*) the scab sloughs off as the epidermal layers are replaced.

Wound Healing

The skin effectively protects against many abrasions, but if a wound does happen, a sequential chain of events promotes rapid healing. The process of wound healing depends on the extent and severity of the injury. Trauma to the epidermal layers stimulates an increased mitotic activity in the stratum germinativum, whereas injuries that extend to the dermis or subcutaneous layer elicit activity throughout the body as well as within the wound itself. General body responses include a temporary elevation of temperature and pulse rate.

In an open wound (fig. 8.23), blood vessels are broken and bleeding occurs. Through the action of **blood platelets** *(plāt'letz)* and protein molecules, called **fibrinogen** *(fibrin'o-jen),* a clot forms and soon blocks the flow of blood. A scab forms and covers and protects the damaged area. Mechanisms are activated to destroy bacteria, dispose of dead or injured cells, and isolate the injured area. These responses are collectively referred to as *inflammation* and are characterized by redness, heat, edema, and pain. Inflammation is a response that confines the injury and promotes healing.

The next step in healing is the differentiation of binding **fibroblasts** from connective tissue at the wound margins. Together with new branches from surrounding blood vessels, **granulation tissue** is formed. Phagocytic cells migrate into the wound and ingest dead cells and foreign debris. Eventually the damaged area is repaired, and the protective scab is sloughed off.

If the wound is severe enough, the granulation tissue may develop into **scar tissue.** Scar tissue differs from normal skin in that its collagen fibers are more dense and it has no stratified squamous or epidermal layer. Scar tissue has fewer blood vessels and may lack hair, glands, and sensory receptors. The closer the edges of a wound, the less granulation tissue develops and the less obvious a scar. This is one reason for suturing a large break in the skin.

Important Clinical Terminology

acne   An inflammatory condition of sebaceous glands. Acne is effected by gonadal hormones and is therefore more common during puberty and adolescence. Pimples and blackheads on the face, chest, and back are expressions of this condition.

albinism   A congenital, genetic deficiency of the pigment of the skin, hair, and eyes due to a metabolic block in the synthesis of melanin.

alopecia   Loss of hair, baldness. Baldness is usually due to genetic factors and cannot be treated. Baldness may signify anatomical maturity.

athlete's foot *(tinea pedis)*   A skin fungus disease of the foot.

blister   A collection of fluid between the epidermis and dermis, caused by excessive friction or a burn.

boil (furuncle)   A localized bacterial infection originating in a hair follicle or skin gland.

callus   A localized buildup of the stratum corneum layer of the epidermis due to excessive friction.

carbuncle   A bacterial infection similar to a boil except that it infects the subcutaneous tissues.

comedo   A plug of sebum and epithelial debris in the hair follicle and excretory duct of the sebaceous gland. Also called a blackhead or whitehead.

corn   A type of callus that is localized on the foot, usually over toe joints.

dandruff   Common dandruff is the continual shedding of epidermal cells of the scalp, which can be removed by normal washing and brushing of the hair. Abnormal dandruff may be caused by certain skin diseases such as seborrhea, psoriasis, and so on.

decubitus ulcer   A bedsore. An exposed ulcer caused by a continual pressure that restricts dermal blood flow to a localized portion of the skin.

dermatitis   An inflammation of the skin.

dermatology   A specialty of medicine concerned with the study of the skin, its anatomy, physiology, histopathology, and the relationship of cutaneous lesions to systemic disease.

dermatome   An instrument for cutting thin slices of skin for grafting or excising small lesions.

eczema *(eg′ze-mah)*   A noncontagious inflammatory condition of the skin producing red, itching, vesicular lesions, which may be crusty or scaly.

erythema *(er″i-the′mah)*   Redness of the skin caused by vasodilation from skin injury, infection, or inflammation.

furuncle   A boil; a localized abscess resulting from an infected hair follicle.

gangrene   Necrosis of tissue due to the obstruction of blood flow. It may be localized or extensive and may secondarily be infected with anaerobic microorganisms.

keratosis   Any abnormal growth and hardening of the stratum corneum layer of the skin.

melanoma   A cancerous tumor originating from proliferating melanocytes within the epidermis of the skin.

nevus *(ne′vus)*   A mole or birthmark; congenital pigmentation of a certain area of the skin.

papilloma   A benign epithelial neoplasm, such as a wart or a corn.

papule   A small inflamed elevation of the skin, such as a pimple.

psoriasis *(so-ri′ah-sis)*   An inherited inflammatory skin disease usually expressed as circular scaly patches of skin.

pustule   A small, localized elevation of the skin containing pus.

seborrhea *(seb″o-re′ah)*   A disease characterized by an excessive activity of the sebaceous glands and accompanied by oily skin and dandruff.

ulcer   An open sore. A decubitus ulcer denotes a bedsore.

urticaria (hives)   A skin eruption of reddish weals, usually with extreme itching. It may be caused by an allergic reaction, stress, or contact with some external or internal precipitating factor.

wart   A horny projection of epidermal cells caused by a virus.

Chapter Summary

I. The Integument as an Organ
The skin is the largest body organ, and together with the epidermal structures (hair, glands, and nails) it constitutes the integumentary system.
  A. The skin is considered an organ because it consists of several kinds of tissues that are structurally arranged to function together.
  B. The appearance of the skin is clinically important because it provides clues to certain body conditions or dysfunctions.
II. Development of the Integumentary System
The integumentary system derives from ectoderm and mesoderm during embryonic development.
  A. Epidermal layers, hair, nails, and cutaneous glands, including mammary glands, derive from ectoderm that specializes into a mitotically active germinal layer.
    1. Hair and sebaceous glands develop together during the formation of a hair follicle.
    2. Sweat glands develop independently from the germinal layer of the epidermis.
    3. Mammary glands are modified sweat glands that develop along a mammary ridge.
    4. Nails begin development at ten weeks at the nail fields.
    5. During early fetal development, melanoblasts migrate to the germinal layer and give rise to melanocytes, which produce melanin.
  B. The dermis and hypodermis derive from mesoderm. Specialized mesoderm, called mesenchyme, gives rise to specific dermal structures.
III. Layers of the Integument
The outer epidermis consists of four or five layers, the dermis consists of two layers, and the hypodermis is a single layer.

  A. The stratified squamous epithelium of the epidermis is divisible into five structural and functional layers: stratum germinativum, stratum spinosum, stratum granulosum, stratum lucidum, and stratum corneum.
    1. Normal skin color is due to a combination of melanin and carotene in the epidermis and hemoglobin in the blood of the dermis and hypodermis.
    2. Fingerprints on the surface of the epidermis are individually unique; flexion creases and flexion lines are acquired.
  B. The thick dermis of the skin is composed of fibrous connective tissue interlaced with elastic fibers. The two layers of the dermis are the upper papillary layer and the deeper reticular layer.
  C. The hypodermis, composed of adipose and fibrous connective tissue, binds the dermis to underlying organs.
IV. Physiology of the Integument
The integument protects the body from pathogens and injury and is important for body homeostasis.
  A. Physical features of the skin provide protection of the body from disease and external injury.
    1. Keratin and an acidic oily secretion on the surface protect the skin from water and microorganisms.
    2. Cornification of the skin protects against abrasion.
    3. Melanin is a barrier to UV light.
  B. The skin regulates body fluids and temperatures.
    1. Fluid loss is minimal due to keratinization and cornification.
    2. Temperature regulation is maintained by radiation, convection, and the antagonistic effects of sweating and shivering.
  C. The skin permits the absorption of UV light, respiratory gases, steroids, fat-soluble vitamins, and certain toxins and pesticides.
  D. The integument synthesizes melanin and keratin, which remain in the skin, and vitamin D, which is used elsewhere in the body.

  E. Sensory reception in the skin is provided through cutaneous receptors throughout the dermis and hypodermis. Cutaneous receptors respond to precise sensory stimuli and are more sensitive in thin skin.
  F. Certain emotions are communicated through the integument.
V. Epidermal Derivatives
Hair, nails, and integumentary glands form from the epidermal layer and are ectodermal in origin. Integumentary glands are important in body defense and maintaining homeostasis.
  A. Hair is characteristic of all mammals but its distribution, function, density, and texture varies among species of different mammals.
    1. Each hair consists of a shaft, root, and bulb, which is positioned in a hair follicle.
    2. The three hair layers are the medulla, cortex, and cuticle.
    3. Lanugo, angora, and definitive are the three kinds of human hair.
  B. Hardened, keratinized nails are found on the distal dorsum of each digit, where they protect the digits; fingernails aid in grasping and picking up small objects.
    1. Each nail consists of a body, free edge, and a root.
    2. The hyponychium, eponychium, and nail fold support the nail on the nail bed.
  C. Integumentary glands are exocrine since they either secrete or excrete substances through ducts.
    1. Sebaceous glands secrete sebum onto the shaft of the hair.
    2. The two types of sudoriferous (sweat) glands are eccrine and apocrine.
    3. Mammary glands are specialized sudoriferous glands that secrete milk during lactation.
    4. Ceruminous glands secrete cerumen (earwax).

Review Activities

Objective Questions

1. Hair, nails, integumentary glands, and the epidermis of the skin are derived from embryonic
   - (a) ectoderm
   - (b) mesoderm
   - (c) endoderm
   - (d) mesenchyme

2. Spoon-shaped nails may result when a person has a dietary deficiency of
   - (a) zinc
   - (b) iron
   - (c) niacin
   - (d) vitamin B_{12}

3. The epidermal layer *not* present in the thin skin of the face is the stratum
   - (a) granulosum
   - (b) lucidum
   - (c) spinosum
   - (d) corneum

4. Which of the following does *not* contribute to skin color?
   - (a) keratin
   - (b) melanin
   - (c) carotene
   - (d) hemoglobin

5. Integumentary cells synthesize vitamin D in the presence of UV light and
   - (a) keratin
   - (b) cortisol
   - (c) trichosiderin
   - (d) dehydrocholesterol

6. Integumentary glands that empty their secretions into hair follicles are
   - (a) sebaceous glands
   - (b) sudoriferous glands
   - (c) eccrine glands
   - (d) ceruminous glands

7. Fetal hair that is present during the last trimester of development is referred to as
   - (a) angora
   - (b) definitive
   - (c) lanugo
   - (d) replacement

8. Which of these conditions is potentially life threatening?
   - (a) acne
   - (b) melanoma
   - (c) eczema
   - (d) seborrhea

9. The skin of a burn victim has been severely damaged through the epidermis and into the dermis; integumentary regeneration will be slow with some scarring but complete. What degree burn is it?
   - (a) first
   - (b) second
   - (c) third

10. The technical name for a blackhead or whitehead is
    - (a) carbuncle
    - (b) melanoma
    - (c) nevus
    - (d) comedo

Essay Questions

1. Discuss the development of the skin and its associated hair, glands, and nails. What role do the ectoderm and mesoderm play in integumentary development?

2. What are some physical and chemical features of the skin that make it an effective protective organ?

3. Of what practical value is it for the outer layers of the epidermis and hair to be composed of dead cells?

4. Contrast the structure and function of sebaceous, sudoriferous, mammary, and ceruminous glands.

5. Discuss what is meant by an inflammatory lesion. What are some frequent causes of skin lesions?

Selected Readings

Bloom, W., and D. W. Fawcett. 1975. *A textbook of histology.* 10th ed. Philadelphia: W. B. Saunders Co.

Bluefarb, S. M. 1974. *Dermatology.* Kalamazoo, Mich.: The Upjohn Co.

Elden, H. R., ed. 1971. *Biophysical properties of the skin.* New York: John Wiley and Sons.

Edelson, R. L., and J. M. Fink. June 1985. The immunologic function of skin. *Scientific American.*

Helwig, E. B., and F. K. Mostofi, eds. 1971. *The skin.* Baltimore: The Williams and Wilkins Co.

Loomis, W. F. December 1970. Rickets. *Scientific American.*

Marples, M. J. 1965. *The ecology of the human skin.* Springfield, Ill.: Charles C. Thomas, Publisher.

Marples, M. J. January 1979. Life on the human skin. *Scientific American.*

Moncrief, J. A. 1973. Burns. *New Engl. J. Med.* 228:444.

Montagna, W. 1962. *The structure and function of the skin.* 2d ed. New York: Academic Press.

Montagna, W. June 1969. The skin. *Scientific American.*

Pawelek, J. M., and A. M. Korner. March–April 1982. The biosynthesis of mammalian melanin. *American Scientist.*

Penrose, L. S. December 1969. Dermatoglyphics. *Scientific American.*

Ross, R. June 1969. Wound healing. *Scientific American.*

Rushmer, R. L., et al. 1966. The skin. *Science* 154:343.

Tregear, R. T. 1966. *Physical functions of skin.* New York: Oxford University Press.

9

Skeletal System

Concepts

The axial and appendicular components of the skeletal system of an adult human consists of 206 individual bones arranged into a strong, flexible body framework.

The bones of the skeleton perform the mechanical functions of support, protection, and body movement and the metabolic functions of hemopoiesis and mineral storage.

All bones derive from specialized mesenchymal mesoderm through endochondral ossification or through intramembranous ossification. Ossification begins by the fourth week of prenatal development and is not completed in certain bones until about age twenty-five or thirty.

Although bones have commonalities of histological structure, each bone has a characteristic pattern of ossification and growth, a characteristic shape, and diagnostic surface features that indicate its functional relationship to other bones, muscles, and to the body structure as a whole.

The skull consists of diagnostic cranial and facial bones that contain several cavities and house the brain.

The twenty-six vertebrae of the vertebral column, separated by intervertebral discs, lend flexibility to the axial skeleton and absorb the stress of movement.

The flexible rib cage, consisting of the thoracic vertebrae, twelve paired ribs, costal cartilages, and the sternum, encloses and protects the thoracic viscera and is involved in breathing.

The pectoral girdle and upper extremities are adapted for freedom of movement and extensive muscle attachment.

The pelvic girdle and lower extremities are adapted for support and locomotion.

Organization of the Skeletal System

The axial and appendicular components of the skeletal system of an adult human consists of 206 individual bones arranged into a strong, flexible body framework.

> *Objective 1.* Describe the structural organization of the skeletal system, and list the bones of the axial and appendicular portions.

Osteology is the science concerned with the study of bones. Each bone is an organ that plays a part in the total functioning of the skeletal system. The skeletal system of an adult human is composed of approximately 206 bones. Actually, the number of bones differs from person to person depending on age and genetic variations. At birth, the skeleton consists of approximately 270 bones. As further bone development (ossification) occurs during infancy, the number increases. During adolescence, however, the number of bones decreases, as there is a gradual fusion of separate bones.

Some adults have extra bones within the joints (sutures) of the skull called **sutural** *(soo'cher-ahl),* or **Wormian, bones.** Additional bones may develop in tendons in response to stress as the tendons repeatedly move across a joint. Bones formed this way are called **sesamoid** *(ses'ah-moid)* **bones.** Sesamoid bones, like the sutural bones, vary in number. The patellae (kneecaps) are two sesamoid bones present in all people.

For the convenience of study, the skeleton is divided into axial and appendicular portions. Anterior and posterior views of the skeleton are shown in figure 9.1. Table 9.1 lists the divisions of the skeleton and the number of bones in each portion.

The **axial skeleton** consists of the bones that form the axis of the body and that support and protect the organs of the head, neck, and torso:

1. **Skull.** The skull consists of two sets of bones: the cranial bones that form the cranium, or braincase, and the facial bones that support the eyes, nose, and jaws.
2. **Auditory ossicles.** Three auditory ossicles are present in the middle ear chamber of each ear and serve to transmit sound impulses.
3. **Hyoid** *(hi'oid)* **bone.** The hyoid bone is located above the larynx and below the lower jaw. It supports the tongue and assists in swallowing.

---

sesamoid: Gk. *sesamon,* like a sesame seed
ossicle: L. *ossiculum,* little bone

Figure 9.1.   (*a*) an anterior view and (*b*) a posterior view of the human skeleton. The bones of the axial skeleton are shaded to differentiate them from the bones of the appendicular skeleton.

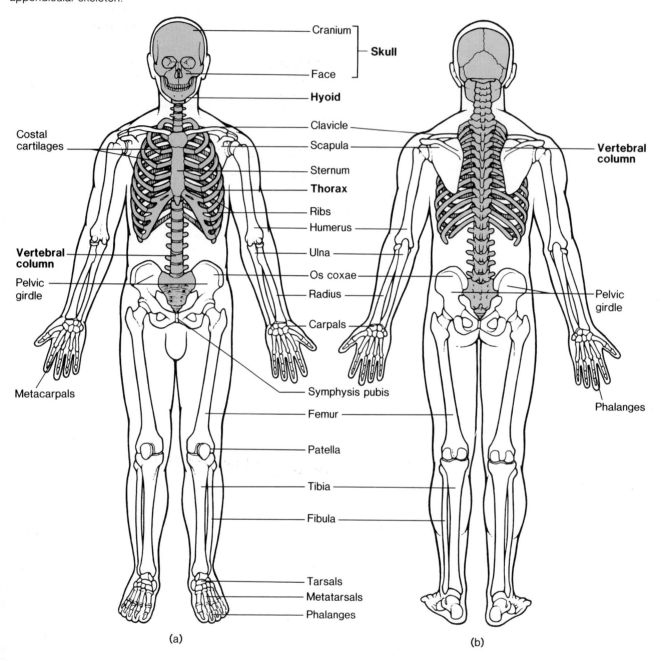

(a)   (b)

4. **Vertebral column.** The vertebral column (backbone) consists of twenty-six individual vertebrae separated by cartilaginous intervertebral discs. In the pelvic region several vertebrae are fused to form the sacrum, which provides attachment for the pelvic girdle. A few terminal vertebrae are fused to form the coccyx, the so-called tailbone.

5. **Rib cage.** The rib cage, or thoracic cage, forms the bony and cartilaginous framework of the thorax. The rib cage articulates posteriorly with the thoracic vertebrae and includes the twelve pairs of **ribs,** the flattened **sternum,** and the **costal cartilages,** which connect the ribs to the sternum on the anterior side.

Table 9.1 Classification of the bones of the adult skeleton

I. Axial skeleton		
a. Skull		22 bones
	8 cranial bones:	
	frontal 1	
	parietal 2	
	occipital 1	
	temporal 2	
	sphenoid 1	
	ethmoid 1	
	13 facial bones:	
	maxilla 2	
	palatine 2	
	zygomatic 2	
	lacrimal 2	
	nasal 2	
	vomer 1	
	inferior nasal concha 2	
	1 mandible	
b. Middle ear ossicles		6 bones
	malleus 2	
	incus 2	
	stapes 2	
c. Hyoid		1 bone
	hyoid bone 1	
d. Vertebral column		26 bones
	cervical vertebra 7	
	thoracic vertebra 12	
	lumbar vertebra 5	
	sacrum 1 (5 fused bones)	
	coccyx 1 (3–5 fused bones)	
e. Rib cage		25 bones
	rib 24	
	sternum 1	
II. Appendicular skeleton		
a. Pectoral girdle		4 bones
	scapula 2	
	clavicle 2	
b. Upper extremities		60 bones
	humerus 2	
	radius 2	
	ulna 2	
	carpal 16	
	metacarpal 10	
	phalanx 28	
c. Pelvic girdle		2 bones
	os coxa 2 (each os coxa contains 3 fused bones)	
d. Lower extremities		60 bones
	femur 2	
	tibia 2	
	fibula 2	
	patella 2	
	tarsal 14	
	metatarsal 10	
	phalanx 28	
	Total	**206 bones**

Source: Van De Graaff, Kent M., *Human Anatomy.* © 1984 Wm. C. Brown Publishers, Dubuque, Iowa. All Rights Reserved. Reprinted by permission.

The **appendicular skeleton** is composed of the bones of the upper and lower extremities and the bony girdles, which anchor the appendages to the axial skeleton:

1. **Pectoral girdle.** The paired **scapulae** and **clavicles** constitute the pectoral girdle. It is not a complete girdle, having only an anterior attachment to the skeleton at the sternum via the clavicles. The primary function of the pectoral girdle is to provide attachment for the muscles that move the brachium and forearm.

2. **Upper extremities.** Each upper extremity consists of a proximal **humerus** within the brachium, an **ulna** and **radius** within the forearm, the **carpal** bones of the wrist, and the **metacarpal** and **phalangeal** (*fah-lan' je-al*) bones of the hand.

3. **Pelvic girdle.** The pelvic girdle is formed by two **ossa coxae** (hipbones) united anteriorly by the **symphysis** (*sim' fĭ-sis*) **pubis** and posteriorly by the sacrum and coccyx of the vertebral column. The pelvic girdle supports the weight of the body through the vertebral column and protects the lower viscera within the pelvic cavity.

4. **Lower extremities.** Each lower extremity consists of a proximal **femur** within the thigh, a **tibia** and **fibula** within the lower leg, the **tarsal** bones of the ankle, and the **metatarsal** and **phalangeal** bones of the foot. In addition, the **patella** (*pah-tel'ah*) is located on the anterior surface of the knee joint between the thigh and lower leg regions.

---

1. List the bones of the body which you can palpate on yourself. Indicate which are of the axial skeleton and which are of the appendicular skeleton.
2. What are sesamoid bones and where are they found?
3. Describe the locations and functions of the pectoral and pelvic girdles.

Functions of the Skeletal System

The bones of the skeleton perform the mechanical functions of support, protection, and body movement and the metabolic functions of hemopoiesis and mineral storage.

Objective 2. Discuss the principal functions of the skeletal system, and identify the body systems that these functions serve.

The strength of bone comes from its inorganic components, which resist decomposition even after death. Much of what we know of prehistoric animals, including man, has been determined from preserved skeletal remains. Frequently when we think of bone, we think of a hard, dry structure. In fact, the term *skeleton* comes from a Greek word meaning "dried up." Living bone is not inert material, however; it is dynamic and adaptable in performing many body functions, including support, protection, body movement, hemopoiesis, and mineral storage.

1. **Support.** The skeleton forms a rigid framework to which are attached the softer tissues and organs of the body.

2. **Protection.** The skull and vertebral column enclose the central nervous system; the rib cage protects the heart, lungs, great vessels, liver, and spleen; and the pelvic cavity supports and protects the pelvic viscera. Even the site where red blood cells are produced is protected within the central, hollow portion of the bone.

3. **Body movement.** Bones serve as anchoring attachments for most skeletal muscles. In this capacity, the bones act as levers with the joints functioning as pivots when muscles contract to cause body movement.

4. **Hemopoiesis** *(he″mo-poi-e′sis).* The red bone marrow of an adult produces white blood cells, red blood cells, and platelets. In an infant, the spleen and liver produce red blood cells, but as the bones mature, the bone marrow assumes the performance of this formidable task. It is estimated that an average of one million red blood cells are produced every second by the bone marrow to replace those that are worn out and destroyed by the liver.

5. **Mineral storage.** The inorganic matrix of bone is composed primarily of the minerals calcium and phosphorus. These minerals give bone its rigidity and account for approximately two-thirds of the weight of bone. About 95% of the calcium and 90% of the phosphorus within the body are deposited in the bones and teeth. Although the concentration of these organic salts within the blood is kept within narrow limits, both of these mineral salts are essential for other body functions. Calcium is necessary for muscle contraction, blood clotting, and the movement of ions and nutrients across plasma membranes of cells. Phosphorus is required for the activities of the nucleic acids, DNA and RNA, as well as for ATP utilization. If mineral salts are not present in the diet in sufficient amounts, they may be withdrawn from the bones until they are replenished through proper nutrition. In addition to calcium and phosphorus, lesser amounts of magnesium and sodium salts are stored in bone tissue.

Vitamin D assists in the absorption of calcium and phosphorus from the intestine into the blood. As the bones develop in a child, it is immensely important that its diet contains an adequate amount of these two minerals and vitamin D. If the diet is deficient in these essentials, the blood level falls below that necessary for calcification and a condition known as **rickets** develops. Rickets is characterized by soft bones that may result in bowlegs and malformation of the head, chest, and pelvic girdle.

In summary, the skeletal system is not an isolated body system. It is closely associated with the muscle system since it stores the calcium needed for muscular contraction and provides an attachment for muscles as they span the movable joints. The skeletal system serves the circulatory system by producing blood cells in protected sites. Also, many of the vessels of the circulatory system are named according to the bones they parallel. The skeletal system supports and protects, to varying degrees, all of the systems of the body.

1. List the functions of the skeletal system.
2. Discuss two ways the skeletal system serves the circulatory system in the production of blood. What two ways does it serve the muscular system?

Development of the Skeletal System

All bones derive from specialized mesenchymal mesoderm through endochondral ossification or through intramembranous ossification. Ossification begins by the fourth week of prenatal development and is not completed in certain bones until about age twenty-five or thirty.

Objective 3. Distinguish between endochondral ossification and intramembranous ossification.
Objective 4. Explain the processes of axial and appendicular skeletal development.
Objective 5. Describe the fontanels of a fetal skull, and explain their importance.

Bone formation, or ossification, begins about the fourth week of embryonic development, but ossification centers cannot be observed until about the eighth week (fig. 9.2). Bone tissue derives from specialized migratory cells known as *mesenchyme.* Some of the embryonic mesenchymal cells will transform into *chondroblasts (kon′dro-blastz)* and develop a cartilage matrix that is later replaced by bone in a process known as **endochondral** *(en″do-kon′dral)* **ossification.** Most of the skeleton is formed in this fashion—first it goes through a hyaline cartilage stage and then ossifies as bone.

A smaller number of mesenchymal cells develop directly into bone without first going through a cartilage stage. This type of bone formation process is referred to as **intramembranous** *(in″trah-mem′brah-nus)* **ossification.** Facial bones and certain bones of the cranium are formed this way. **Sesamoid bones** are specialized intramembranous bones that develop in tendons.

Figure 9.2. Ossification centers of the skeleton of a 10-week-old fetus. (a) the diagram depicts endochondral ossification in solid dark colors and membranous ossification in a stippled pattern. (b) the photograph shows the ossification centers stained with an indicator dye.

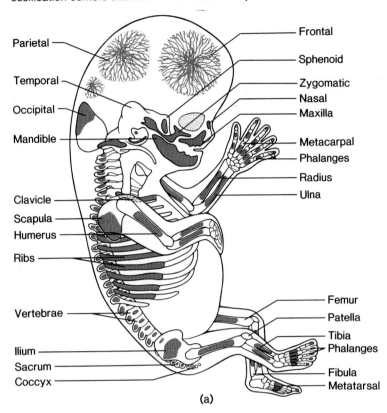

Parietal
Temporal
Occipital
Mandible
Clavicle
Scapula
Humerus
Ribs
Vertebrae
Ilium
Sacrum
Coccyx

Frontal
Sphenoid
Zygomatic
Nasal
Maxilla
Metacarpal
Phalanges
Radius
Ulna
Femur
Patella
Tibia
Phalanges
Fibula
Metatarsal

(a)

(b)

Development of the Axial Skeleton

Development of the Vertebral Column and Rib Cage.

The development of the vertebral column is a sequential process involving embryonic somites and the notochord. **Somites** are paired blocklike condensations of mesoderm that form on both sides of the neural tube toward the end of the third week of embryonic development (fig. 9.3). The **notochord** is a flexible rod of mesodermal tissue that extends the length of the back of an embryo (see fig. 1.8).

During the fourth week of development the somites differentiate into three separate components (fig. 9.4), which give rise to specific structures: the dorsolateral **dermatome** gives rise to the dermis of the skin; the medial **myotome** forms most of the skeletal muscle of the body; and the ventromedial **sclerotome** *(skle′rah-tōm)* forms the **vertebral column** and certain other bones of the body.

At about twenty-eight days of development, the sclerotome portion of the somites begins migrating in three main directions. (1) Mesenchymal cells migrate ventromedially to surround the notochord and form the bodies

dermatome: Gk. *dermia*, skin; *tome*, a cutting
myotome: Gk. *myos*, muscle; *tome*, a cutting
sclerotome: Gk. *scleros*, hard; *tome*, a cutting

Figure 9.3. A dorsal view of a 28-day-old embryo showing the position of the somites lateral to the neural tube forming along the neural groove.

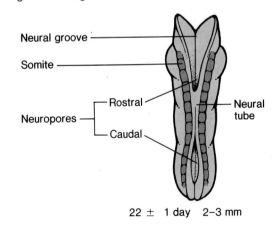

Neural groove
Somite
Neuropores
Rostral
Caudal
Neural tube

22 ± 1 day   2–3 mm

Figure 9.4. A transverse view through the dorsal aspect of a 28-day-old embryo showing the dermatome, myotome, and sclerotome of a somite. The arrows represent the direction of migration of the mesenchymal cells of the sclerotome, which give rise to the vertebral column.

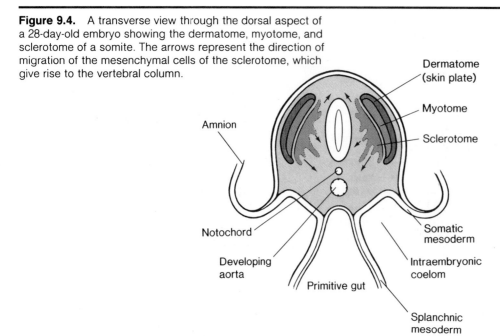

Figure 9.5. The endochondral development of the sternum. (a) an anterior view of the head and thorax of an embryo at thirty-five days showing the paired mesenchymal bands. (b) fusion of the mesenchymal bands and differentiation of mesenchymal cells into chondroblasts forms an unsegmented cartilaginous sternum. (c) the sternum becomes segmented as endochondral ossification progresses in specific ossification sites.

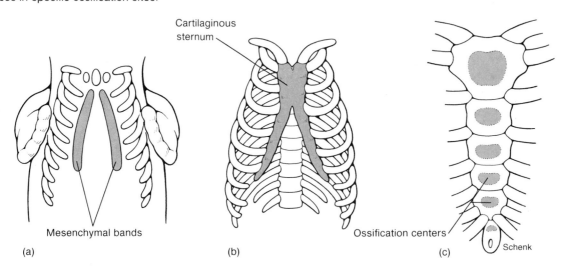

of the vertebrae (fig. 9.4). The notochord eventually degenerates as the vertebrae ossify. Only small portions of the notochord persist as gelatinous centers of the intervertebral discs called the **nucleus pulposus** *(pul-po'sus).* (2) Mesenchymal cells migrate dorsally to form the **vertebral,** or **neural, arches,** covering the neural tube. (3) A group of mesenchymal cells pass ventrolaterally as **costal processes** into the developing body wall to form the ribs and costal cartilage of the rib cage.

The development of the sternum is independent of the ribs and is first apparent at about thirty-five days as a pair of mesenchymal bands form lateral to the ventral midline in the thoracic region (fig. 9.5). Gradually the mesenchymal bands migrate together and fuse as the mesenchymal cells become cartilaginous. Segmentation of the sternum occurs secondarily as ossification centers develop along the cartilaginous sternum.

Figure 9.6.   The embryological development of the skull.
(*a–c*) superior views showing cartilaginous development and
fusion in the formation of the chondrocranium. (*d*) lateral
view at about 20 weeks. (*a*) a view of an embryo at 6 weeks,
(*b*) at 7 weeks, and (*c*) at 12 weeks.

(a)
- Trabeculae cranii
- Ala orbitalis
- Ala temporalis
- Hypophyseal cartilages
- Otic capsule
- Parachordal cartilage
- Occipital sclerotomes
- Notochord

(b)
- Nasal capsule
- Trabeculae fused
- Hypophyseal cartilage
- Internal auditory meatus
- Occipital cartilage
- Former site of notochord

(c)
- Foramen magnum
- Ethmoid
- Lesser and greater wings and body of sphenoid
- Nasal bone
- Maxilla
- Malleus
- Mandible
- Occipital bone
- Petrous portion of temporal bone

(d)
- Parietal bone
- Frontal bone
- Petrous portion of temporal bone
- Squama temporalis
- Occipital bone
- Stapes
- Incus
- Hyoid
- Thyroid cartilage
- Meckel's cartilage

Development of the Skull.   Two distinct portions of mesodermal germ cells contribute to the formation of the skull
around the developing brain. The **neurocranium** forms a
protective case that surrounds the brain, and the **viscerocranium** *(vis″er-o-kra′ne-um)* (splanchnocranium) portion forms the ear ossicles, hyoid bone, laryngeal and
tracheal cartilages, and specific processes of the skull.

One portion of the neurocranium called the **chondrocranium** *(kon′dro-kra″ne-um)* undergoes endochondral
ossification to form the bones supporting the brain, and
another portion called the **dermatocranium** develops
through membranous ossification to form the bones covering the brain and facial region (fig. 9.6).

During fetal development and infancy, the bones of
the neurocranium covering the brain are separated by fibrous sutures. There are also six large membranous areas
of the skull that provide spaces between the developing
bones (fig. 9.7). These areas are called **fontanels** *(fon″tah-
nel)* ("soft spots") and permit the skull to undergo changes

of shape, called *molding,* during parturition. The fontanels also allow for rapid growth of the brain during infancy. Ossification of the fontanels is normally complete
by twenty to twenty-four months of age. A description of
the six fontanels follows:

1. **Anterior (frontal) fontanel.** The anterior fontanel is
   diamond-shaped and is the most prominent of the
   six. It is located on the anteromedian portion of the
   skull.
2. **Posterior (occipital) fontanel.** The posterior fontanel
   is positioned at the back of the skull on the median
   line.
3. **Anterolateral (sphenoidal) fontanels.** The paired
   anterolateral fontanels are found on either side of
   the skull directly below the anterior fontanel.
4. **Posterolateral (mastoid) fontanels.** The paired
   posterolateral fontanels are located on the
   posterolateral sides of the skull.

neurocranium: Gk. *neuron,* nerve; *kranion,* skull
viscerocranium: L. *viscera,* soft parts; Gk. *kranion,* skull
fontanel: Fr. *fontaine,* little fountain

Figure 9.7. The fetal skull showing the six fontanels and the sutures. (*a*) a right lateral view; (*b*) a superior view.

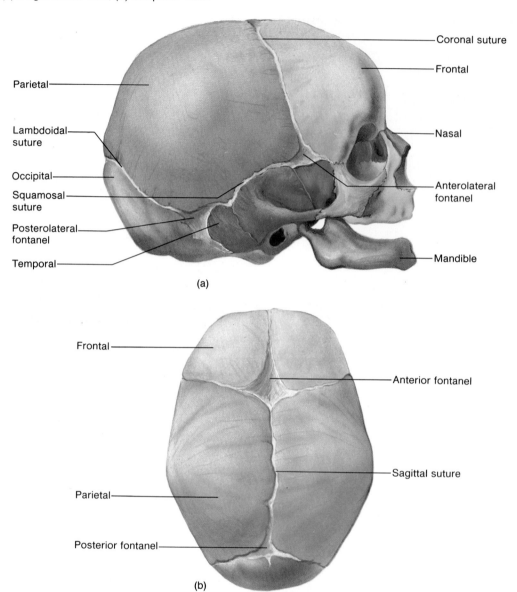

Coronal suture

Frontal

Parietal

Nasal

Lambdoidal suture

Occipital

Anterolateral fontanel

Squamosal suture

Posterolateral fontanel

Temporal

Mandible

(a)

Frontal

Anterior fontanel

Parietal

Sagittal suture

Posterior fontanel

(b)

A prominent **sagittal suture** extends the anteroposterior median length of the skull between the anterior and posterior fontanels. A coronal suture extends from the anterior fontanel to the anterolateral fontanel. A **lambdoidal suture** extends from the posterior fontanel to the posterolateral fontanel. A **squamosal suture** connects the posterolateral fontanel to the anterolateral fontanel.

The viscerocranium is the portion of the skull that forms from the **branchial** *(brang'ke-al),* or **pharyngeal, arches,** which are apparent in the neck region of an embryo during the fourth week of development. In aquatic vertebrates (fish and amphibians), the branchial arches participate in the formation of gills. In humans and other mammals, these arches form the jaws and ear ossicles of the skull and the hyoid bone and larynx within the neck (fig. 9.6). The viscerocranium, like the neurocranium, develops through both endochondral and intramembranous ossification.

Development of the Appendicular Skeleton

The development of the upper and lower extremities is initiated toward the end of the fourth week as four small elevations called **limb buds** appear (fig. 9.8). The anterior pair are the arm buds, which precede the development of the posterior pair of leg buds by a few days. Each limb bud consists of a mass of undifferentiated mesoderm partially covered with a layer of ectoderm, called the **apical** *(ap'e-kal)* **ectodermal ridge,** which promotes bone and muscle development.

lambdoidal: Gk. *lambda,* letter (λ) in Greek alphabet
branchial: L. *branchion,* gill

Figure 9.8. The development of the appendicular skeleton. (a) limb buds are apparent in an embryo by twenty-eight days, and (b) an ectodermal ridge is the percursor of the skeletal and muscular structures. (c) mesenchymal primordial cells are present at thirty-three days. (d) hyaline cartilaginous models of individual bones develop early in the sixth week. (e) later in the sixth week the cartilaginous skeleton of the upper extremity is well formed.

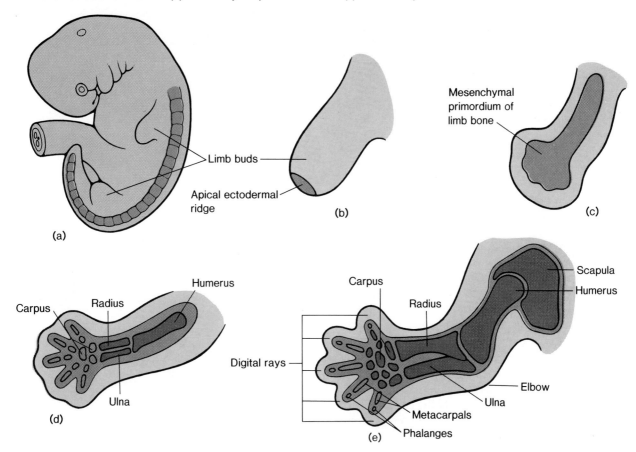

As the limb buds elongate, migrating mesenchymal tissues differentiate into specific cartilaginous bones. Soon primary ossification centers form in each bone, and the hyaline cartilage tissue is gradually replaced by bony tissue in the process of endochondral ossification.

Initially, the developing limbs are directed caudally, but later there is a lateral rotation in the upper extremity and a medial rotation in the lower extremity. As a result, the elbows are directed backward and the knees directed forward.

Digital rays that will form the hands and feet are apparent by the fifth week, and the individual digits separate by the end of the sixth week.

1. Discuss how endochondral ossification differs from intramembranous ossification.
2. What are somites, and how do they participate in the formation of the vertebral column? How does development of the axial skeleton differ from that of the appendicular skeleton?
3. Referring to the neurocranium, viscerocranium, chondrocranium, and dermatocranium, describe the formation of the skull, ear ossicles, and hyoid bone.
4. Describe the fontanels and list their two functions.
5. During which weeks of pregnancy is an embryo most susceptible to deformities of the appendages from the maternal ingestion of teratogenic drugs?

A large number of limb deformities occurred between 1957 and 1962 as a result of mothers ingesting thalidomide during early pregnancy to relieve "morning sickness." It is estimated that 7,000 infants were malformed by thalidomide. The malformations ranged from *micromelia* (short limbs) to *amelia* (absence of limbs).

Bone Structure and Growth

Although bones have commonalities of histological structure, each bone has a characteristic pattern of ossification and growth, a characteristic shape, and diagnostic surface features that indicate its functional relationship to other bones, muscles, and to the body structure as a whole.

Objective 6. Classify bones according to their shapes, and give an example of each type.

Objective 7. Describe the various markings on the surfaces of bones.

Objective 8. Describe the gross features of a typical long bone, and list the functions of each structure.

Objective 9. Describe the process of endochondral ossification as related to bone growth.

Bone Structure

Each bone of the skeleton is an organ since it consists of several types of tissue. Osseous tissue is the principal tissue, but nervous, vascular, and cartilaginous tissues also contribute to the structure and function of each bone.

The shape and surface features of each bone indicate its functional role in the skeleton. Bones that are long, for example, function as levers during body movement. Bones that support the body are massive and have large articular surfaces and processes for muscle attachment. Roughened areas on these bones may serve for the attachment of ligaments, tendons, or muscles. A flattened surface provides placement for a large muscle or may serve for protection. Grooves around an articular end of a bone are where a tendon or nerve passes, and openings through a bone permit the passage of nerves or vessels.

Shapes of Bones. The bones of the skeleton are classified into four principal types on the basis of shape rather than size. The four classes are long bones, short bones, flat bones, and irregular bones (fig. 9.9).

1. **Long bones.** Long bones are longer than they are wide and function as levers. Most of the bones of the upper and lower extremities are of this type (e.g., the humerus, radius, ulna, metacarpals, femur, tibia, fibula, metatarsals, and phalanges).
2. **Short bones.** Short bones are somewhat cube-shaped and are found in confined spaces where they transfer forces (e.g., the wrist and ankle).
3. **Flat bones.** Flat bones have a broad, dense surface for muscle attachment or protection of underlying organs (e.g., the cranium, ribs, and bones of the shoulder girdle).
4. **Irregular bones.** Irregular bones have varied shapes and have many surface markings for muscle attachment or articulation (e.g., the vertebrae and certain bones of the skull).

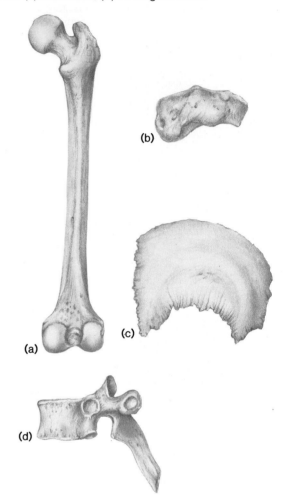

Figure 9.9. Shapes of bones. (*a*) a long bone; (*b*) a short bone; (*c*) a flat bone; (*d*) an irregular bone.

Surface Features of Bone. The following are the various descriptive terms (with examples) used to identify the surface features of bone:

Articulating surfaces
condyle A large, rounded, articulating knob (the occipital condyle of the occipital bone)
head A prominent, rounded, articulating proximal end of a bone (the head of the femur)
facet *(fas'et)* A flattened or shallow articulating surface (the costal facet of a thoracic vertebra)
Nonarticulating prominences
process Any marked bony prominence (the mastoid process of the temporal bone)
tubercle *(tu'ber-kl)* A small rounded process (the greater tubercle of the humerus)
tuberosity A large roughened process (the radial tuberosity of the radius)
trochanter *(tro-kan'ter)* A massive process found only on the femur (the greater trochanter of the femur)

facet: Fr. *facette*, little face
tuberosity: L. *tuberosus*, lump
trochanter: Gk. *trochanter*, runner

Figure 9.10. A section through the skull showing diploe. Diploe is a layer of cancellous bone sandwiched between two surface layers of compact bone.

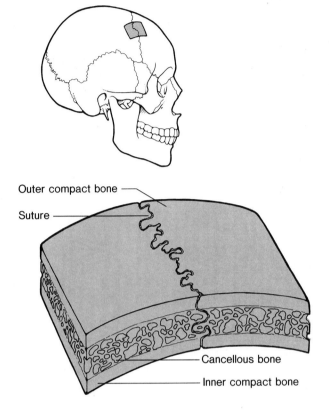

Figure 9.11. A diagram of a long bone shown in longitudinal section.

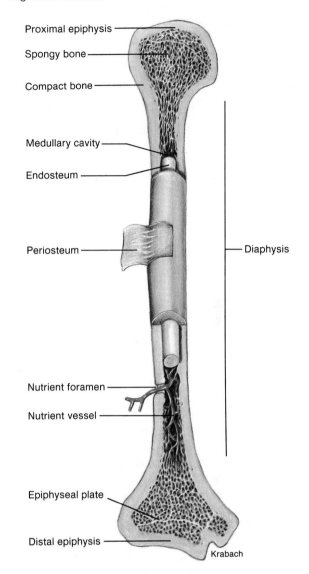

spine A sharp, slender process (the spine of the scapula)

crest A narrow, ridgelike projection (the iliac crest of the os coxa)

epicondyle A projection above a condyle (the medial epicondyle of the femur)

Depressions and openings

fossa A flattened or shallow surface (the mandibular fossa of the temporal bone)

sulcus *(sul'kus)* A groove that accommodates a vessel, nerve, or tendon (the intertubercular sulcus of the humerus)

fissure A narrow, slitlike opening (the superior orbital fissure of the sphenoid bone)

meatus *(me-a'tus),* or **canal** A tubelike passageway through a bone (the external auditory meatus of the temporal bone)

alveolus *(al-ve'o-lus)* A deep pit or socket (the alveoli for teeth in the maxilla)

foramen *(fo-ra'men),* pl. **foramina** A rounded opening through a bone (the foramen magnum of the occipital bone)

sinus A cavity or hollow space in a bone (the frontal sinus of the frontal bone)

fovea A small pit or depression (the fovea capitis of the femur)

Gross Structure of Bone. The arrangement of **osteocytes,** or bone cells, within **Haversian systems,** which form **osseous tissue,** was described in chapter 6. Two types of osseous tissue are found in bone: a dense **compact** type and a **spongy,** or **cancellous,** type. In a flat intramembranous bone of the skull, the cancellous bone is sandwiched between compact bone and is called **diploe** *(dip'lo-e)* (fig. 9.10). In a long bone from an appendage, the bone shaft, or **diaphysis** *(di-af'ĭ-sis),* consists of compact bone forming a cylinder that surrounds a central cavity called the **medullary cavity** (fig. 9.11). The medullary cavity is lined with a thin layer of connective tissue called the **endosteum** *(en-dos'te-um)* and contains **yellow bone marrow,** so named because of the large amounts of fat it contains. On each end of the diaphysis is an **epiphysis** *(ě-pif'ĭ-sis),* consisting of cancellous bone surrounded

diaphysis: Gk. *dia,* throughout; *physis,* growth
epiphysis: Gk. *epi,* upon; *physis,* growth

Figure 9.12. The process of endochondral ossification, beginning with (*a*) the cartilaginous model as it occurs in a fetus. The bone develops (*b–e*) through intermediate stages to (*f*) adult bone. (*g*) the activity of ossification is shown in a photomicrograph from an epiphyseal region (63×).

by a layer of compact bone. **Red bone marrow** is found within the porous chambers of spongy bone. In an adult, erythropoiesis, the production of red blood cells, occurs in the red bone marrow, especially that of the sternum, vertebrae, portions of the ossa coxae, and the proximal epiphyses of the femora and humeri. The red bone marrow is also responsible for the formation of certain white blood cells and platelets and for the phagocytosis of worn-out red blood cells. **Articular cartilage,** which is composed of thin hyaline cartilage, caps each epiphysis and facilitates joint movement. Along the diaphysis are **nutrient foramina,** small openings into the bone that allow passage of nutrient vessels into the bone for nourishment of the living tissue.

Between the diaphysis and epiphysis is an **epiphyseal** *(ep″ĭ-fĭz′e-al)* **plate,** a region of mitotic activity that is responsible for elongation of bone. As bone growth is completed, an **epiphyseal line** replaces the plates and ossification occurs between the epiphysis and the diaphysis. A **periosteum** *(per″e-os′te-um)* of dense, white fibrous tissue covers the surface of the bone. This highly vascular layer serves as a place for a tendon-muscle attachment and is responsible for diametric (width) bone growth.

periosteum: Gk. *peri*, around; *osteon*, bone

Bone Growth

The development of bone from embryonic to adult size depends on the orderly processes of mitotic divisions, growth, and the structural remodeling determined by genetics, hormonal secretions, and nutritional supply. In most bone development, a cartilaginous model is gradually replaced by osseous tissue during endochondral bone formation (fig. 9.12). As the cartilage model grows, the chondrocytes (cartilage cells) in the center of the shaft hypertrophy, and minerals are deposited within the matrix in a process called *calcification*. Calcification restricts the passage of nutrients to the chondrocytes, causing them to die. At the same time, some cells of the perichondrium differentiate into primordial bone cells called **osteoblasts,** which secrete **osteoid,** the organic component of bone. As the perichondrium calcifies, it gives rise to a thin plate of compact bone called the **periosteal bone collar** that is surrounded by the periosteum.

A **periosteal bud,** consisting of osteoblasts and blood vessels, invades the disintegrating center of the cartilage model from the periosteum. Once in the center, the osteoblasts secrete osteoid, and a **primary ossification center** is established from which ossification expands into deteriorating cartilage. This process is repeated in both the proximal and distal epiphyses, forming **secondary ossification centers.**

osteoblast: Gk. *osteon*, bone; *blastos*, offspring or germ

Zone of reserve cartilage
Zone of proliferating cartilage
Zone of hypertrophic cartilage
Zone of calcified matrix

Chondrocytes
Bone
Red bone marrow

(g)

Table 9.2 Average age of completion of bone ossification

Bone	Chronological age of fusion
Scapula	18–20
Clavicle	23–31
Bones of upper extremity (brachium, forearm, hand)	17–20
Os coxa	18–23
Bones of lower extremity (thigh, leg, foot)	18–22
Vertebra	25
Sacrum	23–25
Sternum (body)	23
Sternum (manubrium, xiphoid)	after 30

Source: Van De Graaff, Kent M., *Human Anatomy.* © 1984 Wm. C. Brown Publishers, Dubuque, Iowa. All Rights Reserved. Reprinted by permission.

Bone growth continues as long as the cartilage cells at the epiphyseal plate between the two ossification centers continue to divide (fig. 9.13). When mitosis of the chondrocytes gradually decreases, ossification finally occurs within the plate, thus prohibiting the further lengthwise growth of the bone. The time when epiphyseal plates ossify varies greatly from bone to bone, but generally it occurs between the ages of eighteen and twenty within the long bones of most persons (table 9.2).

Radiologists can determine the ages of persons who are still growing by examining X-ray pictures of their bones. Ossification at the epiphyseal plates occurs at predictable periods of development and is therefore a reliable indicator of age. Large discrepancies between bone age and chronological age may indicate a genetic or endocrine dysfunction.

Figure 9.13. The presence of epiphyseal plates, as seen in an X-ray of a child's hand, indicates that bones are still growing in length.

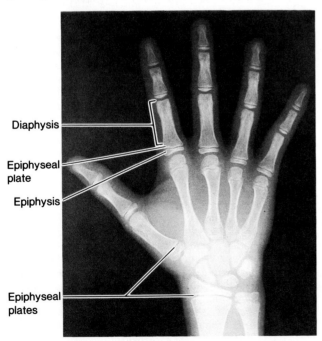

Diaphysis
Epiphyseal plate
Epiphysis
Epiphyseal plates

Bone remodeling is a continual process throughout a person's life. The diagnostic processes on the surface of bones develop as stress is applied to the periosteum, resulting in the osteoblastic secretion of osteoid and the formation of new bone tissue. These processes may continue to change somewhat in persons who are athletically active even though they have stopped growing in height.

Specialized bone cells called **osteoclasts** can enzymatically cause bone resorption. As new bone layers are deposited on the outside surface of the bone, osteoclasts dissolve bone tissue adjacent to the medullary cavity. In this way, the size of the cavity keeps pace with the increased growth of the bone.

Normal bone growth and the maintenance of healthy bones is dependent on certain physiological and even physical factors. The physiological factors and their effect on bone growth and function are summarized in table 9.3.

osteoclast: Gk. *osteon*, bone; *klastos*, broken

Table **9.3** Factors affecting bone physiology	
Substance	**Effect**
Growth hormone	Stimulates osteoblast activity and collagen synthesis
Thyroid hormones	Stimulate osteoblast activity, collagen synthesis, and formation of ossification centers
Sex hormones (especially androgens)	Stimulate osteoblast activity and bone growth
Adrenocorticoid hormones	Stimulate osteoclast activity
Vitamin A	Promotes chondrocyte function; synthesis of lysosomal enzymes for osteoclast activity
Vitamin C	Promotes synthesis of collagen

A normal hormonal balance maintained by feedback mechanisms and a continual source of specific essential minerals and vitamins are necessary for healthy bones. Consider, for example, that **hypothyroidism** results in decreased bone resorption and slowed ossification, and **hyperthyroidism** results in increased bone resorption and premature calcification of the epiphyseal plates. The decrease in sex hormones that accompanies aging causes an increase in bone resorption and loss of bone mass **(osteoporosis** [*os″te-o-po-ro′sis*]). Excessive glucocorticoids also results in osteoporosis, as evidenced in **Cushing's syndrome.** Although vitamin A is needed for chondrocyte and osteoclastic activities, an excess of vitamin A is more serious than a deficiency because it increases demineralization, making bones brittle and easily fractured.

The size, shape, and processes of bones depend on the physical stresses and pressures applied. Initial bone formation, bone growth, and the adaptation of mature bone to units of applied force is reflected in the **stress lines,** or structure of ossification within a bone (fig. 9.14). Stress lines provide maximum strength to the bone in the direction of greatest applied pressure. **Compression lines** are stress lines that are linear and parallel to the applied weight, as in weight-bearing bones like the tibia. **Tension lines** develop in bones subjected to tensile forces such as in the head and neck of the femur, which is positioned on a supportive angle lateral to the os coxa. The diagnostic processes on some bones, such as the greater trochanter of the femur, develop in response to forces of stress applied to the periosteum of the bone where the tendons of muscles attach. In other words, as a muscle is contracted, a pulling stress is applied to the periosteum of the bone where the tendon is attached. This mechanical tension causes an increase in osteoblastic activity and bone deposition, and a bony process is gradually formed.

Bone is highly dynamic and is continually being remodeled in response to mechanical stress or even the absence of stress. The effect of the absence of stress can best

Figure 9.14. Longitudinal section of proximal end of femur showing stress lines.

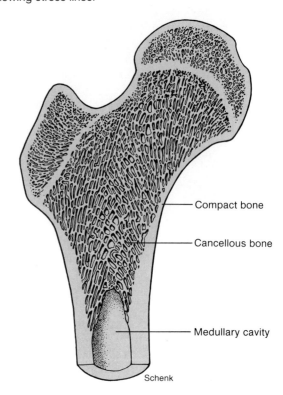

Compact bone

Cancellous bone

Medullary cavity

Schenk

be seen in the bones of inactive patients confined to bed or the bones of persons who are paralyzed. X-ray examination shows the loss of bone mass or even osteoporosis. The absence of gravity that accompanies space flight may result in mineral loss from bones if an exercise program is not maintained.

The movement of teeth in orthodontics involves bone remodeling. The teeth sockets (alveoli) are reshaped through the activity of osteoclast and osteoblast cells as stress is applied through the application of braces. The use of traction in treating certain skeletal disorders has a similar effect.

1. Using examples, discuss the function of each of the four kinds of bones as determined by shape.
2. Define each of the following surface markings on bones: condyle, head, facet, process, crest, epicondyle, fossa, alveolus, foramen, and sinus.
3. Diagram a sagittal view of a typical long bone and label the following: diaphysis, medullary cavity, epiphyses, articular cartilages, nutrient foramen, periosteum, and epiphyseal plates. Explain the function of each of these structures.
4. Define osteocytes, osteoblasts, and osteoclasts, and explain the function of each as related to the processes of endochondral ossification and bone growth.

Table 9.4 Summary of major foramina of the skull

Foramen	Location	Structures transmitted
Carotid canal	Petrous portion of temporal bone	Internal carotid artery and sympathetic nerves
Greater palatine	Palatine bone of hard palate	Greater palatine nerve and descending palatine vessels
Hypoglossal foramen/canal	Anterolateral edge of the occipital condyle	Hypoglossal nerve and branch of ascending pharyngeal artery
Incisive	Anterior region of hard palate, posterior to the incisor teeth	Branches of descending palatine vessels and nasopalatine nerve
Inferior orbital	Between maxilla and greater wing of sphenoid	Maxillary branch of trigeminal nerve, zygomatic nerve, and infraorbital vessels
Infraorbital	Inferior to orbit in maxilla	Infraorbital nerve and artery
Jugular	Between petrous portion of temporal and occipital, posterior to carotid canal	Internal jugular vein; vagus, glossopharyngeal, and accessory nerves
Lacerum	Between petrous portion of temporal and sphenoid	Branch of ascending pharyngeal artery
Lesser palatine	Posterior to greater palatine foramen in hard palate	Lesser palatine nerves
Magnum	Occipital bone	Union of medulla oblongata and spinal cord, meningeal membranes, accessory nerves; vertebral and spinal arteries
Mandibular	Medial surface of ramus of mandible	Inferior alveolar nerve and vessels
Mental	Below the second premolar on the lateral side of mandible	Mental nerve and vessels
Nasolacrimal canal	Lacrimal bone	Nasolacrimal (tear) duct
Olfactory	Cribriform plate of the ethmoid	Olfactory nerves
Optic	Back of orbit in lesser wing of sphenoid	Optic nerve and ophthalmic artery
Ovale	Greater wing of sphenoid	Mandibular branch of trigeminal nerve
Rotundum	Within body of sphenoid	Maxillary branch of trigeminal nerve
Spinosum	Posterior angle of sphenoid	Middle meningeal vessels
Stylomastoid	Between styloid and mastoid processes of temporal	Facial nerve and stylomastoid artery
Superior orbital fissure	Between greater and lesser wings of sphenoid	Four cranial nerves (oculomotor, trochlear, ophthalmic branch of trigeminal and abducens)
Supraorbital	Supraorbital ridge of orbit	Supraorbital nerve and artery
Zygomaticofacial	Anterior surface of zygomatic bone	Zygomaticofacial nerve and vessels

Source: Van De Graaff, Kent M., *Human Anatomy.* © 1984 Wm. C. Brown Publishers, Dubuque, Iowa. All Rights Reserved. Reprinted by permission.

Skull

The human skull, consisting of eight cranial and fourteen facial bones, contains several cavities that house the brain and sensory organs. Each bone of the skull articulates the adjacent bones and has diagnostic and functional processes or other surface features and foramina.

Objective 10. Identify the cranial and facial bones of the skull, and describe their structural characteristics.

Objective 11. Describe the location of each of the bones of the skull, and name the articulations that affix them together.

The skull consists of **cranial bones** and **facial bones.** The eight bones of the cranium articulate firmly with one another to enclose and protect the brain and associated sense organs. The fourteen facial bones form the framework for the facial region and support the teeth. A variation in shape and density of the facial bones is a major contributor to the individuality of each human face. The facial bones, with the exception of the bone within the lower jaw, are also firmly interlocked with one another and the cranial bones.

The skull has several cavities. The **cranial cavity** is the largest, with a capacity of about 1,300–1,350 cc. The **nasal cavity** is formed by both cranial and facial bones and is partitioned into two chambers, or **nasal fossae,** by a **nasal septum** of bone and cartilage. Four sets of **paranasal sinuses** are located within the bones surrounding the nasal area and communicate via ducts into the nasal cavity. **Middle** and **inner ear chambers** are positioned inferior to the cranial cavity and house the organs of hearing and balance. The two **orbits** for the eyeballs are formed by facial and cranial bones. The **oral,** or **buccal, cavity** (mouth), which is only partially formed by bone, is completely within the facial region.

The bones of the skull contain numerous foramina to accommodate nerves, vessels, and other structures. Table 9.4 summarizes the foramina of the skull. Figures 9.15, 9.16, and 9.17 show anterior, lateral, and inferior views of the skull; figure 9.18 shows a midsagittal view, and figure 9.19 views the interior of the braincase. X rays (roentgenograms) of the skull are shown in figure 9.20.

roentgenogram: from Wilhelm K. Roentgen, German physicist, 1845–1923

Figure 9.15.   An anterior view of the skull.

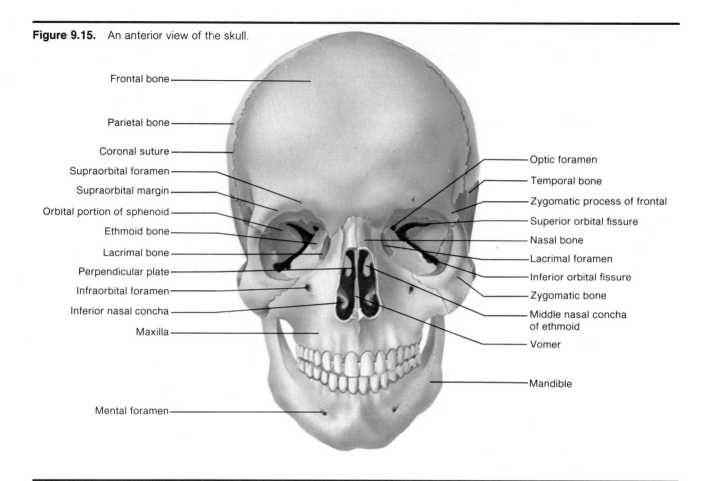

Frontal bone

Parietal bone

Coronal suture

Supraorbital foramen

Supraorbital margin

Orbital portion of sphenoid

Ethmoid bone

Lacrimal bone

Perpendicular plate

Infraorbital foramen

Inferior nasal concha

Maxilla

Mental foramen

Optic foramen

Temporal bone

Zygomatic process of frontal

Superior orbital fissure

Nasal bone

Lacrimal foramen

Inferior orbital fissure

Zygomatic bone

Middle nasal concha of ethmoid

Vomer

Mandible

Figure 9.16.   A lateral view of the skull.

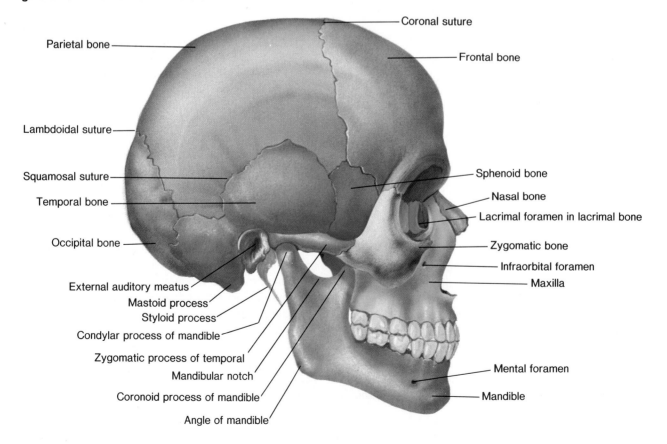

Parietal bone

Lambdoidal suture

Squamosal suture

Temporal bone

Occipital bone

External auditory meatus

Mastoid process

Styloid process

Condylar process of mandible

Zygomatic process of temporal

Mandibular notch

Coronoid process of mandible

Angle of mandible

Coronal suture

Frontal bone

Sphenoid bone

Nasal bone

Lacrimal foramen in lacrimal bone

Zygomatic bone

Infraorbital foramen

Maxilla

Mental foramen

Mandible

Figure 9.17.   An inferior view of the skull.

Figure 9.18.   A midsagittal view of the skull.

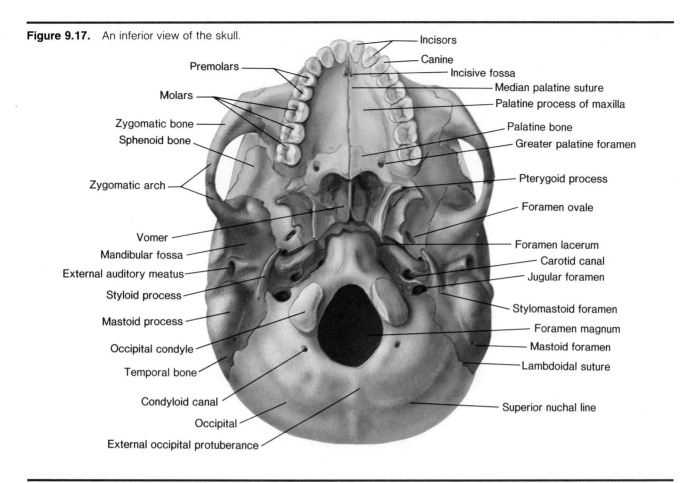

Figure 9.19. The floor of the cranial cavity.

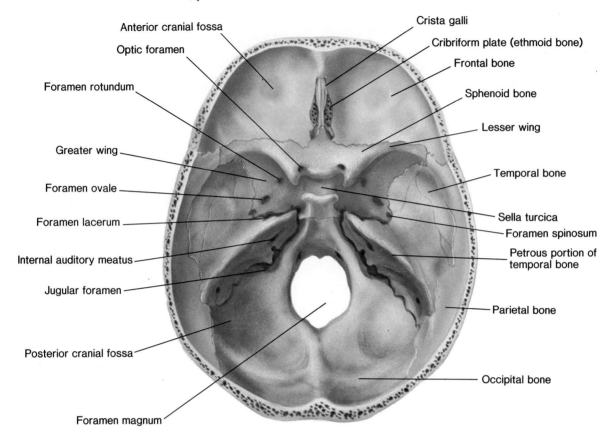

Figure 9.20. X rays of the skull showing paranasal sinuses. (*a*) an anteroposterior view; (*b*) a right lateral view.

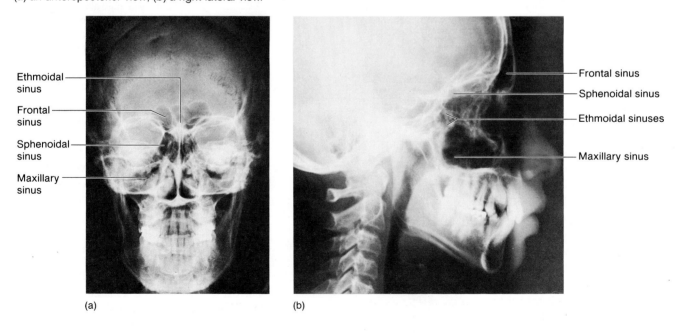

(a)   (b)

Figure 9.21.   A lateral view of the right temporal bone.

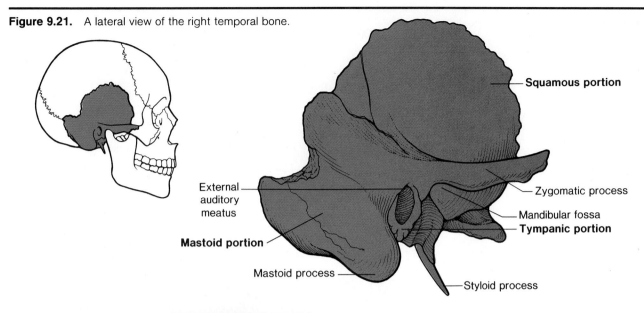

Squamous portion

Zygomatic process

Mandibular fossa

Tympanic portion

External auditory meatus

Mastoid portion

Mastoid process

Styloid process

Table **9.5** Bones forming the orbit	
Region of the orbit	**Contributing bones**
Roof (superior)	Frontal; lesser wing of sphenoid
Floor (inferior)	Maxilla; zygomatic
Lateral wall	Zygomatic
Posterior wall	Greater wing of sphenoid
Medial wall	Maxilla; lacrimal; ethmoid
Superior rim	Frontal
Lateral rim	Zygomatic
Medial rim	Maxilla

Source: Van De Graaff, Kent M., *Human Anatomy.* © 1984 Wm. C. Brown Publishers, Dubuque, Iowa. All Rights Reserved. Reprinted by permission.

Although the hyoid bone and the three paired ear ossicles are not considered part of the skull, they are within the axial skeleton and will be described immediately following the discussion of the skull.

Cranium

The cranial bones enclose the brain and consist of one frontal, two parietals, two temporals, one occipital, one sphenoid, and one ethmoid.

Frontal.   The frontal bone forms the anterior roof of the cranium, the forehead, the roof of the nasal cavity, and the superior arch of the **bony orbits,** which contain the eyeballs. Table 9.5 summarizes the bones of the orbit. The frontal bone develops in two halves that grow together and are usually completely fused by age five or six. Occasionally a complete suture persists between these two portions beyond age six and is referred to as a **metopic *(me-top'ik)* suture.** The **supraorbital margin** is a prominent bony ridge over the orbit. Openings along this ridge, called supraorbital foramina, allow passage of small nerves and vessels.

The frontal bone contains a **frontal sinus,** which is connected to the nasal cavity (fig 9.20). This sinus, along with the other paranasal sinuses, lessens the weight of the skull and acts as a sound chamber for voice resonance.

Parietal.   The **coronal *(ko-ro'nal)* suture** separates the frontal bone from the parietals, and the **sagittal *(saj'ĭ-tal)* suture** along the dorsal midline separates the right and left parietal. The parietal bones form the upper sides and roof of the cranium. The inner concave surface of the parietal bone, as well as the inner concave surfaces of other cranial bones, is marked by shallow impressions from convolutions of the brain and vessels serving the brain.

Temporal.   The two temporal bones form the lower sides of the cranium (figs. 9.16, 9.17, and 9.21). Each temporal is joined to its adjacent parietal bone by the **squamosal *(skwa-mo'sal)* suture.** Structurally, each temporal has four parts:

1. **Squamous portion.** The squamous portion is the flattened plate of bone at the sides of the skull. Projecting forward is a **zygomatic process** that forms the posterior portion of the **zygomatic arch.** On the inferior surface of the squamous portion is the **mandibular fossa,** which receives the articular condyle of the mandible. This articulation is referred to as the **temporomandibular *(tem"po-ro-man-dib'u-lar)* joint.**

2. **Tympanic portion.** The tympanic portion of the temporal bone contains the **external auditory meatus,** or ear canal, located immediately posterior to the mandibular fossa. A thin, pointed **styloid process** (figs. 9.16 and 9.21) projects downward from the tympanic portion.

---

cranium: Gk. *kranion,* skull
metopic suture: Gk. *metopon,* forehead; L. *sutura,* sew

zygomatic: Gk. *zygoma,* yolk
styloid: Gk. *stylos,* pillar

Figure 9.22. The sphenoid bone. (a) a superior view; (b) a posterior view.

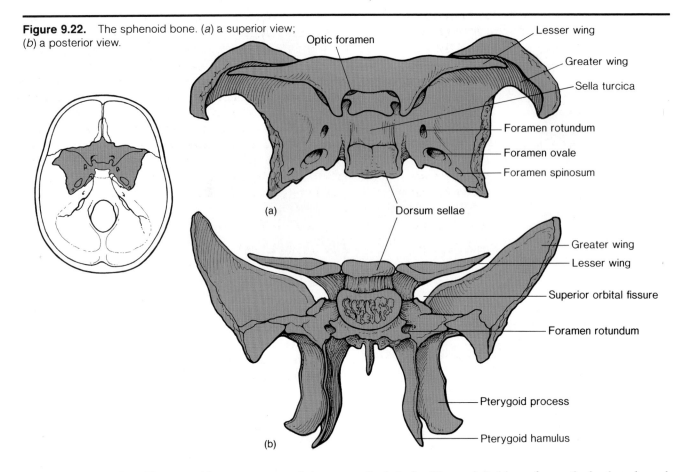

3. **Mastoid portion.** The **mastoid process,** a rounded projection posterior to the external auditory meatus, accounts for the mass of the mastoid portion. The **mastoid foramen** (fig. 9.17) is directly posterior to the mastoid process. The **stylomastoid foramen,** located between the mastoid and styloid processes, is the passage for part of the facial nerve.

4. **Petrous** *(pet′rus)* **portion.** The petrous portion is viewed inferiorly (figs. 9.17 and 9.19). The structures of the middle and inner ear are housed in this dense, bony portion. The **carotid** *(kah-rot′id)* **canal** and the **jugular foramen** border on the medial side of the petrous portion. The carotid canal allows blood into the brain via the internal carotid artery, and the jugular foramen lets blood drain from the brain via the internal jugular vein.

The mastoid process of the temporal bone can be easily palpated as a bony knob immediately behind the earlobe. This process contains a number of small sinuses that are clinically important because they can become infected in mastoiditis. A tubular communication from the mastoid sinuses to the middle ear chamber may permit prolonged ear infections to spread to this region.

Occipital. The occipital bone forms the back and much of the base of the skull. It is fastened to the parietal bones by the **lambdoidal suture.** The **foramen magnum** is the large hole in the occipital bone through which the spinal cord attaches to the brain. On each side of the foramen magnum are the **occipital condyles** *(ok-sip′i-tal kon′dīlz)* (fig. 9.17), which articulate with the atlas of the vertebral column. At the anterolateral edge of the occipital condyle is the **hypoglossal foramen,** through which the hypoglossal nerve passes. A **condyloid** *(kon′dī-loid)* **canal** lies posterior to the occipital condyle. The **external occipital protuberance** is a prominent posterior projection on the occipital bone that can be felt as a definite bump just under the skin. The **superior nuchal** *(nu′kal)* **line** is a ridge of bone extending laterally from the occipital protuberance to the mastoid portion of the temporal bone. Sutural, or Wormian, bones are small clusters of irregularly shaped bones that may be found between the joints of certain cranial bones but generally occur along the lambdoidal suture.

Sphenoid. The sphenoid *(sfe′noid)* bone forms the anterior base of the cranium and can be viewed laterally and inferiorly (figs. 9.16, 9.19). This bone resembles a bat with outstretched wings (fig. 9.22). It consists of a **body** with

---

mastoid: Gk. *mastos,* breast
petrous: Gk. *petra,* rock

magnum: L. *magnum,* great
nuchal: Fr. *nuque,* nape of neck
sphenoid: Gk. *sphenoeides,* wedgelike

Figure 9.23.   An anterior view of the ethmoid bone.

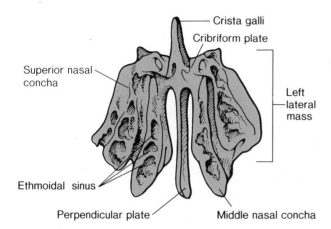

Crista galli
Cribriform plate
Superior nasal concha
Left lateral mass
Ethmoidal sinus
Perpendicular plate
Middle nasal concha

Table **9.6**	Bones forming the nasal cavity
Region of nasal cavity	**Contributing bones**
Roof (superior)	Ethmoid (cribriform plate); frontal
Floor (inferior)	Maxilla; palatine
Lateral wall	Maxilla; palatine
Nasal septum (medial)	Ethmoid (perpendicular plate); vomer; nasal
Bridge	Nasal
Conchae	Ethmoid (superior and middle); inferior nasal conchae

Source: Van De Graaff, Kent M., *Human Anatomy.* © 1984 Wm. C. Brown Publishers, Dubuque, Iowa. All Rights Reserved. Reprinted by permission.

laterally projecting **greater** and **lesser wings,** which form part of the bony orbit. The body is a wedgelike central portion that contains the **sphenoidal** *(sfe-noi'dal)* **sinuses** and a prominent depression called the **sella turcica** *(sel'ah tur'si-kah),* which supports the pituitary gland. The sella turcica (meaning "Turk's saddle") is seen on the floor of the cranium. A pair of **pterygoid** *(ter'i-goid)* **processes** project inferiorly from the sphenoid bone to help form the lateral walls of the nasal cavity. Several foramina (fig. 9.19) are associated with the sphenoid bone:

1. **Optic foramen.** A large opening through the lesser wing into the back of the orbit for passage of the optic nerve and the ophthalmic artery.
2. **Superior orbital fissure.** A triangular opening between the wings of the sphenoid for passage of four cranial nerves (oculomotor, trochlear, ophthalmic branch of the trigeminal, and abducens).
3. **Foramen ovale.** An opening at the base of the pterygoid process through which passes the mandibular branch of the trigeminal nerve.
4. **Foramen spinosum.** A small opening at the posterior angle of the sphenoid for passage of the middle meningeal vessels.
5. **Foramen lacerum** *(las'er-um).* An opening between the sphenoid and the petrous portion of the temporal bone through which pass the internal carotid artery and the meningeal branch of the ascending pharyngeal artery.
6. **Foramen rotundum.** An opening located just posterior to the superior orbital fissure at the junction of the anterior and medial portions of the sphenoid bone. The maxillary branch of the trigeminal nerve passes through this foramen.

Ethmoid.   The ethmoid bone is located in the anterior portion of the floor of the cranium between the orbits, where it forms the roof of the nasal cavity (figs. 9.18, 9.19, 9.23). An inferior projection of the ethmoid, called the **perpendicular plate,** contributes in part to the nasal septum that separates the nasal cavity into two chambers referred to as **nasal fossae.** A spine of the perpendicular plate, the **crista galli,** projects superiorly into the cranial cavity and serves as an attachment for the meninges covering the brain. On both lateral walls of the nasal cavity are two scroll-shaped plates of the ethmoid, called the **superior** and **middle nasal conchae.** At right angles to the perpendicular plate, within the floor of the cranium, is the **cribriform plate,** which has numerous perforations for the passage of olfactory nerves from the nasal cavity. Table 9.6 summarizes the bones of the nasal cavity.

The moist, warm vascular lining within the nasal cavity is susceptible to infections, particularly if a person is not in good health. Infections of the nasal cavity can spread to several surrounding areas. The paranasal sinuses connect to the nasal cavity and are especially prone to infection. The eyes may become reddened and swollen during a nasal infection because of the connection of the nasolacrimal duct through which tears drain from the orbit to the nasal cavity. Organisms may spread via the auditory canal from the nasopharynx to the middle ear. With prolonged nasal infections, organisms may even ascend to the meninges covering the brain, along the sheaths of the olfactory nerves and through the cribriform plate to produce meningitis.

conchae: L. *conchae,* shells
cribriform: L. *cribrum,* sieve; *forma,* like

Figure 9.24.   A medial view of the left maxilla.

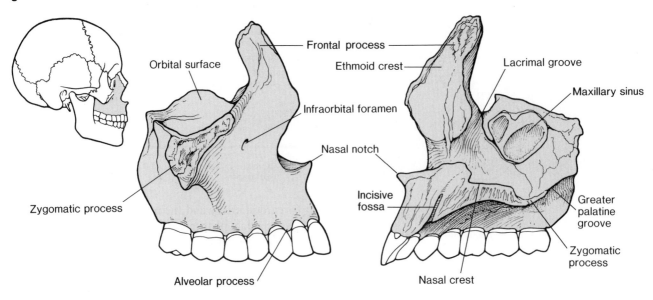

Figure 9.25.   (a) the right palatine bone as it articulates anteriorly with the maxilla. (b) the two palatine bones viewed posteriorly. The two palatine bones form the posterior portion of the hard palate where they articulate at the nasal crest.

(a)

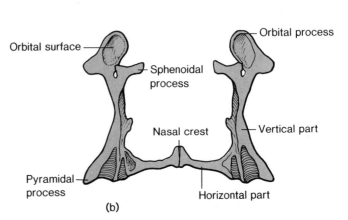

(b)

Facial Bones

The fourteen bones of the skull not in contact with the brain are called facial bones. These bones, together with certain cranial bones (frontal and portions of the ethmoid and temporals), provide the basic shape of the face. Facial bones also support the teeth and provide attachments for various muscles that move the jaw and cause facial expressions. All the facial bones are paired except the vomer and mandible. The articulated facial bones can be seen in figures 9.15, 9.16, 9.17, and 9.18.

Maxilla.   The two maxillae unite at the midline to form the upper jaw, which supports the upper teeth. **Incisors** *(in-si′zor)*, **canines** *(ka′nīn)*, **premolars,** and **molars** are contained in sockets, or **alveoli,** within the **alveolar** *(al-ve′o-lar)* **process** of the maxilla (fig. 9.24). The palatine process, a horizontal plate of the maxilla, forms the greater portion of the **hard palate** *(pal′at)*, or roof of the mouth.

incisor: L. *incidere*, to cut
canine: L. *canis*, dog
molar: L. *mola*, millstone
alveolus: L. *alveus*, little cavity

Figure 9.26.   A lateral view of the bony and cartilaginous structure of the external nose.

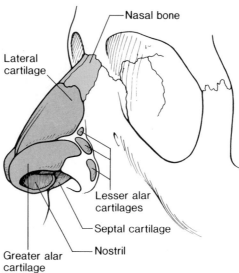

The **incisive fossa** (fig. 9.17) is located in the anterior region of the hard palate behind the incisor teeth. An **infraorbital foramen** is located under each orbit and serves as a passageway for the infraorbital nerve and artery to the nose. A final opening within the maxilla is the **inferior orbital fissure.** It is located between the maxilla and the greater wing of the sphenoid (figs. 9.15 and 9.24) and is the opening for the maxillary branch of the trigeminal nerve and infraorbital vessels. The large **maxillary sinus** located within the maxilla is one of the four paranasal sinuses.

> If the two palatine processes fail to join during early prenatal development (about twelve weeks), a cleft palate results. A cleft palate may be accompanied by a cleft lip (harelip) lateral to the midline. These conditions can be surgically and cosmetically treated with excellent success. A more immediate problem, however, is that a newborn with a cleft palate may have a difficult time swallowing while nursing because it is unable to create the necessary suction within the oral cavity.

Palatine.   The L-shaped palatine bones form the posterior third of the hard palate, a portion of the orbits, and a part of the nasal cavity. The **horizontal plates** of the palatines contribute to the formation of the hard palate (fig. 9.25). On the hard palate of each palatine bone is a large **greater palatine foramen,** which permits the passage of the greater palatine nerve and descending palatine vessels (fig. 9.17). Two or more smaller **lesser palatine foramina** are positioned posterior to the greater palatine foramina. Branches of the lesser palatine nerve pass through these openings.

Zygomatic (Malar).   The two zygomatic bones form the cheekbones of the face. A posteriorly extending **zygomatic process** of this bone unites with that of the temporal bone to form the **zygomatic arch** (fig. 9.17). The zygomatic bone also forms the lateral margin of the orbit. A small **zygomaticofacial** *(zi″go-mat″ĭ-ko-fa′shal)* **foramen,** located on the anterolateral surface of this bone, allows passage of the zygomatic nerves and vessels.

Lacrimal.   The small lacrimals *(lak′rĭ-mal)* are thin bones that form the anterior part of the medial wall of each orbit. Each has a **lacrimal sulcus,** a groove that helps to form the **nasolacrimal canal.** This opening permits the tears of the eye to drain into the nasal cavity (fig. 9.15).

Nasal.   The small, rectangular nasal bones join in the midline to form the bridge of the nose (fig. 9.26). The nasal bones support the **lateral cartilages,** which are a part of the framework of the nose. Fractures of the nasal bones or fragmentation of the associated cartilage are common facial injuries.

Inferior Nasal Concha.   The two inferior nasal conchae are fragile, scroll-like bones that project horizontally and medially from the lateral walls of the nasal cavity (fig. 9.15). They extend into the nasal cavity just below the superior and middle nasal conchae, which are part of the ethmoid bone (fig. 9.23). The inferior nasal conchae are the largest of the three conchae, and, like the other two, they are covered with mucous membranes to warm, moisten, and cleanse inhaled air.

Figure 9.27.   The bones and cartilages that form the nasal septum.

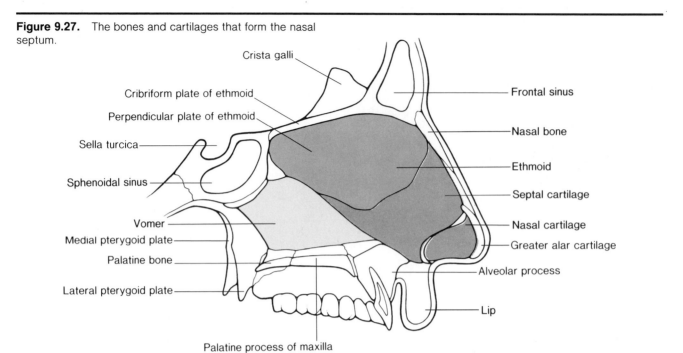

Crista galli

Cribriform plate of ethmoid

Perpendicular plate of ethmoid

Sella turcica

Sphenoidal sinus

Vomer

Medial pterygoid plate

Palatine bone

Lateral pterygoid plate

Frontal sinus

Nasal bone

Ethmoid

Septal cartilage

Nasal cartilage

Greater alar cartilage

Alveolar process

Lip

Palatine process of maxilla

Figure 9.28.   The mandible. (*a*) lateral view; (*b*) posterior view.

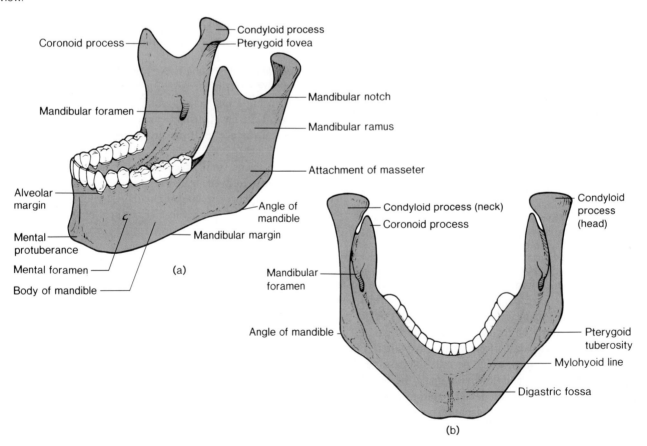

Coronoid process

Condyloid process

Pterygoid fovea

Mandibular foramen

Mandibular notch

Mandibular ramus

Attachment of masseter

Alveolar margin

Angle of mandible

Mental protuberance

Mandibular margin

Mental foramen

Body of mandible

(a)

Condyloid process (neck)

Coronoid process

Mandibular foramen

Angle of mandible

Condyloid process (head)

Pterygoid tuberosity

Mylohyoid line

Digastric fossa

(b)

Figure 9.29. An anterior view of the hyoid bone.

Greater cornu

Lesser cornu

Body

Hyoid bone

Hyoid bone

Larynx

Vomer. The vomer *(vo'mer)* is a thin, plowshare-shaped bone that forms the lower part of the nasal septum (fig. 9.17, 9.27). The vomer, along with the perpendicular plate of the ethmoid, supports the septal cartilage to complete the nasal septum.

Mandible. The mandible, or lower jawbone, is attached to the skull by a temporomandibular articulation and is the only movable bone of the skull. Several muscles that close the jaw extend from the skull to the mandible and are discussed in chapter 11. The mandible of an adult supports sixteen teeth within alveoli, which occlude with those of the maxilla.

The horseshoe-shaped front and horizontal lateral sides of the mandible are referred to as the **body** (fig. 9.28). Extending vertically from the posterior portion of the body are two **rami** *(ra'mi)* (singular, *ramus*). Each ramus has a knoblike **condyloid process,** which articulates with the mandibular fossa of the temporal bone, and a pointed **coronoid process** for the attachment of the temporalis muscle. The depressed area between these two processes is the **mandibular notch.** The angle of the mandible is where the horizontal body and vertical ramus meet at the corner of the jaw.

Two sets of foramina are found on the mandible: the **mental foramen** on the lateral side below the first molar, and the **mandibular foramen** on the medial surface of the

ramus. The mental nerve and vessels pass through the mental foramen, and the inferior alveolar nerve and vessels are transmitted through the mandibular foramen.

Dentists use bony landmarks of the facial region to locate the nerves that traverse the foramina so that anesthetics can be injected. For example, the trigeminal nerve is composed of three large branches, the lower two of which convey sensations from the teeth, gums, and jaws. The mandibular teeth can be desensitized by an injection near the mandibular foramen, called a third division, or lower, nerve block. An injection near the foramen rotundum of the skull, called a second division nerve block, desensitizes all the upper teeth on one side of the maxilla.

Hyoid. The hyoid is a U-shaped bone located in the neck just superior to the larynx (voice box). The hyoid is unique in that it does not attach directly to any other bone but is suspended from the styloid processes of the skull by the **stylohyoid** *(sti"lo-hi'oid)* **ligaments.** The hyoid has a **body,** two **lesser cornua** extending anteriorly, and two **greater cornua,** which project posteriorly to the stylohyoid ligaments (fig. 9.29). Several neck and tongue muscles attach to the hyoid bone. The hyoid may be palpated by placing a thumb and a finger on either side of the upper neck under the lateral portions of the mandible and firmly squeezing medially.

vomer: L. *vomer,* plowshare
mandible: L. *mandere,* to chew
ramus: L. *ramus,* branch
condyloid: L. *condylus,* knucklelike
coronoid: Gk. *korone,* like a crow's beak

Figure 9.30. The three ear ossicles within the middle ear cavity. What is the embryonic origin of the ear ossicles?

Figure 9.31. The vertebral column of an adult has four curves named according to the region in which they occur. The vertebrae are separated by intervertebral discs, which allow flexibility.

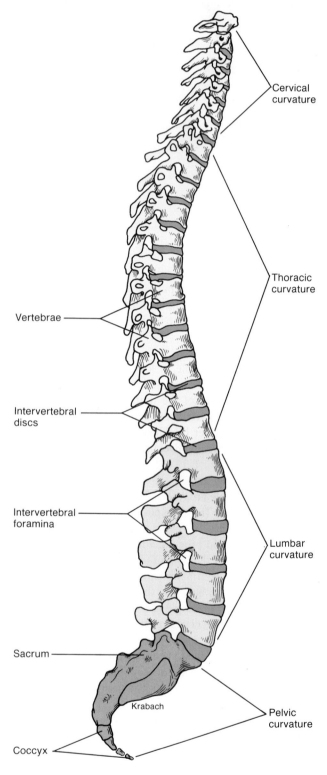

Ear Ossicles. Three small, paired ear ossicles, the **malleus, incus,** and **stapes,** are located within the middle ear chambers in the petrous portion of the temporal bones (fig. 9.30). These bones transfer and amplify sound impulses through the middle ear.

1. List which facial and cranial bones of the skull are paired and which are unpaired. Also, indicate at least two structural features for each of the bones of the skull.
2. Describe the location of each of the bones of the skull, and indicate the sutures that join them.
3. What is the function of the sella turcica, foramen magnum, petrous bone, crista galli, and nasal conchae?
4. Which facial bones support the teeth?

cornu: L. *cornu,* horn
malleus: L. *malleus,* hammer
incus: L. *incus,* anvil
stapes: L. *stapes,* stirrup

Figure 9.32. The development of the vertebral curvatures. An infant is born with the two primary curves and does not develop the secondary curves until it begins sitting upright and walking. Note the differences in curvature between the sexes. (Richard J. Harrison, William Montagna, *Man,* © 1973, p. 158. Reprinted by permission of Prentice-Hall, Inc., Englewood Cliffs, NJ.)

Vertebral Column

The supporting vertebral column consists of twenty-six vertebrae separated by fibrocartilaginous intervertebral discs, which lend flexibility and absorb the stress of movement. Vertebrae enclose and protect the spinal cord, support and permit movement of the skull, articulate with the rib cage, and provide for the attachment of trunk muscles.

> *Objective 12.* Identify the bones of the five regions of the vertebral column, and describe the characteristic curves of each region.
>
> *Objective 13.* Describe the structure of a typical vertebra.

The vertebral column is composed of thirty-three individual vertebrae. There are seven **cervical,** twelve **thoracic** *(tho-ras'ik),* five **lumbar,** five fused **sacral,** and four fused **coccygeal** *(kok-sij'e-al)* vertebrae; thus, the vertebral column is composed of a total of twenty-six movable parts. Vertebrae are separated by fibrocartilaginous intervertebral discs and are secured to one another by interlocking processes and binding ligaments. This structural arrangement provides limited movements between vertebrae but extensive movements for the entire vertebral column. Between the vertebrae are openings called **intervertebral** *(in''ter-ver'te-bral)* **foramina** that permit passage of spinal nerves.

Four curvatures of the vertebral column of an adult can be identified and viewed from the side (fig. 9.31). The **cervical, thoracic,** and **lumbar curves** are identified by the type of vertebrae they include. The **pelvic curve** is formed by the shape of the sacrum and coccyx *(kok'siks).* The curves of the vertebral column play an important functional role in increasing the strength and maintaining the balance of the upper portion of the body and also make possible a bipedal stance.

The four vertebral curves are not present in a newborn baby. Instead, the vertebral column is somewhat anteriorly concave, and except for the cervical region, it remains this way even as an infant learns to crawl (fig. 9.32). The cervical curve begins to develop at about three months of age as a baby begins holding up its head and becomes more pronounced as the baby learns to sit up. The lumbar curve develops as a child begins to walk. The thoracic and pelvic curves are called **primary curves** because they retain the anteriorly concave shape of the fetus. The cervical and lumbar curves are called **secondary curves** because they are modifications of the fetal shape.

Figure 9.33.   Cervical vertebrae. (a) an X-ray of the cervical region. (b) a typical cervical vertebrae. (c) the atlas and axis as they articulate.

(b)

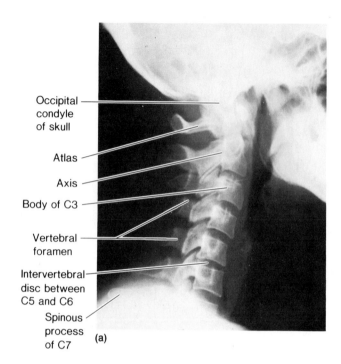

Occipital condyle of skull

Atlas

Axis

Body of C3

Vertebral foramen

Intervertebral disc between C5 and C6

Spinous process of C7   (a)

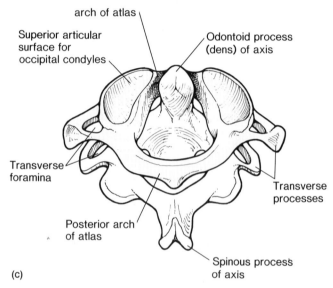

(c)

The vertebral column is commonly called the "backbone" and together with the spinal cord of the nervous system constitutes the spinal column. The vertebral column has three basic functions:

1. To support the head and upper extremities while permitting freedom of movement
2. To provide attachment for various muscles, ribs, and visceral structures
3. To protect the spinal cord and permit passage of the spinal nerves

General Structure of Vertebrae

Vertebrae show similarities in their general structure from one region to another. Figure 9.33 illustrates a typical vertebra. A vertebra is usually composed of an anterior drum-shaped **body (centrum)** adapted to withstand compression.

The body is in contact with **intervertebral discs** on each end. The **vertebral,** or **neural, arch** is affixed to the posterior surface of the body and is composed of two supporting **pedicles** *(ped'i-k'l)* and two arched **laminae.** The hollow space formed by the vertebral arch and body is the **vertebral foramen,** through which the spinal cord passes. Between the pedicles of adjacent vertebrae are the **intervertebral foramina,** through which spinal nerves emerge as they branch off from the spinal cord.

Seven processes arise from the vertebral arch: the **spinous** *(spi'nus)* **process,** two **transverse processes,** two **superior articular processes,** and two **inferior articular processes** (fig. 9.34). The first two processes serve for muscle attachment, and the latter two pairs limit twisting of the vertebral column. The spinous process protrudes posteriorly and inferiorly from the vertebral arch. The transverse process extends laterally from each side of a vertebra at the point where the lamina and pedicle join.

pedicle: L. *pediculus,* small foot
lamina: L. *lamina,* thin layer

Figure 9.34. Thoracic vertebrae. (*a*) representative vertebrae in a right lateral view; (*b*) a superior view.

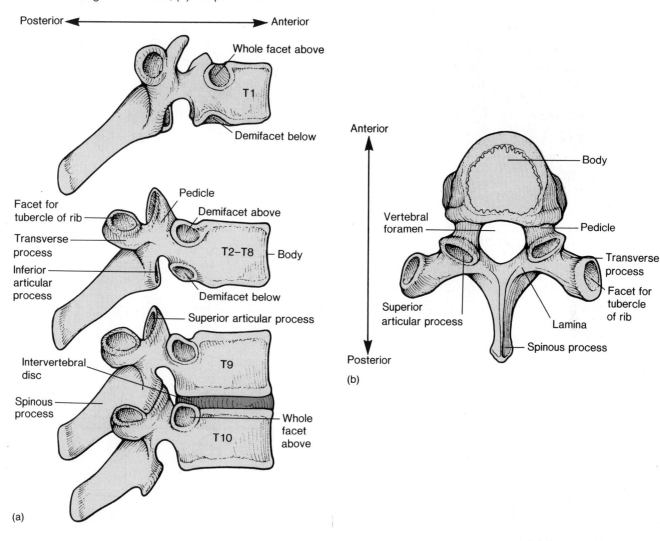

(a)

(b)

The superior articular processes of a vertebra have interlocking articulations with the inferior articular processes of the adjacent anterior vertebra.

Regional Characteristics of Vertebrae

Cervical Vertebrae. The seven cervical vertebrae form the flexible framework of the neck region and support the head. The osseous tissue of cervical vertebrae is more dense than that of the other vertebral regions, and, except for the coccygeal region, the cervical vertebrae are smallest. Cervical vertebrae are distinguished by the presence of a **transverse foramen** in the transverse process (fig. 9.33). The vertebral vessels pass through this opening as they transfer blood to and from the brain. The spinous processes of the third, fourth, and fifth cervical vertebrae are **bifid** *(bi'fid),* or notched, for the attachment of the strong **nuchal ligament,** which attaches to the back of the skull for added support.

The first cervical vertebra, the **atlas,** is adapted to articulate with the occipital condyles of the skull while supporting the head. The atlas has concave **superior articular surfaces** to articulate with the oval-shaped occipital condyles. This joint permits the nodding of the head in a "yes" movement. The atlas lacks a body. It has a short, bifurcated spinous process.

The second cervical vertebra is called the **axis** and is easily identified by the presence of a peglike projection called the **odontoid** *(o-don'toid)* **process,** or **dens.** This process extends superiorly to provide a pivot for rotation with the atlas. This joint permits rotation, or the turning of the head to the side, as in a "no" movement.

---

atlas: from Gk. mythology, Atlas—the Titan who supported the heavens
axis: L. *axis,* axle
odontoid: Gk. *odontos,* tooth

Figure 9.35. Lumbar vertebrae. (*a*) an X-ray; (*b*) a superior view; (*c*) a right lateral view.

(a)

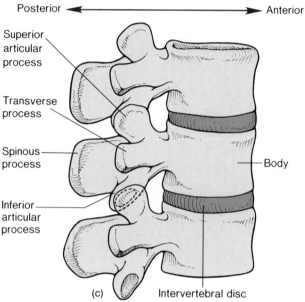

(b)

(c)

Thoracic Vertebrae. The thoracic vertebrae serve as attachments of the ribs to form the posterior anchor of the rib cage. Thoracic vertebrae are larger than cervical vertebrae and increase in size from superior (T1) to inferior (T12). Each thoracic vertebra has a long spinous process that slopes obliquely downward and facets or demifacets for articulation with the ribs (fig. 9.34). The first thoracic vertebra has a superior facet and an inferior demifacet on both sides. The second through eighth have two demifacets on each side so that each head of a rib articulates with two vertebrae. The ninth vertebra has a single demifacet on each side, and the tenth through the twelfth have facets. The first ten vertebrae also have facets on their transverse processes for articulation with the tubercles of the ribs.

Lumbar Vertebrae. The five lumbar vertebrae are easily identified by their heavy bodies and thick, blunt spinous processes (fig. 9.35) for attachment of powerful back muscles. They are the largest vertebrae of the vertebral column. Their articular processes are also distinctive in that the superior articular processes are directed medially instead of superiorly, and the inferior articular processes are directed laterally instead of inferiorly.

lumbar: L. *lumbus*, loin

Figure 9.36. The sacrum and coccyx. (*a*) an anterior view; (*b*) a posterior view.

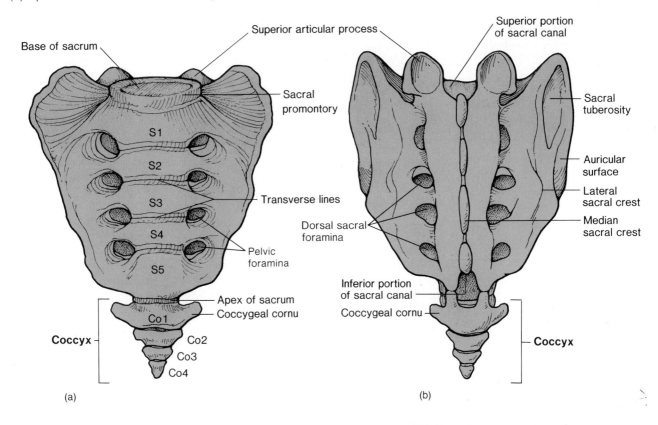

(a)

(b)

Sacrum. The wedge-shaped sacrum (fig. 9.36) consists of five sacral vertebrae, which become fused after age twenty-six. The sacrum is functionally adapted to provide a strong foundation for the pelvic girdle. The sacrum has an extensive **auricular surface** on each side for the formation of an immovable **sacroiliac** *(sa''kro-il'e-ak)* **joint** with the os coxa. A **median sacral crest** is formed along the dorsal surface by the fusion of the spinous processes. **Dorsal sacral foramina** on either side of the crest allow the passage of nerves from the spinal cord. The **sacral canal** is the tubular cavity within the sacrum that is continuous with the vertebral canal. Paired **superior articular processes,** which articulate with the fifth lumbar vertebra, arise from the roughened **sacral tuberosity** along the dorsal surface.

The smooth anterior surface of the sacrum forms the posterior surface of the pelvic cavity. It has four **transverse lines** denoting the fusion of the vertebral bodies. On either side of the transverse lines are the paired **pelvic foramina.** The superior border of the anterior surface of the sacrum, called the **sacral promontory** *(prom'on-to''re),* is an important obstetrical landmark for pelvic measurements.

Coccyx. The coccyx *(kok'siks)* is the so-called tailbone. It is composed of four or five fused coccygeal vertebrae, which form a triangular-shaped structure. The first vertebra of the fused coccyx has two long **coccygeal cornua,**

Table **9.7**	Regions of the vertebral column	
Region	**Number of bones**	**Diagnostic features**
Cervical	7	Transverse foramina; superior facets of atlas articulate with occipital condyle; odontoid process of axis; spinous processes of 3d through 5th vertebrae are bifid
Thoracic	12	Long spinous processes that slope obliquely inferiorly; facets and demifacets for articulation with ribs
Lumbar	5	Large bodies; prominent transverse processes; short, thick spinous processes
Sacrum	5 fused vertebrae	Extensive auricular surface; median sacral crest; dorsal sacral foramina; sacral promontory; sacral canal
Coccyx	4 fused vertebrae	Small, triangular; coccygeal cornua

Source: Van De Graaff, Kent M., *Human Anatomy.* © 1984 Wm. C. Brown Publishers, Dubuque, Iowa. All Rights Reserved. Reprinted by permission.

which are attached by ligaments to the sacrum (fig. 9.36). Lateral to the cornua are the transverse processes.

The regions of the vertebral column are summarized in table 9.7.

sacrum: L. *sacris,* sacred
coccyx: Gk. *kokkyx,* like a cuckoo's beak

Figure 9.37.   The rib cage.

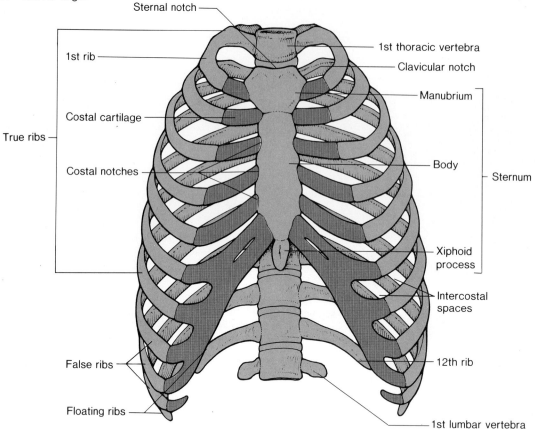

When a person sits, the coccyx flexes anteriorly somewhat, acting as a shock absorber. An abrupt fall on the coccyx, however, may cause a painful subperiosteal bruising, fracture, or fracture-dislocation of the sacrococcygeal joint. An especially difficult childbirth can even injure the coccyx of the mother. Coccygeal trauma is painful and may require months to heal.

1. *Which are the primary curves of the vertebral column, and which are the secondary curves? Describe the characteristic curves of each region.*
2. *What is the function of the transverse foramina of the cervical vertebrae?*
3. *Describe the diagnostic differences between a thoracic and a lumbar vertebra. Which structures are similar and could therefore be characteristic of a typical vertebra?*

Rib Cage

The cone-shaped and flexible rib cage consists of the thoracic vertebrae, twelve paired ribs, costal cartilages, and the sternum. It encloses and protects the thoracic viscera and is directly involved in the mechanics of breathing.

Objective 14.   Identify the parts of the rib cage, and distinguish between the various types of ribs.

The **sternum, ribs, costal cartilages,** and the previously described thoracic vertebrae form the rib cage, or **thoracic cage,** of the thorax (fig. 9.37). The rib cage is anteroposteriorly compressed and more narrow superiorly than inferiorly. It supports the pectoral girdle and upper extremities, protects and supports the thoracic and upper abdominal viscera, and plays a major role in breathing. Certain bones of the rib cage contain active sites for blood cell production in the red bone marrow.

Sternum.   The sternum (breastbone) is an elongated, flattened bony plate consisting of three separate bones: the upper **manubrium,** the central **body,** and the lower **xiphoid** *(zi'foid)* **process.** On the lateral sides of the sternum are **costal notches** where the costal cartilages attach. A **sternal notch** is formed at the superior end of the manubrium, and a **clavicular** *(klah-vik'u-lar)* **notch** for articulation with the clavicle is present on both sides of the sternal notch. The manubrium articulates with the costal cartilages of the first and second ribs. The body of the sternum attaches to

sternum: Gk. *sternon*, chest
manubrium: L. *manubrium*, a handle
xiphoid: Gk. *xiphos*, sword
costal: L. *costa*, rib

Figure 9.38.   The structure of a rib.

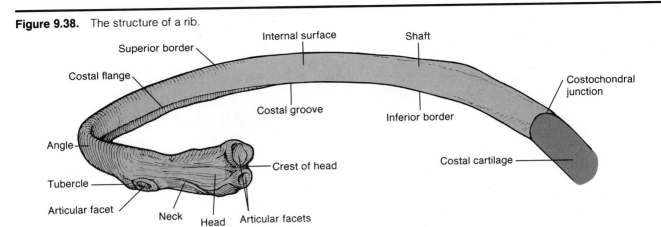

the costal cartilages of the second through the tenth ribs. The xiphoid process does not attach to ribs but is an attachment for abdominal muscles. The costal cartilages of the eighth, ninth, and tenth ribs fuse to form the costal margin. A costal angle is formed where the two costal margins come together at the xiphoid process. The **sternal angle** (angle of Louis) may be palpated as a depression between the manubrium and body of the sternum at the level of the second rib. The costal angle, costal margins, and sternal angle are important surface landmarks of the chest and abdomen.

Ribs.   There are twelve pairs of ribs, each pair being attached posteriorly to a thoracic vertebra. Anteriorly, the first seven pairs are anchored to the sternum by individual costal cartilages and are called **true ribs.** Ribs 8, 9, and 10 are attached to the costal cartilage of rib 7 and are termed **false ribs.** The remaining two paired ribs do not attach to the sternum and are referred to as **floating ribs.** They are attached to the muscles of the body wall.

Although the structure of ribs varies, each of the first ten pairs has a **head** and a **tubercle** for articulation with a vertebra. The last two have a head but no tubercle. In addition, each of the twelve pairs has a **neck, angle,** and **shaft** (fig. 9.38). The head of a rib projects posteriorly and articulates with the body of a thoracic vertebra. The tubercle is a knoblike process just lateral to the head. It articulates with the facet of the transverse process. The neck is the constricted area between the head and the tubercle. The **shaft,** or **body,** is the main, curved part of the rib. Along the inner surface of the shaft is a depressed canal called the **costal groove,** which protects the costal vessels and nerve. Spaces between the ribs are called **intercostal spaces** and are occupied by the intercostal muscles.

Fractures of the ribs are relatively common injuries, and most frequently occur between ribs 3 and 10. The first two pairs of ribs are protected by the clavicles, and the last two pairs move freely and will give with an impact. Little can be done to assist the healing of broken ribs other than binding them tightly to restrict movement.

1. *Describe the rib cage, and list its functions. What determines if a rib is designated as true, false, or floating?*
2. *Define the costal margin and the costal angle.*

Pectoral Girdle and Upper Extremity

The structure of the pectoral girdle and upper extremities is adaptive for freedom of movement and extensive muscle attachment.

> *Objective 15.*   Describe the bones of the pectoral girdle and the positions of articulations.
>
> *Objective 16.*   Identify the bones of the upper extremity, and list the diagnostic features of each bone.

Pectoral Girdle

The two scapulae and two clavicles make up the pectoral (shoulder) girdle. It is not a complete girdle, having only an anterior attachment to the axial skeleton at the sternum. The primary function of the pectoral girdle is to provide attachment for the numerous muscles that move the brachium and forearm. The pectoral girdle is not weight-bearing and is therefore more delicate in structure than the pelvic girdle.

Clavicle.   The slender S-shaped clavicle (collarbone) binds the shoulder to the axial skeleton and positions the shoulder joint away from the trunk for freedom of movement. The articulation of the medial **sternal end** of the clavicle to the manubrium is referred to as the **sternoclavicular joint** (fig. 9.39). The lateral **acromial** *(ah-kro'me-al)* **end** of the clavicle articulates with the acromion process of the scapula. This articulation is referred to as the **acromioclavicular joint.** A **conoid tubercle** is present on the inferior surface of the lateral end, and a **costal tuberosity** is present on the inner surface of the medial end. Both processes serve as points of attachment for ligaments.

Figure 9.39.   The right clavicle. (*a*) a superior view;
(*b*) an inferior view.

(a)

(b)

Figure 9.40.   The right scapula. (*a*) an anterior view;
(*b*) a posterior view.

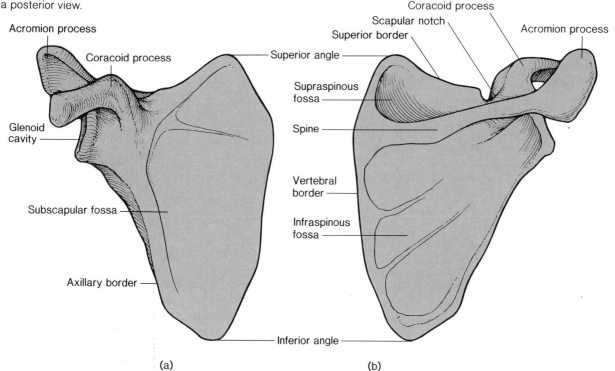

(a)   (b)

Scapula.   The scapula (shoulder blade) is a large triangular flat bone positioned on the posterior aspect of the rib cage, overlying ribs 2 to 7. The *spine* of the scapula is a prominent diagonal bony ridge seen on the posterior surface (fig. 9.40). Above the spine is the **supraspinous** *(su″pra-spi′nus)* **fossa,** and below the spine is the **infraspinous fossa.** The spine broadens toward the shoulder as the **acromion** *(ah-kro′me-on)* **process.** This process serves

for the attachment of several muscles as well as articulation with the clavicle. Inferior to the acromion is a shallow depression, the **glenoid cavity,** into which the head of the humerus fits. The **coracoid process** is a thick, upward projection that lies superior and anterior to the glenoid cavity. On the anterior surface of the scapula is a slightly concave area known as the **subscapular fossa.**

The scapula has three borders separated by two angles. The superior edge is called the **superior border.** The

---

scapula: L. *scapula,* shoulder

---

glenoid: Gk. *glenoeides,* shallow form

Figure 9.41.   The right humerus. (*a*) an anterior view; (*b*) a posterior view.

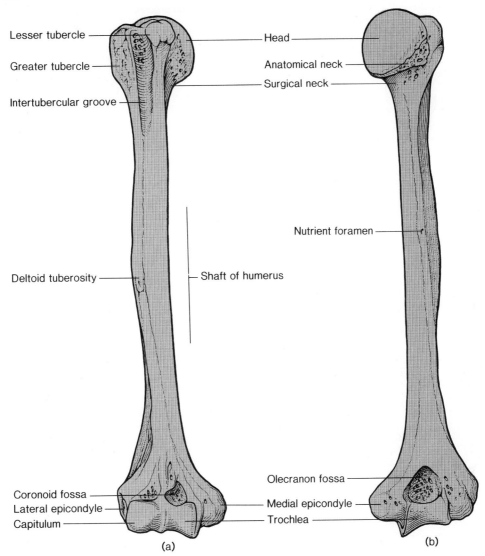

Lesser tubercle

Greater tubercle

Intertubercular groove

Head

Anatomical neck

Surgical neck

Nutrient foramen

Deltoid tuberosity

Shaft of humerus

Coronoid fossa
Lateral epicondyle
Capitulum

Olecranon fossa

Medial epicondyle

Trochlea

(a)

(b)

vertebral, or **medial, border** is nearest the vertebral column positioned about 5 cm (2 in.) away. The **axillary,** or **lateral, border** is directed toward the arm. The **superior angle** is between the superior and vertebral borders, while the **inferior angle** is between the vertebral and axillary borders. Along the superior border is a distinct depression called the **scapular notch.**

The anatomy of the scapula is important to know because some fifteen muscles attach to its processes and fossae. Clinically, the pectoral girdle is significant because the clavicle and acromion process of the scapula are frequently fractured in trying to break a fall. The acromion process is palpated when locating the proper site for an intramuscular injection of the arm.

Brachium

The brachium is the upper arm and contains a single bone, the humerus.

Humerus.   The humerus (fig. 9.41) is the longest bone of the upper extremity. It consists of a proximal **head,** which articulates with the glenoid cavity of the scapula, a **shaft** (body), and a distal end, which is modified to receive the two bones of the forearm. Surrounding the margin of the head is a slightly indented groove denoting the **anatomical neck.** The region where the shaft begins to taper is referred to as the **surgical neck,** a frequent site of fractures. Lateral to the head is a large eminence, the **greater tubercle.** The **lesser tubercle** is slightly anterior to the greater and is separated from the greater by an **intertubercular (bicipital) groove,** through which passes the tendon from the biceps brachii muscle.

Along the lateral midregion of the shaft is a roughened area, the **deltoid tuberosity;** this is important for the attachment of the deltoid muscle, which elevates the arm horizontally to the side. Small openings in the bone along the shaft are called **nutrient foramina.**

Figure 9.42. An anterior view of the right radius and ulna. A cross section shows the binding interosseous ligament.

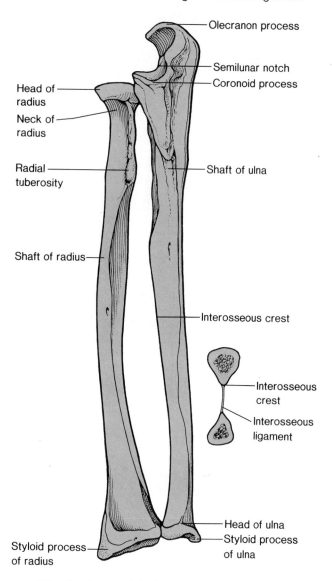

Figure 9.43. A posterior view of the right radius and ulna.

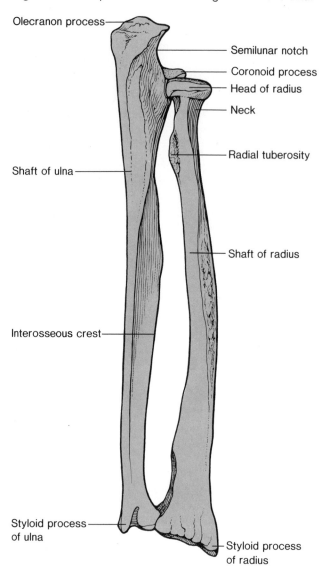

The distal end of the humerus has two rounded condyloid articular surfaces. The **capitulum** *(kah-pit'u-lum)* is the lateral rounded condyle that receives the radius. The **trochlea** *(trok'le-ah)* is the pullylike medial surface that articulates with the ulna. On either side above the condyles are the **lateral** and **medial epicondyles.** The large medial epicondyle protects the ulnar nerve that passes through the ulnar sulcus (see fig. 12.19). The **coronoid fossa** is a depression above the trochlea on the anterior surface. The **olecranon** *(o-lek'rah-non)* **fossa** is a depression on the distal posterior surface. Both fossae are adapted to receive parts of the ulna during movement of the forearm.

Forearm

The skeletal structures of the forearm are the ulna on the medial side and the radius on the lateral (thumb) side (figs. 9.42, 9.43). The ulna is more firmly connected to the humerus than the radius and is longer. The radius, however, contributes more significantly at the wrist joint than does the ulna.

Ulna. The proximal end of the ulna articulates with the humerus and radius. A distinct depression, the **semilunar (trochlear) notch,** articulates with the trochlea of the humerus. The **coronoid process** forms the anterior lip of the semilunar notch, while the **olecranon process** forms the posterior portion, or elbow. Lateral and inferior to the coronoid process is the **radial notch,** which accommodates the head of the radius.

Along the anterolateral surface of the shaft is a sharp ridge called the **interosseous crest.** A ligament extends from this crest to the radius to bind these two bones together (fig. 9.42).

Figure 9.44. A posterior view of the right wrist and hand. (*a*) a drawing; (*b*) a photograph.

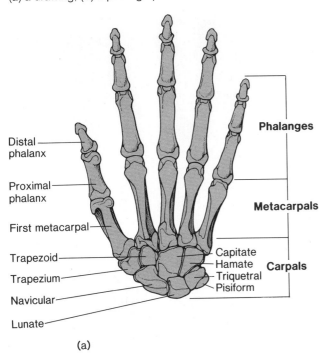

Distal phalanx
Proximal phalanx
First metacarpal
Trapezoid
Trapezium
Navicular
Lunate
Phalanges
Metacarpals
Capitate
Hamate
Triquetral
Pisiform
Carpals

(a)

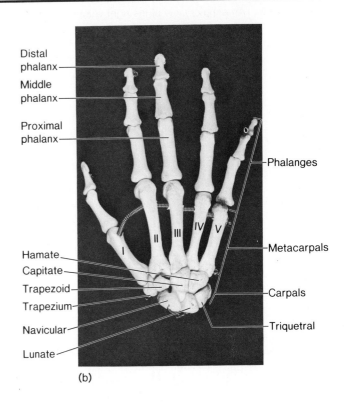

Distal phalanx
Middle phalanx
Proximal phalanx
Hamate
Capitate
Trapezoid
Trapezium
Navicular
Lunate
Phalanges
Metacarpals
Carpals
Triquetral

(b)

The tapered distal end of the ulna has a knobbed portion, the **head,** and a posteromedial projection, the **styloid process.** The ulna articulates with the radius distally as well as proximally.

Radius.   The radius consists of a **shaft** that has a small proximal end and a large distal end. A proximal disc-shaped **head** articulates with the capitulum of the humerus and the radial notch of the ulna. The prominent **radial tuberosity,** for attachment of the biceps muscle, is on the medial side of the shaft just below the head. The distal end of the radius has a double-faceted articular surface: a lateral **navicular facet** and a medial **lunate facet.** The distal end of the radius also has the **styloid process** on the lateral tip and the **ulnar notch** on the medial side for receiving the distal end of the ulna. The styloid processes on the ulna and radius provide lateral and medial support for articulation at the wrist.

When a person falls, the natural tendency is to extend the hand to break the fall. This reflexive movement frequently results in fractured bones. Common fractures of the radius include a fracture of the head as it is driven forcefully against the capitulum, a fracture of the neck, or a fracture of the distal end (Colles' fracture) caused by landing on an outstretched hand.

When falling, it is less traumatic to the body to withdraw the appendages, bend the knees, and let the entire body hit the surface. Athletes learn that this is the safe way to fall.

Wrist and Hand

The wrist and hand contain twenty-seven bones arranged into the carpus, metacarpus, and phalanges (figs. 9.44, 9.45, 9.46).

Carpus.   The carpus, or wrist, consists of eight carpal bones arranged into two transverse rows of four bones each. The proximal row, naming from the lateral (thumb) to medial side, consists of the **navicular** (scaphoid), **lunate, triquetral** *(tri-kwe'tral)* (triangular), and **pisiform** *(pi'si-form).* The pisiform forms in a tendon as a sesamoid bone. The distal row, from lateral to medial, consists of the **trapezium** (greater multangular), **trapezoid** (lesser multangular), **capitate,** and **hamate** *(ham'at).* The navicular and lunate of the proximal row articulate with the distal end of the radius.

Metacarpus.   The metacarpus, or palm of the hand, is composed of five metacarpal bones. Each metacarpal bone consists of a proximal **base,** a **shaft,** and a distal **head,** which is rounded for articulation with the base of each proximal phalanx. The heads of the metacarpals are the knuckles of a clinched fist. The metacarpal bones are numbered from one to five, the lateral, or thumb, side being one.

carpus: Gk. *karpos,* wrist
navicular: L. *navicula,* small ship
lunate: L. *lunare,* crescent or moon-shaped
triquetral: L. *triquetrus,* three-cornered
pisiform: Gk. *pisos,* pea
trapezium: Gk. *trapesion,* small table
capitate: L. *capitatus,* head
hamate: L. *hamatus,* hook

Figure 9.45.   An anterior view of the right wrist and hand.

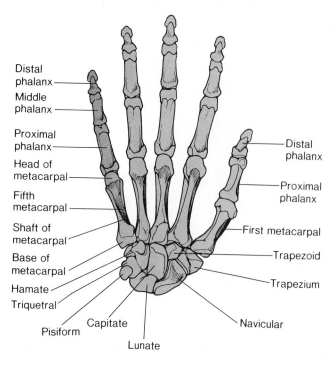

Figure 9.46.   An X-ray of the right wrist and hand shown in an anteroposterior projection.

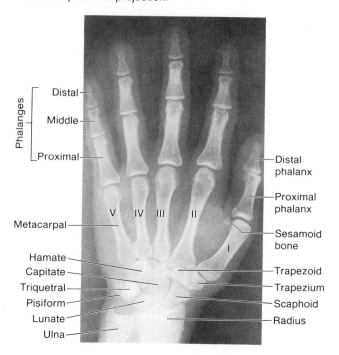

Table 9.8   Bones of the pectoral girdles and upper extremities

Name and number	Location	Diagnostic features
Clavicle (2)	Anterior base of neck, between sternum and scapula	S-shaped; sternal and acromial ends
Scapula (2)	Upper back forming part of the shoulder	Triangular-shaped; spine; acromion and corocoid processes
Humerus (2)	Brachium between scapula and elbow	Longest bone of upper extremity; greater and lesser tubercles; surgical neck; deltoid tuberosity; capitulum; trochlea; coronoid and olecranon fossae
Ulna (2)	Medial side of forearm	Semilunar notch; olecranon and styloid processes
Radius (2)	Lateral side of forearm	Head; radial tuberosity; styloid process
Carpal (16)	Wrist	Short bones arranged in two rows of four bones each
Metacarpal (10)	Palm of hand	Long bones arranged one in line with each digit
Phalanx (28)	Digits	Three in each finger, except two in thumb

Source: Van De Graaff, Kent M., *Human Anatomy.* © 1984 Wm. C. Brown Publishers, Dubuque, Iowa. All Rights Reserved. Reprinted by permission.

Phalanges.   The fourteen phalanges are the skeletal elements of the fingers. A single finger bone is called a **phalanx** *(fa'lanks).* The phalanges of the fingers are arranged into a proximal row, a middle row, and a distal row. The thumb has only a proximal and a distal phalanx. A summary of the bones of the upper extremities is presented in table 9.8.

phalanx: Gk. *phalanx,* finger bone

The hand is a marvel of structural organization, which despite its complexity, is able to sustain considerable abuse. Other than sprained fingers and dislocation, the most common osseous injury is a fracture to the navicular bone of the wrist (about 70% of carpal fractures occur here). When immobilizing the wrist joint with a plaster cast, the wrist is positioned in the plane of relaxed function. This is the position in which the hand is about to grasp an object between the thumb and index finger.

Figure 9.47. An anterior view of the pelvic girdle.

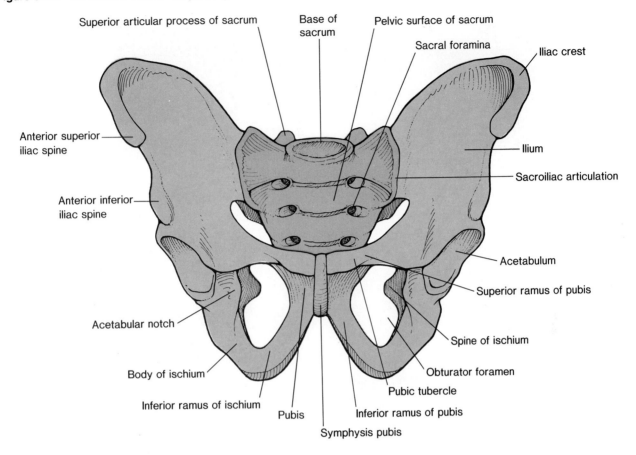

1. Describe the structure of the pectoral girdle. Why is the pectoral girdle considered an incomplete girdle?
2. Name the fossae found on the scapula.
3. Describe each of the long bones of the upper extremity.
4. Where are the styloid processes of the wrist area? What are their functions?
5. Name the bones in the proximal row of the carpus. Which of these bones articulate with the radius?

Pelvic Girdle and Lower Extremity

The structure of the pelvic girdle and lower extremities is adaptive for support and locomotion. There are extensive processes and surface features on certain bones of the pelvic girdle and lower extremities that accommodate massive muscles for posture and locomotion.

Objective 17. Describe the structure of the pelvic girdle, and list its functions.
Objective 18. Describe the structural differences between the male and female pelvis.
Objective 19. Identify the bones of the lower extremity, and list the diagnostic features of each bone.
Objective 20. Describe the structural features and functions of the arches of the foot.

Pelvic Girdle

The pelvic girdle, or pelvis, is formed by two **ossa coxae** (hipbones) united anteriorly by the **symphysis pubis** (fig. 9.47). It is attached posteriorly to the sacrum and coccyx of the vertebral column. The pelvic girdle and its associated ligaments support the weight of the body through the vertebral column. The pelvic girdle also supports and protects the lower viscera, including the urinary bladder, the reproductive organs, and in a pregnant woman, the developing fetus.

Clinically speaking, the basinlike pelvis is frequently divided into a **greater,** or **false, pelvis** and a **lesser,** or **true, pelvis** (see fig. 9.50). These two components are divided by the **pelvic brim,** a curved bony rim passing inferiorly from the sacral promontory to the upper margin of the symphysis pubis. The greater pelvis is the expanded portion of the pelvis superior to the pelvic brim. The pelvic

coxae: L. *coxae,* hips

Figure 9.48. An X-ray of the pelvic girdle and the articulating femora.

Fifth lumbar vertebra
Ilium
Sacrum
Coccyx
Pelvic inlet
Anterior inferior iliac spine
Acetabulum
Pelvic brim
Pubis

Sacral promontory
Sacroiliac joint
Head of femur
Neck of femur
Greater trochanter
Obturator foramen
Lesser trochanter
Ischium
Symphysis pubis

brim not only divides the two portions but surrounds the **pelvic inlet** of the lesser pelvis. The lower circumference of the lesser pelvis bounds the pelvic outlet. During parturition, a child must pass through its mother's lesser pelvis for a natural delivery. *Pelvimetry* measures the dimension of the lesser pelvis to determine if a cesarean delivery might be necessary. Diameters may be determined by vaginal palpation or by X-ray measurements (fig. 9.48).

Each os coxa actually consists of three separate bones: the **ilium,** the **ischium** *(is'ke-um),* and the **pubis** (fig. 9.49). These bones are fused together in the adult. On the lateral surface of the os coxa where the three bones ossify together is a large circular depression, the **acetabulum** *(as''ĕ-tab'u-lum),* which receives the head of the femur. Although in the adult both ossa coxae are single bones, the three components are considered separately for descriptive purposes.

Ilium. The ilium is the largest and uppermost of the three pelvic bones. The ilium presents a crest and four angles, or spines, which serve for muscle attachment and are important surface landmarks. The **iliac crest** forms the prominence of the hip. This crest terminates anteriorly as

the **anterior superior iliac spine.** Just below this spine is the **anterior inferior iliac spine.** The posterior termination of the iliac crest is the **posterior superior iliac spine,** and just below this is the **posterior inferior iliac spine.**

Below the posterior inferior iliac spine is the **greater sciatic** *(si-at'ik)* **notch.** On the medial surface of the ilium is the roughened **auricular surface** that articulates with the sacrum. The **iliac fossa** is the smooth, concave surface on the anterior portion of the ilium. The iliacus muscle originates from this fossa. The **iliac tuberosity,** for the attachment of the sacroiliac ligament, is positioned posterior to the iliac fossa. Three roughened ridges are present on the **gluteal surface** of the posterior aspect of the ilium. These ridges serve to attach the gluteal muscles and are the **inferior, anterior,** and **posterior gluteal lines.**

Ischium. The ischium is the posterior-inferior component of the os coxa. This bone has several significant features. The **spine of the ischium** is the projection immediately posterior to the greater sciatic notch of the ilium and ischium. Inferior to this spine is the **lesser sciatic notch** of the ischium. The **ischial tuberosity** is the bony projection that supports the weight of the body in the sitting position. A deep **acetabular notch** is present on the inferior portion of the acetabulum. The large **obturator foramen** is formed by the **ramus** of the ischium together with the pubis.

ilium: L. *ilia,* loin
ischium: Gk. *ischion,* hip joint
pubis: L. *pubis,* genital area
acetabulum: L. *acetabulum,* vinegar cup

Figure 9.49. The lateral aspect of the right os coxa.

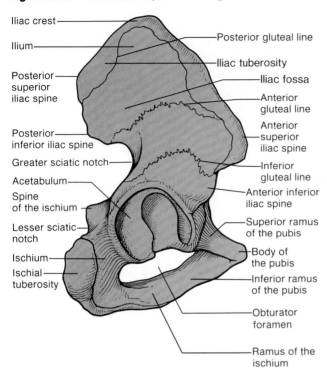

Iliac crest
Ilium
Posterior superior iliac spine
Posterior inferior iliac spine
Greater sciatic notch
Acetabulum
Spine of the ischium
Lesser sciatic notch
Ischium
Ischial tuberosity

Posterior gluteal line
Iliac tuberosity
Iliac fossa
Anterior gluteal line
Anterior superior iliac spine
Inferior gluteal line
Anterior inferior iliac spine
Superior ramus of the pubis
Body of the pubis
Inferior ramus of the pubis
Obturator foramen
Ramus of the ischium

Figure 9.50. A comparison of the (a) male and (b) female pelvic girdle.

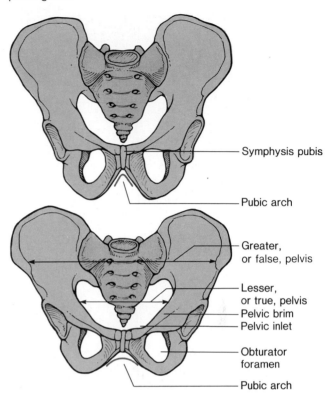

Symphysis pubis
Pubic arch
Greater, or false, pelvis
Lesser, or true, pelvis
Pelvic brim
Pelvic inlet
Obturator foramen
Pubic arch

Pubis. The pubis is the anterior component of the os coxa. This bone consists of a **superior ramus** and an **inferior ramus** that supports the **body** of the pubis. The body contributes to the formation of the symphysis pubis—the joint between the two ossa coxae.

The structure of the human pelvis, in its attachment to the vertebral column, permits an upright posture and locomotion on two legs (bipedal) rather than on four legs like other mammals. Although these structures are well adapted for bipedal locomotion, an upright posture may cause problems. The sacroiliac joint may weaken with age, causing lower back pains. The weight of the viscera may weaken the walls of the lower abdominal area and cause hernias. Some of the problems of childbirth are related to the structure of the mother's pelvis. The hip joint tends to deteriorate with age, and many older persons suffer from fractured hips.

There are structural differences between the pelvis of an adult male and that of an adult female (fig. 9.50) that reflect the female's role in pregnancy and childbirth. The following are characteristics of the female pelvis that differ from that of the male:

1. The entire pelvic girdle is tilted forward.
2. The bones are thinner and lighter.
3. The distance between the anterior superior iliac spines (greater, or false, pelvis) is wider.

4. The true pelvis (space posterior to the pelvic brim) is spherical and wider.
5. The outline of the obturator foramen is more triangular.
6. The symphysis pubis is shallower.
7. The pubic angle is wider and more rounded.

In addition to these osseous differences, the symphysis pubis and sacroiliac joints stretch during pregnancy and parturition.

Thigh
The patella will be considered along with the femur in the discussion of the thigh, even though the femur is the only bone of the thigh.

Femur. The femur (thighbone) is the longest, heaviest, and strongest bone in the body (fig. 9.51). The proximal rounded **head** of the femur articulates with the acetabulum of the os coxa. A roughened, shallow pit, called the **fovea capitis,** is in the lower center of the head of the femur. The fovea capitis provides the point of attachment for the ligamentum teres, which helps to support the head of the femur against the acetabulum. The constricted region supporting the head is called the **neck** and is a common site for fractures in aged persons.

femur: L. *femur,* thigh

Figure 9.51.   The right femur. (a) an anterior view; (b) a posterior view.

Head of femur
Fovea capitus
Greater trochanter
Neck of femur
Intertrochanteric line
Intertrochanteric crest
Lesser trochanter
Linea aspera
Shaft of femur
Lateral epicondyle
Intercondylar fossa
Patellar surface
Medial epicondyle
Medial and lateral condyles

(a)   (b)

The **shaft** of the femur has a slight medial bow so that it converges with the femur of the opposite thigh and brings the knee joints more in line with the body's plane of gravity. The degree of convergence is even greater in the female because of the wide pelvis. The shaft has several important structures for muscle attachment. On the proximal-lateral side of the shaft is the **greater trochanter,** and on the medial side is the **lesser trochanter.** On the anterior side between the trochanters is the **intertrochanteric line.** On the posterior side between the trochanters is the **intertrochanteric crest. The linea aspera** is a roughened vertical ridge on the posterior surface of the shaft.

The distal end of the femur is expanded for articulation with the tibia. The **medial** and **lateral condyles** (*kon'dīl*) are the articular processes for this joint. The

linea aspera: L. *linea*, line; *asperare*, rough

Figure 9.52.   A lateral X-ray of the right knee region.

Femur

Lateral epicondyle of femur

Patella

Head of tibia

Tibia

Fibula

Figure 9.53.   The right tibia, fibula, and patella. (*a*) an anterior view; (*b*) a posterior view.

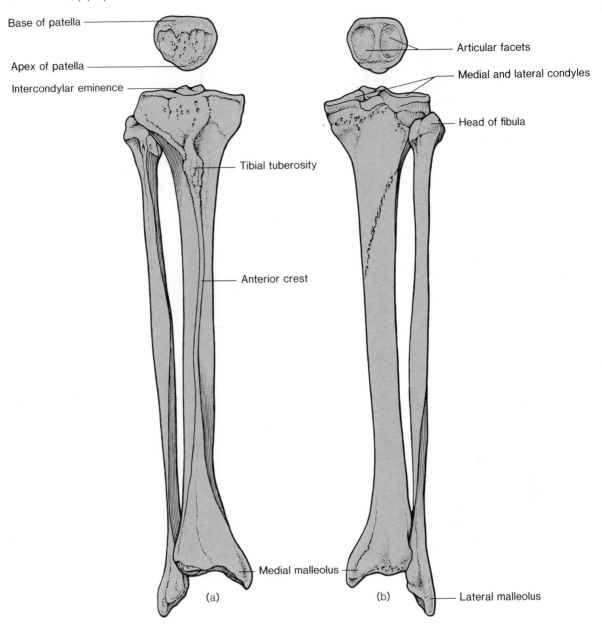

Base of patella

Apex of patella

Intercondylar eminence

Tibial tuberosity

Anterior crest

Medial malleolus

(a)

Articular facets

Medial and lateral condyles

Head of fibula

Medial malleolus

Lateral malleolus

(b)

depression between the condyles on the posterior aspect is called the **intercondylar fossa.** The **patellar surface** is between the condyles on the anterior side. Above the condyles on the lateral and medial sides are the epicondyles, which serve for ligament and tendon attachment.

Patella.   The patella, or kneecap, is a sesamoid bone positioned on the anterior side of the knee joint (figs. 9.52 and 9.53). It develops in response to stress in the tendon of the quadriceps femoris muscle. The patella is a triangular bone with a broad **base** and an inferiorly pointed **apex. Articular facets** on the posterior surface of this bone articulate with the medial and lateral condyles of the femur.

The functions of the patella are to protect the knee joint and to strengthen the quadriceps tendon. It also increases the leverage of the quadriceps femoris muscle as it straightens (extends) the leg.

The patella can be fractured by a direct blow. It usually does not fragment, however, because it is confined within the tendon. Dislocations of the patella may result from injury or may be congenital due to underdevelopment of the lateral condyle of the femur.

Figure 9.54. (a) a superior view of the right foot. (b) a medial X-ray of the right foot and ankle. (c) a superior view of the bones of the right ankle and foot. (d) an inferior view of the bones of the right ankle and foot.

Figure 9.55. The arches of the foot. (*a*) a medial view of the left foot showing both arches. (*b*) a transverse view through the bases of the metatarsals showing a portion of the transverse arch.

(a)

(b)

Lower Leg

The tibia and fibula are the skeletal elements of the lower leg. The tibia is the larger and more medial of the two bones. Figure 9.53 illustrates the skeletal structure of the lower leg.

Tibia. The tibia (shinbone) articulates proximally with the femur at the knee joint to bear the weight of the body. On the distal end, the tibia articulates with the talus of the ankle. Two slightly concave surfaces on the proximal end of the tibia, the **medial** and **lateral condyles,** articulate with the condyles of the femur. Between the condyles is a slight upward projection called the **intercondylar eminence.** The **tibial tuberosity,** for attachment of the patellar ligament, is located on the proximal-anterior portion of the shaft. There is a sharp ridge along the anterior surface of the shaft called the **anterior crest.**

The **medial malleolus** *(mal-le'o-lus)* is a prominent medial knob of bone located on the distal end of the tibia. A **fibular notch,** for articulation with the fibula, is located on the distal-lateral end.

Fibula. The fibula is a long, narrow bone that is more important for muscle attachment than for support. The **head** of the fibula articulates with the lateral-proximal end of the tibia. The distal end has a prominent knob called the **lateral malleolus.**

tibia: L. *tibia,* shinbone, pipe, flute
fibula: L. *fibula,* clasp or brooch
malleolus: L. *malleolus,* small hammer

The lateral and medial malleoli are positioned on either side of the talus of the ankle joint, which they help stabilize. Both processes can be seen as prominent surface features and are easily palpated. Fractures to either or both malleoli are common in skiers. These fractures, clinically referred to as Pott's fractures, result from a shearing force occurring at a vulnerable spot on the lower leg.

Ankle and Foot

The ankle and foot contain twenty-six bones, arranged into the tarsus, metatarsus, and phalanges (figs. 9.54, 9.55). The bones of the ankle and foot are basically similar to those of the wrist and hand. They do, however, have distinct structural differences in order to provide weight support and leverage during walking.

Tarsus. There are seven tarsal bones. The **talus** is the tarsal bone that articulates with the tibia to form the ankle joint. The **calcaneus** *(kal-ka'ne-us)* is the largest of the tarsal bones and forms the heel of the foot. It has a large posterior extension, called the **tuberosity of the calcaneus,** for the attachment of the calf muscles. Anterior to the talus is the block-shaped **navicular** *(nah-vik'u-lar)* bone. The remaining four tarsal bones form a distal series that articulate with the metatarsals. They are, from the medial to lateral side, the **first, second,** and **third cuneiforms** *(ku-ne'i-form)* and the **cuboid.**

tarsus: G. *tarsos,* flat of the foot
talus: L. *talus,* ankle
calcaneus: L. *calcis,* heel

Table 9.9 Bones of the pelvic girdle and the lower extremities		
Name and number	**Location**	**Diagnostic features**
Os coxa (2)	Hip, part of the pelvic girdle; composed of 3 fused bones	Iliac crest, acetabulum, anterior superior iliac spine, ischial tuberosity, obturator foramen
Femur (2)	Bone of the thigh between hip and knee	Head, fovea capitis, neck, greater and lesser trochanters, lateral and medial condyles
Patella (2)	Anterior surface of knee	Triangular sesamoid bone
Tibia (2)	Medial side of leg between knee and ankle	Medial and lateral condyles, tibial crest, medial malleolus
Fibula (2)	Lateral side of leg between knee and ankle	Head, lateral malleolus
Tarsal (14)	Ankle	Large talus and calcaneus to receive the weight of leg; five other wedge-shaped bones to help form arches of foot
Metatarsal (10)	Sole of foot	Long bones, one in line with each digit
Phalanx (28)	Digits	Three in each toe, two in big toe

Source: Van De Graaff, Kent M., *Human Anatomy.* © 1984 Wm. C. Brown Publishers, Dubuque, Iowa. All Rights Reserved. Reprinted by permission.

Metatarsus.   The metatarsus, or sole of the foot, is composed of five metatarsal bones. The metatarsals are numbered from one to five, with the medial, or big toe, side being one. The first metatarsal is larger than the others because of its weight-bearing function.

The metatarsals are long bones, each having a **base, shaft,** and **head.** The proximal bases of the first, second, and third metatarsals articulate proximally with the cuneiforms. The heads of the metatarsals articulate distally with the proximal phalanges. The proximal joints are called **tarsometatarsal joints,** and the distal joints are called **metatarsophalangeal joints.** The ball of the foot is formed by the heads of the first two metatarsal bones.

Arches of the Foot.   The foot has two arches that support the weight of the body and provide leverage when walking. These arches are formed by the structure and arrangement of the bones held in place by ligaments and tendons. The arches are not rigid; they yield when weight is placed on the foot and spring back as the weight is lifted.

The **longitudinal arch** is divided into medial and lateral portions. The medial, or inner, portion is the larger of the two. It is supported by the calcaneus proximally and by the heads of the first three metatarsals distally. The wedge, or "keystone," of this portion of the longitudinal arch is the talus. The shallower lateral portion consists of the calcaneus, cuboid, and fourth and fifth metatarsals. The cuboid is the "keystone" of this portion.

The **transverse arch** extends across the width of the foot and is formed by the distal portion of the calcaneus, the navicular, the cuboid, and the proximal portions of all five metatarsals.

A weakening of the ligaments and tendons of the foot decreases the height of the longitudinal arch in a condition called **pes planus,** or flatfoot.

Table 9.9 contains a summary of the bones of the lower extremities.

1. Describe the structure of the pelvic girdle which reflects its weight-bearing role.
2. How can female and male pelvises be distinguished? What is the clinical importance of the lesser pelvis in females?
3. Describe the structure of each of the long bones of the lower extremity and the structure of each of the tarsal bones.
4. Which bones of the foot participate in the formation of the arches of the foot? What are the functions of the arches?

Clinical Considerations

Bone is a dynamic living tissue that is susceptible to hormonal or nutritional deficiency, diseases, and changes brought on by age. Since the development of bone is genetically governed, congenital conditions can occur. The hardness of bone gives it strength, yet it lacks the resiliency to avoid fracture if it undergoes excessive trauma. All of these aspects of bone provide some important and interesting clinical considerations.

Developmental Disorders

Congenital malformations account for several types of skeletal deformities. Certain bones may fail to form during osteogenesis, or they may form abnormally. **Cleft palate** and **cleft lip** are malformations of the palate and face. Cleft palates vary in severity and seem to have a mixed genetic and environmental cause. **Spina bifida** *(spi'nah bif'ĭ-dah)* is a congenital defect of the vertebral column resulting from a failure of the laminae of the vertebrae to fuse, exposing the spinal cord. The lumbar area is mainly affected, and frequently only a single vertebra is involved.

Minor defects of the extremities are relatively common malformations. Extra digits, a condition called **polydactyly** *(pol''e-dak'tĭ-le),* is the most common limb deformity. Usually an extra digit is incompletely formed and does not function. **Syndactyly,** or webbed digits, is likewise a relatively common limb malformation. Polydactyly is inherited as a dominant trait, whereas syndactyly is a simple dominant or simple recessive trait.

Talipes *(tal'i-pēz),* or clubfoot, is a congenital malformation in which the sole of the foot is twisted medially. It is not certain if abnormal positioning or restricted movement *in utero* causes this condition, but both genetics and environmental conditions are involved in most cases.

Nutritional and Hormonal Disorders

Several bone disorders result from nutritional deficiencies or from excessive or deficient amounts of the hormones that regulate bone development and growth. Vitamin D has a tremendous influence on proper bone structure and function. When there is a deficiency of this vitamin, the body is unable to metabolize calcium and phosphorus. Vitamin D deficiency in children causes **rickets.** The bones of a child with rickets remain soft and are deformed from the weight of the body.

A deficiency in vitamin D in the adult causes the bones to demineralize, or to give up stored calcium and phosphorus. This condition is called **osteomalacia** *(os''te-o-mah-la'she-ah).* Osteomalacia is prevalent in women who have repeated pregnancies and poor diets and are seldom exposed to the sun.

The consequences of endocrine disorders are described in chapter 25. Since the impact of hormones on bone development is great, however, a few endocrine disorders will be briefly mentioned. Hypersecretion of the growth hormone from the pituitary gland leads to **gigantism** in young persons if it starts before ossification of their epiphyseal plates and to **acromegaly** *(ak''ro-meg'ah-le)* in adults. Acromegaly is characterized by hypertrophy of the bones of the face, hands, and feet. In contrast, a hyposecretion of the growth hormone can lead to **dwarfism.**

Osteoporosis is a bone disorder characterized by marked demineralization, which weakens bone. The causes of osteoporosis include aging, prolonged inactivity, malnutrition, and an unbalanced secretion of hormones. It is most common in postmenopausal women.

Paget's disease is a disease of disorganized metabolic processes within bone. The activity of osteoblasts and osteoclasts becomes irregular, producing areas with thickened osseous deposits and other areas where too much bone is removed. The etiology of the disease is unknown, but it is a relatively common affliction in persons over age fifty, and it occurs more frequently in males than in females.

Trauma and Injury

There are a variety of types of trauma to the skeletal system, ranging from injury to the bone itself to damage of the joints in the form of sprains or dislocations. Joint injuries will be discussed in the following chapter.

The most common type of bone injury is a **fracture.** A fracture is the cracking or breaking of a bone. In *radiology,* X rays are used to diagnose the position and extent of a fracture. Fractures may be classified in several ways, and the type and severity of the fracture varies with the age and the general health of the body. **Spontaneous,** or **pathologic, fractures,** for example, result from diseases that weaken the bones. Most fractures, however, are called **traumatic fractures** because they are caused by injuries. The following are descriptions of several kinds of traumatic fractures (see fig. 9.56).

1. **Simple,** or **closed.** The fractured bone does not break through the skin.
2. **Compound,** or **open.** The fractured bone is exposed to the outside through an opening in the skin.
3. **Partial.** The bone is incompletely broken.
4. **Complete.** The fracture has separated the bone into two portions.
5. **Capillary.** A hairlike crack occurs within the bone.
6. **Comminuted** *(kom'i-nut''ed).* The bone is splintered into small fragments.
7. **Spiral.** The bone is twisted as it is broken.
8. **Greenstick.** In this incomplete break one side of the bone is broken and the other side is bowed.
9. **Impacted.** One broken end of a bone is driven into the other.
10. **Transverse.** A fracture occurs across the bone at right angles to the shaft.
11. **Colles'.** The styloid process of the radius is fractured.
12. **Pott's.** The lateral malleolus of the fibula is fractured.
13. **Avulsion.** In this fracture a structure, a finger for example, is torn off.
14. **Depressed.** The broken portion of the bone is driven inwards, as in certain skull fractures.
15. **Displaced.** In this fracture the bone fragments are not in anatomical alignment.
16. **Nondisplaced.** In this fracture the bone fragments are in anatomical alignment.

When a bone fractures, medical treatment involves realigning the broken ends and then immobilizing them until new bone tissue is formed and the fracture is healed. The site and severity of the fracture and the age of the patient will determine the type of immobilization. The methods of immobilization include tape, splints, casts, straps, wires, and steel pins. Even with these various methods of treatment, certain fractures heal poorly. Ongoing research into these problems is producing promising results. It has been found, for example, that applying weak electrical currents to fractured bones promotes healing and reduces the time of immobilization by half. Furthermore,

Figure 9.56.   Examples of types of traumatic fractures.

A *greenstick* fracture is incomplete, and the break occurs on the convex surface of the bend in the bone

A *fissured* fracture involves an incomplete longitudinal break

A *comminuted* fracture is complete and results in several bony fragments

A *transverse* fracture is complete, and the break occurs at a right angle to the axis of the bone

An *oblique* fracture occurs at an angle other than a right angle to the axis of the bone

A *spiral* fracture is caused by twisting a bone excessively

Figure 9.57.   (*a–d*) stages of the repair of a fracture. (*e*) an X-ray of a healing fracture.

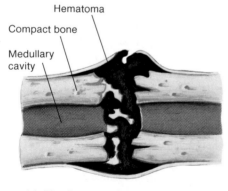

(a) Blood escapes from ruptured blood vessels and forms a hematoma

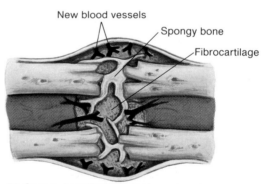

(b) Spongy bone forms in regions close to developing blood vessels, and fibrocartilage forms in more distant regions

(c) Fibrocartilage is replaced by a bony callus

(d) Osteoclasts remove excess bony tissue, making new bone structure much like the original new glasslike material called "bioglass" has been used to bond broken bones. This may be especially useful in elderly patients because their bones are slow in healing.

Physicians can realign and immobilize a fracture, but the ultimate repair of the bone occurs naturally within the bone itself. Several steps occur in the repair of a fracture (fig. 9.57).

1. When a bone is fractured, the surrounding periosteum is usually torn and blood vessels in both tissues are ruptured. A blood clot called a **fracture hematoma** *(hem-ah-to′mah)* soon forms throughout the damaged area. A disrupted blood supply to osteocytes and periosteal cells at the fracture site causes localized cellular death. This is followed by swelling and inflammation.

2. The traumatized area is "cleaned up" by the activity of phagocytic cells within the blood and osteoclasts that resorb bone fragments. As the debris is removed, fibrocartilage fills the gap within the fragmented bone, and a cartilaginous mass called a **callus** is formed. The callus becomes the precursor to the formation of bone in much the same way that hyaline cartilage is the precursor to developing bone.

3. The remodeling of the callus is the final step in the healing process. The cartilaginous callus is broken down, a new vascular supply is established, and compact bone develops around the periphery of the fracture. A healed fracture line is frequently undetectable by X-ray, except that the bone in this area is usually slightly thicker.

Neoplasms of Bone

Malignant bone tumors are three times more common than benign tumors. Pain is the usual symptom of either type of osseous neoplasm, although benign tumors may not have associated pain.

Two types of benign bone tumors are **osteoma,** which is the more frequent and often involves the skull, and **osteoid osteoma,** which is a painful neoplasm of the long bones, usually in children.

Osteogenic sarcoma is the most virulent type of bone cancer and frequently metastasizes through the blood to the lungs. This disease usually originates in the long bones and is accompanied by aching and persistent pain.

A **bone scan** (fig. 9.58) is a diagnostic procedure frequently done on a person who has had a malignancy elsewhere in the body that may have metastasized to the bone. The patient receiving a bone scan may be injected with a radioactive substance that accumulates more rapidly in malignant tissue than normal tissue. Entire body radiographs show malignant bone areas as intensely dark dots.

Figure 9.58. A bone scan.

(e)

Important Clinical Terminology

achondroplasia *(ah-kon''dro-pla'ze-ah)*   A genetic defect resulting in the retarded formation of cartilaginous bone during fetal development.

craniotomy   Surgical cutting into the cranium to provide access to the brain.

epiphysiolysis *(ep''ĭ-fĭz''e-ol'ĭ-sis)*   A weakening and separation of the epiphysis from the diaphysis of a long bone.

laminectomy   The surgical removal of the posterior arch of a vertebra, usually to repair a herniated intervertebral disc.

orthopedics   A branch of medicine concerned with the diagnosis and treatment of trauma, diseases, and abnormalities involving the skeletal and muscular systems.

osteitis   An inflammation of bone tissue.

osteoblastoma   A benign tumor produced from bone-forming cells most frequently in the vertebrae of young children.

osteochondritis   An inflammation of bone and cartilage tissues.

osteomyelitis   An inflammation of bone marrow caused by bacteria or fungi.

osteonecrosis   The death of bone tissue, usually caused by obstructed arteries.

osteopathology   The study of bone diseases.

osteosarcoma   A malignant tumor of bone tissue.

osteotomy   The surgical removal of a bone.

Chapter Summary

I. Organization of the Skeletal System
The axial and appendicular components of the skeletal system provide a strong, flexible body framework.
  A. The axial skeleton consists of the skull, auditory ossicles, hyoid bone, vertebral column, and rib cage.
  B. The appendicular skeleton consists of the pectoral girdle, upper extremities, pelvic girdle, and lower extremities.

II. Functions of the Skeletal System
The functions of the bones of the skeleton include support, protection, body movement, hemopoiesis, and mineral storage.
  A. The mechanical functions of bones include the support and protection of softer body tissues and organs, and certain bones function as levers during body movement.
  B. The metabolic functions of bones include hemopoiesis and mineral storage.

III. Development of the Skeletal System
Bones derive from mesenchymal mesoderm through endochondral ossification or intramembranous ossification.
  A. The development of the axial skeleton involves the notochord, somites, and sclerotomes in the vertebral column and rib cage, and the neurocranium, viscerocranium, and chondrocranium in the skull.
  B. The development of the appendicular skeleton progresses as the limb buds follow the growth of apical ectodermal ridges.

IV. Bone Structure and Growth
Each bone has a characteristic pattern of ossification and growth, a characteristic shape, and diagnostic surface features that indicate functionality.

  A. Bone structure includes the shape and surface features of each bone as well as gross internal components.
    1. Structurally speaking, bones may be classified as long, short, flat, or irregular.
    2. The surface features of bones can be broadly classified into articulating surfaces, nonarticulating prominences, and depressions and openings.
    3. A typical long bone has a diaphysis filled with marrow in the medullary cavity, epiphyses, epiphyseal plates for linear growth, and a covering of periosteum for diametric growth and the attachment of ligaments and tendons.
  B. Bone growth from embryonic to adult size is an orderly process determined by genetics, hormonal secretions, and nutritional supply.
    1. Most bones develop through endochondral ossification.
    2. Bone remodeling is a continual process involving osteoclasts in bone resorption and osteoblasts forming new osseous tissue.

V. Skull
The skull consists of cranial and facial bones that contain several cavities and house the brain.
  A. The cranium encloses and protects the brain and provides for the attachment of muscles.
    1. Sutures are immovable joints between cranial bones.
    2. The eight cranial bones include the frontal, parietals, temporals, occipital, sphenoid, and ethmoid.
  B. Facial bones form the basic shape of the face, support the teeth, and provide for the attachment of the facial muscles.

    1. The fourteen facial bones are the nasals, maxillae, zygomatics, mandible, lacrimals, palatines, inferior nasal conchae, and vomer.
    2. The hyoid bone is located in the neck between the mandible and the larynx.
    3. The three paired ear ossicles are located within the middle ear chambers of the petrous portion of the temporal bones.

VI. Vertebral Column
The twenty-six vertebrae of the vertebral column, separated by intervertebral discs, lend flexibility to the axial skeleton and absorb the stress of movement.
  A. The vertebral column consists of seven cervical, twelve thoracic, five lumbar, five fused sacral, and four fused coccygeal vertebrae.
  B. Cervical vertebrae have transverse foramina; thoracic vertebrae have facets and demifacets for articulation with ribs; lumbar vertebrae have heavy bodies; sacral vertebrae are triangularly fused and articulate with the pelvic girdle; the coccygeal vertebrae form a small triangular bone.

VII. Rib Cage
The flexible rib cage, consisting of the thoracic vertebrae, twelve paired ribs, costal cartilages, and the sternum, encloses and protects the thoracic viscera and is involved in breathing.
  A. The sternum consists of a manubrium, body, and xiphoid process.
  B. There are seven pairs of true ribs, three pairs of false ribs, and two pairs of floating ribs.

VIII. Pectoral Girdle and Upper Extremity
The pectoral girdle and upper extremities are adapted for freedom of movement and extensive muscle attachment.
   A. The pectoral girdle is composed of two scapulae and two clavicles. The clavicles attach the pectoral girdle to the axial skeleton at the sternum.
   B. The brachium contains the humerus, which extends from the scapula to the elbow.
   C. The forearm contains the medial ulna and the lateral radius.

   D. The wrist and hand contain twenty-seven bones arranged into the carpus, metacarpus, and phalanges.
IX. Pelvic Girdle and Lower Extremity
The pelvic girdle and lower extremities are adapted for support and locomotion.
   A. The pelvic girdle is formed by two ossa coxae united anteriorly by the symphysis pubis.
      1. The pelvis is divided into a greater pelvis, which helps to support the pelvic viscera, and a lesser pelvis, which forms the walls of the birth canal.

      2. Each os coxae consists of an ilium, ischium, and pubis.
   B. The thigh contains the femur, which extends from the hip to the knee, where it articulates with the tibia and the patella.
   C. The lower leg contains the medial tibia and the lateral fibula.
   D. The ankle and foot contain twenty-six bones arranged into the tarsus, metatarsus, and phalanges.

Review Activities

Objective Questions

1. A bone is considered to be a(n)
   (a) tissue
   (b) cell
   (c) organ
   (d) system
2. Which of the following statements is *false?*
   (a) Bones are important in the synthesis of vitamin D.
   (b) Bones and teeth contain about 99% of the body's calcium.
   (c) Red bone marrow is the primary site for hemopoiesis.
   (d) Most bones develop through endochondral ossification.
3. Bone tissue derives from specialized migratory mesodermal cells called
   (a) dermatomes
   (b) mesenchyme
   (c) myotomes
   (d) somites
4. Small portions of the embryonic notochord persist in the adult as the
   (a) primary curves
   (b) nucleus pulposus
   (c) pedicles
   (d) centrum

5. Which of the following is *not* a long bone? The
   (a) talus
   (b) proximal phalanx
   (c) metatarsal
   (d) fibula
6. Specialized bone cells that enzymatically reabsorb bone tissue are
   (a) osteoblasts
   (b) osteocytes
   (c) osteons
   (d) osteoclasts
7. The mandibular fossa is located in which structural portion of the temporal bone? The
   (a) squamous
   (b) tympanic
   (c) mastoid
   (d) petrous
8. The crista galli is a structural feature of which bone? The
   (a) sphenoid
   (b) ethmoid
   (c) palatine
   (d) temporal
9. Transverse foramina are characteristic of
   (a) lumbar vertebrae
   (b) sacral vertebrae
   (c) thoracic vertebrae
   (d) cervical vertebrae
10. The bone disorder common in aged persons particularly if they have prolonged inactivity, malnutrition, or an unbalanced secretion of hormones is
   (a) osteitis
   (b) osteonecrosis
   (c) osteoporosis
   (d) osteomalacia

Essay Questions

1. Explain the significance of each of the following structures involved in bone development: mesenchyme, somites, costal processes, neurocranium, viscerocranium, branchial arches, limb buds, and apical ectodermal ridge.
2. List the bones of the skull that are paired. Which are unpaired? Identify the bones of the skull that can be palpated.
3. Diagram a typical long bone, and label the epiphyses, diaphysis, epiphyseal plates, medullary cavity, nutrient foramina, periosteum, and articular cartilages.
4. Compare the pectoral and pelvic girdles in structure, articulation to the axial skeleton, and function.
5. Explain why a proper balance of vitamins, hormones, and minerals is essential in maintaining healthy osseous tissue. Give examples of diseases or skeletal conditions that may occur if there is an imbalance of any of these three essential substances.

Selected Readings

Burnstein, A. H. 1976. Aging of bone tissue: mechanical properties. *J. Bone Joint Surgery.* 58:82.
Evans, F. G., ed. 1966. *Studies in the anatomy and function of bones and joints.* New York: Springer-Verlag.
Hall, B. K. 1970. Cellular differentiation in skeletal tissue. *Biol. Rev.* 45:455.
Harris, W. H., and R. P. Heaney. 1970. *Skeletal renewal and metabolic bone disease.* Boston: Little, Brown, and Co.

Loomis, W. F. December 1970. Rickets. *Scientific American.*
Moore, K. L. 1977. *Before we are born: basic embryology and birth defects.* Rev. ed. Philadelphia: W. B. Saunders Co.
Napier, J. April 1967. The antiquity of human walking. *Scientific American.*
Platzer, W. 1978. *Locomotor system.* Vol. 1 of *Color atlas and textbook of human anatomy.* Chicago: Year Book Medical Publishers.
Prichard, J. J. 1977. Skeletal development. *Postgrad. Med. J.* 53:429.

Sharpe, W. D. 1979. Age changes in human bones: an overview. *Bull. N.Y. Acad. Med.* 55:757.
Snell, R. S. 1978. *Atlas of clinical anatomy.* Boston: Little, Brown and Co.
Trueta, J. 1968. *Studies in the development and decay of the human frame.* Philadelphia: W. B. Saunders Co.
Vaughan, J. M. 1981. *The physiology of bone.* 3d ed. New York: Oxford University Press.

10

Articulations

Concepts

The articulations between the bones of the skeleton are classified by arthrologists into three types according to structure, or into three types according to function, or the degree of movement permitted.

Diarthroses form by the third month as the epiphyses of adjacent endochondral bones develop distinct configurations and as the muscles that move the joints undergo contractions.

Articulating bones in synarthroses are tightly connected by either fibrous tissue or cartilage. Synarthrotic joints are rigid and immovable and are of two types: sutures and synchondroses.

Amphiarthroses allow limited motion in response to twisting, compression, or stress. The two types of amphiarthroses are symphyses and syndesmoses.

Diarthroses are freely movable joints that contain synovial fluid and are enclosed by joint capsules. Types of diarthroses include gliding, hinge, pivot, condyloid, saddle, and ball-and-socket.

Movements at diarthroses are produced by the contraction of skeletal muscles spanning the joints and attaching to or near the bones forming the articulations. In these actions, the bones act as levers, the muscles provide the force, and the joints are the fulcra, or pivots.

Of the numerous joints of the body, some have special structural features that enable them to perform particular functions. Furthermore, these joints are somewhat vulnerable to trauma and are therefore clinically important.

Selected Terms

articulation (ar-tik″u-la′shun)
synovial (sĭ-no′ve-al)
suture (su′chur)
synchondrosis (sin″kon-dro′sis)
symphysis (sim′fi-sis)
syndesmosis (sin″des-mo′sis)
articular cartilage (ar-tik′u-lar kar′tĭ-lij)
meniscus (mĕ-nis′kus)
bursa (ber′sah)
flexion (flek′shun)
extension (ek-sten′shun)
supination (su″pĭ-na′shun)
pronation (pro-na′shun)

Classification of Joints

The articulations between the bones of the skeleton are classified by arthrologists into three types according to structure or into three types according to function, or the degree of movement permitted.

Objective 1. Define the terms *arthrology* and *kinesiology*.
Objective 2. Distinguish between a structural and a functional classification of joints.

One of the functions of the skeletal system is to permit body movement. It is not the rigid bones that allow movement but the **articulations** *(ar-tik″u-la′shun),* or **joints,** between the bones. The structure of a joint determines the range of movement it permits. When we think of joints, we usually visualize movement. Not all joints are flexible, however, and as one part of the body moves, other joints remain rigid to stabilize the body and maintain balance. The coordinated activity of all the joints permits the sinuous, elegant movements of a gymnast or ballet dancer, as well as the mundane activities of walking, eating, writing, and speaking.

Arthrology is the science concerned with the study of joints. Generally speaking, an arthrologist is interested in the structure, classification, and function within joints as well as any dysfunction that may develop. In the more practical and dynamic science of **kinesiology** *(ki-ne″se-ol′o-je),* a kinesiologist studies the functional relationship, or biomechanics, of the skeleton, joints, muscles, and innervation as they work together to produce coordinated movement.

As you study the joints, you should adopt a kinetic approach and be able to demonstrate the various movements permitted at each of the movable joints and understand the adaptive advantage as well as the limitations of each type of movement.

The joints of the body may be classified according to structure or function. Structural classification is based on the presence or absence of a joint cavity and the kind of supportive connective tissue surrounding the joint. Classifying by structure, there are three types of joints:

1. **Fibrous** *(fi′brus).* A fibrous joint lacks a joint cavity, and fibrous connective tissue connects articulating bones.
2. **Cartilaginous** *(kar″tĭ-laj′ĭ-nus).* A cartilaginous joint lacks a joint cavity, and cartilage binds articulating bones.
3. **Synovial** *(sĭ-no′ve-al).* A synovial joint has a joint cavity, and ligaments help to support the articulating bones.

---

arthrology: Gk. *arthron,* joint; *logos,* study
kinesiology: Gk. *kinesis,* movement; *logos,* study

The functional classification of joints is based on the degree of movement permitted within the joint. Using this type of classification, there are three kinds of articulations:

1. **Synarthroses.** Immovable joints.
2. **Amphiarthroses.** Slightly movable joints.
3. **Diarthroses.** Freely movable joints.

This book uses the functional classification scheme in discussing the various kinds of joints of the body. But continual reference will be made to the structural classification as the specific anatomy of a particular joint is presented.

---

1. Explain the statement that kinesiology is applied arthrology.
2. Using both structural and functional classifications, how would a joint be classified that is tightly bound with hyaline cartilage so that only slight movement is possible?
3. List the three types of functional joints, and speculate as to which provides the greatest support and which is the most vulnerable to trauma.

Development of Freely Movable Joints

Diarthroses form by the third month as the epiphyses of adjacent endochondral bones develop distinct configurations and as the muscles that move the joints undergo contractions.

Objective 3. Describe the development of diarthroses, and discuss the importance of fetal movement to the normal development of freely movable joints.

The sites of developing diarthroses *(di″ar-thro′sēz)* (freely movable joints) are discernible at six weeks as mesenchyme becomes concentrated in the areas where precartilage cells differentiate (fig. 10.1). The future joints at this stage appear as intervals of less concentrated mesenchymal cells. As cartilage cells develop within a forming bone, a thin flattened sheet of cells forms around the cartilaginous model to become the perichondrium. These same cells are continuous across the gap between the adjacent developing bone. Surrounding the gap, the flattened mesenchymal cells differentiate to become the **joint capsule.**

During the early part of the third month of development, the mesenchymal cells still remaining within the joint capsule begin migrating toward the epiphyses of the adjacent developing bones. The cleft eventually enlarges

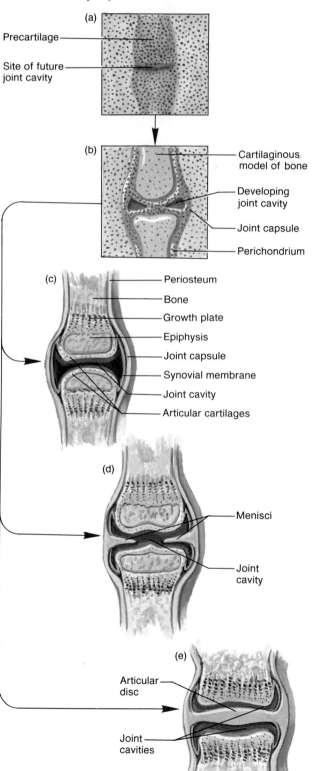

Figure 10.1. Development of diarthroses. (*a*) at six weeks, different densities of mesenchyme denote where the bones and joints will form. (*b*) at nine weeks, a basic diarthrotic model is present. At twelve weeks, the diarthroses are formed and have either (*c*) a free joint cavity (e.g., interphalangeal joint); (*d*) a cavity containing menisci (e.g., knee joint); or (*e*) a cavity with a complete articular disc (e.g., sternoclavicular joint).

(a)
Precartilage
Site of future joint cavity

(b)
Cartilaginous model of bone
Developing joint cavity
Joint capsule
Perichondrium

(c)
Periosteum
Bone
Growth plate
Epiphysis
Joint capsule
Synovial membrane
Joint cavity
Articular cartilages

(d)
Menisci
Joint cavity

(e)
Articular disc
Joint cavities

Figure 10.2. A section across the skull showing a suture.

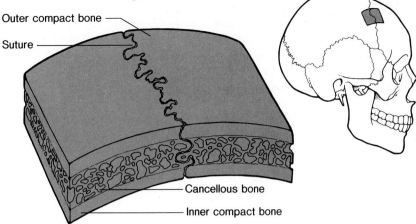

Outer compact bone
Suture
Cancellous bone
Inner compact bone

to become the **joint cavity.** Thin pads of hyaline cartilage develop on the surfaces of the epiphyses that contact the joint cavity. These pads become the **articular cartilages** of the functional joint. As the joint continues to develop, a highly vascular **synovial membrane** forms on the inside of the joint capsule and begins secreting a watery *synovial fluid* into the joint cavity.

In certain developing diarthroses, the mesenchymal cells do not migrate away from the center of the joint cavity but, rather, give rise to cartilaginous wedges, called **menisci,** as in the knee joint (fig. 10.1), or to complete cartilaginous pads, called **articular discs,** as in the sternoclavicular joint.

The formation of most diarthroses is completed by the end of the third month. Shortly after this time, there are fetal muscle contractions, known as *quickening,* that cause movement at these joints. Joint movement enhances the nutrition of the articular cartilage and prevents the fusion of connective tissues within the joint.

---

1. *Describe the development of a diarthrotic joint, and give the approximate prenatal date of each event.*
2. *Describe the variation of cartilaginous separations between the articular cartilages of diarthroses.*
3. *Explain the importance of quickening to the health of a newly formed diarthrotic joint.*

Synarthroses

Articulating bones in synarthroses *(sin''ar-thro'sēz)* are tightly connected by either fibrous tissue or cartilage. Synarthrotic joints are rigid and immovable and are of two types: sutures and synchondroses.

---

suture: L. *sutura,* sew

Objective 4. Describe the structure of a suture, and indicate where sutures are located.

Objective 5. Describe the structure of a synchondrosis; indicate where synchondroses are located and how they change as a person ages.

Sutures

Sutures *(su'chur)* are found only within the skull and are characterized by a thin layer of dense fibrous connective tissue that binds the articulating bones (fig. 10.2). Sutures form at about eighteen months of age and replace the pliable fontanels of an infant's skull (see fig. 9.7).

There are several types of sutures, which can be distinguished on the basis of the appearance of the articulating margin of bone. A **serrate suture** is characterized by serrated, interlocking articulations. This is the most common type of suture, an example being the sagittal suture between the two parietal bones. In a **lap (squamous) suture,** the margin of one bone overlaps that of the articulating bone. The squamous suture formed between the temporal and parietal bones is an example. In a **plane (butt) suture,** the margins of the articulating bones are fairly smooth. An example is the maxillary suture, where the two maxillary bones articulate to form the hard palate. A **synostosis** *(sin''os-to'sis),* a unique type of suture, is present during growth of the skull but becomes totally ossified in the adult. The union between the right and left portions of the frontal bone is an example of synostosis.

Fractures of the skull are fairly common in an adult but much less so in a child. The skull of a child is resilient to blows because of the nature of the bone and the layer of fibrous connective tissue within the sutures. The skull of an adult is much like an eggshell in its lack of resilience and will frequently splinter on impact.

---

synostosis: Gk. *syn,* together; *osteon,* bone

Synchondroses

Synchondroses *(sin″kon-dro′sēs)* are synarthrotic joints that have hyaline cartilage between the bone segments. These joints are typically temporary, forming the growth lines (epiphyseal plates) between the diaphyses and epiphyses in the long bones of children (fig. 10.3). When growth is complete, the synchondrotic joints ossify. A totally ossified synchondrosis can also be referred to as a synostosis. A synchondrosis can be clearly seen in the X-ray of a long bone of a child in figure 10.3.

Considering that there are two synchondroses for each long bone and many sutures within the skull, it is obvious that synarthroses are the most common type of joint.

1. Distinguish between the three principal kinds of sutures, and give an example and the location of each. What do each have in common?
2. Explain why synchondroses are found only on bones that have epiphyses and a diaphysis.

Amphiarthroses

Amphiarthroses allow limited motion in response to twisting, compression, or stress. The two types of amphiarthroses are symphyses and syndesmoses.

Objective 6. Describe the structure of a symphysis, and indicate where symphyses occur.

Objective 7. Describe the structure of a syndesmosis, and indicate where syndesmoses occur.

Symphyses

The adjoining bones of a symphysis *(sim′fĭ-sis)* joint are separated by a pad of fibrocartilage. This pad cushions the joint and allows limited movement. The symphysis pubis and the intervertebral joints (fig. 10.4) at the intervertebral discs are examples of symphyses. Though only limited motion is possible at each intervertebral joint, the combined movement of all the joints of the vertebral column results in extensive spinal action.

Syndesmoses

In a syndesmotic joint, adjacent bones are held together by collagenous fibers or interosseous ligaments. A syndesmosis *(sin″des-mo′sis)* is characteristic of the distal ends of the tibia-fibula and the radius-ulna (fig. 10.5).

---

synchondrosis: Gk. *syn*, together; *chondros*, cartilage
symphysis: Gk. *symphysis*, growing together
syndesmosis: Gk. *syndesmos*, binding together

Figure 10.3.   An X-ray of the left humerus of a ten-year-old child, showing a synchondrotic joint. In a long bone, this type of joint occurs at both the proximal and distal epiphyseal plates. The mitotic activity at a synchondrotic joint is responsible for bone growth in length.

Proximal epiphysis

Proximal epiphyseal plate (site of synchondrotic joint)

Diaphysis

Distal epiphyseal plate

Distal epiphysis

Hormonal action in a pregnant woman causes the symphysis pubis, sacroiliac, and sacrococcygeal joints to soften and become more flexible. This relaxation of the pelvic joints increases the potential capacity of the pelvic cavity and reduces trauma during childbirth by increasing the diameter of the pelvic outlet. These joints will strengthen postpartum, but following the birth of a mother's first child, the diameter of her pelvis will generally be slightly wider than before.

1. Discuss the function of the pad of fibrocartilage in a symphysis joint, and give two examples of symphyses.
2. What structural feature is characteristic to all syndesmoses?

Figure 10.4.   Both (*a*) the symphysis pubis and (*b*) the intervertebral joints are amphiarthrotic joints.

(a)   Fibrocartilaginous disc

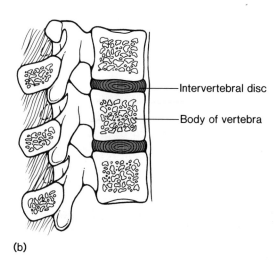

Intervertebral disc

Body of vertebra

(b)

Figure 10.5.   The distal articulation of the ulna and radius forms a syndesmotic joint. Interosseous ligaments tightly bind these bones and permit only slight movement between them.

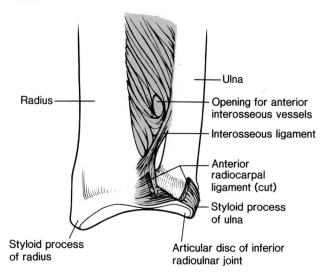

Radius

Styloid process of radius

Ulna

Opening for anterior interosseous vessels

Interosseous ligament

Anterior radiocarpal ligament (cut)

Styloid process of ulna

Articular disc of inferior radioulnar joint

Diarthroses

Diarthroses are freely movable joints that contain synovial fluid and are enclosed by joint capsules. Types of diarthroses include gliding, hinge, pivot, condyloid, saddle, and ball-and-socket.

Objective 8. Describe the structure of a diarthrotic joint, and indicate where diarthroses occur.

Objective 9. List and discuss the various kinds of diarthroses, where they occur, and the movements they permit.

The most obvious type of articulation in the body is the freely movable diarthrosis. The dual function of diarthrotic joints is to provide a wide range of precise, smooth movements and yet maintain stability, strength, and, in certain aspects, rigidity in the body.

Diarthroses are the most complex and varied of the three major types of joints. A diarthrotic joint's range of movement is limited by three factors: (1) the structure of the bones participating in the articulation (certain processes on bones, for example, the oleocranon process of the ulna, actually limit the range of motion and "lock" the articulation to prevent overextension of the joint); (2) the strength and tautness of the associated ligaments, tendons, and joint capsule; (3) the size, arrangement, and action of the muscles that span the joint. There is tremendous individual variation in joint motility, most of which is related to body conditioning. "Double-jointed" is a misnomer because such a joint is not double although it does permit extreme maneuverability.

Structure of a Diarthrotic Joint

In structural terms, diarthroses are synovial joints because they are enclosed by a fibroelastic joint capsule, which is filled with lubricating **synovial fluid,** or **synovium** (fig. 10.6). Synovial fluid is secreted by a thin **synovial membrane** that lines the inside of the capsule. Synovial fluid is similar to interstitial fluid (fluid surrounding cells of a tissue), which has a high concentration of hyaluronic

Figure 10.6. A diarthrotic joint is represented by this diagrammatic lateral view of the knee joint.

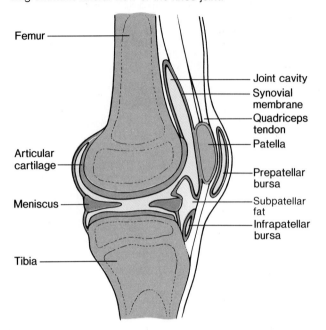

Femur

Joint cavity
Synovial membrane
Quadriceps tendon
Patella
Prepatellar bursa
Subpatellar fat
Infrapatellar bursa

Articular cartilage

Meniscus

Tibia

Figure 10.7. A tendon sheath facilitates the gliding of a tendon as it traverses a fibrous or bony tunnel. Tendon sheaths are closed sacs, one layer of the synovial membrane lining the tunnel, the other folding over the surface of the tendon.

Retinaculum
Tendon sheath
Tendon
Bone
Periosteum

acid, a lubricating substance. The bones that articulate in a diarthrosis are capped with a smooth **articular cartilage** *(ar-tik'u-lar kar'tĭ-lij)*. The avascular articular cartilage is only about 2 mm thick and depends upon the alternating compression and decompression during joint activity for the exchange of nutrients and waste products with the synovial fluid. Ligaments help to bind a diarthrosis and may be located within the joint cavity or on the outside of the capsule. Tough, fibrous, cartilaginous pads called **menisci**—sing., *meniscus (mě-nis'kus)*—are located within the capsule of certain diarthroses (e.g., the knee joint) and serve to cushion as well as to guide the articulating bones.

Articulating bones of diarthroses do not come in contact with one another. Articular cartilage caps the articular surface of each bone, and synovial fluid circulates through the joint during movement. Both of these joint structures minimize friction as well as cushion the articulating bones. Trauma or disease may render either of these two joint structures nonfunctional, and the two articulating bones will come in contact. Bony deposits will then form, and a type of arthritis will develop within the joint.

Located near certain diarthrotic joints are flattened, pouchlike sacs called **bursae**—sing., *bursa (ber'sah)*—which are filled with synovial fluid (fig. 10.6). These closed sacs are commonly located between muscles or within an

meniscus: Gk. *meniskos,* small moon
bursa: Gk. *byrsa,* bag or purse

area where a tendon passes over a bone. The functions of bursae are to cushion certain muscles and to facilitate the movement of tendons or muscles over bony or ligamentous surfaces.

A **tendon sheath** is a modified bursa that surrounds and lubricates the tendons of certain muscles, particularly those that cross the wrist and ankle joints (fig. 10.7). The tendons that are surrounded by sheaths are strong and cordlike and are usually associated with important flexor muscles.

Kinds of Diarthroses

Diarthroses are classified into six main types according to their structure and the kinds of motion they permit. The six types are gliding, hinge, pivot, condyloid, saddle, and ball-and-socket.

Gliding. Gliding joints allow only side-to-side and back-and-forth movements. This is the simplest type of joint movement. The articulating surfaces can be nearly flat or one may be slightly concave and the other slightly convex. The intercarpal and intertarsal joints, as well as the sternoclavicular joint and the joint between the articular processes of adjacent vertebrae, are examples of gliding joints (fig. 10.8).

Hinge. The structure of a hinge joint permits bending in only one plane, much like the hinge of a door. In this type of articulation the surface of one bone is always concave, the other convex (fig. 10.9). Hinge joints are the most common type of diarthroses and include such specific joints as the knee, the elbow (humeroulnar articulation), and the joints of the phalanges.

Figure 10.8. The locations of both gliding and condyloid joints are depicted. Gliding joints are located between the proximal and distal carpals and between the individual carpals. There is a condyloid joint between the radius and the proximal row of carpals. Note the diagrammatic representation of a gliding joint.

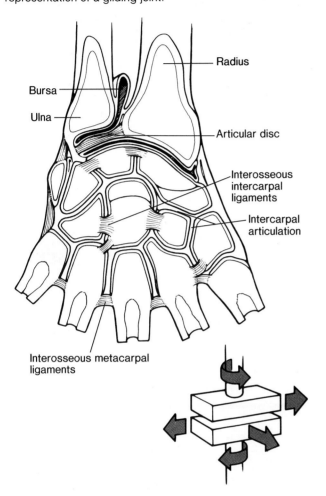

Figure 10.9. A hinge joint permits only a bending movement (flexion and extension). The hinge joint of the elbow involves the distal end of the humerus articulating with the proximal end of the ulna. Note the diagrammatic representation of this joint.

Figure 10.10. The atlas articulating with the axis forms a pivot joint that permits a rotational movement in one axis. Note the diagrammatic representation. Refer to figure 10.9 and determine which articulating bones of the elbow region form a pivot joint.

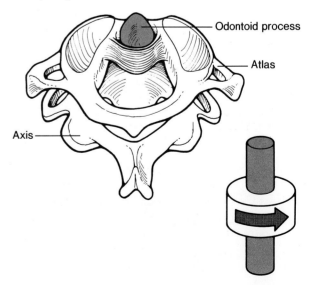

Pivot. The movement in a pivot joint is limited to rotation about a central axis. In this type of articulation the articular surface on one bone is conical or rounded and fits into a depression on another bone. Examples are the proximal articulation of the radius and ulna for rotation of the forearm, as in turning a doorknob, and the articulation between the atlas and axis that makes rotational movement of the head possible (fig. 10.10).

Condyloid (Ellipsoid). A condyloid *(kon'di-loid)* articulation is structured so that an oval, convex articular surface of one bone fits into an elliptical, concave depression on another bone. This permits angular movement in two directions (biaxial), as in an up-and-down and side-to-side motion. A condyloid joint does not permit rotational movement. The radiocarpal joint of the wrist is an example (see fig. 10.8).

Figure 10.11. A saddle joint is formed as the trapezium articulates with the base of the first metacarpal bone. Note the diagrammatic representation of this joint.

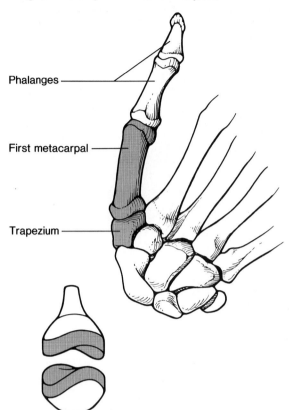

Phalanges

First metacarpal

Trapezium

Figure 10.12. A ball-and-socket articulation illustrated by the hip joint. Note the diagrammatic representation.

Os coxae

Head of femur into acetabulum

Femur

Table **10.1** Types of articulations			
Type	**Structure**	**Movements**	**Example**
Synarthroses	Articulating bones in close contact, bound by a thin layer of fibrous tissue or cartilage	None	Suture between bones of the skull; epiphyseal plate
Amphiarthroses	Articulating bones separated by fibrocartilaginous discs or bound by interosseous ligament	Slightly movable	Intervertebral joints; symphysis pubis and sacroiliac joint; between tibia-fibula and radius-ulna
Diarthroses	Joint capsule containing synovial membrane and synovial fluid	Freely movable	
A. Gliding	Flattened or slightly curved articulating surfaces	Sliding	Intercarpal and intertarsal joints
B. Hinge	Concave surface of one bone articulates with convex surface of another	Bending motion in one plane	Knee; elbow; joints of phalanges
C. Pivot	Conical surface of one bone articulates with a depression of another	Rotation about a central axis	Atlantoaxial joint; proximal radioulnar joint
D. Condyloid	Oval condyle of one bone articulates with elliptical cavity of another	Biaxial movement	Radiocarpal joint
E. Saddle	Concave and convex surface on each articulating bone	Wide range of movements	Carpometacarpal joint of the thumb
F. Ball-and-Socket	Rounded convex surface of one bone articulates with cuplike socket of another	Movement in all planes and rotation	Shoulder and hip joints

Figure 10.13. Examples of flexion and extension. Contraction of the posterior thigh muscles with the femur held rigid results in flexion at the knee joint as the lower leg is moved posteriorly. Contraction of the anterior thigh muscles with the entire leg held rigid results in flexion at the hip joint as the leg is moved anteriorly. Contraction of the posterior thigh muscles with the entire leg held rigid results in extension at the hip joint as the leg is moved posteriorly.

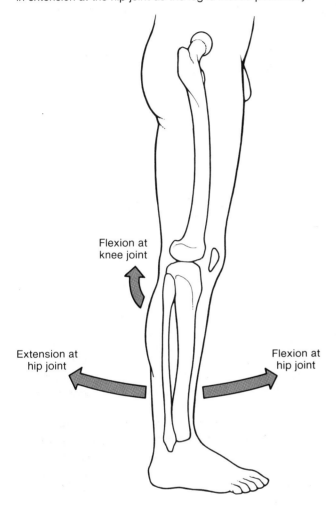

Flexion at knee joint

Extension at hip joint

Flexion at hip joint

Synovial fluid within the diarthrotic joints serves to lubricate the articular surfaces and provide nourishment to the articular cartilage. Trauma to the joint causes the excessive production of synovial fluid in an attempt to cushion and immobilize the joint. This leads to swelling and discomfort to the joint. The most frequent type of joint injury is a sprain, in which the supporting ligaments or joint capsule are damaged to varying degrees.

1. *List the structures of a diarthrosis, and explain the function of each.*
2. *What three factors limit the range of movement within diarthroses?*
3. *Give an example of each of the six kinds of diarthroses, and describe the range of movement possible at each.*

Movements at Diarthroses

Movements at diarthroses are produced by the contraction of skeletal muscles spanning the joints and attaching to or near the bones forming the articulations. In these actions, the bones act as levers, the muscles provide the force, and the joints are the fulcra, or pivots.

Objective 10. List and discuss the various kinds of movements possible within diarthrotic joints.

Objective 11. Describe the components of a lever, and explain the role of diarthrotic joints in lever systems.

Objective 12. Compare the structure of first-, second-, and third-class levers.

As previously mentioned, the range of movement is determined by the structure of the individual joint and the arrangement of the associated muscle and bone. The movement at a hinge joint, for example, occurs in only one plane, whereas the structure of a ball-and-socket joint permits movement around many axes. Movements within joints are broadly classified as **angular** and **circular.** Within each of these categories are specific types of movements, and certain special movements may involve several of the specific types.

Angular

Angular movements increase or decrease the joint angle produced by the articulating bones. The four types of angular movements are flexion, extension, abduction, and adduction.

Saddle. Each articular process of a saddle-shaped joint has a concave surface in one direction and a convex surface in another. This unique articulation is a modified condyloid joint that allows a wide range of movement and is associated only with the thumb. Specifically, the saddle joint is found at the articulation of the trapezium of the carpus with the first metacarpal bone (fig. 10.11).

Ball-and-Socket. Ball-and-socket joints are formed by the articulation of a rounded convex surface with a cuplike cavity (fig. 10.12). This type of articulation provides the greatest range of movement of all the diarthrotic joints. Examples are the hip and shoulder joints.

A summary of the various types of joints within the body is presented in table 10.1.

Flexion. Flexion *(flek'shun)* is a movement that decreases the joint angle on an anterior-posterior plane (fig. 10.13). Examples of flexion are the bending of the elbow or knee. Flexion of the elbow is a forward movement,

flexion: L. *flectere,* to bend

Figure 10.14.   Dorsiflexion and plantar flexion of the foot at the ankle joint.

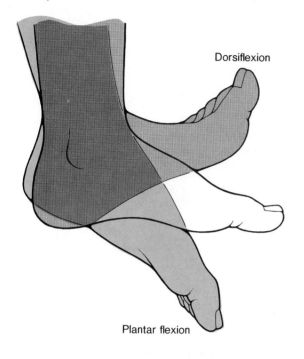

Figure 10.15.   Abduction and adduction of the lower extremities in reference to the main axis of the body.

Figure 10.16.   Supination and pronation of the hand. Note the relative position of the ulna and radius in both positions.

whereas flexion of the knee is a backward movement. Flexion in most joints is simple to understand, such as flexion of the head as it bends forward or the flexing of a digit, but flexion of the ankle and shoulder joints needs further explanation. In the ankle joint, flexion occurs as the dorsum of the foot is elevated. This movement is frequently called dorsiflexion (fig. 10.14). Pressing the foot downward (as in rising on the toes) is **plantar flexion.** The shoulder joint is flexed when the arm is brought forward, thus decreasing the joint angle.

Extension.   In extension *(ek-sten'shun),* which is the reverse of flexion, the joint angle is increased (fig. 10.13). Extension returns a body part to the anatomical position. In an extended joint, the angle between the articulating bones is 180°. The exception to this is the ankle joint, in which a 90° angle exists between the foot and the lower leg in the anatomical position. Examples of extension are the straightening of the elbow or knee joints from flexion positions. **Hyperextension** occurs when a portion of the body is extended beyond the anatomical position so that the joint angle is greater than 180°. An example of hyperextension is bending the head backward.

extension: L. *ex,* out, away from; *tendere,* stretch

Figure 10.17.   Circumduction of the shoulder joint.

Circumduction

Figure 10.18.   Inversion and eversion of the foot on the intertarsal joints.

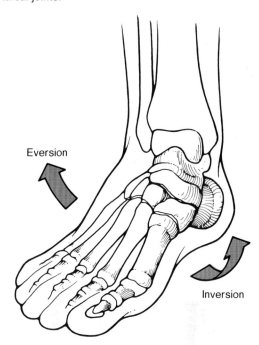

Eversion

Inversion

Abduction.   Abduction is the movement of a body part away from the main axis of the body, or away from the midsagittal plane, in a lateral direction (fig. 10.15). This term usually applies to the arm or leg but can also apply to the fingers or toes, in which case the line of reference is the longitudinal axis of the limb. Examples of abduction are moving the arms upward and away from the body or spreading the fingers apart.

Adduction.   Adduction, the opposite of abduction, is the movement of a body part toward the main axis of the body (fig. 10.15). In the anatomical position, the arms and legs have been adducted toward the midplane of the body.

Circular
Joints that permit circular movement are composed of a bone with a rounded or oval surface articulating with a corresponding cup or depression on another bone. The two basic types of circular movements are rotation and circumduction.

Rotation.   Rotation is the movement of a bone around its own axis (see fig. 10.10). There is no lateral displacement during this movement. Examples are turning the head from side to side in a "no" motion and moving the forearm from a palm-up position to a palm-down position.
    Supination *(su''pĭ-na'shun)* is a specialized rotation of the forearm that results in the palm of the hand being turned forward (anteriorly). In the supine position, the ulna and radius of the forearm are parallel and in the anatomical position. **Pronation** *(pro-na'shun)* is the opposite

of supination (fig. 10.16). It is a rotational movement of the forearm that results in the palm of the hand being directed backward (posteriorly).

Circumduction.   Circumduction is the circular, conelike movement of a body segment. The distal extremity forms the circular movement, while the proximal attachment forms the pivot (fig. 10.17). This type of motion is possible at the shoulder, wrist, trunk, hip, and ankle joints.

Special Movements
Because the terms used to describe generalized movements around axes do not apply to the structure of certain joints, other terms must be used to describe the motion of such joints.

Inversion.   Inversion is the movement of the sole of the foot inward or medially (fig. 10.18). The pivot axis is at the ankle joint.

Eversion.   Eversion is the opposite of inversion and is the movement of the sole of the foot outward or laterally (fig. 10.18). Both inversion and eversion are clinical terms usually used to describe developmental abnormalities.

Protraction.   Protraction is the movement of part of the body forward on a plane parallel to the ground. Examples are thrusting out the lower jaw (fig. 10.19) or movement of the clavicle forward.

abduction: L. *abducere,* lead away
adduction: L. *adductus,* bring to
rotation: L. *rotare,* a wheel

Figure 10.19. Protraction and retraction of the lower jaw.

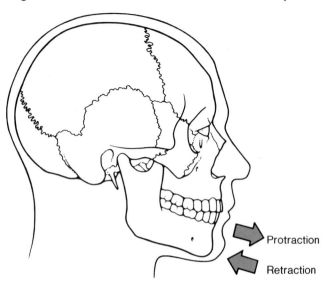

Protraction

Retraction

Retraction.   Retraction is the pulling back of a protracted part of the body on a plane parallel to the ground (fig. 10.19). Retraction of the mandible brings the lower jaw back in alignment with the upper jaw so that the teeth occlude.

Elevation.   Elevation is a movement that results in a portion of the body being lifted upward. Examples of elevation include elevating the mandible to close the mouth or lifting the shoulders to shrug.

Depression.   Depression is the movement opposite of elevation. Both the mandible and shoulders are depressed when moved downward.

Figure 10.20 presents a visual summary of many of the movements permitted at diarthrotic joints.

Biomechanics of Body Movement

A lever is any rigid structure that turns about a fulcrum when force is applied. Because bones are rigid structures that can be moved at diarthrotic joints in response to applied forces, they fit the criteria of levers. There are four

Figure 10.20 Photographic summary of joint movements. (*a*) Adduction of shoulder, hip, and carpophalangeal joints; (*b*) abduction of shoulder, hip, and carpophalangeal joints; (*c*) lateral rotation of vertebral column; (*d*) lateral flexion of vertebral column; (*e*) flexion of vertebral column; (*f*) hyperextension of vertebral column; (*g*) flexion of shoulder, hip, and knee joints of right side of body; extension of elbow and wrist joints; dorsiflexion of right ankle joint; (*h*) extension of shoulder and hip joints on right side of body; plantar flexion of right ankle joint.

(a)   (b)   (c)   (d) basic components to a lever: (*a*) a rigid bar or other such structure; (*b*) a pivot or fulcrum; (*c*) an object or resistance that is moved; and (*d*) a force that is applied to one portion of the rigid structure.

Levers are generally associated with machines but can equally apply to other mechanical structures such as the human body. Diarthrotic joints are always the fulcra (*F*), the muscles provide the force, or effort (*E*), and the bones are the rigid structures that move the resisting object (*R*).

There are three kinds of levers, determined by the arrangement of their parts (fig. 10.21).

1. In a **first-class lever,** the fulcrum is positioned between the effort and the resistance. The parts in a first-class lever are much like those of a seesaw—a sequence of resistance-pivot-effort. Scissors and hemostats are mechanical examples of the first-class levers. In the body, the head at the atlantooccipital *(at-lan″to-ok-sip′ĭ-tal)* joint is a first-class lever. The weight of the skull and facial portion of the head is the resistance, and the posterior neck muscles that contract to maintain the balance of the head on the joint are the effort.

2. In a **second-class lever,** the resistance is positioned between the fulcrum and the effort. The sequence of arrangement is pivot-resistance-effort, as in a wheelbarrow or the action of a crowbar when one end is placed under a rock and the other end lifted. Contracting the calf muscles (*E*) to elevate the body (*R*) on the toes, with the ball of the foot acting as the fulcrum, is another example.

3. In a **third-class lever,** the effort lies between the fulcrum and the resistance. The sequence of the parts is pivot-effort-resistance. An example is using a pair of forceps to grasp an object. The third-class lever is the most common type within the body. The flexion of the forearm at the elbow is an example. The effort occurs as the biceps muscle is contracted to move the resistance of the forearm with the elbow joint forming the fulcrum.

Each skeletal-muscular interaction at a diarthrotic joint forms some kind of lever system. The specific kind of lever is not always easy to identify. Certain joints are adapted for power at the expense of speed, whereas others

(e)

(f)

(g)

(h)

Figure 10.21.   The three classes of levers. (*a*) in a first-class lever, the fulcrum (F) is positioned between the resistance (R) and the effort (E). (*b*) in a second-class lever, the resistance is between the fulcrum and the effort. (*c*) in a third-class lever, the effort is between the fulcrum and the resistance.

Figure 10.22.   The position of a joint (fulcrum) relative to the length of a long bone (lever arm) and the point of attachment of a muscle (force) determines the mechanical advantage when movement occurs. (*a*) the elbow joint and extensor muscles of a human; (*b*) the elbow joint and extensor muscles of an armadillo.

(a)

(b)

are clearly adapted for speed (fig. 10.22). The specific attachment of muscles that span a joint plays an extremely important role in determining the mechanical advantage. The position of the insertion of a muscle relative to the joint is an important factor in the biomechanics of the contraction. An insertion close to the joint (fulcrum), for example, will produce a faster and greater range of movement than an insertion far away from the joint. An attachment far away from the joint takes advantage of the lever arm of the bone and increases power at the sacrifice of speed and range of movement.

1. Describe the structure of a joint that permits rotational movement.
2. What types of joints are involved in lever systems within the body?
3. Which is the most common type of lever in the body?
4. Considering the type of diarthrosis at the hip joint and the location of the gluteal muscles of the buttock, explain why this type of lever system is adapted for rapid, wide-ranging movements.

Figure 10.23. The temporomandibular joint. (*a*) a lateral view; (*b*) a medial view; (*c*) a parasagittal section.

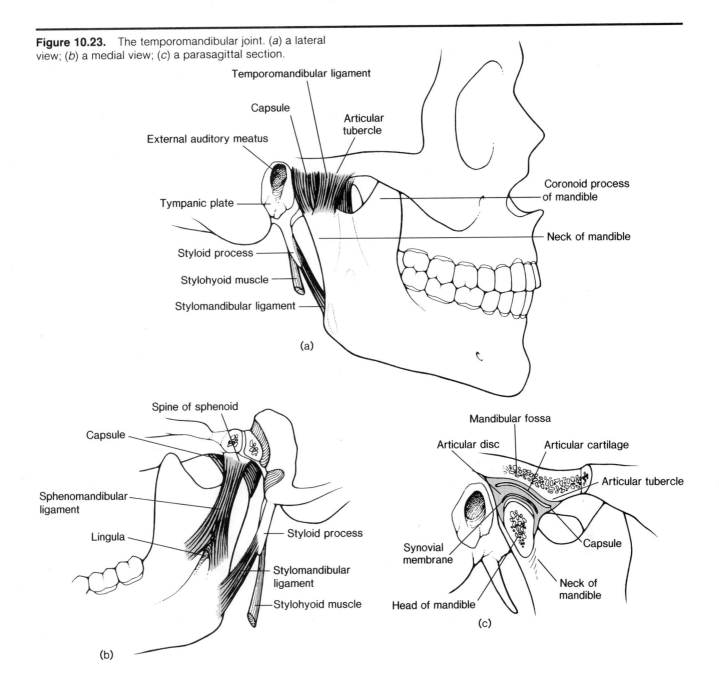

Specific Joints of the Body

Of the numerous joints in the body, some have special structural features that enable them to perform particular functions. Furthermore, these joints are somewhat vulnerable to trauma and are therefore clinically important.

Objective 13. Describe the structure, function, and possible clinical importance of the following joints: temporomandibular, sternoclavicular, humeroscapular, elbow, metacarpophalangeal, interphalangeal, coxal, tibiofemoral, and talocrural.

Temporomandibular Joint

The temporomandibular *(tem″po-ro-man-dib′u-lar)* joint is the only diarthrotic joint in the skull and is a unique combination of a hinge joint and a gliding joint (fig. 10.23). It is formed by the mandibular condyle and the mandibular fossa and the articular tubercle of the temporal bone. An **articular disc** separates the joint cavity into superior and inferior compartments.

Three ligaments support and reinforce the temporomandibular joint. The **temporomandibular ligament** is positioned on the lateral side of the joint capsule and is covered by the parotid salivary gland. This ligament prevents the head of the mandible from being displaced posteriorly and fracturing the tympanic plate when the chin

Figure 10.24. The sternoclavicular joint and associated ligaments. (*a*) an anterior view showing a frontal section; (*b*) a posterior view.

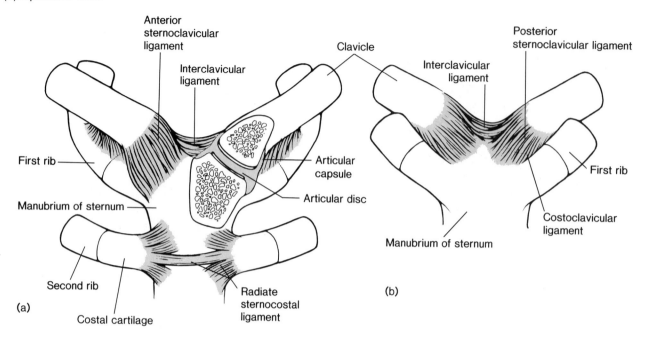

suffers a severe blow. The stylomandibular ligament is not directly associated with the joint but extends inferiorly and anteriorly from the styloid process to the posterior border of the ramus of the mandible. **A sphenomandibular** *(sfe''no-man-dib'u-lar)* **ligament** crosses on the medial side of the joint from the spine of the sphenoid bone to the ramus of the mandible.

The movements of the temporomandibular joint include depression and elevation of the mouth as a hinge joint, protraction and retraction of the jaw as a gliding joint, and lateral rotatory movements. The lateral motion is made possible by the articular disc.

The temporomandibular joint can be easily palpated by applying firm pressure to the area in front of your auricle and opening and closing your mouth. This joint is most vulnerable to dislocation when the mandible is completely depressed, as in yawning. Relocating the jaw is usually a simple task, however, and is accomplished by pressing downward on the molar teeth while pushing the jaw backward.

Temporomandibular joint (TMJ) syndrome is a recently recognized ailment that may afflict an estimated 75 million Americans. The apparent cause of TMJ syndrome is a malalignment of one or both temporomandibular joints. The symptoms of the condition vary from moderate and intermittent pain to intense and continuous pain of the head, neck, shoulders, or back. Some vertigo (disorientation of coordination) and tinnitus (ringing in ear) may be experienced.

Sternoclavicular Joint

The sternoclavicular *(ster''no-klah-vik'u-lar)* joint is formed where the clavicle articulates with the manubrium of the sternum (fig. 10.24). Although a gliding joint, the sternoclavicular joint has a good range of movement because of the presence of an articular disc within the joint capsule.

Four ligaments support and give flexibility to the sternoclavicular joint. **An anterior sternoclavicular ligament** covers the anterior surface of the joint, and a **posterior sternoclavicular ligament** covers the posterior surface. Both ligaments extend from the sternal end of the clavicle to the manubrium. **An interclavicular ligament** extends between the sternal ends of both clavicles, binding them together. The **costoclavicular ligament** extends from the costal cartilage of the first rib to the costal tuberosity of the clavicle.

Of all the joints associated with the rib cage, the sternoclavicular joint is most frequently dislocated. Excessive force along the long axis of the clavicle may displace the clavicle forward and inferiorly. Injury to the costal cartilages is painful and is caused most frequently by a forceful, direct blow onto the costal cartilages.

Humeroscapular (Shoulder) Joint

The shoulder joint is formed by the articulation of the head of the humerus with the glenoid fossa of the scapula (fig. 10.25). It is a ball-and-socket joint and the most freely

Figure 10.25.   The humeroscapular joint. (*a*) an anterior view; (*b*) a posterior view; (*c*) a lateral view with the humerus removed.

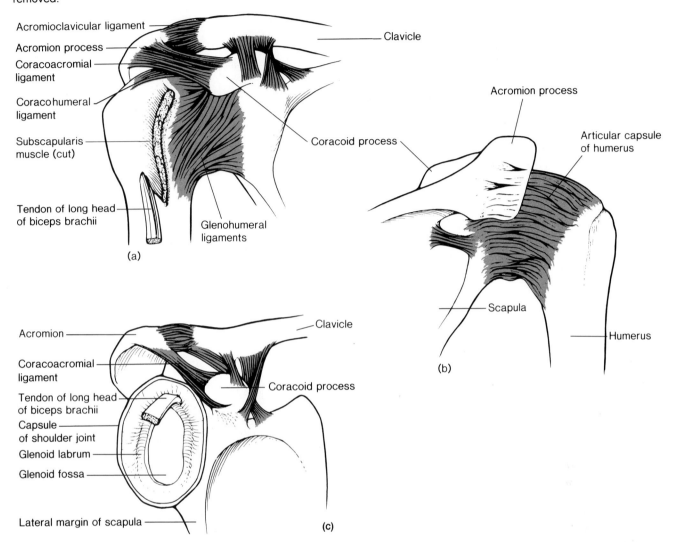

movable joint in the body. A circular band of fibrocartilage called the **glenoid** *(gle'noid)* **labrum** passes around the rim of the shoulder joint to deepen the concavity of the glenoid fossa (fig. 10.25). The shoulder joint is protected from above by an arch formed by the coracoid and acromion processes of the scapula and by the clavicle.

Although three ligaments surround and support the shoulder joint, most of the stability of this joint depends on the powerful muscles and tendons that cross over it. Thus it is an extremely mobile joint, in which stability has been sacrificed for motility. The **coracohumeral** *(kor''ah-ko-hu'mer-al)* **ligament** extends from the coracoid process of the scapula to the greater tubercle of the humerus. The joint capsule is reinforced with three ligamentous bands called the **glenohumeral ligaments.** The final ligament of the shoulder joint is the **transverse humeral ligament** (not illustrated), a thin band that extends from the greater tubercle to the lesser tubercle of the humerus.

Two major and two minor bursae are associated with the shoulder joint (not illustrated). The larger bursae are the **subdeltoid bursa** between the deltoid muscle and the joint capsule and the **subacromial bursa** between the acromion and joint capsule. The **subcoracoid bursa** lies between the coracoid process and the joint capsule and is frequently considered an extension of the subacromial bursa. A small **subscapular bursa** is located between the tendon of the subscapularis muscle and the joint capsule.

The shoulder joint is vulnerable to dislocations from sudden jerks of the arm, especially in children before strong shoulder muscles have developed. Because of the weakness of this joint in children, parents should be careful not to force a child to follow by yanking on the arm. Dislocation of the shoulder is extremely painful and may cause permanent damage or perhaps muscle atrophy due to disuse.

labrum: L. *labrum,* lip

Figure 10.26. The right elbow region. (*a*) an anterior view; (*b*) a posterior view; (*c*) a midsagittal section; (*d*) a lateral view; (*e*) a medial view.

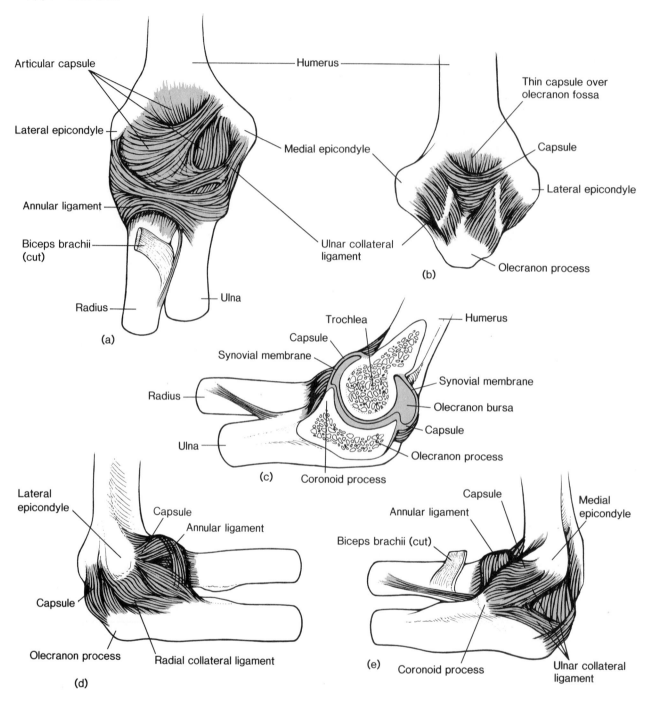

Figure 10.27. Metacarpophalangeal and interphalangeal joints. (*a*) a lateral view; (*b*) a posterior view; (*c*) a palmar view.

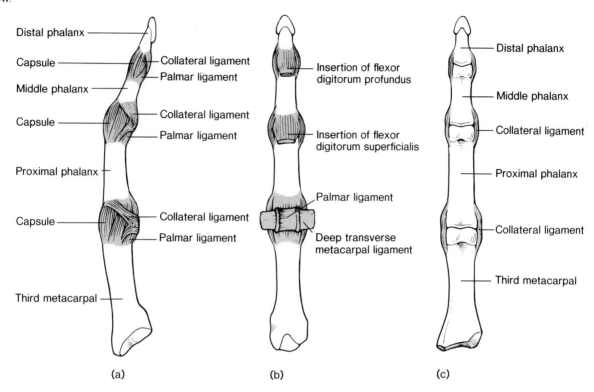

(a)   (b)   (c)

Elbow Joint

There are three sets of articulations in the elbow region, two of which constitute the elbow joint (fig. 10.26). The elbow joint is a hinge joint, formed by the trochlea of the humerus articulating with the trochlear notch of the ulna and the capitulum of the humerus articulating with the head of the radius. Although there are two sets of articulations at the elbow joint, there is only one joint capsule and a large **olecranon bursa** to lubricate this area. A **radial (lateral) collateral ligament** reinforces the elbow joint on the lateral side, and an **ulnar (medial) collateral ligament** strengthens the medial side.

Because so many muscles originate or insert near the elbow, it is a common site of localized tenderness, inflammation, and pain. **Tennis elbow** is a general term for musculotendinous soreness in this area. The structures most generally strained are the tendons attached to the lateral epicondyle of the humerus. The strain is caused by repeated extension of the wrist against some force, as occurs during the backhand stroke in tennis.

Metacarpophalangeal Joints and Interphalangeal Joints

Both metacarpophalangeal and interphalangeal joints are hinge diarthroses. These joints are formed as the heads of the metacarpals articulate with the proximal phalanges and as the phalanges articulate with one another (fig. 10.27). Each joint in both joint types has three ligaments.

A **palmar ligament** spans each joint on the palmar, or anterior, side of the joint capsule. Each joint also has two **collateral ligaments,** one on the lateral side and one on the medial side to further reinforce the joint capsule. There are no supporting ligaments on the posterior sides.

Athletes frequently "jam a finger." It occurs when a ball forcefully strikes a distal phalanx as the fingers are extended, causing a sharp flexion at the joint between the middle and distal phalanges. No ligaments support the joint on the posterior side, but there is a tendon from the digital extensor muscles of the forearm. It is this tendon that is damaged when the finger is jammed. Treatment involves splinting the finger for a period of time. If splinting is not effective, however, surgery is necessary or the person will suffer a permanent crook in the finger.

Coxal (Hip) Joint

The ball-and-socket hip joint is formed by the head of the femur articulating with the acetabulum of the os coxa (fig. 10.28). It bears the weight of the body and is therefore much stronger and more stable than the shoulder joint. The hip joint is secured by a strong fibrous joint capsule, several ligaments, and a number of powerful muscles.

The primary ligaments of the hip joint are the anterior **iliofemoral** *(il″e-o-fem′or-al)* and **pubofemoral ligaments** and the posterior **ischiofemoral** *(is″ke-o-fem′or-al)*

Figure 10.28.   The right coxal (hip) joint. (*a*) an anterior view; (*b*) a posterior view; (*c*) a frontal section.

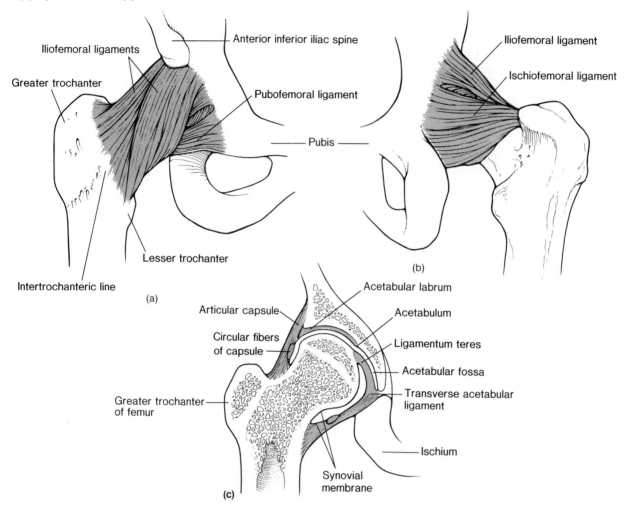

ligament. The **ligamentum teres** is located within the articular capsule and attaches the head of the femur to the acetabulum. The **transverse acetabular** *(as''ĕ-tab'u-lar)* **ligament** crosses the acetabular notch and connects to the joint capsule and the ligamentum teres. The **acetabular labrum,** a fibrocartilaginous rim that rings the head of the femur as it articulates with the acetabulum, is attached to the margin of the acetabulum.

Osteoarthritis is a degenerative disease of the articular cartilage of diarthrotic joints accompanied by the formation of bony spurs in the joint cavities. Immobility results if the hip joint is severely afflicted with this disease. Fortunately, the entire hip joint can be replaced in a procedure called hip arthroplasty. During this surgery, the acetabulum is replaced by a low-friction polyethylene socket, fit into the os coxa using bone cement. The head of the femur is replaced by a stainless steel, ball-shaped prosthesis (see fig. 10.31).

---

prosthesis: Gk. *pros,* in addition to; *thesis,* a setting down

Tibiofemoral (Knee) Joint

The knee joint is the largest, most complex, and probably the most vulnerable joint in the body. The knee joint is formed as the femur and the tibia articulate. It is a complex hinge diarthrosis, which in addition to flexion and extension, permits limited rolling and gliding movements. On the anterior side, the knee joint is stabilized and protected by the patella and the **patellar** *(pah-tel'ar)* **ligament,** which forms a gliding **patellofemoral joint.**

Because of the complexity of the knee joint, only the relative positions of the ligaments, menisci, and bursae will be presented. Although the detailed attachments and functions will not be discussed, the locations of these structures can be seen in figure 10.29.

In addition to the patella and the patellar ligament on the anterior surface, the tendinous insertion of the quadriceps femoris muscle forms two supportive bands called the **lateral** and **medial patellar retinacula** *(ret''ĭ-nak'u-lah).* Four bursae are associated with the anterior aspect of the knee (not illustrated): the **superficial infrapatellar** *(in''frah-pah-tel'ar)* **bursa,** the **suprapatellar bursa,** the **prepatellar bursa,** and the **deep infrapatellar bursa.**

Figure 10.29. The right tibiofemoral (knee) joint. (*a*) an anterior view; (*b*) a superficial posterior view; (*c*) a deep posterior view; (*d*) an anterior view with the knee flexed and the patella removed.

(a)

Femur

Lateral collateral ligament

Patella

Lateral patellar retinaculum

Infrapatellar fat pad

Fibular ligament (anterior head)

Fibula

Medial collateral ligament

Medial patellar retinaculum

Articular capsule

Patellar ligament

Tibia

(b)

Capsule

Oblique popliteal ligament

Lateral collateral ligament

Arcuate popliteal ligament

Popliteus muscle (cut)

Tibia

Fibula

(c)

Femur

Medial condyle of femur

Medial meniscus

Posterior cruciate ligament

Tibia

Anterior cruciate ligament

Lateral condyle of femur

Lateral collateral ligament

Lateral meniscus

Lateral tibial condyle

Fibular head ligaments

Fibula

(d)

Posterior cruciate ligament

Medial condyle of femur

Medial meniscus

Transverse ligament

Medial collateral ligament

Tibia

The posterior aspect of the knee is referred to as the **popliteal region.** The broad **oblique popliteal ligament** and the **arcuate** *(ar'cu-āt)* **popliteal ligament** are superficial in position, whereas the **anterior** and **posterior cruciate** *(kroo'she-āt)* **ligaments** are deep within the joint. The **popliteal bursa** and the **semimembranosus bursa** (not illustrated) are the two bursae associated with the back of the knee.

Strong **collateral ligaments** support both the medial and lateral sides of the knee joint. Two fibrocartilaginous discs called the **lateral** and **medial menisci** are located within the knee joint interposed between the distal femoral and proximal tibial condyles. The two menisci are connected by a **transverse ligament.** Several other bursae are associated with the knee joint. In addition to the four on the anterior side and the two on the posterior side, there are seven bursae on the lateral and medial sides, making a total of thirteen.

cruciate: L. *crucis*, cross

Figure 10.30.   The talocrural (ankle) joint. (*a*) an anterior view; (*b*) a posterior view.

(a)

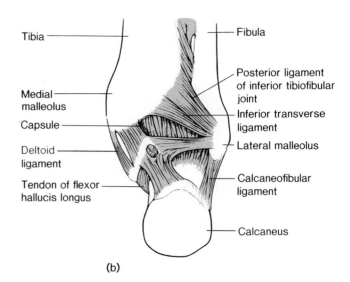

(b)

During normal walking, running, and supporting of the body, the knee joint functions superbly. It can tolerate even moderate stress without tissue damage. However, the knee lacks bony support to withstand sudden forceful stresses, such as commonly occur among professional athletes. Knee injuries frequently require surgery and heal with difficulty due to the avascularity of the cartilaginous tissue. Because this joint is potentially so vulnerable, it is important to realize its limitations by understanding its anatomy.

Ankle sprains are a common type of locomotor injury. They vary extensively in seriousness but generally occur in certain locations. The most common cause of ankle sprain is excessive inversion of the foot, resulting in partial tearing of the anterior talofibular ligament and the calcaneofibular ligament. Less commonly, the deltoid ligament is injured by excessive eversion of the foot. Torn ligaments are extremely painful and are accompanied by immediate local swelling. Reducing the swelling and immobilizing the joint are about the only treatments for moderate sprains. Extreme sprains may require surgery and casting of the joint to facilitate healing.

Talocrural (Ankle) Joint

There are actually two articulations within the ankle joint, both of which are hinge diarthroses. One articulation is formed as the distal end of the tibia and its medial malleolus articulate within the talus, and the other is formed as the lateral malleolus of the fibula articulates with the talus (fig. 10.30).

One joint capsule surrounds the articulations of three bones, and four ligaments support the ankle joint on the outside of the capsule. The strong **deltoid ligament** is associated with the tibia, whereas the **anterior talofibular** *(ta''lo-fib'u-lar)* **ligament, posterior talofibular ligament** (not visible in fig. 10.30), and **calcaneofibular** *(kal-ka''ne-o-fib'u-lar)* **ligament** are associated with the fibula.

The malleoli form a cap over the upper surface of the talus that prohibits side-to-side movement at the ankle joint. Unlike the condyloid joint at the wrist, the movements of the ankle are limited to flexion and extension. Dorsiflexion of the ankle is checked primarily by the Achilles tendon, whereas plantar flexion, or ankle extension, is checked by the tension of the extensor tendons on the front of the joint and the anterior portion of the joint capsule.

Table 10.2 presents a summary of the principal joints of the body and their movement.

1. Which diarthrotic joints have menisci?
2. What are the two types of joints found in the shoulder region? Why is the shoulder joint so vulnerable?
3. Which joints are reinforced with muscles that span the joint?
4. Describe the structure of the knee joint, and indicate which structures protect and reinforce its anterior surface.

Clinical Considerations

A diarthrotic joint is a remarkable biologic system, which acts as a self-lubricating bearing surface, able to move with almost frictionless precision under tremendous loads and impacts. Under normal circumstances and in most people, the many joints of the body perform without problems throughout life. Joints are not indestructible, however, and

Table 10.2 Principal articulations

Joint	Type	Movement
Most skull joints	Synarthrosis (suture)	Immovable
Temporomandibular	Diarthrosis (hinge; gliding)	Elevation, depression; protraction, retraction
Atlantooccipital	Diarthrosis (condyloid)	Flexion, extension, circumduction
Atlantoaxial	Diarthrosis (pivot)	Rotation
Intervertebral		
bodies of vertebrae	Amphiarthrosis (symphysis)	Slight movement
articular processes	Diarthrosis (gliding)	Flexion, extension, slight rotation
Sacroiliac	Amphiarthrosis (synchondrosis); diarthrosis (gliding)	Slight gliding movement; may fuse in adults
Costovertebral	Diarthrosis (gliding)	Slight movement during breathing
Sternocostal	Amphiarthrosis (synchondrosis); diarthrosis (gliding)	Slight movement during breathing
Sternoclavicular	Diarthrosis (gliding)	Slight movement when shrugging shoulders
Sternal	Amphiarthrosis (symphysis)	Slight movement during breathing
Acromioclavicular	Diarthrosis (gliding)	Protraction, retraction; elevation, depression
Humeroscapular (shoulder)	Diarthrosis (ball-and-socket)	Flexion, extension; adduction, abduction; rotation, circumduction
Elbow	Diarthrosis (hinge)	Flexion, extension
Proximal radioulnar	Diarthrosis (pivot)	Rotation
Distal radioulnar	Amphiarthrosis (syndesmosis)	Slight movement
Radiocarpal (wrist)	Diarthrosis (condyloid)	Flexion, extension; adduction, abduction; circumduction
Intercarpal	Diarthrosis (gliding)	Slight movement
Carpometacarpal		
fingers	Diarthrosis (condyloid)	Flexion, extension; adduction, abduction
thumb	Diarthrosis (saddle)	Flexion, extension; adduction, abduction
Metacarpophalangeal and interphalangeal	Diarthrosis (hinge)	Flexion, extension
Symphysis pubis	Amphiarthrosis (symphysis)	Slight movement
Coxal (hip)	Diarthrosis (ball-and-socket)	Flexion, extension; adduction, abduction; rotation; circumduction
Tibiofemoral (knee)	Diarthrosis (hinge)	Flexion, extension; slight rotation when flexed
Proximal tibiofibular	Diarthrosis (gliding)	Slight movement
Distal tibiofibular	Amphiarthrosis (syndesmosis)	Slight movement
Talocrural (ankle)	Diarthrosis (hinge)	Dorsiflexion, plantar flexion; slight circumduction
Intertarsal	Diarthrosis (gliding)	Inversion, eversion
Tarsometatarsal	Diarthrosis (gliding)	Flexion, extension; adduction, abduction

are subject to various forms of trauma and disease. Although not all the diseases of joints are fully understood, medical science has made remarkable progress in the treatment of arthrological problems.

Trauma to Joints

Joints are well adapted to withstand compression and tension forces. Torsion or sudden impact to the side of a joint, however, can be devastating. These types of injuries frequently occur in athletes.

In a **strained joint,** unusual or excessive exertion stretches the tendons spanning a joint or the surrounding muscles but causes no serious damage. Strains frequently result from not "warming up" or activating joints or muscles prior to strenuous use. A **sprain** is a tearing of the ligaments or tendons that surround a joint. There are various grades of sprains, and the severity will determine the treatment. Severe sprains of the knee joint are frequently accompanied by damage to articular cartilages, which generally requires surgery. Sprains are usually accompanied by **synovitis** *(sin''o-vi'tis),* an inflammation of the joint capsule.

Luxation, or **joint dislocation,** is derangement of the articulating bones that compose the joint. Joint dislocation is more serious than a sprain and is usually accompanied by sprains. The shoulder and knee joints are the most vulnerable to dislocation. Self-healing of a dislocated joint, like the knee joint, may be incomplete, leaving the person with a "trick knee" that may unexpectedly give way.

Subluxation is partial dislocation of a joint. Subluxation of the hip joint is a common type of birth defect, which can be treated by bracing or casting the hip joints to promote suitable bone development.

---

luxation: L. *luxus,* out of place

Bursitis *(ber-si'tis)* is an inflammation of the bursa associated with a joint. Because of its close proximity to the joint, bursitis may affect the joint capsule as well. Bursitis may be caused by excessive stress on the bursa from overexertion, or it may be a local or systemic inflammatory process. As the bursa swells, the surrounding muscles become sore and stiff. **Tendonitis** involves the tendon, in or out of a tendon sheath, and usually has the same causes as bursitis.

The flexible vertebral column is a marvel of mechanical engineering. Not only do the individual vertebrae articulate one with another, but together they form the portion of the axial skeleton with which the head, ribs, and pectoral and pelvic girdles articulate. The vertebral column also encloses the spinal cord and provides exits for thirty-one pairs of spinal nerves. With all the articulations within the vertebral column and the physical abuse it receives, it is no wonder that back ailments are second only to headaches as the most common physical complaint. Our way of life causes many of the problems associated with the vertebral column. Improper shoes, athletic exertion, sudden stops in vehicles, or improper lifting can all cause the back to go awry. Body weight, age, and general body condition influence a person's susceptibility to back problems.

The most common cause of back pain is strained muscles, generally the result of overexertion. The second most frequent back ailment is a herniated disc. The dislodged nucleus pulposus of a disc may push against a spinal nerve and cause excruciating pain. The third most frequent back problem is a dislocated articular facet between two vertebrae, caused by sudden torsion of the vertebral column. The treatment of back ailments varies from bed rest to spinal manipulation to extensive surgery.

Curvature disorders are another problem of the vertebral column. **Kyphosis** *(ki-fo'sis)* (hunchback) is an exaggeration of the thoracic curve. **Lordosis** (swayback) is an abnormal anterior convexity of the lumbar curve. **Scoliosis** (crookedness) is an abnormal lateral curvature of the vertebral column, which may be caused by one leg being longer than the other and uneven muscular development on the two sides of the vertebral column.

Diseases of Joints

Arthritis is a generalized term for over fifty different joint diseases, all of which have the symptoms of edema, inflammation, and pain. The causes of arthritis are unknown, but certain types follow joint trauma or bacterial infections. There is evidence that some types of arthritis

kyphosis: Gk. *kyphos,* hunched
lordosis: Gk. *lordos,* curving forward
scoliosis: Gk. *skoliosis,* crookedness

are the result of hormonal or metabolic disorders. The most common forms are rheumatoid arthritis, osteoarthritis, and gouty arthritis.

In **rheumatoid** *(roo'mah-toid)* **arthritis,** the synovial membrane thickens and becomes tender and synovial fluid accumulates. This change is usually followed by an invasion of fibrous tissue and deterioration of the articular cartilage. When the cartilage is destroyed, the exposed osseous tissue is joined by the fibrous tissue and instigates ossification of the joint. It is the joint ossification that causes the crippling effect of rheumatoid arthritis. This disease affects females slightly more than males, and occurs most commonly between the ages of thirty and fifty.

Osteoarthritis is a degenerative joint disease that results from aging and irritation of the joints. Although osteoarthritis is far more common than rheumatoid arthritis, is it usually less damaging. Osteoarthritis is a slow, progressive disease in which the articular cartilages gradually soften and disintegrate. The affected joints seldom swell, and the synovial membrane is rarely damaged. As the articular cartilage deteriorates, ossified spurs are deposited on the exposed bone, causing pain and restricting the movement of articulating bones. Osteoarthritis most frequently affects the knee, hip, and intervertebral joints.

Gouty arthritis results from a metabolic disorder in which an abnormal amount of uric acid is retained in the blood and sodium urate crystals are deposited in the joints. The salt crystals irritate the articular cartilage and synovial membrane, causing swelling, tissue deterioration, and pain. If gout is not treated, the affected joint fuses. Males have a greater incidence of gout than females, and apparently the disease is genetically determined.

Treatment of Joint Disorders

Arthroscopy *(ar-thros'ko-pe)* is widely used in diagnosing and, to a limited extent, treating joint disorders. Arthroscopic inspection involves making a local incision through the skin and into the joint capsule through which the tube-like arthroscopic instrument is threaded. In arthroscopy of the knee, the articular cartilage, synovial membrane, menisci, and cruciate ligaments can be observed. Samples can be extracted, and pictures taken for further evaluation.

Remarkable advancements have been made in the last fifteen years in **joint prostheses** *(pros-the'sēz)* (fig. 10.31). Joint prostheses, or artificial articulations, do not take the place of normal, healthy joints; but they are a valuable option for chronically disabled arthritis patients. Two important advancements have been the development of high-density polyethylene to articulate with stainless steel, and an improved polymethyl methacrylate cement for securing the prosthesis.

rheumatoid: Gk. *rheuma,* a flowing
gout: L. *gutta,* a drop (thought to be caused by "drops of viscous humors")

Figure 10.31.   Two examples of joint prostheses. (*a,b*) the
hip joint; (*c,d*) the knee joint.

(a)

(b)

(c)

(d)

Chapter Summary

I. Classification of Joints
The articulations between the bones of the skeleton are classified by arthrologists into three types according to structure or into three types according to function, or the degree of movement permitted.
  A. Joints are formed as adjacent bones articulate together. Arthrology is the science concerned with the study of joints, and kinesiology is the study of movements involving certain joints.
  B. Joints can be classified according to structure or function.
    1. The structural classification divides joints into fibrous, cartilaginous, and synovial types.
    2. The functional classification divides joints into synarthroses, amphiarthroses, and diarthroses.

II. Development of Freely Movable Joints
Diarthroses form by the third month as the epiphyses of adjacent endochondral bones develop distinct configurations and as the muscles that move the joints undergo contractions.
  A. The development of each movable joint is initiated during the sixth week as the perichondrium from adjacent developing bones spans the future joint cavity and gives rise to the joint capsule.
  B. Articular cartilages form on the epiphyses, and the synovial membrane secretes synovial fluid into the joint cavity.

III. Synarthroses
Articulating bones in synarthroses are tightly connected by either fibrous tissue or cartilage. Synarthrotic joints are rigid and immovable and are of two types: sutures and synchondroses.
  A. Sutures are found only in the skull and are classified as serrate, lap, plane, or synostosis.
  B. Synchondroses are temporary joints formed in the growth lines (epiphyseal plates) between the diaphyses and epiphyses in the long bones of children.

IV. Amphiarthroses
Amphiarthroses allow limited motion in response to twisting, compression, or stress. The two types of amphiarthroses are symphyses and syndesmoses.
  A. The articulating bones of amphiarthroses are connected by ligaments or fibrocartilaginous pads.
  B. The symphysis pubis and the intervertebral joints are examples of symphyses; the distal articulations of the tibia-fibula and the ulna-radius are examples of syndesmoses.

V. Diarthroses
Diarthroses are freely movable joints that contain synovial fluid and are enclosed by joint capsules. Types of diarthroses include gliding, hinge, pivot, condyloid, saddle, and ball-and-socket.
  A. Diarthroses contain a joint cavity, synovial membrane, and articular cartilage. Some also contain articular discs, accessory ligaments, and associated bursae.
  B. The range of movement of a diarthrosis is determined by the structure of the articulating bones, the ligaments and tendons, and the muscles that act on the joint.

VI. Movements at Diarthroses
Movements at diarthroses are produced by the contraction of the skeletal muscles spanning the joints and attaching to or near the bones forming the articulations. In these actions, the bones act as levers, the muscles provide the force, and the joints are the fulcra, or pivots.
  A. Angular movements increase or decrease the joint angle produced by the articulating bones.
    1. Flexion decreases the joint angle on an anterior-posterior plane; extension increases the same joint angle.
    2. Abduction is the movement of a body part away from the main axis of the body; adduction is the movement toward the main axis of the body.
  B. Circular movements can occur only in joints that are composed of a bone with a rounded surface articulating with a corresponding depression on another bone.
    1. Rotation is the movement of a bone around its own axis.
    2. Circumduction is a conelike movement of a body segment.

C. Special joint movements include inversion and eversion, protraction and retraction, and elevation and depression.
D. Diarthroses can be classified as first-, second-, or third-class levers.
    1. A lever consists of a rigid structure, a pivot, a resistance, and a force.
    2. The position of the insertion of a muscle relative to the joint is an important factor in the biomechanics of the contraction.

VII. Specific Joints of the Body
Of the numerous joints in the body, some have special structural features that enable them to perform particular functions. Furthermore, these joints are somewhat vulnerable to trauma and are therefore clinically important.
  A. The temporomandibular joint, a combined hinge and gliding joint, is of clinical importance because of temporomandibular joint (TMJ) syndrome.
  B. The sternoclavicular joint, a gliding joint, is frequently dislocated in injuries to the rib cage.
  C. The humeroscapular (shoulder) joint, a ball-and-socket joint, is vulnerable to dislocations from sudden jerks of the arm, especially in children before strong shoulder muscles have developed.
  D. There are two hinge joints at the elbow as the distal end of the humerus articulates with the proximal ends of the ulna and radius. Strain on the elbow joint is common during certain sports.
  E. Both metacarpophalangeal and interphalangeal joints are hinge diarthroses that may be traumatized as an athlete "jams a finger."
  F. The ball-and-socket coxal (hip) joint is subject to osteoarthritis in elderly people.
  G. The hinged tibiofemoral (knee) joint is the largest, most complex, and most vulnerable joint in the body.
  H. There are two hinged articulations within the talocrural (ankle) joint. Ankle sprains are common injuries of this joint.

Review Activities

Objective Questions

1. Which statement regarding joints is *false*?
   - (a) Joints are the locations where two or more bones articulate.
   - (b) The structural classification of joints includes fibrous, membranous, and cartilaginous types.
   - (c) Arthrology is the study of joints; kinesiology is the study of the biomechanics of joint movement.

2. Synchondroses are a type of
   - (a) synarthrosis
   - (b) diarthrosis
   - (c) amphiarthrosis
   - (d) syndesmosis

3. A synostosis is a unique type of
   - (a) synchondrosis
   - (b) syndesmosis
   - (c) symphysis
   - (d) suture

4. Which of the following joint type-function word pairs is *incorrect*?
   - (a) synchondrosis—growth at the epiphyseal plate
   - (b) symphysis—movement at the intervertebral joint
   - (c) suture—strength and stability in the skull
   - (d) syndesmosis—movement of the jaw

5. Structurally, diarthroses are
   - (a) synovial joints
   - (b) fibrous joints
   - (c) membranous joints
   - (d) cartilaginous joints

6. Which of the following is *not* characteristic of all diarthroses?
   - (a) articular cartilage
   - (b) synovial fluid
   - (c) joint capsule
   - (d) meniscus

7. In a second-class lever, the
   - (a) fulcrum is positioned between the effort and the resistance
   - (b) effort lies between the fulcrum and the resistance
   - (c) resistance is positioned between the fulcrum and the effort
   - (d) effort and resistance are the same

8. Which of the following joints can be readily and comfortably hyperextended?
   - (a) interphalangeal
   - (b) coxal
   - (c) tibiofemoral
   - (d) sternocostal

9. Which of the following is most vulnerable to luxation? The
   - (a) elbow
   - (b) humeroscapular
   - (c) coxal
   - (d) tibiofemoral

10. A thickening and tenderness of the synovial membrane and the accumulation of synovial fluid are signs of the development of
    - (a) arthroscopitis
    - (b) gouty arthritis
    - (c) osteoarthritis
    - (d) rheumatoid arthritis

Essay Questions

1. What is meant by a structural classification of joints, as compared to a functional classification?
2. Why is the anatomical position so important in explaining the movements that are possible at joints?
3. What are the structural components of a diarthrotic joint that determine or limit the range of movement at that joint?
4. What is a bursa? What is the difference between a bursa and a tendon sheath? What are their functions?
5. Identify four types of diarthrotic joints found in the wrist and hand region, and state the types of movement permitted by each.
6. What is meant by a sprained ankle? How does a sprain differ from a strain or a luxation?
7. What occurs within the joint capsule in rheumatoid arthritis? How does rheumatoid arthritis differ from osteoarthritis?
8. Describe the use of an arthroscopic instrument in diagnosing and treating a traumatized joint.

Selected Readings

Aufranco, D. E., and R. H. Turner. October 1971. Total replacement of the arthritic hip. *Hosp. Prac.*

Evans, F. G., ed. 1966. *Studies in the anatomy and function of bones and joints.* New York: Springer-Verlag.

Napier, J. April 1967. The antiquity of human walking. *Scientific American.*

Platzer, W. 1978. *Locomotor system.* Vol. 1 of Kahle, W., H. Leonhardt, and W. Platzer. *Color atlas and textbook of human anatomy.* Chicago: Year Book Medical Publishers.

Rasche, P. J., and R. K. Burke. 1978. *Kinesiology and applied anatomy: The science of human movement.* 6th ed. Philadelphia: Lea and Febiger.

Rosse, C., and D. K. Clawson. 1980. *The musculoskeletal system in health and disease.* Philadelphia: Harper and Row Publishers.

Simon, W. H., ed. 1978. *The human joint in health and disease.* Philadelphia: Univ. of Pennsylvania Press.

Sonstegard, D. A., L. S. Matthews, and H. Kaufer. January 1978. The surgical replacement of the human knee joint. *Scientific American.*

11

Muscular System

Outline

Concepts

The muscles of the skeletal system are adapted to contract in order to carry out the functions of motion, heat production, and posture and body support.

Skeletal muscles begin development at four weeks after conception from blocks of mesoderm, called myotomes, in the trunk area and from loosely organized mesenchyme in the head area. Each muscle fiber is derived from several embryonic myoblast cells.

Skeletal muscle tissue and its binding connective tissue are arranged in a highly organized pattern so that the forces of the contracting muscle fibers are united and directed onto the structure being moved.

Muscle fiber contraction results from a sliding movement within the myofibrils in which the length of the sarcomeres is reduced.

The recruitment of motor units when a sufficient amount of Ca^{++} is bound to tropin results in muscle contraction. The response of a motor unit is determined by the intensity and duration of the stimulus.

Aerobic cellular respiration is required for the production of ATP needed for the cross-bridge activity of the various types of skeletal muscle fibers.

Skeletal muscles are named on the basis of shape, location, attachment, orientation of fibers, relative position, and function.

The muscles of the axial skeleton include those of facial expression, mastication, eye movement, tongue movement, neck movement, respiration, the abdominal wall, the pelvic outlet, and movement of the vertebral column.

The muscles of the appendicular skeleton include those of the pectoral girdle, humerus, forearm, wrist, hand, and fingers, and those of the pelvic girdle, thigh, lower leg, ankle, foot, and toes.

Selected Terms

myoblasts   (mi′o-blasts″)
myotomes   (mi′o-tōms)
fascia   (fash′e-ah)
synergistic   (sin″er-jis′tik)
antagonistic   (an-tag′o-nis′tik)
sarcomeres   (sar′ko-mērs)
myofibrils   (mi″o-fi′brils)
actin   (ak′tin)
myosin   (mi′o-sin)
neurotransmitter   (nu″ro-trans′mit-er)
troponin   (tro′po-nin)
tetanus   (tet′a-nus)
phosphocreatine   (fos″fo-kre′ah-tin)
myoglobin   (mi″o-glo′bin)

Organization and General Functions

The muscles of the skeletal system are adapted to contract in order to carry out the functions of motion, heat production, and posture and body support.

Objective 1. Define the term *myology,* and describe the three principal functions of muscles.

Objective 2. Explain how muscles are described according to their anatomical location and cooperative function.

Myology is the study of muscles. Myology by itself does not have much meaning, however, except as it relates to the skeletal system and joints in the performance of body movements and to the nervous system in the performance of motor control. Muscles are dynamic, and an understanding of them requires an applied approach.

Functions of Muscles

Muscle cells (fibers) are adapted to contract when stimulated by nerve impulses. The stimulation of several fibers is not enough to cause a noticeable effect within a muscle, but these types of isolated fiber contractions are important and occur continuously within a muscle. When many fibers of a skeletal muscle are activated, the muscle contracts and causes body movement. Muscles perform three general functions: (1) motion, (2) heat production, and (3) posture and body support.

1. **Motion.** The most obvious type of motion performed by the skeletal muscles is to move the body and/or its appendages, as in walking, running, writing, chewing, and swallowing. The contraction of skeletal muscle is equally important in breathing and in moving body fluids. The stimulation of isolated fibers within the muscles maintains muscle tonus, which is important in the movement of venous blood and lymphatic fluid. The eye and even the ear ossicles have associated skeletal muscles responsible for various movements.

   The primary impetus for blood flow is the contraction of cardiac muscles within the heart. All of the involuntary body systems (urinary, digestive, respiratory, circulatory, etc.) contain smooth muscles for the involuntary movement of materials through the body. The act of speaking involves several different muscle groups that control air volume, tense the vocal cords, move the tongue, and purse the lips. This is a highly coordinated effort of several muscle groups controlled by several different nerve centers.

2. **Heat production.** Body temperature is remarkably consistent. Metabolism within the cells releases heat as an end product. Since muscles constitute nearly one-half of the body weight and are in a continuous state of fiber activity, they are very important in the production of heat. The rate of heat production increases immensely as a person exercises strenuously.

muscle: L. *muris,* mouse (from the appearance of certain muscles)
myology: Gk. *myos,* muscle; *logos,* study of

Figure 11.1. An anterior view of some of the superficial skeletal muscles.

Frontalis
Orbicularis oculi
Zygomaticus
Orbicularis oris
Masseter
Sternocleidomastoid
Trapezius
Pectoralis major
Deltoid
Serratus anterior
Biceps brachii
Brachialis
External oblique
Brachioradialis
Rectus abdominis
Tensor fasciae latae
Rectus femoris
Vastus lateralis
Gracilis
Vastus medialis
Adductor longus
Sartorius
Peroneus longus
Gastrocnemius
Tibialis anterior
Soleus
Extensor digitorum longus

Nelson

3. **Posture and body support.** The skeletal system gives form and stability to the body, but skeletal muscles maintain posture and support around the flexible joints. Certain muscles are active postural muscles whose primary function is to work in opposition to gravity. Some postural muscles are working even when you think you are relaxed. As you are sitting, for example, the weight of your head is balanced at the atlantooccipital joint through the efforts of the muscles located at the back of the neck. If you start to get sleepy, your head will suddenly nod forward as the postural muscles relax and the weight (resistance) overcomes the effort.

There are three types of muscle tissue in the body: smooth, cardiac, and skeletal. Although these three types differ in structure and function, all muscle tissue possesses some basic characteristic properties:

1. **Irritability.** A muscle tissue receives and responds to a stimulus from a nerve impulse.
2. **Contractility.** A muscle tissue responds to a stimulus by contracting, or shortening, its length.
3. **Extensibility.** Once a stimulus has subsided and the fibers within muscle tissue are shortened and relaxed, they may be stretched back to their original length by the contraction of an opposing muscle. The fibers are then stretched in readiness for another contraction.

Figure 11.2. A posterior view of some of the superficial skeletal muscles.

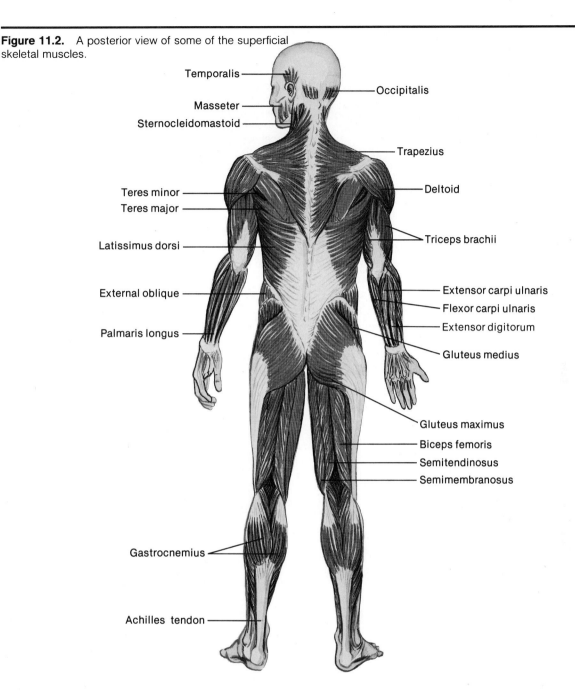

Temporalis

Occipitalis

Masseter

Sternocleidomastoid

Trapezius

Deltoid

Teres minor

Teres major

Triceps brachii

Latissimus dorsi

External oblique

Extensor carpi ulnaris

Flexor carpi ulnaris

Extensor digitorum

Palmaris longus

Gluteus medius

Gluteus maximus

Biceps femoris

Semitendinosus

Semimembranosus

Gastrocnemius

Achilles tendon

4. **Elasticity.** A muscle tissue has an innate tension that causes it to assume a desired shape regardless of how it might be stretched.

A histological description of each of the three muscle types was presented in chapter 6 and should be reviewed at this time. Smooth muscle and cardiac muscle are discussed in chapter 16, as they are innervated by the autonomic nervous system. The remaining information presented in this chapter pertains only to skeletal muscle and the skeletal muscular system of the body.

Skeletal muscle constitutes a body system by itself and accounts for approximately 40% of body weight. Over six hundred individual muscles comprise the skeletal muscular system. The principal superficial muscles are shown in figures 11.1 and 11.2. Muscles are usually described in

groups according to their anatomical location and cooperative function. The *muscles of the axial skeleton* have their attachments to the bones of the axial skeleton and include facial muscles, neck muscles, and anterior and posterior trunk muscles. The *muscles of the appendicular skeleton* include those that act on the pectoral and pelvic girdles and those that move the segments of the appendages.

1. Discuss how the functions of muscles aid in maintaining body homeostasis.
2. What is meant by a postural muscle?
3. Distinguish between the axial and the appendicular muscles.

Development of Skeletal Muscles

Skeletal muscles begin development at four weeks after conception from blocks of mesoderm, called myotomes, in the trunk area, and from loosely organized mesenchyme in the head and appendage areas. Each muscle fiber is derived from several embryonic myoblast cells.

Objective 3. Describe the formation of a muscle fiber from myoblast cells.

Objective 4. Describe the development of skeletal muscles from myotomes.

The formation of skeletal muscles begins during the fourth week of embryonic development as specialized mesodermal cells, called **myoblasts** *(mi'o-blasts''),* begin rapid mitotic division (fig. 11.3). The proliferation of new cells continues while the myoblast cells fuse together into **syncytial** *(sin-sish'al)* **myotubes.** At nine weeks, primitive myofilaments course through the myotubes and the nuclei of the contributing myoblasts are centrally located. Growth in length continues through the addition of myoblasts.

It is not certain when skeletal muscle is sufficiently developed to sustain contractions, but by the seventeenth week the fetal movements known as **quickening** are strong enough to be recognized by the mother. At this time, the individual muscle fibers have thickened, the nuclei have moved peripherally, and the filaments (myofilaments) have differentiated into alternating thin *(actin)* and thick *(myosin)* bands. Growth in length still continues through addition of myoblasts.

Shortly before a baby is born, the formation of myoblast cells ceases, and all the muscle cells of a person have been determined. Differences in strength, endurance, and coordination are somewhat genetically determined but are primarily a result of individual body conditioning. Muscle coordination is an ongoing process of achieving a fine neural control of muscle fibers. A mastery of muscle movement is comparatively slow in humans. It is several months before a newborn has the coordination to crawl and about a year before it achieves bipedal posture and locomotion. By contrast, most mammals can walk and run within a few hours after they are born. Refined, precise movements of the hands for delicate work, such as playing a violin or performing surgery, may be learned only after many hours of disciplined practice; and some persons may not be able to develop these movements at all.

The process of muscle fiber development occurs within specialized mesodermal masses, called **myotomes** *(mi'o-tōms),* in the embryonic trunk area and from loosely organized masses of mesoderm in the head and appendage areas. At six weeks, the trunk of an embryo is segmented into distinct myotomes (fig. 11.4) that are associated dorsally with specific sclerotomes. Sclerotomes give rise to

Figure 11.3. The development of skeletal muscle fibers. (*a*) at four weeks, mesodermal cells differentiate into myoblasts that aggregate into a developing syncytial tube. (*b*) at five weeks, the syncytial myotube is formed as individual plasma membranes are broken down and longitudinal myofilaments appear. Myotubes grow in length by incorporating additional myoblasts; each adds an additional nucleus. (*c*) muscle fibers are distinct at nine weeks, but the nuclei are still centrally located, and growth in length continues through the addition of myoblasts. (*d*) at five months, thin (actin) and thick (myosin) myofilaments are present, and moderate growth in length still continues. (*e*) by birth, the striated myofilaments have aggregated into bundles, the fiber has thickened, and the nuclei have shifted to the periphery. Myoblast activity ceases, and all the muscle fibers a person will have are formed.

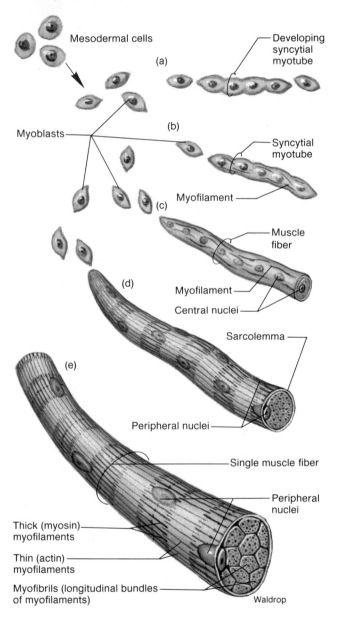

myoblast: Gk. *myos,* muscle; *blastos,* germ
syncytial: Gk. *syn,* with; *cyto,* cell

Figure 11.4. The development of skeletal muscles. (*a*) the distribution of embryonic myotomes at six weeks. Segmental myotomes give rise to the muscles of the trunk area and girdles. Loosely organized masses of mesoderm form the muscles of the head and extremities. (*b*) the arrangement of skeletal muscles at eight weeks. The development of muscles is influenced by the preceding cartilaginous models of bones. The innervation of muscles corresponds to the development of spinal nerves and dermatome arrangement.

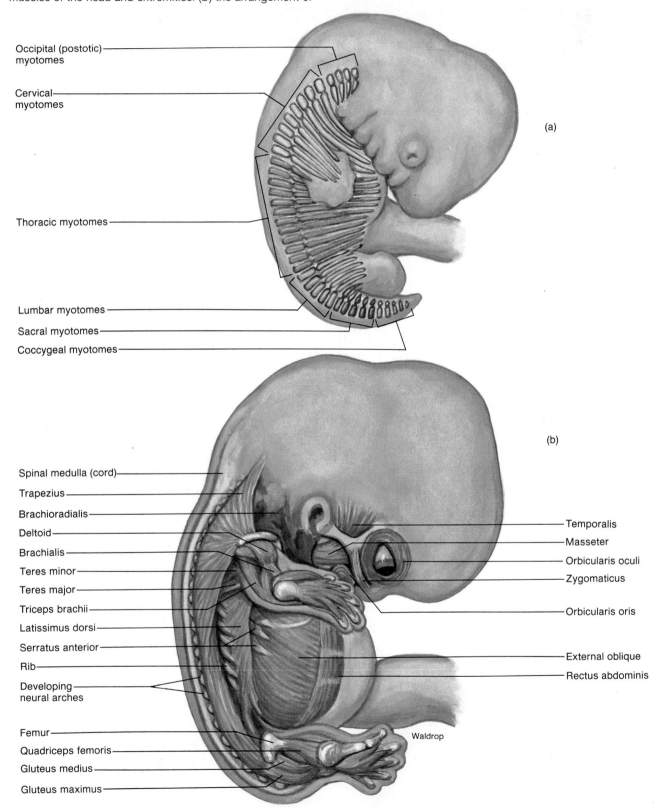

Occipital (postotic) myotomes

Cervical myotomes

Thoracic myotomes

Lumbar myotomes

Sacral myotomes

Coccygeal myotomes

(a)

(b)

Spinal medulla (cord)

Trapezius

Brachioradialis

Deltoid

Brachialis

Teres minor

Teres major

Triceps brachii

Latissimus dorsi

Serratus anterior

Rib

Developing neural arches

Femur

Quadriceps femoris

Gluteus medius

Gluteus maximus

Temporalis

Masseter

Orbicularis oculi

Zygomaticus

Orbicularis oris

External oblique

Rectus abdominis

Waldrop

vertebrae. As will be explained in chapter 15, spinal nerves arise from the spinal cord and exit between vertebrae to innervate (serve with motor and sensory nerve branches) developing muscles in the adjacent myotomes. As myotomes develop, they elongate anteriorly toward the midline of the body, or distally into the developing limbs. The muscles of the entire muscular system are differentiated and in correct position by the eighth week (fig. 11.4). The orientation of the developing muscles is preceded and influenced by cartilaginous models of bones.

1. Describe the process by which muscle fibers become multinucleated.
2. Distinguish between the terms myoblasts, myotubes, and muscle fibers.
3. Briefly describe the embryonic events that occur between weeks six and eight in the formation of specific muscles.

Structure of Skeletal Muscles

Skeletal muscle tissue and its binding connective tissue are arranged in a highly organized pattern so that the forces of the contracting muscle fibers are united and directed onto the structure being moved.

Objective 5. Distinguish between the various binding connective tissues associated with skeletal muscles.

Objective 6. Explain what is meant by synergistic and antagonistic muscle groups.

Objective 7. Describe the various types of muscle fiber architecture, and discuss the biomechanical advantage of each.

Attachments

Skeletal muscles are usually long and narrow, span a joint, and are attached to a bone at either end by a tendon (fig. 11.5). As the muscle contracts, one of the bones moves relative to the other around the joint. The more fixed, or stationary, attachment is designated as the **origin** of a muscle, whereas the movable end is its **insertion.** In muscles associated with the girdles and appendages, the origin is generally the proximal attachment and the insertion the distal attachment. The fleshy, thickened portion of a muscle is referred to as its **belly,** or **gaster.** Generally the belly of a muscle is located on the bone proximal to the bone that is to be moved. The joint is spanned by a **tendon** from the muscle. A tendon is toughened, dense fibrous connective tissue that connects a muscle to the periosteum of a bone. It functions to transfer the force of contraction

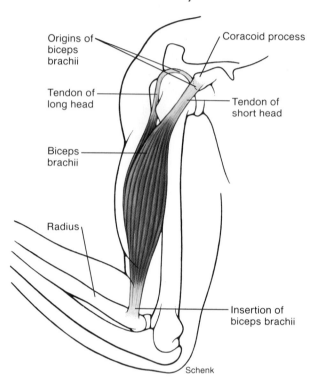

Figure 11.5. The skeletomuscular relationship. The more proximal and fixed point of muscle attachment is the origin, whereas the distal, maneuverable point of attachment is the insertion. The contraction of muscle fibers causes one bone to move relative to another around a joint.

from the muscle across the joint and onto the bone that is to be moved. Flattened, sheetlike tendons are called **aponeuroses** *(ap''o-nu-ro'sēz)* and occur where attachment is over a broad line. In certain places, especially in the wrist and ankle, the tendons are not only enclosed by protective **tendon sheaths,** which lubricate the tendons with synovial fluid, but the entire group of tendons is covered by a thin band of connective tissue called a **retinaculum** *(ret''i-nak'u-lum)* (see fig. 11.47).

Associated Connective Tissue

Contracting muscle fibers would not be effective if they worked as isolated units. Each fiber is bound to adjacent fibers to form bundles, and the bundles in turn are bound to other bundles. With this arrangement, the contraction in one area of a muscle works in conjunction with contracting fibers elsewhere in the muscle. The binding substances within muscles are the associated connective tissues.

Connective tissue is structurally arranged within muscle to protect, strengthen, and bind muscle fibers into bundles and bind the bundles together (fig. 11.6). The individual fibers of skeletal muscles are surrounded by fine

---

aponeurosis: Gk. *aponeurosis,* change into a tendon
retinaculum: L. *retinere,* to hold back (retain)

Figure 11.6. Connective tissue associated with the skeletal muscle.

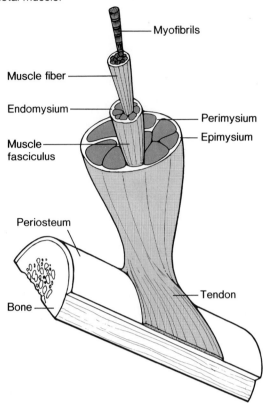

Figure 11.7. Examples of synergistic and antagonistic muscles. The two heads of the biceps are synergistic to each other, as are the three heads of the triceps. The biceps is antagonistic to the triceps, and the triceps is antagonistic to the biceps. When one antagonistic group contracts, the other one must relax, or movement does not occur.

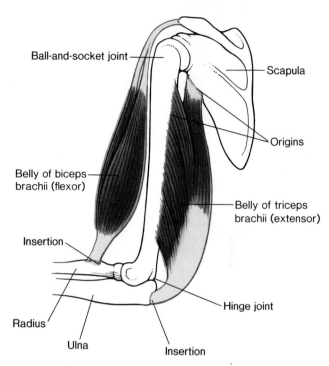

sheaths of connective tissue called **endomysium** *(en''do-mis'e-um)*. The endomysium binds adjacent fibers and supports capillaries and nerve endings serving the muscle. Another connective tissue, called **perimysium,** binds groups of fibers together into bundles called **fasciculi** *(fah-sik'u-li)*. The perimysium supports blood vessels and nerve fibers serving the various fasciculi. The entire muscle is covered by the **epimysium,** which in turn is continuous with a tendon.

A loose fibrous connective tissue, called **fascia** *(fash'e-ah)*, is continuous with the epimysium and binds separate muscles together. Fascia may be divided into **superficial fascia** and **deep fascia.** Superficial fascia is a packinglike tissue that secures the hypodermis of the skin to the underlying muscles. It varies in thickness throughout the body. For example, superficial fascia over the buttocks and anterior abdominal wall is thick and laced with adipose tissue. The superficial fascia under the skin of the dorsum of the hand and facial region is thin. Deep fascia lacks adipose tissue, occurs between individual muscle, and also surrounds adjacent muscles to bind them into functional groups.

Muscle Groups

Just as individual muscle fibers seldom contract independently, muscles generally do not contract separately but work as functional groups. Muscles that contract together and are coordinated in accomplishing a particular movement are said to be *synergistic* (fig. 11.7). *Antagonistic* muscles perform opposite functions and are generally located on the opposite sides of the limb. The two heads of the biceps muscle, for example, contract together to *flex* the elbow joint. The triceps, the antagonists to the biceps, *extend* the elbow as they are contracted.

Seldom does a single muscle cause a movement at a joint. A division of labor is achieved by having several synergistic muscles rather than one massive muscle. One muscle may be an important postural muscle, for example, whereas another may be adapted for rapid, powerful contraction. When total output of the synergistic muscles is required, the contraction of several smaller muscles provides more strength than the contraction of one large one.

endomysium: Gk. *endon*, within; *myos*, muscle
fasciculus: L. *fascis*, bundle
epimysium: Gk. *epi*, upon; *myos*, muscle
fascia: L. *fascia*, a band or girdle
perimysium: Gk. *peri*, around; *myos*, muscle

synergistic: Gk. *synergein*, cooperate
antagonistic: Gk. *antagonistes*, struggle against

Figure 11.8. Skeletal muscle architecture.

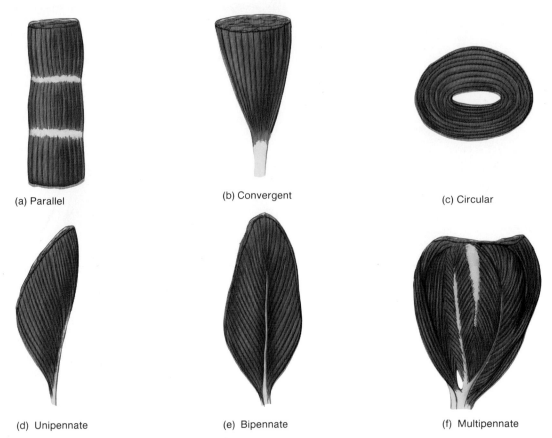

(a) Parallel   (b) Convergent   (c) Circular

(d) Unipennate   (e) Bipennate   (f) Multipennate

Muscle Architecture

Skeletal muscles may be classified on the basis of fiber arrangement (fig. 11.8). Certain advantages are inherent in each of the types of muscles. The following are the major types of fiber arrangement:

1. **Parallel, or fusiform.** Parallel muscles have relatively long excursions (contract over a great distance) and good endurance but are not especially strong. They are long, straplike muscles such as the sartorius and rectus abdominis muscles.
2. **Convergent.** Convergent-fibered muscles are so named because the fibers converge at the insertion point to maximize contraction. They are fan-shaped muscles such as the deltoid and pectoralis major muscles.
3. **Pennate (pinnate).** Pennate-fibered muscles have many fibers per unit area and hence are strong. They provide dexterity. They have short excursions but generally tire quickly. There are three kinds of pennate-fibered muscles: *unipennate*, *bipennate*, and *multipennate*. Examples of each of the pennate types can be found in the forearm.

4. **Circular.** Circular-fibered muscles surround a body opening, or **orifice**, and act as a sphincter when contracted. Examples are the orbicularis oculi around the eye and orbicularis oris surrounding the mouth.

> Muscle-fiber architecture can be observed on a cadaver or other dissection specimen. If you have the opportunity to learn the muscles of the body from a cadaver, observe the fiber architecture of specific muscles, and try to determine the advantages afforded each muscle by its location and action.

Blood and Nerve Supply to Skeletal Muscle

Muscle cells have a high rate of metabolic activity and therefore require extensive vascularity to receive nutrients and oxygen and to eliminate waste products. Smaller muscles generally have a single artery supplying blood and perhaps two veins returning blood (fig. 11.9). Large muscles may have several arteries and veins. The microscopic capillary exchange between arteries and veins occurs throughout the endomysium that surrounds individual fibers.

---

pennate: L. *pennatus*, feather

---

orifice: L. *orificium*, mouth; *facere*, to make

Figure 11.9. The relationship of blood vessels and nerves to skeletal muscles of the axillary region. Note the close proximity of the nerves and vessels as they pass between muscle masses.

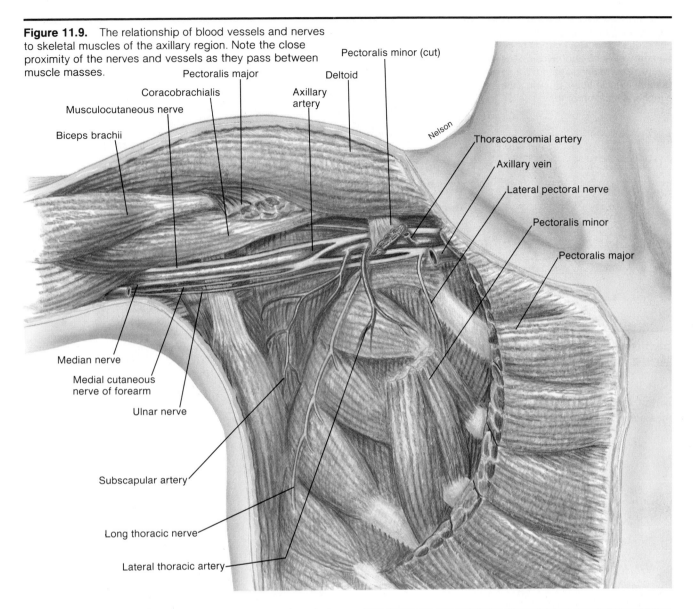

A skeletal muscle fiber cannot contract unless it is stimulated by a nerve impulse. This means that there must be extensive innervation to a muscle to ensure the connection of each muscle fiber to a nerve cell. Actually there are two nerve pathways associated with each muscle. A **motor (efferent) neuron** is a nerve cell that conducts nerve impulses to the muscle fiber to stimulate it to contract. A **sensory (afferent) neuron** conducts nerve impulses away from the muscle fiber to the central nervous system. Muscle fibers will atrophy if they are not periodically stimulated to contract. The names of both the nerves and vessels usually refer to the muscle they serve.

The feeling of stiffness the day after strenuous exercise is due to the accumulation of lactic acid within the muscle. Lactic acid is a waste product of metabolism and is removed by the venous flow from the muscle. If a person is in good physical condition, the vascularity can readily remove the metabolic wastes, and stiffness is not as apt to occur. Being in good physical condition not only improves vascularity, but enlarges muscle fibers and allows them to work more efficiently over a longer duration.

1. Contrast the following terms: endomysium and epimysium; fascia and tendon; aponeurosis and retinaculum.
2. Which type of architecture provides dexterity and strength?

Figure 11.10. (a) a skeletal muscle fiber is composed of numerous threadlike strands of myofibrils that contain the filaments of actin and myosin. A skeletal muscle fiber is striated and multinucleated. (b) an electron micrograph of skeletal muscle fibers showing the striations and the peripheral location of the nuclei.

(a)

Skeletal Muscle Fibers and Mode of Contraction

Muscle fiber contraction results from a sliding movement within the myofibrils in which the length of the sarcomeres is reduced.

Objective 8. Identify the major components of a muscle fiber, and discuss the function of each part.

Objective 9. Discuss the sliding filament theory of muscle contraction.

Objective 10. Describe the length-tension relationship of muscle contraction, and explain how the relationship is produced.

Objective 11. Contrast isotonic and isometric contractions.

Skeletal Muscle Fibers

A skeletal muscle fiber is an elongated, cylindrical cell that may reach a length of 30 cm (12 in.) and have a diameter of 10 to 100 μm. Despite their unusual shape, muscle fibers have the same organelles that are present in other cells: mitochondria, intracellular membranes, glycogen granules, and others. Unlike other cells in the body, however, skeletal muscle fibers are multinucleated and striated (fig. 11.10). Each fiber is surrounded by a plasma membrane, called a **sarcolemma,** and the cytoplasm within the cell is called **sarcoplasm.** A network of membranous channels, called the **sarcoplasmic reticulum,** extends throughout the sarcoplasm (fig. 11.11). A system of **transverse tubules**

sarcolemma: Gk. *sarkos,* flesh; *lemma,* to peel, husk

Figure 11.11. The structural relationship of the myofibrils of a muscle fiber to the sarcolemma, transverse tubules, and sarcoplasmic reticulum. Note the position of the mitochondria.

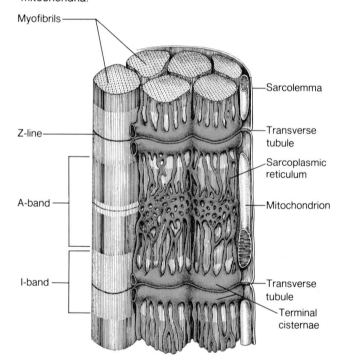

(T-tubules) runs perpendicular to the sarcoplasmic reticulum and opens to the outside through the sarcolemma. Also embedded in the sarcoplasm and extending the entire length of the fiber are many parallel, threadlike structures called **myofibrils** *(mi''o-fi'brils)* (fig. 11.12). Each myofibril is composed of even smaller protein strands called **filaments.** *Thin filaments* are about 50–60 Å thick

Nuclei

Muscle fiber

(b)

Figure 11.12.   (a) a skeletal muscle fiber contains many threadlike structures called myofibrils, each arranged (b) into compartments called sarcomeres. (c) the characteristic striations of a sarcomere are due to the arrangement of thin and thick filaments composed of actin and myosin, respectively.

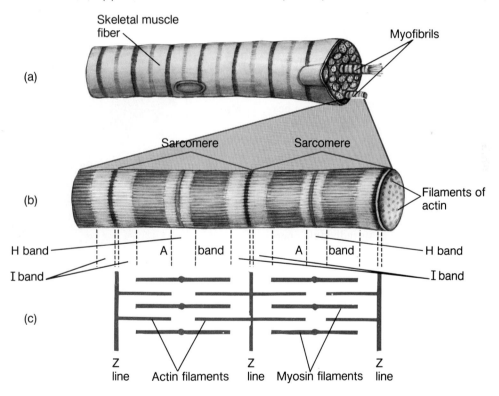

Skeletal muscle fiber

Myofibrils

(a)

Sarcomere   Sarcomere

(b)

Filaments of actin

H band   A   band   A   band   H band

I band   I band

(c)

Z line   Actin filaments   Z line   Myosin filaments   Z line

(1 Å = 10^{-10} m) in diameter and are composed of the protein **actin** *(ak'tin). Thick filaments* are about 110 Å in diameter and are composed of the protein **myosin** *(mi'o-sin).*

The characteristic dark and light striations of skeletal muscle myofibrils are due to the arrangement of these filaments. The dark crossbands, produced by overlapping myosin filaments, are able to polarize visible light (are, thus, anisotropic) and have therefore been named *A bands.* The light crossbands, composed of thin actin filaments, do not polarize light (are, thus, isotropic) and are therefore called *I bands.*

anisotropic: Gk. *anisos,* uneven; *tropos,* a turning
isotropic: Gk. *isos,* equal; *tropos,* a turning

actin: L. *actus,* motion, doing
myosin: L. *myosin,* within muscle

The I bands within a myofibril extend from the edge of one stack of thick myosin filaments to the edge of the next stack of thick filaments. They are light bands because they contain only thin actin filaments. Each thin filament, however, extends partway into the A bands on each side (between the thick filaments of each stack). Since thick and thin filaments overlap at the edges of each A band, the edges are darker in appearance than the central region of the A band. These central lighter regions of the A bands are called the *H bands* (for *helle,* a German word for bright). The central H bands thus contain only myosin that is not overlapped with thin filaments.

At high magnification, thin dark lines can be seen in the middle of the I bands. These are labeled *Z lines* (for *Zwischenscheibe,* a German word meaning "between disc"). The arrangement of thick and thin filaments between a pair of Z lines forms a repeating pattern that serves as the basic subunit of striated muscle contraction. These subunits, from Z line to Z line, are known as **sarcomeres** *(sar′ko-mērs).* A longitudinal section of a myofibril thus presents a side view of successive sarcomeres (fig. 11.13).

This side view is, in a sense, misleading; there are numerous sarcomeres within each myofibril that are out of the plane of the section (and out of the picture). A better appreciation of the three-dimensional structure of a myofibril can be obtained by viewing the myofibril in cross section. In this view, it can be seen that the Z lines are in reality disc shaped and that the thin filaments that penetrate these Z discs surround the thick filaments in a hexagonal arrangement (see fig. 11.13c). If one concentrates on a single row of dark thick filaments in this cross section, the alternating pattern of thick and thin filaments seen in longitudinal section becomes apparent.

Sliding Filament Theory of Contraction

When a muscle is stimulated to contract, it decreases in length as a result of the shortening of its individual fibers. Shortening of the muscle fibers, in turn, is produced by shortening of their myofibrils, which occurs as a result of the shortening of the distance from Z line to Z line (fig. 11.14). As the sarcomeres shorten in length, however, the A bands do *not* shorten but instead move closer together. The I bands—which represent the distance between A bands of successive sarcomeres—decrease in length.

Close examination reveals that the thick myosin and thin actin filaments remain the same length during muscle contraction. Shortening of the sarcomeres is produced, not by shortening of the filaments, but rather by the *sliding* of thin filaments over thick filaments. In the process of contraction, the thin filaments on either side of each A band extend deeper and deeper toward the center, producing increasing amounts of overlap with the thick filaments. The central H bands thus get shorter and shorter during contraction.

Figure 11.13. Electron micrographs of myofibrils of a muscle fiber. (a) at low power (1,600×), a single muscle fiber containing numerous myofibrils. (b) at high power (53,000×), myofibrils in longitudinal section. Notice the sarcomeres and overlapping thick and thin filaments. (c) the hexagonal arrangement of thick and thin filaments as seen in cross section (arrows point to cross-bridges; SR = sarcoplasmic reticulum).

Nucleus

Muscle fiber

(a)

Myofibril

(b)

Sarcomere

Myofibril

SR

(c)

Kessel and Kardon

Figure 11.14.   The sliding filament model of contraction. As the filaments slide, the Z lines are brought closer together. The A bands remain the same length during contraction, but the I and H bands get progressively narrower and may eventually become obliterated.

Figure 11.15. Myosin cross-bridges are oriented in opposite directions on either side of a sarcomere.

Cross-bridges. Sliding of the filaments is produced by the action of numerous **cross-bridges** that extend out from the myosin toward the actin. These cross-bridges are part of the myosin proteins that extend from the axis of the thick filaments to form "arms" that terminate in globular "heads" (fig. 11.15). The orientation of cross-bridges on one side of a sarcomere is opposite to that on the other side, so that when they attach to actin on each side of the sarcomere they can pull the actin from each side toward the center.

Relaxed muscles are easily stretched (although this is opposed *in vivo* by reflex contraction), demonstrating that the myosin cross-bridges are not attached to actin when the muscle is at rest. Biochemical evidence has shown that the globular heads of the cross-bridges contain an **ATPase** that binds ATP when the muscle is at rest but that is incapable of splitting ATP under these conditions. Activation of this ATPase occurs when the myosin cross-bridge attaches to actin.

The attachment of the cross-bridge to actin activates the myosin ATPase; the resultant splitting of ATP provides the energy needed for the *power stroke,* which pulls the thin filaments toward the center of the A bands. The power stroke may be produced by movement of the cross-bridge arm (with the angle of the head constant) or by movement of the head with the arm position fixed. These possible mechanisms are illustrated in figure 11.16.

A single contraction cycle and power stroke of all the cross-bridges in a muscle would shorten the muscle by only about 1% of its resting length. Since muscles can shorten up to 60% of their resting length, it is obvious that the contraction cycles must be repeated many times. In order for this to occur, the cross-bridges must detach from the actin at the end of a power stroke, reassume their resting orientation, and then reattach to the actin to repeat the cycle.

The detachment of a cross-bridge from actin at the end of a power stroke requires the attachment of a "fresh" ATP to the myosin ATPase. The importance of this process is illustrated by the muscular contracture called *rigor mortis* that is observed in a dead person and occurs due to lack of ATP when the muscle dies. This results in the formation of "rigor complexes" between myosin and actin that cannot detach.

> In rigor mortis, all of the cross-bridges are attached to actin at the same time. During normal contraction, however, only about 50% of the cross-bridges are attached at any given time. The power strokes are thus not in synchrony, as the strokes of a competitive rowing team are. Rather, they are like the actions of a team engaged in a tug-of-war, in which the pulling action of the members is asynchronous. Some cross-bridges are engaged in power strokes at all times during the contraction.

Table 11.1 summarizes the events of the sliding filament theory of muscle contraction.

Length-Tension Relationship

The strength of a muscle's contraction is affected by a number of factors. These include the number of muscle fibers within the muscle that are stimulated to contract,

Figure 11.16.   Two possible mechanisms by which cross-bridges could produce sliding of the filaments. (*a*) the cross-bridge arms are presumed to move on a "hinge" while the heads are fixed in position; (*b*) the cross-bridge heads are presumed to swivel on "hinges" from a 90° angle to a 45° angle while the cross-bridge arms are stationary.

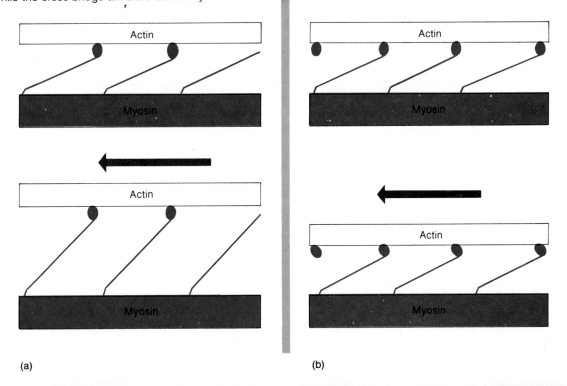

(a)   (b)

Table 11.1 Summary of the sliding filament theory of contraction
1. A myofiber, together with all its myofibrils, shortens by movement of the insertion toward the origin of the muscle.
2. Shortening of the myofibrils is caused by shortening of the sarcomeres—the distance between Z lines (or discs) is reduced.
3. Shortening of the sarcomeres is accomplished by sliding of the myofilaments—each filament remains the same length during contraction.
4. Sliding of the filaments is produced by asynchronous power strokes of myosin cross-bridges, which pull the thin filaments (actin) over the thick filaments (myosin).
5. The A bands remain the same length during contraction, but are pulled toward the origin of the muscle.
6. Adjacent A bands are pulled closer together as the I bands between them shorten.
7. The H bands shorten during contraction as the thin filaments from each end of the sarcomeres are pulled toward the middle.

Source: Fox, Stuart Ira, *Human Physiology.* © 1984 Wm. C. Brown Publishers, Dubuque, Iowa. All Rights Reserved. Reprinted by permission.

the thickness of each muscle fiber (thicker fibers have more myofibrils and thus can exert more power), the muscle architecture, and the initial length of the muscle fibers when they are at rest.

There is an "ideal" resting length of muscle fibers. When the resting length is more than this ideal, the overlap between actin and myosin is so little that few cross-bridges can attach. When the muscle is stretched to the point that there is no overlap of actin with myosin, no cross-bridges can attach to the thin filaments and the muscle cannot contract. When the muscle is shortened to about 60% of its resting length, the Z lines abut against the thick filaments so that further contraction cannot occur.

The strength of a muscle's contraction can be measured by the force required to prevent it from shortening. Under these isometric conditions, the strength of contraction, or *tension,* can be measured when the muscle length at rest is varied (fig. 11.17). Maximum tension is produced when the muscle is at its normal resting length *in vivo* (within the body). This resting length is maintained by reflex contraction in response to passive stretching.

Figure 11.17.   The length-tension relationship in skeletal muscles. Maximum relative tension (1.0) is achieved when the muscle is 100%–120% of its resting length (sarcomere lengths from 2.0 to 2.25 μm). Increases or decreases in muscle (and sarcomere) lengths result in rapid decreases in tension.

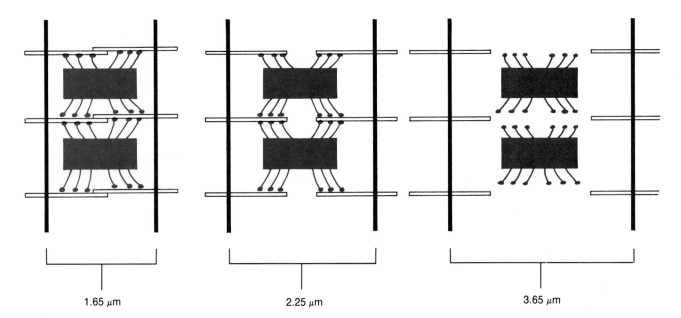

1.65 μm   2.25 μm   3.65 μm

Isotonic and Isometric Contractions

In order for muscle fibers to shorten when they contract, they must generate a force that is greater than the opposing forces that act to prevent movement of the muscle's insertion. Flexion of the forearm, for example, occurs against the force of gravity and the weight of the objects being lifted. The tension produced by the contraction of each muscle fiber separately is insufficient to overcome these opposing forces, but the combined contractions of large numbers of muscle fibers may be sufficient to overcome the opposing force and flex the forearm as the muscle fibers shorten in length.

The contraction that results in muscle shortening is called *isotonic contraction,* so called because the force of contraction remains relatively constant throughout the shortening process (fig. 11.18). In an *isometric contraction,* the length of a muscle remains constant because the antagonist force equals the force in the muscle being contracted. Isometric contractions occur when a person supports an object in a fixed position. An isometric contraction is converted to an isotonic contraction when increased force is generated within the muscle overcoming the resistance and resulting in muscle movement.

isotonic: Gk. *isos*, equal; *tonos*, tension
isometric: Gk. *isos*, equal; *metron*, measure

Figure 11.18. Types of muscle contraction. (*a*) in an isometric contraction, the muscle develops tension but does not shorten. (*b*) in an isotonic contraction, the muscle shortens while under a constant load.

(a)   Time

(b)   Time   Schenk

1. Draw three successive sarcomeres in a myofibril of a resting muscle fiber. Label the myofibril, sarcomeres, A bands, I bands, H bands, and Z lines.
2. Why do the A bands appear darker than the I bands?
3. Draw three successive sarcomeres in a myofibril of a contracted fiber. Indicate which bands get shorter during contraction, and explain how this occurs.
4. Explain why the maximum relative tension in a muscle is achieved when it is 100%–120% of its resting length.
5. Describe how the antagonistic muscles in the brachium can be exercised through both isotonic and isometric contractions.

Stimulation, Regulation, and the Mechanics of Contraction

The recruitment of motor units when a sufficient amount of Ca^{++} is bound to tropin results in muscle contraction. The response of a motor unit is determined by the intensity and duration of the stimulus.

Objective 12. Define a motor unit, and discuss the role of motor units in muscular contraction.

Objective 13. Discuss the events that occur during excitation-contraction coupling.

Objective 14. Define the all-or-none principle of muscular contraction.

Objective 15. Describe the various types of contractions performed by skeletal muscle.

Neuromuscular Junction

A nerve serving a muscle is composed of both motor and sensory neurons. As the motor portion of the nerve penetrates a muscle, it splays into a number of branching neuron processes called **axons.** The terminal ends of axons contact the sarcolemma of muscle fibers by means of **motor end plates** (fig. 11.19). Each axon may split into enough branches to serve dozens of muscle fibers. The area consisting of the motor end plate and the sarcolemma of a muscle fiber is known as the **neuromuscular (myoneural) junction.**

Acetylcholine is a *neurotransmitter chemical* stored in **synaptic vesicles** at the terminal ends of the axons. A nerve impulse reaching the terminal end of an axon causes the release of acetylcholine into the *neuromuscular cleft* of the neuromuscular junction. As this chemical mediator contacts the sarcolemma, it initiates physiological activity within the muscle fiber resulting in contraction.

Motor Unit

A motor unit consists of a single motor neuron and the aggregation of muscle fibers innervated by the motor neuron (see fig. 11.19b). When a nerve impulse travels through a motor unit, all the fibers served by it contract simultaneously to their maximum. Most muscles have an innervation ratio of one motor neuron per 100–150 muscle fibers. Muscles that are capable of precise, dexterous movements, such as an eye muscle, may have an innervation ratio of 1:10. Massive muscles that are responsible for gross body movements, like those of the thigh, may have an innervation ratio exceeding 1:500.

All of the motor units controlling a particular muscle, however, are not the same size. Innervation ratios in a large thigh muscle may vary from 1:100 to 1:2000. Neurons that innervate smaller numbers of muscle fibers have smaller cell bodies and axon diameters than neurons that have larger innervation ratios. The smaller neurons also are stimulated by lower levels of excitatory input (EPSPs). The small motor units, as a result, are the ones that are used most often. The larger motor units are only activated when very forceful contractions are required.

axon: Gk. *axon*, axis

Figure 11.19. A motor end plate at the neuromuscular junction. (*a*) a neuromuscular junction is the site where the nerve fiber and muscle fiber meet. The motor end plate is the specialized portion of the sarcolemma of a muscle fiber surrounding the terminal end of the axon. Note the slight gap between the membrane of the axon and that of the muscle fiber. (*b*) a photomicrograph of muscle fibers and motor end plates. A motor neuron and the muscle fibers it innervates constitute a motor unit.

Skeletal muscles are voluntary in that they can be consciously contracted. The magnitude of the task determines the number of motor units that are activated. Performing a light task, such as lifting a book, requires few motor units, whereas lifting a table requires many. Muscles with pennate architecture have many motor units, are strong and dexterous, but generally fatigue more readily than muscles with fewer motor units. Being mentally "psyched up" to accomplish an athletic feat involves voluntary activation of more motor units within the muscles. Seldom does a person utilize all the motor units within a muscle, but the secretion of *epinephrine (ep''i-nef'rin)* from the adrenal gland does promote an increase in the force that can be produced when a given number of motor units are activated.

Regulation of Contraction

When the cross-bridges attach to actin, the myosin ATPase enzymes are activated resulting in power strokes and muscle contraction. In order for muscle relaxation to occur, the attachment of the myosin cross-bridges to actin must be prevented. The regulation of cross-bridge attachment to actin is a function of two proteins found associated with actin in the thin filaments.

Figure 11.20. The relationship of troponin and tropomyosin to actin in the thin filaments. The tropomyosin is attached to actin, whereas the troponin complex of three subunits is attached to tropomyosin (not directly to actin).

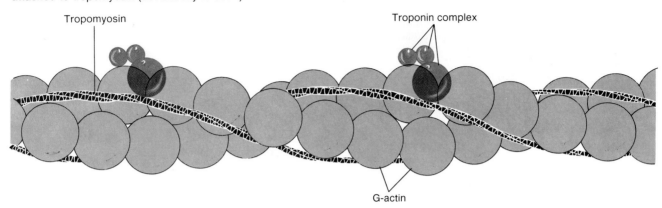

Tropomyosin

Troponin complex

G-actin

The actin filament—or *F-actin*—is a polymer formed of three hundred to four hundred globular subunits (*G-actin*) arranged in a double row and twisted to form a helix. A different type of protein, known as **tropomyosin,** *(tro''po-mi'o-sin),* lies within the groove between the double row of G-actin (fig. 11.20). There are forty to sixty tropomyosin molecules per thin filament, with each tropomyosin spanning a distance of approximately seven actin subunits.

Within the thin filaments, attached to tropomyosin, rather than directly to actin, is a third type of protein, called **troponin** *(tro'po-nin).* Troponin and tropomyosin work together to regulate the attachment of the cross-bridges to actin and thus serve as a switch for muscle contraction and relaxation. In a relaxed muscle tropomyosin is in a position that physically blocks attachment of the cross-bridges to actin. In order for the myosin cross-bridges to attach to actin tropomyosin must be moved.

Role of Ca⁺⁺ in Muscle Contraction. In a relaxed muscle, when tropomyosin blocks the attachment of the cross-bridges to actin, the concentration of Ca^{++} in the sarcoplasm (cytoplasm of muscle cells) is very low. When the muscle cell is stimulated to contract, the concentration of Ca^{++} in the sarcoplasm quickly rises above 10^{-6} molar. Some of this Ca^{++} attaches to a subunit of troponin, causing a conformational change that moves troponin and its attached tropomyosin out of the way so that the cross-bridges can attach to actin (fig. 11.21).

The position of the troponin-tropomyosin complexes in the thin filaments is thus adjustable. When Ca^{++} is not attached to troponin, tropomyosin is in a position that inhibits the attachment of the cross-bridges to actin; muscle contraction is thus prevented. When Ca^{++} attaches to troponin, the troponin-tropomyosin complexes shift position. This allows the cross-bridges to attach to actin; the myosin ATPase is thus activated, ATP is split, the cross-bridges produce a power stroke, and muscle contraction begins.

The cross-bridges can then bond to new ATP, detach from actin, return to their previous orientation, and continue these contraction cycles as long as Ca^{++} is attached to troponin.

Excitation-Contraction Coupling. Muscle contraction is "turned on" when sufficient amounts of Ca^{++} bind to troponin. This occurs when the Ca^{++} concentration of the sarcoplasm rises above 10^{-6} molar. In order for muscle relaxation to occur, therefore, the Ca^{++} concentration of the sarcoplasm must be lowered below this level. Muscle relaxation is produced by the active transport of Ca^{++} out of the sarcoplasm into the sarcoplasmic reticulum (see fig. 11.11).

Most of the Ca^{++} is stored within expanded portions of the sarcoplasmic reticulum known as **terminal cisternae** (fig. 11.11). These terminal cisternae are separated by only a very narrow gap from the transverse tubules, which are continuous with the sarcolemma and which thus open to the extracellular environment through pores in the cell surface.

The release of acetylcholine from the axon terminals of somatic motor neurons at the neuromuscular junctions, as previously described, causes an electrical activation of the skeletal muscle fibers. End-plate potentials are produced that generate action potentials. Muscle action potentials, like the action potentials of axons, are all-or-none events that are regenerated along the cell membrane. It must be remembered that action potentials involve the flow of ions between the extracellular and intracellular environments across a cell membrane that separates these two compartments. In muscle cells, therefore, action potentials can be conducted into the interior of the fiber across the membrane of the transverse tubules.

Figure 11.21. The effect of Ca++ on myosin cross-bridge attachment to actin. Colored G-actin subunits represent attachment sites for cross-bridges. At rest (1) these sites are blocked by tropomyosin. The attachment of Ca++ to troponin (*dark structures*) moves the troponin-tropomyosin so that cross-bridges can attach to G-actin (2). This is followed by the hydrolysis of ATP and a power stroke (3). Detachment of the cross-bridges from actin (4) allows the contraction cycle to be repeated.

Figure 11.22. The setup for observing the contractile behavior of the isolated gastrocnemius muscle of a frog.

Table **11.2** Summary of events that occur during excitation-contraction coupling

1. Action potentials in a somatic motor nerve cause the release of acetylcholine neurotransmitter at the myoneural junction (one myoneural junction per myofiber).

2. Acetylcholine, through its interaction with receptors in the muscle cell membrane (sarcolemma), produces action potentials that are regenerated across the sarcolemma.

3. The membranes of the transverse tubules (T tubules) are continuous with the sarcolemma, and conduct action potentials deep into the muscle fiber.

4. Action potentials in the T tubules, by a mechanism that is poorly understood, stimulate the release of Ca^{++} from the terminal cisternae of the sarcoplasmic reticulum.

5. Ca^{++} released into the sarcoplasm attaches to troponin, causing a change in its structure.

6. The shape change in troponin causes its attached tropomyosin to shift position in the actin filament, thus exposing bonding sites for the myosin cross-bridges.

7. The attachment of the myosin cross-bridges to the exposed sites on actin activates the myosin ATPase, thus hydrolyzing ATP.

8. The hydrolysis of ATP provides energy for the power stroke by which the cross-bridges pull the thin filaments over the thick filaments.

9. The attachment of fresh ATP allows the cross-bridges to detach from actin and repeat the contraction cycle as long as Ca^{++} remains attached to troponin.

10. When action potentials stop being produced, the sarcoplasmic reticulum actively accumulates Ca^{++} and tropomyosin moves again to its inhibitory position.

Source: Fox, Stuart Ira, *Human Physiology.* © 1984 Wm. C. Brown Publishers, Dubuque, Iowa. All Rights Reserved. Reprinted by permission.

Action potentials in the transverse tubules cause the release of Ca^{++} from the sarcoplasmic reticulum, which, in turn, stimulates muscle contraction. As long as action potentials continue to be produced—which is as long as the neural innervation of the muscle continues to be active—Ca^{++} will remain attached to troponin, and the cross-bridges will be able to undergo contraction cycles. When action potentials cease, the sarcoplasmic reticulum actively accumulates Ca^{++} and muscle relaxation occurs.

The neural regulation of skeletal muscle contraction is thus mediated by calcium ions. The mechanisms of electrical excitation and the mechanisms of muscle contraction (sliding of the filaments) are "coupled" through adjustments of the sarcoplasmic Ca^{++} concentration. The events that occur in this *excitation-contraction coupling* are summarized in table 11.2.

Twitch, Summation, and Tetanus

The contractile behavior of skeletal muscles is more easily studied *in vitro* (outside the body) than *in vivo*. In these studies a muscle such as the gastrocnemius (calf muscle) of a frog is usually mounted so that one end is fixed and the other is movable. In the classic laboratory studies, the movable end of the muscle directly produces deflections of a pen, which writes on a rotating drum recorder (fig. 11.22). When more modern equipment is used, the mechanical force of muscle contraction is transduced into an electric current, which can be amplified and displayed as pen deflections in a multi-channel recorder (a physiograph or polygraph—fig. 11.23). Sensitive equipment used in research allows even the contraction of isolated muscle fibers to be recorded.

Figure 11.23. The physiograph Mark III recorder. This is one of many types of electronic recording devices that receive electrical signals produced by transducers and record these signals on moving paper.

When a single muscle fiber is stimulated by an electric shock of a given voltage, it may contract. The minimum voltage that stimulates contraction is the *threshold stimulus.* If the stimulus is below threshold, no response will occur; if a threshold shock is given, the muscle fiber will contract and rapidly relax. This response is called a **twitch** (fig. 11.24). Raising the stimulus voltage above threshold does not increase the strength of the muscle fiber's contraction. A fiber either will contract maximally or not at all—this is the **all-or-none law** of muscle physiology.

The responses of a whole muscle, such as the gastrocnemius, are quite different. As the stimulus strength is increased above threshold, the contraction strength likewise increases, up to a maximum. The contractions of a whole muscle are therefore **graded** rather than all-or-none. Graded contractions are obviously needed for proper control of skeletal movements.

Repeated shocks to an isolated muscle fiber produce only repeated, all-or-none, twitches. Repeated shocks to a whole muscle, however, can result in a **summation of twitches.** This occurs because some fibers in the muscle are stimulated to begin contraction while other fibers relax from their twitch caused by a previous stimulus. In this way the twitches "ride piggyback" on each other and the total contraction strength is increased. If the repeated shocks are delivered very rapidly, summation can cause the twitches to fuse into a smooth, sustained contraction known as **tetanus** *(tet′a-nus).*

When a muscle *in vitro* is kept in tetanus by continuous stimulation, it gradually loses its ability to maintain the contraction; this is known as muscle fatigue. Muscle fatigue in this case is caused by depletion of ATP. Such depletion of ATP in the body would be presaged by feelings of pain, caused by production of lactic acid within the muscle. True fatigue of the muscle would thus rarely occur *in vivo.*

Even when muscles are at rest, they respond to continued stimuli from the nervous system. This results in muscle **tonus** *(to′nus),* or a state of partial, sustained contraction, which keeps the muscles slightly taut in readiness for activity. Tonus is also extremely important in generating body heat. If muscle fibers are not periodically stimulated, they lose tonus, become flaccid, atrophy, and soon die.

1. *Explain why motor units are considered the basic functional units of muscle contraction.*
2. *Draw a figure of the molecular structure of the thick and thin filaments, showing the cross-bridges, G-actin, troponin, and tropomyosin. Describe the roles of Ca^{++}, troponin, and tropomyosin in muscle contraction.*
3. *List, in order of occurrence, the events that occur during the contraction cycles of the cross-bridges. Draw a flow chart (using arrows) of the events that occur between release of ACh at the neuromuscular junction and the attachment of Ca^{++} to troponin.*
4. *Draw a myogram of a twitch and one of tetanus. Explain why a sustained contraction would eventually fatigue.*

Figure 11.24. Myograms of frog muscle responses to stimuli. (*a*) time course of the muscle twitch. (*b*) warmup, or "treppe," in which strength of the twitch is increased to constant stimuli. (*c*) successive stimuli that are rapidly delivered produce summation of twitches and, ultimately, fusion of the twitches into a sustained contraction, or, tetanus.

Figure 11.25. The production and utilization of phosphocreatine in muscles.

Energy Usage by Skeletal Muscle

Aerobic cellular respiration is required for the production of ATP needed for the cross-bridge activity of the various types of skeletal muscle fibers.

Objective 16. Discuss the relationship between ATP and phosphocreatine in muscle physiology.

Objective 17. Describe the differences between slow and fast twitch muscle fibers.

Skeletal muscles at rest obtain most of their energy from the aerobic respiration of fatty acids. During exercise, muscle glycogen and blood glucose are also used as energy sources. Energy thus obtained by cell respiration is used to make ATP, which serves as the immediate source of energy for movement of the cross-bridges and muscle contraction.

During sustained muscle contraction, the utilization of ATP may occur at a faster rate than the rate of ATP production through cell respiration. At these times the rapid renewal of ATP is accomplished by the combination of ADP with phosphate, derived from another "high-energy phosphate" compound called **phosphocreatine** *(fos″fo-kre′ah-tin)*, or **creatine phosphate.**

The phosphocreatine concentration within muscle cells is greater than three times the concentration of ATP and represents a ready reserve of high-energy phosphate that can be donated directly to ADP. During times of rest, the depleted reserve of phosphocreatine can be restored by the reverse reaction—the phosphorylation of creatine with phosphate derived from ATP (fig. 11.25).

The enzyme that transfers phosphate between creatine and ATP is called **creatine kinase** *(ki′nās)*. Skeletal muscle and heart muscle have two different forms of this enzyme (they have different isoenzymes).

The skeletal muscle isoenzyme is found to be elevated in the blood of people with **muscular dystrophy** (degenerative disease of skeletal muscles). The plasma concentration of the isoenzyme characteristic of heart muscle is elevated as a result of **myocardial infarction,** and measurements of this enzyme are thus used as a means of diagnosing this condition (fig. 11.26).

Figure 11.26. Creatine kinase in plasma is separated from other proteins and identified by the rate of its movement in an electric field. The height of each peak indicates concentration, and the position of each peak indicates the isoenzyme form. As can be seen in the *lower figure,* the plasma concentration of creatine kinase is elevated in both heart disease (myocardial infarction) and skeletal muscle disease, but the two types of disease may be distinguished in this test by the appearance of a different isoenzyme in myocardial infarction than in muscle disease.

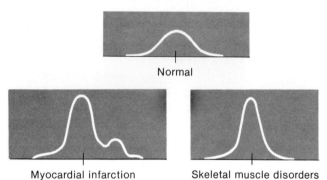

Normal

Myocardial infarction   Skeletal muscle disorders

Fast-, Slow-, and Intermediate-Twitch Fibers

When the cells of a muscle such as the gastrocnemius (calf muscle) are treated with chemicals that stain the fibers in accordance with their content of aerobic respiratory enzymes, three types of fibers are observed (fig. 11.27). The majority of the fibers are thick and pale, indicating relatively low amounts of enzyme activity. These types of fibers are surrounded by relatively few capillaries. Scattered within the muscle are fewer thin, dark-staining fibers, which have higher concentrations of respiratory enzymes and which are surrounded by more capillaries. Fibers that are intermediate between these two extremes are also observed when the correct staining technique is used.

The thinner fibers, which are richly endowed with respiratory enzymes and capillaries, also contain higher concentrations of the pigment *myoglobin (mi"o-glo'bin)* than do the thicker fibers. Myoglobin is a red pigment—hence the term **red fibers** for these cells—which is related to the red hemoglobin pigment in blood. Like hemoglobin, myoglobin can combine with oxygen. The high concentration of myoglobin, high content of aerobic respiratory enzymes, and rich capillary blood supply provide the thinner red fibers with a high capacity for aerobic respiration and with a high resistance to fatigue.

The thicker, pale or **white, fibers,** in contrast, have a limited capacity for aerobic respiration but are rich in stored glycogen and glycolytic enzymes, which increase their ability to respire anaerobically. These fibers are thus less resistant to fatigue than the thin fibers. The thicker white fibers, however, contain more myofibrils than the thinner fibers and have higher concentrations of myosin

Figure 11.27. A cross section of muscle showing white, intermediate, and red (the darker) fibers.

Table **11.3** Characteristics of red, intermediate, and white muscle fibers			
	Red	**Intermediate**	**White**
Diameter	Small	Intermediate	Large
Z-line thickness	Wide	Intermediate	Narrow
Glycogen content	Low	Intermediate	High
Resistance to fatigue	High	Intermediate	Low
Capillaries	Many	Many	Few
Myoglobin content	High	High	Low
Respiration type	Aerobic	Aerobic	Anaerobic
Twitch rate	Slow	Fast	Fast
Myosin ATPase content	Low	High	High

Source: Fox, Stuart Ira, *Human Physiology.* © 1984 Wm. C. Brown Publishers, Dubuque, Iowa. All Rights Reserved. Reprinted by permission.

ATPase (the enzyme found in the cross-bridges). These fibers, thus, can contract more quickly and develop more power than the thinner fibers.

The characteristics of the three types of skeletal muscle fibers are presented in table 11.3.

Because of the different amounts of myosin ATPase in the red, white, and intermediate fibers, each contracts at a different speed. Red fibers have a **slow twitch,** whereas white fibers have a **fast twitch.** The extraocular muscles that position the eyes, for example, contain a high proportion of white fibers and reach maximum force in about 7.3 msec (milliseconds are thousandths of a second). The soleus muscle of the leg, in contrast, which has a high proportion of red slow-twitch fibers, requires about 100 msec

Figure 11.28.   A comparison of the rates with which maximum tension is developed in three muscles. These include the relatively fast-twitch extraocular (*a*) and gastrocnemius (*b*) muscles and the slow-twitch soleus muscle (*c*).

Time (msec) ———➤

Table **11.4** Summary of the effects of endurance training (long-distance running, swimming, bicycling, etc.) on skeletal muscles
1. Improved ability to obtain ATP from oxidative phosphorylation
2. Increased size and number of mitochondria
3. Less lactic acid produced per given amount of exercise
4. Increased myoglobin content
5. Increased intramuscular triglyceride content
6. Increased lipoprotein lipase (enzyme needed to utilize lipids from blood)
7. Increased proportion of energy derived from fat, less from carbohydrates
8. Lower rate of glycogen depletion during exercise
9. Improved efficiency in extracting oxygen from blood

Source: Fox, Stuart Ira, *Human Physiology.* © 1984 Wm. C. Brown Publishers, Dubuque, Iowa. All Rights Reserved. Reprinted by permission.

Figure 11.29.   Biopsies of gastrocnemius muscles treated in such a way that muscle fibers with a high content of aerobic respiratory enzymes stain most darkly. Sample (*a*) came from an endurance-trained athlete, whereas the sample (*b*) was taken from a sprinter.

(a)

(b)

to reach maximum tension (fig. 11.28). Muscles like the soleus that are *postural muscles* are well suited for the sustained contractions needed to maintain an upright position; muscles like the gastrocnemius, which contain a high amount of white fast-twitch fibers, are well adapted to produce a great amount of power quickly.

Interestingly, the conduction rate of motor neurons that innervate fast-twitch white muscles is faster (80–90 meters per second) than the conduction rate of nerve fibers to slow-twitch red muscles (60–70 meters per second). The myofiber type (fast or slow twitch) may indeed be determined by the motor neuron that innervates the muscle. When the motor nerves to red and white fibers are switched in experimental animals, the previously fast-twitch fibers become slow and the slow-twitch fibers become fast. As expected from these observations, all of the muscle fibers innervated by the same motor neuron (that are part of the same motor unit) are of the same type.

Adaptations to Exercise

It is currently believed that the fiber types are determined by their innervation; one fiber type cannot, therefore, change into another as a result of exercise training. All three fiber types, however, do adapt to training (table 11.4). Endurance training, for example, causes an increase in the content of myoglobin and aerobic respiratory enzymes of all three types of fibers. Red and intermediate fibers, however, show the greatest changes under these conditions (fig. 11.29).

Endurance training does not increase the size of muscles. Muscle enlargement is produced only by frequent bouts of muscle contraction against a high resistance—as in weight lifting. As a result of this latter type of training, muscle fibers become thicker and the muscle therefore grows by hypertrophy (an increase in cell size). It has, in fact, long been believed that adult skeletal muscles could only grow by hypertrophy and not by hyperplasia (an increase in cell number). Recent experimental evidence, however, suggests that muscle growth during weight lifting may also be partly due to muscle fiber "splitting," that is, to a longitudinal division of muscle fibers and consequent hyperplasia. Since other recent experiments support the more traditional view that cell numbers remain constant, the proposed ability of skeletal muscles to grow by hyperplasia as a result of weight training is currently very controversial.

---

1. Draw a figure illustrating the relationship between ATP and phosphocreatine.
2. Describe the three different types of muscle fibers in terms of their size, enzyme content, and rates of twitch. Illustrate the differences in rates of twitch with a figure. Describe the functional significance of the different ratios of these fibers in different muscles.
3. Describe the effects of endurance training and strength training on the fiber characteristics of muscles.

Naming of Muscles

Skeletal muscles are named on the basis of shape, location, attachment, orientation of fibers, relative position, and function.

> *Objective 18.* Describe, using examples, the various ways in which muscles are named.

One of the tasks of a myologist is to learn the names of the more than six hundred skeletal muscles within the body. This task is somewhat simplified by the fact that most of the muscles are paired; that is, the right side is the mirror image of the left. Further simplifying the task are the descriptive names of muscles. Only the principal muscles of the body are described in the remaining pages of this chapter. If you desire greater detail, refer to the detailed atlases that are listed at the end of the chapter.

As you learn the muscles of the body, understand how they are named and be able to identify them on yourself. Learn them as functional groups so that you will better understand their actions. If you can identify a muscle on yourself, you will be able to contract the muscle and describe its action. Learning anatomy this way will be easier, more meaningful, and improve retention.

Figure 11.30. Superficial facial muscles involved in facial expression.

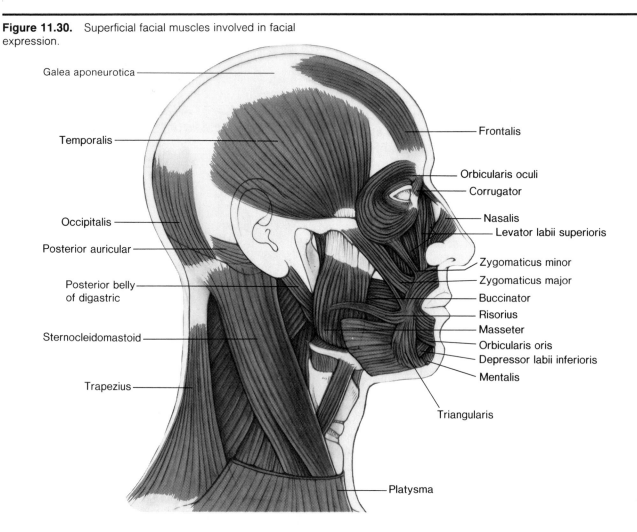

Galea aponeurotica

Temporalis

Occipitalis

Posterior auricular

Posterior belly of digastric

Sternocleidomastoid

Trapezius

Frontalis

Orbicularis oculi

Corrugator

Nasalis

Levator labii superioris

Zygomaticus minor

Zygomaticus major

Buccinator

Risorius

Masseter

Orbicularis oris

Depressor labii inferioris

Mentalis

Triangularis

Platysma

The following are some ways in which the names of muscles have been logically derived, with examples of each given:

1. **Shape:** rhomboideus (like a rhomboid); trapezius (like a trapezoid); or denoting the number of heads of origin: triceps (three heads), biceps (two heads)
2. **Location:** pectoralis (in the chest, or pectus); intercostal (between ribs); brachii (upper arm)
3. **Attachment:** many facial muscles (zygomaticus, temporalis, nasalis); sternocleidomastoid (sternum, clavicle, and mastoid process of the skull)
4. **Size:** maximus, or larger (largest); minimus, or smaller (smallest); longus (long); brevis (short)
5. **Orientation of fibers:** rectus (straight); transverse (across); oblique (in an oblique direction)
6. **Relative position:** lateral, medial, abdominal, internal, and external
7. **Function:** adductor, flexor, extensor, pronator, and levator (lifter)

---

1. Refer to chapter 1, and review the location of the following body regions: cervical, pectoral, abdominal, gluteal, perineal, brachial, antebrachial, inguinal, thigh, and popliteal.
2. Refer to chapter 10, and review the movements permitted at diarthrotic joints.

Muscles of the Axial Skeleton

Muscles of the axial skeleton include those of facial expression, mastication, eye movement, tongue movement, neck movement, respiration, the abdominal wall, the pelvic outlet, and movement of the vertebral column.

Objective 19. Locate the major muscles of the axial skeleton; identify synergistic and antagonistic muscles, and describe the action of each.

Muscles of Facial Expression
Humans have a well-developed facial musculature that provides complex facial expressions as a means of social communication. Indeed, many messages can be conveyed by a person without speaking a word. The following muscles produce facial movements (fig. 11.30 and 11.31).

Epicranius. The broad, extensive epicranius is divisible into two portions: the **frontalis** *(frun-ta'lis),* over the frontal bone, or forehead; and the **occipitalis** *(ok-sip''ĭ-ta'lis),* covering the occipital region of the skull. These muscles are united by an aponeurosis covering the skull called the **galea aponeurotica.** The frontalis originates on the galea aponeurotica, inserts on the soft tissue of the eyebrow, and

functions to elevate the eyebrow and wrinkle the forehead. The occipitalis originates on the occipital bone and mastoid process, inserts on the galea aponeurotica, and functions to retract the scalp. The sustained contraction of this muscle frequently causes headaches.

Corrugator. The corrugator muscle originates on the fleshy tissue above the eyebrow and inserts over the root of the nose. It causes a wrinkling between the eyes, or "frown lines."

Orbicularis Oculi. The orbicularis oculi muscle is a sphincter muscle that encircles the eye. It originates on the bones of the orbit (frontal and maxillary) and inserts on the tissue of the eyelids (palpebrae). The orbicularis oculi permits blinking, winking, and squinting. The contraction of this muscle also compresses the **lacrimal** (tear) **gland,** keeping the eyeball continuously moistened.

Nasalis. The nasalis extends over the bridge of the nose and acts to compress the nostrils.

Orbicularis Oris. The orbicularis oris is a sphincter muscle encircling the lips of the mouth. Its origin is the surrounding tissue, and it inserts on the mucosa of the lips. The orbicularis oris closes the mouth, forms words, and puckers or purses the mouth, as in kissing.

Levator Labii Superioris. The levator labii superioris is a broad, flat muscle originating via three heads from the upper maxilla and a portion of the zygomatic bone. It inserts on the superior margin of the orbicularis oris muscle so that when contracted it elevates the upper lip.

Zygomaticus. The zygomaticus is so named because it originates on the zygomatic bone. It inserts on the superior corner of the orbicularis oris muscle. When contracted, the zygomaticus elevates the corner of the mouth, as in smiling and laughing.

Risorius. The risorius muscle originates on fascia on the side of the face and inserts on the orbicularis oris muscle at the angle of the mouth. It draws the angle of the mouth laterally.

Triangularis. The triangular muscle below either corner of the mouth is the triangularis. It originates on the mandible and inserts on the orbicularis oris muscle just below where the risorius inserts. The triangularis depresses the corner of the mouth, as in frowning.

Depressor Labii Inferioris. The depressor labii inferioris is medial to the triangularis. It originates on the mandible lateral to the midline and inserts on the orbicularis oris muscle. It depresses the bottom lip.

Figure 11.31.   Muscles of facial expression. The actions of each of the muscles of facial expression can be easily identified as a person contracts them. In each of these photographs, identify the muscles that are being contracted.

(a)   (b)   (c)

(d)   (e)   (f)

(g)   (h)

Table 11.5 Muscles of facial expression

Muscle	Derivation	Origin	Insertion	Action	Innervation
Epicranius	*epi* = above; *crani* = skull	Galea aponeurotica and occipital bone	Skin of eyebrow and galea aponeurotica	Wrinkles forehead and moves scalp	Facial
Frontalis	Referring to frontal bone	Galea aponeurotica	Skin of eyebrow	Wrinkles forehead and elevates eyebrow	Facial
Occipitalis	Referring to occipital bone	Occipital bone and mastoid process	Galea aponeurotica	Moves scalp backward	Facial
Corrugator	*ruga* = a wrinkle	Fascia above eyebrow	Root of nose	Draws eyebrows toward midline	Facial
Orbicularis oculi	*orbis* = orbit, circular; *oculus* = eye	Bones of orbit	Tissue of eyelid	Closes eye	Facial
Nasalis	*nasus* = nose	Maxilla and nasal cartilage	Aponeurosis of nose	Compresses nostrils	Facial
Orbicularis oris	*orb* = circular; *or* = mouth	Fascia surrounding lips	Mucosa of lips	Closes and purses lips	Facial
Levator labii superioris	*levator* = lifter; *labii* = lip; *superior* = upper	Upper maxilla and zygomatic bone	Orbicularis oris and skin above lips	Elevates upper lip	Facial
Zygomaticus	Referring to zygomatic bone	Zygomatic bone	Superior corner of orbicularis oris	Elevates corner of mouth	Facial
Risorius	*risor* = laughter	Fascia of cheek	Oribcularis oris at corner of mouth	Draws angle of mouth laterally	Facial
Triangularis	Referring to triangular shape	Mandible	Inferior corner of orbicularis oris	Depresses corner of mouth	Facial
Depressor labii inferioris	*depressor* = depress; *labii* = lip; *inferior* = lower	Mandible	Orbicularis oris	Depresses lower lip	Facial
Mentalis	*mentum* = chin	Mandible (chin)	Orbicularis oris	Elevates and protrudes lower lip	Facial
Platysma	*platy* = broad	Fascia of neck and chest	Inferior border of mandible	Depresses lower lip	Facial
Buccinator	*bucca* = cheek	Maxilla	Orbicularis oris	Compresses cheek	Facial

Source: Van De Graaff, Kent M., *Human Anatomy.* © 1984 Wm. C. Brown Publishers, Dubuque, Iowa. All Rights Reserved. Reprinted by permission.

Mentalis. The mentalis originates on the chin and inserts on the orbicularis oris muscle. The mentalis elevates the lower lip and protrudes it, as in pouting.

Platysma. The platysma is a thin, superficial, sheetlike muscle extending from its origin on the fascia of the pectoralis and deltoid muscles to the lower border of the mandible. The platysma draws the corners of the mouth backward, widening the mouth, as in an expression of horror. It may also assist in opening the mouth.

Buccinator. The buccinator *(buk'si-na''tor)* is a deep cheek muscle that originates on the maxilla and crosses horizontally to insert on the orbicularis oris muscle. It compresses the cheek, and so aids the manipulation of food.

The location and action of the facial muscles are summarized in table 11.5. All the muscles of facial expression are innervated by the facial nerve (VII) and receive their blood supply from branches of the external carotid artery.

Muscles of Mastication

Four pairs of muscles are responsible for the biting and grinding movements of the lower jaw during mastication (figs. 11.30, 11.32). Each extends from the skull to the mandible, is innervated by the mandibular division of the trigeminal nerve (V), and is served blood from branches of the external carotid artery.

Temporalis. The temporalis is the largest of the four pairs. It is a convergent-fibered, fan-shaped muscle that originates on the temporal fossa, passes medially to the zygomatic arch, and inserts on the coronoid process of the mandible. The temporalis closes the jaw.

Masseter. The masseter *(mas-se'ter)* is positioned at the angle of the jaw. It originates on the zygomatic arch and inserts on the lateral surface of the ramus of the mandible. It acts synergistically with the temporalis in closing the jaw.

Medial Pterygoid. Both pterygoid muscles are positioned on the medial (internal) side of the mandible (fig. 11.32). The medial pterygoid originates on the pterygoid process of the sphenoid bone and inserts on the ramus of the mandible. It elevates the jaw and provides lateral jaw movement.

Lateral Pterygoid. The lateral pterygoid muscle also originates on the pterygoid process of the sphenoid. It inserts near the condyle of the mandible and acts to protract the mandible and provide limited lateral jaw movement.

A summary of the muscles of mastication is presented in table 11.6.

Figure 11.32. Muscles of mastication. (*a*) a superficial view; (*b*) a deep view.

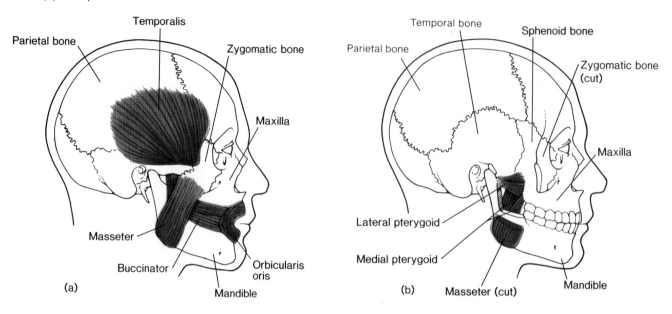

(a)

(b)

Table **11.6** Muscles of mastication

Muscle	Derivation	Origin	Insertion	Action	Innervation
Temporalis	*templus* = time, temple	Temporal fossa	Coronoid process of mandible	Elevates jaw	Trigeminal
Masseter	*maseter* = chew	Zygomatic arch	Lateral ramus of mandible	Elevates jaw	Trigeminal
Medial pterygoid	*medial* = toward midline; *pteryg* = winglike (referring to process of sphenoid bone)	Sphenoid bone	Medial ramus of mandible	Elevates jaw and moves jaw laterally	Trigeminal
Lateral pterygoid	*latus* = side	Sphenoid bone	Anterior side of mandibular condyle	Protracts jaw	Trigeminal

Source: Van De Graaff, Kent M., *Human Anatomy.* © 1984 Wm. C. Brown Publishers, Dubuque, Iowa. All Rights Reserved. Reprinted by permission.

Tetanus is a bacterial disease caused by anaerobic *Clostridium tetani* being introduced to the body, usually from a puncture wound. The bacteria produce a neurotoxin that is carried to the spinal cord by sensory nerves. The motor impulses relayed back cause certain muscles to contract continuously (tetany). The jaw muscles are affected first, which has given the disease the common name of **lockjaw.**

Ocular Muscles

The movements of the eyeball are controlled by six extrinsic eye muscles called the extrinsic ocular muscles (fig. 11.33). These muscles arise from the margin of the optic foramen at the back of the orbital cavity and insert on the outer layer (sclera) of the eyeball. Four **rectus muscles** maneuver the eyeball in the direction indicated by their names (**superior, inferior, lateral,** and **medial**), and two **oblique muscles (superior** and **inferior)** rotate the eyeball on its axis. The medial rectus on one side contracts with the medial rectus of the opposite eye when focusing on close objects. When looking to the side, the lateral rectus of one eyeball works with the medial rectus of the opposite eyeball to keep both eyes focused together. The superior oblique muscle passes through a pulleylike cartilaginous loop, the **trochlea,** before attaching to the eyeball. The ocular muscles are innervated by three cranial nerves (table 11.7).

Another muscle, the **levator palpebrae** *(le-va'tor pal'pĕ-bre)* **superioris,** is located in the ocular region but is not attached to the eyeball. It extends into the upper eyelid and raises the eyelid when contracted.

Figure 11.33. Extrinsic ocular muscles of the left eyeball. (*a*) a lateral view; (*b*) a superior view.

Superior rectus
Trochlea
Superior oblique
Lateral rectus
Optic foramen
Medial rectus
Circle of Zinn
Inferior rectus
Inferior oblique
(a)

Trochlea
Superior oblique
Levator palpebrae superioris (cut)
Superior rectus
Medial rectus
Lateral rectus
Inferior rectus
Inferior oblique
(b)

Table 11.7 Ocular muscles

Muscle	Cranial nerve innervation	Movement of eyeball
Lateral rectus	Abducens	Lateral
Medial rectus	Oculomotor	Medial
Superior rectus	Oculomotor	Superior and medial
Inferior rectus	Oculomotor	Inferior and medial
Inferior oblique	Oculomotor	Superior and lateral
Superior oblique	Trochlear	Inferior and lateral

Source: Van De Graaff, Kent M., *Human Anatomy.* © 1984 Wm. C. Brown Publishers, Dubuque, Iowa. All Rights Reserved. Reprinted by permission.

Muscles That Move the Tongue

The tongue is a highly specialized muscular organ that functions in speaking, manipulating food during mastication, cleansing the teeth, and swallowing. Two groups of muscles are responsible for tongue movement: intrinsic and extrinsic. The *intrinsic muscles* are confined within the tongue and are responsible for its motility and changes of shape. This group includes numerous unnamed bundles. *Extrinsic tongue muscles* are those that originate on structures away from the tongue and insert onto it to cause gross tongue movement (see fig. 11.34 and table 11.8). The four paired extrinsic muscles are the genioglossus, styloglossus, hyoglossus, and stylohyoid.

Genioglossus. The thick genioglossus (*je''ne-o-glos'us*) muscle arises from the mandible on the inner side of the chin and fans out to insert upon the underside of the tongue. When the anterior portion of the genioglossus is contracted, the tongue is depressed and thrust forward. If both genioglossus muscles are contracted together along their entire lengths, the dorsal surface of the tongue becomes transversely concave. This muscle is extremely important in infants for sucking; the tongue is positioned around the nipple with a concave groove channeled toward the pharynx.

Styloglossus. The straplike styloglossus muscle originates from the styloid process of the temporal bone and inserts on the side and undersurface of the tongue. It is an antagonist to the genioglossus in that it retracts and elevates the tongue.

Hyoglossus. The short hyoglossus muscle originates from the hyoid bone and inserts on the side of the tongue. This muscle depresses the tongue, especially along the sides.

These three extrinsic tongue muscles (genioglossus, styloglossus, hyoglossus) are innervated with motor fibers by the hypoglossal nerve.

Stylohyoid. The stylohyoid muscle does not attach directly to the tongue but has an indirect effect on the tongue as it elevates the hyoid bone where it inserts. The stylohyoid originates from the styloid process of the temporal bone and is innervated by the facial nerve.

Figure 11.34. Extrinsic muscles of the tongue.

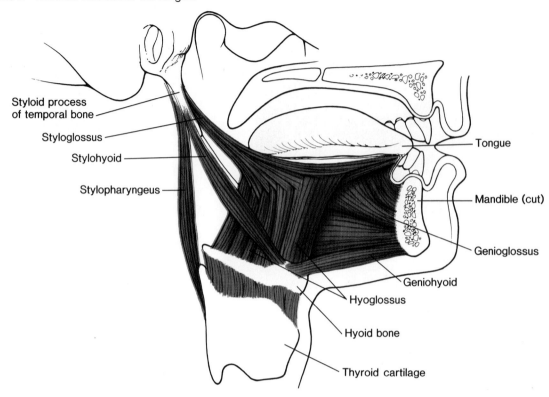

Table **11.8** Extrinsic tongue muscles

Muscle	Derivation	Origin	Insertion	Action	Innervation
Genioglossus	*geneion* = chin; *glossus* = tongue	Mental spine of mandible	Undersurface of tongue	Depresses and protracts tongue	Hypoglossal
Styloglossus	*stylo* = referring to styloid process of skull	Styloid process of temporal bone	Lateral side and undersurface of tongue	Elevates and retracts tongue	Hypoglossal
Hyoglossus	*hyo* = referring to hyoid bone	Body of hyoid	Side of tongue	Depresses sides of tongue	Hypoglossal
Stylohyoid		Styloid process of temporal bone	Body of hyoid	Elevates and retracts hyoid bone and tongue	Facial

Source: Van De Graaff, Kent M., *Human Anatomy.* © 1984 Wm. C. Brown Publishers, Dubuque, Iowa. All Rights Reserved. Reprinted by permission.

Muscles of the Neck

Muscles of the neck either support and move the head or are associated with structures within the neck region, such as the hyoid bone and larynx. Only the more obvious neck muscles will be considered in this chapter.

Several of the muscles in this section and the sections to follow can be observed on yourself. Refer to chapter 12 to determine which muscles are important surface landmarks. These muscles are illustrated in figures 11.35 and 11.36 and are summarized in table 11.9.

Posterior Muscles. The posterior muscles include the sternocleidomastoid, trapezius, splenius capitis, semispinalis capitis, and longissimus capitis.

As the name implies, the **sternocleidomastoid** *(ster″no-kli″do-mas′toid)* muscle originates on the sternum and clavicle and inserts on the mastoid process of the skull (fig. 11.36). When contracted on one side, it turns the head sideways. If both sternocleidomastoid muscles are contracted, the head is pulled forward and down. The sternocleidomastoid is covered by the platysma, which has already been described (see fig. 11.30) and is innervated by the accessory nerve.

Although a portion of the **trapezius** muscle extends over the posterior neck region, it is primarily a superficial muscle of the back and will be described later. The trapezius is served by the accessory nerve.

The **splenius capitis** is a broad muscle positioned deep to the trapezius. It originates on the ligamentum nuchae and the spinous processes of the seventh cervical and first

Figure 11.35. Deep muscles in the back of the neck.

Splenius capitis

Longissimus capitis

Longissimus cervicis

Splenius capitis (cut)

Longissimus capitis (cut)

Semispinalis capitis

Schenk

Figure 11.36. Muscles of the anterior and lateral neck regions.

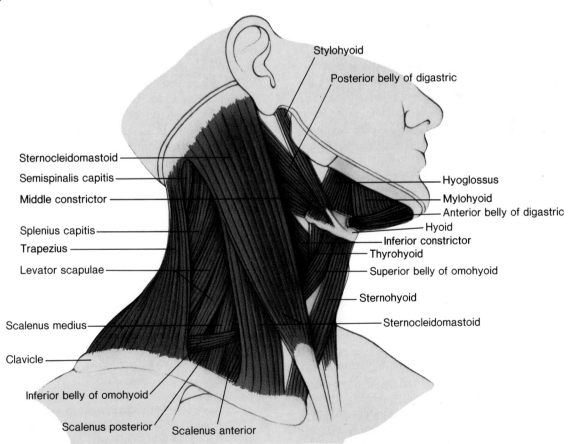

Stylohyoid

Posterior belly of digastric

Sternocleidomastoid

Semispinalis capitis

Middle constrictor

Splenius capitis

Trapezius

Levator scapulae

Scalenus medius

Clavicle

Inferior belly of omohyoid

Scalenus posterior   Scalenus anterior

Hyoglossus

Mylohyoid

Anterior belly of digastric

Hyoid

Inferior constrictor

Thyrohyoid

Superior belly of omohyoid

Sternohyoid

Sternocleidomastoid

Table **11.9** Muscles of the neck

Muscle	Derivation	Origin	Insertion	Action	Innervation
Sternocleidomastoid	*sternum* = referring to sternal bone; *cleido* = clavicle; *mastoid* = mastoid process of skull	Sternum; clavicle	Mastoid process of temporal bone	Turns head to side; flexes neck and head	Accessory
Digastric	*di* = two; *gaster* = belly	Inferior border of mandible; mastoid groove	Hyoid bone	Opens jaw; elevates hyoid	Trigeminal (ant. belly); facial (post. belly)
Mylohyoid	*mylos* = akin to; *hyo*=hyoid bone	Inferior border of mandible	Body of hyoid and median raphe	Elevates hyoid bone and floor of mouth	Trigeminal
Stylohyoid	*stylo* = referring to styloid process	Styloid process of temporal bone	Body of hyoid	Elevates and retracts tongue	Facial
Hyoglossus	*hyo* = hyoid bone; *glossus*=tongue	Body of hyoid bone	Side of tongue	Depresses side of tongue	Hypoglossal
Sternohyoid		Manubrium	Body of hyoid	Depresses hyoid	Hypoglossal
Sternothyroid	*thyro* = shield shaped	Manubrium	Thyroid cartilage	Depresses thyroid cartilage	Hypoglossal
Thyrohyoid		Thyroid cartilage	Great cornu of hyoid	Depresses hyoid; elevates thyroid	Hypoglossal
Omohyoid	*omo* = shoulder	Superior border of scapula	Clavicle; body of hyoid	Depresses hyoid	Hypoglossal

three thoracic vertebrae. It inserts on the back of the skull below the superior nuchal line and the mastoid process of the temporal bone. When the splenius capitis contracts on one side, the head rotates and extends to one side. Contracted together, these muscles extend the head. Further contraction causes hyperextension of the neck and head.

The broad, sheetlike **semispinalis capitis** muscle extends upward from the seventh cervical and first six thoracic vertebrae to insert on the occipital bone. When the two semispinalis capitis muscles contract together, they function to extend the head along with the splenius capitis muscle. If one of the muscles acts alone, the head is rotated to the side.

The narrow, straplike **longissimus capitis** muscle ascends from processes of the lower four cervical and upper five thoracic vertebrae and inserts on the mastoid process of the temporal bone. This muscle extends the head, bends it to the one side, or rotates it slightly.

Suprahyoid Muscles. The group of suprahyoid muscles located above the hyoid bone includes the digastric, mylohyoid, stylohyoid, and hyoglossus (see fig. 11.36).

The **digastric** is a two-bellied muscle of double origin that inserts on the hyoid bone. The anterior origin is on the mandible at the point of the chin, and the posterior origin is near the mastoid process of the skull. The digastric can open the mouth or elevate the hyoid.

The **mylohyoid** forms the floor of the mouth. It originates on the inferior border of the mandible and inserts on the median raphe and body of the hyoid. As this muscle contracts, the floor of the mouth is elevated. It aids swallowing by forcing the food toward the back of the mouth.

The slender **stylohyoid** muscle extends from the styloid process of the skull to the hyoid bone, which it elevates as it contracts. The secondary effect of this muscle on tongue movement has already been described.

Infrahyoid Muscles. Infrahyoid muscles are paired, thin, straplike muscles located below the hyoid bone. They are individually named on the basis of their origin and insertion and include the sternohyoid, sternothyroid, thyrohyoid, and omohyoid (see fig. 11.36).

The **sternohyoid** muscle originates on the manubrium of the sternum and inserts on the hyoid. It depresses the hyoid bone as it contracts.

The **sternothyroid** (not illustrated) muscle also originates on the manubrium but inserts on the thyroid cartilage of the larynx. When this muscle contracts, the larynx is pulled downward.

The short **thyrohyoid** muscle extends from the thyroid cartilage to the hyoid bone. Its actions are to elevate the larynx and lower the hyoid bone.

The long, thin **omohyoid** muscle originates on the superior border of the scapula and inserts on the hyoid bone. It acts to depress the hyoid bone.

The coordinated movements of the hyoid bone and the larynx are impressive. The hyoid bone does not articulate with any other bone, yet it has seven paired muscles and one unpaired muscle attached to it. Two paired muscles involve tongue movement, one paired muscle lowers the jaw, one muscle elevates the floor of the mouth, and four paired muscles depress the hyoid or elevate the thyroid.

Figure 11.37. Muscles of respiration.

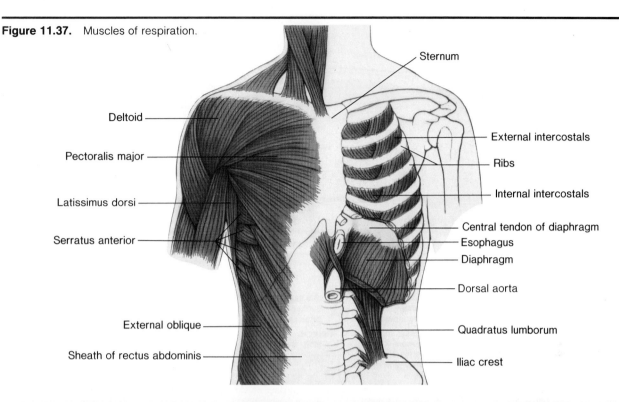

| Sternum |
| Deltoid |
| Pectoralis major |
| Latissimus dorsi |
| Serratus anterior |
| External oblique |
| Sheath of rectus abdominis |
| External intercostals |
| Ribs |
| Internal intercostals |
| Central tendon of diaphragm |
| Esophagus |
| Diaphragm |
| Dorsal aorta |
| Quadratus lumborum |
| Iliac crest |

Table 11.10 Muscles of respiration

Muscle	Derivation	Origin	Insertion	Action	Innervation
Diaphragm	*dia* = across; *phragm* = wall	Xiphoid process, costal cartilages of last six ribs, and lumbar vertebrae	Central tendon	Pulls central tendon inferiorly and increases vertical dimension of thorax	Phrenic
External intercostals	*inter* = between; *costa* = rib	Inferior border of a rib	Superior border of a rib below	Elevates and draws ribs together	Intercostal
Internal intercostals		Superior border of a rib	Inferior border of a rib above	Draws ribs together and depresses thorax	Intercostal

Source: Van De Graaff, Kent M., *Human Anatomy.* © 1984 Wm. C. Brown Publishers, Dubuque, Iowa. All Rights Reserved. Reprinted by permission.

Muscles of Respiration

The muscles of respiration are skeletal muscles that continually, rhythmically, and usually involuntarily contract. Breathing, or *pulmonary ventilation,* is divided into two phases: inspiration (inhalation) and expiration (exhalation).

During normal, relaxed inspiration, the important muscles are the **diaphragm** *(di'ah-fram)* and the **external intercostal** muscles (fig. 11.37). A contraction of the dome-shaped diaphragm downward causes a vertical increase in thoracic volume. A simultaneous contraction of the external intercostals produces an increase in the lateral dimension of the thorax. The external intercostal muscles arise from the lower portion of the first eleven ribs, and the fibers extend downward and forward to insert upon the superior borders of the ribs below. The external intercostal muscles are innervated by the intercostal nerves, and the diaphragm receives its stimuli through the phrenic nerves.

Expiration is primarily a passive process, occurring as the muscles of inspiration are relaxed and the rib cage recoils to its original position. During forced expiration, the **internal intercostal** muscles contract, causing the rib cage to be depressed. The internal intercostals lie under the external intercostals, and their fibers are directed downward and backward. They also have intercostal nerve innervation. The abdominal muscles may also contract during forced expiration, which increases pressure within the abdominal cavity and forces the diaphragm superiorly, squeezing additional air out of the lungs.

The muscles of respiration are summarized in table 11.10.

Muscles of the Abdominal Wall

The anterolateral abdominal wall is composed of four pairs of flat, sheetlike muscles: the **external oblique, internal oblique, transversus abdominis,** and **rectus abdominis** (figs. 11.38 and 11.43). These muscles support and protect the

Figure 11.38.   Muscles of the anterior abdominal wall.

Table **11.11**   Muscles of the abdominal wall

Muscle	Derivation	Origin	Insertion	Action	Innervation
External oblique	*external* = closer to surface; *oblique* = diagonal	Lower eight ribs	Iliac crest, linea alba	Compresses abdomen; lateral rotation	Intercostals, iliohypogastric, and ilioinguinal
Internal oblique	*internal* = deep	Iliac crest, inguinal ligament, and lumbodorsal fascia	Linea alba, costal cartilage of last three or four ribs	Compresses abdomen; lateral rotation	Intercostals, iliohypogastric, and ilioinguinal
Transversus abdominis	*transverse* = horizontal; *abdomino* = belly	Iliac crest, inguinal ligament, lumbar fascia, and costal cartilage of last six ribs	Xiphoid process, linea alba, and pubis	Compresses abdomen	Intercostals, iliohypogastric, and ilioinguinal
Rectus abdominis	*rectus* = straplike; straight	Pubic crest and symphysis pubis	Costal cartilage of fifth to seventh ribs and xiphoid process	Flexes vertebral column	Intercostals

organs of the abdominal cavity and aid in breathing. When they contract, the pressure in the abdominal cavity increases, which can aid in defecation and in stabilizing the spine during heavy lifting.

External Oblique.   The external oblique muscle is the strongest and most superficial of the three layered muscles of the lateral abdominal wall. Its fibers are directed inferiorly and medially. The external oblique originates on the inferior surfaces of the lower eight ribs in contact with the serratus anterior muscle. Its insertion is a broad, thin aponeurosis that unites the external oblique of one side with its counterpart on the opposite side. The aponeurosis forms the **linea alba** ("white line"), a prominent band of connective tissue on the midline of the abdomen, and the **inguinal ligament** (see fig. 11.51), which extends from the anterior superior spine of the ilium to the pubic tubercle. This muscle has extensive innervation from branches of intercostal nerves eight through twelve and from iliohypogastric and ilioinguinal nerves.

Internal Oblique.   The internal oblique muscle is located between the external oblique and the transversus abdominis. Its fibers run at right angles to those of the external oblique. The internal oblique originates on the

inguinal ligament, the iliac crest, and the lumbodorsal fascia of the back. Its primary insertion is on the linea alba and the pubic crest. An anterior portion of this muscle inserts upon the costal cartilages of the lower three ribs. The internal oblique is innervated by branches of the same nerves that serve the external oblique.

Transversus Abdominis.   The deepest of the abdominal muscles is the transversus abdominis. It has an extensive origin on the inguinal ligament, the iliac crest, the lumbodorsal fascia, and the last five ribs. The fibers of the transversus abdominis muscle pass directly medially around the abdominal wall to insert upon the linea alba and the pubic crest. The transversus abdominis is innervated by branches of intercostal nerves seven through twelve and by iliohypogastric and ilioinguinal nerves.

Rectus Abdominis.   The long, straplike rectus abdominis muscle (fig. 11.43) extends from the pubis to the rib cage on both sides of the linea alba. It is entirely enclosed in a fibrous sheath formed from the aponeuroses of the other three abdominal muscles. Contraction of the rectus abdominis results in flexion of the lumbar portion of the vertebral column. The rectus abdominis muscle is served by branches of intercostal nerves seven through twelve.

Refer to table 11.11 for a summary of the muscles of the abdominal wall.

Figure 11.39. Muscles of the pelvic outlet. (*a*) male;
(*b*) female.

(a)

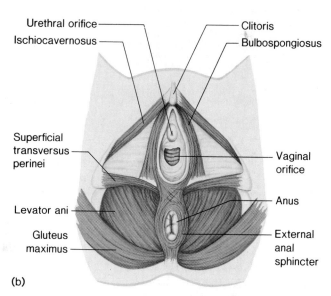

(b)

Muscles of the Pelvic Outlet

The pelvic outlet, or floor of the pelvis, is formed by two groups of muscular sheets and the associated fascia. The more posterior group is called the **pelvic diaphragm** and consists of the levator ani and coccygeus muscles. The anterior group, associated with the genitalia, is called the **urogenital diaphragm** and consists of the superficial transversus perinei, the bulbospongiosus, and the ischiocavernosus (fig. 11.39). The pelvic diaphragm is similar in the male and female, but the urogenital diaphragm of each sex differs markedly. The muscles of the pelvic outlet are innervated by the pudendal nerve or perineal branches of the pudendal nerve.

Levator Ani. The levator ani muscle arises from the pubic bone and the spine of the ischium and inserts at the midline on the coccyx and on the anterior fibers of the same muscle of the opposite side. The two levator ani muscles together form a thin sheet of muscle that helps to support the pelvic viscera and constrict the lower part of the rectum, pulling it forward and aiding defecation.

Coccygeus. The fan-shaped coccygeus muscle arises from the spine of the ischium and inserts on the sacrum and the coccyx (not shown in the illustration). The coccygeus muscle aids the levator ani in its functions. Either or both of these muscles are occasionally stretched and torn during parturition.

Superficial Transversus Perinei. The superficial transversus perinei is a small straplike muscle that arises from the ischial tuberosity, passes below the genitalia, and inserts on the central tendon of the midline. It assists other muscles in supporting the pelvic viscera.

Bulbospongiosus (Bulbocavernosus). In males, the bulbospongiosus muscle of one side unites with that of the opposite side to form a muscular constriction surrounding the base of the penis. When contracted, the two muscles constrict the urethral canal and assist in emptying the urethra. In females, these muscles are separated by the vaginal orifice, which they constrict as they contract.

Ischiocavernosus. The ischiocavernosus muscle originates from the tuberosity of the ischium and inserts onto the crus of the penis in the male and the clitoris of the female. This muscle aids the erection of the penis in the male and the clitoris in the female.

External Anal Sphincter. Although the external anal sphincter is not generally considered part of the pelvic outlet, it is in this region and is a skeletal muscle. This funnel-shaped constrictor muscle surrounds the anal canal and functions to keep the anal canal closed.

The muscles of the pelvic outlet are summarized in table 11.12.

Table 11.12 Muscles of the pelvic outlet

Muscle	Derivation	Origin	Insertion	Action	Innervation
Levator ani	*anus* = ring	Spine of ischium and pubic bone	Coccyx	Supports pelvic viscera; aids in defecation	Pudendal plexus
Coccygeus	*coccyx* = like a cuckoo's bill	Ischial spine	Sacrum and coccyx	Supports pelvic viscera; aids in defecation	Perineal branch of pudendal nerve
Superficial transversus perinei	*perineum* = referring to area between genitals and anus	Ischial tuberosity	Central tendon	Supports pelvic viscera	Pudendal plexus
Bulbospongiosus	*bulbus* = little bulb; *spongia* = spongelike	Central tendon	Males: base of penis; females: root of clitoris	Constricts urethral canal; constricts vagina	Perineal branch of pudendal nerve
Ischiocavernosus	*ischio* = ischium; *caverna* = hollow spaces	Ischial tuberosity	Pubic arch	Aids erection of penis or clitoris	Perineal branch of pudendal nerve

Source: Van De Graaff, Kent M., *Human Anatomy.* © 1984 Wm. C. Brown Publishers, Dubuque, Iowa. All Rights Reserved. Reprinted by permission.

Muscles of the Vertebral Column

The muscles that move the vertebral column are strong and complex because they have to provide support and movement in resistance to the effect of gravity.

The vertebral column can be flexed, extended, abducted, and rotated. The muscle that flexes the vertebral column, the rectus abdominis, has already been described as a paired, straplike muscle of the anterior abdominal wall. The extensor muscles located on the posterior side of the vertebral column have to be stronger than the flexors because extension (such as lifting an object) is in opposition to gravity. The extensor muscles consist of a superficial group and a deep group. Only the superficial muscles of the vertebral column will be described.

Quadratus Lumborum. The quadratus lumborum muscle is shown in figures 11.40 and 11.51. It originates on the iliac crest and the lower three lumbar vertebrae. It inserts on the transverse processes of the first four lumbar vertebrae and the inferior margin of the twelfth rib. When the right and left quadratus lumborum contract together, the vertebral column in the lumbar region extends. Separate contraction causes lateral flexion or abduction of the spine.

Erector Spinae (sacrospinalis). The erector spinae is a massive muscle group that extends from the sacrum to the skull. It actually consists of three groups of muscles: **iliocostalis, longissimus,** and **spinalis** (fig. 11.40). Each of these groups, in turn, consists of overlapping slips of muscle. The iliocostalis is the most lateral group, the longissimus is intermediate in position, and the spinalis medially is positioned in contact with the spinous processes of the vertebrae.

The erector spinae muscles are innervated by dorsal rami of spinal nerves. A summary of these muscles is presented in table 11.13.

The erector spinae muscles are frequently strained through improper lifting. A heavy object should not be lifted with the vertebral column flexed; instead, the thighs and knees should be flexed so that the pelvic and leg muscles can aid in the task.

The erector spinae muscles are also frequently strained in women during pregnancy. Pregnant women will try to counterbalance the effect of the protruding abdomen by hyperextending the vertebral column. This causes an exaggerated lumbar curvature, strained muscles, and a peculiar gait.

1. *Identify the facial muscles that do the following: (a) wrinkle the forehead; (b) purse the lips; (c) protrude the lower lip; (d) produce frown lines; (e) smile; (f) frown; (g) wink; and (h) elevate the upper lip to show the teeth.*
2. *Describe the actions of the extrinsic muscles that move the tongue.*
3. *Which muscles of the neck either originate from or insert on the hyoid bone?*
4. *Describe the actions of the muscles of inspiration. Which muscles participate in forced expiration?*
5. *Which muscles of the pelvic outlet support the floor of the pelvic cavity, and which are associated with the genitalia? What is the extent and the compartmentalization of the sacrospinalis muscle?*

Muscles of the Appendicular Skeleton

The muscles of the appendicular skeleton include those of the pectoral girdle, humerus, forearm, wrist, hand, and fingers, and those of the pelvic girdle, thigh, lower leg, ankle, foot, and toes.

> *Objective 20.* Locate the major muscles of the appendicular skeleton; identify synergistic and antagonistic muscles, and describe the action of each.

Figure 11.40.   Muscles of the vertebral column.

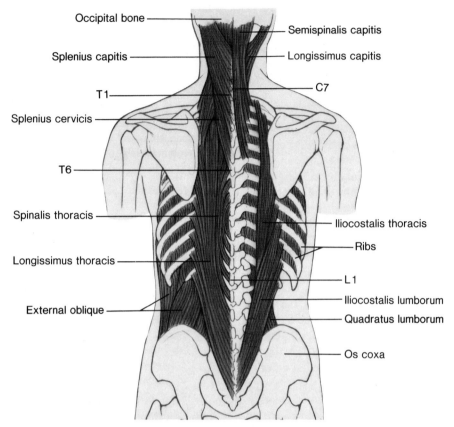

Occipital bone
Splenius capitis
T1
Splenius cervicis
T6
Spinalis thoracis
Longissimus thoracis
External oblique

Semispinalis capitis
Longissimus capitis
C7
Iliocostalis thoracis
Ribs
L1
Iliocostalis lumborum
Quadratus lumborum
Os coxa

Table 11.13   Muscles of the vertebral column

Muscle	Derivation	Origin	Insertion	Action	Innervation
Quadratus lumborum	*quad* = four; *lumb* = referring to lumbar region	Iliac crest and lower three lumbar vertebrae	Twelfth rib and upper four lumbar vertebrae	Extends lumbar region; lateral flexion of vertebral column	T12, L1
Erector spinae	Consists of three groups of muscles: iliocostalis, longissimus, and spinalis. The iliocostalis and longissimus are further subdivided into three groups on the basis of location along the vertebral column.				
Iliocostalis lumborum	*ilium* = flank	Crest of ilium	Lower six ribs	Extends lumbar region	Dorsal rami of lumbar nerves
Iliocostalis thoracis	*thorax* = chest	Lower six ribs	Upper six ribs	Extends thoracic region	Dorsal rami of thoracic nerves
Iliocostalis cervicis	*cervix* = neck	Angles of third to sixth rib	Transverse processes of fourth to sixth cervical vertebrae	Extends cervical region	Dorsal rami of cervical nerves
Longissimus thoracis		Transverse processes of lumbar vertebrae	Transverse processes of all the thoracic vertebrae and lower nine ribs	Extends thoracic region	Dorsal rami of spinal nerves
Longissimus cervicis		Transverse processes of upper four or five thoracic vertebrae	Transverse processes of second to sixth cervical vertebrae	Extends cervical region and lateral flexion	Dorsal rami of spinal nerves
Longissimus capitis	(See the muscles of the neck in table 11.9.)				Dorsal rami of middle and lower cervicals
Spinalis thoracis		Spinous processes of upper lumbar and lower thoracic vertebrae	Spinous processes of upper thoracic vertebrae	Extends vertebral column	Dorsal rami of spinal nerves

Muscles That Act on the Pectoral Girdle

The shoulder is attached to the axial skeleton only at the sternoclavicular joint, so strong, straplike muscles are necessary. Furthermore, muscles that move the brachium originate on the scapula, and during brachial movement the scapula has to be held fixed. The muscles that act on the pectoral girdle originate on the axial skeleton and can be divided into anterior and posterior groups.

Anterior.   The anterior group of muscles that act on the pectoral girdle includes the serratus anterior, pectoralis minor, and subclavius (fig. 11.41).

The **serratus** *(ser-ra'tus)* **anterior** muscle originates on the lateral surface of the first eight or nine ribs and inserts on the anterior portion of the vertebral border of the scapula. This muscle pulls the scapula forward and downward, holding it against the rib cage. The serratus anterior muscle is innervated by the long thoracic nerve.

The **pectoralis minor** muscle is deep to the pectoralis major on the anterior thoracic wall. It is a flat, triangular muscle originating on the anterior surface of the third, fourth, and fifth ribs and inserting on the coracoid process of the scapula. The pectoralis minor has several actions; the major one is to depress the shoulder while drawing the scapula medially. It is innervated by the medial anterior thoracic nerve.

The **subclavius** is a thin, straplike muscle that originates on the anterior surface of the first rib and inserts on the subclavian groove of the clavicle. This muscle depresses the clavicle and hence draws the shoulder downward.

> Treatment of advanced stages of breast cancer requires the surgical removal of both pectoralis major and pectoralis minor muscles in a procedure called a radical mastectomy. Postoperative physical therapy is primarily geared toward strengthening the synergistic muscles of this area. As you learn the muscles that act on the brachium, determine which are synergists with the pectoralis major.

Posterior.   The posterior group of muscles that act on the pectoral girdle includes the trapezius, levator scapulae, and rhomboideus (fig. 11.42).

The **trapezius** is a large, superficial, triangular muscle of the neck and upper back. It arises on the occipital bone, the spinous process of the seventh cervical, and all of the thoracic vertebrae. It inserts on the clavicle, the acromion process of the scapula, and the spine of the scapula. The trapezius can adduct the scapula, elevate the scapula (as in shrugging the shoulder), and hyperextend the head. The trapezius is innervated by the accessory nerve.

Figure 11.41.   Deep anterior thoracic muscles that move the pectoral girdle.

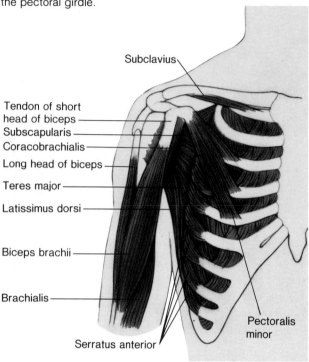

The long, thin, straplike **levator scapulae** muscle originates on the fourth and fifth cervical vertebrae and attaches to the upper vertebral border of the scapula. When contracted, it elevates the scapula. This muscle is served by the dorsal scapular nerve.

The **major** and **minor rhomboids** are two separate but continuous muscles. They both extend from the spinous processes of the vertebrae to the vertebral border of the scapula. The rhomboideus major muscle is larger and positioned inferior to the minor muscle. Working together, they elevate and adduct the scapula. They are both served by the dorsal scapular nerve.

Table 11.14 summarizes the muscles that act on the pectoral girdle.

Muscles That Move the Humerus (Brachium)

Of the nine muscles that span the shoulder joint to insert on the humerus, only two of them, the pectoralis major and latissimus dorsi, do not originate on the scapula. These two are designated as axial muscles, whereas the remaining seven are scapular muscles. Figures 11.42 and 11.43 show the muscles of this region, and figure 11.44 shows the attachments of all the muscles that either originate or insert on the scapula.

> In terms of their development, the pectoralis major and the latissimus dorsi muscles are not axial muscles at all. They develop in the forelimb and extend to the trunk secondarily. They are considered axial muscles only because their origins are on the axial skeleton.

Figure 11.42. Posterior muscles of the neck, thorax, and abdomen. (Superficial muscles removed from right side.)

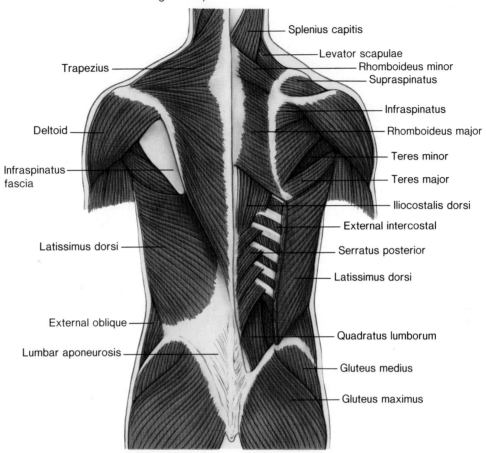

Trapezius	Splenius capitis
Deltoid	Levator scapulae
Infraspinatus fascia	Rhomboideus minor
Latissimus dorsi	Supraspinatus
External oblique	Infraspinatus
Lumbar aponeurosis	Rhomboideus major

(labels, left side top to bottom: Trapezius, Deltoid, Infraspinatus fascia, Latissimus dorsi, External oblique, Lumbar aponeurosis)

(labels, right side top to bottom: Splenius capitis, Levator scapulae, Rhomboideus minor, Supraspinatus, Infraspinatus, Rhomboideus major, Teres minor, Teres major, Iliocostalis dorsi, External intercostal, Serratus posterior, Latissimus dorsi, Quadratus lumborum, Gluteus medius, Gluteus maximus)

Table 11.14 Muscles that act on the pectoral girdle

Muscle	Derivation	Origin	Insertion	Action	Innervation
Serratus anterior	*serratus* = serrated; *anterior* = front	Upper eight or nine ribs	Anterior vertebral border of scapula	Pulls scapula forward and downward	Long thoracic
Pectoralis minor	*pectus* = chest; *minor* = lesser	Sternal ends of third, fourth, and fifth ribs	Coracoid process of scapula	Pulls scapula forward and downward	Medial anterior thoracic
Subclavius	*sub* = below, under	First rib	Subclavian groove of clavicle	Draws clavicle downward	C5, C6
Trapezius	*trapezoeides* = trapezoid shaped	Occipital bone and spines of cervical and thoracic vertebrae	Clavicle, spine of scapula, and acromion process	Elevates scapula, draws head back, adducts scapula, braces shoulder	Accessory
Levator scapulae	*levator* = elevator; *scapulae* = scapula	Fourth and fifth cervical vertebrae	Vertebral border of scapula	Elevates scapula	Dorsal scapular
Rhomboideus major	*rhomboides* = rhomboid shaped	Spines of second to fifth thoracic vertebrae	Vertebral border of scapula	Elevates and retracts scapula	Dorsal scapular
Rhomboideus minor		Seventh cervical and first thoracic vertebrae	Vertebral border of scapula	Elevates and retracts scapula	Dorsal scapular

Figure 11.43. Anterior muscles of the neck, thorax, and abdomen. (Superior muscles removed from the right side.)

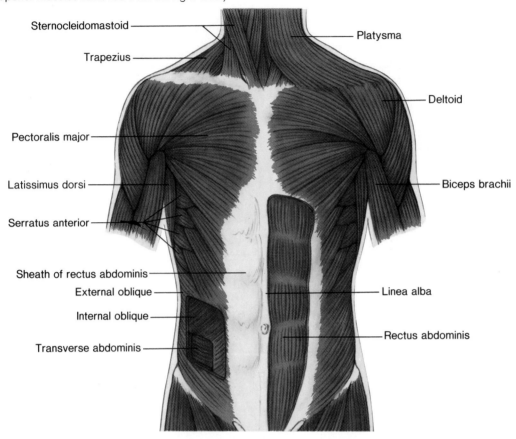

Table **11.15** Muscles that move the humerus

Muscle	Derivation	Origin	Insertion	Action	Innervation
Pectoralis major	*pectus* = chest; *major* = greater	Clavicle, sternum, costal cartilages of second to sixth rib	Greater tubercle of humerus	Flexes, adducts, and rotates arm medially	Medial and lateral anterior thoracic
Latissimus dorsi	*latissimus* = widest; *dorsum* = back	Spines of sacral, lumbar, and lower thoracic vertebrae; iliac crest and lower ribs	Intertubercular groove of humerus	Extends, adducts, and rotates humerus medially; retracts shoulder	Thoracodorsal
Deltoid	*delta* = triangular	Clavicle, acromion process; spine of scapula	Deltoid tuberosity of humerus	Abducts arm; extends or flexes humerus	Axillary
Supraspinatus	*supra* = above; *spina* = spine of scapula	Fossa—superior to spine of scapula	Greater tubercle of humerus	Abducts and laterally rotates humerus	Suprascapular
Infraspinatus	*infra* = below	Fossa—inferior to spine of scapula	Greater tubercle of humerus	Rotates arm laterally	Suprascapular
Teres major		Inferior angle of scapula	Intertubercular groove of humerus	Extends humerus, or adducts and rotates arm medially	Lower subscapular
Teres minor	*teres* = rounded; *minor* = lesser	Axillary border of scapula	Greater tubercle and groove of humerus	Rotates arm laterally	Axillary
Subscapularis	*sub* = below, under; *scapula* = scapula	Subscapular fossa	Lesser tubercle of humerus	Rotates arm medially	Subscapular
Coracobrachialis	*coraco* = referring to coracoid process; *brachium* = arm	Coracoid process of scapula	Shaft of humerus	Flexes and adducts shoulder joint	Musculocutaneous

Figure 11.44. A posterior view of the scapula and humerus, showing the areas of attachment of the associated muscles.

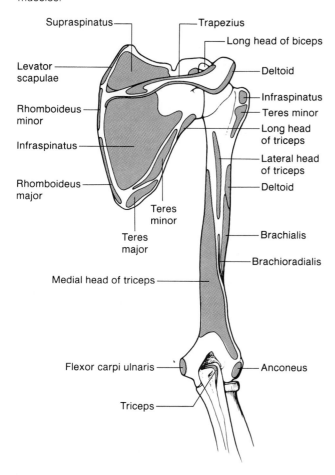

Axial Muscles. The axial muscles include the pectoralis major and latissimus dorsi.

The **pectoralis major** is a large chest muscle that originates on the clavicle, sternum, costal cartilages, and rectus sheath (fig. 11.43). It inserts on the greater tubercle of the humerus. Its primary functions are to adduct, flex, and rotate the brachium medially. The pectoralis major is innervated by the medial and lateral anterior thoracic nerves.

The large, flat, triangular **latissimus dorsi** muscle covers the lower half of the thoracic region of the back (fig. 11.42). It arises from the broad lumbar aponeurosis that is attached to the spines of the lower thoracic, lumbar, and sacral vertebrae, and from the iliac crest. The latissimus dorsi is frequently called the "swimmer's muscle" because it powerfully adducts the arm, drawing it downward and backward while it rotates medially. It is innervated by the thoracodorsal nerve.

Scapular Muscles. The scapular muscles include the deltoid, supraspinatus, infraspinatus, teres major, teres minor, subscapularis, and coracobrachialis.

The **deltoid** is a thick, powerful muscle that caps the shoulder joint. It originates on the clavicle, acromion, and

spine of the scapula and inserts on the deltoid tuberosity of the humerus (figs. 11.43 and 11.44). It is innervated by the axillary nerve, and its function is to abduct the brachium. The deltoid muscle is a common site for intramuscular injections.

The **supraspinatus** arises from the supraspinous fossa of the scapula, passes over the shoulder joint, and inserts on the greater tubercle of the humerus (see figs. 11.42 and 11.44). In addition to stabilizing the shoulder joint, it laterally rotates the humerus and is synergistic with the deltoid in abducting the brachium. The supraspinatus is innervated by the suprascapular nerve.

The **infraspinatus** muscle originates on the infraspinous fossa of the scapula and inserts on the greater tubercle of the humerus (see figs. 11.42 and 11.44). Its function is to rotate the arm laterally, and its innervation is the suprascapular nerve.

The **teres major** arises from the inferior angle of the scapula and attaches to the crest of the lesser tuberosity immediately above where the latissimus dorsi inserts (figs. 11.42 and 11.44). The action of the teres major is similar to that of the latissimus dorsi, adducting and medially rotating the humerus. The teres major is innervated by the lower subscapular nerve.

The **teres minor** arises from the axillary border of the scapula and inserts on the greater tubercle of the humerus (figs. 11.42 and 11.44). This muscle works with the infraspinatus in laterally rotating the brachium. It is served by the axillary nerve.

The **subscapularis** is a deep muscle of the shoulder, located between the serratus anterior and the anterior surface of the scapula (see fig. 11.41). It originates on the subscapular fossa and inserts on the lesser tubercle of the humerus. The subscapularis is a strong stabilizer of the shoulder and also aids in medially rotating the brachium. It is innervated by the subscapular nerve.

The mass of the **coracobrachialis** is seen on the upper medial side of the brachium (see fig. 11.41). It has a tendinous origin on the coracoid process of the scapula and inserts on the medial surface of the humerus, about two-thirds of the way down the brachium. It is innervated by the musculocutaneous nerve and functions to flex and adduct the shoulder joint.

A summary of the muscles that act on the humerus is presented in table 11.15. The sites of attachments of the muscles of this region are shown in figure 11.44.

Muscles That Act on the Forearm

The powerful muscles of the brachium are responsible for movement of the forearm at the elbow and at the radioulnar joint. These muscles are the biceps brachii, brachialis, brachioradialis, and triceps brachii (fig. 11.45). A transverse section through the brachium in figure 11.46 provides a different perspective of the brachial region.

Figure 11.45. The right shoulder and brachium. (*a*) an anterior view; (*b*) a posterior view.

(a)

(b)

Figure 11.46.   The axillary region and a cross section through the brachium.

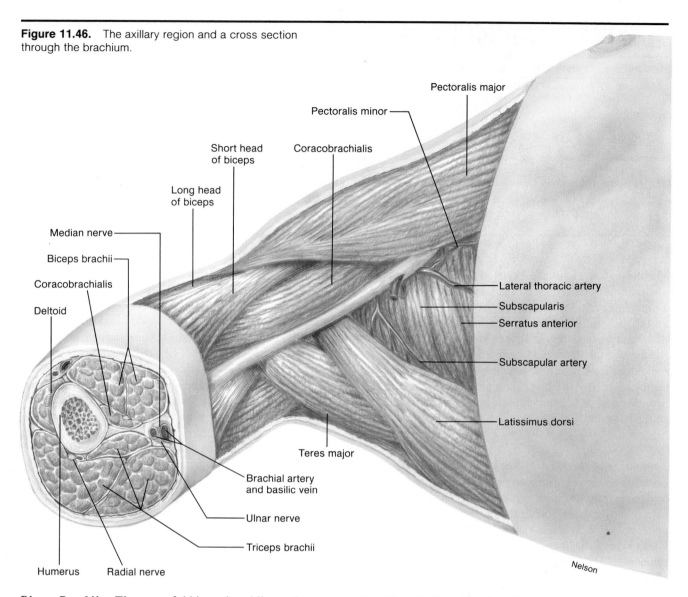

Pectoralis major

Pectoralis minor

Coracobrachialis

Short head of biceps

Long head of biceps

Median nerve

Biceps brachii

Coracobrachialis

Deltoid

Lateral thoracic artery

Subscapularis

Serratus anterior

Subscapular artery

Latissimus dorsi

Teres major

Brachial artery and basilic vein

Ulnar nerve

Triceps brachii

Humerus   Radial nerve

Nelson

Biceps Brachii.   The powerful biceps brachii muscle, positioned on the anterior surface of the humerus, is the most familiar muscle of the arm; yet it has no attachments on the humerus. The biceps has a dual origin: a medial tendinous head, the **short head,** arises from the coracoid process of the scapula, and the **long head** originates on the superior tuberosity of the glenoid fossa and passes through the intertubercular groove on the humerus. Both heads of the biceps insert on the radial tuberosity and are served by the musculocutaneous nerve. When contracted, the biceps brachii flexes the forearm and also assists supination of the hand.

Brachialis.   The brachialis is located on the distal anterior half of the humerus, deep to the biceps brachii. It originates directly from the periosteum of the humerus and inserts on the coronoid process of the ulna. The brachialis works synergistically with the biceps in flexing the forearm. It is innervated by the musculocutaneous, radial, and median nerves.

Brachioradialis.   The brachioradialis is the prominent muscle positioned along the lateral (radial) surface of the forearm. It originates proximal to the lateral epicondyle of the humerus and inserts on the lateral surface of the radius immediately above the styloid process. The brachioradialis is innervated by the radial nerve.

Triceps Brachii.   The triceps brachii muscle, located on the posterior surface of the brachium is the muscle antagonistic to the biceps brachii. It has three heads, or origins. Two of the three, the **lateral head** and **medial head,** arise from the humerus, whereas the **long head** arises from the infraglenoid tuberosity of the scapula. A common tendinous insertion attaches the triceps brachii muscle to the olecranon process of the ulna. The triceps extends the forearm as it is stimulated through the radial nerve.

Table 11.16 summarizes the muscles that act on the forearm.

Table **11.16** Muscles that act on the forearm

Muscle	Derivation	Origin	Insertion	Action	Innervation
Biceps brachii	*biceps* = two heads; *brachium* = upper arm	Coracoid process; tuberosity above glenoid cavity of scapula	Radial tuberosity	Flexes and supinates forearm	Musculocutaneous
Brachialis		Anterior shaft of humerus	Coronoid process of ulna	Flexes forearm	Musculocutaneous, median, and radial
Brachioradialis	*radialis* = pertaining to radius	Supracondyloid ridge of humerus	Proximal to styloid process of radius	Flexes forearm	Radial
Triceps brachii	*triceps* = three heads	Tuberosity below glenoid cavity; lateral and medial surfaces of humerus	Olecranon process of ulna	Extends forearm	Radial

Source: Van De Graaff, Kent M., *Human Anatomy.* © 1984 Wm. C. Brown Publishers, Dubuque, Iowa. All Rights Reserved. Reprinted by permission.

Muscles That Move the Wrist, Hand, and Fingers

The muscles that cause wrist, hand, and most finger movements are positioned along the forearm (figs. 11.47, 11.48). Several of these muscles act on two joints, the elbow and wrist. Others act on the joints of the wrist, hand, and digits. Still others produce rotational movement at the radioulnar joint. The precise actions of these muscles are complex, so only the basic movements will be described here. Most of these muscles perform four primary actions on the hand and digits: supination, pronation, flexion, and extension.

Supination of the Hand. The **supinator** muscle is positioned around the upper posterior portion of the radius. It originates from the lateral epicondyle of the humerus and inserts on the lateral edge of the radial tuberosity and upper posterior surface of the radius. The supinator muscle works synergistically with the biceps brachii to supinate the hand. The supinator is innervated by the radial nerve (fig. 11.48).

Pronation of the Hand. Two muscles are responsible for pronating the hand, the pronator teres and pronator quadratus.

The **pronator teres** muscle is located on the upper medial side of the forearm (fig. 11.48). It arises on the medial epicondyle of the humerus and inserts on the upper lateral shaft of the radius. It acts to pronate the hand as it is stimulated by the median nerve.

The deep, anteriorly positioned **pronator quadratus** muscle extends between the ulna and radius on the distal fourth of the forearm (fig. 11.48). It originates on the ulna and inserts on the distal lateral portion of the radius. This muscle works synergistically with the pronator teres to rotate the palm of the hand posteriorly and position the thumb medially. It is served by the volar interosseous branch of the median nerve.

Flexion of the Wrist, Hand, and Fingers. Six muscles that flex the wrist, hand, and fingers will be described (fig. 11.47). They will be discussed from lateral to medial and from superficial to deep. The brachioradialis, already described, is an obvious reference muscle for locating the muscles of the forearm that flex the wrist, hand, and fingers.

The **flexor carpi radialis** muscle arises from the medial epicondyle of the humerus, extends diagonally across the anterior surface of the forearm, and inserts onto the base of the second and third metacarpal bones (fig. 11.48). Three joints spanned by this muscle function to flex and abduct the hand. The flexor carpi radialis muscle itself flexes and abducts the hand. It is innervated by the median nerve.

The narrow **palmaris longus** muscle arises from the medial epicondyle of the humerus and has a long, slender tendon that attaches to the palmar aponeurosis and assists in flexing the hand as it is stimulated by the median nerve.

The **flexor carpi ulnaris** muscle is positioned on the medial anterior side of the forearm (figs. 11.47 and 11.48). It originates on the medial epicondyle of the humerus and the olecranon process, and it inserts on the lateral carpal bones and the base of the fifth metacarpal. The flexor carpi ulnaris muscle assists in flexing and adducting the hand, and it is innervated by the ulnar nerve.

The broad **flexor digitorum superficialis (sublimis)** muscle lies directly beneath the three flexors just described (see figs. 11.47, 11.48). It has an extensive origin, involving all three of the long bones of the arm: the medial epicondyle of the humerus, the olecranon process of the ulna, and the anterior border of the radius. The tendon at the distal end of this muscle is united across the wrist joint but then splits to attach to the middle phalanx of digits two through five. This muscle flexes the wrist, hand, and the second, third, fourth, and fifth fingers. It is innervated by the median nerve.

Figure 11.47. Muscles of the right forearm that move the wrist, hand, and digits. (*a*) an anterior view; (*b*) a posterior view.

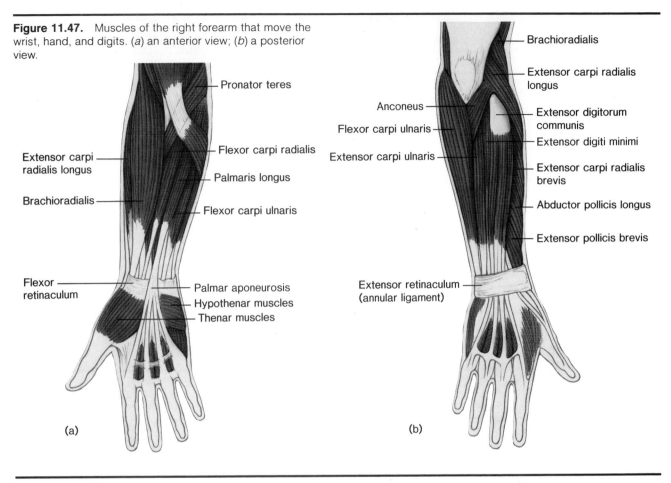

(a)

Pronator teres

Extensor carpi radialis longus

Flexor carpi radialis

Brachioradialis

Palmaris longus

Flexor carpi ulnaris

Flexor retinaculum

Palmar aponeurosis

Hypothenar muscles

Thenar muscles

(b)

Brachioradialis

Extensor carpi radialis longus

Anconeus

Flexor carpi ulnaris

Extensor carpi ulnaris

Extensor digitorum communis

Extensor digiti minimi

Extensor carpi radialis brevis

Abductor pollicis longus

Extensor pollicis brevis

Extensor retinaculum (annular ligament)

Figure 11.48. Deep muscles of the forearm. (*a*) rotators; (*b*) flexors; (*c*) extensors.

(a)

Supinator

Pronator teres

Pronator quadratus

(b)

Palmaris longus (cut distally to display deep tendons)

Flexor carpi ulnaris

Flexor carpi radialis

Flexor digitorum superficialis

Flexor pollicis longus

Lumbricals (from tendons of flexor digitorum profundus)

Tendons of flexor digitorum superficialis

Tendons of flexor digitorum profundus

(c)

Extensor carpi radialis longus

Extensor digitorum

Extensor carpi ulnaris

Abductor pollicis longus

Extensor pollicis longus

Extensor indicis

Table 11.17 Muscles of the forearm that act on the wrist, hand, and digits

Muscle	Derivation	Origin	Insertion	Action	Innervation
Supinator	*supin* = bend back; palms up	Lateral epicondyle of humerus and crest of ulna	Lateral surface of radius	Supinates forearm	Radial
Pronator teres	*pron* = bend forward; palms down; *teres* = long, round	Medial epicondyle of humerus	Lateral surface of radius	Pronates forearm	Median
Pronator quadratus	*quad* = four (square)	Distal fourth of ulna	Distal fourth of radius	Pronates hand	Median
Flexor carpi radialis	*flexor* = decrease angle; *carpus* = wrist; *radial* = radius bone	Medial epicondyle of humerus	Base of second and third metacarpals	Flexes and abducts hand	Median
Palmaris longus	*palma* = flat of hand; *longus* = long	Medial epicondyle of humerus	Palmar aponeurosis	Flexes hand	Median
Flexor carpi ulnaris		Medial epicondyle and olecranon process	Carpal and metacarpal bones	Flexes and adducts wrist	Ulnar
Flexor digitorum superficialis	*digitis* = finger or toe; *superficialis* = superficial	Medial epicondyle, olecranon process, and anterior border of radius	Middle phalanges of digits	Flexes forearm, wrist, hand, and digits	Median
Flexor digitorum profundus	*profundus* = deep	Proximal two-thirds of ulna and interosseous membrane	Distal phalanges	Flexes wrist, hand, and digits	Median and ulnar

Source: Van De Graaff, Kent M., *Human Anatomy.* © 1984 Wm. C. Brown Publishers, Dubuque, Iowa. All Rights Reserved. Reprinted by permission.

The **flexor digitorum profundus** muscle (not illustrated) lies deep to the flexor digitorum superficialis muscle just described. It is a bipennate muscle arising from the anterior, proximal two-thirds of the ulna and the interosseous ligament between the ulna and radius. It inserts on the bases of the second through fifth distal phalanges (fig. 11.48). This muscle flexes the wrist, hand, and the distal phalanges of all digits except the first (thumb), and it is innervated by both the median and ulnar nerves.

The **flexor pollicis longus** is a deep, lateral muscle originating on the anterior surface of the radius, its adjacent interosseous ligament, and the coronoid process of the ulna (fig. 11.48). Its tendinous insertion is on the base of the distal phalanx of the thumb. The flexor pollicis longus muscle flexes the thumb, assisting the grasping mechanism of the hand. It is innervated by the volar interosseous branch of the median nerve.

> The tendons of the muscles that flex the hand can be seen on your wrist as you make a fist. These tendons are securely positioned by the flexor retinaculum (fig. 11.47), which crosses the wrist area transversely.

Extension of the Hand. The muscles that extend the hand are located on the posterior side of the forearm. Most of the primary extensors of the hand can be seen superficially in fig. 11.47 and will be discussed from lateral to medial.

The long, tapered **extensor carpi radialis longus** muscle is located posterior to the brachioradialis. It arises from the lateral supracondylar ridge of the humerus and inserts on the base of the second metacarpal. The extensor carpi radialis longus muscle extends and abducts the hand. It is innervated by the radial nerve.

The **extensor carpi radialis brevis** muscle is positioned immediately medial to the extensor carpi radialis longus muscle and performs approximately the same functions. Its origin and insertion, however, are different. The extensor carpi radialis brevis muscle originates on the lateral epicondyle of the humerus and inserts on the base of the third metacarpal. It too is innervated by the radial nerve.

The bipennate-fibered **extensor digitorum communis** muscle is positioned in the center of the forearm along the posterior surface. It originates on the lateral epicondyle of the humerus. Its tendon of insertion divides at the wrist, beneath the extensor retinaculum, into four tendons that attach to the distal tip of the medial phalanges of digits two through five. The function of this muscle is to extend the wrist and phalanges, and it is innervated by the radial nerve.

The **extensor digiti minimi** is a long, narrow muscle located on the ulnar side of the extensor digitorum communis muscle. It arises from the lateral epicondyle of the humerus. The tendinous insertion of this muscle fuses with the tendon of the extensor digitorum communis going to the fifth digit. It is innervated by the radial nerve and assists in extending the wrist and little finger.

The **extensor carpi ulnaris** muscle is the most medial muscle on the posterior surface of the forearm. It arises

Muscle	Derivation	Origin	Insertion	Action	Innervation
Flexor pollicis longus	*pollex* = thumb	Shaft of radius, interosseous membrane, and coronoid process of ulna	Distal phalanx of thumb	Flexes thumb	Median
Extensor carpi radialis longus	*extensor* = increase joint angle	Lateral supracondylar ridge of humerus	Second metacarpal	Extends and adducts hand	Radial
Extensor carpi radialis brevis	*brevis* = short	Lateral epicondyle of humerus	Third metacarpal	Extends and adducts hand	Radial
Extensor digitorum communis	*communis* = common	Lateral epicondyle of humerus	Posterior surfaces of phalanges II–V	Extends wrist and phalanges	Radial
Extensor digiti minimi	*minimus* = small	Lateral epicondyle of humerus	Extensor aponeurosis of fifth digit	Extends fifth digit and wrist	Radial
Extensor carpi ulnaris		Lateral epicondyle of humerus and olecranon process	Base of fifth metacarpal	Extends and adducts wrist	Radial
Extensor pollicis longus		Distal shaft of ulna, lateral side	Base of distal phalanx of thumb	Extends thumb; abducts hand	Radial
Extensor pollicis brevis		Distal shaft of radius and interosseous membrane	Base of first phalanx of thumb	Extends thumb; abducts hand	Radial
Abductor pollicis longus		Distal radius and ulna and interosseous membrane	Base of first metacarpal bone	Abducts first digit and hand	Radial

on the lateral epicondyle of the humerus and portions of the ulna. It inserts on the base of the fifth metacarpal bone where it functions to extend and abduct the hand. It too is served by the radial nerve.

The **extensor pollicis longus** muscle extends from the mid-ulnar region, across the lower two-thirds of the forearm, and inserts onto the base of the distal phalanx of the thumb (see fig. 11.48). It extends the thumb and abducts the hand and is innervated by the radial nerve.

The **extensor pollicis brevis** muscle arises from the lower midportion of the radius and inserts on the base of the proximal phalanx of the thumb (fig. 11.47). The action of this muscle is similar to that of the extensor pollicis longus, and it is served by the same nerve.

As its name implies, the **abductor pollicis longus** muscle abducts the thumb. It originates on the interosseous ligament, between the ulna and radius, and inserts on the base of the first metacarpal bone. The radial nerve innervates this muscle as well.

The muscles that act on the wrist, hand, and digits are summarized in table 11.17.

Notice that your hand is partially contracted even when relaxed. The muscles that extend your hand are not as strong as the muscles that flex it. This is why persons who receive strong electrical shocks through the arms will tightly flex their hands and hold on. All the muscles of the arm are stimulated to contract, but the flexors, being stronger, cause the hands to close tightly.

Muscles of the Hand

The hand is a marvelously complex structure adapted to permit an array of intricate movements. Flexion and extension movements of the hand and phalanges are accomplished by the muscles of the forearm just described. Precise finger movements that require coordinating abduction and adduction with flexion and extension are the function of the small intrinsic muscles of the hand. These muscles and associated structures of the hand are depicted in figure 11.49. The position and actions of the muscles of the hand are listed in table 11.18.

The muscles of the hand are divided into **thenar** *(the'nar)*, **hypothenar** *(hi-poth'ĕ-nar)*, and **intermediate** groups. The **thenar eminence** is the fleshy base of the thumb and is formed by three muscles: the **abductor pollicis brevis**, the **flexor pollicis brevis**, and the **opponens pollicis.** All three muscles are innervated by the median nerve. The most important of the thenar muscles is the opponens pollicis, which opposes the thumb to the palm of the hand.

The **hypothenar eminence** is the elongated, fleshy bulge at the base of the little finger. It also is formed by three muscles: the **abductor digiti minimi,** the **flexor digiti minimi,** and the **opponens digiti minimi.** All of these muscles are supplied by branches of the ulnar nerve.

Muscles of the intermediate group are positioned between the metacarpal bones in the region of the palm. This group includes the **adductor pollicis,** the **lumbricales** *(lum'bri-kal),* and the **palmar** and **dorsal interossei.** All the muscles of the intermediate group, except two lumbricales, are innervated by branches of the ulnar nerve. The first and second lumbricales receive branches from the median nerve.

Figure 11.49. Intrinsic muscles of the hand and associated structures. (*a*) an anterior view; (*b*) a cross section.

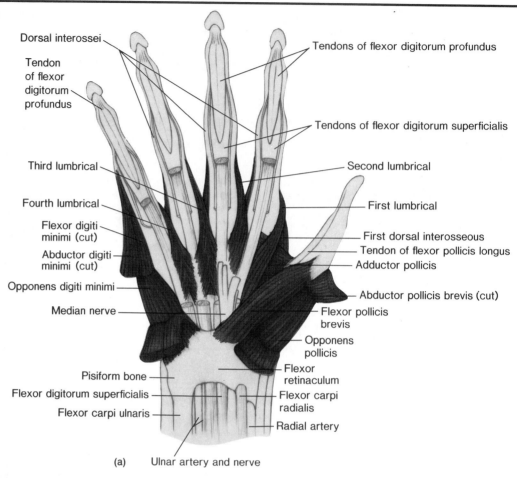

(a) Ulnar artery and nerve

Table 11.18 Intrinsic muscles of the hand

Muscle	Derivation	Origin	Insertion
Thenar Muscles			
Abductor pollicis brevis	*pollex* = thumb; *brevis* = short	Flexor retinaculum, scaphoid, and trapezium	Proximal phalanx of thumb
Flexor pollicis brevis		Transverse carpal ligament and trapezium	Proximal phalanx of thumb
Opponens pollicis	*opponens* = against	Trapezium and transverse carpal ligament	First metacarpal
Intermediate Muscles			
Adductor pollicis (oblique and transverse heads)		Oblique head, capitate; transverse head, second and third metacarpals	Proximal phalanx of thumb
Lumbricales (4)	*lumbricalis* = referring to digits	Tendons of flexor digitorum profundus	Extensor expansions of digits
Palmar interossei (3)	*palma* = palm	Medial side of second metacarpal; lateral sides of fourth and fifth metacarpals	Proximal phalanges of index, ring, and little fingers and extensor digitorum communis
Dorsal interossei (4)		Adjacent sides of metacarpals	Proximal phalanges of index and middle fingers (lateral sides) plus proximal phalanges of middle and ring fingers (medial sides) and extensor digitorum communis
Hypothenar Muscles			
Abductor digiti minimi	*minimus* = very small	Pisiform and tendon of flexor carpi ulnaris	Proximal phalanx of digit 5
Flexor digiti minimi		Flexor retinaculum and hook of hamate	Proximal phalanx of digit 5
Opponens digiti minimi		Flexor retinaculum and hook of hamate	Fifth metacarpal

(b)

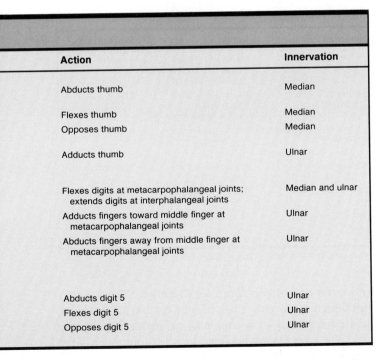

Action	Innervation
Abducts thumb	Median
Flexes thumb	Median
Opposes thumb	Median
Adducts thumb	Ulnar
Flexes digits at metacarpophalangeal joints; extends digits at interphalangeal joints	Median and ulnar
Adducts fingers toward middle finger at metacarpophalangeal joints	Ulnar
Abducts fingers away from middle finger at metacarpophalangeal joints	Ulnar
Abducts digit 5	Ulnar
Flexes digit 5	Ulnar
Opposes digit 5	Ulnar

Figure 11.50. Anterior pelvic muscles.

Muscles That Move the Thigh

The muscles that move the thigh originate from the pelvic girdle and insert on various places on the femur. These muscles stabilize a highly movable hip joint and provide support for the body during bipedal stance and locomotion. Found in this region are the most massive muscles of the body as well as some extremely small muscles. The muscles that move the thigh are divided into anterior, posterior, and medial groups.

Anterior Muscles.   The anterior muscles that move the thigh are the iliacus and psoas major (figs. 11.50, 11.51).

The triangular **iliacus** muscle arises from the iliac fossa and inserts on the lesser trochanter of the femur. It is innervated by the femoral nerve.

Figure 11.51. Anterior pelvic muscles that move the hip.

Figure 11.52. Deep gluteal muscles.

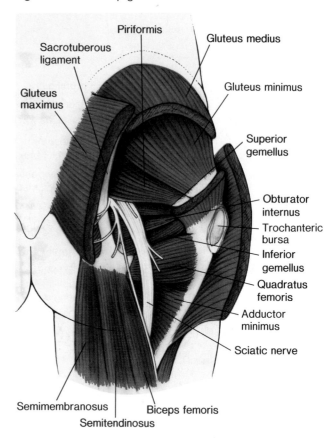

Table **11.19** Anterior and posterior muscles that move the thigh					
Muscle	**Derivation**	**Origin**	**Insertion**	**Action**	**Innervation**
Iliacus	*iliacus* = pertaining to ilium (hip)	Iliac fossa	Lesser trochanter of femur	Flexes and rotates thigh laterally; flexes vertebral column	Femoral
Psoas major	*psoa* = muscle of loin	Transverse processes of all lumbar vertebrae	Lesser trochanter with iliacus	Flexes and rotates thigh laterally; flexes vertebral column	L2, L3
Gluteus maximus	*gloutos* = rump; *maximus* = greatest	Iliac crest, sacrum, coccyx, and aponeurosis of back	Gluteal tuberosity and iliotibial tract	Extends and rotates thigh laterally	Inferior gluteal
Gluteus medius	*medius* = middle	Lateral surface of ilium	Greater trochanter	Abducts and rotates thigh medially	Superior gluteal
Gluteus minimus	*minimus* = smallest	Lateral surface of lower half of ilium	Greater trochanter	Abducts and rotates thigh medially	Superior gluteal
Tensor fasciae latae	*tensor* = tightener; *fascia* = band; *latus* = wide	Anterior border of ilium and iliac crest	Iliotibial tract	Abducts thigh	Superior gluteal

Source: Van De Graaff, Kent M., *Human Anatomy.* © 1984 Wm. C. Brown Publishers, Dubuque, Iowa. All Rights Reserved. Reprinted by permission.

The long, thick **psoas major** muscle originates on the transverse processes of the lumbar vertebrae and inserts, along with the iliacus, on the lesser trochanter. The psoas major muscle is innervated by the spinal nerves L2 and L3. The psoas major muscle and the iliacus work synergistically in flexing and rotating the thigh and flexing the vertebral column. They are frequently referred to as a single muscle, the **iliopsoas** *(il″e-o-so′as).*

Posterior and Lateral (Buttock) Muscles. The posterior muscles that move the thigh include the gluteus maximus, gluteus medius, gluteus minimus, and tensor fasciae latae.

The large **gluteus maximus** muscle forms much of the prominence of the buttock (fig. 11.52). It is a powerful extensor muscle of the thigh and is very important for bipedal stance and locomotion. The gluteus maximus originates on the ilium, sacrum, coccyx, and aponeurosis of the back. It inserts on the gluteal tuberosity of the femur and the **iliotibial tract,** a tendinous band extending down

Figure 11.53. Adductor muscles of the right thigh.

the thigh (see fig. 11.54). The gluteus maximus is served by the inferior gluteal nerve. The mass of this muscle is of clinical significance as a site for intramuscular injections.

The **gluteus medius** muscle is located immediately deep to the gluteus maximus (fig. 11.52). It originates on the lateral surface of the ilium and inserts on the greater trochanter of the femur. The gluteus medius abducts and medially rotates the thigh. It is innervated by the superior gluteal nerve.

The **gluteus minimus** is the smallest and deepest of the gluteal muscles. It also arises from the lateral surface of the ilium and inserts on the lateral surface of the greater trochanter, where it acts synergistically with the gluteus medius to abduct and medially rotate the thigh. It too is innervated by the superior gluteal nerve.

The quadrangular **tensor fasciae latae** muscle is positioned superficially on the lateral surface of the hip (see figs. 11.50, 11.54). It originates on the iliac crest and inserts on a broad lateral fascia of the thigh called the **fascia lata.** The fascia lata is continuous with the iliotibial tract. The tensor fasciae latae is innervated by the superior gluteal nerve and, when stimulated, causes abduction of the thigh.

A deep group of six lateral rotators of the thigh is positioned directly over the posterior aspect of the hip joint. These muscles are not discussed but are identified in figures 11.52 and 11.53 from superior to inferior as the **piriformis, superior gemellus, obturator externus, inferior gemellus, obturator internus,** and **quadratus femoris.**

The anterior and posterior group of muscles that move the thigh are summarized in table 11.19.

Medial, or Adductor, Muscles.   The medial muscles that move the thigh include the gracilis, pectineus, adductor longus, adductor brevis, and adductor magnus (figs. 11.53, 11.54).

The long, thin **gracilis** *(gras′il-is)* muscle is the most superficial of the medial thigh muscles. It arises from the lower ramus of the pubis immediately below the symphysis pubis and inserts on the proximal medial surface of the tibia along with the sartorius. The gracilis is a two-joint muscle and can adduct the thigh or flex the leg. It is innervated by the obturator nerve.

The **pectineus** is the uppermost of the medial muscles that move the thigh. It is a flat, quadrangular muscle that

Figure 11.54. Muscles of the thigh. (a) an anterior view; (b) a posterior view.

Gluteus medius

Sartorius

Tensor fasciae latae

Rectus femoris

Vastus lateralis

Patellar tendon

Inguinal ligament

Iliopsoas

Pectineus

Adductor longus

Gracilis

Vastus medialis

(a)

Anterior inferior iliac spine

Greater trochanter

Intertrochanteric line

Rectus femoris

Vastus lateralis

Patella

Tibia

Patellar ligament

Anterior superior iliac spine

Intertrochanteric line

Vastus intermedius

Sartorius

Vastus medialis

Patella

Tibia

Patella

Gluteus maximus

Iliotibial tract

Adductor magnus

Gracilis

Semitendinosus

Semimembranosus

Sartorius

Biceps femoris

(b)

Ischial tuberosity

Semi-tendinosus

Semi-membranosus

Head of fibula

Ischial tuberosity

Long head of biceps femoris

Linea aspera

Short head of biceps femoris

Fibula

Schenk

Table **11.20** Medial muscles that move the thigh

Muscle	Derivation	Origin	Insertion	Action	Innervation
Gracilis	*gracilis* = slender	Inferior edge of symphysis pubis	Proximal medial surface of tibia	Adducts thigh; flexes and rotates leg at knee	Obturator
Pectineus	*pecten* = comb	Pectineal line of pubis	Distal to lesser trochanter of femur	Adducts and flexes thigh	Femoral
Adductor longus	*ad* = toward; *longus* = long	Pubis—below pubic crest	Linea aspera of femur	Adducts, flexes, and laterally rotates thigh	Obturator
Adductor brevis	*brevis* = short	Inferior ramus of pubis	Linea aspera of femur	Adducts, flexes, and laterally rotates thigh	Obturator
Adductor magnus	*magnus* = large	Inferior ramus of ischium and pubis	Linea aspera and medial epicondyle of femur	Adducts, flexes, and laterally rotates thigh	Obturator

Source: Van De Graaff, Kent M., *Human Anatomy.* © 1984 Wm. C. Brown Publishers, Dubuque, Iowa. All Rights Reserved. Reprinted by permission.

originates on the pubis and inserts just below the lesser trochanter of the femur. The pectineus flexes and adducts the thigh and is served by the femoral nerve.

The **adductor longus** muscle is immediately medial to the gracilis on the upper third of the thigh. It originates near the symphysis pubis and inserts on the linea aspera of the femur. The adductor longus and the following two adductors are synergistic in adducting, flexing, and rotating the thigh. They are all innervated by the obturator nerve.

The thick, triangular **adductor brevis** muscle is located behind the adductor longus. It arises from the inferior ramus of the pubis and inserts on the linea aspera of the femur.

As the name implies, the **adductor magnus** is a large, thick muscle, somewhat triangular in shape. It has an extensive origin along the inferior ramus of the ischium and pubis. It has a dual insertion along the linea aspera of the medial epicondyle of the femur.

Table 11.20 summarizes the muscles that adduct the thigh.

Muscles of the Thigh That Move the Lower Leg

The muscles that move the lower leg originate on the thigh and are surrounded and compartmentalized by tough fascial sheets, which are a continuation of the fascia lata and iliotibial tract. They are divided according to function and position into two groups: anterior extensors and posterior flexors.

Anterior, or Extensor, Muscles. The anterior muscles that move the lower leg are the sartorius and quadriceps femoris (fig. 11.54).

The long, straplike **sartorius** muscle obliquely crosses the anterior aspect of the thigh. It arises from the anterior superior spine of the ilium and extends medially to insert on the proximal medial surface of the tibia. The sartorius can act on both the hip and knee joints to flex and rotate the thigh laterally and also to assist in flexing the leg. The sartorius muscle is the longest muscle of the body and is frequently called the "tailor's muscle" because it enables one to assume a cross-legged sitting position. It is innervated by the femoral nerve.

The **quadriceps femoris** is actually a composite of four distinct muscles that have separate origins but a common insertion on the patella and **patellar tendon.** These muscles function synergistically to extend the leg, as in kicking a football. The four muscles of the quadriceps femoris muscle are the rectus femoris, vastus lateralis, vastus medialis, and vastus intermedius. All four are innervated by the femoral nerve.

The **rectus femoris** occupies a superficial position and is the only one of the four quadriceps that contracts over both the hip and knee joints. It has a dual origin on the anterior inferior spine of the ilium and the lip of the acetabulum.

The laterally positioned **vastus lateralis** is the largest muscle of the quadriceps femoris. It arises from the greater trochanter and linea aspera of the femur.

The **vastus medialis** occupies a medial position along the thigh. It originates on the linea aspera.

The **vastus intermedius** (not illustrated) lies deep to the rectus femoris and originates on the anterior surface of the shaft of the femur.

The anterior thigh muscles that move the lower leg are summarized in table 11.21.

Posterior, or Flexor, Muscles. There are three posterior thigh muscles antagonistic to the quadriceps femoris muscles in flexing the leg. These muscles are known as the **hamstrings** (fig. 11.54) and are each innervated by the tibial nerve that branches from the sciatic nerve.

The **biceps femoris** muscle occupies the posterior lateral aspect of the thigh. It has a superficial long head and a deep short head. The long head arises from the ischial tuberosity, and the short head arises from the linea aspera. The biceps femoris inserts on the head of the fibula

Table 11.21 Anterior thigh muscles that move the lower leg

Muscle	Derivation	Origin	Insertion	Action	Innervation
Sartorius	*sartor* = tailor	Spine of ilium	Medial surface of tibia	Flexes leg and thigh, abducts thigh, rotates thigh laterally, and rotates leg medially	Femoral
Quadriceps femoris	*quad* = four; *cephalic* = head; *femoris* = femur		Patella by common tendon, which continues as patellar tendon to tibial tuberosity	Extends leg at knee	Femoral
Rectus femoris	*rectus* = straight fibers	Spine of ilium and lip of acetabulum			
Vastus lateralis	*vastus* = large; *lateralis* = lateral	Greater trochanter and linea aspera of femur			
Vastus medialis	*medialis* = medial	Medial surface of femur			
Vastus intermedius	*intermedius* = middle	Anterior and lateral surfaces of femur			

Source: Van De Graaff, Kent M., *Human Anatomy*. © 1984 Wm. C. Brown Publishers, Dubuque, Iowa. All Rights Reserved. Reprinted by permission.

Table 11.22 Posterior thigh muscles that move the lower leg

Muscle	Derivation	Origin	Insertion	Action	Innervation
Biceps femoris	*bi* = two; *cephal* = head; *femoris* = femur	Long head—ischial tuberosity; short head—linea aspera of femur	Head of fibula and lateral condyle of tibia	Flexes leg and extends thigh	Tibial
Semitendinosus	*semi* = half; *tendo* = tendon	Ischial tuberosity	Proximal portion of medial surface of body of tibia	Flexes leg and extends thigh	Tibial
Semimembranosus	*membran* = membrane	Ischial tuberosity	Medial condyle of tibia	Flexes leg and extends thigh	Tibial

Source: Van De Graaff, Kent M., *Human Anatomy*. © 1984 Wm. C. Brown Publishers, Dubuque, Iowa. All Rights Reserved. Reprinted by permission.

and the lateral condyle of the tibia. The action of this muscle is a bit complicated because it works over both the hip and knee joints. It flexes the knee as well as extends, laterally rotates, and adducts the thigh.

The superficial **semitendinosus** muscle is fusiform and is located on the posterior medial aspect of the thigh. It originates on the ischial tuberosity from a common tendon with the biceps femoris. The long and tapering tendon inserts onto the proximal lateral portion of the tibia. The semitendinosus muscle also works over two joints, providing extension and medial rotation of the thigh and flexion of the leg.

The **semimembranosus** lies deep to the semitendinosus on the posterior medial aspect of the thigh. It is a flat muscle that arises from the ischial tuberosity and inserts on the medial condyle of the tibia. The action of the semimembranosus is similar to that of the semitendinosus.

The posterior thigh muscles that move the lower leg are summarized in table 11.22. The relationship of the muscles of the thigh is illustrated in figure 11.55.

Hamstring injuries are a common occurrence in some sports. The injury usually occurs when sudden lateral or medial stress to the knee joint tears the muscles or tendons. Because of its structure and the stress applied to it in competition, the knee joint is highly susceptible to injury. To reduce the number of knee injuries, the rules in contact sports should be altered or additional support and protection should be provided for this vulnerable joint.

Muscles of the Lower Leg That Move the Ankle, Foot, and Toes

The muscles of the leg, the **crural muscles,** are responsible for the movements of the foot. There are three groups of crural muscles: anterior, lateral, and posterior. The anteromedial aspect of the leg along the shaft of the tibia lacks muscle attachment. Muscles in the anterior compartment are innervated by the deep peroneal nerve, those in the lateral compartment by the superficial peroneal nerve, and those in the posterior compartment by the tibial nerve.

Figure 11.55.   A cross section of the right thigh as seen from above. Note the position of the vessels and nerves.

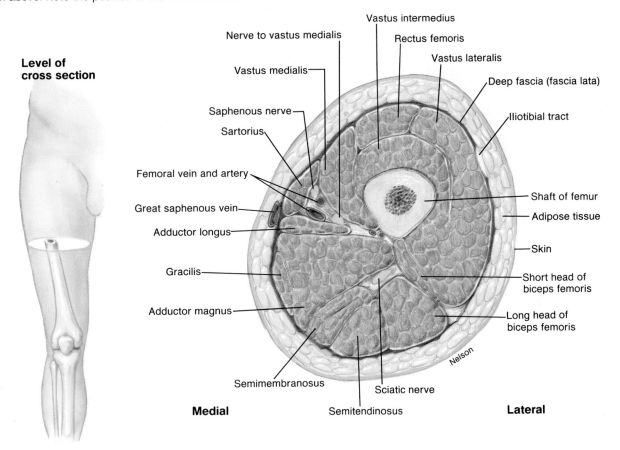

Level of cross section

Nerve to vastus medialis
Vastus intermedius
Rectus femoris
Vastus lateralis
Vastus medialis
Deep fascia (fascia lata)
Saphenous nerve
Iliotibial tract
Sartorius
Femoral vein and artery
Shaft of femur
Adipose tissue
Great saphenous vein
Adductor longus
Skin
Gracilis
Short head of biceps femoris
Adductor magnus
Long head of biceps femoris
Nelson
Semimembranosus
Sciatic nerve
Semitendinosus
Medial
Lateral

Anterior Crural Muscles.   The anterior crural muscles include the tibialis anterior, extensor digitorum longus, extensor hallucis longus, and peroneus tertius (figs. 11.56, 11.57).

The large, superficial **tibialis anterior** muscle can be easily palpated on the anterior lateral portion of the tibia (fig. 11.62). It arises from the lateral condyle and lateral surface of the proximal half of the tibia. It has a long tendon of insertion that traverses to the medial plantar surface of the foot and attaches to the first cuneiform and base of the first metatarsal. In this position, the tendon of the tibialis anterior aids in supporting the longitudinal arch of the foot. The tibialis anterior flexes and inverts the foot.

The **extensor digitorum longus** muscle lies lateral to the tibialis anterior. It originates on the lateral condyle of the tibia and the proximal portion of the fibula. Its tendon of insertion branches into four parts that attach to the superior surface of the second, third, fourth, and fifth toes (fig. 11.62). The extensor digitorum longus is responsible for the dorsiflexion of the ankle and the extension of the toes.

The **extensor hallucis longus** muscle is positioned deep between the tibialis anterior and the extensor digitorum longus (fig. 11.62). It originates on the middle portion of the fibula and the interosseous membrane. The tendinous insertion of this muscle is on the dorsal surface of the distal phalanx of the great toe. As the name implies, the extensor hallucis longus extends the big toe and aids in dorsiflexion of the foot.

A portion of the **peroneus tertius** muscle can be seen in figures 11.57 and 11.62. It is a small muscle that arises from the anterior surface of the distal one-fourth of the fibula and interosseous membrane. Its tendinous insertion is on the dorsal surface of the base of the fifth metatarsal. The peroneus tertius works over the ankle joint, assisting in dorsiflexion and eversion of the foot.

Lateral Crural Muscles.   The lateral crural muscles are the peroneus longus and peroneus brevis (figs. 11.56, 11.57).

Figure 11.56.   Anterior crural muscles.

Figure 11.57.   Lateral crural muscles.

The long, flat **peroneus longus** muscle originates from the proximal two-thirds of the fibula and a portion of the lateral condyle of the tibia. On the distal end, the long tendon of this muscle passes behind the lateral malleolus and under the arch of the foot to insert on the first metatarsal and first cuneiform (fig. 11.61c and d). The peroneus longus is responsible for plantar flexion and eversion of the foot.

The **peroneus brevis** lies deep to the peroneus longus and is positioned closer to the foot. It originates on the distal two-thirds of the fibula. It inserts on the base of the fifth metatarsal by means of a long tendon that passes behind the lateral malleolus. The action of the peroneus brevis is synergistic with the peroneus longus in plantar flexion and eversion of the foot.

Posterior Crural Muscles.   The seven posterior crural muscles can be grouped into a superficial and a deep group. The superficial group is composed of the gastrocnemius, soleus, and plantaris muscles (fig. 11.58). The four deep posterior crural muscles are the popliteus, flexor hallucis longus, flexor digitorum longus, and tibialis posterior (fig. 11.59).

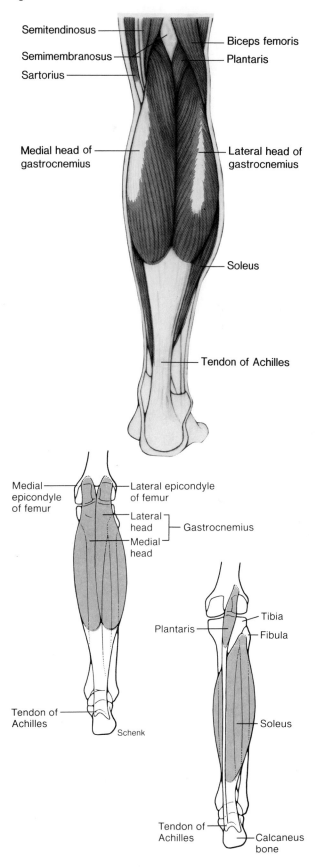

Figure 11.58. Posterior crural muscles and popliteal region.

The **gastrocnemius** *(gas″trok-ne′me-us)* is a large superficial muscle that forms the major portion of the calf of the leg. It consists of two distinct heads, which arise from the posterior surfaces of the medial and lateral condyles of the femur. This muscle and the deeper soleus muscle insert onto the calcaneus bone via the common **tendon of Achilles (tendo calcaneus).** The gastrocnemius acts over two joints to cause flexion of the knee and plantar flexion of the foot.

The **soleus** lies deep to the gastrocnemius. These two muscles are frequently referred to as a single muscle, the **triceps surae.** The soleus originates on the proximal heads of the fibula and tibia. It has a common insertion with the gastrocnemius but acts on only one joint, the ankle, in plantar flexing the foot.

The small **plantaris** muscle arises just above the origin of the lateral head of the gastrocnemius on the lateral condyle of the femur. It has a very long, slender tendon of insertion onto the calcaneus bone. The tendon of this muscle is frequently mistaken for a nerve by those dissecting it for the first time. The plantaris is a weak muscle with limited ability to flex the leg and plantar flex the foot.

Figure 11.59.   Deep posterior and crural muscles.

Figure 11.60.   Muscles that surround the popliteal fossa.

Table **11.23** Muscles of the lower leg that move the ankle, foot, and toes

Muscle	Derivation	Origin	Insertion	Action	Innervation
Tibialis anterior	*tibialis* = tibia; *ante* = in front	Lateral condyle and body of tibia	First metatarsal and first cuneiform	Dorsiflexion and inversion of foot	Deep peroneal
Extensor digitorum longus	*extensor* = increase joint angle; *digit* = finger or toe; *longus* = long	Lateral condyle of tibia and anterior surface of fibula	Extensor expansions of digits II–V	Extends digits II–V and dorsiflexes foot	Deep peroneal
Extensor hallucis longus	*hallux* = great toe	Anterior surface of fibula and interosseous membrane	Distal phalanx of digit I	Extends big toe and assists dorsiflexion of foot	Deep peroneal
Peroneus tertius	*perone* = fibula; *tertius* = third	Anterior surface of fibula and interosseous membrane	Dorsal surface of fifth metatarsal	Dorsiflexes and everts foot	Deep peroneal
Peroneus longus	*longus* = long	Lateral condyle of tibia and head and shaft of fibula	First cuneiform and metatarsal I	Plantar flexes and everts foot	Superficial peroneal
Peroneus brevis	*brevis* = short	Lower aspect of fibula	Metatarsal V	Plantar flexes and everts foot	Superficial peroneal
Gastrocnemius	*gaster* = belly; *kneme* = leg	Lateral and medial condyle of femur	Posterior surface of calcaneus	Plantar flexes foot; flexes knee	Tibial
Soleus	*soleus* = sole of foot	Posterior aspect of fibula and tibia	Calcaneus	Plantar flexes foot	Tibial
Plantaris	*planta* = plantar surface of foot	Lateral supracondylar ridge of femur	Calcaneus	Plantar flexes foot	Tibial
Popliteus	*poples* = ham of the knee	Lateral condyle of femur	Upper posterior aspect of tibia	Flexes and medially rotates leg	Tibial
Flexor hallucis longus	*flexor* = decrease angle	Posterior aspect of fibula	Distal phalanx of big toe	Flexes distal phalanx of big toe	Tibial
Flexor digitorum longus		Posterior surface of tibia	Distal phalanges of digits II–V	Flexes distal phalanges of digits II–V	Tibial
Tibialis posterior	*posterior* = back	Tibia and fibula and interosseous membrane	Navicular, cuneiforms, cuboid, and metatarsals II–IV	Plantar flexes and inverts foot; supports arches	Tibial

Source: Van De Graaff, Kent M., *Human Anatomy.* © 1984 Wm. C. Brown Publishers, Dubuque, Iowa. All Rights Reserved. Reprinted by permission.

The thin, triangular **popliteus** muscle is situated deep to the heads of the gastrocnemius where it forms part of the floor of the popliteal fossa. The **popliteal fossa** is the depression on the back side of the knee joint (fig. 11.60). The popliteus arises from the lateral condyle of the femur and inserts on the posterior surface of the proximal portion of the tibia (fig. 11.59).

The **flexor hallucis longus** muscle lies deep to the soleus on the posterolateral side of the leg. It is a bipennate muscle that arises from the distal two-thirds of the fibula. Its long tendon of attachment passes posterior to the medial malleolus and along the lateral sole of the foot to insert upon the hallux (the big toe). The flexor hallucis longus flexes the big toe and assists in plantar flexing and inverting the foot.

The **flexor digitorum longus** muscle also lies deep to the soleus and parallels the flexor hallucis longus on the medial side of the leg. It originates on the posterior surface of the shaft of the tibia. Its distal tendon passes behind the medial malleolus and continues along the plantar surface of the foot where it branches into four tendinous slips. These four slips attach to the bases of the terminal phalanges of the second, third, fourth, and fifth toes (fig. 11.61b). The flexor digitorum longus works over several joints, flexing four of the digits and assisting in plantar flexing and inverting the foot.

The **tibialis posterior** muscle is located deep to the soleus, directly on the interosseous membrane on the posterior surface of the leg. It originates on the interosseous membrane and the adjacent proximal surfaces of the tibia and fibula. The distal tendon of the tibialis posterior passes behind the medial malleolus and inserts on the plantar surfaces of the navicular, the cuneiforms, the cuboid, and the second, third, and fourth metatarsals (fig. 11.61d). The tibialis posterior plantar flexes and inverts the foot and gives support to the arches of the foot.

The crural muscles are summarized in table 11.23.

Muscles of the Foot

With the exception of one additional intrinsic muscle, the **extensor digitorum brevis,** the muscles of the foot are similar in number and nomenclature to those of the hand. The

Figure 11.61.   The four musculotendinous layers of the plantar aspect of the foot.

functions of the muscles of the foot are different, however, because the foot is adapted to provide support while bearing body weight rather than for grasping.

The muscles of the foot are topographically arranged into four layers (fig. 11.61) that are difficult to dissociate even in dissection. The muscles function either to move the toes or to support the arches of the foot through their contraction. Because of their complexity, the muscles of the foot will be presented only in illustrations. (See figs. 11.61 and 11.62.)

1. *List all the muscles that either originate from or insert on the scapula.*
2. *On the basis of function, list the muscles of the upper extremity into the categories of flexors, extensors, abductors, adductors, and rotators. (Each muscle may fit into two or more categories.)*
3. *Which muscles of the lower extremity span two joints and therefore have two different actions?*

Figure 11.62.   An anterior view of the dorsum of the foot.

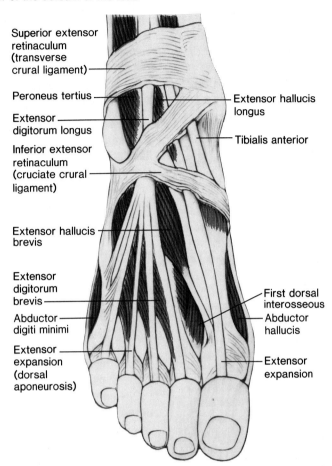

Superior extensor retinaculum (transverse crural ligament)

Peroneus tertius

Extensor digitorum longus

Inferior extensor retinaculum (cruciate crural ligament)

Extensor hallucis brevis

Extensor digitorum brevis

Abductor digiti minimi

Extensor expansion (dorsal aponeurosis)

Extensor hallucis longus

Tibialis anterior

First dorsal interosseous

Abductor hallucis

Extensor expansion

Clinical Considerations

Compared to the other systems of the body, the muscular system has few clinical afflictions. If properly exercised, the muscles of the body adequately serve a person for a lifetime. Muscles are capable of doing incredible amounts of work and through conditioning can become even stronger.

There are, however, several clinical considerations of muscles, including the diagnosis of muscle conditions, physiological conditions in muscles, and other diseases of muscles.

Diagnosis of Muscle Condition

The clinical symptoms of muscle diseases are usually weakness, loss of muscle mass (atrophy), and pain. The diagnosis and extent of the disease depends on the severity of the symptoms.

Several standard procedures are used to assess skeletal muscle dysfunction and disease. The most obvious is a clinical examination of the patient. Following this, it may be necessary to test the muscle function using **electromyography (EMG)** to measure conduction rates and motor unit activity within a muscle. Laboratory tests may be conducted, such as serum enzyme assays that reflect muscle destruction or muscle biopsy examinations under the microscope. Muscle biopsy is perhaps the definitive source of diagnostic information. Progressive atrophy, polymyositis, and metabolic diseases of muscles can be determined through a biopsy.

Physiological Conditions in Muscles

Muscles depend on systematic, periodic contraction to maintain optimal health. Obviously overuse or disease will cause a change in muscle tissue. The immediate effect of overexertion of muscle tissue is the accumulation of lactic acid, resulting in fatigue and soreness. Excessive contraction of a muscle can also damage the fibers or associated connective tissue, resulting in a **strained muscle.**

When skeletal muscles are not contracted, either because the motor nerve supply is blocked or because the limb is immobilized (as when a broken bone is in a cast), the muscle fibers **atrophy,** or diminish in size. Atrophy is reversible if exercise is resumed, as after a healed fracture, but tissue death is inevitable if the nerves cannot be stimulated.

The fibers in healthy muscle tissue increase in size, or **hypertrophy,** if a muscle is systematically exercised. This increase in muscle size and strength is due not to an increase in the number of muscle cells but to the production of more myofibrils accompanied by a strengthening of the associated connective tissue.

A **cramp** within a muscle is an involuntary, painful, and prolonged contraction. Cramps can occur while muscles are in use or at rest. The precise cause of cramps is unknown, but evidence indicates that cramps may be related to conditions within the muscle (e.g., calcium or oxygen deficiencies) or to stimulation of the motor neurons.

A condition called **rigor mortis** (rigidity of death) affects skeletal muscle tissue several hours after death as depletion of ATP within the fibers causes a state of muscle contracture and stiffness of the joint. In a few days, however, as the muscle proteins decompose, the rigidity of the corpse disappears.

Diseases of Muscles

Fibromyositis is an inflammation of both skeletal muscular tissue and the associated connective tissue. Its causes are not fully understood. Fibromyositis frequently occurs in the extensor muscles of the lumbar region of the vertebral column where extensive aponeuroses exist. Fibromyositis of this region is called **lumbago,** or **rheumatism.**

Muscular dystrophy is a genetic disease characterized by a gradual atrophy and weakening of muscle tissue. There are several kinds of muscular dystrophy, none of whose etiology is completely understood. The most frequent type affects children and is sex-linked to the male child. As muscular dystrophy progresses, the muscle fibers atrophy and are replaced by adipose tissue. Most children who have muscular dystrophy die before the age of twenty.

The disease **myasthenia gravis** is characterized by extreme muscle weakness and low endurance. There is a defective transmission of impulses at the neuromuscular junction. Myasthenia gravis is believed to be an autoimmune disease, and it typically affects women between the ages of twenty and forty.

Poliomyelitis (polio) is actually a viral disease of the nervous system that causes a paralysis of muscles. The viruses are usually localized in the ventral (anterior) horn of the spinal cord where they affect the motor nerve impulses to skeletal muscles.

Neoplasms of muscle are rare, but when they do occur they are usually malignant. **Rhabdomyosarcoma** *(rab″do-mi″o-sar-ko′mah)* is a malignant tumor of skeletal muscle. It can arise in any skeletal muscle, and most often afflicts young children and elderly persons.

Additional Clinical Terms

convulsion   An involuntary, spasmodic contraction of skeletal muscle.

fibrillation *(fi-bri-la′shun)*   A series of rapid, uncoordinated, and spontaneous contractions of individual motor units within a muscle.

hernia   The rupture or protrusion of a portion of the underlying viscera through muscle tissue. The most common hernias occur in the normally weak places of the abdominal wall. There are four basic types of hernias:

1. **femoral**—viscera passing through the femoral ring
2. **hiatal**—the superior portion of the stomach protruding through the esophageal opening of the diaphragm
3. **inguinal**—viscera protruding through the inguinal canal
4. **umbilical**—a hernia occurring at the navel

intramuscular injection   A hypodermic injection at certain heavily muscled areas to avoid damaging nerves. The most common site is the buttock.

myalgia   Pain within a muscle resulting from any muscular disorder or disease.

myokymia *(mi″o-kim′e-ah)*   Continual quivering of a muscle.

myoma   A tumor of muscle tissue.

myopathy   Any muscular disease.

myotomy *(mi-ot′o-me)*   Surgical cutting of muscle tissue.

myotonia   A prolonged muscular spasm.

paralysis   The loss of nervous control of a muscle.

shin splints   Tenderness and pain on the anterior surface of the lower leg, caused by straining the flexor digitorum longus muscle.

torticollis (wryneck)   A persistent contraction of a sternocleidomastoid muscle, drawing the head to one side and distorting the face. Torticollis may be acquired or congenital.

---

lumbago: L. *lumbus*, loin
myasthenia: Gk. *myos*, muscle; *astheneia*, weakness
poliomyelitis: Gk. *polios*, gray; *myolos*, marrow

---

rhabdomyosarcoma: Gk. *rhabdos*, rod; *myos*, muscle; *oma*, a growth

Chapter Summary

I. Organization and General Functions
The muscles of the skeletal system are adapted to contract in order to carry out the functions of motion, heat production, and posture and body support.
   A. The contraction of skeletal muscle fibers results in body motion, heat production, and the maintenance of posture and body support.
   B. The four basic properties characteristic of all muscle tissue are irritability, contractility, extensibility, and elasticity.
   C. Axial muscles include facial muscles, neck muscles, and trunk muscles; appendicular muscles include those that act on the girdles and those that move the segments of the appendages.

II. Development of Skeletal Muscles
Skeletal muscles begin development at four weeks after conception from blocks of mesoderm, called myotomes, in the trunk area and from loosely organized mesenchyme in the head area. Each muscle fiber is derived from several embryonic myoblast cells.
   A. The union of myoblasts forms a syncytial myotube, which becomes a skeletal muscle fiber as contractile filaments develop.
   B. The development of muscle fibers occurs within myotomes within the trunk region and loosely organized mesenchyme areas within the head region.

III. Structure of Skeletal Muscles
Skeletal muscle tissue and its binding connective tissue are arranged in a highly organized pattern so that the forces of the contracting muscle fibers are united and directed onto the structures being moved.
   A. The origin of a muscle is the more stationary attachment. The insertion is the more movable attachment.
   B. Individual muscle fibers are covered by endomysium. Fasciculi are covered by perimysium. The entire muscle is covered by epimysium.
   C. Synergistic muscles contract together. Antagonistic muscles perform in opposition to a particular group of muscles.
   D. Muscles may be classified according to fiber arrangement as parallel, convergent, pennate, or circular.
   E. Each muscle has motor and sensory innervation.

IV. Skeletal Muscle Fibers and Mode of Contraction
Muscle fiber contraction results from a sliding movement within the myofibrils in which the length of the sarcomeres is reduced.
   A. Each skeletal muscle fiber is a multinucleated, striated cell composed of myofibrils and enclosed by a sarcolemma.
      1. Myofibrils are characterized by alternating A and I bands; each I band is bisected by a Z line, and the portion between two Z lines is the sarcomere.
      2. Extending through the sarcoplasm is a network of membranous channels called the sarcoplasmic reticulum and a system of transverse tubules (T tubules).
   B. The activity of the cross-bridges causes sliding of the filaments.
      1. The actin on each side of the A bands is pulled toward the center.
      2. The H bands thus get shorter as more actin overlaps with myosin.
      3. The I bands also get shorter as adjacent A bands are pulled closer together.
      4. The A bands stay the same length because the filaments (both thick and thin) do not shorten during muscle contraction.
   C. Maximum tension in a muscle is produced when the muscle is at its normal resting length *in vivo*.
   D. When a muscle exerts tension without shortening, the contraction is termed isometric; when shortening does occur, the contraction is isotonic.

V. Stimulation, Regulation, and the Mechanics of Contraction
The recruitment of motor units when a sufficient amount of Ca^{++} is bound to tropin results in muscle contraction. The response of a motor unit is determined by the intensity and duration of the stimulus.
   A. The neuromuscular junction is the area consisting of the motor end plate and the sarcolemma of a muscle fiber. In response to a nerve impulse, the synaptic vesicles of the terminal end of an axon secretes a neurotransmitter, which diffuses across the neuromuscular cleft of the neuromuscular junction and stimulates the muscle fiber.
   B. The motor neuron and the muscle fibers innervated by the motor neuron is called a motor unit.
      1. When a muscle is composed of many motor units (such as in the hand), there is a fine control of muscle contraction.
      2. The large muscles of the leg have relatively few motor units, which are correspondingly large.
      3. Graded contractions are produced by the asynchronous stimulation of different motor units.
   C. Calcium ions serve to couple electrical excitation of the muscle fiber with contraction.
      1. When a muscle is at rest, the Ca^{++} concentration of the sarcoplasm is very low and cross-bridges are prevented from attaching to actin by complexes of troponin and tropomyosin in the thin filaments.
      2. Action potentials are conducted by transverse tubules into the muscle fiber, where they stimulate the release of Ca^{++} from the sarcoplasmic reticulum.
      3. The binding of Ca^{++} to troponin causes movement of tropomyosin, which allows cross-bridges to attach and muscle contraction to occur.
      4. When action potentials cease, Ca^{++} is removed from the sarcoplasm and stored in the sarcoplasmic reticulum.
   D. The contraction of separate muscle fibers is all-or-none.
      1. The contraction of separate muscle fibers is called a twitch.
      2. A whole muscle also produces a twitch in response to a single electrical pulse *in vitro*.
      3. A summation of fiber twitches can occur so rapidly that the muscle produces a smooth, sustained contraction known as tetanus.

VI. Energy Usage by Skeletal Muscle
Aerobic cellular respiration is required for the production of ATP needed for the cross-bridge activity of the various types of skeletal muscle fibers.

A. ATP is produced from the combination of ADP with phosphate derived from phosphocreatine.
 1. The phosphocreatine represents a ready reserve of high-energy phosphate during sustained muscle contractions.
 2. The phosphocreatine is produced at rest from creatine, and phosphate is derived from ATP.
B. There are three types of muscle fibers.
 1. Slow-twitch, thin fibers are adapted for aerobic respiration and are resistant to fatigue.
 2. Fast-twitch, thicker fibers are adapted for anaerobic respiration.

 3. Intermediate-twitch fibers are the third type of muscle fibers.
C. Physical training affects the enzyme content and characteristics of the muscle fibers.
 1. Strength training increases the enzymes of anaerobic respiration and makes the fast and intermediate fibers larger and stronger.
 2. Endurance training increases the capacity for aerobic respiration of the muscle fibers.
VII. Naming of Muscles
Skeletal muscles are named on the basis of shape, location, attachment, orientation of fibers, relative position, and function.
 A. Most of the muscles are paired; that is, the right side of the body is an image of the left.

 B. Muscles should be learned as functional groups.
VIII. Muscles of the Axial Skeleton
The muscles of the axial skeleton include those of facial expression, mastication, eye movement, tongue movement, neck movement, respiration, the abdominal wall, the pelvic outlet, and movement of the vertebral column.
IX. Muscles of the Appendicular Skeleton
The muscles of the appendicular skeleton include those of the pectoral girdle, humerus, forearm, wrist, hand, and fingers, and those of the pelvic girdle, thigh, lower leg, ankle, foot, and toes.

Review Activities

Objective Questions

1. Which of the following is *not* used as a means of naming muscles?
 (a) location
 (b) action
 (c) shape
 (d) attachment
 (e) strength of contraction
2. A graded muscle contraction is produced by variations in
 (a) the strength of the fiber's contraction
 (b) the number of fibers that are contracting
 (c) both of the above
 (d) neither of the above
3. Sustained muscle contraction (tetanus) is produced by
 (a) a sustained contraction of muscle fibers
 (b) asynchronous twitches of the muscle fibers
 (c) both of the above
 (d) neither of the above
4. A flexor of the shoulder joint is the
 (a) pectoralis major
 (b) supraspinatus
 (c) teres major
 (d) trapezius
 (e) latissimus dorsi
5. Which of the following muscles have motor units with the lowest innervation ratio?
 (a) brachial muscles
 (b) muscles of the forearm
 (c) thigh muscles
 (d) abdominal muscles

6. Neurotransmitters are stored in synaptic vesicles within
 (a) motor end plates
 (b) motor units
 (c) myofibrils
 (d) terminal ends of axons
7. When a skeletal muscle shortens during contraction, which of the following statements is *false*?
 (a) The A bands shorten.
 (b) The H bands shorten.
 (c) The I bands shorten.
 (d) The sarcomeres shorten.
8. Electrical excitation of a muscle fiber most directly causes
 (a) movement of tropomyosin
 (b) attachment of the cross-bridges to actin
 (c) the release of Ca^{++} from the sarcoplasmic reticulum
 (d) the splitting of ATP
9. The energy for muscle contraction is most directly obtained from
 (a) phosphocreatine
 (b) ATP
 (c) anaerobic respiration
 (d) aerobic respiration
10. Which of the following muscles does *not* have either an origin or insertion upon the humerus? The
 (a) teres minor
 (b) biceps brachii
 (c) supraspinatus
 (d) brachialis
 (e) pectoralis major

Essay Questions

1. Describe how muscle fibers are formed, and explain why the fibers are multinucleated.
2. Discuss the physical features and physiological properties of skeletal muscle that enables it to contract.
3. What is a motor unit, and what is its role in muscle contraction?
4. Discuss the position of flexor and extensor muscles relative to the shoulder, elbow, and wrist joints.
5. Describe exercises that strengthen the following muscles: (a) the pectoralis major; (b) the deltoid; (c) the triceps; (d) the pronator teres; (e) the rhomboideus major; (f) the trapezius; (g) the serratus anterior; and (h) the latissimus dorsi.
6. Give three examples of synergistic muscle groups within the lower extremity, and identify the antagonistic muscle group for each.

Selected Readings

Anderson, J. E. 1978. *Grant's atlas of anatomy.* 7th ed. Baltimore: The Williams and Wilkins Co.

Basmajian, J. V. 1980. *Grant's method of anatomy.* 10th ed. Baltimore: The Williams and Wilkins Co.

————. 1978. *Muscles alive.* 4th ed. Baltimore: The Williams and Wilkins Co.

Bendall, J. R. 1969. *Muscles, molecules, and movement.* New York: Elsevier Publishers.

Chapman, C. B., and J. H. Mitchell. May 1965. The physiology of exercise. *Scientific American.*

Clemente, C. D. 1981. *Anatomy: A regional atlas of the human body.* 2d ed. Philadelphia: Lea and Febiger.

Close, R. I. 1972. Dynamic properties of mammalian skeletal muscles. *Physiol. Rev.* 52:129.

Cohen, C. November 1975. The protein switch of muscle contraction. *Scientific American.*

Eisenberg, E., and L. E. Greene. 1980. The relation of muscle biochemistry to muscle physiology. *Ann. Rev. Physiol.* 42:293.

Grinnell, A. D., and M. A. B. Brazier, eds. 1981. *Regulation of muscle contraction: Excitation-contraction coupling.* New York: Academic Press.

Hinson, M. M. 1981. *Kinesiology.* 2d ed. Dubuque, IA: Wm. C. Brown Company Publishers.

Hoyle, G. April 1970. How a muscle is turned off and on. *Scientific American.*

Huxley, H. E. November 1968. The contraction of muscle. *Scientific American.*

————. 1969. The mechanism of muscular contraction. *Science* 164:1356.

Junge, D. 1981. *Nerve and muscle excitation.* 2d ed. Sunderland, MA.: Sinauer Associates, Inc.

Last, R. J. 1977. *Anatomy, regional and applied.* 6th ed. Boston: Churchill Livingstone.

Margaria, R. March 1972. The sources of muscular energy. *Scientific American.*

Merton, P. A. May 1972. How we control the contraction of our muscles. *Scientific American.*

Murray, J. H., and A. Weber. February 1974. The cooperative action of muscle proteins. *Scientific American.*

Nadel, E. R. 1985. Physiological adaptations to aerobic training. *Amer. Scientist* 73:334.

Peterson, B. W. 1979. Reticulospinal projections to spinal motor nuclei. *Ann. Rev. Physiol.* 41:127.

Porter, K. R., and C. Franzini-Armstrong. March 1965. The sarcoplasmic reticulum. *Scientific American.*

Rasch, P. J., and R. K. Burke. 1978. *Kinesiology and applied anatomy: The science of human movement.* 6th ed. Philadelphia: Lea and Febiger.

Rosse, C., and D. K. Clawson. 1980. *The musculoskeletal system in health and disease.* Philadelphia: Harper and Row Publishers.

Sandow, A. 1970. Skeletal muscle. *Physiol. Rev.* 32:87.

Snell, R. S. 1978. *Atlas of clinical anatomy.* Boston: Little, Brown, and Co.

Sobotta, J., and F. Figge. 1974. *Atlas of human anatomy.* 9th ed. New York: Hafner Press.

Stefani, E., and D. J. Chiarandini. 1982. Ion channels in skeletal muscles. *Ann. Rev. Physiol.* 44:357.

Stein, R. B., ed. 1980. *Nerve and muscle: Membranes, cells, and systems.* New York: Plenum Publishing Corp.

Wilkie, D. R. 1968. *Muscle.* New York: St. Martin's Press.

12

Surface Anatomy and

Concepts

Surface anatomy is a branch of gross anatomy concerned with identifying body structures through observation and palpation. The study of surface anatomy has tremendous application to physical fitness and medical diagnosis and treatment.

The surface anatomy of a neonatal infant differs from that of an adult because it represents an early stage of human development. Certain aspects of the surface anatomy of a neonate are of clinical importance in ascertaining the degree of physical development, general health, and possible congenital abnormalities.

The surface anatomy of the head is of concern because the head contains vital sense organs and provides openings into the respiratory and digestive systems. Of further concern are esthetic appearances and the many medical problems that involve this part of the body including trauma, diseases, and dysfunctions.

Major organs are located within the flexible neck, and several structures (the spinal cord, vessels, trachea, and esophagus) that are essential for body sustenance pass through the neck.

The locations of vital visceral organs in the cavities of the torso make the surface anatomy of this body region especially important.

The surface features of the pelvic region are important primarily to identify reproductive organs and clinical problems with these organs.

The surface anatomy of the shoulder and upper extremity is of clinical importance because of frequent trauma to these body regions. Vessels of the upper extremity are also used as pressure points and for intravenous injections or blood withdrawal.

The massive bones and muscles of the buttock and lower extremity are important as weight bearers and locomotors. Many of the surface features of these regions are important in relation to locomotion or locomotor dysfunction.

Body Topography

Introduction to Surface Anatomy

Surface anatomy is a branch of gross anatomy concerned with identifying body structures through observation and palpation. The study of surface anatomy has tremendous application to physical fitness and medical diagnosis and treatment.

Objective 1. Describe the value of surface anatomy in learning internal anatomical structures.

Objective 2. Discuss how surface anatomy is important in diagnosing and treating various diseases or conditions of the body.

Objective 3. Describe subcutaneous differences in adult males and females that may have a bearing on their surface anatomy.

It is amazing how much anatomical information can be learned by studying the surface anatomy of one's own body. Surface anatomy is the study of the structure and markings of the surface of the body that can be identified through observation or **palpation** *(pal-pa'shun)*. Surface features can be readily identified through visual **observation,** and anatomical features beneath the skin can be located by **palpation** (feeling with firm pressure).

With the exception of certain cranial bones, the bones of the entire skeleton can be palpated. Once the position, shape, and processes of these bones are identified, these skeletal features can serve as landmarks for other anatomical structures. One can observe the location of many skeletal muscles and their tendinous attachments as they are contracted and made to bulge. The location and range of movement of each of the joints of the body can be determined as the articulating bones are moved by muscle contractions. On some persons, one can locate the positions of superficial veins and trace their courses. Even the location and function of the valves within the veins can be demonstrated on the surface of the skin (fig. 12.1). The locations of some of the arteries can be seen as they pulsate beneath the skin. The locations of these arterial pressure points are an important clinical aspect of surface anatomy. Other structures can be identified from the surface, including certain nerves, lymph nodes, glands, and other internal organs.

Surface anatomy is an essential part of the study of anatomy and physiology. Knowing the location of muscles and muscle groups as observed in surface anatomy can be extremely important in physical fitness and body conditioning. In many medical and paramedical professions the surface anatomy of a patient is of immeasurable value in diagnosis and treatment. Knowing where to record a pulse, insert needles and tubes, listen to the functioning of internal organs, take X rays, and perform physical therapy requires a knowledge of surface body landmarks.

The effectiveness of observation and palpation in studying a person's surface anatomy is dependent on the amount of subcutaneous adipose tissue present (fig. 12.2). In examining an obese person, it may be extremely difficult to ob-

Figure 12.1. A demonstration of the presence and function of valves within the veins of the forearm, conducted by the great English anatomist William Harvey. (After William Harvey, *On the Motion of the Heart and Blood in Animals,* 1628.) Surface anatomy was extremely important to understanding the concept of a closed circulatory system (e.g., blood contained within vessels).

serve or palpate certain internal structures that are readily discernible in a thin person. The hypodermis and subcutaneous tissue in a woman is naturally thicker than in a man (fig. 12.3). This tends to smooth the surface contour of a woman and obscure the internal features, such as muscles, veins, and bony prominences, that are apparent in men.

Make anatomy relevant and applicable by identifying your surface anatomy. Use yourself as a model from which to learn and review, and anatomy as a science will take on a new meaning. As you learn about a bone or a process on a bone, palpate that part of your body. Contract the muscles you are studying so that you better understand their locations, attachments, and actions. Do this and you will become better acquainted with your body and anatomy will become more enjoyable and easier to learn. Your body is one "crib sheet" you can take with you to exams.

Figure 12.2. Subcutaneous adipose tissue. (*a*) an anterior view of the left brachial region. (*b*) a posterior view of the left lower leg.

(a)

(b)   Nelson

This chapter will be of great value to you in reviewing the bones of the skeleton, articulations, and muscles you have already studied. Refer back to this chapter as you learn the anatomy of the remaining body systems. Reviewing this way will reinforce your external perspective of where the various organs and structures are located.

1. *Define the terms* observation *and* palpation, *and explain the value of surface anatomy in learning the location of internal structures.*
2. *Discuss the medical importance of surface anatomy.*
3. *Describe the subcutaneous sexual difference that has a bearing on surface anatomy.*

Surface Anatomy of the Newborn

The surface anatomy of a neonatal infant differs from that of an adult because it represents an early stage of human development. Certain aspects of the surface anatomy of a neonate are of clinical importance in ascertaining the degree of physical development, general health, and possible congenital abnormalities.

Objective 4. Describe the surface anatomy of a normal, full-term neonate.
Objective 5. List some internal anatomical structures that can be palpated in a neonate.

The birth of a baby is a dramatic culmination of a nine-month gestation during which the miraculous development of the fetus prepares it for extrauterine life. The normal, full-term neonate is physiologically prepared for life but is totally dependent on parental care. The physical assessment of the neonate is extremely important to insure its survival. Much of the assessment is performed through inspection and palpation of its surface anatomy. The surface anatomy of a neonate is obviously different than that of an adult because of the transitional stage of development from fetus to infant.

Figure 12.3. Principal areas of adipose deposition of a female. (*a*) an anterior view. (*b*) a lateral view. The outline of a male is superposed in the lateral view. There is significantly more adipose tissue interlaced in the fascia covering the muscles, vessels, and nerves in a female than in a male. The hypodermis layer of the skin is also approximately 10 percent thicker in a female than in a male.

(a)

(b)   Sims

Table **12.1** Surface anatomy of the neonate		
Body structure	**Normal conditions**	**Common variations**
General posture	Flexion of vertebral column and extremities	Extended legs and neck; abducted and rotated thighs (breech birth)
Skin	Red or pink skin with vernix caseosa and lanugo; edematous face, extremities, and genitalia	Neonatal jaundice; integumentary blisters; Mongolian spots
Skull	Fontanels large, flat, and firm but soft to touch	Molded skull, bulging fontanels; cephalhematoma
Eyes	Lids edematous; color—gray, dark blue, or brown; absence of tears; corneal, pupillary, and blink reflexes	Conjunctivitis, subconjunctival hemorrhage
Ears	Auricle flexible, cartilage present; top of auricle positioned on horizontal line with outer canthus of eye	Auricle flat against head
Neck	Short and thick surrounded by neck folds	Torticollis
Chest	Equal anterioposterior and lateral dimensions; xiphoid process evident; breast enlargement	Funnel or pigeon chest; additional nipples (polythelia); secretions from breast (witch's milk)
Abdomen	Cylindric in shape; liver and kidneys palpable	Umbilical hernia
Genitalia	(♂ and ♀) Edematous and darkly pigmented; (♂) testes palpable in scrotum; periodic erection of penis	(♀) Blood-tinged discharge (pseudomenstruation); hymenal tag; (♂) testes palpable in inguinal canal; inability to retract prepuce; inguinal hernia
Extremities	Symmetry of extremities; 10 fingers and toes; soles flat with moderate to deep creases	Partial syndactyly; asymmetric length of toes

Although the surface anatomy of a neonate is discussed at this point in the book, prenatal development is discussed in chapter 29 and body growth with its accompanying physiological changes are discussed in chapter 30. A summary of the surface anatomy of the neonate is presented in table 12.1.

General Appearance

As a result of *in utero* position, the posture of the full-term neonate is one of flexion (fig. 12.4). The neonate born vertex (head first) keeps the neck and vertebral column flexed, with the chin resting on the upper chest. The hands are flexed (clenched), as are the arms, and held toward the chest. The legs are flexed at the knees, and the hips are flexed bringing the thighs toward the abdomen. The feet are dorsiflexed.

The skin is the one organ of the neonate that is completely visible and is therefore a source of considerable information concerning state of the development and the clinical condition of the newborn. At birth, the skin is covered with a grayish, cheeselike substance called **vernix caseosa.** If it is not washed away during bathing, it will dry and disappear within a couple of days. Fine, silklike hair, called **milia** *(mil'e-ah),* may appear as tiny white papules on the nose, cheeks, and chin. Skin color depends on genetic background although certain areas, such as the gen-nose, cheeks, and chin. Skin color depends on genetic background although certain areas, such as the genitalia,

---

neonate: Gk. *neos*, new; L. *natalis*, birth
vernix caseosa: L. *vernix*, varnish; *caseus*, cheese
milia: L. *miliarius*, relating to millet

areola, and linea alba may appear darker pigmented because of a response to the maternal and placental hormones that enter the fetal circulation. **Mongolian spots** occur in about 90% of newborn Negroes, Orientals, and American Indians. These blue-gray pigmented areas vary in size and are usually located in the lumbosacral region. Mongolian spots generally fade within the first year or two.

Abnormal skin color is clinically important in the physical assessment of the neonate. **Cyanosis** (bluish discoloration) is usually due to a pulmonary disease such as atelectasis or pneumonia or to congenital heart disease. Although **jaundice** (yellowish discoloration) is common in infants and is usually of no concern, it may indicate liver or bone marrow problems. **Pallor** (paleness) may indicate anemia, edema, or shock.

The appearance of the nails and nail beds is especially valuable in determining body dysfunctions, certain genetical conditions, and even normal gestations. Cyanosis, pallor, and capillary pulsations are best observed at the nails. Yellow nails are found in postmature neonates.

Local edema (swelling) is not uncommon particularly in the skin of the face, legs, hands, feet, and genitalia. Creases on the palms of the hands and soles of the feet should be prominent; the absence of creases accompanies prematurity (fig. 12.5). The nose is usually flattened after birth, and bruises here or on other areas of the face are not uncommon. The auricle of the ear is flexible with the top edge positioned on a horizontal line with the outer canthus (corner) of the eye.

The neck of a neonate is short and thick and is surrounded by neck folds. The chest is rounded in cross section, and the abdomen is cylindrical. The abdomen may bulge in the upper right quadrant due to the large liver.

Figure 12.4.   (*a* and *b*) a flexion position of a neonate is an indication of a healthy gestation and a normal delivery.

(a)

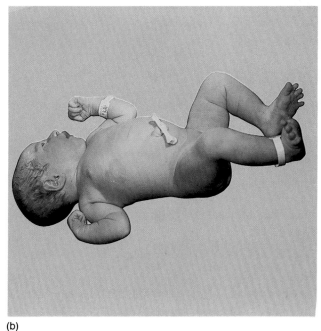

(b)

If the newborn is thin, peristaltic intestinal waves may be observed. At birth, the **umbilical** *(um-bil′i-kal)* **cord** appears a bluish white and moist. After clamping it begins to dry and appears yellowish brown. It progressively shrivels and becomes greenish black prior to its falling off by the second week.

The genitalia of both sexes may appear darkly pigmented due to maternal hormonal influences. In a female neonate, a **hymenal** *(hi′men-al)* **tag** is frequently present and is visible from the posterior opening of the vagina. It is comprised of tissue from the hymen and labia minora. The hymenal tag usually disappears by the end of the first month.

Figure 12.5.   Sole creases at different ages of gestation as seen from footprints of two premature babies (*a* and *b*) and a full-term baby (*c*). (*a*) at 26 weeks, only an anterior transverse crease is present; (*b*) by 33 weeks, creases develop along the medial instep; and (*c*) the entire sole is creased by 38 weeks.

(a)   (b)   (c)

Palpable Structures

The six **fontanels** can be lightly palpated as the "soft spots" on the infant's head. The liver is palpable 2–3 cm (1 in.) below the right costal margin. A physician doing a physical examination of a neonate will palpate both kidneys soon after delivery before the intestines fill with air. The suprapubic area is also palpated for an abnormal distended urinary bladder. The newborn should void urine within the first twenty-four hours after birth.

The testes of a male child should always be palpated in the scrotum. If the neonate is small or premature, the testes may be palpable in the inguinal canals. An examination for inguinal hernias is facilitated by the crying of an infant, which creates abdominal pressure.

> 1. Describe the appearance of each of the following in a normal neonate: skin, head, thorax, abdomen, genitalia, and extremities. What is meant by the normal flexion position of a neonate?
> 2. Which internal body organs are palpable in a neonate?

Head

The surface anatomy of the head is of concern because the head contains vital sense organs and provides openings into the respiratory and digestive systems. Of further concern are esthetic appearances and the many medical problems that involve this part of the body including trauma, diseases, and dysfunctions.

Figure 12.6.   The surface anatomy of the facial region: (a) an anterior view; (b) a lateral view.

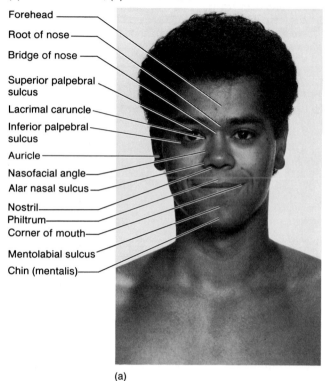

Forehead
Root of nose
Bridge of nose
Superior palpebral sulcus
Lacrimal caruncle
Inferior palpebral sulcus
Auricle
Nasofacial angle
Alar nasal sulcus
Nostril
Philtrum
Corner of mouth
Mentolabial sulcus
Chin (mentalis)

(a)

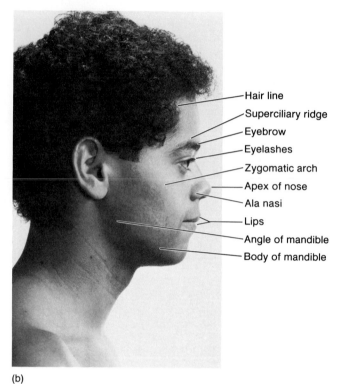

Hair line
Superciliary ridge
Eyebrow
Eyelashes
Zygomatic arch
Apex of nose
Ala nasi
Lips
Angle of mandible
Body of mandible

(b)

Objective 6. Identify by observation or palpation various surface features of the cranial and facial regions.

The head, or caput, contains the brain and sense organs. It also provides openings into the respiratory and digestive systems. The head is structurally and developmentally divided into the cranium and the face.

Cranium

The **cranium** *(kra'ne-um),* also known as the **braincase,** is covered by the *scalp.* The scalp is attached anteriorly, at the level of the **eyebrows,** to the **supraorbital ridges.** The scalp continues posteriorly over the area commonly called the forehead and across the crown (vertex) of the top of the head to the **superior nuchal** *(nu'kal)* **line,** a ridge on the back of the skull. Both the supraorbital ridge above the **orbit,** or socket of the eye, and the superior nuchal line at the back of the skull can be easily palpated. Laterally, the scalp covers the **temporal region** and terminates at the fleshy portion of the ear called the **auricle,** or **pinna.** The temporal region is the attachment for the **temporalis muscle,** which can be palpated as you repeatedly clench your jaw. This region is clinically important since it is a point of entrance to the cranial cavity in many surgical procedures.

Only a portion of the scalp is covered with hair, and the variable **hairline** is genetically determined. The scalp is clinically important because of the dense connective tissue layer beneath the skin that supports nerves and vessels. When the scalp is cut, the wound is held together by the connective tissue, but at the same time the vessels are held open, resulting in profuse bleeding.

Face

The face, or **facies** *(fa'she-ēz),* (fig. 12.6) is composed of four regions: the **ocular region,** which includes the eye and associated structures; the **auricular region,** which includes the ear; the **nasal region,** which includes the external and internal structures of the nose; and the **oral region,** which includes the mouth and associated structures.

The skin of the face is relatively thin and contains many sensory receptors, particularly in the oral region. Certain facial regions also have numerous **sweat glands** and **sebaceous** *(se-ba'shus)* (oil-secreting) **glands.** Acne is a serious facial dermatological problem for many teenagers. Facial hair appears over most of the facial region in males after they go through puberty; unwanted facial hair may occur sparsely on some females and can become a social problem.

The muscles of facial expression are important for the way they affect surface features. Various emotions are

cranium: Gk. *kranion,* skull
nuchal: L. *nucha,* nape of neck

Figure 12.7. The surface anatomy of the ocular region.

<table>
<tr><td colspan="2">Table 12.2 Surface anatomy of the ocular region</td><td colspan="2"></td></tr>
</table>

Structure	Comments	Structure	Comments
Eyebrow	Ridge of hair that superiorly arches the eye. It protects the eye against sunlight and is important in facial expression.	Iris	Circular, colored, muscular portion of the eyeball that surrounds the pupil. It reflexly regulates the amount of incoming light.
Eyelids	Movable folds of skin and muscle that cover the eyeball anteriorly. They assist in lubricating the anterior surface of the eyeball and reflexly close to protect the eyeball.	Pupil	Opening in the center of the iris through which light enters the eyeball
		Palpebral fissure	Space between the eyelids when they are open
Eyelashes	Row of hairs on the margin of eyelid. They prevent airborne substances from contacting eyeball.	Subtarsal sulcus	Groove beneath the eyelid that parallels the margin of the lid. It traps small foreign particles that contact the conjunctiva.
Conjunctiva	Thin mucous membrane that covers the anterior surface of the eyeball and lines the undersurface of the eyelids. It aids in reducing friction during blinking.	Medial commissure	Medial corner of the eye where the upper and lower eyelids come together
		Lateral commissure	Lateral corner of the eye where the upper and lower eyelids come together
Sclera	Outer fibrous layer of eyeball; the "white" of the eye that gives form to the eyeball	Lacrimal caruncle	Fleshy, pinkish elevation at the medial commissure. It contains sebaceous and sweat glands.
Cornea	Transparent anterior portion of eyeball. It is slightly convex to refract incoming light waves.		

Source: Van De Graaff, Kent M., *Human Anatomy.* © 1984 Wm. C. Brown Publishers, Dubuque, Iowa. All Rights Reserved. Reprinted by permission.

conveyed when these muscles are contracted. These muscles originate on the different facial bones and insert into the dermis (second major layer) of the skin. Excessive contraction of these muscles may cause permanent crease lines in the skin.

Because the organs of the facial region are so complex and specialized, there are professional fields of speciality associated with the various regions. Optometry and ophthalmology are concerned with the structure and function of the eye. Dentistry is entirely devoted to the health and functional and cosmetic problems of the oral region, particularly the teeth. An otorhinolaryngologist is an ear, nose, and throat specialist.

The *ocular region* includes the eyeball and associated structures. Most of the surface features of the ocular region protect the eye. **Eyebrows** protect against potentially damaging sunlight and mechanical blows; **eyelids** reflexly close to protect against objects or visual stimuli; **eyelashes** prevent airborne objects from contacting the eyeball; and the **lacrimal** *(lak'ri-mal)* **secretions** (tears) wash away chemical or foreign materials and prevent the surface of the eyeball from drying. Figure 12.7 depicts many of the surface features of the ocular region, and table 12.2 describes these structures.

The *auricular region* includes the visible surface structures as well as internal organs that function in hearing and in maintaining equilibrium. The fleshy **auricle,** or **pinna** *(pin'nah),* and the tubular opening into the middle ear, called the **external auditory meatus** *(me-a'tus),* are the only observable surface features of the auricular region. The rim of the auricle is called the **helix,** and the inferior portion is referred to as the **lobule.** The lobule is composed primarily of connective and fatty tissue and therefore can be easily pierced. For this reason, it is sometimes used when obtaining blood for a blood count. The **tragus** *(tra'gus)* is a small, posteriorly directed, projection partially covering and protecting the external auditory meatus. Further protection is provided by the many fine hairs that surround the opening into this canal. The head of the mandible can be palpated at the opening of the external auditory meatus by placing your little finger in the opening and then vigorously moving your jaw. The **mastoid process** can be palpated as a bony knob on the skull immediately posterior to the lobule of the auricle. This process serves as the point of attachment for the muscle that turns the head. Refer to figure 12.8 and table 12.3 for an illustration and discussion of other surface features of the auricular region.

The inspection of some of the internal structures of the ear is part of a routine physical examination and is performed using an otoscope. Earwax may accumulate in the canal, but this is a protective substance. It waterproofs the eardrum and because of its bitter taste is thought to be an insect repellent.

A few structural features of the *nasal region* are apparent from its surface anatomy (fig. 12.9 and table 12.4). The principal function of the nose is associated with the respiratory system, and the need for a permanent body opening to permit gaseous ventilation accounts for its surface features. The **root** (nasion) is located at about the level of the eyebrows where the nose begins to protrude from the forehead. The firm, narrow portion between the eyes is the **bridge** of the nose and is formed by the union of the nasal bones. The nose below this level has a pliable cartilaginous framework that maintains an opening. The tip of the nose is called the **apex,** and the region between the bridge and the apex is the **dorsum nasi.** As the lateral portion of the nose blends with the face, it is called the **nasofacial angle.** The **external nares** *(na'rēz),* or **nostrils,** are the paired openings into the nose. The **nasal septum** forms a partition between the nostrils, and the **ala** *(a'lah)* forms the flaired outer margin of each nostril.

---

meatus: L. *meatus,* path
helix: Gk. *helix,* to turn or roll
tragus: Gk. *tragos,* goat (because of the tuft of hairs on it)
ala: L. *ala,* winglike

Figure 12.8.   The surface anatomy of the auricular region.

Table 12.3 Surface anatomy of the auricular region

Structure	Comments
Auricle (pinna)	Expanded, fleshy portion of the ear projecting from the side of the head that funnels sound waves into the external auditory meatus.
Helix	Outer rim of auricle. It gives form and shape to the pinna.
Lobule	Inferior, fleshy portion of auricle
Tragus	Posterior cartilaginous projection of auricle. It partially covers and protects the external auditory meatus.
Antitragus	Small, cartilaginous anterior projection opposite the tragus
Antihelix	Semicircular ridge anterior to the greater portion of the helix
Concha	Depressed hollow of auricle. It funnels sound waves.
External auditory meatus	Canal extending inward to the tympanic membrane. It is slightly S-shaped and contains glands that secrete earwax for protection.
Triangular fossa	Triangular depression in superior portion of antihelix

Figure 12.9. The surface anatomy of the nasal and oral regions: (*a*) the nose and lips; (*b*) the teeth.

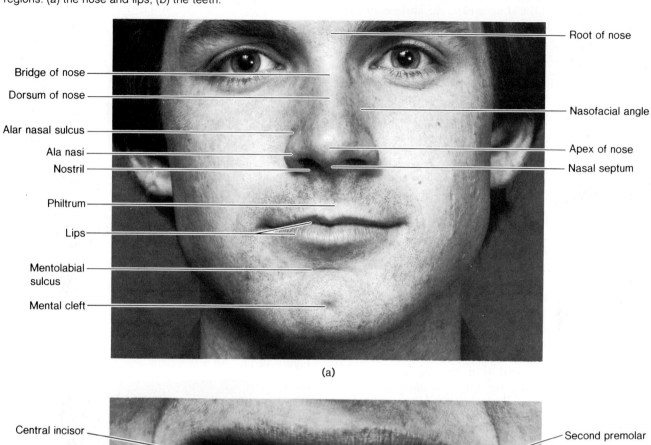

Root of nose

Bridge of nose

Dorsum of nose

Nasofacial angle

Alar nasal sulcus

Ala nasi

Apex of nose

Nostril

Nasal septum

Philtrum

Lips

Mentolabial sulcus

Mental cleft

(a)

Central incisor

Second premolar

Lateral incisor

First premolar

Canine

Second premolar

Central incisor

First premolar

Lateral incisor

Canine

(b)

Table 12.4 Surface anatomy of the nasal and oral regions

Structure	Comments	Structure	Comments
Root (nasion)	Superior attachment of nose to the cranium	External nares	External openings into nasal cavity
Bridge	Anterior bony framework of nose formed by union of nasal bones	Nasal septum	Partition between external nares
Dorsum nasi	Anterior ridge of nose	Ala	Laterally expanded border of external nare
Nasofacial angle	Anteriolateral boundary of nose as it blends with tissues of the face	Philtrum	Vertical depression in medial portion of upper lip
		Lips	Upper and lower anterior borders of the mouth
Apex	Terminal tip of nose	Chin (mental)	Anterior portion of lower jaw

Structures of the *oral region* that are important in surface anatomy include the fleshy upper and lower **labia** *(la'be-ah)*, lips, and the structures of the **oral cavity** that can be observed when the mouth is open. The lips are shown in figure 12.9, and the structures of the oral cavity are seen in figure 12.10.

The color of the lips and other mucous membranes of the oral cavity are diagnostic of certain body dysfunctions. The lips may appear pale in patients with severe anemia and bluish or cyanotic in persons lacking in oxygen. Pernicious anemia may cause a lemon yellow tint to the lips, as does jaundice.

1. *What are the boundaries of the cranial region, and why is this region clinically important?*
2. *Why do scalp wounds bleed so freely? How might this relate to infections?*
3. *What are the subdivisions of the facial region?*

Neck

Major organs are located within the flexible neck and several structures (the spinal cord, vessels, trachea, and esophagus) that are essential for body sustenance pass through the neck.

Objective 7. Discuss the functions of the neck.

Objective 8. List by name and location the triangles of the neck and the structures contained within these triangles.

The neck, or **collum,** is a complex region of the body that connects the head to the thorax. The spinal cord, digestive and respiratory tracts, and major vessels traverse this highly flexible area. Several organs and glands are located here as well. Remarkable musculature in the neck produces an array of movements. Because of this complexity, the neck is a clinically important area. The surface features provide landmarks for determining the location of internal structures.

Surface Features

The neck is divided into four regions: 1) *an anterior region,* called the **cervix,** which contains the digestive and respiratory tracts, the **larynx** *(lar'inks)* (voice box), vessels to and from the head, nerves, and the **thyroid** and **parathyroid glands;** 2) right and 3) left *lateral regions,* each composed of major neck muscles and **cervical** *(ser'vi-kal)* **lymph nodes;** and 4) a *posterior region,* referred to as the **nucha** *(nu'kah),* which includes the spinal cord, cervical vertebrae, and associated structures.

cervix: L. *cervix,* neck
larynx: Gk. *larynx,* upper windpipe

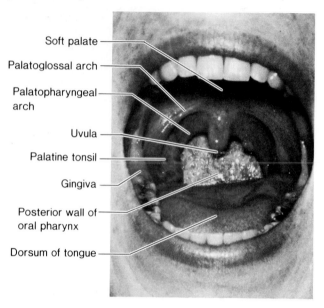

Figure 12.10.   Surface structures seen in the oral cavity with the mouth open. (Note an inflammation of the right palatine tonsil.)

Soft palate
Palatoglossal arch
Palatopharyngeal arch
Uvula
Palatine tonsil
Gingiva
Posterior wall of oral pharynx
Dorsum of tongue

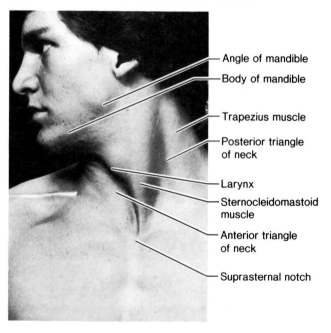

Figure 12.11.   An anterolateral view of the neck.

Angle of mandible
Body of mandible
Trapezius muscle
Posterior triangle of neck
Larynx
Sternocleidomastoid muscle
Anterior triangle of neck
Suprasternal notch

The most prominent structure of the cervix of the neck is the **thyroid cartilage** (Adam's apple) of the larynx (fig. 12.11). The thyroid cartilage is associated with the vocal cords and is larger in sexually mature males than in females, which accounts for the male's deeper voice. The **hyoid bone** can be palpated just above the larynx. Both of these structures are elevated during swallowing, which is

hyoid: Gk. *hyoeides,* U-shaped

Table 12.5 Location and contents of the triangles of the neck

Triangle	Boundaries	Contents
Anterior	Sternocleidomastoid muscle, median line of neck, inferior border of mandible	Four lesser triangles, salivary glands, larynx, trachea, thyroid glands, various vessels and nerves
Carotid	Sternocleidomastoid, posterior digastric, and omohyoid muscles	Carotid arteries, internal jugular vein, vagus nerve
Submandibular	Digastric muscle (both heads), inferior border of mandible	Salivary glands
Suprahyoid	Digastric muscle, hyoid bone (This is the only unpaired triangle of the neck.)	Muscles of the floor of the mouth, salivary glands and ducts
Muscular	Sternocleidomastoid and omohyoid muscles, midline of neck	Larynx, trachea, thyroid gland, carotid sheath
Posterior	Sternocleidomastoid and trapezius muscles, clavicle	Nerves and vessels
Occipital	Sternocleidomastoid, trapezius, and omohyoid muscles	Cervical plexus, accessory nerve
Subclavian	Sternocleidomastoid and omohyoid muscles; clavicle	Brachial plexus, subclavian artery

Source: Van De Graaff, Kent M., *Human Anatomy.* © 1984 Wm. C. Brown Publishers, Dubuque, Iowa. All Rights Reserved. Reprinted by permission.

one of the actions that directs food and fluid into the esophagus. You can note this action on yourself by gently cupping your fingers on the larynx, then swallowing. Below the thyroid cartilage is the **cricoid** *(kri'koid)* **cartilage** followed by the **trachea** *(tra'ke-ah)* (windpipe), both of which can be palpated. The trachea is clinically important in case a respiratory tube has to be inserted during a *tracheotomy.* The thyroid gland can be palpated on either side just below the level of the larynx. Pulsations of the **common carotid artery** can be observed and palpated on either side of the neck just lateral and a bit superior to the level of the larynx.

The **suprasternal notch** is the depression in the midline of the cervix just superior to the sternum. The two **clavicles** are obvious in all persons because they are subcutaneous (just under the skin).

The **sternocleidomastoid** and **trapezius muscles** are the prominent structures of each lateral region (fig. 12.11). The sternocleidomastoid muscle can be palpated its entire length when the head is turned to the side. The tendon of this muscle is especially prominent to the side of the suprasternal notch. The trapezius muscle can be felt if the shoulders are shrugged. An inflammation of the trapezius muscle causes a "stiff neck." If a person is angry or if a shirt collar is too tight, the **external jugular vein** can be seen as it courses obliquely across the sternocleidomastoid muscle. **Cervical lymph nodes** of the lateral neck region may become swollen and painful from infectious diseases of the oral or pharyngeal regions.

Most of the structures of the nuchal region are too deep to be of importance in surface anatomy. The spines of the lower cervical vertebrae, however, can be observed and palpated when the neck is flexed. In this same position, a firm ridge is raised, called the **ligamentum nuchae,** that extends anteriorly along the midline. Clinically, this region is tremendously important because of the debilitating damage to it from whiplash injury or a broken neck.

Triangles of the Neck

The triangles of the neck, created by the arrangement of specific muscles and bones, are clinically important because of the specific structures included in each. The structures of the neck that are important in surface anatomy have already been identified, however, so only a summary of the two major and six minor triangles will be presented in table 12.5 and depicted in figure 12.12. The sternocleidomastoid muscle obliquely transects the neck, dividing it into an **anterior triangle** and a **posterior triangle.** The apex of the anterior triangle is directed inferiorly. The median line of the neck forms the anterior boundary of the anterior triangle, the inferior border of the mandible forms its superior boundary, and the sternocleidomastoid muscle forms its posterior boundary. The posterior triangle is formed by the sternocleidomastoid muscle anteriorly and the trapezius muscle posteriorly, and the clavicle forms its base inferiorly.

Three structures traversing the neck are extremely important and potentially vulnerable. These structures are the common carotid artery, which carries blood to the head, the internal jugular vein, which drains blood from the head, and the vagus nerve, which conducts nerve impulses to visceral organs. These structures are protected in the neck by their deep position near the sternocleidomastoid muscle and by being enclosed in a tough connective tissue sheath called the **carotid sheath.**

1. List four functions of the neck, and state which body systems are located, in part, within the neck.
2. What are the structural regions of the neck? What structures are included in each region?
3. Using your knowledge of the triangles of the neck, describe where you would palpate to feel a pulse, the trachea, cervical lymph nodes, and the thyroid gland.

Figure 12.12. Triangles of the neck. (a) the two large triangular divisions; (b) the six lesser triangular subdivisions; (c) the detailed muscular anatomy of the neck.

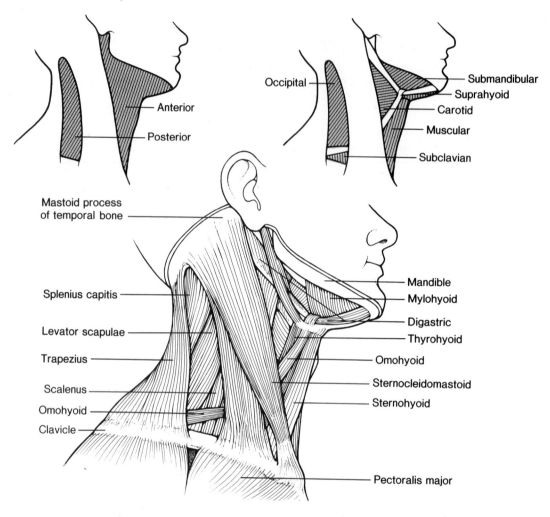

Anterior

Posterior

Occipital

Submandibular

Suprahyoid

Carotid

Muscular

Subclavian

Mastoid process of temporal bone

Splenius capitis

Levator scapulae

Trapezius

Scalenus

Omohyoid

Clavicle

Mandible

Mylohyoid

Digastric

Thyrohyoid

Omohyoid

Sternocleidomastoid

Sternohyoid

Pectoralis major

Torso

The locations of vital visceral organs in the cavities of the torso make the surface anatomy of this body region especially important.

> *Objective 9.* Identify by observation or palpation various surface features of the torso.
> *Objective 10.* List the auscultation sites of the thorax and abdomen.

The **torso** *(tor'so),* or **trunk,** is divided into the **back** (dorsum), **thorax** *(tho'raks)* (chest), **abdomen** *(ab'do-men)* (venter), and **pelvis.** A region called the **perineum** *(per''i-ne'um)* forms the floor of the pelvis and includes the external genitalia. The surface anatomy of these regions is particularly important in determining the location and condition of the visceral organs. The usability of some of these surface features varies depending on age, sex, and body weight.

Back

No matter how obese a person is, a **median furrow** can be seen on the back along with some of the **vertebral spines** (fig. 12.13). The entire series of vertebral spines can be observed if the vertebral column is flexed. This position is important in determining vertebral-column defects (see clinical comments in chap. 14). The back of the **scapula** presents other important surface landmarks. The base of the spine of the scapula is level with the third thoracic vertebra, and the inferior angle of the scapula is even with the seventh thoracic vertebra. Several muscles of the

Figure 12.13. The surface anatomy of the back. (*a*) flexion of the upper extremities; (*b*) extension of the upper extremities. (Note the three distinct heads of the deltoid muscle: posterior, middle, and anterior.)

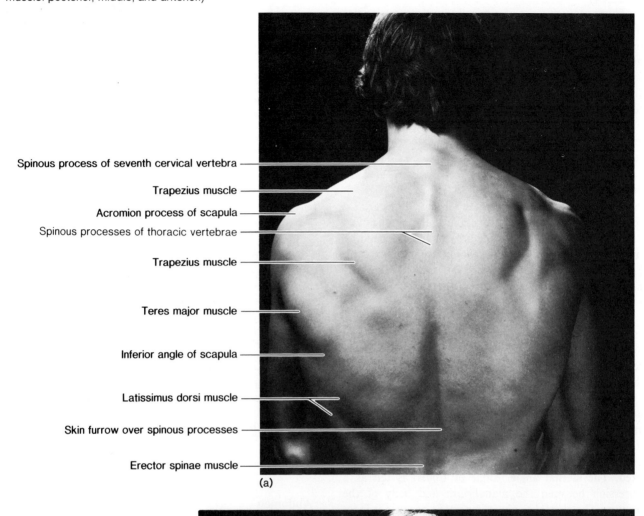

Spinous process of seventh cervical vertebra

Trapezius muscle

Acromion process of scapula

Spinous processes of thoracic vertebrae

Trapezius muscle

Teres major muscle

Inferior angle of scapula

Latissimus dorsi muscle

Skin furrow over spinous processes

Erector spinae muscle

(a)

Posterior portion of deltoid muscle

Trapezius muscle

Infraspinatus muscle

Teres major muscle

Triangle of auscultation

Latissimus dorsi muscle

(b)

scapula can be observed on a lean, muscular person and are identified in figure 12.13. Many of the ribs and muscles that attach to the ribs can be seen in a lateral view (fig. 12.14).

There are two pairs of clinically important triangles on the back. The **triangle of auscultation** (fig. 12.13) is bound by the trapezius muscle, the latissimus dorsi muscle, and the medial (vertebral) border of the scapula. Because there is a space between the superficial back muscles in this area, heart and respiratory sounds are not muffled by the muscles when a stethoscope is placed here. The **lumbar triangle** (not illustrated) is bound medially by the latissimus dorsi muscle, laterally by the external oblique muscle, and inferiorly by the iliac crest.

Thorax

The leading causes of death in the United States are associated with disease or dysfunction of the thoracic organs. With the exception of the breasts and surrounding lymph nodes, the organs of the thorax are within the rib cage. Because of the location of the thoracic visceral organs and their clinical importance, the surface anatomy of the thorax is extremely important.

The flexible rib cage presents several bony landmarks that can be observed or palpated (fig. 12.15). The paired clavicles and the suprasternal notch have already been identified as being important surface features of the neck. These structures are also important in the thoracic region as reference points for counting the ribs. Many of the ribs can be seen on a thin person. All of the ribs except the first, and at times the twelfth, can be palpated. The sternum is composed of three separate bones (manubrium, body, xiphoid process), each of which can be palpated. The **sternal angle** (angle of Louis) is felt as a depression between the manubrium and body of the sternum. The sternal angle is important because it is located at the level of the second rib. The articulation between the body of the sternum and the xiphoid process, called the **xiphisternal** *(zif″ĭ-ster′nal)* **joint,** is positioned over the lower border of the heart and the diaphragm. The **costal margin** of the rib cage is the lower oblique boundary and can be easily identified when a person inhales and holds his breath.

The nipples in the male are located at the fourth intercostal spaces (the area between the fourth and fifth ribs) and about 10 cm (4 in.) from the midline. They vary in position in sexually mature women according to age, size, and the pendulousness of the breasts. The position of the left nipple in males is an important landmark for knowing where to listen to various heart sounds and for determining if the heart is enlarged. For diagnostic purposes, an imaginary line, called the **midclavicular line,** can be extended vertically from the middle of the clavicle through the nipple. Several superficial chest muscles can be observed or palpated and are therefore important surface features. These muscles and the structures described above are depicted in figure 12.15.

Figure 12.14.   An anterolateral view of the torso and axilla.

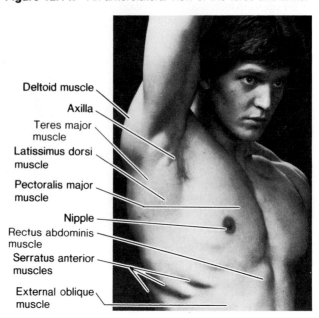

Deltoid muscle
Axilla
Teres major muscle
Latissimus dorsi muscle
Pectoralis major muscle
Nipple
Rectus abdominis muscle
Serratus anterior muscles
External oblique muscle

In addition to helping one know where to listen with a stethoscope to heart sounds, surface features of the thorax are important for auscultations of the lungs, X rays, tissue biopsies, sternal taps for bone marrow studies, or thoracic surgery. Although the anatomical features of the rib cage are quite consistent, slight deformities and asymmetries do occur. These generally cause no disability and require no treatment. Most of the abnormalities are congenital and include conditions such as a projecting sternum ("pigeon breast") or a receding sternum ("funnel chest").

Abdomen

The abdomen is the portion of the body between the diaphragm and the pelvis. The abdomen does not have a bony framework as does the thorax, so the surface anatomical features are not as well defined. Bony landmarks of both the thorax and pelvis are used when referring to abdominal structures (fig. 12.16). The right costal margin of the rib cage is located over the liver on the right side, and the left costal margin is positioned over the stomach on the left. The xiphoid process is important because from this point a tendinous, midventral raphe, called the **linea alba,** extends the length of the abdomen to attach to the **symphysis** *(sim′fĭ-sis)* **pubis.** The symphysis pubis can be palpated at the anterior union of the two halves of the pelvic girdle. The **navel,** or **umbilicus,** is the former site of attachment of the fetal umbilical cord and is located along the linea alba. The linea alba separates the paired, straplike **rectus abdominis** muscles, which can be seen when a person flexes the abdomen (such as when doing sit-ups).

linea alba: L. *linea,* line; *alba,* white
umbilicus: L. *umbilicus,* navel
navel: O. E. *nafela,* umbilicus

Figure 12.15.   The surface anatomy of the anterior thoracic region of the male.

Trapezius muscle
Clavicle
Acromion process of scapula
Pectoralis major muscle
Areola
Nipple
Costal margin
Serratus anterior muscle

Suprasternal notch
Supraclavicular fossa
Deltoid muscle
Body of sternum
Xiphoid process

Figure 12.16.   The surface anatomy of the anterior abdominal region.

Serratus anterior muscle
Linea semilunaris
Umbilicus
McBurney's point
Iliac crest
Anterior superior iliac spine
Site of inguinal ligament

Xiphoid process
Costal margin
Linea alba
External oblique muscle
Tendinous intersections of rectus abdominis muscle

Clinically, the linea alba is a favored site for abdominal surgery because an incision made along this line severs no muscles and few vessels and nerves. The linea alba also heals readily (it has been said that only a zipper would provide a more convenient entry to the abdominal cavity).

The lateral margin of the rectus abdominis muscle can be observed on some persons, and the surface line produced is called the **linea semilunaris.** The **external oblique** muscle is the superficial layer of the muscular abdominal wall. The **iliac crest** is subcutaneous and can be palpated along its entire length. The highest point of the crest lies opposite the body of the fourth lumbar vertebra, an important level in spinal anesthesia. Another important landmark is **McBurney's point,** located one-third of the distance from the right anterior-superior iliac crest to the navel (see fig. 12.16). This point overlies the appendix of the intestinal tract.

The abdominal region is frequently divided into nine regions or four quadrants in order to describe the location of internal organs and to clinically identify the sites of various pains or conditions. These regions have been adequately described in chapter 1 (see figs. 1.17, 1.18).

Although the position of the umbilicus is relatively consistent in all persons, its shape and health is not. Embryological remains of the umbilicus may cause clinical problems such as an opening to the outside, called a **fistula,** or herniation of some of the abdominal contents. Acquired umbilical hernias may develop in children who have a weak abdominal wall in this area, or umbilical hernias may develop in pregnant women because of the extra pressure exerted at this time. A depressed umbilicus on an obese person is difficult to keep clean, and so various types of infections may occur there.

1. Which structures of the torso can be readily observed? Which can be palpated?
2. Describe the location of the common auscultation sites of the torso.
3. Define linea alba, costal margin, linea semilunaris, and McBurney's point.

Pelvis and Perineum

The surface features of the pelvic region are important primarily to identify reproductive organs and clinical problems of these organs.

Objective 11. Describe the location of the perineum and list the organs of the pelvic and perineal regions.

The important bony structures of the **pelvis** *(pel"vis)* include the crest of the ilium and symphysis pubis, located anteriorly, and the ischium and os coccyx, which are palpable posteriorly. An **inguinal ligament** extends from the crest of the ilium to the symphysis pubis and is clinically important because hernias occur along it. Although the inguinal ligament cannot be seen, an oblique groove overlying the ligament is an apparent surface feature.

The **perineum** *(per"i-ne'um)* (see fig. 1.19) is the region that contains the external sex organs and the anal opening. The surface features of this region are further discussed in chapters 27 and 28. The surface anatomy of the perineum of a female becomes particularly important during parturition (childbirth).

1. Define the term perineum. What structures are located within the perineum?
2. List three body systems that have openings within the pelvic region.

Shoulder and Upper Extremity

The surface anatomy of the shoulder and upper extremity is of clinical importance because of frequent trauma to these body regions. Vessels of the upper extremity are also used as pressure points and for intravenous injections or blood withdrawal.

Objective 12. Identify by observation or palpation various surface features of the shoulder and upper extremity.
Objective 13. Discuss the clinical importance of the axilla, cubital fossa, and wrist.

Shoulder

The scapula, clavicle, and proximal portion of the humerus form the shoulder, and portions of each of these bones are important surface landmarks in this region. Posteriorly, the spine of the scapula and acromion process are subcutaneous and easily located. The **angle of acromion** *(ah-kro'me-on)* is frequently used by clinicians to make measurements down the brachium (fig. 12.17).

The acromion process and the clavicle, as well as several large shoulder muscles, can be seen anteriorly (fig. 12.18). The rounded curve of the shoulder is formed by the deltoid muscle covering the greater tuberosity of the humerus. The deltoid muscle is frequently a site for intramuscular injections. The large pectoralis major muscle is prominent as it crosses the shoulder joint and attaches to the humerus. A small depression, called the **deltopectoral triangle,** is situated below the outer third of the clavicle and is bounded on either side by the deltoid and pectoralis major muscles.

inguinal: L. *inguinalis,* groin
acromion: Gk. *akros,* extreme, tip; *omion,* small shoulder

Figure 12.17. A posterior view of the shoulder showing the spine of the scapula, acromion, and angle of acromion. The angle of acromion is used as a reference for locating certain brachial structures such as the surgical neck of the humerus or the axillary nerve, which is positioned against the humerus, 5 cm below the angle of acromion.

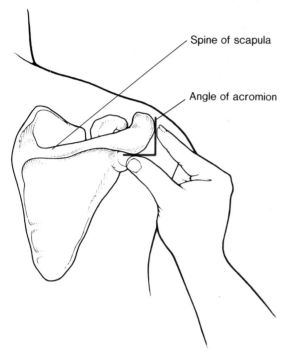

Spine of scapula

Angle of acromion

Figure 12.18. An anterior view of the shoulder region.

Supraclavicular fossa

Trapezius muscle

Acromion process of scapula

Deltoid muscle

Pectoralis major muscle

Insertion of deltoid muscle

Triceps brachii muscle

Biceps brachii muscle

Axilla

The **axilla** is commonly called the armpit. This depressed region of the shoulder supports axillary hair in sexually mature individuals. The axilla is clinically important because of the subcutaneous position of vessels, nerves, and lymph nodes here. Two muscles form the anterior and posterior borders of this region (fig. 12.19). The **anterior axillary** *(ak′sĭ-lar″e)* **fold** is formed by the pectoralis major muscle, and the **posterior axillary fold** consists primarily of the latissimus dorsi muscle extending from the lumbar vertebrae to the humerus. Axillary lymph nodes are palpable in some persons.

In sexually mature females, the axillary tail of the mammary gland, which is positioned on the pectoralis major muscle, extends partially into the axilla. In doing a breast self-examination (see chapter 29), a woman should palpate the axillary area as well as the entire breast because the lymphatic drainage pathway is toward the axilla.

Brachium

Several muscles are clearly visible in the brachial region (figs. 12.19, 12.20). The belly of the biceps brachii muscle becomes prominent when the elbow is flexed with the palm upward. While the arm is in this position, the deltoid muscle can be traced as it inserts upon the humerus. The

triceps brachii muscle forms the bulk of the posterior surface of the brachium. A groove forms on the medial side of the brachium between the biceps and triceps muscles where pulsations of the brachial artery may be felt as it carries blood toward the forearm. This region is clinically important because it is where arterial blood pressure is taken with a sphygmomanometer. It is also the place to apply pressure in case of severe arterial hemorrhage in the forearm or hand.

Three bony prominences can be located in the region of the elbow (fig. 12.21). The medial and lateral epicondyles are processes on the humerus, while the olecranon is a proximal process of the ulna. When the elbow is extended, these prominences lie on the same straight plane; when the elbow is flexed, these points form a triangle. The ulnar nerve can be palpated in the groove behind the medial epicondyle. This area is commonly known as the "funny bone" or "crazy bone."

The **cubital** *(ku′bĭ-tal)* **fossa** is the triangular depression on the anterior surface of the elbow region where the median cubital vein links the cephalic and basilic veins. These veins are subcutaneous and become more conspicuous if a proximal compression is applied. For this reason, these veins, particularly the median cubital, are an important location for the removal of venous blood for analyses and transfusions or for intravenous therapy (fig. 12.22).

Figure 12.19. An anterior view of the upper extremity.

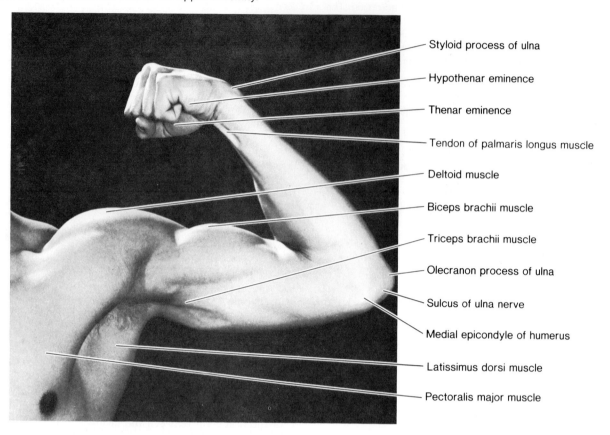

- Styloid process of ulna
- Hypothenar eminence
- Thenar eminence
- Tendon of palmaris longus muscle
- Deltoid muscle
- Biceps brachii muscle
- Triceps brachii muscle
- Olecranon process of ulna
- Sulcus of ulna nerve
- Medial epicondyle of humerus
- Latissimus dorsi muscle
- Pectoralis major muscle

Figure 12.20. A lateral view of the upper extremity.

- Deltoid muscle
- Biceps brachii muscle
- Long head of triceps brachii muscle
- Lateral head of triceps brachii muscle
- Brachioradialis muscle
- Lateral epicondyle of humerus
- Olecranon process of ulna

Figure 12.21.   A posterior view of the elbow (joint slightly extended).

Lateral epicondyle

Extensor carpi radialis longus muscle

Site of palpation of head of radius

Olecranon process of ulna

Medial epicondyle

Basilic vein

Cephalic vein

Figure 12.22.   An anterior view of the forearm and hand.

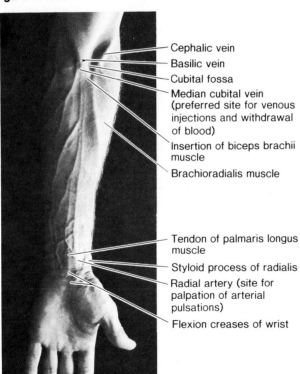

Cephalic vein

Basilic vein

Cubital fossa

Median cubital vein (preferred site for venous injections and withdrawal of blood)

Insertion of biceps brachii muscle

Brachioradialis muscle

Tendon of palmaris longus muscle

Styloid process of radialis

Radial artery (site for palpation of arterial pulsations)

Flexion creases of wrist

Figure 12.23.   A posterior view of the forearm and hand.

Triceps brachii muscle

Olecranon process of ulna

Styloid process of ulna

Brachioradialis muscle

Extensor carpi radialis longus and brevis muscles

First dorsal interosseous muscle

Forearm

Contained within the forearm are two parallel bones (the ulna and radius) and the muscles that control the movements of the hand. The forearm tapers distally toward the wrist, where the muscles give way to tendinous cords that attach to various bones of the hand. Several muscles of the forearm can be identified as surface features and are depicted in figures 12.22 and 12.23.

Figure 12.24. Palpation of the styloid processes of the ulna and radius. The radial styloid is always about 1 cm lower than that of the ulna, an important fact for a physician to know when treating a fractured bone of the forearm.

Radius   Ulna

Because of the frequency of fractures involving the forearm, bony landmarks are clinically important when setting broken bones. The ulna can be palpated its entire length from the olecranon to the distal styloid process. The distal one-half of the radius is palpable as the forearm is rotated, and its styloid process can be located using a technique illustrated in figure 12.24.

Nerves, tendons, and vessels are close to the surface at the wrist, making cuts to this area potentially dangerous. Tendons from four flexor muscles can be observed as surface features if the anterior forearm muscles are strongly contracted while making a fist. The tendons that can be observed along this surface, from lateral to medial, are from the following muscles: flexor carpi radialis, palmaris longus, flexor digitorum superficialis, and flexor carpi ulnaris. The median nerve going to the hand is located under the tendon of the palmaris longus muscle (see fig. 12.22), and the ulnar nerve is lateral to the tendon of the flexor carpi ulnaris. The radial artery lies along the surface of the radius immediately medial to the tendon of the flexor carpi radialis. This is the artery commonly used when taking a pulse. By careful palpation, pulsations can also be detected in the ulnar artery lateral to the tendon of the flexor carpi ulnaris.

Figure 12.25. The anatomical "snuffbox."

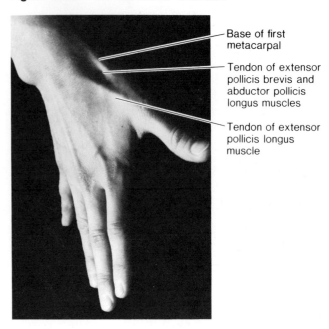

Base of first metacarpal

Tendon of extensor pollicis brevis and abductor pollicis longus muscles

Tendon of extensor pollicis longus muscle

Two tendons that attach to the thumb can be seen on the posterior surface of the wrist as the thumb is extended backwards. The tendon of the extensor pollicis brevis is positioned anterolaterally along the thumb, and the tendon of the extensor pollicis longus lies posteromedially (fig. 12.25). The depression created between these two tendons as they are pulled taut is referred to as the "anatomical snuffbox." Pulsations of the radial artery can be detected in this depression.

> The median nerve, which serves the thumb, is the nerve in the forearm most commonly injured by stab wounds or the penetration of glass. If this nerve is severed, the muscles of the thumb are paralyzed and waste away, resulting in an inability to oppose the thumb in grasping.

Hand

Much of the surface anatomy of the hand, such as flexion creases, fingerprints, and fingernails, includes features of the skin that are discussed in chapter 8. Other surface features are the extensor tendons from the extensor digitorum muscle, which can be seen going to each of the fingers on the back side of the hand as the hand is extended (fig. 12.26). The "knuckles" of the hand are the distal ends of the second through the fifth metacarpal bones. Each of the joints of the fingers and the individual phalanges can be palpated. The **thenar** *(the'nar)* **eminence** is the thickened, muscular portion of the hand forming the base of the thumb (fig. 12.27).

thenar: Gk. *thenar*, palm

Figure 12.26.   A posterior view of the hand.

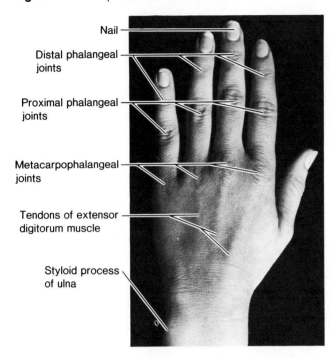

Nail

Distal phalangeal joints

Proximal phalangeal joints

Metacarpophalangeal joints

Tendons of extensor digitorum muscle

Styloid process of ulna

Figure 12.27.   An anterior view of the hand.

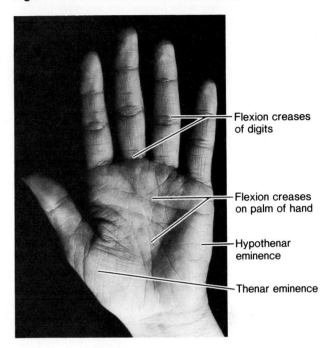

Flexion creases of digits

Flexion creases on palm of hand

Hypothenar eminence

Thenar eminence

---

1. List the clinically important structures that can be observed or palpated in the shoulder and upper extremity.
2. Describe the locations of the axilla, brachium, cubital fossa, and wrist.
3. Bumping the ulnar nerve causes a "tingling" sensation along the medial portion of the forearm to the little finger of the hand. What does this tell you about its distribution?
4. Which of the two bones of the forearm is more stationary as the arm is rotated?

Buttock and Lower Extremity

The massive bones and muscles of the buttock and lower extremity are important as weight bearers and locomotors. Many of the surface features of these regions are important in relation to locomotion or locomotor dysfunction.

Objective 14. Identify by observation or palpation various surface features of the buttock and lower extremity.

Objective 15. Discuss the clinical importance of the buttock, femoral triangle, popliteal space, ankle, and arches of the foot.

Buttock

The superior borders of the **buttocks** *(but'oks),* or **gluteal** *(gloo'te-al)* **region,** are formed by the iliac crests (fig. 12.28). Each crest can be palpated medially to the level of the second sacral vertebra. From this point, the **natal cleft** extends vertically to separate the buttocks into two prominences, each formed by pads of fat as well as by the massive gluteal muscles. An ischial tuberosity can be palpated in the lower portion of each buttock. In a sitting position, the ischial tuberosities support the weight of the body; but when standing, these processes are covered by the gluteal muscles. The sciatic nerve, which is the major nerve to the lower extremity, lies under the gluteus maximus muscle. The inferior border of the gluteus maximus muscle forms the **fold of the buttock.**

Because of the thickness of the gluteal muscles and the rich blood supply, the buttock is a preferred site for intramuscular injections. Care must be taken, however, not to inject into the sciatic nerve. For this reason, the surface landmark of the iliac crest is important. The injection is usually administered 5–7 cm (2–3 in.) below the iliac crest in what is known as the upper lateral quadrant of the buttock.

---

buttock: O. E. *buttuc,* end or rump

Figure 12.28. The buttocks and the posterior aspect of the thigh. (Note the relation of the angle of the elbow joint to the pelvic region, which is characteristic of females.)

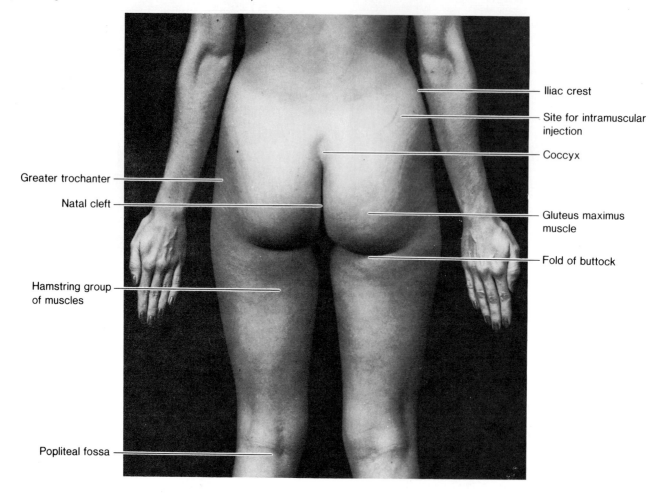

Iliac crest

Site for intramuscular injection

Coccyx

Gluteus maximus muscle

Fold of buttock

Greater trochanter

Natal cleft

Hamstring group of muscles

Popliteal fossa

Thigh

The femur is the only bone of the thigh, but there are three groups of thigh muscles. The anterior group of muscles, referred to as the **quadriceps,** extend the leg when they are contracted (fig. 12.29). The medial muscles are the **adductors,** and when contracted they draw the thigh medially. The **hamstrings** are positioned on the posterior aspect of the thigh (see fig. 12.28) and serve to extend the hip as well as flex the leg when they are contracted. The tendinous attachments of the hamstrings can be palpated along the posterior aspect of the knee joint when the leg is flexed. The hamstrings or their attachments are often injured in athletic events.

The **femoral** *(fem'or-al)* **triangle** is an extremely important part of the surface anatomy of the thigh. It can be seen as a depression inferior to the location of the inguinal *(ing'gwĭ-nal)* ligament on the anterior surface in the upper part of the thigh (see fig. 19.40). The major

vessels of the leg as well as the femoral nerve traverse through this region. Hernias are frequent in this area. More importantly, the femoral triangle is an arterial pressure point where it is vital to apply pressure in the case of uncontrolled hemorrhage of the lower extremity.

The greater trochanter of the femur can be palpated on the lateral, upper surface of the thigh. At the knee, the lateral and medial condyles of the femur and tibia can be identified. The patella (kneecap) can be easily located within the patellar tendon anterior to the knee joint. Stress or injury to this joint may cause swelling, commonly called "water on the knee."

The depression on the posterior aspect of the knee joint is referred to as the **popliteal** *(pop''li-te'al)* **space** (fig. 12.30). This area becomes more clinically important in elderly persons who suffer degenerative conditions. Aneurysms of the popliteal artery are common, as are popliteal abscesses due to infected lymph nodes. The small saphenous vein as it traverses the popliteal space may become varicose in elderly persons.

Figure 12.29.   An anterior view of the thigh and knee.

Quadriceps femoris group of muscles

Adductor muscles

Vastus lateralis muscle

Vastus medialis muscle

Lateral condyle of femur

Medial condyle of femur

Patella

Patellar ligament

Tibial tuberosity

Anterior border of tibia

Figure 12.30.   (a) lateral and (b) posterior surfaces of the right leg.

Iliotibial tract

Vastus lateralis muscle

Biceps femoris muscle

Patella

Biceps femoris tendon of insertion

Head of fibula

Gastrocnemius muscle

Anterior tibialis muscle

Peroneus muscle group

Tendon of biceps femoris muscle

Popliteal space

Gastrocnemius muscle

(a)

(b)

Figure 12.31. The right leg and foot: (*a*) a lateral view; (*b*) a medial view; (*c*) an anterior view; (*d*) a posterior view.

Tendon of peroneus longus
Lateral malleolus
Extensor digitorum brevis muscle
Tuberosity of the base of the fifth metatarsal
Tendons of extensor digitorum longus

(a)

Gastrocnemius muscle
Soleus muscle
Great saphenous vein
Medial malleolus
Calcaneus
Tubercle of navicular
Longitudinal arch

(b)

Peroneus longus
Lateral malleolus
Medial malleolus
Muscle belly of extensor digitorum brevis
Tendons of extensor digitorum longus
Tendon of extensor hallicus longus
First metatarsal phalangeal joint

(c)

Gastrocnemius muscle
Medial malleolus
Lateral malleolus
Posterior tibial artery
Achilles tendon
Tuberosity of base of fifth metatarsal

(d)

Leg

Portions of the tibia and fibula, the bones of the leg, can be observed as surface features. The medial surface and anterior border (commonly called the shin) of the tibia are subcutaneous and are palpable throughout their length. At the ankle, the medial malleolus of the tibia and the lateral malleolus of the fibula are easy to observe as prominent eminences (fig. 12.31). Of clinical importance in setting fractures of the leg is knowing that the top of the medial malleolus lies about 1.3 cm (0.6 in.) proximal to the level of the tip of the lateral malleolus.

The heel is not part of the leg but the posterior portion of the calcaneus bone. It needs to be mentioned with the leg, however, because of its functional relationship to it. The tendo calcaneus (Achilles tendon) is the strong, cordlike tendon that attaches to the calcaneus from the calf of the leg. The muscle forming the "belly" of the calf is the gastrocnemius. Pulsations from the posterior tibial artery can be detected by palpating between the medial malleolus and the calcaneus.

The superficial veins of the leg can be observed on many persons. The great saphenous vein can be seen subcutaneously along the medial aspect of the leg. The less conspicuous, small saphenous vein drains the lateral surface of the leg. If these veins become excessively enlarged, they are called varicose veins.

> Leg injuries are common among athletes. Shin splints, probably the result of a stress fracture or periosteum damage of the tibia, is a common condition in runners. A fracture of one or both malleoli is caused by a severe twisting of the ankle region. Skiing fractures are generally caused by strong torsion forces on the shafts of the tibia or fibula.

Foot

The feet are adapted to support the weight of the body, to maintain balance, and to function mechanically during locomotion. The structural features and surface anatomy of the foot are indicative of these functions. The **arch of the foot** is located on the medial portion of the plantar surface (fig. 12.31) and provides a spring effect when locomoting. The head of the first metatarsal bone forms the medial "ball" of the foot just proximal to the hallux (big toe). The fifth metatarsal bone forms much of the lateral border of the plantar surface of the foot. The tendons of the extensor digitorum longus muscle can be seen along the dorsal surface of the foot, especially if the toes are elevated. Pulsations of the dorsal pedis artery can be palpated on the dorsal surface of the foot between the first and second metatarsal bones. The individual phalanges of the toes, the joints between these bones, and the toenails are obvious surface landmarks.

> 1. What surface features form the boundaries of a buttock?
> 2. List the clinically important structures that can be observed or palpated in the buttock and lower extremity.
> 3. Describe the anatomic location where each of the following could be observed or palpated: the distal tendinous attachments of the hamstring muscles; the greater trochanter; the greater and lesser saphenous veins; the femoral, posterior tibial, and dorsal pedis arteries; and the medial malleolus.

Chapter Summary

I. Introduction to Surface Anatomy
   Surface anatomy is concerned with identifying body structures through observation and palpation. Surface anatomy has tremendous application to physical fitness and medical diagnosis and treatment.
   A. Most of the bones of the skeleton are palpable and provide landmarks for other anatomical structures.
   B. The effectiveness of observation and palpation in studying a person's surface anatomy is dependent on the thickness of the hypodermis and is due to the amount of subcutaneous adipose tissue present.

II. Surface Anatomy of the Newborn
   Certain aspects of the surface anatomy of a neonate are of clinical importance in ascertaining the degree of physical development, general health, and possible congenital abnormalities.
   A. The posture of a full-term, normal neonate is one of flexion.
   B. Portions of the skin and subcutaneous tissues are edematous. Vernix caseosa covers the body of a neonate, and lanugo may be present on the head, neck, and back.
   C. The fontanels, liver, kidneys, and testes of a male should be palpable.

III. Head
   The surface anatomy of the head is of concern because the head contains vital sense organs and provides openings into the respiratory and digestive systems. Of further concern are the esthetic appearances, trauma, disease, and dysfunctions that involve this part of the body.
   A. Surface features of the cranium include the forehead, the crown, the temporalis muscles, and the hair and hair line.
   B. The face is composed of the ocular region that surrounds the eye, the auricular region of the ear, the nasal region serving the respiratory system, and the oral region of the mouth.

IV. Neck
Major organs are located within the flexible neck, and structures that are essential for body sustenance pass through the neck.
   A. The neck consists of a cervix, right and left lateral regions, and a nucha.
   B. Two major and six minor triangles, which contain specific structures, are located on both sides of the neck.

V. Torso
Vital visceral organs in the torso make the surface anatomy of this region especially important.
   A. The median furrow is observable and the vertebral spines and scapulae are palpable on the back.
   B. Palpable structures of the thorax include the sternum, the ribs, and the costal margins.
   C. The important surface anatomy of the abdomen includes the linea alba, umbilicus, costal margins, and iliac crests.

VI. Pelvis and Perineum
The surface features of the pelvic region are important to identify the reproductive organs and their clinical problems.

   A. The crest of the ilium, symphysis pubis, and inguinal ligament are important pelvic landmarks.
   B. The perineum is the region that contains the external genitalia and the anal opening.

VII. Shoulder and Upper Extremity
The surface anatomy of the shoulder and upper extremity is important because of frequent trauma to these regions. Vessels of the upper extremity are also used as pressure points and for intravenous injections or blood withdrawal.
   A. The scapula, clavicle, and humerus are palpable in the shoulder.
   B. The axilla is clinically important because of the vessels, nerves, and lymph nodes located there.
   C. The brachial artery is an important pressure point in the brachium, and the median cubital vein is important for the removal of blood or for intravenous therapy.
   D. The ulna, radius, and their processes are palpable landmarks of the forearm.
   E. The knuckles, fingernails, and tendons for the extensor muscles of the forearm can be observed on the posterior aspect of the hand.

   F. Flexion creases, fingernails, and the thenar eminence are important surface features on the anterior surface.

VIII. Buttock and Lower Extremity
The massive bones and muscles in the buttock and lower extremity serve as weight bearers and locomotors. Many of the surface features of these regions are important in relation to locomotion or locomotor dysfunction.
   A. The prominences of the buttocks are formed by the gluteal muscles and are separated by the natal cleft.
   B. The thigh has three muscle groups: quadriceps, adductors, and hamstrings. The femoral triangle and popliteal spaces are clinically important surface landmarks.
   C. The structures of the leg include the tibia and fibula, the muscles of the calf, and the saphenous veins.
   D. The surface anatomy of the foot includes structures adapted to support the weight of the body, maintain balance, and function during locomotion.

Review Activities

Objective Questions

1. Eyebrows are located on the
   (a) palpebral fissure
   (b) subtarsal sulcus
   (c) scalp
   (d) supraorbital ridges
   (e) c and d
2. Which of the following structures is *not* part of the auricle (pinna) of the ear? The
   (a) tragus
   (b) ala
   (c) lobule
   (d) helix
3. Which of the following clinical-structural word pairs is *incorrectly* matched?
   (a) cleft lip—philtrum
   (b) broken nose—nasion
   (c) pierced ears—lobule
   (d) "black eye"—concha
4. The conjunctiva
   (a) covers the entire eyeball
   (b) is a thick nonmucous membrane
   (c) secretes tears
   (d) none of the above
5. Which of the following could *not* be palpated within the cervix of the neck?
   (a) larynx
   (b) hyoid bone
   (c) trachea
   (d) cervical vertebrae

6. Palpation of a pulse to the head is best accomplished within the
   (a) carotid triangle
   (b) occipital triangle
   (c) suprahyoid triangle
   (d) submandibular triangle
   (e) muscular triangle
7. Which nerve traverses behind the medial epicondyle of the humerus?
   (a) ulnar
   (b) median
   (c) radial
   (d) brachial
   (e) cephalic
8. Which of the following surface features could *not* be observed on obese people?
   (a) suprasternal notch
   (b) scapular muscles
   (c) clavicles
   (d) vertebral spines
   (e) natal cleft
9. Which pairs of muscles form the anterior and posterior borders of the axilla?
   (a) deltoid—pectoralis minor
   (b) biceps brachii—triceps brachii
   (c) latissimus dorsi—pectoralis major
   (d) triceps brachii—pectoralis major
   (e) latissimus dorsi—deltoid
10. Varicose veins occur when which of the following becomes excessively enlarged?
   (a) saphenous veins
   (b) tibial veins
   (c) external iliac vein
   (d) popliteal vein
   (e) all of the above

Essay Questions

1. List four surface features of the cranium, and explain the relationship of the scalp to the cranium.
2. Identify the four regions of the face, and list at least two surface features of each region.
3. Which surface features can be observed on the torso on any person, regardless of how obese he or she is?
4. Identify the two major triangles of the neck and list the associated six minor triangles. Discuss the importance of knowing the contents and boundaries of these triangles.
5. Describe the locations of the arteries that can be palpated as they pulsate in the following regions: neck, brachial, forearm, thigh, and ankle. Which of these are considered clinical pressure points?
6. Describe the locations and clinical importance of the following regions: cubital fossa, femoral triangle, axilla, perineum, and popliteal space.

Selected Readings

Anderson, J. E. 1978. *Grant's atlas of anatomy.* 7th ed. Baltimore: The Williams and Wilkins Co.

Basmajian, J. V. 1980. *Grant's method of anatomy.* 10th ed. Baltimore: The Williams and Wilkins Co.

————. 1977. *Surface anatomy: an instruction manual.* Baltimore: The Williams and Wilkins Co.

Clemente, C. D. 1981. *Anatomy: a regional atlas of the human body.* 2d ed. Baltimore: Urban and Schwarzenberg.

Hall-Craggs, E. C. B. 1985. *Anatomy as a basis for clinical medicine.* Baltimore: Urban and Schwarzenberg.

Last, R. J. 1977. *Anatomy, regional and applied.* Boston: Little, Brown and Co.

Royce, J. 1973. *Surface anatomy.* Philadelphia: F. A. Davis Co.

Snell, R. S. 1978. *Atlas of clinical anatomy.* Boston: Little, Brown and Co.

Sobotta, J., and F. Figge. 1974. *Atlas of human anatomy.* 9th ed. New York: Hafner Press.

Van De Graaff, K. M. 1984. *Human Anatomy.* Dubuque: Wm. C. Brown Co.

Woodburne, R. T. 1978. *Essentials of human anatomy.* 6th ed. New York: Oxford Univ. Press.

Reference Figures

Regional Anatomy

Figure 8. Anterior view of neck and torso.

Inferior border of mandible

Mylohyoid m.

Submandibular gland

Internal jugular vein

Sternohyoid m.

Omohyoid m.

Sternocleidomastoid m.

Heart

Lung

Pericardium (cut)

Laryngeal prominence

Platysma m.

Deltoid m.

Pectoralis minor m.
(rib cage removed)

Pectoralis major m.
(rib cage removed)

Cephalic vein

Triceps brachii m.

Biceps brachii m.

Diaphragm

Liver

Gallbladder

Transverse colon

Greater omentum

Small intestine

Anterior sigmoid colon
(normal variation)

Figure 9. Viscera of thorax.

Pectoralis major
and minor mm.

Right lung

Pericardium (cut)

Right phrenic
arteries and veins

Diaphragm

Pulmonary trunk

Anterior cardiac
arteries and veins

Left lung

Heart

Left coronary
artery and
cardiac vein
(branches)

Cardiac notch

Apex of heart

Figure 10. Viscera of abdomen.

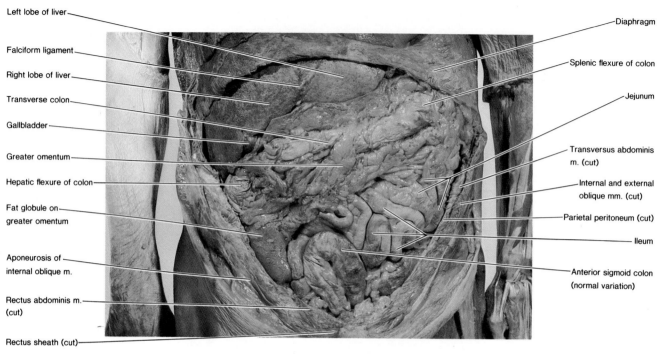

Left lobe of liver

Falciform ligament

Right lobe of liver

Transverse colon

Gallbladder

Greater omentum

Hepatic flexure of colon

Fat globule on greater omentum

Aponeurosis of internal oblique m.

Rectus abdominis m. (cut)

Rectus sheath (cut)

Diaphragm

Splenic flexure of colon

Jejunum

Transversus abdominis m. (cut)

Internal and external oblique mm. (cut)

Parietal peritoneum (cut)

Ileum

Anterior sigmoid colon (normal variation)

Figure 11. Posterior view of shoulder and thoracic musculature.

Trapezius m.

Infraspinatus m.

Posterior deltoid m.

Teres minor m.

Teres major m.

Rhomboideus m.

Triceps brachii m. (long head)

Triangle of auscultation

Latissimus dorsi m.

Levator scapulae m.

Rhomboideus minor m.

Supraspinatus m.

Clavicle

Spine of scapula

Infraspinatus m.

Teres minor m.

Teres major m.

Rhomboideus major m.

Figure 12.   Musculature of lower back.

Latissimus dorsi m.

Spinalis thoracis m.

Longissimus thoracis m.

External intercostal m.

Iliocostalis lumborum m.

External oblique m.

Iliac crest

Gluteus medius m.

Figure 13.   Axillary region.

Deltoid m.

Musculocutaneous nerve

Coracobrachialis m.

Biceps brachii m. (short head)

Median nerve

Ulnar nerve

Latissimus dorsi m.

Circumflex scapular artery

Circumflex scapular vein

Pectoralis major and minor mm. (reflected)

Axillary vein

Brachial veins

Subscapular vein

Posterior humeral circumflex vein

Thoracodorsal vein

Thoracodorsal artery

Serratus anterior m.

Figure 14. Posterior view of left forearm and hand.

Brachioradialis m.
Extensor carpi radialis longus tendon
Extensor carpi radialis brevis m.
Extensor digitorum m.
Abductor pollicis longus m.
Extensor pollicis brevis m.
Extensor pollicis longus m.
Radius
Extensor retinaculum
Extensor carpi radialis longus tendon
Extensor pollicis longus tendon
Extensor pollicis brevis tendon
First dorsal interosseous m.

Extensor carpi ulnaris m.
Extensor digiti minimi m.
Ulna
Extensor carpi radialis brevis tendon
Extensor indicis tendon
Extensor digiti minimi tendon
Extensor digitorum tendons
Interteninous connections

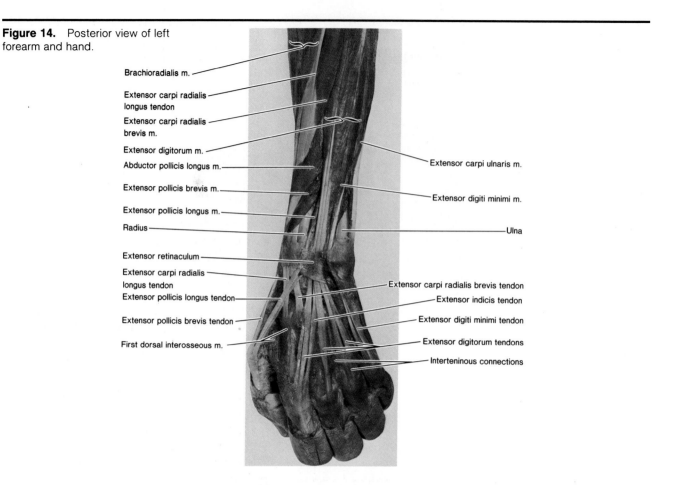

Figure 15. Anterior view of left forearm and hand.

Flexor carpi ulnaris m.
Extensor carpi ulnaris m.
Flexor digitorum superficialis m.
Pisiform bone
Abductor digiti minimi m.
Flexor digiti minimi m.
Opponens digiti minimi m.
Lumbrical m.
Flexor digitorum superficialis tendon
Flexor digitorum profundus tendon
Fibrous digital sheath

Palmaris longus tendon
Flexor carpi radialis tendon
Pronator quadratus m.
Extensor pollicis brevis tendon
Extensor pollicis longus tendon
Abductor pollicis brevis m.
Flexor pollicis brevis m.
Adductor pollicis m. (oblique head)
Adductor pollicis m. (transverse head)
Opponens pollicis m.

Figure 16. Posterior view of right
buttock and upper thigh.

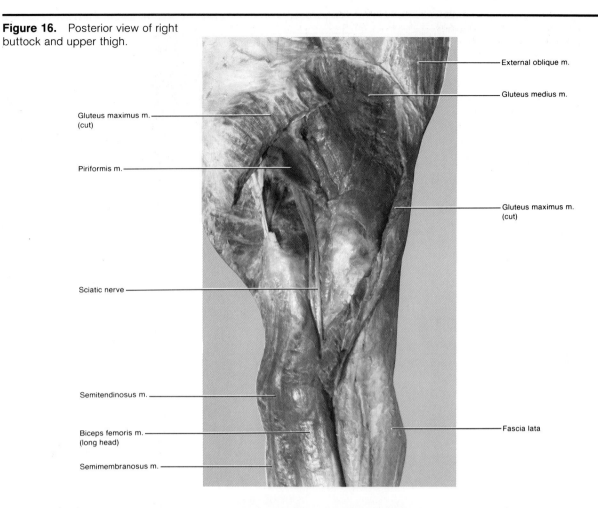

External oblique m.

Gluteus medius m.

Gluteus maximus m.
(cut)

Piriformis m.

Gluteus maximus m.
(cut)

Sciatic nerve

Semitendinosus m.

Biceps femoris m.
(long head)

Fascia lata

Semimembranosus m.

Figure 17. Anterior view of left thigh.

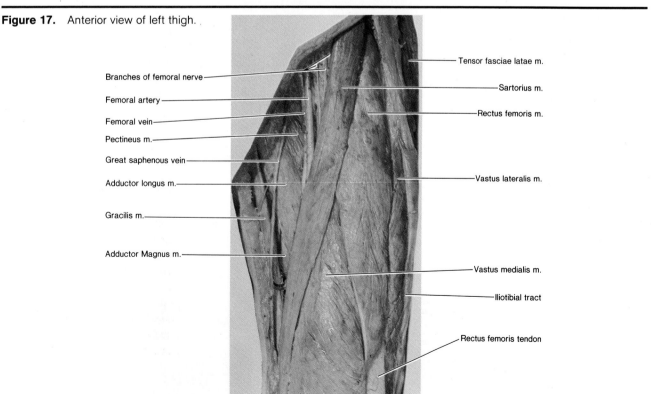

Branches of femoral nerve

Femoral artery

Femoral vein

Pectineus m.

Great saphenous vein

Adductor longus m.

Gracilis m.

Adductor Magnus m.

Tensor fasciae latae m.

Sartorius m.

Rectus femoris m.

Vastus lateralis m.

Vastus medialis m.

Iliotibial tract

Rectus femoris tendon

Figure 18. Medial view of right knee and surrounding musculature.

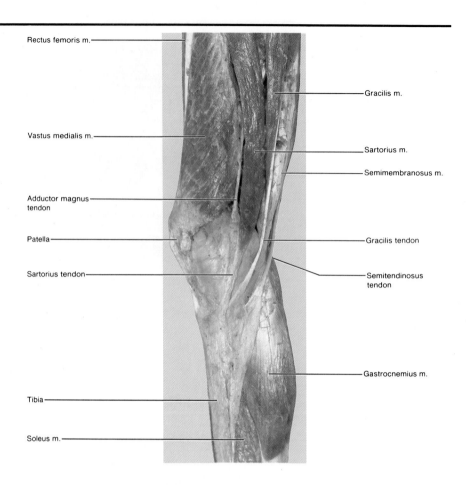

Rectus femoris m.

Gracilis m.

Vastus medialis m.

Sartorius m.

Semimembranosus m.

Adductor magnus tendon

Patella

Gracilis tendon

Sartorius tendon

Semitendinosus tendon

Gastrocnemius m.

Tibia

Soleus m.

Figure 19. Postero-lateral view of right knee musculature and popliteal space.

Semitendinosus m.

Biceps femoris m. (long head)

Sciatic nerve

Fascia lata

Semimembranosus m.

Popliteal artery

Semitendinosus tendon

Common peroneal nerve

Tibial nerve

Medial head of gastrocnemius m.

Lateral head of gastrocnemius m.

Peroneus longus m.

Figure 20. Anterior view of right ankle and foot.

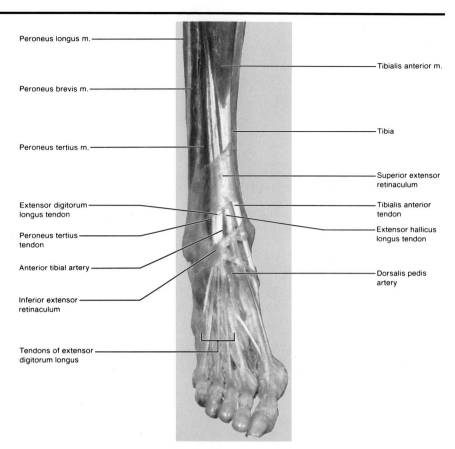

Peroneus longus m.

Peroneus brevis m.

Peroneus tertius m.

Extensor digitorum longus tendon

Peroneus tertius tendon

Anterior tibial artery

Inferior extensor retinaculum

Tendons of extensor digitorum longus

Tibialis anterior m.

Tibia

Superior extensor retinaculum

Tibialis anterior tendon

Extensor hallicus longus tendon

Dorsalis pedis artery

Figure 21. Medial view of right ankle and foot.

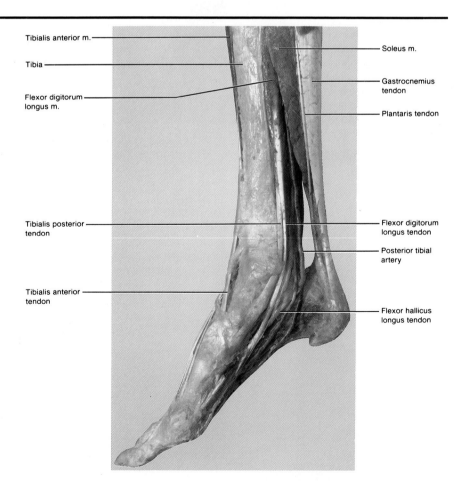

Tibialis anterior m.

Tibia

Flexor digitorum longus m.

Tibialis posterior tendon

Tibialis anterior tendon

Soleus m.

Gastrocnemius tendon

Plantaris tendon

Flexor digitorum longus tendon

Posterior tibial artery

Flexor hallicus longus tendon

Unit 3

Integration and Control

The chapters of Unit III are concerned with those body systems that regulate the activities of the other body systems. The nervous system and its sensory receptor organs are discussed along with the endocrine system. These chapters describe how specific organs help a person to respond appropriately to external and internal stimuli and to maintain homeostasis.

The chapters and principal concepts of Unit III follow:

Chapter 13—Functional Organization of Neural Tissue

Neurons and neuroglial cells comprise the nervous system, which is specialized to transmit nerve impulses along neurons and to activate precise chemical responses at synapses.

Chapter 14—Central Nervous System

The brain and spinal cord comprise the central nervous system (CNS), which functions in the body's orientation and coordination, assimilation of experiences, and programming of instinctual behavior.

Systems of the Human Body

13 Functional Organization of

Concepts

The central nervous system, including the brain and spinal cord, forms from the embryonic neural tube. The peripheral nervous system, including the cranial and spinal nerves, is formed predominately from the neural crest.

The dendrites of a neuron receive stimulation, and the axon conducts nerve impulses. Neuroglial cells provide structural and functional support for the activities of neurons.

The permeability of the axon membrane to specific ions is regulated by gates. An increase in permeability results in ion currents across the membrane that produce electrical changes, which can be propagated from one region to another.

Synaptic transmission may be either electrical or chemical. The predominant type are chemical synapses, in which neurotransmitters are released from the endings of nerve fibers in response to stimulation by action potentials.

Acetylcholine can stimulate depolarization of skeletal muscle cells and neurons by opening the chemically regulated gates. This depolarization serves as a stimulus for the production of action potentials in the postsynaptic cell.

The actions of catecholamine neurotransmitters is mediated by cyclic AMP produced by the postsynaptic cell. There are also many putative neurotransmitters.

Excitatory inputs can summate with each other and with inhibitory postsynaptic potentials. The release of neurotransmitters can be modified by experience and other factors.

Neural Tissue

Organization and Development of the Nervous System

The embryonic neural tube forms the brain and spinal cord, which together constitute the central nervous system. The cranial and spinal nerves, formed predominately from neural crest, comprise the peripheral nervous system.

> *Objective 1.* Describe the divisions of the nervous system.
> *Objective 2.* List the five developmental regions of the brain, and discuss the embryonic formation of these regions.
> *Objective 3.* Describe the development of the spinal cord.
> *Objective 4.* Define a dermatome, and describe the clinical significance of the pattern of dermatome innervation in the body.

The nervous system is divided into the **central nervous system (CNS),** which includes the **brain** and **spinal cord,** and the **peripheral** *(pĕ-rif'er-al)* **nervous system (PNS),** which includes the **cranial nerves** arising from the brain and the **spinal nerves** arising from the spinal cord (fig. 13.1).

The **autonomic nervous system (ANS)** is a functional subdivision of the entire nervous system. The controlling centers of the ANS are within the brain and are considered part of the CNS, and the nerve portions of the ANS are subdivided into the **sympathetic** and **parasympathetic** divisions.

The first indication of nervous tissue development occurs about seventeen days following conception when a thickening appears along the entire dorsal length of the embryo. This thickening, called the **neural plate** (fig. 13.2), differentiates and eventually gives rise to all of the neurons and to most of the **neuroglial** *(nu-rog'le-al)* **cells,** which are the supporting cells of the nervous system.

As development progresses, the midline of the neural plate invaginates to become the **neural groove.** At the same time, there is a proliferation of cells along the lateral margins of the neural plate, which become the thickened **neural folds.** The neural groove continues to deepen as the neural folds elevate. By the twentieth day, the neural folds meet and fuse at the midline, and the neural groove becomes a **neural tube.** The neural tube, for a short time, is open both cranially and caudally. These openings, called **neurophores** (fig. 13.2), close during the fourth week. Once formed, the neural tube separates from the surface ectoderm and eventually develops into the CNS (brain and spinal cord).

The **neural crest** forms from the neural folds as they fuse longitudinally along the dorsal midline. The neural crest is positioned between the surface ectoderm and the neural tube. Most of the PNS (cranial and spinal nerves) forms from the neural crest. Some neural crest cells break away from the main tissue mass and migrate to other locations where they differentiate into motor nerve cells of the sympathetic nervous system or into Schwann cells, which are a type of neuroglial cell important in the PNS.

Figure 13.1. The nervous system consists of the central nervous system (brain and spinal cord) and the peripheral nervous system (cranial nerves and spinal nerves). The nerves within the extremities are also part of the peripheral nervous system.

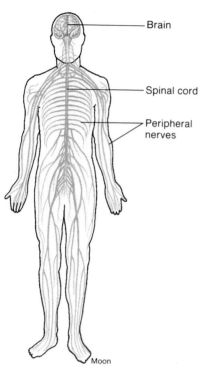

Nervous system
Function: regulation of
all body activities:
learning and memory

Figure 13.2. The early development of the nervous system from embryonic ectoderm. (a) a dorsal view of an 18-day-old embryo showing the formation of the neural plate. (b) a transverse section through an 18-day-old embryo showing differentiation of the neural plate into the neural groove and

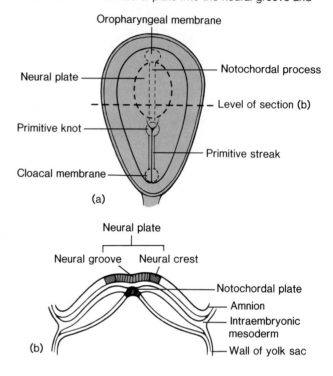

Figure 13.3. The developmental sequence of the brain. During the fourth week, the three principal regions of the brain are formed. During the fifth week, a five-regioned brain develops, and specific structures begin to form.

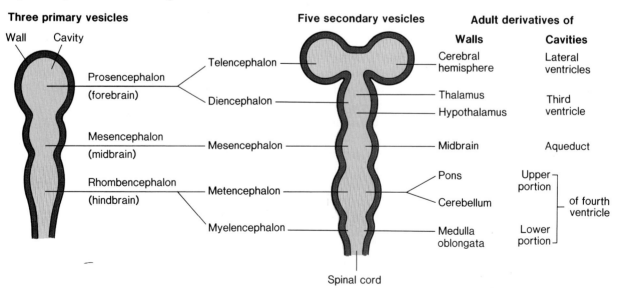

paired neural folds. (*c*) a dorsal view of a 22-day-old embryo showing the rostral and caudal neuropores. (*d–f*) transverse sections through a 22-day-old embryo from the locations indicated in (*c*). Note the amount of fusion of the neural tube at the various levels of the 22-day-old embryo. Note also the relationship of the notochord to the neural tube.

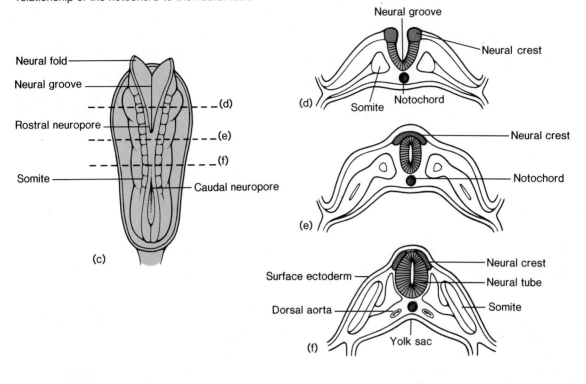

Figure 13.4. Pronounced flexures form in the brain by the end of the fifth week.

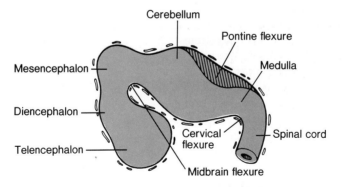

Development of the Brain

The brain begins its embryonic development as the cephalic end of the neural tube starts to grow rapidly and to differentiate (fig. 13.3). By the middle of the fourth week, three distinct swellings are evident: the **prosencephalon** *(pros''en-sef'ah-lon)* (forebrain), the **mesencephalon** (midbrain), and the **rhombencephalon** (hindbrain). Further development during the fifth week results in the formation of five specific regions: the **telencephalon** and the **diencephalon** *(di''en-sef'ah-lon)* derive from the forebrain; the mesencephalon remains unchanged; and the **metencephalon** and **myelencephalon** form from the hindbrain.

The rapid growth of the brain during the fourth and fifth weeks causes marked bendings or flexures to appear (fig. 13.4). The most prominent of these is the **midbrain flexure** along the ventral aspect of the mesencephalon. The **cervical flexure** occurs at the junction of the myelencephalon and the spinal cord. The less prominent **pontine** *(pon'tēn)* **flexure** occurs between the midbrain and the cervical flexure in the roof of the hindbrain.

During the developmental transformation of the brain, hollow chambers known as **ventricles** form within each region. The **first** and **second ventricles** (lateral ventricles)

. the forebrain, a narrow **third ventricle** forms in . midbrain, and the cavity of the hindbrain becomes the **fourth ventricle.** The ventricles of the brain are continuous with each other and with the **central canal** of the spinal cord.

Once the five principal regions of the brain are developed, rapid differentiation occurs in each of the regions, and shortly after all of the major structures of the brain are recognizable. The greatest amount of growth occurs within the telencephalon as its dorsal portions rapidly expand to form the two **hemispheres** of the **cerebrum** (fig. 13.5). The swelling of the cerebrum causes it to cover the diencephalon, the midbrain, and a portion of the hindbrain. By the thirty-fifth week, the lobes of the cerebrum are formed, and the surface of the cerebrum is distinctly convoluted.

Within the diencephalon three swellings form in the lateral walls of the third ventricle and become the **epithalamus** *(ep''ĭ-thal'ah-mus)*, the **thalamus,** and the **hypothalamus** *(hi''po-thal'ah-mus)*. The **pineal** *(pin'e-al)* **gland** also forms in the diencephalon as a midline diverticulum develops from the roof of the third ventricle. A portion of the **pituitary gland** develops from the ventral portion of the diencephalon.

Few developmental changes occur within the mesencephalon. The most obvious structures to develop are the four aggregates of neurons that become the paired **superior** and **inferior colliculi,** which are concerned with visual and auditory functions respectively. Fibers from the cerebrum extend through the midbrain to form two tracts called the **cerebral peduncles** *(pe-dung'k'l)*.

The walls of the metencephalon greatly expand to form the **cerebellum** dorsally and the **pons** ventrally. The cerebellum actually forms from the joining of a pair of dorsal swellings along the midline. The pons develops from numerous bands of nerve fibers. A vascular capillary network called the **choroid plexus** *(ko'roid plek'sus)* develops in the roof of the fourth ventricle. Similar plexuses form in the roof of the third ventricle and the medial walls of the lateral ventricles. These plexuses secrete **cerebrospinal** *(ser''ĕ-bro-spi'nal)* **fluid,** which circulates throughout the cavities of the CNS and within the **subarachnoid** *(sub''ah-rak'noid)* **space** outside the brain.

The caudal portion of the myelencephalon is continuous with and resembles the spinal cord. The entire adult myelencephalon is known as the **medulla oblongata** and forms from an aggregate of nerve fibers. The ventral **pyramids** of the myelencephalon form from the nerve fibers that extend from the developing cerebral cortex.

Development of the Spinal Cord

The spinal cord, like the brain, develops as the neural tube undergoes differentiation and specialization. Throughout the developmental process, the hollow central canal persists while the specialized white and gray matter forms

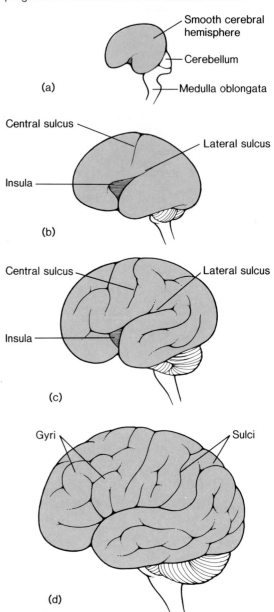

Figure 13.5. The sequential development of the cerebrum within the telencephalon. (*a*) 13 weeks, (*b*) 26 weeks, (*c*) 35 weeks, (*d*) newborn. Note the gradual formation of the cerebral lobes, including the internally positioned insula, and the progressive convolutions of the cerebral cortex.

(fig. 13.6). Changes in the neural tube become apparent during the sixth week as the lateral walls thicken to form a groove, called the **sulcus limitans,** along each lateral wall of the central canal. A pair of **alar** *(a'lar)* **plates** forms dorsal to the sulcus limitans, and a pair of **basal plates** forms ventrally. By the ninth week, the alar plates have specialized to become the **dorsal horns,** containing **afferent cell bodies,** and the basal plates have specialized to form the **ventral** and **lateral horns,** containing efferent cell bodies. Afferent fibers of spinal nerves conduct impulses toward the spinal cord, whereas efferent fibers conduct impulses away from the spinal cord.

Figure 13.6. The development of the spinal cord. (*a*) a transverse section through the neural tube of an embryo at twenty-three days. (*b*) the formation of the alar and basal plates is evident in a transverse section through the spinal cord at six weeks. (*c*) the central canal has reduced in size, and functional dorsal and ventral horns have formed at nine weeks.

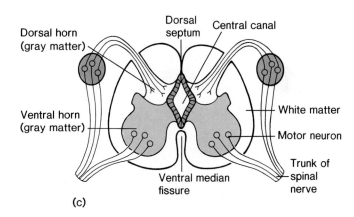

Mitotic activity within nervous tissue is completed during prenatal development. Thus, a person is born with all the neurons he is capable of producing. However, nervous tissue continues to grow and specialize after a person is born, particularly in the first several years of postnatal life.

The pattern of dermatome innervation is of clinical importance when a physician desires to anesthetize a particular portion of the body. Because adjacent dermatomes overlap in the trunk, at least three spinal nerves must be blocked to produce complete anesthesia in that region. Abnormally functioning dermatomes provide clues about injury to the spinal cord or specific spinal nerves. If a dermatome is stimulated but no sensation is perceived, the physician can infer that the injury involves the innervation to that dermatome.

1. List the five regions of the brain that develop during the fifth week.
2. Describe the locations of the ventricles within the brain.
3. Describe the embryonic origin of afferent and efferent cell bodies of spinal nerves.
4. Describe the dermatomes, and explain their clinical significance.

Development of the Peripheral Nervous System

Development of the peripheral nervous system produces the pattern of dermatomes within the body (fig. 13.7). A **dermatome** *(der'mah-tōm)* is an area of the skin innervated by all the cutaneous neurons of a certain spinal or cranial nerve. Most of the scalp and face is innervated by sensory fibers from the trigeminal nerve. With the exception of the first cervical nerve (Cl), all the spinal nerves are associated with specific dermatomes. Dermatomes are consecutive in the neck and thoracic regions. In the trunk, however, adjacent dermatome innervations overlap. The apparently uneven dermatome arrangement in the appendages is due to the uneven rate of nerve growth into the limb buds. Actually the limbs are segmented, and dermatomes overlap only slightly.

---

dermatome: Gk. *derma*, skin; *tomia*, a cutting

Figure 13.7. The pattern of dermatomes and the peripheral distribution of spinal nerves. (*a*) an anterior view; (*b*) a posterior view.

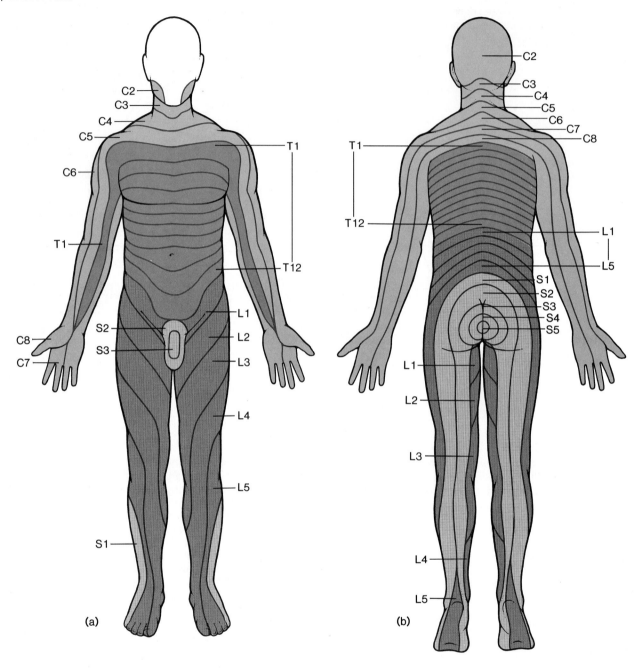

Histology of the Nervous System

Neurons have many forms but all contain dendrites for reception and an axon for the conduction of nerve impulses. Different types of neuroglial cells support the neurons, both structurally and functionally. Neurons contain axons that may be sensory and/or motor in function.

Objective 5. Describe the microscopic structure of a neuron.

Objective 6. Describe how a neurilemmal sheath and a myelin sheath are formed.

Objective 7. List the types of neuroglial cells, and describe their functions.

Objective 8. Describe the functions and locations of afferent and efferent nerve fibers.

The highly specialized and complex nervous system is composed of only two principal categories of cells, neurons and neuroglia. **Neurons** are the basic structural and functional units of the nervous system. They are specialized to respond to physical and chemical stimuli, conduct impulses, and release specific chemical regulators. Through these activities, neurons perform functions such as storing memory, thinking, and regulating other organs and glands. Neurons cannot divide mitotically, although some neurons can regenerate a severed portion or sprout small new branches under some conditions.

Neuroglia, or **glial** *(gli'al)* **cells,** are supportive cells in the nervous system that aid the function of neurons. Glial cells are about five times more abundant than neurons and have limited mitotic abilities.

Neurons

Although neurons vary considerably in size and shape, they generally have three principal components: (1) a cell body; (2) dendrites; and (3) an axon (figs. 13.8 and 13.9).

The **cell body,** or **perikaryon** *(per''i-kar'e-on),* is the enlarged portion of the neuron that contains the nucleus and nucleolus surrounded by cytoplasm. Besides containing organelles typically found in cells, the cytoplasm of neurons is characterized by the presence of **Nissl bodies** and filamentous strands of protein called **neurofibrils** *(nu''ro-fi'bril).* Nissl bodies are specialized layers of granular (rough) endoplasmic reticulum that synthesize fibrils and minute microtubules, which appear to be involved in transporting material within the cell. The cell bodies within the CNS are frequently clustered into regions called **nuclei** (not to be confused with the nucleus of a cell). Cell bodies in the PNS generally occur in clusters called **ganglia** (table 13.1).

Nissl body: from Franz Nissl, German neuroanatomist, 1860–1919
ganglion: Gk. *ganglion,* swelling

Dendrites are branched processes that extend from the cytoplasm of the cell body. Dendrites function to receive a stimulus and conduct impulses to the cell body. Some dendrites are covered with minute **dendritic spines,** which greatly increase their surface area and provide contact points for other neurons. The volume occupied by dendrites is referred to as the **dendritic zone** of a neuron.

The **axon,** or **nerve fiber,** is the second type of cytoplasmic extension from the cell body. An axon is a relatively long, cylindrical process that conducts impulses away from the cell body. The conical tapering region of the axon, where it originates from the cell body, is referred to as the **axon hillock.** Axons vary in length from a few millimeters in the CNS to over a meter between the distal portions of the extremities and the spinal cord. Side branches called **axon collaterals** extend a short distance from the axon. The cytoplasm of an axon contains many mitochondria, microtubules, and neurofibrils.

Proteins and other molecules are transported through the axon at faster rates than could be achieved by simple diffusion. This rapid movement is produced by two different mechanisms: axoplasmic flow and axonal transport (table 13.2). **Axoplasmic flow,** the slower of the two, results from rhythmic waves of contraction that push the cytoplasm from the axon hillock to the nerve endings. **Axonal transport,** which is more rapid and more selective, may occur in a retrograde as well as a forward direction. Indeed, such retrograde transport may be responsible for the movement of herpes virus, rabies virus, and tetanus toxin from nerve terminals into cell bodies.

Neuroglia

Unlike other organs that are "packaged" in connective tissue derived from mesoderm, the supporting **neuroglial (glial) cells** of the nervous system are derived from the same ectoderm that produce neurons. There are six categories of neuroglial cells: (1) **Schwann cells,** which form myelin sheaths around peripheral axons; (2) **oligodendrocytes** *(ol''i-go-den'dro-sīt),* which form myelin sheaths around axons in the CNS; (3) **microglia** *(mi-krog'le-ah),* which are phagocytic cells that migrate through the CNS and remove foreign and degenerated material; (4) **astrocytes,** which help regulate the passage of molecules from the blood to the brain; (5) **ependyma** *(ĕ-pen'di-mah),* which line the ventricles of the brain and the central canal of the spinal cord; and (6) **satellite cells,** which support neuron cell bodies within the ganglia of the PNS (table 13.3).

---

dendrite: Gk. *dendron,* tree branch
axon: Gk. *axon,* axis
neuroglia: Gk. *neuron,* nerve; *glia,* glue
Schwann cell: from Theodor Schwann, German histologist, 1810–1882
oligodendrocyte: Gk. *oligos,* few; L. *dens,* tooth; Gk. *kytos,* hollow (cell)
microglia: Gk. *mikros,* small; *glia,* glue
ependyma: Gk. *ependyma,* upper garment
satellite: L. *satelles,* attendant

Figure 13.8. The structure of two kinds of neurons. (*a*) a motor neuron; (*b*) a sensory neuron.

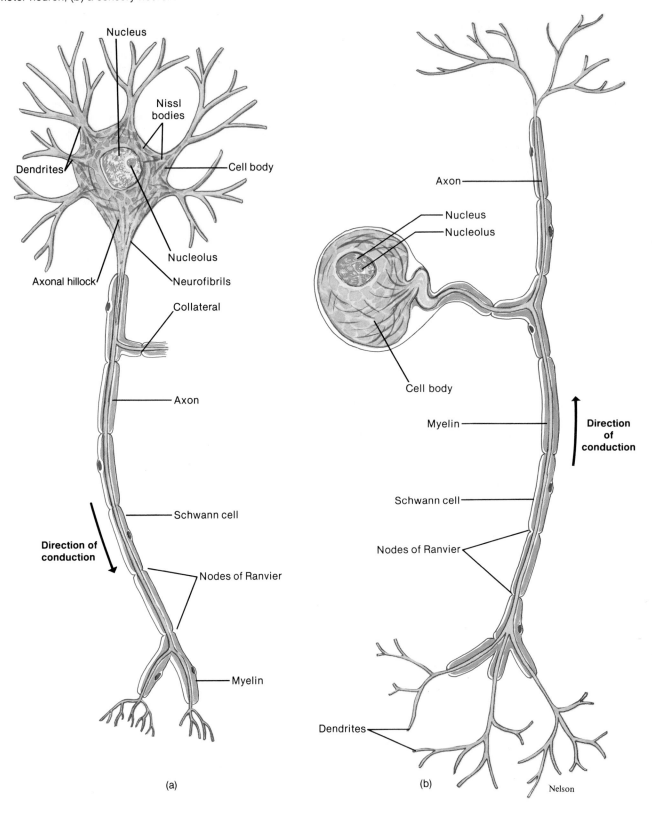

Nucleus

Nissl bodies

Dendrites

Cell body

Nucleolus

Axonal hillock

Neurofibrils

Collateral

Axon

Schwann cell

Direction of conduction

Nodes of Ranvier

Myelin

(a)

Axon

Nucleus

Nucleolus

Cell body

Myelin

Direction of conduction

Schwann cell

Nodes of Ranvier

Dendrites

(b)

Nelson

Figure 13.9. A photomicrograph of neurons from the anterior column of gray matter of the spinal cord (120X).

Table 13.1 Anatomical terms used in describing the nervous system

Term	Definition
Central nervous system (CNS)	Brain and spinal cord
Peripheral nervous system (PNS)	Nerves and ganglia
Sensory nerve fiber	Nerve that transmits impulses from a sensory receptor into CNS (an afferent fiber)
Motor nerve fiber	Nerve that transmits impulses from CNS to an effector organ, e.g., muscle (efferent)
Nerve	Cablelike collection of nerve fibers; may be "mixed" (contain both sensory and motor fibers)
Somatic motor nerve	Nerve that stimulates contraction of skeletal muscles
Autonomic motor nerve	Nerve that stimulates contraction (or inhibits contraction) of smooth muscle and cardiac muscles, and stimulates secretion of glands
Ganglion	Collection of neuron cell bodies located outside the CNS
Nucleus	Groupings of neuron cell bodies within the CNS
Tract	Collections of nerve fibers that interconnect regions of the CNS

Source: Fox, Stuart Ira, *Human Physiology.* © 1984 Wm. C. Brown Publishers, Dubuque, Iowa. All Rights Reserved. Reprinted by permission.

Table 13.2 Comparison of axoplasmic flow with axonal transport

Axoplasmic flow	Axonal transport
Transport rate comparatively slow (1–2 mm/day)	Transport rate comparatively fast (200–400 mm/day)
Molecules transported only from cell body	Molecules transported from cell body to axon endings and in reverse direction
Bulk movement of proteins in axoplasm, including microfilaments and tubules	Transport of specific proteins, mainly of membrane proteins and acetylcholinesterase
Transport accompanied by peristaltic waves of axon membrane	Transport dependent on cagelike microtubule structure within axon, and on actin and Ca^{++}

Source: Fox, Stuart Ira, *Human Physiology.* © 1984 Wm. C. Brown Publishers, Dubuque, Iowa. All Rights Reserved. Reprinted by permission.

Table 13.3 Structure and function of neuroglia

Type	Structure	Function
Astrocytes	Stellate with numerous processes	Form structural support between capillaries and neurons within the CNS; blood-brain barrier
Oligodendrocytes	Similar to astrocytes but with shorter and fewer processes	Form myelin in CNS; guide development of neurons within CNS
Microglia	Minute cells with few short processes	Phagocytize pathogens and cellular debris within CNS
Ependyma	Columnar cells that may have ciliated free surfaces	Line ventricles and central canal within CNS where cerebrospinal fluid is circulated by ciliary motion
Satellite cells	Small, flattened cells	Support ganglia within PNS
Schwann cells	Flattened cells arranged in series around axons or dendrites	Form myelin within the PNS

Source: Van De Graaff, Kent M., *Human Anatomy.* © 1984 Wm. C. Brown Publishers, Dubuque, Iowa. All Rights Reserved. Reprinted by permission.

Neurilemmal and Myelin Sheaths.   Some types of neuroglia form sheaths around axons. In the peripheral nervous system this function is served by Schwann cells; in the central nervous system oligodendrocytes serve a similar function. Axons that are surrounded by only a simple sheath of these neuroglial cells are said to be *unmyelinated.* The living sheath of neuroglial cells around an axon is called a **neurilemma** *(nu''ril-lem'mah).* Axons that are additionally surrounded by successive wrappings of the plasma membrane of these cells are said to be *myelinated* (fig. 13.10).

Axons that are less than two micrometers in diameter are usually unmyelinated. Larger axons are generally surrounded by a **myelin** *(mi'e-lin)* **sheath** in addition to their neurilemmal sheath. In the process of myelin sheath formation, Schwann cells (in the PNS) or oligodendrocytes (in the CNS) wrap around the axon so that their cytoplasm becomes squeezed toward the outermost region, like toothpaste rolled from the bottom of a tube (fig. 13.11). The neurilemmal sheath is thus outside the myelin sheath in myelinated axons. Nerves often contain a mixture of both myelinated and unmyelinated nerve fibers (fig. 13.12).

Astrocytes and the Blood-Brain Barrier.   Astrocytes are large, stellate cells with numerous cytoplasmic processes that radiate outward. These are the most abundant neuroglia in the CNS, and in some locations in the brain they constitute 90% of the nervous tissue.

Capillaries in the brain, unlike those of most other organs, do not have pores between adjacent endothelial cells. Molecules within these capillaries must thus be moved through the endothelial cells by active transport, endocytosis, and exocytosis. Astrocytes within the brain have numerous extensions, called **perivascular feet,** which surround most of the outer surface of the brain capillaries (fig. 13.13). Before molecules in the blood can enter neurons in the CNS, they may have to pass through both the endothelial cells and the astrocytes. Astrocytes, therefore, contribute to the **blood-brain barrier,** which is highly selective; some molecules are permitted to pass, whereas closely related molecules may not be allowed to cross the barrier.

The blood-brain barrier presents difficulties in the chemotherapy of brain diseases because drugs that could enter other organs may not be able to enter the brain. In the treatment of Parkinson's disease, for example, patients who need a chemical called dopamine in the brain must be given a precursor molecule called levodopa (L-dopa). This is because dopamine cannot cross the blood-brain barrier, whereas L-dopa can enter neurons and be changed to dopamine in the brain.

myelin: Gk. *myelos,* marrow
astrocyte: Gk. *aster,* star; *kytos,* hollow (cell)

Figure 13.10.   A myelinated neuron. (*a*) a myelin sheath forms to support and protect each peripheral neuron as Schwann cells tightly wrap around the axon. (*b*) myelin is composed of a lipid-protein substance within the Schwann cell. The nucleus of the Schwann cell remains toward the outside of the myelin, embedded within the membranous neurilemma.

Dendrite

Neuron nucleus

(a)

Neuron cell body

Axon

Unmyelinated region

Myelinated region

Schwann cell nucleus

(b)

Neurilemmal sheath

Axon

Neurofibrils

Myelin

Figure 13.11. The formation of a myelin sheath in a peripheral axon. The myelin sheath is formed by successive wrappings of the Schwann cell membranes, leaving most of the Schwann cell cytoplasm outside the myelin. The neurilemmal sheath of Schwann cells is thus located outside the myelin sheath.

Figure 13.12. An electron micrograph of unmyelinated (*left*) and myelinated axons (*right*).

Figure 13.13. A diagram showing the perivascular feet of astrocytes (a type of neuroglial cell), which cover most of the surface area of brain capillaries.

Figure 13.14.   The process of neuron regeneration. (*a*) if a neuron is severed through a myelinated axon, the proximal portion may survive, but (*b*) the distal portion degenerates. The myelin sheath provides a pathway for (*c*) the regeneration of an axon and (*d*) innervation is restored.

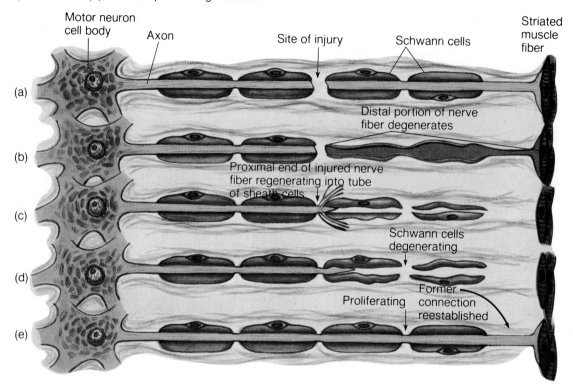

Figure 13.15.   The relationship between sensory and motor fibers of the peripheral nervous system (*PNS*) and the central nervous system (*CNS*).

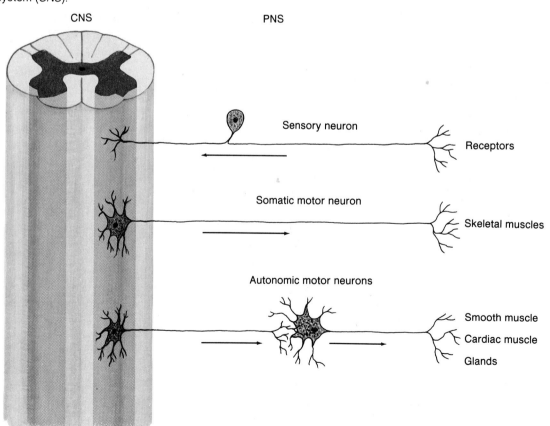

Figure 13.16. Three different types of neurons.

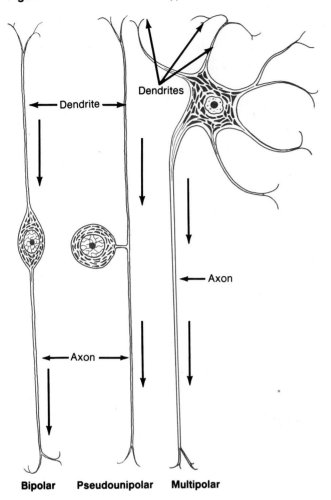

Bipolar   Pseudounipolar   Multipolar

Figure 13.17. The structure of a nerve.

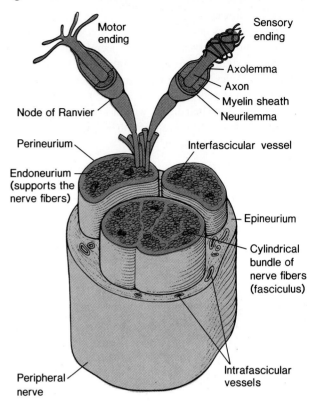

Regeneration of a Cut Axon. When an axon in a peripheral nerve is cut, the distal region of the axon that was severed from the cell body degenerates and is phagocytosed by Schwann cells. The Schwann cells then form a *regeneration tube* (fig. 13.14), as the part of the axon that is connected to the cell body begins to grow and exhibit ameboid movement. The Schwann cells of the regeneration tube are believed to secrete chemicals that attract the growing axon tip, and the tube helps to guide the regenerating axon to its proper destination. Even a severed major nerve may be surgically reconnected and the function of the nerve reestablished if the surgery is performed before tissue death. The CNS lacks Schwann cells, and central axons are generally believed to have a much more limited ability to regenerate than peripheral axons.

Classification of Neurons and Nerves

Neurons may be classified according to structure or function. The functional classification is based on the direction of conducted impulses. Sensory impulses originate in sensory receptors and are conducted by **sensory,** or **afferent, neurons** to the CNS. Motor impulses originate in the CNS and are conducted by **motor,** or **efferent, neurons** to a muscle or gland (fig. 13.15). **Association neurons,** or **interneurons,** are located between sensory and motor neurons and are found within the spinal cord and brain.

The structural classification of neurons is based upon the number of processes that extend from the cell body of the neuron (fig. 13.16). The spindle-shaped **bipolar neuron** has a process at either end; this type occurs in the retina of the eye. **Multipolar neurons** are the most common type and are characterized by several dendrites and one axon extending from the cell body. Motor neurons are a good example of this type. A **pseudounipolar neuron** has a single process that divides into two. Sensory neurons are pseudounipolar and have their cell bodies located in dorsal root ganglia of spinal and cranial nerves.

A **nerve** is a collection of nerve fibers outside the CNS. Fibers within a nerve are held together and strengthened by loose fibrous connective tissue (fig. 13.17). Each individual nerve fiber is enclosed in a connective tissue sheath called the **endoneurium** *(en''do-nu're-um).* A group of fibers, called a **fasciculus** *(fah-sik'u-lus),* is surrounded by a connective tissue sheath called a **perineurium.** The entire nerve is surrounded and supported by connective tissue called the **epineurium,** which contains tiny blood vessels and often adipose cells. Perhaps less than a quarter of the bulk of a nerve consists of nerve fibers: more than half is associated connective tissue, and approximately a quarter is the myelin that surrounds the nerve fibers.

---

fasciculus: L. diminutive of *fascis,* bundle

Most nerves are composed of both motor and sensory fibers and thus are called **mixed nerves.** Some of the cranial nerves, however, are composed either of sensory neurons only (**sensory nerves**) or of motor neurons only (**motor nerves**). Sensory nerves serve the special senses, such as taste, smell, sight, and hearing. Motor nerves conduct impulses to muscles, causing them to contract, or to glands, causing them to secrete.

Neurons and their fibers within nerves may be classified according to the area of innervation into the following scheme (fig. 13.18):

1. **Somatic afferent.** Sensory receptors within the skin, muscles, and joints receive stimuli and convey nerve impulses through somatic afferent fibers to the CNS for interpretation.
2. **Somatic efferent.** Impulses from the CNS through somatic efferent fibers cause the contraction of skeletal muscles.
3. **Visceral afferent.** Visceral afferent fibers convey impulses from visceral organs and blood vessels to the CNS for interpretation.
4. **Visceral efferent.** Visceral efferent fibers, also called **autonomic fibers,** are part of the autonomic nervous system. They originate in the CNS and innervate cardiac muscle, glands, and smooth muscle within the visceral organs.

---

1. *Briefly describe the functions of dendrites, cell bodies, and axons, and describe the differences between bipolar, pseudounipolar, and multipolar neurons.*
2. *Distinguish between myelinated and unmyelinated axons, and describe how a myelin sheath is formed.*
3. *Explain the nature of the blood-brain barrier, and describe its structure.*
4. *Explain the differences between a neuron, nerve fiber, and nerve. Describe how nerves are classified.*

Electrical Activity in Nerve Fibers

The permeability of a cell membrane to specific ions is regulated by structures in the membrane known as "gates." Increases in the permeability to these ions at specific locations in the membrane allows the ions to move by diffusion down their concentration gradients. Such ion currents create characteristic electrical changes across the cell membrane, which can be propagated from one region to another.

Objective 9. Describe how voltage-regulated ion gates function and how the diffusion of Na^+ and K^+ affects the membrane potential.

Figure 13.18. The classification of nerve fibers by origin and function.

Objective 10. Describe the properties of an action potential, and explain how it is produced.
Objective 11. Define the absolute and relative refractory periods, and explain the all-or-none law of action potentials.
Objective 12. Explain how action potentials are conducted in unmyelinated and myelinated nerve fibers.

Although all cells in the body maintain a potential difference across their cell membranes, only nerve and muscle cells can alter their membrane potential in response to appropriate stimulation. These alterations are achieved by varying the membrane permeability to different ions. Tissues that can respond to stimuli in this manner are said to be *irritable.*

An increase in membrane permeability to a specific ion results in the diffusion of that ion down its concentration gradient, either into or out of the cell. These *ion currents* occur only across limited patches of membrane (located fractions of a millimeter apart), where specific ion channels are located. Changes in the potential difference across the membrane at these points can be measured by the voltage developed between two electrodes: one placed inside the cell, the other placed outside the cell

Figure 13.19.   The difference in potential (in millivolts) between an intracellular and extracellular recording electrode is displayed on an oscilloscope screen. The resting membrane potential (*rmp*) of the axon may be reduced (depolarization) or increased (hyperpolarization).

membrane at the region being recorded. The voltage between these two *recording electrodes* can be visualized by connecting these electrodes to an oscilloscope (fig. 13.19).

In an oscilloscope, electrons from a cathode ray "gun" are sprayed across a fluorescent screen, producing a line of light. Changes in the potential difference between the two recording electrodes produces a deflection of this line. The oscilloscope can be calibrated in such a way that an upward deflection of this line indicates that the inside of the membrane has become less negative (or more positive) compared to the outside of the membrane. A downwards deflection of the line, conversely, indicates that the inside of the cell has become more negative.

If both recording electrodes were placed outside of the cell, the potential difference between the two would be zero (there is no charge separation). When one of the two electrodes penetrates the cell membrane, the oscilloscope shows that the intracellular electrode is electrically negative with respect to the extracellular electrode. If appropriate stimulation causes positive charges to flow into the cell, the line would deflect upward; this change is called **depolarization** *(de-po''lar-i-za'shun),* or hypopolarization, because the potential difference between the two recording electrodes is reduced. If the inside of the membrane becomes more negative as a result of stimulation, the line on the oscilloscope would deflect downwards; this is called **hyperpolarization.**

Ion "Gating" in Nerve Fibers
As discussed in chapter 5, the intracellular potassium concentration is much higher than the concentration of potassium in the extracellular fluid. The sodium concentration, conversely, is about twenty times higher in extracellular fluid than it is within the cell. These concentration differences are a result of (1) electrical attraction by negatively charged proteins and organic phosphates that are "fixed" within the cell; (2) the greater permeability of the resting cell membrane to K^+ than to Na^+; and (3) active

transport by the Na^+/K^+ pump. Because of this ion distribution, the inside of a neuron is about sixty-five millivolts negative (-65 mV) compared to the extracellular fluid (the exact value is different in different cells).

The permeability of the membrane to Na^+, K^+, and other ions is regulated by proteins within the membrane that function as **gates.** At the *resting membrane potential* of an axon (equal to about -65 mV), the Na^+ gates are almost—but not completely—closed. The K^+ gates, however, are slightly open, so that the membrane is more permeable to K^+ than to Na^+.

Depolarization of a small region of an axon can be experimentally induced by a pair of stimulating electrodes. If a pair of recording electrodes are placed in the same region (one electrode within the axon and one outside), an upward deflection of the oscilloscope line will be observed as a result of this depolarization. If a certain level of depolarization is achieved (from -65 mV to -55 mV) by this artificial stimulation, a sudden and very rapid change in the membrane potential will be observed. This is due to the fact that depolarization to a threshold level causes the Na^+ *gates* to open. Sodium, as a result, diffuses down its concentration gradient and into the cell.

A fraction of a second after the Na^+ gates open, they close again. At this time, the depolarization stimulus causes the K^+ *gates* to open. Membrane permeability to K^+, in other words, greatly increases, and K^+ diffuses out of the cell down its concentration gradient. The Na^+ and K^+ gates in an axon membrane are thus **voltage regulated**—they are relatively closed at the resting membrane potential and open (sooner for Na^+, later for K^+) when the membrane voltage is depolarized to a threshold value.

Action Potentials
When the axon membrane has been depolarized to a threshold level—by stimulating electrodes, for example—the Na^+ gates open as previously described. This permits Na^+ to enter the axon by diffusion, which further depolarizes the membrane (makes the inside less negative, or

oscilloscope: L. *oscillatio,* a swing; Gk. *skopos,* watch, look at
depolarize: L. *de,* away; *polus,* the end of an axis (pole)

Figure 13.20. Depolarization of an axon has two effects: (1) Na⁺ gates open and Na⁺ diffuses into the cell and (2) after a brief period, K⁺ gates open and K⁺ diffuses out of the cell. An inward diffusion of Na⁺ causes further depolarization—this causes further opening of Na⁺ gates in a positive feedback (+) fashion. The opening of K⁺ gates and outward diffusion of K⁺ make the inside of the cell more negative and thus have a negative feedback effect (−) on the initial depolarization.

more positive). Since the Na⁺ gates of the axon are voltage regulated, this depolarization causes a further increase in Na⁺ permeability. A positive feedback loop is thus created as increased Na⁺ entry into the cell causes more depolarization, and more depolarization causes increased Na⁺ diffusion into the cell.

The effect of depolarization of the axon membrane on K⁺ permeability has the opposite result. Depolarization—with a slight time delay—causes an increase in K⁺ permeability (opens the K⁺ gates). Potassium, as a result, moves out of the cell and makes the inside of the axon membrane more negative. This process is called **repolarization** and helps to bring the membrane potential back toward the resting level of −65 mV (fig. 13.20).

The stimulus of depolarization thus has two effects on the axon: (1) Na⁺ gates open and Na⁺ diffuses into the cell—this makes the inside of the cell more positively charged compared to the outside of the cell (up to about +40 mV); and (2) the K⁺ gates open and K⁺ diffuses out of the cell, helping to reestablish the resting membrane potential of −65 mV. Depolarization thus causes a transient increase in Na⁺ permeability and more depolarization, followed by a transient increase in K⁺ permeability and repolarization. These events produce an **action potential,** or **nerve impulse** (fig. 13.21).

All-or-None Law. Once a region of axon membrane has been depolarized to a threshold value, the positive feedback effect of depolarization on Na⁺ permeability and of Na⁺ permeability on depolarization causes the membrane potential to shoot toward about +40 mV. It does not normally become more positive because of the fact that the

Na⁺ gates quickly close and the K⁺ gates open. The length of time that the Na⁺ and K⁺ gates stay open is independent of the strength of the depolarization stimulus.

The amplitude of action potentials is therefore **all-or-none.** When depolarization is below a threshold value, the voltage-regulated gates are closed; when depolarization reaches threshold, a maximum potential change (the action potential) is produced. Since the change from −65 mV to +40 mV and back to −65 mV lasts only about three msec (milliseconds, or thousandths of a second), the image of an action potential on an oscilloscope screen looks like a spike. Action potentials are therefore sometimes called *spike potentials.*

Both the rising (depolarization) and falling (repolarization) phases of an action potential are, in summary, produced by gating the *diffusion* of ions down their opposing concentration gradients. Since the gates are open for a fixed period of time, the duration of each action potential is about the same. Likewise, since the concentration gradient for Na⁺ is relatively constant, the amplitude of each action potential is about the same in all axons at all times (from −65 mV to +40 mV, or about 100 mV in total amplitude).

Coding for Stimulus Intensity. If one depolarization stimulus is greater than another, the greater stimulus strength is not coded by a greater amplitude of action potentials. The code for stimulus strength in the nervous system is not **AM** (amplitude modulated). When a greater stimulus strength is applied to a neuron, identical action potentials are produced more frequently (more are produced per minute). Therefore, the code for stimulus strength in the nervous system is frequency modulated (**FM**). This is shown in figure 13.22.

Figure 13.21. An action potential (*upper figure*) is produced by an increase in sodium conductance followed, with a short time delay, by an increase in potassium conductance (*lower figure*). This drives the membrane potential first toward the sodium equilibrium potential and then toward the potassium equilibrium potential.

Figure 13.22. Recordings from a single sensory fiber of a sciatic nerve of a frog stimulated by varying degrees of stretch of the gastrocnemius muscle. Note that increasing amounts of stretch (indicated by increasing weights attached to the muscle) result in an increased frequency of action potentials.

When an entire collection of axons (in a nerve) is stimulated, different axons will be stimulated at different stimulus intensities. A low-intensity stimulus will only activate those few fibers with low thresholds, whereas high-intensity stimuli can activate fibers with higher thresholds. As the intensity of stimulation increases, more and more fibers will become activated. This process, called **recruitment,** represents another mechanism by which the nervous system can code for stimulus strength.

Refractory Periods.   If a stimulus of a given intensity is maintained at one point of an axon and depolarizes it to threshold, action potentials will be produced at that point at a given frequency (number per minute). As the stimulus strength is increased, the frequency of action potentials produced at that point will increase accordingly. As action potentials are produced with increasing frequency, the time between successive action potentials will decrease—but only up to a minimum time interval. The interval between successive action potentials will never get so short that one action potential is produced before the preceding one has finished.

During the time that a patch of axon membrane is producing an action potential, it is incapable of responding—or *refractory (re-frak'to-re)*—to further stimulation. If a second stimulus is applied, for example, while the Na^+ gates are open in response to a first stimulus, the second stimulus cannot have any effect (the gates are already open). During the time that the Na^+ gates are open, therefore, the membrane is in an **absolute refractory period** and cannot respond to any subsequent stimulus. If a second stimulus is applied while the K^+ gates are open (and the membrane is in the process of repolarizing), the membrane is in a **relative refractory period.** During this time only a very strong stimulus can depolarize the membrane and produce a second action potential (fig. 13.23).

As a result of the fact that the cell membrane is refractory during the time it is producing an action potential, each action potential remains a separate, all-or-none event. In this way, as a continuously applied stimulus increases in intensity, its strength can be coded strictly by the frequency of the action potential it produces at each point of the axon membrane.

One might think that as an axon produces a large number of action potentials, the relative concentrations of Na^+ and K^+ would be changed in the extracellular and intracellular compartments. This is not true. In a typical mammalian axon that is one micrometer in diameter, for example, only one intracellular K^+ ion in three thousand

Figure 13.23. The absolute and relative refractory periods. While a segment of axon is producing an action potential, the membrane is absolutely or relatively resistant (refractory) to further stimulation.

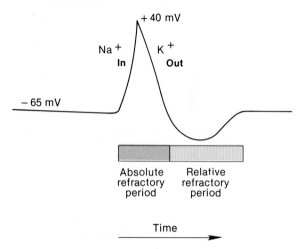

Figure 13.24. The conduction of a nerve impulse (action potential) in an unmyelinated nerve fiber (axon). Each action potential "injects" positive charges that spread to adjacent regions. The region that has previously produced an action potential is refractory. The previously unstimulated region is partially depolarized. As a result, its voltage-regulated Na$^+$ gates open, repeating the process.

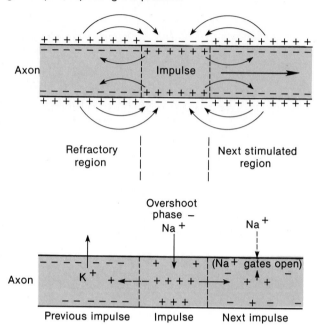

would be exchanged for a Na$^+$. Since a typical neuron has about one million Na$^+$/K$^+$ pumps, able to transport nearly 200 million ions per second, these small changes can be quickly corrected.

Cable Properties of Neurons. If a pair of stimulating electrodes produces a depolarization that is too weak to cause the opening of voltage-regulated Na$^+$ gates—that is, if the depolarization is below threshold (about -55 mV)—the change in membrane potential will be *localized* to within one to two millimeters of the point of stimulation. For example, if the stimulus causes depolarization from -65 mV to -60 mV at one point, and the recording electrodes are placed only three millimeters away from the stimulus, the membrane potential recorded will remain at -65 mV (the resting potential). The axon is thus a very poor conductor compared to metal wires.

The ability of a neuron to transmit charges through its cytoplasm is known as its *cable properties*. These cable properties are quite poor because there is a high internal resistance to the spread of charges and because many charges leak out of the axon through its membrane. If an axon had to conduct only through its cable properties, therefore, no axon could be more than a millimeter in length. The fact that some axons are a meter or more in length suggests that the conduction of nerve impulses does not rely on the cable properties of the axon.

Conduction of Nerve Impulses

When stimulating electrodes artificially depolarize one point of an axon membrane to a threshold level, voltage-regulated gates open and an action potential is produced at that small region of axon membrane containing those gates. For about the first millisecond of the action potential, when the membrane voltage changes from -65 mV to $+40$ mV, a current of Na$^+$ enters the cell by diffusion

due to the opening of the Na$^+$ gates. Each action potential thus "injects" positive charges (Na$^+$ ions) into the axon.

These positively charged Na$^+$ ions are conducted, by the cable properties of the axon, to an adjacent region that still has a membrane potential of -65 mV. Within the limits of the cable properties of the axon (one to two millimeters), this helps to depolarize the adjacent region of axon membrane. When this adjacent region of membrane reaches a threshold level of depolarization, it too produces an action potential as its voltage-regulated gates open.

Each action potential thus acts as a stimulus for the production of another action potential at the next region of membrane that contains voltage-regulated gates. In the previous description of action potentials, the stimulus for their production was artificial—depolarization produced by a pair of stimulating electrodes. Now it can be seen that an action potential at one point along an axon is normally produced by depolarization that results from the production of a preceding action potential. This line of reasoning leads to the production of the first action potential, generally at the beginning of the axon (the axon hillock). The first action potential usually results from depolarization induced by neurotransmitter chemicals.

Conduction in an Unmyelinated Nerve Fiber. In an unmyelinated axon, every patch of membrane that contains Na$^+$ and K$^+$ gates can produce an action potential. Action potentials are thus produced at locations only a fraction of a micrometer apart all along the length of the axon.

Figure 13.25. The conduction of the nerve impulse in a myelinated nerve fiber. Since the myelin sheath prevents inward Na⁺ current, action potentials can only be produced at the interruptions in the myelin sheath, or nodes of Ranvier. This "leaping" of the action potential from node to node is known as saltatory conduction.

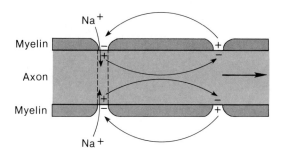

Table 13.4 Examples of conduction velocities and functions of mammalian nerves of different diameters		
Diameter (μm)	Conduction velocity (m/sec)	Examples of functions served
12–22	70–120	Somatic motor fibers
5–13	30–70	Sensory: muscle position pressure
3–8	15–40	Sensory: touch, pressure
1–5	12–30	Sensory: pain, temperature
1–3	3–15	Autonomic fibers to ganglia
0.3–1.3	0.7–2.2	Autonomic fibers to muscles

The cablelike spread of depolarization induced by Na⁺ influx during one action potential helps to depolarize the adjacent regions of membrane. This process is also aided by movements of ions on the outer surface of the axon membrane (fig. 13.24). This process would depolarize the adjacent membranes on each side of the region producing an action potential, but the area which had previously produced an action potential cannot produce a new one at this time because it is still in its refractory period. In this way, action potentials are passed in one direction only along the axon.

Notice that action potentials are not really conducted, although it is convenient to use that word. Each action potential is a separate, complete event (flow of Na⁺ and K⁺ across the membrane to produce an all-or-none change in potential at a limited region of membrane) that is repeated, or *regenerated,* along the axon's length. The action potential produced at the end of the axon is thus a completely new event that was produced in response to depolarization from the previous action potential. The last action potential has the same amplitude as the first. Action potentials are thus said to be **conducted without decrement** (without decreasing in amplitude).

Since action potentials are produced at every fraction of a micrometer of axon membrane in an unmyelinated fiber, the conduction rate is relatively slow. This conduction rate is greater when the axon is thicker, because the ability of fibers to conduct charges by cable properties improves with increasing diameter. The conduction rate is also faster when the fiber is myelinated.

Conduction in a Myelinated Nerve Fiber. The myelin sheath provides insulation for the axon, preventing movements of Na⁺ and K⁺ through the membrane. If the myelin sheath were continuous, therefore, action potentials could not be produced. Fortunately, there are interruptions in the myelin known as **nodes of Ranvier** *(rahn-ve-a').*

Since action potentials can only be produced at the nodes of Ranvier and since the cable properties of axons can only conduct depolarization a very short distance (1 to 2 mm), it follows that the nodes must be no farther than about one millimeter apart. Action potentials thus seem to "leap" from node to node; this is called **saltatory** *(sal'tah-to''re)* **conduction** (fig. 13.25).

Saltatory conduction in a myelinated nerve fiber allows a *faster rate of conduction* than is possible in an unmyelinated fiber. Conduction rate, as previously discussed, is also related to fiber diameter. For a fiber of a given diameter, the presence of a myelin sheath (and thus of saltatory conduction) increases conduction rate by a factor of about twenty-five. Conduction rate in the human nervous system varies from 1.0 m/sec (for thin, unmyelinated fibers that mediate slow, visceral responses) to greater than 100 m/sec (225 miles per hr) for thick myelinated fibers involved in quick stretch reflexes of skeletal muscles (table 13.4).

1. Define the meaning of the terms depolarization and repolarization, *and illustrate these processes with a figure.*
2. Describe how the permeability of the axon membrane to Na⁺ and K⁺ is regulated, and how changes in permeability to these ions affects the membrane potential.
3. Describe how gating of Na⁺ and K⁺ in the axon membrane results in the production of an action potential.
4. Define the all-or-none law of action potentials, and describe the effect of increased stimulus strength on action potential production. Explain how the refractory periods affect the maximum frequency of action potential production.
5. Describe how action potentials are conducted by unmyelinated nerve fibers, and explain why saltatory conduction in myelinated fibers is more rapid.

saltatory: L. *saltatio,* to leap

nodes of Ranvier: from Louis A. Ranvier, French pathologist, 1835–1922

Figure 13.26. Different types of synapses: (*a*) axodendrite, (*b*) axoaxonic, (*c*) dendrodendritic, and (*d*) axosomatic.

Synaptic Transmission

Synaptic transmission may be either electrical or chemical. The most predominant type are chemical synapses, in which neurotransmitters are released from the endings of nerve fibers in response to stimulation by action potentials.

Objective 13. Describe the different types of synapses.

Objective 14. Describe the structure and function of gap junctions.

Objective 15. Describe the structure of presynaptic nerve endings of chemical synapses, and explain how neurotransmitters are released.

A **synapse** *(sin'aps)* is the functional connection between a neuron and a second cell. In the CNS this other cell is also a neuron. In the PNS the other cell may be a neuron or an *effector cell* within either a muscle or a gland. Although the physiology of neuron-neuron synapses and neuron-muscle synapses is similar, the latter synapses are often distinguished by the names **myoneural** *(mi"o-nu'ral)* or **neuromuscular junctions.**

Neuron-neuron synapses usually involve a connection between the axon of one neuron and the dendrites, cell body, or axon of a second neuron. These are called, respectively, **axodendritic, axosomatic,** and **axoaxonic synapses** (fig. 13.26). In almost all synapses, transmission is in one direction only—from the axon of the first (or presynaptic) cell to the second (or postsynaptic) cell. **Dendrodendritic synapses** do not fit this classic pattern; in these synapses, two dendrites from different neurons make reciprocal innervations—some of these synapses conduct in one direction while others conduct in the opposite direction.

In the early part of the twentieth century, most physiologists believed that synaptic transmission was *electrical*—that is, that action potentials were conducted directly from one cell to the next. This belief was reasonable in view of the facts that nerve endings appeared to touch the postsynaptic cells and that the delay in synaptic conduction was extremely short (about 0.5 msec). Improved histological techniques, however, revealed tiny gaps in the synapses, and experiments demonstrated that the actions of autonomic nerves could be duplicated by certain chemicals. This led to the suspicion that synaptic transmission might be *chemical*—that the presynaptic nerve endings might release chemical **neurotransmitters** that stimulated action potentials in the postsynaptic cells.

synapse: Gk. *syn,* together; *haptein,* to fasten

Figure 13.27. Gap junctions are shown in electron micrographs, (a) and (b). (a) the photograph shows that cell membranes of two cells are fused together in the gap junction; (b) a surface view of a gap junction is seen. (c) the information presented in these and other electron micrographs is interpreted by the illustration.

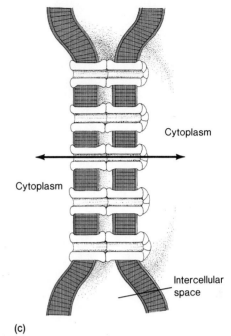

(a)   (b)   (c)

In 1921, a physiologist named Otto Loewi published the results of an experiment suggesting that synaptic transmission was chemical, at least at the junction between a branch of the parasympathetic vagus nerve and the heart. He had isolated the heart of a frog and—while stimulating the branch of the vagus that innervates the heart—perfused the heart with an isotonic salt solution. Stimulation of this nerve slowed the heart rate, as expected. More importantly, application of this salt solution to the heart of a second frog caused the second heart to slow its rate of beat.

Loewi concluded that the nerve endings of the vagus must have released a chemical—which he called *vagusstoff (va'gus-tof)*—that inhibited the heart rate. This chemical was subsequently identified as **acetylcholine** *(as''ĕ-til-ko'lēn)*, or ACh. In the decades following Loewi's discovery, many other examples of chemical synapses were discovered, and the theory of electrical synaptic transmission fell into disrepute. More recent evidence, ironically, has shown that electrical synapses do exist (though they are the exception) in the nervous system, within smooth muscles, and between cardiac cells in the heart.

Electrical Synapses: Gap Junctions

In order for two cells to be electrically coupled they must be approximately equal in size and must be joined by areas of contact with low electrical resistance. In this way, impulses can be regenerated from one cell to the next without interruption—and without Frankenstein-like sparks between cells.

Adjacent cells that are electrically coupled are joined together by **gap junctions.** In gap junctions, the membranes of the two cells are separated by only two nanometers (10^{-9} meter). When gap junctions are observed in the electron microscope from a surface view, hexagonal arrays of particles are seen that are believed to be channels through which ions and molecules may pass from one cell to the next (fig. 13.27).

Gap junctions are present in smooth and cardiac muscle, where they allow excitation and rhythmic contraction of large masses of muscle cells. Gap junctions have also been observed in various regions of the brain. Although their functional significance here is unknown, it has been speculated that they may allow a two-way transmission of impulses (in contrast to chemical synapses, which are always one-way). Gap junctions have also been observed between neuroglial cells, which do not produce electrical impulses; perhaps these act as channels for the passage of informational molecules between cells. It is interesting in this regard that many embryonic tissues have gap junctions and that these gap junctions disappear as the tissue becomes more specialized.

Otto Loewi: German pharmacologist (1873–1961)

Figure 13.28. An electron micrograph of a chemical synapse, showing synaptic vesicles at the end of an axon.

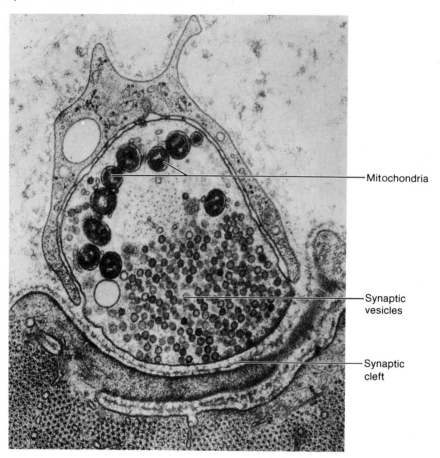

— Mitochondria

— Synaptic vesicles

— Synaptic cleft

Chemical Synapses

Transmission across the majority of synapses in the nervous system is one-way and occurs through the release of chemical neurotransmitters from presynaptic axon endings. These presynaptic endings, which are called **terminal boutons** *(boo-tanz′)* because of their swollen appearance, are separated from the postsynaptic cell by a **synaptic cleft** so narrow that it can only be seen clearly with an electron microscope (fig. 13.28).

Neurotransmitter molecules within the presynaptic neuron endings are contained within many small, membrane-enclosed **synaptic vesicles.** In order for the neurotransmitter within these vesicles to be released into the synaptic cleft, the vesicle membrane must fuse with the axon membrane in the process of *exocytosis (eks″o-si-to′sis).* The neurotransmitter is released in *quanta*—that is, in multiples of the amount contained in one vesicle. The number of vesicles that undergo exocytosis is directly related to the frequency of action potentials produced at the presynaptic axon ending.

The release of neurotransmitters following electrical excitation of the presynaptic endings accounts for most of the 0.5 msec time delay in synaptic transmission. During this time interval there is a sudden, transient inflow of Ca^{++} into the presynaptic endings. This inflow of Ca^{++} is apparently due to opening of Ca^{++} gates in response to electrical excitation and is required for the release of neurotransmitters. The Ca^{++} is believed to activate previously inactive enzymes within the axon terminals. This process produces a change in the ratio of phospholipids in the axon membrane, which may change the fluidity of the membrane and thus allow exocytosis to occur.

1. Define a synapse, and describe the structures that form synapses.
2. Describe the structure, locations, and functions of gap junctions.
3. Explain how neurotransmitter chemicals are stored and released by presynaptic neuron endings.

---

bouton: Fr. *bouton,* button
exocytosis: Gk. *exo,* outside; *kytos,* cell; *osis,* a condition

Figure 13.29. (*a* and *b*) the binding of acetylcholine to receptor proteins causes the opening of chemically regulated gates in the postsynaptic membrane. (*c*) this results in the increased diffusion of Na$^+$ and K$^+$ through the membrane.

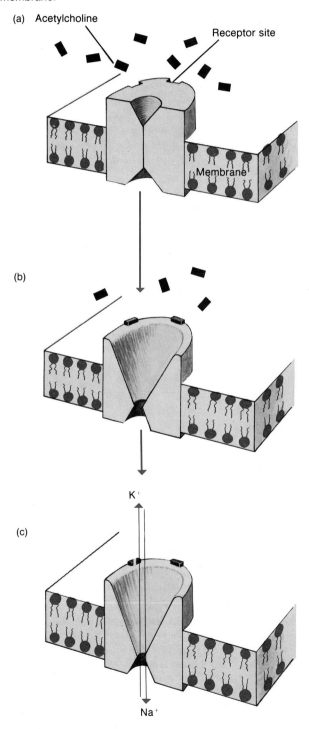

(a) Acetylcholine

Receptor site

Membrane

(b)

(c)

K$^+$

Na$^+$

Acetylcholine as a Neurotransmitter

Acetylcholine can stimulate depolarization of skeletal muscle cells and neurons by stimulating the opening of chemically regulated Na$^+$ and K$^+$ gates. This depolarization serves as a stimulus for the production of action potentials in the postsynaptic cell.

Objective 16. Describe how acetylcholine produces end plate potentials in skeletal muscle fibers.

Objective 17. Explain the function of acetylcholinesterase.

Objective 18. Describe the characteristics of excitatory postsynaptic potentials (EPSPs), and explain how EPSPs differ from action potentials.

Acetylcholine (ACh) is used as a neurotransmitter by some neurons in the CNS, by somatic motor neurons at the neuromuscular junction, and by parasympathetic nerve endings. The effects of this chemical are excitatory in the first two synapses and inhibitory in the third. The excitatory effects of ACh will be discussed in this section; its inhibitory effects are considered in a later section on synaptic integration.

Acetylcholine at the Neuromuscular Junction

The neuromuscular junction—the synapse between a somatic motor neuron and a skeletal muscle fiber—is the most accessible synapse to study and thus the best understood. The presynaptic neuron endings have synaptic vesicles, which each contain about ten thousand molecules of ACh. Once these molecules are released by exocytosis, they quickly diffuse across the narrow synaptic cleft to the membrane of the postsynaptic cell. Here they chemically bond to **receptor proteins** that are built into the postsynaptic membrane. These receptor proteins combine with ACh in a specific manner, analogous to the specific interaction between transport proteins and their substrates.

Acetylcholine is not, however, transported into the postsynaptic cell. Instead, the interaction between ACh and its receptor proteins causes changes in membrane structure that result in the opening of the Na$^+$ and K$^+$ gates. These **chemically regulated gates** contrast with the voltage-regulated gates previously discussed in the axon, which open in response to depolarization.

In contrast to voltage-regulated gates, where the outward flow of K$^+$ occurs after the inward flow of Na$^+$, chemically regulated gates allow the simultaneous diffusion of Na$^+$ and K$^+$ (fig. 13.29). The depolarizing effect of Na$^+$ diffusion predominates because the electrochemical gradient for Na$^+$ is greater than for K$^+$. The outflow

Figure 13.30. Mechanisms of the release of acetylcholine (*ACh*) from presynaptic nerve endings and the binding of ACh to receptor proteins (*R*) in the postsynaptic membrane. Acetylcholine that combines with acetylcholinesterase (*AChE*) in the postsynaptic membrane is hydrolyzed and thus inactivated.

of K⁺, however, does prevent the "overshoot" characteristic of action potentials—the membrane potential can reach 0 mV but cannot reverse polarity (as occurs in action potentials). Because of the characteristics of chemically regulated gates, neurotransmitters do not directly produce action potentials. They can only produce depolarization, which may stimulate the opening of voltage-regulated gates and production of action potentials a short distance away from the site of the synapse.

Acetylcholinesterase.   The bond between ACh and its receptor protein exists for only a brief instant. The ACh-receptor complex quickly dissociates but can be quickly reformed as long as free ACh is in the vicinity. In order

for activity in the postsynaptic cell to be controlled (in this case, for control of skeletal muscle contraction), free ACh must be inactivated very soon after it is released. The inactivation of ACh is achieved by means of an enzyme called **acetylcholinesterase** *(as″ĕ-til-ko″lin-es′ter-ās),* **AChE,** which is built into the cell membrane near the ACh receptor proteins (fig. 13.30).

Nerve gas exerts its odious effects by inhibiting AChE in skeletal muscles. Since ACh is not degraded it can continue to combine with receptor proteins and can continue to stimulate the postsynaptic cell, leading to spastic paralysis. Clinically, cholinesterase inhibitors (such as **neostigmine** *[ne″o-stig′min]*) are used to enhance the effects of ACh on muscle contraction when neuromuscular transmission is weak, as in the disease myasthenia gravis.

Table 13.5 Derivation and effect of various drugs that affect neural control of skeletal muscles

Drug	Origin	Effects
Botulinus toxin	Produced by *Clostridium botulinum* (bacteria)	Inhibits release of acetylcholine (ACh)
Curare	Resin from a South American tree	Prevents interaction of ACh with the postsynaptic receptor protein
α-bungarotoxin	Venom of *Bungarus* snakes	Binds to ACh receptor proteins
Saxitoxin	Red tide *(Gonyaulax)*	Blocks Na^+ channels
Tetrodoxin	Pufferfish	Blocks Na^+ channels
Nerve gas	Artificial	Inhibits acetylcholinesterase in postsynaptic cell
Prostigmine	Nigerian bean	Inhibits acetylcholinesterase in postsynaptic cell
Strychnine	Seeds of an Asian tree	Prevents IPSPs in spinal cord that inhibit contraction of antagonistic muscles

Source: Fox, Stuart Ira, *Human Physiology.* © 1984 Wm. C. Brown Publishers, Dubuque, Iowa. All Rights Reserved. Reprinted by permission.

End Plate Potential.   The interaction of ACh with its receptors in the muscle cell membrane opens chemically regulated gates and depolarizes that region of membrane. The area of the muscle cell membrane innervated by somatic motor nerve endings and affected in this way by ACh is called a **motor end plate.** Depolarizations of this membrane in response to ACh are known as **end plate potentials.** The depolarizations of the end plate potentials do not overshoot 0 mV, for reasons previously discussed, and differ from action potentials in a number of other respects.

Unlike action potentials, end plate potentials have *no threshold;* a single quantum of ACh (released from a single synaptic vesicle) produces a tiny depolarization of the end plate. When more quanta of ACh are released, the depolarization of the motor end plate is correspondingly greater. End plate potentials are therefore *graded* in magnitude, unlike all-or-none action potentials. Since end plate potentials can be graded, they are capable of *summation.* This is quite different from action potentials, which are prevented from summating by their all-or-none nature and by the presence of refractory periods.

When the end plate becomes sufficiently depolarized, the depolarization serves as a stimulus for the opening of voltage-regulated gates in adjacent areas of the skeletal muscle cell membrane. The opening of voltage-regulated gates causes action potentials to be produced. These action potentials in turn serve as stimuli for the production of other action potentials in the next area of membrane. Skeletal muscle fibers conduct action potentials in the same manner as unmyelinated axons. Electrical excitation of a muscle fiber in this manner stimulates muscle contraction.

If any stage in the process of neuromuscular transmission is blocked, muscle weakness—sometimes leading to paralysis and death—may result. The drug *curare (koo-rah're),* for example, competes with ACh for attachment to the receptor proteins and thus reduces the size of the end plate potentials. This drug was first used on poison darts by South American Indians because it produced flaccid paralysis of their victims. Clinically, curare is used as a muscle relaxant during anesthesia and to prevent muscle damage during electroconvulsive shock therapy.

Muscle weakness in the disease **myasthenia** *(mi''as-the'ne-ah)* **gravis** is due to the fact that ACh receptors are blocked by antibodies secreted by the immune system of the affected person. Paralysis in people who eat shellfish poisoned with saxitoxin, produced by unicellular organisms that cause the red tides, results from inhibition of the chemically regulated gates. The effects of these and other poisons on neuromuscular transmission are summarized in table 13.5.

Acetylcholine in Neuron-Neuron Synapses

Within the nervous system, the axon terminals of one neuron typically synapse with the dendrites or cell body of another. The dendrites and cell body thus serve as the receptive area of the neuron, and it is in these regions that receptor proteins for neurotransmitters and chemically regulated gates are located. The first voltage-regulated gates are located at the beginning of the axon, at the axon hillock. It is here that action potentials are first produced (fig. 13.31).

Release of ACh at these synapses results, through opening of chemically regulated gates, in the production of graded depolarizations in the dendrites and cell body. These graded depolarizations are known as **excitatory postsynaptic potentials (EPSPs).** A comparison of EPSPs and action potentials is provided in table 13.6.

---

myasthenia: Gk. *myos,* muscle; *asthenia,* weakness

Figure 13.31. A diagram illustrating the functional specialization of different regions in a "typical" neuron.

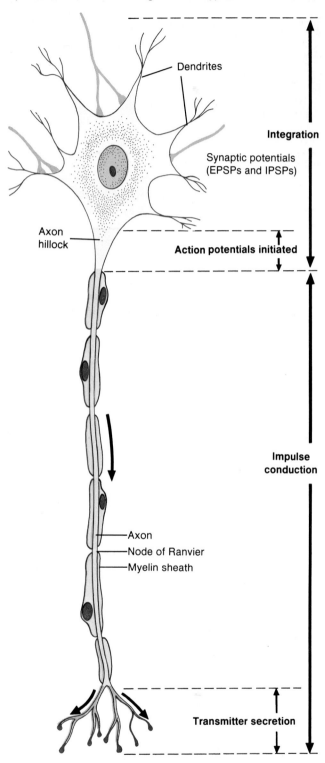

Dendrites

Integration

Synaptic potentials (EPSPs and IPSPs)

Axon hillock

Action potentials initiated

Impulse conduction

Axon
Node of Ranvier
Myelin sheath

Transmitter secretion

Table 13.6 Comparison of action potentials with synaptic potentials at the neuromuscular junction

Characteristic	Action potential	Excitatory postsynaptic potential
Stimulus for opening of ionic gates	Depolarization	Acetylcholine (ACh)
Initial effect of stimulus	Na^+ gates open	Na^+ and K^+ gates open
Production of repolarization	Opening of K^+ gates	Loss of intracellular positive charges with time and distance
Conduction distance	Not conducted—regenerated over length of axon	1–2mm; a localized potential
Positive feedback between depolarization and opening of Na^+ gates	Present	Absent
Maximum depolarization	+40 mV	Close to zero
Summation	No summation—is all-or-none	Summation of EPSPs, producing graded depolarizations
Refractory period	Present	Absent
Effect of drugs	Inhibited by tetrodoxin, not by curare	Inhibited by curare, not by tetrodoxin

Since voltage-regulated gates are absent in the dendrites and cell body, depolarizations produced in these regions must be conducted by cable properties to the axon hillock. This causes the amplitude of the EPSP to decrease with distance from its point of origin. If the depolarization is at a threshold level by the time the EPSP reaches the axon hillock, it will cause opening of voltage-regulated gates and the production of action potentials. If an inadequate amount of ACh is released from the presynaptic nerve ending, the EPSP produced at the axon hillock of the postsynaptic cell may be insufficient to generate action potentials (fig. 13.32).

A skeletal muscle fiber receives synaptic input from only one neuron. A neuron, in contrast, may receive synaptic inputs from as many as one thousand other neurons. So, whereas the synaptic input from a single axon must be sufficient to excite a muscle cell, a neuron may require the summation of many EPSPs before a threshold depolarization is produced at the axon hillock. Summation, which can only occur between graded synaptic potentials (action potentials can't summate), allows the postsynaptic neuron to "weigh" different combinations of synaptic inputs and respond appropriately.

Figure 13.32. The graded nature of excitatory postsynaptic potentials is shown, in which stimuli of increasing strength produce increasing amounts of depolarization. When a threshold level of depolarization is produced, action potentials are generated in the axon.

1. Describe the events that occur between the release of ACh and the production of end plate potentials in a skeletal muscle fiber.
2. Describe the function of acetylcholinesterase and its physiological significance.
3. Compare the properties of EPSPs and action potentials, and describe where these events occur in a postsynaptic neuron.
4. Explain how EPSPs produce action potentials in the postsynaptic neuron.

Catecholamine and Other Neurotransmitters

The actions of catecholamine neurotransmitters are mediated by cyclic AMP produced by the postsynaptic cell. In addition to acetylcholine and the catecholamines, there are many putative neurotransmitters.

Objective 19. Describe how catecholamines are inactivated at the synapse.

Objective 20. Explain how catecholamines influence the activity of postsynaptic cells.

Objective 21. Identify some of the putative neurotransmitters, and explain the possible significance of the endorphins.

Catecholamines *(kat″ĕ-kol′ah-mēn)* are a group of regulatory molecules that include dopamine, epinephrine, and norepinephrine. All three types function as neurotransmitters in the CNS; epinephrine and norepinephrine additionally serve as hormones secreted by the adrenal medulla gland.

Like ACh, catecholamine neurotransmitters are released by exocytosis from presynaptic vesicles and diffuse across the synaptic cleft to interact with specific receptor proteins in the membrane of the postsynaptic cell. The stimulatory effects of these catecholamines, like those of ACh, are quickly inhibited. The inhibition of ACh stimulation is accomplished by acetylcholinesterase; the inhibition of catecholamine action is due to (1) *reuptake* of catecholamines into the presynaptic neuron endings; (2) enzymatic degradation of catecholamines in the presynaptic neuron endings by *monoamine (mon″o-am′en) oxidase (MAO);* and (3) their enzymatic degradation in the postsynaptic neuron by *catecholamine-O-methyltransferase (COMT)*. This is shown in figure 13.33. Drugs that inhibit MAO and COMT thus promote the effects of catecholamine action.

The interaction of catecholamines with their receptors does not directly open chemically regulated Na^+ and K^+ gates. Instead, these neurotransmitters activate an enzyme in the postsynaptic cell membrane known as *adenyl cyclase (ad′ĕ-nil si′klās)*. This enzyme converts ATP to **cyclic AMP (cAMP)** and pyrophosphate (two inorganic phosphates) within the postsynaptic cell cytoplasm. Cyclic AMP in turn activates other enzymes that open the chemically regulated ionic gates in the postsynaptic membrane. Cyclic AMP is thus a **second messenger** in the action of these neurotransmitters (fig. 13.34).

In addition to promoting the opening of ionic gates and the production of electrical impulses, cyclic AMP may also promote more long-term modifications of the postsynaptic neuron. These long-term modifications may involve cAMP-mediated changes in genetic expression.

Figure 13.33. A diagram showing the production, release, and reuptake of catecholamine neurotransmitters from presynaptic nerve endings. The transmitters combine with receptor proteins (*R*) in the postsynaptic membrane. Although most of the catecholamines are recaptured by the presynaptic nerve ending, some of the transmitters are inactivated within the postsynaptic cell. (COMT = catecholamine O-methyltransferase.)

Putative Neurotransmitters

Many other compounds in addition to acetylcholine and catecholamines may function as neurotransmitters. Since complete experimental proof of their transmitter function is not currently available, these compounds are identified as putative neurotransmitters. These include amino acids, such as glycine and glutamic acid, and many polypeptides (table 13.7). Interestingly, some of the polypeptides that function as hormones secreted by the intestine and other endocrine glands are also produced in the brain and may function there as neurotransmitters.

Figure 13.34.   Catecholamine neurotransmitters induce the intracellular production of cyclic AMP (cAMP) in the postsynaptic cell. The cAMP, in turn, stimulates the opening of ionic gates (short-term effects) and genetic expression (long-term effects) through the activation of previously inactive enzymes.

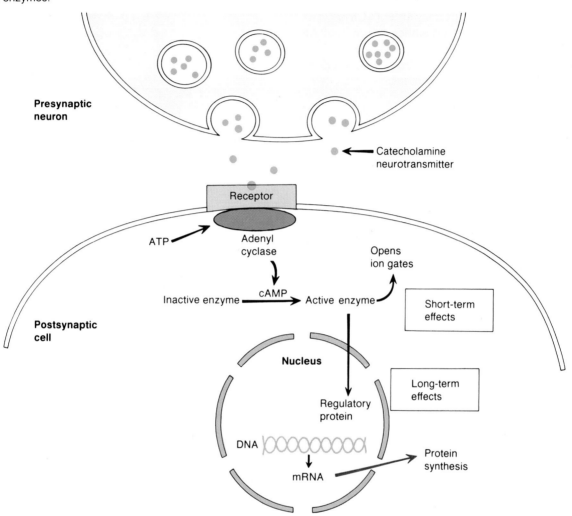

Table **13.7**	Examples of chemicals that are either proven or putative neurotransmitters		
Category	**Chemicals**	**Category**	**Chemicals**
Amines	Acetylcholine GABA (gamma-aminobutyric acid) Histamine Serotonin	Gut-brain polypeptides	Glucagon Insulin Somatostatin Substance P ACTH (adrenocorticotrophic hormone)
Catecholamines	Dopamine Epinephrine Norepinephrine		Angiotensin II Endorphins
Amino acids	Aspartic acid Glutamic acid Glycine	Polypeptides (other)	LHRH (luteinizing hormone-releasing hormone) TRH (thyrotrophin-releasing hormone) Vasopressin

Source: Fox, Stuart Ira, *Human Physiology.* © 1984 Wm. C. Brown Publishers, Dubuque, Iowa. All Rights Reserved. Reprinted by permission.

Although some of these molecules may function as neurotransmitters in the traditional sense—by stimulating the opening of ionic gates and causing changes in the membrane potential—others may have more subtle and poorly understood effects. An exciting recent discovery is that some neurons in both the PNS and CNS produce both a classical neurotransmitter (ACh or a catecholamine) and a putative polypeptide transmitter. Additionally, there is evidence to suggest that these neurons may secrete one or the other of their transmitters depending upon environmental conditions.

Endorphins. The ability of opium and its analogues, the opioids, to relieve pain has been known for centuries. Morphine, for example, has long been used for this purpose. The discovery in 1973 of opioid receptor proteins in the brain suggested that the effects of these drugs might be due to the stimulation of specific neuron pathways. This implied that opioids might resemble neurotransmitters produced by the brain—as do LSD, mescaline, and other mind-altering drugs. Other experiments suggested that the brain might be producing its own morphinelike compounds.

These morphinelike compounds have been identified as a family of chemicals called **endorphins** (for "endogenously produced morphinelike compounds"). They include a thirty-one-amino-acid polypeptide called **β-endorphin** and a group of five-amino-acid peptides called the **enkephalins** *(en-kef'ah-lin)*. There is a large body of evidence supporting the concept that endorphins function as endogenous analgesics (pain-suppressants) and some evidence that these compounds may function as a type of pleasure reward. The precise physiology of endorphins, however, is at present incompletely understood.

---

1. *Describe three ways that catecholamine neurotransmitters are inactivated at the synapse.*
2. *Explain how catecholamines stimulate the production of cyclic AMP, and why cAMP is called a second messenger in the action of these neurotransmitters.*

Synaptic Integration

Interaction occurs between many synaptic inputs. Excitatory inputs can summate with each other and with inhibitory postsynaptic potentials. The release of neurotransmitters can be modified by experience and other factors, and synapses exhibit plasticity at a molecular and cellular level.

Objective 22. Describe how spatial and temporal summation occurs, and explain the functional significance of these processes.

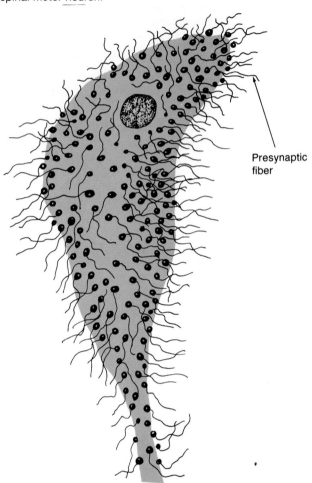

Figure 13.35. A diagram illustrating the convergence of large numbers of presynaptic fibers on the cell body of a spinal motor neuron.

Presynaptic fiber

Objective 23. Explain how inhibitory postsynaptic potentials (IPSPs) are produced, and give examples of their physiological significance.

Objective 24. Explain how presynaptic inhibition is produced.

Objective 25. Explain post-tetanic potentiation.

Since voltage-regulated Na^+ and K^+ gates are absent in the dendrites and cell bodies, changes in membrane potential induced by neurotransmitters in these areas do not have the all-or-none characteristics of action potentials. Synaptic potentials are graded and can add together, or summate. **Spatial summation** occurs because many presynaptic nerve fibers converge on a single postsynaptic neuron (fig. 13.35). In spatial summation, synaptic depolarizations (EPSPs) produced at different synapses may summate in the postsynaptic dendrites and cell body. In **temporal summation,** the successive activity of presynaptic axon terminals, causing successive waves of transmitter release, may result in the summation of EPSPs in the postsynaptic neuron. The summation of EPSPs helps

Figure 13.36. Excitatory postsynaptic potentials (EPSPs) can summate over distance (spatial summation) and time (temporal summation). When summation results in a threshold level of depolarization at the axon hillock, voltage-regulated Na^+ gates are opened and an action potential is produced.

to insure that the depolarization that reaches the axon hillock will be sufficient to generate new action potentials in the postsynaptic nerve fiber (fig. 13.36).

When a presynaptic neuron is stimulated continuously, even for as short a time as a few seconds, its ability to excite a postsynaptic neuron is enhanced—or potentiated—when this neuron pathway is subsequently stimulated. This potentiation of synaptic transmission may last for hours or even weeks following the tetanic (continuous) stimulation of the presynaptic neuron. This phenomenon is called **post-tetanic** *(tĕ-tan'ik)* **potentiation.**

Post-tetanic potentiation is due to the increased release of neurotransmitter as a result of previous tetanic stimulation of the presynaptic neuron. Experiments suggest that there is increased Ca^{++} concentration in the axon terminals following tetanic stimulation. Since Ca^{++} mediates exocytosis of the synaptic vesicles in response to action potentials, this observation may explain the improved synaptic efficiency that results. Post-tetanic potentiation may favor transmission along frequently used neural pathways and thus may represent a mechanism of neural "learning."

Synaptic Inhibition

Although most neurotransmitters depolarize the postsynaptic membrane (produce EPSPs), some transmitters have the opposite effect. These inhibitory neurotransmitters cause *hyperpolarization* of the postsynaptic membrane: they make the inside of the membrane more negative than it is at rest. Since hyperpolarization (from -65 mV to, for example, -85 mV) takes the membrane potential farther away from the threshold depolarization required to

Figure 13.37. An inhibitory postsynaptic potential (IPSP) makes the inside of the postsynaptic membrane more negative than the resting potential—it hyperpolarizes the membrane. Subsequent or simultaneous excitatory postsynaptic potentials (EPSPs), which are depolarizations, must thus be stronger to reach the threshold required to generate action potentials at the axon hillock.

stimulate action potentials, such hyperpolarization inhibits the activity of the postsynaptic neuron. Hyperpolarizations produced by neurotransmitters are therefore called **inhibitory postsynaptic potentials (IPSPs).** The inhibition produced in this way is called **postsynaptic inhibition.**

Excitatory and inhibitory inputs (EPSPs and IPSPs) to a postsynaptic neuron can summate in an algebraic fashion (fig. 13.37). The effects of IPSPs in this way reduce, or may even abolish, the ability of EPSPs to generate action potentials in the postsynaptic cell. Considering that the nervous system contains approximately 10^{12} neurons and that a given neuron may receive as many as 1,000 presynaptic inputs, the possibilities for synaptic integration in this way are awesome.

In addition to serving an integrative function in the CNS, IPSPs also have more specialized roles. Although ACh produces depolarization at the neuromuscular junction and stimulates skeletal muscle contraction, for example, the same transmitter causes hyperpolarization in the heart. This is due to the fact that the combination of ACh with its receptor protein in the myocardial cells causes opening of only K^+ gates, and the outward diffusion of K^+ causes hyperpolarization. The baseline membrane potential is thus lowered, so that a longer time is required to reach threshold. The parasympathetic fibers that innervate the heart in this way cause a slowing of the heart rate.

The neurotransmitter called *gamma-aminobutyric acid (GABA)* is exclusively inhibitory. This neurotransmitter is believed to hyperpolarize central neurons by opening chloride gates. Chloride ion (Cl^-), as a result, diffuses into the neuron and makes the membrane potential more negative on the inside. The amino acid *glycine* is also believed to function as a neurotransmitter. This transmitter may help muscular coordination by inhibiting those neurons in the spinal cord that activate antagonistic muscles.

In **presynaptic inhibition,** the release of neurotransmitter from the axon of one neuron is reduced by a second neuron, which makes an excitatory axoaxonic synapse with the first. Depolarization of the first neuron's axon produced by this axoaxonic synapse puts it in a partial refractory period. This reduces the frequency of action potentials in the first neuron's axon, thus reducing the amount of neurotransmitter it is stimulated to release.

Recent evidence indicates that synapses are more changeable than was previously believed. This has been called **synaptic plasticity** *(plas-tis'i-te),* and it exists at both the molecular and cellular levels. At the molecular level, it has been shown that some neurons produce both a classic neurotransmitter and a putative polypeptide neurotransmitter, as previously discussed. At the cellular level, there is evidence that axonal sprouting over short distances, even in the CNS, produces a turnover of synapses in the mature nervous system. This breakdown and reforming of synapses can occur within hours. The full physiological significance of these interesting discoveries is not currently known.

1. Define spatial summation, temporal summation, and post-tetanic potentiation. Explain their functional importance.
2. Explain how postsynaptic inhibition is produced and how IPSPs and EPSPs can interact.
3. Explain the mechanism of presynaptic inhibition.

Clinical Considerations

The clinical aspects of the nervous system are extensive and usually complex. Numerous diseases and developmental problems directly involve the nervous system, and the nervous system is indirectly involved with most of the diseases that afflict the body because of the perception of pain. Pain receptors are free nerve endings that elicit pain in disease or trauma; such pain is important in diagnosing specific diseases or dysfunctions.

Only some of the diseases involving developmental problems of the nervous system and clinical aspects of nerve conduction and synaptic transmission will be considered in this section. Other clinical aspects of neurology will be discussed in later chapters.

Developmental Problems

Congenital malformations of the CNS are common and frequently involve overlying bone, muscle, and connective tissue. The more severe abnormalities make life impossible, and the less severe malformations frequently result in functional disability.

Most congenital malformations of the nervous system occur during the sensitive embryonic period. Neurological malformations are generally caused by genetic abnormalities but may result from environmental factors such as anoxia, infectious agents, drugs, and ionizing radiation.

Spina bifida is a defective fusion of the vertebral elements and may or may not involve the spinal cord. **Spina bifida occulta** is the most common and least serious type of spina bifida. This defect usually involves few vertebrae, is not externally apparent except for perhaps a pigmented spot with a tuft of hair, and usually does not cause neurological disturbances. **Spina bifida cystica,** a severe type of spina bifida, is a saclike protrusion of skin and underlying meninges, which may contain portions of the spinal cord and nerve roots. Spina bifida cystica is most common in the lower thoracic, lumbar, and sacral regions. The position and extent of the defect determines the degree of neurological impairment.

Anencephalia *(an"en-sĕ-fa'le-ah)* is a markedly defective development of the brain and the surrounding cranial bones. Anencephalia occurs in one per thousand births, and makes sustained extrauterine life impossible. This congenital defect apparently results from the failure of the neural folds at the cranial portion of the neural plate to fuse and form the prosencephalon.

Microcephaly *(mi"kro-sef'ah-le)* is an uncommon condition in which brain development is not completed. If enough neurological tissue is present, the infant will survive but will be severely mentally retarded.

Defective skull development frequently causes **cranial encephalocele** *(en-sef'ah-lo-se'le).* This condition occurs approximately once per two thousand births. Cranial encephalocele is characterized by a herniated portion of the brain and meninges through a defect in the skull, usually in the occipital region. Occasionally the herniation contains fluid within the meninges but not the brain tissue. In this case, it is referred to as a **cranial meningocele** *(mĕ-ning′ go-sēl).*

Hydrocephalus is the abnormal accumulation of cerebrovascular fluid in the ventricles and subarachnoid or subdural space. Hydrocephalus may be caused by the excessive production or blocked flow of cerebrospinal fluid. It may also be associated with other congenital problems such as spina bifida cystica or encephalocele. Hydrocephalus frequently causes cranial bones to thin and the cerebral cortex to atrophy.

Many congenital disorders cause an impairment of intelligence, known as **mental retardation.** Chromosomal abnormalities, maternal and fetal infections such as syphilis and German measles, and excessive irradiation of the fetus are common causes of mental retardation.

Diseases of the Myelin Sheath

Multiple sclerosis (MS) is a relatively common neurological disease in persons between the ages of twenty and forty. MS is a chronic, degenerating, remitting, and relapsing disease that progressively destroys the myelin sheaths of neurons in multiple areas of the CNS. Initially, lesions form on the myelin sheaths and soon develop into hardened *scleroses,* or scars (hence the name). Destruction of the myelin sheaths prohibits the normal conduction of impulses, resulting in a progressive loss of functions. Because myelin degeneration is widely distributed, MS has a greater variety of symptoms than any other neurological disease. This characteristic, coupled with remissions, frequently causes misdiagnosis of this disease.

During the early stages of MS, many patients are considered neurotic because of the wide variety and temporary nature of their symptoms. As the disease progresses, the symptoms may include double vision (diplopia), spots in the visual field, blindness, tremor, numbness of appendages, and locomotor difficulty. Eventually the patient is bedridden, and death may occur anytime from seven to thirty years after the first symptoms appear.

In **Tay-Sachs disease** the myelin sheaths are destroyed by the excessive accumulation of one of the lipid components of the myelin. This results from an enzyme defect due to the inheritance of genes that are carried by the parents in a recessive state. Tay-Sachs disease, which

is inherited primarily in Jewish families, appears when the infant is under one year of age. It causes blindness, loss of mental and motor ability, and ultimately death by the age of three. Potential parents can tell if they are carriers for this condition by the use of a special blood test for the defective enzyme.

Diseases Involving Neurotransmitters

Parkinson's disease, or **paralysis agitans,** is a major cause of neurological disability in people over sixty years of age. It is a progressively degenerative disease of unknown cause. Nerve cells within the substantia nigra, an area within the basal ganglia of the brain, are destroyed. This causes muscle tremors, muscular rigidity, speech defects, and other severe problems. The symptoms of this disease can be partially treated by altering the neurotransmitter status of the brain. These patients are given L-dopa, to increase the production of dopamine in the brain, and may also be given anticholinergic drugs to decrease the production of acetylcholine.

Alzheimer's disease is the most common cause of dementia, often beginning in middle age and producing progressive mental deterioration. The cause of Alzheimer's disease is unknown, but evidence suggests that it is associated with the decreased ability of the brain to produce acetylcholine. Attempts to increase ACh production by increased ingestion of precursor molecules (choline or lecithin) have thus far not been successful. Drugs that block AChE and thus enhance the action of endogenously produced ACh offer promise but are so far still in the experimental stage.

At least some psychiatric disorders may be produced by dysfunction of neurotransmitters. There is evidence that **schizophrenia** *(skiz″o-fre′ne-ah)* is associated with hyperactivity of the neurons that use dopamine as a neurotransmitter; drugs that are effective in the treatment of schizophrenia (e.g., chlorpromazine) act by blocking dopamine receptor proteins. **Depression** is associated with decreased activity of the neurons that use monoamines—norepinephrine and serotonin—as neurotransmitters; antidepressant drugs enhance the action of these neurotransmitters. In a similar way, barbiturates and benzodiazepine (e.g., Valium), which decrease **anxiety,** act by enhancing the action of the neurotransmitter GABA in the central nervous system.

---

multiple sclerosis: L. *multiplus,* many parts; Gk. *skleros,*
    hardened
Tay-Sachs disease: from Warren Tay, English physician,
    1843–1927, and Bernard Sachs, U.S. neurologist,
    1858–1944

Parkinson's disease: from James Parkinson, English physician,
    1755–1824
Alzheimer's disease: from Alois Alzheimer, German
    neurologist, 1864–1915

Chapter Summary

I. Organization and Development of the Nervous System
The central nervous system, including the brain and spinal cord, forms from the embryonic neural tube. The peripheral nervous system, including the cranial and spinal nerves, is formed predominately from neural crest tissue.
A. The prosencephalon, mesencephalon, and rhombencephalon develop from the neural tube.
1. The telencephalon and diencephalon develop from the forebrain (prosencephalon).
2. The metencephalon and myelencephalon develop from the hindbrain (rhombencephalon).
3. The diencephalon contains the epithalamus, thalamus, hypothalamus, pineal gland, and pituitary gland.
B. As the spinal cord develops, afferent and efferent neurons are formed that conduct into and out of the spinal cord, respectively.
C. The development of the peripheral nervous system produces a pattern of dermatomes in the skin.

II. Histology of the Nervous System
The dendrites of a neuron receive stimulation, and the axon conducts nerve impulses. Neuroglial cells provide structural and functional support for the activities of neurons.
A. A neuron contains dendrites, a cell body, and an axon.
1. The cell body contains the nucleus, Nissl bodies, neurofibrils, and other organelles.
2. Dendrites receive stimuli, and the axon conducts nerve impulses away from the cell body.
B. There are six different categories of neuroglial cells.
1. The glial cells that surround an axon form an outer covering called a neurilemma; this is produced by Schwann cells in the PNS.
2. Some neurons are surrounded by successive wrappings of glial cell membranes called a myelin sheath; this is produced by Schwann cells in the PNS and oligodendrocytes in the CNS.
3. Astrocytes in the CNS contribute to the blood-brain barrier.
C. A nerve is a collection of axons in the PNS.
1. Sensory neurons are afferent and pseudounipolar.
2. Motor neurons are efferent and multipolar.
3. Interneurons, or association neurons, are located entirely within the CNS.
4. Somatic motor nerves innervate skeletal muscle; visceral motor, or autonomic, nerves innervate smooth muscle, cardiac muscle, and glands.

III. Electrical Activity in Nerve Fibers
The permeability of the axon membrane to specific ions is regulated by gates. An increase in permeability results in ion currents across the membrane that produce electrical changes, which can be propagated from one region to another.
A. The permeability of the axon membrane to Na^+ and K^+ is regulated by gates.
1. At the resting membrane potential of -65 mV, the membrane is relatively impermeable to Na^+ and only slightly permeable to K^+.
2. These gates are voltage regulated; the Na^+ and K^+ gates open in response to the stimulus of depolarization.
3. When the membrane is depolarized to a threshold level, the Na^+ gates open first, followed quickly by opening of the K^+ gates.
B. The opening of voltage-regulated gates produces an action potential.
1. The opening of Na^+ gates in response to depolarization allows Na^+ to diffuse into the axon, thus further depolarizing the membrane in a positive feedback fashion.
2. The inward diffusion of Na^+ causes a reversal of the membrane potential from -65 mV to $+40$ mV.
3. The opening of K^+ gates and outward diffusion of K^+ causes the reestablishment of the resting membrane potential; this is called repolarization.
4. Action potentials are all-or-none events.
5. The refractory periods of an axon membrane prevent action potentials from running together.
6. Stronger stimuli produce action potentials at greater frequency.
C. One action potential serves as the depolarization stimulus for production of the next action potential in the axon.
1. In unmyelinated axons, action potentials are produced fractions of a micrometer apart.
2. In myelinated actions, action potentials are only produced at the nodes of Ranvier; this saltatory conduction is faster than conduction in an unmyelinated nerve fiber.

IV. Synaptic Transmission
Synaptic transmission may be either electrical or chemical. The predominant type are chemical synapses, in which neurotransmitters are released from the endings of nerve fibers in response to stimulation by action potentials.
A. Gap junctions are electrical synapses, found in cardiac muscle, smooth muscle, and some synapses in the CNS.
B. In chemical synapses, neurotransmitters are packaged in synaptic vesicles and released by exocytosis into the synaptic cleft.

V. Acetylcholine as a Neurotransmitter
Acetylcholine can stimulate depolarization of skeletal muscle fibers and neurons by opening the chemically regulated gates. This depolarization serves as a stimulus for the production of action potentials in the postsynaptic cell.
A. The combination of ACh with its receptor protein in the postsynaptic membrane causes the chemically regulated gates to open and produces depolarizations called end plate potentials.
1. ACh is inactivated by acetylcholinesterase.
2. Depolarizations called end plate potentials are produced by ACh in skeletal muscle cells.
3. End plate potentials are graded depolarizations that stimulate the production of action potentials in skeletal muscle fibers.
B. Depolarization produced by neurotransmitters in neurons is called an excitatory postsynaptic potential (EPSP).
1. EPSPs are graded and capable of summation, and they decrease in amplitude as they are conducted.
2. EPSPs produced at synapses in the dendrites or cell body travel to the axon hillock, stimulate opening of voltage-regulated gates, and generate action potentials in the axon.

VI. Catecholamine and Other Neurotransmitters
The actions of catecholamine neurotransmitters are mediated by cyclic AMP produced by the postsynaptic cell. There are also many putative neurotransmitters.
A. Catecholamines include dopamine, norepinephrine, and epinephrine.
1. These neurotransmitters are inactivated after being released primarily by reuptake into the presynaptic nerve endings.
2. Catecholamines activate adenyl cyclase in the postsynaptic cell, which catalyzes the formation of cyclic AMP.
3. Cyclic AMP, formed in the postsynaptic cell, produces the effects of the neurotransmitter.

B. In addition to the classical neurotransmitters, neurons produce a large number of other chemicals that are believed to have a neurotransmitter function.

VII. Synaptic Integration

Excitatory inputs can summate with each other and with inhibitory postsynaptic potentials. The release of neurotransmitters can be modified by experience and other factors.

A. Spatial and temporal summation of EPSPs allows a sufficient depolarization to be produced to cause the stimulation of action potentials in the postsynaptic neuron.

B. Neurotransmitters that cause hyperpolarization of the postsynaptic membrane produce inhibitory postsynaptic potentials (IPSPs).

1. IPSPs and EPSPs from different synaptic inputs can summate.
2. The production of IPSPs is called postsynaptic inhibition.
3. Presynaptic inhibition occurs in an axoaxonic synapse and reduces the amount of neurotransmitter released by the inhibited neuron.

Review Activities

Objective Questions

Match the following structures of the brain to the region in which they are located:
1. Cerebellum
2. Cerebral cortex
3. Medulla oblongata
   (a) telencephalon
   (b) diencephalon
   (c) mesencephalon
   (d) metencephalon
   (e) myelencephalon
4. The neuroglial cells that form myelin sheaths in the peripheral nervous system are
   (a) oligodendrocytes
   (b) satellite cells
   (c) Schwann cells
   (d) astrocytes
   (e) microglia
5. A collection of neuron cell bodies located outside the CNS is called a
   (a) tract
   (b) nerve
   (c) nucleus
   (d) ganglion
6. Which of the following neurons is pseudounipolar?
   (a) sensory neurons
   (b) somatic motor neurons
   (c) neurons in the retina
   (d) autonomic motor neurons
7. Depolarization of an axon is produced by the
   (a) inward diffusion of Na^+
   (b) active extrusion of K^+
   (c) outward diffusion of K^+
   (d) inward active transport of Na^+
8. Repolarization of an axon during an action potential is produced by the
   (a) inward diffusion of Na^+
   (b) active extrusion of K^+
   (c) outward diffusion of K^+
   (d) inward active transport of Na^+

9. As the strength of a depolarizing stimulus to an axon is increased,
   (a) the amplitude of action potentials increases
   (b) the duration of action potentials increases
   (c) the speed with which action potentials are conducted increases
   (d) the frequency with which action potentials are produced increases
10. The conduction of action potentials in a myelinated nerve fiber is
    (a) saltatory
    (b) without decrement
    (c) faster than in an unmyelinated fiber
    (d) all of the above
11. Which of the following is *not* a characteristic of synaptic potentials?
    (a) They are all-or-none in amplitude.
    (b) They decrease in amplitude with distance.
    (c) They are produced in dendrites and cell bodies.
    (d) They are graded in amplitude.
    (e) They are produced by chemically regulated gates.
12. Which of the following is *not* a characteristic of action potentials?
    (a) They are produced by voltage-regulated gates.
    (b) They are conducted without decrement.
    (c) Na^+ and K^+ gates open at the same time.
    (d) The membrane potential reverses polarity during depolarization.
13. A drug that inactivates acetylcholinesterase
    (a) inhibits the release of ACh from presynaptic endings

(b) inhibits the attachment of ACh to its receptor protein
(c) increases the ability of ACh to stimulate muscle contraction
(d) all of the above
14. Postsynaptic inhibition is produced by
   (a) depolarization of the postsynaptic membrane
   (b) hyperpolarization of the postsynaptic membrane
   (c) axoaxonic synapses
   (d) post-tetanic potentiation

Essay Questions

1. Describe the formation of the neural crest, and explain how derivatives of neural crest tissue can be located in different parts of the body.
2. Compare the characteristics of action potentials with those of synaptic potentials.
3. Explain how voltage-regulated gates produce an all-or-none action potential.
4. Explain how action potentials are regenerated along an axon.
5. Explain why conduction in a myelinated axon is faster than in an unmyelinated axon.
6. Trace the course of events between the production of an EPSP and the generation of action potentials at the axon hillock. Explain the effect of spatial and temporal summation on this process.
7. Explain how an IPSP is produced and how IPSPs can inhibit activity of the postsynaptic neuron.

Selected Readings

Axelrod, J. June 1974. Neurotransmitters. *Scientific American.*

Barchas, J. D. et al. 1978. Behavioral neurochemistry: Neuroregulators and behavioral states. *Science* 200:964.

Bartus, R. T. et al. 1982. The cholinergic hypothesis of geriatric memory dysfunction. *Science* 217:408.

Block, I. B. et al. 1984. Neurotransmitter plasticity at the molecular level. *Science* 225:1266.

Blusztajn, J. K., and R. J. Wurtman. 1983. Choline and cholinergic neurons. *Science* 221:614.

Cottman, C. W., and M. Nieto-Sampedro. 1984. Cell biology of synaptic plasticity. *Science* 225:1287.

Coyle, J. T., D. L. Prince, and M. R. DeLong. 1983. Alzheimer's disease: A disorder of cortical cholinergic innervation. *Science* 2B 219:1184.

Jacobs, B. L., and M. E. Trulson. 1979. Mechanism of action of LSD. *American Scientist* 67:396.

Kandel, E. R. 1979. Psychotherapy and the single synapse. *New Engl. J. Med.* 301:1028.

Katz, B. November 1952. The nerve impulse. *Scientific American.*

Keynes, R. D. March 1979. Ion channels in the nerve cell membrane. *Scientific American.*

Krieger, D. T. 1983. Brain peptides: What, where, and why? *Science* 222:975.

Kuffler, S. W., and J. G. Nicholls. 1976. *From neuron to brain: A cellular approach to the function of the nervous system.* Sunderland, Mass.: Sinauer Associates, Inc., Publishers.

Lester, H. A. February 1977. The response of acetylcholine. *Scientific American.*

Morrell, P., and W. Norton. May 1980. Myelin. *Scientific American.*

Nathanson, J. A., and P. Greengard. August 1977. "Second messengers" in the brain. *Scientific American.*

Nicol, S. E., and I. I. Gottesman. 1983. Clues to the genetics and neurobiology of schizophrenia. *American Scientist* 71:398.

Schwartz, J. H. April 1980. The transport of substances in nerve cells. *Scientific American.*

Snyder, S. H. 1984. Drugs and neurotransmitter receptors in the brain. *Science* 224:22.

Snyder, S. H. 1980. Brain peptides as neurotransmitters. *Science* 209:976.

Stevens, C. F. September 1979. The neuron. *Scientific American.*

Wurtman, R. J. April 1982. Nutrients that modify brain function. *Scientific American.*

14 | Central Nervous System

Outline

General Features of the Central Nervous System
Functions of the Central Nervous System
Characteristics of the Central Nervous System
Cerebrum
Structure of the Cerebrum
Lobes of the Cerebrum
White Matter of the Cerebrum
Basal Ganglia
Diencephalon
Thalamus
Hypothalamus
Epithalamus
Pituitary Gland
Mesencephalon
Metencephalon
Pons
Cerebellum
Myelencephalon
Medulla Oblongata
Reticular Formation
Meninges of the Central Nervous System
Dura Mater
Arachnoid Membrane
Pia Mater
Ventricles and Cerebrospinal Fluid
Ventricles of the Brain
Cerebrospinal Fluid
Brain Metabolism
Blood-Brain Barrier
Neurotransmitters
Spinal Cord
Structure of Spinal Cord
Spinal Cord Tracts
Clinical Considerations
Neurological Assessment and Drugs
Developmental Problems
Injuries
Diseases and Infections
Degenerative Disorders of the Nervous System
Chapter Summary
Review Activities
Selected Readings

Concepts

The central nervous system (CNS), consisting of the brain and spinal cord, is covered with meninges, is bathed in cerebrospinal fluid, and contains gray and white matter.

The cerebrum, consisting of five paired lobes within two convoluted hemispheres, is concerned with higher brain functions, such as the perception of sensory impulses, the instigation of voluntary movement, the storage of memory, thought, and reasoning.

The diencephalon is a major autonomic region of the brain that consists of vital structures such as the thalamus, hypothalamus, epithalamus, and pituitary gland.

The mesencephalon contains the corpora quadrigemina, concerned with visual and auditory reflexes, and the cerebral peduncles, composed of fiber tracts. It also contains specialized nuclei that help to control posture and movement.

The metencephalon contains the pons, which relays impulses, and the cerebellum, which coordinates skeletal muscle contractions.

The medulla oblongata, contained within the myelencephalon, connects to the spinal cord and contains nuclei for the cranial nerves and vital autonomic functions.

The CNS is covered by protective meninges, consisting of a dura mater, an arachnoid membrane, and a pia mater.

The ventricles, central canal, and subarachnoid space contain cerebrospinal fluid, formed by the active transport of substances from blood plasma in the choroid plexus.

The brain has a high rate of metabolic activity and a unique internal environment because of the blood-brain barrier and the numerous specific chemicals synthesized and secreted by the cells of the brain.

The spinal cord consists of centrally located gray matter involved in reflexes, and peripherally located ascending and descending tracts of white matter, which conduct impulses to and from the brain.

General Features of the Central Nervous System

The central nervous system (CNS), consisting of the brain and spinal cord, is covered with meninges, is bathed in cerebrospinal fluid, and contains gray and white matter.

Objective 1. Define the term *neurology.*
Objective 2. List the functions of the central nervous system.
Objective 3. Describe the general characteristics of the brain and spinal cord.

The immensely complex brain and spinal cord constitute the central nervous system (CNS). The CNS, along with the endocrine system, regulates the functions of other body systems. The brain, however, does much more than that, and its potential is perhaps greatly underestimated. It is incomprehensible that one's personality, thoughts, and aspirations result from the functioning of a body organ. Plato referred to the brain as "the divinest part of us." The thought processes of this organ have devised the technology for launching space shuttles, curing diseases, and splitting atoms. But with all of these achievements, the brain still remains amazingly ignorant of its own workings.

Neurology *(nu-rol'o-je),* or the study of the nervous system, has been referred to as "the last frontier of anatomy and physiology." Some basic questions concerning the functioning of the nervous system remain unanswered: How do nerve cells store and retrieve memory? What are the roles of the many chemical compounds within the brain? What causes mental illness or senility? Scientists are still developing the technology and skills necessary to understand the functional complexity of the nervous system. The next few decades should be very interesting as scientists come to better understand the nervous system.

Functions of the Central Nervous System

The nervous system, in general, is specialized to respond to stimuli and relay nerve impulses to the brain. The interpretation of nerve impulses is referred to as *perception.* Perception is made possible by the highly specialized neurons. The nervous system is present throughout the body as it functions with the endocrine system to coordinate closely the activities of the other body systems. In addition to integrating body activities, the nervous system has the ability to store experiences (*memory*) and to establish patterns of response based on prior experiences (*learning*).

The specific functions of the CNS are

1. orientating the body to internal and external environments;
2. coordinating and controlling bodily activities;
3. assimilating experiences requisite to memory, learning, and intelligence; and
4. programming instinctual behavior (apparently more important in vertebrates other than humans).

---

neurology: Gk. *neuron,* nerve; L. *logos,* study of

Figure 14.1. The CNS consists of the brain and the spinal cord, both of which are covered with meninges and bathed in cerebrospinal fluid.

Characteristics of the Central Nervous System

The entire delicate CNS is protected by a bony encasement—the cranium surrounding the brain and the vertebral column surrounding the spinal cord. The **meninges** *(mĕ-nin'jēz)* form a protective membrane between the bone and the soft tissue of the CNS. The CNS is bathed in **cerebrospinal** *(ser''ĕ-bro-spi'nal)* **fluid,** which circulates within the hollow **ventricles** of the brain, the **central canal** of the spinal cord (fig. 14.1), and the **subarachnoid** *(sub''ah-rak'noid)* **space** surrounding the entire CNS.

The CNS is composed of gray and white matter. **Gray matter** consists of either nerve cell bodies and dendrites or bundles of unmyelinated axons and neuroglia. The gray matter of the brain exists as the outer convoluted **cortex layer** of the cerebrum and cerebellum. There are also specialized gray matter clusters of nerve cells, called **nuclei,** deep within the white matter. **White matter** forms the tracts within the CNS and consists of aggregations of myelinated axons and associated neuroglia.

The brain of an adult weighs nearly 1.5 kg (3–3.5 lbs) and is composed of an estimated 100 billion (10^{11}) neurons. Neurons communicate with one another by means of innumerable synapses between the axons and dendrites within the brain. Neurotransmission within the brain is regulated by specialized neurotransmitter chemicals which are thought to account for specific mental functions.

1. List some specific studies that would fit within the realm of neurology.
2. Using examples, explain the functions of the central nervous system.
3. What characteristics do the brain and spinal cord have in common? Describe the general features of each.

Cerebrum

The cerebrum, consisting of five paired lobes within two convoluted hemispheres, is concerned with higher brain functions, such as the perception of sensory impulses, the instigation of voluntary movement, the storage of memory, thought processes, and reasoning ability.

Objective 4. Describe the structure of the cerebrum, and list the functions of the cerebral lobes.

Objective 5. Define the term *electroencephalogram,* and discuss its clinical importance.

Objective 6. Describe the fiber tracts within the cerebrum.

Structure of the Cerebrum

The **cerebrum** *(ser'ĕ-brum),* located in the region of the telencephalon, is the largest and most obvious portion of the brain. It accounts for about 80% of the mass of the

cerebrum: L. *cerebrum,* brain

Figure 14.2.   The cerebrum. (a) a lateral view; (b) a superior view.

brain and is responsible for the higher mental functions, including memory and reason. The cerebrum consists of **right** and **left hemispheres,** which are incompletely separated by a **longitudinal fissure** (fig. 14.2). Portions of the two hemispheres are connected internally by the **corpus callosum** *(kah-lo′sum),* a large tract of white matter (see fig. 14.1). A portion of the meninges, called the **falx** *(falks)* **cerebri,** extends into the longitudinal fissure. The cerebrum contains a central cavity called the **lateral ventricle** (fig. 14.3), which is lined with ependymal cells and filled with cerebrospinal fluid.

The two cerebral hemispheres carry out different functions. In most people, the left hemisphere controls analytical and verbal skills such as reading, writing, and mathematics. The right hemisphere is the source of spatial and artistic kinds of intelligence. The corpus callosum unifies attention and awareness between the two hemispheres and permits a sharing of learning and memory. Severing the corpus callosum is a radical treatment to control severe epileptic seizures. Although this surgery is successful, it results in the cerebral hemispheres functioning as separate structures, each with its own memories and thoughts, competing for control. A more recent and effective technique of controlling epileptic seizures is a precise laser treatment of the corpus callosum.

The cerebrum consists of two layers. The surface layer, referred to as the **cerebral cortex,** is composed of gray matter that is 2–4 mm (0.08–0.16 in.) thick (fig. 14.3). Beneath the cerebral cortex is the thick **white matter** of the cerebrum, which constitutes the second layer. The cerebral cortex is characterized by numerous folds and grooves called **convolutions** *(kon-vo-lu′shuns).* Convolutions form during early fetal development when brain size increases rapidly and the cortex enlarges out of proportion to the underlying white matter. The elevated folds of the convolutions are the **gyri** (singular, *gyrus*), and the depressed grooves are the **sulci** *(sul′si)* (singular, *sulcus*). The convolutions greatly increase the area of the gray matter, which is comprised of nerve cell bodies.

Lobes of the Cerebrum

Each cerebral hemisphere is subdivided by deep sulci, or fissures, into five lobes, four of which appear on the surface of the cerebrum and are named according to the overlying cranial bones (fig. 14.4). The reasons for the separate cerebral lobes, as well as two cerebral hemispheres, have to do with specificity of function (table 14.1).

gyrus: Gk. *gyros,* circle
sulcus: L. *sulcus,* a furrow or ditch

Figure 14.3.   Sections through the cerebrum and
diencephalon. (*a*) a coronal section. (*b*) a transverse section.

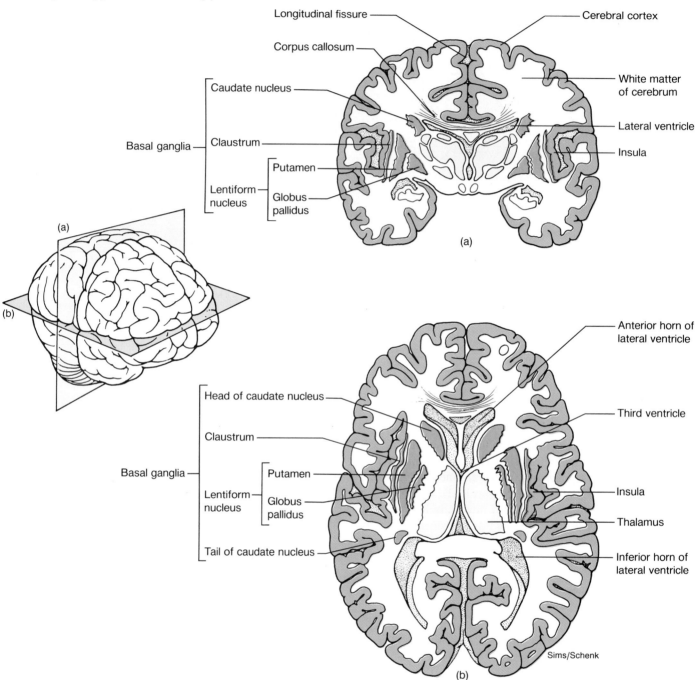

Table **14.1**	Functions of the cerebral lobes		
Lobe	**Functions**	**Lobe**	**Functions**
Frontal	Voluntary motor control of skeletal muscles; personality; higher intellectual processes (e.g., concentration, planning, and decision making); verbal communication	Temporal	Interpretation of auditory sensations; storage (memory) of auditory and visual experiences
Parietal	Somatesthetic interpretation (e.g., cutaneous and muscular sensations); understanding speech and formulating words to express thoughts and emotions	Occipital	Integrates movements in focusing the eye; correlating visual images with previous visual experiences and other sensory stimuli; conscious perception of vision
		Insula	Memory; integrates other cerebral activities

Figure 14.4. The lobes of the left cerebral hemisphere showing the principal motor and sensory areas of the cerebral cortex. Dotted lines show the division of the lobes.

Frontal Lobe. The frontal lobe forms the anterior portion of each cerebral hemisphere (fig. 14.4). A prominent deep furrow, called the **central sulcus (fissure of Rolando),** separates the frontal lobe from the parietal lobe. The central sulcus extends at right angles from the longitudinal fissure to the lateral sulcus. The **lateral sulcus (fissure of Sylvius)** extends laterally from the inferior surface of the cerebrum to separate the frontal and temporal lobes. The **precentral gyrus** (see figs. 14.2, 14.4), an important motor area, is positioned immediately in front of the central sulcus. The frontal lobe's functions include initiating voluntary motor impulses for the movement of skeletal muscles, analyzing sensory experiences, and providing responses relating to personality. The frontal lobes also involve responses related to memory, emotions, reasoning, judgment, planning, and verbal communication.

Broca's area, or the **motor speech** area, is a highly specialized portion of the frontal lobe, generally located in the left hemisphere in both right-handed and left-handed persons, immediately superior to the lateral cerebral sulcus

(fig. 14.4). Mental activity in Broca's area causes selective stimulation of motor impulses in motor centers elsewhere in the frontal lobe, which in turn, causes coordinated skeletal muscle movement in the pharynx and larynx. At the same time, motor impulses are sent to the respiratory muscles to regulate air movement across the vocal cords. The combined muscular stimulation translates thought patterns into speech.

Parietal Lobe. The parietal lobe is posterior to the frontal lobe. Except for the central sulcus along its anterior border, the parietal lobe lacks distinct boundaries. An important sensory area called the **postcentral gyrus** (see figs. 14.2, 14.4) is positioned immediately behind the central sulcus. The postcentral gyrus is designated as a somatesthetic area because it responds to stimuli from cutaneous and muscular receptors throughout the body.

The size of the portions of the precentral gyrus responsible for motor movement and the size of the portions of the postcentral gyrus that respond to sensory stimuli do not correspond to the size of the part of the body being served but to the number of motor units activated or the

fissure of Rolando: from Luigi Rolando, Italian anatomist, 1773–1831
fissure of Sylvius: from Franciscus Sylvius de la Boe, Dutch anatomist, 1614–72
Broca's area: from Pierre P. Broca, French neurologist, 1824–80

Figure 14.5. Motor and sensory areas of the cerebral cortex. (*a*) motor areas that control skeletal muscles. (*b*) sensory areas that receive somatesthetic sensations.

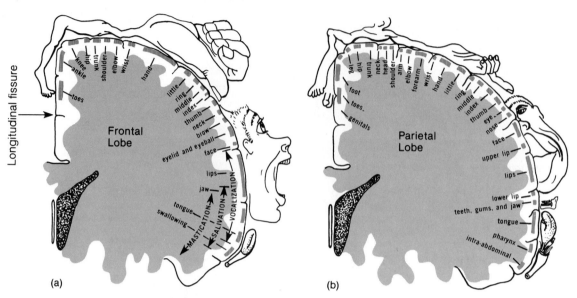

density of receptors (fig. 14.5). For example, because the hand has many motor units and sensory receptors, larger portions of the precentral and postcentral gyri serve it than serve the thorax, which is much larger in size.

In addition to responding to somatesthetic stimuli, the parietal lobe functions in the understanding of speech and in verbal articulation of thoughts and emotions. The parietal lobe also interprets the textures and shapes of objects as they are handled.

Temporal Lobe. The temporal lobe is located below the parietal lobe and the posterior portion of the frontal lobe. It is separated from both by the lateral sulcus. The temporal lobe contains auditory centers that receive sensory fibers from the cochlea of the ear. This lobe also interprets some sensory experiences and stores memories of both auditory and visual experiences.

Occipital Lobe. The occipital lobe forms the posterior portion of the cerebrum and has no distinct separation from the temporal and parietal lobes. The occipital lobe is superior to the cerebellum and is separated from it by a shell-like infolding of the meningeal layer called the **tentorium cerebelli** (see fig. 14.1). The principal functions of the occipital lobe concern vision. The occipital lobe integrates eye movements by directing and focusing the eye. It is also responsible for visual association, correlating visual images with previous visual experiences and other sensory stimuli.

Insula. The insula is a deep lobe of the cerebrum that cannot be viewed on the surface (see fig. 14.3). It is deep to the lateral sulcus and is covered by portions of the frontal, parietal, and temporal lobes. Little is known of the function of the insula, though it primarily integrates other cerebral activities and may have some function in memory.

Because of its size and position, portions of the cerebrum frequently suffer brain trauma. A concussion to the brain may cause a temporary or permanent impairment of cerebral functions; a stroke usually affects cerebral function. Much of what is known about cerebral function comes from observing body dysfunctions when specific regions of the cerebrum are traumatized.

Brain Waves. Neurons within the cerebral cortex continuously generate electrical activity, which can be recorded by electrodes attached to precise locations on the scalp, producing an **electroencephalogram (EEG).** An EEG pattern, commonly called *brain waves,* is the collective expression of millions of action potentials from neurons.

Brain waves are first emitted from a developing brain during early fetal development and continue throughout a person's life. The cessation of brain-wave patterns may be a decisive factor in the legal determination of death.

Certain distinct EEG patterns signify healthy mental functions. Deviations from these patterns are of clinical significance in diagnosing trauma, mental depression, hematomas, and various diseases, such as tumors, infections, and epilepsy. Normally, there are four kinds of EEG patterns (fig. 14.6).

insula: L. *insula,* island

Figure 14.6. Brain waves. (*a*) electrodes attached to scalp. (*b*) types of electroencephalograms.

(a)

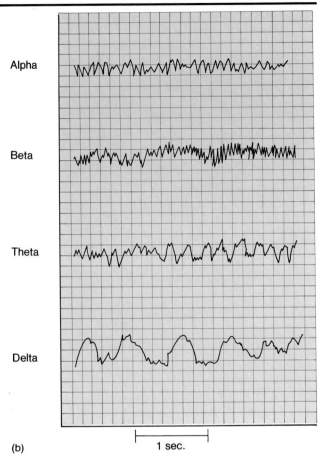

(b)

Alpha

Beta

Theta

Delta

1 sec.

1. **Alpha waves** are best recorded from the parietal and occipital regions while a person is awake and relaxed but with the eyes closed. These waves are rhythmic oscillations of about 10–12 cycles/second. The alpha rhythm of a child younger than eight years old occurs at a slightly lower frequency of 4–7 cycles/second.

2. **Beta waves** are strongest from the frontal lobes, especially the area near the precentral gyrus. These waves are sensory evoked and respond to visual and mental activity. Because they respond to stimuli from receptors and are superimposed on the continuous activity patterns, they constitute *evoked activity*. The frequency of beta waves is 13–25 cycles/second.

3. **Theta waves** are emitted from the temporal and occipital lobes. They have a frequency of 5–8 cycles/second and are common in newborn infants. The recording of theta waves in adults generally indicates severe emotional stress and can be a forewarning of a nervous breakdown.

4. **Delta waves** are seemingly emitted in a general pattern from the cerebral cortex. These waves have a frequency of 1–5 cycles/second and are common during sleep and in an awake infant. The presence of delta waves in an awake adult indicates brain damage.

White Matter of the Cerebrum

The thick white matter of the cerebrum is deep to the cortex (see fig. 14.3) and consists of dendrites and myelinated axons. These fibers form the billions of connections within the brain by which information in the form of electrical impulses is transmitted to the appropriate places. There are three types of fiber tracts within the white matter, which are named according to location and the direction in which they conduct impulses (fig. 14.7).

1. **Association fibers** are confined to a given cerebral hemisphere and conduct impulses between neurons within that hemisphere.

2. **Commissural** *(kom-mis'u-ral)* **fibers** connect the neurons and gyri of one hemisphere with those of the other. The **corpus callosum** and **anterior commissure** *(kom'ĭ-shŭr)* (fig. 14.8) are composed of commissural fibers.

3. **Projection fibers** form the ascending and descending tracts that transmit impulses from the cerebrum to other parts of the brain and spinal cord and from the spinal cord and other parts of the brain to the cerebrum.

Figure 14.7. Types of fiber tracts within the white matter associated with the cerebrum. (a) association fibers of a given hemisphere. (b) commissural fibers connecting the hemispheres and projection fibers connecting the hemispheres with other structures of the CNS. Note the decussation (crossing over) of projection fibers within the medulla oblongata.

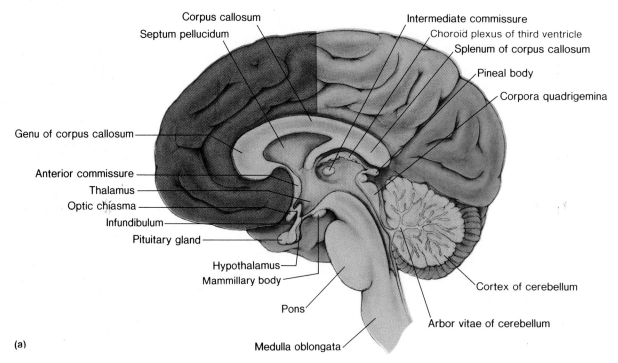

Figure 14.8. A midsagittal section through the brain. (a) a diagram. (b) a photograph.

(a)

Basal Ganglia

The **basal ganglia** are specialized paired masses of gray matter located deep within the white matter of the cerebrum (see fig. 14.3). The most prominent of the basal ganglia is the **corpus striatum,** so named because of its striped appearance. The corpus striatum is composed of several masses of nuclei. The **caudate nucleus** is the upper mass. A thick band of white matter lies between the caudate nucleus and the next lower two masses, collectively called the **lentiform nucleus.** The lentiform nucleus consists of a lateral portion, called the **putamen** *(pu-ta'men),* and a medial portion, called the **globus pallidus.**

The basal ganglia are associated with other structures of the brain, particularly within the mesencephalon. The caudate nucleus and putamen of the basal ganglia control unconscious contractions of certain skeletal muscles, such as those of the upper extremities involved in involuntary arm movements during walking. The globus pallidus regulates the muscle tone necessary for specific, intentional body movements. Neural diseases or physical trauma to the basal ganglia generally causes a variety of motor movement dysfunctions, including rigidity, tremor, and rapid and aimless movements.

corpus striatum: L. *corpus,* body; *striare,* striped
lentiform: L. *lentis,* elongated
putamen: L. *putare,* to cut, prune
globus pallidus: L. *globus,* sphere; *pallidus,* pale

1. *Diagram a lateral view of the cerebrum, and label the four superficial lobes and the fissures that separate them.*
2. *List the functions of each of the paired cerebral lobes.*
3. *What is a brain-wave pattern? How are these patterns monitored clinically?*
4. *Describe the arrangement of the fiber tracts within the cerebrum.*
5. *Define basal ganglia, and list their functions.*

Diencephalon

The diencephalon is a major autonomic region of the brain that consists of vital structures such as the thalamus, hypothalamus, epithalamus, and pituitary gland.

Objective 7. List the autonomic functions of the thalamus and the hypothalamus.

Objective 8. Describe the location and structure of the pituitary gland.

The **diencephalon** *(di''en-sef'ah-lon)* is the second subdivision of the forebrain and is almost completely surrounded by the cerebral hemispheres of the telencephalon. The third ventricle (see fig. 14.18) forms a midplane cavity within the diencephalon. The most important structures of the diencephalon are the thalamus, hypothalamus, epithalamus, and pituitary gland.

(b)

Thalamus

The **thalamus** *(thal'ah-mus)* is a large ovoid mass of gray matter, constituting nearly four-fifths of the diencephalon. It is actually a paired organ, with each portion positioned immediately below the lateral ventricle of its respective cerebral hemisphere (see figs. 14.3, 14.8). The principal function of the thalamus is to act as a relay center for all sensory impulses, except smell, to the cerebral cortex. Specialized masses of nuclei relay the incoming impulses to precise locations within the cerebral lobes for interpretation.

The thalamus also performs some sensory interpretation. The cerebral cortex discriminates pain and other tactile stimuli, but the thalamus responds to general sensory stimuli and provides crude awareness. The thalamus probably plays a role in the initial autonomic response of the body to intense pain and is, therefore, partially responsible for the physiological shock that frequently follows serious trauma.

Hypothalamus

The **hypothalamus** *(hi''po-thal'ah-mus)* is a small portion of the diencephalon located below the thalamus, where it forms the floor and part of the lateral walls of the third ventricle (fig. 14.8). The hypothalamus consists of several masses of nuclei interconnected with other parts of the nervous system. Despite its small size, the hypothalamus performs numerous vital functions, most of which relate directly or indirectly to the regulation of visceral activities. It also performs emotional and instinctual functions.

The hypothalamus has been described as an autonomic nervous center because of its role in accelerating or decreasing certain body functions. Another major function of the hypothalamus is to regulate the release of hormones from the pituitary gland. The principal autonomic and limbic functions of the hypothalamus are described below:

1. **Cardiovascular regulation.** Although the heart has an innate pattern of contraction, impulses from the hypothalamus cause autonomic acceleration or deceleration of the heart. Impulses from the posterior hypothalamus produce a rise in arterial blood pressure and an increase of the heart rate. Impulses from the anterior portion have an opposite effect. The impulses from these regions do not travel directly to the heart but pass first to the cardiovascular centers of the medulla oblongata.

2. **Body-temperature regulation.** Specialized nuclei within the anterior portion of the hypothalamus are sensitive to changes in body temperature. If the arterial blood flowing through this portion of the hypothalamus is above normal temperature, the hypothalamus initiates impulses that cause heat loss through sweating and vasodilation of cutaneous vessels of the skin. A blood temperature below normal causes the hypothalamus to relay impulses that result in heat production and retention through the initiation of shivering, the contraction of cutaneous blood vessels, and the cessation of sweating.

3. **Regulation of water and electrolyte balance.** Specialized *osmoreceptors* in the hypothalamus continuously monitor the osmotic concentration of the blood. An increased osmotic concentration due to lack of water causes antidiuretic hormone (ADH) to be produced by the hypothalamus and released from the posterior pituitary gland. At the same time, a *thirst center* within the hypothalamus produces feelings of thirst.

4. **Regulation of hunger and control of gastrointestinal activity.** The *feeding center* is a specialized portion of the lateral hypothalamus that monitors the blood glucose, fatty-acid, and amino-acid levels. Low levels of these substances in the blood are partially responsible for a sensation of hunger elicited from the hypothalamus. When sufficient amounts of food have been ingested, the *satiety (sah-ti'e-te) center* in the midportion of the hypothalamus inhibits the feeding center. The hypothalamus also receives sensory impulses from the abdominal viscera and regulates glandular secretions and the peristaltic movements of the digestive tract.

5. **Regulation of sleeping and wakefulness.** The hypothalamus has both a *sleep center* and a *wakefulness center* that function with other parts of the brain to determine the level of conscious alertness.

6. **Sexual response.** Specialized *sexual center* nuclei within the dorsal portion of the hypothalamus respond to sexual stimulation of the tactile receptors within the genital organs. The experience of orgasm involves neural activity within the sexual center of the hypothalamus.

7. **Emotions.** Found within the hypothalamus are a number of nuclei associated with specific emotional responses such as anger, fear, pain, and pleasure.

8. **Control of endocrine functions.** The hypothalamus produces neurosecretory chemicals that stimulate the anterior pituitary to release various hormones.

Obviously the hypothalamus is one of the most vital structures of the body. Dysfunction of the hypothalamus may seriously affect the autonomic, somatic, or psychic body functions. Unsurprisingly, the hypothalamus is implicated as a principal factor in psychosomatic illness. Insomnia, peptic ulcers, palpitation of the heart, diarrhea, and constipation are a few symptoms of psychosomatic problems.

---

thalamus: L. *thalamus,* inner room

Epithalamus

The **epithalamus** is the dorsal portion of the diencephalon that includes a thin roof over the third ventricle. The inside lining of the roof consists of a vascular **choroid plexus** where cerebrospinal fluid is produced. A small cone-shaped mass, called the **pineal gland (epiphysis),** which has a neuroendocrine function, extends outward from the posterior end of the epithalamus (see fig. 14.8). The posterior commissure, located ventral to the pineal gland, is a tract of commissural fibers that connects the superior colliculi.

Pituitary Gland

The **pituitary** *(pĭ-tu'ĭ-tār"e)* **gland,** or **hypophysis** *(hi-pof'ĭ-sis),* is positioned on the inferior aspect of the diencephalon and is attached to the hypothalamus by a stalk-like structure called the **infundibulum** *(in"fun-dib'u-lum)* (see fig. 14.8). The pituitary is a rounded, pea-shaped gland measuring about 1.3 cm (0.5 in.) in diameter. It is covered by the dura mater and is supported by the sella turcica of the sphenoid bone. The circle of Willis surrounds the highly vascular pituitary gland, providing it with a rich blood exchange. The pituitary, which has an endocrine function, is structurally and functionally divided into an anterior portion, called the **adenohypophysis** *(ad"ĕ-no-hi-pof'ĭ-sis),* and a posterior portion, called the **neurohypophysis.**

1. *Discuss what is meant by the statement that the thalamus is the pain center of the brain.*
2. *List the body systems and the functions over which the hypothalamus has some control.*
3. *Describe the location of the pituitary gland relative to the rest of the diencephalon and the rest of the brain. What are the two portions of the pituitary gland and how are they supported?*

Mesencephalon

The mesencephalon contains the corpora quadrigemina, concerned with visual and auditory reflexes, and the cerebral peduncles, composed of fiber tracts. It also contains specialized nuclei that help to control posture and movement.

Objective 9. List the structures of the mesencephalon, and explain their functions.

The mesencephalon *(mes"en-sef'ah-lon),* or midbrain, is a short section of the brain stem between the diencephalon and the pons (see fig. 14.11). Within the midbrain is the **cerebral aqueduct (aqueduct of Sylvius)** (see fig. 14.18), which connects the third and fourth ventricle.

The midbrain also contains the corpora quadrigemina (fig. 14.8), cerebral peduncle (see figs. 14.7, 14.11), red nucleus, and substantia nigra.

The **corpora quadrigemina** *(kwod"rĭ-jem'ĭ-nah)* are the four rounded elevations on the dorsal portion of the midbrain. The two upper eminences, the **superior colliculi,** are concerned with visual reflexes. The two posterior eminences, the **inferior colliculi,** are responsible for auditory reflexes. The **cerebral peduncles** *(pe-dung'k'l)* are a pair of cylindrical structures composed of ascending and descending projection fiber tracts that support and connect the cerebrum to the other regions of the brain.

The **red nucleus** is deep within the midbrain between the cerebral peduncle and the cerebral aqueduct. The red nucleus is gray matter that connects the cerebral hemispheres and the cerebellum and functions in reflexes concerned with motor coordination and maintaining posture. Another nucleus called the **substantia nigra** is ventral to the red nucleus. The substantia nigra is thought to inhibit forced involuntary movements.

1. *Discuss what possible symptoms might be apparent as a result of a tumor in the midbrain.*
2. *Which structures of the midbrain function with the cerebellum in controlling posture and movement?*

Metencephalon

The metencephalon contains the pons, which relays impulses, and the cerebellum, which coordinates skeletal muscle contractions.

Objective 10. Describe the location and structure of the pons and cerebellum, and list their functions.

The metencephalon *(met"en-sef'ah-lon)* is the most superior portion of the hindbrain. Two vital structures of the metencephalon are the pons and cerebellum. The cerebral aqueduct of the mesencephalon enlarges to become the **fourth ventricle** (see fig. 14.18) within the metencephalon.

Pons

The **pons** can be observed as a rounded bulge on the underside of the brain, between the midbrain and the medulla oblongata (fig. 14.9). The pons consists of white fiber tracts that course in two principal directions. The surface fibers extend transversely to connect with the cerebellum through the middle cerebral peduncles. The deeper longitudinal fibers are part of the motor and sensory tracts that connect the medulla with the tracts of the midbrain.

---

pineal: L. *pinea,* pine cone
pituitary: L. *pituita,* phlegm (this gland was originally thought to secrete mucus into nasal cavity)
infundibulum: L. *infundibulum,* funnel
aqueduct of Sylvius: from Jacobus Sylvius, French anatomist, 1478–1555

---

corpora quadrigemina: L. *corpus,* body; *quadri,* four; *geminus,* twin
colliculus: L. *colliculus,* small mound
pons: L. *pons,* bridge

Figure 14.9. Nuclei within the pons and medulla oblongata that constitute the respiratory center.

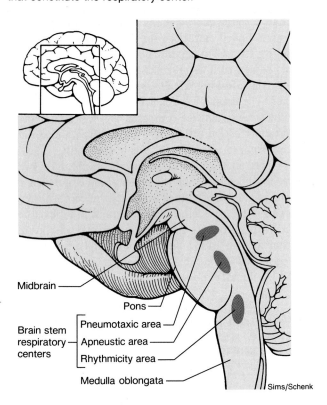

Midbrain

Pons

Brain stem respiratory centers { Pneumotaxic area / Apneustic area / Rhythmicity area

Medulla oblongata

Sims/Schenk

Figure 14.10. The structure of the cerebellum. (*a*) a superior view; (*b*) an inferior view.

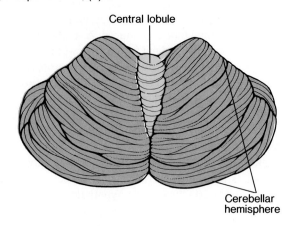

Central lobule

Cerebellar hemisphere

(a)

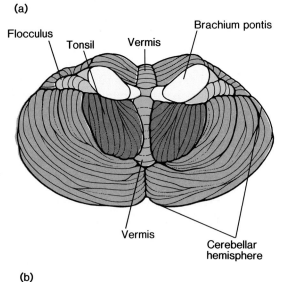

Flocculus
Tonsil
Vermis
Brachium pontis
Vermis
Cerebellar hemisphere

(b)

Scattered throughout the pons are several nuclei associated with specific cranial nerves. The cranial nerves that have nuclei within the pons include the trigeminal (V), which transmits impulses for chewing and sensory sensations from the head; the abducens (VI), which controls certain movements of the eyeballs; the facial (VII), which transmits impulses for facial movements and sensory sensations from the taste buds; and the vestibular branches of the vestibulocochlear (VIII), which maintain equilibrium and convey impulses of sound.

Other nuclei of the pons function with nuclei of the medulla oblongata to regulate the rate and depth of breathing. The two respiratory centers of the pons are called the **apneustic** and the **pneumotaxic areas** (fig. 14.9).

Cerebellum

The **cerebellum** *(ser"ĕ-bel'um)* is the second largest structure of the brain. It is located in the metencephalon and occupies the inferior and posterior aspect of the cranial cavity. The cerebellum is separated from the overlying cerebrum by a **transverse fissure** (see fig. 14.1). A portion of the meninges called the **tentorium cerebelli** extends into the transverse fissure. The cerebellum consists of two **hemispheres** and a central constricted area called the **vermis** (fig. 14.10). The **falx cerebelli** is the portion of the meninges that partially extends between the hemispheres.

Like the cerebrum, the cerebellum has a thin, outer layer of gray matter, called the **cerebellar cortex,** and a thick, deeper layer of white matter. The cerebellum is convoluted into a series of slender, parallel **gyri.** The tracts of white matter within the cerebellum have a distinctive branching pattern called the **arbor vitae,** which can be seen in a sagittal view (see fig. 14.8).

Three paired bundles of nerve fibers, called **cerebellar peduncles,** support the cerebellum and provide it with tracts for communicating with the rest of the brain (fig. 14.11). The cerebellar peduncles are as follows:

1. **Superior cerebellar peduncles** connect the cerebellum with the midbrain. The fibers within these peduncles originate primarily from specialized **dentate nuclei** within the cerebellum and pass through the red nucleus to the thalamus and then to the motor areas of the cerebral cortex. Impulses through the fibers of these peduncles provide feedback to the cerebrum.

---

cerebellum: L. diminutive of *cerebrum,* brain
vermis: L. *vermis,* worm

arbor vitae: L. *arbor,* tree; *vitae,* life
peduncle: L. diminutive of *pes,* foot

Figure 14.11. The cerebellar peduncles can be seen when the cerebellum has been removed from its attachment to the brain stem.

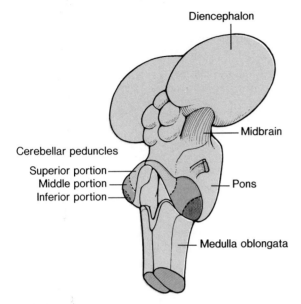

Diencephalon

Midbrain

Cerebellar peduncles
Superior portion
Middle portion
Inferior portion

Pons

Medulla oblongata

2. **Middle cerebellar peduncles** convey impulses of voluntary movement from the cerebrum through the pons and to the cerebellum.
3. **Inferior cerebellar peduncles** connect the cerebellum with the medulla oblongata and the spinal cord. They contain both incoming vestibular and proprioceptive fibers and outgoing motor fibers.

The principal function of the cerebellum is coordinating skeletal muscle contractions. The cerebellum does this by recruiting precise motor units within the muscles voluntarily, subconsciously, or reflexively. Impulses for voluntary muscular movement originate in the cerebral cortex and are coordinated by the cerebellum. The cerebellum constantly initiates impulses to selective motor units for maintaining posture and muscle tone. The cerebellum also adjusts to incoming impulses from **proprioceptors** *(pro''pre-o-sep'tor)* within muscles, tendons, joints, and special sense organs.

Trauma or diseases of the cerebellum, such as cerebral palsy or a stroke, frequently causes an impairment of skeletal muscle function. Movements become jerky and uncoordinated in a condition known as **ataxia.** There is also a loss of equilibrium, resulting in a disturbance of gait. Alcohol intoxication causes similar uncoordinated body movements.

1. *Describe the locations and relative sizes of the pons and the cerebellum.*
2. *Which cranial nerves have nuclei located within the pons?*
3. *Define the terms* tentorium cerebelli, vermis, arbor vitae, *and cerebellar peduncles.*
4. *List the functions of the pons and the cerebellum.*

Myelencephalon

The medulla oblongata, contained within the myelencephalon *(mi''ĕ-len-sef-ah-lon),* connects to the spinal cord and contains nuclei for the cranial nerves and vital autonomic functions.

> *Objective 11.* Describe the location and structure of the medulla oblongata, and list its functions.
> *Objective 12.* Define the term *reticular formation,* and explain its function.

Medulla Oblongata

The **medulla oblongata** *(mĕ-dul'ah ob''long-gah'tah),* or simply **medulla,** is a bulbous structure about 3 cm (1 in.) long that is continuous with the pons anteriorly and the spinal cord posteriorly at the level of the foramen magnum (see figs. 14.7, 14.8). Externally, the medulla resembles the spinal cord, except for the two triangular, elevated structures, called **pyramids,** on the ventral side and an oval enlargement, called the **olive,** on each lateral surface. The **fourth ventricle** is the space within the medulla. It is continuous posteriorly with the central canal of the spinal cord and anteriorly with the cerebral aqueduct.

The medulla is composed of vital nuclei and white matter that form all of the descending and ascending tracts communicating between the spinal cord and various parts of the brain. Most of the fibers within these tracts cross over to the opposite side through the pyramidal region of the medulla, permitting one side of the brain to receive information from and send information to the opposite side of the body.

The gray matter of the medulla consists of several important nuclei for the cranial nerves, sensory relay, and for autonomic functions (fig. 14.12). The **nucleus ambiguus** and the **hypoglossal nucleus** are the centers from which arise the vestibulocochlear (VIII), glossopharyngeal (IX), accessory (XI), and hypoglossal (XII) nerves. The vagus nerves (X) arise from **vagus nuclei,** one on each lateral side of the medulla adjacent to the fourth ventricle. The **nucleus gracilis** and the **nucleus cuneatus** relay sensory information to the thalamus and then to the cerebral cortex. The **inferior olivary nuclei** and the **accessory olivary nuclei** of the olive mediate impulses passing from the forebrain and midbrain through the inferior cerebellar peduncles to the cerebellum.

Three other nuclei within the medulla function as autonomic centers for controlling vital visceral functions.

1. **Cardiac center.** Both *inhibitory* and *accelerator fibers* arise from nuclei of the cardiac center. Inhibitory impulses constantly travel through the vagus nerves to slow the heartbeat. Accelerator

medulla: L. *medulla,* marrow

Figure 14.12. A sagittal section of the medulla oblongata and pons showing the cranial nerve nuclei of gray matter.

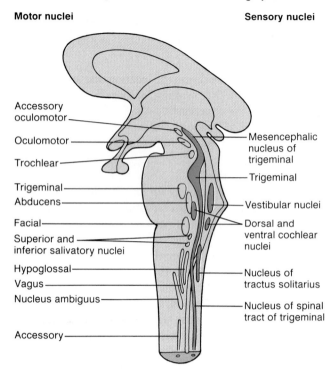

Motor nuclei | Sensory nuclei

- Accessory oculomotor
- Oculomotor
- Trochlear
- Trigeminal
- Abducens
- Facial
- Superior and inferior salivatory nuclei
- Hypoglossal
- Vagus
- Nucleus ambiguus
- Accessory

- Mesencephalic nucleus of trigeminal
- Trigeminal
- Vestibular nuclei
- Dorsal and ventral cochlear nuclei
- Nucleus of tractus solitarius
- Nucleus of spinal tract of trigeminal

impulses travel through the spinal cord and eventually innervate the heart through fibers within spinal nerves T-1–T5.

2. **Vasomotor center.** Nuclei of the vasomotor center send impulses via the spinal cord and spinal nerves to the smooth muscles of arteriole walls, causing them to constrict and elevate arterial blood pressure.

3. **Respiratory center.** The respiratory center of the medulla controls the rate and depth of breathing and functions with nuclei of the pons (see fig. 14.9) to produce rhythmic breathing.

Other nuclei of the medulla function as centers for nonvital respiratory movements such as sneezing, coughing, swallowing, and vomiting. Some of these activities, such as swallowing, may be initiated voluntarily, but once they progress to a certain point they become involuntary and cannot be stopped.

Reticular Formation

The **reticular formation** is a complex network of nuclei and nerve fibers within the brain stem that functions as the *reticular activating system (RAS),* which arouses the cerebrum. Portions of the reticular formation are located in the spinal cord, pons, midbrain, and parts of the thalamus and hypothalamus (fig. 14.13). The reticular formation contains ascending and descending fibers from most of the structures within the brain.

Nuclei within the reticular formation generate a continuous flow of impulses unless they are inhibited by other

Figure 14.13. The reticular activating system. The arrows indicate the direction of impulses along nerve pathways that connect with the RAS.

- Thalamus
- Midbrain
- Pons
- Reticular formation
- Medulla

parts of the brain. The principal functions of the RAS are to keep the cerebrum in a state of alert consciousness and to monitor selectively the afferent impulses perceived by the cerebrum. The RAS also helps the cerebellum activate selected motor units to maintain muscle tonus and produce smooth, coordinated contractions of skeletal muscles.

The RAS is sensitive to changes in and trauma to the brain. The sleep response is thought to occur because of a decrease in activity within the RAS, perhaps due to the secretion of specific neurotransmitters. A blow to the head or certain drugs and diseases may damage the RAS, causing unconsciousness. A coma is a state of unconsciousness and inactivity of the RAS that even the most powerful external stimuli cannot disturb.

1. When a nurse prepares a case study, the location, structure, function, and possible dysfunction of an afflicted structure or organ is reported. Prepare a brief case study of a patient who has suffered severe trauma to the medulla oblongata from a blow to the back of the skull.
2. Which cranial nerves arise from the medulla oblongata?
3. Describe the location of the reticular formation. What are two functions of the RAS?

Meninges of the Central Nervous System

The CNS is covered by protective meninges, consisting of a dura mater, an arachnoid membrane, and a pia mater.

Objective 13. Describe the position of the meninges as they protect the CNS.

meninges: L. plural form of *meninx,* membrane

Figure 14.14.   The meninges of the brain.

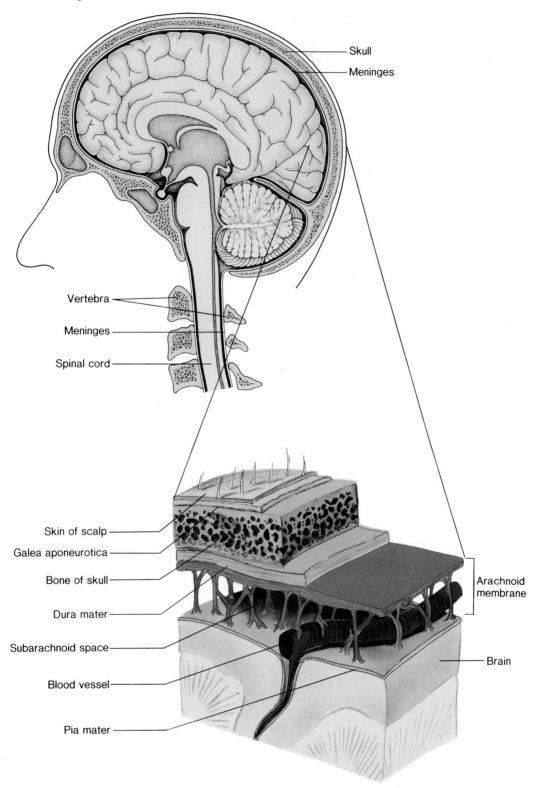

The entire delicate CNS is protected by a bony encasement: the cranium, surrounding the brain, and the vertebral column, surrounding the spinal cord. It is also protected by three connective tissue membranous coverings called the meninges. See figures 14.14, 14.15, 14.16, 14.17. Individually, from the outside in, they are known as the dura mater, the arachnoid membrane, and the pia mater.

Figure 14.15.   The dural sinuses.

Dural sinus

Skin of scalp

Arachnoid granulations

Galea aponeurotica

Bone of skull

Periosteal layer
of the dura mater

Dura mater

Meningeal layer
of the dura mater

Arachnoid membrane

Subarachnoid space

Pia mater

Cortex of brain

Partition of meninges

Figure 14.16.   A cross section showing the relationship of
the spinal cord and meninges to the protective epidural
space.

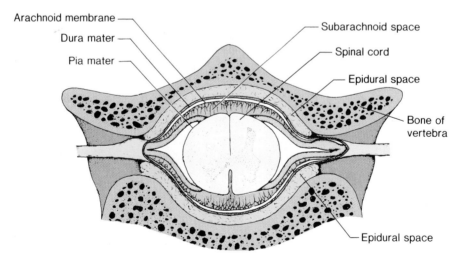

Arachnoid membrane

Subarachnoid space

Dura mater

Spinal cord

Pia mater

Epidural space

Bone of
vertebra

Epidural space

Figure 14.17.   The spinal cord and the meninges.

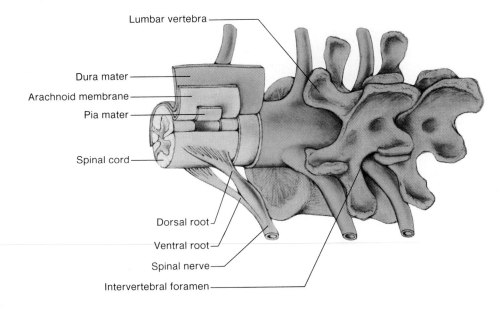

Lumbar vertebra

Dura mater

Arachnoid membrane

Pia mater

Spinal cord

Dorsal root

Ventral root

Spinal nerve

Intervertebral foramen

Dura Mater

The **dura mater** is in contact with the bone and is composed primarily of tough, white fibrous connective tissue. The **cranial dura mater** is a double-layered structure. The thicker, outer **periosteal** *(per''e-os'te-al)* **layer** adheres lightly to the cranium where it is the periosteum. The thinner, inner **meningeal layer** follows the general contour of the brain. The **spinal dura mater** is not double layered but is similar to the meningeal layer of the cranial dura mater.

The two layers of the cranial dura mater are fused over most of the brain. In certain regions, however, the layers are separated, enclosing **dural sinuses** (see fig. 14.15), which collect venous blood and drain it to the internal jugular veins of the neck.

In four locations, the meningeal layer of the cranial dura forms distinct septa to partition major structures on the surface of the brain and anchor the brain to the inside of the cranial case. These septa have been previously identified in this chapter and are reviewed in table 14.2.

The spinal dura mater forms a tough, tubular **dural sheath** that continues into the vertebral canal and surrounds the spinal cord. There is no connection between the dural sheath and the vertebrae forming the vertebral canal, but instead there is a potential cavity called the **epidural space** (see fig. 14.16). The epidural space is highly vascular and contains areolar and adipose connective tissue, which form a protective pad around the spinal cord.

---

dura mater: L. *dura*, hard; *mater,* mother

Table **14.2** Septa of the cranial dura mater	
Septa	**Location**
Falx cerebri	Extends downward into the longitudinal fissure to partition the right and left cerebral hemispheres; anchored anteriorly to the crista galli of the ethmoid bone and posteriorly to the tentorium
Tentorium cerebelli	Separates the occipital lobes of the cerebrum from the cerebellum; anchored to the tentorium, petrous bones, and occipital bone
Falx cerebelli	Partitions the right and left cerebellar hemispheres; anchored to occipital crest
Diaphragma sellae	Forms the roof of the sella turcica

Arachnoid Membrane

The **arachnoid membrane** is the middle of the three meninges. This delicate, netlike membrane spreads over the CNS but generally does not extend into the sulci or fissures of the brain. The **subarachnoid space,** located between the arachnoid membrane and the deepest meninx, the pia mater, contains cerebrospinal fluid. The subarachnoid space is maintained by delicate, web-like strands that connect the arachnoid membrane and pia mater (see fig. 14.15).

---

arachnoid: L. *arachnoides,* like a cobweb

Figure 14.18. The ventricles of the brain. (*a*) an anterior view; (*b*) a lateral view.

Lateral ventricle

Interventricular foramen

Third ventricle

Cerebral aqueduct

Fourth ventricle

To central canal of spinal cord

(a)

Lateral ventricle

Cerebral aqueduct

Fourth ventricle

Interventricular foramen

Third ventricle

To central canal of spinal cord

(b)

Pia Mater

The thin **pia mater** *(pi'ah ma'ter)* is attached to the surfaces of the CNS and follows the irregular contours of the brain and spinal cord. The pia mater is composed of modified areolar connective tissue. It is highly vascular and functions to support the vessels that nourish the underlying cells of the brain and spinal cord. The pia mater is specialized over the roofs of the ventricles, where along with the arachnoid membrane, it contributes to the formation of the choroid plexuses. Lateral extensions of the pia mater along the spinal cord form the **ligamentum denticulatum,** which attaches the cord to the dura mater.

pia mater: L. *pia,* soft or tender; *mater,* mother

Meningitis is an inflammation of the meninges, usually caused by certain bacteria or viruses. The arachnoid and the pia mater are the two meninges most frequently affected. Meningitis is accompanied by high fever and severe headache. Complications may cause sensory impairment, paralysis, or mental retardation. Untreated meningitis generally results in coma and death.

1. Contrast cranial and spinal dura mater.
2. Explain how the meninges support and protect the CNS.
3. Describe the location of the dural sinuses and the epidural space.

Figure 14.19. CT scans of the brain showing (*a*) a normal configuration of the ventricles and (*b*) an abnormal configuration of the ventricles due to hydrocephalism.

—Ventricles

(a)

—Ventricles

(b)

Ventricles and Cerebrospinal Fluid

The ventricles, central canal, and subarachnoid space contain cerebrospinal fluid, formed by the active transport of substances from blood plasma in the choroid plexuses.

> *Objective 14.* Discuss the formation, function, and flow of cerebrospinal fluid.

Cerebrospinal fluid *(ser''ĕ-bro-spi'nal floo'id), CSF,* is a clear, lymphlike fluid that forms a protective cushion around and within the CNS. The fluid also buoys the brain. CSF circulates through the various **ventricles** of the brain, the **central canal** of the spinal cord, and the **subarachnoid space** around the entire CNS. The cerebrospinal fluid returns to the circulatory system by draining through the walls of the **arachnoid villi.**

Ventricles of the Brain

The ventricles of the brain are connected to one another and to the central canal of the spinal cord (figs. 14.18, 14.19). Each of the two **lateral ventricles** (first and second ventricles) is located in each hemisphere of the cerebrum, inferior to the corpus callosum. The **third ventricle** is located in the diencephalon between the thalami. Each lateral ventricle is connected to the third ventricle by a narrow, oval opening called the **interventricular foramen** (foramen of Monro). The **fourth ventricle** is located in the

brain stem, within the pons, cerebellum, and medulla oblongata. The **cerebral aqueduct** (aqueduct of Sylvius) passes through the midbrain to link the third and fourth ventricles. The fourth ventricle also communicates posteriorly with the central canal. Cerebrospinal fluid exits from the fourth ventricle into the subarachnoid space (fig. 14.20) through three foramina: the **foramen of Magendie,** a medial opening, and two lateral **foramina of Luschka** (not illustrated). Cerebrospinal fluid returns to the venous blood through the arachnoid villi.

Internal hydrocephalus is a condition in which cerebrospinal fluid builds up within the ventricles of the brain (fig. 14.19b). It is more common in infants whose cranial sutures have not yet strengthened or ossified. If the pressure is excessive, the condition may have to be treated surgically.

External hydrocephalus, an accumulation of fluid within the subarachnoid space, usually results from an obstruction of drainage at the arachnoid villi.

A *pneumoencephalogram* is an X-ray picture of the skull and the brain, which is obtained after injecting air or oxygen into the cerebrospinal fluid surrounding the spinal cord and waiting for it to move to the ventricles and subarachnoid spaces of the brain (see fig. 14.21). Brain lesions and tumors may be identified using this technique.

---

foramen of Magendie: from Francois Magendie, French physiologist, 1783–1855
foramen of Luschka: from Hubert Luschka, German anatomist, 1820–75

---

foramen of Monro: from Alexander Monro, Jr., Scottish anatomist, 1733–1817

Figure 14.20. The flow of cerebrospinal fluid.

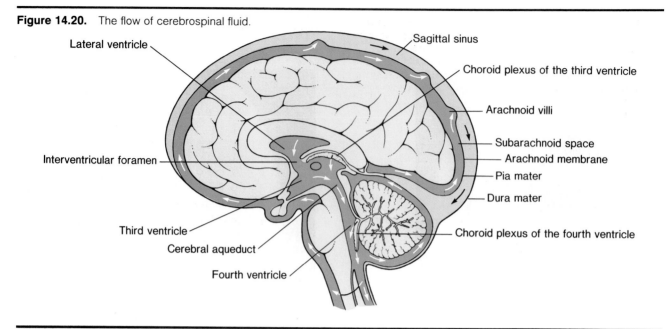

Figure 14.21. Pneumoencephalograms showing the ventricles of the brain of a 12-year-old child. (*a*) Lateral view; (*b*) anterior view.

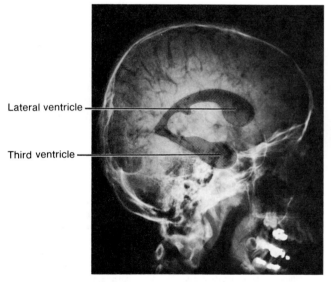

Cerebrospinal Fluid

CSF buoys the CNS and protects it from mechanical injury. CSF has a specific gravity of 1.007, which is a density close to that of brain tissue. The brain weight is about 1,500 grams, but suspended in CSF its buoyed weight is about 50 grams. This means that the brain has a near neutral buoyancy and can therefore function effectively as a relatively heavy organ. At a true neutral buoyancy, an object does not float or sink but is suspended in its fluid environment.

In addition to buoying the CNS, CSF reduces the damaging effect of an impact to the head by spreading the force over a larger area. CSF also helps to remove metabolic wastes from nervous tissue. Since the CNS lacks lymphatic circulation, the CSF moves cellular wastes into the venous return at its places of drainage.

The clear, watery CSF is continuously produced from materials within the blood by masses of specialized capillaries called **choroid plexuses** *(ko'roid plek'sus-es)* and, to a lesser extent, by secretions of the ependymal cells. The ciliated ependymal cells cover the choroid plexuses as well as line the central canal and presumably aid the movement of the CSF.

CSF is formed mainly by the active transport and ultrafiltration of substances within the blood plasma. CSF has more sodium, chloride, magnesium, and hydrogen ions than blood plasma and less calcium, potassium, and glucose. In addition, CSF contains some proteins, urea, and white blood cells.

As much as 800 ml of cerebrospinal fluid are produced each day, although only 140–200 ml are bathing

Figure 14.22. Positron emission tomographic (PET) brain scan of a schizophrenic patient receiving shocks to the right forearm while the eyes are closed. Patient was not medicated. Red areas indicate high glucose use (uptake of 18-F-deoxyglucose). The scan shows highest glucose uptake in the posterior region, where the brain's visual center is located.

the CNS at any moment. A person lying in a horizontal position has a slow but continuous circulation of cerebrospinal fluid, with a fluid pressure of about 10 mm Hg.

The homeostatic consistency of the CSF composition is critical and a chemical unbalance may have marked effects on CNS functions. An increase in amino acid glycine concentration, for example, produces hypothermia and hypotension as temperature and blood pressure regulatory mechanisms are disrupted. A slight change in pH may affect the respiratory rate and depth.

1. Describe the location of the ventricles within the brain.
2. What are the functions of cerebrospinal fluid?
3. Where is cerebrospinal fluid produced, and where does it drain?

Brain Metabolism

The brain has a high rate of metabolic activity and a unique internal environment because of the blood-brain barrier and the numerous specific chemicals synthesized and secreted by the cells of the brain.

Objective 15. Explain the importance of the blood-brain barrier in maintaining homeostasis within the brain.

Objective 16. Define the term *neurotransmitter,* and list the common neurotransmitters of the brain and their functions.

The metabolic rate of nervous tissue is very high and constant. The blood flow to the brain through the paired internal carotid arteries and vertebral arteries is approximately 20% of the cardiac output at rest. This continuous flow is so crucial that a failure of cerebral circulation for as short as ten seconds causes unconsciousness. The brain is composed of perhaps the most sensitive tissue of the body. Not only does it have to have continuous oxygen, but because of its high metabolic rate, it also has to have a continuous nutrient supply and rapid removal of wastes. The brain is also very sensitive to certain toxins and drugs. The CSF aids the metabolic needs of the brain through the distribution of nutrients and the removal of wastes. CSF also maintains a protective homeostatic environment within the brain. The blood-brain barrier and the secretory activities of neural tissue also helps to maintain homeostasis.

There are measurable increases in regional blood flow within the brain and in glucose and oxygen metabolism that accompany mental functions, including perception and emotion. These metabolic changes can be assessed through the use of **positron-emission tomography (PET).** The technique of a PET scan (fig. 14.22) is based on injecting radioactive tracer molecules labeled with carbon-11, fluorine-18, and oxygen-15 into the blood stream and photographing the gamma rays that are emitted from the patient's brain through the skull. PET scans are of value in studying neurotransmitters and neuroreceptors as well as the substrate metabolism of the brain. They are also of value in the diagnosis and treatment of several neurological disorders including schizophrenia and Parkinson's disease.

Blood-Brain Barrier

The **blood-brain barrier** (**BBB**) is a structural arrangement of capillaries, surrounding connective tissue, and specialized neuroglial cells, called astrocytes (see fig. 13.13), that selectively determine which substances can move from the plasma of the blood to the extracellular fluid of the brain. Certain substances such as H_2O, O_2, CO_2, glucose, and lipid-soluble compounds (alcohol, for example) pass readily through the BBB. The inorganic ions Na^+, K^+, and Cl^- pass more slowly so that the concentrations of these ions are different in the brain than in the plasma. Other substances, such as proteins, lipids, creatine, urea, inulin, certain toxins, and most antibiotics, are restricted in passage. The BBB is an important factor to consider when planning drug therapy for neurological disorders.

Neurotransmitters

Brain function depends on the presence and perhaps a delicate balance of **neurotransmitter** *(nu''ro-trans-mit'er)* chemicals. Neurotransmitter chemicals are secreted by the neurons that synthesize them. It is estimated that there are over two hundred specific neurotransmitter chemicals within the brain, each of which has a precise function.

Some of the excitatory neurotransmitters include acetylcholine, epinephrine, norepinephrine, serotonin, and dopamine. *Acetylcholine (as''ē-til-ko'lēn)* released at the synapses within the brain facilitates the transmission of nerve impulses. *Epinephrine (ep''ĭ-nef'rin)* and *norepinephrine* are thought to cause arousal and to maintain alertness. *Serotonin* is released by neurons in the mesencephalon and is thought to play a role in temperature regulation, sensory perception, and the onset of sleep. *Dopamine* is released by neurons in the substantia nigra and is concerned with motor control. A deficiency of dopamine is thought to cause Parkinson's disease, and an excess of dopamine is thought to be involved in schizophrenia.

Two common inhibitory neurotransmitters are gamma-aminobutyric acid (GABA) and glycine. *GABA* is released by specific cells of the cerebellum to inhibit certain motor neurons, and thus it is involved in coordination. *Glycine* is released by neurons in the spinal cord to inhibit transmission along certain spinal cord tracts.

Neuropeptides *(nu''ro-pep'tīd)* are a group of neurotransmitters composed of relatively short chains of amino acids. Many of them are better known as hormones in nonneural tissue. These include such proteins as somatostatin, vasopressin, oxytocin, angiotensin, and cholecystokinin.

Enkephalins and *endorphins* are two recently discovered neuropeptides that are involved in the transmission and perception of pain. Enkephalins are found in the spinal cord, limbic system (see fig. 16.12), and thalamus. Endorphins are found in the pituitary gland. These neuropeptides have an effect similar to the narcotic drug morphine in that they interact with the receptor sites that block pain. Enkephalins and endorphins are considered the body's natural pain killers.

Substance P is neuropeptide in the brain that aids in transmission of pain by exciting spinal neurons that impulse from pain receptors. The secretion of enkephalins and endorphins apparently produces an analgesic effect by inhibiting the release of Substance P in the spinal cord.

---

1. *Construct a flow chart depicting a substance that can flow freely across the BBB from the plasma to the nervous tissue. Indicate also a substance that passes more slowly and one that is restricted in its passage. Why is the BBB functionally important?*
2. *Distinguish between a neurotransmitter and a neuropeptide. Construct a table that lists the common neurotransmitters of the brain and their functions.*

Spinal Cord

The spinal cord consists of centrally located gray matter involved in reflexes, and peripherally located ascending and descending tracts of white matter, which conduct impulses to and from the brain.

> *Objective 17.* Describe the structure of the spinal cord.
>
> *Objective 18.* Describe the arrangement of ascending and descending tracts within the spinal cord.

The spinal cord is the portion of the CNS that extends through the neural canal of the vertebral column (fig. 14.23). It is continuous with the brain through the foramen magnum of the skull. The spinal cord has two principal functions:

1. It provides a means of neural communication to and from the brain through tracts of white matter. **Ascending tracts** conduct impulses from the peripheral sensory receptors of the body to the brain. **Descending tracts** conduct motor impulses from the brain to the muscles and glands.
2. It serves as a center for spinal reflexes. Specific nerve pathways enable some movements to be reflexive rather than initiated voluntarily by the brain. Movements of this type are not confined to skeletal muscles; reflexive movements of cardiac and smooth muscles control heart rate, breathing rate, blood pressure, and digestive activities. Spinal nerve pathways are also involved in swallowing, coughing, sneezing, and vomiting.

Structure of Spinal Cord

The **spinal cord** extends inferiorly from the position of the foramen magnum of the occipital bone to the level of the first lumbar vertebra (L1). The spinal cord is somewhat flattened dorsoventrally, making it oval in cross-section. Two prominent enlargements can be seen in a dorsal view (see fig. 14.23). The **cervical enlargement** is located between the third cervical and the second thoracic vertebrae. Nerves emerging from this region serve the upper extremities. The **lumbar enlargement** lies between the ninth and twelfth thoracic vertebrae. Nerves from the lumbar enlargement supply the lower extremities.

The embryonic spinal cord develops more slowly than the associated vertebral column; thus, in the adult, the cord does not extend beyond L1. The tapering, terminal portion of the spinal cord is called the **conus medullaris.** The **filum terminale,** a fibrous strand composed mostly of pia mater, extends inferiorly from the conus medullaris at the level of L1 to the coccyx. Nerves also radiate inferiorly

---

filum terminalis: L. *filum,* filament; *terminus,* end

Figure 14.23. The spinal cord and plexuses.

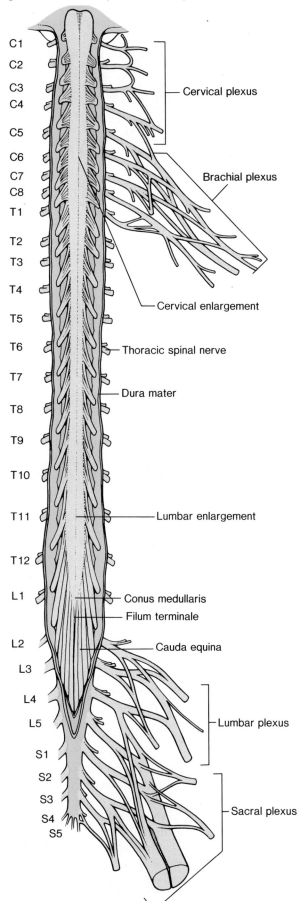

C1
C2
C3
C4
C5
C6
C7
C8
T1
T2
T3
T4
T5
T6
T7
T8
T9
T10
T11
T12
L1
L2
L3
L4
L5
S1
S2
S3
S4
S5

Cervical plexus

Brachial plexus

Cervical enlargement

Thoracic spinal nerve

Dura mater

Lumbar enlargement

Conus medullaris
Filum terminale

Cauda equina

Lumbar plexus

Sacral plexus

from the conus medullaris through the vertebral canal. These nerves are collectively referred to as the **cauda equina** because they resemble a horse's tail.

The spinal cord develops as thirty-one segments, each of which gives rise to a pair of **spinal nerves** that emerge from the cord through the intervertebral foramina. Two grooves, an **anterior median fissure** and a **posterior median sulcus,** extend the length of the spinal cord and partially divide the cord into right and left portions. The spinal cord, like the brain, is protected by three distinct meninges and is cushioned by cerebrospinal fluid. The pia mater contains an extensive vascular network.

The **gray matter** of the spinal cord is centrally located and surrounded by white matter. It is composed of nerve cell bodies, neuroglia, and unmyelinated internuncial neurons. The **white matter** consists of bundles, or tracts, of myelinated fibers of sensory and motor neurons.

The relative size and shape of the gray and white matter varies throughout the spinal cord. The amount of white matter increases toward the brain as the nerve tracts become thicker. More gray matter exists in the cervical and lumbar enlargements where innervations from the upper and lower extremities respectively make connections.

The core of gray matter roughly resembles the letter H (fig. 14.24). Projections of the gray matter within the spinal cord are called horns and are named according to the direction in which they project. The paired **posterior (dorsal) horns** extend posteriorly, and the paired **anterior (ventral) horns** project anteriorly. A pair of short **lateral horns** extend to the sides and are located between the other two pairs. Lateral horns are prominent only in the thoracic and upper lumbar regions. The transverse bar of gray matter that connects the paired horns across the center of the spinal cord is called the **gray commissure.** Contained within the gray commissure is the **central canal,** which is continuous with the ventricles of the brain and filled with cerebrospinal fluid.

Spinal Cord Tracts

Impulses are conducted through the ascending and descending tracts of the spinal cord within columns of white matter. The spinal cord has six columns of white matter called **funiculi** *(fu-nik'u-li),* which are named according to their relative position within the cord. The two **anterior (ventral) funiculi** are located between the two anterior horns of gray matter to either side of the anterior median fissure (fig. 14.24). The two **posterior (dorsal) funiculi** are located between the two posterior horns of gray matter to either side of the posterior median sulcus. Two **lateral funiculi** are located between the anterior and posterior horns of gray matter.

cauda equina: L. *cauda,* tail; *equus,* horse
commissure: L. *commissura,* a joining
funiculus: L. diminutive of *funis,* cord, rope

Figure 14.24. The (a) spinal cord in cross section. (b) a photograph of the spinal cord and nerve.

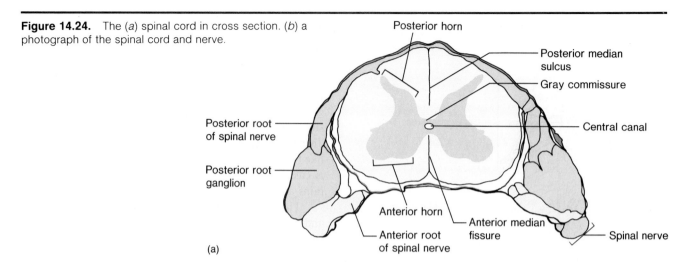

Posterior horn

Posterior median sulcus

Gray commissure

Central canal

Posterior root of spinal nerve

Posterior root ganglion

Anterior horn

Anterior median fissure

Anterior root of spinal nerve

Spinal nerve

(a)

(b)

Each funiculus consists of both ascending and descending tracts. The nerve fibers within the tracts are generally myelinated and have specific sites of origin and termination. In fact, the names of the various tracts reflect their origin and termination. The fibers of the tracts either remain on the same side of the brain and spinal cord or cross over within the medulla or the spinal cord. The crossing over of nerve tracts is referred to as *decussation (de''kus-sa'shun)*. Figure 14.25 illustrates a descending tract that decussates within the medulla, and figure 14.26 illustrates an ascending tract that decussates within the medulla.

---

decussation: L. *decussare,* to form an X intersection

The principal ascending and descending tracts within the funiculi are presented with their functions in table 14.3 and are illustrated in figure 14.27.

Descending tracts are grouped as either corticospinal or extrapyramidal. **Corticospinal (pyramidal) tracts** descend directly, without synaptic interruption, from the cerebral cortex to the lower motor neurons. The cell bodies of the neurons that contribute fibers to these tracts are located primarily in the precentral gyrus of the frontal lobe. Most (about 85%) of the corticospinal fibers decussate in the pyramids of the medulla (see fig. 14.7). The fibers that cross comprise the **lateral corticospinal tracts** and the remaining uncrossed fibers comprise the **anterior corticospinal tracts.** Because of the crossing of fibers from higher motor neurons in the pyramids, the right hemisphere primarily controls the musculature on the left side of the body, whereas the left hemisphere controls the right musculature.

Figure 14.25. Descending tracts composed of motor fibers that cross over within the medulla oblongata.

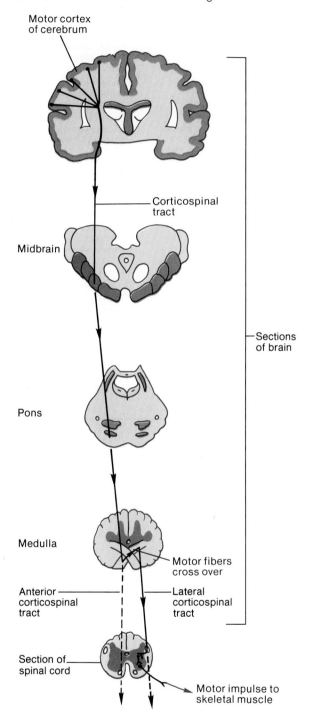

Motor cortex
of cerebrum

Corticospinal
tract

Midbrain

Sections
of brain

Pons

Medulla

Motor fibers
cross over

Anterior
corticospinal
tract

Lateral
corticospinal
tract

Section of
spinal cord

Motor impulse to
skeletal muscle

Figure 14.26. Ascending tracts composed of sensory fibers that cross over within the medulla.

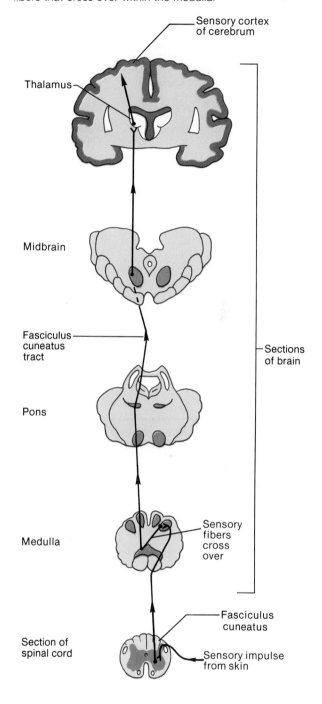

Sensory cortex
of cerebrum

Thalamus

Midbrain

Fasciculus
cuneatus
tract

Sections
of brain

Pons

Medulla

Sensory
fibers
cross
over

Section of
spinal cord

Fasciculus
cuneatus

Sensory impulse
from skin

Table 14.3 Principal ascending and descending tracts of spinal cord

Tract	Funiculus	Origin	Termination	Function
Ascending tracts				
Anterior spinothalamic	Anterior	Posterior horn on one side of cord but crosses to opposite side	Thalamus, then cerebral cortex	Conducts sensory impulses for crude touch and pressure
Lateral spinothalamic	Lateral	Posterior horn on one side of cord but crosses to opposite side	Thalamus, then cerebral cortex	Conducts pain and temperature impulses that are interpreted within cerebral cortex
Fasciculus gracilis and fasciculus cuneatus	Posterior	Peripheral afferent neurons; does not cross over	Nucleus gracilis and nucleus cuneatus of medulla; eventually thalamus, then cerebral cortex	Conducts sensory impulses from skin, muscles, tendons, and joints, which are interpreted as sensations of fine touch, precise pressures, and body movements.
Posterior spinocerebellar	Lateral	Posterior horn; does not cross over	Cerebellum	Conducts sensory impulses from one side of body to same side of cerebellum for subconscious proprioception necessary for coordinated muscular contractions
Anterior spinocerebellar	Lateral	Posterior horn; some fibers cross, others do not	Cerebellum	Conducts sensory impulses from both sides of body to cerebellum for subconscious proprioception necessary for coordinated muscular contractions

Source: Van De Graaff, Kent M., *Human Anatomy.* © 1984 Wm. C. Brown Publishers, Dubuque, Iowa. All Rights Reserved. Reprinted by permission.

Figure 14.27. A cross section showing the principal ascending and descending tracts within the spinal cord.

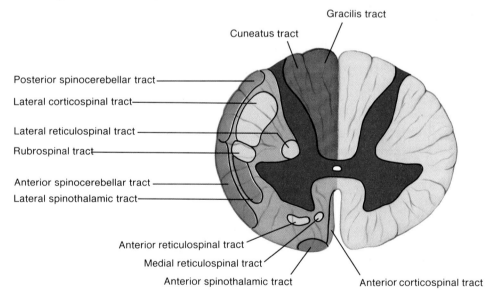

Tract	Funiculus	Origin	Termination	Function
Descending tracts				
Anterior corticospinal	Anterior	Cerebral cortex on one side of brain; crosses to opposite side of cord	Anterior horn	Conducts motor impulses from cerebrum to spinal nerves and outward to cells of anterior horns for coordinated, precise voluntary movements of skeletal muscle
Lateral corticospinal	Lateral	Cerebral cortex on one side of brain; crosses in base of medulla to opposite side of cord	Anterior horn	Conducts motor impulses from cerebrum to spinal nerves and outward to cells of anterior horns for coordinated, precise voluntary movements
Tectospinal	Anterior	Midbrain; crosses to opposite side of cord	Anterior horn	Conducts motor impulses to cells of anterior horns and eventually to muscles that move the head in response to visual, auditory, or cutaneous stimuli
Rubrospinal	Lateral	Midbrain (red nucleus); crosses to opposite side of cord	Anterior horn	Conducts motor impulses concerned with muscle tone and posture
Vestibulospinal	Anterior	Medulla; does not cross over	Anterior horn	Conducts motor impulses that regulate body tone and posture (equilibrium) in response to movements of head
Anterior and medial reticulospinal	Anterior	Reticular formation of brain stem	Anterior horn	Conducts motor impulses that control muscle tone and sweat gland activity
Lateral reticulospinal	Lateral	Reticular formation of brain stem	Anterior horn	Conducts motor impulses that control muscle tone and sweat gland activity

---

The corticospinal tracts appear to be particularly important in voluntary movements that require complex interactions between the motor cortex and sensory input. Speech, for example, is impaired when the corticospinal tracts are damaged in the thoracic region of the spinal cord, whereas involuntary breathing continues. Damage to the pyramidal motor system can be detected clinically by a positive **Babinski's reflex,** in which stimulation of the sole of the foot causes extension (upward movement) of the toes. This positive Babinski reflex is normal in infants because neural control is not yet fully developed.

The remaining descending tracts are extrapyramidal tracts which originate in the midbrain and brain stem regions (fig. 14.28). Electrical stimulation of the cerebral cortex, the cerebellum, and the basal ganglia indirectly evokes movements because of synaptic connections within extrapyramidal tracts.

The **reticulospinal** *(rĕ-tik″u-lo-spi′nal)* **tracts** are the major descending pathways of the extrapyramidal system. These tracts originate in the reticular formation of the brain stem. Neurostimulation of the reticular formation by the cerebrum or cerebellum either facilitates or inhibits (depending on the area stimulated) the activity of lower motor neurons (fig. 14.29).

There are no descending tracts from the cerebellum. The cerebellum can only influence motor activity indirectly, through the vestibular nuclei, red nucleus, and basal ganglia. These structures, in turn, affect lower motor neurons via the **vestibulospinal tracts, rubrospinal tracts,** and reticulospinal tracts. Damage to the cerebellum interferes with the coordination of movements with spatial judgment. Underreaching or overreaching for an object may occur, followed by *intention tremor,* in which the limb moves back and forth in a pendulumlike motion.

The basal ganglia, acting through synapses in the reticular formation particularly, appear normally to exert an inhibitory influence on the activity of lower motor neurons. Damage to the basal ganglia thus results in increased muscle tone. People with such damage display *akinesia (ah″ki-ne′ze-ah)* (lack of desire to use the affected limb) and *chorea*—sudden and uncontrolled random movements.

Babinski's reflex: from Joseph F. Babinski, French neurologist, 1857–1932

akinesia: Gk. *a*, without; *kinesis,* movement
chorea: Fr. *choros,* a dance

Figure 14.28. Areas of the brain containing neurons involved in the control of skeletal muscles (higher motor neurons). The thalamus is a relay center between the motor cortex and other brain areas.

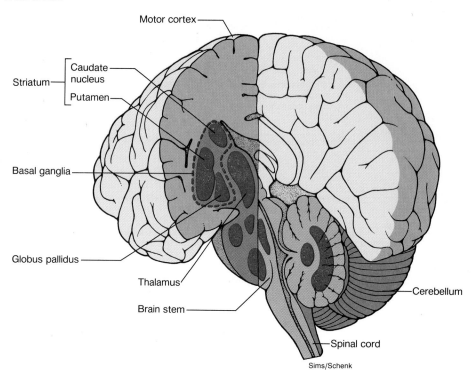

Sims/Schenk

Paralysis agitans, better known as **Parkinson's disease,** is a disorder of the basal ganglia involving the degeneration of fibers from the substantia nigra. These fibers, which use dopamine as a neurotransmitter, are required to antagonize the effects of other fibers that use acetylcholine (ACh) as a transmitter. The relative deficiency of dopamine compared to ACh is believed to produce the symptoms of Parkinson's disease, including *resting tremor.* This "shaking" of the limbs tends to disappear during voluntary movements and then reappear when the limb is again at rest.

Parkinson's disease is treated with drugs that block the effects of ACh and by the administration of L-Dopa, which can be converted to dopamine in the brain (dopamine cannot be given directly because it does not cross the blood-brain barrier).

1. *Diagram a cross section of the spinal cord, and label the structures of the gray matter and the white matter. Describe the location of the spinal cord.*
2. *List the structures and function of the corticospinal tracts. Make a similar list for the extrapyramidal tracts.*
3. *Explain why damage to the right side of the brain primarily affects motor activities on the left side of the body.*

Parkinson's disease: from James Parkinson, English phy 1755–1824

Clinical Considerations

The clinical aspects of the central nervous system are extensive and usually complex. Numerous diseases and developmental problems directly involve the nervous system, and the nervous system is indirectly involved with most of the diseases that afflict the body because of the location and activity of sensory pain receptors. Pain receptors are free nerve endings that are present throughout living tissue. The pain sensations elicited by disease or trauma are important for localizing and diagnosing specific diseases or dysfunctions.

Only a few of the many clinical considerations of the central nervous system will be discussed, including neurological assessment and drugs, developmental problems, injuries, infections and diseases, and degenerative disorders.

Neurological Assessment and Drugs

Neurological assessment has become exceedingly sophisticated and accurate in the past few years. In most physical examinations, only the basic aspects such as reflexes and sensory functions are assessed. But if the physician suspects abnormalities involving the nervous system, further neurological tests are done, employing the following techniques.

Figure 14.29.   Pathways involved in the higher motor
neuron control of skeletal muscles.

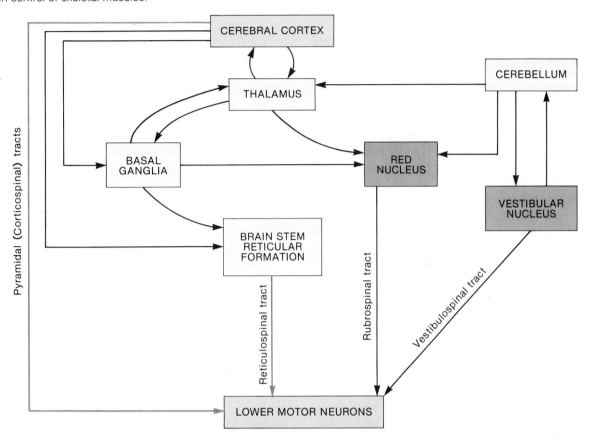

A **lumbar puncture** is performed by inserting a fine
needle between the third and fourth lumbar vertebrae and
withdrawing a sample of cerebrospinal fluid from the subarachnoid space (fig. 14.30). A **cisternal puncture** is similar to a lumbar puncture except that the cerebrospinal
fluid is withdrawn from the cisterna magna near the foramen magnum of the skull. The pressure of the cerebrospinal fluid, which is normally about 10 mm of mercury,
is measured with a *manometer*. Samples of cerebrospinal
fluid may also be examined for abnormal constituents.
Also, excessive fluids, accumulated as a result of disease
or trauma, may be drained.

Examining the brain and the ventricles for lesions,
aneurysms, tumors, or certain diseases may require several techniques. A **pneumoencephalogram** (see fig. 14.21)
provides an X-ray picture of the size, position, and shape
of the ventricles and is therefore an excellent method for
detecting lesions of the brain. To obtain a pneumoencephalogram, a lumbar puncture is performed, a given
amount of cerebrospinal fluid is withdrawn, and air or another gas is injected into the subarachnoid space where it
rises and provides a contrast medium in the ventricles.

The condition of the arteries of the brain can be determined through a **cerebral angiogram.** In this technique,

Figure 14.30.   (*a*) a lumbar puncture is performed by
inserting a needle between the third and fourth lumbar
vertebrae (L3-L4) and (*b*) withdrawing cerebrospinal fluid
from the subarachnoid space.

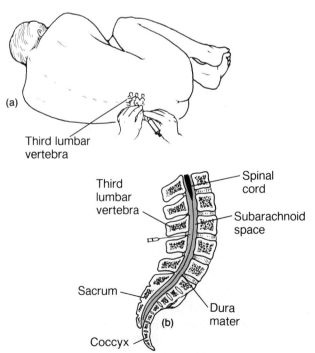

a radiopaque substance is injected into the common ca- rotid arteries and allowed to disperse through the cerebral vessels. Aneurysms and vascular constrictions or displace- ments by tumors may then be revealed on X rays.

The development of the **CT scanner,** or **Computerized Axial Tomographic scanner,** has revolutionized the diag- nosis of brain disorders. The CT scanner projects a sharply focused, detailed tomogram, or cross section, of a pa- tient's brain onto a television screen. The versatile CT scanner allows quick and accurate diagnoses of tumors, aneurysms, blood clots, and hemorrhage. The CT scanner may also be used to detect certain types of birth defects, brain damage, scar tissue, and old or recent strokes.

A machine with even greater potential than the CT scanner is the **DSR,** or **dynamic spatial reconstructor.** Like the CT scanner, the DSR is computerized to transform X- ray pictures into composite video images. The advantage of the DSR over the CT is the three-dimensional view it presents and the speed in which the image is portrayed. DSR can produce 75,000 cross-section images in five sec- onds, whereas CT can produce only one. At that speed, body functions as well as structures may be studied. Blood flow through blood vessels of the brain can be observed. This type of data is important in detecting early symp- toms of a stroke or other disorders.

Certain disorders of the brain may be diagnosed more simply by examining brain-wave patterns using an **elec- troencephalogram.** Sensitive electrodes placed on the scalp record particular EEG patterns being emitted from evoked cerebral activity. EEG recordings are used to monitor epi- leptic patients to predict seizures and determine proper drug therapy and also to monitor comatose patients

The fact that the nervous system is extremely sensi- tive to various drugs is both fortunate and potentially di- sastrous to a person. *Drug abuse* is a major clinical concern because of the addictive and devastating effect that cer- tain drugs have on the nervous system. Much has been written on drug abuse, and it is beyond the scope of this text to elaborate on the effects of drugs. A positive aspect of drugs is their administration in medicine to temporarily interrupt the passage or perception of sensory impulses. Injecting an anesthetic drug near a nerve, as in dentistry, desensitizes a specific area and causes a *nerve block.* Nerve blocks of a limited extent occur if an appendage is cooled or if a nerve is compressed for a period of time. Before the discovery of pharmacological drugs, physicians would fre- quently cool an affected appendage with ice or snow be- fore performing surgery. **General anesthetics** affect the brain and render a person unconscious. A **local anesthetic** causes a nerve block by desensitizing a specific area.

Developmental Problems

Congenital malformations of the CNS are common and frequently involve overlying bone, muscle, and connective tissue. The more severe abnormalities make life impos- sible, and the less severe malformations frequently result in functional disability.

Most congenital malformations of the nervous system occur during the sensitive embryonic period. Neurological malformations are generally caused by genetic abnor- malities but may result from environmental factors such as anoxia, infectious agents, drugs, and ionizing radiation.

Spina bifida is a defective fusion of the vertebral ele- ments and may or may not involve the spinal cord. **Spina bifida occulta** is the most common and least serious type of spina bifida. This defect usually involves few vertebrae, is not externally apparent except for perhaps a pigmented spot with a tuft of hair, and usually does not cause neu- rological disturbances. **Spina bifida cystica,** a severe type of spina bifida, is a saclike protrusion of skin and under- lying meninges that may contain portions of the spinal cord and nerve roots. Spina bifida cystica is most common in the lower thoracic, lumbar, and sacral regions. The posi- tion and extent of the defect determines the degree of neu- rological impairment.

Anencephalia *(an''en-sĕ-fa'le-ah)* is a markedly de- fective development of the brain and the surrounding cra- nial bones. Anencephalia occurs once per thousand births and makes sustained extrauterine life impossible. This congenital defect apparently results from the failure of the neural folds at the cranial portion of the neural plate to fuse and form the prosencephalon.

Microcephaly is an uncommon condition in which brain development is not completed. If enough neurolog- ical tissue is present, the infant will survive but will be severely mentally retarded.

Defective skull development frequently causes **cra- nial encephalocele** *(en-sef'ah-lo-sēl).* This condition oc- curs approximately once per two thousand births. Cranial encephalocele is characterized by a herniated portion of the brain and meninges through a defect in the skull, usu- ally in the occipital region. Occasionally the herniation contains fluid with the meninges and not the brain tissue. In this case, it is referred to as a **cranial meningocele** *(mĕ- ning'go-sēl).*

Hydrocephalus is the abnormal accumulation of cere- brospinal fluid in the ventricles and subarachnoid or sub- dural space. Hydrocephalus may be caused by the excessive production of or blocked flow of cerebrospinal fluid. It may also be associated with other congenital problems such as spina bifida cystica or encephalocele. Hydrocephalus frequently causes the cranial bones to thin and the cerebral cortex to atrophy.

Many congenital disorders cause an impairment of intelligence, known as **mental retardation.** Chromosomal abnormalities, maternal and fetal infections such as syph- ilis and German measles, and excessive irradiation of the fetus are all commonly associated with mental retarda- tion.

Injuries

Although the brain and spinal cord seem to be well pro- tected within a bony encasement, they are sensitive or- gans highly susceptible to injury.

Certain symptomatic terms are used when determining possible trauma within the CNS. **Headaches** are the most common ailment of the CNS. Most headaches are due to dilated blood vessels within the meninges of the brain. Headaches are generally asymptomatic of brain disorders, associated rather with physiological stress, eyestrain, or fatigue. Persistent and intense headaches may indicate a more serious problem such as a brain tumor. A **migraine** is a specific type of headache commonly preceded or accompanied by visual impairments and gastrointestinal unrest. It is not known why only 5%–10% of the population periodically suffer from migraines or why they are more common in women. Fatigue, allergy, or emotional stress tends to trigger migraines.

Fainting is a brief loss of consciousness that may result from a rapid pooling of blood in the lower extremities. It may occur when a person rapidly arises from a reclined position, receives a blow to the head, or experiences an intense psychologic stimulus, such as viewing a cadaver for the first time. Fainting is of more concern when it is symptomatic of a particular disease.

A **concussion** is a sudden movement of the brain caused by a violent blow to the head, which may or may not fracture bones of the skull. A concussion usually results in a brief period of unconsciousness followed by mild **delirium,** in which the patient is in a state of confusion. **Amnesia** is a more intense disorientation in which the patient suffers various degrees of memory loss.

A person who survives a severe head injury may be **comatose** for a short or an extended period of time. A coma is a state of unconsciousness from which the patient cannot be aroused by even the most intense external stimuli. The area of the brain most likely to cause a coma from trauma is the reticular activating system. Although a head injury is the most common cause of coma, certain diseases, drugs, poisons, infections, and some chemical imbalances (e.g., diabetes) may also be responsible.

The flexibility of the vertebral column is essential for body movements, but because of this flexibility the spinal cord and spinal nerves are somewhat vulnerable to trauma. Falls or severe blows to the back are a common cause of injury. A skeletal injury, such as a fracture, dislocation, or compression of the vertebrae, usually traumatizes nervous tissue as well. Other frequent causes of trauma to the spinal cord include gunshot wounds, stabbings, herniated discs, and birth injuries. The consequences of the trauma depend upon the severity and location of the injury and the medical treatment the patient receives. If nerve fibers of the spinal cord are severed, motor or sensory functions will be permanently lost.

Paralysis is a permanent loss of motor control, usually resulting from disease or a lesion of the spinal cord or specific nerves. Paralysis of both lower extremities is

amnesia: L. *amnesia,* forgetfulness
comatose: Gk. *koma,* deep sleep
paralysis: Gk. *paralysis,* loosening

Figure 14.31. Whiplash varies in severity from muscle and ligament strains to dislocation of the vertebrae and compression of the spinal cord. Injuries such as this may cause permanent loss of some or all of the spinal cord functions.

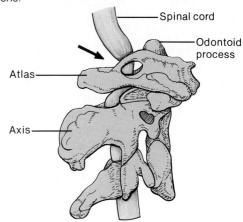

called **paraplegia.** Paralysis of both the upper and lower extremity on the same side is called **hemiplegia,** and paralysis of all four extremities is **quadriplegia.** Paralysis may be flaccid or spastic. **Flaccid** *(flak'sid)* **paralysis** generally results from a lesion of the anterior-horn cells and is characterized by noncontractible muscles that atrophy. **Spastic paralysis** results from lesions of the corticospinal tracts of the spinal cord and is characterized by hypertonicity of the skeletal muscles.

Whiplash is a sudden hyperextension and flexion of the cervical vertebrae (fig. 14.31) such as may occur during a rear-end automobile collision. Recovery of a minor whiplash (muscle and ligament strains) is generally complete but requires a long time. Severe whiplash (spinal cord compression) may cause permanent paralysis to the structures below the level of injury.

Diseases and Infections
Mental illness is a major clinical consideration of the nervous system and is perhaps the least understood. The two principal categories of mental disorders are neurosis and psychosis. In *neurosis,* a maladjustment to certain aspects of life interferes with normal functioning, but contact with reality is maintained. An irrational fear *(phobia)* is an example of neurosis. Neurosis frequently causes intense anxiety or abnormal distress that brings about increased sympathetic stimulation. *Psychosis,* a more serious mental condition, is typified by a withdrawal from reality and is usually socially unacceptable. The more common forms of psychosis include *schizophrenia,* in which a person withdraws into a world of fantasy; *paranoia,* in which a person has systematized delusions often of a persecutory nature; and *manic-depressive psychosis,* in which a person's moods swing widely from intense elation to deepest despair.

paraplegia: Gk. *para,* beside; *plessein,* to strike

Epilepsy is a relatively common brain disorder with a strong hereditary basis but is also caused by head injuries, tumors, childhood infectious diseases, or it can be idiopathic (without demonstrable cause). A person with epilepsy may periodically suffer from an *epileptic seizure,* which has various symptoms depending on the type of epilepsy.

The most common kinds of epilepsy are petit mal, psychomotor epilepsy, and grand mal. **Petit mal** *(pĕ-te'mahl')* occurs almost exclusively in children between the ages of three and twelve. A child experiencing a petit mal seizure loses contact with reality for 5–30 seconds but does not lose consciousness or display convulsions. There may, however, be slight uncontrollable facial gestures or eye movements, and the child stares as if in a daydream. During a petit mal seizure, the thalamus and hypothalamus produce an extremely slow EEG pattern of 3 waves per second. Children with petit mal usually outgrow the condition by age nine or ten and generally require no medication.

Psychomotor epilepsy is often confused with mental illness because of the symptoms characteristic of the seizure. During such a seizure, EEG activity accelerates in the temporal lobes, causing a person to become disoriented and lose contact with reality. Occasionally during a seizure, specific cerebral motor areas cause a person to smack his lips or clap his hands involuntarily. If motor areas in the brain are not stimulated, a person having a psychomotor epileptic seizure may wander aimlessly until the seizure subsides.

Grand mal is a more serious form of epilepsy characterized by periodic convulsive seizures that generally render a person unconscious. Grand mal epileptic seizures are accompanied by rapid EEG patterns of 25–30 waves per second. This sudden increase of EEG patterns from the normal of about 10 waves per second may cause an extensive stimulation of motor units and, therefore, uncontrollable skeletal-muscle activity. During a grand mal seizure, a person loses consciousness, convulses, and may lose urinary and bowel control. The unconsciousness and convulsions usually last a few minutes, after which the muscles relax and the person awakes but remains disoriented for a short time.

Epilepsy almost never affects intelligence and can be effectively treated with drugs in about 85% of the patients.

Cerebral Palsy. Cerebral palsy is a condition of motor disorders characterized by paresis (partial paralysis) and lack of muscular contraction. It is caused by damage to the motor areas of the brain during prenatal development, birth, or infancy. During neural development within an

embryo, radiation or bacterial toxins (such as from German measles), transferred through the placenta of a pregnant mother, may cause cerebral palsy. Oxygen deprivation due to complications at birth and hydrocephalus in a newborn may also cause cerebral palsy. The three areas of the brain most severely affected by this disease are the cerebral cortex, the basal ganglia, and the cerebellum. The type of cerebral palsy is determined by which region of the brain is affected.

Some degree of mental retardation occurs in 60%–70% of cerebral-palsy victims. Partial blindness, deafness, and speech problems frequently accompany this disease. Cerebral palsy is nonprogressive (i.e., these impairments do not worsen as a person ages), but neither are there organic improvements.

Neoplasms of the CNS. Neoplasms of the CNS are either **intracranial tumors,** which affect cells within or associated with the brain, or they are **intravertebral (intraspinal) tumors,** which affect cells within or near the spinal cord. **Primary neoplasms** develop within the CNS. Approximately one-half are benign, but they may become lethal because of the pressure they apply upon vital centers as they grow. Patients with **secondary,** or **metastatic, neoplasms** within the brain have a poor prognosis because the cancer has already established itself in another body organ, frequently the liver, lung, or breast, and has only secondarily spread to the brain. The symptoms of a brain tumor include headache, convulsions, pain, paralysis, or a change in behavior.

Neoplasms of the CNS are classified according to the tissues in which the cancer occurs. Tumors arising in neuroglia cells are called **gliomas** *(gli-o'mah)* and account for about one-half of all primary neoplasms within the brain. Gliomas are frequently spread throughout cerebral tissue, develop rapidly, and usually cause death within a year after diagnosis. **Astrocytoma** *(as''tro-si-to'mah),* **oligodendroglioma** *(ol''i-go-den''dro-gli-o'mah),* and **ependymoma** *(ĕ-pen''di-mo'mah)* are common types of gliomas.

Meningiomas arise from meningeal coverings of the brain and account for about 15% of primary intracranial tumors. Meningiomas are usually benign if they can be treated readily.

Intravertebral tumors are classified as **extramedullary** when they develop on the outside of the spinal cord and as **intramedullary** when they develop within the substance of the spinal cord. Extramedullary neoplasms may cause pain and numbness in body structures away from the tumor as the growing tumor compresses the spinal cord. An intramedullary neoplasm causes a gradual loss of sensory, temperature, and motor function below the spinal-segmental level of the affliction.

The methods of detecting and treating cancers within the CNS have greatly improved in the last few years. Early detection and competent treatment have lessened the likelihood of death from this disease and reduced the probability of physical impairment.

---

epilepsy: Gk. *epi,* upon; *lepsis,* seize
petit mal: L. *pitinnus,* small child; *malus,* bad

Dyslexia. Dyslexia is a defect in the language center within the brain. In dyslexia, otherwise intelligent people reverse the order of letters in syllables, of syllables in words, and of words in sentences. The sentence: "The man saw a red dog," for example might be read by the dyslexic as "A red god was the man." Dyslexia is believed to result from the failure of one cerebral hemisphere to respond to written language, perhaps due to structural defects.

Meningitis. The nervous system is vulnerable to a variety of organisms and viruses that may cause abscesses or infections. Meningitis is an infection of the meninges, which may be confined to the spinal cord, in which case it is referred to as **spinal meningitis,** or it may involve the brain and associated meninges, in which case it is known as **encephalitis** or **encephalomyelitis** *(en-sef''ah-lo-mi''ĕ-li'tis),* respectively. The microorganisms that most commonly cause meningitis are meningococci, streptococci, pneumococci, and tubercle bacilli. Viral meningitis is more serious than bacterial meningitis. Nearly 20% of viral encephalitides are fatal. The organisms that cause meningitis probably enter the body through respiratory passageways.

Poliomyelitis. Poliomyelitis, or infantile paralysis, is primarily a childhood disease caused by a virus that destroys nerve cell bodies within the anterior horn of the spinal cord, especially those within the cervical and lumbar enlargements. This degenerative disease is characterized by fever, severe headache, stiffness and pain, and the loss of certain somatic reflexes. Muscle paralysis follows within several weeks and eventually the muscles atrophy. Death results if the virus invades the vasomotor and respiratory nuclei within the medulla oblongata or anterior horn cells controlling respiratory muscles. Poliomyelitis has been effectively controlled with immunization.

Syphilis. Syphilis is a venereal disease that, if untreated, progressively destroys body organs. When syphilis causes organ degeneration, it is said to be in the *tertiary stage* (ten to twenty years after the primary infection). The organs of the nervous system are frequently infected, causing a condition called **neurosyphilis.** Neurosyphilis is classified according to the tissue involved, and the symptoms vary correspondingly. If the meninges are infected, the condition is termed **chronic meningitis. Tabes dorsalis** is a form of neurosyphilis in which there is a progressive degeneration of the posterior funiculi of the spinal cord and dorsal roots of spinal nerves. Motor control is gradually lost and patients eventually become bedridden, unable even to feed themselves.

Degenerative Disorders of the Nervous System
Degenerative diseases of the CNS are characterized by a progressive, symmetrical deterioration of vital structures of the brain or spinal cord. The etiologies of these diseases are poorly understood, but it is thought that most of them are genetic.

Cerebrovascular Accident (CVA). Cerebrovascular accident is the most common disease of the nervous system. It is the third highest cause of death in the United States and perhaps the major cause of disability. The term **stroke** is frequently used synonymously with CVA, but actually a stroke refers to the sudden and dramatic appearance of a neurological defect. Cerebral thrombosis, in which a thrombus, or clot, forms in an artery of the brain, is the most common cause of CVA. Other causes of CVA include intracerebral hemorrhages, aneurysms, atherosclerosis, and arteriosclerosis of the cerebral arteries.

Patients who recover from CVA frequently suffer partial paralysis and mental disorders, such as loss of language skills. The dysfunction depends upon the severity of the CVA and the regions of the brain that were injured. Patients surviving a CVA can often be rehabilitated, but approximately two-thirds die within three years of the initial damage.

Multiple Sclerosis. Multiple sclerosis (MS) is a relatively common neurological disease in persons between the ages of twenty and forty. MS is a chronic, degenerating, remitting, and relapsing disease that progressively destroys the myelin sheaths of neurons in multiple areas of the CNS. Initially, lesions form on the myelin sheaths and soon develop into hardened *scleroses (skle-ro'sēz),* or scars (hence the name). The destruction of myelin sheaths prohibits the normal conduction of impulses, resulting in a progressive loss of functions. Because myelin degeneration is widely distributed, MS has a greater variety of symptoms than any other neurologic disease. This characteristic, coupled with remission, frequently causes misdiagnosis of this disease.

During the early stages of MS, many patients are considered neurotic because of the wide variety and temporary nature of their symptoms. As the disease progresses, the symptoms may include double vision (diplopia), spots in the visual field, blindness, tremor, numbness of appendages, and locomotor difficulty. Eventually the patient is bedridden, and death may occur anytime from seven to thirty years after the first symptoms appear.

Syringomyelia. Syringomyelia *(si-ring''go-mi-e'le-ah)* is a relatively uncommon condition characterized by the appearance of cystlike cavities, called *syringes,* within the gray matter of the spinal cord. These syringes progressively destroy the cord from the inside out. Syringomyelia is a chronic, slow progressing disease of unknown cause. As the spinal cord deteriorates, the patient experiences muscular weakness and atrophy and sensory loss, particularly the senses of pain and temperature.

---

dyslexia: Gk. *dys,* bad; *lexis,* speech

Chapter Summary

I. General Features of the Central Nervous System

The central nervous system (CNS), consisting of the brain and spinal cord, is covered with meninges, is bathed in cerebrospinal fluid, and contains gray and white matter.
  A. Neurology is the study of the nervous system. It is a complex science that is still undergoing intense and exciting investigation.
  B. The functions of the nervous system are orientation, coordination, assimilation, and instinctual behavior.
  C. The CNS, composed of gray matter and white matter, is protected by meninges and is bathed in cerebrospinal fluid.

II. Cerebrum

The cerebrum, consisting of five paired lobes within two convoluted hemispheres, is concerned with higher brain functions, such as the perception of sensory impulses, the instigation of voluntary movement, the storage of memory, thought processes, and reasoning ability.
  A. The cerebral cortex of the two cerebral hemispheres is convoluted with gyri and sulci.
  B. Each cerebral hemisphere contains frontal, parietal, temporal, occipital, and insula lobes.
  C. Brain waves generated by the cerebral cortex are recorded as an electroencephalogram and may provide valuable diagnostic information.
  D. The white matter of the cerebrum consists of association, commissural, and projection fibers.
  E. The basal ganglia are specialized masses of gray matter located within the white matter of the cerebrum.

III. Diencephalon

The diencephalon is a major autonomic region of the brain that consists of vital structures such as the thalamus, hypothalamus, epithalamus, and pituitary gland.
  A. The thalamus is an ovoid mass of gray matter that functions as a relay center for sensory impulses and responds to pain.
  B. The hypothalamus is an aggregation of specialized nuclei that regulate many visceral activities. It also performs emotional and instinctual functions.
  C. The epithalamus contains the pineal gland and the vascular choroid plexus over the roof of the third ventricle.

IV. Mesencephalon

The mesencephalon contains the corpora quadrigemina, concerned with visual and auditory reflexes, and the cerebral peduncles, composed of fiber tracts. It also contains specialized nuclei that help to control posture and movement.
  A. The superior colliculi are concerned with visual reflexes and the inferior colliculi are concerned with auditory reflexes.
  B. The red nucleus and the substantia nigra are concerned with motor activities.

V. Metencephalon

The metencephalon contains the pons, which relays impulses, and the cerebellum, which coordinates skeletal muscle contractions.
  A. The pons consists of fiber tracts connecting the cerebellum and medulla to other structures of the brain. The pons also contains nuclei for certain cranial nerves and the regulation of respiration.
  B. The cerebellum consists of two hemispheres connected by the vermis and supported by three paired peduncles.
    1. It is composed of white matter surrounded by a thin convoluted cortex of gray matter.
    2. The cerebellum is concerned with coordinated contractions of skeletal muscle.

VI. Myelencephalon

The medulla oblongata, contained within the myelencephalon, connects to the spinal cord and contains nuclei for the cranial nerves and vital autonomic functions.
  A. The medulla oblongata is composed of the ascending and descending tracts of the spinal cord and contains nuclei for several autonomic functions.
  B. The reticular formation functions as the reticular activating system in arousing the cerebrum.

VII. Meninges of the Central Nervous System

The CNS is covered by protective meninges, consisting of a dura mater, an arachnoid membrane, and a pia mater.
  A. The cranial dura mater consists of an outer periosteal layer and an inner meningeal layer. The spinal dura mater is a single layer that is surrounded by the vascular epidural space.
  B. The arachnoid membrane is a netlike meninx surrounding the subarachnoid space, which contains cerebrospinal fluid.

C. The thin pia mater adheres to the contour of the CNS.

VIII. Ventricles and Cerebrospinal Fluid

The ventricles, central canal, and subarachnoid space contain cerebrospinal fluid, formed by the active transport of substances from blood plasma in the choroid plexuses.
  A. The lateral, third, and fourth ventricles are interconnected chambers within the brain that are continuous with the central canal of the spinal cord.
  B. These chambers are filled with cerebrospinal fluid, which also flows throughout the subarachnoid space.
  C. Cerebrospinal fluid is continuously secreted by the choroid plexuses and is absorbed into the blood at the arachnoid villi.

IX. Brain Metabolism

The brain has a high rate of metabolic activity and a unique internal environment because of the blood-brain barrier and the numerous specific chemicals synthesized and secreted by the cells of the brain.
  A. The blood-brain barrier determines which substances within blood plasma can enter the extracellular fluid of the brain.
  B. Neurotransmitter chemicals within the brain are excitatory or inhibitory.
  C. Neuropeptides are specific neurotransmitters composed of short chains of amino acids.

X. Spinal Cord

The spinal cord consists of centrally located gray matter, involved in reflexes, and peripherally located ascending and descending tracts of white matter, which conduct impulses to and from the brain.
  A. The spinal cord is composed of thirty-one segments, each of which gives rise to a pair of spinal nerves.
    1. It is characterized by a cervical enlargement, a lumbar enlargement, and two longitudinal grooves, which divide it into right and left halves.
    2. The conus medullaris is the terminal portion of the spinal cord, and the cauda equina radiates inferiorly from that point.
  B. Ascending and descending spinal cord tracts are referred to as funiculi.
    1. Descending tracts are grouped as either corticospinal or extrapyramidal.
    2. Many of the fibers in the funiculi decussate (cross over) in the spinal cord or in the medulla oblongata.

Review Activities

Objective Questions

1. Which of the following is *not* a lobe of the cerebrum?
   (a) parietal
   (b) sphenoid
   (c) temporal
   (d) insula
   (e) occipital

2. The principal connection between the cerebral hemispheres is the
   (a) corpus callosum
   (b) pons
   (c) intermediate mass
   (d) vermis
   (e) precentral gyrus

3. The structure of the brain that is most directly involved in the autonomic response to pain is the
   (a) pons
   (b) hypothalamus
   (c) medulla oblongata
   (d) thalamus

4. Which statement is *false* concerning the basal ganglia?
   (a) They are located within the cerebrum.
   (b) They regulate the basal metabolic rate.
   (c) They consist of the caudate nucleus, lentiform nucleus, putamen and globus pallidus.
   (d) They indirectly exert an inhibitory influence on lower motor neurons.

5. The corpora quadrigemina, red nucleus, and substantia nigra are structures of the
   (a) diencephalon
   (b) metencephalon
   (c) mesencephalon
   (d) myelencephalon

6. The fourth ventricle is contained within the
   (a) cerebrum
   (b) cerebellum
   (c) midbrain
   (d) metencephalon

7. The right cerebral cortex controls voluntary movements on the left side of the body because
   (a) most persons are right-handed
   (b) the right hemisphere dominates
   (c) many of the fibers in the funiculi decussate in the medulla oblongata
   (d) of distinct cerebral specialization of hemispheres

8. A patient experiencing a fluctuating body temperature, lack of hunger and thirst, and psychosomatic disorders, may have a malfunctioning
   (a) hypothalamus
   (b) midbrain
   (c) cerebellum
   (d) medulla oblongata

9. Spinal cord tracts that descend from the cerebral cortex to the lower motor neurons without synaptic interruption are called
   (a) reticulospinal tracts
   (b) corticospinal tracts
   (c) rubrospinal tracts
   (d) vestibulospinal tracts

10. The disease characterized by the destruction of the myelin sheaths of neurons in the CNS and the formation of plaques is
    (a) syringomyelia
    (b) neurosyphilis
    (c) poliomyelitis
    (d) multiple sclerosis

Essay Questions

1. List the types of brain waves recorded on an electroencephalogram, and explain the diagnostic value of each.

2. List the functions of the hypothalamus. Why is the hypothalamus considered a major part of the autonomic nervous system?

3. What structures are within the midbrain? List the nuclei located in the midbrain, and give the function of each.

4. Describe the location and structure of the medulla oblongata. List the nuclei found within this structure. What are the functions of the medulla oblongata?

5. What is cerebrospinal fluid? Where is it produced, and what is its pathway of circulation?

6. Define the following abbreviations: EEG, ANS, CSF, PNS, RAS, CT scan, MS, DSR, and CVA.

7. Describe the various techniques available for conducting a neurological assessment.

8. List and define the various psychological terms used to describe mental illness.

9. What is epilepsy? What causes it, and how is it controlled?

10. What do meningitis, poliomyelitis, neurosyphilis, and herpes zoster have in common, and how do these conditions differ?

Selected Readings

Angevine, J. R., Jr., and C. W. Cotman. 1981. *Principles of neuroanatomy.* New York: Oxford Univ. Press.

Axelrod, J. June 1974. Neurotransmitters. *Scientific American.*

Bloom, F. E. October 1981. Neuropeptides. *Scientific American.*

Cowen, W. M. September 1979. The development of the brain. *Scientific American.*

Decara, L. V. January 1970. Learning in the autonomic nervous system. *Scientific American.*

Evarts, E. V. September 1979. Brain mechanisms of movement. *Scientific American.*

Ferstrom, J. D., and R. J. Wurtman. February 1974. Nutrition and the brain. *Scientific American.*

Fincher, J. 1981. *The brain: mystery of matter and mind.* Washington, D.C.: U.S. News Books.

Geschwind, N. September 1979. Specializations of the human brain. *Scientific American.*

Hubel, D. H. September 1979. The brain. *Scientific American.*

Hubel, D. H. 1984. The brain. *Scientific American Offprint.*

Hubel, D. H., and T. N. Wiesel. September 1979. Brain mechanisms of vision. *Scientific American.*

Iversen, L. L. September 1979. The chemistry of the brain. *Scientific American.*

Jacobs, B. L., and M. E. Trulson. 1979. Mechanisms of action of LSD. *American Scientist* 67:396.

Kandel, E. R. September 1979. Small systems of neurons. *Scientific American.*

Kety, S. S. September 1979. Disorders of the human brain. *Scientific American.*

Kimura, D. March 1973. The asymmetry of the human brain. *Scientific American.*

Lester, H. A. February 1977. The response to acetylcholine. *Scientific American.*

Llinas, R. R. January 1975. The cortex of the cerebellum. *Scientific American.*

Nathanson, J. A., and P. Greengard. August 1977. Second messenger in the brain. *Scientific American.*

Nauta, W. J. H., and M. Feirtag. September 1979. The organization of the brain. *Scientific American.*

Norman, D. A. 1982. *Learning and memory.* San Francisco: W. H. Freeman.

Norton, W. T., and P. Morell. May 1980. Myelin. *Scientific American.*

Schwartz, J. H. April 1980. The transport of substances in nerve cells. *Scientific American.*

Shashoua, V. E. July–August 1985. The role of extracellular proteins in learning and memory. *American Scientist.*

Shepherd, G. M. February 1978. Microcircuits in the nervous system. *Scientific American.*

Siegal, R. K. October 1977. Hallucinations. *Scientific American.*

Stevens, C. F. September 1979. The neuron. *Scientific American.*

Wurtman, R. J. April 1982. Nutrients that modify brain function. *Scientific American.*

15

Peripheral

Concepts

The peripheral nervous system consists of nerves that convey impulses to and from the central nervous system.

Twelve pairs of cranial nerves emerge from the inferior surface of the brain and pass through foramina of the skull to innervate structures in the head, neck, and visceral organs of the torso.

Each of the thirty-one pairs of spinal nerves is formed by the union of a dorsal and a ventral spinal root, which emerges from the spinal cord through an intervertebral foramen to innervate a body dermatome.

Except in the thoracic nerves T2–T12, the ventral rami of the spinal nerves combine and then split again as a network of nerves referred to as a plexus. There are four plexuses of spinal nerves: the cervical, the brachial, the lumbar, and the sacral. Nerves emerge from the plexuses and are named according to the structures they innervate or the general course they take.

The conduction pathway of a reflex arc consists of a receptor, a sensory neuron, a motor neuron and its innervation, in the PNS, and an association neuron, in the CNS. The reflex arc provides the mechanisms for a rapid, automatic response to a potentially threatening stimulus.

Nervous System

Introduction to the Peripheral Nervous System

The peripheral nervous system consists of nerves that convey impulses to and from the central nervous system.

> *Objective 1.* Define the peripheral nervous system, and distinguish between sensory, motor, and mixed nerves.

The peripheral nervous system (PNS) is that portion of the nervous system outside of the central nervous system. The PNS functions to convey impulses to and from the brain or spinal cord. Sensory receptors within the sensory organs, neurons, nerves, ganglia, and plexuses are all part of the PNS, which serves virtually every part of the body (fig. 15.1). The sensory receptors are discussed in chapter 17.

The nerves of the PNS are classified as cranial nerves or spinal nerves, depending on whether they arise from the brain or the spinal cord. The terms *sensory nerve, motor nerve,* and *mixed nerve* relate to the direction in which the nerve impulses are being conducted. **Sensory nerves** consist of sensory (afferent) fibers that convey impulses toward the CNS. **Motor nerves** consist of motor (efferent) fibers that convey impulses away from the CNS. **Mixed nerves** are composed of both sensory and motor fibers and therefore convey impulses in both directions. Reflexes that are considered in this chapter involve sensory and motor nerves of the PNS and specific portions of the CNS.

1. *Make a quick sketch of the brain and the tongue to depict the relationship between the CNS and the PNS. Indicate diagrammatically with lines and arrows the sensory and motor innervation of the tongue. Define a mixed nerve. What kinds of sensory stimulation arise from the tongue? What type of response is caused by motor stimulation to the tongue?*
2. *List the structures of the nervous system that are considered a part of the PNS.*

Figure 15.1.   The peripheral nervous system is composed of cranial nerves, spinal nerves, plexuses, and the nerves that arise from the plexuses.

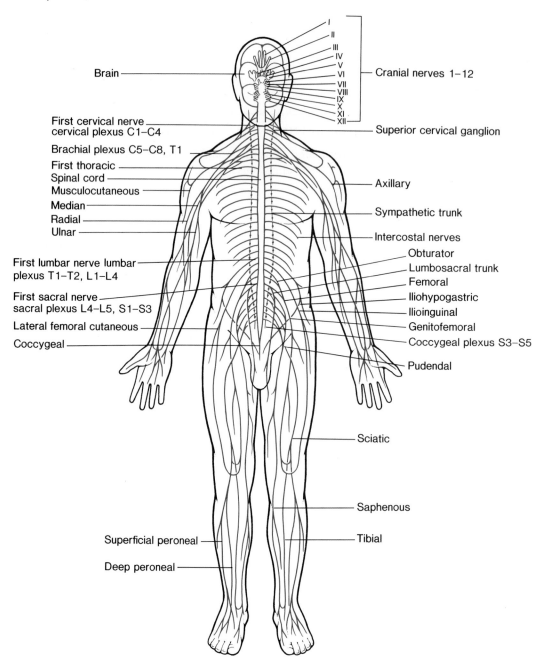

Brain

Cranial nerves 1–12

I
II
III
IV
V
VI
VII
VIII
IX
X
XI
XII

First cervical nerve cervical plexus C1–C4

Brachial plexus C5–C8, T1

First thoracic

Spinal cord

Musculocutaneous

Median

Radial

Ulnar

First lumbar nerve lumbar plexus T1–T2, L1–L4

First sacral nerve sacral plexus L4–L5, S1–S3

Lateral femoral cutaneous

Coccygeal

Superior cervical ganglion

Axillary

Sympathetic trunk

Intercostal nerves

Obturator

Lumbosacral trunk

Femoral

Iliohypogastric

Ilioinguinal

Genitofemoral

Coccygeal plexus S3–S5

Pudendal

Sciatic

Saphenous

Superficial peroneal

Deep peroneal

Tibial

Figure 15.2. Cranial nerves.

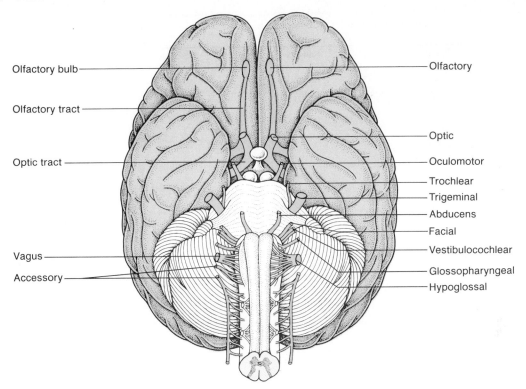

Olfactory bulb

Olfactory tract

Optic tract

Vagus

Accessory

Olfactory

Optic

Oculomotor

Trochlear

Trigeminal

Abducens

Facial

Vestibulocochlear

Glossopharyngeal

Hypoglossal

Cranial Nerves

Twelve pairs of cranial nerves emerge from the inferior surface of the brain and pass through foramina of the skull to innervate structures in the head, neck, and visceral organs of the torso.

> *Objective 2.* List the twelve pairs of cranial nerves, and describe the location and function of each.
>
> *Objective 3.* Describe the clinical methods for determining cranial nerve dysfunction.

Structure and Function of the Cranial Nerves

Of the twelve pairs of cranial nerves, two pairs arise from the forebrain and ten pairs arise from the midbrain and brain stem (fig. 15.2). The cranial nerves are designated by Roman numerals and with names. The Roman numerals refer to the order in which the nerves are positioned from the front of the brain to the back. The names indicate the structures innervated or the principal functions of the nerves. A summary of the cranial nerves is presented in table 15.1, and the locations of the nuclei from which they arise are illustrated in figure 14.12.

Although most cranial nerves are mixed, some are associated with special senses and consist of sensory fibers only. The cell bodies of sensory fibers are located in ganglia outside the brain.

Generations of anatomy students have used a mnemonic device to help them remember the order in which the cranial nerves emerge from the brain: "On old Olympus' towering top, a Finn and German viewed a hop." The initial letter of each word in this jingle corresponds to the initial letter of each pair of cranial nerves.

Table 15.1 Summary of cranial nerves

Number and name	Foramen transmitting	Composition	Location of cell bodies	Function
I Olfactory	Foramina in ethmoidal cribriform plate	Sensory	Bipolar cells in nasal mucosa	Olfaction
II Optic	Optic canal	Sensory	Ganglion cells of retina	Vision
III Oculomotor	Superior orbital fissure	Motor	Oculomotor nucleus	Motor impulses to levator palpebrae superioris and extrinsic eye muscles except superior oblique and lateral rectus; innervation to muscles that regulate amount of light entering eye and that focus the lens
		Sensory: proprioception		Proprioception from muscles innervated with motor fibers
IV Trochlear	Superior orbital fissure	Motor	Trochlear nucleus	Motor impulses to superior oblique muscle of eyeball
		Sensory: proprioception		Proprioception from superior oblique muscle of eyeball
V Trigeminal Ophthalmic division	Superior orbital fissure	Sensory	Semilunar ganglion	Sensory impulses from cornea, skin of nose, forehead, and scalp
Maxillary division	Foramen rotundum	Sensory	Semilunar ganglion	Sensory impulses from nasal mucosa, upper teeth and gums, palate, upper lip, and skin of cheek
Mandibular division	Foramen ovale	Sensory	Semilunar ganglion	Sensory impulses from temporal region, tongue, lower teeth and gum, and skin of chin and lower jaw
		Sensory: proprioception		Proprioception from muscles of mastication
		Motor	Motor trigeminal nucleus	Motor impulses to muscles of mastication and muscle that tenses tympanum
VI Abducens	Superior orbital fissure	Motor	Abducens nucleus	Motor impulses to lateral rectus muscle of eyeball
		Sensory: proprioception		Proprioception from lateral rectus muscle of eyeball
VII Facial	Stylomastoid foramen	Motor	Motor facial nucleus	Motor impulses to muscles of facial expression and muscle that tenses the stapes
		Motor: parasympathetic	Superior salivatory nucleus	Secretion of tears from lacrimal gland and salivation from sublingual and submandibular salivary glands
		Sensory	Geniculate ganglion	Sensory impulses from taste buds on anterior two-thirds of tongue; nasal and palatal sensation
		Sensory: proprioception		Proprioception from muscles of facial expression

Number and name	Foramen transmitting	Composition	Location of cell bodies	Function
VIII Vestibulocochlear	Internal acoustic meatus	Sensory	Vestibular ganglion	Sensory impulses associated with equilibrium
			Spiral ganglion	Sensory impulses associated with hearing
IX Glossopharyngeal	Jugular foramen	Motor	Nucleus ambiguus	Motor impulses to muscles of pharynx used in swallowing
		Sensory: proprioception	Petrosal ganglion	Proprioception from muscles of pharynx
		Sensory	Petrosal ganglion	Sensory impulses from taste buds on posterior one-third of tongue; pharynx, middle-ear cavity, carotid sinus
		Parasympathetic	Inferior salivatory nucleus	Salivation from parotid salivary gland
X Vagus	Jugular foramen	Motor	Nucleus ambiguus	Contraction of muscles of pharynx (swallowing) and larynx (phonation)
		Sensory: proprioception		Proprioception from muscles of pharynx and larynx
		Sensory	Nodose ganglion	Sensory impulses from taste buds on rear of tongue; sensations from auricle of ear; general visceral sensations
		Motor: parasympathetic	Dorsal motor nucleus	Regulate visceral motility
XI Accessory	Jugular foramen	Motor	Nucleus ambiguus	Laryngeal movement
			Spinal accessory nucleus	Motor impulses to trapezius and sternocleidomastoid muscles for movement of head, neck, and shoulders
		Sensory: proprioception		Proprioception from muscles that move head, neck, and shoulders
XII Hypoglossal	Hypoglossal canal	Motor	Hypoglossal nucleus	Motor impulses to intrinsic and extrinsic muscles of tongue and infrahyoid muscle
		Sensory: proprioception		Proprioception from muscles of tongue

Figure 15.3.   The olfactory nerve.

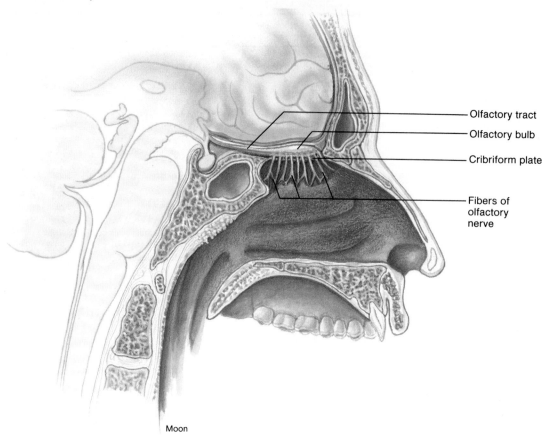

Olfactory tract

Olfactory bulb

Cribriform plate

Fibers of olfactory nerve

Moon

Figure 15.4.   The optic nerve.

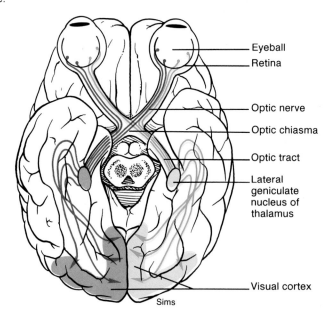

Eyeball

Retina

Optic nerve

Optic chiasma

Optic tract

Lateral geniculate nucleus of thalamus

Visual cortex

Sims

Figure 15.5.   The oculomotor, trochlear, and abducens cranial nerves.

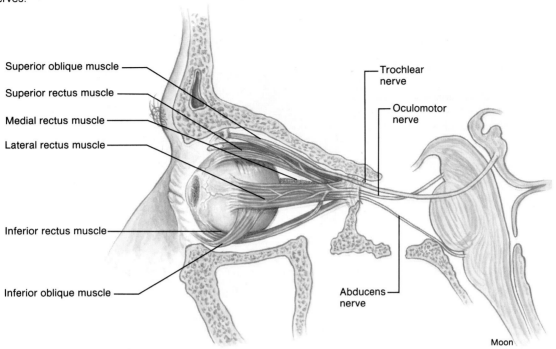

Superior oblique muscle

Superior rectus muscle

Medial rectus muscle

Lateral rectus muscle

Inferior rectus muscle

Inferior oblique muscle

Trochlear nerve

Oculomotor nerve

Abducens nerve

Moon

I Olfactory Nerve.   Actually, numerous olfactory nerves relay sensory impulses of smell from the mucous membranes of the nasal cavity (fig. 15.3). Olfactory nerves are composed of bipolar neurons that function as *chemoreceptors,* responding to volatile chemical particles breathed into the nasal cavity. The dendrites and cell bodies of olfactory neurons are positioned within the mucosa, primarily that which covers the superior nasal conchae and adjacent nasal septum. The axons of these neurons pass through the cribriform plate of the ethmoid to the **olfactory bulb** where synapses are made, and the sensory impulses are passed through the **olfactory tract** to the primary olfactory area in the cerebral cortex.

II Optic Nerve.   The optic nerve, another sensory nerve, conducts impulses from the *photoreceptors* (rods and cones) in the retina of the eye. Each optic nerve is composed of an estimated 1.25 million nerve fibers that converge at the back of the eyeball and enter the cranial cavity through the optic foramen. The two optic nerves unite on the floor of the diencephalon to form the **optic chiasma** *(ki-as'mah)* (fig. 15.4). Nerve fibers that arise from the medial half of each retina cross at the chiasma to the opposite side of the brain, whereas fibers arising from the lateral half remain on the same side of the brain. The optic nerve fibers pass posteriorly from the chiasma in the **optic tracts.** The optic tracts lead to the thalami, where a majority of the fibers terminate within certain thalamic nuclei. A few of the ganglion-cell axons that reach the thalamic nuclei have collaterals that convey impulses to

the superior colliculi. Synapses within the thalamic nuclei, however, permit impulses to pass through neurons to the **visual cortex** within the occipital lobes. Other synapses permit impulses to reach the nuclei for the oculomotor, trochlear, and abducens nerves, which regulate intrinsic (internal) and extrinsic (from orbit to eyeball) eye muscles. The visual pathway into the eyeball functions reflexively to produce motor responses to light stimuli. If an optic nerve is damaged, the eyeball served by that nerve is blinded.

III Oculomotor.   The oculomotor nerve produces certain extrinsic and intrinsic movements of the eyeball. The oculomotor is primarily a motor nerve that arises from nuclei within the midbrain. The oculomotor divides into superior and inferior branches as it passes through the superior orbital fissure in the orbit (fig. 15.5). The superior branch innervates the **superior rectus** muscle, which moves the eyeball superiorly, and the **levator palpebrae** *(le-va'tor pal'pe-bre)* **superioris** muscle, which raises the upper eyelid. The inferior branch innervates the **medial rectus, inferior rectus,** and **inferior oblique** eye muscles for medial, inferior, and superior and lateral movement of the eyeball, respectively. In addition, fibers from the inferior branch of the oculomotor enter the eyeball to supply motor innervation to the intrinsic smooth muscles of the iris for pupil dilation and to the muscles within the ciliary body for lens accommodation.

oculomotor: L. *oculus,* eye; *movere,* move

olfactory: L. *olfacere,* smell out
optic: L. *optica,* see
chiasma: Gk. *chiasma,* an X-shaped arrangement

Figure 15.6. The trigeminal nerve.

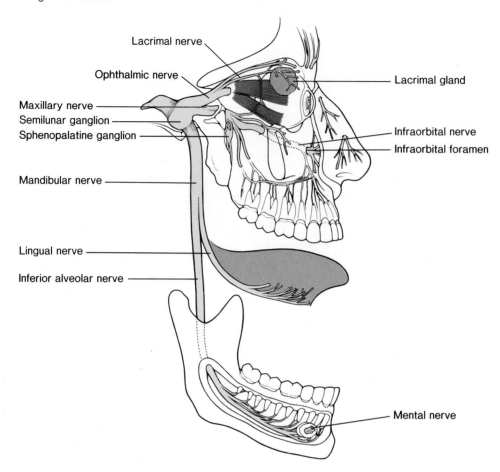

Lacrimal nerve

Ophthalmic nerve

Maxillary nerve

Semilunar ganglion

Sphenopalatine ganglion

Mandibular nerve

Lingual nerve

Inferior alveolar nerve

Lacrimal gland

Infraorbital nerve

Infraorbital foramen

Mental nerve

A few sensory fibers of the oculomotor originate from proprioceptors within the intrinsic muscles of the eyeball. These fibers convey impulses that affect the position and activity of the muscles they serve. A person whose oculomotor nerve is damaged may have a drooping upper eyelid or a dilated pupil or be unable to move the eyeball in the directions permitted by the four extrinsic muscles innervated by this nerve.

IV Trochlear. The trochlear is a very small mixed nerve that emerges from a nucleus within the midbrain and passes from the cranium through the superior orbital fissure of the orbit. The trochlear innervates the **superior oblique** muscle of the eyeball with both motor and sensory fibers (fig. 15.5). Motor impulses to the superior oblique cause the eyeball to rotate downward and away from the midline. Sensory impulses originate in proprioceptors of the superior oblique muscle and provide information about its position and activity. Damage to the trochlear nerve impairs movement in the direction permitted by the superior oblique eye muscle.

V Trigeminal. The large trigeminal nerve is a mixed nerve with motor functions originating from the nuclei within the pons and sensory functions terminating in nuclei within the midbrain, pons, and medulla. Two roots of the trigeminal are apparent as they emerge from the ventrolateral side of the pons (see fig. 15.2). The larger, **sensory root** immediately enlarges into a swelling called the **semilunar (Gasserian) ganglion,** located in a bony depression on the inner surface of the petrous portion of the temporal bone. Three large branches arise from the semilunar ganglion (fig. 15.6): the **ophthalmic branch** enters the orbit through the superior orbital fissure; the **maxillary branch** extends through the foramen rotundum; and the **mandibular branch** passes through the foramen ovale. The motor fibers of the trigeminal nerve accompany the mandibular branch through the foramen ovale and innervate the muscles of mastication and certain muscles in the floor of the mouth. Impulses through the motor branch of the mandibular portion of the trigeminal nerve stimulate contraction of the muscles involved in chewing, including the medial and lateral pterygoids, masseter, temporalis, mylohyoid, and the anterior belly of the digastric muscle.

trochlear: Gk. *trochos,* a wheel

trigeminal: L. *trigeminus,* three born together
Gasserian ganglion: from Johann L. Gasser, Vienna anatomist, 18th century
ophthalmic: L. *ophthalmia,* region of the eye

Although the trigeminal is a mixed nerve, its sensory functions are much more extensive than its motor functions. The three sensory branches of the trigeminal respond to touch, temperature, and pain sensations from the face. More specifically, the ophthalmic branch consists of sensory fibers from the anterior half of the scalp, skin of the forehead, upper eyelid, surface of the eyeball, lacrimal (tear) gland, side of the nose, and upper mucosa of the nasal cavity. The maxillary branch is composed of sensory fibers from the lower eyelid, lateral and inferior mucosa of the nasal cavity, palate and portions of the pharynx, teeth and gums of the upper jaw, upper lip, and skin of the cheek. Sensory fibers of the mandibular branch transmit impulses from the teeth and gums of the lower jaw, anterior two-thirds of the tongue (not taste), mucosa of the mouth, auricle of the ear, and lower part of the face. Trauma to the trigeminal nerve causes a lack of sensation from specific facial structures, whereas damage to the mandibular branch impairs chewing.

> The trigeminal nerve is the principal nerve relating to the practice of dentistry. Before teeth are filled or extracted, anesthetic is injected into the appropriate nerve to block sensation. A maxillary nerve block, or second-division nerve block, performed by injecting the sphenopalatine (pterygopalatine) ganglion (see fig. 15.6), desensitizes the teeth in the upper jaw. A mandibular, or third-division, nerve block desensitizes the lower teeth. This is performed by injecting anesthetic into the inferior alveolar branch of the mandibular nerve where it enters the mandible through the mandibular foramen.

VI Abducens.   The small abducens nerve originates from a nucleus within the pons and emerges from the lower portion of the pons and the anterior border of the medulla oblongata. It is a mixed nerve that traverses the superior orbital fissure of the orbit to innervate the **lateral rectus** eye muscle (see fig. 15.5). Impulses through the motor fibers of the abducens nerve cause the lateral rectus eye muscle to contract and the eyeball to rotate laterally away from the midline. Sensory impulses through the abducens nerve originate in proprioceptors in the lateral rectus muscle and are conveyed to the pons where muscle contraction is mediated. If the abducens nerve were damaged, not only would the patient be unable to turn the eyeball laterally, but because of the lack of muscle tonus to the lateral rectus, the eyeball would be pulled medially.

VII Facial.   The facial nerve arises from nuclei within the lower portion of the pons, traverses the petrous portion of the temporal bone, and emerges on the side of the face near the parotid salivary gland. The facial nerve is mixed. Impulses through the motor fibers cause contraction of the posterior belly of the digastric muscle and the muscles of

abducens: L. *abducere,* lead away

Figure 15.7.   Facial nerve branches.

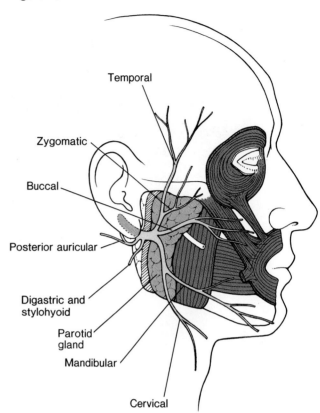

Temporal

Zygomatic

Buccal

Posterior auricular

Digastric and stylohyoid

Parotid gland

Mandibular

Cervical

facial expression, including the scalp and platysma muscles (fig. 15.7). The submandibular and sublingual salivary glands also receive some autonomic motor innervation from the facial nerve.

Sensory fibers of the facial nerve arise from taste buds on the anterior two-thirds of the tongue. Taste buds function as *chemoreceptors* because they respond to specific chemical stimuli. The **geniculate ganglion** *(gang'gle-on)* is the enlargement of the facial nerve prior to the entry of the sensory portion into the pons. Sensory sensations of taste are conveyed into nuclei within the medulla, through the thalamus, and finally to the gustatory (taste) area of the cerebral cortex of the insula.

Trauma to the facial nerve destroys the ability to contract facial muscles on the affected side of the face and distorts taste perception, particularly of sweets. The affected side of the face tends to sag since muscle tonus is lost. *Bell's palsy* is a functional disorder (probably of viral origin) of the facial nerve.

VIII Vestibulocochlear (Auditory).   The vestibulocochlear nerve is also referred to as the **auditory, acoustic,** or **statoacoustic** nerve. It is purely sensory and is composed of two branches that arise within the inner ear (fig. 15.8).

Bell's palsy: from Sir Charles Bell, Scottish physician, 1774–1842
vestibulocochlear: L. *vestibulum,* chamber; *cochlea,* snail shell

Figure 15.8.   The vestibulocochlear nerve.

Figure 15.9.   The glossopharyngeal nerve.

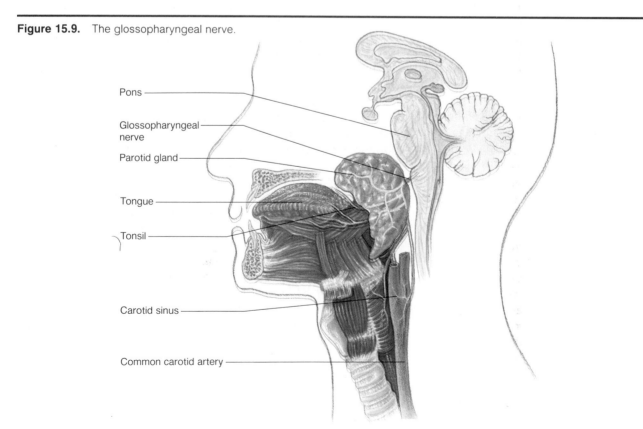

Figure 15.10.   The distribution of the vagus nerve.

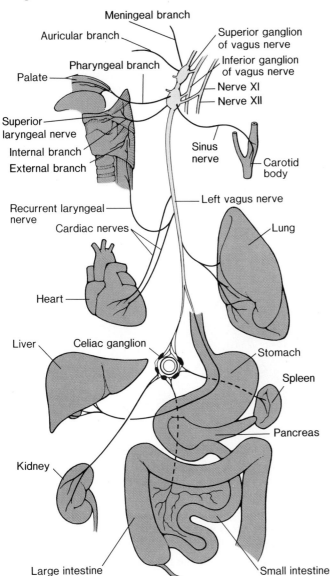

The **vestibular branch** arises from the **vestibular organs** of equilibrium and balance. Bipolar neurons from the vestibular organs (saccule, utricle, and semicircular canals) extend to the **vestibular ganglion,** where cell bodies are contained. From there, fibers convey impulses to the **vestibular nuclei** *(nu′kle-i)* within the pons and medulla oblongata. Fibers from there extend to the thalamus and the cerebellum.

The **cochlear branch** arises from the **organ of Corti** within the cochlea and is associated with hearing. The cochlear branch is composed of bipolar neurons that convey impulses through the **spiral ganglion** to the cochlear nuclei within the medulla oblongata. From there,

organ of Corti: from Alfonso Corti, Italian anatomist, 1822–88

fibers extend to the thalamus and synapse with neurons that convey the impulses to the auditory areas of the cerebral cortex.

Injury to the cochlear portion of the vestibulocochlear nerve results in perception deafness, whereas damage to the vestibular portion causes dizziness and an inability to maintain balance.

IX Glossopharyngeal.   The glossopharyngeal nerve is a mixed nerve that innervates part of the tongue and pharynx (fig. 15.9). The motor fibers of this nerve originate in a nucleus within the medulla oblongata and pass through the jugular foramen. The motor fibers innervate the muscles of the pharynx and the parotid salivary gland to stimulate the swallowing reflex and the secretion of saliva.

The sensory fibers of the glossopharyngeal nerve arise from the pharyngeal region, the parotid salivary gland, and the taste buds on the posterior one-third of the tongue. These taste buds, like those innervated by the facial nerve, are *chemoreceptors*. Some sensory fibers also arise from sensory receptors within the carotid sinus of the neck and help regulate blood pressure. Impulses from the glossopharyngeal nerve travel through the medulla and into the thalamus where they synapse with fibers that convey the impulses to the gustatory area of the cerebral cortex.

Damage to the glossopharyngeal nerve results in the loss of perception of bitter and sour taste from taste buds on the posterior portion of the tongue. If the motor portion of this nerve is damaged, swallowing becomes difficult.

X Vagus.   The vagus nerve has motor and sensory fibers that innervate visceral organs of the thoracic and abdominal cavities (fig. 15.10). The vagus nerve passes through the jugular foramen. The motor portion of the vagus arises from the **nucleus ambiguus** and **dorsal motor nucleus** of the vagus within the medulla. Through various branches, it innervates the muscles of the pharynx, larynx, respiratory tract, lungs, heart, esophagus, and abdominal viscera with the exception of the lower portion of the large intestine. One motor branch of the vagus nerve, the **recurrent laryngeal** nerve, innervates the larynx, enabling one to speak.

Sensory fibers of the vagus convey impulses from essentially the same organs served by motor fibers. Impulses through the sensory fibers relay specific sensations such as hunger pangs, distension, intestinal discomfort, or laryngeal movements. Sensory fibers also arise from proprioceptors in the muscles innervated by the motor fibers of this nerve.

If both vagus nerves are damaged, death ensues rapidly because vital autonomic functions stop, but the injury of one nerve causes vocal impairment, difficulty in swallowing, or other visceral disturbances.

glossopharyngeal: L. *glossa*, tongue; Gk. *pharynx*, throat
vagus: L. *vagus*, wandering

Figure 15.11. The accessory and hypoglossal nerves.

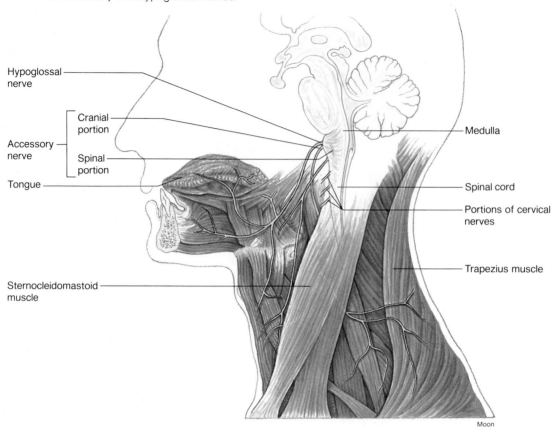

Hypoglossal nerve

Cranial portion

Accessory nerve

Spinal portion

Tongue

Sternocleidomastoid muscle

Medulla

Spinal cord

Portions of cervical nerves

Trapezius muscle

Moon

XI Accessory. The accessory nerve is principally a motor nerve, but it does contain some sensory fibers from proprioceptors within the muscles it innervates. The accessory nerve is unique in that it arises from both the brain and the spinal cord (fig. 15.11). The **cranial motor component,** also called the **bulbar (medullary) portion,** arises from nuclei within the medulla (ambiguus and spinal accessory), passes through the jugular foramen with the vagus nerve, and innervates the skeletal muscles of the soft palate, pharynx, and larynx, which contract reflexively during swallowing. The **spinal motor component** arises from the first five segments of the cervical portion of the spinal cord, passes cranially through the foramen magnum to join with the bulbar portion, and passes through the jugular foramen. The spinal motor component of the accessory nerve innervates the sternocleidomastoid and the trapezius muscles that move the head, neck, and shoulders. Damage to an accessory nerve makes it difficult to move the head or shrug the shoulders.

XII Hypoglossal. The hypoglossal nerve is a mixed nerve. The motor fibers arise from the hypoglossal nucleus within the medulla oblongata and pass through the hypoglossal canal of the skull to innervate both the extrinsic

and intrinsic muscles of the tongue (see fig. 15.11). Motor impulses along these fibers account for the coordinated contraction of the tongue muscles necessary for such activities as food manipulation, swallowing, and speech.

The sensory portion of the hypoglossal nerve arises from proprioceptors within the same tongue muscles and conveys impulses to the medulla oblongata regarding the position and function of the muscles.

If a hypoglossal nerve were damaged, a person would have difficulty in speaking, swallowing, and protruding the tongue.

Neurological Assessment of Cranial Nerves
Head injuries and brain concussions are common occurrences in automobile accidents. The cranial nerves would seem to be well protected on the inferior side of the brain. But the brain, immersed in and filled with cerebrospinal fluid, is like a water-sodden log; a blow to the top of the head can cause a serious rebound of the brain from the floor of the cranium. Routine neurological examinations involve testing for cranial-nerve dysfunction.

Commonly used clinical methods for determining cranial-nerve dysfunction are presented in table 15.2.

---

hypoglossal: Gk. *hypo,* under; L. *glossa,* tongue

Table 15.2 Methods of determining cranial-nerve dysfunction

Nerve	Techniques of examination	Comments
Olfactory	With eyes closed, patient differentiates odors (tobacco, coffee, soap, etc.).	Nasal passages must be patent and tested separately by occluding the opposite side.
Optic	Examine the optic fundi with ophthalmoscope; test visual acuity by reading eye charts.	Visual acuity must be determined with lenses on, if patient wears them.
Oculomotor	Have patient follow examiner's finger movement with eyes—especially cross-eyed movement; observe pupillary change by shining light into each eye separately.	Examiner should note rate of pupillary change and coordinated constriction of pupils. Light in one eye should cause a similar pupillary change in other eye but to a lesser degree.
Trochlear	Have patient follow examiner's finger movement with eyes—especially lateral and downward movement	
Trigeminal	Motor portion: examiner palpates temporalis and masseter muscles as patient clenches teeth; patient is asked to open mouth against resistance applied by examiner.	Muscles of both sides of the jaw should show equal contractile strength.
	Sensory portion: tactile and pain receptors are tested by lightly touching entire face with cotton and then with pin stimulus.	Patient's eyes should be closed and innervation areas for all three trigeminal branches should be tested.
Abducens	Have patient follow examiner's finger movement—especially lateral movement.	Motor functioning of cranial nerves III, IV, and VI may be tested simultaneously through selective movements of eyeball.
Facial	Motor portion: have patient raise eyebrows, frown, tightly constrict eyelids, smile, puff out cheeks, and whistle.	Examiner should note lack of tonus expressed by sagging regions of face.
	Sensory portion: examiner should place sugar on each side of tip of tongue.	Not reliable test for specific facial-nerve dysfunction because of tendency to stimulate taste buds on both sides of tip of tongue.
Vestibulocochlear	Vestibular portion: patient asked to walk a straight line.	Not usually tested unless patient complains of dizziness or balance problems.
	Cochlear portion: test with tuning fork.	Note ability to discriminate sounds.
Glossopharyngeal and vagus	Motor: examiner should note disturbances in swallowing, talking, and movement of soft palate; stimulate back of throat and note gag reflex.	Visceral innervation of vagus cannot be examined except for innervation to larynx, which is also served by glossopharyngeal.
Accessory	Patient is asked to shrug shoulders against resistance of examiner's hand and to rotate head against resistance.	Sides should show uniformity of strength.
Hypoglossal	Patient is requested to protrude tongue; tongue thrust may be resisted with tongue blade.	Tongue should protrude straight; deviation to side indicates ipsilateral-nerve dysfunction; asymmetry, atrophy, or lack of strength should be noted.

Source: Van De Graaff, Kent M., *Human Anatomy.* © 1984 Wm. C. Brown Publishers, Dubuque, Iowa. All Rights Reserved. Reprinted by permission.

1. *Which cranial nerves consist of sensory fibers only?*
2. *Which cranial nerves pass through the superior orbital fissure? through the jugular foramen?*
3. *Which cranial nerves have anything to do with tasting, chewing and manipulating food, and swallowing?*
4. *Which cranial nerves have anything to do with the structure, function, or movement of the eyeball?*
5. *List the cranial nerves, and indicate how each would be tested (both motor and sensory) for possible dysfunction.*

Objective 4. Discuss how the spinal nerves are named.

Objective 5. Describe the general distribution of a spinal nerve.

The thirty-one pairs of spinal nerves (fig. 15.12) are grouped as follows: eight cervical, twelve thoracic, five lumbar, five sacral, and one coccygeal. With the exception of the first cervical nerve, the spinal nerves leave the spinal cord and vertebral canal through intervertebral foramina. The first pair of cervical nerves emerges between the occipital bone of the skull and the atlas vertebra. The second through the seventh pairs of cervical nerves emerge above the vertebrae for which they are named, whereas the eighth pair of cervical nerves passes between the seventh cervical and first thoracic vertebrae. The remaining pairs of spinal nerves emerge below the vertebrae for which they are named.

Spinal Nerves

Each of the thirty-one pairs of spinal nerves is formed by the union of a dorsal and a ventral spinal root, which emerges from the spinal cord through an intervertebral foramen to innervate a body dermatome.

Figure 15.12. The distribution of the spinal nerves.

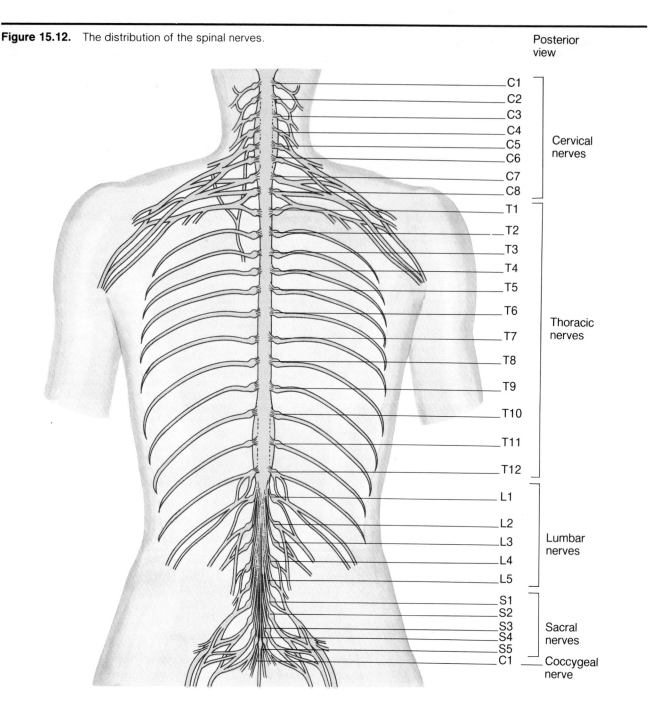

Posterior view

C1
C2
C3
C4
C5
C6
C7
C8
} Cervical nerves

T1
T2
T3
T4
T5
T6
T7
T8
T9
T10
T11
T12
} Thoracic nerves

L1
L2
L3
L4
L5
} Lumbar nerves

S1
S2
S3
S4
S5
} Sacral nerves

C1 — Coccygeal nerve

A **spinal nerve** is a mixed nerve attached to the spinal cord by a **dorsal (posterior) root** composed of sensory fibers and a **ventral (anterior) root** composed of motor fibers (fig. 15.13). The dorsal root contains an enlargement called the **dorsal-root ganglion,** where the cell bodies of sensory neurons are located. The axons of sensory neurons convey sensory impulses through the dorsal root and into the spinal cord where synapses occur with dendrites of other neurons. The ventral root consists of axons of motor neurons that convey motor impulses away from the CNS. A spinal nerve is formed as the fibers from the dorsal and ventral roots converge and emerge through an intervertebral foramen.

The disease **herpes zoster,** also known as shingles, is a viral infection of the dorsal-root ganglia. Herpes zoster causes painful, often unilateral, clusters of fluid-filled vesicles in the skin along the paths of the affected peripheral sensory neurons. This disease requires no special treatment, as the lesions gradually heal, and is usually not serious except in elderly debilitated patients, who may die from exhaustion.

Figure 15.13. A cross section of the spinal cord and the distribution of a spinal nerve.

A spinal nerve divides into several branches immediately after it emerges through the intervertebral foramen. The small **meningeal branch** reenters the vertebral canal to innervate the meninges, vertebrae, and vertebral ligaments. A larger branch, called the **dorsal ramus,** innervates the muscles, joints, and skin of the back along the vertebral column (fig. 15.13). A **ventral ramus** of a spinal nerve innervates the muscles and skin on the lateral and ventral (anterior) side of the torso.

The **rami communicantes** are two branches off of the spinal nerve that connect to a **sympathetic ganglion,** which is part of the autonomic nervous system. The rami communicantes are composed of a **gray ramus,** containing unmyelinated postganglionic fibers, and a **white ramus,** containing myelinated preganglionic fibers. All spinal nerves have a gray ramus, but only the spinal nerves T1 through T3 have both gray and white rami.

---

1. List the number of nerves in each of the five regions of the vertebral column.
2. What are the four principal branches, or rami, from a spinal nerve, and what does each innervate?

Nerve Plexuses

Except in the thoracic nerves T2–T12, the ventral rami of the spinal nerves combine and then split again as a network of nerves referred to as a plexus. There are four plexuses of spinal nerves: the cervical, the brachial, the lumbar, and the sacral. Nerves emerge from the plexuses and are named according to the structures they innervate or the general course they take.

Objective 6. List the spinal nerve composition of each of the plexuses arising from the spinal cord.

Objective 7. List the principal nerves that emerge from the plexuses, and describe their general innervation.

Cervical Plexus. The cervical plexus *(plek′sus)* is positioned deep on the side of the neck, lateral to the first four cervical vertebrae (fig. 15.14). It is formed by the ventral rami of the first four cervical nerves (C1–C4) and a portion of C5. Branches from the cervical plexus innervate the skin and muscles of the neck and portions of the head and shoulders. Branches of the cervical plexus also combine with the accessory and hypoglossal cranial nerves to

---

plexus: L. *plectere*, to twine or braid

Figure 15.14. The cervical plexus.

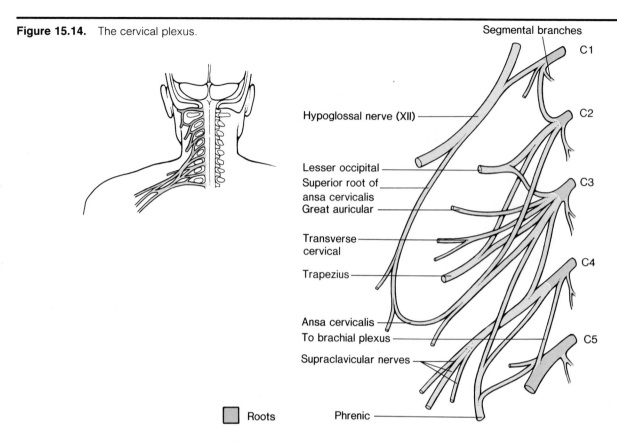

Segmental branches

C1

Hypoglossal nerve (XII)

C2

Lesser occipital

Superior root of ansa cervicalis

Great auricular

C3

Transverse cervical

Trapezius

C4

Ansa cervicalis

To brachial plexus

C5

Supraclavicular nerves

Phrenic

Roots

Table **15.3**, Branches of cervical plexus		
Nerve	**Spinal component**	**Innervation**
Superficial Cutaneous Branches		
Lesser occipital	C2,C3	Skin of scalp above and behind ear
Greater auricular	C2,C3	Skin in front of, above, and below ear
Transverse cervical	C2,C3	Skin of anterior aspect of neck
Supraclavicular	C3,C4	Skin of upper portion of chest and shoulder
Deep Motor Branches		
Ansa cervicalis Superior root	C1,C2	Geniohyoid, thyrohyoid, and infrahyoid muscles of neck
Inferior root	C3,C4	Omohyoid, sternohyoid, and sternothyroid muscles of neck
Phrenic	C3,C4,C5	Diaphragm
Segmental branches	C1–C5	Deep muscles of neck (levator scapulae ventralis, trapezius, scalenus, and sternocleidomastoid)

Source: Van De Graaff, Kent M., *Human Anatomy*. © 1984 Wm. C. Brown Publishers, Dubuque, Iowa. All Rights Reserved. Reprinted by permission.

supply dual innervation to some specific neck and pharyngeal muscles. Fibers from the third, fourth, and fifth cervical nerves unite to become the **phrenic** *(fren'ik)* **nerve,** which innervates the diaphragm. Motor impulses through the paired phrenic nerves cause the diaphragm to contract, inspiring air into the lungs.

The nerves of the cervical plexus are summarized in table 15.3.

Brachial Plexus. The brachial plexus is positioned to the side of the last four cervical and first thoracic vertebrae. It is formed by the ventral rami of C5 through T1 and some fibers from C4 and T2. From its emergence, the brachial plexus extends downward and laterally, passes over the first rib behind the clavicle, and enters the axilla. Each brachial plexus innervates the entire upper extremity of one side plus a number of shoulder and neck muscles.

Structurally, the brachial plexus is divided into *roots, trunks, divisions,* and *cords* (fig. 15.15). The roots of the brachial plexus are simply continuations of the ventral rami of the cervical nerves. The roots of C5 and C6 converge to become the **upper trunk,** C7 becomes the **middle trunk,** and C8 and T1 converge to become the **lower trunk.** Each of the three trunks immediately divides into an **anterior division** and a **posterior division.** The divisions then converge to form three cords. The **posterior cord** is formed by the convergence of the posterior divisions of the upper, middle, and lower trunks and hence contains fibers from C5 through C8. The **medial cord** is a continuation of the

phrenic: Gk. *phren,* diaphragm or midriff

Figure 15.15. The brachial plexus.

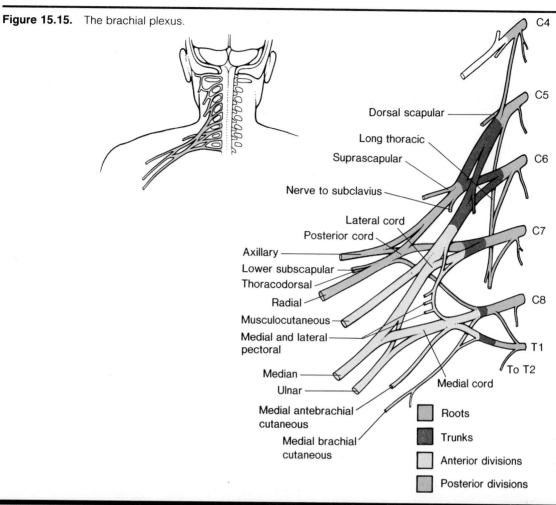

Labels on figure:
C4
C5
Dorsal scapular
Long thoracic
Suprascapular
Nerve to subclavius
Lateral cord
Posterior cord
Axillary
Lower subscapular
Thoracodorsal
Radial
Musculocutaneous
Medial and lateral pectoral
Median
Ulnar
Medial antebrachial cutaneous
Medial brachial cutaneous
C6
C7
C8
T1
To T2
Medial cord

Roots
Trunks
Anterior divisions
Posterior divisions

Table **15.4** Branches of brachial plexus		
Nerve	**Cord and spinal components**	**Innervation**
Axillary	Posterior cord (C5,C6)	Skin of shoulder; shoulder joint, deltoid and teres minor muscles
Radial	Posterior cord (C5–C8,T1)	Skin of posterior lateral surface of arm, forearm, and hand; extensor muscles of posterior upper arm and forearm (triceps brachii, supinator, anconeus, brachioradialis, extensor carpi radialis brevis, extensor carpi radialis longus, extensor carpi ulnaris) and several intrinsic muscles of the hand
Musculocutaneous	Lateral cord (C5,C6,C7)	Skin of lateral surface of forearm; muscles of anterior upper arm (coracobrachialis, biceps brachii, brachialis)
Ulnar	Medial cord (C8,T1)	Skin of medial third of hand; flexor muscles of anterior forearm, medial palm, and intrinsic flexor muscles of hand (flexor carpi ulnaris, flexor digitorum profundus, third and fourth lumbricales)
Median	Medial cord (C6,C7,C8,T1)	Skin of lateral two-thirds of hand; flexor muscles of anterior forearm, lateral palm, and first and second lumbricales

Source: Van De Graaff, Kent M., *Human Anatomy.* © 1984 Wm. C. Brown Publishers, Dubuque, Iowa. All Rights Reserved. Reprinted by permission.

anterior division of the lower trunk and primarily contains fibers from C8 and T1. The **lateral cord** is formed by the convergence of the anterior division of the upper and middle trunk and consists of fibers from C5 through C7.

Trauma to the brachial plexus is not uncommon, especially if the clavicle, upper ribs, or lower cervical vertebrae are seriously fractured. Occasionally, the brachial plexus of a newborn is severely strained during a difficult delivery. In such cases, the arm of the injured side is paralyzed and withers as the muscles atrophy in relation to the extent of the injury.

In summary, the brachial plexus is composed of nerve fibers from spinal nerves C5 through T1 and a few fibers from C4 and T2. Roots are continuations of the ventral rami. The roots converge to form trunks, and the trunks branch into divisions. The divisions in turn form cords, and the nerves of the upper extremity arise from the cords.

Five major nerves and several smaller ones arise from the three cords of the brachial plexus. The principal nerves of the brachial plexus are summarized in table 15.4.

Figure 15.16.   The lumbar plexus.

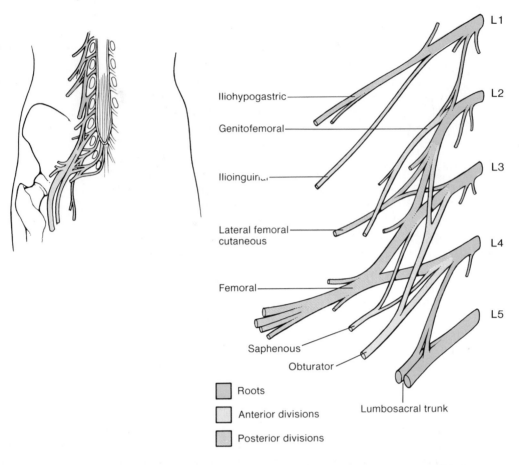

Iliohypogastric

Genitofemoral

Ilioinguinal

Lateral femoral cutaneous

Femoral

Saphenous

Obturator

L1

L2

L3

L4

L5

Roots

Anterior divisions

Posterior divisions

Lumbosacral trunk

Table 15.5	Branches of lumbar plexus	
Nerve	**Spinal components**	**Innervation**
Iliohypogastric	T12–L1	Skin of lower abdomen and buttock; muscles of anterolateral abdominal wall (external oblique, internal oblique, transversus abdominis)
Ilioinguinal	L1	Skin of upper median thigh, scrotum and root of penis in male, and labia majora in female; muscles of anterolateral abdominal wall with iliohypogastric nerve
Genitofemoral	L1,L2	Skin of middle anterior surface of thigh, scrotum in male, and labia majora in female; cremaster muscle in male
Lateral femoral cutaneous	L2,L3	Skin of anterior, lateral, and posterior aspects of thigh
Femoral	L2–L4	Skin of anterior and medial aspect of thigh and medial aspect of leg and foot; anterior muscles of thigh (iliacus, psoas major, pectineus, rectus femoris, sartorius) and extensor muscles of leg (rectus femoris, vastus lateralis, vastus medialis, vastus intermedius)
Obturator	L2–L4	Skin of medial aspect of thigh; adductor muscles of leg (obturatorius externus, pectineus, adductor longus, adductor brevis, adductor magnus, gracilis)
Saphenous	L2–L4	Skin of medial aspect of leg

Source: Van De Graaff, Kent M., *Human Anatomy.* © 1984 Wm. C. Brown Publishers, Dubuque, Iowa. All Rights Reserved. Reprinted by permission.

saphenous: L. *saphena,* the hidden one

Figure 15.17. The sacral plexus.

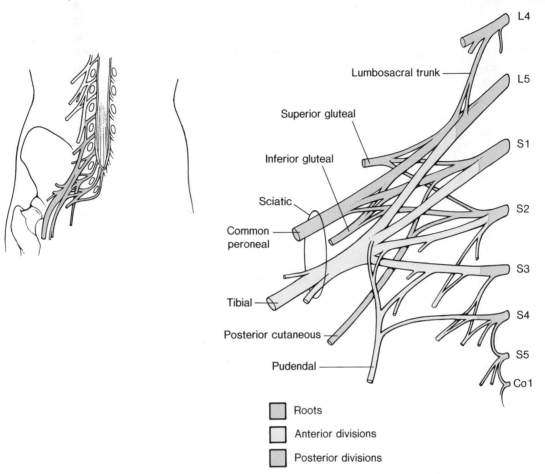

Roots

Anterior divisions

Posterior divisions

> You become keenly aware of the nerves in your body when you accidentally bump one. When you hit your "funny bone" (the back of the elbow) on the corner of a table, you are actually hitting your ulnar nerve. It tingles into your little finger because this is its area of innervation.
>
> Likewise, when your leg or arm "goes to sleep" it is due to compression of the major nerves of that area, not, as many people believe, to the blood supply being cut off.

Lumbar Plexus. The lumbar plexus is positioned to the side of the first four lumbar vertebrae. It is formed by the ventral rami of spinal nerves L1–L4 and some fibers from T12 (fig. 15.16). The nerves that arise from the lumbar plexus innervate structures of the lower abdomen and anterior and medial portions of the lower extremity. The lumbar plexus is not as complex as the brachial plexus, having only roots and divisions rather than the roots, trunks, divisions, and cords as in the brachial plexus.

Structurally, the **posterior division** of the lumbar plexus passes obliquely outward, deep to the psoas major muscle, whereas the **anterior division** is superficial to the

---

lumbar: L. *lumbus,* loin

---

quadratus lumborum muscle. From these two divisions arise the large **femoral** *(fem'or-al)* **nerve,** which innervates the anterior muscles of the thigh and the **obturator nerve,** which innervates the adductor muscles of the leg. Several smaller nerves also arise from the lumbar plexus and are summarized in table 15.5.

> A herniated intervertebral disc in the lumbar region may cause a condition called **sciatica** *(si-at'i-kah).* This condition is characterized by a sharp pain in the gluteal region and by pains that extend down the posterior side of the thigh, caused by the disc compressing the spinal roots that form the sciatic nerve.

Sacral Plexus. The sacral plexus is positioned immediately caudal to the lumbar plexus. It is formed by the ventral rami of spinal nerves L4, L5, and S1–S4 (fig. 15.17). The nerves arising from the lumbar plexus innervate the lower back, pelvis, perineum, posterior surface of

---

sacral: L. *sacris,* sacred

Table 15.6 Branches of sacral plexus

Nerve	Spinal components	Innervation
Superior gluteal	L4,L5,S1	Abductor muscles of thigh (gluteus minimus, gluteus medius, tensor fasciae latae)
Inferior gluteal	L5–S2	Extensor muscle of thigh (gluteus maximus)
Nerve to piriformis	S1,S2	Abductor and rotator of thigh (piriformis)
Nerve to quadratus femoris	L4,L5,S1	Rotators of thigh (gemellus inferior, quadratus femoris)
Nerve to obturator internus	L5–S2	Rotators of thigh (gemellus superior, obturatorius internus)
Perforating cutaneous	S2,S3	Skin over lower medial surface of buttock
Posterior cutaneous	S1–S3	Skin over lower lateral surface of buttock, anal region, upper posterior surface of thigh, upper aspect of calf, scrotum in male, and labia majora in female
Sciatic	L4–S3	Composed of two nerves (tibial and common peroneal) within sciatic sheath; splits into two portions at popliteal fossa; branches from sciatic in thigh region to hamstring muscles (biceps femoris, semitendinosus, semimembranosus) and adductor magnus
Tibial (sural, medial and lateral plantar)	L4–S3	Skin of posterior surface of leg and sole of foot; muscle innervation includes gastrocnemius, soleus, flexor digitorum longus, flexor hallucis longus, tibialis posterior, popliteus, and intrinsic muscles of the foot
Common peroneal (superficial and deep peroneal)	L4–S2	Skin of anterior surface of the leg and dorsum of foot; muscle innervation includes peroneus tertius, peroneus brevis, peroneus longus, tibialis anterior, extensor hallucis longus, extensor digitorum longus, extensor digitorum brevis
Pudendal	S2–S4	Skin of penis and scrotum in male and skin of clitoris, labia majora, labia minora, and lower vagina in female; muscles of perineum

Source: Van De Graaff, Kent M., *Human Anatomy.* © 1984 Wm. C. Brown Publishers, Dubuque, Iowa. All Rights Reserved. Reprinted by permission.

the thigh and leg, and the dorsal and ventral surfaces of the foot. Like the lumbar plexus, the sacral plexus consists of **roots** and **anterior** and **posterior divisions** from which nerves arise. Because some of the nerves of the sacral plexus also contain fibers from the nerves of the lumbar plexus through the **lumbosacral trunk,** these two plexuses are frequently described collectively as the **lumbosacral plexus.**

The **sciatic** *(si-at'ik)* **nerve** is the largest nerve arising from the sacral plexus and, in fact, is the largest nerve in the body. The sciatic nerve passes from the pelvis through the greater sciatic notch of the os coxa and extends down the posterior aspect of the thigh. The sciatic nerve is actually composed of two nerves, the common peroneal and tibial nerves, wrapped in a common sciatic sheath.

A summary of the nerves of the sacral plexus and their distribution is presented in table 15.6.

---

sciatic: L. *sciaticus,* hip joint
peroneal: Gk. *perone,* pin, fibula

The sciatic nerve in the buttock lies deep to the gluteus maximus muscle, midway between the greater trochanter and the ischial tuberosity. Because of its position, the sciatic nerve is of tremendous clinical importance. A posterior dislocation of the hip joint will generally injure the sciatic nerve. A herniated disc (slipped disc), pressure from the uterus during pregnancy, or an improperly administered injection into the buttock may damage the roots leading to the sciatic nerve or the sciatic nerve itself.

1. Define a nerve plexus. What are the four plexuses, and which spinal nerves contribute to each?
2. Which of the spinal nerves are not involved in a plexus?
3. Distinguish between a dorsal ramus and a ventral ramus. Which is involved in the formation of plexuses?
4. Construct a table that lists the plexus of origin and the general region of innervation for the following nerves: pudendal, phrenic, femoral, ulnar, median, sciatic, saphenous, axillary, radial, ansa cervicalis, tibial, and common peroneal.

Figure 15.18. The reflex arc.

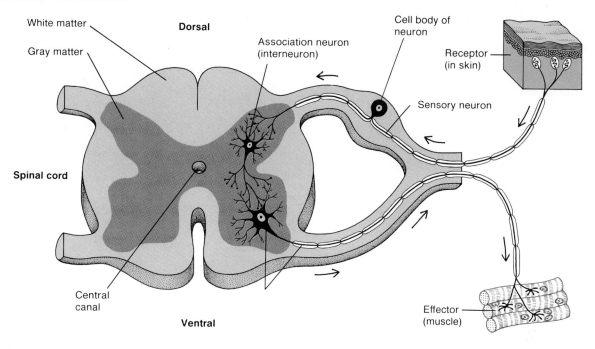

Reflex Arcs and Reflexes

The conduction pathway of a reflex arc consists of a receptor, a sensory neuron, a motor neuron and its innervation, in the PNS, and an association neuron, in the CNS. The reflex arc provides the mechanism for a rapid, automatic response to a potentially threatening stimulus.

Objective 8. Define a reflex arc, and list its five components.

Objective 9. Distinguish between the various kinds of reflexes.

Specific **nerve pathways** provide routes by which impulses travel through the nervous system. Frequently, a nerve pathway begins with impulses being conducted to the CNS through sensory receptors and sensory neurons of the PNS. Once within the CNS, impulses may immediately travel back through motor portions of the PNS to activate specific skeletal muscles, glands, or smooth muscles, and also be sent simultaneously to other parts of the CNS through ascending tracts within the spinal cord.

Components of the Reflex Arc

The simplest type of nerve pathway is a **reflex arc** (fig. 15.18). A reflex arc implies an automatic, unconscious, protective response to a situation in an attempt to maintain body homeostasis. A reflex arc leads by a short route from sensory to motor neurons and includes only two or three neurons. The five components of a reflex arc are the receptor, sensory neuron, center, motor neuron, and effector. The **receptor** includes the dendrite of a sensory neuron and the place where the electrical impulse is initiated. The sensory neuron relays the impulse through the posterior root to the CNS. The **center** is within the CNS and usually involves an association neuron (interneuron). It is here that the arc is made and other impulses are sent through synapses to other parts of the body. The **motor neuron** conducts the impulse to an effector organ (generally a skeletal muscle) that responds. The result of an impulse through a reflex arc is called a reflex action or, simply, a *reflex.*

Figure 15.19. The knee-jerk reflex is an ipsilateral reflex. The receptor and effector organs are on the same side of the spinal cord. The knee-jerk reflex is also a monosynaptic reflex because it involves only two neurons and one synapse.

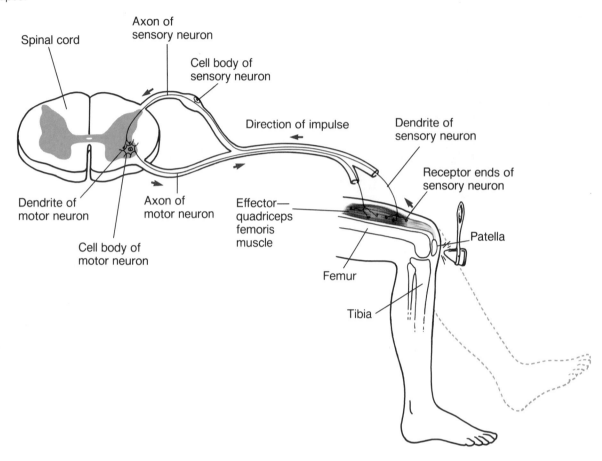

Kinds of Reflexes

Visceral Reflexes. Reflexes that cause smooth or cardiac muscle to contract or glands to secrete are **visceral** *(vis′er-al)* **(autonomic) reflexes.** Visceral reflexes help control the body's many involuntary processes such as heart rate, respiratory rate, blood pressure, and digestion. Swallowing, sneezing, coughing, and vomiting may also be reflexive, although they involve skeletal muscles that function involuntarily.

Somatic Reflexes. Somatic reflexes are those that result in the contraction of skeletal muscles. There are three principal kinds of somatic reflexes, named according to the response they produce.

The **stretch reflex** involves only two neurons and one synapse in the pathway and is therefore called a *monosynaptic reflex arc.* Slight stretching of neuromuscular spindle receptors within a muscle initiates an impulse along an afferent neuron to the spinal cord. A synapse with a motor neuron occurs in the anterior gray horn and a motor unit is activated, causing specific muscle fibers to contract. Since the receptor and effector organs of the stretch reflex involve structures on the same side of the spinal cord, the reflex arc is an *ipsilateral reflex arc*. The knee-jerk reflex is an ipsilateral reflex (fig. 15.19), as are all monosynaptic reflex arcs.

A **flexor reflex,** or **withdrawal reflex,** consists of a *polysynaptic reflex arc* (fig. 15.20). Flexor reflexes involve association neurons (interneurons) in addition to the sensory and motor neurons. A flexor reflex is initiated as a person encounters a painful stimulus such as a hot or sharp object. As a receptor organ is stimulated, afferent neurons transmit the impulse to the spinal cord where association neurons (interneurons) are activated. There the impulses are directed through motor neurons to flexor muscles that contract in response. Simultaneously, antagonistic muscles are inhibited (relaxed) so that the traumatized extremity can be quickly withdrawn from the harmful source of stimulation.

Figure 15.20. The flexor, or withdrawal, reflex is a polysynaptic reflex because it involves association neurons (interneurons) in addition to sensory and motor neurons.

Several additional reflexes may be activated while a flexor reflex is in progress. In an *intersegmental reflex arc,* motor units from several segments of the spinal cord are activated by impulses coming in from the receptor organ. An intersegmental reflex arc produces stimulation of more than one effector organ. Frequently, afferent impulses from a receptor organ cross over through the spinal cord to activate effector organs within the opposite (*contralateral*) limb. This type of reflex is called a **crossed extensor reflex** (fig. 15.21) and is important for maintaining body balance while a flexor reflex is in progress. The reflexive inhibition of certain muscles to contract, called *reciprocal inhibition,* also helps maintain balance while either flexor or crossed extensor reflexes are in progress.

Certain reflexes are important for physiological functions, while others are important for avoiding injury. Some common reflexes are presented in table 15.7 and illustrated in figure 15.22.

Part of a routine physical examination involves testing a person's reflexes. The condition of the nervous system, particularly the functioning of the synapses, may be determined by examining reflexes. In case of injury to some portion of the nervous system, testing certain reflexes may indicate the location and extent of the injury. Also, an anesthesiologist may try to initiate a reflex to ascertain the effect of an anesthetic.

1. How are reflexes important in maintaining body homeostasis?
2. List the five components of a reflex arc.
3. Define or describe the following: visceral reflex, somatic reflex, stretch reflex, flexor reflex, crossed extensor reflex, ipsilateral, and contralateral.

anesthesia: Gk. *an,* without; *aisthesis,* sensation

Figure 15.21.   A crossed extensor reflex causes a reciprocal inhibition of muscles of the opposite appendage. This type of reflex inhibition is important in maintaining balance.

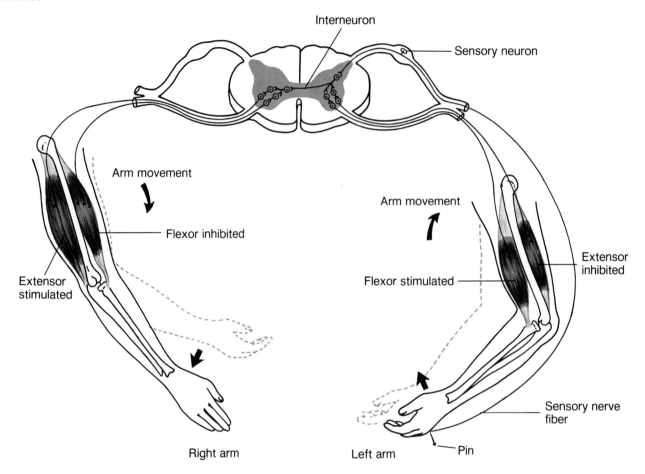

Table 15.7   Some clinically important reflexes

Reflex	Spinal segment	Site of receptor stimulation	Effector action
Biceps reflex	C5, C6	Biceps tendon near attachment on radial tuberosity	Contracts biceps brachii to flex elbow
Triceps reflex	C7, C8	Triceps tendon near attachment on olecranon process	Contracts triceps brachii to extend elbow
Supinator or brachioradialis reflex	C5, C6	Radial attachment of supinator and brachioradialis muscles	Supinates forearm and hand
Knee reflex	L2, L3, L4	Patellar tendon just below patella	Contracts quadriceps to extend the knee
Ankle reflex	S1, S2	Achilles tendon near attachment on calcaneus	Plantar flexes ankle
Plantar reflex	L4, L5, S1, S2	Lateral aspect of sole from heel to ball of foot	Plantar flexes foot and flexes toes
Babinski reflex*	L4, L5, S1, S2	Lateral aspect of sole from heel to ball of foot	Dorsiflexes great toe and fans other toes
Abdominal reflexes	T8, T9, T10 above umbilicus and T10, T11, T12 below umbilicus	Sides of abdomen above and below level of umbilicus	Contracts abdominal muscles and deviates umbilicus toward stimulus
Cremasteric reflex	L1, L2	Stroking upper inside of thigh in males	Contracts cremasteric muscle and elevates testis on same side of stimulation

*If the Babinski reflex rather than the plantar reflex occurs as the sole of the foot is stimulated, it may indicate damage to the corticospinal tract within the spinal cord. However, the Babinski reflex is present in infants up to 12 months of age because of the immaturity of their corticospinal tracts.

Source: Van De Graaff, Kent M., *Human Anatomy*. © 1984 Wm. C. Brown Publishers, Dubuque, Iowa. All Rights Reserved. Reprinted by permission.

Babinski's reflex: from Joseph F. Babinski, French neurologist, 1857–1932

Figure 15.22.   Some reflexes of clinical importance.

Glabellar reflex

Ankle (Achilles) reflex

Babinski reflex

Abdominal reflex

Plantar reflex

Knee (patellar) reflex

Biceps reflex

Supinator (brachioradialis) reflex

Triceps reflex

Chapter Summary

I. Introduction to the Peripheral Nervous System
The peripheral nervous system consists of nerves that convey impulses to and from the central nervous system.
  A. The cranial nerves arise from the brain, and the spinal nerves arise from the spinal cord.
  B. Sensory (afferent) nerves convey impulses toward the CNS, motor (efferent) nerves convey impulses away from the CNS, and mixed nerves are composed of both sensory and motor fibers.

II. Cranial Nerves
Twelve pairs of cranial nerves emerge from the inferior surface of the brain and pass through foramina of the skull to innervate structures in the head, neck, and visceral organs of the torso.
  A. The names of the cranial nerves indicate their primary function or the general distribution of their fibers.
  B. Most of the cranial nerves are mixed, some consist of sensory fibers only, and others are primarily motor.
  C. Some of the cranial nerve fibers are somatic, and others are autonomic.
  D. A test for cranial-nerve dysfunction is clinically important in a neurological examination.

III. Spinal Nerves
Each of the thirty-one pairs of spinal nerves is formed by the union of a dorsal and ventral spinal root, which emerges from the spinal cord through an intervertebral foramen to innervate a body dermatome.

  A. The spinal nerves are grouped according to the levels of the spinal column from which they arise, and they are numbered in sequence.
  B. Each spinal nerve is a mixed nerve consisting of a dorsal root of sensory fibers and a ventral root of motor fibers.
  C. Just beyond its intervertebral foramen, each spinal nerve divides into several branches.

IV. Nerve Plexuses
Except in the thoracic nerves T2–T12, the ventral rami of the spinal nerves combine and then split again as a network of nerves referred to as a plexus. There are four plexuses of spinal nerves: the cervical, the brachial, the lumbar, and the sacral. Nerves emerge from the plexuses and are named according to the structures they innervate or the general course they take.
  A. The cervical plexus is formed by the ventral rami of C1–C4 and by a portion of C5.
  B. The brachial plexus is formed by the ventral rami of C5–T1 and by some fibers from C4 and T2.
    1. The brachial plexus is divided into roots, trunks, divisions, and cords.
    2. The axillary, radial, musculocutaneous, ulnar, and median are the principal nerves arising from the brachial plexus.
  C. The lumbar plexus is formed by the ventral rami of L1–L4 and by some fibers from T12.

  1. The lumbar plexus is divided into roots and divisions.
  2. The femoral and obturator are the principal nerves arising from the lumbar plexus.
  D. The sacral plexus is formed by the ventral rami of L4, L5, and S1–S4.
    1. The sacral plexus is divided into roots and divisions.
    2. The sciatic nerve, composed of the common peroneal and tibial nerve, arises from the sacral plexus.

V. Reflex Arcs and Reflexes
The conduction pathway of a reflex arc consists of a receptor, a sensory neuron, a motor neuron and its innervation, in the PNS, and an association neuron, in the CNS. The reflex arc provides the mechanisms for a rapid, automatic response to a potentially threatening stimulus.
  A. A reflex arc is the simplest type of nerve pathway.
  B. Visceral reflexes cause smooth or cardiac muscle to contract or glands to secrete.
  C. Somatic reflexes cause skeletal muscles to contract.
    1. The stretch reflex is a monosynaptic reflex arc.
    2. The flexor reflex is a polysynaptic reflex arc.

Review Activities

Objective Questions

1. Which of the following is a *false* statement concerning the peripheral nervous system?
  (a) It consists of cranial and spinal nerves only.
  (b) It contains components of the autonomic nervous system.
  (c) Sensory receptors, neurons, nerves, ganglia, and plexuses are all part of the PNS.

2. An inability to look cross-eyed would most likely indicate a problem with which cranial nerve? The
  (a) optic
  (b) oculomotor
  (c) abducens
  (d) facial

3. Which cranial nerve innervates the muscle that raises the upper eyelid? The
  (a) trochlear
  (b) oculomotor
  (c) abducens
  (d) facial

4. The inability to walk a straight line may indicate damage to which cranial nerve? The
  (a) trigeminal
  (b) facial
  (c) vestibulocochlear
  (d) vagus

5. Which cranial nerve passes through the stylomastoid foramen? The
  (a) facial
  (b) glossopharyngeal
  (c) vagus
  (d) hypoglossal

6. Which of the following cranial nerves does not contain parasympathetic fibers? The
  (a) oculomotor
  (b) accessory
  (c) vagus
  (d) facial

7. Which of the following is not a spinal nerve plexus? The
  (a) cervical
  (b) brachial
  (c) thoracic
  (d) lumbar
  (e) sacral

8. Roots, trunks, divisions, and cords are characteristic of the
   (a) sacral plexus
   (b) thoracic plexus
   (c) lumbar plexus
   (d) brachial plexus
9. Which of the following nerve-plexus association is incorrect?
   (a) median—sacral
   (b) phrenic—cervical
   (c) axillary—brachial
   (d) femoral—lumbar

10. Extending the leg when the patellar tendon is tapped is an example of a(n)
    (a) visceral reflex
    (b) flexor reflex
    (c) ipsilateral reflex
    (d) crossed extensor reflex

Essay Questions

1. Explain the structural and functional relationship between the central nervous system, the autonomic nervous system, and the peripheral nervous system.
2. List the cranial nerves, and describe the major function(s) of each. How is each cranial nerve tested for dysfunction?
3. Describe the structure of a spinal nerve.
4. List the roots of each of the spinal plexuses, and describe where each is located and the nerves which originate from them.
5. Distinguish between monosynaptic, polysynaptic, ipsilateral, stretch, and flexor reflexes.

Selected Readings

Barr, M. L. 1979. *The human nervous system, an anatomical viewpoint.* 3rd ed. New York: Harper and Row Publishers.

Carpenter, M. B. 1978. *Core text of neuroanatomy.* Baltimore: Williams and Wilkins Co.

Clark, R. G. 1979. *Essentials of clinical neuroanatomy and neurophysiology.* Philadelphia: F. A. Davis.

Kandel, E. R., and E. R. Schwartz, eds. 1981. *Principles of neural science.* New York: Elsevier North-Holland.

Morell, P., and W. T. Norton. May 1980. Myelin. *Scientific American.*

Waxman, S. G. 1982. Membrane, myelin, and the pathophysiology of multiple sclerosis. *N. Engl. J. Med.* 306:1529.

16

Autonomic

Concepts

The involuntary effectors, including smooth muscle, cardiac muscle, and glands, are regulated by the autonomic nervous system. The nature of this regulation is in many ways different from the neural regulation of voluntary striated muscles.

The autonomic system consists of preganglionic neurons that originate in the CNS and postganglionic neurons that originate in ganglia. The specific origin of the preganglionic fibers and the location of the ganglia, however, are different in the sympathetic and parasympathetic divisions of the autonomic system.

The sympathetic and parasympathetic divisions of the autonomic system may act in a complementary or antagonistic way to help maintain homeostasis. The sympathetic division activates the body to "fight or flight" through adrenergic effects; the parasympathetic division often has antagonistic actions through cholinergic effects.

Brain centers regulate the activity of descending pathways to preganglionic autonomic neurons. The neural centers that control the activity of autonomic nerves are influenced by higher brain areas as well as by sensory input.

Nervous System

The Autonomic Nervous System and Its Visceral Effectors

The action of effectors—muscles and glands—is controlled to a large degree by motor neurons. Skeletal muscles, which are the voluntary effectors, are regulated by somatic motor neurons. The involuntary effectors, including smooth muscle, cardiac muscle, and glands, are regulated by autonomic motor neurons.

Objective 1. Define the terms *preganglionic neuron* and *postganglionic neuron* as you explain the difference between the efferent pathways of the somatic motor and autonomic motor systems.

Objective 2. Describe how the neural regulation of visceral effector organs can differ from the neural regulation of skeletal muscles.

Objective 3. Compare the structure and regulation of single-unit smooth muscle with that of multi-unit smooth muscle.

Autonomic motor nerves innervate organs whose functions are not usually under voluntary control. The effectors that respond to autonomic regulation include **cardiac muscle** (the heart), **smooth** (visceral) **muscles,** and **glands.** These are part of the organs of the *viscera* and of blood vessels and specialized structures within other organs. The involuntary effects of autonomic innervation contrast with the voluntary control of skeletal muscles by way of somatic motor innervation.

The Autonomic Nervous System

Unlike somatic motor neurons, which conduct impulses along a single axon from the spinal cord to the neuromuscular junction, the autonomic *(aw''to-nom'ik)* motor pathway involves two neurons in the efferent flow of impulses (table 16.1).

Table 16.1 Comparisons of the somatic motor system with the autonomic motor system

Feature	Somatic motor	Autonomic motor
Effector organs	Skeletal muscles	Cardiac muscle, smooth muscle, and glands
Presence of ganglia	No ganglia outside CNS	Cell bodies of postganglionic autonomic fibers located in paravertebral, prevertebral (collateral), and terminal ganglia
Number of neurons from CNS to effector	One	Two
Type of neuromuscular junction	Specialized motor end plate	No specialization of postsynaptic membrane; all areas of smooth muscle cells contain receptor proteins for neurotransmitters
Effect of nerve impulse on muscle	Excitatory only	Either excitatory or inhibitory
Type of nerve fibers	Fast-conducting, thick (9–13 μm), and myelinated	Slow conducting; preganglionic fibers, lightly myelinated but thin (3 μm); postganglionic fibers, unmyelinated and very thin (about 1.0 μm)
Effect of denervation	Flaccid paralysis and atrophy	Much muscle tone and function remains; target cells show denervation hypersensitivity

Source: Fox, Stuart Ira, *Human Physiology.* © 1984 Wm. C. Brown Publishers, Dubuque, Iowa. All Rights Reserved. Reprinted by permission.

viscus, pl. viscera: L. *viscera,* internal organs
autonomic: Gk. *auto,* self; *nomos,* law

Figure 16.1.   A comparison of a somatic motor reflex with an autonomic motor reflex. Although, for the sake of clarity, each is shown on different sides of the spinal cord, both visceral and somatic sensory neurons are found bilaterally in the dorsal roots of spinal nerves and somatic and autonomic motor fibers are found bilaterally in the ventral roots.

The first of these neurons has its cell body in the gray matter of the brain or spinal cord. The axon of this neuron does not directly innervate the effector organ but instead synapses with a second neuron within an *autonomic ganglion* (a ganglion is a collection of cell bodies outside the CNS). The first neuron is thus called a **preganglionic, or presynaptic, neuron.** The second neuron in this pathway, called a **postganglionic, or postsynaptic, neuron,** produces an axon that leaves the autonomic ganglion and synapses with the cells of an effector organ (fig. 16.1).

Preganglionic autonomic fibers originate in the midbrain and hindbrain and from the upper thoracic to the fourth sacral levels of the spinal cord. Autonomic ganglia are located in the head, neck, and abdomen; chains of autonomic ganglia also parallel the right and left sides of the spinal cord. The origin of the preganglionic fibers and the location of the autonomic ganglia help to differentiate the **sympathetic** and **parasympathetic** subdivisions of the autonomic system, which will be discussed in later sections of this chapter.

ganglion: Gk. *ganglion,* a swelling or knot

Visceral Effector Organs

Unlike skeletal muscles, which enter a state of flaccid paralysis when their motor nerves are severed, the involuntary effectors are somewhat independent of their innervation. Smooth muscles maintain a resting tone (tension) in the absence of nerve stimulation, for example. Damage to an autonomic nerve, in fact, makes its target muscle more sensitive than normal to stimulating agents. This phenomenon is called **denervation** *(de″ner-va′shun)* **hypersensitivity.**

In addition to their intrinsic ("built-in") muscle tone, many smooth muscles and the cardiac muscle take their autonomy a step further. These muscles can contract rhythmically, even in the absence of nerve stimulation, in response to electrical waves of depolarization initiated by the muscles themselves. Autonomic nerves also maintain a resting "tone," in the sense that they maintain a baseline firing rate that can be either increased or decreased. Changes in tonic neural activity produce changes in the intrinsic activity of the effector organ. A decrease in the excitatory input to the heart, for example, will slow its rate of beat.

Unlike somatic motor neurons—where release of the neurotransmitter (ACh) always stimulates the effector organ (skeletal muscles)—some autonomic nerves release transmitters that inhibit the activity of their effectors. An increase in the activity of such an inhibitory nerve to the heart, for example, will slow the heart rate.

Cardiac Muscle. Like skeletal muscle cells, cardiac muscle cells, or *myocardial cells,* are striated; they contain actin and myosin filaments arranged in the form of sarcomeres, and they contract by means of the sliding filament mechanism. The long, fibrous skeletal muscle cells, however, are structurally and functionally separated from each other, whereas the myocardial cells are short, branched, and interconnected. Adjacent myocardial cells are joined by electrical synapses, or **gap junctions.** Gap junctions have an affinity for stain that gives them the appearance of **intercalated** *(in-ter' kah-lat-ed)* **discs** in a light microscope.

Electrical impulses that originate at any point in the mass of myocardial cells, called the **myocardium** *(mi''o-kar' de-um),* can spread to all cells in the mass that are joined by gap junctions. Because all cells in a myocardium are electrically joined, the myocardium behaves as a single functional unit, or a *functional syncytium (sin-sish'e-um).* Unlike skeletal muscles, which can produce graded contractions with a strength that depends on the number of cells stimulated, the heart contracts with an **all-or-none** contraction.

Unlike skeletal muscles, which require external stimulation by somatic motor nerves before they can produce action potentials and contract, cardiac muscle is able to produce action potentials automatically. Cardiac action potentials normally originate in a specialized group of cells called the *pacemaker.* However, the rate of this spontaneous depolarization and, thus, the rate of the heartbeat is regulated by autonomic innervation.

Smooth Muscles. Smooth (visceral) muscles are arranged in circular layers around the walls of blood vessels, bronchioles (small air passages in the lungs), and in the sphincter muscles of the digestive tract. However, both circular and longitudinal smooth muscle layers are found in the tubular digestive tract, the ureters (which transport urine), and the ductus deferens (which transport sperm), and the uterine tubes (which transport ova). The alternate contraction of circular and longitudinal smooth muscle layers produces **peristaltic waves,** which propel the contents of these tubes in one direction.

Smooth muscle cells do not contain sarcomeres (which produce striations in skeletal and cardiac muscle). Smooth muscle cells do, however, contain a great amount of actin and some myosin, which produces a ratio of thin-to-thick filaments of about 16:1 (in striated muscles the ratio is 2:1). The myosin filaments in smooth muscle cells are quite long (fig. 16.2).

Figure 16.2. An electron micrograph of the thick and thin filaments of smooth muscle. A longitudinal section of complete long filament is shown between the arrows (32,000×).

(a)   (b)

The long length of myosin filaments and the fact that they are not organized into sarcomeres may be advantageous for the function of smooth muscles. Smooth muscles must be able to exert tension even when greatly stretched—in the urinary bladder, for example, the smooth muscle cells may be stretched up to two and a half times their resting length. Skeletal muscles, in contrast, lose their ability to contract when the sarcomeres are stretched to the point where actin and myosin no longer overlap.

As in striated muscles, the contraction of smooth muscles is triggered by a sharp rise in the Ca^{++} concentration within the cytoplasm of the muscle cells. Despite their structural differences, it is currently believed that both smooth muscles and striated muscles contract by means of a *sliding filament* mechanism. The details of this mechanism in smooth muscles, however, are not well understood. Some comparisons of skeletal muscle, cardiac muscle, and smooth muscle are given in table 16.2

Single-Unit and Multi-Unit Smooth Muscles. Smooth muscles are often grouped into two functional categories: **single-unit** and **multi-unit.** Single-unit smooth muscles have numerous gap junctions (electrical synapses) between adjacent cells that weld them together electrically; they thus behave as a single unit. Multi-unit smooth muscles have

Table 16.2 Some comparisons of skeletal, cardiac, and smooth muscles

Skeletal muscle	Cardiac muscle	Smooth muscle
Striated; actin and myosin arranged in sarcomeres	Striated; actin and myosin arranged in sarcomeres	Not striated; more actin than myosin; actin inserts into dense bodies and cell membrane
Well-developed sarcoplasmic reticulum and transverse tubules	Moderately developed sarcoplasmic reticulum and transverse tubules	Sarcoplasmic reticulum poorly developed; no transverse tubules
Contains troponin in the thin filaments	Contains troponin in the thin filaments	Contains a Ca^{++} binding protein; may be located in thick filaments
Ca^{++} released into cytoplasm from sarcoplasmic reticulum	Ca^{++} enters cytoplasm from sarcoplasmic reticulum and extracellular fluid	Ca^{++} enters cytoplasm from extracellular fluid, sarcoplasmic reticulum, and perhaps mitochondria
Cannot contract without nerve stimulation; denervation results in muscle atrophy	Can contract without nerve stimulation; action potentials originate in pacemaker cells of heart	Maintains tone in absence of nerve stimulation; visceral smooth muscle produces pacemaker potentials; denervation results in hypersensitivity to stimulation
Muscle fibers stimulated independently; no gap junctions	Gap junctions present as intercalated discs	Gap junctions present in most smooth muscles

Source: Fox, Stuart Ira, *Human Physiology.* © 1984 Wm. C. Brown Publishers, Dubuque, Iowa. All Rights Reserved. Reprinted by permission.

Table 16.3 Some comparisons between single-unit and multi-unit smooth muscles

	Single-unit muscle	Multi-unit muscle
Location	Gastrointestinal tract; uterus; ureter; small arteries (arterioles)	Arrector pili muscles of hair follicles; ciliary muscle (attached to lens); iris; vas deferens; large arteries
Origin of electrical activity	Spontaneous activity by pacemakers (myogenic)	Not spontaneously active; potentials are neurogenic
Type of potentials	Action potentials	Graded depolarizations
Response to stretch	Responds to stretch by contraction; not dependent on nerve stimulation	No inherent response to stretch
Presence of gap junctions	Numerous gap junctions join all cells together electrically	Few (if any) gap junctions
Type of contraction	Slow, sustained contractions	Slow, sustained contractions

Source: Fox, Stuart Ira, *Human Physiology.* © 1984 Wm. C. Brown Publishers, Dubuque, Iowa. All Rights Reserved. Reprinted by permission.

few, if any, gap junctions; the individual cells must thus be stimulated separately by impulses through nerve endings. This is similar to the numerous motor units required for the control of skeletal muscles.

Single-unit smooth muscles display *pacemaker* activity in which certain cells stimulate others in the mass. Single-unit smooth muscles also display intrinsic, or *myogenic (mi''o-jen'ik)* electrical activity and contraction in response to stretch. For example, the stretch induced by an increase in the luminal contents of a small artery or a section of the gastrointestinal tract can stimulate myogenic contraction. Such contraction does not require stimulation by autonomic nerves. Contraction of multi-unit smooth muscles, in contrast, requires nerve stimulation. Comparisons of single-unit and multi-unit smooth muscles are given in table 16.3.

Autonomic Innervation of Smooth Muscles.
There are significant differences between the neural control of skeletal muscles and that of smooth muscles. A skeletal muscle fiber has only one junction with a somatic nerve fiber, and the receptors for the neurotransmitter are localized at the neuromuscular junction in the membrane of the skeletal muscle fiber. In contrast, the entire surface of smooth muscle cells contains transmitter receptor proteins. Neurotransmitter molecules are released along a stretch of an autonomic nerve fiber that is located some distance from the smooth muscle cells. The regions of the autonomic fiber that release transmitters appear as bulges, or *varicosities,* and the neurotransmitters released from these varicosities stimulate a number of smooth muscle cells. Since there are a number of varicosities along a stretch of an autonomic nerve ending, they form synapses "in passing"—or *synapses en passant*—with the smooth muscle cells.

---

1. Describe how the regulation of the contraction of cardiac and smooth muscle cells differs from that of skeletal muscle cells and how these muscles are affected by the experimental removal of their innervation.
2. Define the terms preganglionic and postganglionic neurons in the autonomic system, and use a diagram to illustrate the difference in efferent outflow between somatic and autonomic nerves.
3. Compare the structure of single-unit and multiunit smooth muscles, and explain how these two types differ in the way they are regulated by autonomic nerves.
4. Explain how synapses between autonomic nerves and smooth muscles differ from the neuromuscular junctions of somatic motor nerves with skeletal muscle fibers.

---

myogenic: Gk. *mys,* muscle; *genesis,* origin

Figure 16.3. The sympathetic chain of paravertebral ganglia, showing its relationship to the vertebral column and the spinal cord.

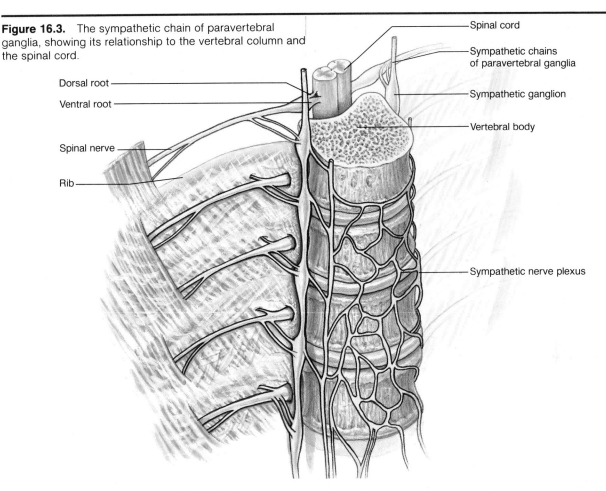

Dorsal root

Ventral root

Spinal nerve

Rib

Spinal cord

Sympathetic chains of paravertebral ganglia

Sympathetic ganglion

Vertebral body

Sympathetic nerve plexus

Structure of the Autonomic System

The sympathetic and parasympathetic divisions of the autonomic system share structural similarities: both consist of preganglionic neurons that originate in the CNS and postganglionic neurons that originate outside of the CNS in ganglia. The specific origin of the preganglionic fibers and the location of the ganglia, however, are different in the two subdivisions of the autonomic system.

Objective 4. Describe the origin of preganglionic sympathetic neurons and the location of sympathetic ganglia.

Objective 5. Explain the relationship between the sympathetic system and the adrenal medulla.

Objective 6. Describe the origin of the preganglionic parasympathetic neurons and the location of the parasympathetic ganglia.

Objective 7. Describe the distribution of the vagus nerve, and explain its significance within the parasympathetic division.

Sympathetic (Thoracolumbar) Division

The sympathetic system is also called the thoracolumbar division of the autonomic system because its preganglionic fibers exit the spinal cord from the first thoracic (T1) to the second lumbar (L2) levels. Most sympathetic nerve fibers, however, separate from the somatic motor fibers and synapse with postganglionic neurons within a double chain of sympathetic ganglia, or **paravertebral ganglia,** located on either side of the spinal cord (fig. 16.3).

Since the preganglionic sympathetic fibers are myelinated and thus appear white, they are called **white rami communicantes.** Some of these preganglionic sympathetic fibers synapse with postganglionic neurons located at their same level in the chain of sympathetic ganglia. Other preganglionic fibers travel up or down within the sympathetic chain before synapsing with postganglionic neurons. Since the fibers of the postganglionic sympathetic neurons are unmyelinated and thus appear gray, they form the **gray rami communicantes.** Postganglionic axons in the gray rami communicantes extend directly back to the ventral roots of the spinal nerves and travel distally within the spinal nerves to innervate their effector organs (fig. 16.4).

ramus: L. *ramus,* a branch

Figure 16.4. Sympathetic chain ganglia, the sympathetic chain, and rami communicantes of the sympathetic division of the ANS.

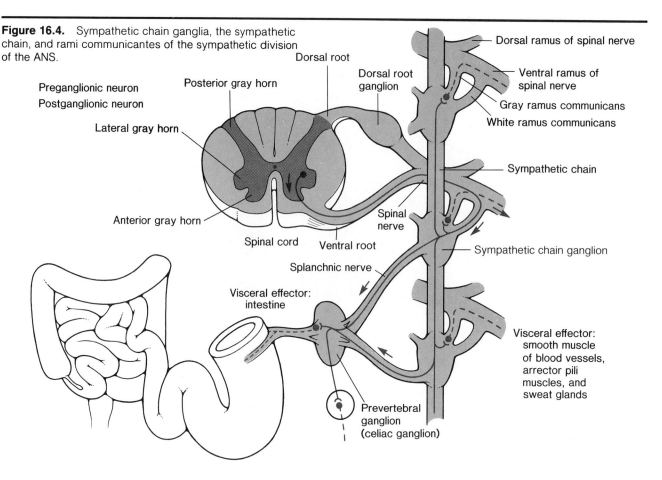

Within the chain of paravertebral ganglia, *divergence* is seen when preganglionic fibers branch to synapse with many postganglionic neurons located at different levels in the chain. *Convergence* is also seen when a given postganglionic neuron receives synaptic input from a large number of preganglionic fibers. The divergence of impulses from the spinal cord to the ganglia and the convergence of impulses within the ganglia usually results in the **mass activation** of almost all of the postganglionic fibers. This explains why the sympathetic system is usually activated as a unit and affects all of its effector organs at the same time.

Many preganglionic fibers that exit the spinal cord in the upper thoracic level travel into the neck, where they synapse in cervical sympathetic ganglia (fig. 16.5). Postganglionic fibers from here innervate the smooth muscles and glands of the head and neck.

Collateral Ganglia. Many preganglionic fibers that exit the spinal cord below the level of the diaphragm pass through the sympathetic chain of ganglia without synapsing. Beyond the sympathetic chain, these preganglionic fibers form **splanchnic** *(splank'nik)* **nerves** (fig. 16.4). Preganglionic fibers in the splanchnic nerves synapse in **collateral ganglia,** which are also called **prevertebral ganglia.** These include the **celiac** *(se'le-ak),* **superior mesenteric** *(mes''en-ter'ik),* and **inferior mesenteric** ganglia (figs. 16.6 and 16.7).

The *greater splanchnic nerve* arises from preganglionic sympathetic fibers T4–T9 and synapses in the celiac ganglion. These fibers contribute to the *celiac (solar) plexus.* Postganglionic fibers from the celiac ganglion innervate the stomach, spleen, liver, small intestine, and kidneys. The *lesser splanchnic nerve* terminates in the superior mesenteric ganglion. Postganglionic fibers from here innervate the small intestine and colon. The *lumbar splanchnic nerve* synapses in the inferior mesenteric ganglion, and the post ganglionic fibers innervate the distal colon and rectum, urinary bladder, and genital organs.

Adrenal Glands. The paired adrenal glands are located above each kidney. Each adrenal is composed of two parts: an outer **cortex** and an inner **medulla.** These two parts are really two functionally different glands with different embryonic origins, different hormones, and different regulatory mechanisms. The adrenal cortex secretes steroid hormones; the adrenal medulla secretes the hormone **epinephrine** (adrenalin) and, to a lesser degree, **norepinephrine,** when it is stimulated by the sympathetic system.

The adrenal medulla is a modified sympathetic ganglion whose cells are derived from postganglionic sympathetic neurons. The cells of the adrenal medulla are

splanchnic: Gk. *splanchno-,* relating to viscera
adrenal: L. *ad,* to; *renes,* kidney
cortex: L. *cortex,* bark
medulla: L. *medulla,* marrow

Figure 16.5. The cervical sympathetic ganglia.

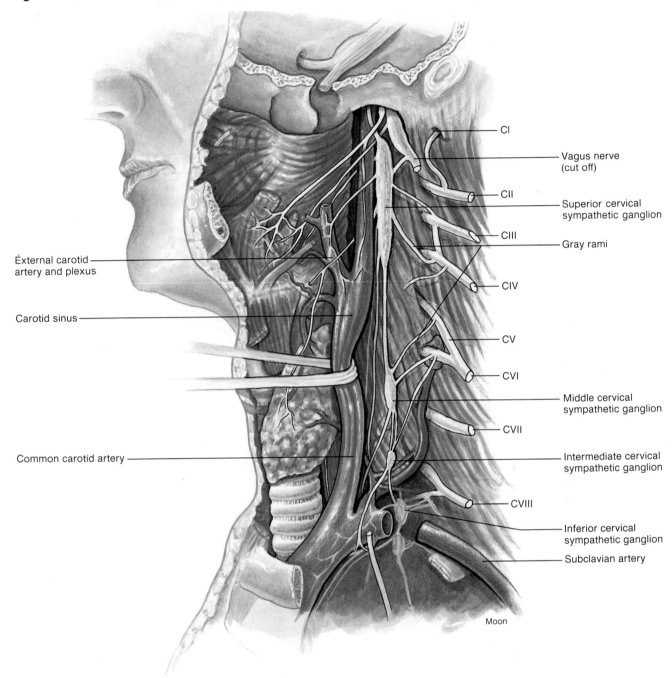

CI

Vagus nerve (cut off)

CII

Superior cervical sympathetic ganglion

CIII

Gray rami

External carotid artery and plexus

Carotid sinus

CIV

CV

CVI

Middle cervical sympathetic ganglion

CVII

Common carotid artery

Intermediate cervical sympathetic ganglion

CVIII

Inferior cervical sympathetic ganglion

Subclavian artery

Moon

innervated by preganglionic sympathetic fibers originating in the thoracic level of the spinal cord; they secrete epinephrine into the blood in response to this neural stimulation. The effects of epinephrine are complementary to those of the neurotransmitter norepinephrine, which is released from postganglionic sympathetic nerve endings. For this reason and because the adrenal medulla is stimulated as part of the mass activation of the sympathetic system, the two are often grouped together as the **sympathoadrenal system.**

Parasympathetic (Craniosacral) Division

The parasympathetic system is also known as the *craniosacral division* of the autonomic system. This is because its preganglionic fibers originate in the brain (specifically, the midbrain and the medulla oblongata of the brain stem) and in the second through fourth sacral levels of the spinal cord. These preganglionic parasympathetic fibers synapse in ganglia that are located next to—or actually within—the organs innervated. These parasympathetic ganglia, which are called **terminal ganglia,** supply the postganglionic fibers that synapse with the effector cells. Tables 16.4 and 16.5 show the comparative structures of the sympathetic and parasympathetic divisions. It should be

Figure 16.6. The collateral sympathetic ganglia: the celiac, superior, and inferior mesenteric ganglia.

Celiac ganglion

Superior mesenteric ganglion

First lumbar sympathetic ganglion

Renal plexus

Aortic plexus

Inferior mesenteric ganglion and plexus

Pelvic sympathetic chain

Moon

Table 16.4 The sympathetic (thoracolumbar) system

Parts of body innervated	Spinal origin of preganglionic fibers	Origin of postganglionic fibers
Eye	C–8 and T–1	Cervical ganglia
Head and neck	T–1 to T–4	Cervical ganglia
Heart and lungs	T–1 to T–5	Upper thoracic (paravertebral) ganglia
Upper extremity	T–2 to T–9	Lower cervical and upper thoracic (paravertebral) ganglia
Upper abdominal viscera	T–4 to T–9	Celiac and superior mesenteric (collateral) ganglia
Adrenal	T–10 and T–11	Adrenal medulla
Urinary and reproductive systems	T–12 to L–2	Celiac and inferior mesenteric (collateral) ganglia
Lower extremities	T–9 to L–2	Lumbar and upper sacral (paravertebral) ganglia

Source: Fox, Stuart Ira, *Human Physiology.* © 1984 Wm. C. Brown Publishers, Dubuque, Iowa. All Rights Reserved. Reprinted by permission.

Figure 16.7. The autonomic nervous system. The sympathetic division is shown in black, the parasympathetic in red. The solid lines indicate preganglionic fibers, and the dotted lines indicate postganglionic fibers.

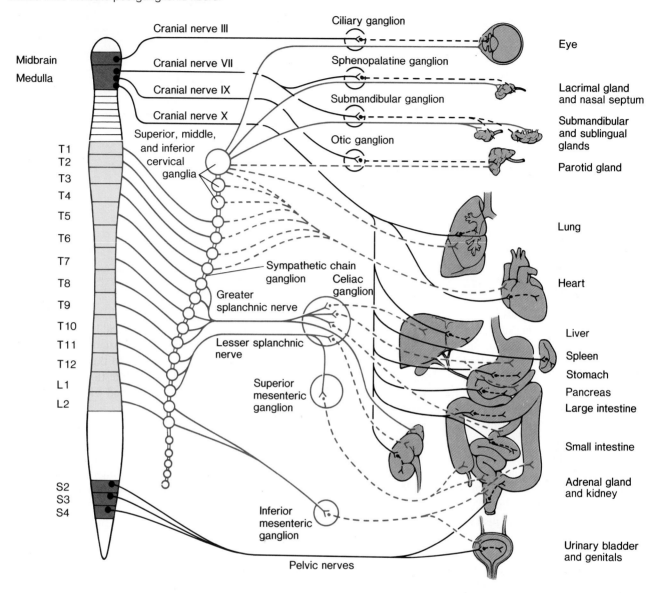

Table **16.5** The parasympathetic (craniosacral) system			
Effector organs	**Origin of preganglionic fibers**	**Nerve**	**Location of terminal ganglia**
Eye (ciliary and iris muscles)	Midbrain (cranial)	Oculomotor (third cranial) nerve	Ciliary ganglion
Lacrimal, mucus, and salivary glands in head	Medulla oblongata (cranial)	Facial (seventh cranial) nerve	Sphenopalatine and submandibular ganglia
Parotid (salivary) gland	Medulla oblongata (cranial)	Glossopharyngeal (ninth cranial) nerve	Otic ganglion
Heart, lungs, gastrointestinal tract, liver, pancreas	Medulla oblongata (cranial)	Vagus (tenth cranial) nerve	Terminal ganglia in or near organ
Lower half of large intestine, rectum, urinary bladder, and reproductive organs	S-2 to S-4 (sacral)	Through pelvic spinal nerves	Terminal ganglia near organs

Source: Fox, Stuart Ira, *Human Physiology*. © 1984 Wm. C. Brown Publishers, Dubuque, Iowa. All Rights Reserved. Reprinted by permission.

noted that, unlike sympathetic fibers, most parasympathetic fibers do not travel within spinal nerves. Cutaneous effectors (blood vessels, sweat glands, and arrector pili muscles) and blood vessels in skeletal muscles thus receive sympathetic but not parasympathetic innervation.

Four of the twelve pairs of cranial nerves contain preganglionic parasympathetic fibers. These are the oculomotor (III), facial (VII), glossopharyngeal (IX), and vagus (X) nerves. Parasympathetic fibers within the first three of these cranial nerves synapse in ganglia located in the head; fibers in the vagus nerve synapse in terminal ganglia located in many regions of the body.

The oculomotor nerve contains somatic motor and parasympathetic fibers that originate in the oculomotor nuclei of the midbrain. These parasympathetic fibers synapse in the **ciliary ganglion,** whose postganglionic fibers innervate the ciliary muscle and constrictor fibers in the iris of the eye. Preganglionic fibers that originate in the medulla oblongata travel in the facial nerve to the **pterygopalatine** *(ter''i-go-pal'ah-tēn)* **ganglion,** also called the sphenopalatine ganglion, which sends postganglionic fibers to the nasal mucosa, pharynx, palate, and lacrimal glands. Another group of fibers in the facial nerve terminate in the **submandibular ganglion,** which sends postganglionic fibers to the submandibular and sublingual salivary glands. Preganglionic fibers of the glossopharyngeal nerve synapse in the **otic** *(o'tik)* **ganglion,** which sends postganglionic fibers to innervate the parotid salivary gland.

Other nuclei in the medulla oblongata contribute preganglionic fibers to the very long vagus nerves, which provide the most extensive parasympathetic innervation in the body. As the paired vagus nerves pass through the thorax they contribute to the *pulmonary plexuses* within the mediastinum. Branches of the pulmonary plexuses accompany blood vessels and bronchi into the lungs. Below the pulmonary plexuses, the vagus nerves merge to form the *esophageal (ē-sof''ah-je'al) plexuses.*

At the lower end of the esophagus, vagal fibers collect to form an **anterior** and **posterior vagal** *(va'gal)* **trunk,** each composed of fibers from both vagus nerves. The vagal trunks enter the abdominal cavity through the esophageal opening in the diaphragm. Fibers from the vagal trunks innervate the stomach on the anterior and posterior sides. Branches of the vagus nerves within the abdominal cavity also contribute to the *celiac plexus* and *plexuses of the abdominal aorta.*

The preganglionic fibers in the vagus synapse with postganglionic neurons that are actually located *within* the innervated organs. These preganglionic fibers are thus quite long, and provide parasympathetic innervation to the heart, lungs, esophagus, stomach, pancreas, liver, small intestine, and upper half of the large intestine. Postganglionic parasympathetic fibers arise from terminal ganglia within these organs and synapse with effector cells.

vagus: L. *vagus,* wandering

Table 16.6 Some comparisons between the structure of the sympathetic and parasympathetic systems

	Sympathetic	Parasympathetic
Origin of preganglionic outflow	Thoracolumbar levels of spinal cord	Midbrain, hindbrain, and sacral levels of spinal cord
Location of ganglia	Chain of paravertebral ganglia and prevertebral (collateral) ganglia	Terminal ganglia in or near effector organs
Distribution of postganglionic fibers	Throughout the body	Mainly limited to the head and the viscera of the chest, abdomen, and pelvis
Divergence of impulses from pre- to postganglionic fibers	Great divergence (one preganglionic may activate twenty postganglionic fibers)	Little divergence (one preganglionic only activates a few postganglionic fibers)
Mass discharge of system as a whole	Yes	Not normally

Source: Fox, Stuart Ira, *Human Physiology.* © 1984 Wm. C. Brown Publishers, Dubuque, Iowa. All Rights Reserved. Reprinted by permission.

Preganglionic fibers from the sacral levels of the spinal cord provide parasympathetic innervation to the lower half of the large intestine, rectum, and to the urinary and reproductive systems. These fibers, like those of the vagus, synapse with terminal ganglia located within the effector organs.

Parasympathetic nerves to the visceral organs thus consist of preganglionic fibers, whereas sympathetic nerves to these organs contain postganglionic fibers. A composite view of the sympathetic and parasympathetic systems is provided in figure 16.7, and these comparisons are summarized in table 16.6.

1. Compare the origin of preganglionic sympathetic and parasympathetic fibers and the location of sympathetic and parasympathetic ganglia.
2. Using a simple line drawing, illustrate the sympathetic pathway from the spinal cord to the heart. Label the preganglionic neuron, postganglionic neuron, and the ganglion.
3. Use a simple line diagram to show the parasympathetic innervation of the heart. Label the preganglionic and postganglionic neurons, the name of the nerve involved, and the terminal ganglion.
4. Describe the distribution of the vagus nerve and its physiological significance.
5. Define the terms white ramus communicantes and gray ramus communicantes, and explain why blood vessels in the skin and skeletal muscles receive sympathetic but not parasympathetic innervation.
6. Describe the structure of the adrenal gland, and explain its relationship to the sympathetic nervous system.

Table 16.7 Effects of autonomic nerve stimulation on various visceral effector organs

Effector organ	Sympathetic effect	Parasympathetic effect
Eye		
Iris (radial muscle)	Dilates pupil	—
Iris (sphincter muscle)	—	Constricts pupil
Ciliary muscle	Relaxes (for far vision)	Contracts (for near vision)
Glands		
Lacrimal (tear)	—	Stimulates secretion
Sweat	Stimulates secretion	—
Salivary	Decreases secretion; saliva becomes thick	Increases secretion; saliva becomes thin
Stomach	—	Stimulates secretion
Intestine	—	Stimulates secretion
Adrenal medulla	Stimulates secretion of hormones	—
Heart		
Rate	Increases	Decreases
Conduction	Increases rate	Decreases rate
Strength	Increases	—
Blood vessels	Mostly constricts; affects all organs	Dilates in a few organs (e.g., penis)
Lungs		
Bronchioles (tubes)	Dilates	Constricts
Mucous glands	Inhibits secretion	Stimulates secretion
Gastrointestinal tract		
Motility	Inhibits movement	Stimulates movement
Sphincters	Stimulates closing	Inhibits closing
Liver	Stimulates hydrolysis of glycogen	—
Adipose (fat) cells	Stimulates hydrolysis of fat	—
Pancreas	Inhibits exocrine secretions	Stimulates exocrine secretions
Spleen	Stimulates contraction	—
Urinary Bladder	Helps set muscle tone	Stimulates contraction
Arrector muscles	Stimulates erection of hair and "goosebumps"	—
Uterus	If pregnant: contraction pili if not pregnant: relaxation	—
Penis	Erection; ejaculation	Erection (due to vasodilation)

Source: Fox, Stuart Ira, *Human Physiology.* © 1984 Wm. C. Brown Publishers, Dubuque, Iowa. All Rights Reserved. Reprinted by permission.

Functions of the Autonomic Nervous System

The actions of the sympathetic and parasympathetic divisions of the autonomic nervous system, together with the actions of hormones, help to maintain a state of dynamic constancy in the internal environment. The sympathetic division activates the body to "fight or flight" through adrenergic effects; the parasympathetic division often has antagonistic actions through cholinergic effects. Homeostasis thus depends, in large part, on the complementary and often antagonistic effects of sympathetic and parasympathetic innervation.

Objective 8. List the neurotransmitters of the preganglionic and postganglionic neurons of the sympathetic and parasympathetic systems.

Objective 9. Describe the actions of the sympathoadrenal system, and explain the significance of alpha-adrenergic, beta-adrenergic, and cholinergic stimulation.

Objective 10. Describe the effects of acetylcholine released by postganglionic parasympathetic nerve fibers.

Objective 11. Explain the antagonistic, complementary, and cooperative effects of sympathetic and parasympathetic innervation.

The sympathetic and parasympathetic divisions of the autonomic system (fig. 16.7) affect the visceral organs in different ways. Mass activation of the *sympathetic system* prepares the body for intense physical activity in emergencies; the heart rate increases, blood glucose rises, and blood is diverted to the skeletal muscles (away from the visceral organs and skin). These and other effects are listed in table 16.7. The "theme" of the sympathetic system has been aptly summarized in a phrase: **fight or flight.**

The effects of *parasympathetic nerve* stimulation are in many ways opposite to the effects of sympathetic stimulation. The parasympathetic system, however, is not normally activated as a whole. Stimulation of separate parasympathetic nerves can result in slowing of the heart, dilation of visceral blood vessels, and an increased activity of the digestive tract (table 16.7). The different responses of visceral organs to sympathetic and parasympathetic nerve activity is due to the fact that the postganglionic fibers of these two divisions release different neurotransmitters.

Neurotransmitters of the Autonomic System

Acetylcholine *(as″ĕ-til-kō'lēn) (ACh)* is the neurotransmitter of all preganglionic fibers (both sympathetic and parasympathetic). Acetylcholine is also the transmitter released by all parasympathetic postganglionic fibers at their synapses with effector cells (fig. 16.8). Transmission at the autonomic ganglia and at synapses of postganglionic nerve fibers is thus said to be **cholinergic** *(ko″lin-er'jik).*

The neurotransmitter released by most postganglionic sympathetic nerve fibers is **norepinephrine** *(nor-adrenalin).* Transmission at these synapses is thus said to be **adrenergic** *(ad″ren-er'jik).* There are a few exceptions to this rule: some sympathetic fibers that innervate blood vessels in skeletal muscles, as well as sympathetic fibers to sweat glands, release ACh (are cholinergic).

cholinergic: Gk. chole, bile; ergon, work

Figure 16.8.   Neurotransmitters of the autonomic motor system. ACh = acetylcholine; NE = norepinephrine; E = epinephrine. Those nerves that release ACh are called cholinergic; those nerves that release NE are called adrenergic. The adrenal medulla secretes both epinephrine (85%) and norepinephrine (15%) as hormones into the blood.

Figure 16.9.   The structure of the catecholamines norepinephrine and epinephrine.

In view of the fact that the cells of the adrenal medulla are derived from postganglionic sympathetic neurons, it is not surprising that the hormones they secrete (normally about 85% epinephrine and 15% norepinephrine) are similar to the transmitter of postganglionic sympathetic neurons. Epinephrine differs from norepinephrine only in the presence of an additional methyl (CH₃) group, as shown in figure 16.9. Epinephrine, norepinephrine, and dopamine (a transmitter within the CNS) are collectively termed **catecholamines** *(kat''ě-kol-am'in)*. The catecholamines, together with serotonin (another transmitter within the CNS), are often referred to as *monoamines,* or *biogenic amines.*

Responses to Adrenergic Stimulation

Adrenergic stimulation—by epinephrine in the blood and by norepinephrine released from sympathetic nerve endings—has both excitatory and inhibitory effects. The heart, dilatory muscles of the iris, and the smooth muscles of many blood vessels are stimulated to contract. The smooth muscles of the bronchioles and of some blood vessels, however, are inhibited from contracting; adrenergic chemicals, therefore, cause these structures to dilate.

Since excitatory and inhibitory effects can be produced in different tissues by the same chemical, the responses clearly depend on the biochemistry of the tissue cells rather than on the intrinsic properties of the chemical. Included in the biochemical differences among the

Table 16.8 Effects of epinephrine and norepinephrine on adrenergic receptors in various organ systems

Organ (or organ system)	Receptor	Response	Relative order of catecholamine potency	
			NE	E
Heart	Beta	Increased heart rate; increased contractility; increased conduction velocity	3	2
Blood vessels	Alpha	Constriction	2	1
	Beta†	Dilatation	3	2
Bronchi	Beta	Relaxation	3	2
Intestine				
Muscle	Alpha, beta	Inhibition	2	1
Sphincters	Alpha	Contraction	2	1
Bladder				
Detrusor muscle	Beta	Relaxation	3	2
Trigone muscle of sphincter	Alpha	Contraction	2	1
Uterus	Alpha, beta	Variable§	2	1
Eye				
Radial muscle	Alpha	Contraction	2	1
Ciliary muscle	Beta	Relaxation	3	2

*NE = norepinephrine; E = epinephrine.
†Only demonstrated in skeletal muscle and visceral muscle vessels.
‡Probably not active in intestine.
§Response is dependent upon stage of menstrual cycle and pregnancy.
Source: Oliver Carrier, Jr., *Pharmacology of the Peripheral Nervous System* (Chicago: Year Book Medical Publisher, Inc. 1972).

target tissues for catecholamines are differences in the *membrane receptor proteins* for these chemical agents. There are two major classes of these receptor proteins, designated **alpha** (α) and **beta** (β) **receptors.**

Alpha-adrenergic effects are usually excitatory and beta-adrenergic effects are usually inhibitory in the effector organs. There are exceptions to this rule, however. As shown in table 16.8, stimulation of both alpha and beta receptors in the intestine inhibits intestinal activity, and adrenergic stimulation of the heart is mediated by beta receptors.

Responses to Cholinergic Stimulation

Somatic motor neurons, all preganglionic autonomic neurons, and all postganglionic parasympathetic neurons are cholinergic—they use acetylcholine as a neurotransmitter. The cholinergic effects of somatic motor neurons and preganglionic autonomic neurons are always excitatory. The cholinergic effects of postganglionic parasympathetic fibers are usually excitatory, with some notable exceptions; the parasympathetic fibers innervating the heart, for example, cause slowing of the heart rate.

The drug *muscarine (mus'kar-in),* derived from some poisonous mushrooms, blocks the cholinergic effects of parasympathetic nerves in the heart, smooth muscles, and glands. This drug, however, does not block the effects of somatic motor nerves or the cholinergic transmission in autonomic ganglia. The acetylcholine receptors in visceral organs are, therefore, said to promote the *muscarinic effects* of ACh. These effects are identical to those produced by parasympathetic nerve stimulation and are believed to be mediated by different receptors than those that mediate other cholinergic effects.

The muscarinic effects of ACh are specifically inhibited by the drug **atropine,** derived from the deadly nightshade plant *(Atropa belladonna).* Indeed, extracts of this plant were used by women during the middle ages to dilate their pupils (atropine inhibits parasympathetic stimulation of the iris). This was done to enhance their beauty (belladonna—beautiful woman). Atropine is used clinically today to dilate pupils during eye examinations, to dry mucous membranes of the respiratory tract prior to general anesthesia, and to inhibit spasmodic contractions of the lower digestive tract.

Organs with Dual Innervation

Many organs receive a dual innervation—they are innervated by both sympathetic and parasympathetic fibers. When this occurs, the effects of these two divisions may be antagonistic, complementary, or cooperative.

Antagonistic Effects. The effects of sympathetic and parasympathetic innervation of the pacemaker region of the heart is the best example of the antagonism of these two systems. In this case sympathetic and parasympathetic fibers innervate the same cells. Adrenergic stimulation from sympathetic fibers increases the heart rate, and cholinergic stimulation from parasympathetic fibers inhibits the pacemaker cells and, thus, decreases the heart rate. Antagonism is also seen in the digestive tract, where sympathetic nerves inhibit and parasympathetic nerves stimulate intestinal movements and secretions.

Figure 16.10. Reciprocal innervation of the iris muscles by the sympathetic and parasympathetic systems. Stimulation of sympathetic nerves produces contraction of the dilator (radial) muscles, which enlarges the pupil. Stimulation of the parasympathetic nerves produces contraction of the constrictor (circular) muscle layer, which makes the pupil smaller.

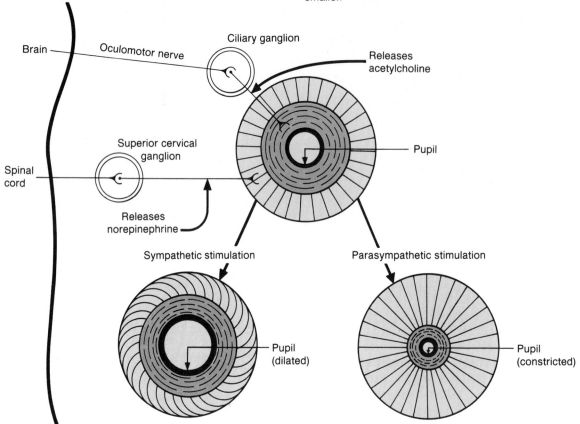

The effects of sympathetic and parasympathetic stimulation on the diameter of the pupil of the eye is analogous to the reciprocal innervation of flexor and extensor skeletal muscles by somatic motor neurons. This is because the iris contains antagonistic muscle layers. Contraction of the radial muscles, which is stimulated by sympathetic nerves, causes dilation; contraction of the circular muscles, which are innervated by parasympathetic nerve endings, causes constriction of the pupils (fig. 16.10).

Complementary Effects. The effects of sympathetic and parasympathetic stimulation on salivary gland secretion are complementary. The secretion of watery saliva is stimulated through parasympathetic nerves, which also stimulate the secretion of other exocrine glands in the digestive tract. Impulses through sympathetic nerves stimulate the constriction of blood vessels throughout the digestive tract. The resultant decrease in blood flow to the salivary glands causes the production of a thicker, more viscous saliva.

Cooperative Effects. The effects of sympathetic and parasympathetic stimulation on the urinary and reproductive systems are cooperative. Erection of the penis, for example, is due to vasodilation resulting from parasympathetic nerve stimulation; ejaculation is due to stimulation through sympathetic nerves. Although the contraction of the urinary bladder is myogenic (independent of nerve stimulation), it is promoted in part by the action of parasympathetic nerves. This *micturition (mik″tu-rish′un),* or urination, urge and reflex is also enhanced by sympathetic nerve activity, which increases the tone of the bladder muscles. Emotional states that are accompanied by high sympathetic nerve activity may thus result in reflex urination at bladder volumes that are normally too low to trigger this reflex.

Organs without Dual Innervation

Although most organs are innervated by both sympathetic and parasympathetic nerves, some—including the adrenal medulla, arrector pili muscles, sweat glands, and most blood vessels—receive only sympathetic innervation. In these cases regulation is achieved by increases or decreases in the "tone" (firing rate) of the sympathetic fibers. Constriction of blood vessels, for example, is produced by increased sympathetic activity, which stimulates alpha-adrenergic receptors, and vasodilation results from decreased sympathetic nerve stimulation.

micturition: L. *micturire,* to urinate

The sympathoadrenal system is required for *nonshivering thermogenesis:* animals deprived of their sympathetic system and adrenals cannot tolerate cold stress. The sympathetic system itself is required for proper thermoregulatory responses to heat. In a hot room, for example, decreased sympathetic stimulation produces dilation of the blood vessels in the surface of the skin, which increases cutaneous blood flow and provides better heat radiation. During exercise, in contrast, there is increased sympathetic activity, which causes constriction of the blood vessels in the skin of the limbs and stimulation of sweat glands in the trunk.

The eccrine sweat glands in the trunk secrete a watery fluid in response to sympathetic stimulation. Evaporation of this dilute sweat helps to cool the body. The eccrine sweat glands also secrete a chemical called **bradykinin** *(brad''e-ki'nin)* in response to sympathetic stimulation. Bradykinin stimulates dilation of the surface blood vessels near the sweat glands, helping to radiate heat. At the conclusion of exercise, sympathetic stimulation is reduced and blood flow to the surface of the limbs is increased, which aids in the elimination of metabolic heat. Notice that all of these thermoregulatory responses are achieved without the direct involvement of the parasympathetic system.

---

1. *Define the meaning of the terms* adrenergic *and* cholinergic, *and use these terms to describe the neurotransmitters of different autonomic nerve fibers.*
2. *List the effects of sympathoadrenal stimulation on different effector organs, and indicate which effects are due to alpha or beta receptor stimulation.*
3. *Describe the effects of the drug* atropine, *and explain these effects in terms of the actions of the parasympathetic system.*
4. *Explain how the sympathetic and parasympathetic systems can have antagonistic, cooperative, and complementary effects, and give examples.*

Control of the Autonomic System by Higher Brain Centers

Visceral functions are regulated, to a large degree, by autonomic reflexes. In most autonomic reflexes, sensory input is directed to brain centers, which in turn regulate the activity of descending pathways to preganglionic autonomic neurons. The neural centers that directly control the activity of autonomic nerves are influenced by higher brain areas as well as by sensory input.

Objective 12. Describe the area of the brain that most directly controls the activity of autonomic nerves, and describe the higher brain areas that influence autonomic activity.

Table 16.9 Some reflexes stimulated by sensory input from afferent fibers in the vagus that are transmitted to centers in the medulla oblongata

Organs	Type of receptors	Reflex effects
Lungs	Stretch receptors	Inhibits further inhalation; stimulates an increase in cardiac rate and vasodilation
	Type J receptors	Stimulated by pulmonary congestion—produces feelings of breathlessness, and causes a reflex fall in cardiac rate and blood pressure
Aorta	Chemoreceptors	Stimulated by rise in CO_2 and fall in O_2—produces increased rate of breathing, fall in heart rate, and vasoconstriction
	Baroreceptors	Stimulated by increased blood pressure—produces a reflex decrease in heart rate
Heart	Atrial stretch receptors	Inhibits antidiuretic hormone secretion, thus increasing the volume of urine excreted
	Stretch receptors in ventricles	Produces a reflex decrease in heart rate and vasodilation
Gastrointestinal tract	Stretch receptors	Feelings of satiety, discomfort, and pain

Source: Fox, Stuart Ira, *Human Physiology.* © 1984 Wm. C. Brown Publishers, Dubuque, Iowa. All Rights Reserved. Reprinted by permission.

Objective 13. Explain, in terms of the structures involved, how the activity of the autonomic system and the activity of the endocrine system can be coordinated.

Objective 14. Explain, in terms of the structures involved, how autonomic functions can be affected by emotions.

The **medulla oblongata** of the brain stem is the area that most directly controls the activity of the autonomic system. Almost all autonomic responses can be elicited by experimental stimulation of the medulla, which contains centers for the control of the cardiovascular, pulmonary, urinary, reproductive, and digestive systems. Much of the sensory input to these centers travels in the afferent fibers of the vagus nerve, which is a mixed nerve containing both sensory and motor fibers. These reflexes are listed in table 16.9.

The Hypothalamus

The hypothalamus (fig. 16.11) is an extremely important brain region located just above (superior to) the pituitary gland, which in turn is located above the posterior portion of the roof of the mouth. By means of efferent fibers to the brain stem and posterior pituitary and by means of hormones that regulate the anterior pituitary, the hypothalamus serves to orchestrate somatic, autonomic, and endocrine responses during various behavioral states.

Experimental stimulation of different areas of the hypothalamus can evoke the autonomic responses characteristic of aggression, sexual behavior, eating, or satiety.

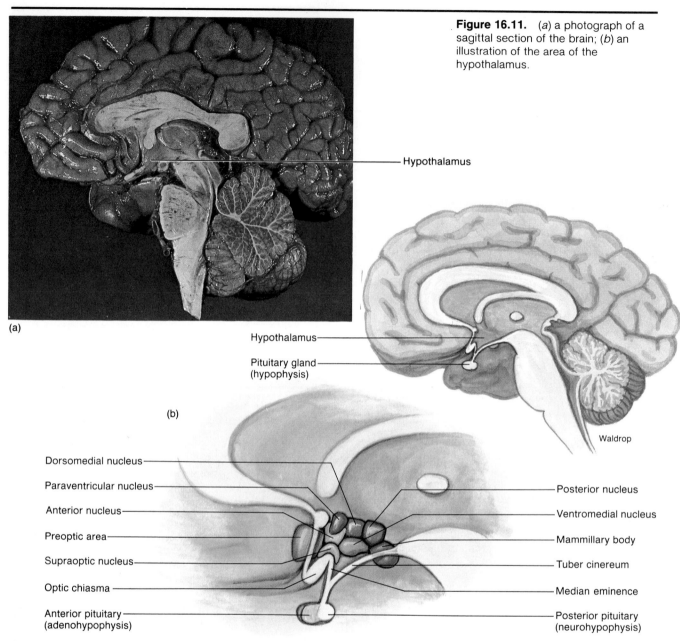

Figure 16.11. (a) a photograph of a sagittal section of the brain; (b) an illustration of the area of the hypothalamus.

Hypothalamus

(a)

Hypothalamus

Pituitary gland (hypophysis)

Waldrop

(b)

Dorsomedial nucleus

Paraventricular nucleus

Anterior nucleus

Preoptic area

Supraoptic nucleus

Optic chiasma

Anterior pituitary (adenohypophysis)

Posterior nucleus

Ventromedial nucleus

Mammillary body

Tuber cinereum

Median eminence

Posterior pituitary (neurohypophysis)

Chronic stimulation of the lateral hypothalamus, for example, can make an animal eat and become obese, whereas stimulation of the medial hypothalamus inhibits eating. Other areas contain osmoreceptors that stimulate thirst and the secretion of antidiuretic hormone (ADH) from the posterior pituitary.

The hypothalamus is also where the body's "thermostat" is located. Experimental cooling of the preoptic-anterior hypothalamus causes shivering (a somatic response) and nonshivering thermogenesis (sympathetic responses). Experimental heating of this hypothalamic area results in hyperventilation (stimulated by somatic motor nerves), vasodilation, salivation, and sweat gland secretion (stimulated by autonomic nerves).

The coordination of sympathetic and parasympathetic reflexes by the medulla oblongata is thus integrated with the control of somatic and endocrine responses by the hypothalamus. The activities of the hypothalamus are in turn influenced by higher brain centers.

Limbic System, Cerebellum, and Cerebrum

The limbic system is a group of fiber tracts and nuclei that form a ring (a limbus) around the brain stem. It includes the cingulate gyrus of the cerebral cortex, the hypothalamus, the fornix (a fiber tract), the hippocampus, and the amygdaloid nucleus (fig. 16.12). These structures, which were derived early in the course of vertebrate evolution, were once called the *rhinencephalon (ri''nen-sef'ah-lon),* or "smell brain," because of their importance in the central processing of olfactory information.

limbic: L. *limbus*, edge or border

Figure 16.12. The limbic system and the pathways that interconnect the structures of the limbic system (note: the left temporal lobe of the cerebral cortex has been removed).

Corpus callosum
Cingulate gyrus
Septal area
Preoptic area
Olfactory bulb
Olfactory tract
Hypothalamus
Amygdaloid nucleus
Cortex of right hemisphere

Fornix
Thalamic nucleus
Mammillothalamic tract
Mammillary body
Hippocampus
Cortex of left hemisphere

Waldrop

In higher vertebrates, these structures are now recognized as centers involved in basic emotional drives—such as anger, fear, sex, and hunger—and in short-term memory. Complex circuits between the hypothalamus and other parts of the limbic system (illustrated in figure 16.12) contribute visceral responses to emotions.

The autonomic correlates of motion sickness—nausea, sweating, and cardiovascular changes—are abolished by cutting the efferent tracts of the cerebellum. This demonstrates that impulses from the cerebellum to the medulla influence activity of the autonomic system. Experimental and clinical observations have also demonstrated that the frontal and temporal lobes of the cerebral cortex influence lower brain areas as part of their involvement in emotion and personality.

One of the most dramatic examples of the role of higher brain areas in personality and emotion is the famous crowbar accident, which occurred in 1848. A twenty-five-year-old railroad foreman, Phineas P. Gage, was tamping gunpowder into a hole in a rock when it exploded. The rod—three feet, seven inches long and one and one-fourth inches thick—was driven through his left eye, passed through his brain, and emerged in the back of his skull.

After a few minutes of convulsions, Gage got up, rode a horse three-quarters of a mile into town, and walked up a long flight of stairs to see a doctor. He recovered well, with no noticeable sensory or motor deficits. His associates, however, noted striking personality changes. Before the accident Gage was a responsible, capable, and financially prudent man. Afterwards, he appeared to lose social inhibitions, engaged in gross profanity (which he did not do before the accident), and seemed to be tossed about by chance whims. He was eventually fired from his job, and his previous friends remarked that he was "no longer Gage."

1. Describe the role of the medulla oblongata in the regulation of the autonomic system.
2. Describe the role of the hypothalamus in the regulation of the autonomic and endocrine systems.
3. Explain, in terms of the neural pathways involved, how emotional states can be associated with particular autonomic responses (such as blushing).

Clinical Considerations

Autonomic Dysreflexia

Autonomic dysreflexia, a serious condition producing rapid elevations in blood pressure that can lead to stroke (cerebrovascular accident), occurs in 85% of people with quadriplegia and others with spinal cord lesions above the sixth thoracic level. Lesions to the spinal cord first produce the symptoms of spinal shock, characterized by the loss of both skeletal muscle and autonomic reflexes. After a period of time both types of reflexes return in an exaggerated state; the skeletal muscles may become spastic, due to absence of higher inhibitory influences, and the visceral organs experience denervation hypersensitivity. Patients in this state have difficulty emptying their urinary bladders and must often be catheterized.

Noxious stimuli, such as overdistension of the urinary bladder, can result in reflex activation of the sympathetic nerves below the spinal cord lesion. This produces goose bumps, cold skin, and vasoconstriction in the regions served by the spinal cord below the level of the lesion. The rise in blood pressure resulting from this vasoconstriction activates pressure receptors that transmit impulses along sensory nerve fibers to the medulla. In response to this sensory input, the medulla directs a reflex slowing of the heart and vasodilation. Since descending impulses are blocked by the spinal lesion, however, the skin is warm and moist (due to vasodilation and sweat gland secretion) above the lesion, but cold below the level of spinal cord damage.

Pharmacology of the Autonomic System

Epinephrine, norepinephrine, acetylcholine, and some chemicals that are not normally found in the body can either enhance or block the physiological effects of the autonomic system. Drugs that enhance a particular effect are called **agonists,** and those that block an effect are called **antagonists.** Drugs that affect the functions of the autonomic system can be further divided into those that enhance or block adrenergic or cholinergic effects.

Adrenergic Drugs. Epinephrine, norepinephrine, and artificially produced chemicals that are similar to these endogenous catecholamines promote sympathetic nerve effects by stimulating adrenergic receptors. The study of these drugs, as well as of others that block adrenergic effects, have led pharmacologists to the discovery that there are two main classes of adrenergic receptors, designated *alpha* and *beta*. It was further learned that there are two subclasses of alpha receptors (alpha$_1$ and alpha$_2$) and two subclasses of beta receptors (beta$_1$ and beta$_2$). The location and effects produced by stimulation of these receptor subtypes are summarized in table 16.10. The use of specific drugs allows clinicians to more or less selectively produce effects such as slowing of the heart rate or bronchodilation.

Cholinergic Drugs. Acetylcholine is used as a neurotransmitter by somatic motor neurons, preganglionic autonomic fibers, all postganglionic parasympathetic nerve

Table 16.10 Adrenergic receptor subtypes: examples of their location and action

Adrenergic receptor	Tissue location	Effect
Alpha$_1$	Vascular smooth muscle	Vasoconstriction
	Uterus	Contraction
Alpha$_2$	Platelets	Aggregation
Beta$_1$	Heart	Faster and stronger beat
	Adipose tissue	Lipolysis
Beta$_2$	Vascular smooth muscle	Vasodilation
	Bronchiolar smooth muscle	Bronchodilation

fibers, and some sympathetic nerve fibers. Drugs that are similar in structure and action to acetylcholine (methacholine and bethanecholine) thus can promote the effects of these nerves. Drugs that inhibit the action of acetylcholinesterase (an enzyme that degrades ACh)—such as physostigmine and neostigmine—enhance the action of ACh. The effects of these drugs on the neuromuscular junction are used in the treatment of myasthenia gravis and other problems associated with impaired activation of the skeletal muscles. Some of these drugs are also used as *parasympathomimetics (par''ah-sim''pah-tho-mi-met'ik)*—drugs that duplicate the action of parasympathetic nerves—in the treatment of glaucoma. Atropine and related drugs block the action of ACh released by postganglionic parasympathetic neurons and thus serve to block the actions of the parasympathetic system.

Chapter Summary

I. The Autonomic Nervous System and its Visceral Effectors
   The involuntary effectors, including smooth muscle, cardiac muscle, and glands, are regulated by the autonomic nervous system. The nature of this regulation is in many ways different from the neural regulation of voluntary striated muscles.
   A. Preganglionic autonomic neurons originate from the brain or spinal cord; postganglionic neurons originate from ganglia located outside the CNS.
   B. Smooth muscle, cardiac muscle, and glands receive autonomic innervation.
      1. The involuntary effectors are somewhat independent of their innervation and become hypersensitive when their innervation is removed.
      2. Myocardial cells are interconnected by electrical synapses, or gap junctions, to form a functional syncytium with independent pacemaker activity.
      3. Single-unit smooth muscles have gap junctions and pacemaker

activity; multi-unit smooth muscles lack gap junctions and pacemaker activity.

II. Structure of the Autonomic System
   The autonomic system consists of preganglionic neurons that originate in the CNS and postganglionic neurons that originate in ganglia. The specific origin of the preganglionic fibers and the location of the ganglia, however, are different in the sympathetic and parasympathetic divisions of the autonomic system.
   A. Preganglionic neurons of the sympathetic division originate in the spinal cord between the thoracic and lumbar levels.
      1. Many of these fibers synapse with postganglionic neurons whose cell bodies are located in a double chain of sympathetic (paravertebral) ganglia outside the spinal cord.
      2. Some preganglionic fibers synapse in collateral ganglia; these are the celiac, superior mesenteric, and the inferior mesenteric ganglia.
      3. Some preganglionic fibers innervate the adrenal medulla,

which secretes epinephrine (and some norepinephrine) into the blood in response to this stimulation.
   B. Preganglionic parasympathetic fibers originate in the brain and in the sacral levels of the spinal cord.
      1. Preganglionic parasympathetic fibers contribute to cranial nerves III, VII, IX, and X.
      2. Preganglionic fibers of the vagus (X) nerve are very long and synapse in terminal ganglia located next to or within the innervated organ; short postganglionic fibers then innervate the effector cells.
      3. The vagus provides sympathetic innervation to the heart, lungs, esophagus, stomach, liver, small intestine, and upper half of the large intestine.
      4. Parasympathetic outflow from the sacral levels of the spinal cord innervates terminal ganglia in the lower half of the large intestine, the rectum, and the urinary and reproductive systems.

III. Functions of the Autonomic Nervous System
The sympathetic and parasympathetic divisions of the autonomic system may act in a complementary or antagonistic way to help maintain homeostasis. The sympathetic division activates the body to "fight or flight" through adrenergic effects; the parasympathetic division often has antagonistic actions through cholinergic effects.
  A. All preganglionic autonomic nerve fibers are cholinergic (use ACh as a neurotransmitter).
    1. All postganglionic parasympathetic fibers are cholinergic.
    2. Most postganglionic sympathetic fibers are adrenergic (use norepinephrine as a neurotransmitter).
    3. Sympathetic fibers that innervate sweat glands and those that innervate blood vessels in skeletal muscles are cholinergic.
  B. Adrenergic effects include stimulation of the heart, vasoconstriction in the viscera and skin, bronchodilation, and glycogenolysis in the liver.

  1. There are two main groups of adrenergic receptor proteins: alpha and beta receptors.
  2. Different organs have only alpha or beta receptors, and some organs (such as the heart) have both receptors.
  C. Cholinergic effects of parasympathetic nerves are promoted by the drug muscarine and inhibited by atropine.
  D. In organs with dual innervation, the actions of the sympathetic and parasympathetic divisions can be antagonistic, complementary, or cooperative.
    1. The effects are antagonistic in the heart and pupils.
    2. The actions are complementary in the regulation of salivary gland secretion and are cooperative in the regulation of the reproductive and urinary systems.
  E. In organs without dual innervation (such as most blood vessels), regulation is achieved by variations in sympathetic nerve activity.
IV. Control of the Autonomic System by Higher Brain Centers
Visceral sensory input to the brain may

result in the activity of the descending pathways to the preganglionic autonomic neurons. The centers in the brain that control autonomic activity are influenced by higher brain areas as well as by sensory input.
  A. The medulla oblongata of the brain stem is the area that most directly controls the activity of the autonomic system.
    1. The medulla oblongata is in turn influenced by sensory input and by input from the hypothalamus.
    2. The hypothalamus orchestrates somatic, autonomic, and endocrine responses during various behavioral states.
  B. The activity of the hypothalamus is influenced by input from the limbic system, cerebellum, and cerebrum; these interconnections provide an autonomic component to changes in body position, emotion, and various expressions of personality.

Review Activities

Objective Questions

1. Which of the following statements about the superior mesenteric ganglion is true?
   (a) It is a parasympathetic ganglion.
   (b) It is a paravertebral sympathetic ganglion.
   (c) It is located in the head.
   (d) It contains postganglionic sympathetic neurons.
2. The pterygopalatine, ciliary, submandibular, and otic ganglia are
   (a) collateral sympathetic ganglia
   (b) cervical sympathetic ganglia
   (c) parasympathetic ganglia that receive fibers from the vagus
   (d) parasympathetic ganglia that receive fibers from the third, seventh, and ninth cranial nerves
3. Parasympathetic ganglia are located
   (a) in a chain parallel to the spinal cord
   (b) in the dorsal roots of spinal nerves
   (c) next to or within the organs innervated
   (d) in the brain
4. The neurotransmitter of preganglionic sympathetic fibers is
   (a) norepinephrine
   (b) epinephrine
   (c) acetylcholine
   (d) dopamine

5. Which of the following results from stimulation of alpha-adrenergic receptors?
   (a) constriction of blood vessels
   (b) dilation of bronchioles
   (c) increased heart rate
   (d) sweat gland secretion
6. Which of the following fibers release norepinephrine?
   (a) preganglionic parasympathetic fibers
   (b) postganglionic parasympathetic fibers
   (c) postganglionic sympathetic fibers in the heart
   (d) postganglionic parasympathetic fibers in sweat glands
   (e) all of these
7. The actions of sympathetic and parasympathetic fibers are cooperative in the
   (a) heart
   (b) reproductive system
   (c) digestive system
   (d) eyes
8. Propanolol is a "beta-blocker." It would therefore cause
   (a) vasodilation
   (b) slowing of the heart rate
   (c) increased blood pressure
   (d) secretion of saliva

9. Atropine blocks parasympathetic nerve effects. It would therefore cause
   (a) dilation of the pupils
   (b) decreased mucous secretion
   (c) decreased movements of the digestive tract
   (d) increased heart rate
   (e) all of these
10. The area of the brain that is most directly involved in the reflex control of the autonomic system is the
   (a) hypothalamus
   (b) cerebral cortex
   (c) medulla oblongata
   (d) cerebellum

Essay Questions

1. Compare the sympathetic and parasympathetic systems in terms of the location of their ganglia and the distribution of their nerves.
2. Explain the anatomical and physiological relationship between the sympathetic nervous system and the adrenal glands.
3. Compare the effects of adrenergic and cholinergic stimulation on the cardiovascular and digestive systems.
4. Explain how effectors that receive only sympathetic innervation are regulated by the autonomic system.
5. Explain why a person may sweat more profusely immediately after exercise than during exercise.

Selected Readings

Angevine, J. B., Jr., and C. W. Cottman. 1981. *Principles of neuroanatomy.* New York: Oxford University Press.
Decara, L. V. January 1970. Learning in the autonomic nervous system. *Scientific American.*
Hoffman, B. B., and R. J. Lefkowitz. 1980. Alpha-adrenergic receptor subtypes. *New Engl. J. Med.* 302:1390.
Lefkowitz, B. B. 1976. Beta-adrenergic receptors: Recognition and regulation. *New Engl. J. Med.* 295:323.
Motulsky, J. H., and P. A. Insel. 1982. Adrenergic receptors in man. *New Engl. J. Med.* 307:18.
Noback, C. E., and R. J. Demerest. 1975. *The human nervous system: Basic principles of neurobiology.* 2d ed. New York: McGraw-Hill Book Company.

17

Sensory Organs

Outline

Concepts

Sensory organs are specialized
    extensions of the nervous system that
    contain sensory neurons adapted to
    respond to specific stimuli and
    conduct nerve impulses.

The development of the eye and ear
    from embryonic germ layers begins
    during the fourth week. The sensory
    components of the eye and ear form
    by twenty weeks, and the accessory
    structures form by thirty-two weeks.

The senses are classified according to
    the complexity and location of
    receptors or on the basis of the
    stimulus energy they transduce.

A sensory receptor responds to stimulus
    energy and initiates the nerve
    impulse that produces a particular
    modality of sensation.

The perception of somatic sensations is
    determined by the density of the
    receptors in the receptive field and
    the intensity of the sensation, as well
    as by accommodation, afterimage,
    and projection.

Olfactory receptors of the olfactory
    nerve respond to chemical stimuli
    and transmit the sensation of
    olfaction to the cerebral cortex.

Taste receptors in taste buds respond to
    chemical stimuli and transmit
    impulses through the
    glossopharyngeal or facial cranial
    nerve to the cerebral cortex.

Rods and cones within the eyeball are
    sensitive to light waves and transmit
    impulses through ganglion fibers of
    the optic nerve to the visual cortex of
    the occipital cerebral lobe.

Structures of the outer, middle, and
    inner ear are involved in hearing.
    The inner ear also provides for a
    sense of balance, or equilibrium.

Overview of Sensory Perception

Sensory organs are highly specialized extensions of the nervous system in that they contain sensory neurons adapted to respond to specific stimuli and conduct nerve impulses to the brain for interpretation.

Objective 1. Explain what is necessary for perceiving a sensation.
Objective 2. Discuss the selectivity of sensory receptors to specific stimuli.

Your sense organs are actually extensions of the nervous system that allow you to perceive changes in your internal and external environments. These sense organs have been described as "windows for the brain" because it is through them that you are aware of your environment. You must receive a stimulus before you can interpret the sensation and make the necessary body adjustments dictated by your central nervous system. Your sense organs permit daily experiences of pleasure and are vital to your daily survival by allowing you to hear sounds, see dangers, taste undesirable foods, and perceive pain, hunger pangs, or intestinal aches.

A **sensation** *(sen-sa′shun)* is the arrival of a sensory impulse to the brain. The interpretation of a sensation is referred to as **perception** *(per-sep′shun).* In other words, you feel, see, hear, taste, and smell with your brain. In order to perceive a sensation, the following four conditions are necessary.

1. A **stimulus** sufficient to initiate a response in the nervous system must be present.
2. A **receptor** must convert the stimulus to a nerve impulse. A receptor is a specialized, peripheral dendritic ending of an afferent nerve fiber or the specialized receptor cell associated with it.
3. The **conduction of the nerve impulse** must occur from the receptor to the brain along a nervous pathway.
4. The **interpretation of the impulse** in the form of a perception must occur within a specific portion of the brain.

Only impulses reaching the cerebral cortex of the brain are consciously interpreted as sensations. If impulses terminate in the spinal cord or brain stem, they initiate a reflexive motor activity response rather than a conscious sensation. Impulses reaching the higher brain centers travel through nerve fibers composing *sensory,* or *ascending, tracts.* Clusters of neuron cell bodies, called **nuclei,** are synaptic sites along sensory tracts within the CNS. The nuclei that sensory impulses pass through before reaching the cerebral cortex are located in the spinal cord, medulla oblongata, pons, and thalamus.

Through the use of scientific instruments, we know that our senses act as energy filters that allow us to perceive only a narrow range of energy. Vision, for example, is limited to light waves in the visible spectrum. Other types of waves of the same type of energy as visible light, such as X rays, radio waves, and ultraviolet and infrared light, cannot normally excite the sensory receptors in the eyes. Because there is no such thing as cold—only varying degrees of heat—the perception of cold is entirely a function of the nervous system. The perception of cold has obvious survival value. Though filtered and distorted by the limitations of sensory functions, our perceptions allow us to interact effectively with the environment.

---

1. *Distinguish between sensation and perception.*
2. *List the four conditions necessary for perception, and identify which of the four must always involve consciousness in order for perception to occur.*
3. *Using examples, explain the statement that each of our senses acts as a filter allowing us to perceive only a narrow range of energy.*

Development of the Sensory Organs

The development of the eye and ear begins during the fourth week and involves all three embryonic germ layers—ectoderm, mesoderm, and endoderm. The sensory components of the eye and ear form by twenty weeks, and the accessory structures form by thirty-two weeks.

Objective 3. Describe the development of the eye and the ear in the human embryo.

The sensory organs are derived from ectoderm for the most part, as is the rest of the nervous system. The sensory organs develop simultaneously, beginning early in the fourth week and continuing through the second trimester. The sensory receptors located in the skin, visceral organs, and muscles and tendons develop along with those organs that they serve. Because the development of these organs has been described elsewhere in this text, only the development of the eye and ear will be discussed.

Development of the Eye

The development of the eye is a complex, rapid process that involves the precise interaction of neuroectoderm, surface ectoderm, and mesoderm. Differentiation of these germ layers is evident early in the fourth week. The initial differentiation involves neuroectoderm forming a lateral diverticulum on either side of the prosencephalon. As the diverticulum increases in size, the distal portion dilates to become the **optic** *(op'tik)* **vesicle,** and the proximal portion constricts to become the **optic stalk** (fig. 17.1). Once the optic vesicle is formed, the overlying surface ectoderm thickens and invaginates. The thickened portion is the **lens placode** *(plak'ōd),* and the invagination is the **lens pit.**

During the fifth week, the lens placode is depressed and eventually cut off from the surface ectoderm, causing the formation of the **lens vesicle.** Simultaneously with the formation of the lens vesicle, the optic vesicle invaginates and differentiates into the two-layered **optic cup.** A groove called the **optic fissure** appears along the inferior surface of the optic cup and is continuous with a depression along the optic stalk. The **hyaloid artery** and **hyaloid vein** traverse the optic fissure longitudinally to serve the developing eyeball. The walls of the optic fissure eventually close so that the hyaloid vessels are within the tissue of the optic stalk. They become the **central vessels of the retina** *(ret'i-nah)* of the mature eye. The optic stalk eventually becomes the **optic nerve,** composed of sensory axons from the retina.

By the end of the sixth week and the early part of the seventh week, the optic cup has differentiated into two sheets of epithelial tissue that become the sensory and pigmented layers of the **retina.** Both of these layers also line the entire vascular coat, including the **ciliary body, iris,** and the **choroid** *(ko'roid).* A proliferation of cells in the lens vesicle leads to the formation of the **lens.**

To this point of eye development, the tissue mass surrounding the differentiating ectoderm is undifferentiated mesoderm. Once the primordial ectodermal structures are formed, however, the mesodermal germ tissue begins to differentiate. The **lens capsule** (fig. 17.34) forms from the mesoderm surrounding the lens. The mesoderm between the lens and retina, through which the hyaloid vessels pass, is transformed into the **vitreous humor** (figs. 17.25 and 17.36). Mesoderm surrounding the optic cup differentiates into two distinct layers of the developing eyeball. The inner layer of mesoderm becomes the vascular **choroid,** and the outer layer becomes the toughened **sclera** *(skle'rah)* posteriorly and the transparent **cornea** anteriorly. The anterior chamber and the posterior chamber are the fluid-filled spaces that form between the cornea and the iris and between the iris and the lens, respectively. Once the cornea has formed, additional surface ectoderm gives rise to the thin **conjunctiva** covering the anterior surface of the eyeball. Epithelium of the **eyelids** and the **lacrimal glands** and **ducts** develop from surface ectoderm, whereas the **extrinsic eye muscles** and all connective tissue associated with the eye develop from mesoderm. These accessory structures of the eye gradually develop during the embryonic period and into the fetal period as late as the fifth month.

---

optic: L. *optica,* see
hyaloid: Gk. *hyalos,* glass; *eiodos,* form

Figure 17.1. Diagrams of successive developmental stages of the eye. (*a*) a dorsal view of the cranial end of a 22-day-old embryo and the formation of the optic vesicle from the neuroectoderm of the prosencephalon (forebrain). (*b*) the development of the optic stalk and optic cup from the optic vesicle and the formation of the lens pit within the lens placode of surface ectoderm. This, and the other diagrams, is shown in sagittal view. (*c, d*) successive stages of the development of the eye during the fifth week. Note the enclosure of the hyaloid vessels within the optic stalk. (*e, f*) the successive development of the eye at six weeks and twenty weeks, respectively. Note the role of mesoderm during these stages in the formation of the internal structures of the eye. (*g*) the eye of the newborn.

(a)

(b)

(c)

(d)

(e)

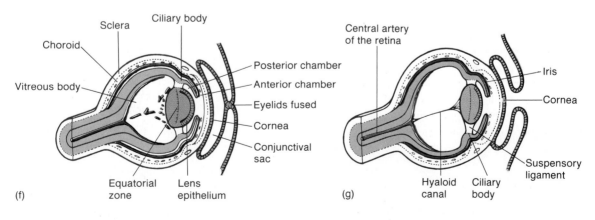

(f)

(g)

Figure 17.2. Diagrams of successive developmental stages of the inner ear. (*a*) a dorsal view of a 22-day-old embryo and the formation of the otic placode from surface ectoderm. (*b, c, e, f*) successive stages of development of the otocyst in the 22-day-old to the 30-day-old embryo. Schematic sections shown are coronal views. (*d*) lateral view of the cranial region of 28-day-old embryo, showing the position and external appearance of the developing otocyst. (*g–k*) lateral views of the developing otocyst into the cochlea and semicircular canals from the fifth to eighth week.

Figure 17.3. The formation of the cochlea and the organ of Corti from the otic capsule. (a–d) successive stages of development of the perilymphatic space and the organ of Corti from the eighth to the twentieth week.

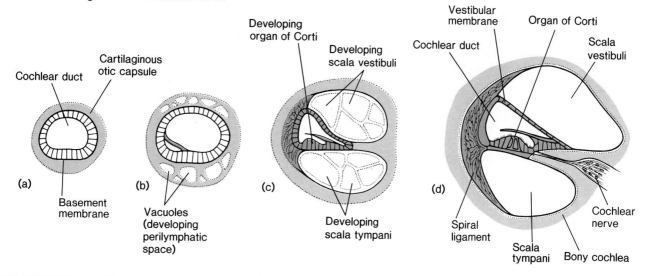

The developing eye is extremely sensitive to and may be impaired by teratogenic agents, particularly rubella (German measles). If a pregnant woman contracts rubella, there is a 90% chance the embryo or fetus will contract it also. An embryo afflicted with rubella has more than a 30% chance of being aborted, stillborn, or congenitally deformed. Rubella interferes with the mitotic process and thus causes underdeveloped organs. An embryo with rubella may suffer from a number of physical deformities, cataracts and glaucoma being common deformities of the eye.

Development of the Ear

The ear begins to develop at the same time as the eye, early during the fourth week. All three embryonic germ layers—ectoderm, mesoderm, and endoderm—are involved in the formation of the ear. Both types of ectoderm (neuroectoderm and surface ectoderm) play a role.

The ear of an adult is structurally and functionally divided into an **external ear,** a **middle ear,** and an **inner ear,** each of which has a separate embryonic origin. The inner ear does not develop from deep embryonic tissue as one might expect but rather begins to form early in the fourth week when a plate of surface ectoderm called the **otic** (o'tik) **placode** appears lateral to the developing hindbrain (fig. 17.2). The otic placode soon invaginates and forms an **otic pit.** Toward the end of the fourth week, the outer edges of the invaginated otic pit come together and fuse to form an **otocyst (otic vesicle).** The otocyst soon pinches off and separates from the surface ectoderm. The otocyst further differentiates to form a dorsal **utricular portion** and

a ventral **saccular portion.** Three separate diverticula extend outward from the utricular portion and develop into the **semicircular canals,** which later function in balance and equilibrium. A tubular diverticulum called the **cochlear duct** extends in a coiled fashion from the saccular portion and forms the membranous portion of the **cochlea** of the ear (fig. 17.2). The **organ of Corti,** which is the functional portion of the cochlea, differentiates from cells along the wall of the cochlear duct (fig. 17.3). The sensory nerves that innervate the inner ear are derived from neuroectoderm from the developing brain.

The differentiating otocyst is surrounded by mesodermal tissue that soon forms a cartilaginous **otic capsule** (fig. 17.3). As the otocyst and surrounding otic capsule grow in size, vacuoles containing the fluid **perilymph** form within the otic capsule. The vacuoles soon enlarge and coalesce to form the **perilymphatic space,** which divides into the **scala tympani** and the **scala vestibuli.** Eventually, the cartilaginous otic capsule ossifies to form the **bony (osseous) labyrinth** of the inner ear. The chamber of the middle ear derives from endoderm, whereas the bones, or **ossicles,** of the middle ear derive from the neural crest. The middle ear chamber is referred to as the **tympanic cavity** and derives from the first pharyngeal pouch (see figs. 17.2, 17.4). The ossicles, which amplify incoming sound waves, derive from the first and second pharyngeal arch cartilages. As the tympanic cavity enlarges, it surrounds and encloses the developing ossicles (fig. 17.4). The connection of the tympanic cavity to the pharynx gradually elongates to develop into the **auditory (Eustachian) tube,** which remains patent throughout life and is important in maintaining an equilibrium of air pressure between the pharyngeal and tympanic cavities.

otic: Gk. *otikos,* ear

Figure 17.4. The coronal plane showing the sequential development of the middle ear and the ear ossicles. (*a*) the embryo at four weeks illustrating the relation of the otocyst to the first branchial groove and the first pharyngeal pouch. (*b*) at five weeks, the derivatives of the first and second pharyngeal arch cartilages are apparent. (*c*) by the sixth week, the pharyngeal arch cartilages have differentiated into the ear ossicles, which are enveloped by the tubotympanic recess (future tympanic cavity). (*d*) by the end of the eighth week, the ear ossicles are formed and enclosed within the tympanic cavity.

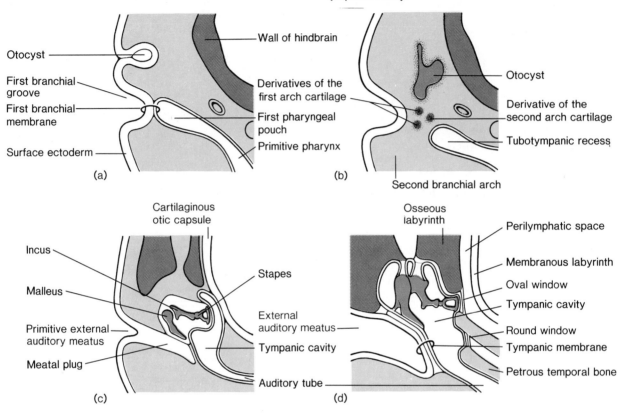

Figure 17.5. The progressive development of the auricle of the ear: (*a*) 5 weeks; (*b*) 6 weeks; (*c*) 8 weeks; (*d*) 32 weeks. The ears develop from structures of the neck region but gradually migrate to the side of the head.

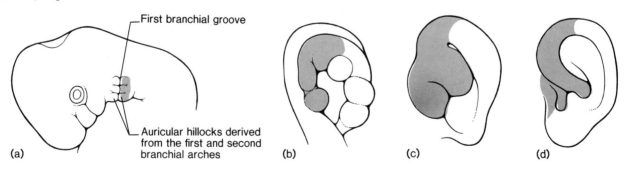

The external ear includes the fleshy **auricle** attached to the side of the head and the tubular **external auditory meatus** that extends into the temporal bone of the skull. The external auditory meatus forms from the surface ectoderm that covers the dorsal end of the first branchial groove (see fig. 17.4). A solid epithelial plate called the **meatal plug** soon develops at the bottom of the funnel-shaped branchial groove. The meatal plug is involved in the formation of the inner wall of the external auditory meatus and contributes to the **tympanic membrane** (eardrum). The tympanic membrane has a dual origin from surface ectoderm and the endoderm lining the first pharyngeal pouch (see fig. 17.4).

The auricle develops from six separate swellings of surface ectoderm called **auricular hillocks** that form around the first branchial groove (fig. 17.5). As the auricular hillocks enlarge, they coalesce and are supported by mesodermal mesenchyme that differentiates into fibrous tissue. The auricles migrate superiorly as the face of the embryo and fetus forms, so that by thirty-two weeks the auricles have ascended to the dorsoventral level of the eyes.

---

1. *List four structures of the eye that develop from the neuroectoderm.*
2. *Name the layers of the eyeball that are derived from mesoderm.*
3. *Which structures of the ear are derived from surface ectoderm?*
4. *Describe the development of the tympanic cavity and the auditory tube.*

Classification of the Senses

The senses are classified as general or special according to the simplicity or complexity of the receptors and neural pathways. They are also classified as somatic or visceral according to the location of the receptors. A functional classification groups sensory receptors according to the type of stimulus energy they transduce.

Objective 4. Compare and contrast somatic, visceral, and special senses.

Objective 5. Define and give examples of exteroceptors, visceroceptors, and proprioceptors.

Objective 6. Distinguish between phasic and tonic receptors.

Structurally, the sensory receptor can be the dendrites of sensory neurons, which are either free (such as those in the skin that mediate pain and temperature) or are encapsuled within non-neural structures, such as pacinian corpuscles, or pressure receptors in the skin (see fig. 17.9). Other receptors, including taste buds, photoreceptors in the eyes, and hair cells in the inner ears derive from epithelial cells that synapse with sensory dendrites.

The senses of the body can be classified as general or special according to the simplicity or complexity of the receptors and neural pathways involved. **General senses** are widespread through the body and are simple in structure. Examples are touch, pressure, cold-heat, and pain. **Special senses** are localized in complex receptor organs and have extensive neural pathways. Among the special senses are taste, smell, sight, hearing, and balance.

The senses can also be classified as somatic or visceral according to the location of the receptors. **Somatic senses** are those in which the receptors are localized within the body wall. These include the cutaneous receptors and those within muscles, tendons, and joints. **Visceral senses** are those in which the receptors are located within visceral organs. These classes of senses intersect in certain organs. Hearing, for example, is a special somatic sense, whereas pain from the digestive tract is a general visceral sense.

Senses are also classified according to the location of the receptors and the types of stimuli to which they respond. There are three basic kinds of receptors: exteroceptors, visceroceptors, and proprioceptors.

Exteroceptors. Exteroceptors *(eks″ter-o-sep′tor)* are located near the surface of the body where they respond to stimuli from the external environment. They include the following:

1. Cones and rods in the retina of the eye—*photoreceptors (fo″to-re-sep′tor).*
2. Hair cells in the organ of Corti within the inner ear—*pressure receptors.*
3. Olfactory receptors in the nasal epithelium of the nasal cavity—*chemoreceptors (ke″mo-re-sep′tor).*
4. Taste receptors on the tongue—*chemoreceptors.*
5. Skin receptors within the dermis—*tactile receptors* for touch; *pressure receptors,* or *mechanoreceptors (mek″ah-no-re-sep′tor),* for pressure; *thermoreceptors* for temperature; and *nocioreceptors* (pain receptors).

Pain receptors are stimulated by chemicals released from damaged tissue cells and thus are a type of chemoreceptor. Although there are specific pain receptors, nearly all types of receptors when intensely stimulated transmit impulses that are perceived as pain. Pain receptors exist throughout the body but only those located within the skin are classified as exteroceptors.

Visceroceptors (Enteroceptors). As the name implies, visceroceptors *(vis″er-o-sep′tor)* produce sensations arising from the viscera, such as internal pain, hunger, thirst, fatigue, or nausea. Specialized visceroceptors located within the circulatory system are sensitive to changes in blood pressure (baroreceptors).

---

somatic: Gk. *somatikos,* body
visceral: L. *viscera,* body organs

Figure 17.6. Tonic receptors (*a*) continue to fire at a relatively constant rate as long as the stimulus is maintained. These produce slowly adapting sensations. Phasic receptors (*b*) respond with a burst of action potentials when the stimulus is first applied but quickly reduce their rate of firing when the stimulus is maintained. This produces rapidly adapting sensations.

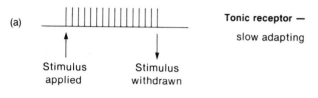

(a)   **Tonic receptor —**
slow adapting

Stimulus applied   Stimulus withdrawn

(b)   **Phasic receptor —**
rapidly adapting

Stimulus applied   Stimulus withdrawn

Proprioceptors. Proprioceptors *(pro″pre-o-sep′tor)* relay information about body position, equilibrium, and movement. They are located in the inner ear, joints, tendons, and muscles.

The three types of receptors may be further classified on the basis of sensory adaptation. Some receptors respond with a burst of activity when a stimulus is first applied but quickly decrease their firing rate as they adapt to the stimulus even though the stimulus is maintained. Receptors with this response pattern are called **phasic receptors.** Receptors that produce a relatively constant rate of firing as long as the stimulus is maintained are known as **tonic receptors** (fig. 17.6).

Phasic receptors alert us to changes in sensory stimuli and are in part responsible for the fact that we can cease paying attention to constant stimuli. This ability is called *sensory adaptation.* Odor and touch, for example, adapt rapidly; bath water feels hotter when we first enter it. Sensations of pain, in contrast, adapt little if at all.

---

1. *Using examples, explain how sensory receptors can be classified according to complexity, location, structure, and the type of stimuli to which they respond.*
2. *Describe sensory adaptation in olfactory and pain receptors. Using a line drawing, relate sensory adaptation to the responses of phasic and tonic receptors.*

---

proprioceptor: L. *proprius,* one's own; *ceptus,* taken

Physiology of Sensory Receptors

A sensory receptor responds to stimulus energy and initiates the nerve impulse in the sensory neuron. Stimulation of a particular sensory receptor produces a particular modality of sensation.

> *Objective 7.* Define the law of specific nerve energies.
>
> *Objective 8.* Explain how action potentials are generated in sensory neurons in response to sensory stimuli.

Law of Specific Nerve Energies

Stimulation of a sensory nerve fiber produces only one sensation—touch, cold, pain, and so on. According to the **law of specific nerve energies,** the sensation characteristic of each type of sensory neuron is that produced by its normal, or *adequate stimulus.* The adequate stimulus for the photoreceptors of the eye, for example, is light. If these receptors are stimulated by some other means—such as by pressure produced by a punch to the eye—a flash of light (the adequate stimulus) may be perceived.

Paradoxical cold provides another example of the law of specific nerve energies. First, a receptor for cold is located by touching the tip of a cold metal rod to the skin. Sensation gradually disappears as the rod warms to body temperature. Applying the tip of a rod heated to 45° C to the same spot, however, causes the sensation of cold to reappear. This paradoxical cold is produced because the heat slightly damages receptor endings, and by this means produces an "injury current" that stimulates the receptor.

Regardless of how a sensory neuron is stimulated, therefore, only one sensory modality will be perceived. This specificity is due to the synaptic pathways within the brain that are activated by the sensory neuron. The ability of receptors to function as sensory filters and be stimulated normally by only one type of stimulus (the adequate stimulus) allows usually the brain to perceive the stimulus accurately.

Sensory Neurons and Receptor Potentials

Sensory neurons are *pseudounipolar* (unipolar). Neurons of this type have a *peripheral branch,* comparable to a dendrite, and a *central branch,* comparable to the axon (fig. 17.7). Sensory neurons thus form a continuous pathway from the sensory endings, past the dorsal root ganglia (which contain the cell bodies of sensory neurons), to the gray matter of the spinal cord. Excluding the sensory endings, the entire length of a sensory neuron contains voltage regulated gates that produce action potentials.

The electrical behavior of the sensory endings is similar to that of the dendrites of other neurons. Upon stimulation, dendrites exhibit a local, graded depolarization

Figure 17.7. The conduction of sensory impulses by a pseudounipolar neuron into the gray matter of the spinal cord.

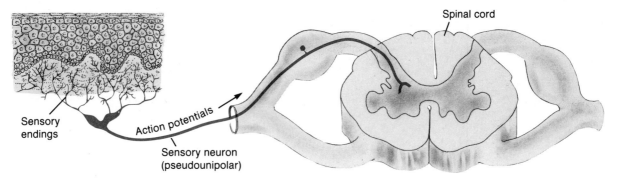

Sensory endings

Action potentials

Sensory neuron (pseudounipolar)

Spinal cord

that increases in proportion to the stimulus intensity. This potential is analogous to an excitatory postsynaptic potential (EPSP). In sensory endings, however, it is known as a **receptor potential,** or a **generator potential,** because it stimulates opening of voltage-regulated gates and thus serves to generate action potentials (fig. 17.8) in response to sensory stimuli.

---

1. *Our perceptions are products of our brains; they are incompletely and inconstantly related to physical reality. Explain this statement, using examples of vision and perceptions of cold.*
2. *Describe how the magnitude of a sensory stimulus is transduced into a receptor potential and how the magnitude of the receptor potential is coded in the sensory nerve process.*

Somatic Senses

Somatic senses include cutaneous receptors and proprioceptors. The perception of somatic sensations is determined by the density of the receptors in the stimulated receptive field and the intensity of the sensation. Accommodation, afterimage, and projection are also factors of somatic perception.

Objective 9. Describe the structure, function, and location of the various tactile receptors and the neural pathway for somatic sensation.

Objective 10. Explain the purpose of pain, the receptors that respond to pain, and the impulse pathway to the brain.

Objective 11. Define and give examples of referred pain and phantom pain.

Objective 12. Describe the physiological factors involved in the perception of somatic sensations.

Figure 17.8. Sensory stimuli result in the production of local, graded potential changes known as the receptor, or generator potential (number 1 through 4). If the receptor potential reaches a threshold value of depolarization, it generates action potentials (number 5) in the sensory neuron.

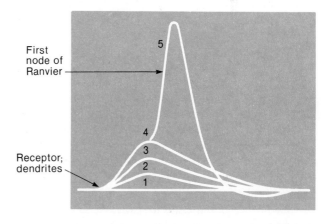

First node of Ranvier

Receptor; dendrites

The **somatic senses,** or **somasthetic senses,** include the receptors that are located in the skin and cause cutaneous sensations such as touch, tickling, pressure, cold, heat, and pain. The somatic senses also include proprioceptors that are located in the inner ear, joints, tendons, and muscles, and relay information about body position, equilibrium, and movement.

Tactile and Pressure Receptors

Both tactile receptors and pressure receptors are sensitive to mechanical forces that distort or displace the tissue in which they are located. Tactile receptors respond to *fine,* or *light,* touch and are located primarily in the dermis and hypodermis of the skin. Pressure receptors respond to *deep pressure* and are commonly found in the hypodermis of the skin and in the tendons and ligaments of joints. There are three kinds of tactile receptors and one kind of pressure receptor (fig. 17.9).

Figure 17.9.   A diagrammatic section of the skin showing the general location and magnified structure of cutaneous receptors.

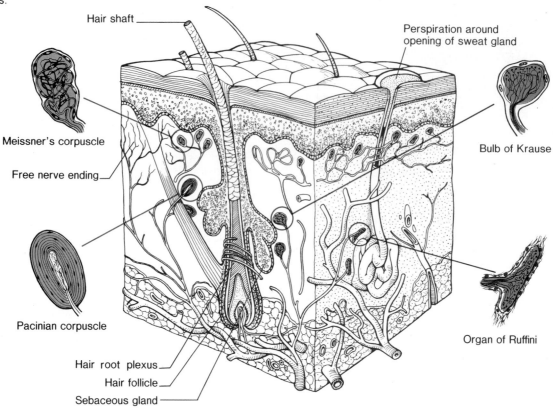

Meissner's Corpuscles.   A Meissner's corpuscle *(kor' pus'l)* is an oval receptor composed of a mass of dendritic endings from two or three nerve fibers enclosed by connective tissue sheaths. These corpuscles are numerous in the hairless portions of the body, such as the eyelids, lips, tip of the tongue, fingertips, palms of the hands, nipples, and external genital organs. Meissner's corpuscles lie within the papillary layer of the dermis where they are especially sensitive to the motion of objects that barely contact the skin. Sensations of fine or light touch are perceived as these receptors are stimulated. They also function when a person touches an object to determine its texture.

Free Nerve Endings.   Free nerve endings are the least modified and the most superficial of the tactile receptors. These receptors extend into the lower layers of the epidermis where they end as knobs between the epithelial cells. Free nerve endings are most important as pain receptors, although they also respond to objects that are in continuous contact with the skin, such as clothing.

Meissner's corpuscle: from George Meissner, German histologist, 1829–1905
corpuscle: L. *corpusculum,* diminutive of corpus, body

Hair Root Plexuses.   Hair root plexuses are a specialized type of free nerve ending. They are coiled around hair follicles where they respond to movement of the hair.

Pacinian Corpuscles.   Pacinian corpuscles are large, onion-shaped receptors composed of the dendritic endings of several sensory nerve fibers enclosed by connective tissue layers. They are commonly found within the synovial membranes of diarthrotic joints, in the perimysium of muscle tissue, and in certain visceral organs. Pacinian corpuscles are also abundant in the skin of the palms and fingers of the hand, soles of the feet, external genital organs, and breasts.

Pacinian corpuscles respond to heavy pressures, generally those that are constantly applied. They can also detect vibrations in tissues and organs.

Thermoreceptors

Thermoreceptors *(ther''mo-re-sep'tor)* are widely distributed throughout the dermis of the skin but are especially abundant in the lips and the mucous membranes of the mouth and anal regions. There are two kinds of thermoreceptors—one that responds to heat and the other to cold (fig. 17.9).

pacinian corpuscle: from Filippo Pacini, Italian anatomist, 1812–1883

Table **17.1** Cutaneous receptors

Type	Location	Function	Sensation
Meissner's corpuscles (mechanoreceptors)	Papillae of dermis; numerous in hairless portions of body (eyelids, fingertips, lips, nipples, external genitalia)	Detect light motion against surface of skin	Fine touch; texture
Free nerve endings (pressure receptors; pain receptors)	Lower layers of epidermis	Detect changes in pressure; detect tissue damage	Touch, pressure; pain
Hair root plexuses (tactile receptors)	Around hair follicles	Detect movement of hair	Touch
Pacinian corpuscles (mechanoreceptors)	Hypodermis; synovial membranes; perimysium; certain visceral organs	Detect changes in pressure	Deep pressure; vibrations
Organs of Ruffini (thermoreceptors)	Lower layers of dermis	Detect changes in temperature	Heat
Bulbs of Krause (thermoreceptors)	Dermis	Detect changes in temperature	Cold

Source: Van De Graaff, Kent M., *Human Anatomy*. © 1984 Wm. C. Brown Publishers, Dubuque, Iowa. All Rights Reserved. Reprinted by permission.

Organs of Ruffini. The organs of Ruffini are heat receptors located deep within the dermis. Heat receptors are elongated, oval structures that are most sensitive to temperatures from 25° C (77° F). Temperatures above 45° C (113° F) elicit impulses through the organs of Ruffini that are perceived as painful, burning sensations.

Bulbs of Krause. The bulbs of Krause are receptors for the sensation of cold. They are more abundant than heat receptors and are closer to the surface of the skin. The bulbs of Krause are most sensitive to temperatures below 20° C (68° F). Temperatures below 10° C elicit responses through the bulbs of Krause that are perceived as painful, freezing sensations.

Pain Receptors

The principal receptors for pain are the free nerve endings. Several million free nerve endings are distributed throughout the skin and internal tissues. Pain receptors are sparse in certain visceral organs and absent within the nervous tissue of the brain. Although the free nerve endings are specialized to respond to tissue damage, all of the cutaneous receptors will relay impulses that are interpreted as pain if stimulated excessively. The cutaneous receptors are summarized in table 17.1.

The protective value of pain receptors is obvious. Unlike other cutaneous receptors, free nerve endings have little accommodation, so impulses are relayed continuously to the CNS as long as the irritating stimulus is present. Although pain receptors can be activated by all types of stimuli, they are particularly sensitive to chemical stimulation. Muscle spasms, muscle fatigue, or an inadequate supply of blood to an organ may also cause pain.

Sensory impulses for pain are conducted to the spinal cord through afferent nerve fibers. Impulses are then conducted to the thalamus along the *lateral spinothalamic tract* of the spinal cord and from there to the somatesthetic area of the cerebral cortex. Although an awareness of pain occurs in the thalamus, the type and intensity of pain is interpreted in the specialized areas of the cerebral cortex.

The sensation of pain can be clinically classified as **somatic pain** and **visceral pain.** Stimulation of the cutaneous pain receptors results in the perception of superficial somatic pain. Deep somatic pain comes from stimulation of receptors in skeletal muscles, joints, and tendons.

Stimulation of the receptors within the viscera causes the perception of visceral pain. Through precise neural pathways, the brain is able to perceive the area of stimulation and project the pain sensation back to that area. The sensation of pain from certain visceral organs, however, may not be perceived as arising from those organs but from other somatic locations. This phenomenon is known as **referred pain** (fig. 17.10). The sensation of referred pain is consistent from one person to another and is clinically important in diagnosing organ dysfunctions. The pain of a heart attack, for example, may be perceived subcutaneously over the heart and down the medial side of the left arm. Ulcers of the stomach may cause pain that is perceived as coming from the upper central (epigastric) region of the torso. Pain from problems of the liver or gallbladder may be perceived as localized visceral pain or perceived as referred pain arising from the right neck and shoulder regions.

Referred pain is not totally understood but seems to be related to the development of the tracts within the spinal cord. It is thought that there are some *common nerve pathways* used by sensory impulses coming from both the cutaneous areas and from visceral organs (fig. 17.11). Consequently, impulses along these pathways may be incorrectly interpreted as arising cutaneously rather than from within a visceral organ.

---

organs of Ruffini: from Angelo Ruffini, Italian anatomist, 1864–1929
bulbs of Krause: from Wilhelm J. F. Krause, German anatomist, 1833–1910

Figure 17.10. Sites of referred pain are perceived cutaneously but actually originate from specific visceral organs.

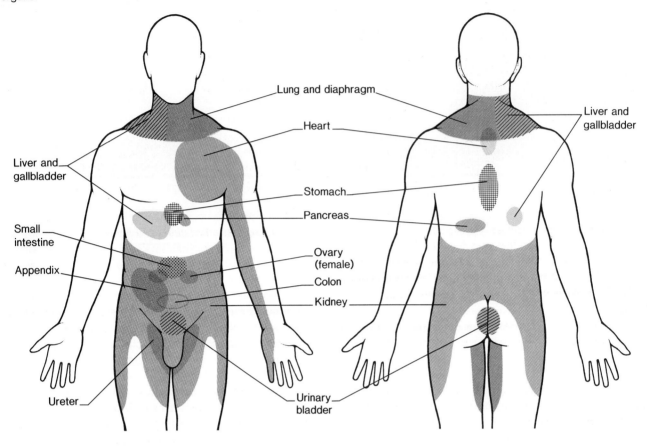

Figure 17.11. An explanation of referred pain. Pain originating from the myocardium of the heart may be perceived as coming from the skin of the left arm because sensory impulses from these two organs are conducted through common nerve pathways to the brain.

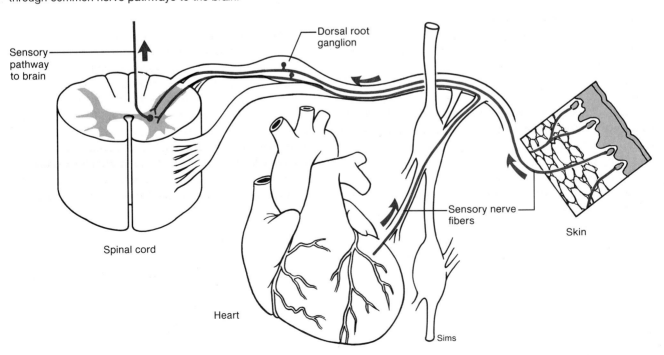

Figure 17.12. Proprioreceptors. (*a*) a neuromuscular spindle; (*b*) a neurotendinous receptor (Golgi tendon organ).

Muscle fibers

Tendon

(a)

(b)

Phantom pain is frequently experienced by an amputee who continues to feel pain from the body part that was amputated. After amputation, the severed sensory neurons heal and function in the remaining portion of the appendage. Although it is not known why impulses that are interpreted as pain are sent periodically through these neurons, the sensations evoked in the brain are projected to the amputated region, resulting in phantom pain.

Proprioceptors

Proprioceptors monitor internal conditions, frequently those under voluntary control. They are located within skeletal muscle tissue, tendons, and the synovial membranes and connective tissue surrounding joints. Proprioceptors are sensitive to changes in stretch and tension. Some of the sensory impulses from proprioceptors reach the level of consciousness as the **kinesthetic sense,** by which the position of the body parts is perceived. Other proprioceptor information is not consciously interpreted and is used to adjust the intensity and timing of muscle contractions to provide coordinated movements. With the kinesthetic sense, we can determine the position and movement of our limbs without visual sensations, such as when we dress or walk in the dark. The kinesthetic sense, along with hearing, becomes keenly developed in a blind person.

High-speed transmission is a vital characteristic of the kinesthetic sense, since rapid feedback to various body parts is essential for quick, smooth, coordinated body movements.

There are three types of proprioceptors: joint kinesthetic receptors, neuromuscular spindles, and neurotendinous receptors.

Joint Kinesthetic Receptors. Joint kinesthetic receptors are located in the connective tissue capsule in diarthrotic joints where they are stimulated by changes in position caused by movement at the joints.

Neuromuscular Spindles. Neuromuscular spindles are located in skeletal muscle, particularly in the muscles of the extremities. They consist of the endings of sensory neurons spiraled around individual muscle fibers (fig. 17.12). Neuromuscular spindles are stimulated by an increase in muscle tension due to the lengthening or stretching of the individual spindles and thus provide information about the length of the muscle and the rate of change of length (i.e., velocity of contraction).

Neurotendinous Receptors. Neurotendinous receptors, also called *Golgi (gol′ je) tendon organs,* are located where a muscle attaches to a tendon (fig. 17.12). They are stimulated by the tension produced in a tendon when the attached muscle is either stretched or contracted.

Receptive Fields and Sensory Acuity

The *receptive field* of a neuron serving cutaneous sensation is the area of the skin whose stimulation results in changes in the firing rate of the neuron. The area of each receptive field in the skin varies inversely with the *density of the receptors* in the region. Cutaneous receptors are abundant in parts of the face, palms and fingers of the hands, soles of the feet, and the genitalia. They are less abundant in the skin of the torso, particularly along the back and on the back of the neck, and are sparse in the skin over joints, especially the elbow. Generally speaking, the thinner the skin, the greater the sensitivity. The *intensity of a sensation* depends not only on the density of the receptors but also on the number of receptors stimulated in a given amount of time.

> The high tactile acuity of the fingertips is exploited in the reading of Braille. Braille symbols consist of dots that are raised 1 mm up from the page and separated from each other by 2.5 mm. Experienced Braille readers can scan words at about the same speed that a sighted person can read aloud—a rate of about 100 words per minute.

When a blunt object touches the skin a number of receptive fields may be stimulated. Those receptive fields in the center areas where the touch is strongest will be stimulated more than in neighboring fields where the touch is lighter. We do not usually feel a subtle sensory experience of light touch when it surrounds a center of stronger touch, however. Instead, only a single touch is felt, which is somewhat sharper than the actual shape of the blunt object. This sharpening of sensation is due to a process called *lateral inhibition* (fig. 17.13).

Lateral inhibition and the sharpening of sensation that results from it occur within the central nervous system. Those sensory neurons whose receptive fields are stimulated most strongly inhibit—via interneurons that pass "laterally" within the CNS—sensory neurons that serve neighboring receptive fields. Lateral inhibition similarly plays a prominent role in pitch discrimination.

When the intensity of a given stimulus does not change once the sensation has been perceived in the CNS, *accommodation* may occur. This means that the sensation is not perceived continuously at the same level even though the stimulus is continuous. Accommodation occurs, for example, when you get into a tub of hot water and perceive a burning sensation. As the sensory neurons accommodate, the stimulus becomes one of comfortable warmth, even though the temperature of the water has not changed.

Afterimage is the reverse of accommodation, such as when a sensation continues to be perceived even though the stimulus is no longer present. The sensation of a hat on your head after you have removed it is an example of afterimage.

As the brain perceives a sensation, it interprets or identifies the site from which the stimulus is being applied. This process is called *projection* because the brain projects the sensation back to the apparent source of the sensation. Because of projection we attribute the function of sight to the eyes, hearing to the ears, feeling to the skin, and so forth, when actually these perceptions are functions of the brain.

Braille: from Louis Braille, French teacher of the blind, 1809–1852

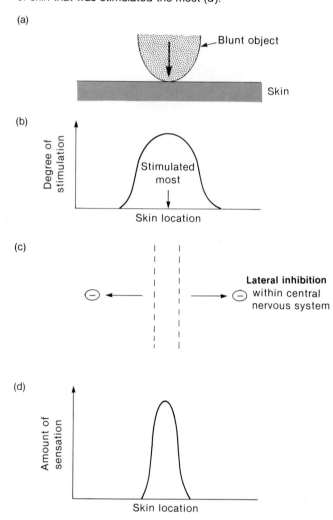

Figure 17.13.   When an object touches the skin (*a*), receptors in the center of the touched skin are stimulated more than neighboring receptors (*b*). As a result of lateral inhibition within the central nervous system (*c*), input from these neighboring sensory neurons is reduced. Sensation, as a result, is more sharply localized to the area of skin that was stimulated the most (*d*).

> One method of determining sensitivity of the skin is the **two-point touch threshold** test (fig. 17.14). In this test, the points of two pins are applied to the skin. The distance between the points of stimulation is then consistently narrowed until the minimum distance is found at which two distinct sites of stimulation can be distinguished. For example, the receptors in the tip of the tongue can discriminate two points of stimulation at a minimum distance of 1.4 mm (0.06 in.). If the test is performed on the back of the neck, the two points are perceived until a minimum distance of 36.2 mm (1.43 in.). At a distance less than this, a single point of stimulation is perceived. The two-point touch threshold is thus an indication of **tactile acuity,** or the sharpness of touch perception. Table 17.2 indicates the two-point threshold for different regions of the body.

acuity: L. *acuo,* sharpen

Figure 17.14.   The two-point touch threshold test. If each point touches the receptive fields of different sensory neurons, two separate points of touch will be felt. If both caliper points touch the receptive field of one sensory neuron, only one point of touch will be felt.

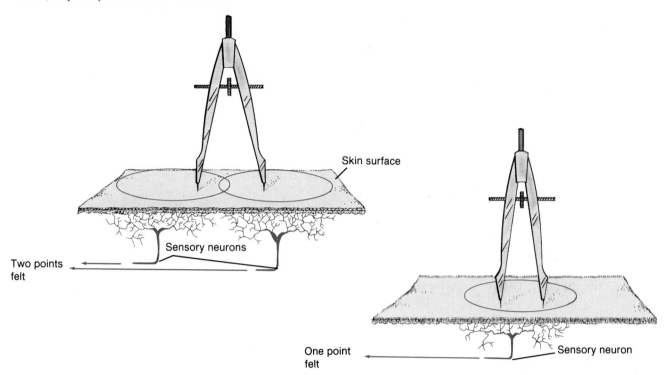

Skin surface

Sensory neurons

Two points felt

One point felt

Sensory neuron

Table **17.2** The two-point touch threshold for different regions of the body	
Body region	**Two-point touch threshold (mm)**
Big toe	10
Sole of foot	22
Calf	48
Thigh	46
Back	42
Abdomen	36
Upper arm	47
Forehead	18
Palm of hand	13
Thumb	3
First finger	2

Source: From Weinstein, S., and D. R. Kenshalo (editor), The Skin Senses, 1968. Courtesy of Charles C Thomas, Publisher, Springfield, Illinois.

Neural Pathways for Somatic Sensation

The conduction pathways for the somatic senses are shown in figure 17.15. Proprioception and pressure are carried by large, myelinated nerve fibers that ascend in the dorsal columns of the spinal cord on the ipsilateral (same) side. These fibers do not synapse until they reach the medulla oblongata of the brain stem; fibers that carry these sensations from the feet are thus incredibly long. After synapsing in the medulla with other, second-order sensory neurons, information in the latter neurons crosses over to the contralateral (opposite) side as it ascends via a fiber tract, called the **medial lemniscus,** to the thalamus. Third-order sensory neurons in the thalamus that receive this input in turn project to the **postcentral gyrus.**

Sensations of hot, cold, and pain are carried by thin, unmyelinated sensory neurons into the spinal cord. These synapse within the spinal cord with second-order interneurons, which cross over to the contralateral side and ascend to the brain in the **lateral spinothalamic tract.** Fibers that mediate touch and pressure ascend in the **ventral spinothalamic tract.** Fibers of both spinothalamic tracts synapse in the thalamus with third-order neurons, which in turn project to the postcentral gyrus of the parietal cerebral lobe. Note that, in all cases, somatic information is carried to the postcentral gyrus in third-order neurons. Also, because of crossing over, somatic information from each side of the body is projected to the postcentral gyrus of the contralateral cerebral hemisphere.

All somatic information from the same area of the body projects to the same area of the postcentral gyrus, so that a *sensory homunculus* "map" of the body can be

Figure 17.15. Pathways that lead from the cutaneous receptors and proprioreceptors into the postcentral gyrus in the cerebral cortex.

Postcentral gyrus

Ventro-lateral nucleus of the thalamus

Midbrain

Pain, hot, and cold

Touch and pressure

Lower medulla

Lateral spinothalamic tract

Ventral spinothalamic tract

Proprioception

Spinal cord

Waldrop

Figure 17.16. A sensory homunculus map depicts the areas of the postcentral gyrus of the cerebral cortex devoted to sensation in different parts of the body. Notice that a disproportionately large area of the cortex is devoted to the fingers and lips.

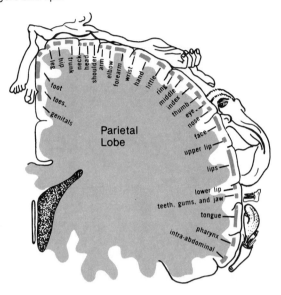

drawn on the postcentral gyrus to represent sensory projection points (fig. 17.16). This map is greatly distorted, however, because it shows larger areas of cerebral cortex devoted to sensation in the face and hands than in other areas of the body. This disproportionately larger area of the cortex devoted to the face and hands reflects the fact that there is a higher density of sensory receptors in these regions.

---

1. List the types of tactile receptors, and state where they are located. What portion of the brain interprets tactile sensations?
2. Discuss the importance of pain. List the receptors that respond to pain and the structures of the brain which are particularly important in the perception of pain sensation.
3. Using examples, distinguish between referred pain and phantom pain. Discuss why it is important for a physician to know the referred pain sites.
4. Describe the meaning of sensory acuity, and explain how acuity is affected by the density of receptive fields and by the process of lateral inhibition.
5. Using a flow diagram, describe the neural pathways leading from cutaneous pain and pressure receptors to the postcentral gyrus. Indicate where crossing over occurs.

Olfactory Sense

Olfactory receptors are the dendritic endings of the olfactory (first cranial) nerve that respond to chemical stimuli and transmit the sensation of olfaction directly to the olfactory portion of the cerebral cortex.

Objective 13. Describe the afferent pathway of olfaction.

Olfactory reception in humans is not highly developed compared to that of certain other vertebrates. Because we do not rely on smell for communicating or for finding food, the olfactory sense is probably the least important of the senses; it is more important in detecting the presence of an odor rather than its intensity. Accommodation occurs relatively rapidly with this sense. Olfaction *(ol-fak'shun)* functions closely with gustation (taste) in that the receptors of both are chemoreceptors and require dissolved substances for stimuli.

Olfactory receptor cells are located in the nasal epithelium within the roof of the nasal cavity on both sides of the nasal septum in an area referred to as the **nasal cleft** (fig. 17.17). Olfactory cells are moistened and stabilized by the surrounding glandular goblet cells and the supporting columnar epithelial cells. The olfactory cells are actually bipolar neurons whose cell bodies lie between the supporting columnar cells. The free end of each olfactory cell contains several dendritic endings called **olfactory hairs,** which are the sensitive portion of the receptor cell.

The afferent pathway for olfaction consists of several neural segments. The unmyelinated axons of the olfactory cells unite to form the **olfactory nerves,** which traverse the foramina of the cribriform plate and terminate in the paired masses of gray and white matter called the **olfactory bulbs.** The olfactory bulbs lie on both sides of the crista galli of the ethmoid bone beneath the frontal lobes of the cerebrum. The neurons of the olfactory nerves synapse within the olfactory bulb with dendrites of the **olfactory tract.** Sensory impulses are conveyed along the olfactory tract and into the olfactory portion of the cerebral cortex where they are interpreted as odor and cause the sensation of smell.

Unlike taste, which is divisible into only four modalities, thousands of distinct odors can be distinguished by people who are trained in this capacity (as in the perfume industry). The molecular basis of olfaction is not understood, but it is known that a single odorant molecule is sufficient to excite an olfactory receptor.

Figure 17.17. The olfactory receptive area. (a) a frontal section through the nose showing the relationship of the nasal cavity and the nasal septum to the olfactory sensory structures within the nasal cleft. (b) a sagittal section of the forebrain and the nasal cavity showing the location of olfactory receptors.

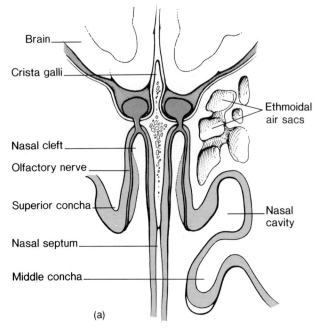

Brain

Crista galli

Nasal cleft

Olfactory nerve

Superior concha

Nasal septum

Middle concha

Ethmoidal air sacs

Nasal cavity

(a)

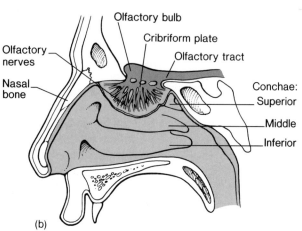

Olfactory bulb

Cribriform plate

Olfactory tract

Olfactory nerves

Nasal bone

Conchae:
Superior
Middle
Inferior

(b)

Only about 2% of the air you inhale comes in contact with the olfactory receptors, which are positioned above the main stream of air flow. Olfactory sensitivity can be increased by forceful sniffing, which draws the air into contact with the receptors.

Certain chemicals activate the trigeminal (fifth) as well as the olfactory (first) cranial nerves and cause reactions. Pepper, for example, may cause sneezing; onions cause the eyes to water; and smelling salts (ammonium salts) initiate respiratory reflexes and are used to revive unconscious persons.

1. What are olfactory hairs? Where are they located?
2. Trace the pathway of an olfactory stimulation from the olfactory hairs to the cerebrum, where interpretation occurs.

Gustatory Sense

Taste receptors are specialized epithelial cells clustered together into taste buds that respond to chemical stimuli and transmit the sense of taste through the glossopharyngeal (ninth cranial) nerve or the facial (seventh cranial) nerve to the cortex of the parietal cerebral lobe for interpretation.

Objective 14. List the three types of papillae, and explain how they function in the perception of taste.

Objective 15. Identify the cranial nerves and the afferent pathways of gustation.

The receptors of taste are located in the **taste buds.** Taste buds are specialized sensory organs that are most numerous on the dorsum of the tongue but are also present on the soft palate and oropharynx. The cylindrical taste bud is composed of many sensory **gustatory cells** that are encapsuled by **supporting cells** (fig. 17.18). Each gustatory cell contains a dendritic ending called a **gustatory hair** that projects to the surface through an opening in the taste bud called the **taste pore.** The gustatory hairs are the sensitive portion of the receptor cells. Saliva provides a moistened environment necessary for a chemical stimulus to activate the gustatory cells.

Taste buds are elevated by surrounding connective tissue and epithelium to form *papillae (pah-pil'ah)* (fig. 17.18). Three types of papillae can be identified:

1. **Circumvallate** *(ser''kum-val'āt)* **papillae.** The largest but least numerous are the circumvallate papillae, which are arranged in an inverted V-shape pattern on the back of the tongue.
2. **Fungiform papillae.** Knoblike fungiform papillae are present on the tip and sides of the tongue.
3. **Filiform papillae.** Short, thickened, threadlike filiform papillae are located on the anterior two-thirds of the tongue.

There are only four basic modalities of taste, known as gustation *(gus-ta'shun)* which are sensed most acutely on particular regions of the tongue (fig. 17.19). These are *sweet* (tip of tongue), *sour* (sides of tongue), *bitter* (back of tongue), and *salty* (over most of the tongue).

gustatory: L. *gustare*, to taste
papilla: L. *papilla*, nipple

Figure 17.18. Papillae of the tongue and associated taste buds. (*a*) numerous taste buds are positioned within each papilla. (*b*) each gustatory cell and its associated gustatory hair is encapsuled by supporting cells. (*c*) a photomicrograph of some taste buds.

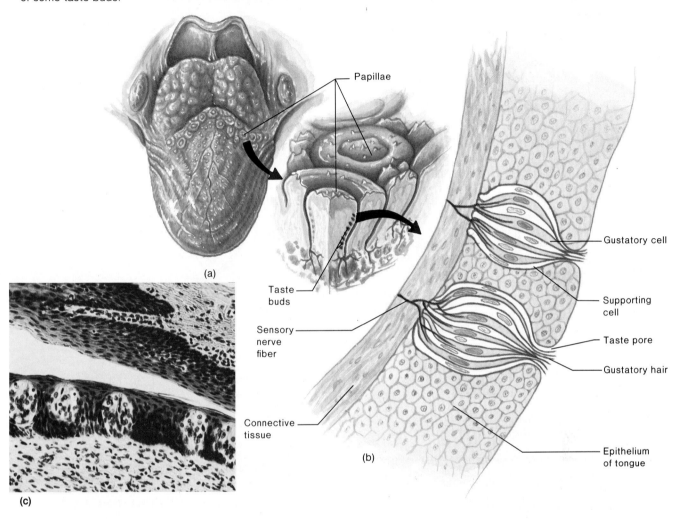

Figure 17.19. Patterns of taste receptor distribution on the dorsum of the tongue. (*a*) sweet receptors; (*b*) sour receptors; (*c*) salt receptors; (*d*) bitter receptors.

Sour taste is produced by hydrogen ions (H^+); all acids therefore taste sour. Most organic molecules, particularly sugars, taste sweet to varying degrees. Only pure table salt (NaCl) has a pure salty taste. Other salts, such as KCl (commonly used in place of NaCl by people with hypertension), taste salty but have bitter overtones. Bitter taste is evoked by quinine and seemingly unrelated molecules.

The afferent pathway of taste receptors to the brain involves mainly two cranial nerves (fig. 17.20). Taste buds on the posterior one-third of the tongue have a sensory pathway through the glossopharyngeal nerve, whereas the anterior two-thirds of the tongue is served by the chorda tympani branch of the facial nerve.

Taste sensations passing through the nerves just mentioned are conveyed through the medulla and thalamus to the parietal lobe of the cerebral cortex where they are interpreted.

Because taste and smell are both chemoreceptors, they complement each other. We often confuse a substance's smell with its taste; and if we have a head cold or hold our nose while eating, food seems to lose its flavor.

1. Distinguish between papillae, taste buds, and gustatory cells. Discuss the function of each as related to taste.
2. List the three types of papillae.
3. Which cranial nerves have sensory innervation associated with taste? What are the afferent pathways to the brain where perception of taste occurs?

Visual Sense

Rods and cones are the photoreceptors within the eyeball that are sensitive to light waves and are stimulated to transmit nerve impulses through the visual cortex of the occipital lobes where interpretation of vision occurs.

Objective 16. Describe the associated structures of the eye and the structure of the eyeball.

Objective 17. Trace the path of light waves through the eye, and explain how they are focused on distant and near objects.

Objective 18. Describe the neural pathway of a visual impulse, and discuss the neural processing of visual information.

The eyes are organs that refract (bend) and focus the incoming light waves onto the sensitive photoreceptors at the back of each eye. Nerve impulses from the stimulated photoreceptors are conveyed through visual pathways within the brain to the occipital lobes of the cerebrum where the sense of vision is perceived. The specialized

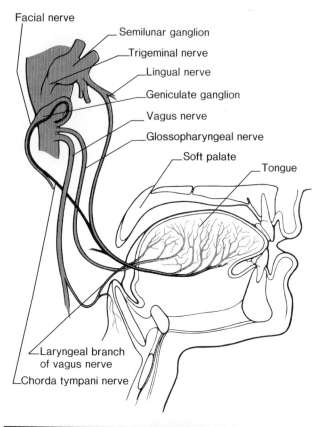

Figure 17.20. The principal innervation of the tongue. Viscerosensory innervation (taste) is provided by the facial and glossopharyngeal nerves and motor innervation by the glossopharyngeal nerve. Branches from the vagus nerves and the trigeminal nerves also provide some sensory innervation.

Facial nerve
Semilunar ganglion
Trigeminal nerve
Lingual nerve
Geniculate ganglion
Vagus nerve
Glossopharyngeal nerve
Soft palate
Tongue
Laryngeal branch of vagus nerve
Chorda tympani nerve

Table 17.3 Comparisons between eye structures and analogous structures in a camera

Eye structure	Analogous camera structure
Cornea and lens	Lens system
Iris and pupil	Variable aperture system
Eyelid	Lens cap
Sclera	Camera frame
Pigment epithelium and choroid	Black interior of camera
Retina*	Film

*Since neural processing begins in the retina, this eye structure may also be considered analogous (in part) to the photographer!

Source: Fox, Stuart Ira, *Human Physiology.* © 1984 Wm. C. Brown Publishers, Dubuque, Iowa. All Rights Reserved. Reprinted by permission.

photoreceptor cells can respond to an incredible one billion different stimuli each second. Further, these cells are sensitive to about 10 million gradations of light intensity and 7 million different shades of color.

The eyes are anteriorly positioned on the skull and set just far enough apart to achieve *binocular (stereoscopic) vision* when focusing on an object. This three-dimensional perspective allows a person to assess depth. Often likened to a camera (table 17.3) the eyes are responsible for approximately 80% of all knowledge that is assimilated.

Figure 17.21. Accessory structures of the eyeball as seen in a sagittal section of the eyelids and an anterior portion of the eyeball within the orbit.

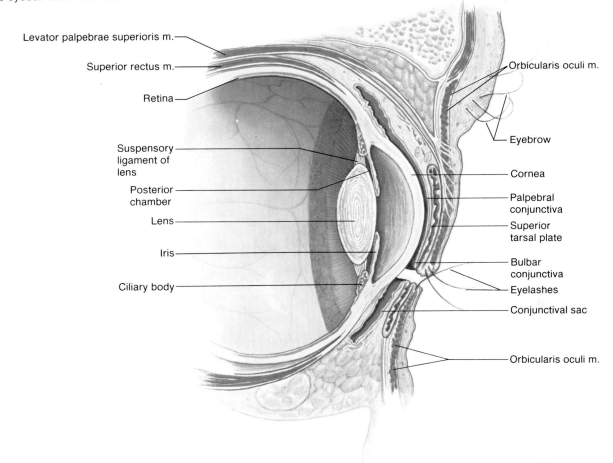

Levator palpebrae superioris m.
Superior rectus m.
Retina
Suspensory ligament of lens
Posterior chamber
Lens
Iris
Ciliary body

Orbicularis oculi m.
Eyebrow
Cornea
Palpebral conjunctiva
Superior tarsal plate
Bulbar conjunctiva
Eyelashes
Conjunctival sac
Orbicularis oculi m.

The eyes of other vertebrates are basically similar to yours. Certain species, however, have adaptive modifications. Consider, for example, the extremely keen eyesight of a hawk, which soars high in the sky searching for food, or the eyesight of a nocturnal animal such as an owl, which feeds only at night. Note how the location of the eyes on the head corresponds to behavior. Predatory species, like cats, have eyes that are directed forward, allowing depth perception. Prey species, like deer, have eyes positioned high on the sides of their heads, allowing panoramic vision to detect distant threatening movements, even while grazing.

Associated Structures of the Eye

Associated structures of the eye either protect the eyeball or provide eye movement. Protective structures include the bony orbit, eyebrow, facial muscles, eyelids, eyelashes, conjunctiva, and the lacrimal apparatus that produces tears. Eyeball movements are made possible by the actions of the extrinsic ocular eye muscles that arise from the orbit and insert on the outer layer of the eyeball.

Orbit. Each eyeball is positioned in a bony depression in the skull called the orbit (see fig. 9.15 and table 9.5). Seven bones of the skull (frontal, lacrimal, ethmoid, zygomatic, maxilla, sphenoid, and palatine) form the walls of the orbit that support and protect the eye.

Eyebrow. Eyebrows consist of short, thick hair positioned transversely above both eyes along the superior orbital ridges of the skull (see figs. 12.6 and 17.21). Located here they effectively shade the eyes from the sun and prevent perspiration or falling objects from getting into the eyes. Underneath the skin of each eyebrow is the orbital portion of the orbicularis oculi muscle and a portion of the corrugator muscle (see fig. 11.30). Contraction of either of these muscles causes the eyebrow to move, often reflexively, to protect the eye.

Eyelids and Eyelashes. Eyelids (**palpebrae**) develop as reinforced folds of skin with attached skeletal muscle so that they are movable. In addition to the orbicularis oculi

palpebra: L. *palpebra,* eyelid (related to *palpare,* to pat gently) muscle attached to the skin that surrounds the front of the eye, the **levator palpebrae** (le-va'tor pal'pĕ-bre) **superioris** muscle attaches along the upper eyelid and provides it with greater movability than the lower eyelid. Contraction of the orbicularis oculi muscle closes the eyelids over the eye, and contraction of the levator palpebrae superioris muscle elevates the upper eyelid to expose the eye. The eyelids protect the eyeball from desiccation by reflexively blinking about every seven seconds and moving fluid across the anterior surface of the eyeball. To avoid a blurred image, the eyelid will blink when the eyeball moves to a new position of fixation.

The **palpebral fissure** (see fig. 12.7) is the interval between the upper and lower eyelids. The **commissures (canthi)** of the eye are the medial and lateral angles where the eyelids come together. The **medial commissure** is broader than the **lateral commissure** and is characterized by a small, reddish, fleshy elevation called the **lacrimal caruncle** (kar'ung-kl) (see fig. 17.22), which contains sebaceous and sudoriferous glands. The shape of the palpebral fissure is elliptical when the eyes are open.

Each eyelid supports a row of numerous **eyelashes** that protect the eye from airborne objects. The shaft of each eyelash is surrounded by a root hair plexus that provides the hair with the sensitivity necessary to elicit a reflexive closure of the lids. Eyelashes of the upper lid are long and turn upward, whereas those of the lower lid are short and turn downward.

In addition to the layers of the skin and the underlying connective tissue and orbicularis oculi muscle fibers, each eyelid contains a tarsal plate, tarsal glands, and conjunctiva. The **tarsal plates,** composed of dense fibrous connective tissue, are important in maintaining the shape of the eyelids (fig. 17.21). Specialized sebaceous glands called **tarsal meibomian** (mi-bo'me-an) **glands** are embedded within the tarsal plates along the exposed inner surfaces of the eyelids. The ducts of the tarsal glands open onto the edges of the eyelids, and their oily secretions help keep the eyelids from adhering to each other. A *chalazion* (kah-la'ze-on) is a tumor or cyst on the eyelid that results from an infection of the tarsal glands. Modified sweat glands called **ciliary glands** are also located within the eyelids along with additional sebaceous glands at the bases of the hair follicles of the eyelashes. An infection of these sebaceous glands is referred to as a *sty.*

commissure: L. *commissura,* a joining
caruncle: L. *caruncula,* diminutive of caro, flesh
tarsal: Gk. *tarsos,* flat basket
meibomian glands: from Heinrich Meibom, German anatomist, 1638–1700
chalazion: Gk. *chalazion,* hail; a small tubercle

Figure 17.22. The lacrimal apparatus. The lacrimal gland produces lacrimal fluid, which moistens and cleanses the conjunctiva that lines the underside of the eyelids and covers the anterior surface of the eyeball.

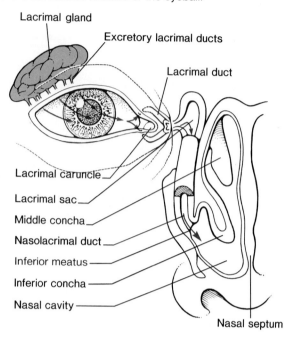

Lacrimal gland
Excretory lacrimal ducts
Lacrimal duct
Lacrimal caruncle
Lacrimal sac
Middle concha
Nasolacrimal duct
Inferior meatus
Inferior concha
Nasal cavity
Nasal septum

Conjunctiva. The conjunctiva is a thin mucus-secreting epithelial membrane that lines the interior surface of each eyelid and exposed anterior surface of the eyeball (fig. 17.21). It consists of stratified squamous epithelium that varies in thickness in different regions. The **palpebral conjunctiva** is thick and adheres to the tarsal plates of the eyelids. As the conjunctiva reflects onto the anterior surface of the eyeball, it is known as the **bulbar conjunctiva.** This portion is transparent and especially thin where it covers the cornea. Because the conjunctiva is continuous and reflects from the eyelids to the anterior surface of the eyeball, a space called the **conjunctival sac** exists when the eyelids are closed. The conjunctival sac protects the eyeball by preventing objects (including a contact lens) from passing beyond the confines of the sac. The conjunctiva can repair itself rapidly if it is scratched.

Lacrimal Apparatus. The lacrimal apparatus consists of the lacrimal gland, which secretes the *lacrimal fluid* (tears), and a series of ducts that drain the secretion into the nasal cavity (fig. 17.22). The **lacrimal gland,** which is about the size and shape of an almond, is located in the superolateral portion of the orbit. It is a compound tubuloacinar gland that secretes lacrimal fluid through several excretory lacrimal ducts into the conjunctival sac of the upper eyelid. With each blink of the eyelids, lacrimal

Figure 17.23.   The extrinsic ocular muscles of the left eyeball.

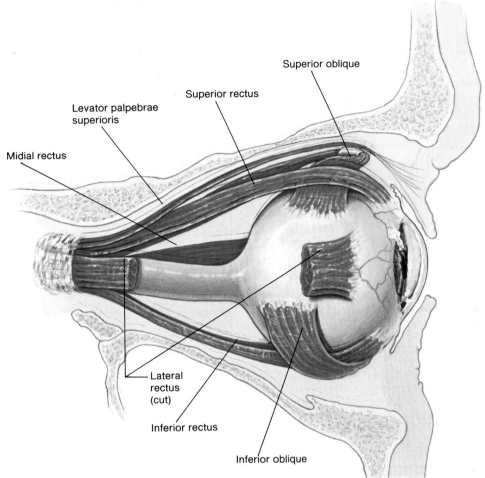

fluid passes medially and downward and drains into two small openings, called **puncta lacrimalia,** on both sides of the lacrimal caruncle. From here the lacrimal fluid drains through the lacrimal duct into the lacrimal sac and continues through the nasolacrimal duct to the inferior meatus of the nasal cavity.

Lacrimal fluid is an aqueous, mucus secretion that contains a bactericidal enzyme called *lysozyme.* Lacrimal fluid not only moistens and lubricates the conjunctival sac but also reduces the chance of eye infections. Normally about one milliliter of lacrimal fluid is produced each day by the lacrimal gland of each eye. If irritating substances, such as particles of sand or chemicals from onions, make contact with the conjunctiva, the lacrimal glands are stimulated to oversecrete. The extra lacrimal fluid protects the eye by diluting and washing away the irritating substance.

Humans are the only animals known to weep in response to emotional stress. While crying, the volume of lacrimal secretion is so great that the tears may spill over the edges of the eyelids and the nasal cavity fill with fluid. The crying response is an effective means of communicating one's emotions and results from stimulation of the lacrimal glands with parasympathetic motor impulses.

Extrinsic Eye Muscles.   The movements of the eyeball are controlled by six extrinsic eye muscles called the **extrinsic ocular muscles** (fig. 17.23). Each extrinsic ocular muscle originates from the bony orbit and inserts by a tendinous attachment to the tough outer tunic of the eyeball.

Four **recti muscles** maneuver the eyeball in the direction indicated by their names (**superior, inferior, lateral,** and **medial**), and two **oblique muscles** (**superior** and **inferior**) rotate the eyeball on its axis. One of the extrinsic ocular muscles, the superior oblique, passes through a pulleylike cartilaginous loop, called the **trochlea** *(trok'le-ah),* before attaching to the eyeball. Although stimulation of each muscle causes a precise movement of the eyeball, most of the movements involve the combined contraction of usually two muscles.

The motor units of the intrinsic ocular muscles are the smallest in the body. This means that a single motor neuron serves about ten muscle fibers, resulting in precision movements. The eyes move in synchrony by contracting synergistic muscles while relaxing antagonistic muscles.

The extrinsic ocular muscles are innervated by three cranial nerves (table 17.4). Innervation of the other skeletal and smooth muscles that serve the eye is also indicated in table 17.4.

A physical examination may include an eye movement test. As the patient's eyes follow the movement of a physician's finger, the physician can assess weaknesses in specific muscles or dysfunctions of specific cranial nerves. The patient experiencing double vision (diplopia) when moving his eyes may be suffering from muscle weakness. Looking laterally tests the abducens cranial nerve; looking inferiorly and medially tests the trochlear cranial nerve; and looking cross-eyed tests the oculomotor and trochlear cranial nerves of both eyes.

Structure of the Eyeball

The eyeball of an adult is essentially spherical and is approximately 25 mm (1 in.) in diameter. About four-fifths of the eyeball is positioned within the orbit of the skull. The eyeball consists of three basic layers: the fibrous tunic, the vascular tunic, and the internal tunic, or retina (fig. 17.24).

Fibrous Tunic.   The fibrous tunic is the outer layer of the eyeball. It is divided into two regions: the posterior five-sixths is the opaque sclera, and the anterior one-sixth is the transparent cornea.

The toughened **sclera** is the "white of the eye." It is composed of tightly bound elastic and collagenous fibers that give shape to the eyeball and protect its inner structures. The sclera is avascular but does contain sensory receptors for pain. The large **optic nerve** exits through the sclera at the posterior portion of the eyeball.

---

trochlea: Gk. *trochos,* a wheel
sclera: Gk. *skleros,* hard

Table **17.4** Muscles of the eye		
Extrinsic muscles (striated)		
Superior rectus	Oculomotor nerve (III)	Rotates eye upward and toward midline
Inferior rectus	Oculomotor nerve II/ (III)	Rotates eye downward and toward midline
Medial rectus	Oculomotor nerve (III)	Rotates eye toward midline
Lateral rectus	Abducens nerve (VI)	Rotates eye away from midline
Superior oblique	Trochlear nerve (IV)	Rotates eye downward and away from midline
Inferior oblique	Oculomotor nerve (III)	Rotates eye upward and away from midline
Intrinsic muscles (smooth)		
Ciliary muscles	Oculomotor nerve (III) parasympathetic fibers	Causes suspensory ligaments to relax
Iris, circular muscles	Oculomotor nerve (III) parasympathetic fibers	Causes size of pupil to decrease
Iris, radial muscles	Sympathetic fibers	Causes size of pupil to increase

Source: Fox, Stuart Ira, *Human Physiology.* © 1984 Wm. C. Brown Publishers, Dubuque, Iowa. All Rights Reserved. Reprinted by permission.

The **cornea** is transparent and convex to permit the passage and cause the refraction (bending) of incoming light waves. The transparency of the cornea is due to tightly packed, avascular, dense connective tissue. Also, the relatively few cells that are present in the cornea are arranged in unusually regular patterns. The circumferential edge of the cornea is continuous structurally with the sclera. The outer surface of the cornea is covered with a thin, nonkeratinized, stratified squamous epithelial layer called the **bulbar (corneal) epithelium,** which is actually a continuation of the conjunctiva of the sclera (fig. 17.21).

Vascular Tunic.   The vascular tunic, or **uvea** *(u've-ah)* of the eyeball consists of the choroid, the ciliary body, and the iris (fig. 17.24).

The **choroid** is a thin, highly vascular layer that lines most of the internal surface of the sclera. The choroid contains numerous pigment-producing melanocytes, which give it a dark brownish color to prevent light waves from being reflected out of the eyeball. There is an opening in the choroid at the back of the eyeball where the optic nerve is located.

---

cornea: L. *cornu,* horn
uvea: L. *uva,* grape
choroid: Gk. *chorion,* membrane

Figure 17.24. The internal anatomy of the eyeball.

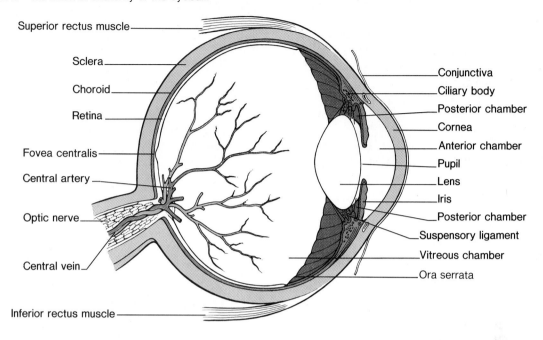

Superior rectus muscle

Sclera

Choroid

Retina

Fovea centralis

Central artery

Optic nerve

Central vein

Inferior rectus muscle

Conjunctiva

Ciliary body

Posterior chamber

Cornea

Anterior chamber

Pupil

Lens

Iris

Posterior chamber

Suspensory ligament

Vitreous chamber

Ora serrata

Figure 17.25. The detailed structure of the anterior portion of the eyeball.

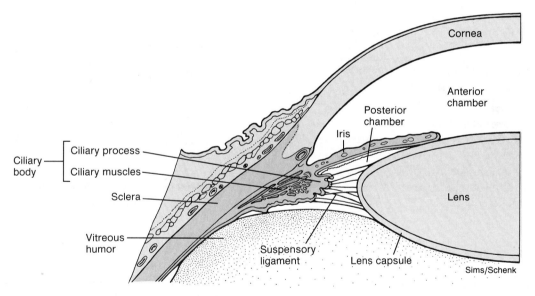

Ciliary body

Ciliary process

Ciliary muscles

Sclera

Vitreous humor

Cornea

Anterior chamber

Posterior chamber

Iris

Lens

Suspensory ligament

Lens capsule

Sims/Schenk

The **ciliary body** is the thickened, anterior portion of the vascular tunic that forms an internal muscular ring toward the front of the eyeball (figs. 17.25, 17.26). Three distinct planes of smooth muscle fibers called **ciliary muscles** are found within the ciliary body.

Numerous extensions of the ciliary body called **ciliary processes** attach to the **zonular fibers,** which in turn attach to the lens. Collectively, the zonular fibers constitute the **suspensory ligament.** The transparent **lens** consists of tight layers of protein fibers arranged like the layers of an onion. A thin, clear **lens capsule** encloses the lens and provides it with a constant environment. The suspensory ligament attaches to the margin of the lens capsule.

---

zonular fibers: L. *zona*, a girdle

Figure 17.26.   The ciliary body, suspensory ligaments, and the lens viewed from posterior to anterior.

Ciliary processes of ciliary body

Lens

Ora serrata

Suspensory ligaments

Figure 17.27.   Changes in the shape of the lens. (a) the lens is flattened for distant vision when the ciliary muscle fibers are relaxed and the suspensory ligaments are taut. (b) the lens is more spherical for closeup vision when the ciliary muscle fibers are contracted and the suspensory ligaments are relaxed.

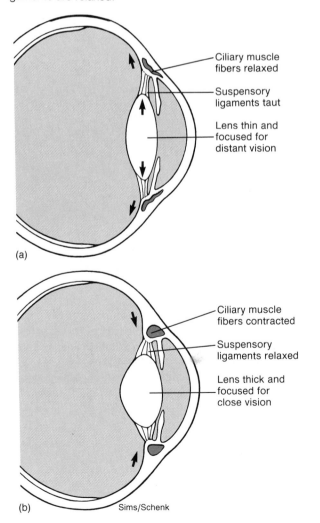

Ciliary muscle fibers relaxed

Suspensory ligaments taut

Lens thin and focused for distant vision

(a)

Ciliary muscle fibers contracted

Suspensory ligaments relaxed

Lens thick and focused for close vision

(b)   Sims/Schenk

Figure 17.28.   Dilation and constriction of the pupil. In dim light, the radially arranged smooth muscle fibers are stimulated to contract by sympathetic stimulation, dilating the pupil. In bright light, the circularly arranged smooth muscle fibers are stimulated to contract by parasympathetic stimulation, constricting the pupil.

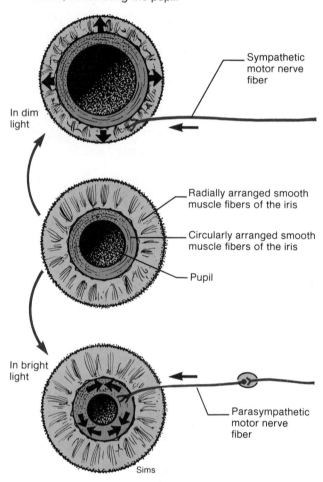

In dim light

Sympathetic motor nerve fiber

Radially arranged smooth muscle fibers of the iris

Circularly arranged smooth muscle fibers of the iris

Pupil

In bright light

Parasympathetic motor nerve fiber

Sims

The shape of the lens determines how much the light waves that pass through will be refracted. Constant tension of the suspensory ligament, when the ciliary muscles are relaxed, flattens the lens somewhat (fig. 17.27). Contraction of the ciliary muscles relaxes the suspensory ligament and makes the lens more spherical. The constant tension within the lens capsule results in the surface of the lens becoming more convex when the suspensory ligament is not taut. A flattened lens facilitates viewing a distant object, whereas a rounded lens permits viewing a close object.

The **iris** is the anterior portion of the vascular tunic and is continuous with the choroid. The iris is viewed from the outside as the colored portion of the eyeball (figs. 17.24, 17.25). It consists of smooth muscle fibers arranged in a circular and a radial pattern. The contraction of the smooth muscle fibers regulates the diameter of the **pupil** (table 17.4 and fig. 17.28), an opening in the center of the iris. Contraction of the circularly arranged fibers of the iris, stimulated by bright light, constricts the pupil and

Figure 17.29. The layers of the retina. (*a*) a schematic view (the arrow indicates the direction through which light travels to stimulate the photoreceptors); (*b*) a micrograph of the retina.

(b)

diminishes the amount of light entering the eyeball. Contraction of the radially arranged fibers, in response to dim light, dilates the pupil and permits more light to enter.

Retina. The retina covers the choroid as the innermost layer of the eyeball. It consists of a thin, outer **pigmented layer,** or visual portion. The retina is principally in the posterior portion of the eyeball (fig. 17.24). The visual layer of the retina terminates in a jagged margin near the ciliary body called the **ora serrata.** The pigmented layer extends anteriorly over the back of the ciliary body and iris.

> The pigmented layer and nervous layer of the retina are not attached to each other except surrounding the optic nerve and at the ora serrata. Because of this loose connection, the two layers may become separated as a detached retina. Such a separation can be corrected by fusing the layers with a laser.

The nervous layer of the retina is composed of three principal layers of neurons. Listing them in the order in which they conduct impulses, they are rod and cone cells, bipolar neurons, and ganglion neurons (fig. 17.29).

ora serrata: L. *ora*, margin; *serra*, saw

Rods and **cones** are photoreceptors. Rods number over 100 million per eye and are more slender and elongated than cones (fig. 17.30). Rods are positioned on the peripheral parts of the retina where they respond to dim light for black-and-white vision. They also respond to form and movement but provide poor visual acuity. Cones number about 7 million per eye. They provide daylight color vision and are responsible for visual acuity.

Cones are concentrated in a depression near the center of the retina called the **fovea centralis,** which is the area of keenest vision (figs. 17.24, 17.31). Surrounding the fovea centralis is the yellowish **macula lutea,** which also has an abundance of cones (fig. 17.32). There are no photoreceptors at the point where the optic nerve is attached. This area is a "blind spot" and is referred to as the **optic disc.** Normally a person is unaware of the blind spot because the eyes continually move about and because an object is viewed from a different angle with each eye. The blind spot can easily be demonstrated as described in figure 17.33.

fovea: L. *fovea*, small pit
macula lutea: L. *macula*, spot; *luteus*, yellow

Figure 17.30. Receptor cells of the eyeball.

Rod cell   Cone cell

Mitochondria

Nuclei

Synaptic endings

Figure 17.31. A diagram of the retina sectioned through the fovea centralis.

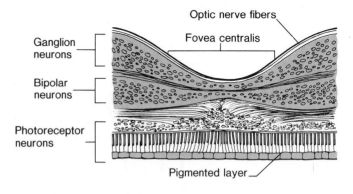

Optic nerve fibers

Fovea centralis

Ganglion neurons

Bipolar neurons

Photoreceptor neurons

Pigmented layer

Figure 17.32. A view of the retina as seen with an ophthalmoscope. Optic nerve fibers leave the eyeball at the optic disc to form the optic nerve. Note the blood vessels that can be seen entering the eyeball at the optic disc.

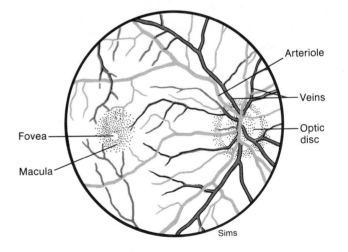

Arteriole

Veins

Optic disc

Fovea

Macula

Sims

Figure 17.33. The blind spot. Hold the drawing about twenty inches from your face, with your left eye closed and your right eye focused on the circle. Slowly move the drawing closer to your face until the cross disappears. This occurs because the image of the cross is focused on the optic disc, where photoreceptors are absent.

Figure 17.34. Aqueous humor maintains the intraocular pressure within the anterior and posterior chambers. It is secreted into the posterior chamber, flows through the pupil into the anterior chamber, and drains from the eyeball through the venous sinus.

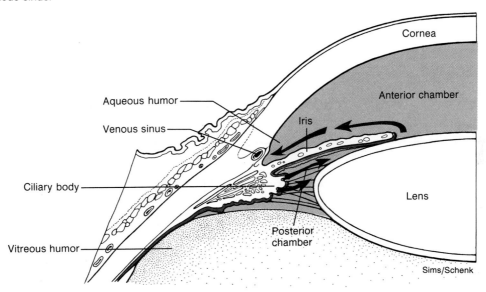

Sims/Schenk

Blood Supply to the Eyeball. Both the choroid and the retina are richly supplied with blood. Two **ciliary arteries** pierce the sclera at the posterior aspect of the eyeball and traverse the choroid to the ciliary body and base of the iris. Although the ciliary arteries enter the eyeball independently, they anastomose (connect) extensively throughout the choroid.

The **central artery** (**central retinal artery**) branches from the ophthalmic artery and enters the eyeball in contact with the optic nerve. As the central artery passes through the optic disc, it divides into superior and inferior branches, each of which then divides into temporal and nasal branches to serve the inner layers of the retina (fig. 17.24). The **central vein** drains blood from the eyeball through the optic disc. The branches of the central artery can be observed within the eyeball through an ophthalmoscope (fig. 17.32).

An internal examination of the eyeball with an ophthalmoscope is frequently part of a routine physical examination. Capillary vessels, which can be seen within the eyeball, can reveal certain diseases or body dysfunctions. Diseases such as arteriosclerosis, diabetes, cataracts, or glaucoma can be detected.

Chambers of the Eyeball. The interior of the eyeball is separated by the lens into two main cavities (see fig. 17.24). The cavity anterior to the lens is subdivided by the iris into an **anterior chamber** and a **posterior chamber.** The anterior chamber is located in front of the iris and behind the cornea. The posterior chamber is located between the iris and the suspensory ligaments and the lens. The anterior and posterior chambers connect through the pupil and are filled with a watery fluid called **aqueous humor.** The constant production of aqueous humor maintains an *intraocular (in''trah-ok'u-lar) pressure* of about 24 mm of mercury within the anterior and posterior chambers. Aqueous humor also provides nutrients and oxygen to the avascular lens and cornea. An estimated 5–6 ml of aqueous humor is secreted each day from the vascular epithelium of the ciliary body (fig. 17.34). From its site of secretion within the posterior chamber, the aqueous humor passes through the pupil into the anterior chamber. From here, it drains from the eyeball through the **venous sinus** *(ve'nus si'nus)* (**canal of Schlemm**) into the bloodstream. The venous sinus is located at the junction of the cornea and iris.

canal of Schlemm: from Friedrich S. Schlemm, German anatomist, 1795–1858

Table 17.5 Location and functions of the structures of the eyeball

Tunic and structure	Location	Composition	Function
Fibrous tunic	Outer layer of eyeball	Avascular connective tissue	Provides shape of eyeball
Sclera	Posterior, outer layer; "white of the eye"	Tightly bound elastic and collagen fibers	Supports and protects eyeball
Cornea	Anterior surface of eyeball	Tightly packed dense connective tissue—transparent and convex	Transmits and refracts light
Vascular tunic (uvea)	Middle layer of eyeball	Highly vascular pigmented tissue	Supplies blood; prevents reflection
Choroid	Middle layer in posterior portion of eyeball	Vascular layer	Supplies blood to eyeball
Ciliary body	Anterior portion of vascular tunic	Smooth muscle fibers and glandular epithelium	Supports the lens through suspensory ligaments and determines its shape; secretes aqueous humor
Iris	Anterior portion of vascular tunic continuous with ciliary body	Pigment cells and smooth muscle fibers	Regulates the diameter of the pupil and hence the amount of light entering the vitreous chamber
Retina	Inner layer in posterior portion of eyeball	Photoreceptor neurons (rods and cones), bipolar neurons, and ganglion neurons	Photoreception; transmits impulses
Lens (not part of any tunic)	Between posterior and vitreous chambers; supported by suspensory ligaments of ciliary body	Tightly arranged protein fibers; transparent	Refracts light and focuses onto fovea centralis

Source: Van De Graaff, Kent M., *Human Anatomy.* © 1984 Wm. C. Brown Publishers, Dubuque, Iowa. All Rights Reserved. Reprinted by permission.

Between the lens and retina is the large **vitreous chamber,** which is filled with a transparent jellylike **vitreous humor.** Vitreous humor also contributes to the intraocular pressure to maintain the shape of the eyeball and to hold the retina against the choroid. Unlike aqueous humor, vitreous humor is not continuously produced but formed prenatally.

The location and functions of the structures within the eyeball are summarized in table 17.5.

> Puncture wounds to the eyeball are especially dangerous and frequently cause blindness. Protective equipment such as goggles, shields, and shatterproof lenses should be used in hazardous occupations and certain sports. If the eye is punctured, the principal thing to remember is to **not remove the object** if it is still impaling the eyeball. Removal may allow the fluids to drain from the eyeball and cause the loss of the intraocular pressure. This is particularly serious if the nonreplaceable vitreous humor drains from the eyeball.

Physiology of Vision

The focusing of light waves and stimulation of photoreceptors of the retina require five basic processes:

1. The transmission of light waves through transparent media of the eyeball.
2. The refraction of light waves through media of different densities.
3. Accommodation of the lens to focus the light waves.

---

vitreous: L. *vitreus,* glassy

4. Constriction of the pupil by the iris to regulate the amount of light entering the vitreous chamber.
5. Convergence of the eyeballs so that visual acuity is maintained.

Visual impairment may exist if one or more of these processes does not function properly (see clinical considerations).

Transmission of Light Waves. Light waves entering the eyeball pass through four transparent media before they stimulate the photoreceptors. In sequence, the media through which light waves pass are the cornea, aqueous humor, lens, and vitreous humor. The cornea and lens are solid media composed of tightly packed, avascular protein fibers. An additional thin, transparent membranous continuation of the conjunctiva covers the outer surface of the cornea. The aqueous humor is a low viscosity fluid and the vitreous humor is jellylike in consistency.

Refraction of Light Waves. Refraction (re-frak'shun) is the bending of light waves. Refraction occurs as light waves pass at an oblique angle from a medium of one optical density to a medium of another optical density (fig. 17.35). The convex cornea is the principal refractive medium; the fluids within the various chambers produce minimal refraction. The lens is particularly important for refining and altering refraction. Of the refractive media, only the lens can be altered in shape to achieve precise refraction.

The refraction of light waves is so extensive that the visual image is formed upside down on the retina (fig. 17.36). Nerve impulses of the image in this position are relayed to the visual cortex of the occipital lobe where the inverted image is interpreted as right side up.

Figure 17.35. The shape of a lens determines the refraction of the light waves that pass through. (*a*) a lens with a convex surface causes light waves to converge to a focal point. (*b*) a lens with a concave surface causes light waves to diverge.

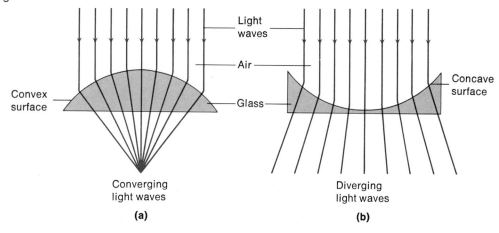

(a)   **(b)**

Figure 17.36. The refraction of light waves within the eyeball causes the image of an object to be inverted on the retina.

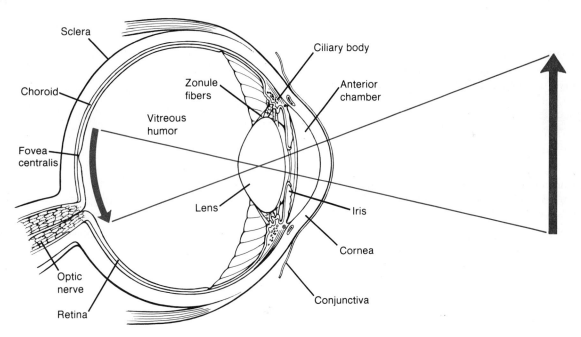

Accommodation of the Lens. Accommodation *(ah-kom"o-da'shun)* is the automatic adjustment of the curvature of the lens by contraction of ciliary muscles to bring light waves into sharp focus on the retina. The lens of the eyeball is biconvex. When an object is viewed closer than about twenty feet, the lens must make an adjustment, or accommodation, if a clear focus is to appear on the retina. Contraction of the smooth muscle fibers of the ciliary body causes the suspensory ligament to relax and the lens to become thicker (see fig. 17.27). A thicker, more convex lens causes a greater convergence of the refracted light waves (see fig. 17.35) necessary for viewing close objects.

Constriction of the Pupil. Constriction of the pupil through parasympathic stimulation that causes contraction of the circular muscle fibers of the iris (see fig. 17.28) is important for two reasons. One is the reduction in the amount of light that enters the vitreous chamber. A reflexive pupil constriction protects the retina from sudden or intense bright light. Another more important function is that a reduced pupil diameter prevents light waves from entering the vitreous chamber through the periphery of the lens. Light waves refracted from the periphery would not be brought into focus on the retina and would cause blurred vision. Autonomic constriction of the pupil and accommodation of the lens occur simultaneously.

Figure 17.37. The electromagnetic spectrum (*top*) is shown in Angstrom units (1Å = 10^{-10} meter). The visible spectrum comprises only a small range of this spectrum (*bottom*), shown in nanometer units (1 nm = 10^{-9} meter).

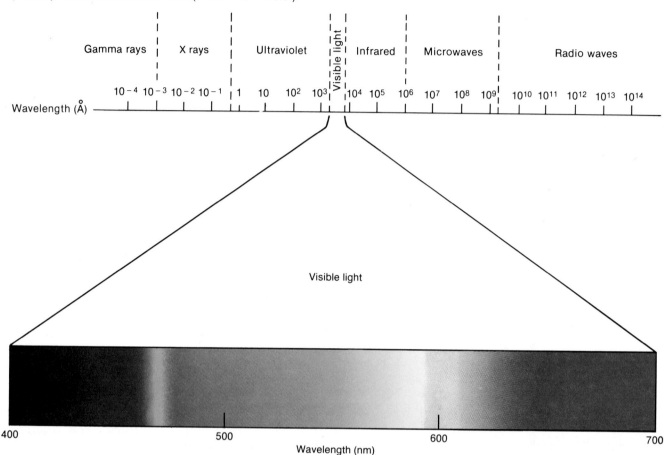

Convergence of the Eyeballs.

Convergence of the eyeballs means that the eyes rotate more medially when viewing closer objects. In fact, focusing on an object close to the tip of the nose causes a person to appear cross-eyed. The eyeballs must converge when viewing close objects because only then can the light waves focus on the same portions in both retinas.

Visual Spectrum.

The eyes transduce energy in the *electromagnetic spectrum* (fig. 17.37) into nerve impulses. Only a limited part of this spectrum can excite the photoreceptors. Electromagnetic energy with wavelengths between 400 and 700 nanometers (nm) comprise *visible light*. Light of longer wavelengths, which are in the infrared regions of the spectrum, do not have sufficient energy to excite photoreceptors but are perceived as heat. Ultraviolet light, which has shorter wavelengths and more energy than visible light, is filtered out by the yellow color of the eyes' lens. Certain insects, such as honeybees, and people who have had their lens removed, can see light in the ultraviolet range.

Visual Pigments.

Rods and cones are activated when light produces a chemical change in the molecules of pigment contained within the receptor cells. Rods, for example, contain a purple pigment known as **rhodopsin** *(ro'dop'sin)*. This pigment appears purple (a combination of red and blue) because it transmits light in the red and blue regions of the spectrum and absorbs light energy in the green region. The wavelength of the light that is absorbed best—the *absorption maximum*—is about 500 nm (corresponding to yellowish-green).

Cars and other objects that are yellowish green in color are seen more easily at night (when rods are used for vision) than are red objects. This is because red light is not well absorbed by rhodopsin, and only absorbed light can produce the photochemical reactions that result in vision. In response to absorbed light, rhodopsin dissociates into its two component parts: a pigment called **retinaldehyde** (or retinene) and a protein called **opsin.**

Retinaldehyde can exist in two possible configurations (shapes)—one known as the all-*trans* form and one called the ll-*cis* form (fig. 17.38). The all-*trans* form is the most stable, but only the ll-*cis* form is found attached to opsin. In response to absorbed light energy, the ll-*cis* retinaldehyde is converted to its all-*trans* form, causing it

---

rhodopsin: Gk. *rhodon*, rose; *ops*, eye
opsin: Gk. *ops*, eye

Figure 17.38. The photopigment rhodopsin consists of the protein opsin combined with 11–*cis* retinaldehyde. In response to light the retinaldehyde is converted to a different form, called all-*trans,* and dissociates from the opsin. This photochemical reaction induces changes in ionic permeability that ultimately results in stimulation of ganglion cells in the retina.

to dissociate from the opsin. This dissociation reaction in response to light initiates changes in the ionic permeability of the rod cell membrane and ultimately results in the production of nerve impulses in the ganglion cells.

The photochemical dissociation of rhodopsin is called the **bleaching reaction.** Rods are thus bleached by light; in the dark, new rhodopsin is made, by using either the all-*trans* retinaldehyde left from the bleaching reaction or *vitamin A,* which is a precursor of retinaldehyde. This is why eating carrots (which are rich in vitamin A) can improve *night vision* in people who are vitamin A deficient. In the *dark-adapted eye,* there is a higher content of rhodopsin in the rods, and sight is possible under very dim illumination (as occurs after a few minutes in a dark theater). Under these conditions of low illumination, rods provide black-and-white vision.

Cones are less sensitive than rods to light, but provide *color vision* and greater visual acuity. During the day, therefore, the high light intensity bleaches out the rods, and color vision with high acuity is provided by the cones. There appear to be three types of cones: blue cones, green cones, and red cones. Each type contains a different pigment protein than the other types of cones, all of which differ from the opsin protein contained in rods. These differences in pigment proteins produce differences in the absorption maximum of each pigment (fig. 17.39). Green

Figure 17.39. There are three types of cones. Each type of cone contains retinaldehyde combined with a different type of protein, producing a pigment that absorbs light maximally at a different wavelength. Color vision, according to the trichromatic theory, is produced by activity of these blue cones, green cones, and red cones.

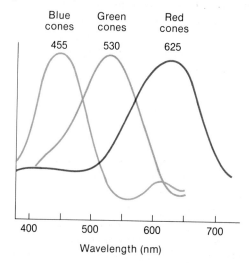

cones, for example, have an absorption maximum in the green region of the spectrum (at 530 nm), and red cones have an absorption maximum at 625 nm in the red region.

Since rods do not absorb red light but red cones do, a red light in a photographic darkroom allows vision (because of the red cones), but does not cause bleaching of the rods. When the light is turned off, therefore, the eyes will still be dark adapted and the person will be able to see (because of the rods).

Our ability to see a multitude of colors with only three types of cones is called the **trichromatic** *(tri″kro-mat′ik)* **theory** of color vision. *Color blindness* is due to a congenital lack of one or more types of cones. In red-green color blindness, for example, red and green appear the same because one of these two types of cones is missing. People with this condition must rely on other cues, such as the relative intensity of light and position, to tell the differences between red and green traffic signals.

Neural Pathways of Vision, Eye Movements, and Processing Visual Information

The photoreceptor neurons, rods and cones, are the functional units of sight in that they respond to light waves and produce nerve impulses. Nerve impulses from the rods and cones pass through bipolar neurons to ganglion neurons (see fig. 17.29). The optic nerve consists of axons of aggregated ganglion neurons that emerge through the posterior aspect of the eyeball. The two optic nerves (one from each eyeball) converge at the **optic chiasma** *(ki-as′mah)* (fig. 17.40). The fibers partially cross at the optic chiasma so that all the fibers arising from the medial (nasal) half of each retina cross to the opposite side. The fibers of the optic nerve that arise from the lateral (temporal) half of the retina do not cross. The **optic tract** is a continuation of optic nerve fibers from the optic chiasma and is composed of fibers arising from the retinas of both eyeballs.

As the optic tracts enter the brain, some of the fibers in the tracts terminate in the **superior colliculi** *(ko-lik′u-li)*. These fibers and the motor pathways they activate constitute the **tectal system,** which is responsible for body-eye coordination.

Approximately 70%–80% of the fibers in the optic tract pass to the **lateral geniculate** *(jĕ-nik′u-lat)* **body** of the thalamus (fig. 17.41). Here the fibers synapse with neurons whose axons constitute a pathway called the **optic radiation.** The optic radiation transmits impulses to the **striate** *(stri′āt)* **cortex** area of the occipital cerebral lobe. This entire arrangement of visual fibers is known as the **geniculostriate system,** which is responsible for perception of the visual field.

Figure 17.40. Visual fields of the eyes and neural pathways for vision. An overlapping of the visual field of each eye provides binocular vision, which is the ability to perceive depth.

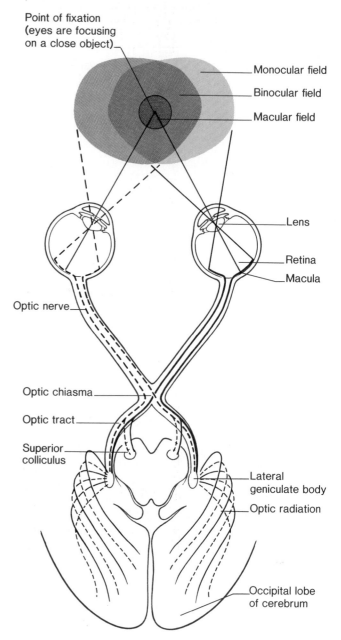

Point of fixation (eyes are focusing on a close object)

Monocular field

Binocular field

Macular field

Lens

Retina

Macula

Optic nerve

Optic chiasma

Optic tract

Superior colliculus

Lateral geniculate body

Optic radiation

Occipital lobe of cerebrum

The nerve fibers that cross at the optic chiasma arise from the retinas in the medial portions of the eyeballs. The photoreceptors of these fibers are stimulated by light entering the eyeball from the periphery. If the optic chiasma were severed longitudinally, peripheral vision would be lost and leave only "tunnel vision." If an optic tract were severed, both eyes would be partially blind, the lateral field of vision lost for one eye and the medial field of vision lost for the other.

Figure 17.41. The lateral geniculate body. Each lateral geniculate consists of six layers (numbered 1 through 6 in this figure). Each of these layers receives input from only one eye, with right and left eyes alternating. An arrow through these six layers (see figure) of the left lateral geniculate, for example, encounters corresponding projections from a part of the right visual field in right and left eyes, alternatively, as it passes from the outer to the inner layers.

The Superior Colliculi and Eye Movements. Neural pathways from the superior colliculi to motor neurons in the spinal cord help mediate the "startle" response to the sight of an unexpected intruder. Other nerve fibers from the superior colliculi stimulate the **extrinsic eye muscles** (table 17.4), which are the striated muscles that move the eyes.

There are two types of eye movements coordinated by the superior colliculi. *Smooth pursuit movements* track moving objects and keep the image focused on the fovea centralis. *Saccadic (sah-kad'ik) eye movements* are short (lasting 20 to 50 msec), jerky movements that occur while the eyes appear to be still. These saccadic movements are believed to be important in maintaining visual acuity.

The tectal system is also involved in the control of the **intrinsic eye muscles**—the iris and the muscles of the ciliary body. Shining a light into one eye stimulates the *pupillary reflex* in which both pupils constrict. This is caused by activation of parasympathetic neurons by fibers from the superior colliculi. Postganglionic neurons in the ciliary ganglia behind the eyes, in turn, stimulate constrictor fibers in the iris. Contraction of the ciliary body during *accommodation* also involves stimulation of the superior colliculi.

Processing of Visual Information. In order for visual information to have meaning it must be associated with past experience and integrated with information from other senses. Some of this higher processing occurs in the inferior temporal lobes of the cortex. Experimental removal of these areas from monkeys impairs their ability

to remember visual tasks that they previously learned and hinders their ability to associate visual images with the significance of the object. Monkeys with their inferior temporal lobes removed, for example, will handle a snake without fear. The symptoms produced by loss of the inferior temporal lobes are known as the *Klüver-Bucy syndrome.*

In an attempt to reduce the symptoms of severe epilepsy, surgeons at one time cut the corpus callosum in some patients. This fiber tract, as previously described, transmits impulses between the right and left cerebral hemispheres. The right cerebral hemisphere of patients with such *split brains,* therefore, receives sensory information from only the left half of the external world. The left hemisphere, similarly cut off from communication with the right hemisphere, receives sensory information from only the right half of the external world.

Experiments with split-brain patients have revealed that the functions of the two hemispheres are different. This is true even though each hemisphere would normally receive input from both halves of the external world through the corpus callosum. If the sensory image of an object, such as a key, is delivered to only the left hemisphere (by showing it to only the right visual field), the object can be named. If the object is presented to the right cerebral cortex, the person knows what the object is but cannot name it. Experiments such as this suggest that (in right-handed people) the left hemisphere is needed for language and the right hemisphere is responsible for pattern recognition (fig. 17.42).

1. List the associated structures of the eye that either cause the eye to move or protect it within the bony orbit.
2. Diagram the structure of the eye, and label the following: sclera, cornea, choroid, retina, fovea centralis, iris, pupil, lens, and ciliary body. What are the principal cells or tissues in each of the three layers of the eye?
3. Trace the path of light through the three chambers of the eye, and explain the mechanism of light refraction and how the eye is focused for viewing distant and near objects.
4. List the different layers of the retina, and describe the path of light and of nerve activity through these layers. Continue tracing the path of a visual impulse to the cerebral cortex, and list in order the structures traversed.
5. Describe the photochemical reaction that leads to the activation of rods and the role of vitamin A in vision.

Figure 17.42. Different functions of the right and left cerebral hemispheres, as revealed by experiments with people who have had the tract connecting the two hemispheres (the corpus callosum) surgically split.

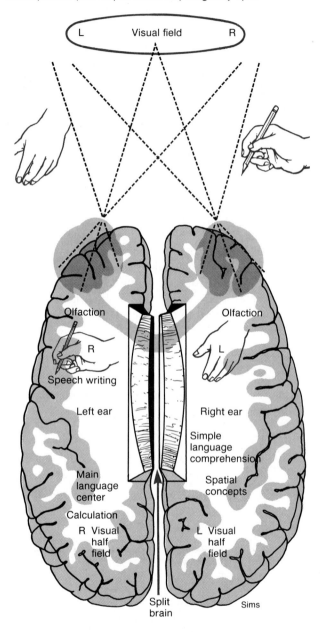

Olfaction   Olfaction
R
Speech writing
Left ear   Right ear
Simple language comprehension
Main language center   Spatial concepts
Calculation
R Visual half field   L Visual half field
Split brain   Sims
L   Visual field   R

Senses of Hearing and Balance

Structures of the outer, middle, and inner ear are involved in the sense of hearing. The inner ear also contains structures that provide for a sense of balance, or equilibrium.

Objective 19. Describe the structures of the ear that relate to hearing, and give their locations and functions.

Objective 20. Trace the path of sound waves through the ear, and describe how they are transmitted and converted to nerve impulses.

Objective 21. Explain the mechanisms by which equilibrium is maintained.

Structure of the Ear

The ear is the organ of hearing and equilibrium. It contains receptors that respond to movements of the head and receptors that convert sound waves into nerve impulses. Impulses from both receptor types are transmitted through the vestibulocochlear (VIII) cranial nerve to the brain for interpretation. The ear consists of three principal regions: the external ear, the middle ear, and the inner ear (fig. 17.43).

External Ear. The external ear consists of the auricle, or pinna, and the external auditory meatus. The **auricle** *(aw'rĕ-kl)* is the visible fleshy appendage attached to the side of the head. It consists of a cartilaginous framework of elastic connective tissue covered with skin. The rim of the auricle is called the **helix,** and the inferior fleshy portion is the **lobe** (see fig. 12.8 for the surface features of the auricle). The lobe is the only portion of the auricle that is not supported with cartilage. The auricle has a ligamentous attachment to the skull and poorly developed auricular muscles inserting superiorly and posteriorly within it. The blood supply to the auricle is from the posterior auricular and occipital arteries, which branch from the external carotid and superficial temporal arteries, respectively. The structure of the auricle directs sound waves into the external auditory meatus.

The **external auditory meatus** is a slightly S-shaped canal about 2.5 cm (1 in.) in length, extending slightly upward from the auricle to the tympanic membrane (fig. 17.43). The skin that lines the meatus contains fine hairs and sebaceous glands near the entrance. Specialized wax-secreting glands called **ceruminous** *(sĕ-roo'mĭ-nus)* **glands** are located in the skin, deep within the meatus. *Cerumen* (earwax) secreted from ceruminous glands keeps the tympanic membrane soft and waterproof. Cerumen and the hairs also help to prevent small foreign objects from reaching the tympanic membrane. The bitter cerumen is probably an insect repellent as well.

The **tympanic** *(tim-pan'ik)* **membrane** (eardrum) is a thin, double layered, epithelial partition between the external auditory meatus and the middle ear. It is approximately 1 cm in diameter and is composed of an outer concave layer of stratified squamous epithelium and an inner convex layer of low columnar epithelium. The tympanic membrane is extremely sensitive to pain and is innervated by the auriculotemporal nerve and the auricular branch of the vagus.

Figure 17.43.   The ear. Note the external, middle, and inner regions of the ear.

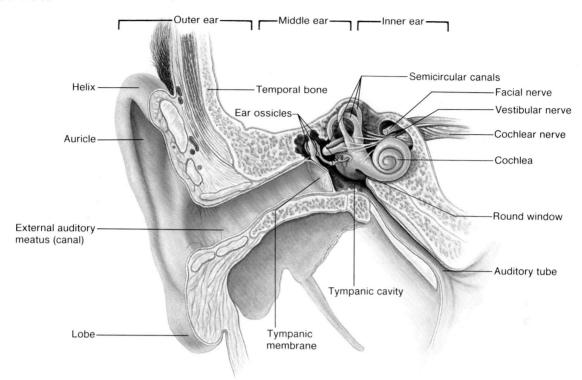

Outer ear — Middle ear — Inner ear

Helix

Auricle

External auditory meatus (canal)

Lobe

Temporal bone

Ear ossicles

Tympanic membrane

Tympanic cavity

Semicircular canals

Facial nerve

Vestibular nerve

Cochlear nerve

Cochlea

Round window

Auditory tube

Middle Ear.   The laterally compressed middle ear is an air-filled chamber called the **tympanic cavity** in the petrous portion of the temporal bone (figs. 17.43, 17.45). The tympanic membrane separates the middle ear from the external auditory meatus of the outer ear. A bony partition containing the **oval window (fenestra vestibuli)** and the **round window (fenestra cochlea)** separates the middle ear from the inner ear.

> Inspecting the tympanic membrane with an otoscope during a physical examination provides significant information about the condition of the middle ear. A diagram of the normal tympanic membrane is presented in figure 17.44. The color, curvature, presence of lesions, or position of the malleus of the middle ear are features of particular importance. If ruptured, the tympanic membrane can generally regenerate and readily heal itself.

There are two openings into the tympanic cavity. The **epitympanic recess** in the posterior wall connects the tympanic cavity to the **mastoidal air cells** within the mastoid process of the temporal bone. The **auditory** *(aw'di-to"re)*, or **Eustachian, tube** connects the tympanic cavity anteriorly with the nasopharynx and equalizes air pressure on both sides of the tympanic membrane.

otoscope: Gk. *otikos*, ear; *skopein*, to examine

Figure 17.44.   The right tympanic membrane as seen with an auriscope (a form of otoscope). A small depression, the *umbo*, occurs at the tip of the handle of the malleus. When the tympanic membrane is illuminated with an auriscope, the concavity of the umbo produces a "cone of light" that radiates anteriorly and inferiorly.

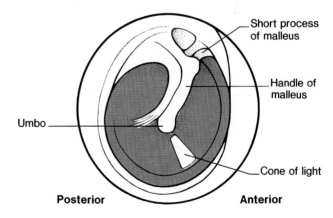

Short process of malleus

Handle of malleus

Umbo

Cone of light

Posterior

Anterior

auriscope: L. *auris*, ear; Gk. *skopein*, to examine

Figure 17.45.   The position of the ear ossicles within the tympanic cavity.

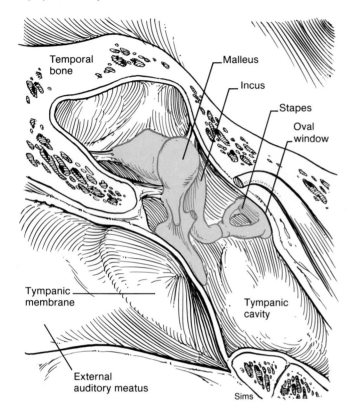

Figure 17.46.   A medial view of the middle ear showing the position of the auditory muscles.

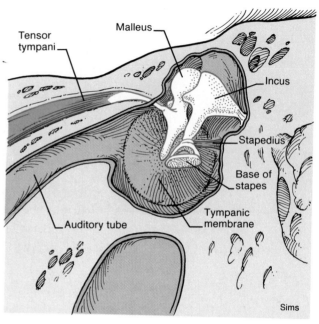

The mucous membranes that line the tympanic cavity, the mastoidal air cells, and the auditory tube are continuous with those of the nasopharynx. For this reason, infections of the nose or throat may spread to the tympanic cavity and cause a middle ear infection or even to the mastoidal air cells and cause mastoiditis. Forcefully blowing the nose advances the spread of the infection.

An equalization of air pressure on both sides of the tympanic membrane is of functional importance to hearing. When atmospheric pressure is reduced, as occurs at higher altitudes, the tympanic membrane bulges outward in response to the greater air pressure within the tympanic cavity. The bulging is painful and may impair hearing by reducing flexibility. The auditory tube, which is collapsed most of the time in adults, opens during swallowing or yawning and allows the air pressures to equalize.

A series of three **ear ossicles** crosses the tympanic cavity from the tympanic membrane to the oval window (fig. 17.45). These tiny bones, from outer to inner, are the **malleus** *(mal'e-us)* (hammer), **incus** (anvil), and **stapes** (stirrup). The ear ossicles are attached to the wall of the tympanic cavity by ligaments. Vibrations of the tympanic membrane cause the ear ossicles to move and transmit sound waves across the tympanic cavity to the oval window.

Vibration of the oval window moves a fluid within the inner ear and stimulates the receptors of hearing.

As the ear ossicles transmit vibrations, they act as a lever system to increase the force of vibration. In addition, the force of vibration is intensified as it is transmitted from the relatively large surface of the tympanic membrane to the smaller surface area of the oval window. The combined effect increases the amplification of sound by about twenty times.

Two small skeletal muscles, the **tensor tympani** and the **stapedius** *(sta-pe'de-us)* (fig. 17.46), attach to the malleus and stapes respectively and contract reflexively to protect the inner ear against loud noises. When contracted, the tensor tympani pulls the malleus inward and the stapedius pulls the stapes outward. This combined action reduces the force of vibration of the ossicles.

Inner Ear.   The inner ear, or **labyrinth,** contains the functional organs for hearing and equilibrium. The labyrinth consists of two parts: an outer **bony labyrinth** *(lab'i-rinth)* and a **membranous** *(mem'brah-nus)* **labyrinth** contained within the bony labyrinth (fig. 17.47). The space between the bony labyrinth and the membranous labyrinth is filled with a fluid, called *perilymph,* secreted by cells lining the bony canals. Within the tubular chambers of the membranous labyrinth is another fluid called *endolymph.* These two fluids provide a liquid-conducting medium for the vibrations involved in hearing and the maintenance of equilibrium.

The bony labyrinth is structurally and functionally divided into three areas: vestibule, semicircular canals, and cochlea.

Figure 17.47.   The labyrinths of the inner ear. The
membranous labyrinth is contained within the bony labyrinth.

Semicircular canals:
Anterior
Posterior
Lateral

Semicircular ducts

Utricle

Saccule

Vestibule

Cochlear nerve

Cochlea

Cochlear duct

Membranous ampullae:
Anterior
Lateral
Posterior

Connection to
cochlear duct

Apex of cochlea

Vestibule.   The vestibule is the central portion of the bony labyrinth that contains the oval window into which the stapes fits and the round window on the opposite end (fig. 17.47).

The membranous labyrinth within the vestibule consists of two connected sacs called the **utricle** *(u'tre-k'l)* and the **saccule** *(sak'ūl.)* The utricle is larger than the saccule and lies in the upper back portion of the vestibule. Both the utricle and saccule contain receptors that are sensitive to gravity and linear movement of the head.

Semicircular Canals.   The three bony semicircular canals of each ear are at right angles to each other and are positioned posterior to the vestibule. The thinner **semicircular ducts** form the membranous labyrinth within the semicircular canals (fig. 17.47). Each of the three semicircular ducts has a **membranous ampulla** *(am-pul'lah)* at one end and connects with the upper back part of the utricle. Receptors within the semicircular ducts are sensitive to angular acceleration and deceleration of the head, as in rotational movement.

Cochlea.   The snail-shaped cochlea *(kok'le-ah)* is coiled two and a half times around a central axis of bone. There are three chambers in the cochlea (fig. 17.48). The upper

chamber, the **scala** *(ska'lah)* **vestibuli,** begins at the oval window and is continuous with the vestibule. The lower chamber, the **scala tympani,** terminates at the round window. The **cochlear duct** is the triangular middle chamber. The roof of the cochlear duct is called the **vestibular membrane,** and the floor is called the **basilar** *(bas'i-lar)* **membrane.** The scala vestibuli and the scala tympani contain perilymph and are completely separated, except at the narrow apex of the cochlea, called the **helicotrema,** *(hel''i-ko-tre'mah)* where they are continuous (fig. 17.50). The cochlear duct is filled with endolymph and ends at the round window.

The **organ of Corti** is contained within the cochlear duct of the cochlea. The sound receptors that transform mechanical vibrations into nerve impulses are located along the basilar membrane of this structure, making it the functional unit of hearing. The epithelium of the organ of Corti consists of supporting cells and hair cells (fig. 17.49). The bases of the hair cells are anchored in the basilar membrane, and their tips are embedded in the tectorial membrane that forms a gelatinous canopy over them.

scala: Gk. *scala,* staircase
helicotrema: Gk. *helix,* a spiral; *trema,* a hole
organ of Corti: from Alfonso Corti, Italian anatomist, 1822–1888

cochlea: L. *cochlea,* snail shell

Figure 17.48. The organ of Corti (*a*) is located within the cochlear duct of the cochlea (*b*).

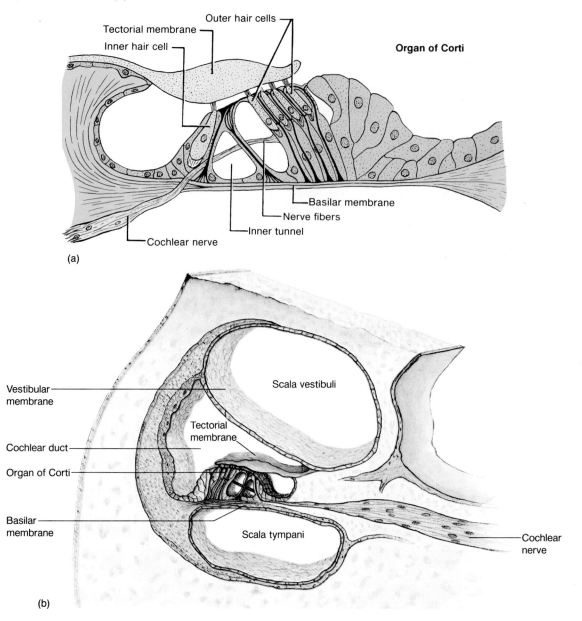

(a)

(b)

Figure 17.49. A scanning electron micrograph of the hair cells of the organ of Corti.

Figure 17.50. (*a*) the cochlea is illustrated as a straightened structure to show that the perilymph in the scala vestibuli and scala tympani is continuous at the helicotrema. These two chambers are separated by the blind-ending cochlear duct (scala media). The basilar membrane of the cochlear duct is narrower at the base and wider at the apex. (*b*) sounds of different frequencies (pitches) produces maximum displacement of the basilar membrane in these regions.

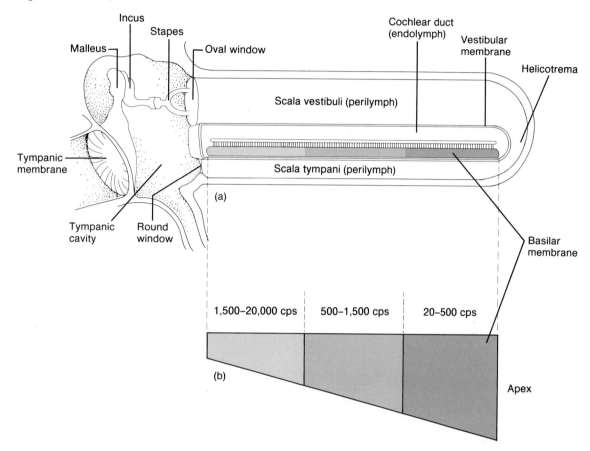

Sound Waves and Neural Pathways for Hearing

Sound Waves. Sound waves travel in all directions from their source, like ripples in a pond after a stone is dropped. These waves are characterized by their frequency and their intensity. The **frequency,** or distances between crests of the sound waves, is measured in *hertz* (*Hz*), which is the modern designation for *cycles per second* (*cps*). The *pitch* of a sound is directly related to its frequency—the higher the frequency of a sound, the higher its pitch.

The **intensity,** or loudness of a sound, is directly related to the amplitude of the sound waves. This is measured in units known as *decibels* (*db*). A sound that is barely audible—at the threshold of hearing—has an intensity of zero decibels. Every 10 decibels indicates a tenfold increase in sound intensity; a sound is ten times higher than threshold at 10 db, 100 times higher at 20 db, a million times higher at 60 db and ten billion times higher at 100 db.

The ear of a trained, young individual can hear sound over a frequency range of 20,000–30,000 Hz, yet can distinguish between two pitches that have only a 0.3% difference in frequency. The human ear can detect differences in sound intensities of only 0.1 to 0.5 db, whereas the range of audible intensities covers twelve orders of magnitude (10^{12}), from the barely audible to the limits of painful loudness.

Sound waves funneled through the external auditory meatus produce extremely small vibrations of the tympanic membrane. Movements of the tympanum during speech (average intensity of 60 db) are estimated to be equal to the diameter of a molecule of hydrogen.

Sound waves passing through the solid medium of the ear ossicles are amplified by about twenty times as they reach the footplate of the stapes against the oval window. As the oval window is displaced, pressure waves pass through the fluid medium of the scala vestibuli (fig. 17.50), which pass around the helicotrema to the scala tympani.

Figure 17.51. Sounds of low frequency cause pressure waves of perilymph to pass through the helicotrema. Sounds of higher frequency cause pressure waves to "short cut" through the cochlear duct. This sets up traveling waves in the basilar membrane.

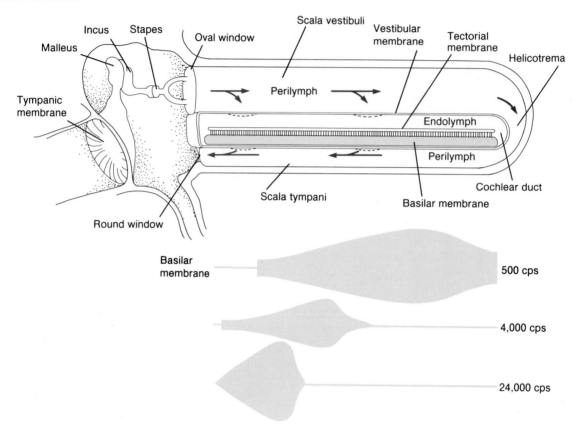

Movements of perilymph within the scala tympani, in turn, displace the round window into the tympanic cavity.

When the sound frequency (pitch) is sufficiently low, there is adequate time for the pressure waves of perilymph within the upper scala vestibuli to travel through the helicotrema to the scala tympani. As the sound frequency increases, however, pressure waves of perilymph within the scala vestibuli do not have time to travel all the way to the apex of the cochlea. Instead, they are transmitted through the *vestibular membrane,* which separates the scala vestibuli from the cochlear duct, and through the *basilar membrane,* which separates the cochlear duct from the scala tympani, to the perilymph of the scala tympani. The distance that these pressure waves travel, therefore, decreases as the sound frequency increases.

Sounds of low pitch (with frequencies below about 50 Hz) cause movements of the entire length of the basilar membrane—from the base to the apex. Higher sound frequencies result in maximum displacement of the basilar membrane closer to its base, as illustrated in figure 17.51.

Displacement of the basilar membrane and hair cells by movements of perilymph causes the microvilli that are embedded in the tectorial membrane to bend. This stimulation excites the sensory cells, which causes the release of an unknown neurotransmitter that excites sensory endings of the cochlear (eighth cranial) nerve.

Neural Pathways for Hearing. Cochlear sensory neurons in the eighth cranial nerve synapse with neurons in the medulla (fig. 17.52), which project to the inferior colliculi of the midbrain. Neurons in this area in turn project to the thalamus, which sends axons to the auditory cortex of the temporal lobe. By means of this pathway, neurons

Figure 17.52.   Neural pathways from the spiral ganglia of the cochlea of the auditory cortex.

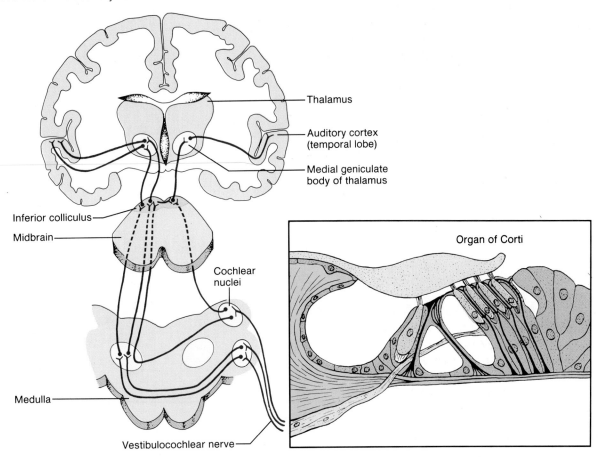

Thalamus

Auditory cortex (temporal lobe)

Medial geniculate body of thalamus

Inferior colliculus

Midbrain

Cochlear nuclei

Organ of Corti

Medulla

Vestibulocochlear nerve

in different regions of the basilar membrane stimulate neurons in corresponding areas of the auditory cortex; each area of this cortex thus represents a different part of the basilar membrane and a different pitch (fig. 17.53).

Mechanics of Equilibrium

Maintaining equilibrium *(e"kwĭ-lib're-um)* is a complex process that depends on constant input from the vestibular organs of both inner ears. Although the vestibular organs are the principal source of sensory information for equilibrium, visual input from the eyes, tactile receptors within the skin, and proprioceptors of tendons, muscles, and joints also provide sensory input that is vital for maintaining equilibrium (fig. 17.54).

The vestibular organs provide the CNS with two kinds of receptor information. One kind is provided by receptors within the saccule and utricle that are sensitive to gravity and to linear acceleration and deceleration of the head such as in riding in a car. The other is provided by receptors within the semicircular canals that are sensitive to angular acceleration and deceleration such as when turning the head, spinning, or tumbling.

Figure 17.53. Sounds of different frequencies (pitches) excite different sensory neurons in the cochlea; these in turn send their input to different regions of the auditory cortex.

Correspondence between cochlea and acoustic area of cortex:

Blue—low tones

Red—medium tones

Yellow—high tones

BECK

Figure 17.54. Neural processing involved in maintenance of equilibrium and balance.

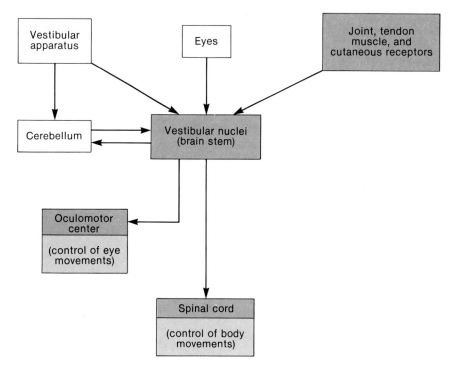

Figure 17.55. A scanning electron microscope of hairs and kinocilium within the vestibular apparatus.

The receptor hair cells of the vestibular organs contain twenty to fifty microvilli and one cilium called a **kinocilium** *(ki''no-sil'e-um)* (fig. 17.55). When the hair cells are displaced in the direction of the kinocilium, the cell membrane is depressed and becomes depolarized (fig. 17.56). When the hair cells are displaced in the opposite direction, the membrane becomes hyperpolarized.

Saccule and Utricle. Receptor hair cells of the saccule and utricle are located in a small, thickened area of the walls of these organs called the **macula** (fig. 17.57). Cytoplasmic extensions of the hair cells project into a gelatinous material that supports microscopic crystals of calcium carbonate called **otoliths.** The otoliths increase the weight of the gelatinous mass, which results in a higher *inertia* (resistance to change in movement).

When a person is upright, the hairs of the utricle are oriented vertically, whereas those of the saccule are oriented horizontally into the otolith membrane. During forward acceleration, the otolith membrane lags behind the

kinocilium: Gk. *kineo*, to move; L. *cilium*, eyelash
macula: L. *macula*, spot
otoliths: Gk. *otos*, ear; *lithos*, stone

Figure 17.56. (*a*) sensory hair cells in the vestibular apparatus contain hairs (microvilli) and one kinocilium. (*b*) when hair cells are bent in the direction of the kinocilium, the cell membrane is depressed (see arrow) and the sensory neuron innervating the hair cell is stimulated. (*c*) when the hairs are bent in a direction opposite to the kinocilium the sensory neuron is inhibited.

(b)

Kinocilium

(a)   At rest

(c)

Figure 17.57. The macula of the utricle. (*a*) when the head is in an upright position, the weight of the otoliths applies direct pressure to the sensitive cytoplasmic extensions of the hair cells. (*b*) as the head is tilted forward, the extensions of the hair cells bend in response to gravitational force and cause the hair cells to be stimulated.

(a)

Macula

Otoliths

Hair cells

Sensory nerve fibers

Supporting cells

HEAD UPRIGHT

(b)

Gelatinous material sags

Cytoplasmic extensions of hair cells bent

Gravitational force

HEAD BENT FORWARD

hair cells, so the hairs are pushed backward. This is similar to the backward thrust of the body when a car accelerates rapidly forward. The inertia of the otolith membrane similarly causes the hairs of the utricle to be pushed upward when a person jumps from a raised platform. These effects, and the opposite ones that occur when a person accelerates backward or upward, allow us to maintain our equilibrium with respect to gravity during linear acceleration.

Sensory impulses from the vestibular organs are conveyed to the brain via the vestibular portion of the vestibulocochlear nerve.

Semicircular Canals. Receptors of the semicircular canals are contained within the ampulla at the base of each semicircular duct. An elevated area of the ampulla called the **crista ampullaris** *(am''pu-lar'is)* contains numerous hair cells and supporting cells (fig. 17.58). Like the saccule and utricle, the hair cells have cytoplasmic extensions that project into a dome-shaped gelatinous mass called the **cupula** *(ku'pu-lah)*. When the hair cells within the cupula are bent by rapid displacement of the fluid within the semicircular ducts, as in spinning around, sensory impulses travel to the brain via the vestibular portion of the vestibulocochlear nerve.

Neural Pathways. Stimulation of the hair cells in the vestibular apparatus activates the sensory neurons of the *vestibulocochlear (eighth cranial)* nerve. These fibers transmit impulses to the cerebellum and to the vestibular nuclei of the medulla oblongata. The vestibular nuclei, in turn, send fibers to the oculomotor center of the brain stem and to the spinal cord. Neurons in the oculomotor center control eye movements, and neurons in the spinal cord stimulate movements of the head, neck, and limbs. Movements of the eyes and body produced by these pathways serve to maintain balance and "track" the visual field during rotation.

1. List the structures of the outer ear, middle ear, and inner ear, and explain the function of each as related to hearing.
2. Use a flow chart to describe how sound waves in air within the external auditory meatus are transduced into movements of the basilar membrane.
3. Explain how movements of the basilar membrane of the organ of Corti can code for different sound frequencies (pitches).
4. Explain how the vestibular organs maintain a sense of balance and equilibrium.

cupula: L. *cupula,* cup shaped

Figure 17.58.   (a) a crista ampullaris within an ampulla. (b) movement of the endolymph during rotation causes the cupula to displace, thus stimulating the hair cells.

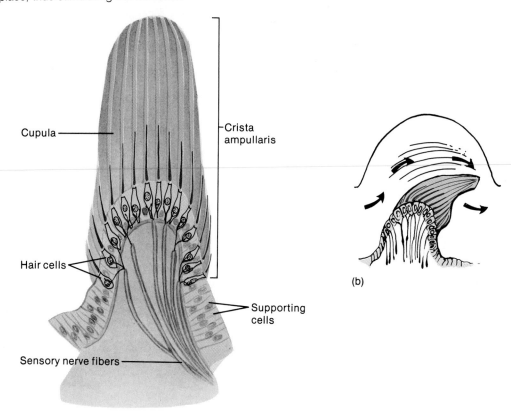

Cupula

Crista ampullaris

Hair cells

Supporting cells

Sensory nerve fibers

(b)

Clinical Considerations

Numerous disorders and diseases afflict the sensory organs. Some of these occur during the sensitive period of prenatal development and have a congenital effect. Other sensory impairments, some of which are avoidable, can occur at any time of life. Still others are the result of changes during the natural aging process. The loss of a sense is frequently a traumatic adjustment. Fortunately, however, when a sensory function is impaired or lost, the other senses seem to become more keen to lessen the extent of the handicap. A blind person, for example, compensates somewhat for the loss of sight by developing a remarkable hearing ability.

Many books have been written on diseases and dysfunctions of the sensory organs, and entire specialities within medicine are devoted to treating the disorders of the specific sensory organs. It is beyond the scope of this book to attempt a comprehensive discussion of the numerous aspects of the sensory organs. Some general comments will be made, however, on the diagnosis of sensory disorders and on developmental problems that can affect the eyes or ears. In addition, the more common diseases and dysfunctions of these organs will be discussed.

Diagnosis of Sensory Organs

Eye.   There are two distinct professional specialities concerned with the structure and function of the eye. *Optometry* is the paramedical profession concerned wth assessing vision and treating visual problems. An *optometrist* is not a physician and does not treat eye diseases but prescribes lenses or visual training. *Ophthalmology (of"thal-mol'o-je)* is the speciality of medicine concerned with diagnosing and treating eye diseases.

Although the eyeball is an extremely complex organ, it is quite accessible to examination. Various techniques and instruments are used during an eye examination, but the following are used most frequently: (1) a *cycloplegic drug,* which is instilled into the eyes to dilate the pupils and temporarily inactivate the ciliary muscles; (2) a *Snellen's chart,* which is used to determine the visual acuity of a person standing 20 feet from the chart (a reading of 20/20 is considered normal for the test); (3) an *ophthalmoscope,* which contains a light, mirrors, and lenses to illuminate and magnify the interior of the eyeball so that the structures within may be examined; and (4) a *tonometer,* used to measure ocular tension, which is important in detecting glaucoma.

Ear. Otorhinolaryngology *(o''to-ri''no-lar''in-gol'o-je)* is the speciality of medicine dealing with the diagnosis and treatment of diseases or conditions of the ear, nose, and throat. *Audiology* is the study of hearing, particularly assessing the ear and its functioning.

There are three common instruments or techniques used to examine the ears to determine auditory function: (1) an *otoscope* is an instrument used to examine the tympanic membrane of the ear; abnormalities of this membrane are informative when diagnosing specific auditory problems, including middle ear infections; (2) *tuning fork tests* are useful in determining hearing acuity and especially for discriminating the various kinds of hearing loss; (3) *audiometry* is a functional examination for hearing sensitivity and speech discrimination.

Developmental Problems of the Eyes and Ears

Although there are many congenital abnormalities of the eyes and ears, most of them are rare. The sensitive period of development of these organs is from about twenty-four to forty-five days after conception. Most congenital disorders of the eyes and ears are caused by genetic factors or intrauterine infections such as rubella virus.

Eye. Most **congenital cataracts** are hereditary, but they may also be caused by maternal rubella infection during the critical fourth to sixth week of eye development. In this condition, the lens is opaque and frequently appears grayish white.

Cyclopia is a rare condition in which the eyes are partially fused into a median eye enclosed by a single orbit. Other severe malformations, which are incompatible with life, are generally expressed with this condition.

Ear. **Congenital deafness** is generally caused by an autosomal recessive gene but may also be caused by a maternal rubella infection. The actual functional impairment is generally either a defective set of ossicles or improper development of the neurosensory structures of the inner ear.

Although the shape of the auricle varies widely, **auricular abnormalities** are not uncommon, especially in infants with chromosomal syndromes causing mental deficiencies. In addition, the external auditory meatus frequently does not develop in these children, producing a condition called **atresia** *(ah-tre'ze-ah)* of the external auditory meatus.

Functional Impairments of the Eye

Few people have perfect vision. Slight variations in the shape of the eyeball or curvature of the cornea or lens cause an imperfect focal point of light waves onto the retina. Most variations are slight, however, and the error of refraction goes unnoticed. Severe deviations that are not corrected may cause blurred vision, fatigue, chronic headaches, and depression.

There are four principal clinical considerations caused by defects in the refractory structures or general shapes of the eyeball. **Myopia** (nearsightedness) is an elongation of the eyeball that causes light waves to focus at a point in the vitreous humor in front of the retina (fig. 17.59). Only light waves from close objects can be focused clearly on the retina; distant objects appear blurred. **Hyperopia** (farsightedness) is a condition in which the eyeball is too short, which causes light waves to have a focal point behind the retina. Although visual accommodation aids a hyperopic person, it generally does not help enough for the person to view clearly very close or distant objects. **Presbyopia** is a condition in which the lens tends to lose its elasticity and ability to accommodate. It is relatively common in elderly people. In order to read print on a page, a person with presbyopia must hold the page farther from the eyes. **Astigmatism** is a condition in which an irregular curvature of the cornea or lens of the eye distorts the refraction of light waves. If a person with astigmatism views a circle, the image will not appear clear in all 360 degrees; the parts of the circle that appear blurred can thus be used to map the astigmatism.

Various glass or plastic lenses can generally aid persons with the above visual impairments. Myopia may be corrected with a biconcave lens; hyperopia with a biconvex lens; and presbyopia with bifocals, or a combination of two lens adjusted for near and distant vision. Correction for astigmatism requires a careful assessment of the irregularities and a prescription of a specially ground corrective lens.

Cataracts. A cataract is a clouding of the lens that leads to a gradual blurring of vision and the eventual loss of sight. A cataract is not a growth within or upon the eye but a chemical change in the protein of the lens caused by injury, poisons, infections, or age degeneration.

---

myopia: Gk. *myein*, to shut; *ops*, eye
hyperopia: Gk. *hyper*, over; *ops*, eye
presbyopia: Gk. *presbys*, old man; *ops*, eye
astigmatism: Gk. *a*, without; *stigma*, point
cataract: Gk. *katarrhegnynai*, to break down

Figure 17.59. In a normal eye (a), parallel rays of light are brought to focus on the retina by refraction in the cornea and lens. If the eye is too long, as in myopia (b), the focus is in front of the retina. This can be corrected by a concave lens. If the eye is too short (c), as in hyperopia, the focus is behind the retina. This is corrected by a convex lens.

no correction necessary

(a)

Concave lens   Myopia

(b)

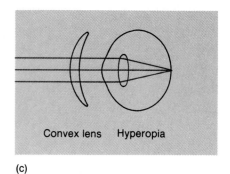

Convex lens   Hyperopia

(c)

Cataracts are the leading cause of blindness. A cataract can be removed surgically, however, and vision restored by implanting a tiny intraocular lens that either clips to the iris or is secured into the vacant lens capsule. Special contact lenses or thick lenses for glasses are other options.

Detachment of the Retina. Retinal detachment is a separation of the retina from the underlying pigment epithelium. It generally begins as a minute tear in the retina that gradually extends as vitreous fluid accumulates between the layers. Retinal detachment may result from hemorrhage, a tumor, degeneration, or trauma from a violent blow to the eye. A detached retina may be repaired by using laser beams, cryoprobes, or intense heat to destroy the tissue beneath the tear and rejoin the layers.

Strabismus. Strabismus is a condition in which both eyes do not focus upon the same axis of vision. This prevents stereoscopic vision, and the persons afflicted will have varied visual impairments. Strabismus is usually caused by a weakened extrinsic eye muscle.

Strabismus is assessed while the patient attempts to look straight ahead. If the afflicted eye is turned toward the nose, the condition is called convergent strabismus (**esotropia**). If the eye is turned outward, it is called divergent strabismus (**exotropia**). Disuse of the afflicted eye causes a visual impairment called **amblyopia.** Visual input from the normal eye and the eye with strabismus results in **diplopia,** or double vision. A normal, healthy person who has excessively consumed alcohol may experience diplopia.

Infections and Diseases of the Eye

Infections. Infections and inflammation can occur in any of the associated structures of the eye and in structures within or on the eyeball itself. The causes of infections are usually microorganisms, mechanical irritation, or sensitivity to particular substances.

Conjunctivitis (inflammation of the conjunctiva) may result from sunburn or an infection by organisms such as staphylococci, viruses, or streptococci. Bacterial conjunctivitis is commonly called "pinkeye."

Keratitis (inflammation of the cornea) may develop secondarily from conjunctivitis or be caused by diseases such as tuberculosis, syphilis, mumps, or measles. Keratitis is painful and may cause blindness if untreated.

A **chalazion** *(kah-la'ze-on)* is a cyst caused by an infection and a subsequent blockage in the ducts of the sebaceous glands along the free edge of the eyelids.

Styes (hordeola) are relatively common, but mild, infections of the follicle of an eyelash or the sebaceous gland of the follicle. Styes may spread readily from one eyelash to another if untreated. Poor hygiene and the excessive use of cosmetics may contribute to development of styes.

Diseases. **Trachoma** *(trah-ko'mah)* is a highly contagious bacterial disease of the conjunctiva and cornea. Although rare in the United States, it is estimated that over 500 million people are afflicted by this disease. Trachoma may be treated readily with sulfonamides and some antibiotics, but if untreated, it will spread progressively until it covers the cornea. At this stage, vision is lost and the eye undergoes degenerative changes.

Figure 17.60. The development of otosclerosis in the middle ear immobilizes the stapes, leading to conduction deafness.

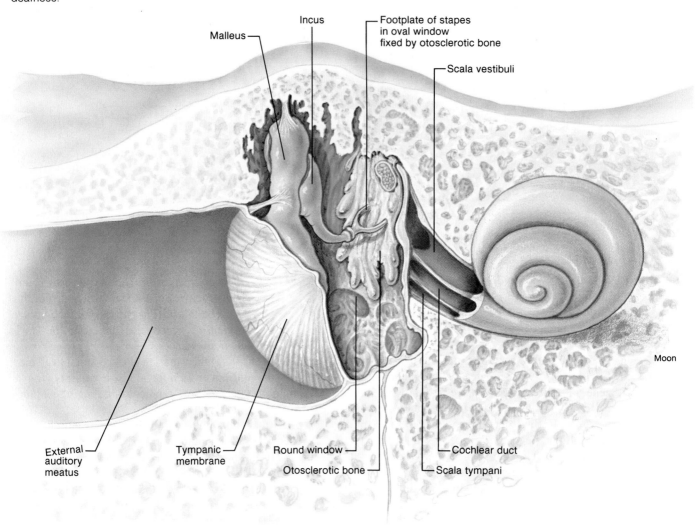

Glaucoma is the second leading cause of blindness and is particularly common in underdeveloped countries. Although it can afflict persons of any age, 95% of the cases involve persons over the age of forty. Glaucoma is an abnormal increase in the intraocular pressure of the eyeball. Aqueous humor does not drain through the venous sinus as quickly as it is produced. Glaucoma can be caused by several things, including malnutrition. Glaucoma causes compression of the blood vessels in the eyeball and compression of the optic nerve. Retinal cells die and the optic nerve may atrophy, producing blindness.

glaucoma: Gk. *glaukos,* gray

Infections, Diseases, and Functional Impairments of the Ear

Disorders of the ear are common and may affect both hearing and the vestibular functions. The ear is afflicted by numerous diseases as well as environmental factors—some of which can be prevented.

Infections and Diseases. **External otitis** is a general term for infections of the external ear. The causes of external otitis range from dermatitis to fungal and bacterial infections.

Acute purulent otitis media is a middle ear infection. Pathogens of this disease usually enter through the auditory tube, most often following a cold or tonsillitis. Children frequently have middle ear infections because of their susceptibility to infections and their short and straight auditory tubes. As a middle ear infection progresses to the inflammatory stage, the auditory tube is closed and drainage prohibited. An intense earache is a common symptom of a middle ear infection. The pressure from the inflammation may eventually rupture the tympanic membrane to permit drainage.

Repeated middle ear infections, particularly in children, usually necessitate an incision of the tympanic membrane, called a **myringotomy** *(mir''in-got'o-me),* and the implantation of a tiny tube within the tympanic membrane to assist the patency of the auditory tube. The tubes, which are eventually sloughed out of the ear, permit the infection to heal and help prohibit further infections by keeping the auditory tube open.

Perforation of the tympanic membrane may occur as the result of infections or trauma. Sudden, intense noise might rupture the membrane. Spontaneous perforation of the membrane usually heals rapidly, but scar tissue may form and lessen the sensitivity to sound vibrations.

Otosclerosis is a progressive deterioration of the normal bone of the bony labyrinth and its replacement with vascular spongy bone (fig. 17.60). This frequently causes hearing loss as the ear ossicles are immobilized. Surgical scraping of the bone growth and replacing the stapes with a prosthesis is the most frequent treatment of otosclerosis.

Meniere's disease afflicts the inner ear and may cause hearing loss as well as equilibrium disturbance. The causes of Meniere's disease vary and are not completely understood but are thought to be related to a dysfunction of the autonomic nervous system that causes a vasoconstriction within the inner ear. The disease is characterized by recurrent periods of **vertigo** (dizziness and a sensation of rotation), **tinnitus** *(ti-ni'tus)* (ringing in the ear), and progressive deafness in the affected ear. Meniere's disease is chronic and affects both sexes equally. It is more common in middle-aged and elderly persons.

Nystagmus and Vertigo. When a person first begins to spin, the inertia of the endolymph within the semicircular canals causes the cupula to bend in the opposite direction.

Meniere's disease: from Prosper Meniere, French physician, 1799–1862
vertigo: L. *vertigo,* dizziness
tinnitus: L. *tinnitus,* ring or tingle

As the spin continues, however, the inertia of the endolymph is overcome and the cupula straightens. At this time the endolymph and the cupula are moving in the same direction and at the same speed. If movement is suddenly stopped, the greater inertia of the endolymph causes it to continue moving in the previous direction of spin and to bend the cupula in that direction.

Bending of the cupula after movement has stopped affects muscular control of the eyes and body. The eyes slowly drift in the direction of the previous spin and are then rapidly jerked back to the midline position, producing involuntary oscillations. These movements are called post-rotatory **vestibular nystagmus** *(nis-tag'mus).* People at this time may feel that they, or the room, are spinning. The loss of equilibrium that results is called **vertigo.** If the vertigo is sufficiently severe or the person particularly susceptible, the autonomic system may become involved. This can produce dizziness, pallor, sweating, and nausea.

Auditory Impairment. Loss of hearing results from disease, trauma, or developmental problems involving any portion of the auditory apparatus, cochlear nerve and auditory pathway, or areas of auditory perception within the brain. Hearing impairment varies from slight disablement, which may or may not worsen, to total deafness. Some types of hearing impairment, including deafness, can be mitigated through hearing aids or surgery.

There are two principal types of auditory impairments. **Conduction deafness** is caused by an interference with the sound waves through the external or middle ear. Conduction problems include impacted cerumen (wax), a ruptured tympanic membrane, a severe middle ear infection, or adhesions of one or more ear ossicles *(otosclerosis).* Conductive deafness can be successfully treated.

Perceptive (sensorineural) deafness results from disorders that affect the inner ear, the cochlear nerve or nerve pathway, or auditory centers within the brain. Perceptive impairment ranges in severity from the inability to hear certain frequencies to deafness. Such deafness may be caused by a number of factors, including diseases, trauma, and genetic or developmental problems. Hearing impairments of this type, which are common in elderly persons, are related to age. The ability to perceive high-frequency sounds is generally affected first. Hearing aids may help patients with perceptive deafness. This type of deafness is permanent, however, because it involves sensory structures that cannot heal themselves through regeneration or replication.

Chapter Summary

I. Overview of Sensory Perception
Sensory organs are specialized extensions of the nervous system that contain sensory neurons adapted to respond to specific stimuli and conduct nerve impulses.
   A. A stimulus to a receptor that conducts an impulse to the brain is necessary for perception.
   B. Senses act as energy filters that permit perception of only a narrow range of energy.

II. Development of the Sensory Organs
The development of the eye and ear from embryonic germ layers begins during the fourth week. The sensory components of the eye and ear form by twenty weeks, and the accessory structures form by thirty-two weeks.
   A. The primordial ectodermal structures are developed by the seventh week followed by development of the mesodermal structures of the eyeball and accessory features.
   B. Surface ectoderm gives rise to the majority of the structures of the middle and inner ear. The auricle derives from surface ectoderm, and the tympanic membrane derives from surface ectoderm and endoderm.

III. Classification of the Senses
The senses are classified according to the complexity and location of receptors or on the basis of the stimulus energy they transduce.
   A. General senses are widespread throughout the body and are simple in structure. Special senses are localized in complex receptor organs and have extensive neural pathways.
   B. Somatic senses are localized within the body wall, and visceral senses are located within the visceral organs.
   C. Phasic receptors respond quickly to a stimulus but then adapt and decrease firing rate. Tonic receptors produce a constant rate of firing.

IV. Physiology of Sensory Receptors
A sensory receptor responds to stimulus energy and initiates the nerve impulse that produces a particular modality of sensation.
   A. According to the law of specific nerve energies, the sensation characteristic of each type of sensory neuron is that produced by its adequate stimulus.
   B. The entire length of a sensory neuron contains voltage regulated gates that produce action potentials.

V. Somatic Senses
The perception of somatic sensations is determined by the density of the receptors in the receptive field and the intensity of the sensation, as well as by accommodation, afterimage, and projection.
   A. Meissner's corpuscles, free nerve endings, and hair root plexuses are tactile receptors, responding to light touch.
   B. Pacinian corpuscles are pressure receptors in the hypodermis.
   C. Thermoreceptors include the organs of Ruffini (heat) and bulbs of Krause (cold).
   D. Free nerve endings are the principal pain receptors, serving to protect the body from injury.
   E. Joint kinesthetic receptors, neuromuscular spindles, and neurotendinous receptors are proprioceptors that are sensitive to changes in stretch and tension.
   F. The acuity of sensation depends on receptor density and lateral inhibition.

VI. Olfactory Sense
Olfactory receptors of the olfactory nerve respond to chemical stimuli and transmit the sensation of olfaction to the cerebral cortex.
   A. Olfaction functions closely with gustation in that the receptors of both are chemoreceptors, requiring dissolved substances for stimuli.
   B. Olfactory receptors are located in the nasal epithelium of the nasal cavity.

VII. Gustatory Sense
Taste receptors in taste buds respond to chemical stimuli and transmit impulses through the glossopharyngeal or facial cranial nerve to the cerebral cortex.
   A. Gustatory cells compose the taste buds located in circumvallate, fungiform, and filiform papillae.
   B. The four modalities of taste sensation are sweet, salty, sour, and bitter.

VIII. Visual Sense
Rods and cones within the eyeball are sensitive to light waves and transmit impulses through ganglion fibers of the optic nerve to the visual cortex of the occipital cerebral lobe.
   A. Accessory structures of the eye (the orbit, eyebrow, eyelids, eyelashes, conjunctiva, lacrimal gland, and ocular muscles) either protect the eyeball or provide eye movement.
   B. The eyeball consists of the fibrous tunic, which is divided into the sclera and cornea; the vascular tunic, which consists of the choroid, the ciliary body, and the iris; and the internal tunic, or retina, which consists of an outer pigmented layer and an inner nervous layer.
   C. Rods and cones, which are the photoreceptors in the nervous layer of the retina, respond to dim and colored light respectively. Cones are concentrated in the fovea centralis, the area of keenest vision.
   D. The anterior and posterior chambers contain aqueous humor, and the vitreous chamber contains vitreous humor.
   E. The visual process includes the transmission and refraction of light waves, accommodation of the lens, constriction of the pupil, and convergence of the eyes.
   F. Rods and cones contain specific pigments that provide sensitivity to different light waves.
   G. Neural pathways from the retina to the superior colliculus help regulate eye and body movements. Most fibers from the retina project to the lateral geniculate body and from there to the striate cortex.

IX. Senses of Hearing and Balance
Structures of the outer, middle, and inner ear are involved in hearing. The inner ear also provides for a sense of balance, or equilibrium.
   A. The external ear consists of the auricle and the external auditory meatus.
   B. The middle ear, bounded by the tympanic membrane and the oval and round windows, contains the ear ossicles (malleus, incus, and stapes) and the auditory muscles (tensor tympani and stapedius).
   C. The vestibule of the middle ear connects to the pharynx through the auditory tube.
   D. The inner ear, or labyrinth, contains the functional organs for hearing and equilibrium: the organ of Corti in the cochlea and the semicircular canals, saccule, and utricle.

Review Activities

Objective Questions

1. Which of the following must occur for the perception of a sensation to take place?
   - (a) stimulus present
   - (b) nerve impulse conduction
   - (c) receptor activated
   - (d) all of the above
2. The cutaneous receptor sensitized for deep pressure detection is the
   - (a) hair root plexus
   - (b) pacinian corpuscle
   - (c) organ of Ruffini
   - (d) free nerve endings
3. Cutaneous receptive fields are smallest in the
   - (a) back
   - (b) fingertips
   - (c) thighs
   - (d) arms
4. The sensation of visceral pain perceived as arising from another somatic location is known as
   - (a) related pain
   - (b) phantom pain
   - (c) referred pain
   - (d) parietal pain
5. When a person with normal vision views an object from a distance of at least 20 feet;
   - (a) the ciliary muscles are relaxed
   - (b) the suspensory ligament is taut
   - (c) the lens is flat, having the least convex shape
   - (d) all of the above
6. Which of the following is an avascular ocular tissue?
   - (a) sclera
   - (b) cornea
   - (c) ciliary body
   - (d) iris
7. A bar of light in a specific part of the retina, with a particular length and orientation, is the most effective stimulus for
   - (a) ganglion cells
   - (b) simple cortical cells
   - (c) lateral geniculate cells
   - (d) complex cortical cells
8. The stimulation of hair cells in the semicircular canals results from the movement of
   - (a) endolymph
   - (b) perilymph
   - (c) the otolith membrane
9. The middle ear is separated from the inner ear by the
   - (a) round window
   - (b) tympanic membrane
   - (c) oval window
   - (d) both a and c
10. Glasses with concave lenses help to correct
    - (a) presbyopia
    - (b) myopia
    - (c) hyperopia
    - (d) astigmatism

Essay Questions

1. Outline the major events in the development of the eye and ear. When are congenital deformities most likely to occur?
2. Name the six senses of the body. Differentiate between the special and somatic senses. What are the similarities between these two senses?
3. List the functions of proprioceptors, and differentiate between the three types. What role do proprioceptors play in the kinesthetic sense?
4. Diagram the structure of the eyeball, and label the sclera, cornea, choroid, macula lutea, ciliary body, suspensory ligaments, lens, iris, pupil, retina, optic disc, and fovea centralis.
5. Outline and explain the process of focusing light waves onto the fovea centralis.
6. Trace a sound wave through the structures of the ear, and explain how the mechanism of hearing functions.

Selected Readings

Brindley, G. S. 1970. Central pathways of vision. *Ann. Rev. Physiol.* 32:259.
Casey, K. L. 1973. Pain: a current view of neural mechanisms. *American Scientist* 61:194.
Daw, N. W. 1973. Neurophysiology of color vision. *Physiol. Rev.* 53:571.
Durrant, J. D., and J. H. Lovrinic. 1977. *Bases of hearing science.* Baltimore: The Williams and Wilkins Co.
Freese, A. S. 1977. *The miracle of vision.* New York: Harper and Row Publishers.
Gombrich, E. H. September 1972. The visual image. *Scientific American.*
Green, D. M. 1976. *An introduction to hearing.* New York: Halsted Press.

Harris, J. D. 1972. Audition. *Ann. Rev. Psych.* 23:313.
Hubel, D. H., and T. N. Weisel. September 1979. Brain mechanisms of vision. *Scientific American.*
Kaufman, H. E. July 1973. Corneal transplantation, a progress report. *Hosp. Prac.*
Lim, R. K. S. 1970. Pain. *Ann. Rev. Physiol.* 32:269.
Loeb, G. E. February, 1985. The functional replacement of the ear. *Scientific American.*
Pettigrew, J. D. August 1972. The neurophysiology of binocular vision. *Scientific American.*
Ross, J. March 1976. The resources of binocular perception. *Scientific American.*
Rubenstein, E. March 1980. Diseases caused by impaired communication among cells. *Scientific American.*
Rushton, W. A. H. March 1975. Visual pigments and color blindness. *Scientific American.*

Snyder, S. H. March 1977. Opiate receptors and internal opiates. *Scientific American.*
Toates, F. M. 1972. Accommodation function of the human eye. *Physiol. Rev.* 52:828.
Van Heyninger, R. December 1975. What happens to the human lens in cataract? *Scientific American.*
Von Bekesy, G. August 1975. The ear. *Scientific American.*
Werblin, F. S. January 1973. The control of sensitivity in the retina. *Scientific American.*
Wertenbaker, L. 1981. *The eye: window to the world.* Washington, D.C.: U.S. New Books.
Young, R. W. October 1970. Visual cells. *Scientific American.*

18

Endocrine System

Outline

Concepts

Hormones are regulatory molecules
     secreted into the blood by endocrine
     glands. The action of a hormone on
     its target tissue is affected by a
     number of variables.

Hormones that are chemically similar
     have similar mechanisms of action.
     These similarities involve the location
     of cellular receptor proteins and the
     events that occur in the target cell
     after hormone-receptor interaction.

The embryonic development of the
     endocrine glands establishes the
     structural and functional
     relationships of the endocrine system.

The hormones secreted by the posterior
     pituitary are produced by the
     hypothalamus. The anterior pituitary
     produces its own hormones. Secretion
     of both parts of the pituitary gland is
     controlled by the hypothalamus.

The adrenal cortex and medulla secrete
     different hormones, which have
     different functions in the body and
     are under different control
     mechanisms.

The thyroid gland secretes hormones
     that help regulate energy
     metabolism, and the parathyroid
     glands secrete a hormone that
     contributes to the regulation of
     calcium and phosphate balance.

The pancreas contains the islets of
     Langerhans, which secrete insulin
     and glucagon. Many other organs
     also secrete hormones that help
     regulate body functions.

Selected Terms

precursor (pre'kur-sor)
synergistic (sin''er-jis'tik)
adenyl cyclase (ad''ē-nil si'klās)
calmodulin (kal''mod-u'lin)
phosphodiesterase
  (fos''fo-di-es'ter-ās)
cecum (se'kum)
diverticulum (di''ver-tik'u-lum)
infundibulum (in''fun-dib'u-lum)
adenohypophysis (ad''ē-no-hi-pof'ĭ-sis)
corticotrophin (kor''ti-ko-tro'fin)
hypophyseal (hi''po-fiz'e-al)
interstitial (in''ter-stish'al)
adrenal (ah-dre'nal)
chromaffin (kro-maf'in)
corticosteroids (kor''ti-ko-ste'roids)
tetraiodothyronine
  (tet''rah-i''o-do-thi'ro-nēn)
thyroxine (thi-rok'sin)
triiodothyronine (tri''i-o''do-thi'ro-nēn)
glucagon (gloo'kah-gon)
islets of Langerhans
  (i'lets of Lahng'er-hanz)
thymus (thi'mus)

Hormones: Actions and Interactions

Hormones are regulatory molecules that are secreted into the blood by endocrine glands, which may be specialized organs or may be organs that perform endocrine as well as other functions. The effect of a given hormone on a given target tissue is relatively specific but is affected by the hormone's concentration and by the concentration of other hormones in the blood.

Objective 1. Define the terms *prohormone* and *prehormone,* and give examples of each.

Objective 2. Describe how the concentration of a given polypeptide hormone can affect the response of its target tissue to subsequent exposures to the same hormone.

Objective 3. Explain the different ways that one hormone can affect the response of a target tissue to a different hormone.

Endocrine *(en'do-krin)* **glands** (fig. 18.1), which lack the ducts present in exocrine glands, secrete biologically active chemicals called **hormones** into the blood (table 18.1). Hormones may be steroids, derivatives of amino acids, polypeptides, or glycoproteins. Whatever their chemical nature, hormones combine with specific receptor proteins in their target cells and modify the production or action of specific cellular enzymes and other proteins.

Hormones affect the metabolism of their target organs and, by this means, help regulate (1) total body metabolism, (2) growth, and (3) reproduction. The effects of hormones on body metabolism and growth are discussed in chapter 25; the regulation of reproductive functions by hormones is included in chapters 27 and 28.

Organs such as the adrenals, thyroid, parathyroids, pituitary, and pineal glands appear to be exclusively endocrine in function. Other organs of the endocrine system secrete hormones while they serve additional functions. The stomach and small intestine, for examples, secrete hormones that help regulate their digestive functions. The pancreas is both an exocrine and an endocrine gland—pancreatic juice is an exocrine secretion that enters the pancreatic duct, whereas the hormones insulin and glucagon are secreted into the blood by groups of cells known as the islets of Langerhans in the pancreas (fig. 18.1). Likewise, the skin, thymus, liver, kidneys, and brain secrete hormones and perform other specialized functions.

---

endocrine: Gk. *endon,* within; *krinein,* to separate
exocrine: Gk. *exo,* outside, *krinein,* to separate
hormone: Gk. *hormon,* to set in motion

Figure 18.1. (a) the anatomy of some of the endocrine glands. (b) the islets of Langerhans within the pancreas.

(b)

Activation and Inactivation of Hormones

Hormone molecules that affect the metabolism of target cells are often derived from less active "parent," or **precursor** *(pre'kur-sor)* molecules. When cells of the islets of Langerhans produce insulin and when cells of the anterior pituitary produce ACTH (adrenocorticotrophic hormone), for example, these hormones are derived from larger, less active polypeptides within the endocrine cells. These larger peptides are called **prohormones.** Under normal conditions, prohormones are either not secreted at all or are only secreted in very small amounts.

In the case of insulin and ACTH, the conversion of the less active precursor molecules into the fully active hormones occurs in the endocrine glands prior to secretion. Thyroxine (T$_4$), secreted by the thyroid, testosterone (a sex steroid from the testes), and vitamin D$_3$ (secreted by the skin), however, are converted into more active molecules within their target cells (table 18.2). Since the molecules that are secreted by some endocrine glands are inactive until changed by their target cells, the secreted forms are sometimes called **prehormones.**

The concentration of hormones (or prehormones) in the blood primarily reflects the rate of secretion by the endocrine glands—hormones do not generally accumulate in the blood. This is because the *half-life* of hormones in the blood is quite short, ranging from under two minutes to a few hours for most hormones. Hormones are rapidly removed from the blood by target organs and by the liver. Hormones removed from the blood by the liver are converted by enzymatic reactions to less active products. Steroids, for example, are converted to more polar derivatives. These less active, more water-soluble polar derivatives are released into the blood and are excreted in the urine and bile.

Effects of Hormone Concentrations on Tissue Response

The effects of hormones are very concentration dependent. Normal tissue responses are only produced when the hormones are present within their normal, or *physiological,* range of concentrations. When some hormones are taken in abnormally high, or *pharmacological,* concentrations (as when they are taken as drugs), they may produce noncharacteristic effects. This may be partly due to

Table 18.1 A partial list of the endocrine glands

Endocrine gland	Major hormones	Primary target organs	Primary effects
Adrenal cortex	Cortisol Aldosterone	Liver, muscles Kidneys	Glucose metabolism; Na$^+$ retention, K$^+$ excretion
Adrenal medulla	Epinephrine	Heart, bronchioles, blood vessels	Adrenergic stimulation
Hypothalamus	Releasing and inhibiting hormones	Anterior pituitary	Regulates secretion of anterior pituitary hormones
Intestine	Secretin and cholecystokinin	Stomach, liver, and pancreas	Inhibits gastric motility; stimulates bile and pancreatic juice secretion
Islets of Langerhans (pancreas)	Insulin Glucagon	Many organs Liver and adipose tissue	Insulin promotes cellular uptake of glucose and formation of glycogen and fat; glucagon stimulates hydrolysis of glycogen and fat
Ovaries	Estradiol-17β and progesterone	Female genital tract and mammary glands	Maintains structure of genital tract; promotes secondary sexual characteristics
Parathyroids	Parathyroid hormone	Bone, intestine, and kidneys	Increases Ca^{++} concentration in blood
Pineal	Melatonin	Hypothalamus and anterior pituitary	Affects secretion of gonadotrophic hormones
Pituitary, anterior	Trophic hormones	Endocrine glands and other organs	Stimulates growth and development of target organs; stimulates secretion of other hormones
Pituitary, posterior	Antidiuretic hormone Oxytocin	Kidneys, blood vessels Uterus, mammary glands	Antidiuretic hormone promotes water retention and vasoconstriction. Oxytocin stimulates contraction of uterus and mammary secretory units
Stomach	Gastrin	Stomach	Stimulates acid secretion
Testes	Testosterone	Prostate, seminal vesicles, other organs	Stimulates secondary sexual development
Thymus	Thymosin	Lymph nodes	Stimulates white blood cell production
Thyroid	Thyroxine (T_4) and triiodothyronine (T_3)	Most organs	Growth and development; stimulates basal rate of cell respiration (basal metabolic rate or BMR)

Source: Fox, Stuart Ira, *Human Physiology.* © 1984 Wm. C. Brown Publishers, Dubuque, Iowa. All Rights Reserved. Reprinted by permission.

Table 18.2 Conversion of some prehormones into biologically active derivatives

Endocrine gland	Prehormone	Active products	Comments
Skin	Vitamin D_3	1,25-dihydroxy vitamin D_3	Hydroxylation reactions occur in the liver and kidneys.
Testes	Testosterone	Dihydrotestosterone (DHT)	DHT and other 5α-reduced androgens are formed in most androgen-dependent tissue.
		Estradiol-17β (E_2)	E_2 is formed in the brain from testosterone, where it is believed to affect both endocrine function and behavior; small amounts of E_2 are also produced in the testes.
Thyroid	Thyroxine (T_4)	Triiodothyronine (T_3)	Conversion of T_4 to T_3 occurs in almost all tissues.

Source: Fox, Stuart Ira, *Human Physiology.* © 1984 Wm. C. Brown Publishers, Dubuque, Iowa. All Rights Reserved. Reprinted by permission.

the fact that abnormally high concentrations of a hormone may cause the hormone to bond to tissue receptor proteins of different but related hormones. Also, since some steroid hormones can be converted by their target cells into products with other biological effects (such as the conversion of androgens into estrogens), the administration of large amounts of one steroid can result in the production of a significant amount of other steroids with different effects.

Pharmacological doses of hormones, particularly of steroids, can thus have widespread and often damaging "side effects." People with inflammatory diseases who are treated with high doses of cortisone over long periods of time, for example, may develop characteristic changes in bone and soft tissue structure. Contraceptive pills, which contain sex steroids, have a number of potential side effects that could not have been predicted at the time the pill was first introduced.

Desensitization and Downregulation.
Prolonged exposure to high concentrations of polypeptide hormones has been found to *desensitize* the target cells. Subsequent exposure to the same concentration of the same hormone thus produces less of a target tissue response. This desensitization may be partially due to the fact that high concentration of these hormones cause a decrease in the number of receptor proteins in their target cells, a phenomenon called *downregulation*. Such desensitization and downregulation of receptors has been shown to occur, for example, in adipose cells exposed to high concentration of insulin and in testicular cells exposed to high concentrations of luteinizing hormone (LH).

Hormone Interactions

The responsiveness of a target tissue to a particular hormone is affected not only by the concentration of that hormone, but also by the effects of other hormones on that tissue. One hormone can modify the response of a target cell to another hormone by (1) altering the sensitivity of the target cell to the other hormone, either by changing the number of receptor proteins or by causing other changes; (2) altering the activity of enzymes needed to activate the other hormone (if the other hormone functions as a prehormone); and (3) altering the activity of target cell enzymes that function to inactivate the other hormone.

Permissive and Synergistic Effects.
A hormone is said to have a **permissive effect** when it enhances the response of a target organ to another hormone or when it increases the activity of another hormone. Prior exposure of the

Table 18.3 Types of interactions between the effects of different hormones on target cells

Interactions	Description	Example
Permissive effects	The effects of one hormone on a target organ require previous or simultaneous exposure of the organ to other hormones.	Estrogen stimulates the production of progesterone receptors in the uterus, thus having a permissive effect on the response of the uterus to progesterone.
Synergistic effects	The normal response of a target organ depends on the sum of the effects of different hormones.	Follicle-stimulating hormone (FSH) is required for the beginning and final stages of sperm production in testes; androgen is required for the intermediate stages of spermatogenesis.
Antagonistic effects	The effects of one hormone block or are opposite to the effects of another in a target organ.	Estrogen inhibits the formation of prolactin receptors in mammary glands, thus antagonizing the ability of prolactin to stimulate milk production.

Source: Fox, Stuart Ira, *Human Physiology.* © 1984 Wm. C. Brown Publishers, Dubuque, Iowa. All Rights Reserved. Reprinted by permission.

uterus to estrogen, for example, induces the formation of receptor proteins for progesterone, which increases the response of the uterus when it is subsequently exposed to progesterone. Estrogen thus has a permissive effect on the responsiveness of the uterus to progesterone. Prior exposure of the gonads to FSH, as another example, stimulates the production of LH receptors, which enhances the responsiveness of the testes and ovaries to LH.

Vitamin D_3 is a prehormone that must first be converted by enzymes in the kidneys and liver into the active hormone 1,25-dihydroxy vitamin D_3, which helps to raise blood calcium levels. Parathyroid hormone (PTH) has a permissive effect on the actions of 1,25-dihydroxy vitamin D_3 because it stimulates the activity of these hydroxylating enzymes. Conversely, 1,25-dihydroxy vitamin D_3 also has a permissive effect on the ability of PTH to stimulate the intestinal absorption of Ca^{++}.

Two or more hormones are said to exert **synergistic** *(sin''er-jis'tik)* **effects** when their effects are complementary, so that the sum of their actions cannot be reproduced by any one hormone separately. The development of sperm, for example, requires the synergistic action of both FSH and androgens during puberty (table 18.3). Likewise, the ability of mammary glands to produce and secrete milk requires the synergistic action of many hormones—estrogen, cortisol, prolactin, oxytocin, and others.

Table 18.4 Functional categories of hormones, based on the location of their receptor proteins and the mechanisms of their action

Types of hormones	Secreted by	Location of receptors	Effects of hormone–receptor interaction
Catecholamines, polypeptides, glycoproteins	All glands except adrenal cortex, gonads, and thyroid	Outer surface of cell membrane	Stimulates production of intracellular "second messenger," which activates previously inactive enzymes
Steroids	Adrenal cortex, testes, ovaries	Cytoplasm of target cells	Stimulates translocation of hormone–receptor complex to nucleus and activation of specific genes
Thyroxine (T_4)	Thyroid	Nucleus of target cells	After conversion to triiodothyronine (T_3), activates specific genes

Antagonistic Effects.   In some situations the actions of one hormone antagonize the effects of another. Lactation during pregnancy, for example, is prevented because the high concentration of estrogen in the blood inhibits the secretion and action of prolactin. Another example of antagonism is the action of insulin and glucagon (two hormones from the islets of Langerhans) on adipose tissue; the formation of fat is promoted by insulin, whereas glucagon promotes fat breakdown.

1. *Define the terms* prohormone *and* prehormone, *and give examples of each.*
2. *Distinguish between the physiological effect and the pharmacological effect of a hormone.*
3. *Explain how the response of the body to a given hormone can be affected by the concentration of that hormone in the blood.*
4. *Explain how the effects of one hormone can influence the response of a target organ to other hormones.*

Mechanisms of Hormone Action

Although each hormone has its own characteristic effects on specific target cells, hormones that are in the same chemical category have similar mechanisms of action. These similarities involve the location of cellular receptor proteins and the events in the target cells after hormone-receptor interaction has occurred.

Objective 4. Describe the differences in cellular location of the receptor proteins for different hormones.

Objective 5. Explain how steroid hormones and thyroxine exert their effects on target cells.

Objective 6. Describe what is meant by a "second messenger" in hormone action, and explain the sequence of events that occurs after a hormone bonds to a receptor when cyclic AMP functions as a second messenger.

Objective 7. Explain how calcium may function as a second messenger in the action of some hormones.

Hormones are delivered by the blood to every cell in the body, but only the **target cells** are able to respond to each hormone. In order to respond to a hormone, a target cell must have specific receptor proteins for that hormone. Receptor protein–hormone interaction is highly specific, much like the interaction of an enzyme with its substrate. Receptor proteins are not enzymes, however, and since they cannot be detected by the techniques used to assay enzymes, other methods must be used. These methods are based on the observation that hormones bond to receptors with a *high affinity* (high bond strength, as compared to the nonspecific binding of hormones) and with a *low capacity*. The latter characteristic refers to the fact that there are only a limited number of receptors per target cell (usually a few thousand), so that it is possible to saturate receptors with hormones. Notice that the characteristics of specific bonding and saturation of receptor proteins are similar to the characteristics of enzyme and transport carrier proteins discussed in previous chapters.

The location of a hormone's receptor proteins in its target cells depends on the chemical nature of the hormone. Based on the location of the receptor proteins, hormones can be grouped into three categories: (1) receptor proteins within the nucleus of target cells—*thyroid hormones;* (2) receptor proteins within the cytoplasm of target cells—*steroid hormones* and (3) receptor proteins in the outer surface of the target cell membrane—*amine, polypeptide, and glycoprotein hormones.* This information is summarized in table 18.4.

Figure 18.2. The mechanism of the action of a steroid hormone (*H*) on the target cells.

Mechanisms of Steroid and Thyroid Hormone Action

Steroid and thyroid hormones are similar in size and in the fact that they are nonpolar and thus are not very water-soluble. Unlike other hormones, therefore, steroids and thyroid hormones (primarily thyroxine) do not travel dissolved in the aqueous plasma but instead are transported to their target cells attached to carrier proteins. These hormones then dissociate from the carrier proteins in the blood and easily pass through the lipid component of the target cell's membrane.

Steroid Hormones. Once through the cell membrane, steroid hormones attach to *cytoplasmic receptor proteins* in the target cells. The steroid hormone–receptor protein complex then *translocates* to the nucleus and attaches by means of the receptor proteins to the chromatin. The sites of attachment, termed *acceptor sites,* are specific for the target tissue. This specificity is believed to be determined by acidic (nonhistone) proteins in the chromatin. According to one theory, part of the receptor bonds to an acidic protein while a different part of the receptor bonds to DNA.

The attachment of the receptor protein–steroid complex to the acceptor site "turns on genes." Specific genes become activated by this process and produce nuclear RNA, which is then processed into messenger RNA (mRNA). This new mRNA enters ribosomes and codes for the production of new proteins. Since some of these newly synthesized proteins may be enzymes, the metabolism of the target cell is changed in a specific manner (fig. 18.2). Steroid hormones, in short, affect their target cells by stimulating genetic transcription (RNA synthesis) followed by genetic translation (protein synthesis).

Thyroxine. The major hormone secreted by the thyroid gland is thyroxine, or **tetraiodothyronine** *(tet″rah-i″o-do-thi′ro-nēn)* (**T₄**). Like steroid hormones, thyroxine travels

Figure 18.3. The mechanism of the action of T_3 (triiodothyronine) on the target cells.

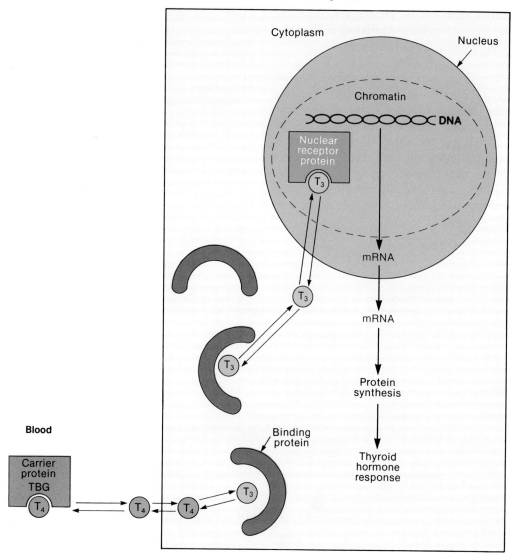

in the blood attached to carrier proteins (primarily attached to *thyroxine-binding globulin,* or *TBG*). The thyroid also secretes a small amount of **triiodothyronine,** or T_3. The carrier proteins have a higher affinity for T_4 than for T_3, however, and as a result, the amount of unbound (or "free") T_3 is about ten times greater than the amount of free T_4 in the plasma.

Approximately 99.96% of the thyroxine in the blood is attached to carrier proteins; the rest is free. Only the free thyroxine can enter target cells; the protein-bound thyroxine serves as a "reservoir" of this hormone in the blood (this is why it takes a couple of weeks after surgical removal of the thyroid for the symptoms of hypothyroidism to develop). Once the free thyroxine passes into the target cell cytoplasm, it is enzymatically converted into T_3. Thyroxine, therefore, is not the chemical form of the hormone that is active in the target cells.

Inactive T_3 receptor proteins are already in the nucleus attached to chromatin. These receptors are inactive until T_3 enters the nucleus from the cytoplasm. The attachment of T_3 to the chromatin-bound receptor proteins activates genes and results in the production of new mRNA and new proteins. This sequence of events is summarized in figure 18.3.

Membrane Receptor Proteins and Second Messengers

Amine, polypeptide, and glycoprotein hormones cannot pass through the lipid barrier of the target cell membrane. Although some of these hormones may enter the cell by pinocytosis, most of the effects of these hormones are believed to result from interaction of these hormones with receptor proteins in the outer surface of the target cell membrane. Since these hormones do not have to enter the target cells to exert their effects, other molecules must mediate the actions of these hormones within the target cells. If you think of hormones as "messengers" from the endocrine glands, the intracellular mediators of the hormone's action can be called **second messengers.**

Cyclic AMP. Cyclic adenosine monophosphate (abbreviated cAMP) was the first "second messenger" to be discovered and is the best understood. The hormonal effects of epinephrine (adrenalin) and norepinephrine are due to the actions of cAMP, and the effects of norepinephrine released as a neurotransmitter by sympathetic nerve endings are also due to cAMP production. It was later discovered that the effects of many (but not all—see table 18.5) polypeptide and glycoprotein hormones are also mediated by cAMP.

Table 18.5 Hormones that use cAMP as a second messenger and hormones that use other second messengers

Hormones that stimulate adenyl cyclase	Hormones that do not stimulate adenyl cyclase
Adrenocorticotrophic hormone (ACTH)	Catecholamines (α-adrenergic)
Calcitonin	Growth hormone (GH)
Epinephrine (β-adrenergic)	Insulin
Follicle-stimulating hormone (FSH)	Oxytocin
Glucagon	Prolactin
Luteinizing hormone (LH)	Somatomedin
Parathyroid hormone	Somatostatin
Thyrotrophin-releasing hormone (TRH)	
Thyroid-stimulating hormone (TSH)	
Antidiuretic hormone	

Source: Fox, Stuart Ira, *Human Physiology.* © 1984 Wm. C. Brown Publishers, Dubuque, Iowa. All Rights Reserved. Reprinted by permission.

Figure 18.4. Cyclic AMP (cAMP) as a second messenger in the action of catecholamine, polypeptide, and glycoprotein hormones.

Table 18.6 Sequence of events that occurs with cyclic AMP as a second messenger

1. The hormones combine with their receptors on the outer surface of target cell membranes.
2. Hormone-receptor interaction stimulates activation of adenyl cyclase on the cytoplasmic side of the membranes.
3. Activated adenyl cyclase catalyzes the conversion of ATP to cyclic AMP (cAMP) within the cytoplasm.
4. Cyclic AMP activates protein kinase enzymes that were already present in the cytoplasm in an inactive state.
5. Activated cAMP-dependent protein kinase transfers phosphate groups to (phosphorylates) other enzymes in the cytoplasm.
6. The activity of specific enzymes is either increased or inhibited by phosphorylation.
7. Altered enzyme activity mediates the target cell's response to the hormone.

Source: Fox, Stuart Ira, *Human Physiology*. © 1984 Wm. C. Brown Publishers, Dubuque, Iowa. All Rights Reserved. Reprinted by permission.

The bonding of these hormones to their membrane receptor proteins activates an enzyme called **adenyl cyclase** *(ad"ĕ-nil si'klas)*. This enzyme is built into the cell membrane and when activated, it catalyzes the following reaction:

$$ATP \rightarrow cAMP + PP_i$$

Adenosine triphosphate (ATP) is thus converted into cAMP plus two inorganic phosphates (*pyrophosphate,* abbreviated PP_i). As a result of the interaction of the hormone with its receptor and the activation of adenyl cyclase, therefore, the intracellular concentration of cAMP is increased. Cyclic AMP activates a previously inactive enzyme in the cytoplasm called **protein kinase.** The inactive form of this enzyme consists of two subunits: a catalytic subunit and an inhibitory subunit. The enzyme is produced in an inactive form and becomes active only when cAMP attaches to the inhibitory subunit and causes it to dissociate from the catalytic subunit, which then becomes active (fig. 18.4). The hormone, in summary—acting through an increase in cAMP production—causes an increase in protein kinase enzyme activity within its target cells.

Active protein kinase catalyzes the attachment of phosphate groups to different proteins in the target cells. This causes some enzymes to become activated, and others to become inactivated. Cyclic AMP, acting through protein kinase, thus modulates the activity of enzymes that are already present in the target cell. This alters the metabolism of the target tissue in a manner characteristic of the actions of that specific hormone (table 18.6).

Like all biologically active molecules, cAMP must be rapidly inactivated for its regulation to be effective. This function is served by an enzyme called **phosphodiesterase** *(fos"fo-di-es'ter-ās)* within the target cells, which hydrolyzes cAMP into inactive fragments.

Drugs that inhibit the activity of phosphodiesterase thus prevent the breakdown of cAMP and result in increased concentrations of cAMP within the target cells. The drug **theophylline** *(the'o-fil'in),* for example, is used clinically to raise cAMP levels within bronchiolar smooth muscle. This duplicates the effect of epinephrine on the bronchioles (producing dilation) in people who suffer from asthma.

Cyclic GMP and Ca⁺⁺ as Second Messengers. Liver cells respond to epinephrine by an increase in cAMP production and a breakdown of stored glycogen; the action of insulin, in contrast, produces an increase in the production of glycogen. Clearly, since the action of epinephrine is mediated by cAMP, the action of insulin in liver cells must be mediated by a different second messenger.

Some of the hormones that do not use cAMP as a second messenger may act via production of **cyclic guanosine** *(gwan'o-sin)* **monophosphate (cGMP).** Indeed, cAMP and cGMP produced in response to the action of different hormones may act antagonistically within the target cells. The control of mitotic cell division and the cell cycle, for example, appears to be related to the ratio of cAMP to cGMP. In addition to these two "cyclic nucleotides," **calcium ion (Ca^{++})** may also function as a second messenger in the action of certain hormones.

Active transport carriers help to maintain a very steep concentration gradient for Ca^{++}, which is at a much higher concentration in the extracellular fluid than it is in the cytoplasm of cells. Transient inhibition of this Ca^{++} pump produces a sudden inflow of Ca^{++} that can serve as a second messenger. Analogous events occur in muscles and nerves, where calcium stimulates contraction and the release of neurotransmitters, respectively, in response to electrical stimulation.

In some tissue cells, Ca^{++} binds to and activates a cellular protein called **calmodulin** *(kal"mod-u'lin).* The activation of calmodulin by a sudden influx of Ca^{++} is analogous to the activation of protein kinase by cAMP. Recent experiments, for example, suggest that insulin may exert at least some of its effects via a transient influx of Ca^{++} that activates calmodulin.

Activated calmodulin, like activated protein kinase, can affect various target cell enzymes. Figure 18.5 illustrates how a single process—the hydrolysis of glycogen—can be stimulated in one case by the hormonal activation of the cAMP/protein kinase regulatory system and in another case by the neural stimulation that activates the Ca^{++}/calmodulin system. Glycogen breakdown in the liver

Figure 18.5. The hydrolysis of glycogen (glycogenolysis) in the liver and the skeletal muscles in response to cyclic AMP (cAMP) and Ca^{++} as second messengers. Both protein kinase and calmodulin are proteins activated by the second messengers.

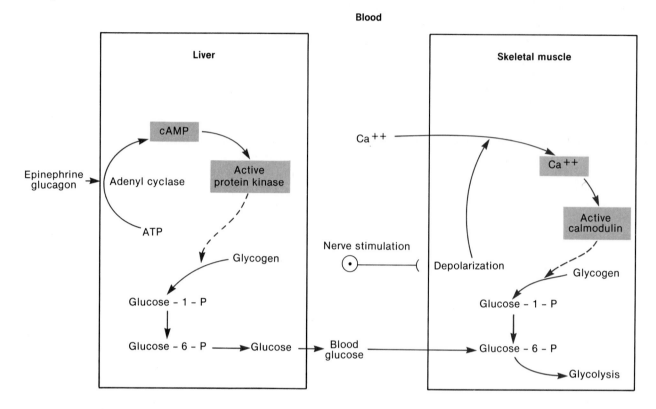

in response to epinephrine is mediated by cAMP and protein kinase; glycogen breakdown in the skeletal muscles during nerve stimulation appears to be mediated by Ca^{++} inflow and the activation of calmodulin. The effects of cAMP and of Ca^{++} in different tissues are summarized in table 18.7.

A given cell may respond to a variety of different hormones that produce similar as well as antagonistic effects. The response of a target cell to these different hormones may depend upon interactions between different second messengers. Calmodulin, for example, has been found to stimulate adenyl cyclase (and thus to raise cAMP levels in target cells) under some circumstances and to stimulate phosphodiesterase activity (and thus to lower cAMP) in other instances. These effects, together with the antagonism between cAMP and cGMP, suggest that the effects of various hormones in a target cell may depend on complex and poorly understood interactions between different second messengers.

1. *Using diagrams, describe how steroid hormones and thyroxine exert their effects on their target cells.*
2. *Using a diagram, describe how cyclic AMP is produced within a target cell in response to hormone stimulation. List three hormones that use cAMP as a second messenger.*
3. *Explain how cAMP functions as a second messenger in the action of some hormones.*
4. *Explain how Ca^{++} can have regulatory functions in different tissues and why it may also be considered to be a second messenger in the action of some hormones.*

Development of the Endocrine System

The embryonic development of the endocrine glands establishes the anatomical relationships between the glands and associated structures. A knowledge of the origin and nature of these structural relationships aids the understanding of endocrine function.

Table 18.7 Comparison of the effects of cyclic AMP (cAMP) and Ca^{++} as second messengers in different tissues

Second messenger	Mechanism	Effects: nerves	Effects: muscles	Effects: hormones
cAMP	Activation of cAMP-dependent protein kinase	Opens ion channels in postsynaptic membrane	Phosphorylation of myosin in skeletal muscle in response to depolarization (significance controversial)	Activation or inactivation of target cell enzymes in response to many polypeptide and glycoprotein hormones
Ca^{++}	Activation of Ca^{++}-dependent regulatory protein (calmodulin)	Mediates release of neurotransmitter in response to depolarization	Phosphorylation of myosin activates actomyosin ATPase in smooth muscles and stimulates contraction	May mediate the effects of insulin on target cells; may increase cAMP by stimulation of adenyl cyclase or decrease cAMP by stimulation of phosphodiesterase

Source: Fox, Stuart Ira, *Human Physiology.* © 1984 Wm. C. Brown Publishers, Dubuque, Iowa. All Rights Reserved. Reprinted by permission.

Objective 8. Describe the structure and embryonic development of the pituitary gland.

Objective 9. Describe the embryonic development of the thyroid gland and pancreas.

Objective 10. Describe the embryonic development of the adrenal cortex and adrenal medulla.

The endocrine system is the only body system whose organs are not structurally connected. Because these organs are isolated from each other and are distributed throughout the body, each endocrine organ has a separate and independent development. All three embryonic germ layers (endoderm, mesoderm, and ectoderm) contribute to the development of the endocrine system.

The development of the pituitary, thyroid, pancreas, and adrenal glands will be discussed separately. The development of the testes and ovaries is discussed in chapters 27 and 28, respectively.

Pituitary Gland

Although the pituitary gland is a single organ, it is actually composed of two distinct types of tissues that have different embryonic origins, that secrete different hormones, and that are under different control systems. The pituitary gland, or **hypophysis** *(hi-pof'ĭ-sis),* has two parts. The anterior portion, called the **adenohypophysis** *(ad"ĕ-no-hi-pof'ĭ-sis),* develops from ectoderm that lines the primitive oral cavity. The posterior portion, called the **neurohypophysis,** develops from neuroectoderm of the developing brain (fig. 18.6).

The adenohypophysis begins to develop during the third week as a diverticulum *(di"ver-tik'u-lum),* a pouch-like extension, called **Rathke's pouch.** It arises from the roof of the primitive oral cavity and grows toward the brain. At the same time, another diverticulum called the

infundibulum *(in"fun-dib'u-lum)* forms from the diencephalon on the inferior aspect of the brain. As the two diverticula come in contact, Rathke's pouch loses its connection with the oral cavity and the primordial tissue of the adenohypophysis is formed. The fully developed adenohypophysis includes the **pars distalis, pars tuberalis,** and **pars intermedia.**

The neurohypophysis develops as the infundibulum extends inferiorly from the diencephalon to come in contact with the developing adenohypophysis. The fully formed neurohypophysis consists of the infundibulum and the **pars nervosa.** Specialized nerve fibers that connect the hypothalamus with the pars nervosa develop within the infundibulum.

Notice that the neurohypophysis is essentially an extension of the brain, and indeed the entire pituitary gland, like the brain, is surrounded by the meninges. The adenohypophysis, in contrast, is derived from nonneural tissue (the same embryonic tissue that will form the epithelium over the roof of the mouth). These different embryologies have important consequences with regards to the physiology of the two parts of the pituitary gland.

Thyroid Gland

The thyroid gland is derived from endoderm and begins its development during the third week as a thickening in the floor of the primitive pharynx. The thickening soon out-pouches downward as the **thyroid diverticulum** (fig. 18.7). A narrow **thyroglossal duct** connects the descending primordial thyroid tissue to the pharynx. As the descent continues, the tongue starts to develop, and the opening into the thyroglossal duct, called the **foramen cecum** *(se'kum),* pierces through the base of the tongue.

---

pituitary: L. *pituita,* phlegm (this gland was originally thought to secrete mucus into nasal cavity)

adenohypophysis: Gk. *adeno,* gland; *hypo,* under; *physis,* a growing

Rathke's pouch: from Martin H. Rathke, German anatomist, 1793–1860

---

infundibulum: L. *infundibulum,* a funnel

foramen cecum: L. *foramen,* open passage; *caecum,* blind tube

Figure 18.6. The embryonic development of the pituitary gland. (*a*) a midsagittal section through the head of an embryo at four weeks shows the neuroectoderm of the developing diencephalon and the ectoderm of the oral cavity, which both contribute in forming the pituitary gland. (*b–d*) successive stages of the developing pituitary gland. (*e*) by eight weeks, Rathke's pouch is no longer connected to the oral cavity. (*f*) the development of the pituitary gland is completed during the fetal stage.

Figure 18.7. The embryonic development of the thyroid gland. (*a*) at four weeks, a thyroid diverticulum begins to form in the floor of the pharynx at the level of the second brachial arch. (*b,c*) the thyroid diverticulum extends downward during the fifth and sixth weeks. (*d*) a sagittal section through the head and neck of an adult shows the path of development and final position of the thyroid gland.

(d) Regressing stalk of Rathke's pouch

(e) Former site of stalk of Rathke's pouch

Anterior lobe
Pars intermedia

Optic chiasma
Pars tuberalis
Pars distalis (anterior lobe)
Developing sphenoid bone
Pharyngeal roof

Median eminence
Infundibular stem
Pars intermedia
Pars nervosa (posterior lobe)
Colloid-containing vesicles

(f)

By the seventh week, the thyroid gland occupies a position immediately inferior to the larynx and around the front and lateral sides of the trachea. Once in position, the thyroglossal duct disappears and the foramen cecum regresses in size to a vestigial pit that persists throughout life.

Pancreas

The pancreas begins development during the fifth week, as dorsal and ventral pancreatic buds of endoderm arise from the caudal portion of the foregut. These primordial pancreatic tissues continue to grow independently until the ventral bud is carried dorsally and fuses with the dorsal bud as the duodenum rotates to the right. The fusion of the two portions of the pancreas occurs during the seventh week.

The pancreas thus develops from and maintains connections with (via the pancreatic duct) the intestine. The exocrine products of the pancreas, contained in **pancreatic juice,** are channeled through the pancreatic duct to the intestine. The endocrine parts of the pancreas—the **islets of Langerhans**—however, secrete their products (insulin and glucagon) into the blood. The precise origin of the endocrine cells of the pancreas remains uncertain. It is believed that some of the endocrine cells develop as buds from pancreatic ductules and that others arise from neuroectodermal cells that migrate to the pancreas to form these endocrine cells and the autonomic innervation of the pancreas.

pancreas: Gk. *pan*, all; *kreas*, flesh

Adrenal Glands

The adrenal glands begin development during the fifth week from two different germ layers. Each adrenal gland has an outer part, or **cortex,** which develops from mesoderm, and an inner part, or **medulla,** which develops from neuroectoderm. The mesodermal ridge that forms the adrenal cortex is in the same region from which the gonads develop.

The neuroectodermal cells that form the adrenal medulla are derived from the neural crest of the neural tube. The developing adrenal medulla is gradually encapsulated by the adrenal cortex, a process that continues into the fetal stage. The formation of the adrenal gland is not completed until the end of the third year of age.

Notice that the adrenal gland, like the pituitary, has a dual origin; part is neural and part is not. Like the pituitary, the adrenal cortex and medulla are really two different endocrine tissues, which are located in the same organ but secrete different hormones and are regulated by different control systems.

1. Describe the structure and embryonic origins of the adenohypophysis and neurohypophysis.
2. Name two endocrine glands that develop from endoderm, and describe the developmental process shared in common by these two glands.
3. Describe the embryonic formation of the adrenal glands, and explain how this is similar to the formation of the pituitary gland.

adrenal: L. *ad*, to; *renes*, kidney

Figure 18.8. The structure of the pituitary gland as seen in the sagittal view.

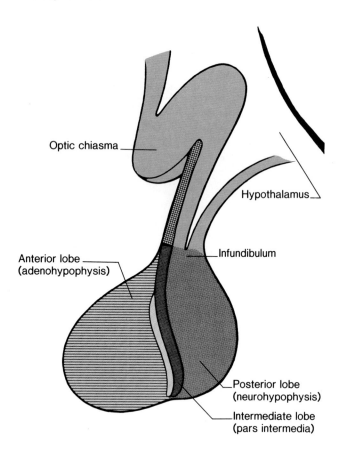

Optic chiasma

Hypothalamus

Infundibulum

Anterior lobe (adenohypophysis)

Posterior lobe (neurohypophysis)

Intermediate lobe (pars intermedia)

Pituitary Gland

The neurohypophysis secretes hormones that are produced by the hypothalamus, whereas the adenohypophysis secretes its own hormones in response to regulation from hypothalamic hormones. The secretions of the pituitary gland are thus controlled by the hypothalamus, as well as by negative feedback influences from the target glands.

Objective 11. List the hormones secreted by the adenohypophysis and neurohypophysis.

Objective 12. Describe, in a general way, the actions of anterior pituitary hormones, and explain why the term "master gland" has been used in the past to describe the adenohypophysis.

Objective 13. Describe how the secretions of anterior and posterior pituitary hormones are controlled by the hypothalamus.

Objective 14. Explain how the secretion of anterior pituitary hormones is regulated by negative feedback.

The pituitary gland, or hypophysis, is located on the inferior aspect of the brain in the region of the diencephalon, and is attached to the hypothalamus by a stalklike structure called the *infundibulum* (fig. 18.8). The pituitary is a rounded, pea-shaped gland measuring about 1.3 cm (0.5 in.) in diameter. It is covered by the dura mater and is supported by the sella turcica of the sphenoid bone. The circle of Willis surrounds the highly vascular pituitary gland, providing it with a rich blood supply.

The pituitary gland is structurally and functionally divided into an anterior lobe, or **adenohypophysis** *(ad''ē-no-hi-pof'i-sis)*, and a posterior lobe called the **neurohypophysis.** The adenohypophysis consists of three parts: (1) the **pars distalis** is the bulbar portion; (2) the **pars tuberalis** is the thin extension in contact with the infundibulum; and (3) the **pars intermedia** is between the anterior and posterior parts of the pituitary. These parts are illustrated in figure 18.8.

The neurohypophysis is the neural part of the pituitary gland. It consists of the bulbar **pars nervosa,** which is in contact with the pars intermedia and the pars distalis of the adenohypophysis, and the **infundibulum,** which is the connecting stalk to the hypothalamus. Nerve fibers extend through the infundibulum along with minute neuroglia-like cells, called **pituicites** *(pǐ-tu'ǐ-sīt).*

The pituitary is the structure of the brain perhaps most subject to neoplasms. A tumor of the pituitary is generally easily detected as it begins to grow and interfere with hormonal activity. If a tumor of the pituitary grows superiorly, it may exert sufficient pressure on the optic chiasma to cause bitemporal hemianopia, which is blindness in the temporal field of vision of both eyes, or "tunnel vision." Surgical removal of a neoplasm of the pituitary gland (hypophysectomy) may be performed transcranially through the frontal bone or through the nasal cavity and sphenoidal air sinus.

Pituitary Hormones

There are eight important hormones secreted by the pituitary gland. The first six in the following list are secreted by the anterior pituitary (the pars distalis of the adenohypophysis). The last two hormones in the list are produced in the hypothalamus, transported through axons in the infundibulum to the posterior pituitary, and secreted by the posterior pituitary (the pars nervosa of the neurohypophysis) into the blood.

The hormones secreted by the pars distalis are called **trophic** *(trof'ik)* **hormones.** The term *trophic* means "food." This term is used because high amounts of the anterior pituitary hormones make their target glands hypertrophy. These hormones are secreted by the pars distalis:

1. **Growth hormone (GH, or somatotrophin [*so-mah-to-tro'fin*]).** This hormone stimulates growth in all body organs. It also promotes the movement of

amino acids into tissue cells and the incorporation of these amino acids into tissue proteins.

2. **Thyroid-stimulating hormone (TSH, or thyrotrophin).** This hormone stimulates the thyroid gland to produce and secrete thyroxine (tetraiodothyronine, or T_4).

3. **Adrenocorticotrophic hormone (ACTH, or corticotrophin).** This hormone stimulates the adrenal cortex to secrete the glucocorticoids, such as hydrocortisone (cortisol).

4. **Follicle-stimulating hormone (FSH, or folliculotrophin).** This hormone stimulates the growth and secretion of ovarian follicles in the ovaries of females and the production of sperm in the testes of males.

5. **Luteinizing hormone (LH, or luteotrophin).** This hormone and FSH are collectively called **gonadotrophic** *(gon''ah-do-tro'fik)* **hormones.** In females, LH stimulates ovulation and the conversion of the ovulated ovarian follicle into an endocrine structure called a corpus luteum. In males, LH (which is sometimes also called interstitial *[in''ter-stish'al]* cell-stimulating hormone, or ICSH) stimulates the secretion of male sex hormones (mainly testosterone) from the interstitial cells of Leydig in the testes.

6. **Prolactin.** This hormone is secreted by both males and females, but its function is only well understood in females where it stimulates the production of milk in the mammary glands of women after the birth of their babies.

The posterior pituitary, or pars nervosa, secretes only two hormones, both of which are produced in the hypothalamus and merely stored in the posterior lobe of the pituitary:

1. **Antidiuretic hormone (ADH, or vasopressin).** Antidiuretic *(an''ti-di-u-ret'ik)* hormone stimulates the kidneys to retain water so that less water is excreted in the urine and more water is retained in the blood. This hormone also causes vasoconstriction in experimental animals, although the significance of this effect in humans is controversial.

2. **Oxytocin.** This hormone has no known function in males, but in females it is known to stimulate contractions of the uterus during labor and contractions of the mammary gland alveoli and ducts, which result in the milk-ejection reflex during lactation.

Oxytocin, like ADH, is produced by neurons in the hypothalamus and travels down axons in the infundibulum to the pars nervosa, where it is stored and secreted in response to nerve impulses from the hypothalamus.

Figure 18.9. The posterior pituitary, or neurohypophysis, stores and secretes hormones (vasopressin and oxytocin) produced in neuron cell bodies within the supraoptic and paraventricular nuclei of the hypothalamus. These hormones are transported to the posterior pituitary by nerve fibers of the hypothalamo-hypophyseal tract.

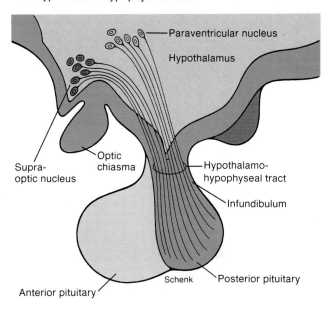

Injections of oxytocin may be given to a woman during labor if she is having difficulties in parturition. Increased amounts of oxytocin assist uterine contractions and generally speed up delivery. Oxytocin administration after parturition causes the uterus to regress in size and squeezes the blood vessels, thus minimizing the danger of hemorrhage.

Hypothalamic Control of the Neurohypophysis

The posterior pituitary (pars nervosa of the neurohypophysis) secretes two hormones: antidiuretic hormone (ADH) and oxytocin. These two hormones, however, are actually produced in neuron cell bodies of the **supraoptic** *(su''prah-op'tik)* **nuclei** and **paraventricular nuclei** of the hypothalamus. These nuclei within the hypothalamus are thus endocrine glands; the hormones they produce are transported along axons of the **hypothalamo-hypophyseal** *(hi''po-fiz'e-al)* **tract** (fig. 18.9) to the posterior pituitary, which stores and later secretes these hormones. The posterior pituitary is thus more a storage organ than a true gland.

The secretion of ADH and oxytocin from the posterior pituitary is controlled by **neuroendocrine** *(nu''ro-en'do-krin)* **reflexes.** In nursing mothers, for example, the stimulus of sucking acts via sensory nerve impulses to the hypothalamus to stimulate the reflex secretion of oxytocin.

Figure 18.10. The hormones secreted by the anterior pituitary and the target organs for those hormones.

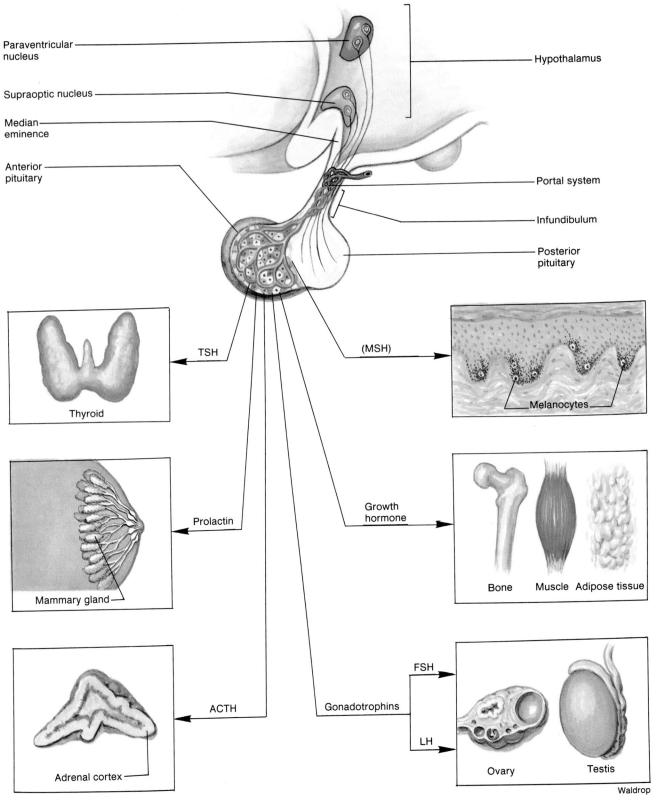

Paraventricular nucleus

Supraoptic nucleus

Median eminence

Anterior pituitary

Hypothalamus

Portal system

Infundibulum

Posterior pituitary

TSH

Thyroid

(MSH)

Melanocytes

Prolactin

Mammary gland

Growth hormone

Bone   Muscle   Adipose tissue

ACTH

Adrenal cortex

Gonadotrophins

FSH

LH

Ovary   Testis

Waldrop

Table **18.8** Hormones secreted by the anterior pituitary

Hormone	Target tissue	Stimulated by hormone	Regulation of secretion
ACTH (adrenocorticotrophic hormone)	Adrenal cortex	Secretion of glucocorticoids	Stimulated by CRH (corticotrophin-releasing hormone); inhibited by glucocorticoids
TSH (thyroid-stimulating hormone)	Thyroid gland	Secretion of thyroid hormones	Stimulated by TRH (thyrotrophin-releasing hormone); inhibited by thyroid hormones
GH (growth hormone)	Most tissue	Protein synthesis and growth; lipolysis and increased blood glucose	Inhibited by somatostatin; stimulated by growth hormone-releasing hormone
FSH (follicle-stimulating hormone) and LH (luteinizing hormone)	Gonads	Gamete production and sex steroid hormone secretion	Stimulated by GnRH (gonadotrophin-releasing hormone); inhibited by sex steroids
Prolactin	Mammary glands and other sex accessory organs	Milk production Controversial actions in other organs	Inhibited by PIH (prolactin-inhibiting hormone)
LH (luteinizing hormone)	Gonads	Sex hormone secretion: ovulation and corpus luteum formation	Stimulated by GnRH

Source: Fox, Stuart Ira, *Human Physiology.* © 1984 Wm. C. Brown Publishers, Dubuque, Iowa. All Rights Reserved. Reprinted by permission.

The secretion of ADH is stimulated by osmoreceptor neurons in the hypothalamus in response to a rise in blood osmotic pressure; its secretion is inhibited by sensory impulses from stretch receptors in the left atrium of the heart in response to a rise in blood volume.

Hypothalamic Control of the Adenohypophysis

At one time the anterior pituitary was called the "master gland" because it secretes hormones that regulate some other endocrine glands (see fig. 18.10 and table 18.8). Adrenocorticotrophic hormone (ACTH), thyroid-stimulating hormone (TSH), and the gonadotrophic hormones (FSH and LH) stimulate the adrenal cortex, thyroid, and gonads, respectively, to secrete their hormones. The anterior pituitary hormones also have a "trophic" effect on their target glands, in that the structure and health of the target glands depend on adequate stimulation by anterior pituitary hormones. The anterior pituitary, however, is not really the master gland, because secretion of its hormones is in turn controlled by hormones secreted by the hypothalamus.

Releasing and Inhibiting Hormones.   Since axons do not enter the anterior pituitary, hypothalamic control of the anterior pituitary is achieved through hormonal rather than neural regulation. Neurons in the hypothalamus produce releasing and inhibiting hormones, which are transported to axon endings in the basal portion of the hypothalamus. This region, known as the **median eminence,** contains blood capillaries that are drained by venules in the stalk of the pituitary.

The venules that drain the median eminence deliver blood to a second capillary bed in the anterior pituitary. Since this second capillary bed receives venous rather than arterial blood, the vascular link between the median eminence and the anterior pituitary comprises a portal system. (This is analogous to the hepatic portal system that delivers venous blood from the intestine to the liver.) The vascular link between the hypothalamus and the anterior pituitary is thus called the **hypothalamo-hypophyseal portal system.**

Neurons of the hypothalamus secrete short polypeptide hormones into this portal system which regulate the secretions of the anterior pituitary (table 18.9). Thyrotrophin-releasing hormone (**TRH**) stimulates the secretion of TSH, and corticotrophin *(kor″ti-ko-tro′fin)*- releasing hormone (**CRH**) stimulates the secretion of ACTH from the anterior pituitary. A single releasing hormone, gonadotrophin-releasing hormone, or **GnRH,** appears to stimulate the secretion of both gonadotrophic hormones (FSH and LH) from the anterior pituitary. The secretion of prolactin and of growth hormone from the anterior pituitary is regulated by hypothalamic-inhibitory hormones, known as **PIH** (prolactin-inhibiting hormone) and **somatostatin,** respectively. Recently, a specific hypothalamic-releasing hormone that stimulates growth hormone secretion has also been identified. The secretion of growth hormone from the anterior pituitary thus seems to be unique in that it is regulated by both a releasing and an inhibiting hormone from the hypothalamus. (The existence of a prolactin-releasing hormone from the hypothalamus has been suggested but not yet experimentally proven).

Table 18.9 Some hypothalamic hormones involved in the control of the anterior pituitary

Hypothalamic hormone	Structure	Effect on anterior pituitary	Action of anterior pituitary hormone
Corticotrophin-releasing hormone (CRH)	Presently unknown	Stimulates secretion of adrenocorticotrophic hormone (ACTH)	Stimulates secretions of adrenal cortex
Gonadotrophin-releasing hormone (GnRH)	10 amino acids	Stimulates secretion of follicle-stimulating hormone (FSH) and luteinizing hormone (LH)	Stimulates gonads to produce gametes (sperm and ova) and secrete sex steroids
Prolactin-inhibiting hormone (PIH)	Presently unknown	Inhibits prolactin secretion	Stimulates production of milk in mammary glands
Somatostatin	14 amino acids	Inhibits secretion of growth hormone	Stimulates anabolism and growth in many organs
Thyrotrophin-releasing hormone (TRH)	3 amino acids	Stimulates secretion of thyroid-stimulating hormone (TSH)	Stimulates secretion of thyroid gland

Source: Fox, Stuart Ira, *Human Physiology.* © 1984 Wm. C. Brown Publishers, Dubuque, Iowa. All Rights Reserved. Reprinted by permission.

Feedback Control of the Adenohypophysis

In view of its secretion of releasing and inhibiting hormones, the hypothalamus might be considered the "master gland." The chain of command, however, is not linear; the hypothalamus and anterior pituitary are controlled by the effects of their own actions. In the endocrine system, to use an analogy, the general takes orders from the private. The hypothalamus and anterior pituitary are not master glands because their secretions are controlled by the target glands they regulate.

Anterior pituitary secretion of ACTH, TSH, and the gonadotrophins (FSH and LH) is controlled by **negative feedback inhibition** from the target gland hormones. Secretion of ACTH is inhibited by a rise in corticosteroid secretion, for example, and TSH is inhibited by a rise in the secretion of thyroxine from the thyroid. These negative feedback relationships are easily demonstrated by removal of the target glands. Castration (surgical removal of the gonads), for example, produces a rise in the secretion of FSH and LH. In a similar manner, removal of the adrenals or the thyroid would result in an abnormal increase in ACTH or TSH secretion from the anterior pituitary.

These effects demonstrate that under normal conditions the target glands exert an inhibitory effect on the anterior pituitary. This inhibitory effect can occur at two levels: (1) the target gland hormones could act on the hypothalamus and inhibit the secretion of releasing hormones, and (2) the target gland hormones could act on the anterior pituitary and inhibit its response to the releasing hormones. Thyroxine, for example, appears to inhibit the response of the anterior pituitary to TRH and thus acts to reduce TSH secretion (fig. 18.11). Sex steroids, in contrast, reduce the secretion of gonadotrophins by inhibiting both GnRH secretion and the ability of the anterior pituitary to respond to stimulation by GnRH (see fig. 18.12).

Recent evidence suggests that there may be retrograde transport of blood from the anterior pituitary to the hypothalamus. This may permit a **short feedback loop** where a particular trophic hormone inhibits the secretion of its releasing hormone from the hypothalamus. A high secretion of TSH, for example, may inhibit further secretion of TRH by this means.

In addition to negative feedback control of the anterior pituitary, there is one known example of *positive feedback* in which a target gland hormone stimulates the secretion of a pituitary hormone. During the middle phase of the menstrual cycle the ovaries secrete increasing amounts of a sex steroid called estradiol-17β, which stimulates a "surge" of LH secretion from the anterior pituitary. This positive feedback effect is unique for this phase of the menstrual cycle; at other times estradiol-17β inhibits LH secretion.

Higher Brain Function and Pituitary Secretion

The positive feedback effect of estradiol-17β on LH secretion is believed to result from activation of neurons in the amygdaloid nucleus in the limbic system of the brain. This illustrates the fact that the activity of the hypothalamus and, thus, of the anterior pituitary and its target glands, is influenced by higher brain functions. As a result of the effects of higher brain activity on the hypothalamus and pituitary, emotional disturbances can affect ovarian secretion and the menstrual cycle. Stress, as described later in this chapter, produces an increase in CRH secretion from the hypothalamus, which in turn results in elevated ACTH and corticosteroid secretion. Growth hormone is secreted in increased amounts during sleep, and the secretion of LH likewise follows a *circadian (ser''kah-de'an)* ("about a day") *rhythm* as a result of the influence of higher brain centers on the hypothalamus.

Figure 18.11. The secretion of thyroxine from the thyroid is stimulated by the thyroid-stimulating hormone (TSH) from the anterior pituitary. The secretion of TSH is stimulated by the thyrotrophin-releasing hormone (TRH) secreted from the hypothalamus into the hypothalamo-hypophyseal portal system. This stimulation is balanced by the negative feedback inhibition of thyroxine, which decreases the responsiveness of the anterior pituitary to stimulation by TRH.

Figure 18.12. Negative feedback control of gonadotrophin secretion. The possible short negative feedback loop involving the inhibition of GnRH secretion by the retrograde transport of gonadotrophins is shown by a dotted arrow.

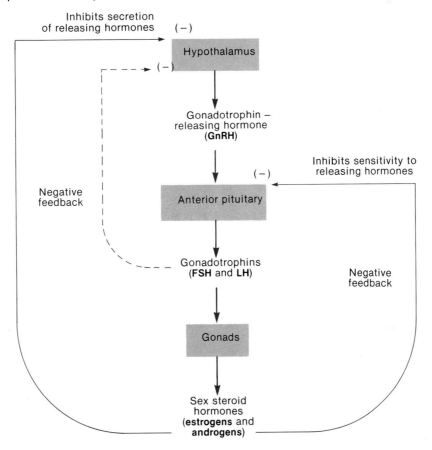

1. Describe the embryonic origins of the adenohypophysis and neurohypophysis, and list the parts of each. Indicate which of these parts are also called the ''anterior pituitary'' and ''posterior pituitary.''
2. List the hormones secreted by the posterior pituitary. Describe the site of origin of these hormones and the mechanisms by which their secretions are regulated.
3. List the hormones secreted by the anterior pituitary, and describe how the hypothalamus controls the secretion of each hormone.
4. Draw a negative feedback loop showing the control of ACTH secretion. Explain how this system would be affected by (a) an injection of ACTH; (b) surgical removal of the pituitary; (c) an injection of corticosteroids; and (d) surgical removal of the adrenal glands.

Adrenal Glands

The adrenal cortex and adrenal medulla are structurally and functionally different. The adrenal medulla secretes catecholamine hormones, which complement the action of the sympathetic nervous system. The adrenal cortex secretes corticosteroids, which participate in the regulation of mineral balance, energy balance, and reproductive function.

Objective 15. Describe the action of epinephrine and norepinephrine, and explain how the secretion of these hormones is regulated.

Objective 16. Describe the different categories of corticosteroids, identify the parts of the adrenal cortex that secretes each, and explain how the secretion of these hormones is regulated.

Objective 17. Explain the relationship between stress and the secretions of the adrenal cortex.

The adrenal *(ah-dre'nal)* glands (also called **suprarenal** [*su''prah-re'nal*] **glands**) are paired organs that cap the superior borders of the kidneys (fig. 18.13). The adrenal glands, along with the kidneys, are retroperitoneal and are embedded against the muscles of the back in a protective pad of fat.

Each adrenal gland is about 50 mm (2 in.) in length, 30 mm (1.1 in.) in width, and 10 mm (0.4 in.) in depth, and is generally pyramidal in appearance. Each adrenal gland consists of an outer cortex and inner medulla (fig. 18.13), which function as separate glands.

The **adrenal cortex** makes up the bulk of the gland and is histologically subdivided into three zones: an outer **zona glomerulosa** *(glo-mer''u-lo'sah),* an intermediate

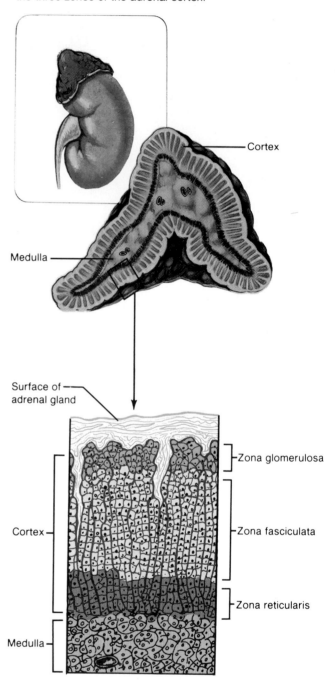

Figure 18.13. The structure of the adrenal gland showing the three zones of the adrenal cortex.

Cortex

Medulla

Surface of adrenal gland

Cortex

Medulla

Zona glomerulosa

Zona fasciculata

Zona reticularis

zona fasciculata *(fah-sik''u-la'tah),* and an inner **zona reticularis.** The **adrenal medulla** is composed of tightly packed clusters of **chromaffin** *(kro-maf'in)* **cells,** which are arranged around blood vessels. Each cluster of chromaffin cells receives direct autonomic innervation.

Like other endocrine glands, the adrenal glands are highly vascular. Three separate suprarenal arteries supply blood to each adrenal gland. One arises from the inferior

Figure 18.14.   Simplified pathways for the synthesis of steroid hormones in the adrenal cortex. The adrenal cortex produces steroids that regulate Na^+ and K^+ balance (mineralocorticoids), steroids that regulate glucose balance (glucocorticoids), and small amounts of sex steroid hormones.

phrenic artery, another from the aorta, and a third is a branch of the renal artery. The venous drainage goes through a suprarenal vein into the inferior vena cava for the right adrenal gland and through the suprarenal vein into the left renal vein for the left adrenal gland.

The adrenal glands are innervated by preganglionic fibers of the splanchnic nerves and by fibers of the celiac and associated sympathetic plexuses.

Functions of the Adrenal Cortex

The adrenal cortex secretes steroid hormones called **corticosteroids** *(kor″ti-ko-ste′roids),* or **corticoids,** for short. There are three functional categories of corticosteroids: (1) **mineralocorticoids,** which regulate Na^+ and K^+ balance; (2) **glucocorticoids,** which regulate the metabolism of glucose and other organic molecules; and (3) **sex steroids,** which are weak androgens (and lesser amounts of estrogens) that supplement the sex steroids secreted by the gonads.

Aldosterone is the most potent mineralocorticoid. The mineralocorticoids are produced in the zona glomerulosa (fig. 18.14). The predominant glucocorticoid in humans is cortisol (hydrocortisone), which is secreted by the zona fasciculata and perhaps also by the zona reticularis. The secretion of aldosterone by the zona glomerulosa is stimulated by angiotensin II and by blood K^+. The secretion of cortisol by the zona fasciculata is stimulated by ACTH (table 18.10 and fig. 18.15).

Table 18.10   Hormones of the adrenal cortex

Hormone	Action	Source of regulation
Mineralocorticoids	Regulate the concentration of extracellular electrolytes, especially sodium and potassium	Electrolyte concentration in blood
Glucocorticoids	Influence the metabolism of carbohydrates, proteins, and fats; promote vasoconstriction; anti-inflammatory compounds	ACTH from the adenohypophysis of the pituitary gland in response to stress
Gonadocorticoids	Supplement the sex hormones from the gonads	

Figure 18.15. The activation of the pituitary-adrenal axis by nonspecific stress.

Cortisol and other glucocorticoids suppress inflammation and inhibit the immune response. These compounds are thus often used to treat patients suffering from arthritis and various allergies (which are produced by the immune system). They are also frequently used in sports medicine to treat traumatized joints and are used in organ transplantation and tissue grafts to keep the host from rejecting donor tissues. Unfortunately, cortisol inhibits the regeneration of connective tissue and also has numerous other undesirable side effects, so that it must be used sparingly and only when necessary.

The General Adaptation Syndrome. In 1936, Hans Selye discovered that injections of cattle ovaries into rats (1) stimulated growth of the adrenal cortex; (2) caused atrophy of the lymphoid tissue of the spleen, lymph nodes, and thymus; and (3) produced bleeding peptic ulcers. At first he thought that these ovarian extracts contained a specific hormone that caused these effects. He later discovered that injections of a variety of substances, including foreign chemicals such as formaldehyde, could produce the same effects. Indeed, the same pattern of effects occurred when he placed rats in cold environments or when he dropped them into water and made them swim until they were exhausted.

The specific pattern of effects produced by these procedures suggested that these effects were the result of something that the procedures shared in common. Selye reasoned that all of the procedures were *stressful* and that the pattern of changes he observed represented a specific response to any stressful agent. He later discovered that

Table 18.11 Comparison of the hormones from the adrenal medulla

Epinephrine	Norepinephrine
Elevates blood pressure because of increased cardiac output and peripheral vasoconstriction	Elevates blood pressure because of generalized vasoconstriction
Accelerates respiratory rate and dilates respiratory passageways	Similar effect but to a lesser degree
Increases efficiency of muscular contraction	Similar effect but to a lesser degree
Increases rate of glycogen breakdown into glucose, so level of blood glucose rises	Similar effect but to a lesser degree
Increases rate of fatty acid released from fat, so level of blood fatty acids rises	Similar effect but to a lesser degree
Increases release of ACTH and TSH from the adenohypophysis of the pituitary gland	No effect

Source: Van De Graaff, Kent M., *Human Anatomy.* © 1984 Wm. C. Brown Publishers, Dubuque, Iowa. All Rights Reserved. Reprinted by permission.

all forms of stress produce these effects because all stressors stimulate the pituitary-adrenal axis. Under stressful conditions, there is increased secretion of ACTH and thus increased secretion of corticosteroids from the adrenal cortex.

Since stress is so very difficult to define, many people prefer to define stress operationally as any stimulus that activates the pituitary-adrenal axis. Using this criterion, it has been found that pleasant changes in one's life—such as marriage, a recent promotion, and so on—can be as stressful as unpleasant changes. On this basis, Selye has attempted to define stress as "a nonspecific response of the body to readjust itself following any demand made upon it."

Selye termed this nonspecific response the **general adaptation syndrome (GAS)**. Stress, in other words, produces GAS. There are three stages in the response to stress: (1) the *alarm reaction,* when the adrenal glands are activated; (2) the *stage of resistance,* in which readjustment occurs; and (3) if the readjustment is not complete, the *stage of exhaustion* may follow, leading to sickness and possibly death.

Functions of the Adrenal Medulla

The chromaffin cells of the adrenal medulla secrete **epinephrine** and **norepinephrine** in an approximate ratio of 4:1, respectively. These hormones are classified as amines (more specifically, as catecholamines), and are derived from the amino acid tyrosine.

The effects of these hormones are similar to those caused by stimulation of the sympathetic nervous system, except that the hormonal effect lasts about ten times longer. The hormones from the adrenal medulla increase cardiac output and heart rate, dilate coronary blood vessels, increase mental alertness, increase the respiratory

Figure 18.16. Stress is the physiological change in the body that prepares it for fight or flight. These changes are elicited by neural impulses from the hypothalamus that directly influence body tissue and by impulses to the adrenal medulla that secondarily influence the same body tissues through the release of epinephrine.

rate, and elevate metabolic rate. A comparison of the effects of epinephrine and norepinephrine are presented in table 18.11.

The adrenal medulla is innervated by sympathetic nerve fibers. The impulses are initiated from the hypothalamus via the spinal cord in response to stress. Stress therefore activates the adrenal medulla as well as the adrenal cortex. Activation of the adrenal medulla prepares the body for greater physical performance—the *fight or flight* response (see fig. 18.16).

Excessive stimulation of the adrenal medulla can result in depletion of the body's energy reserves, and high levels of corticosteroid secretion from the adrenal cortex can significantly impair the immune system. It is reasonable to expect, therefore, that prolonged stress can result in increased susceptibility to disease. Indeed, many studies show that prolonged stress results in an increased incidence of cancer and other diseases.

1. List the categories of corticosteroids and the zones of the adrenal cortex that secrete these hormones.
2. List the hormones of the adrenal medulla, and describe their effects.
3. Explain how the secretions of the adrenal cortex and adrenal medulla are regulated.
4. Describe how stress affects the secretions of the adrenal cortex and medulla, and explain how adrenal hormones can result in increased susceptibility to disease.

Thyroid and Parathyroids

The thyroid gland secretes thyroxine and triiodothyronine, which participate in the regulation of energy metabolism and are critically important for proper growth and development. The thyroid also secretes calcitonin, which may antagonize the action of parathyroid hormone in the regulation of calcium and phosphate balance.

> *Objective 18.* Describe the structure of the thyroid follicles and how these follicles produce the thyroid hormones.
>
> *Objective 19.* Describe the metabolic effects of thyroxine and triiodothyronine, and explain how the secretion of these hormones is regulated.
>
> *Objective 20.* Describe the structure and location of the parathyroid glands, and list the actions of parathyroid hormone.

The thyroid gland is positioned just below the larynx (fig. 18.17). This gland consists of two lobes that lie on either side of the trachea and are connected anteriorly by a broad **isthmus.** The thyroid is the largest of the endocrine glands, weighing between 20 and 25 g. It receives an abundant blood supply (80–120 ml/minute) through the paired superior thyroid branches of the external carotid arteries and the paired inferior thyroid branches of the subclavian arteries. The venous return is through the paired superior and middle thyroid veins that pass into the internal jugular veins and through the inferior thyroid veins that empty into the brachiocephalic veins.

On a microscopic level, the thyroid gland consists of many spherical hollow sacs called **thyroid follicles** (fig. 18.18). These follicles are lined with a simple cuboidal epithelium composed of **principal cells,** which synthesize the principal thyroid hormones (see table 18.12). The interior of the follicles contains **colloid** *(kol'oid),* which is a protein-rich fluid. Between the follicles are epithelial cells called **parafollicular cells,** which produce a hormone called calcitonin (or thyrocalcitonin).

Production and Action of Thyroid Hormones

The thyroid follicles actively accumulate iodide (I⁻) from the blood and secrete it into the colloid. Once the iodide is in the colloid it is oxidized to iodine and attached to specific amino acids (tyrosines) within the polypeptide chain of a protein called **thyroglobulin** *(thi-ro-glob'u-lin).* The attachment of one iodine to tyrosine produces **monoiodotyrosine** *(mon''o-i-o''do-ti'ro-sēn)* (**MIT**); the attachment of two iodines produces **diiodotyrosine** (**DIT**).

Within the colloid, enzymes modify the structure of MIT and DIT and couple them together (fig. 18.19). When

---

Figure 18.17. The thyroid gland. (*a*) its relationship to the larynx and trachea. (*b*) a scan of the thyroid gland twenty-four hours after the intake of radioactive iodine.

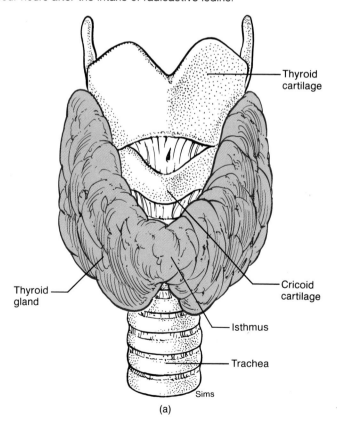

Thyroid cartilage

Thyroid gland

Cricoid cartilage

Isthmus

Trachea

Sims

(a)

(b)

Figure 18.18. A thyroid gland, showing numerous thyroid follicles. Each follicle consists of follicular cells surrounding the fluid known as colloid. (The gaps between the follicular cells and the colloid are artifacts made in the preparation of the slide.)

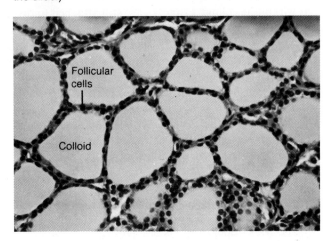

Follicular cells

Colloid

Table **18.12** Hormones of the thyroid gland		
Hormone	Action	Source of regulation
Thyroxine (T₄)	Increases rate of protein synthesis and rate of energy release from carbohydrates; regulates rate of growth; stimulates maturity of nervous system	Hypothalamus and release of TSH from adenohypophysis of the pituitary gland
Triiodothyronine (T₃)	Same as above	Same as above
Calcitonin (thyrocalcitonin)	Lowers blood calcium by inhibiting the release of calcium from bone tissue	Calcium levels in blood

Source: Van De Graaff, Kent M., *Human Anatomy.* © 1984 Wm. C. Brown Publishers, Dubuque, Iowa. All Rights Reserved. Reprinted by permission.

Figure 18.19. Stages in the formation and secretion of thyroid hormones. Iodide is actively accumulated by the follicular cells. In the colloid it is converted into iodine and attached to tyrosine amino acids within the thyroglobulin protein. Pinocytosis of iodinated thyroglobulin, coupling of MIT and DIT, and the release of thyroid hormones are stimulated by TSH from the anterior pituitary.

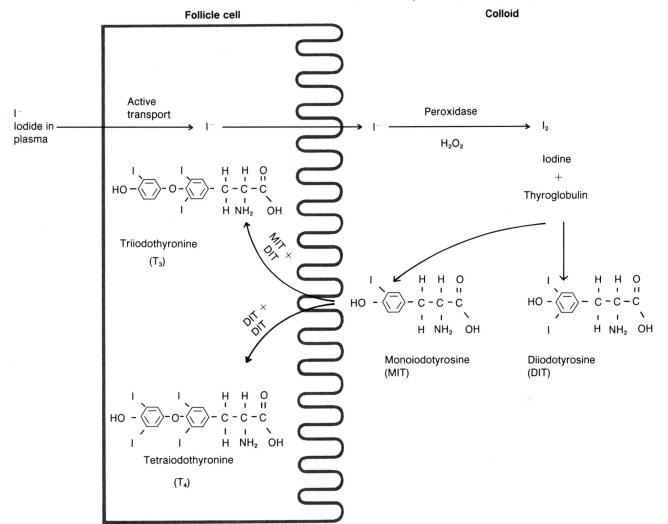

Figure 18.20. The mechanism of goiter formation in iodine deficiency.

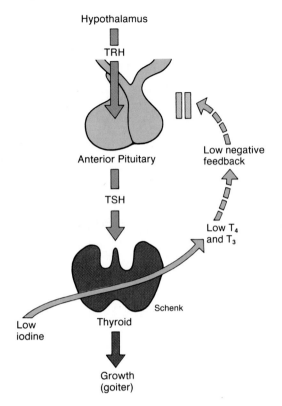

Hypothalamus

TRH

Anterior Pituitary

TSH

Low negative feedback

Low T₄ and T₃

Low iodine

Thyroid

Schenk

Growth (goiter)

two DIT molecules that are appropriately modified are coupled together, a molecule of **tetraiodothyronine** *(tet″rah-i″o-di-thi′ro-nēn)*, **T₄,** or **thyroxine** *(thi-rok′sin)*, is produced. The combination of one MIT with one DIT forms **triiodothyronine** *(tri″i-o″do-thi′ro-nēn)*, **T₃.** Note that within the colloid T₄ and T₃ are still attached to thyroglobulin.

As previously described, most of the thyroxine in blood is attached to carrier proteins. Only the very small percentage of thyroxine that is free in the plasma can enter the target cells, where it is converted to triiodothyronine and attached to nuclear receptor proteins. Through activation of genes, thyroid hormones stimulate protein synthesis, promote maturation of the nervous system, and increase the rate of energy utilization by the body.

> Thyroid-stimulating hormone (TSH) from the anterior pituitary stimulates the thyroid to secrete thyroxine and exerts a trophic effect on the thyroid gland. This trophic effect is dramatically revealed in people who develop an **iodine-deficiency (endemic) goiter.** In the absence of sufficient dietary iodine the thyroid cannot produce adequate amounts of T₄ and T₃. The resulting lack of negative feedback inhibition causes abnormally high levels of TSH secretion, which in turn stimulate the abnormal growth of the thyroid (a goiter). These events are summarized in figure 18.20.

Figure 18.21. A posterior view of the parathyroid glands.

Pharynx

Superior parathyroid

Inferior parathyroid

Thyroid gland

Esophagus

Trachea

The parafollicular cells of the thyroid secrete a hormone called **calcitonin** *(kal''si-to'nin)* (also called **thyrocalcitonin**). Under certain conditions, calcitonin promotes a decrease in blood calcium levels and thus antagonizes the effects of parathyroid hormone and vitamin D_3.

Parathyroid Glands

The small, flattened parathyroid glands are embedded in the posterior surfaces of the lateral lobes of the thyroid gland (see fig. 18.21). There are usually four parathyroid glands: a **superior** and an **inferior** pair. Each parathyroid gland is a small yellowish brown body measuring 3–8 mm (0.1–0.3 in.) in length, 2–5 mm (.07–0.2 in.) in width, and about 1.5 mm (0.05 in.) in depth.

On a microscopic level, the parathyroids are composed of two types of epithelial cells (fig. 18.22). The cells that synthesize parathyroid hormone are called **chief cells** and are scattered among **oxyphil** *(ok'se-fil)* **cells.** Oxyphil cells support the chief cells and are believed to produce reserve quantities of parathyroid hormone.

The blood supply and drainage of the parathyroid glands is similar to that of the thyroid gland. The innervation of the parathyroids, however, is a bit different. The parathyroids receive fibers from the pharyngeal branches of the vagus nerves as well as fibers arising from the cervical sympathetic ganglia.

Figure 18.22. The histology of the parathyroid gland.

Oxyphil and chief cells

Connective tissue septa

The parathyroid glands secrete one hormone called **parathyroid hormone (PTH).** This hormone promotes a rise in blood calcium levels by acting on the bones, kidneys, and intestine (fig. 18.23).

1. *Describe the structure of the thyroid gland, and list the effect of thyroid hormones.*
2. *Describe how thyroid hormones are produced and how their secretion is regulated, and explain the consequences of an inadequate dietary intake of iodine.*
3. *Describe the structure of the parathyroid glands, and explain the actions of parathyroid hormone.*

Pancreas and Other Endocrine Glands

The islets of Langerhans in the pancreas secrete two hormones, insulin and glucagon, which are critically involved in the regulation of metabolic balance in the body. Additionally, many other organs secrete hormones that help regulate digestion, metabolism, growth, immune function, and reproduction.

Figure 18.23. Actions of parathyroid hormone. An increased level of parathyroid hormone causes the bones to release calcium, the kidneys to conserve calcium loss through the urine, and the absorption of calcium through the intestinal wall. Negative feedback of increased calcium levels in the blood inhibits the secretion of this hormone.

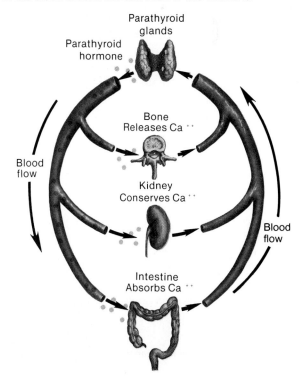

Parathyroid glands

Parathyroid hormone

Bone Releases Ca⁺⁺

Blood flow

Kidney Conserves Ca⁺⁺

Blood flow

Intestine Absorbs Ca⁺⁺

Figure 18.24.   The pancreas and the associated islets of Langerhans.

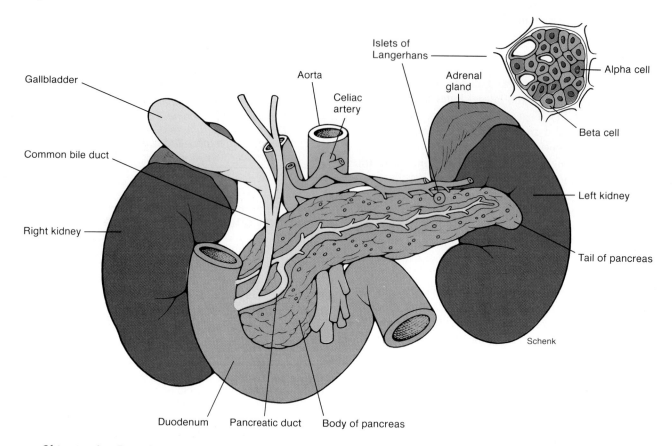

Objective 21.   Describe the structure of the endocrine portion of the pancreas and the origin of insulin and glucagon.

Objective 22.   Describe the actions of insulin and glucagon.

Objective 23.   Describe the structure and anatomic location of the pineal and thymus glands, and indicate their endocrine functions.

The pancreas is both an endocrine and an exocrine gland. The gross structure of this gland and its exocrine functions in digestion are described in chapter 24. The endocrine portion of the pancreas consists of scattered clusters of cells called the **islets of Langerhans** *(i'lets of Lahng'er-hanz).* These endocrine structures are most common in the body and tail of the pancreas (fig. 18.24).

Histologically, the most conspicuous cells in the islets are the *alpha* and *beta* cells (fig. 18.25). The alpha cells secrete the hormone **glucagon** *(gloo'kah-gon),* and the beta cells secrete **insulin.**

_____

islets of Langerhans: from Paul Langerhans, German
        anatomist, 1847–1888
insulin: L. *insula,* island

Alpha cells secrete glucagon in response to a fall in the blood glucose concentrations. Glucagon stimulates the liver to convert glycogen to glucose (**glycogenolysis** [*gli"ko-je-nol'i-sis*]), which causes the blood glucose level to rise. This effect represents the completion of a negative feedback loop. Glucagon also stimulates the hydrolysis of stored fat (**lipolysis**) and the consequent release of free fatty acids into the blood. This effect helps to provide energy substrates for the body during fasting, when blood glucose levels decrease. Glucagon, together with other hormones, also stimulates the conversion of fatty acids to ketone bodies, which can be secreted by the liver into the blood and used by many organs as an energy source.

Beta cells secrete insulin in response to a rise in the blood glucose concentrations. Insulin promotes the entry of glucose into tissue cells, and the conversion of this glucose into energy storage molecules of glycogen and fat. Insulin also aids the entry of amino acids into cells and the production of cellular protein. The actions of insulin and glucagon are thus antagonistic (table 18.13). After a meal, insulin secretion is increased and glucagon secretion is decreased; fasting, in contrast, causes a rise in glucagon and a fall in insulin secretion.

Figure 18.25.   The histology of the islets of Langerhans.

Figure 18.26.   The thymus is a bilobed organ within the mediastinum of the thorax.

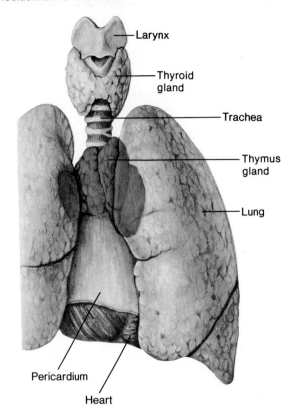

Table **18.13** Hormones of the pancreas		
Hormone	**Action**	**Source of regulation**
Glucagon	Stimulates the liver to convert glycogen into glucose, causing the blood glucose level to rise	Blood glucose level through negative feedback in pancreas
Insulin	Promotes movement of glucose through plasma membranes; stimulates liver to convert glucose into glycogen; promotes the transport of amino acids into cells; assists in synthesis of proteins and fats	Blood glucose level through negative feedback in pancreas

Source: Van De Graaff, Kent M., *Human Anatomy.* © 1984 Wm. C. Brown Publishers, Dubuque, Iowa. All Rights Reserved. Reprinted by permission.

Pineal Gland

The small, cone-shaped pineal gland, or **epiphysis cerebri,** is located in the roof of the third ventricle near the corpora quadrigemina, where it is encapsulated by the meninges covering the brain. The pineal gland of a child weighs about 0.2 g and is 5–8 mm (0.2–0.3 in.) long and 9 mm wide. The gland begins to regress in size at about age seven and in the adult appears as a thickened strand of fibrous tissue. Histologically, the pineal gland consists of specialized parenchymal and neuroglial cells. Although the pineal gland lacks direct nervous connections to the rest of the brain, it is highly innervated by the sympathetic nervous system from the superior cervical ganglion.

The principal hormone of the pineal is **melatonin** *(mel″ah-to′nin).* The secretion of melatonin is inhibited by light and is therefore maximal at night. It has long been suspected that this hormone inhibits the pituitary-gonad axis. Melatonin secretion is highest in children of ages one to five and decreases thereafter, reaching its lowest levels at the end of puberty, where concentrations are 75% lower than during early childhood. The secretion of melatonin is, therefore, believed by some researchers to play an important role in the onset of puberty.

Thymus

The thymus *(thi′mus)* is a bilobed organ positioned in the upper mediastinum, in front of the aorta and behind the manubrium of the sternum (fig. 18.26). Although the size of the thymus varies considerably from person to person, it is relatively large in newborns and children and sharply regresses in size after puberty. Besides decreasing in size, the thymus of adults becomes infiltrated with strands of fibrous and fatty connective tissue.

pineal: L. *pinea,* pine cone

thymus: Gk. *thymos,* a warty excrescence

Table 18.14 Summary of the physiological effects of gastrointestinal hormones

Secreted by	Hormone	Effects
Stomach	Gastrin	Stimulates parietal cells to secrete HCl Stimulates chief cells to secrete pepsinogen Maintains structure of gastric mucosa
Small intestine	Secretin	Stimulates water and bicarbonate secretion in pancreatic juice Potentiates actions of cholecystokinin on pancreas
Small intestine	Cholecystokinin (CCK)	Stimulates contraction of the gallbladder Stimulates secretion of pancreatic juice enzymes Potentiates action of secretin on pancreas Maintains structure of exocrine pancreas (acini)
Small intestine	Gastric inhibitory peptide (GIP)	Inhibits gastric emptying Inhibits gastric acid secretion Stimulates secretion of insulin from endocrine pancreas (islets of Langerhans)

Source: Fox, Stuart Ira, *A Laboratory Guide to Human Physiology, Concepts and Clinical Applications*, 3d ed. © 1976, 1980, 1984 Wm. C. Brown Publishers, Dubuque, Iowa. All Rights Reserved. Reprinted by permission.

The thymus serves as the site of production of **T cells (thymus-dependent cells),** which are the lymphocytes involved in cell-mediated immunity. In addition to providing T cells, the thymus secretes a number of hormones that are believed to stimulate T cells after they leave the thymus.

Gastrointestinal Tract

The stomach and small intestine secrete a number of hormones that act on the gastrointestinal tract itself and on the pancreas and gallbladder (table 18.14). The effects of these hormones act together with regulation by the autonomic nervous system to coordinate the activities of different regions of the digestive tract and the secretions of pancreatic juice and bile.

Gonads and Placenta

The gonads (**testes** and **ovaries**) secrete **sex steroids.** These include male sex hormones, or **androgens** *(an'dro-jen),* and female sex hormones—**estrogens** and **progestogens.** The principal hormones in each of these categories are **testosterone** *(tes-tos'tĕ-rōn),* **estradiol-17β,** and **progesterone,** respectively.

The testes consist of two compartments: **seminiferous tubules,** which produce sperm, and **interstitial tissue** between the convolutions of the tubules. Within the interstitial tissue are **Leydig cells,** which secrete testosterone and lesser amounts of weaker androgens (as well as small amounts of estradiol-17β). Testosterone is needed for the development and maintenance of the male genitalia (penis and scrotum) and male sex accessory organs (prostate, seminal vesicles, epididymides, and ductus deferens), as well as for the development of male secondary sexual characteristics.

During the first half of the menstrual cycle, estrogen is secreted by many small structures within the ovary called **ovarian follicles.** These follicles contain the egg cell, or **ovum,** and **granulosa cells** that secrete estrogen. By about mid-cycle, one of these follicles grows very large and, in the process of ovulation, extrudes its ovum from the ovary. The empty follicle, under the influence of luteinizing hormone (LH) from the anterior pituitary, then becomes a new endocrine structure called a **corpus luteum.** The corpus luteum secretes progesterone as well as estradiol-17β.

The **placenta** is the organ responsible for nutrient and waste exchange between the fetus and mother. The placenta is also an endocrine gland; it secretes large amounts of estrogens and progesterone, as well as a number of polypeptide and protein hormones that are similar to some hormones secreted by the anterior pituitary gland. These latter hormones include **human chorionic gonadotrophin (hCG),** which is similar to LH, and **somatomammotrophin,** which is similar in action to growth hormone and prolactin.

1. Describe the structure of the endocrine pancreas, and indicate the sites of origin of insulin and glucagon.
2. Describe how insulin and glucagon secretion is affected by eating and by fasting, and explain the actions of these two hormones.
3. Describe the location of the pineal gland and the possible function of melatonin.
4. Describe the location and function of the thymus gland.

placenta: L. *placenta,* flat cake

Leydig cells: from Franz von Leydig, German anatomist, 1821–1908

Clinical Considerations

Disorders of the Pituitary Gland

Panhypopituitarism.   A reduction in the activity of the pituitary gland is called **hypopituitarism** *(hi″po-pi-tu′i-tah-rizm″)* and can result from intracranial hemorrhage, a blood clot, prolonged steroid treatments, or a tumor. Total pituitary impairment, termed **panhypopituitarism,** brings about a progressive and general loss of hormonal activity. For example, the gonads stop functioning and the person suffers from amenorrhea (lack of menstruation) or aspermia (no sperm production) and loss of pubic and axillary hair. The thyroid and adrenals also eventually stop functioning. People with this condition, and those who have had their pituitary surgically removed (a procedure called **hypophysectomy**) receive thyroxine, cortisone, growth hormone, and gonadal hormones throughout life to maintain normal body function.

Abnormal Growth Hormone Secretion.   Inadequate growth hormone secretion during childhood causes **pituitary dwarfism.** Hyposecretion of growth hormone in an adult produces a rare condition called **pituitary cachexia** *(kah-kek′se-ah)* **(Simmonds' disease).** One of the symptoms of this disease is premature aging caused by tissue atrophy. Oversecretion of growth hormone during childhood, in contrast, causes **gigantism.** Excessive growth hormone secretion in an adult does not cause further growth in length because the epiphyseal discs have ossified. Hypersecretion of growth hormone in an adult causes **acromegaly** *(ak″ro-meg′ah-le),* in which the person's appearance gradually changes as a result of thickening of bones and growth of soft tissues, particularly in the face, hands, and feet.

Inadequate ADH Secretion.   A dysfunction of the neurohypophysis results in a deficiency in ADH secretion, causing a condition called **diabetes insipidus.** Symptoms of this disease include polyuria (excessive urination), polydipsia (drinking large amounts of water), and severe ionic imbalances. Diabetes insipidus is treated by injections of ADH.

Disorders of the Adrenal Glands

Tumors of the Adrenal Medulla.   Tumors of the chromaffin cells of the adrenal medulla are referred to as **pheochromocytomas** *(fe-o-kro″mo-si-to′mah).* These tumors cause hypersecretion of epinephrine and norepinephrine, which produce an effect similar to continuous sympathetic nervous stimulation. The symptoms of this condition are hypertension, elevated metabolism, hyperglycemia and sugar in the urine, nervousness, digestive problems, and sweating. It does not take long for the body to become totally fatigued under these conditions, making the patient susceptible to other diseases.

Addison's Disease.   This disease is caused by inadequate secretion of both glucocorticoids and mineralocorticoids, which results in hypoglycemia, sodium and potassium imbalance, dehydration, hypotension, rapid weight loss, and generalized weakness. A person with this condition who is not treated with corticosteroids will die within a few days because of the severe electrolyte imbalance and dehydration. Another symptom of this disease is darkening of the skin. This is caused by high secretion of ACTH and possibly MSH (because MSH is derived from the same parent molecule as ACTH), which is a result of lack of negative feedback inhibition of the pituitary by corticosteroids.

Cushing's Syndrome.   Hypersecretion of corticosteroids results in Cushing's syndrome. This is generally caused by a tumor of the adrenal cortex or by oversecretion of ACTH from the adenohypophysis. Cushing's syndrome is characterized by changes in carbohydrate and protein metabolism, hyperglycemia, hypertension, and muscular weakness. Metabolic problems give the body a puffy appearance and can cause structural changes characterized as "buffalo hump" and "moon face." Similar effects are also seen when people with chronic inflammatory diseases receive prolonged treatment with corticosteroids, which are given to reduce inflammation and inhibit the immune response.

Adrenogenital Syndrome.   Usually associated with Cushing's syndrome, this condition is caused by hypersecretion of adrenal sex hormones, particularly the androgens. Adrenogenital syndrome in young children causes premature puberty and enlarged genitals, especially the penis in a male and the clitoris in a female. An increase in body hair and a deeper voice are other characteristics. This condition in a mature woman can cause growth of a beard.

Addison's disease: from Thomas Addison, English physician, 1793–1860

Cushing's syndrome: from Harvey Cushing, U.S. physician, 1869–1939

Simmonds' disease: from Morris Simmonds, German physician, 1855–1925

acromegaly: Gk. *akron,* extremity; *megas,* large

diabetes: Gk. *diabetes,* to pass through a siphon

pheochromocytomas: Gk. *phaios,* dusky; *chroma,* color; *oma,* tumor

Disorders of the Thyroid and Parathyroids

Hypothyroidism. The infantile form of hypothyroidism is known as **cretinism** *(kre'tin-izm)* (fig. 18.27). An affected child usually appears normal at birth because thyroxine is received from the mother through the placenta. The clinical symptoms of cretinism are stunted growth, thickened facial features, abnormal bone development, mental retardation, low body temperature, and general lethargy. If cretinism is diagnosed early, it can be successfully treated by administering thyroxine.

Myxedema. Hypothyroidism in an adult causes myxedema *(mik''se-de'mah).* This disorder affects body fluids, causing edema and increasing blood volume, hence increasing blood pressure. A person with myxedema has a low metabolic rate, lethargy, and a tendency to gain weight. This condition is treated with thyroxine or with triiodothyronine, which are taken orally (as pills).

Endemic Goiter. A goiter is an abnormal growth of the thyroid gland. When this is a result of inadequate dietary intake of iodine, the condition is called endemic goiter (fig. 18.28). In this case, growth of the thyroid is due to excessive TSH secretion, which results from low levels of thyroxine secretion. Endemic goiter is thus associated with hypothyroidism.

Graves' Disease. Graves' disease, also called **toxic goiter,** involves growth of the thyroid associated with hypersecretion of thyroxine. This hyperthyroidism is produced by antibodies that act like TSH and stimulate the thyroid; it is an autoimmune disease. As a consequence of high levels of thyroxine secretion, the metabolic rate and heart rate increase, there is loss of weight, and the autonomic nervous system induces excessive sweating. In about half of the cases, **exophthalmos** *(ek''sof-thal'mos)* (bulging of the eyes) also develops (fig. 18.29) because of edema in the tissues of the eye sockets and swelling of the extrinsic eye muscles.

Disorders of the Parathyroid Glands. Surgical removal of the parathyroids sometimes unintentionally occurs when the thyroid is removed because of a tumor or the presence of Graves' disease. The resulting fall in parathyroid hormone (PTH) causes a decrease in plasma calcium levels, which can lead to severe muscle tetany. Hyperparathyroidism is usually caused by a tumor that secretes excessive amounts of PTH. This stimulates demineralization of bone, which makes the bones soft and raises the blood

myxedema: Gk. *myxa*, mucus; *oidema*, swelling
Graves' disease: from Robert James Graves, Irish physician, 1796–1853
exophthalmos: Gk. *ex*, out; *opthalmos*, eyeball

Figure 18.27. Cretinism is a disease of infancy caused by an underactive thyroid gland.

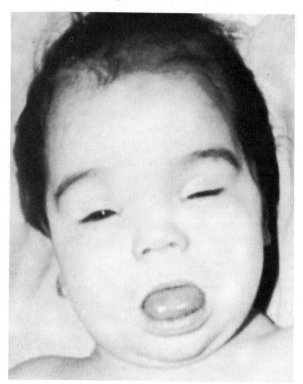

levels of calcium and phosphate. As a result of these changes, bones are subject to deformity and fracture, and stones composed of calcium phosphate are likely to develop in the urinary tract.

Disorders of the Islets of Langerhans

Diabetes Mellitus. Diabetes mellitus is characterized by fasting hyperglycemia and the presence of glucose in the urine. There are two forms of this disease. **Type I,** or **juvenile-onset** diabetes mellitus, is caused by destruction of the beta cells and the resulting lack of insulin secretion. **Type II,** or **maturity-onset** diabetes mellitus (which is the more common form), is caused by decreased tissue sensitivity to the effects of insulin, so that larger amounts of insulin are required to produce a normal effect. Both types of diabetes mellitus are also associated with abnormally high levels of glucagon secretion.

Reactive Hypoglycemia. People who have a genetic predisposition for type II diabetes mellitus often first develop reactive hypoglycemia. In this condition, the rise in blood glucose that follows the ingestion of carbohydrates stimulates excessive secretion of insulin, which in turn causes the blood glucose levels to fall below the normal range. This can result in weakness, changes in personality, and mental disorientation.

Figure 18.28. A simple or endemic goiter is caused by insufficient iodine in the diet.

Figure 18.29. Hyperthyroidism is characterized by an increased metabolic rate, weight loss, muscular weakness, and nervousness. The eyes may also protrude.

Chapter Summary

I. Hormones: Actions and Interactions
Hormones are regulatory molecules secreted into the blood by endocrine glands. The action of a hormone on its target tissue is affected by a number of variables.
   A. Precursors of active hormones may be called either prohormones or prehormones.
      1. Prohormones are relatively inactive precursor molecules made in the endocrine cells.
      2. Prehormones are the normal secretions of an endocrine gland that must be converted to other derivatives by target cells to be active.
   B. The effects of a hormone in the body depend on its concentration.
      1. Abnormally high amounts of a hormone can result in effects that are not normally found.
      2. Target tissues can become desensitized by high hormone concentrations.
   C. Hormones can interact in permissive, synergistic, or antagonistic ways.

II. Mechanisms of Hormone Action
Hormones that are chemically similar have similar mechanisms of action. These similarities involve the location of cellular receptor proteins and the events that occur in the target cell after hormone-receptor interaction.
   A. Steroid and thyroid hormones enter their target cells.
      1. Thyroid hormones attach to chromatin-bound receptors located in the nucleus.
      2. Steroid hormones bond to cytoplasmic receptor proteins and translocate to the nucleus.
      3. Attachment of the hormone-receptor protein complex to the chromatin activates genes and thereby stimulates RNA and protein synthesis.
   B. Amine, polypeptide, and glycoprotein hormones bond to receptor proteins on the outer surface of the target cell membrane.
      1. In many cases, this leads to the intracellular production of cyclic AMP, which serves as a second messenger in the action of these hormones.
      2. In other cases, cyclic GMP or Ca^{++} may serve as a second messenger in the action of the hormone.

III. Development of the Endocrine System
The embryonic development of the endocrine glands establishes the structural and functional relationships of the endocrine system.
   A. The pituitary consists of an adenohypophysis and a neurohypophysis.
      1. The adenohypophysis is derived from an oral diverticulum and includes the pars distalis, pars tuberalis, and pars intermedia.
      2. The neurohypophysis develops as a downgrowth of the brain and consists of the infundibulum and the pars nervosa.
   B. The thyroid is derived from endoderm in the primitive pharynx.
   C. The pancreas is derived from endoderm in the foregut.
   D. The adrenal cortex develops from mesoderm; the adrenal medulla develops from neural crest cells of the neural tube.

IV. Pituitary Gland

The hormones secreted by the posterior pituitary are produced by the hypothalamus. The anterior pituitary produces its own hormones. Secretion of both parts of the pituitary gland is controlled by the hypothalamus.

A. The pituitary secretes eight hormones.
   1. The anterior pituitary secretes growth hormone, thyroid stimulating hormone, adrenocorticotrophic hormone, follicle-stimulating hormone, luteinizing hormone, and prolactin.
   2. The posterior pituitary secretes antidiuretic hormone (also called vasopressin) and oxytocin.

B. Secretions of the posterior pituitary are controlled by the hypothalamo-hypophyseal nerve tract.

C. Secretions of the anterior pituitary are controlled by hypothalamic hormones that stimulate or inhibit secretions of the anterior pituitary.
   1. Hypothalamic hormones include TRH, CRH, GnRH, PIH, somatostatin, and a growth hormone–releasing hormone.
   2. These hormones are carried to the anterior pituitary by the hypothalamo-hypophyseal portal system.

D. Secretions of the anterior pituitary are also regulated by the feedback (usually negative feedback) of hormones from the target glands.

E. Higher brain centers, acting through the pituitary, can influence pituitary secretion.

V. Adrenal Glands

The adrenal cortex and medulla secrete different hormones, which have different functions in the body and are under different control mechanisms.

A. The adrenal cortex secretes mineralocorticoids (mainly aldosterone), glucocorticoids (mainly hydrocortisone), and sex steroids (primarily weak androgens).
   1. The glucocorticoids help regulate energy balance; they also can inhibit inflammation and suppress immune function.
   2. The pituitary-adrenal axis is stimulated by stress as part of the general adaptation syndrome.

B. The adrenal medulla secretes epinephrine and lesser amounts of norepinephrine, which complement the action of the sympathetic nervous system.

VI. Thyroid and Parathyroids

The thyroid gland secretes hormones that help regulate energy metabolism, and the parathyroid glands secrete a hormone that contributes to the regulation of calcium and phosphate balance.

A. The thyroid follicles secrete tetraiodothyronine (T_4, or thyroxine) and lesser amounts of triiodothyronine (T_3).
   1. These hormones are formed within the colloid of the thyroid follicles.
   2. The parafollicular cells of the thyroid secrete the hormone calcitonin, which may act to lower blood calcium levels.

B. The parathyroids are small structures embedded within the thyroid gland; the parathyroids secrete a hormone that promotes a rise in blood calcium levels.

VII. Pancreas and Other Endocrine Glands

The pancreas contains the islets of Langerhans, which secrete insulin and glucagon. Many other organs also secrete hormones that help regulate body functions.

A. Beta cells in the islets secrete insulin; alpha cells secrete glucagon.
   1. Insulin lowers blood glucose and stimulates the production of glycogen, fat, and protein.
   2. Glucagon raises blood glucose by stimulating the breakdown of liver glycogen; glucagon also promotes lipolysis and the formation of ketone bodies.
   3. The secretion of insulin is stimulated by a rise in blood glucose following meals; the secretion of glucagon is stimulated by a fall in blood glucose during periods of fasting.

B. The pineal gland, located on the roof of the third ventricle of the brain, secretes melatonin; this hormone may play a role in regulating reproductive function.

C. The thymus is located in front of the aorta in the mediastinum; it is the site of the production of T cell lymphocytes and secretes a number of hormones that may help regulate the immune system.

D. The gastrointestinal tract secretes a number of hormones that help regulate functions of the digestive system.

E. The gonads secrete sex steroid hormones.
   1. Leydig cells in the interstitial tissue of the testes secrete testosterone and other androgens.
   2. Granulosa cells of the ovarian follicles secrete estrogen.
   3. The corpus luteum of the ovaries secretes progesterone as well as estrogen.

F. The placenta secretes estrogen, progesterone, and a variety of polypeptide hormones that have actions similar to some anterior pituitary hormones.

Review Activities

Objective Questions

Match the gland to its embryonic origin.
1. Adenohypophysis
2. Neurohypophysis
3. Adrenal medulla
4. Pancreas
5. Thyroid gland

(a) endoderm of pharynx
(b) diverticulum from brain
(c) endoderm of foregut
(d) neural crest ectoderm
(e) Rathke's pouch

6. Hypothalamic releasing hormones
   (a) are secreted into capillaries in the median eminence
   (b) are transported by portal veins to the anterior pituitary
   (c) stimulate the secretion of specific hormones from the anterior pituitary
   (d) all of the above

7. The hormone primarily responsible for setting the basal metabolic rate and for promoting the maturation of the brain is
   (a) cortisol
   (b) ACTH
   (c) TSH
   (d) thyroxine

8. Which of the following statements about the adrenal cortex is *true*?
   (a) It is developed from mesoderm.
   (b) It secretes some androgens.
   (c) The zona granulosa secretes aldosterone.
   (d) The zona fasciculata is stimulated by ACTH.
   (e) All of the above are true.

9. The hormone insulin
   (a) is secreted by alpha cells in the islets of Langerhans
   (b) is secreted in response to a rise in blood glucose
   (c) stimulates the production of glycogen and fat
   (d) both a and b
   (e) both b and c

Match the hormone with the primary agent that stimulates its secretion.

10. Epinephrine
11. Thyroxine
12. Corticosteroids
13. ACTH

(a) TSH
(b) ACTH
(c) Growth hormone
(d) Sympathetic nerves
(e) CRH

14. Steroid hormones are secreted by
   (a) the adrenal cortex
   (b) the gonads
   (c) the thyroid
   (d) both a and b
   (e) both b and c

15. The secretion of which of the following hormones would be *increased* in a person with endemic goiter?
   (a) TSH
   (b) thyroxine
   (c) triiodothyronine
   (d) all of the above

Essay Questions

1. Explain how the functions of the neurohypophysis and adrenal medulla are related to their embryonic origins.
2. Compare steroid and polypeptide hormones in terms of their mechanism of action in target organs.
3. Explain the significance of the term *trophic* in regard to the actions of anterior pituitary hormones.
4. Suppose a drug blocks the conversion of T_4 to T_3. Explain what the effects of this drug would be on (a) TSH secretion, (b) thyroxine secretion, and (c) the size of the thyroid gland.
5. Explain why the phrase "master gland" is sometimes used to describe the anterior pituitary, and why this term is misleading.
6. Suppose a person's immune system made antibodies against insulin receptor proteins. Describe the possible effect of this condition on carbohydrate and fat metabolism.

Selected Readings

Austin, L. A., and H. Heath, III. 1981. Calcitonin: Physiology and pathophysiology. *New Engl. J. Med.* 304:269.

Axelrod, J., and T. D. Reisine. 1984. Stress hormones: Their interaction and regulation. *Science* 224:452.

Baxter, J. D., and W. J. Funder. 1979. Hormone receptors. *New Engl. J. Med.* 300:117.

Brownstein, M. J., et. al. 1980. Synthesis, transport and release of posterior pituitary hormones. *Science* 207:373.

Carmichael, S. W., and H. Winkler. August, 1985. The adrenal chromaffin cell. *Scientific American.*

Frohman, L. A. 1975. Neurotransmitters as regulators of endocrine functions. *Hosp. Prac.* 10:54.

Ganong, W. F., L. C. Alpert, and T. C. Lee. 1974. ACTH and the regulation of adrenocortical secretion. *New Engl. J. Med.* 290:1006.

Gillie, R. B. June 1971. Endemic goiter. *Scientific American.*

Goldsmith, R. S. 1969. Hyperparathyroidism. *New Engl. J. Med.* 281:367.

Guillemin, R., and R. Burgus. November 1972. The hormones of the hypothalamus. *Scientific American.*

Katzenellenbogen, B. S. 1980. Dynamics of steroid hormone receptor action. *Ann. Rev. Physiol.* 42:17.

Krieger, D. T. 1984. Brain peptides: what, where, and why? *Science* 222:975.

McEwen, B. S. July 1976. Interactions between hormones and nerve tissue. *Scientific American.*

O'Malley, B., and W. T. Shrader. February 1976. The receptors of steroid hormones. *Scientific American.*

Rasmussen, H., and M. M. Pechet. October 1970. Calcitonin. *Scientific American.*

Roth, J. 1980. Insulin receptors in diabetes. *Hosp. Prac.* 15:98.

Roth, J., and S. I. Taylor. 1982. Receptors for peptide hormones: alterations in diseases of humans. *Ann. Rev. of Physiol.* 44:639.

Schally, A. V. 1978. Aspects of the hypothalamic control of the pituitary gland. *Science* 202:18.

Selye, H. 1973. The evolution of the stress concept. *American Scientist* 61:692.

Turner, C. D. 1976. *General endocrinology.* 6th ed. Philadelphia: W. B. Saunders Co.

Ville, D. B. 1975. *Human endocrinology: A developmental approach.* Philadelphia: W. B. Saunders Co.

Williams, R. 1974. *Textbook of endocrinology.* 5th ed. Philadelphia: W. B. Saunders Co.

Unit 4 Regulation and Maintenance

Unit IV is concerned with the anatomy and physiology of those body systems whose principal function is to maintain homeostasis. The blood of the circulatory system transports materials for each of these systems. Oxygen and nutrients are made available to the living cells of the body through the blood. The blood also removes metabolic wastes from the cells and carries them to the kidneys, where the wastes are filtered out and eliminated from the body. The regulatory body systems maintain stable concentrations of various substances in the body's fluids. Homeostasis is further maintained by the immune system, which acts to protect the body from many diseases.

The chapters and principal concepts of Unit IV follow:

Chapter 19—Circulatory System
The continuously pumping heart propels blood through the vessels of the body, bringing life-sustaining oxygen and nutrients to the cells and carrying away their metabolic wastes.

Chapter 20—Cardiac Output, Blood Flow, and Blood Pressure
The parameters of cardiac output, blood flow, and blood pressure are determined by intrinsic and extrinsic factors and are continuously adjusted to keep pace with the changing needs of the body.

Chapter 21—Blood Composition and Clotting Mechanisms
The formed elements—erythrocytes, leukocytes, and thrombocytes—and the blood plasma transport respiratory gases, hormones, nutrients, and metabolic wastes and provide clotting capabilities as well as immunity.

of the Human Body

Chapter 22—Immune System
Nonspecific and specific defenses
protect the body from the invasion
of disease-causing agents. B-
lymphocytes provide antibody-
mediated immunity, and T-
lymphocytes provide cell-mediated
immunity.

Chapter 23—Respiratory System
The respiratory system takes in
oxygen, eliminates carbon dioxide,
and helps maintain the acid-base
balance of the body. Respiratory
function involves ventilation and
the transport of respiratory gases
and depends upon the proper
regulation of these processes.

Chapter 24—Digestive System
The digestive system prepares
food for cellular utilization through
the processes of ingestion,
mastication, deglutition,
peristalsis, chemical digestion,
and absorption.

Chapter 25—Regulation of
Metabolism
The utilization and storage of food
energy is regulated by a variety of
hormones and is also dependent
upon specific vitamins, minerals,
amino acids, and fatty acids. The
energy needs of the body are met
by the caloric value of food.

Chapter 26—Urinary System
The kidneys of the urinary system
clear the blood plasma of
metabolic wastes, especially urea,
in the process of forming urine
and return water, electrolytes, and
nutrients to the blood. The kidneys
also regulate the volume, acid-
base balance, and chemical
composition of the extracellular
fluids.

19

Circulatory System

Concepts

The circulatory system consists of components that provide functions essential to the life of complex, multicellular organisms.

The development of the heart and blood vessels reveals evolutionary relationships between different vertebrates, and a knowledge of these developmental events helps to explain some congenital defects in the cardiovascular system.

The structure of the heart and the action of its valves allow it to function as a double pump.

The cardiac cycle of systole and diastole can be monitored by the electrical activity that produces the cycle and through sounds created by the closing of the heart valves.

Arteries and veins are structured to carry blood to and from the heart. Capillaries allow the exchange of molecules between blood and tissue fluid.

The branches of the aorta provide oxygenated blood to all cells in the body.

Systemic veins of ever increasing diameters return oxygen-depleted blood to the right atrium via the superior and inferior vena cavae.

There are structural adaptations in the fetal cardiovascular system to the fact that oxygen and nutrients are obtained from the placenta rather than from the fetal lungs and digestive tract.

The lymphatic system functions to return tissue fluid to the venous system and to protect the body from disease.

Functions and Major Components of the Circulatory System

The circulatory system consists of a number of components that provide functions essential to the life of complex, multicellular organisms.

Objective 1. Describe the functions of the circulatory system.

Objective 2. Describe the major components of the circulatory system.

A unicellular organism can provide for its own maintenance and continuity by performing the wide variety of functions needed for life. By contrast, the complex human body is composed of trillions of specialized cells that demonstrate a division of labor. Cells of a multicellular organism depend on one another for the very basics of their existence. The majority of the cells of the body are firmly implanted in tissues and are incapable of procuring food and oxygen or even moving away from their own wastes. Therefore, a highly specialized and effective means of transporting materials within the body is needed.

The blood serves this transportation function. An estimated 60,000 miles of vessels throughout the body of an adult ensure that continued sustenance reaches each of the trillions of living cells. The blood can also, however, serve to transport disease-causing viruses, bacteria and their toxins. To ensure against this, the circulatory system has protective mechanisms: the white blood cells and lymphatic system. In order to perform its various functions, the circulatory system works together with the respiratory, urinary, digestive, endocrine, and integumentary systems in maintaining homeostasis.

Functions of the Circulatory System

The functions of the circulatory system can be divided into three areas: transportation, regulation, and protection.

1. **Transportation.** All of the substances essential for cellular metabolism are transported by the circulatory system. These substances can be categorized as follows:
   a. Respiratory. Red blood cells, called **erythrocytes** *(ē-rith'ro-sīt)*, transport oxygen to the tissue cells. In the lungs, oxygen from the inhaled air attaches to hemoglobin molecules within the erythrocytes and is transported to the cells for aerobic respiration. Carbon dioxide produced by cell respiration is carried by the blood to the lungs for elimination in the exhaled air.
   b. Nutritive. The digestive system is responsible for the mechanical and chemical breakdown of food so that it can be absorbed through the intestinal wall and into the blood vessels of the circulatory system. The blood then carries these absorbed products of digestion through the liver and to the cells of the body (figure 19.1).
   c. Excretory. Metabolic wastes, excessive water and ions, as well as other molecules in plasma (the fluid portion of blood) are filtered through the capillaries of the kidneys and excreted in urine.

Figure 19.1.   A schematic diagram of the circulatory system.

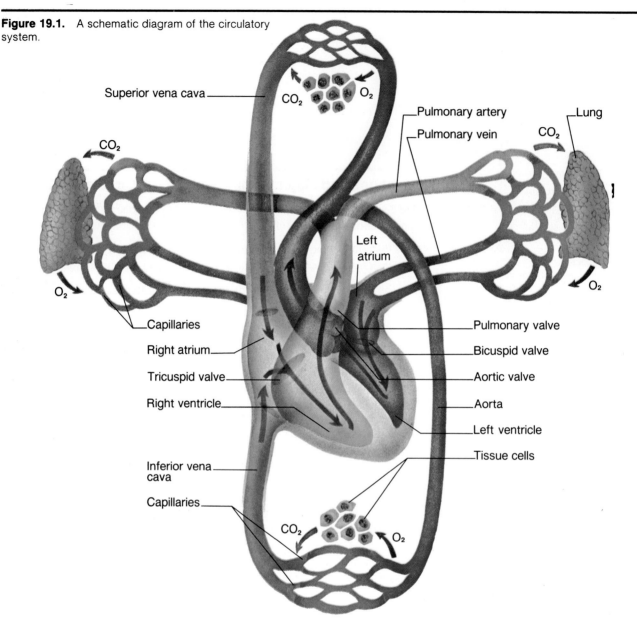

2. **Regulation.** The blood carries hormones and other regulatory molecules from their site of origin to distant target tissues.

3. **Protection.** The circulatory system protects against injury and foreign microbes or toxins introduced into the body. The clotting mechanism protects against blood loss when vessels are damaged, and **leukocytes** *(lu'ko-sīt)* (white blood cells) provide immunity to many disease-causing agents.

Major Components of the Circulatory System

The circulatory system is frequently divided into the **cardiovascular system,** which consists of the heart and blood vessels, and the **lymphatic system,** which consists of lymph vessels and lymph nodes.

The **heart** is a four-chambered, double pump. Its pumping action creates the pressure head needed to push blood in vessels to the lungs and body cells. At rest, the heart of an adult pumps about 5.0 L/min of blood. It takes about one minute for blood to be circulated to the most distal extremity and back to the heart.

Blood vessels form a tubular network that permits blood to flow from the heart to all the living cells of the body and then back to the heart. *Arteries* carry blood away from the heart, while *veins* return blood to the heart. Arteries and veins are continuous with each other through smaller blood vessels.

Arteries branch extensively to form a "tree" of ever smaller vessels. Those that are microscopic in diameter are called *arterioles.* Conversely, microscopic-sized veins, called *venules (ven'ūl),* deliver blood into ever larger vessels that empty into the large veins. Blood passes from the arterial to the venous system in *capillaries (kap'ĭ-lar''e),* which are the thinnest and most numerous blood vessels. All exchanges of fluid, nutrients, and wastes between the blood and tissues occur across the walls of capillaries.

Fluid derived from plasma that passes out of capillary walls into the surrounding tissues is called *tissue fluid,* or *interstitial (in''ter-stish'al) fluid.* Some of this fluid returns directly to capillaries, and some enters into **lymph vessels** located in the connective tissues around the blood vessels. Fluid in lymph vessels is called *lymph (limf).* This fluid is returned to the venous blood at particular sites. **Lymph nodes,** positioned along the way, cleanse the lymph prior to its return to the venous blood.

Endothermic (warm-blooded) animals, including humans, need an efficient circulatory system to transport oxygen-rich blood rapidly to all parts of the body. Of the vertebrates, only birds and mammals with their consistently warm body temperatures are considered endothermic, and only birds, mammals, and a few reptiles (crocodiles and alligators) have a four-chambered heart. This is far more efficient than the three-chambered hearts of other reptiles and amphibians or the two-chambered hearts of fishes.

---

1. Name the components of the circulatory system that function in oxygen transport, in the transport of nutrients from the digestive system, and in protection.
2. Define and describe the function of arteries, veins, and capillaries.
3. Define the terms interstitial fluid *and* lymph. Describe their relationships to blood plasma.

Development of the Cardiovascular System

The heart and blood vessels develop from mesoderm. The pattern of this developmental process reflects evolutionary relationships among different vertebrate groups, and a knowledge of this developmental process helps to explain certain congenital defects sometimes found in the human cardiovascular system.

Objective 3. Describe the events in the development of the embryonic heart from the eighteenth through the twenty-fifth day.

Objective 4. Trace the formation and fate of the six aortic arches during embryonic development.

The cardiovascular system is one of the first systems to form in the embryo, delivering nutrients to the mitotically active cells and disposing of waste products through its association with the maternal blood vessels in the placenta. Blood is formed and is circulated through vessels by the pumping action of the heart on about the twenty-fifth day after conception.

Throughout pregnancy, the fetus is dependent on the mother's circulatory system for nutrient, gaseous, and waste exchange. Therefore, some unique structures of the fetal circulatory system develop to accommodate the placental arrangement. This section of the chapter is concerned with the development of the fetal heart and associated major vessels. The circulation of blood through the fetal cardiovascular system is discussed at the end of the chapter after some basic concepts of the circulatory system are introduced.

Development of the Heart

The remarkable development of the heart requires only six to seven days. Heart development is first apparent at the eighteenth or nineteenth day in the *cardiogenic (kar''de-o-jen'ik) area* of the mesoderm layer. A small paired mass of specialized cells called *heart cords* form here. Shortly after, a hollow center develops in each heart cord, and the structure is then referred to as a *heart tube.* The heart tubes begin to migrate toward each other during the twenty-first day and soon fuse to form a single median *endocardial heart tube.* During this time the heart tubes undergo dilations and constrictions so that when fusion is completed during the fourth week, five distinct regions of the heart can be identified. These are the **truncus arteriosus, bulbus cordis, ventricle, atrium,** and **sinus venosus** (fig. 19.2).

After the fusion of the heart tubes and the formation of distinct dilations, the heart begins to pump blood. Partitioning of the heart chambers begins during the middle of the fourth week and is completed by the end of the fifth week. It is during this crucial time that many congenital heart problems develop.

Major changes occur in each of the five primitive dilations of the developing heart during the week-and-a-half embryonic period beginning in the middle of the fourth week. The truncus arteriosus differentiates to form a partition between the aorta and the pulmonary trunk. The bulbus cordis is incorporated in the formation of the walls of the ventricles. The sinus venosus forms the **coronary sinus** and a portion of the wall of the right atrium. The ventricle is divided into the right and left chambers by the growth of the **interventricular septum.** The atrium is partially partitioned into right and left chambers. An opening between the two atria called the **foramen ovale** persists throughout fetal development. This opening is covered by a flexible valve, which permits blood to pass from the right to the left side of the heart.

The development of the heart is summarized in table 19.1.

Congenital defects of the cardiac septa are a relatively common form of birth abnormality. Approximately 0.7% of live births and 2.7% of stillbirths show cardiac abnormalities. Ventricular septal defects are the most common of the cardiac defects. An infant with a congenital cardiac defect may suffer from inadequate oxygenation of blood and is termed a "blue baby."

---

cardiogenic: Gk. *kardia,* heart; *genesis,* be born (origin)

Figure 19.2. The formation of the heart chambers during the fourth and fifth weeks. (*a, b*) the heart tubes fuse during days 21 and 22, and (*c, d*) the heart bends during days 23 and 24.

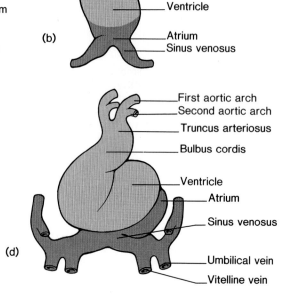

Table **19.1** Structural changes in the development of the heart	
Primitive cardiac dilation	**Developmental fate**
Truncus arteriosus	Aorticopulmonary septum, which forms the partition between the ascending aorta and the pulmonary trunk
Bulbus cordis	Incorporated into the walls of the ventricles
Sinus venosus	Differentiates to form the coronary sinus and a portion of the wall of the right atrium
Ventricle	Development of the interventricular septum to form the right and left ventricles
Atrium	Development of the septum secundum to partially partition the right and left atria, and formation of the foramen ovale with the valve of the foramen ovale

Source: Van De Graaff, Kent M., *Human Anatomy.* © 1984 Wm. C. Brown Publishers, Dubuque, Iowa. All Rights Reserved. Reprinted by permission.

Development of the Major Arteries

The formation of the major arteries occurs simultaneously with the development of the heart. The most complex and fascinating vascular formation is the development of the **aortic arches** associated with the pharyngeal pouches and branchial arches in the neck region (fig. 19.3). These aortic arches arise from the truncus arteriosus. Although six pairs of aortic arches develop, they are not all present at the same time, and none of them persists in entirety through fetal development.

The transformation of the six aortic arches into the basic adult arterial arrangement occurs during the sixth to the eighth week of embryonic development. The first and second pairs of aortic arches disappear before the formation of the sixth. The third pair of aortic arches form the common carotid arteries and the internal carotid arteries. The right fourth aortic arch forms the base of the right subclavian artery, and the left fourth arch contributes to part of the arch of the aorta. The first, second, and fifth pairs of aortic arches have no derivatives. The right sixth aortic arch forms the proximal portion of the right pulmonary artery. The left sixth arch forms the proximal portion of the left pulmonary artery, and the distal portion of this arch persists as a shunt between the pulmonary trunk and the aorta called the **ductus arteriosus.**

The derivatives of the aortic arches are summarized in table 19.2.

Although the embryonic vessels that develop from the truncus arteriosus are called aortic arches, they should not be confused with the aortic arch that is the major systemic artery leaving the heart in the adult. Only the left fourth arch participates in the formation of the adult aortic arch.

An examination of the persisting aortic arches in the adults of different classes of vertebrates reveals interesting evolutionary relationships. Fish have six aortic arches persisting in the gill region. Reptiles have only one pair, a branch to the left and one to the right. All birds have a single right aortic arch, and all mammals have a single left aortic arch.

Figure 19.3. The formation of the aortic arch and major arteries of the thoracic region. (*a*) a schematic lateral view of the pharyngeal pouches and aortic arches within the neck region of a 28-day-old embryo. (*b*) the aortic arches at six weeks. Note that by this time arches 1, 2, and 5 have largely disappeared. (*c*) the aortic arches at seven weeks. (*d*) arterial configuration at eight weeks.

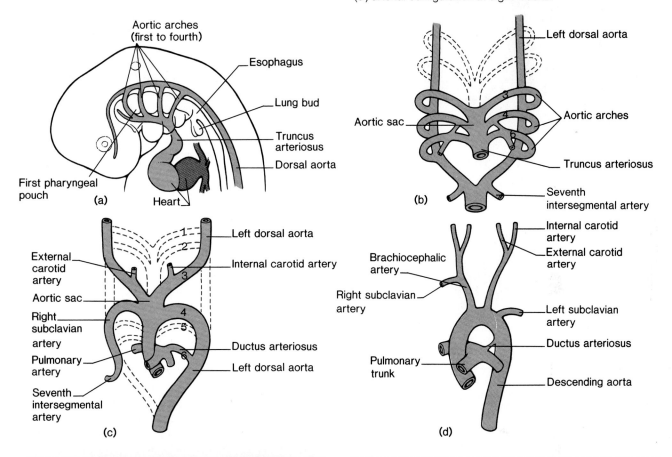

Table **19.2** Derivatives of the aortic arches	
Aortic arch	**Derivative**
First pair	No derivative
Second pair	No derivative
Third pair	Proximal portions form the common carotid arteries. Distal portions form the internal carotid arteries.
Fourth pair	Right arch forms the base of the right subclavian artery. Left arch forms a portion of the arch of the aorta.
Fifth pair	No derivative
Sixth pair	Right arch: proximal portion persists as proximal part of the right pulmonary artery; distal portion degenerates. Left arch: proximal portion forms the proximal part of the left pulmonary artery; distal part persists as the ductus arteriosus

Source: Van De Graaff, Kent M., *Human Anatomy.* © 1984 Wm. C. Brown Publishers, Dubuque, Iowa. All Rights Reserved. Reprinted by permission.

1. When does the heart begin development, and when does it begin to pump blood?
2. List the five regions of the embryonic heart tube, and indicate which regions of the heart they help form.
3. Describe the fate of the six embryonic aortic arches in humans.

Structure of the Heart

The structure of the heart muscle and the action of its valves allow it to function as a double pump. In this way, the heart pumps blood low in oxygen to the lungs and pumps oxygen-rich blood to the body.

Objective 5. Describe the location of the heart in relation to other organs of the thoracic cavity and their associated serous membranes.

Objective 6. Describe the structure and functions of the three layers of the heart wall.

Objective 7. Identify the chambers, valves, and associated vessels of the heart.

Objective 8. Trace the flow of blood through the heart, and distinguish between the pulmonary and systemic circulations.

Objective 9. Describe the location of the parts of the conduction system of the heart, and trace the pathway of impulse initiation and conduction.

Figure 19.4.   An X-ray of the heart situated within the mediastinum between the lungs.

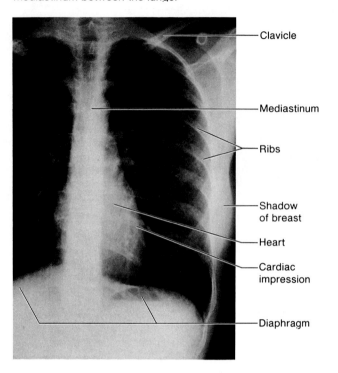

- Clavicle
- Mediastinum
- Ribs
- Shadow of breast
- Heart
- Cardiac impression
- Diaphragm

The heart is a hollow, four-chambered muscular organ, which is roughly the size of your clenched fist and averages 255 grams in adult females and 310 grams in adult males. It is estimated that the heart contracts some 42 million times and ejects 700,000 gallons of blood during a year.

The heart is located within the thoracic cavity between the lungs in the mediastinum (fig. 19.4). About two-thirds of the heart is located left of the midline, with its *apex,* or cone-shaped end, pointing downward in contact with the diaphragm. The *base* of the heart is the broad superior end where the large vessels attach.

The heart is enclosed and protected by a loose-fitting, serous sac with dense fibrous connective tissue called the **parietal pericardium** *(per″i-kar′de-um),* or **pericardial sac** (fig. 19.5). The parietal pericardium separates the heart from the other thoracic organs and forms the wall of the *pericardial cavity,* which contains a watery, lubricating *pericardial fluid.* The parietal pericardium is actually composed of an outer *fibrous layer* and an inner *serous layer.* The serous layer secretes the pericardial fluid.

---

pericardium: Gk. *peri,* around; *kardia,* heart

Figure 19.5.   The position of the heart and associated serous membranes within the thoracic cavity.

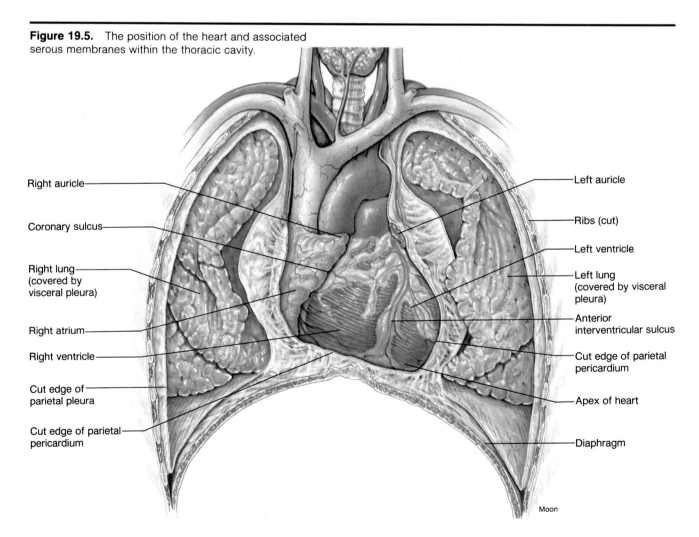

- Right auricle
- Coronary sulcus
- Right lung (covered by visceral pleura)
- Right atrium
- Right ventricle
- Cut edge of parietal pleura
- Cut edge of parietal pericardium
- Left auricle
- Ribs (cut)
- Left ventricle
- Left lung (covered by visceral pleura)
- Anterior interventricular sulcus
- Cut edge of parietal pericardium
- Apex of heart
- Diaphragm

Moon

Pericarditis is an inflammation of the parietal pericardium associated with increases in the secretion of fluid into the pericardial cavity. Because the tough, fibrous portion of the parietal pericardium is inelastic, an increase in fluid pressure impairs the movement of blood into and out of the chambers of the heart. Some of the pericardial fluid may be withdrawn for analysis by injecting a needle to the left of the xiphoid process to pierce the parietal pericardium.

Heart Wall

The wall of the heart is composed of three distinct layers (fig. 19.6). The outer layer is the **epicardium,** also called the *visceral pericardium.* The space between this layer and the parietal pericardium is the pericardial cavity just described. The thick middle layer of the heart is called the **myocardium.** It is composed of cardiac muscle and arranged in such a way that the contraction of the muscle bundles results in squeezing or wringing of the heart chambers (fig. 19.7). The thickness of the myocardium varies depending on the force needed to eject blood from the particular chambers. The thickest portion of the myocardium, therefore, surrounds the left ventricle. The atrial walls are relatively thin. The inner layer, called the **endocardium,** is continuous with the endothelium of blood vessels. The endocardium also forms part of the valves of the heart. Inflammation of the endocardium is called *endocarditis.*

Table 19.3 summarizes the characteristics of the three layers of the heart wall.

Table **19.3** Layers of the heart wall		
Layer	**Characteristics**	**Function**
Epicardium (visceral pericardium)	Serous membrane of connective tissue covered with epithelium and including blood capillaries, lymph capillaries, and nerve fibers	Lubricative outer covering
Myocardium	Cardiac muscle tissue separated by connective tissues and including blood capillaries, lymph capillaries, and nerve fibers	Muscular contractions that eject blood from the heart chambers
Endocardium	Serous membrane of epithelium and connective tissues, including elastic and collagenous fibers, blood vessels, and specialized muscle fibers	Lubricative inner lining of the chambers and valves

Source: Van De Graaff, Kent M., *Human Anatomy.* © 1984 Wm. C. Brown Publishers, Dubuque, Iowa. All Rights Reserved. Reprinted by permission.

Figure 19.6. The structure of the heart wall and the pericardium.

Fibrous pericardium

Serous pericardium

Pericardial cavity

Epicardium

Myocardium

Endocardium

Coronary vessel

Trabeculae carneae

Figure 19.7. The structural arrangement of the cardiac muscles forming the myocardium of the heart. The fiber arrangement around both ventricles (*a*) and (*b*) is a whorled pattern.

Myocardial muscle fibers

(a)

(b)

Figure 19.8.   The structure of the heart. (*a*) an anterior view; (*b*) a posterior view; (*c*) an internal view.

Right pulmonary arteries

Right pulmonary veins

Aorta

Superior vena cava

Right atrium

Right coronary artery

Right ventricle

Inferior vena cava

Pulmonary trunk

Left pulmonary arteries

Left pulmonary veins

Left atrium

Left coronary artery

Circumflex artery

Anterior interventricular artery

Left ventricle

(a)

Left common carotid artery

Left subclavian artery

Arch of the aorta

Descending aorta

Left pulmonary arteries

Left pulmonary veins

Left atrium

Posterior cardiac vein

Coronary sinus

Left ventricle

Brachiocephalic artery (innominate)

Superior vena cava

Azygos vein

Right pulmonary arteries

Oblique sinus

Right pulmonary veins

Right atrium

Inferior vena cava

(b)

Right ventricle

Middle cardiac vein

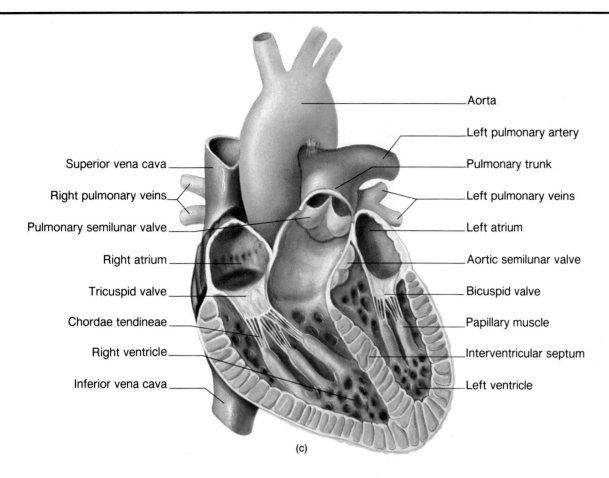

Aorta

Left pulmonary artery

Superior vena cava

Pulmonary trunk

Right pulmonary veins

Left pulmonary veins

Pulmonary semilunar valve

Left atrium

Right atrium

Aortic semilunar valve

Tricuspid valve

Bicuspid valve

Chordae tendineae

Papillary muscle

Right ventricle

Interventricular septum

Inferior vena cava

Left ventricle

(c)

Chambers and Valves

The heart is a four-chambered, double pump, consisting of upper right and left **atria** and lower right and left **ventricles;** the atria contract and empty simultaneously into the ventricles (fig. 19.8), which also contract in unison. The atria are separated by the thin muscular *interatrial septum,* while the ventricles are separated by the thick muscular *interventricular septum.* **Atrioventricular** *(a''tre-o-ven-trik'u-lar)* **valves** *(bicuspid and tricuspid)* are located between the atria and ventricles, and **semilunar valves** are present at the bases of the two large vessels leaving the heart. These valves maintain the flow of blood in one direction.

Grooved depressions on the surface of the heart indicate the partitions between the chambers and also contain **coronary vessels** that supply blood to the wall of the heart. The most prominent groove is the *atrioventricular (coronary) sulcus* that encircles the heart and marks the division between the atria and ventricles. The partition between right and left ventricles is denoted by two (anterior and posterior) *interventricular sulci.*

The following discussion follows the sequence in which blood flows through the structures of the atria, ventricles, and valves.

Right Atrium. The right atrium *(a'tre-um)* receives systemic venous blood from the **superior vena cava,** which drains the upper portion of the body, and from the **inferior vena cava,** which drains the lower portion (fig. 19.8). There is an additional venous opening into the right atrium from the *coronary sinus,* which returns venous blood from the myocardium of the heart itself.

Right Ventricle. Blood from the right atrium passes through the **tricuspid valve** to fill the right ventricle *(ven'tri-k'l).* The tricuspid valve is an atrioventricular (AV) valve that gets its name from its three valve leaflets, or cusps. Each cusp is held in position by strong tendinous cords called **chordae tendineae,** which are secured to the ventricular wall by cone-shaped **papillary muscles.** These structures prevent the valves from everting, like an umbrella in a strong wind, when the ventricles contract.

Ventricular contraction causes the tricuspid valve to close and the blood to leave the right ventricle through the **pulmonary trunk** and to enter the lungs via the right and left **pulmonary arteries.** The **pulmonary semilunar** ("half-moon shaped") **valve** lies at the base of the pulmonary trunk as it leaves the right ventricle.

---

atrium: L. *atrium,* chamber
ventricle: L. *ventriculus,* diminutive of *venter,* belly

---

vena cava: L. *vena,* vein; *cava,* empty
chordae tendineae: L. *chorda,* string; *tendere,* to stretch

Figure 19.9. A superior view of the heart valves. (*a*) a photograph of the semilunar valves. (*b*) a diagram of the relative position of the valves and the supporting fibrous connective tissue.

(a)

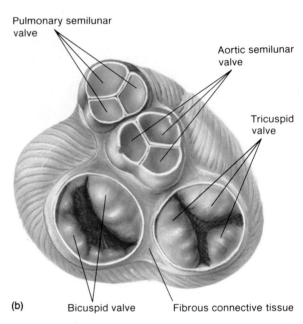

Pulmonary semilunar valve

Aortic semilunar valve

Tricuspid valve

(b)   Bicuspid valve   Fibrous connective tissue

Table **19.4** Valves of the heart		
Valve	**Location**	**Comments**
Tricuspid valve	Between right atrium and right ventricle, surrounding atrioventricular orifice	Composed of three cusps that prevent a backflow of blood from right ventricle into right atrium during ventricular contraction
Pulmonary semilunar valve	Entrance to pulmonary trunk	Composed of three half-moon-shaped flaps that prevent a backflow of blood from pulmonary trunk into right ventricle during ventricular relaxation
Bicuspid (mitral) valve	Between left atrium and left ventricle, surrounding atrioventricular orifice	Composed of two cusps that prevent a backflow of blood from left ventricle to left atrium during ventricular contraction
Aortic semilunar valve	Entrance to ascending aorta	Composed of three half-moon-shaped flaps that prevent a backflow of blood from aorta into left ventricle during ventricular relaxation

Left Atrium. After gas exchange has occurred within the capillaries of the lungs, oxygenated blood is transported to the left atrium through four **pulmonary veins,** two from each lung.

Left Ventricle. The left ventricle receives blood from the left atrium. These two chambers are separated by the **bicuspid,** or **mitral** *(mi′tral),* valve. When the left ventricle is relaxed, the valve is open and allows blood to flow from the atrium to the ventricle; when the left ventricle contracts, the valve closes. Closing of the valve during ventricular contraction prevents backflow of blood into the atrium.

The walls of the left ventricle are thicker than those of the right ventricle. The endocardium of both chambers has distinct ridges called **trabeculae** *(trah-bek′u-le)* **carneae** (fig. 19.6). Oxygenated blood leaves the left ventricle through the **ascending aorta** *(a-or′tah).* The **aortic semilunar valve,** located at the base of the ascending aorta, closes as a result of the pressure of the blood when the left ventricle relaxes and thus prevents backflow of blood into the relaxed ventricle.

The appearance of the valves is shown in figure 19.9, and their actions are summarized in table 19.4.

bicuspid: L. *bi,* two; *cuspis,* tooth point or spike
mitral: L. *mitra,* like a bishop's mitre
trabeculae carneae: L. *trabecula,* small beams; *carneus,* flesh

Figure 19.10. Coronary circulation. (*a*) a plastic cast of the coronary arteries and their major branches. (*b*) an anterior view of the arterial supply to the heart. (*c*) an anterior view of the venous drainage.

(b)

(c)

(a)

Circulatory Routes

Figure 19.1 illustrates the circulatory routes of the blood. There are two principal divisions of the circulatory blood flow: the pulmonary and systemic circulations.

The **pulmonary circulation** consists of blood vessels that transport blood to the lungs for gas exchange and then back to the heart. It consists of the right ventricle that ejects the blood, the pulmonary trunk with its semilunar valve, the pulmonary arteries that transport blood to the lungs, the pulmonary capillaries within each lung, four pulmonary veins that transport oxygenated blood back to the heart, and the left atrium that receives the blood from the pulmonary veins.

The **systemic** *(sis-tem'ik)* **circulation** is composed of all the remaining vessels of the body that are not part of the pulmonary circulation. This includes the right atrium, the left ventricle, the aorta with its semilunar valve, all of the branches of the aorta, all capillaries other than those in the lungs, and all veins other than the pulmonary veins. The right atrium receives all the venous return of oxygen-depleted blood from the systemic veins.

Coronary Circulation. The myocardium is supplied with blood by the right and left **coronary arteries** (fig. 19.10). These two vessels arise from the aorta immediately beyond the aortic semilunar valve. The coronaries encircle the heart within the atrioventricular sulcus, the depression between the atria and ventricles. Two branches arise from both the right and left coronaries to serve the atrial and ventricular walls. The left coronary artery gives rise to the **anterior interventricular artery,** which courses within the anterior interventricular sulcus to serve both ventricles, and the **circumflex artery,** which supplies oxygenated blood to the walls of the left atrium and left ventricle. The right coronary artery gives rise to the **posterior interventricular artery,** which courses through the posterior interventricular sulcus to serve the two ventricles, and a **marginal artery,** which serves the walls of the right atrium

---

circumflex: L. *circum,* around; *flectere,* to bend

and right ventricle. The main trunks of the right and left coronaries anastomose (join together) on the posterior surface of the heart.

From the capillaries in the myocardium, the blood enters the **coronary veins.** The course of these vessels parallels that of the coronary arteries. The coronary veins, however, have thinner walls and are more superficial than the arteries. The two principal coronary veins are the **great cardiac vein,** which returns blood from the anterior aspect of the heart, and the **middle cardiac vein,** which drains the posterior aspect of the heart. These coronary veins converge into a large venous channel on the posterior surface of the heart called the **coronary sinus** (fig. 19.10). The coronary venous blood then enters the heart through an opening into the right atrium.

Heart attacks are the most common cause of death in the United States. The most common type of heart attack involves an occlusion of a coronary artery, which reduces the delivery of oxygen to the myocardium. There are several things you can do to minimize the risk of heart attack: (1) reduce excess weight; (2) avoid hypertension through proper diet, stress reduction, or medication if necessary; (3) avoid smoking; and (4) exercise regularly.

Conduction System of the Heart

Contractions of the atria and ventricles occur automatically as a result of electrical impulses that originate in and are conducted by heart tissue. The rhythm of contraction (**systole**) and relaxation (**diastole**) is thus intrinsic or built-in, although the rate of this rhythm is modified by stimulation through autonomic nerves and the hormone epinephrine.

The **sinoatrial** *(si″no-a′tre-al)* **node,** or **SA node** (the *pacemaker*), is located in the posterior wall of the right atrium (fig. 19.11). The SA node initiates the cardiac cycle of systole and diastole by producing an electrical impulse that spreads over both atria, causing them to contract simultaneously. The impulse then passes to the **atrioventricular node,** or **AV node,** which is located on the inferior portion of the interatrial septum. From here, the impulse continues through the **atrioventricular bundle,** or **bundle of His,** located at the top of the interventricular septum. The atrioventricular bundle divides into right and left branches, which are continuous with the **Purkinje** *(pur-kin′je)* **fibers** within the ventricular walls. Stimulation of these fibers causes the ventricles to contract simultaneously and eject blood into the pulmonary and systemic circulations.

---

systole: Gk. *systole*, contraction
diastole: Gk. *diastole*, prolonged or expansion
bundle of His: from Wilhelm His, Jr., Swiss physician, 1863–1934
Purkinje fibers: from Johannes E. von Purkinje, Bohemian anatomist, 1787–1869

Figure 19.11.   The conduction system of the heart.

1. List the three layers of the heart wall, and describe the structures associated with each layer.
2. List the valves that control blood flow through the heart, and describe their locations and functions.
3. Trace the flow of a drop of blood from the lungs, through the heart, and back to the lungs.
4. Trace the flow of electrical impulses through the conduction system of the heart.

The Cardiac Cycle, the Heart Sounds, and the Electrocardiogram

The cycle of contraction and relaxation of the heart chambers results from electrical activity, which can be monitored with an electrocardiogram. The cardiac cycle produces changes in intraventricular pressure, which can be followed through the sounds created by the closing of the AV and semilunar valves.

Objective 10.   Describe the cycle of contraction and relaxation of the atria and ventricles and how these events are related in time.

Objective 11.   Describe how pressure changes in the heart affect the AV valves and semilunar valves and how these events produce the heart sounds.

Figure 19.12. The cardiac cycle of ventricular systole and diastole. Contraction of the atria occurs in the last 0.1 second of ventricular diastole. Relaxation of the atria occurs during ventricular systole. The durations of systole and diastole given are accurate for a cardiac rate of seventy-five beats per minute.

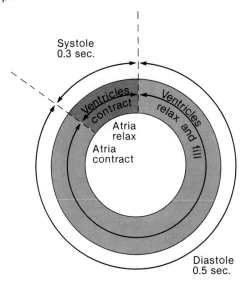

Objective 12. Describe the electrical activity of the SA node, and explain why the SA node functions as the normal pacemaker.

Objective 13. Explain the correlation between the first and second heart sounds and the waves of the ECG.

The cardiac cycle refers to the repeating pattern of contraction and relaxation of the heart. The phase of contraction is called **systole** *(sis'to-le),* and the phase of relaxation is called **diastole** *(di-as'to-le).* When these terms are used alone they refer to contraction and relaxation of the ventricles. It should be noted, however, that the atria also contract and relax. There is an atrial systole and diastole. Atrial contraction occurs when the ventricles are relaxed in diastole; when the ventricles contract during systole, the atria are relaxed.

The heart thus has a two-step pumping action. The right and left atria contract almost simultaneously, followed about 0.1–0.2 seconds later by contraction of the right and left ventricles. During the time when both the atria and ventricles are relaxed the venous return of blood fills the atria and (because the AV valves are open) fills the ventricles. It has been estimated that the ventricles are about 80 percent filled with blood even before the atria contract. Contraction of the atria adds the final one-fifth to the *end-diastolic volume* of blood in the ventricles.

Contraction of the ventricles in systole ejects about two-thirds of the blood in the ventricles, leaving one-third of the initial amount as the *end-systolic volume.* The ventricles then fill with blood during the next cycle. At an average **cardiac rate** of seventy-five beats per minute, each cycle lasts 0.8 second; 0.5 second is spent in diastole, and systole takes 0.3 second (fig. 19.12).

Figure 19.13. The relationship between the heart sounds and the intraventricular pressure and volume. Closing of the A-V valves occurs during the early part of contraction, when the intraventricular pressure rises prior to ejection of blood. Closing of the semilunar valves occurs at the beginning of ventricular relaxation, just prior to filling. The first and second sounds thus appear during the stages of isovolumic contraction and isovolumic relaxation (iso=same).

Pressure Changes During the Cardiac Cycle

When the heart is in diastole, pressure in the systemic arteries averages about 80 mm Hg. The following events in the cardiac cycle then occur:

1. *Isovolumic contraction* of the ventricles occurs when the inflow valves (atrioventricular) and outflow valves (semilunar) are closed.
2. When the pressure in the left ventricle becomes greater than the pressure in the aorta the phase of *ejection* begins as the semilunar valves open. The pressure in the left ventricle and aorta rises to about 120 mm Hg (fig. 19.13) as the ventricular volume decreases.
3. As the pressure in the left ventricle falls below the pressure in the aorta, the back pressure closes the semilunar valves. The pressure in the aorta falls to 80 mm Hg while pressure in the left ventricle falls to 0 mm Hg.
4. During *isolumic relaxation* the AV and semilunar valves are closed. This phase lasts until the pressure in the ventricles falls below pressure in the atria.

5. When the pressure in the ventricles falls below the pressure in the atria, a phase of *rapid filling* occurs.

6. *Atrial contraction* (*systole*) empties the final amount of blood into the ventricles immediately prior to the next phase of isolumic contraction of the ventricles.

Heart Sounds

Closing of the AV and semilunar valves produces sounds that can be heard at the surface of the chest with a stethoscope. These sounds are often described phonetically as *lub-dub.* The "lub," or **first sound,** is produced by closing of the AV valves. The "dub," or **second sound,** is produced by closing of the semilunar valves. The first sound is thus heard when the ventricles contract at systole, and the second sound is heard when the ventricles relax at the beginning of diastole.

> The first sound may, particularly during inhalation, be heard to split into tricuspid and mitral components. Closing of the tricuspid is best heard at the fourth right intercostal space (between the ribs), and closing of the mitral valve is best heard at the apex of the heart (fig. 19.14). The second sound may also be split under certain conditions. Closing of the pulmonary and aortic semilunar valves is best heard at the second left and right intercostal spaces, respectively.

Heart Murmurs. Murmurs are abnormal heart sounds produced by abnormal patterns of blood flow in the heart. Many murmurs are caused by defective heart valves. Defective heart valves may be congenital, or more commonly, they can occur as a result of *rheumatic endocarditis,* associated with rheumatic fever. In this disease the valves become damaged by antibodies made in response to an infection caused by streptococcus bacteria (the same bacteria that produce strep throat).

In *mitral stenosis,* for example, the mitral valve becomes thickened and calcified. This can impair the blood flow from the left atrium to the left ventricle. An accumulation of blood in the left atrium may cause a rise in left atrial and pulmonary vein pressure, resulting in pulmonary hypertension. As a compensation for the increased pulmonary pressure, the right ventricle grows thicker and stronger.

Valves are said to be incompetent when they do not close properly, and murmurs may be produced as blood regurgitates through the valve flaps. One important cause of incompetent AV valves is damage to the papillary muscles. When this occurs, the tension in the chordae tendineae may not be sufficient to prevent the valve from everting as pressure in the ventricle rises during systole.

---

stenosis: Gk. *stenosis,* a narrowing

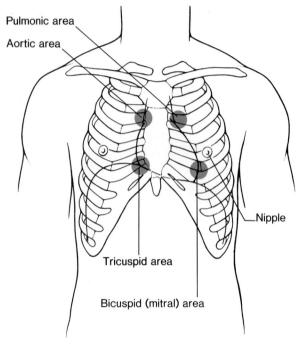

Figure 19.14. Routine stethoscope positions for listening to the heart sounds.

Murmurs can also be produced by the flow of blood through *septal defects*—holes in the septum between the right and left sides of the heart. These are usually congenital and may occur either in the interatrial or interventricular septum (fig. 19.15).

> The lungs of a fetus are collapsed, and blood is routed away from the pulmonary circulation by an opening in the interatrial septum called the foramen ovale and by a connection between the pulmonary trunk and aorta called the ductus arteriosus (fig. 19.16). These shunts normally close after birth, but when they remain open (are **patent**) murmurs can result. If a significant amount of oxygen-depleted blood from the right side of the heart enters the systemic circulation through these defects the baby may be **cyanotic** (blue).

Electrical Activity of the Heart

Most myocardial cells have the potential for spontaneous electrical activity; they have an intrinsic rhythm. In the normal heart, however, spontaneous electrical activity is limited to a small region (about $2 \times 5 \times 15$ mm) in the right atrium near the opening of the superior vena cava. This region is the **sinoatrial,** or **SA, node** and serves as the normal *pacemaker* of the heart.

Cells of the SA node are never in an electrically resting state; as soon as they have repolarized from a previous action potential they begin to depolarize. This depolarization, resulting from spontaneous changes in membrane permeability to cations, is called a *pacemaker potential.* When the pacemaker potential produces a depolarization

Figure 19.15. Abnormal patterns of blood flow due to septal defects. Left-to-right shunting of blood is shown *(shaded circles)* because the left pump is at a higher pressure than the right pump. Under abnormal conditions, however, the pressure in the right atrium may exceed that of the left, causing right-to-left shunting of blood through a septal defect in the atria.

Septal defect
in atria

Septal defect
in ventricles

Figure 19.16. The flow of blood through a patent (open) ductus arteriosus.

Figure 19.17. Pacemaker potentials and action potentials in the SA node.

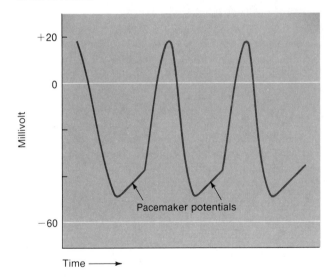

to about -55 mV, an action potential is stimulated. As in action potentials produced by nerve and muscle fibers, the membrane potential first reverses polarity (to about $+20$ mV) due to the inflow of Na^+ and then moves toward about -60 mV due to the outflow of K^+ and the action

of the Na^+/K^+ pump. Permeability to K^+ gradually diminishes as the action potential ends, but instead of stabilizing at a resting potential, the membrane gradually depolarizes again as a new pacemaker potential is produced by the inward leakage of Na^+. This pattern is repeated for the lifetime of a normal heart (fig. 19.17).

Some other regions of the heart, including the area around the SA node and the atrioventricular bundle, can potentially produce pacemaker potentials. The rate of spontaneous depolarization of these cells, however, is slower than that of the SA node. The potential pacemaker cells are therefore stimulated by action potentials from the SA node before they can depolarize spontaneously. If action potentials from the SA node are prevented from reaching these areas, they will generate their own pacemaker potentials and serve as sites for the origin of action potentials; they will function as pacemakers. A pacemaker other than the SA node is called an *ectopic (ektop'ik) pacemaker* (also called an *ectopic focus*). From this discussion, it is clear that the rhythm set by such an ectopic pacemaker is usually slower than that normally set by the SA node.

Ordinary myocardial cells stimulated by action potentials from the SA node produce and conduct their own action potentials. The ordinary myocardial cells have resting potentials of about -90 mV. When stimulated by action potentials from a pacemaker region, these cells depolarize to $+20$ mV as a result of the inward diffusion of Na^+. Following the rapid reversal of the membrane polarity, the voltage quickly declines to about -10 to -20 mV. Unlike the action potentials of other cells, however, this level of depolarization is maintained for 200–300 msec (milliseconds) before repolarization (fig. 19.18). This *plateau phase* results from a slow inward diffusion of Ca^{++}, which balances a slow outward diffusion of cations. Rapid repolarization at the end of the plateau phase is achieved, as in other cells, by the opening of K^+ gates and the rapid outward diffusion of K^+ that results.

Conduction of the Impulse. Action potentials from the SA node spread very quickly—at a rate of 0.8 to 1.0 meter (m) per second—across the myocardial cells of both atria. These impulses, however, cannot spread directly across into the ventricles. They must instead be conducted by specialized tissue, derived from heart muscle, that traverses the connective tissue sheath between the atria and ventricles. This specialized conductive tissue includes the AV node, atrioventricular bundle, and Purkinje fibers.

Slow conduction of impulses (0.03 to 0.05 m per second) through the AV node accounts for over half of the time delay between excitation of the atria and ventricles. After the impulses spread through the AV node, the conduction rate increases greatly in the atrioventricular bundle and reaches very high velocities (5 m per second) in the Purkinje fibers. As a result of this rapid conduction of impulses, ventricular contraction begins about 0.1–0.2 seconds after contraction of the atria.

Unlike skeletal muscles, the heart cannot sustain a contraction by the summation of twitches. This is because the atria and ventricles behave as if they were composed of only one muscle cell; the entire myocardium is electrically stimulated as a single unit and contracts as a unit. This contraction, lasting almost 300 msec, is equivalent to a twitch produced by a single skeletal muscle fiber (which only lasts 20–100 msec in comparison). The heart

Figure 19.18. An action potential in a myocardial cell from the ventricles. The plateau phase of the action potential is maintained by a slow inward diffusion of CA^{++}. The cardiac action potential, as a result, has a duration that is about one hundred times longer than the "spike potential" of an axon.

normally cannot be stimulated again until after it has relaxed from its previous contraction because myocardial cells have *long refractory periods* (fig. 19.19), which correspond to the long duration of their action potentials. The rhythmic pumping action of the heart is thus insured.

The Electrocardiogram

A pair of surface electrodes placed directly on the heart will record a repeating pattern of potential changes. As action potentials spread from the atria to the ventricles, the voltage measured between these two electrodes will vary in a way that provides a "picture" of the electrical activity of the heart. The nature of this picture can be varied by changing the position of the recording electrodes; different positions provide different perspectives, enabling an observer to gain a more complete picture of the electrical events.

The body is a good conductor of electricity because tissue fluids contain a high concentration of ions that can move (creating a current) in response to potential differences. Potential differences generated by the heart are thus conducted to the body surface where they can be recorded as an **electrocardiogram** (**ECG,** or **EKG**). The ECG waves are shown in figure 19.20.

There are two types of ECG recording electrodes, or "leads." The *bipolar limb leads* record the voltage between electrodes placed on the wrists and legs. These bipolar leads include lead I (right arm to left arm), lead II (right arm to left leg), and lead III (left arm to left leg). In the *unipolar leads,* voltage is recorded between a single "exploratory electrode" placed on the body and an electrode that is built into the electrocardiograph and maintained at zero potential (ground).

Figure 19.19. The time course for the myocardial action potential (A) is compared with the duration of contraction (B). Notice that the long action potential results in a correspondingly long absolute refractory period (ARP) and relative refractory period (RRP). These refractory periods last almost as long as the twitch, so that the myocardial cells cannot be stimulated a second time until they have finished their contraction from the first stimulus.

Figure 19.20. The electrocardiogram indicates the conduction of electrical impulses through the heart (a) and measures and records both the intensity of this electrical activity (in millivolts) and the time intervals involved (b).

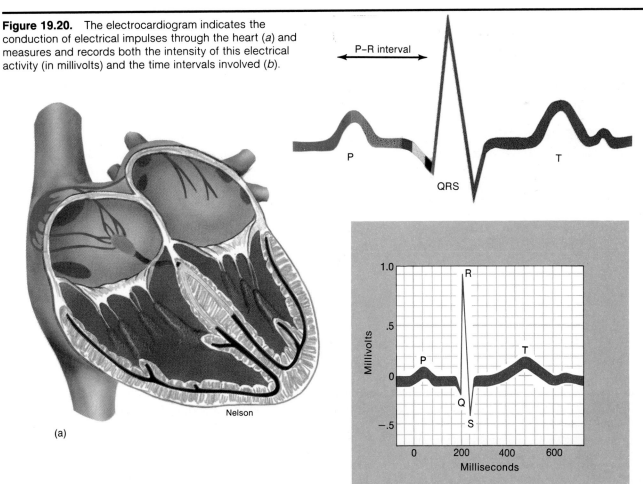

Figure 19.21. The placement of the bipolar limb leads and the exploratory electrode for the unipolar chest leads in an electrocardiogram (ECG).

Table **19.5** The electrocardiograph (ECG) leads	
Name of lead	**Placement of electrodes**
Bipolar limb leads	
I	Right arm and left arm
II	Right arm and left leg
III	Left arm and left leg
Unipolar limb leads	
AVR	Right arm
AVL	Left arm
AVF	Left leg
Unipolar chest leads	
V_1	4th intercostal space right of sternum
V_2	4th intercostal space left of sternum
V_3	5th intercostal space to the left of the sternum
V_4	5th intercostal space in line with the middle of the clavicle
V_5	5th intercostal space to the left of V_4
V_6	5th intercostal space in line with the middle of the axilla

The unipolar limb leads are placed on the right arm, left arm, and left leg, and are abbreviated AVR, AVL, and AVF, respectively. The unipolar chest leads are labeled one through six, starting from the midline position (fig. 19.21). There are thus a total of twelve standard ECG leads that "view" the changing pattern of the heart's electrical activity from different perspectives (table 19.5). This is important because certain abnormalities are best seen with particular leads and may not be visible at all with other leads.

Each cardiac cycle produces three distinct ECG waves, the first, or **P wave,** occurs as a result of atrial depolarization. The second, or **QRS wave,** represents depolarization of the ventricles. Atrial repolarization occurs during this time but is hidden by the larger electrical changes in the ventricles. The last ECG wave, the **T wave,** is produced by repolarization of the ventricles (fig. 19.22).

Correlation of the ECG with Heart Sounds. Depolarization of the ventricles, as indicated by the QRS wave, stimulates contraction by promoting the uptake of Ca^{++} into the regions of the sarcomeres. The QRS wave is thus

Figure 19.22.   The conduction of electrical impulses in the heart, as indicated by the electrocardiogram (ECG). The direction of the arrows in (e) indicates that depolarization of the ventricles occurs from the inside (endocardium) out (to the epicardium), whereas the arrows in (g) indicate that repolarization of the ventricles occurs in the opposite direction.

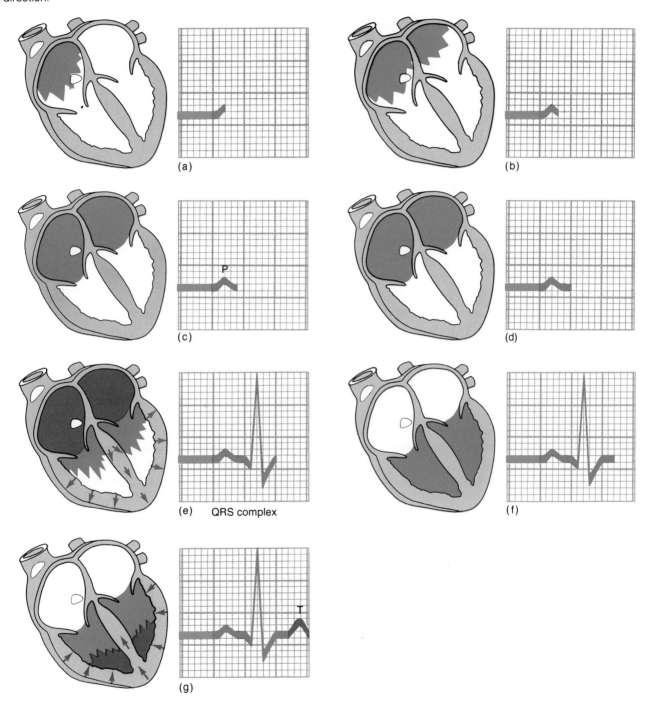

(a)

(b)

(c)   P

(d)

(e)   QRS complex

(f)

(g)   T

seen to occur at the beginning systole. The rise in intraventricular pressure that results causes the AV valves to close, so that the first heart sound (S_1, or lub) is produced immediately after the QRS wave (fig. 19.23).

Repolarization of the ventricles, as indicated by the T wave, occurs at the same time that the ventricles relax at the beginning of diastole. The resulting fall in intraventricular pressure causes the aortic and pulmonary semilunar valves to close, so that the second heart sound (S_2, or dub) is produced shortly after the T wave in an electrocardiogram begins.

---

1. Describe how action potentials are produced by the SA node, and explain why the SA node functions as the normal pacemaker.
2. Explain why the heart does not sustain a contraction.
3. Describe how the heart sounds are produced, and explain the relationship between these sounds and the waves of the ECG.

Blood Vessels

The structure of arteries and veins allows them to function in the transport of blood from the heart to the capillaries and from the capillaries back to the heart. The structure of capillaries provides the mechanism for the exchange of plasma fluid and dissolved molecules between the blood and the surrounding tissues.

Objective 14. Describe the structures of muscular and elastic arteries, and explain how these structures aid their functions.

Objective 15. Describe the structure of veins, and explain how this structure aids their function.

Objective 16. Describe the structure of different types of capillaries, and explain how these structures relate to their function.

Blood vessels form a tubular network throughout the body that permits blood to flow from the heart to all the living cells of the body and then back to the heart. Blood leaving the heart passes through vessels of progressively smaller diameters referred to as **arteries, arterioles,** and **capillaries.** Capillaries are microscopic vessels that join the arterial flow to the venous flow. Blood returning to the heart from the capillaries passes through vessels of progressively larger diameters called **venules** and **veins.**

Figure 19.23. The relationship between changes in intraventricular pressure and the electrocardiogram during the cardiac cycle. The QRS wave (representing depolarization of the ventricles) occurs at the beginning of systole, while the T wave (representing repolarization of the ventricles) occurs at the beginning of diastole.

The walls of arteries and veins are composed of three coats, or "tunics." The outermost layer is the **tunica** *(tu'ni-kah)* **externa,** the middle layer is the **tunica media,** and the inner layer is the **tunica intima.** The tunica externa is composed of connective tissue. The tunica media is composed primarily of smooth muscle. The tunica intima consists of two parts: (1) an innermost simple squamous epithelium, the *endothelium (en''do-the'le-um),* which lines the lumina of all blood vessels and (2) a layer of elastic fibers, or *elastin.*

Although both arteries and veins have the same basic structure (fig. 19.24), there are some important differences between the two types. Arteries have more muscle for their diameters than do comparably sized veins. Also, arteries appear more round in cross section, whereas veins are usually partially collapsed. This is due to the fact that veins are not usually filled to their capacity; they can stretch when they receive more blood and thus function as reservoirs or "capacitance" vessels. In addition, many veins have valves, which are absent in arteries.

---

tunica: L. *tunica,* covering or coat
lumen: L. *lumen,* opening

Figure 19.24.   The structure of a medium-sized artery and vein showing the relative thickness and composition of the tunicas.

Artery   **Vein**

Tunica intima

Valve

Tunica media

Tunica externa

Serosa

Arteries

The aorta and other large arteries contain numerous layers of elastin fibers between smooth muscle cells in the tunica media. These large arteries expand when the pressure of the blood rises as a result of the heart's contraction; they recoil, like a stretched rubber band, when the blood pressure falls during relaxation of the heart. This elastic recoil helps to produce a smoother, less pulsatile flow of blood through the smaller arteries and arterioles.

The small arteries and arterioles are less elastic than the larger arteries and have a thicker layer of smooth muscle for their diameters. Unlike the larger *elastic arteries,* therefore, the smaller *muscular arteries* retain almost the same diameter as the pressure of the blood rises and falls during the heart's pumping activity. Since arterioles and small muscular arteries have narrow lumina, they provide the greatest resistance to blood flow through the arterial system.

Small muscular arteries that are 100 μm or less in diameter branch to form smaller arterioles (20–30 μm in diameter). In some tissues, blood from the arterioles can enter the venules directly through **arteriovenous anastomoses** *(ah-nas''to-mo'sis).* In most cases, however, blood

from arterioles passes into capillaries (fig. 19.25). Capillaries are the narrowest of blood vessels (7–10 μm in diameter), and serve as the "business end" of the circulatory system in which exchanges of gases and nutrients between the blood and the tissues occur.

Capillaries

The arterial system branches extensively (table 19.6) to deliver blood to over forty billion capillaries in the body. The extensiveness of these branchings is indicated by the fact that all tissue cells are located within a distance of only 60–80 μm of a capillary and by the fact that capillaries provide a total surface area of 1,000 square miles for exchanges between blood and tissue fluid.

Despite their large number, capillaries contain only about 250 ml of blood at any time, out of a total blood volume of about 5,000 ml (most is contained in the venous system). The amount of blood flowing through a particular capillary bed is determined in part by the action of the **precapillary sphincter muscles.** These muscles allow only 5%–10% of the capillary beds in skeletal muscles, for example, to be open at rest. Blood flow to an organ is regulated by the action of these precapillary sphincters and

anastomosis: Gk. *anastomoo,* to furnish with a mouth (coming together)

Figure 19.25. The microcirculation. Metarterioles (arteriovenous anastomoses) provide a path of least resistance between arterioles and venules. Precapillary sphincter muscles regulate the flow of blood through the capillaries.

Table **19.6**	Characteristics of the vascular supply to the mesenteries in a dog				
Kind of vessel	Diameter (mm)	Number	Total cross-sectional area (cm²)	Length (cm)	Total volume (cm³)
Aorta	10	1	0.8	40	30
Large arteries	3	40	3.0	20	60
Main artery branches	1	600	5.0	10	50
Terminal branches	0.6	1,800	5.0	1	25
Arterioles	0.02	40,000,000	125	0.2	25
Capillaries	0.008	1,200,000,000	600	0.1	60
Venules	0.03	80,000,000	570	0.2	110
Terminal veins	1.5	1,800	30	1	30
Main venous branches	2.4	600	27	10	270
Large veins	6.0	40	11	20	220
Vena cava	12.5	1	1.2	40	50
					930

Source: Reprinted with permission of Macmillan Publishing Company from *Animal Physiology: Principles and Adaptations*, 3d ed., by Malcolm S. Gordon. Copyright © 1977 by Malcolm S. Gordon.

Figure 19.26. An electron micrograph of a capillary in a coronary vessel. Notice the thin intercellular channel and the fact that the capillary wall is only one cell thick. Arrows show some of the many pinocytic vesicles.

Nucleus

Capillary lumen

by the degree of resistance to blood flow (due to constriction or dilation) provided by the small arteries and arterioles in the organ.

Unlike the vessels of the arterial and venous systems, the walls of capillaries are composed of only one cell layer—a simple squamous epithelium, or endothelium (fig. 19.26). The absence of smooth muscle and connective tissue layers permits a more rapid rate of transport of materials between the blood and the tissues.

Types of Capillaries. Different organs have different types of capillaries, which are distinguished by significant differences in structure. In terms of their endothelial lining, these capillary types include those that are *continuous,* those that are *discontinuous,* and those that are *fenestrated (fen'es-trāt''ed).*

Continuous capillaries are those in which adjacent endothelial cells are closely joined together. These are found in muscles, lungs, adipose tissue, and in the central nervous system. Continuous capillaries in the CNS lack intercellular channels; this fact contributes to the blood-brain barrier. Continuous capillaries in other organs have narrow intercellular channels (about 40–45 A in width) that allow the passage of molecules other than protein between the capillary blood and tissue fluid (fig. 19.27).

Examination of endothelial cells with an electron microscope has revealed the presence of pinocytotic vesicles, which suggests that the intracellular transport of material may occur across the capillary walls. This type of transport appears to be the only mechanism of capillary exchange available within the central nervous system and may account, in part, for the selective nature of the blood-brain barrier.

Figure 19.27. Diagrams of continuous, fenestrated, and discontinuous capillaries as they appear in the electron microscope. This classification is derived from the continuity of the endothelial layer.

Continuous

Fenestrated

Discontinuous

Figure 19.28. The action of the one-way venous valves. Contraction of skeletal muscles helps to pump blood toward the heart, but is prevented from pushing blood away from the heart by closure of the venous valves.

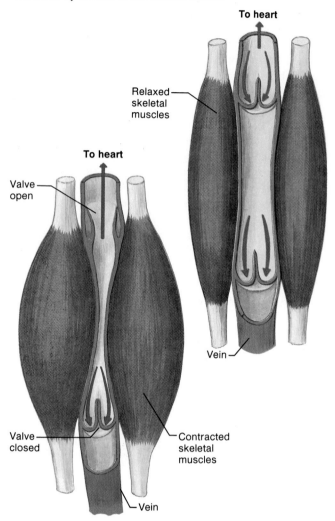

Figure 19.29. A classical demonstration by William Harvey of the existence of venous valves that prevent the flow of blood away from the heart.

the veins are squeezed by contracting skeletal muscles, a one-way flow of blood to the heart is insured by the presence of **venous valves.** The ability of these valves to prevent the flow of blood away from the heart was demonstrated in the seventeenth century by William Harvey (fig. 19.29). After applying a tourniquet to a subject's arm, Harvey found that he could push the blood in a bulging vein toward the heart but not in the reverse direction.

The effect of the massaging action of skeletal muscles on venous blood flow is often described as the **skeletal muscle pump.** The rate of venous return to the heart is dependent, in large part, on the action of skeletal muscle pumps. When these pumps are less active, as when a person stands still or is bedridden, blood accumulates in the veins and causes them to bulge. When a person is more active, blood returns to the heart at a faster rate and less is left in the venous system.

The accumulation of blood in the veins of the legs over a long period of time, as may occur in people with occupations that require standing still all day, can cause the veins to stretch to the point where the venous valves are no longer efficient. This can produce **varicose** *(var'i-kōs)* **veins.** During walking the movements of the foot activate the soleus muscle pump. This effect can be produced in bedridden people by upward and downward manipulations of the feet.

1. Describe the basic structural pattern of arteries and veins. Describe how arteries and veins differ in structure and how these differences contribute to the resistance function of arteries and the capacitance function of veins.
2. Describe the functional significance of the "skeletal muscle pump," and explain the action of venous valves.
3. Explain the functions of capillaries, and describe the structural differences between capillaries in different organs.

The kidneys, endocrine glands, and intestines have *fenestrated capillaries,* characterized by wide intercellular pores (800–1,000 Å) that are covered by a layer of mucoprotein, which may serve as a diaphragm. In the bone marrow, liver, and spleen, the distance between endothelial cells is so great that these *discontinuous* capillaries appear as little cavities (*sinusoids*) in the organ.

Veins

The average pressure in the veins is only 2 mm Hg (millimeters of mercury), compared to a much higher average arterial pressure of about 100 mm Hg. These pressures represent the hydrostatic pressure that the blood exerts on the walls of the vessels, and the numbers indicate the differences from atmospheric pressure.

The low venous pressure is insufficient to return blood to the heart, particularly from the lower limbs. Veins, however, pass between skeletal muscle groups that produce a massaging action as they contract (fig. 19.28). As

varicose: L. *varicosus,* dilated vein

Figure 19.30. Principal arteries of the body (a. = artery; aa. = arteries).

External carotid a.

Right common carotid a.

Vertebral a.

Right subclavian a.

Axillary

Brachiocephalic a.

Aorta

Hepatic a.

Superior mesenteric a.

Ulnar a.

Radial a.

External iliac a.

Palmar arch a.

Deep femoral a.

Femoral a.

Posterior tibial a.

Anterior tibial a.

Internal carotid a.

Left common carotid a.

Left subclavian a.

Arch of the Aorta

Coronary aa.

Celiac a.

Brachial a.

Left gastric a.

Splenic a.

Renal aa.

Testicular/ovarian (gonadal) aa.

Inferior mesenteric a.

Common iliac a.

Internal iliac a.

Popliteal a.

Moon

Principal Arteries of the Body

The aorta leaves the left ventricle, ascends to a position just above the heart, and then descends through the thorax and abdomen. Branches of the aorta carry oxygenated blood to all of the cells of the body.

Objective 17. In the form of a flow chart, list the arterial branches of the ascending aorta and aortic arch.

Objective 18. Describe the arterial supply to the brain.

Objective 19. Describe the arterial pathways that supply the upper extremities.

Objective 20. Describe the major arteries to the thorax, abdomen, and lower extremities.

Contraction of the left ventricle forces oxygenated blood into the arteries of the systemic circulation. The principal arteries of the body are shown in figure 19.30. They will

Figure 19.31. The structural relationship between the major arteries and veins to and from the heart (v. = vein; vv. = veins).

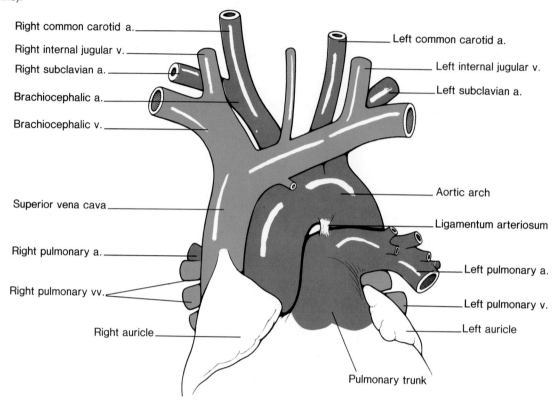

Right common carotid a.

Right internal jugular v.

Right subclavian a.

Brachiocephalic a.

Brachiocephalic v.

Superior vena cava

Right pulmonary a.

Right pulmonary vv.

Right auricle

Left common carotid a.

Left internal jugular v.

Left subclavian a.

Aortic arch

Ligamentum arteriosum

Left pulmonary a.

Left pulmonary v.

Left auricle

Pulmonary trunk

be described by region and identified in order from largest to smallest, or as the blood flows through the system. The major systemic artery is the **aorta,** from which all the primary arteries arise.

Arch of the Aorta

The systemic vessel that ascends from the left ventricle of the heart is called the **ascending aorta.** The right and left **coronary arteries** serving the myocardium of the heart with blood are the only branches that arise off the ascending aorta. The aorta arches to the left and posteriorly over the pulmonary arteries as the **aortic arch** (fig. 19.31). Three vessels arise from the aortic arch: the **brachiocephalic** *(brak″e-o-sĕ-fal′ik)* **(innominate),** the **left common carotid** *(kah-rot′id),* and the **left subclavian.**

The brachiocephalic is the first vessel to branch from the aortic arch and, as its name suggests, supplies blood to the tissue of the arm and head on the right side of the body. It is a short vessel rising superiorly through the mediastinum to a point near the junction of the sternum and the right clavicle. There it bifurcates into the **right common carotid,** which extends to the right side of the neck and head, and the **right subclavian,** which carries blood to the right upper extremity.

The remaining two branches from the arch of the aorta are the left common carotid and the left subclavian arteries. The left common carotid transports blood to the left side of the neck and head, and the left subclavian supplies the left upper extremity.

Arteries of the Neck and Head

The common carotids course upward in the neck along either side of the trachea (fig. 19.32). Several small vessels arise from the common carotid to supply blood to the larynx, thyroid, anterior neck muscles, and lymph glands of the neck. The common carotid bifurcates into the **internal** and **external carotid arteries** slightly below the angle of the mandible. By pressing gently in this area you can detect your pulse. At the base of the internal carotid, near the bifurcation, is a slight dilation called the **carotid sinus.** The carotid sinus contains *baroreceptors,* which monitor blood pressure, and *chemoreceptors (ke″mo-re-sep′tor)* within the **carotid body,** which respond to chemical changes in the blood.

Blood Supply to the Brain. The brain is supplied with arterial blood that arrives through four vessels that eventually unite on the inferior aspect of the brain surrounding the pituitary gland (fig. 19.33). The four vessels are the paired internal carotid arteries and the paired vertebral arteries. The value of having four separate vessels that

carotid: Gk. *karotikos,* stupefying (a state that can be induced by finger pressure in a region of the carotid called the carotid sinus)

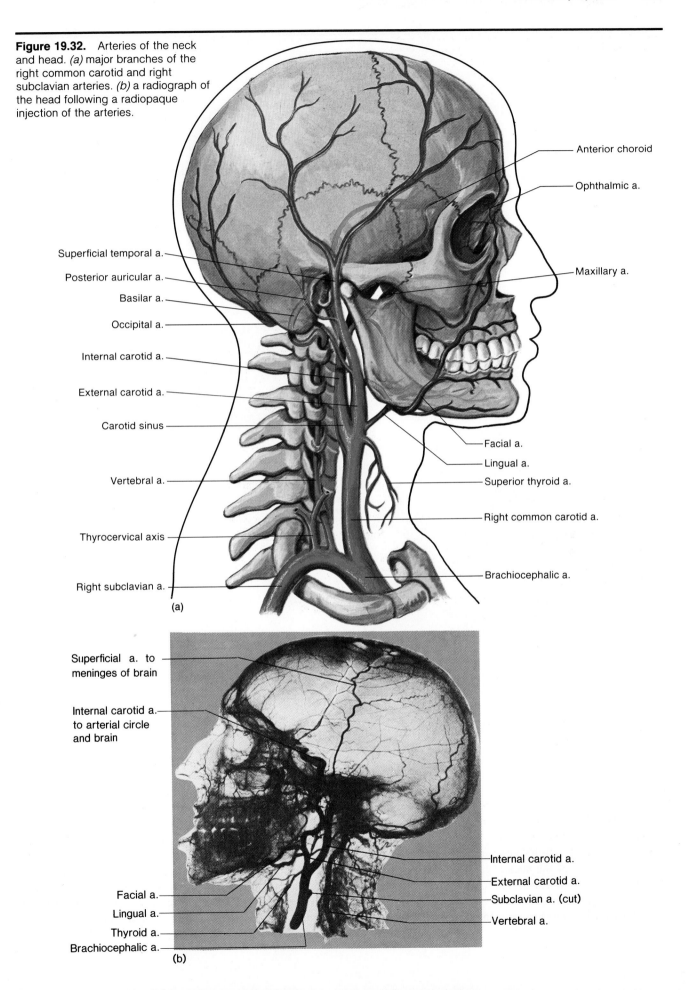

Figure 19.32. Arteries of the neck and head. *(a)* major branches of the right common carotid and right subclavian arteries. *(b)* a radiograph of the head following a radiopaque injection of the arteries.

Anterior choroid

Ophthalmic a.

Superficial temporal a.

Posterior auricular a.

Basilar a.

Occipital a.

Internal carotid a.

External carotid a.

Carotid sinus

Vertebral a.

Thyrocervical axis

Right subclavian a.

Maxillary a.

Facial a.

Lingual a.

Superior thyroid a.

Right common carotid a.

Brachiocephalic a.

(a)

Superficial a. to meninges of brain

Internal carotid a. to arterial circle and brain

Facial a.

Lingual a.

Thyroid a.

Brachiocephalic a.

Internal carotid a.

External carotid a.

Subclavian a. (cut)

Vertebral a.

(b)

Figure 19.33. Arteries that supply blood to the brain.

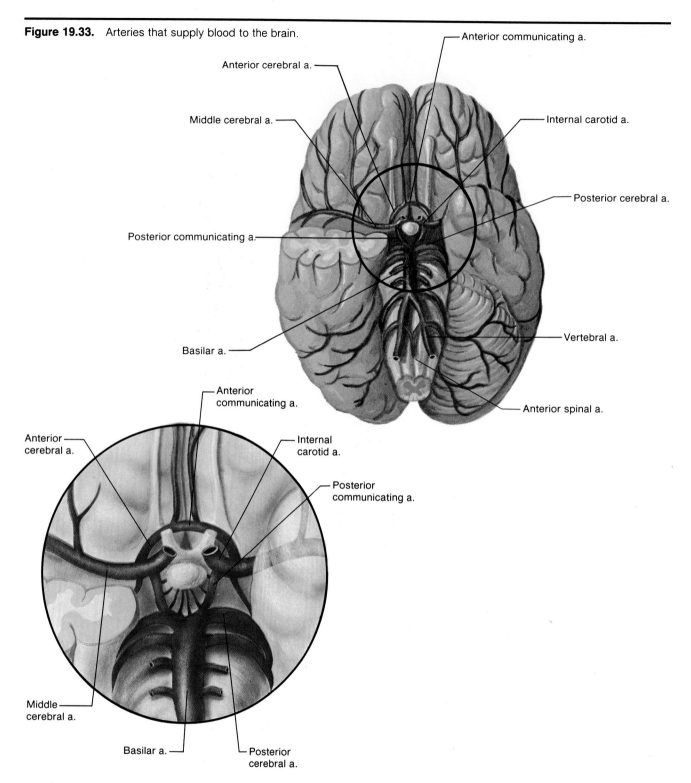

anastomose at one location is that if one becomes occluded, the three alternate routes may insure an adequate blood supply to the brain.

The **vertebral arteries** arise from the subclavian arteries at the base of the neck (fig. 19.32). They pass superiorly through the transverse foramina of the cervical vertebrae and enter the skull through the foramen magnum. Within the braincase, the two vertebral arteries unite to form the **basilar artery** at the level of the pons. The basilar ascends along the inferior surface of the brain stem and inner ear; it terminates by forming two **posterior cerebral arteries,** which supply the posterior portion of the cerebrum. The *posterior communicating arteries* are branches that arise from the posterior cerebral arteries and participate in forming the *arterial circle* around the pituitary gland, also called the **circle of Willis.**

Figure 19.34.   Branches of the external carotid artery. Note
that the internal carotid artery ascends to the base of the
brain through the carotid canal of the temporal bone. The right vertebral artery is also shown as it ascends to the brain
through the transverse foramina of the cervical vertebrae.

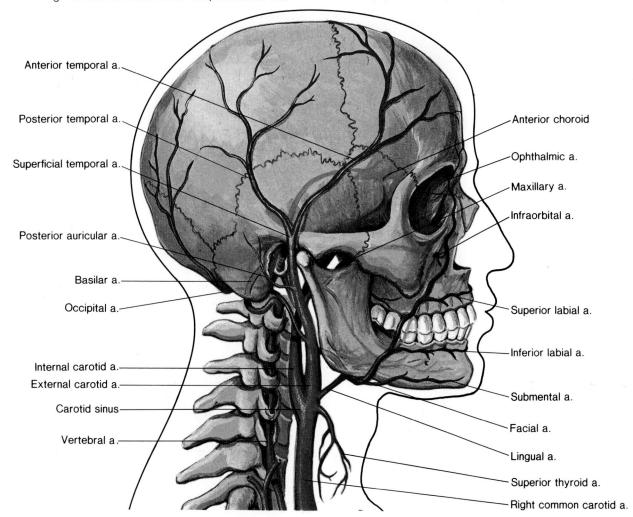

Anterior temporal a.

Posterior temporal a.

Superficial temporal a.

Posterior auricular a.

Basilar a.

Occipital a.

Internal carotid a.
External carotid a.
Carotid sinus
Vertebral a.

Anterior choroid
Ophthalmic a.
Maxillary a.
Infraorbital a.

Superior labial a.

Inferior labial a.

Submental a.

Facial a.

Lingual a.

Superior thyroid a.

Right common carotid a.

The **internal carotid artery** branches from the common
carotid artery and ascends in the neck until it reaches the
base of the skull, where it enters the carotid canal of the
temporal bone. Several branches arise from the internal
carotid once it is on the inferior surface of the brain.
Three of the more important ones are the **ophthalmic**
(of'thal'mik) **artery** (not seen in the illustration), which
supplies the eye and associated structures, and the **anterior** and **middle cerebral arteries,** which provide blood to
the cerebrum. The internal carotids are connected to the
posterior cerebral arteries at the circle of Willis.

Capillaries within the pituitary gland receive both
arterial and venous blood. The venous blood arrives
from venules immediately superior to the pituitary,
which drain capillaries in the hypothalamus of the
brain. This arrangement of two capillary beds in
series—where the second capillary bed receives
venous blood from the first—is called a **portal
system.** The venous blood that travels from the
hypothalamus to the pituitary contains hormones
from the hypothalamus that help regulate pituitary
gland hormone secretion.

External Carotid Artery.   The external carotid artery
gives off several branches as it extends upward along the
side of the neck and head (fig. 19.34). The names of these
branches are determined by the areas or structures that
they serve. The principal vessels that arise from the external carotid are the following:

1. **Superior thyroid artery,** which serves the muscles of
   the hyoid region, the larynx and vocal cords, and
   the thyroid gland.
2. **Ascending pharyngeal artery** (not seen in
   illustration), which serves the pharyngeal area and
   various lymph nodes.
3. **Lingual artery,** which provides extensive
   vascularization of the tongue and sublingual
   salivary gland.
4. **Facial artery,** which traverses a notch on the
   inferior margin of the mandible to serve the
   pharyngeal area, palate, chin, lips, and nasal region.
   The facial artery is an important vessel to apply
   pressure when controlling bleeding from the face.

Figure 19.35. Arteries of the upper extremity. (*a*) a radiograph of the forearm and hand following a radiopaque injection of the arteries. (*b*) an anterior view of major arteries.

Anastomoses between arteries at fingertips

Phalanges

Branches to fingers

Metacarpal of index finger

Branch to thumb and index finger

Radial a. joining deep arch of hand

Carpal bones of wrist

Ulnar artery dividing to form deep and superficial arches of hand

Radial a.

Ulnar a.

Radius

Ulna

(a)

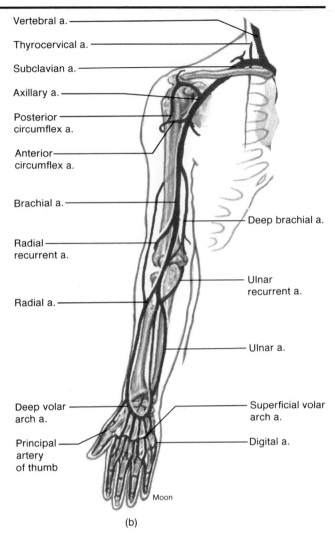

Vertebral a.

Thyrocervical a.

Subclavian a.

Axillary a.

Posterior circumflex a.

Anterior circumflex a.

Brachial a.

Radial recurrent a.

Radial a.

Deep volar arch a.

Principal artery of thumb

Deep brachial a.

Ulnar recurrent a.

Ulnar a.

Superficial volar arch a.

Digital a.

Moon

(b)

5. **Occipital artery,** which serves the posterior portion of the scalp, the meninges over the brain, the mastoid process, and certain posterior neck muscles.
6. **Posterior auricular artery,** which serves the ear and scalp over the ear.

The external carotid artery terminates at a level near the mandibular condyle by dividing into **maxillary** and **superficial temporal arteries.** The maxillary artery gives off branches to the teeth and gums, the muscles of mastication, nasal cavity, eyelids, and meninges. The superficial temporal artery supplies blood to the parotid salivary gland and to the superficial structures on the side of the head. Pulsations through the temporal artery can be easily detected by placing your fingertips immediately in front of the ear at the level of the eye. This vessel is frequently used by anesthesiologists to check a patient's pulse rate during surgery.

Headaches are usually caused by vascular pressure on the sensitive meninges covering the brain. The two principal vessels serving the meninges are the occipital and maxillary arteries. Vasodilation of these vessels creates excessive pressure on the sensory receptors within the meninges, resulting in a headache.

Arteries of the Upper Extremity

The right subclavian artery branches off the brachiocephalic, and the left subclavian artery arises directly from the arch of the aorta (fig. 19.30). The **subclavian** *(sub-cla've-an)* **artery** passes laterally deep to the clavicle, carrying blood toward the arm (fig. 19.35). You can detect the pulsations of your subclavian artery by pressing firmly on the tissue just above the medial portion of your clavicle. From each subclavian artery arises a **vertebral artery** that carries blood to the brain (already described), a short **thyrocervical trunk** that serves the thyroid gland, trachea, and larynx, and an **internal thoracic (mammary) artery** that

Figure 19.36.   Arteries that serve the thoracic wall.

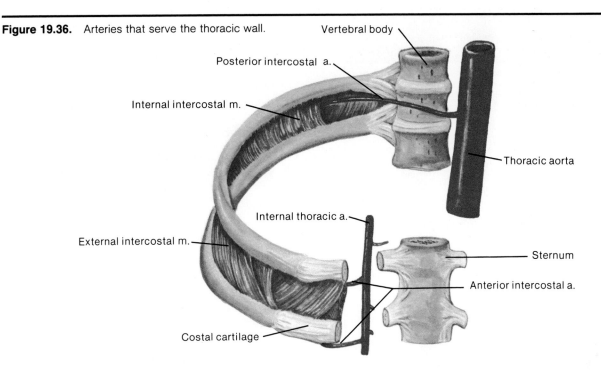

Vertebral body

Posterior intercostal a.

Internal intercostal m.

Thoracic aorta

Internal thoracic a.

External intercostal m.

Sternum

Anterior intercostal a.

Costal cartilage

descends into the thorax to serve the thoracic wall, thymus, and pericardium. A branch of the internal thoracic artery supplies blood to the muscles and tissues (mammary glands) of the anterior thorax.

The subclavian artery becomes the **axillary** *(ak′si-lar″e)* **artery** as it passes into the axillary region. The axillary artery is that portion of the major artery of the upper extremity between the outer border of the first rib and the lower border of the teres major muscle. Several small branches arise from the axillary artery and supply blood to the tissues of the upper thorax and shoulder region.

The **brachial** *(bra′ke-al)* **artery** is the continuation of the axillary artery through the brachial region. The brachial artery courses on the medial side of the humerus where it is a major pressure point and the most common site for determining blood pressure. A **deep brachial artery** branches from the brachial artery and curves posteriorly near the radial nerve to supply the triceps muscle. Two additional branches from the brachial, the **anterior** and **posterior circumflex arteries,** form a continuous ring of vessels around the proximal and distal portions of the arm, respectively.

Just proximal to the cubital fossa, the brachial bifurcates into the **radial** and **ulnar arteries,** which supply blood to the forearm and a portion of the hand and digits. The radial artery courses down the lateral, or radial, side of the arm where it sends numerous small branches to the muscles of the forearm. The **radial recurrent artery** serves the region of the elbow and is the first and largest off of the radial artery. The radial artery is important as a site to record the pulse near the wrist.

The ulnar artery is a bit larger in caliber than the radial artery. It extends down the ulnar side of the forearm and gives off many small branches to the muscles on the medial side of the forearm. It too has an initial large branch, which arises from the proximal portion near the elbow and is called the **ulnar recurrent artery.** At the wrist, the ulnar and radial arteries anastomose to form the **palmar arch** in the hand, from which **digital arteries** extend into the fingers.

Branches of the Thoracic Aorta

The thoracic aorta is a continuation of the arch of the aorta as it descends through the thoracic cavity to the diaphragm. This large vessel gives off branches to the organs and muscles of the thoracic region. These branches include **pericardial arteries,** going to the pericardium of the heart; **bronchial arteries** for systemic circulation to the lungs; **esophageal** *(ē-sof″ah-je′al)* **arteries,** going to the esophagus as it passes through the mediastinum; segmental **intercostal arteries,** serving the intercostal muscles and structures of the wall of the thorax (fig. 19.36); and **phrenic** *(fren′ik)* **arteries,** supplying blood to the diaphragm.

These vessels are summarized according to their location and function in table 19.7.

Branches of the Abdominal Aorta

The **abdominal aorta** is the segment of the aorta between the diaphragm and the level of the fourth lumbar vertebra, where it divides into the right and left **common iliac** *(il′e-ak)* **arteries.** The first branch of the abdominal aorta is called the **celiac** *(se′le-ak)* **artery.** It is a short, thick trunk that arises anteriorly just below the diaphragm. The

Table **19.7** Segments and branches of the aorta

Segment of aorta	Arterial branch	General region or organ served
Ascending aorta	Right and left coronary	Heart
Arch of aorta	Brachiocephalic	
	Right common carotid	Right side of head and neck
	Right subclavian	Right shoulder and right upper extremity
	Left common carotid	Left side of head and neck
	Left subclavian	Left shoulder and left upper extremity
Thoracic aorta	Pericardials	Pericardium of heart
	Intercostals	Intercostal and thoracic muscles, pleurae
	Superior phrenics	Diaphragm
	Bronchials	Bronchi of lungs
	Esophageals	Esophagus
	Inferior phrenics	Inferior surface of diaphragm
Abdominal aorta	Celiac	
	Common hepatic	Liver, upper pancreas, duodenum
	Left gastric	Stomach, esophagus
	Splenic	Spleen, pancreas, stomach
	Superior mesenteric	Small intestine, pancreas, cecum, ascending and transverse colons
	Suprarenals	Suprarenal (adrenal) glands
	Renals	Kidneys
	Gonadals	
	Testiculars	Testes
	Ovarians	Ovaries
	Inferior mesenteric	Transverse, descending, and sigmoid colons; rectum
	Common iliacs	
	External iliacs	Lower extremities
	Internal iliacs (hypogastrics)	Genital organs, gluteal muscles

Source: Van De Graaff, Kent M., *Human Anatomy.* © 1984 Wm. C. Brown Publishers, Dubuque, Iowa. All Rights Reserved. Reprinted by permission.

Figure 19.37. An anterior view of the abdominal aorta and its principal branches. (*a*) a drawing; (*b*) an X-ray.

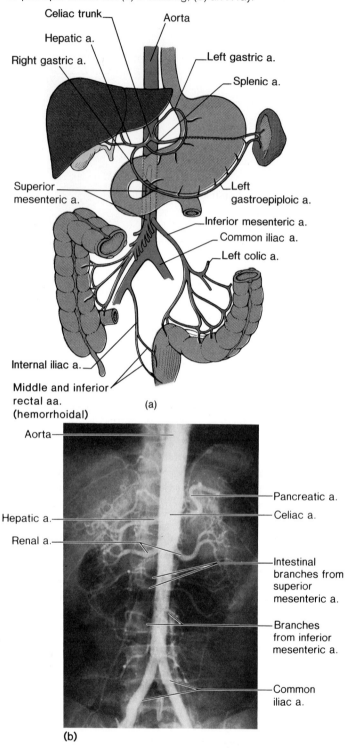

celiac divides immediately into three arteries: the **splenic** *(splen'ik),* going to the spleen, the **left gastric,** going to the stomach, and the **hepatic** *(hĕ-pat'ik),* going to the liver (fig. 19.37).

The **superior mesenteric artery** is another unpaired vessel and arises anteriorly from the abdominal aorta just below the celiac. The superior mesenteric supplies blood to the small intestine (except for a portion of the duodenum), the cecum, the appendix, the ascending colon, and the transverse colon.

The next major vessels to arise from the abdominal aorta are the paired **renal arteries,** which carry blood to the kidneys. Smaller **suprarenal arteries** are located just above the renal arteries and serve the suprarenal (adrenal) glands. The **testicular** (internal spermatic) **arteries** in the male and the **ovarian arteries** in the female are small, paired vessels that arise from the abdominal aorta just below the renals and serve the gonads.

Figure 19.38.   Arteries of the pelvic region.

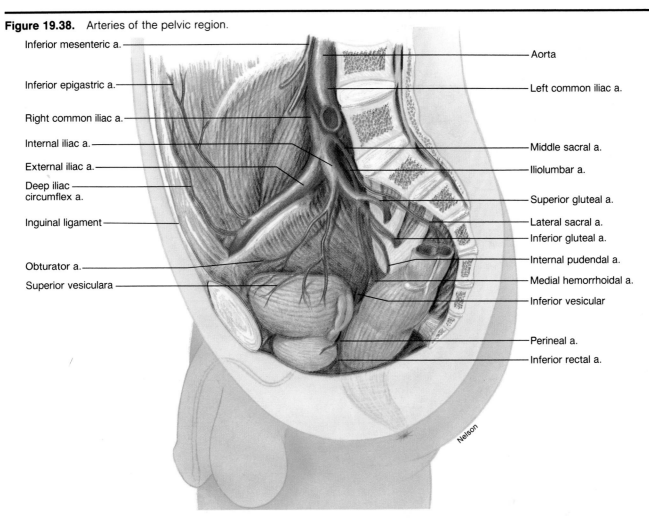

Inferior mesenteric a.

Inferior epigastric a.

Right common iliac a.

Internal iliac a.

External iliac a.

Deep iliac circumflex a.

Inguinal ligament

Obturator a.

Superior vesiculara

Aorta

Left common iliac a.

Middle sacral a.

Iliolumbar a.

Superior gluteal a.

Lateral sacral a.

Inferior gluteal a.

Internal pudendal a.

Medial hemorrhoidal a.

Inferior vesicular

Perineal a.

Inferior rectal a.

Nelson

The **inferior mesenteric artery** is the last major branch of the abdominal aorta. It is an unpaired anterior vessel that arises just before the iliac bifurcation. The inferior mesenteric supplies blood to the descending colon and rectum.

Several **lumbar arteries** branch posteriorly from the abdominal aorta throughout its length and serve the muscles and the spinal cord in the lumbar region. In addition, an unpaired **middle sacral artery** (fig. 19.38) arises from the posterior terminal portion of the abdominal aorta to supply the sacrum and coccyx.

Arteries of the Pelvis and Lower Extremities

The aorta terminates in the posterior pelvic area as it bifurcates into the right and left common iliac arteries. These vessels pass downward approximately 5 cm on their respective sides and terminate by dividing into the **internal** and **external iliac arteries.**

The internal iliac has extensive branches to supply arterial blood to the gluteal muscles and the organs of the pelvic region (fig. 19.38). The wall of the pelvis is served by the **iliolumbar** and **lateral sacral arteries.** The internal visceral organs of the pelvis are served by the **middle rectal** and the **superior, middle** and **inferior vesicular arteries** to the urinary bladder. In addition, in females **uterine** and **vaginal** *(vaj'ĭ-nal)* **arteries** branch from the internal iliacs to serve the reproductive organs. The muscles of the buttock are served by the **superior** and **inferior gluteal arteries.** Some of the upper medial thigh muscles are supplied with blood from the **obturator artery.** The **internal pudendal artery** of the internal iliac serves the external genitalia of male and female and is an important vessel in sexual activity. Erection of the penis in the male and corresponding vascular engorgement of the genitalia of the female are vascular phenomena controlled by the autonomic nervous system.

The external iliac artery passes out of the pelvic cavity beneath the inguinal ligament (fig. 19.38) and becomes the **femoral artery.** Two branches arise from the external iliac, however, before it traverses beneath the inguinal ligament. An **inferior epigastric artery** branches immediately from the external iliac and passes superiorly to supply the skin and muscles of the abdominal wall. The **deep circumflex iliac artery** is a small branch that extends laterally to supply the muscles attached to the iliac fossa.

The femoral artery passes through an area called the **femoral triangle** on the upper medial portion of the thigh (figs. 19.39 and 19.40). At this point, the femoral artery

Figure 19.39. Arteries of the right lower extremity. (*a*) an anterior view; (*b*) a posterior view.

(a) (b)

is close to the surface and not only can be palpated but is an important pressure point. Several vessels arise from the femoral to serve the thigh region. The largest of these, the **deep femoral artery,** passes posteriorly to serve the hamstring muscles. The **lateral** and **medial femoral circumflex arteries** encircle the proximal end of the femur and serve muscles in this region. The femoral artery becomes the **popliteal** *(pop-lit′e-al)* **artery** as it passes across the posterior aspect of the knee.

Hemorrhage can be a serious problem in many accidents. Therefore, one should know the pressure points where the arterial blood flow can be curtailed to prevent a victim from bleeding to death. The pressure points for the appendages are the brachial artery on the medial side of the arm and the femoral artery in the groin. Firmly applied pressure to these regions greatly diminishes the flow of blood to traumatized areas below. A tourniquet may have to be applied if bleeding is severe enough to endanger life.

The popliteal artery supplies small branches to the knee joint and then divides into an **anterior tibial artery** and **posterior tibial artery** (fig. 19.39). These vessels traverse the anterior and posterior aspects of the leg, respectively, providing blood to the muscles of these regions and to the foot.

At the ankle, the anterior tibial artery becomes the **dorsalis pedis artery,** which serves the ankle and dorsum of the foot, after which it contributes to the formation of the **plantar arch** of the foot. Clinically, palpation of the dorsalis pedis artery can provide information about circulation to the foot, but more importantly it can provide information about the circulation in general because these pulses are taken at the most distal portion of the body.

The posterior tibial artery sends a large **peroneal** *(per″o-ne′al)* **artery** to serve the peroneal muscles of the leg. At the ankle, the posterior tibial bifurcates into the **lateral** and **medial plantar arteries,** which supply the muscles and structures on the sole of the foot. The lateral

Figure 19.40.   The femoral triangle.

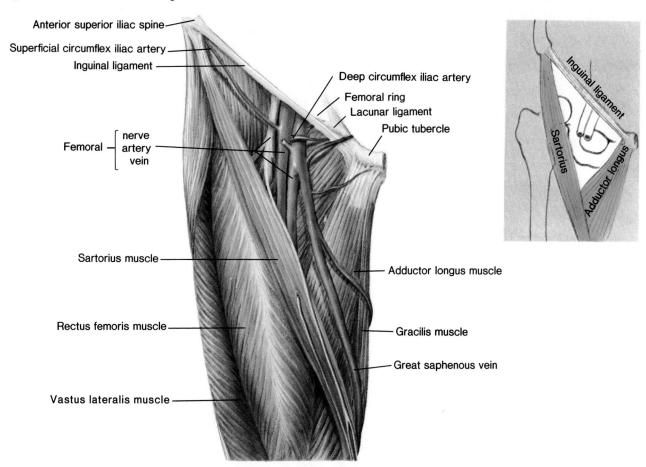

Anterior superior iliac spine

Superficial circumflex iliac artery

Inguinal ligament

Deep circumflex iliac artery

Femoral ring

Lacunar ligament

Pubic tubercle

Femoral { nerve / artery / vein }

Sartorius muscle

Adductor longus muscle

Rectus femoris muscle

Gracilis muscle

Great saphenous vein

Vastus lateralis muscle

Inguinal ligament

Sartorius

Adductor longus

plantar artery anastomoses with the dorsal pedis artery to form the plantar arch, similar to the arterial arrangement in the hand. **Digital arteries** arise from the plantar arch to supply the toes with blood.

> 1. *Describe the blood supply to the brain. Where is the circle of Willis located, and how is it formed?*
> 2. *Describe the clinical significance of the brachial and radial arteries.*
> 3. *Describe the arterial pathway from the subclavian artery to the digital arteries.*
> 4. *List the arteries that supply blood to the lower abdominal wall, the external genitalia, the hamstring muscles, the knee joint, and the dorsum of the foot.*

Principal Veins of the Body

After systemic blood has passed through the tissues and become depleted of oxygen, it returns to the right atrium of the heart in veins of ever increasing diameters.

Objective 21.   Describe the venous drainage of the head, neck, and upper extremities.

Objective 22.   Describe venous drainage of the thorax, lower extremities, and abdominal region.

Objective 23.   Describe the vessels involved in the hepatic portal system.

In the venous portion of the systemic circulation blood flows from smaller vessels into larger ones, so that a vein receives smaller tributaries instead of giving off branches as an artery does. The veins from all parts of the body converge into two major vessels that empty into the right atrium: the **superior** and **inferior vena cavae.** Veins are more numerous than arteries and are both superficial and deep. Superficial veins can generally be seen just beneath the skin and are clinically important in drawing blood and giving injections. Deep veins are close to the principal arteries and are usually similarly named. As with arteries, veins are named according to the region in which they are found or the organ that they serve (when a vein serves an organ it drains blood away from it).

The principal systemic veins of the body are illustrated in figure 19.41.

Figure 19.41. Principal veins of the body. Superficial veins are depicted in the left extremities and deep veins in the right extremities.

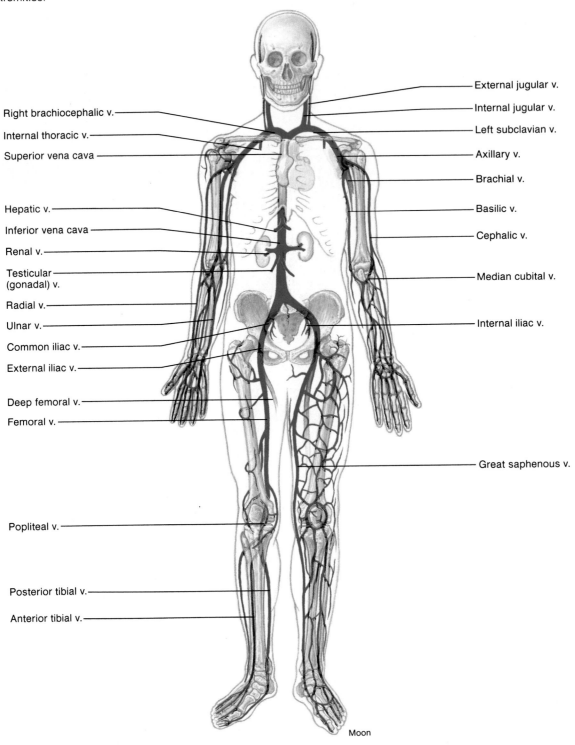

External jugular v.

Internal jugular v.

Left subclavian v.

Right brachiocephalic v.

Internal thoracic v.

Superior vena cava

Axillary v.

Brachial v.

Hepatic v.

Basilic v.

Inferior vena cava

Cephalic v.

Renal v.

Testicular (gonadal) v.

Median cubital v.

Radial v.

Ulnar v.

Common iliac v.

Internal iliac v.

External iliac v.

Deep femoral v.

Femoral v.

Great saphenous v.

Popliteal v.

Posterior tibial v.

Anterior tibial v.

Moon

Figure 19.42.   Veins that drain the head and neck.

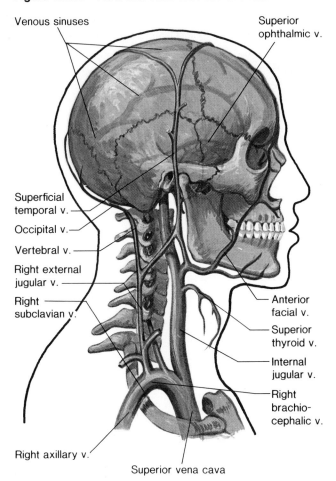

Venous sinuses

Superior
ophthalmic v.

Superficial
temporal v.

Occipital v.

Vertebral v.

Right external
jugular v.

Right
subclavian v.

Right axillary v.

Superior vena cava

Anterior
facial v.

Superior
thyroid v.

Internal
jugular v.

Right
brachio-
cephalic v.

Figure 19.43.   Cranial venous sinuses and the internal
jugular vein.

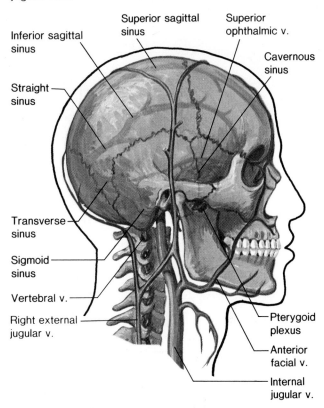

Inferior sagittal
sinus

Superior sagittal
sinus

Superior
ophthalmic v.

Cavernous
sinus

Straight
sinus

Transverse
sinus

Sigmoid
sinus

Vertebral v.

Right external
jugular v.

Pterygoid
plexus

Anterior
facial v.

Internal
jugular v.

Veins Draining the Head and Neck

Blood from the scalp, portions of the face, and the superficial neck regions is drained by the **external jugular veins** (fig. 19.42). These vessels descend on either side of the neck, superficial to the sternocleidomastoid muscle and deep to the platysma. They drain into the right and left **subclavian veins** located just behind the clavicles.

The paired **internal jugular veins** drain blood from the brain, meninges, and deep regions of the face and neck. The internal jugular veins are larger and deeper than the external jugular veins. They arise from numerous cranial **venous sinuses,** which are a series of both paired and unpaired channels positioned between the two layers of dura mater. The venous sinuses, in turn, receive venous blood from the **cerebral,** the **cerebellar,** the **ophthalmic,** and the **meningeal** *(mĕ-nin'je-al)* **veins.** Figure 19.43 illustrates the principal cranial venous sinuses and the formation of the internal jugular vein.

The internal jugular vein passes inferiorly down the neck adjacent to the common carotid artery and the vagus nerve. All three are surrounded by the protective *carotid sheath* and are positioned beneath the sternocleidomastoid muscle. The internal jugular empties into the subclavian vein, and the union of these two vessels forms the large **brachiocephalic (innominate) vein** on each side. The two brachiocephalic veins then merge to form the **superior vena cava,** which drains into the right atrium of the heart (fig. 19.41).

The external jugular vein is the vessel that is seen on the side of the neck when a person is angry or wearing a tight collar. You can voluntarily distend this vein by performing the **Valsalva maneuver.** To do this, take a deep breath and hold it while you forcibly contract your abdominal muscles as in a forced exhalation. This procedure is automatically performed when lifting a heavy object or defecating. The increased thoracic pressure that results compresses the vena cavae and interferes with the return of blood to the right atrium. The Valsalva maneuver, as a result, can be dangerous if performed by people with cardiovascular disease.

jugular: L. *jugulum,* throat or neck

Valsalva maneuver: from Antonio Valsalva, Italian anatomist,
1666–1723

Figure 19.44. Veins that drain the upper extremity.

Right external jugular v.

Right subclavian

Axillary v.

Brachial v.

Cephalic v.

Radial v.

Right internal jugular v.

Left brachiocephalic v.

Right brachiocephalic v.

Superior vena cava

Basilic v.

Median cubital v.

Ulnar v.

Dorsal arch v.

Moon

Veins of the Upper Extremity

The upper extremity has both superficial and deep venous drainage (fig. 19.44). The superficial veins are highly variable and form an extensive network just below the skin. The deep veins accompany the arteries of the same region and are given similar names. The deep veins of the upper extremity will be described first. Both the **radial vein** on the lateral side of the forearm and the **ulnar vein** on the medial side drain blood from the **deep** and **superficial volar arches** of the hand. The radial and ulnar veins join in the cubital fossa to form the **brachial vein,** which continues up the medial side of the brachium.

The main superficial vessels of the upper extremity are the **basilic vein** and the **cephalic vein.** The basilic vein passes on the ulnar side of the forearm and the medial side

of the arm. Near the head of the humerus, the basilic vein merges with the brachial vein and forms the **axillary vein.**

The cephalic vein drains the superficial portion of the hand and forearm on the radial side and then continues up the lateral side of the arm. In the shoulder region, the cephalic vein pierces the fascia and joins the axillary vein. The axillary vein then passes the first rib to form the subclavian vein, which unites with the external jugular to form the brachiocephalic vein of that side.

Superficially, in the cubital fossa of the elbow, the **median cubital vein** ascends from the cephalic vein on the lateral side to connect with the basilic vein on the medial side. The median cubital vein is a frequent site for venipuncture in order to remove a sample of blood or add fluids to the blood.

Figure 19.45.   Veins of the thoracic region. The lungs and heart have been removed.

External jugular v.

Subclavian v.

Superior vena cava

Axillary v.

Brachial v.

Basilic v.

Azygos v.

Internal jugular v.

Brachiocephalic v v.

Cephalic v.

Posterior intercostal v.

Hemiazygos v.

Veins of the Thorax

The **superior vena cava,** formed by the union of the two brachiocephalic veins, empties venous blood from the head, neck, and upper extremities directly into the right atrium of the heart. These large vessels lack the valves that are characteristic of most other veins in the body.

In addition to receiving blood from the brachiocephalic veins, the superior vena cava collects blood from the azygos system of veins arising from the posterior thoracic wall (fig. 19.45). The **azygos** *(az'ĭ-gos)* **vein** extends superiorly along the dorsal abdominal and thoracic walls on the right side of the vertebral column. The azygos vein ascends through the mediastinum to join the superior vena cava at the level of the fourth thoracic vertebra. Tributaries of the azygos vein include the **ascending lumbar veins** that drain the lumbar and sacral regions, **intercostal veins** draining from the intercostal muscles, and the **accessory hemiazygos** and **hemiazygos veins,** which form the major tributaries to the left of the vertebral column.

Veins of the Lower Extremity

The lower extremities, like the upper extremities, have both a deep and a superficial group of veins (fig. 19.46). The deep veins accompany corresponding arteries and have more valves than do the superficial veins. The deep veins will be described first.

The **posterior** and **anterior tibial veins** originate in the foot and course upward behind and in front of the tibia to the back of the knee where they merge to form the **popliteal vein.** The popliteal vein receives blood from the knee region. Just above the knee, this vessel becomes the **femoral vein.** The femoral vein in turn continues up the thigh and receives blood from the **deep femoral vein** near the groin. Just above this, the femoral vein receives blood from the **great saphenous** *(sah-fe'nus)* **vein** and then becomes the **external iliac vein** as it passes under the inguinal ligament. The external iliac curves upward to the level of the sacroiliac joint. There it merges with the **internal iliac (hypogastric) vein** at the pelvic and genital regions to form the **common iliac vein.** At the level of the fifth lumbar vertebra, the right and left common iliacs unite to form the large **inferior vena cava** (fig. 19.46).

The superficial veins of the lower extremity are the **small** and **great saphenous veins.** The small saphenous vein arises from the lateral side of the foot, courses posteriorly along the surface of the calf of the leg, and descends deep to enter the popliteal vein behind the knee. The great saphenous vein is the longest vessel in the body. It originates at the arch of the foot and ascends superiorly along the medial aspect of the leg and thigh before draining into the femoral vein.

saphenous: L. *saphena*, the hidden one

azygos: Gk. *a*, without; *zygon*, yoke

Figure 19.46. Veins of the lower extremity. (*a*) superficial veins, medial and posterior aspects. (*b*) deep veins, medial view.

Superficial epigastric v.

Femoral v.

Great saphenous v.

Popliteal v.

Small saphenous v.

Inferior vena cava

Right common iliac v.

Internal iliac v.

External iliac v.

Lateral femoral circumflex v.

Deep femoral

Femoral

Popliteal v.

Small saphenous v.

Anterior tibial v.

Posterior tibial v.

Dorsal pedis v.

Medial plantar v.

Lateral plantar v.

Moon

(a)

(b)

Figure 19.47. The hepatic portal system.

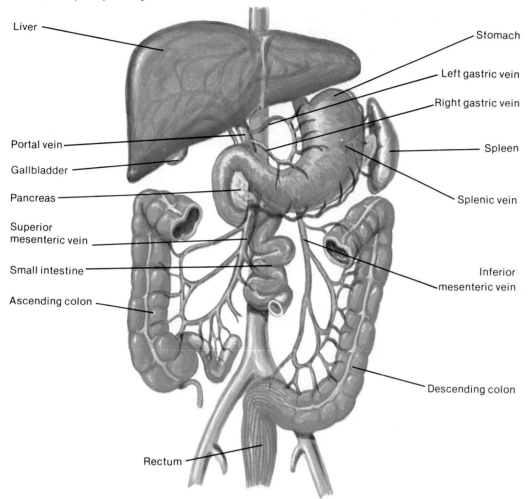

Liver

Portal vein

Gallbladder

Pancreas

Superior
mesenteric vein

Small intestine

Ascending colon

Rectum

Stomach

Left gastric vein

Right gastric vein

Spleen

Splenic vein

Inferior
mesenteric vein

Descending colon

Veins of the Abdominal Region

The **inferior vena cava** parallels the abdominal aorta on the right side as it ascends through the abdominal cavity to penetrate the diaphragm and enter the right atrium (fig. 19.47). It is the largest vessel in diameter in the body and is formed by the union of the two common iliac veins draining the lower extremities. As the inferior vena cava ascends through the abdominal cavity, it receives tributaries from veins that correspond in name and position to arteries previously described.

Four paired **lumbar veins** (not shown) drain the posterior abdominal wall, the vertebral column, and the spinal cord. In the male, the **testicular (internal spermatic) veins** receive blood from the testes. The right testicular vein empties into the left renal vein. The drainage pathway is similar in the female, but the vessels are called the **ovarian veins** and receive blood from the ovaries. The **renal veins** drain blood from the kidneys and ureters into the inferior vena cava. Small **suprarenal veins** drain the suprarenal (adrenal) glands. The right suprarenal vein empties directly into the inferior vena cava, and the left suprarenal connects to the left renal vein. The **inferior phrenic veins** receive blood from the inferior side of the diaphragm and

drain into the inferior vena cava. Right and left **hepatic veins** originate from the capillary sinusoids of the liver and empty into the inferior vena cava immediately below the diaphragm.

Note that the inferior vena cava does not receive blood directly from the digestive tract or the pancreas or the spleen. Instead, the venous outflow from these organs first passes through capillaries in the liver.

Hepatic Portal System

A *portal system* is one in which the veins that drain one group of capillaries deliver blood to another group of capillaries, which in turn are drained by more usual systemic veins that carry blood to the vena cavae and the right atrium of the heart. There are thus two capillary beds in series. The hepatic portal system is composed of veins that drain blood from capillaries in the intestines, pancreas, spleen, stomach, and gallbladder into capillaries in the liver (called *sinusoids*) and of the right and left **hepatic veins** that drain the liver and empty into the inferior vena cava (fig. 19.47). As a result of the hepatic portal system, the absorbed products of digestion must first pass through the liver before entering the general circulation.

Figure 19.48.   Fetal circulation.

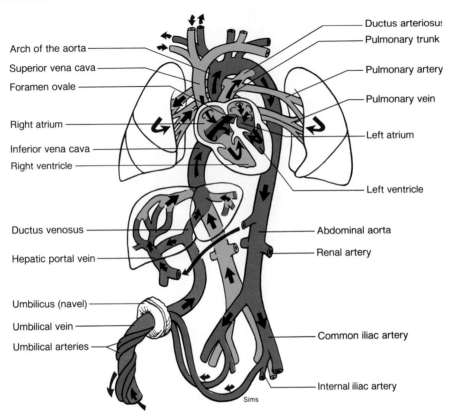

The **hepatic portal vein** is the large vessel that receives blood from the digestive organs. It is formed by a union of the **superior mesenteric vein,** which drains nutrient-rich blood from the small intestine, and the **splenic vein.** The splenic vein drains the spleen but is enlarged because of a convergence of the following four tributaries: (1) the **inferior mesenteric vein** from the large intestine, (2) the **pancreatic vein** from the pancreas, (3) the **left gastroepiploic vein** from the stomach, and (4) the **right gastroepiploic vein,** also from the stomach.

Three additional veins empty into the portal vein. The right and left **gastric veins** drain the lesser curvature of the stomach, and the **cystic vein** drains blood from the gallbladder.

One of the functions of the liver is to detoxify harmful substances, such as alcohol, that are absorbed into the blood from the small intestine. However, excessive quantities of alcohol cannot be processed during a single pass through the liver, and so a person becomes intoxicated. Eventually, the liver is able to process the alcohol as the circulating blood is repeatedly exposed to the liver sinusoids via the hepatic artery. Alcoholics may eventually suffer from cirrhosis of the liver as the normal liver tissue is destroyed.

gastroepiploic: Gk. *gastros,* stomach; *epiplein,* to float on
   (referring to greater omentum)

In summary, it is important to note that the sinusoids of the liver receive blood from two sources. The hepatic artery supplies oxygen-rich blood to the liver, whereas the hepatic portal vein transports nutrient-rich blood from the intestine for processing. These two blood sources become mixed in the liver sinusoids. Liver cells exposed to this blood obtain nourishment from it and are uniquely qualified (because of their anatomical position and enzymatic ability) to modify the chemical nature of the venous blood that enters the general circulation from the digestive tract.

1. Using a flow chart, list the venous drainage from the head and neck to the superior vena cava. Indicate which vein may bulge in the side of the neck when a person performs the Valsalva maneuver and which vein is commonly used as a site for venipuncture.
2. Describe the positions, sources, and drainages of the small and great saphenous veins.
3. Describe the hepatic portal system, and explain the functional importance of this system.

Fetal Circulation

The fetal circulation demonstrates adaptations to the fact that the fetal lungs are nonfunctional and that oxygen and nutrients are obtained from the placenta. These structural adaptations to fetal life are normally eliminated quickly after birth.

Table 19.8 Cardiovascular structures of the fetus and changes in the neonatal infant

Structure	Location	Function	Fate in neonatal infant
Umbilical vein	Connects the placenta to the liver; forms a major portion of umbilical cord	Transports nutrient-rich, oxygenated blood from the placenta to the fetus	Forms the round ligament of the liver
Ductus venosus	Venous shunt within the liver to connect with the inferior vena cava	Transports oxygenated blood directly into the inferior vena cava	Forms the ligamentum venosum, a fibrous cord in the liver
Foramen ovale	Opening between the right and left atria	A shunt to bypass the pulmonary circuitry	Closes at birth and becomes the fossa ovalis, a depression in the interatrial septum
Ductus arteriosus	Between pulmonary trunk and arch of the aorta	A shunt to bypass the pulmonary circuitry	Closes shortly after birth, atrophies, and becomes the ligamentum arteriosum
Umbilical arteries	Arise from internal iliac arteries and associated with umbilical cord	Transports blood from the fetus to the placenta	Atrophies to become the lateral umbilical ligaments

Source: Van De Graaff, Kent M., *Human Anatomy*. © 1984 Wm. C. Brown Publishers, Dubuque, Iowa. All Rights Reserved. Reprinted by permission.

Objective 24. Describe the fetal circulation to and from the placenta.

Objective 25. Describe the structure and function of the foramen ovale, ductus venosus, and ductus arteriosus.

The circulation of blood through a fetus is by necessity different from that of a newborn infant (fig. 19.48). Respiration, the procurement of nutrients, and the elimination of metabolic wastes occur through the maternal blood instead of through the organs of the fetus. The capillary exchange between the maternal and fetal circulation occurs within the **placenta.** This remarkable structure includes part of the uterus of the mother during pregnancy and is discharged following delivery as the afterbirth.

The **umbilical cord** is the connection between the placenta and the fetal umbilicus. It includes one **umbilical vein** and two **umbilical arteries** surrounded by a gelatinous substance. Oxygenated and nutrient-rich blood flows through the umbilical vein toward the inferior surface of the liver. At this point, the umbilical vein bifurcates into one branch that joins with the portal vein, while the other branch, called the **ductus venosus,** enters the inferior vena cava. Thus, oxygenated blood is mixed with venous blood returning from the lower extremities of the fetus before it enters the heart. The umbilical vein is the only vessel of the fetus that carries fully oxygenated blood.

The inferior vena cava empties into the right atrium of the fetal heart. From here, most of the blood is directed through the **foramen ovale,** an opening between the two atria. Here it mixes with a small quantity of blood returning through the pulmonary circulation. The blood then passes into the left ventricle, from which it is pumped into the aorta and through the body of the fetus. Some blood entering the right atrium passes into the right ventricle and out of the heart via the pulmonary trunk. Since the lungs of the fetus are not functional, only a small portion of blood continues through the pulmonary circulation (the resistance to blood flow is very high in the collapsed fetal lungs). Most of the blood in the pulmonary trunk passes through the **ductus arteriosus** into the aortic arch, where it mixes with blood coming from the left ventricle. Blood is returned to the placenta by the two **umbilical arteries** that arise from the internal iliac arteries.

Notice that in the fetus oxygen-rich blood is transported by the inferior vena cava to the heart and via the foramen ovale and ductus arteriosus to the systemic circulation.

Important changes occur in the cardiovascular system at birth. The foramen ovale, ductus arteriosus, ductus venosus, and the umbilical vessels are no longer necessary. The foramen ovale abruptly closes with the first breath of air because the reduced pressure in the right side of the heart causes a flap to cover the opening. This reduced pressure is due to the fact that (1) when the lungs fill with air, the vascular resistance to blood flow in the pulmonary circulation falls far below that of the systemic circulation; and (2) the pressure in the inferior vena cava and right atrium falls as a result of the loss of the placental circulation.

The constriction of the ductus arteriosus occurs gradually over a period of about six weeks after birth as the vascular muscle cells constrict in response to the higher oxygen concentration in the postnatal blood. The remaining structure of the ductus gradually atrophies and becomes nonfunctional as a blood vessel. The fate of the unique fetal cardiovascular structures is summarized in table 19.8.

1. Describe the path of blood from the fetal heart, through the placenta, and back to the fetal heart.
2. Trace the pathway of oxygenated blood through the fetal circulation.
3. Describe the foramen ovale and ductus arteriosus, and explain why blood flows the way it does through these structures in the fetal circulation.

Figure 19.49. A schematic diagram showing the structural relationship of a capillary bed and a lymph capillary.

Lymphatic System

The lymphatic system consists of lymphatic vessels and lymph nodes. It functions to return tissue fluid to the venous system and to help protect the body from diseases.

Objective 26. Describe the pattern of lymph flow from the lymphatic capillaries to the venous system.

Objective 27. Describe the structure and location of the lymph nodes, and list two lymphatic organs.

The lymphatic system has three basic functions: (1) it transports interstitial (tissue) fluid, which was initially formed as a blood filtrate, back to the bloodstream; (2) it serves as the route by which absorbed fat from the intestine is transported to the blood; and (3) its cells—called **lymphocytes** *(lim'fo-sīt)* (produced in lymph nodes and lymphatic organs)—help provide immunological defenses against disease-causing agents (chapter 22).

Lymph and Lymph Capillaries

The smallest vessels of the lymphatic *(lim-fat'ik)* system are the **lymph capillaries** (fig.19.49). Lymph capillaries are microscopic, closed-ended tubes that form vast networks in the intercellular spaces within most tissues. Within the villi of the small intestine, for example, lymph capillaries, called *lacteals (lak'te-al),* transport the products of fat absorption away from the digestive tract.

Because the walls of lymph capillaries are composed of a highly permeable simple squamous epithelium, tissue fluid can readily enter. This fluid is formed as a filtrate of plasma through blood capillaries and is identical in composition to plasma except for a lower protein concentration (proteins are largely held back by the filtration process). Adequate lymphatic drainage is needed to prevent the accumulation of tissue fluid, a condition called *edema (ĕ-de'mah).* Once fluid enters the lymphatic capillaries, it is referred to as **lymph.**

Lymph Ducts

From merging lymph capillaries, the lymph is carried into larger **lymph ducts.** The walls of lymph ducts are similar to those of veins in that they have the same three layers and contain valves to prevent the backflow of lymph. Interconnecting lymph ducts eventually empty into one of the two principal vessels: the **thoracic *(tho-ras'ik)* duct** and the **right lymphatic duct** (fig. 19.50).

The larger thoracic duct drains lymph from the lower extremities, abdomen, left thoracic region, left upper extremity, and left side of the head and neck. The main trunk of this vessel ascends along the spinal column and drains into the left subclavian vein. In the abdominal area is a saclike enlargement of the thoracic duct called the **cisterna chyli** *(sis-ter'nah ki'le).* The shorter right lymphatic duct drains lymph vessels from the right upper extremity, right thoracic region, and right side of the head and neck. The right lymphatic duct empties into the right subclavian vein near the right internal jugular vein.

lymph: L. *lympha,* clear water
lacteal: L. *lacteus,* milk

edema: Gk. *edema,* a swelling
cisterna chyli: L. *cisterna,* box, Gk. *chylos,* juice

Figure 19.50. Lymphatic vessels.
(*a*) a magnified view of the upper right
quadrant showing the lymph drainage
of the right breast.

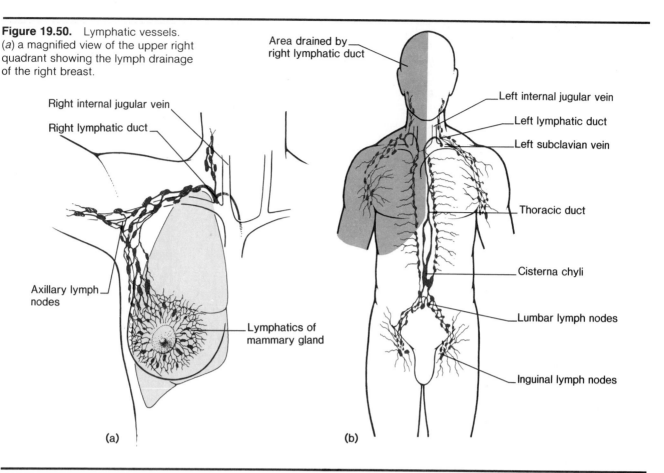

(a)

(b)

Figure 19.51. The structure of a lymph
node. (*a*) a lymph node and associated vessels. (*b*) a photomicrograph of
lymphatic tissue (63X).

(b)

The pressure that pushes lymph through the lymph
ducts comes from the massaging actions produced by
skeletal muscle contractions, intestinal movements, and
other body movements. The abundant valves keep lymph
moving in one direction.

Lymph Nodes

Lymph filters through the reticular tissue of **lymph nodes**
(fig. 19.51), which contain phagocytic cells that help pu-
rify the fluid. Lymph nodes are small, oval bodies enclosed
within fibrous connective tissue *capsules*. Specialized
connective tissue bands called *trabeculae* divide the node.

Afferent lymphatic vessels carry lymph into the node, where it is circulated through sinuses in the *cortical tissue.* Lymph leaves the node through the *efferent lymphatic vessel,* which emerges from the *hilus,* the depression on the concave side. **Germinal centers** within the node are the sites of lymphocyte production.

Lymph nodes usually occur in clusters in specific regions of the body (fig. 19.52). Some of the principal groups of lymph nodes are the **popliteal** and **inguinal nodes** of the lower extremity; the **lumbar nodes** of the pelvic region; the **cubital** and **axillary nodes** of the upper extremity; the **thoracic nodes** of the chest; and the **cervical nodes** of the neck. The submucosa of the small intestine contains numerous scattered lymphocytes and lymphatic nodules and larger aggregations of lymphatic tissue called **Peyer's patches.**

Migrating cancer cells (metastases) are especially dangerous if they enter the lymphatic system, which can disperse them widely. On entering the lymph nodes, the cancer cells can multiply and establish secondary tumors in organs that are distal to the lymphatic drainage of the site of the primary tumor.

Lymphoid Organs

The **spleen** and the **thymus** are lymphoid organs. The spleen is located posterior and lateral to the stomach, from which it is suspended (fig. 19.53). The spleen is not a vital organ in an adult, but it does assist other body organs in producing lymphocytes, filtering the blood, and destroying old erythrocytes. In an infant, it is an important site for the production of erythrocytes.

The thymus is located in the anterior thorax, deep to the manubrium of the sternum (fig. 19.54). It is much larger in a fetus and child than in an adult, because it regresses in size during puberty. The thymus plays a key role in the immune system.

Table 19.9 summarizes the lymphatic organs of the body.

The tonsils, of which there are three pairs, are actually lymphatic organs of the pharyngeal region. The function of the tonsils is to combat infection of the ear, nose, and throat regions. Because of the persistent infections that some children suffer, however, the tonsils may become so overrun with infections that they actually become the source of infections. A tonsillectomy may then have to be performed. This operation is not as common as in the past because of the availability of powerful antibiotics and because the functional value of the tonsils is more greatly appreciated.

Peyer's patches: from Johann K. Peyer, Swiss Anatomist, 1653–1712
spleen: L. *splen,* low spirits (thought to cause melancholy)
thymus: Gk. *thymos,* thyme (compared to the flower of this plant by Galen)

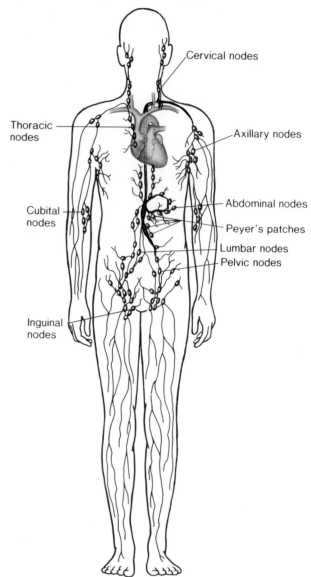

Figure 19.52. Major locations of lymph nodes.

Cervical nodes
Thoracic nodes
Axillary nodes
Cubital nodes
Abdominal nodes
Peyer's patches
Lumbar nodes
Pelvic nodes
Inguinal nodes

1. Describe the relationship between lymph and blood in terms of its origin, composition, and fate.
2. List the two major lymph vessels of the body, and describe their relationships to the vascular system.
3. Explain the functions of lymph nodes and lymphoid organs.

Clinical Considerations

Arrhythmias Detected by the Electrocardiograph

Arrhythmias, or abnormal heart rhythms, can be detected and described by the abnormal ECG patterns they produce. Although the proper clinical interpretation of electrocardiograms requires more knowledge than is presented in this chapter, some knowledge of abnormal rhythms is interesting in itself and is useful in understanding normal physiology.

Figure 19.53.   The structure of the spleen.

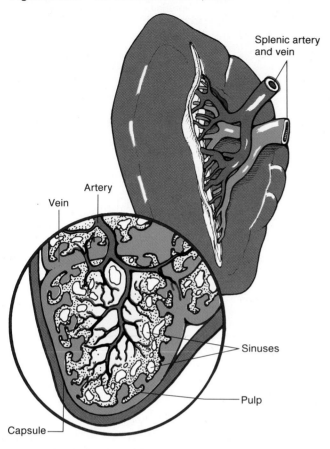

Figure 19.54.   The location of the thymus within the mediastinum between the lungs.

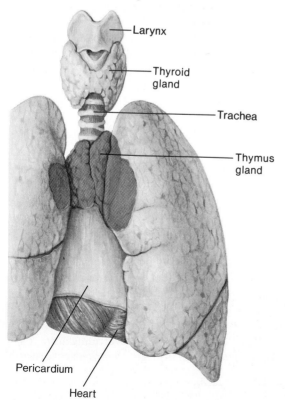

Table **19.9** Lymphatic organs		
Organ	**Location**	**Function**
Lymph nodes	In clusters or chains along the paths of larger lymphatic vessels	Lymphocyte production; house T-lymphocytes and B-lymphocytes that are responsible for immunity; phagocytes filter foreign particles and cellular debris from lymph
Thymus	Within the mediastinum behind the manubrium	Important site of immunity in a child; houses lymphocytes; changes undifferentiated lymphocytes into T-lymphocytes
Spleen	In upper left portion of abdominal cavity beneath the diaphragm and suspended from the stomach	Serves as blood reservoir; phagocytes filter foreign particles, cellular debris, and worn erythrocytes from the blood; houses lymphocytes

Source: Van De Graaff, Kent M., *Human Anatomy.* © 1984 Wm. C. Brown Publishers, Dubuque, Iowa. All Rights Reserved. Reprinted by permission.

Since a heartbeat occurs whenever a normal QRS complex is seen and since the ECG chart paper moves at a known speed, so that its x-axis indicates time, the cardiac rate (beats per minute) can easily be obtained from the ECG recording. A cardiac rate slower than 60 beats per minute indicates **bradycardia** *(brad″e-kar′de-ah);* a rate faster than 100 beats per minute is described as **tachycardia.**

Both bradycardia and tachycardia can occur normally. Endurance-trained athletes, for example, commonly have a slower heart rate than the general population. This *athlete's bradycardia* occurs as a result of higher levels of parasympathetic inhibition of the SA node and is a beneficial adaptation. Activation of the sympathetic system during exercise or emergencies causes a normal tachycardia to occur.

Abnormal tachycardia occurs when a person is at rest. This may result from abnormally fast pacing by the atria, due to drugs or to the development of abnormally fast *ectopic pacemakers*—cells located outside the SA node that assume a pacemaker function. This abnormal atrial tachycardia thus differs from normal "sinus" (SA node) tachycardia. *Ventricular tachycardia* results when abnormally fast ectopic pacemakers in the ventricles cause them to beat rapidly and independently of the atria (fig. 19.55). This is very dangerous because it can quickly degenerate into a lethal condition known as ventricular fibrillation.

Ventricular Fibrillation.   Fibrillation is caused by a continuous recycling of electrical waves through the myocardium. This recycling is normally prevented by the fact

Figure 19.55. In (a) the heartbeat is paced by the normal pacemaker—the S-A node (hence the name sinus rhythm). This can be abnormally slow (bradycardia—46 beats per minute in this example) or fast (tachycardia—136 beats per minute in this example). Compare the pattern of tachycardia in (a) with the tachycardia in (b), which is produced by an ectopic pacemaker in the ventricles. This dangerous condition can quickly lead to ventricular fibrillation, also shown in (b).

Sinus Bradycardia

Sinus Tachycardia

(a)

Ventricular tachycardia

Ventricular fibrillation

(b)

that the myocardium enters a refractory period simultaneously at all regions. If some cells emerge from their refractory periods before others, however, electrical waves can be continuously regenerated and conducted. The recycling of electrical waves along continuously changing pathways produces uncoordinated contraction and an impotent pumping action. These effects can be produced by damage to the myocardium.

Fibrillation can sometimes be stopped by a strong electric shock delivered to the chest, a procedure called **electrical defibrillation.** The electric shock depolarizes all the myocardial cells at the same time, causing them to enter a refractory state. The conduction of random, recirculating impulses thus stops, and—within a short time—the SA node can begin to stimulate contraction in a normal fashion. This does not correct the initial problem that caused the abnormal electrical patterns, but it does keep the person alive long enough to take other corrective measures.

Structural Heart Disorders

Congenital heart problems result from abnormalities in embryonic development and may be attributed to heredity, nutritional problems of the pregnant mother, or viral infections such as rubella. Congenital heart diseases occur in approximately 3 of every 100 births and account for about 50% of early childhood deaths.

Heart murmurs are both congenital and acquired. Nearly 10% of the population have heart murmurs, but most are not clinically significant. A *septal defect* is the most common type of congenital heart problem. An atrial septal defect, or *patent foramen ovale,* is a failure of the foramen ovale to close after birth. A *ventricular septal defect* is caused by abnormal development of the interventricular septum. *Pulmonary stenosis* is a narrowing of

Figure 19.56. The tetralogy of Fallot. The four defects of this anomaly are (1) a ventricular septal defect; (2) an overriding aorta; (3) pulmonary stenosis; and (4) right ventricular hypertrophy.

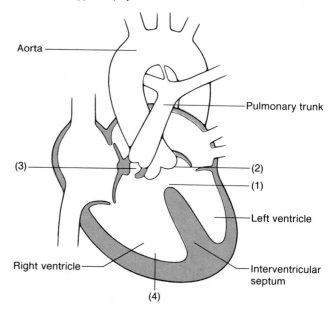

the opening into the pulmonary trunk from the right ventricle. It may lead to pulmonary embolism and is usually recognized by extreme lung congestion.

The **tetralogy of Fallot** is a combination of four defects within a newborn and immediately causes a cyanotic condition, leading to the newborn being called a "blue baby." The four characteristics of this anomaly are (1) a ventricular septal defect, (2) an overriding aorta, (3) pulmonary stenosis, and (4) right ventricular hypertrophy

tetralogy of Fallot: from Etienne-Louis A. Fallot, French
     physician, 1850–1922

Figure 19.57. (a) the lumen (opening) of a human coronary artery is partially occluded by an atheroma and (b) almost completely occluded by a thrombus. (c) close-up view of the cleared left anterior descending coronary artery containing calcified atherosclerotic plaques. The heart is of an 85-year-old female. (d) the structure of an atheroma is diagrammed.

(a)

(b)

Smooth muscle cells   Tunica media   Lumen of vessel

Ulceration

Endothelium

Fat

Cholesterol crystals

Nelson

(c)   (d)

(fig. 19.56). Pulmonary stenosis obstructs blood flow to the lungs and causes hypertrophy of the right ventricle. Open-heart surgery is required to correct this condition, and the overall mortality is about 5%.

Acquired heart diseases include those that result from arterial damage and those that result from bacterial infection. *Bacterial endocarditis* is a disease of the lining of the heart, especially of the valve cusps. It is caused by bacteria that enter the bloodstream. The most common cause of valve defects results from infection by the same organisms that cause rheumatic fever.

Atherosclerosis and Ischemic Heart Disease

Atherosclerosis is the most common form of arteriosclerosis (hardening of the arteries). In atherosclerosis, localized *plaques,* or **atheromas,** protrude into the lumen of the artery and thus reduce blood flow. The plaques additionally serve as sites for *thrombus* (blood clot) formation, which can further occlude the blood supply to an organ (fig. 19.57).

---

atheroma: Gk. *athere*, mush; *oma*, tumor

It is currently believed that atheromas begin as benign tumors of smooth muscle cells that migrate from the tunica media to the intima and proliferate by cell division. In later stages, these cells accumulate cholesterol and other lipids, giving them a "foamy cell" appearance. In fully developed plaques, cholesterol and other lipids accumulate outside the smooth muscle cells, and these accumulations can become calcified. Damage to the endothelium that covers the plaque results in the exposure of subendothelial tissue to the blood. This may contribute to growth of the atheroma and to the formation of blood clots.

Cholesterol and Plasma Lipoproteins. It has long been believed that high blood cholesterol is associated with an increased risk of atherosclerosis. In recent years, however, it has been shown that total blood cholesterol is not as good a predictor of the risk factor for atherosclerosis as is the amount of cholesterol attached to particular plasma protein carriers.

Since cholesterol is a lipid, very little can be dissolved in blood plasma; most travels in the blood attached to protein carriers known as lipoproteins. There are three types of these carriers, based upon their relative densities as indicated by the rate at which they travel to the bottom of a test tube during centrifugation. These are the *very-low-density lipoproteins (VLDL)*, the *low density lipoproteins (LDL)*, and the *high density lipoproteins (HDL)*.

The tendency towards atherosclerosis appears to increase with increasing amounts of cholesterol attached to VLDL and LDL carriers. These carriers are believed to transport cholesterol to the artery walls. The risk of atherosclerosis, however, is *inversely* related to the HDL-cholesterol concentrations; high levels of HDL-cholesterol appear to actually protect the arteries from atherosclerosis.

The concentration of HDL-cholesterol appears to be higher and the risk of atherosclerosis lower in people who exercise regularly. The HDL-cholesterol concentration, for example, is higher in marathon runners than in joggers and is higher in joggers than in inactive men. In addition, before menopause women have higher HDL-cholesterol concentrations and a lower risk of atherosclerosis than men.

Ischemic Heart Disease. A tissue is said to be **ischemic** *(is-kem'ik)* when it receives an inadequate supply of oxygen because of an inadequate blood flow. The most common cause of myocardial ischemia is atherosclerosis of the coronary arteries. The adequacy of blood flow is relative—it depends on the metabolic requirements of the tissue for oxygen. An obstruction in a coronary artery, for example, may allow sufficient blood flow at rest but may produce ischemia when the heart is stressed by exercise or emotional conditions.

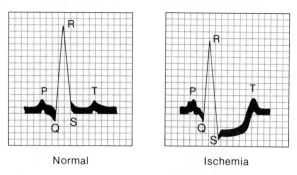

Figure 19.58. Depression of the S-T segment of the electrocardiogram as a result of myocardial ischemia.

Normal   Ischemia

Myocardial ischemia is associated with increased concentrations of blood lactic acid produced by anaerobic respiration of the ischemic tissue. This condition often causes substernal pain, which may also be referred to the left shoulder and arm, as well as other areas. This referred pain is called **angina pectoris.** People with angina frequently take nitroglycerin or related drugs that help relieve the ischemia and pain. These drugs are effective because they stimulate vasodilation, which improves circulation to the heart and decreases the work that the heart must perform to eject blood into the arteries.

Myocardial cells are adapted to respire aerobically and cannot respire anaerobically for more than a few minutes. If ischemia and anaerobic respiration continue for more than a few minutes, *necrosis* (cellular death) may occur in the areas most deprived of oxygen. A sudden, irreversible injury of this kind is called a **myocardial infarction,** or **MI.**

Myocardial ischemia may be detected by characteristic changes in the ECG (fig. 19.58). The diagnosis of myocardial infarction is aided by the blood concentration of various enzymes that are released from the damaged cells (table 19.10).

Other Vascular Disorders

An **aneurism** *(an'u-rizm)* is an expansion or bulging of the heart, aorta, or any other artery. A **coarctation** is a constriction in a segment of a vessel, usually the aorta, and is frequently caused by a remnant of the ductus arteriosus tightening around the vessel. **Varicose veins** are weakened veins that become stretched and swollen. Varicose veins are most common in the legs because the force of gravity tends to weaken the valves and overload the veins. Varicose veins can also occur in the rectum, in which case they are called **hemorrhoids.** *Vein stripping* is the surgical removal of superficial weakened veins. **Phlebitis** *(flĕ-bi'tis)* is inflammation of a vein. It may develop as a result of trauma or an aftermath of surgery. Frequently, however, it appears for no apparent reason. Phlebitis interferes with normal venous circulation.

Table 19.10 Changes in the enzyme activity in plasma following a myocardial infarction

Serum enzyme	Earliest increase (hr)	Maximum concentration (hr)	Return to normal (days)	Amplitude of increase (× normal)
Creatine kinase	3 to 6	24 to 36	3	7
Malate dehydrogenase	4 to 6	24 to 48	5	4
AST	6 to 8	24 to 48	4 to 6	5
Lactate dehydrogenase*	10 to 12	48 to 72	11	3
α-Hydroxybutyrate dehydrogenase	10 to 12	48 to 72	13	3 to 4
Aldolase	6 to 8	24 to 48	4	4
ALT	Usually normal, unless there are other complications			
Isocitrate dehydrogenase	Usually normal			

*Frequently done twice, once before and once after mild heat denaturation.

Source: From Montgomery, Rex, Dryer, Robert L., Conway, Thomas W., and Spector, Arthur A.: Biochemistry: a case-oriented approach, ed. 4, St. Louis, 1983, The C. V. Mosby Co.

Chapter Summary

I. Functions and Major Components of the Circulatory System
The circulatory system consists of components that provide functions essential to the life of complex, multicellular organisms.
A. The blood transports oxygen and nutritive molecules to the tissue cells and carbon dioxide and other wastes away from tissue cells; it also carries hormones and other regulatory molecules to their target organs.
B. Leukocytes and their products serve to protect the body from infection.

II. Development of the Cardiovascular System
The development of the heart and blood vessels reveals evolutionary relationships between different vertebrates, and a knowledge of these developmental events helps to explain some congenital defects in the cardiovascular system.
A. The heart develops from a heart tube that becomes subdivided into regions called the truncus arteriosus, bulbus cordis, ventricle, atrium, and sinus venosus.
   1. The ventricle is divided into right and left chambers by the growth of an interventricular septum.
   2. A septum also forms in the atrium, but it contains an opening called the foramen ovale, which allows communication between the right and left atria.
B. The major arteries are formed from six pairs of aortic arches.
   1. Different members of these aortic arches form or contribute to the common carotid, internal carotid, right subclavian, and pulmonary arteries.

2. Only the left fourth arch contributes to the aortic arch of the adult.
3. A part of the sixth left arch persists as a shunt between the pulmonary trunk and the aorta called the ductus arteriosus.

III. Structure of the Heart
The structure of the heart and the action of its valves allow it to function as a double pump.
A. The heart is enclosed within a pericardial sac; the wall of the heart consists of epicardium, myocardium (the heart muscle), and endocardium.
   1. The right atrium receives blood from the superior and inferior vena cavae, and the right ventricle pumps blood into the pulmonary trunk to the pulmonary arteries.
   2. The left atrium receives blood from the pulmonary veins and pumps blood into the aorta.
   3. Atrioventricular valves (tricuspid and mitral) are located between the atria and ventricles; semilunar valves are located at the base of the pulmonary trunk and aorta.
B. The pulmonary circulation includes the vessels that carry blood from the right ventricle, through the lungs, and from there to the left atrium.
   1. The systemic circulation includes all other arteries and veins in the body, which carry blood from the left ventricle, through the body, and return blood to the right atrium.
   2. The coronary arteries arise from the aorta immediately beyond the semilunar valve and provide arterial blood to the myocardium via several branches; venous blood ultimately enters a coronary sinus that drains into the right atrium.

C. Contraction of the atria and ventricles is produced by action potentials that originate in the pacemaker region—the sinoatrial (SA) node.
   1. These electrical waves spread over the atria and then enter the atrioventricular (AV) node.
   2. From here, the impulses are conducted by the atrioventricular bundle (bundle of His) and Purkinje fibers into the ventricles.

IV. The Cardiac Cycle, the Heart Sounds, and the Electrocardiogram
The cardiac cycle of systole and diastole can be monitored by the electrical activity that produces the cycle and through the sounds created by the closing of the heart valves.
A. During contraction of the ventricles the intraventricular pressure rises and causes the AV valves to close; during relaxation the semilunar valves close as a result of the fact that the pressure is greater in the arteries than in the ventricles.
B. Closing of the AV valves causes the first sound (lub), and closing of the semilunar valves causes the second sound (dub); heart murmurs are commonly caused by abnormal valves or by septal defects.
C. The action potential is longer in heart cells than in other tissues and corresponds in duration to contraction of the heart muscle.
D. The pattern of electrical conduction is observed by an electrocardiogram (ECG).

V. Blood Vessels
Veins and arteries carry blood to and from the heart. Capillaries allow the exchange of molecules between the blood and tissue fluid.
   A. Arteries and veins have a tunica externa, tunica media, and tunica intima.
      1. Arteries have thicker muscle layers for their calibers than do veins because they must withstand a higher blood pressure.
      2. Veins have venous valves that direct blood to the heart when the veins are compressed by the skeletal muscle pumps.
   B. Capillaries are composed only of endothelial cells and may be continuous or discontinuous (fenestrated capillaries and sinusoids).

VI. Principal Arteries of the Body
The branches of the aorta provide oxygenated blood to all cells in the body.
   A. Three arteries arise from the aortic arch: the brachiocephalic, the left common carotid, and the left subclavian; the brachiocephalic divides into the right common carotid and the right subclavian.
   B. The head and neck receive an arterial supply from branches of the internal and external carotid arteries and the vertebral arteries.
      1. The brain receives blood from the paired internal carotid arteries and the paired vertebral arteries, which form the circle of Willis around the pituitary gland.
      2. The external carotid artery gives off many branches that supply the head and neck.

   C. The upper extremity is served by the subclavian artery and its derivatives.
      1. The subclavian artery becomes first the axillary and then the brachial artery as it enters the arm.
      2. The brachial artery bifurcates to form the radial and ulnar arteries, which supply blood to the forearm and a portion of the hand and digits.
   D. The abdominal aorta produces the following branches: the celiac, superior mesenteric, renal, suprarenal, testicular (or ovarian), and inferior mesenteric arteries; the abdominal aorta divides into the right and left iliac arteries.
   E. The common iliac arteries divide into the internal and external iliac arteries, which supply arteries to the pelvis and lower extremities.

VII. Principal Veins of the Body
Systemic veins of ever increasing diameters return oxygen-depleted blood to the right atrium via the superior and inferior vena cavae.
   A. Blood from the head and neck is drained by the external and internal jugular veins; blood from the brain is drained by the internal jugular vein.
   B. The upper extremity is drained by superficial and deep veins.
   C. In the thorax, the superior vena cava is formed by the union of the two brachiocephalic veins and also receives blood from the azygos vein.
   D. The lower extremity is drained by both superficial and deep veins; at the level of the fifth lumbar vertebra, the right and left iliac veins unite to form the inferior vena cava.
   E. Blood from capillaries in the digestive tract is drained via the hepatic portal vein to the liver.

      1. This venous blood then passes through hepatic sinusoids (capillaries) and is drained from the liver in the hepatic vein.
      2. The pattern of circulation in which two capillary beds are in a series is called a portal system.

VIII. Fetal Circulation
There are structural adaptations in the fetal cardiovascular system to the fact that oxygen and nutrients are obtained from the placenta rather than from the fetal lungs and digestive tract.
   A. Fully oxygenated blood is carried only in the umbilical vein, which drains the placenta; this blood is carried via the ductus venosus to the inferior vena cava of the fetus.
   B. Partially oxygenated blood is shunted from the right to the left atrium via the foramen ovale and from the pulmonary trunk to the aorta via the ductus arteriosus.

IX. Lymphatic System
The lymphatic system functions to return tissue fluid to the venous system and to help protect the body from disease.
   A. Lymph capillaries drain tissue fluid, which is formed from plasma; when this fluid enters lymph capillaries, it is called lymph.
   B. Lymph is returned to the venous system via two large lymph ducts—the thoracic duct and the right lymphatic duct.
   C. Lymph filters through lymph nodes, which contain phagocytic cells and germinal centers that produce lymphocytes.
   D. The spleen and thymus are lymphoid organs because they produce lymphocytes.

Review Activities

Objective Questions

1. All arteries in the body contain oxygen-rich blood with the exception of the
   (a) aorta
   (b) pulmonary artery
   (c) renal artery
   (d) coronary arteries
2. Most blood from the coronary circulation directly enters the
   (a) inferior vena cava
   (b) superior vena cava
   (c) right atrium
   (d) left atrium
3. The "lub," or first heart sound, is produced by the closing of the
   (a) aortic semilunar valve
   (b) pulmonary semilunar valve
   (c) tricuspid valve
   (d) bicuspid valve
   (e) (a) and (b)
   (f) (c) and (d)

4. The first heart sound is produced at the
   (a) beginning of systole
   (b) end of systole
   (c) beginning of diastole
   (d) end of diastole
5. The QRS wave of an ECG is produced by
   (a) depolarization of the atria
   (b) repolarization of the atria
   (c) depolarization of the ventricles
   (d) repolarization of the ventricles
6. The second heart sound immediately follows the occurrence of the
   (a) P wave
   (b) QRS wave
   (c) T wave
   (d) U wave

7. The cells that normally have the fastest rate of spontaneous depolarization are located in the
   (a) SA node
   (b) AV node
   (c) atrioventricular bundle
   (d) Purkinje fibers
8. Which of the following arteries do *not* arise from the aortic arch? The
   (a) brachiocephalic
   (b) coronary
   (c) left common carotid
   (d) left subclavian
9. Which of the following arteries do *not* supply blood to the brain? The
   (a) external carotid
   (b) internal carotid
   (c) vertebral
   (d) basilar

10. The maxillary and superficial temporal arteries are derived from the
    (a) external carotid
    (b) internal carotid
    (c) vertebral
    (d) facial
11. Which of the following organs is/are served by a portal vein?
    (a) the liver
    (b) the pituitary
    (c) both (a) and (b)
    (d) neither (a) nor (b)
12. Which of the following fetal blood vessels carries fully oxygenated blood? The
    (a) ductus arteriosus
    (b) ductus venosus
    (c) umbilical vein
    (d) aorta

Essay Questions

1. Describe the functions of the foramen ovale and ductus arteriosus in a fetus, and explain why these can be dangerous if they remain patent after birth.
2. Trace the flow of blood from the left ventricle to the upper teeth.
3. Trace the flow of blood from the intestine, to the heart, and back to the intestine.
4. Define the term portal system, and explain the functional significance of this system in the liver and pituitary gland.
5. Can a defective valve be detected by an ECG? Can a partially damaged AV node be detected by auscultation with a stethoscope? Explain.
6. Compare the durations of the heart's contraction with the myocardial action potential and refractory period. Explain the significance of these relationships.

Selected Readings

Adolph, A. F. March 1967. The heart's pacemaker. *Scientific American*.

Benditt, E. P. February 1977. The origin of atherosclerosis. *Scientific American*.

Dubin, D. 1981. *Rapid interpretation of EKGs*. 3rd ed. Tampa, FL: Cover Publishing Co.

Eder, H. A. May 1983. Lipoproteins and coronary artery disease. *Hosp. Prac.* 18:215.

Ferrans, V. J. July 1983. Morphology of the heart in hypertrophy. *Hosp. Prac.* 18:67.

Fuchs, R., and S. S. Scheidt. May 1983. Prevention of coronary atherosclerosis: part 1. *Cardiovascular Reviews and Reports* 4:671.

Folkow, B., and E. Neill. 1971. *Circulation*. New York: Oxford Univ. Press.

Lehman, J. 1972. Auscultation of heart sounds. *Amer. J. Nurs.* 72:1242.

Levy, M. N., and P. J. Martin. 1981. Neural regulation of the heartbeat. *Ann. Rev. Physiol.* 43:443.

Mayerson, H. S. June 1963. The lymphatic system. *Scientific American*.

Rushmer, R. F. 1976. *Structure and function of the cardiovascular system*. 2d ed. Philadelphia: W. B. Saunders Co.

Spear, J. F., and E. N. Moore. 1982. Mechanisms of cardiac arrhythmias. *Ann. Rev. Physiol.* 44:485.

Whittemore, R. December 1983. Congenital heart disease: Its impact on pregnancy. *Hosp. Prac.* 18:65.

Wood, J. E. January 1968. The venous system. *Scientific American*.

20

Cardiac Output,

Concepts

The cardiac output is a measure of the
    pumping ability of the heart and is
    affected by mechanisms that regulate
    the cardiac rate and stroke volume.

The volume of fluid in the vascular and
    tissue compartments is regulated by
    the kidneys. This is adjusted
    according to the needs of the body by
    antidiuretic hormone and
    aldosterone.

The rate of blood flow is dependent in
    part on vascular resistance.
    Resistance is regulated by
    vasoconstriction and vasodilation,
    produced by extrinsic and intrinsic
    mechanisms.

Blood flow to the heart and skeletal
    muscles is regulated by both intrinsic
    and extrinsic mechanisms that
    increase the rate of flow when the
    metabolic requirements of these
    tissues increase during exercise.

Control mechanisms help to maintain
    relatively constant rates of cerebral
    blood flow. Blood flow to the skin is
    varied according to the needs of the
    body.

The blood pressure rise from diastolic to
    systolic levels provides the driving
    force for blood flow. Blood pressure is
    regulated by a variety of control
    mechanisms and is usually measured
    indirectly by the auscultatory
    method.

Blood Flow, and Blood Pressure

Cardiac Output

The cardiac output is a measure of the pumping ability of the heart and is affected by mechanisms that regulate the cardiac rate and stroke volume.

Objective 1. Describe how the cardiac output is affected by cardiac rate and stroke volume.

Objective 2. Describe the effects of autonomic stimulation of the heart.

Objective 3. Explain the Frank-Starling law of the heart.

Objective 4. Explain how the venous return of blood to the heart is regulated.

The cardiac output is equal to the milliliters (ml) of blood pumped per minute by each ventricle. The average resting **cardiac rate** in an adult is 70 beats per minute; the average **stroke volume** (ml of blood pumped per beat by each ventricle) is 70–80 ml per beat. The product of these two variables gives an average cardiac rate of 5,000 ml per minute:

cardiac output (ml/min) = stroke volume (ml/beat) × cardiac rate (beats/min)

The **total blood volume** is also equal to about 5–6 liters. This means that each ventricle pumps the equivalent of the total blood volume each minute under resting conditions. Put another way, it takes about a minute for a drop of blood to complete the systemic and pulmonary circuits. An increase in cardiac output, as occurs during exercise, must thus be accompanied by an increased rate of blood flow through the circulation. This is accomplished, in part, by factors that regulate the cardiac rate and stroke volume.

Regulation of Cardiac Rate

In the complete absence of neural influences, the heart will continue to beat according to the rhythm set by the SA node. This automatic rhythm is produced by the spontaneous decay of the resting membrane potential to a threshold depolarization, at which point voltage regulated membrane gates are opened, action potentials are produced, and contraction occurs.

Normally, however, sympathetic and vagus (parasympathetic) nerve fibers to the heart are continuously active. Norepinephrine released primarily by sympathetic nerve endings and epinephrine secreted by the adrenal medulla stimulate an increase in the spontaneous rate of firing of the SA node. Acetylcholine released from parasympathetic endings hyperpolarizes the SA node and thus decreases the rate of its spontaneous firing (fig. 20.1). The actual pace set by the SA node at any time depends on the net effect of these antagonistic influences.

Autonomic innervation of the SA node represents the major means by which cardiac rate is regulated. Autonomic nerves do, however, affect cardiac rate by other mechanisms to a lesser degree. Sympathetic endings in the musculature of the atria and ventricles increase the strength of contraction and decrease slightly the time spent in systole when the cardiac rate is high (table 20.1).

Figure 20.1. The effects of sympathetic and parasympathetic activity on the pacemaker function of the SA node. Sympathetic stimulation causes action potentials to be generated more rapidly, and parasympathetic inhibition causes action potentials to be generated more slowly, than would occur in the absence of autonomic influences (*shaded areas*).

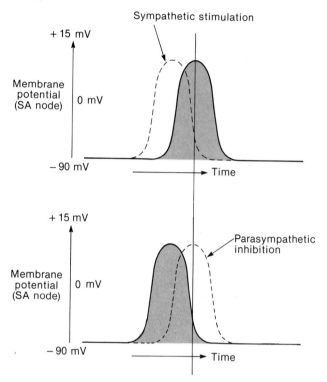

Table **20.1** Effects of autonomic nerve activity on the heart		
Region affected	**Sympathetic nerve effects**	**Parasympathetic nerve effects**
SA node	Increased rate of diastolic depolarization; increased cardiac rate	Decreased rate of diastolic depolarization; decreased cardiac rate
AV node	Increased conduction rate	Decreased conduction rate
Atrial muscle	Increased strength of contraction	Decreased strength of contraction
Ventricular muscle	Increased strength of contraction	No significant effect

Source: Fox, Stuart Ira, *Human Physiology*. © 1984 Wm. C. Brown Publishers, Dubuque, Iowa. All Rights Reserved. Reprinted by permission.

In exercise the cardiac rate increases as a result of decreased vagus nerve inhibition and increased sympathetic nerve stimulation. The resting bradycardia of endurance-trained athletes is due largely to increased vagus nerve activity. These changes are coordinated by *cardiac control centers* in the *medulla oblongata* of the brain stem. There may be separate cardioaccelerator and cardioinhibitory centers (this is currently controversial) in the medulla that control autonomic innervation to the heart. These centers, in turn, are affected by higher brain centers and by sensory feedback from pressure receptors, or *baroreceptors,* in a reflex fashion.

Regulation of Stroke Volume

The stroke volume is regulated by three variables: (1) the **end-diastolic volume (EDV),** which is the volume of blood in the ventricles at the end of diastole; (2) the average, or mean, aortic blood pressure, which is measured as the **mean arterial pressure;** and (3) the **contractility** *(kon''trak-til'ĭ-te),* or strength, of ventricular contraction.

The end-diastolic volume is the amount of blood in the ventricles just prior to contraction. This is a work load imposed on the ventricles prior to contraction and is thus sometimes called a *preload.* The stroke volume is directly proportional to the preload; an increase in EDV results in

an increase in stroke volume. The stroke volume is also directly proportional to contractility; when the ventricles contract more strongly, they pump more blood.

In order to eject blood, the pressure generated by ventricular contraction must be greater than the mean arterial blood pressure. The arterial pressure thus represents an impedance to the ejection of blood from the ventricles, or an *afterload* imposed on the ventricles after contraction has begun. The stroke volume is inversely proportional to the afterload; the greater the mean arterial pressure (other factors being equal), the lower the stroke volume.

The proportion of the EDV that is ejected against a given afterload depends upon the strength of ventricular contraction. Normally, contraction strength is sufficient to eject 70–80 ml of blood out of a total end-diastolic volume of 110–130 ml. The *ejection fraction* is thus about two-thirds. Amazingly, this fraction remains relatively constant as the EDV increases. This implies that the strength of ventricular contraction must increase as the EDV increases.

Intrinsic Control of Contraction Strength. Two physiologists, Frank and Starling, demonstrated that normally the strength of ventricular contraction varies directly with the end-diastolic volume. Even in experiments where the heart is removed from the body (and is thus not subject to neural or hormonal regulation) and where the heart is filled with blood flowing from a reservoir, an increase in EDV within the physiological range results in increased contraction strength and, therefore, in increased stroke volume. This relationship between EDV, contraction strength, and stroke volume is thus a built-in, or **intrinsic,** property of heart muscle.

Frank-Starling Law of the Heart. The intrinsic control of contraction strength and stroke volume is due to variations in the degree to which the myocardium is stretched by the EDV. As the EDV rises within the physiological range, the myocardium is increasingly stretched and, as

Figure 20.2.   The Frank-Starling mechanism (law of the heart). When the heart muscle is subject to increased amounts of stretch, it contracts more strongly. As a result of the increased contraction strength (shown as tension), the time required to reach maximum contraction is the same regardless of the degree of stretch.

Resting sarcomere lengths

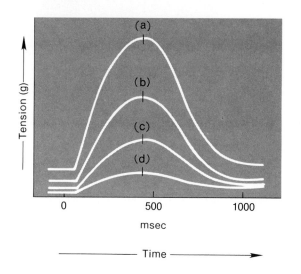

Figure 20.3.   The regulation of cardiac output. Factors that stimulate cardiac output are shown as solid arrows; factors that inhibit cardiac output are shown as dotted arrows.

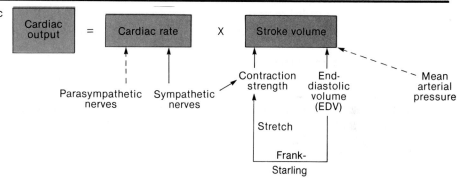

a result, contracts more forcibly. This mechanism that allows the heart to adjust intrinsically to changing demands is known as the **Frank-Starling law of the heart.**

Stretch can also increase the contraction strength of skeletal muscles. The resting length of skeletal muscles, however, is close to ideal, so that significant stretching decreases contraction strength. This is not true of the heart. Prior to filling with blood during diastole, the sarcomere lengths of myocardial cells are only about 1.5 μm. At this length the actin filaments from each side overlap in the middle of the sarcomeres, and the cells can only contract weakly (fig. 20.2).

As the ventricles fill with blood the myocardium stretches, so that the actin filaments overlap with myosin only at the edges of the A bands (fig. 20.2). This allows more force to be developed during contraction. Since this more advantageous overlapping of actin and myosin is produced by stretching of the ventricles and since the de-

gree of stretching is controlled by the degree of filling (the end-diastolic volume), the strength of contraction is intrinsically adjusted so that more blood will be ejected as the EDV rises.

Extrinsic Control of Contractility.   The *contractility* is the contraction strength at any given fiber length. At any given degree of stretch, the strength of ventricular contraction depends on the activity of the sympathoadrenal system. Norepinephrine from sympathetic endings and epinephrine from the adrenal medulla produce an increase in contraction strength. This is often called a **positive inotropic** *(in″o-trop′ik)* **effect.**

The cardiac output is thus affected in two ways by the activity of the sympathoadrenal system: the positive inotropic effect on contractility, and the **positive chronotropic** *(kron″o-trop′ik)* **effect** that increases cardiac rate (fig. 20.3). Stimulation through the parasympathetic nerve to

---

Starling: Ernest Henry Starling, English physiologist, 1866–1927

---

chronotropic: Gk. *chronos,* time; *trope,* turn, change

the heart has a **negative chronotropic effect** (decreases cardiac rate), but does not directly affect the contraction strength of the ventricles.

Venous Return

The end-diastolic volume—and thus the stroke volume and cardiac output—is controlled by factors that affect the *venous return* of blood to the heart. The rate at which the atria are filled with venous blood depends on the total blood volume and the venous pressure, which serves as the driving force for the return of blood to the heart.

Veins have thinner, less muscular walls than do arteries and thus have a higher *compliance.* This means that a given amount of pressure will cause more distension (expansion) in veins than in arteries, so that the veins can hold more blood. Approximately two-thirds of the total blood volume is located in the veins (fig. 20.4). Veins are therefore called *capacitance vessels,* by analogy with capacitors in electronics, which can accumulate electrical charges. Muscular arteries and arterioles expand less under pressure (are less compliant) and thus are called *resistance vessels.*

Although veins contain almost 70% of the total blood volume, the mean venous pressure is only 2 mm Hg, compared to a mean arterial pressure of 90–100 mm Hg. The lower venous pressure is due in part to a pressure drop between arteries and capillaries and in part to the high venous compliance.

The venous pressure is highest in the venules (10 mm Hg) and lowest at the junction of the vena cavae with the right atrium (0 mm Hg). In addition to this pressure difference, the venous return to the heart is aided by (1) sympathetic nerve activity; (2) the skeletal muscle pump, which constricts veins during muscle contraction; and (3) the pressure difference between the thoracic and abdominal cavities.

Sympathetic stimulation of the veins causes some contraction of the smooth muscle walls. This does not cause much constriction of the lumina of veins, in contrast to the effects on arteries; it does, however, make the veins less compliant. This reduced venous compliance helps to raise venous pressure and aid the return of venous blood to the heart.

Contraction of skeletal muscles functions as a "pump" by virtue of its squeezing action on veins. Contraction of the diaphragm (a muscular sheet separating the thoracic and abdominal cavities) during inhalation also improves venous return. This results from the fact that as the diaphragm contracts, it lowers to increase the thoracic volume and decrease the abdominal volume. This creates a partial vacuum in the thoracic cavity and a higher pressure in the abdominal cavity. The pressure difference thus produced favors blood flow from abdominal to thoracic veins (fig. 20.5).

Figure 20.4.   The distribution of blood within the circulatory system at rest.

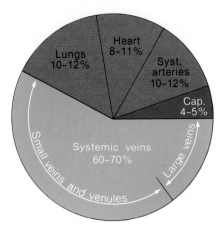

1. Describe how the stroke volume is intrinsically regulated by the end-diastolic volume, and explain how this regulation is accomplished.
2. Describe the effects of autonomic nerve stimulation on the cardiac rate and stroke volume.
3. Define the terms preload and afterload, and explain how these affect the cardiac output.
4. List the factors that affect venous return, and draw a flow chart to show how an increased venous return can result in an increased cardiac output.

Blood Volume

Fluid in the extracellular environment of the body is distributed between the plasma and the tissue fluid compartments. Blood volume is affected by the function of the kidneys. Through their actions on the kidneys, ADH and aldosterone help regulate the blood volume.

Objective 5. Describe how tissue fluid is formed from and returns to the blood capillaries.

Objective 6. Describe the regulation of antidiuretic hormone (ADH) secretion, and the effects of ADH on the blood volume.

Objective 7. Describe the regulation of aldosterone secretion and the effects of aldosterone on blood volume and pressure.

Blood volume represents one part, or "compartment," of the total body water. Approximately two-thirds of the total body water is located within cells—in the **intracellular compartment.** The remaining one-third is in the **extracellular environment.** This extracellular fluid is normally distributed so that about 80% is in the tissues—as tissue or **interstitial** *(in''ter-stish'al)* **fluid**—and 20% is in the blood plasma (fig. 20.6).

The distribution of water between these compartments is determined by a balance between opposing forces.

Figure 20.5. Variables that affect venous return and thus end-diastolic volume. Direct relationships are indicated by solid arrows; inverse relationships are shown with dashed arrows.

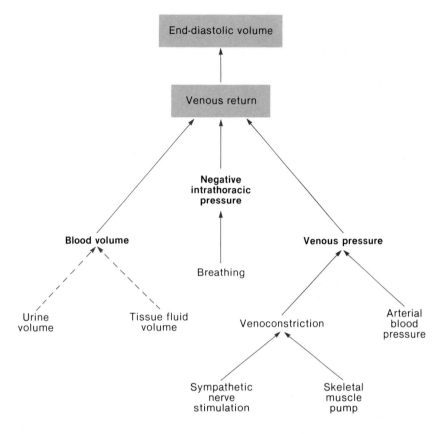

Figure 20.6. The daily intake and excretion of body water and its distribution between different intracellular and extracellular compartments.

Figure 20.7. Tissue or interstitial fluid is formed by filtration as a result of blood pressures at the arteriolar ends of capillaries and returned to venular ends of capillaries by the colloid osmotic pressure of plasma proteins. Solid arrows indicate the net direction of flow.

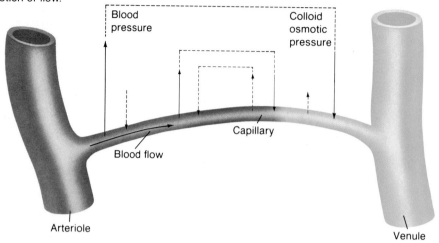

Figure 20.8. The lymphatic system returns excess tissue fluid to the venous system.

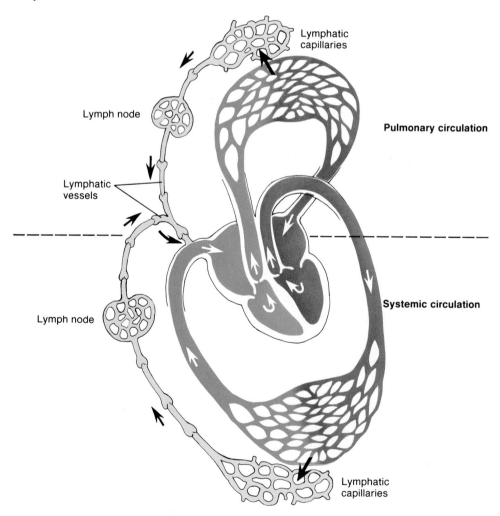

Blood pressure, for example, promotes formation of tissue fluid at the expense of plasma, while osmotic forces draw water from tissues into the vascular system. The total volume of intracellular and extracellular fluid is normally maintained constant by a balance between water loss and water gain. Mechanisms that affect drinking, urine volume, and the distribution of water between plasma and tissue fluid thus help regulate blood volume, and by this means help regulate cardiac output and blood flow.

Exchange of Fluid between Capillaries and Tissues

The distribution of extracellular fluid between the plasma and interstitial compartments is in a state of dynamic equilibrium. Tissue fluid is not normally a "stagnant pond" but is rather a continuously circulating medium, formed from and returning to the vascular system. In this way, the tissue cells receive a continuously fresh supply of glucose and other plasma solutes that are filtered through tiny endothelial channels in the capillary walls.

Filtration results from the blood pressure within the capillaries. This hydrostatic pressure, which is exerted against the inner capillary wall, is equal to about 30 mm Hg at the arteriolar end of systemic capillaries and to 10–15 mm Hg at the venular end of the capillaries. This produces an average capillary pressure (near the middle) of about 17 mm Hg.

The **net filtration pressure** is equal to the hydrostatic pressure of the blood in the capillaries minus the hydrostatic pressure of tissue fluid outside the capillaries. If, as an extreme example, these two values were equal, there would be no filtration. The normal value of tissue hydrostatic pressure is currently controversial. Most researchers believe that this value is actually a *negative pressure,* meaning that it is lower than the pressure of the atmosphere. If the tissue fluid hydrostatic pressure is −6.5 mm Hg, as believed by some authorities, the average net filtration pressure would be 17 − (−6.5) = 23.5 mm Hg.

Glucose, comparably sized organic molecules, inorganic salts, and ions are filtered along with water through the capillary channels. The concentrations of these substances in tissue fluid is thus the same as in plasma. The protein concentration of tissue fluid (2 g/100 ml), however, is less than the protein concentration of plasma (6–8 g/100 ml). This difference is due to the fact that filtration of proteins is restricted by the capillary pores. The osmotic pressure exerted by plasma proteins—called the **colloid osmotic pressure,** or **oncotic pressure,** of the plasma—is, therefore, greater than the colloid osmotic pressure of tissue fluid.

The colloid osmotic pressure of plasma averages 28 mm Hg, and that of tissue fluid averages 5 mm Hg. The difference between these two values (23 mm Hg) represents the net force favoring osmosis of water into the capillaries. If these values are correct, this pressure is lower than the filtration pressure at the arteriolar ends of capillaries and greater than the filtration pressure at the venular ends of capillaries. Fluid and solutes would thus be

Table 20.2 The causes of edema	
Variables that cause edema	**Effects**
Increased blood pressure	Increases filtration pressure so that more tissue fluid is formed at the arteriolar ends of capillaries
Increased tissue protein concentration	Decreases osmosis of water into the venular ends of capillaries. Usually a localized tissue edema due to leakage of plasma proteins through capillaries during inflammation and allergic reactions
Decreased plasma protein concentration	Decreases osmosis of water into the venular ends of capillaries. May be caused by liver disease (which can be associated with insufficient plasma protein production), kidney disease (due to leakage of plasma protein into urine), or protein malnutrition
Obstruction of lymphatic vessels	Infections by filaria roundworms (nematodes) transmitted by a certain species of mosquito block lymphatic drainage, causing edema and tremendous swelling of the affected areas (a condition called elephantiasis)

Source: Fox, Stuart Ira, *Human Physiology.* © 1984 Wm. C. Brown Publishers, Dubuque, Iowa. All Rights Reserved. Reprinted by permission.

filtered out of capillaries at the arteriolar ends, and fluid would be returned by osmosis to the venular ends of capillaries (fig. 20.7).

This "classic" view of capillary dynamics has recently been challenged by some investigators who believe that when capillaries are open, the net filtration force exceeds the force for the osmotic return of water throughout the length of the capillary. They believe that the opposite is true in closed capillaries (capillaries can be opened or closed by the action of precapillary sphincter muscles). Net filtration, in summary, would occur in open capillaries, whereas net absorption of water would occur in closed capillaries.

By either proposed mechanism, plasma and tissue fluid are continuously interchanged through the action of filtration and colloid osmotic pressure forces. The return of fluid to the vascular system by osmosis, however, does not exactly equal the amount filtered. According to some estimates, 20 L of filtrate are formed per day, but only about 18 L are returned to the capillaries by osmosis. The difference—about 2 L per day—is ultimately returned to the vascular system by way of **lymphatic vessels** (fig. 20.8).

Causes of Edema. The excessive accumulation of tissue fluid is called *edema.* This condition is normally prevented by a proper balance between capillary filtration and the osmotic uptake of water and by normal lymphatic drainage. Edema may thus result from (1) high blood pressure; (2) the leakage of plasma proteins into tissue fluid; (3) decreased plasma protein concentration; and (4) the obstruction of lymphatic drainage (table 20.2).

In the tropical disease **filariasis** *(fil''ah-ri'ah-sis)* mosquitoes transmit a nematode worm parasite to humans. The larvae of these worms invade lymphatic vessels and block lymphatic drainage. The edema that results can be so severe that the tissues swell to produce an elephantlike appearance—a condition aptly termed **elephantiasis** *(el''ĕ-fan-ti'ah-sis)* (fig. 20.9).

Figure 20.9.   Parasitic larvae that block lymphatic drainage produce tissue edema, resulting in elephantiasis.

Regulation of Blood Volume by the Kidneys

The formation of urine begins in the same manner as the formation of tissue fluid—by filtration of plasma through capillary channels. The kidneys produce about 180 L per day of blood filtrate, but since there is only 5.5 L of blood in the body, it is clear that most of this filtrate must be returned to the vascular system and recycled. Only about 1–2 L per day of urine is excreted; 98%–99% of the amount filtered is **reabsorbed** back into the vascular system.

The volume of urine excreted can be varied by changes in the reabsorption of filtrate. If 99% of the filtrate is reabsorbed, for example, 1% must be excreted. Decreasing the reabsorption by only 1% thus doubles the urine volume. The percent of the kidney filtrate reabsorbed—and thus the urine and blood volume—is regulated by hormones.

Regulation by Antidiuretic Hormone (ADH).   One of the major hormones involved in the regulation of blood volume is **antidiuretic** *(an''tĭ-di''u-ret'ik)* **hormone (ADH),** also known as *vasopressin (vas''o-pres'in).* This hormone is produced by neurons in the hypothalamus, transported by axons into the neurohypophysis (posterior pituitary), and released from this storage gland in response to hypothalamic stimulation. The secretion of ADH from the posterior pituitary occurs when neurons called **osmoreceptors** in the hypothalamus detect an increase in plasma osmolality (osmotic pressure).

An increase in plasma osmolality occurs when the plasma becomes more concentrated. This can be produced either by *dehydration* or by *excessive salt intake.* Stimulation of osmoreceptors produces sensations of thirst, leading to increased water intake and increased ADH secretion from the posterior pituitary. Through mechanisms that will be discussed in conjunction with kidney physiology in chapter 26, ADH stimulates water reabsorption from the filtrate. A smaller volume of urine is thus excreted as a result of the action of ADH (fig. 20.10).

A person who is dehydrated or who eats excessive amounts of salt thus drinks more and urinates less. This raises the blood volume and, in the process, dilutes the plasma to lower its previously elevated osmolality. The rise in blood volume that results from these mechanisms is extremely important in a dehydrated person with low blood volume and pressure. In a person who eats excessive amounts of salt, however, the increased blood volume could produce an abnormally high blood pressure (hypertension).

Drinking excessive amounts of water without excessive amounts of salt does not result in hypertension. The water does enter the blood from the intestine and momentarily raises the blood volume; at the same time, however, it dilutes the blood, which decreases the plasma osmolality and thus inhibits ADH secretion. With less ADH there is less reabsorption of filtrate in the kidneys—a larger volume of urine is excreted. Water is therefore a **diuretic** *(di''u-ret'ik)*—a substance that promotes urine formation—because it inhibits the secretion of antidiuretic hormone.

Regulation of Aldosterone.   Since Na^+ and Cl^- are easily filtered in the kidneys, a mechanism must exist to promote the reabsorption and retention of salt when the dietary salt intake is too low. Throughout most of human history, salt was in short supply and was thus highly valued. Moorish merchants in the sixth century traded an ounce of salt for an ounce of gold; salt cakes were used as money in Abyssinia; part of a Roman soldier's pay was given in salt—it is from this practice that the word salary *(sal =* salt) is derived. The phrase "worth his salt" is derived from the fact that Greeks and Romans sometimes purchased slaves with salt.

Aldosterone *(al-dos'ter-ōn),* a steroid hormone secreted by the adrenal cortex, stimulates the retention of salt by the kidneys. Aldosterone is thus a "salt-retaining hormone." Retention of salt indirectly promotes retention of water (in part, by the action of ADH, as previously discussed). The action of aldosterone thus causes an increase in blood volume, but—unlike the effects of ADH—does not cause a change in plasma osmolality. The secretion of aldosterone is stimulated during salt deprivation, when the

diuretic: Gk. *dia,* completely; *ouresis,* urination

Figure 20.10. The role of antidiuretic hormone (ADH) in the regulation of blood osmolality and volume. Both dehydration and salt ingestion stimulate drinking and increased water retention. In dehydration, this provides a negative feedback mechanism (shown by dashed arrows) that helps to counter the loss of blood volume. As a result of salt ingestion, this same mechanism produces a rise in blood volume. Numbers indicate the sequence of cause-and-effect steps.

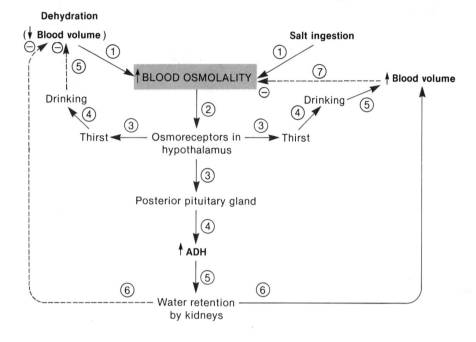

Figure 20.11. The renin-angiotensin-aldosterone system. Negative feedback effects are shown with dashed arrows. Numbers indicate the sequence of cause-and-effect steps.

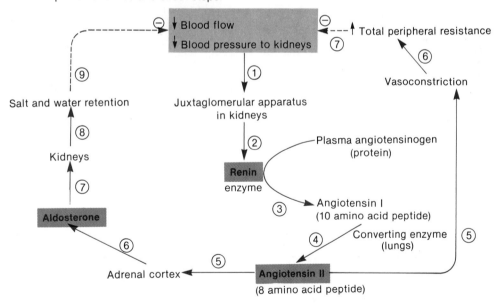

blood volume and pressure are reduced. The adrenal cortex is not directly stimulated by these conditions, however; an intermediate mechanism is required.

Renin-Angiotensin System. When the blood flow and pressure are reduced in the renal artery (as they would be in the low blood volume state of salt deprivation), a group of cells in the kidneys called the **juxtaglomerular** *(juks"tah-glo-mer'u-lar)* **apparatus** secretes the enzyme renin *(re'nin)* into the blood. This enzyme cleaves a ten-amino-acid polypeptide called **angiotensin** *(an"je-o-ten'sin)* **I** from a plasma protein called **angiotensinogen.** As angiotensin I passes through different organs, a *converting enzyme* removes two amino acids. This leaves an eight-amino-acid polypeptide called **angiotensin II** (fig. 20.11). Conditions of salt deprivation, low blood volume, and low blood pressure, in summary, cause increased production of angiotensin II in the blood.

Table 20.3 Estimated distribution of the cardiac output at rest

Organs	Blood flow	
	Milliliters per minute	Percent total
Gastrointestinal tract and liver	1,400	24
Kidneys	1,100	19
Brain	750	13
Heart	250	4
Skeletal muscles	1,200	21
Skin	500	9
Other organs	600	10
Total organs	5,800	100

Source: From Wade, O. L., and J. M. Bishop, *Cardiac Output and Regional Blood Flow.* Copyright © 1962 Blackwell Scientific Publications, Ltd., England. Used with permission.

Angiotensin II has numerous effects that cause a rise in blood pressure. This rise in pressure is due to the constriction of muscular arteries and arterioles (vasoconstriction) and to increased blood volume. Vasoconstriction is produced directly by the effects of angiotensin II on blood vessels. Angiotensin II promotes a rise in blood volume by stimulating the adrenal cortex to secreted aldosterone.

This mechanism can also work in the opposite direction: high salt intake, leading to high blood volume and pressure, normally inhibits renin secretion. With less angiotensin II formation and less aldosterone secretion less salt is retained by the kidneys and more is excreted in the urine. Unfortunately, many people with chronically high blood pressure may have normal or even elevated levels of renin secretion. In these cases, the intake of salt must be lowered to match the impaired ability to excrete salt in the urine.

1. *Describe the composition of tissue fluid. Using a flow diagram, explain how tissue fluid is formed and how it is returned to the vascular system.*
2. *Define the term edema, and describe four different mechanisms that can produce this condition.*
3. *Describe the effects of dehydration on blood and urine volumes, and explain the cause-and-effect mechanism involved.*
4. *Describe how salt deprivation causes increased salt and water retention by the kidneys. Explain how a high-salt diet can be compensated by increased urinary excretion of salt.*

Vascular Resistance and Blood Flow

The rate of blood flow through the vessels of an organ depends largely on the arterial blood pressure on the vascular resistance to blood flow. Resistance to blood flow can be changed by vasoconstriction or vasodilation, induced by extrinsic and intrinsic mechanisms.

Figure 20.12. The flow of blood in the systemic circulation is ultimately dependent on the pressure difference (ΔP) between the origin of the flow (mean pressure of about 100 mm Hg in the aorta) and the end of the circuit—zero mm Hg in the vena cava where it joins the right atrium (*RA*). (LA = left atrium, RV = right ventricle, LV = left ventricle.)

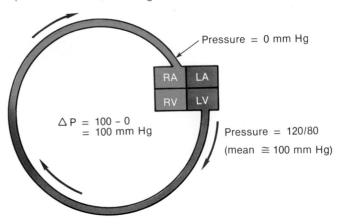

Pressure = 0 mm Hg

$\Delta P = 100 - 0$
$= 100$ mm Hg

Pressure = 120/80
(mean ≅ 100 mm Hg)

Objective 8. Describe how blood flow is affected by blood pressure and vascular resistance.
Objective 9. Describe how vascular resistance is affected by the autonomic system and by other extrinsic agents.
Objective 10. Describe the intrinsic regulation of vascular resistance.

The amount of blood that the heart pumps per minute is equal to the rate of venous return and thus is equal to the rate of blood flow through the entire circulation. The cardiac output of 5–6 L per minute is distributed unequally to the different organs. At rest, blood flow is about 2,500 ml per minute through the liver, kidneys, and gastrointestinal tract, 1,200 ml per minute through the skeletal muscles, 750 ml per minute through the brain, and 250 ml per minute through the coronary arteries of the heart. The balance of the cardiac output (500–1,100 ml per minute) is distributed to the other organs (table 20.3).

Physical Laws Describing Blood Flow

The flow of blood through the vascular system, like the flow of any fluid through a tube, depends in part on the difference in pressure at the two ends of the tube. If the pressure at both ends of the tube is the same there will be no flow. If the pressure at one end is greater than at the other, blood will flow from the region of higher to the region of lower pressure. The rate of blood flow is proportional to the pressure difference ($P_1 - P_2$) between the two ends of the tube. The term **pressure difference** is abbreviated ΔP, in which the Greek letter Δ (*delta*) means "change in" (fig. 20.12).

If the systemic circulation is pictured as a single tube leading from and back to the heart (fig. 20.12), blood flow through this system would occur as a result of the pressure difference between the beginning of the tube (the

Figure 20.13. Relationships between blood flow, vessel radius, and resistance. (*a*) the resistance and blood flow is equally divided between two branches of a vessel. (*b*) a doubling of the radius of one branch and halving of the radius of the other produces a sixteenfold increase in blood flow in the former and a sixteenfold decrease of blood flow in the latter. The total vascular resistance in (*a*) is equal to that of (*b*).

(a)

Radius = 1 mm
Resistance = R
Blood flow = F

← Arterial blood

Radius = 1 mm
Resistance = R
Blood flow = F

Radius = 2 mm
Resistance = 1/16 R
Blood flow = 16F

← Arterial blood

Radius = 1/2 mm
Resistance = 16 R
Blood flow = 1/16 F

(b)

aorta) and the end of the tube (the junction of the venae cavae with the right atrium). The average, or mean, arterial pressure is about 100 mm Hg; the pressure at the right atrium is 0 mm Hg. The "pressure head," or driving force (ΔP), is therefore about $100 - 0 = 100$ mm Hg.

Blood flow is directly proportional to the pressure difference between the two ends of the tube (ΔP) but is *inversely proportional* to the frictional resistance to blood flow through the vessels. Inverse proportionality is expressed by showing one of the factors in the denominator of a fraction, since a fraction decreases when the denominator increases:

$$\text{Blood flow} \propto \frac{\Delta P}{\text{Resistance}}$$

The **resistance** to blood flow through a vessel is directly proportional to the length of the vessel and to the viscosity of the blood (the "thickness," or ability of molecules to "slip over" each other). Vascular resistance is inversely proportional to the fourth power of the radius of the vessel:

$$\text{Resistance} \propto \frac{L\eta}{r^4}$$

where: L = length of vessel

η = viscosity of blood

r = radius of vessel

If one vessel has half the radius of another and if all other factors are the same, the smaller vessel would have sixteen times (2^4) the resistance of the larger vessel. Blood flow through the larger vessel, as a result, would be sixteen times greater than in the smaller vessel (fig. 20.13).

When physical constants are added to this relationship, the rate of blood flow can be calculated according to **Poiseuille's** *(pwah-zuh'yez)* **law:**

$$\text{Blood flow} = \frac{\Delta P\ r^4\ (\pi)}{\eta\ L\ (8)}$$

Since vessel length and blood viscosity do not vary significantly in normal physiology, the major factors affecting blood flow through an organ are the mean arterial pressure and the vascular resistance. At a given mean arterial pressure, blood can be diverted from one organ to another by variations in the degree of vasoconstriction and vasodilation in the arterial supplies of the organs. Since arterioles are the smallest arteries and can become narrower by vasoconstriction, they provide the greatest resistance to blood flow. Blood flow to an organ can be increased by dilation of its arterioles and can be decreased by constriction of its arterioles.

Total Peripheral Resistance. The sum of all the vascular resistances within the systemic circulation is called the *total peripheral resistance.* The arterial supplies to the organs, which contribute to this total, are generally in parallel rather than in series with each other. That is, arterial blood passes through only one set of resistance vessels (arterioles) before returning to the heart (fig. 20.14). Since one organ is not "downstream" from another in terms of its arterial supply, changes in resistance within one organ directly affect blood flow in only that organ.

Poiseuille's law: from Jean Poiseuille, French physiologist, 1799–1869

Figure 20.14. A diagram of the systemic and pulmonary circulations. Notice that with few exceptions (such as the blood flow in the renal circulation), the flow of arterial blood is in parallel rather than in series (arterial blood does not usually flow from one organ to another).

Extrinsic agent	Effect	Comments
Sympathetic nerves		
Alpha-adrenergic	Vasoconstriction	It occurs throughout the body. This is the dominant effect of sympathetic nerve stimulation on the vascular system.
Beta-adrenergic	Vasodilation	There is some activity in arterioles in skeletal muscles and in coronary vessels, but effects are masked by dominant alpha-receptor-mediated constriction.
Cholinergic	Vasodilation	Effects are localized to arterioles in skeletal muscles and are only activated during defense (fight or flight) reaction.
Parasympathetic nerves	Vasodilation	Effects are primarily restricted to the gastrointestinal tract, external genitals, and salivary glands and have little effect on total peripheral resistance.
Angiotensin II	Vasoconstriction	A powerful vasoconstrictor produced as a result of secretion of renin from the kidneys, it may function to help maintain adequate filtration pressure in kidneys when systemic blood flow and pressure is reduced.
ADH (vasopressin)	Vasoconstriction	Although the effects of this hormone on vascular resistance and blood pressure in anesthetized animals are well documented, the importance of these effects in conscious humans is controversial.
Histamine	Vasodilation	It promotes localized vasodilation during inflammation and allergic reactions.
Bradykinins	Vasodilation	Bradykinins are polypeptides secreted by sweat glands that promote local vasodilation.
Prostaglandins	Vasodilation or vasoconstriction	Prostaglandins are cyclic fatty acids that can be produced by most tissues, including blood vessel walls. Prostaglandin I_2 is a vasodilator, whereas thromboxane A_2 is a vasoconstrictor. The physiological significance of these effects is presently controversial.

Table 20.4 Extrinsic control of vascular resistance and blood flow

Vasodilation in a large organ might, however, significantly decrease the total peripheral resistance and, by this means, might decrease the mean arterial pressure. If this is not prevented by compensations, the driving force for blood flow through all organs might be reduced. This is normally prevented by an increase in the cardiac output and by vasoconstriction in other areas. During exercise of large muscles, for example, the arterioles in the exercising muscles are dilated, the cardiac output is increased, and there is constriction of arterioles in the viscera and skin.

Extrinsic Regulation of Blood Flow

The term **extrinsic** *(eks-trin'sik)* **regulation** refers to control by the autonomic nervous system and endocrine system. Angiotensin II, for example, directly stimulates vascular smooth muscle to produce generalized vasoconstriction. Antidiuretic hormone (ADH) also has a vasoconstrictor effect at high concentrations; this is why it is also called **vasopressin.**

Regulation by Sympathetic Nerves. Stimulation of the sympathoadrenal system produces an increase in the cardiac output (as previously discussed) and an increase in total peripheral resistance. The latter effect is due to alpha-adrenergic stimulation of vascular smooth muscle by norepinephrine and, to a lesser degree, by epinephrine. This produces vasoconstriction of arterioles in the viscera and skin.

Even when a person is calm, the sympathoadrenal system is active to a certain degree and helps set the "tone" of vascular smooth muscles. In this case, **adrenergic** *(ad''ren-er'jik)* **sympathetic fibers** (those that release norepinephrine) cause a basal level of vasoconstriction in skeletal muscles and throughout the body. During the fight or flight reaction the activity of adrenergic fibers increases, so that vasoconstriction is produced in the digestive tract, kidneys, and skin.

Arterioles in skeletal muscles receive **cholinergic** *(ko''lin-er'jik)* **sympathetic fibers,** which release acetylcholine as a neurotransmitter. During the fight or flight reaction, the activity of these cholinergic fibers increase. This causes vasodilation. Thus, while the total peripheral resistance and blood pressure are increased by adrenergic fibers that stimulate vasoconstriction in the viscera and skin, vasodilation in skeletal muscles is produced by cholinergic fibers. Blood flow is thus diverted to skeletal muscles during emergency conditions, which may produce an "extra edge" for the skeletal muscles' response to the emergency.

Parasympathetic Control of Blood Flow. Parasympathetic endings in arterioles are always cholinergic and always promote vasodilation. Parasympathetic innervation of blood vessels, however, is limited to the digestive tract, external genitals, and salivary glands. Because of this limited distribution, the parasympathetic system is less important than the sympathetic system in the control of total peripheral resistance. The extrinsic control of blood flow is summarized in table 20.4.

Intrinsic Control of Blood Flow

Extrinsic control mechanisms affect resistance and flow in many regions of the body. In contrast to these more generalized effects, intrinsic ("built-in") mechanisms within individual organs provide a more localized regulation of vascular resistance and blood flow. Intrinsic mechanisms are: (1) **myogenic** *(mi''o-jen'ik)* and (2) **metabolic** *(met''ah-bol'ik).* Some organs, the brain and kidneys in particular, utilize these intrinsic mechanisms to maintain relatively constant flow rates despite wide fluctuations in blood pressure. This ability is termed **autoregulation.**

Myogenic Control Mechanisms. If the arterial blood pressure and flow through an organ is inadequate—if the organ is inadequately **perfused** *(per-fūzd')* with blood—the metabolism of the organ cannot be maintained beyond a limited period of time. Excessively high blood pressure can also be dangerous, particularly in the brain, because this may cause fine blood vessels to burst (causing cerebrovascular accident—CVA, or stroke).

Changes in systemic arterial pressure are compensated for in the brain and some other organs by the appropriate responses of vascular smooth muscle. A decrease in arterial pressure causes cerebral vessels to dilate, so that adequate rates of blood flow can be maintained despite the decreased pressure. High blood pressure, in contrast, causes cerebral vessels to constrict so that finer vessels downstream are protected from the elevated pressure. These responses are myogenic; they are direct responses by the vascular smooth muscle to changes in pressure.

Metabolic Control Mechanisms. Local vasodilation within an organ can occur as a result of the chemical environment created by the organ's metabolism. The localized chemical conditions that promote vasodilation include (1) **decreased oxygen concentrations** that result from increased metabolic rate; (2) **increased carbon dioxide concentrations;** (3) **decreased tissue pH** (due to CO_2, lactic acid, and other metabolic products); and (4) the **release of adenosine or K^+** from the tissue cells.

The vasodilation that occurs in response to tissue metabolism can be demonstrated by constricting the blood supply to an area for a short time and then removing the constriction. The constriction allows metabolic products to accumulate by preventing venous drainage of the area. When the constriction is removed and blood flow resumes, the metabolic products that have accumulated cause vasodilation. The tissue thus appears red. This response is called **reactive hyperemia** *(hi''per-e'me-ah).* A similar increase in blood flow occurs in skeletal muscles and other organs as a result of increased metabolism. This is called **active hyperemia.** Intrinsic control mechanisms are summarized in table 20.5.

1. Describe the relationship between blood flow rate, arterial blood pressure, and vascular resistance.
2. Describe the relationship between vascular resistance and the radius of a vessel. Explain how blood flow can be diverted from one organ to another.
3. Explain how vascular resistance and blood flow is regulated by (1) sympathetic adrenergic fibers, (2) sympathetic cholinergic fibers, and (3) parasympathetic fibers.
4. Define autoregulation, *and explain how this is accomplished through myogenic and metabolic mechanisms.*

Table **20.5** The intrinsic control of vascular resistance and blood flow

Category	Agent (↑ = increase; ↓ = decrease)	Comments
Myogenic	↑ Blood pressure	Stretching of the arterial wall as the blood pressure rises directly stimulates increased smooth muscle tone (vasoconstriction).
Metabolic	↓ Oxygen ↑ Carbon dioxide ↓ pH ↑ Adenosine ↑ K	Local changes in gas and metabolite concentrations act directly on vascular smooth muscle walls to produce vasodilation in the systemic circulation. The importance of different agents varies in different organs.
Myogenic	—	It helps to maintain relatively constant rates of blood flow and pressure within an organ despite changes in systemic arterial pressure (autoregulation).
Metabolic	—	It aids in autoregulation of blood flow and also helps to shunt increased amounts of blood to organs with higher metabolic rates (active hyperemia).

Source: Fox, Stuart Ira, *Human Physiology*. © 1984 Wm. C. Brown Publishers, Dubuque, Iowa. All Rights Reserved. Reprinted by permission.

Blood Flow to the Heart and Skeletal Muscles

Blood flow to the heart and skeletal muscles is regulated by both extrinsic and intrinsic mechanisms. These mechanisms provide increased rates of blood flow when the metabolic requirements of these tissues are increased during exercise.

Objective 11. Describe how coronary blood flow is regulated, and explain how it is affected by the heart's contraction and relaxation.

Objective 12. Describe how skeletal muscle blood flow is regulated, and explain how this flow is increased during exercise.

Objective 13. Explain the mechanisms that contribute to increased cardiac output during exercise.

Survival requires that the heart and brain receive adequate rates of blood flow at all times. The ability of skeletal muscles to respond quickly in emergencies and to maintain continued high levels of activity may also be critically important for survival. During such times, high rates of blood flow to the skeletal muscles must be maintained without compromising blood flow to the heart and brain. This is accomplished by mechanisms that increase the cardiac output and that divert a higher proportion of the cardiac output to the heart, skeletal muscles, and brain and away from the viscera and skin.

Table **20.6** Some vascular and metabolic comparisons between heart and skeletal muscle

	Cardiac muscle	Skeletal muscle
Number of capillaries	4×	1×
Mean blood flow	10–20×	1×
Myolemma (sarcolemma)	Thin, low resistance	Thicker higher resistance
Sarcomere length	1×	1.7×
Glycogen concentration	Maintained	Depleted by fasting, diabetes
Glycolytic enzyme systems	1×	2×, strongly developed
Creatine phosphate concentration	1×	6×
Anaerobic energy production	Beating: 2 min Arrested: 30–90 min	Up to 40% total energy
Lactic acid production	Terminal mechanism	Frequent when incurring O₂ debt
Ability to incure O₂ debt	1×	4×
Increased O₂ requirement met primarily by	Increased flow	Increased extraction
Oxygen consumption	3×	1×
Oxygen extraction at rest	Near maximal	Significant reserve
Increased O₂ consumption with increased work	2×	30×
Myoglobin	Present	Present in red skeletal muscle
Mitochondria	Abundant, giant	Fewer, smaller
Krebs cycle enzymes	2–3×	1×
Cytochrome-C	6×	1×
Myosin ATPase activity	1×	3×

Source: Brachfeld, N. The Physiology of Muscular Exercise, *Primary Cardiology*, June, 1979, p. 112. Reproduced with permission.

Aerobic Requirements of the Heart

The coronary arteries supply an enormous number of capillaries, which are packed within the myocardium at a density of about 2,500–4,000 per cubic millimeter of tissue. Fast-twitch skeletal muscles, in contrast, have a capillary density of 300–400 per cubic millimeter of tissue. Each myocardial cell, as a consequence, is within 10 μm of a capillary (compared to an average distance in other organs of 60–80 μm). The exchange of gases by diffusion between myocardial cells and capillary blood thus occurs very quickly.

Contraction of the myocardium squeezes the coronary arteries. Unlike blood flow in all other organs, flow in the coronary vessels thus decreases in systole and increases during diastole. The myocardium, however, contains large amounts of **myoglobin** *(mi"o-glo'bin)*, a pigment related to hemoglobin, which stores oxygen during diastole and releases its oxygen during systole. In this way,

Figure 20.15. An arteriogram of the left coronary artery in a patient (*a*) when the ECG was normal and (*b*) when the ECG showed evidence of myocardial ischemia. Notice that a coronary artery spasm—see arrow in (*b*)—appears to accompany the ischemia.

(a)

(b)

Figure 20.16. A diagram of coronary artery bypass surgery.

the myocardial cells can receive a continuous supply of oxygen even though coronary blood flow is temporarily reduced during systole.

In addition to containing large amounts of myoglobin, heart muscle contains numerous mitochondria and aerobic respiratory enzymes. This indicates that—even more than slow-twitch skeletal muscles—the heart is extremely specialized for aerobic respiration (table 20.6). The normal heart always respires aerobically, even during heavy exercise when the metabolic demand for oxygen can rise to five times resting levels. This increased oxygen requirement is met by a corresponding increase in coronary blood flow, from about 80 ml at rest to about 400 ml per minute per 100 g tissue during heavy exercise.

Regulation of Coronary Blood Flow

Sympathetic nerve fibers, through stimulation of alpha-adrenergic receptors in the coronary arterioles, produce a relatively high vascular resistance in the coronary circu-

lation at rest. Vasodilation of coronary vessels may be produced in part by sympathetic nerve activation of beta-adrenergic receptors. Most of the vasodilation that occurs during exercise, however, is due to intrinsic metabolic control mechanisms. As the metabolism of the myocardium increases, local accumulation of carbon dioxide and adenosine *(ah-den'o-sēn)* acts directly on the vascular smooth muscles to cause relaxation and vasodilation.

Under abnormal conditions, the blood flow to the myocardium may be inadequate, resulting in myocardial ischemia. This can result from the blockage by atheromas and/or blood clots or from muscular spasm of a coronary artery (fig. 20.15). Occlusion of a coronary artery can be visualized by a technique called **selective coronary arteriography.** In this procedure, a catheter (plastic tube) is inserted into a brachial or femoral artery all the way to the opening of the coronary arteries in the aorta, and radiographic contrast material is injected. The picture thus obtained is called an **angiogram** *(an'je-o-gram'').*

If the occlusion is sufficiently great a coronary bypass operation may be performed. In this procedure a length of blood vessel, usually taken from the saphenous vein in the leg, is sutured to the aorta and to the coronary artery at a location beyond the site of the occlusion (fig. 20.16).

Condition	Blood flow (ml/min)	Mechanism
Rest	1,000	High adrenergic sympathetic stimulation of vascular alpha-receptors causing vasoconstriction
Beginning exercise	Increased	Dilation of arterioles in skeletal muscles due to cholinergic sympathetic nerve activity
Heavy exercise	20,000	Fall in alpha-adrenergic activity Increased sympathetic cholinergic activity Increased metabolic rate of exercising muscles producing intrinsic vasodilation

Table 20.7 Changes in skeletal muscle blood flow under conditions of rest and exercise

Source: Fox, Stuart Ira, *Human Physiology.* © 1984 Wm. C. Brown Publishers, Dubuque, Iowa. All Rights Reserved. Reprinted by permission.

Figure 20.17. The distribution of blood flow (cardiac output) during rest and heavy work. At rest, the cardiac output is 5 L per minute (*bottom of figure*); during heavy work the cardiac output increases to 25 L per minute (*top of figure*). During rest, for example, the brain receives 15% of 5 L per minute (= 750 ml/min.), whereas during exercise it receives 3% to 4% of 25 L per minute (0.03 × 25 = 750 ml/min.). Flow to the skeletal muscles increases more than twentyfold, because the total cardiac output increases (from 5 L/min. to 25 L/min.) and because the percent of the total received by the muscles increases from 15% to 80%.

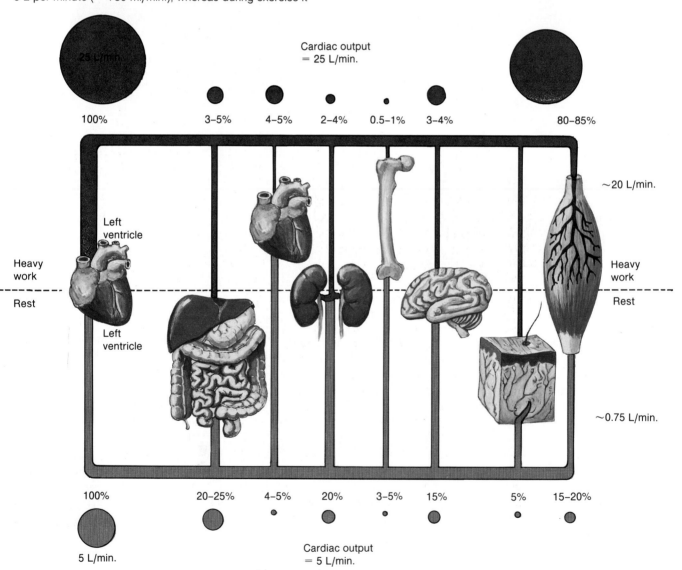

Regulation of Blood Flow through Skeletal Muscles

The arterioles in skeletal muscles, like those of the coronary circulation, have a high vascular resistance at rest as a result of alpha-adrenergic sympathetic stimulation. This produces a relatively low rate of blood flow per tissue weight (4–6 ml per minute per 100 g tissue), but because muscles have such a large mass, this still accounts for 20–25% of the total blood flow in the body at rest.

In addition to adrenergic fibers (those that release norepinephrine), there are also sympathetic cholinergic fibers in skeletal muscles. The release of acetylcholine (ACh) from these fibers stimulates vasodilation just prior to exercise (table 20.7). The increased blood flow that results from this vasodilation may provide an extra edge that improves skeletal muscle performance once exercise begins.

As exercise progresses, the vasodilation and increased skeletal muscle blood flow that occurs is almost entirely due to intrinsic metabolic control. The high metabolic rate of skeletal muscles during exercise causes local changes such as increased carbon dioxide concentrations, decreased pH (due to carbonic acid and lactic acid), decreased oxygen, increased extracellular K^+, and the secretion of adenosine. Like the intrinsic control of the coronary circulation, these changes cause vasodilation of arterioles in skeletal muscles. This decreases the vascular resistance and increases the rate of blood flow. This effect is combined with the recruitment of capillaries by the opening of precapillary sphincter muscles (only 5%–10% of the skeletal muscle capillaries are open at rest). As a result of these changes, skeletal muscles can receive as much as 85% of the total blood flow in the body during maximal exercise.

While the vascular resistance in skeletal muscles decreases during exercise, the resistance to flow through visceral organs and skin increases. This increased resistance occurs because of vasoconstriction stimulated by adrenergic sympathetic fibers and results in decreased rates of blood flow through these organs. During exercise, therefore, the blood flow to skeletal muscles increases because of increased total blood flow (cardiac output) and because blood is diverted away from the viscera and skin. Blood flow to the heart also increases during exercise, whereas blood flow to the brain does not appear to change significantly (fig. 20.17).

Although the skeletal muscles consume large amounts of oxygen during exercise, the oxygen concentration of arterial blood is usually not reduced. This is due to the fact that the rate and depth of breathing is also increased during exercise and to the fact that the pulmonary blood flow increases to match the increased systemic blood flow.

Table 20.8 Relationship between age and maximum cardiac rate

Age	Maximum cardiac rate
20–29	190 beats/min
30–39	160 beats/min
40–49	150 beats/min
50–59	140 beats/min
60 and above	130 beats/min

Source: Fox, Stuart Ira, *Human Physiology.* © 1984 Wm. C. Brown Publishers, Dubuque, Iowa. All Rights Reserved. Reprinted by permission.

The oxygen concentration of venous blood, however, can be reduced to one-half or one-third resting levels by exercising muscles, which indicates that the muscles are extracting a much higher proportion of the oxygen delivered to them in the arterial blood. The ability of muscles to extract oxygen from arterial blood is improved by endurance training.

Changes in Cardiac Output during Exercise

During exercise the cardiac output can increase fivefold, from about 5 L per minute to about 25 L per minute. This is partially due to an increase in cardiac rate. The cardiac rate, however, can only increase to a maximum value (table 20.8) that does not permit a five-fold increase above the resting rate. The stroke volume must, therefore, increase as well, even though at a faster cardiac rate the filling time between ventricular contractions is reduced.

Despite the reduced filling time at high cardiac rates, the end-diastolic volume is not significantly decreased. This is because the rate of venous return is increased by the improved action of the skeletal muscle pumps and by increased respiratory movements during exercise (fig. 20.18). Since the end-diastolic volume is not changed during exercise, the increase in stroke volume that occurs must be due to an increase in the proportion of blood ejected per stroke.

The proportion of the end-diastolic volume ejected per stroke can increase from 67% at rest to as much as 90% during heavy exercise. This increased **ejection fraction** is produced by increased contractility as a result of sympathoadrenal stimulation. There may also be a decrease in total peripheral resistance as a result of vasodilation in the skeletal muscles, which decreases the afterload and thus further augments the increase in stroke volume. The cardiovascular changes that occur during exercise are summarized in table 20.9.

Figure 20.18. Cardiovascular adaptations to exercise.

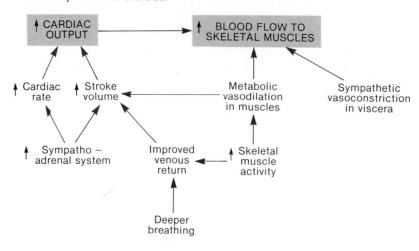

Table **20.9** Cardiovascular changes that occur during moderate exercise		
Variable	**Change**	**Mechanisms**
Cardiac output	Increased	Cardiac rate and stroke volume increased
Cardiac rate	Increased	Increased sympathetic nerve activity; decreased activity of the vagus nerve
Stroke volume	Increased	Increased myocardial contractility due to stimulation by sympathoadrenal system; decreased total peripheral resistance
Total peripheral resistance	Decreased	Vasodilation of arterioles in skeletal muscles (and in skin when thermoregulatory adjustments are needed)
Arterial blood pressure	Increased	Increased systolic and pulse pressure due primarily to increased cardiac output; diastolic pressure rises less due to decreased total peripheral resistance
End-diastolic volume	Unchanged	Decreased filling time at high cardiac rates is compensated by increased venous pressure, increased activity of the skeletal muscle pump, and decreased intrathoracic pressure aiding the venous return
Blood flow to heart and muscles	Increased	Increased muscle metabolism produces intrinsic vasodilation; aided by increased cardiac output and increased vascular resistance in visceral organs
Blood flow to visceral organs	Decreased	Vasoconstriction in digestive tract, liver, and kidneys due to sympathetic nerve stimulation
Blood flow to skin	Increased	Metabolic heat produced by exercising muscles produces reflex (involving hypothalamus) that reduces sympathetic constriction of arteriovenous shunts and arterioles
Blood flow to brain	Unchanged	Autoregulation of cerebral vessels, which maintains constant cerebral blood flow despite increased arterial blood pressure

Source: Fox, Stuart Ira, *Human Physiology.* © 1984 Wm. C. Brown Publishers, Dubuque, Iowa. All Rights Reserved. Reprinted by permission.

1. Describe blood flow and oxygen delivery to the myocardium during systole and diastole.
2. Describe how blood flow to the heart is affected by exercise, and explain how blood flow to the heart is regulated at rest and during exercise.
3. Describe the mechanisms that produce vasodilation of the arterioles in skeletal muscles during exercise, and describe three other mechanisms that increase skeletal muscle blood flow during exercise.
4. Explain how the stroke volume can increase during exercise despite the fact that the filling times are reduced at high cardiac rates.

Blood Flow to the Brain and Skin

Physiological control mechanisms help to maintain relatively constant rates of cerebral blood flow. Blood flow to the skin is varied according to the needs of the body.

Objective 14. Describe the mechanisms that control cerebral blood flow. Explain the significance of the autoregulation of cerebral blood flow.

Objective 15. Describe the cutaneous circulation, and explain how it is affected by ambient temperature, exercise, and emotional states.

Figure 20.19. Computerized picture of blood flow distribution in the brain after injecting the carotid artery with a radioactive isotope. In (*a*), on the left, the subject followed a moving object with his eyes. High activity is seen over the occipital lobe of the brain. In (*a*), on the right, the subject listened to spoken words. Notice that the high activity is seen over the temporal lobe (the auditory cortex). In (*b*), on the left, the subject moved his fingers on the side of the body opposite to the cerebral hemisphere being studied. In (*b*), on the right, the subject counted to twenty. High activity is shown over the mouth area of the motor cortex, the supplementary motor area, and the auditory cortex.

(a)

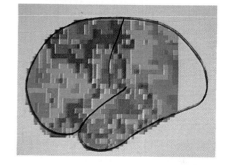

(b)

The examination of cerebral *(ser'ĕ-bral)* and cutaneous *(ku-ta'ne-us)* blood flow is a study in contrasts. Cerebral blood flow is regulated primarily by intrinsic mechanisms; cutaneous blood flow is regulated by extrinsic mechanisms. Cerebral blood flow is relatively constant; cutaneous blood flow exhibits more variation than blood flow in any other organ. The brain is the organ that can least tolerate and the skin is the organ that can most tolerate low rates of blood flow.

Cerebral Circulation

When the brain is deprived of oxygen for a few seconds, the person will lose consciousness; irreversible brain injury may occur after a few minutes. For these reasons, the cerebral blood flow is remarkably constant at about 750 ml per minute. This amounts to about 15% of the total cardiac output at rest.

Unlike the coronary and skeletal muscle blood flow, cerebral blood flow is not normally influenced by sympathetic nerve activity. Only when the mean arterial pressure rises to about 200 mm Hg do sympathetic nerves cause a significant degree of vasoconstriction in the cerebral circulation. This vasoconstriction helps to protect small, thin-walled arterioles from bursting under the pressure and thus helps to prevent cerebrovascular accident (stroke).

In the normal range of arterial pressures, cerebral blood flow is regulated almost exclusively by intrinsic mechanisms. These mechanisms help insure a constant rate of blood flow despite changes in systemic arterial pressure—a process called *autoregulation*. The autoregulation of cerebral blood flow is achieved by both myogenic and metabolic mechanisms.

Myogenic regulation occurs when there is variation in systemic arterial pressure. The cerebral arteries automatically dilate when the blood pressure falls and constrict when the pressure rises. This helps to maintain a constant flow rate during the normal pressure variations that occur during rest, exercise, and emotional states.

The cerebral vessels are also sensitive to the carbon dioxide concentration of arterial blood. When the carbon dioxide concentration rises, as a result of inadequate ventilation (hypoventilation), the cerebral arterioles dilate. This is believed to be due to decreases in the pH of cerebrospinal fluid rather than to a direct effect of CO_2 on the cerebral vessels. Conversely, when the arterial CO_2 falls below normal during hyperventilation, the cerebral vessels constrict. The resulting decrease in cerebral blood flow is responsible for the dizziness that occurs during hyperventilation.

The cerebral arterioles are exquisitely sensitive to local changes in metabolic activity, so that those brain regions with the highest metabolic activity get the most blood. Indeed, areas of the brain that control specific processes have been mapped by the changing patterns of blood flow that result when these areas are activated. Vision and hearing, for example, increase blood flow to the appropriate sensory areas of the cerebral cortex, whereas motor activities such as movements of the eyes, arms, and organs of speech result in different patterns of blood flow (fig. 20.19).

Cutaneous Blood Flow

The skin is the outer covering of the body and as such serves as the first line of defense against invasion by disease-causing organisms. The skin, as the interface between the internal and external environments, also serves

Figure 20.20. Circulation in the skin showing arteriovenous anastomoses, which function as shunts allowing blood to be diverted directly from the arteriole to the venule, thus bypassing superficial capillary loops.

Sims/Schenk

to help maintain a constant deep-body temperature despite changes in the ambient (external) temperature, a process called *thermoregulation.* The small thickness and large size of the skin (1.0–1.5 mm thick; 1.7–1.8 square meters in surface area) make it an effective radiator of heat when the body temperature is greater than the ambient temperature. The transfer of heat from the body to the external environment is aided by the flow of warm blood through capillary loops near the surface of the skin.

Blood flow through the skin is adjusted to maintain deep-body temperature at about 37° C (98.6° F). When the body needs to lose heat, blood flows into superficial capillary loops so that the heat can be radiated to the external environment. When heat must be conserved, unique **arteriovenous anastomoses** (fig. 20.20) in the cutaneous vessels shunt (divert) blood directly from arterioles to deep venules, thus bypassing the superficial capillary loops. Opening of these shunts and constriction of cutaneous arterioles, as a result of sympathetic nerve stimulation, reduces cutaneous blood flow when the ambient temperature is cold.

Skin can tolerate an extremely low blood flow rate in cold weather because its metabolic rate decreases when the ambient temperature decreases. In cold weather, therefore, the skin requires less blood. In very cold weather, however, blood flow to the skin can be so severely restricted that the tissue does die: this is *frostbite.* Blood

flow to the skin can vary from less than 20 ml per minute at maximal vasoconstriction to as much as 3–4 L per minute at maximal vasodilation.

As the temperature warms, cutaneous arterioles in the hands and feet dilate as a result of decreased sympathetic nerve activity. Continued warming causes dilation of arterioles in other areas of the skin. If the resulting increase in cutaneous blood flow is not sufficient to cool the body, secretion of sweat glands may be stimulated. Sweat helps to cool the body as it evaporates from the surface of the skin. Also, sweat glands secrete *bradykinin (brad"e-ki'nin),* a polypeptide that stimulates vasodilation. This increases blood flow to the skin and to the sweat glands, so that larger volumes of more dilute sweat are produced.

Under the usual conditions of ambient temperature, the cutaneous vascular resistance is high and the blood flow is low when a person is not exercising. In the pre-exercise state of "fight or flight," sympathetic nerve activity reduces cutaneous blood flow still further. During exercise, however, the need to maintain a deep-body temperature takes precedence over the need to maintain an adequate systemic blood pressure. As the body temperature rises during exercise, vasodilation in cutaneous vessels occurs together with vasodilation in the exercising muscles. This can produce an even greater lowering of total peripheral resistance. If exercise is performed in hot and

Figure 20.21. A constriction increases blood pressure upstream (analogous to the arterial pressure) and decreases pressure downstream (analogous to capillary and venous pressure).

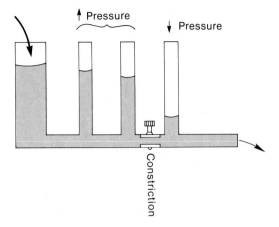

humid weather and if restrictive clothing is worn that increases skin temperature and cutaneous vasodilation, a dangerously low blood pressure may be produced after exercise has ceased and the cardiac output has declined. People have lost consciousness and even died as a result.

Changes in cutaneous blood flow occur as a result of changes in sympathetic nerve activity. Since the activity of the sympathetic nervous system is controlled by the brain, emotional states, acting through control centers in the medulla oblongata, can affect sympathetic activity and cutaneous blood flow. During fear reactions, for example, vasoconstriction in the skin together with activation of sweat glands can produce a pallor and a "cold sweat." Other emotions may cause vasodilation and blushing.

---

1. *Define the term* autoregulation, *and describe how this is accomplished in the cerebral circulation.*
2. *Describe how hyperventilation can cause dizziness.*
3. *Explain how cutaneous blood flow is adjusted to maintain a constant deep-body temperature.*

Blood Pressure

The rise in pressure from diastolic to systolic levels provides the driving force for blood flow through the organs. The arterial pressure is normally maintained within limits by the baroreceptor reflex and other mechanisms and is commonly measured indirectly by the auscultatory method.

Objective 16. Describe how the blood pressure changes as blood passes from large arteries to smaller arteries, arterioles, capillaries, venules, and veins. Explain the reasons for these changes.

Objective 17. Describe the factors that directly affect the arterial blood pressure.

Objective 18. Describe the baroreceptor reflex, and explain its physiological significance.

Objective 19. Describe the auscultatory method of blood pressure measurement, and explain how it works.

Resistance to flow in the arterial system is greatest in the arterioles because these vessels have the smallest diameters. Although the total blood flow through a system of arterioles must be equal to the flow in the larger vessel that gave rise to those arterioles, the narrow diameter of each arteriole reduces the flow rate in each according to Poiseuille's law. Blood flow rate and pressure is thus reduced in the capillaries, which are located downstream of the high resistance imposed by the arterioles. The blood pressure upstream of the arterioles—in the medium and large arteries—is correspondingly increased (fig. 20.21).

The blood pressure and flow rate within the capillaries is further reduced by the fact that their total cross-sectional area is much greater, due to their large number, than the cross-sectional areas of arteries and arterioles (fig. 20.22). Thus, although each capillary is much narrower than each arteriole, the capillary beds served by arterioles do not provide as great a resistance to blood flow as do the arterioles.

Variations in the diameter of arterioles due to vasoconstriction and vasodilation thus simultaneously affect both blood flow through capillaries and the *arterial* (upstream) *blood pressure*. An increase in total peripheral resistance due to vasoconstriction of arterioles can raise arterial blood pressure. Blood pressure can also be raised by an increase in the cardiac output. This may be due to elevations in cardiac rate or stroke volume, which in turn is affected by other factors. The three most important variables affecting blood pressure are the **cardiac rate, blood volume,** and the **total peripheral resistance.** An increase in any of these three, if not compensated by a decrease in another variable, will result in an increased blood pressure.

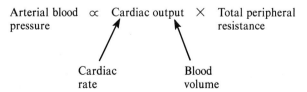

Arterial blood pressure $\propto$ Cardiac output $\times$ Total peripheral resistance

Cardiac rate   Blood volume

Blood pressure can thus be regulated by the kidneys, which control blood volume, and by the sympathoadrenal system. Increased activity of the sympathoadrenal system can raise blood pressure by stimulating vasoconstriction of arterioles (thus raising total peripheral resistance) and by promoting an increased cardiac output. Sympathetic stimulation can also affect blood volume indirectly, by stimulating constriction of renal bood vessels and thus reducing urine output.

Figure 20.22.   As blood passes from the aorta to the smaller arteries, arterioles, and capillaries, the cross-sectional data increases as the pressure decreases.

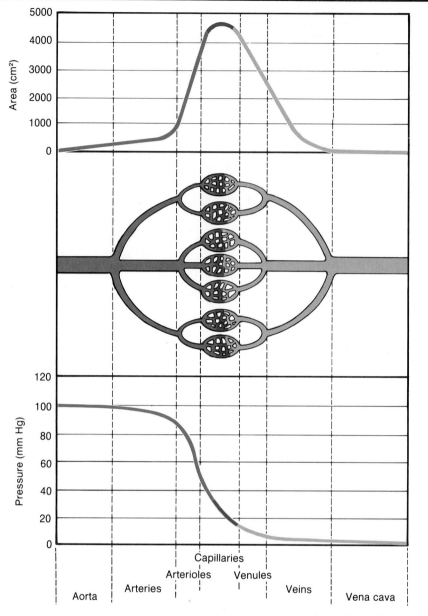

Figure 20.23.   Action potential frequency in sensory nerve fibers from baroreceptors in the carotid sinus and aortic arch. As the blood pressure increases, the baroreceptors become increasingly stretched. This results in an increase in the frequency of action potentials that are transmitted to the cardiac and vasomotor control centers in the medulla oblongata.

Figure 20.24.   The baroreceptor reflex. Sensory stimuli from baroreceptors in the carotid sinus and the aortic arch, acting via control centers in the medulla oblongata, affect the activity of sympathetic and parasympathetic nerve fibers in the heart.

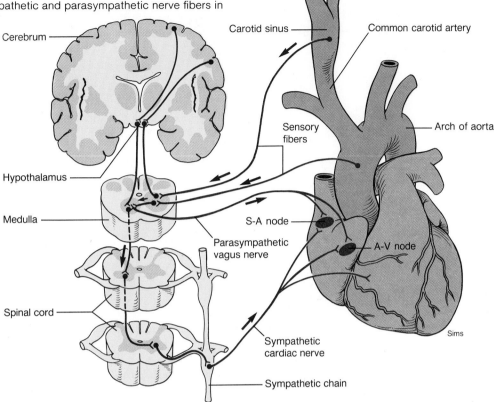

Regulation of Blood Pressure

In order for blood pressure to be maintained within limits, specialized receptors for pressure are needed. These **baroreceptors** *(bar″o-re-sep′tor)* are stretch receptors located in the **aortic arch** and in the **carotid sinuses.** An increase in pressure causes the walls of these arterial regions to stretch and stimulate the activity of sensory nerve endings (fig. 20.23). A fall in pressure below the normal range, in contrast, causes a decrease in the frequency of action potentials produced by these sensory nerve fibers.

Sensory nerve activity from the baroreceptors ascends via the vagus and glossopharyngeal nerves to the medulla oblongata, which directs the autonomic system to respond appropriately. **Vasomotor control centers** in the medulla control vasoconstriction/vasodilation and hence help regulate total peripheral resistance. **Cardiac control centers** (separate cardiostimulatory and cardioinhibitory regions may exist) in the medulla regulate the cardiac rate (fig. 20.24).

Baroreceptor Reflex.   When a person goes from a lying to a standing position, there is a shift of 500–700 ml of blood from the veins of the thoracic cavity to veins in the lower extremities, which expand to contain the extra volume of blood. This pooling of blood reduces the venous return and cardiac output. The resulting fall in blood

pressure is almost immediately compensated by the baroreceptor reflex. A decrease in baroreceptor sensory information to the medulla oblongata inhibits parasympathetic activity and promotes sympathetic nerve activity, resulting in increased cardiac rate and vasoconstriction. These responses help maintain an adequate blood pressure upon standing (fig. 20.25).

An effective baroreceptor reflex helps to maintain adequate blood flow to the brain upon standing. If the baroreceptor sensitivity is reduced, perhaps by atherosclerosis, an uncompensated fall in pressure may occur when a person stands up too rapidly. This condition—called **postural,** or **orthostatic, hypotension** (hypotension = low blood pressure)— can make a person feel dizzy or even faint because of inadequate perfusion of the brain.

The baroreceptor reflex can also mediate the opposite response. When the blood pressure rises above an individual's normal range, the baroreceptor reflex causes a slowing of the cardiac rate and vasodilation. Manual massage of the carotid sinus, a procedure sometimes employed by physicians to reduce tachycardia and lower blood pressure, also evokes this reflex. Such carotid massage should be used cautiously, however, because the intense vagus-nerve-induced slowing of the cardiac rate could cause loss of consciousness (as occurs in emotional fainting).

---

baroreceptor: Gk. *baros,* pressure; L. *receiver,* to receive

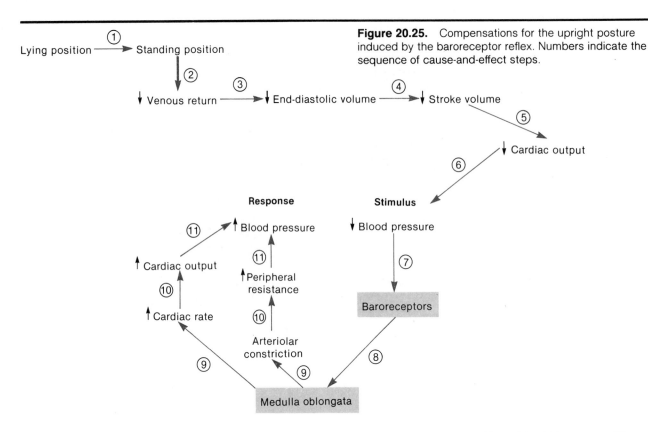

Figure 20.25. Compensations for the upright posture induced by the baroreceptor reflex. Numbers indicate the sequence of cause-and-effect steps.

Figure 20.26. The use of a pressure cuff and sphygmomanometer to measure blood pressure.

Other Reflexes Controlling Blood Pressure. The reflex control of ADH secretion by osmoreceptors in the hypothalamus and the control of angiotensin II production and aldosterone secretion by the juxtaglomerular apparatus of the kidneys have been previously discussed. Antidiuretic hormone and aldosterone increase blood volume, whereas angiotensin II stimulates vasoconstriction.

Another reflex that may be important to blood pressure regulation is initiated by stretch receptors located in the left atrium of the heart. These receptors are activated by increased venous return to the heart, and produce (1) reflex tachycardia, as a result of increased sympathetic nerve activity; and (2) inhibition of ADH secretion, resulting in larger volumes of urine excretion and a lowering of blood volume.

Measurement of Blood Pressure

Stephen Hales accomplished the first documented measurement of blood pressure by inserting a cannula into the artery of a horse and measuring the height to which blood would rise in the vertical tube. Modern clinical blood pressure measurements, fortunately, are less direct. The indirect, or **auscultatory** *(aws-kul'tah-to''re),* **method** of blood pressure measurement is based on the correlation of blood pressure and arterial sounds.

In the auscultatory method, an inflatable rubber bladder within a cloth cuff is wrapped around the upper arm and a stethoscope is applied over the brachial artery

Stephen Hales: English physiologist, 1677–1761
auscultatory: L., *auscultare,* to listen to

Figure 20.27. Korotkoff sounds are produced by the turbulent flow of blood through the partially constricted brachial artery. This occurs when the cuff pressure is greater than the diastolic pressure but less than the systolic pressure.

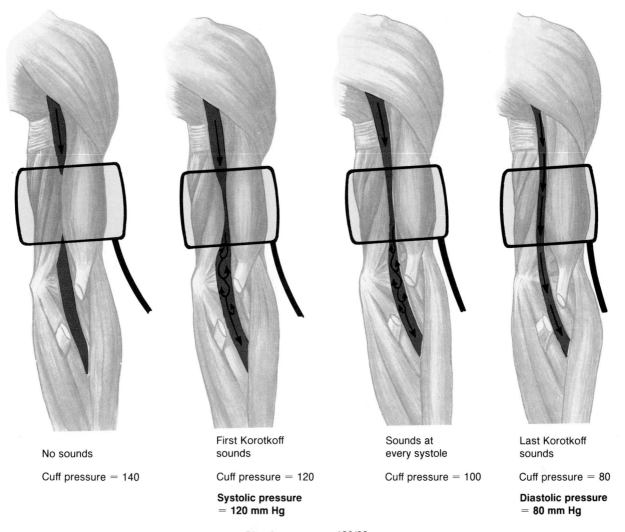

No sounds

Cuff pressure = 140

First Korotkoff sounds

Cuff pressure = 120

Systolic pressure = 120 mm Hg

Sounds at every systole

Cuff pressure = 100

Last Korotkoff sounds

Cuff pressure = 80

Diastolic pressure = 80 mm Hg

Blood pressure = 120/80

(fig. 20.26). The artery is normally silent before inflation of the cuff because blood normally travels in a smooth *laminar flow* through the arteries. The term laminar means layered—blood in the central axial stream moves the fastest, and blood flowing closer to the artery wall moves more slowly. There is little transverse movement between these layers that would produce mixing.

The laminar flow that normally occurs in arteries produces little vibration and is thus silent. When the artery is pinched, however, blood flow through the constriction becomes turbulent, which causes the artery to vibrate and produce sounds (much like the sounds produced by water through a kink in a garden hose). The tendency of the cuff pressure to constrict the artery is opposed by the blood pressure. Thus, in order to constrict the artery, the cuff pressure must be greater than the diastolic blood pressure. If the cuff pressure is also greater than the systolic blood pressure the artery would be pinched off and silent. *Turbulent flow* and sounds produced by vibrations

of the artery as a result of this flow, therefore, only occurs when the cuff pressure is greater than the diastolic and less than the systolic blood pressure.

Suppose that a person has a systolic pressure of 120 mm Hg and a diastolic pressure of 80 mm Hg (the average normal values). When the cuff pressure is between 80 and 120 mm Hg, the artery will be closed during diastole and opened during systole. As the artery begins to open with every systole, turbulent flow of blood through the constriction will create vibrations that are heard as the **sounds of Korotkoff** *(ko-rot′kof),* as shown in figure 20.27. These are usually "tapping" sounds because laminar flow and silence is resumed with every diastole. It should be understood that the sounds of Korotkoff are *not* "lub-dub" sounds produced by closing of the heart valves (those sounds can only be heard on the chest, not on the brachial artery).

Nicolai S. Korotkoff: Russian physician, born 1874

Figure 20.28.   The indirect, or auscultatory, method of blood-pressure measurement. Korotkoff sounds, produced by turbulent blood flow through a constricted artery, occur whenever the cuff pressure is less than the systolic blood pressure and greater than the diastolic blood pressure. As a result, the first Korotkoff sound is heard when the cuff pressure is equal to the systolic blood pressure, and the last sound is heard when the cuff pressure and diastolic blood pressures are equal.

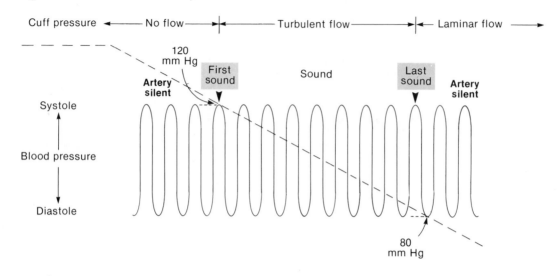

Figure 20.29.   The five phases of blood-pressure measurement.

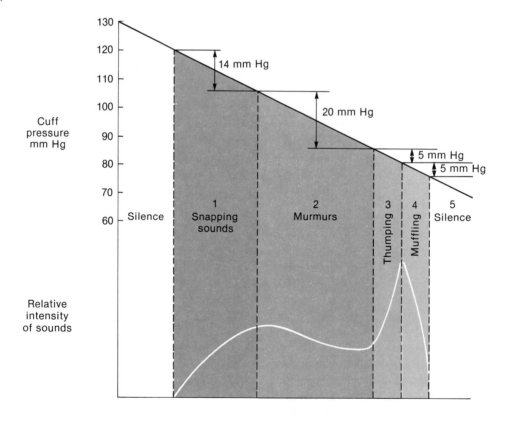

Table 20.10 Normal arterial blood pressure at different ages

Age	Systolic Men	Systolic Women	Diastolic Men	Diastolic Women	Age	Systolic Men	Systolic Women	Diastolic Men	Diastolic Women
1 day	70				16 years	118	116	73	72
3 days	72				17 years	121	116	74	72
9 days	73				18 years	120	116	74	72
3 weeks	77				19 years	122	115	75	71
3 months	86				20–24 years	123	116	76	72
6–12 months	89	93	60	62	25–29 years	125	117	78	74
1 year	96	95	66	65	30–34 years	126	120	79	75
2 years	99	92	64	60	35–39 years	127	124	80	78
3 years	100	100	67	64	40–44 years	129	127	81	80
4 years	99	99	65	66	45–49 years	130	131	82	82
5 years	92	92	62	62	50–54 years	135	137	83	84
6 years	94	94	64	64	55–59 years	138	139	84	84
7 years	97	97	65	66	60–64 years	142	144	85	85
8 years	100	100	67	68	65–69 years	143	154	83	85
9 years	101	101	68	69	70–74 years	145	159	82	85
10 years	103	103	69	70	75–79 years	146	158	81	84
11 years	104	104	70	71	80–84 years	145	157	82	83
12 years	106	106	71	72	85–89 years	145	154	79	82
13 years	108	108	72	73	90–94 years	145	150	78	79
14 years	110	110	73	74	95–106 years	145	149	78	81
15 years	112	112	75	76					

From Diem, K., and Lentner, C., Eds., *Documenta Geigy Scientific Tables,* 7th ed., J. R. Geigy S. A., Basle, Switzerland, 1970. With permission.

Initially, the cuff is usually inflated to produce a pressure greater than the systolic pressure so that the artery is pinched off and silent. The pressure in the cuff is read from an attached meter called a *sphygmomanometer (sfig"mo-mah-nom'e-ter).* A valve is then turned to allow the release of air from the cuff, causing a gradual decrease in cuff pressure. When the cuff pressure is equal to the systolic pressure, the **first sound** of Korotkoff is heard as blood passes in a turbulent flow through the constricted opening of the artery.

Korotkoff sounds will continue to be heard at every systole as long as the cuff pressure remains greater than the diastolic pressure. When the cuff pressure becomes equal to or less than the diastolic pressure, the sounds disappear because the artery remains open (fig. 20.28). The **last sound** of Korotkoff thus occurs when the cuff pressure is equal to the diastolic pressure.

Different phases in the measurement of blood pressure are identified on the basis of the quality of the Korotkoff sounds (fig. 20.29). In some people, the Korotkoff sounds do not disappear even when the cuff pressure is reduced to zero (zero pressure means that it is equal to atmospheric pressure). In these cases—and often routinely—the onset of muffling of the sounds (phase 4 in fig. 20.29) is used rather than the onset of silence (phase 5) as an indication of diastolic pressure. Normal blood pressure values are shown in table 20.10.

Figure 20.30. The pulse pressure provides the driving force for blood flow in the arteries.

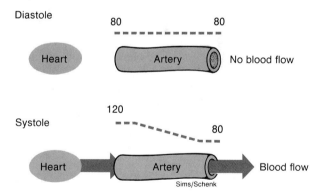

Pulse Pressure

The *pulse pressure* is equal to the difference between the systolic and diastolic pressures. If a person has a blood pressure of 120/80 (systolic/diastolic), for example, the pulse pressure would be 40 mm Hg. This value is significant because it is the pressure difference (ΔP) that drives blood along the arteries to the capillaries. If the pressure were constant in an artery, there would be no blood flow (fig. 20.30).

sphygmomanometer: Gr., *sphygmos,* pulse; *manos,* thin;
   metro, measure

Table 20.11   Some possible causes of secondary hypertension

System involved	Examples	Mechanisms
Kidneys	Kidney disease	Decreased urine formation
	Renal artery disease	Secretion of vasoactive chemicals
Endocrine	Excess catecholamines (tumor of adrenal medulla)	Increased cardiac output and total peripheral resistance
	Excess aldosterone (Conn's disease)	Excess salt and water retention by the kidneys
Nervous	Increased intracranial pressure	Activation of sympathoadrenal system
	Damage to vasomotor center	Activation of sympathoadrenal system
Cardiovascular	Complete heart block; patent ductus arteriosus	Increased stroke volume
	Arteriosclerosis of aorta; coarctation of aorta	Decreased distensibility of aorta

Source: Fox, Stuart Ira, *Human Physiology.* © 1984 Wm. C. Brown Publishers, Dubuque, Iowa. All Rights Reserved. Reprinted by permission.

A decrease in pulse pressure can thus result in dangerously low rates of blood flow through the body organs, even if the absolute pressure values are high. Arteriosclerosis, for example, can increase total peripheral resistance and so raise the diastolic pressure that the heart cannot generate a high enough systolic pressure to adequately perfuse the organs. A blood pressure of 160/145, for example, yields a pulse pressure of only 15 mm Hg, which may not result in adequate rates of blood flow. Such a low pulse pressure reflects the fact that the cardiac output is reduced, which may lead to symptoms of congestive heart failure.

Dehydration or blood loss results in decreased cardiac output and thus in decreased pulse pressure. The opposite conditions occur during exercise, when the cardiac output is greatly increased. The blood pressure, as a result, may rise as high as 200/100. Under these conditions the pulse pressure is large (100 mm Hg in this example), and arterial blood flow is much more rapid than at rest.

---

1. Describe the relationship between blood pressure and the total surface area of arteries, arterioles, and capillaries. Describe how arterioles influence blood flow through capillaries and arterial blood pressure.
2. Describe how the baroreceptor reflex helps to compensate for a fall in blood pressure. Explain how a person who is severely dehydrated would have a rapid pulse.
3. Describe how the sounds of Korotkoff are produced and how these sounds are used to measure blood pressure.
4. Define pulse pressure and explain its physiological significance.

Clinical Considerations

Hypertension

Approximately 20% of all adults in the United States have *hypertension*—blood pressure in excess of the normal range for the person's age and sex. Hypertension that is a result of ("secondary to") known disease processes is logically called **secondary hypertension.** These comprise only about 10% of the hypertensive population. Hypertension that is the result of complex and poorly understood processes is not-so-logically called **primary,** or **essential, hypertension.**

Diseases of the kidneys and arteriosclerosis of the renal arteries can cause secondary hypertension because of high blood volume. More commonly, the reduction of renal blood flow can raise blood pressure by stimulating the secretion of vasoactive chemicals from the kidneys. Experiments in which the renal artery is pinched, for example, produces hypertension that is associated (at least initially) with elevated renin secretion. These and other causes of secondary hypertension are summarized in table 20.11.

Essential Hypertension.   The vast majority of people with hypertension have essential hypertension. An increased total peripheral resistance is a universal characteristic of this condition. Cardiac rate and the cardiac output are elevated in many, but not all, of these cases.

The secretion of renin, which is correlated with angiotensin II production and aldosterone secretion, is likewise variable. Although some people with essential hypertension have low renin secretion, most have either normal or elevated levels of renin secretion. Renin secretion in the normal range is inappropriate for people with hypertension, since high blood pressure should inhibit renin secretion and, through a lowering of aldosterone, result in greater excretion of salt and water. Inappropriately high levels of renin secretion could thus contribute to hypertension by promoting (via stimulation of aldosterone secretion) salt and water retention and high blood volume.

Sustained high stress (acting via the sympathetic nervous system) and high salt intake appear to act synergistically in the development of hypertension. There is some evidence that Na^+ enhances the vascular response to sympathetic stimulation. Further, sympathetic nerves can cause constriction of the renal blood vessels and thus decrease the excretion of salt and water. In one study, hypertensive students given an IQ test had decreased renal blood flow, increased renin secretion, decreased excretion of salt and water, and a greater increase in blood pressure than did students with normal blood pressure. There also

Figure 20.31. The sequence of events that has been proposed as a possible cause of essential hypertension.

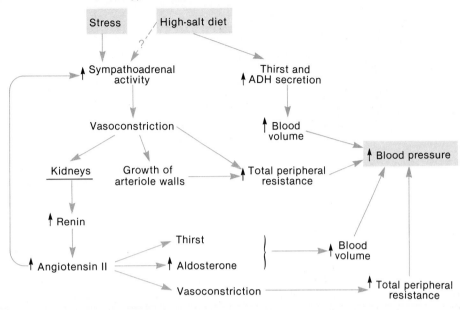

Table **20.12**	Examples and mechanisms of action of some antihypertensive drugs	
Category of drugs	**Examples**	**Mechanisms**
Extracellular fluid volume depletors	Thiazide diuretics	Increase volume of urine excreted, thus lowering blood volume
Sympathoadrenal system inhibitors	Clonidine; alpha-methyldopa	Acts on brain to decrease sympathoadrenal stimulation
	Guanethidine; reserpine	Depletes norepinephrine from sympathetic nerve endings
	Propranolol	Blocks beta-adrenergic receptors, decreasing cardiac output and/or renin secretion
	Phentolamine	Blocks alpha-adrenergic receptors, decreasing sympathetic vasoconstriction
Vasodilators	Hydralazine; sodium nitroprusside	Causes vasodilation by acting directly on vascular smooth muscle

Source: Fox, Stuart Ira, *Human Physiology.* © 1984 Wm. C. Brown Publishers, Dubuque, Iowa. All Rights Reserved. Reprinted by permission.

appears to be a genetic component to these responses; students who were normotensive but who had a family history of hypertension had responses that were intermediate between hypertensive students and students who were normotensive with no family history of hypertension.

The interactions between salt intake, sympathetic nerve activity, cardiovascular responses to sympathetic nerve activity, kidney function, and genetics make it difficult to sort out the cause-and-effect sequence that leads to essential hypertension. Many people have suggested that there is no single cause-and-effect sequence but rather a web of causes and effects (see fig. 20.31). This is currently controversial.

Dangers of Hypertension. If other factors remain constant, blood flow increases as arterial blood pressure increases. People with hypertension thus have adequate perfusion of their organs with blood until the hypertension causes vascular damage. Hypertension, as a result, is usually without symptoms until a dangerous amount of vascular damage is produced.

Hypertension is dangerous because (1) high arterial pressure increases the afterload, making it more difficult for the ventricles to eject blood; this increases the amount of work that the heart must perform and may result in pathological changes in heart structure and function, leading to congestive heart failure; (2) high pressure may damage cerebral blood vessels, leading to cerebrovascular accident (stroke); and (3) hypertension contributes to the development of atherosclerosis, which can itself lead to heart disease and stroke as previously described.

Treatment of Hypertension. Hypertension is usually treated by restricting salt intake and by taking drugs that act in a variety of ways. Most commonly, these drugs are *diuretics* that increase urine volume, thus decreasing blood volume and pressure. Sympathetic-blocking drugs are also often used; drugs that block beta-adrenergic receptors (such as propranolol) decrease cardiac rate. Various vasodilators (see table 20.12) may also be used to decrease total peripheral resistance.

Table 20.13 Signs of shock

	Early sign	Late sign
Blood pressure	Decreased pulse pressure Increased diastolic pressure	Decreased systolic pressure
Urine	Decreased Na$^+$ concentration Increased osmolality	Decreased volume
Blood pH	Increased pH (alkalosis) due to hyperventilation	Decreased pH (acidosis) due to "metabolic" acids
Effects of poor tissue perfusion	Slight restlessness; occasionally warm, dry skin	Cold, clammy skin "Cloudy" senses

Source: Wilson, R. F., ed., *Principles and Techniques of Critical Care*, vol. 1. © 1977 Upjohn Company. Reprinted by permission of F. A. Davis Company.

Table 20.14 Cardiovascular reflexes that help to compensate for circulatory shock

Organ(s)	Compensatory mechanisms
Heart	Sympathoadrenal stimulation produces increased cardiac rate and increased stroke volume, due to "positive inotropic effect" on myocardial contractility
Digestive tract and skin	Decreased blood flow due to vasoconstriction as a result of sympathetic nerve stimulation (alpha-adrenergic effect)
Kidneys	Decreased urine production as a result of sympathetic-nerve-induced constriction of renal arterioles; increased salt and water retention due to increased aldosterone and antidiuretic hormone (ADH) secretion

Source: Fox, Stuart Ira, *Human Physiology.* © 1984 Wm. C. Brown Publishers, Dubuque, Iowa. All Rights Reserved. Reprinted by permission.

Circulatory Shock

Circulatory shock occurs when there is inadequate blood flow and/or oxygen utilization by the tissues. Some of the signs of shock (table 20.13) are a result of inadequate tissue perfusion; other signs of shock are produced by cardiovascular responses that help to compensate for the poor tissue perfusion (table 20.14). When these compensations are effective, they (together with emergency medical care) are able to reestablish adequate tissue perfusion. In some cases, however, and for reasons that are not clearly understood, the shock may progress to an irreversible stage and death may result.

Hypovolemic Shock. The term **hypovolemic** *(hi″po-vo-le′mik)* **shock** refers to circulatory shock due to low blood volume, as might be caused by hemorrhage (bleeding), dehydration, or burns. This is accompanied by decreased blood pressure and decreased cardiac output. In response to these changes, the sympathoadrenal system is activated as a result of the baroreceptor reflex. As a result, tachycardia is produced and vasoconstriction occurs in the skin, digestive tract, kidneys, and muscles (table 20.14). Decreased blood flow through the kidneys stimulates renin secretion and activation of the renin-angiotensin-aldosterone system. A person in hypovolemic shock thus has low blood pressure, rapid pulse, cold, clammy skin, and little urine excretion.

Other Causes of Circulatory Shock. A rapid fall in blood pressure occurs in **anaphylactic** *(an″ah-fĭ-lak′tik)* **shock** as a result of severe allergic reaction (usually to bee stings or penicillin). This occurs because of widespread release of histamine, which causes vasodilation and thus decreases total peripheral resistance. A rapid fall in blood pressure also occurs in **neurogenic shock,** in which sympathetic tone is decreased, usually because of upper spinal cord damage or spinal anesthesia. **Cardiogenic shock** results from cardiac failure, as defined by a cardiac output that is inadequate to maintain tissue perfusion.

Congestive Heart Failure

Cardiac failure occurs when the cardiac output is insufficient to maintain the blood flow required by the body. This may be due to heart disease—resulting from myocardial infarction or congenital defects—or to hypertension, which increases the afterload of the heart. The most common causes of left pump failure are myocardial infarction, aortic valve stenosis, and incompetence of the aortic and bicuspid (mitral) valves. Failure of the right pump is usually caused by prior failure of the left pump.

The term **congestive** is often used in describing heart failure because of the increased venous volume and pressure that results. Failure of the left pump, for example, raises the left atrial pressure and produces pulmonary congestion and edema, which causes shortness of breath and fatigue. Failure of the right pump results in increased right atrial pressure, which produces congestion and edema in the systemic circulation.

The compensatory responses that occur during congestive heart failure are similar to those that occur during hypovolemic shock. Activation of the sympathoadrenal system stimulates cardiac rate, contractility of the ventricles (fig. 20.32), and constriction of arterioles. As in hypovolemic shock, renin secretion is increased and urine output is reduced.

As a result of these compensations, chronically low cardiac output is associated with elevated blood volume and dilation and hypertrophy of the ventricles. These changes can themselves be dangerous. Elevated blood volume places a work overload on the heart, and the enlarged ventricles have a higher metabolic requirement for oxygen. These problems are often treated with drugs that increase myocardial contractility (such as digitalis), drugs that are vasodilators (such as nitroglycerin), and diuretic drugs that lower blood volume by increasing the volume of urine excreted.

Figure 20.32. The relationship between end-diastolic volume and ventricular pressure (a measure of contraction strength). At a given end-diastolic volume the ventricular pressure is decreased during heart failure and increased by sympathetic stimulation. Heart failure may be compensated by sympathetic stimulation.

Chapter Summary

I. Cardiac Output
The cardiac output is a measure of the pumping ability of the heart and is affected by mechanisms that regulate the cardiac rate and stroke volume.
- A. Cardiac rate is increased by sympathoadrenal stimulation and decreased by the effects of parasympathetic fibers that innervate the SA node.
- B. Stroke volume is regulated both extrinsically and intrinsically.
  1. The Frank-Starling law of the heart describes the way the end-diastolic volume, through various degrees of myocardial stretching, influences the contraction strength of the myocardium and thus the stroke volume.
  2. The end-diastolic volume is called the preload; the total peripheral resistance, through its effect on arterial blood pressure, provides an afterload that acts to reduce the stroke volume.
  3. At a given end-diastolic volume, the amount of blood ejected depends on contractility; strength of contraction is increased by sympathoadrenal stimulation.
- C. The venous return of blood to the heart is dependent largely on the total blood volume and mechanisms that improve the flow of blood in veins.
  1. The total blood volume is regulated by the kidneys.
  2. The venous flow of blood to the heart is aided by the action of skeletal muscle pumps and the effects of breathing.

II. Blood Volume
The volume of fluid in the vascular and tissue compartments is regulated by the kidneys. This is adjusted according to the needs of the body by antidiuretic hormone and aldosterone.
- A. Tissue fluid is formed from and returns to the blood.
  1. The hydrostatic pressure of the blood forces fluid from the arteriolar ends of capillaries into the interstitial spaces of the tissues.
  2. Since the colloid osmotic pressure of plasma is greater than tissue fluid, water returns by osmosis to the venular ends of capillaries.
  3. Excess tissue fluid is returned to the venous system by lymphatic vessels.
  4. Edema occurs when there is an accumulation of tissue fluid.
- B. The kidneys control the blood volume by regulating the amount of filtered fluid that will be reabsorbed.
  1. Antidiuretic hormone stimulates reabsorption of water from the kidney filtrate and thus acts to maintain the blood volume.
  2. A decrease in blood flow through the kidneys activates the renin-angiotensin system.
  3. Angiotensin II stimulates vasoconstriction and stimulates the adrenal cortex to secrete aldosterone.
  4. Aldosterone acts on the kidneys to promote the retention of salt and water.

III. Vascular Resistance and Blood Flow
The rate of blood flow is dependent in part on vascular resistance. Resistance is regulated by vasoconstriction and vasodilation, produced by extrinsic and intrinsic mechanisms.
- A. Poiseuille's law describes the fact that blood flow is directly related to the pressure difference between the two ends of a vessel and is inversely related to the resistance to blood flow through the vessel.
- B. Extrinsic regulation of vascular resistance is provided mainly by the sympathetic nervous system, which stimulates vasoconstriction of arterioles in the viscera and skin.
- C. Intrinsic control of vascular resistance allows organs to autoregulate their own blood flow rates.
  1. Myogenic regulation occurs when vessels constrict or dilate as a direct response to a rise or fall in blood pressure.
  2. Metabolic regulation occurs when vessels dilate in response to the local chemical environment within the organ.

IV. Blood Flow to the Heart and Skeletal Muscles
Blood flow to the heart and skeletal muscles is regulated by both intrinsic and extrinsic mechanisms that increase the rate of flow when the metabolic requirements of these tissues increase during exercise.
- A. The heart normally always respires aerobically because of its high capillary supply, myoglobin content, and enzyme content.

B. During exercise, when the heart's metabolism increases, intrinsic metabolic mechanisms stimulate vasodilation of the coronary vessels and thus increase coronary blood flow.

C. Just prior to exercise and at the start of exercise, blood flow through skeletal muscles increases due to vasodilation caused by cholinergic sympathetic nerve fibers; during exercise intrinsic metabolic vasodilation occurs.

D. Since cardiac output can increase by a factor of five or more during exercise, the heart and skeletal muscles receive an increased proportion of a higher total blood flow.
   1. The cardiac rate increases due to lower activity of the vagus nerve and higher activity of the sympathetic nerve.
   2. The venous return is faster because of higher activity of the skeletal muscle pumps and increased breathing.
   3. Increased contractility of the heart, combined with a decrease in total peripheral resistance, can result in a higher stroke volume.

V. Blood Flow to the Brain and Skin
Control mechanisms help to maintain relatively constant rates of cerebral blood flow. Blood flow to the skin is varied according to the needs of the body.
   A. Cerebral blood flow is regulated both myogenically and metabolically.
      1. Cerebral vessels automatically constrict if the systemic blood pressure rises too high.
      2. Metabolic products cause local vessels to dilate and supply more active areas with more blood.
   B. The skin has unique arteriovenous anastomoses, which can shunt the blood away from surface capillary loops.
      1. Sympathetic nerve fibers cause constriction of cutaneous arterioles.
      2. As a thermoregulatory response, there is increased cutaneous blood flow and increased flow through surface capillary loops.

VI. Blood Pressure
The blood pressure rise from diastolic to systolic levels provides the driving force for blood flow. Blood pressure is regulated by a variety of control mechanisms, and it is usually measured indirectly by the auscultatory method.

A. Baroreceptors in the aortic arch and carotid sinuses affect, via the sympathetic nervous system, the cardiac rate and the total peripheral resistance.
   1. The baroreceptor reflex causes pressure to be maintained when an upright posture is assumed and can cause a lowered pressure when the carotid sinuses are massaged.
   2. Other mechanisms that affect blood volume help to regulate blood pressure.

B. Blood pressure is commonly measured indirectly by auscultation of the brachial artery when a pressure cuff is inflated and deflated.
   1. The first sound of Korotkoff, caused by turbulent flow of blood through a constriction in the artery, occurs when the cuff pressure equals the systolic pressure.
   2. The last sound of Korotkoff is heard when the cuff pressure equals the diastolic blood pressure.

C. The pulse pressure, which is the difference between systolic and diastolic pressures, represents the driving force for blood flow through the arterial system.

Review Activities

Objective Questions

1. According to the Frank-Starling law, the strength of ventricular contraction is
   (a) directly proportional to the end-diastolic volume
   (b) inversely proportional to the end-diastolic volume
   (c) independent of the end-diastolic volume

2. In the absence of compensations, the stroke volume will decrease when
   (a) blood volume increases
   (b) venous return increases
   (c) contractility increases
   (d) arterial blood pressure increases

3. Which of the following statements about tissue fluid is *false?*
   (a) It contains the same glucose and salt concentration as plasma.
   (b) It contains a lower protein concentration than plasma.
   (c) Its colloid osmotic pressure is greater than that of plasma.
   (d) Its hydrostatic pressure is less than that of plasma.

4. Edema may be caused by
   (a) high blood pressure
   (b) decreased plasma protein concentration
   (c) leakage of plasma protein into tissue fluid
   (d) blockage of lymphatic vessels
   (e) all of the above

5. Both ADH and aldosterone act to
   (a) increase urine volume
   (b) increase blood flow
   (c) increase total peripheral resistance
   (d) all of the above

6. The greatest resistance to blood flow occurs in the
   (a) large arteries
   (b) medium-sized arteries
   (c) arterioles
   (d) capillaries

7. If a vessel were to dilate to twice its previous radius and if pressure remained constant, blood flow through this vessel would
   (a) increase by a factor of 16
   (b) increase by a factor of 4
   (c) increase by a factor of 2
   (d) decrease by a factor of 2

8. The sounds of Korotkoff are produced by
   (a) closing of the semilunar valves
   (b) closing of the AV valves
   (c) the turbulent flow of blood through an artery
   (d) elastic recoil of the aorta

9. Vasodilation in the heart and skeletal muscles during exercise is primarily due to the effects of
   (a) alpha-adrenergic stimulation
   (b) beta-adrenergic stimulation
   (c) cholinergic stimulation
   (d) products released by the exercising muscle cells

10. Blood flow in the coronary circulation is
    (a) increased during systole
    (b) increased during diastole
    (c) constant throughout the cardiac cycle

11. Blood flow in the cerebral circulation
    (a) varies with systemic arterial pressure
    (b) is regulated primarily by the sympathetic system
    (c) is maintained constant within physiological limits
    (d) is increased during exercise

12. Which of the following organs is able to tolerate the greatest restriction in blood flow? The
    (a) brain
    (b) heart
    (c) skeletal muscles
    (d) skin

13. Arteriovenous shunts in the skin
    (a) divert blood to superficial capillary loops
    (b) are closed when the ambient temperature is very cold
    (c) are closed when the deep-body temperature rises much above 37° C.
    (d) all of the above

Essay Questions

1. Define the terms *contactility*, *preload*, and *afterload*, and explain how these factors affect the cardiac output.
2. Explain, using the Frank-Starling law, how the stroke volume is affected by (a) bradycardia and (b) a "missed beat."
3. Which part of the cardiovascular system contains the most blood? Which part provides the greatest resistance to blood flow? Which part provides the greatest surface area? Explain.
4. Explain how the kidneys regulate blood volume.
5. A person who is dehydrated drinks more and urinates less. Explain the mechanisms involved.
6. Explain, using Poiseuille's law, how arterial blood flow can be diverted from one organ system to another.
7. Describe the mechanisms that increase the cardiac output during exercise and that increase the rate of blood flow to the heart and skeletal muscles.
8. Explain how an anxious person may have a cold, clammy skin and how the skin becomes hot and flushed on a hot, humid day.

Selected Readings

Berne, R. M. and M. N. Levy. 1981. *Cardiovascular Physiology*. 4th ed. St. Louis: C. V. Mosby Co.

Braunwald, E. 1974. Regulation of the circulation. *New Engl. J. Med.* 290: first part, p. 1124; second part, p. 1420.

Brody, M. J., J. R. Haywood, and K. B. Toun. 1980. Neural mechanisms in hypertension. *Ann. Rev. Physiol.* 42:441.

Brown, M. S. and J. L. Goldstein. November 1984. How LDL receptors influence cholesterol and atherosclerosis. *Scientific American*.

Donald, D. E., and J. T. Shepard. 1980. Autonomic regulation of the peripheral circulation. *Ann. Rev. Physiol.* 42:429.

Folkow, B., and E. Neil. 1971. *Circulation*. London: Oxford University Press.

Franciosa, J. A. 1981. Hypertensive left heart failure: Pathogenesis and therapy. *Hosp. Prac.* 16:165.

Granger, D. E., and P. R. Kvietys. 1981. The splanchnic circulation: Intrinsic regulation. *Ann. Rev. Physiol.* 43:409.

Herd, J. A. 1984. Cardiovascular response to stress in man. *Ann. Rev. Physiol.* 46:177.

Hilton, S. M., and K. M. Spyer. 1980. Central nervous regulation of vascular resistance. *Ann. Rev. Physiol.* 42:399.

Kaplan, N. M. 1980. The control of hypertension: A therapeutic breakthrough. *American Scientist* 68:537.

Katz, A. M. 1975. Congestive heart failure. *New Engl. J. Med.* 293:1184.

Kontos, H. A. 1981. Regulation of the cerebral circulation. *Ann. Rev. Physiol.* 43:397.

Light, K. C., J. P. Loepke, P. A. Obrist, and P. W. Willis, IV. 1983. Psychological stress induces sodium and fluid retention in men at high risk for hypertension. *Science* 220:249.

Olsson, R. A. 1981. Local factors regulating cardiac and skeletal muscle blood flow. *Ann. Rev. Physiol.* 43:385.

Ross, J., Jr. 1983. The failing heart and circulation. *Hosp. Prac.* 18:151.

Schatz, I. J. 1983. Orthostatic hypotension: Diagnosis and treatment. *Hosp. Prac.* 18:59.

Stephensen, R. B. 1984. Modification of reflex regulation of blood pressure by behavior. *Ann. Rev. Physiol.* 46:133.

Vatner, S. F., and E. Braunwald. 1975. Cardiovascular control mechanisms in the conscious state. *New Engl. J. Med.* 293:970.

Weber, K. T., J. S. Janicki, and W. Laskey. 1983. The mechanics of ventricular function. *Hosp. Prac.* 18:113.

Zelis, R., S. F. Flaim, A. J. Liedke, and S. H. Nellis. 1981. Cardiovascular dynamics in the normal and failing heart. *Ann. Rev. Physiol.* 43:455.

21 Blood Composition

Concepts

Blood is a specialized connective tissue consisting of formed elements that are suspended and carried in plasma. The constituents of blood function in gas transport, immunity, and blood-clotting mechanisms.

Trauma to a blood vessel initiates a sequence of events that ultimately results in the formation of a clot. Another sequence of events follows the healing of the blood vessel that results in the dissolution of the clot.

and Clotting Mechanisms

Selected Terms

eosinophil (e″o-sin′o-fil)
erythrocyte (ĕ-rith′ro-sīt)
erythropoiesis (ĕ-rith″ro-poi-e′sis)
leukocyte (loo′ko-sīt)
megakaryocyte (meg″ah-kar′e-o-sīt)
platelet (plăt′let)
hemostasis (he-mos′tah-sis)
bradykinin (brad″e-ki′nin)
kallikrein (kal″li-kre′in)
thromboplastin (throm″bo-plas′tin)

Composition of the Blood

Blood, a highly specialized connective tissue, consists of formed elements (erythrocytes, leukocytes, and thrombocytes) that are suspended and carried in plasma. The constituents of blood function in gas transport, immunity, and blood-clotting mechanisms.

> *Objective 1.* List the different types of substances present in plasma.
> *Objective 2.* Describe the origin and function of the different categories of plasma proteins.
> *Objective 3.* List the different types of formed elements of blood, and describe their appearance.
> *Objective 4.* Describe the origin of erythrocytes, leukocytes, and thrombocytes.

The average-sized adult has about 5 liters of blood, which constitute about 8% of the total body weight. Blood leaving the heart is referred to as *arterial blood.* Arterial blood, with the exception of that going to the lungs, is bright red in color due to a high concentration of oxyhemoglobin (the combination of oxygen with hemoglobin) in the erythrocytes. *Venous blood* is blood returning to the heart and, except for the venous blood from the lungs, has a darker color (due to increased amounts of hemoglobin that are no longer combined with oxygen). Blood has a viscosity that ranges between 4.5 and 5.5. This means that it flows thicker than water, which has a viscosity of 1. Blood has a pH range of 7.35 to 7.45 and a temperature within the thorax of the body of about 38° C (100.4° F).

Blood is composed of a cellular portion, called **formed elements** and a fluid portion, called **plasma.** When a blood sample is centrifuged, the heavier formed elements are packed into the bottom of the tube, leaving plasma at the top (fig. 21.1). The formed elements constitute approximately 45% of the total blood volume (the *hematocrit*), and the plasma accounts for the remaining 55%.

Plasma

Plasma is a straw-colored liquid consisting of water and dissolved solutes. Sodium ion is the major solute of plasma in terms of its concentration (Na^+ contributes most to the total osmolality of plasma). In addition to Na^+, plasma contains many other salts and ions, as well as organic molecules such as metabolites, hormones, enzymes, antibodies, and other proteins. The values of some of these constituents of plasma are shown in table 21.1.

Plasma Proteins. Plasma proteins constitute 7%–9% of the plasma. The three types of proteins are albumins, globulins, and fibrinogen. **Albumins** *(al-bu′min)* account for most (60%–80%) of the plasma proteins and are the smallest in size. They are produced by the liver and serve to provide the osmotic pressure

---

hematocrit: Gk. *haima,* blood; *krino,* to separate
plasma: Gk. *plasma,* to form or mold
albumin: L. *albumen,* white

Figure 21.1. Blood cells become packed at the bottom of the test tube when whole blood is centrifuged, leaving the fluid plasma at the top of the tube. Red blood cells are the most abundant of the blood cells—white blood cells and platelets form only a thin, light-colored "buffy coat" at the interface between the packed red blood cells and the plasma.

needed to draw water from the surrounding tissue fluid into the capillaries. This action is needed to maintain blood volume and pressure. **Globulins** *(glob'u-lin)* are divided into three subtypes: **alpha globulins, beta globulins,** and **gamma globulins.** The alpha and beta globulins are produced by the liver and function to transport lipids and fat-soluble vitamins in the blood. Gamma globulins are antibodies produced by lymphocytes (one of the formed elements found in blood and lymphoid tissues) and function in immunity. **Fibrinogen** *(fi-brin'o-jen)* accounts for only about 4% of the protein content of plasma. Fibrinogens, which are synthesized in the liver (table 21.2), are converted to insoluble threads of **fibrin** when blood clots. The fluid from clotted blood, which is called **serum,** is therefore identical to plasma except for the absence of fibrinogen.

The Formed Elements of Blood

The formed elements of blood include two types of blood cells: **erythrocytes** *(ĕ-rith'ro-sīt),* or red blood cells, and **leukocytes** *(loo'ko-sīt),* or white blood cells. Erythrocytes are by far the most numerous of these two types: a cubic millimeter of blood contains 5.1 million to 5.8 million

erythrocytes in males and 4.3 million to 5.2 million erythrocytes in females. The same volume of blood, in contrast, contains only 5,000 to 9,000 leukocytes.

Erythrocytes. Erythrocytes are flattened, biconcave discs, about 7 μm in diameter and 2.2 μm thick. Their unique shape relates to their function of transporting oxygen and provides an increased surface area through which gas can diffuse (fig. 21.2). Erythrocytes lack a nucleus and mitochondria (they get energy from anaerobic respiration). Because of these deficiencies, erythrocytes have a circulating life span of only about 120 days before they are destroyed by phagocytic cells in the liver, spleen, and bone marrow.

Each erythrocyte contains approximately 280 million *hemoglobin* molecules, which give blood its red color. Each hemoglobin molecule consists of a protein, called globin, and an iron-containing pigment, called heme. The iron group of heme is able to combine with oxygen in the lungs and release oxygen in the tissues.

Leukocytes. Leukocytes, or white blood cells, differ from erythrocytes in several ways. Leukocytes contain nuclei and mitochondria and can move in an amoeboid fashion

---

globulin: L. *globulus,* small globe
fibrinogen: L. *fibra,* fibrous
serum: L. *serum,* liquid
erythrocytes: Gk. *erythros,* red; *kytos,* globe

---

hemoglobin: Gk. *haima,* blood; *globus,* globe
leukocytes: Gk. *leukos,* white; *kytos,* hollow (cell)

Table 21.1 Representative normal plasma values

Measurement	Normal range
Blood volume	80–85 ml/kg body weight
Blood osmolality	285–295 mOsm
Blood pH	7.35–7.45
Enzymes	
Creatine phosphokinase (CPK)	Female: 5–35 mU/ml
	Male: 5–55 mU/ml
Lactic dehydrogenase (LDH)	60–120 U/ml
Phosphatase (acid)	Female: 0.01–0.56 Sigma U/ml
	Male: 0.13–0.63 Sigma U/ml
Hematology values	
Hematocrit	Female: 37%–48%
	Male: 45%–52%
Hemoglobin	Female: 12–16 g/100 ml
	Male: 13–18 g/100 ml
Red blood cell count	4.2–5.9 million/mm^3
White blood cell count	4,300–10,880/mm^3
Hormones	
Testosterone	Male: 300–1,100 ng/100 ml
	Female: 25–90 ng/100 ml
Adrenocorticotrophic Hormone (ACTH)	15–70 pg/ml
Growth hormone	Children: over 10 ng/ml
	Adult male: below 5 ng/ml
Insulin	6–26 µU/ml (fasting)
Ions	
Bicarbonate	24–30 mmol/l
Calcium	2.1–2.6 mmol/l
Chloride	100–106 mmol/l
Potassium	3.5–5.0 mmol/l
Sodium	135–145 mmol/l
Organic molecules (other)	
Cholesterol	120–220 mg/100 ml
Glucose	70–110 mg/100 ml (fasting)
Lactic acid	0.6–1.8 mmol/l
Protein (total)	6.0–8.4 g/100 ml
Triglyceride	40–150 mg/100 ml
Urea nitrogen	8–25 mg/100 ml
Uric acid	3–7 mg/100 ml

Source: Reprinted by permission of *The New England Journal of Medicine*, 1980.

Table 21.2 Plasma proteins

Protein	Percentage of total	Origin	Function
Albumin	60	Liver	Gives viscosity to blood and assists in maintaining blood osmotic pressure
Globulin	36		
Alpha globulins		Liver	Transport lipids and fat-soluble vitamins
Beta globulins		Liver	Transport lipids and fat-soluble vitamins
Gamma globulins		Lymphoid tissue	Constitute antibodies of immunity
Fibrinogen	4	Liver	Assist thrombocytes in the formation of clots

Source: Van De Graaff, Kent M., *Human Anatomy.* © 1984 Wm. C. Brown Publishers, Dubuque, Iowa. All Rights Reserved. Reprinted by permission.

Figure 21.2. The structure of an erythrocyte.

Side view

2.0 microns

Top view

7.5 microns

(erythrocytes are not able to move independently). Because of their ameboid ability, leukocytes can squeeze through pores in capillary walls and get to a site of infection, whereas erythrocytes usually remain confined within blood vessels. The movement of leukocytes through capillary walls is called *diapedesis.*

Leukocytes, which are almost invisible under the microscope unless they are stained, are classified according to their stained appearance. Those leukocytes that have granules in their cytoplasm are called **granular leukocytes,** and those that do not are **agranular** (or nongranular) **leukocytes.** The granular leukocytes are also identified by their oddly shaped nuclei, which in some cases are contorted into lobes separated by thin strands. The granular leukocytes are therefore known as **polymorphonuclear** *(pol″e-mor″fo-nu′kle-ar)* **leukocytes** (abbreviated PMN).

The stain used to identify leukocytes is usually a mixture of a pink-to-red stain called eosin and a blue-to-purple stain called a "basic stain." Granular leukocytes with pink-staining granules are therefore called **eosinophils** *(e″o-sin′o-fil),* and those with blue-staining granules are called

basophils. Those with granules that have little affinity for either stain are **neutrophils.** Neutrophils are the most abundant type of leukocyte, comprising 50%–70% of the leukocytes in the blood.

There are two types of agranular leukocytes: lymphocytes and monocytes. **Lymphocytes** are usually the second most numerous type of leukocyte; they are small cells with round nuclei and little cytoplasm. **Monocytes,** in contrast, are the largest of the leukocytes and generally have kidney- or horseshoe-shaped nuclei. In addition to these two cell types, there are smaller numbers of large lymphocytes, or **plasma cells,** that may be difficult to distinguish from monocytes. Plasma cells produce and secrete large amounts of antibodies.

---

diapedesis: Gk. *dia,* through; *pedester,* on foot

Figure 21.3. Types of formed elements.

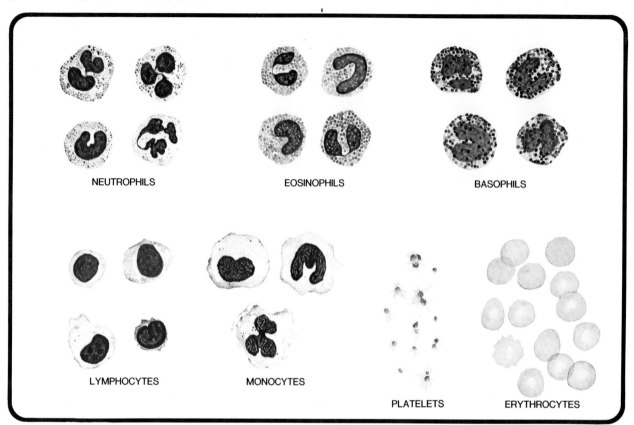

NEUTROPHILS   EOSINOPHILS   BASOPHILS

LYMPHOCYTES   MONOCYTES   PLATELETS   ERYTHROCYTES

Table 21.3 Formed elements of the blood

Component	Description	Number present	Function
Erythrocyte (red blood cell)	Biconcave disc without nucleus; contains hemoglobin; survives 100–120 days	4,000,000 to 6,000,000/mm³	Transports oxygen and carbon dioxide
Leukocytes (white blood cells)		5,000 to 10,000/mm³	Aid in defense against infections by microorganisms
Granulocytes	About twice the size of red blood cells; cytoplasmic granules present; survive 12 hours to 3 days		
1. Neutrophil	Nucleus with 2–5 lobes; cytoplasmic granules stain slightly pink	54%–62% of white cells present	Phagocytic
2. Eosinophil	Nucleus bilobed; cytoplasmic granules stain red in eosin stain	1%–3% of white cells present	Helps to detoxify foreign substances; secretes enzymes that break down clots
3. Basophil	Nucleus lobed; cytoplasmic granules stain blue in hematoxylin stain	Less than 1% of white cells present	Releases anticoagulant heparin
Agranulocytes	Cytoplasmic granules absent; survive 100–300 days		
1. Monocyte	2–3 times larger than red blood cell; nuclear shape varies from round to lobed	3%–9% of white cells present	Phagocytic
2. Lymphocyte	Only slightly larger than red blood cell; nucleus nearly fills cell	25%–33% of white cells present	Provides specific immune response (including antibodies)
Thrombocyte (platelet)	Cytoplasmic fragment; survives 5–9 days	130,000 to 360,000/mm³	Clotting

Thrombocytes. Thrombocytes, or **platelets** *(plāt′let)*, are the smallest of the formed elements and are actually fragments of large cells called **megakaryocytes** *(meg″ah-kar′e-o-sīt)*, found in bone marrow. (This is why the term *formed elements* is used rather than *blood cells* to describe erythrocytes, leukocytes, and thrombocytes). The fragments that enter the circulation as platelets lack nuclei but, like leukocytes, are capable of ameboid movement. The platelet count per cubic millimeter of blood is 130,000 to 360,000. Platelets survive about five to nine days and then are destroyed by the spleen and liver.

Platelets play an important role in blood clotting. They constitute the major portion of the mass of the clot, and phospholipids in their cell membranes serve to activate the clotting factors in plasma that result in threads of fibrin, which reinforce the platelet plug. Platelets that attach together in a blood clot also release a chemical called *serotonin (ser″o-to′nin)*, which stimulates constriction of blood vessels and thus reduces the flow of blood to the injured area.

The appearance of the formed elements of the blood is shown in figure 21.3, and a summary of the characteristics of these formed elements is presented in table 21.3.

> Blood cell counts are an important source of information in determining the health of a person. An abnormal increase in erythrocytes, for example, is termed **polycythemia** *(pol″e-si-the′me-ah)* and is indicative of several dysfunctions. An abnormally low red blood cell count is **anemia.** An elevated leukocyte count, called **leukocytosis,** is often associated with localized infection. A large number of immature leukocytes within a blood sample is diagnostic of the disease **leukemia.**

Hemopoiesis

Blood cells are constantly formed through a process called *hemopoiesis* (fig. 21.4). The term **erythropoiesis** *(ĕ-rith″ro-poi-e′sis)* refers to the formation of erythrocytes, and **leukopoiesis** to the formation of leukocytes. These processes occur in two classes of tissues. **Myeloid** *(mi′ĕ-loid)* **tissue** is the red bone marrow of the humeri, femora, ribs, sternum, pelvis, and portions of the skull that produces erythrocytes, granular leukocytes, and platelets.

Lymphoid tissue—including the lymph nodes, tonsils, spleen, and thymus—produces the agranular leukocytes (monocytes and lymphocytes). During embryonic and fetal development the hemopoietic centers are located in the yolk sac, liver, and spleen. After birth, the liver and spleen serve as sites of blood cell destruction.

Erythropoiesis is an extremely active process. It is estimated that about 2.5 million erythrocytes are produced every second in order to replace the number that are continuously destroyed by the spleen and liver. During the

destruction of erythrocytes, iron is salvaged and returned to the red bone marrow where it is used again in the formation of erythrocytes. The life span of an erythrocyte is approximately 120 days. Agranular leukocytes remain functional for 100–300 days under normal body conditions. Granular leukocytes, in contrast, have an extremely short life span of 12 hours to three days.

Hemopoiesis begins the same in both myeloid and lymphoid tissues (fig. 21.4). Undifferentiated mesenchymal-like cells develop into stem cells called **hemocytoblasts.** These stem cells are able to divide rapidly; some of the daughter cells become new stem cells (so the stem cell population is never depleted), whereas other daughter cells become specialized along different paths of blood cell formation. Hemocytoblasts, for example, may develop into **proerythroblasts** *(pro″ĕ-rith′ro-blast)* that form erythrocytes, **myeloblasts** that form granular leukocytes (neutrophils, eosinophils and basophils), **lymphoblasts** that form lymphocytes, **monoblasts** that form monocytes, or **megakaryoblasts** *(meg″ah-kar′e-o-blast)* that form platelets (thrombocytes).

> 1. List the different classes of plasma proteins, and describe their functions.
> 2. Describe erythropoiesis in terms of its location, rate, and the specific steps that lead to the production of mature erythrocytes.
> 3. Describe the different types of leukocytes, and explain where and how these different types are produced.
> 4. Explain how platelets differ from other formed elements of blood, and describe how they are produced.

Blood Clotting

Trauma to a blood vessel initiates a sequence of events that ultimately results in the formation of a clot. Another sequence of events follows the healing of the blood vessel that results in the dissolution of the clot.

Objective 5. Describe the mechanisms that lead to the formation of a platelet plug and the role of platelets in the activation of clotting factors.

Objective 6. Describe the intrinsic and extrinsic clotting pathways.

Objective 7. Explain how blood clots are dissolved.

When a blood vessel is injured, a number of physiological mechanisms are activated that promote **hemostasis** *(he-mos′tah-sis),* or the cessation of bleeding. Breakage of the endothelial lining of a vessel exposes collagen proteins from

---

hemopoiesis: Gk. *haima*, blood; *poiesis*, production
erythropoiesis: Gk. *erythros*, red; *poiesis*, production
leukopoiesis: Gk. *leukos*, white; *poiesis*, production

---

megakaryoblasts: Gk. *megas*, great; *karyon*, nut; *blastos*, germ
hemostasis: Gk. *haima*, blood; *stasis*, a standing

Figure 21.4. The processes of hemopoiesis. Formed elements begin as hemocytoblasts and differentiate into the various kinds of blood cells depending on the needs of the body.

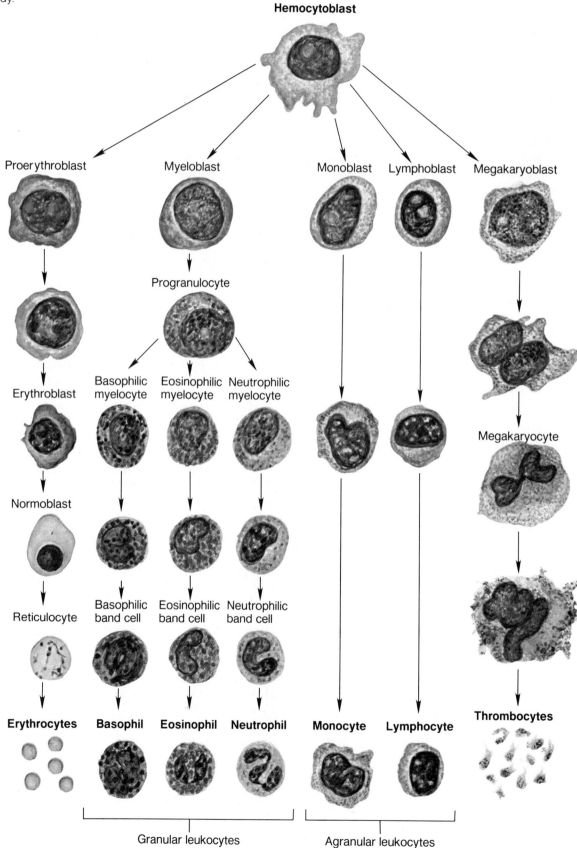

Hemocytoblast

Proerythroblast Myeloblast Monoblast Lymphoblast Megakaryoblast

Progranulocyte

Erythroblast Basophilic myelocyte Eosinophilic myelocyte Neutrophilic myelocyte

Megakaryocyte

Normoblast

Reticulocyte Basophilic band cell Eosinophilic band cell Neutrophilic band cell

Erythrocytes Basophil Eosinophil Neutrophil Monocyte Lymphocyte Thrombocytes

Granular leukocytes   Agranular leukocytes

Figure 21.5. A scanning electron micrograph showing threads of fibrin.

Figure 21.6. The sequence of events leading to platelet aggregation and the formation of a blood clot.

the subendothelial connective tissue to the blood. This initiates three separate, but overlapping, hemostatic mechanisms: (1) vasoconstriction; (2) the formation of a platelet plug; and (3) the production of a web of fibrin proteins around the platelet plug.

Functions of Platelets

In the absence of vessel damage, platelets are repelled from each other and from the endothelial lining of vessels. The repulsion of platelets from an intact endothelium is believed to be due to **prostacyclin,** a derivative of prostaglandins, produced within the endothelium. Mechanisms that prevent platelets from sticking to the blood vessels and to each other are obviously needed to prevent inappropriate blood clotting.

Damage to the endothelium of vessels exposes subendothelial connective tissue to the blood. Platelets are able to stick to exposed collagen proteins, which form the major supporting fibers of connective tissues. Platelets contain secretory granules; when platelets stick to collagen, they *degranulate* as the secretory granules release their products. These products include ADP (adenosine diphosphate), serotonin, prostaglandins, and phospholipids. This event is known as the **platelet release reaction.**

Serotonin and a prostaglandin called **thromboxane A_2** stimulate vasoconstriction, which helps to decrease blood flow to the injured vessel. Phospholipids released from platelets participate in the activation of clotting factors. The release of ADP is essential for the formation of a platelet plug.

The release of ADP from platelets that are stuck to exposed collagen makes other platelets in the vicinity "sticky," so that they adhere to those stuck to the collagen. The second layer of platelets, in turn, undergoes a

platelet release reaction, and the ADP that is secreted causes additional platelets to aggregate at the site of injury. This produces a **platelet plug** in the damaged vessel, which is strengthened by the activation of plasma-clotting factors.

> In order to undergo a release reaction, the production of prostaglandins by the platelets is required. **Aspirin** inhibits the conversion of arachidonic acid (a cyclic fatty acid) into prostaglandins and thus inhibits the release reaction and consequent formation of a platelet plug. The ingestion of excessive amounts of aspirin can thus significantly prolong bleeding time, which is why blood donors and women in the last trimester of pregnancy are advised to avoid aspirin.

Clotting Factors: Formation of Fibrin

The platelet plug is strengthened by a meshwork of insoluble protein fibers known as **fibrin** (fig. 21.5). Blood clots therefore contain platelets, fibrin, and usually trapped red blood cells that give the clot a red color (clots formed in arteries generally lack red blood cells and are grey in color). Finally, contraction of the platelets in the process of **clot retraction** forms a more compact and effective plug (fig. 21.6). Fluid squeezed from the clot as it retracts is called **serum,** which is plasma without fibrinogen (the soluble precursor of fibrin).

There are two pathways that result in the conversion of fibrinogen into fibrin. Blood left in a test tube will clot without the addition of any external chemicals; the pathway that produces this clot is thus called the **intrinsic pathway.** The intrinsic pathway also produces clots in damaged blood vessels when collagen is exposed to plasma.

Damaged tissues, however, release a chemical that initiates a "shortcut" to the formation of fibrin. Since this chemical is not part of blood, the shorter pathway is called the **extrinsic pathway.**

Intrinsic Pathway.   The intrinsic pathway is initiated by the exposure of plasma to negatively charged surfaces, such as that provided by collagen or glass. This activates a plasma protein called **factor XII** (table 21.4), which is a protein-digesting enzyme (protease). Active factor XII in turn activates another plasma protein—**factor XI**—by cleaving part of its inactive precursor. Activated factor XI, in turn, activates **factor IX.**

The next steps in the sequence require the presence of phospholipids, which are provided by platelets, and Ca^{++}. The combination of these with active factor IX and **factor VIII** forms a complex that activates **factor X.** A complex is then formed between active factor X, **factor V,** platelet phospholipids, and Ca^{++} that converts **prothrombin** (inactive **factor II**) to **thrombin.**

Thrombin is a protease that converts the soluble protein **fibrinogen (factor I)** into **fibrin** monomers. These monomers are joined together to form insoluble fibrin polymer by the action of **factor XIII.** The intrinsic clotting sequence is shown on the right side of figure 21.7.

Table 21.4   The plasma clotting factors

Factor	Name	Function	Pathway
I	Fibrinogen	converted to fibrin	common
II	Prothrombin	enzyme	common
III	Tissue thromboplastin	cofactor	extrinsic
IV	Calcium ions (Ca^{++})	cofactor	intrinsic, extrinsic, and common
V	Proaccelerin	cofactor	common
VII	Proconvertin	enzyme	extrinsic
VIII	Antihemophilic factor	cofactor	intrinsic
IX	Plasma thromboplastin component; Christmas factor	enzyme	intrinsic
X	Stuart-Prower factor	enzyme	common
XI	Plasma thromboplastin antecedent	enzyme	intrinsic
XII	Hageman factor	enzyme	intrinsic
XIII	Fibrin stabilizing factor	enzyme	common

Figure 21.7.   The extrinsic and intrinsic clotting pathways that lead to the formation of insoluble fibrin polymers.

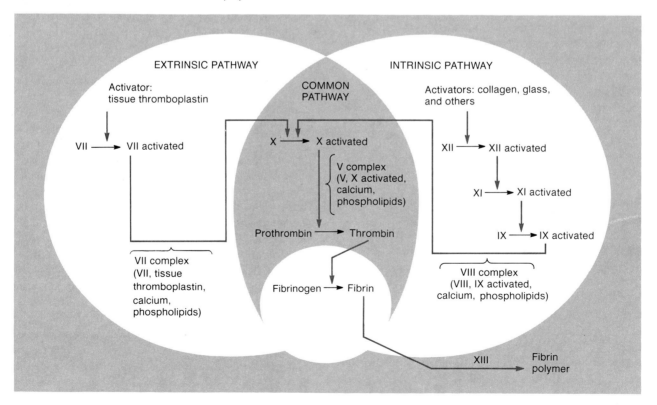

Extrinsic Pathway. The formation of fibrin can occur more rapidly as a result of the release of **tissue thromboplastin** *(throm"bo-plas'tin)*, or **factor III,** from damaged tissue cells. Tissue thromboplastin activates and combines with **factor VII;** factor VII, together with phospholipids and Ca^{++}, forms a complex (the VII complex) that activates factor X. The extrinsic clotting sequence is shown on the left side of figure 21.7.

The extrinsic and intrinsic pathways overlap at this point (see fig. 21.7) into a common pathway. The V complex, formed by active factor X, factor V, phospholipids, and Ca^{++} converts prothrombin to thrombin, which in turn converts fibrinogen to fibrin.

Dissolution of Clots

As the damaged blood vessel wall is repaired, activated factor XII promotes the conversion of another inactive molecule in plasma, **prekallikrein** to the active form called **kallikrein** *(kal"li-kre'in)*. Kallikrein, in turn, catalyzes the conversion of **plasminogen** *(plaz-min'o-jen)* into the active molecule called **plasmin.** Plasmin is an enzyme that digests fibrin into "split products," thus promoting dissolution of the clot (fig. 21.8). Since a clotting factor (factor XII) initiates the pathway leading to dissolution of the clot, the action of plasmin represents the completion of a delayed negative feedback loop.

There are a number of plasminogen activators besides kallikrein. Some of these are endogenous—activators produced by tissues and leukocytes. **Streptokinase** *(strep"to-ki'nās)* is a potent activator of plasminogen but is classified as exogenous because it is made by bacteria. Streptokinase is widely used medically to dissolve clots. It may be injected into the general circulation or injected specifically into a coronary vessel that has become occluded by a thrombus (blood clot). Streptokinase is also used in medical laboratories to measure the plasminogen levels in blood samples.

Figure 21.8. Events that produce dissolution of the blood clot and vasodilation.

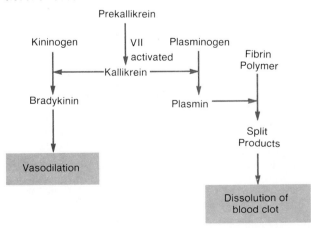

Kallikrein, in addition to its function in the formation of plasmin, also stimulates the formation of a molecule called **bradykinin** *(brad"e-ki'nin)* from its precursor **(kininogen).** Bradykinin stimulates vasodilation (fig. 21.8), which helps to counter the vasoconstrictor effects of prostaglandins and serotonin released by platelets during the formation of a clot.

1. Describe how a platelet plug is formed when a vessel is cut.
2. List the steps shared in common by the intrinsic and extrinsic clotting pathways.
3. Explain the meaning of the terms intrinsic and extrinsic in terms of the clotting pathways, and describe how these pathways differ from each other.
4. Describe the steps that lead to the formation of plasmin, and explain how this might be regarded as a negative feedback mechanism.

Clinical Considerations

Blood Cells

Anemia. Anemia is present when there is an abnormally low hemoglobin concentration and/or red blood cell count. The most common cause of this condition is a deficiency in iron **(iron-deficiency anemia),** which is an essential component of the hemoglobin molecule. In **pernicious anemia,** there is inadequate availability of vitamin B_{12}, which is needed for red blood cell production. This results from atrophy of the glandular mucosa of the stomach, which normally secretes a substance, called **intrinsic factor,** that is needed for absorption of vitamin B_{12} obtained in the diet. **Aplastic** *(ah-plas'tik)* **anemia** is anemia due to destruction of the bone marrow, which may be caused by chemicals (including benzene and arsenic) and by X rays.

Polycythemia. Polycythemia is an abnormally high red blood cell count. This may be a primary condition, of unknown etiology, or a secondary condition, in which low levels of oxygen in the blood stimulate (via secretion of a hormone called **erythropoietin** *[ĕ-rith"ro-poi'ĕ-tin]*) the marrow to produce more erythrocytes. Most commonly, polycythemia is a relative condition due to concentration of the blood caused by dehydration.

Leukocytosis. Leukocytosis is an abnormal increase in the number of leukocytes per cubic millimeter of blood, generally caused by an infection. Unlike leukemia, leukocytosis involves mature forms of leukocytes that are normally seen in blood samples. **Mononucleosis** *(mon"o-nu"kle-o'sis)* is one form of leukocytosis, in which the

bradykinin: Gk. *bradys,* slow; *kinesis,* movement

number of mononuclear (agranular) leukocytes is abnormally raised due to a viral infecton. Mononucleosis is also associated with enlarged lymph nodes and an enlarged spleen.

Leukopenia. Leukopenia is an abnormally low white blood cell count. One common cause of leukopenia is treatment with corticosteroids such as cortisone and its derivatives.

Leukemia. Leukemia is a form of malignancy in which the white blood cell count is extremely high (often a million per cubic millimeter of blood), and in which the leukocytes seen in blood are abnormal, immature forms.

Blood Clotting
Anticoagulants. Clotting of blood in test tubes can be prevented by the addition of **citrate** or **EDTA,** which chelates (binds to) calcium. By this means Ca^{++} levels in the blood that can participate in the clotting sequence are lowered, and clotting is inhibited. A mucoprotein called **heparin** can also be added to the tube to prevent clotting. Heparin activates a plasma protein called antithrombin III, which combines with and inactivates thrombin. Heparin is also given intravenously during certain medical procedures to prevent clotting. Patients may also be given **coumarins** *(koo'mah-rin)* as anticoagulants. The coumarins prevent blood clotting by competing with vitamin K, which is needed for the formation of factors II, VII, IX, and X by the liver. In contrast to the immediate effects of heparin, therefore, coumarin must be given to a patient for several days to be effective.

Hereditary Clotting Disorders. Examples of hereditary clotting disorders include two different genetic defects in factor VIII. A defect in one subunit of factor VIII prevents this factor from participating in the intrinsic clotting pathway. This genetic disease, called **hemophilia A,** is an X-linked recessive trait that is prevalent in the royal families of Europe. A defect in another subunit of factor VIII results in **von Willebrand's disease.** In this disease, rapidly circulating platelets are unable to stick to collagen and a platelet plug cannot be formed. Some acquired and inherited defects in the clotting system are summarized in table 21.5.

von Willebrand's disease: from E. A. von Willebrand, Finnish physician, 1870–1949

Table 21.5 Some acquired and inherited defects in the clotting mechanism

Category	Cause of disorder	Comments
Acquired clotting disorders	Vitamin K deficiency	Inadequate formation of prothrombin and other clotting factors in the liver
	Aspirin	Inhibits prostaglandin production, resulting in defective platelet release reaction
Anticoagulants	Coumarin	Competes with the action of vitamin K
	Heparin	Inhibits activity of thrombin
	Citrate	Combines with Ca^{++} and thus inhibits the activity of many clotting factors
Inherited clotting disorders	Hemophilia A (defective factor VIII$_{AHF}$)	Recessive trait carried on X chromosome; results in delayed formation of fibrin
	Von Willebrand's disease (defective factor VIII$_{VWF}$)	Dominant trait carried on autosomal chromosome; impaired ability of platelets to adhere to collagen in subendothelial connective tissue
	Hemophilia B (defective factor IX), also called Christmas disease	Recessive trait carried on X chromosome; results in delayed formation of fibrin

Source: Fox, Stuart Ira, *Human Physiology.* © 1984 Wm. C. Brown Publishers, Dubuque, Iowa. All Rights Reserved. Reprinted by permission.

Chapter Summary

I. Composition of the Blood
Blood consists of plasma and formed elements, including erythrocytes, leukocytes, and thrombocytes. These formed elements share a common origin but have different functions.
  A. Plasma is the fluid part of the blood.
    1. It consists of water and dissolved solutes, such as salts, small organic molecules, and proteins.
    2. The plasma proteins are divided into classes: albumin, alpha-1-globulin, alpha-2-globulin, beta globulin, and gamma globulin.
    3. Albumins provide the osmotic pressure needed to maintain blood volume; gamma globulins are antibodies.

  B. The formed elements of blood include erythrocytes, leukocytes, and thrombocytes.
    1. Erythrocytes are flattened, biconcave discs that lack nuclei but contain hemoglobin molecules and function in oxygen transport.
    2. Leukocytes contain nuclei and are subdivided into those that are mononuclear (monocytes, lymphocytes) and those that are polymorphonuclear (neutrophils, eosinophils and basophils).
    3. Neutrophils, eosinophils, and basophils are granular; monocytes and lymphocytes are agranular.

    4. Thrombocytes, or blood platelets, are fragments formed from a bone marrow cell called a megakaryocyte.
  C. The formation of blood cells is called hemopoiesis.
    1. Myeloid tissue is the red marrow that forms erythrocytes, granular leukocytes, and thrombocytes.
    2. Lymphoid tissue forms the agranular leukocytes.
    3. All hemopoiesis begins with undifferentiated stem cells called hemocytoblasts, which produce progeny cells that differentiate along different developmental lines.

II. Blood Clotting

Injury of a blood vessel results in the formation of a platelet plug reinforced by threads of fibrin. Such injury also results, after a time delay, in the formation of plasmin which dissolves the clot.

A. Platelets adhere to exposed collagen when a blood vessel is injured.
1. These platelets undergo a release reaction, in which ADP and other molecules are secreted by the stuck platelets.
2. ADP makes other platelets sticky, so that these adhere to the platelets that are stuck to collagen, forming a platelet plug.
3. Platelets also release prostaglandins, which stimulate vasoconstriction, and phospholipids, which aid in the pathways of fibrin production.

B. Fibrin is an insoluble threadlike protein formed by two overlapping pathways.
1. The intrinsic pathway begins when factor XII is activated by contact with collagen or glass.
2. The intrinsic pathway requires the activation and use of factors XII, XI, IX, VIII, calcium, and phospholipids before factor X becomes activated.
3. Active factor X forms a complex with factor II, calcium, and phospholipids that activates factor II (converts prothrombin to thrombin).
4. Thrombin converts fibrinogen to fibrin, which is then—through the action of factor XIII—formed into insoluble fibrin polymers.

5. In the extrinsic pathway, tissue thromboplastin activates factor VII, which forms a complex that activates factor X.
6. The intrinsic and extrinsic pathways overlap once factor X is activated.

C. Activated factor XII also converts prekallikrein to kallikrein, which in turn converts plasminogen into plasmin.
1. Plasmin is an enzyme that digests fibrin and thus promotes the dissolution of the blood clot.
2. Kallikrein also promotes the production of bradykinin, which stimulates vasodilation.

Review Activities

Objective Questions

1. Which of the following statements about erythrocytes is *false*?
   (a) They are produced by myeloid tissue.
   (b) Each erythrocyte lives about 120 days.
   (c) Erythrocytes have nuclei.
   (d) Some erythrocytes are produced within the skull.
2. Which of the following leukocytes is *not* polymorphonuclear?
   (a) neutrophils
   (b) monocytes
   (c) eosinophils
   (d) basophils
3. The most common leukocyte is the
   (a) neutrophil
   (b) monocyte
   (c) eosinophil
   (d) basophil
4. Which of the following is a cell fragment rather than a complete cell?
   (a) erythrocyte
   (b) thrombocyte
   (c) granular leukocyte
   (d) agranular leukocyte
5. Which of the following is *not* formed in bone marrow?
   (a) erythrocytes
   (b) thrombocytes
   (c) granular leukocytes
   (d) agranular leukocytes

6. The activation of factor X is
   (a) part of the intrinsic pathway only
   (b) part of the extrinsic pathway only
   (c) part of both the intrinsic and extrinsic pathways
   (d) not part of either the intrinsic or extrinsic pathways
7. Platelets
   (a) form a plug by sticking to each other
   (b) release chemicals that stimulate vasoconstriction
   (c) provide phospholipids needed for the intrinsic pathway
   (d) all of the above
8. Which of the following statements about collagen is *false*?
   (a) Platelets stick to collagen.
   (b) Collagen is not normally exposed to the blood.
   (c) Collagen can activate factor XII.
   (d) Blood will not clot unless it is exposed to collagen.
9. Plasmin
   (a) is formed after kallikrein is produced
   (b) digests fibrin
   (c) is produced when streptokinase is injected into the blood
   (d) all of the above
10. The ability of platelets to stick together is promoted by
    (a) ATP
    (b) ADP
    (c) cyclic AMP
    (d) adenosine

11. A defect in factor VIII could prevent
    (a) the intrinsic pathway of fibrin formation
    (b) the extrinsic pathway of fibrin formation
    (c) the ability of platelets to stick together
    (d) both (a) and (b)
    (e) both (a) and (c)
12. Vitamin K is needed in blood clotting because it
    (a) is required by the liver for the production of several clotting factors
    (b) is needed for the activation of clotting factors in the plasma
    (c) is required for the platelet release reaction
    (d) converts fibrin monomers into fibrin polymers

Essay Questions

1. Describe the locations and functions of myeloid and lymphoid hemopoietic tissues.
2. Define the term *stem cell,* and explain the role of stem cells in hemopoiesis.
3. Compare and contrast the pathways that lead to the production of erythrocytes, granular leukocytes, agranular leukocytes, and thrombocytes.
4. Describe the roles of platelets in blood clotting.
5. Explain how a cut in the skin initiates both the intrinsic and extrinsic clotting pathways. Which pathway finishes first? Why?

Selected Readings

Arkel, Y. S. January/February 1982. Shedding light on platelet dysfunction. *Diagnostic Med.,* p. 27.

Bleich, H. L., and E. S. Boro. 1975. Emerging concepts of platelet function. *New Eng. J. Med.* 293:970.

Statland, B. E., and R. K. Ito. May/June 1984. How cancer upsets hemostasis. *Diagnostic Med.,* p. 75.

Till, J. E. 1981. Cellular Diversity in the blood-forming system. *American Scientist* 69:522.

Weiss, H. J. 1975. Platelet physiology and abnormalities of platelet function. *New Eng. J. Med.* 293: first part, p. 531; second part, p. 580.

Zucker, M. B. June 1980. The function of blood platelets. *Scientific American.*

22

Immune System

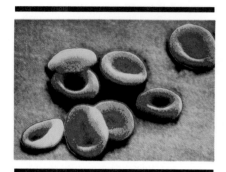

Outline

Defense Mechanisms
Nonspecific Immunity
Specific Immunity
Lymphocytes
Humoral Immunity
Antibodies
The Complement System
Local Inflammation
Cell-Mediated Immunity
Actions of T Lymphocytes
Thymus Gland
Immunizations
Active Immunity and the Clonal Selection
    Theory
Passive Immunity
Monoclonal Antibodies
Tolerance and Non-Self-Recognition
Tolerance
Histocompatibility Antigens
Red Blood Cell Antigens and Blood Typing
Tumor Immunology
Clinical Considerations
Autoimmunity
Immune Complex Diseases
Allergy
Chapter Summary
Review Activities
Selected Readings

Concepts

The immune system includes all of the
    structures and processes that provide
    a defense against potential
    pathogens. These defenses can be
    grouped into nonspecific and specific
    categories.

The structure of antibodies allows them
    to combine with antigens in a specific
    fashion and, by this means, to
    activate other elements of the
    immune system to protect against
    pathogens.

Killer T lymphocytes, acting through
    the secretion of lymphokines, provide
    cell-mediated immunity.
    Subpopulations of helper and
    suppressor T lymphocytes regulate
    the immune response.

Resistance to pathogens can be gained
    by both active and passive
    immunizations. The change from a
    primary to a secondary response in
    active immunity is best explained by
    the clonal selection theory.

The immune system normally can
    tolerate self-antigens while attacking
    non-self-antigens. A system of
    histocompatibility cell antigens
    identifies the cells of each person.

Defense Mechanisms

The immune system includes all of the structures and processes that provide a defense against potential pathogens. These defenses can be grouped into non-specific and specific categories.

Objective 1. Describe some of the mechanisms involved in nonspecific immunity.

Objective 2. Explain what an antigen is, and define the terms *hapten* and *antigenic determinant site.*

Objective 3. Describe the origin and functions of B and T lymphocytes.

Objective 4. Distinguish between humoral and cell-mediated immunity.

Nonspecific defense mechanisms are inherited as part of the structure of each organism. Epithelial membranes that cover the body surfaces, for example, restrict infection by most pathogens. The strong acidity of gastric juice (pH 1–2) also helps to kill many microorganisms before they can invade the body. These external defenses are backed by internal defenses, such as phagocytosis *(fag''o-si-to'sis),* that function in both a specific and nonspecific manner (table 22.1).

Each individual can acquire the ability to defend against specific pathogens by a prior exposure to those pathogens. This specific immune response is a function of lymphocytes. Internal specific and nonspecific defense mechanisms function together to combat infection, with lymphocytes interacting in a coordinated effort with phagocytic cells (listed in table 22.2).

Table 22.1 Structures and defense mechanisms of nonspecific immunity

	Structure	Mechanisms
External	Skin	Anatomic barrier to penetration by pathogens; secretions have lysozyme (enzyme that destroys bacteria)
	Digestive tract	High acidity of stomach Protection by normal bacterial population of colon
	Respiratory tract	Secretion of mucus; movement of mucus by cilia; alveolar macrophages
	Genitourinary tract	Acidity of urine Vaginal lactic acid
Internal	Phagocytic cells	Ingest and destroy bacteria, cellular debris, denatured proteins, and toxins
	Interferons	Inhibit replication of viruses
	Complement proteins	Promote destruction of bacteria and other effects of inflammation
	Endogenous pyrogen	Secreted by leukocytes and other cells; produces fever

Source: Fox, Stuart Ira, *Human Physiology.* © 1984 Wm. C. Brown Publishers, Dubuque, Iowa. All Rights Reserved. Reprinted by permission.

---

pathogen: Gk. *pathema,* suffering; *gen,* to produce

Table **22.2** Phagocytic cells and their location	
Phagocyte	**Location**
Neutrophils	Blood and all tissues
Monocytes	Blood and all tissues
Tissue macrophages (histiocytes)	All tissues (including spleen, lymph nodes, bone marrow)
Kupffer cells	Liver
Alveolar macrophages	Lungs
Microglia	Central nervous system

Source: Fox, Stuart Ira, *Human Physiology.* © 1984 Wm. C. Brown Publishers, Dubuque, Iowa. All Rights Reserved. Reprinted by permission.

Nonspecific Immunity

Invading pathogens that have crossed epithelial barriers enter connective tissues. These invaders—or chemicals called *toxins* secreted from them—may enter blood or lymphatic capillaries and be carried to other areas of the body. The invasion and spread of infection is fought in two stages: (1) nonspecific immunological defenses are employed; if these are sufficiently effective the pathogens may be destroyed without progression to the next step; (2) lymphocytes may be recruited and their specific actions used to reinforce the previously nonspecific immune defenses.

Phagocytosis. There are two major groups of phagocytic cells: (1) neutrophils; and (2) the cells of the *mononuclear phagocyte system*. This latter category includes *monocytes* in the blood, *macrophages* (derived from monocytes) in the connective tissues, and *organ-specific phagocytes* in the liver, spleen, lymph nodes, lungs, and brain (table 22.2).

Connective tissues contain a resident population of all leukocyte types. Neutrophils and monocytes in particular can be highly mobile within connective tissues as they scavenge for invaders and cellular debris. These leukocytes are recruited to the site of an infection by a process known as **chemotaxis** *(ke''mo-tak'sis)*—movement toward chemical attractants. Neutrophils are the first to arrive at the site of an infection; monocytes arrive later and can be transformed into macrophages as the battle progresses.

If the infection is sufficiently large, new phagocytic cells from the blood may join those already in the connective tissue. These new neutrophils and monocytes are able to squeeze through the tiny gaps between adjacent endothelial cells in the capillary wall and enter the connective tissues. This process is called **diapedesis** *(di''ah-pĕ'de'sis)*, which is illustrated in figure 22.1.

toxin: Gk. *toxikon*, poison
phagocytosis: Gk. *phagein*, to eat; *kytos*, a hollow (cell)
chemotaxis: Gk. *chemeia*, alchemy; *taxis*, orderly arrangement
diapedesis: Gk. *dia*, through; *pedesis*, a leaping

Figure 22.1. Diapedesis. White blood cells squeeze through openings between capillary endothelial cells to enter underlying connective tissues.

Blood capillary

Connective tissue   Schenk

Phagocytic cells engulf particles in a manner similar to the way an amoba eats. The particle becomes surrounded by cytoplasmic extensions called pseudopods, which ultimately fuse together. The particle thus becomes surrounded by a membrane derived from the plasma membrane (fig. 22.2).

The ingested particle thus becomes contained within an organelle analogous to a food vacuole in an amoeba. This vacuole then fuses with lysosomes (organelles that contain digestive enzymes), so that the ingested particle and the digestive enzymes remain separated from the cytoplasm by a continuous membrane. Often, however, lysosomal enzymes are released before the food vacuole has completely formed. When this occurs, free lysosomal enzymes may be released into the infected area and contribute to inflammation.

The *Kupffer cells* in the liver, together with phagocytic cells in the spleen and lymph nodes, are **fixed phagocytes.** This term refers to the fact that these cells are immobile ("fixed") in the channels within these organs. As blood flows through the liver and spleen and as lymph percolates through the lymph nodes, foreign chemicals and debris are removed by phagocytosis and chemically inactivated within the phagocytic cells. Invading pathogens are very effectively removed in this manner, so that blood is usually sterile after a few passes through the liver and spleen.

Fever. Fever may be a component of the nonspecific defense system. Body temperature is regulated by the hypothalamus, which contains a thermoregulatory control center (a "thermostat") that coordinates skeletal muscle shivering and the activity of the sympathoadrenal system

pseudopod: Gk. *pseudes*, false; *pous*, foot
Kupffer cell: from Karl W. Kupffer, German anatomist, 1829–1902

Figure 22.2. Phagocytosis by a neutrophil or macrophage. A phagocytic cell extends its pseudopods around the object to be engulfed (such as a bacterium). Dots represent lysosomal enzymes (L = lysosomes). If the pseudopods fuse to form a complete food vacuole (1), lysosomal enzymes are restricted to the organelle formed by the lysosome and food vacuole. If the lysosome fuses with the vacuole before fusion of the pseudopods is complete (2), lysosomal enzymes are released into the infected area of tissue.

to maintain body temperature at about 37° C. This thermostat is reset upwards in response to a chemical called **endogenous pyrogen** *(en-doj'ĕ-nus pi'ro-jen),* secreted by leukocytes. Endogenous pyrogen secretion is stimulated by a chemical called *endotoxin,* which is released by certain bacteria.

Although high fevers are definitely dangerous, many believe that a mild to moderate fever may be a beneficial response that aids recovery from bacterial infections. There is some evidence to support this view, but the mechanisms involved are not clearly understood. One theory is that elevated body temperature may interfere with the nutritional requirements of some bacteria.

Interferons. In 1957, researchers demonstrated that cells infected with a virus produced polypeptides that "interfered with" the ability of a second, unrelated strain of virus to infect other cells in the same culture. These **interferons** *(in''ter-fēr'on)* as they were called, thus produced a nonspecific, short-acting resistance to viral infection. This discovery produced a great deal of excitement, but further research in this area was hindered by the fact that human interferons could only be obtained in very small quantities

and that animal interferons had little effect in humans. In 1980, however, technological breakthroughs allowed researchers to introduce human interferon genes into bacteria—through a technique called *genetic recombination*—so that bacteria could act as interferon factories.

Leukocytes, fibroblasts, and probably many other cells make their own characteristic types of interferons. These polypeptides act as messengers that protect other cells in the vicinity from viral infection. The viruses are still able to penetrate these other cells, but the ability of the viruses to replicate and assemble new virus particles (fig. 22.3) is inhibited.

Lymphocytes called *T lymphocytes* release interferons in response to viral infections and perhaps as part of their immunological surveillance against cancer. Interferons may destroy cancer cells directly and indirectly by the activation of T lymphocytes and *natural killer cells.* Some of the proposed effects of interferons are summarized in table 22.3.

Recent clinical trials have suggested that interferons may be effective in the treatment of such viral infections as viral hepatitis and herpes-induced "cold sores." Current evidence suggests that interferons may not prove to

---

interferon: L. *inter,* between; *ferio,* to strike

Figure 22.3. The sequence of events that can occur when human cells are infected with virus particles.

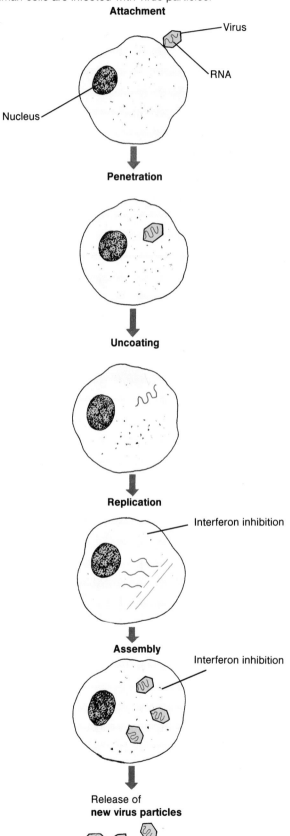

Table **22.3** Some of the proposed effects of interferons

Stimulation	Inhibition
Macrophage phagocytosis	Cell division
Activity of cytotoxic ("killer") T cells	Tumor growth
Activity of natural killer cells	Maturation of adipose cells
Production of antibodies	Maturation of erythrocytes

be a "magic bullet" against cancer but may, when combined with other methods, significantly augment the treatment of some forms of cancer. The emphasis is on treatment rather than prevention here because the protective effects of interferons are of short duration.

Specific Immunity

In 1890, von Behring demonstrated that a guinea pig which had been previously injected with a sublethal dose of diphtheria toxin could survive subsequent injections of otherwise lethal doses of that toxin. Further, von Behring showed that this immunity could be transferred to a second, nonexposed animal by injections of serum from the immunized guinea pig. He concluded that the immunized animal had chemicals in its serum—which he called **antibodies**—that were responsible for the immunity. He also showed that these antibodies conferred immunity only to subsequent diphtheria infections; the antibodies were *specific* in their actions. It was later learned that antibodies are *proteins* produced by a particular type of lymphocyte.

Antigens. Antigens are molecules that stimulate antibody production and combine with specific antibodies. Most antigens are large molecules (such as proteins) with a molecular weight greater than about 10,000, and they are foreign to the blood and other body fluids (although there are exceptions to both descriptions). The ability of a molecule to function as an antigen depends not only on its size but also on the complexity of its structure. Proteins are therefore more antigenic than polysaccharides, which have a simpler structure. Plastics used in artificial implants are composed of large molecules but are not very antigenic because of their simple, repeating structures.

A large, complex foreign molecule can have a number of different **antigenic** *(an-ti-jen'ik)* **determinant sites,** which are areas of the molecule that stimulate production of and combine with different antibodies. Most naturally occurring antigens have many antigenic determinant sites and stimulate the production of different antibodies with specificities for these sites.

Haptens. Many small organic molecules are not antigenic by themselves but can become antigens if they bond to proteins (and thus become antigenic determinant sites on the proteins). This was discovered by Karl Landsteiner, the same man who discovered the ABO blood groups. By bonding these small molecules—which Landsteiner called **haptens** *(hap'ten)*—to proteins in the laboratory, new antigens could be created for research or diagnostic purposes. The bonding of foreign haptens to a person's own proteins can also occur in the body; by this means, derivatives of penicillin, for example, that would otherwise be harmless can produce fatal allergic reactions in susceptible people.

Immunoassays. When the antigen or antibody is attached to the surface of a cell or to particles of latex rubber (in commercial diagnostic tests), the antigen-antibody reaction becomes visible because the particles *agglutinate* (clump) as a result of antigen-antibody bonding (fig. 22.4). These agglutinated particles can be used to assay a variety of antigens, and tests that utilize this procedure are called *immunoassays (im"u-no-as'sa)*. Blood typing and modern pregnancy tests are examples of such immunoassays.

Lymphocytes

Leukocytes, erythrocytes, and blood platelets are all ultimately derived from ("stem from") unspecialized cells in the bone marrow. These *stem cells* produce the specialized blood cells, and they replace themselves by cell division so that the stem cell population is not exhausted. Lymphocytes produced in this manner seed the thymus, spleen, and lymph nodes, producing self-replacing lymphocyte colonies in these organs.

The lymphocytes that become seeded in the thymus become **T lymphocytes.** These cells have surface characteristics and an immunological function that is different from other lymphocytes. The thymus, in turn, seeds other organs; about 65% to 85% of the lymphocytes in blood and most of the lymphocytes in lymph nodes are T lymphocytes. T lymphocytes therefore come from or had an ancestor that came from the thymus gland.

Most of the lymphocytes that are not T lymphocytes are called **B lymphocytes.** The letter *B* is derived from immunological research performed in chickens. Chickens have an organ called the *bursa of Fabricius* at the hind end of their digestive tracts, which is the equivalent of the thymus located at the anterior end. The thymus processes T lymphocytes; the bursa, in a chicken, processes B lymphocytes. Since mammals do not have a bursa and since the site of the processing of B lymphocytes in mammals is not known, the *B* is often translated as the "bursa equivalent" for humans and other mammals.

Both B and T lymphocytes function in specific immunity. The B lymphocytes combat bacterial and some viral infections by secreting antibodies into the blood and

Karl Landsteiner: Austrian pathologist and American immunologist, 1868–1943

bursa of Fabricius: from Hieronymus Fabricius, Italian anatomist, 1533–1619

Figure 22.4. Immunoassay using the agglutination technique. Antibodies against a particular antigen are adsorbed to red blood cells or latex particles. When these particles are mixed with a solution that contains the appropriate antigen, the formation of the antigen-antibody complexes produces clumping (agglutination) of the particles that can be seen with the unaided eye.

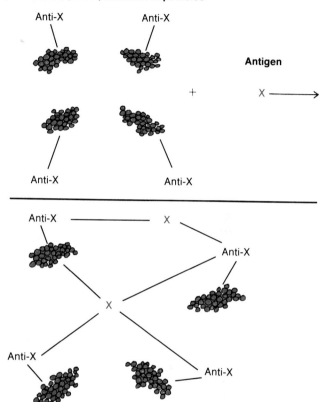

lymph. They are therefore said to provide **humoral immunity.** T lymphocytes attack host cells that have become infected with viruses or fungi, transplanted human cells, and cancerous cells. The T lymphocytes do not secrete antibodies; they must come into close proximity or have actual physical contact with the victim cell in order to destroy it. T lymphocytes are therefore said to provide **cell-mediated immunity** (table 22.4).

1. List the phagocytic cells in blood and lymph, and indicate which organs contain fixed phagocytes.
2. Describe the actions of interferons.
3. List four properties that antigens usually have. Explain why proteins are more antigenic than polysaccharides.
4. Describe the meaning of the term hapten, *and give an example.*
5. Distinguish between B and T lymphocytes in terms of their origins and immune functions.

humor: L. *umor,* liquid

Table 22.4 Comparison of B and T lymphocytes

Characteristic	B lymphocyte	T lymphocyte
Site where processed	Bursa of Fabricius or equivalent in mammals	Thymus
Type of immunity	Humoral (secretes antibodies)	Cell-mediated
Subpopulations	Memory cells and plasma cells	Cytotoxic (killer) T cells, helper cells, suppressor cells
Presence of surface antibodies	Yes—IgM or IgD	Not detectable
Receptors for antigens	Present—are surface antibodies	Present—nature unknown
Life span	Short	Long
Tissue distribution	High in spleen, low in blood	High in blood and lymph
Percent of blood lymphocytes	10%–15%	75%–80%
Transformed by antigens to	Plasma cells	Small lymphocytes
Secretory product	Antibodies	Lymphokines
Immunity to viral infections	Enteroviruses, poliomyelitis	Most others
Immunity to bacterial infections	Streptococcus, staphylococcus, many others	Tuberculosis, leprosy
Immunity to fungal infections	None known	Many
Immunity to parasitic infections	Trypanosomiasis, maybe to malaria	Most others

Source: Fox, Stuart Ira, *Human Physiology.* © 1984 Wm. C. Brown Publishers, Dubuque, Iowa. All Rights Reserved. Reprinted by permission.

Figure 22.5. B lymphocytes have antibodies on their surface that function as receptors for specific antigens. The interaction of antigens and antibodies on the surface stimulates cell division and the maturation of the B cell progeny into memory cells and plasma cells. Plasma cells produce and secrete large amounts of the antibody (note the extensive rough endoplasmic reticulum in these cells).

Humoral Immunity

The structure of antibodies allows them to combine with only very specific antigens. The formation of antigen-antibody complexes activates other elements of the immune system and results in more effective defense against pathogens.

> *Objective 5.* Describe the structure and origin of antibodies.
>
> *Objective 6.* Explain how antibodies promote the destruction of invading pathogens.
>
> *Objective 7.* Describe the complement system and its function.
>
> *Objective 8.* Describe the events that occur in a local inflammation.

Exposure of a B lymphocyte to the appropriate antigens results in cell growth followed by many cell divisions. Some of the progeny become **memory cells,** which are indistinguishable from the original cell; others are transformed into **plasma cells** (fig. 22.5). Plasma cells are protein factories that produce about two thousand antibody proteins per second in their brief (five to seven day) life span.

The antibodies that are produced by plasma cells when B lymphocytes are exposed to a particular antigen react specifically with that antigen. Such antigens may be isolated molecules, or they may be molecules at the surface of an invading foreign cell. The specific bonding of antibodies to antigens serves to identify the enemy and to activate defense mechanisms that lead to the invader's destruction.

Antibodies

Antibody proteins are also known as **immunoglobulins** *(im''u-no-glob'u-lin).* These are found in the gamma globulin class of plasma proteins, as identified by a technique called *electrophoresis (e-lek''tro-fo-re'sis),* in which classes of plasma proteins are separated by their movement in an electric field (fig. 22.6). With this technique, five distinct bands of proteins appear: albumin, alpha-1-globulin, alpha-2-globulin, beta globulin, and gamma globulin.

Figure 22.6. The separation of serum protein by electrophoresis. A = albumin, alpha–1 (α_1), alpha–2 (α_2), beta (β), and gamma (γ) globulin.

Table **22.5** The immunoglobulins	
Immunoglobulin	**Examples of functions**
IgG	Main form of antibodies in circulation: production increased after immunization
IgA	Main antibody type in external secretions, such as saliva and mother's milk
IgE	Responsible for allergic symptoms in immediate hypersensitivity reactions
IgM	Function as antigen receptors on lymphocyte surface prior to immunization; secreted during primary response
IgD	Function as antigen receptors on lymphocyte surface prior to immunization; other functions unknown

Source: Fox, Stuart Ira, *Human Physiology.* © 1984 Wm. C. Brown Publishers, Dubuque, Iowa. All Rights Reserved. Reprinted by permission.

The gamma globulin band is wide and diffuse because it represents a heterogenous class of molecules. Since antibodies are specific in their actions, it could be predicted that different types of antibodies have different structures. An antibody against smallpox, for example, does not confer immunity to poliomyelitis and, therefore, must have a slightly different structure than an antibody against polio. Despite these differences, antibodies are structurally related and form only a few subclasses.

There are five subclasses of immunoglobulins (abbreviated Ig): *IgG, IgA, IgM, IgD,* and *IgE.* Most of the antibodies in serum are in the IgG subclass, whereas most of the antibodies in external secretions (saliva and milk) are IgA (table 22.5). Antibodies in the IgE subclass are involved in allergic reactions.

Antibody Structure. All antibody molecules consist of four interconnected polypeptide chains. Two longer, higher molecular weight chains (the *H chains*) are joined to two shorter, lighter *L chains.* Research has shown that these four chains are arranged in the form of a Y. The stalk of the Y has been called the "crystallizable fragment" (abbreviated F_c), whereas the top of the Y is the "antigen-binding fragment" (F_{ab}). This is shown in figure 22.7.

The amino acid sequences of some antibodies have been determined using antibodies derived from people with multiple myelomas. These lymphocyte tumors are derived from the division of a single B lymphocyte, forming a population of genetically identical cells that secrete identical antibodies. These populations and the antibodies they secrete are different, however, in different patients. Analyses of these antibodies have shown that the F_c regions of different antibodies are the same (are *constant*), whereas the F_{ab} regions are *variable.* Variability of the antigen-binding regions is required for the specificity of antibodies for antigens.

Each immunoglobulin subclass is characterized by its own unique polypeptide in the constant region of the heavy chains. Substituting one of these polypeptides for another, therefore, changes the immunoglobulin subclass. Before a B lymphocyte is stimulated by an antigen, for example, it

Figure 22.7. Antibodies are composed of four polypeptide chains—two are heavy (*H*) and two are light (*L*). In (*a*) these chains are shown as sticks; in (*b*) different regions within these chains are shown separately. The variable regions are abbreviated V, and the constant regions are abbreviated C. Antigens combine with the variable regions. Each antibody molecule is divided into an F_{ab} (antigen-binding) fragment and an F_c (crystallizable) fragment.

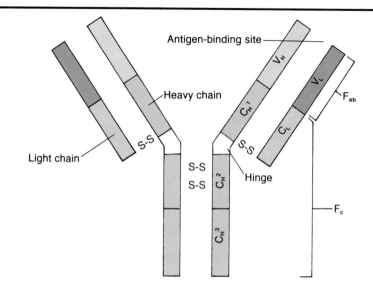

has IgM antibodies on its surface that function as *receptors* for the antigen. After stimulation by antigens, the B lymphocyte becomes transformed into a plasma cell that secretes IgG antibodies instead of IgM antibodies. This replacement involves a change only in the constant part of the heavy chains; since the antigen-binding parts are unaltered, the antibody specificity remains the same when IgM is replaced by IgG.

Diversity of Antibodies. It is estimated that there are about 100 million trillion (10^{20}) antibody molecules in each individual, representing a few million different specificities for different antigens. Considering that antibodies to particular antigens can cross-react to some degree with closely related antigens, this tremendous antibody diversity usually insures that there are some antibodies that can combine with any antigen the person may encounter. These observations evoke a question that has long fasci-

nated scientists: How can a few million different antibodies be produced? A person cannot possibly inherit a correspondingly large number of genes devoted to antibody production.

Two mechanisms have been proposed to explain antibody diversity. First, since different combinations of heavy and light chains can produce different antibody specificities, a person does not have to inherit a million different genes to code for a million different antibodies. If a few hundred genes code for different H chains and a few hundred code for different L chains, different combinations of these polypeptide chains could produce millions of different antibodies. Second, the diversity of antibodies could increase during development if, when some lymphocytes divided, the progeny received antibody genes that have been slightly altered by mutations. This is called *somatic mutation* since the mutation occurs in a body cell rather than in a sperm or ovum. The diversity of antibodies would thus increase as the lymphocyte population increases.

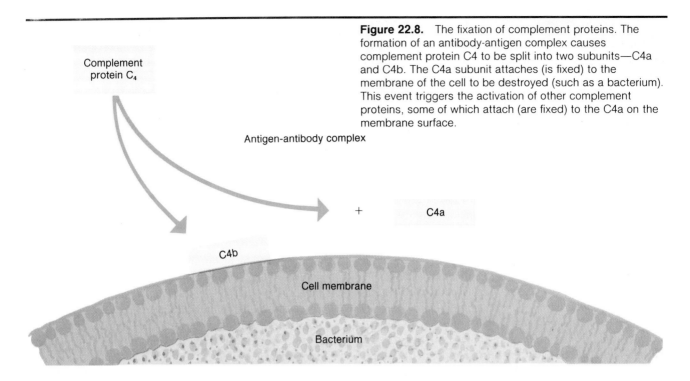

Figure 22.8. The fixation of complement proteins. The formation of an antibody-antigen complex causes complement protein C4 to be split into two subunits—C4a and C4b. The C4a subunit attaches (is fixed) to the membrane of the cell to be destroyed (such as a bacterium). This event triggers the activation of other complement proteins, some of which attach (are fixed) to the C4a on the membrane surface.

Opsonization. The combination of antibodies with antigens does not itself produce destruction of the antigens or of the pathogenic organisms that contain these antigens. Antibodies, rather, serve to identify the targets for immunological attack and to activate nonspecific immune processes that destroy the invader. Bacteria that are buttered with antibodies, for example, are better targets for phagocytosis by neutrophils and macrophages. The ability of antibodies to stimulate phagocytosis is termed **opsonization** *(op''so-ni-za'shun).* Immune destruction of bacteria is also promoted by antibody-induced activation of a system of serum proteins known as *complement.*

The Complement System

It was learned in the early part of the twentieth century that rabbit antibodies to sheep red blood cell antigens could not lyse (destroy) these cells unless certain protein components of serum were present. These proteins, called **complement,** are a nonspecific defense system that is directed against specific invaders that have been identified by antibodies.

There are eleven complement proteins, designated C1 (which has three protein components) through C9. These proteins are present in an inactive state within plasma and other body fluids and become activated by the attachment of antibodies to antigens. In terms of their functions, the complement proteins can be subdivided into three components: (1) recognition (C1); (2) activation (C4, C2, and C3, in that order); and (3) attack (C5–C9). The attack phase consists of **complement fixation,** in which complement proteins attach to the cell membrane and destroy the victim cell.

Antibodies of the IgG and IgM subclasses attach to antigens on the invading cell's membrane, bond to C1, and by this means activate its enzyme activity. Activated C1

Figure 22.9. Complement proteins C5 through C9 (illustrated as a doughnut-shaped ring) puncture the membrane of the cell to which they are attached (fixed). This aids destruction of the cell.

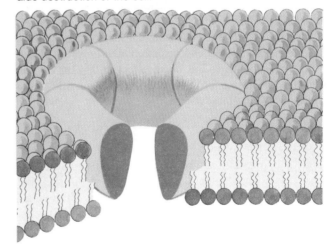

catalyzes the hydrolysis of C4 into two fragments (fig. 22.8), designated C4a and C4b. The C4b fragment bonds to the cell membrane (is "fixed") and becomes an active enzyme that splits C2 into two fragments, C2a and C2b. The C2a becomes attached to C4b and cleaves C3 into C3a and C3b. Fragment C3b becomes attached to the growing complex of complement proteins on the cell membrane. The C3b converts C5 to C5a and C5b. The C5b, and eventually C6 through C9, become fixed to the cell membrane.

Complement proteins C5 through C9 create large pores in the membrane (fig. 22.9). These pores allow the osmotic influx of water, so that the victim cell swells and bursts. Notice that the complement proteins, not the antibodies directly, kill the cell; antibodies only serve as

Table 22.6 Summary of events that occur in a local inflammation when a break in the skin permits entry of bacteria

Category	Events
Nonspecific immunity	Bacteria enter through break in anatomic barrier of skin.
	Resident phagocytic cells—neutrophils and macrophages—engulf bacteria.
	Nonspecific activation of complement proteins occurs.
Specific immunity	B cells are stimulated to produce specific antibodies.
	Phagocytosis is enhanced by antibodies attached to bacterial surface antigens.
	Specific activation of complement proteins occurs, which stimulates phagocytosis, chemotaxis of new phagocytes to the infected area, and secretion of histamine from tissue mast cells.
	Diapedesis allows new phagocytic leukocytes (neutrophils and monocytes) to invade the infected area.
	Vasodilation and increased capillary permeability (as a result of histamine secretion) produce redness and edema.

Source: Fox, Stuart Ira, *Human Physiology*. © 1984 Wm. C. Brown Publishers, Dubuque, Iowa. All Rights Reserved. Reprinted by permission.

activators of this process. Other molecules can also activate the complement system in an alternate nonspecific pathway that bypasses the early phases of the specific pathway described here.

Complement fragments that are not fixed but instead are liberated into the surrounding fluid have a number of effects. These effects include (1) *chemotaxis*—the liberated complement fragments attract phagocytic cells to the site of complement activation; (2) *opsonization*—phagocytic cells have receptors for C3b, so that this fragment may form bridges between the phagocyte and the victim cell that facilitates phagocytosis; and (3) fragments C3a and C5a *stimulate release of histamine (his'tah-mēn)* from mast cells (a connective tissue cell type) and basophils. As a result of histamine release, there is increased blood flow to the infected area due to vasodilation and increased capillary permeability. The latter effect can result in the leakage of plasma proteins into the surrounding tissue fluid, producing local edema.

Local Inflammation

Aspects of the nonspecific and specific immune responses and their interactions are well illustrated by the events that occur when bacteria enter a break in the skin and produce a local inflammation (table 22.6). The inflammatory reaction is initiated by the nonspecific mechanisms of phagocytosis and complement activation. Activated complement further increases this nonspecific response by attracting new phagocytes to the area and by increasing the activity of these phagocytic cells.

After some time, B lymphocytes are stimulated to produce antibodies against specific antigens that are part of the invading bacteria. Attachment of these antibodies to antigens in the bacteria greatly increases the previously nonspecific response. This occurs because of greater activation of complement, which directly destroys the bacteria and which—together with the antibodies themselves—promotes the phagocytic activity of neutrophils, macrophages, and monocytes (fig. 22.10).

As inflammation progresses, the release of lysosomal enzymes from macrophages causes the destruction of leukocytes and other tissue cells. These effects, together with

those produced by histamine and other chemicals released from mast cells, produce the characteristic symptoms of a local inflammation: *redness and warmth* (due to vasodilation); *swelling* (edema); and *pus* (the accumulation of dead leukocytes). If the infection continues, the release of endogenous pyrogen from leukocytes and macrophages may produce a fever.

---

1. Illustrate the structure of an antibody molecule. Label the constant and variable regions, the F_c and F_{ab} parts, and the heavy and light chains.
2. Define opsonization, and name two types of molecules that promote this process.
3. Describe complement fixation, and explain the roles of complement fragments that do not become fixed.
4. Explain how nonspecific and specific immune mechanisms cooperate during a local inflammation.

Cell-Mediated Immunity

Killer T lymphocytes, acting through the secretion of lymphokines, provide cell-mediated immunity. The helper and suppressor subpopulations of T lymphocytes help regulate both cell-mediated and humoral immunity.

Objective 9. Describe the function of killer T lymphocytes in cell-mediated immunity and the composition and actions of lymphokines.

Objective 10. Describe the roles of macrophages in aiding specific immunity.

Objective 11. Explain the functions of helper and suppressor lymphocytes.

Objective 12. Explain the role of the thymus gland in cell-mediated immunity.

The T lymphocytes provide specific immune protection without secreting antibodies. Unlike the B lymphocytes that can kill at a distance by secreting antibodies, T lymphocytes must be near or in actual contact with their victims. This cell-mediated immunity is provided by **killer T lymphocytes,** which defend against viral and fungal infections, as well as some bacterial infections. Killer T

Figure 22.10. The entry of bacteria through a cut in the skin produces a local inflammatory reaction. In this reaction, antigens on the bacterial surface are coated with antibodies and ingested by phagocytic cells. Symptoms of inflammation are produced by the release of lysosomal enzymes and by the secretion of histamine and other chemicals from tissue mast cells.

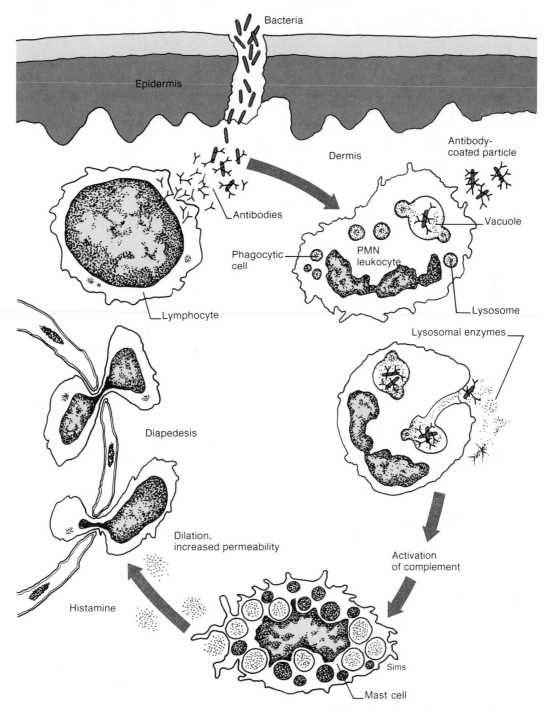

lymphocytes are also responsible for transplant rejection and for immunological surveillance against cancer. Two other subpopulations of T cells, called **helper** and **suppressor T lymphocytes,** don't attack invaders directly but rather help regulate the specific immune system.

Although most bacterial infections are fought by B lymphocytes, as previously described, some are the targets of cell-mediated attack. This is the case with the *tubercle (tu'ber-k'l) bacilli* that cause tuberculosis. Injections

Figure 22.11. The clonal selection theory. In this example, particular antigens interact with specific membrane receptors on the T cell surface. As a result of this selection process, the lymphocyte that is specific for this antigen proliferates to produce a population of genetically identical cells (a clone).

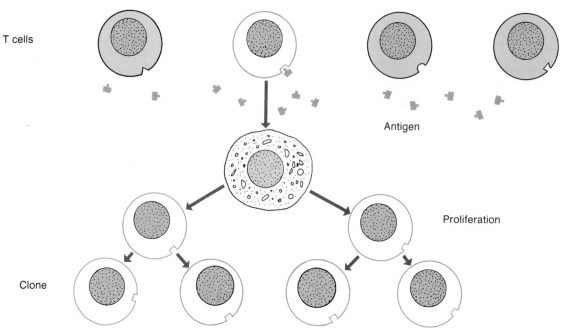

Figure 22.12. Interactions between macrophages and T cells. In this example, the activation of helper T cells occurs after the foreign antigens have been "processed" by the macrophages. The interaction between T cells and macrophages requires that both cells have the same histocompatibility antigens. Once activated, the helper T cells can stimulate cytotoxic ("killer") T cells and B cells.

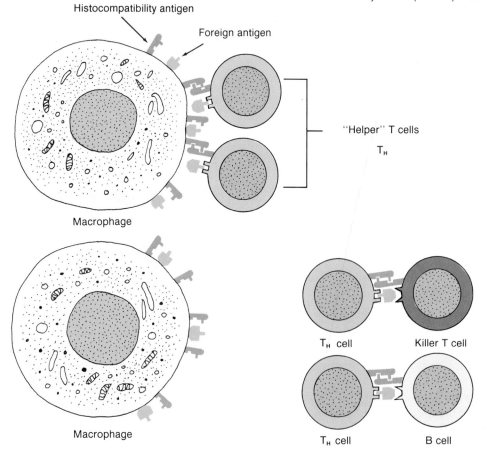

of some of these bacteria under the skin produce inflammation after a latent period of forty-eight to seventy-two hours. This *delayed hypersensitivity (hi''per-sen'si-tiv''i-te)* reaction is cell-mediated rather than humoral; an unexposed guinea pig, for example, can produce this response if it receives lymphocytes from an exposed guinea pig but not if it receives serum from the exposed animal.

Actions of T Lymphocytes

Unlike B cells, T cells do not make antibodies or have antibodies on their surface to function as receptors for specific antigens. The T lymphocytes do, however, have some type of receptor proteins on their membrane surface that interact with specific antigens. This interaction stimulates the T cell to divide many times, creating a population of genetically identical cells—a *clone (klōn)*—which is illustrated in figure 22.11. A similar process of clone selection occurs in B lymphocytes.

Interaction with Macrophages.   In order for both T cell and B cell lymphocytes to be stimulated by many antigens, the antigens must first be processed by macrophages. The nature of this processing is not well understood, but since intimate contact is required between the macrophage *(mak'ro-fāj)* and lymphocyte, it is believed that the macrophage "presents" the antigen to the lymphocyte in a more effective way than could occur in the absence of the macrophage.

Macrophages from one person cannot activate lymphocytes from another. This is due to the fact that the tissue cells from each person have a unique combination of surface molecules called *histocompatibility antigens,* which serve as genetic markers (and which can be used for tissue typing). Macrophage-lymphocyte interaction apparently requires a proper fitting of the surface receptors to these antigens (fig. 22.12).

Lymphokines.   The cell-mediated destruction of victim cells is due to action of killer T lymphocytes, which secrete a family of low-molecular weight polypeptides called **lymphokines** *(lim'fo-kīns)*. The lymphokines include (1) *interferon;* (2) *lymphotoxin,* which directly destroys the victim cells; (3) *macrophage-activating factor,* which promotes phagocytic activity; (4) chemicals that attract leukocytes to the area by chemotaxis; (5) *macrophage-migration-inhibiting factor,* which prevents phagocytic cells from leaving the area; and (6) *interleukin-2,* which stimulates T lymphocyte division and proliferation.

The lymphokines thus promote the destruction of the victim cells directly, by means of interferon and lymphotoxin, and indirectly by factors that increase the concentration and activity of phagocytic cells. The entire response is amplified by interleukin-2-stimulated proliferation of T lymphocytes.

Helper and Suppressor T Cells.   In addition to killer T lymphocytes, there are two other subpopulations of T cells that do not themselves participate in the destruction of the invading pathogens. Instead, they modulate the activity of B cells and thus affect humoral immunity (the secretion of antibodies). The activity of B lymphocytes is increased by **helper T cells** and decreased by **suppressor T cells.** The amount of antibodies secreted in response to antigens is thus affected by the relative numbers of helper to suppressor T cells that develop in response to the same antigen (fig. 22.13).

As a result of advances in recombinant DNA technology (genetic engineering), our knowledge of histocompatibility antigens, the production of monoclonal antibodies, and other technologies, it is now possible for clinical laboratories to distinguish between B lymphocytes, killer T lymphocytes, helper T cells, suppressor T cells, and natural killer cells by means of their surface antigens. Counting the lymphocytes in each of these categories provides far more information about diseases and their causes than was previously available.

> **Acquired immune deficiency syndrome (AIDS)** is a disease that particularly strikes certain high-risk groups. In terms of the number of cases reported, these are homosexual men (71%), intravenous drug abusers (17%), Haitian immigrants (5%), and hemophiliacs (1%). Through the use of the laboratory tests, it has been learned that AIDS is associated with an abnormally low helper T lymphocyte count. This results in decreased immunological function and thus an increased incidence of opportunistic infections. *Pneumocystis carinii pneumonia* and *Kaposi's sarcoma* are two previously rare conditions that account for a high percentage of deaths in AIDS victims. Recent evidence suggests that AIDS is caused by a specific strain of virus that infects human lymphocytes.

Thymus Gland

The thymus gland extends from below the thyroid in the neck into the thoracic cavity. This organ grows during childhood but gradually regresses after puberty. Lymphocytes from the fetal liver and spleen in prenatal life and from the bone marrow postnatally seed the thymus and become transformed into T cells. These lymphocytes in turn enter the blood and seed lymph nodes and other organs, where they divide to produce new T cells when stimulated by antigens.

Small T lymphocytes that have not yet been stimulated by antigens have very long life spans—months or perhaps years. Still, new T cells must be continuously produced to provide efficient cell-mediated immunity. Since the thymus atrophies after puberty, this organ may not be able to provide new T cells in later life. Colonies of T cells in the lymph nodes and other organs are apparently able to produce new T cells under the stimulation of various **thymus hormones.**

Kaposi's sarcoma: from Moritz K. Kaposi, Austrian physician, 1837–1902

Figure 22.13. A given antigen can stimulate the production of both B and T cell clones. The ability to produce B cell clones, however, is also influenced by the relative effects of helper and suppressor T cells.

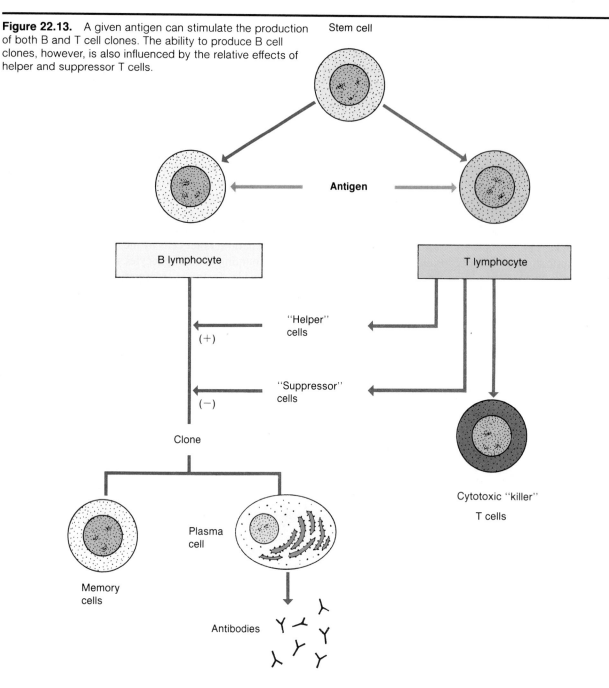

Two hormones that are believed to be secreted by the thymus—*thymopoietin (thi"mo-poi'ě-tin) I* and *thymopoietin II*—may promote the transformation of lymphocytes into T cells. Another thymus hormone, called *thymosin (thi'mo-sin),* may promote the maturation of T lymphocytes.

There is some experimental evidence supporting the notion that the administration of these thymus hormones may be able to restore cell-mediated immunity in some cases where T cell function has declined. This decline occurs in some congenital and acquired diseases as well as

naturally in the course of aging in conjunction with an increased susceptibility to viral infections and cancer. Clinical manipulations of T cell function—through the administration of thymus hormones, interferon, interleukin-2, and other techniques—is an exciting possibility that may result from further experimental research.

1. Define the term lymphokines, *identify their origin, and list the different functions of these molecules.*
2. Describe the interactions between macrophages and lymphocytes and between T lymphocytes and B lymphocytes.
3. Explain the role of the thymus in cell-mediated immunity.

thymus: Gk. *thymos,* sweetbread

Figure 22.14.   Active immunity to a pathogen can be gained by exposure to the fully virulent form or by inoculation with a pathogen whose virulence (ability to cause disease) has been attenuated (reduced) but whose antigens are the same as in the virulent form.

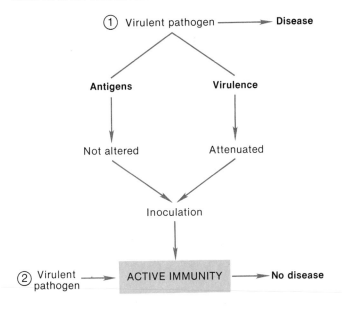

Figure 22.15.   A comparison of antibody production in the primary response (upon first exposure to an antigen) to antibody production in the secondary response (upon subsequent exposure to the antigen). The greater secondary response is believed to be due to the development of lymphocyte clones produced during the primary response.

Immunizations

Resistance to pathogens can be gained by both active and passive immunizations. The change from a primary to a secondary response in active immunity is best explained by the clonal selection theory.

Objective 13. Describe the primary and secondary immune responses.

Objective 14. Explain the clonal selection theory of active immunity, and relate this theory to the mechanism of active immunizations.

Objective 15. Explain how passive immunizations are performed.

Objective 16. Explain how monoclonal antibodies are produced and how they might be used.

It was first known in the mid-eighteenth century that the fatal effects of smallpox could be prevented by inducing mild cases of the disease. This was accomplished at that time by rubbing needles into the pustules of people who had mild forms of smallpox and injecting these needles into healthy people. Understandably, this method of immunization was not widely popular.

Acting on the observation that milkmaids who contracted cowpox—a disease similar to smallpox but less *virulent* (pathogenic)—were immune to smallpox, an English physician, named Edward Jenner, inoculated a healthy boy with cowpox. When the boy recovered, Jenner inoculated him with an otherwise deadly amount of smallpox, from which he also proved to be immune. (This was fortunate for both the boy—who was an orphan—and Jenner; Jenner's fame spread, and as the boy grew into manhood he proudly gave testimonials on Jenner's behalf). This experiment, performed in 1796, began the first widespread immunization program.

A similar, but more sophisticated, demonstration of the effectiveness of immunizations was performed by Louis Pasteur almost a century later. Pasteur isolated the bacteria that cause anthrax and heated them until their ability to cause disease was greatly reduced (their virulence was *attenuated*). He then injected these attenuated bacteria into twenty-five sheep, leaving twenty-five unimmunized. Several weeks later, before a gathering of scientists, he injected all fifty sheep with the completely active anthrax bacteria. All twenty-five of the unimmunized sheep died—all twenty-five of the immunized sheep survived (fig. 22.14).

Active Immunity and the Clonal Selection Theory

When a person is exposed to a particular pathogen for the first time, there is a latent period of five to ten days before measurable amounts of specific antibodies appear in the blood. This sluggish **primary response** may not be sufficient to protect the individual against the disease caused by the pathogen. Antibody concentrations in the blood during this primary response reach a plateau in a few days and decline after a few weeks.

A subsequent exposure of the same individual to the same antigen results in a **secondary response** (fig. 22.15).

Edward Jenner: English physician, 1749–1823
Louis Pasteur: French chemist and bacteriologist, 1822–1895

Figure 22.16.   The greater antibody production in the secondary response compared to the primary response is due mainly to increased amounts of IgG antibodies. Notice that the amount of IgM antibodies is about the same in both primary and secondary responses.

Figure 22.17.   According to the clonal selection theory, exposure to an antigen stimulates the production of lymphocyte clones and the maturation of some members of B cell clones into antibody-secreting plasma cells.

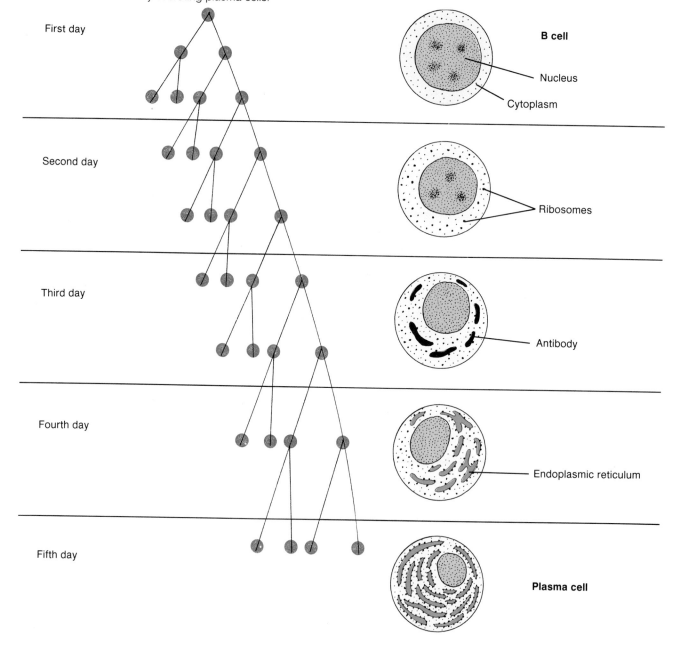

Table 22.7 Summary of the clonal selection theory (with regard to B cells)

Process	Results
Lymphocytes inherit the ability to produce specific antibodies.	Prior to antigen exposure, lymphocytes are already present in the body that can make the appropriate antibodies.
Antigens interact with antibody receptors on the lymphocyte surface.	Antigen-antibody interaction stimulates cell division and the development of lymphocyte clones containing memory cells and plasma cells that secrete antibodies.
Subsequent exposure to the specific antigens produce a more efficient response.	Exposure of lymphocyte clones to specific antigens results in greater and more rapid production of specific antibodies.

Source: Fox, Stuart Ira, *Human Physiology.* © 1984 Wm. C. Brown Publishers, Dubuque, Iowa. All Rights Reserved. Reprinted by permission.

Compared to the primary response, antibody production during the secondary response is much more rapid. Maximum antibody concentrations in the blood are reached in less than two hours and are maintained for a longer time than in the primary response. This rapid rise in antibody production is usually sufficient to prevent the disease.

Before stimulation by a particular antigen, the B cells that make the appropriate antibodies produce mainly IgM antibodies. These IgM antibodies therefore account for a high proportion of the antibodies made during the primary response. In contrast, most antibodies made during the secondary response are in the IgG subclass. This is shown in figure 22.16.

Clonal Selection Theory. The immunization procedures of Jenner and Pasteur were effective because the people who were inoculated produced a secondary rather than a primary response when exposed to the virulent pathogens. This protection is not simply due to accumulations of antibodies in the blood, because secondary responses occur even after antibodies produced by the primary response have disappeared. Immunizations, therefore, seem to produce a type of "learning" in which the ability of the immune system to combat a particular pathogen is improved by prior exposure.

The mechanisms by which secondary responses are produced are not completely understood; the **clonal selection theory,** however, appears to account for most of the evidence. According to this theory, B lymphocytes *inherit* the ability to produce particular antibodies (and T lymphocytes inherit the ability to respond to particular antigens). Some lymphocytes, therefore, can respond to smallpox and produce antibodies against it even if the person has never been previously exposed to this disease.

Exposure to smallpox antigens stimulates these specific lymphocytes to divide many times until a large population of genetically identical cells—a clone—is produced. Some of these cells become plasma cells that secrete antibodies for the primary response; others become memory cells that can be stimulated to secrete antibodies during the secondary response (fig. 22.17).

Notice that according to the clonal selection theory (table 22.7), antigens do not induce lymphocytes to make the appropriate antibodies. Rather, antigens select lymphocytes (through interaction with surface receptors) that are already able to make antibodies against that antigen.

This is analogous to evolution by natural selection. An environmental agent (in this case, antigens) acts on the genetic diversity already present in a population of organisms (lymphocytes) to cause increasing numbers of the individuals that are selected.

Active Immunity. The development of a secondary response provides **active immunity** against the specific pathogens. Development of active immunity requires prior exposure to the specific antigens, during which time a primary response is produced and the person may get sick. Some parents, for example, deliberately expose their children to others who have measles, chicken pox, and mumps so that their children will be immune to these diseases in later life.

Clinical immunization programs induce primary responses by inoculating people with pathogens whose virulence has been attenuated or destroyed (such as Pasteur's heat-inactivated anthrax bacteria) or by using closely related strains of microorganisms that are antigenically similar but less pathogenic (such as Jenner's cowpox inoculations). These procedures cause the development of lymphocyte clones that can combat the virulent pathogens by producing secondary responses.

The first successful polio vaccine (the Salk vaccine) was composed of viruses that had been inactivated by treatment with formaldehyde. These "killed" viruses were injected into the body, in contrast to the currently used oral (Sabin) vaccine. The oral vaccine contains "living" viruses that have attenuated virulence. These viruses invade the epithelial lining of the intestine and multiply but do not invade nerve tissue. The immune system can, therefore, become sensitized to polio antigens and produce a secondary response if polio viruses that attack the nervous system are later encountered.

Passive Immunity

The ability to mount an immune response—called **immunological competence**—does not develop until about a month after birth. The fetus, therefore, cannot immunologically reject its mother. The immune system of the mother is fully competent but does not usually respond to

Salk vaccine: from Jonas Salk, American immunologist, 1914
Sabin vaccine: from Albert B. Sabin, American virologist, 1906

Table 22.8 Comparison of active and passive immunity

Characteristic	Active immunity	Passive immunity
Injection of person with	Antigens	Antibodies
Source of antibodies	The person inoculated	An animal that is inoculated with the antigen; or the mother
Method	Injection with killed or attenuated pathogens or their toxins	Natural—transfer of antibodies across the placenta; artificial—injection with antibodies
Time to develop resistance	5 to 14 days	Immediately after injection
Duration of resistance	Long (perhaps years)	Short (days to weeks)
When used	Before exposure to pathogen	Before or after exposure to pathogen

Source: Fox, Stuart Ira, *Human Physiology.* © 1984 Wm. C. Brown Publishers, Dubuque, Iowa. All Rights Reserved. Reprinted by permission.

fetal antigens, for reasons that are not completely understood. Some IgG antibodies from the mother do cross the placenta and enter the fetal circulation, however, and these serve to confer **passive immunity** to the fetus.

The fetus and the newborn baby are, therefore, immune to the same antigens as the mother. Since the baby did not itself produce the lymphocyte clones needed to form these antibodies, such passive immunity disappears when the infant is about one month old. If the baby is breast-fed it can receive additional antibodies of the IgA subclass in its mother's first milk (the *colostrum*).

Passive immunizations are used clinically to protect people who have been exposed to extremely virulent infections or toxins, such as snake venom, tetanus, and others. In these cases the affected person is injected with **antiserum** (serum containing antibodies), also called **antitoxin** *(an''ti-tok'sin),* from an animal that has been previously exposed to the pathogen. The animal develops the lymphocyte clones and active immunity and thus has a high concentration of antibodies in its blood. Since the person who is injected with these antibodies does not develop active immunity, he or she must again be injected with antitoxin upon subsequent exposures. A comparison of active and passive immunity is shown in table 22.8.

Monoclonal Antibodies

In addition to their use in passive immunity, antibodies are also commercially prepared for use in research and clinical laboratory tests. In the past, antibodies were obtained by first chemically purifying a specific antigen and then injecting this antigen into animals. Since an antigen typically has many different antigenic determinant sites, however, the antibodies obtained by this method were polyclonal; they had different specificities. This decreased their sensitivity to a particular antigenic site and resulted in some degree of crossover with closely related antigen molecules.

In the preparation of monoclonal antibodies, an animal is injected with an antigen and then subsequently killed. B lymphocytes are then obtained from its spleen and placed in thousands of different *in vitro* incubation vessels. These cells soon die, however, unless they are hybridized with cancerous multiple myeloma cells. Cell fusion is promoted by a chemical, polyethylene glycol. The

fusion of a B lymphocyte with a cancerous cell produces a hybrid that undergoes cell division and produces a clone, called a **hybridoma** *(hi''bri-do'mah).* Each hybridoma secretes large amounts of identical, **monoclonal** *(mon''o-klōn'al)* antibodies.

The availability of large quantities of pure monoclonal antibodies has resulted in the development of much more sensitive clinical laboratory tests (of pregnancy, for example). These pure antibodies have also been used to pick one molecule (the specific antigen interferon, for example) out of a solution of many molecules and thus isolate and concentrate it. In the future, monoclonal antibodies against specific tumor antigens may aid the diagnosis of cancer. Even more exciting, cytotoxic drugs that can kill normal as well as cancerous cells might be aimed directly at a tumor by combining these drugs with monoclonal antibodies against specific tumor antigens.

1. Describe two methods used to induce active immunity.
2. Explain the characteristics of the primary and secondary immune responses, and draw graphs to illustrate your discussion.
3. Explain the clonal selection theory and how this theory accounts for the secondary response.
4. Explain passive immunity, and give natural and clinical examples of this type of immunization.

Tolerance and Non-Self-Recognition

The immune system normally can tolerate self-antigens while attacking non-self-antigens. A system of histocompatibility antigens and red blood cell antigens identifies the cells of each person.

Objective 17. Describe how the immune system might become tolerant to self-antigens.

Objective 18. Explain the nature and significance of the histocompatibility antigens.

Objective 19. Describe the red blood cell antigens and explain how blood is typed.

Objective 20. Explain how the normal immune system provides immunological surveillance against cancer.

Figure 22.18.   According to the clonal deletion theory, tolerance to self-antigens may be due to destruction of the lymphocytes that are able to make autoantibodies (antibodies against self-antigens).

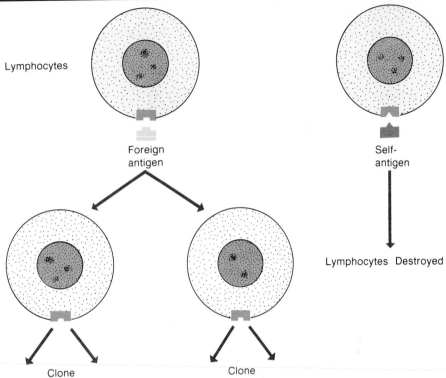

Lymphocytes

Foreign antigen

Self-antigen

Lymphocytes Destroyed

Clone   Clone

The ability to produce antibodies against **non-self**-antigens while tolerating **self**-antigens occurs early in life when immunological competence is established (the first month or so of postnatal life). If a fetal mouse of one strain receives transplanted antigens from a different strain, therefore, it will not recognize tissue transplanted later in life from the other strain as foreign and, as a result, will not immunologically reject the transplant.

Tolerance

The ability of an individual's immune system to recognize and tolerate self-antigens requires continuous exposure of the immune system to those antigens. If this exposure begins when the immune system is weak—such as in fetal and early postnatal life—tolerance is more complete and long lasting than when exposure occurs later in life. Some self-antigens, however, are normally hidden from the blood, such as thyroglobulin within the thyroid gland and lens protein in the eye. An exposure to these self-antigens results in antibody production just as if these proteins were foreign. Antibodies made against self-antigens are called **autoantibodies.**

Two major theories have been proposed to account for immunological tolerance: (1) *clone deletion;* and (2) *immunological suppression.* According to the clone deletion theory, tolerance to self-antigens is achieved by destruction of the lymphocytes that inherit the ability to make autoantibodies (fig. 22.18). It is not known how this might occur.

According to the immunological suppression theory, the lymphocytes that make autoantibodies are present throughout life but are normally inhibited from attacking self-antigens. This is presumably due to the effects of suppressor T lymphocytes. An alteration in the ratio of suppressor to helper T lymphocytes in later life, therefore, might result in the production of autoantibodies.

Histocompatibility Antigens

Tissue that is transplanted from one person to another contains antigens that are foreign to the host. This is because every tissue cell, with the exception of mature red blood cells, is genetically marked with a characteristic combination of **histocompatibility** *(his″to-kom-pat″i-bil′i-te)* **antigens** on its surface. The greater the difference in these antigens between the donor and recipient in a transplant, the greater will be the chance of transplant rejection. Prior to organ transplantation, therefore, the "tissue type" of the recipient is matched to that of potential donors. Since the person's white blood cells are used for this purpose, an alternate name for histocompatibility antigens in humans is **human leukocyte antigens,** abbreviated **HLA.**

The histocompatibility antigens (HLA) are proteins that are coded by four genes located on chromosome number 6 (fig. 22.19). These four genes are labeled A, B, C, and D. Each of these genes can code for only one protein in a given individual, but this protein can be different in different people. Two people, for example, may both have antigen A3, but one might have antigen B17 while the other has antigen B21. The closer two people are related, the more similar their histocompatibility antigens will be.

Figure 22.19. There are four human histocompatibility antigens (or human leukocyte antigens—HLA). Each of these antigens is coded by a gene located on chromosome number 6.

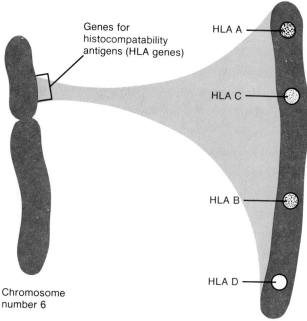

Chromosome number 6

Figure 22.20. (a) an electron micrograph showing contact between a macrophage (*left*) and a lymphocyte (*right*). As illustrated in (b), such contact between a macrophage and a T cell requires that both cells share the same histocompatibility (or HLA) antigens. By presenting antigens to the T cell, the interaction of the macrophage with the T cell may help convert the T cell into a helper or a suppressor or a cytotoxic (killer) T cell.

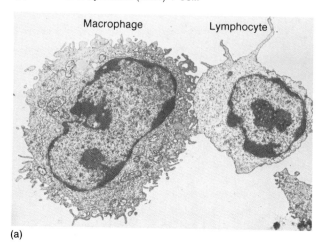

(a)

Interaction between T cells and macrophages, which is often required for the activation of the T cells by antigens, requires that both cells have the same histocompatibility antigens (fig. 22.20). Likewise, the ability of killer T cells to destroy another human cell (which has been infected with a virus) requires that both the T cell and the victim cell have the same histocompatibility antigens (fig. 22.21).

Clinical interest has been generated by the observation that certain diseases are much more common in people who have particular histocompatibility antigens (table 22.9). Ankylosing spondylitis (a type of rheumatoid arthritis), for example, is much more common in people who have antigen B27, and psoriasis (a skin disorder) is three times more common in people with antigen B17 than in the general population. Some other diseases that have a high correlation with particular histocompatibility antigens include Hodgkin's disease (a cancer of the lymph nodes), myasthenia gravis, Graves' disease, and juvenile-onset diabetes.

Red Blood Cell Antigens and Blood Typing

The histocompatibility antigens found on the surface of tissue cells are far more varied than the antigens on red blood cells. There are several groups of red blood cell antigens, but the major group is known as the **ABO system.** In terms of the antigens present on the red blood cell surface, a person may be *type A* (with only A antigens), *type B* (with only B antigens), *type AB* (with both A and B antigens), or *type O* (with neither A nor B antigens).

Figure 22.21. In order for a killer T cell to destroy a tissue cell infected with viruses, the T cell and the target cell must share the same histocompatibility (or HLA for humans) antigens.

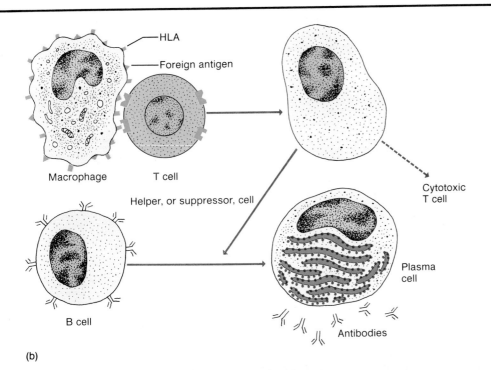

(b)

Table **22.9** Association between particular human histocompatibility antigens and diseases			
Disease	HLA antigen	Frequency in patients (%)	Frequency in controls (%)
Ankylosing spondylitis	B27	90	7
	B13	18	4
Psoriasis	B17	29	8
	B16	15	5
Graves' disease	B8	47	21
Coeliac disease	B8	78	24
Dermatitis herpetiformis	B8	62	27
Myasthenia gravis	B8	52	24
SLE (systemic lupus erythematosus)	B15	33	8
Multiple sclerosis	A3	36	25
	B7	36	25
Acute lymphatic leukemia	A2	63	37
	B35	25	16
Hodgkin's disease	A1	39	32
	B8	26	22
Chronic hepatitis	B8	68	18
Ragweed hay fever	B7	50	19

Source: Bodmer, W., in *Lancet 1*, p. 1269. © 1974 Lancent Ltd. Reprinted by permission.

Table **22.10** The ABO system of antigens on red blood cells	
Antigen on RBCs	Antibody in plasma
A	Anti-B
B	Anti-A
O	Anti-A and anti-B
AB	Neither anti-A nor anti-B

Source: Fox, Stuart Ira, *Human Physiology.* © 1984 Wm. C. Brown Publishers, Dubuque, Iowa. All Rights Reserved. Reprinted by permission.

Each person inherits two genes (one from each parent) that control the production of the ABO antigens. The gene for A or B antigens is dominant to the gene for O, since the latter simply means the absence of A or B. A person who is type A, therefore, may have inherited the A gene from each parent (may have the genotype AA) or the A gene from one parent and the O gene from the other parent (and have the genotype AO). Likewise, a person who is type B may have the genotype BB or BO. It follows that a type O person inherited the O gene from each parent (has genotype OO), whereas a type AB person inherited the A gene from one parent and the B gene from the other (there is no dominance-recessive relationship between A and B).

The immune system is tolerant to its own red blood cell antigens. A person who is type A, for example, does not produce anti-A antibodies. Surprisingly, however, people with type A blood do make antibodies against the B antigen, and conversely, people with type B blood make antibodies against the A antigen. This is believed to result from the fact that antibodies made in response to some common bacteria can cross-react with the A or B antigens. A person who is type A, therefore, acquires antibodies that can react with B antigens by exposure to these bacteria but does not develop antibodies that can react with A antigens because this is prevented by tolerance mechanisms.

People who are type AB develop tolerance to both of these antigens and thus do not produce either anti-A nor anti-B antibodies. Those who are type O, in contrast, do not develop tolerance to either antigen and, therefore, have both anti-A and anti-B antibodies in their plasma (table 22.10).

Figure 22.22.   The agglutination (clumping) of red blood cells occurs when cells with A-type antigens are mixed with anti-A antibodies and when cells with B-type antigens are mixed with anti-B antibodies. No agglutination would occur with type O blood (not shown).

Anti-B   Anti-A

Type A

Type B

Type AB

Transfusion Reactions.   Before transfusions are performed, a *major crossmatch* is made by mixing serum from the recipient with blood cells from the donors. If the types do not match—if the donor is type A, for example, and the recipient is type B—the recipient's antibodies attach to the donor's red blood cells and form bridges that cause the cells to clump together, or **agglutinate** (fig. 22.22). Because of this agglutination *(ah-gloo''tĭ-na'shun)* reaction, the A and B antigens are sometimes called **agglutinogens** *(ag''loo-tin'o-jen)* and the antibodies against them are called **agglutinins** *(ah-gloo'tĭ-nin).* Transfusion errors that result in such agglutination in the blood can produce a blockage of small blood vessels and organ damage.

In emergencies, type O blood has been given to people who are type A, B, AB, or O. Since type O red blood cells lack A or B antigens, the recipient's antibodies cannot cause agglutination of the donor red blood cells. Type O is, therefore, a *universal donor,* but only as long as the volume of plasma donated is small because plasma from a type O person would agglutinate type A, type B, and type AB red blood cells. Likewise, type AB people are *universal recipients* because they lack anti-A and anti-B antibodies, and thus cannot agglutinate donor red blood cells. (Donor plasma could agglutinate recipient red blood cells if the transfusion volume were too large). Because of the dangers involved, use of the universal donor and recipient concept in blood transfusions is strongly discouraged.

Figure 22.23. A killer T cell (*a*) contacts a cancer cell (the larger cell), in a manner that requires specific interaction with antigens on the cancer cell. The killer T cell releases lymphokines, including toxins that cause the death of the cancer cell (*b*).

(a)   Andrejs Liepins

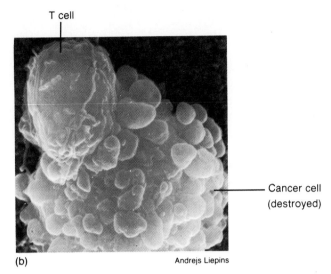

(b)   Andrejs Liepins

Rh Factor. Another important group of antigens found in most red blood cells is the *Rh factor* (Rh stands for Rhesus monkey, in which these antigens were first discovered). People who have these antigens are said to be **Rh positive,** whereas those who do not are **Rh negative.** There are fewer Rh negative people because this condition is recessive to Rh positive. The Rh factor is of particular significance when Rh negative mothers give birth to Rh positive babies.

At the time she gives birth, the immune system of an Rh negative mother may become exposed to Rh antigens from her baby. Since these antigens are foreign to her, she can become sensitized and build up lymphocyte clones against the Rh antigen. This does not affect the first baby, which is already born, but can affect subsequent Rh positive fetuses because antibodies against the Rh factor can cross the placenta. These antibodies cause the destruction of the fetus's erythrocytes, so that the baby is born anemic, with a condition called **erythroblastosis** *(ĕ-rith″ro-blas-to′sis)* **fetalis.**

Erythroblastosis fetalis can be prevented by injecting the Rh negative mother with *antibodies against the Rh factor* (one trade name for this is RhoGAM) within seventy-two hours after the birth of each Rh positive baby. This is a type of passive immunization in which the injected antibodies inactivate the Rh antigens and thus prevent the mother from becoming actively immunized to them.

Tumor Immunology

Oncology (the study of tumors) has revealed that tumor biology is similar to and interrelated with the functions of the immune system. Most tumors appear to be clones of single cells that have become transformed. This is similar to the development of lymphocyte clones in response to specific antibodies. Lymphocyte clones, however, are under complex inhibitory control systems—such as those exerted by suppressor T lymphocytes and negative feedback by antibodies. The division of tumor cells, in contrast, is not effectively controlled by normal inhibitory mechanisms. Tumor cells are also relatively unspecialized—they *dedifferentiate,* which means that they become similar to the less-specialized cells of an embryo.

Tumors are described as *benign* when they are relatively slow growing and limited to a specific location (warts, for example). Benign tumors do not undergo *metastasis (mĕ-tas′tah-sis),* a term that refers to the dispersion of tumor cells and resultant seeding of new tumors in different locations. *Malignant* tumors grow more rapidly and do metastasize. The term **cancer,** as it is usually used, refers to malignant, life-threatening tumors.

As tumors dedifferentiate, they reveal surface antigens that can stimulate the immune destruction of the tumor cells. Consistent with the concept of dedifferentiation, some of these antigens are proteins produced in embryonic or fetal life that are not normally produced postnatally. Since they are absent at the time immunological competence is established they are treated as foreign and fit subjects for immunological attack when they are produced by cancerous cells. The release of two such antigens into the blood has provided the basis for a laboratory diagnosis of some cancers. *Carcinoembryonic (kar″sin-o-em″bre-on′ik) antigen* tests are useful in the diagnosis of colon cancer, for example, and tests for *alpha-fetoprotein* (normally produced only by the fetal liver) help in the diagnosis of liver cancer.

Tumors are attacked by the cell-mediated immune system, although humoral immunity (antibodies) may have a supportive role. Killer T lymphocytes (fig. 22.23)

benign: L. *benignus,* kind
metastasis: Gk. *metastasis,* a removing
cancer: L. *cancer,* a crab

and natural killer cells recognize antigens on the tumor cell surface and normally destroy such cells before a recognizable tumor develops. This function is known as **immunological surveillance** against cancer.

Natural Killer Cells. Researchers observed that a strain of hairless mice, which genetically lacks a thymus and T lymphocytes, does not suffer from a particularly high incidence of tumor production. This surprising observation led to the discovery of **natural killer (NK) cells,** which are lymphocytes that are related to, but distinct from, T lymphocytes. Unlike killer T cells, NK cells destroy tumors in a nonspecific fashion and do not require prior exposure for sensitization to the tumor antigens. The NK cells thus provide a first line of cell-mediated defense, which is subsequently backed up by a specific response mediated by killer T cells. These two cell types interact, however; the activity of NK cells is stimulated by interferon released as one of the lymphokines from T lymphocytes.

Effects of Aging and Stress. Susceptibility to cancer varies greatly. The Epstein-Barr virus that causes Burkitt's lymphoma in a few individuals in Africa, for example, can also be found in healthy people throughout the world. In most cases the virus is harmless; in some cases mononucleosis (involving a limited proliferation of white blood cells) is produced. Only rarely does this virus cause the uncontrolled proliferation of leukocytes occurring in Burkitt's lymphoma. The reasons for these different responses to the Epstein-Barr virus and indeed for the different susceptibilities of people to other forms of cancer are not well understood.

It is known that cancer risk increases with age. According to one theory, this is due to the fact that aging lymphocytes accumulate genetic errors over the years that decrease their effectiveness. The secretion of thymus hormones also decreases with age in parallel with a decrease in cell-mediated immune competence. Both of these changes and perhaps others not yet discovered could increase susceptibility to cancer.

Numerous experiments have demonstrated that tumors grow faster in experimental animals subject to stress than in unstressed control animals. This is generally believed to result from the fact that stressed animals, including humans, have increased secretion of corticosteroid hormones, which act to suppress the immune system (this is why cortisone is given to people who receive organ transplants and to people with chronic inflammatory diseases). Some recent experiments, however, suggest that the stress-induced suppression of the immune system may also be due to other factors that do not involve the adrenal cortex. Future advances in cancer therapy may incorporate methods of strengthening the immune system together with methods that directly destroy tumors.

1. *Define the term* immunological tolerance, *and explain how this might be produced.*
2. *Describe the nature, functions, and clinical applications of the histocompatibility antigens.*
3. *Describe the red blood cell antigens, and explain how the agglutination reactions are produced.*
4. *Describe how erythroblastosis fetalis is produced, and explain the technique used to prevent this condition.*
5. *Explain how T lymphocytes and natural killer cells recognize cancer cells.*

Clinical Considerations

The ability of the normal immune system to tolerate self-antigens while identifying and attacking foreign antigens provides a specific defense against invading pathogens. In every individual, however, this system of defense against invaders at times commits domestic offenses. This can result in diseases that range in severity from the sniffles to sudden death.

Diseases caused by the immune system can be grouped into three interrelated categories: (1) *autoimmune diseases;* (2) *immune complex diseases;* and (3) *allergy,* or *hypersensitivity.* It is important to remember that these diseases are not caused by foreign pathogens but by abnormal responses of the immune system.

Autoimmunity

Autoimmune diseases are those produced by failure of the immune system to recognize and tolerate self-antigens. This failure results in the production of autoantibodies that can cause inflammation and organ damage. Such autoimmune destruction may occur as a result of the following mechanisms.

1. **An antigen that does not normally circulate in the blood may become exposed to the immune system.** Thyroglobulin protein that is normally trapped within the thyroid follicles, for example, can stimulate the production of autoantibodies that cause the destruction of the thyroid (fig. 22.24); this occurs in *Hashimoto's thyroiditis.* Similarly, autoantibodies developed against lens protein in a damaged eye may cause the destruction of a healthy eye (in *sympathetic ophthalmia.*)
2. **A self-antigen, which is otherwise tolerated, may be altered by combining with a foreign hapten.** The disease *thrombocytopenia* (low platelet count), for example, can be caused by the autoimmune destruction of thrombocytes (platelets). This occurs when drugs such as aspirin, sulfonamide, antihistamines, digoxin, and others combine with platelet proteins to produce new antigens. The symptoms of this disease usually stop when the person stops taking these drugs.

Figure 22.24. Autoimmune thyroiditis in a rabbit, induced experimentally by injection with thyroglobulin. Compare the picture of a normal thyroid (*left*) with that of the diseased thyroid (*right*). The grainy appearance of the diseased thyroid is due to the infiltration of large numbers of lymphocytes and macrophages.

Figure 22.25. One proposed mechanism for the development of rheumatoid arthritis. An initial joint inflammation (*1*) results in the release of lysosomal enzymes (*2*) that cause damage to IgG antibodies (*3*). These damaged IgG antibodies act as antigens (*4*) to stimulate B lymphocytes, resulting in the production of IgM antibodies antigens. This, in turn, produces further inflammation.

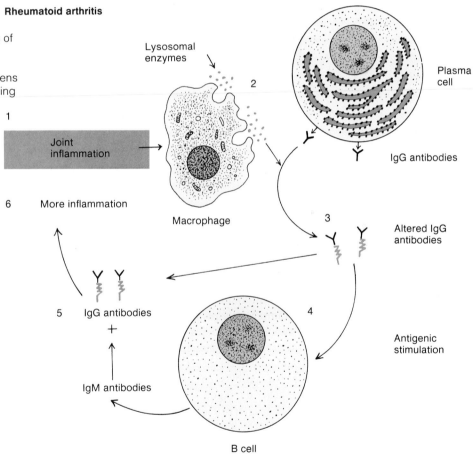

Rheumatoid arthritis

3. **A self-antigen may become damaged and, as a result, may expose new antigenic sites.** This may occur in *rheumatoid arthritis*. An initial inflammation of the joints (perhaps through viral infection) may result in the release of digestive enzymes from lysosomes in phagocytic cells. When these enzymes digest part of IgG antibodies they expose new antigenic sites. This stimulates the production of IgM antibodies directed against the altered IgG proteins (fig. 22.25). The IgM antibodies that a person develops against his own IgG antibodies are called *rheumatoid factor* and can be used to diagnose this condition.

Figure 22.26.   A proposed model to explain the increased frequency of autoimmune diseases with age. Early in life, those lymphocytes that produce autoantibodies may be destroyed. With time, however, some descendants of the remaining lymphocytes may mutate to produce autoantibodies. When this occurs later in life, these cells are not destroyed and may cause autoimmune diseases.

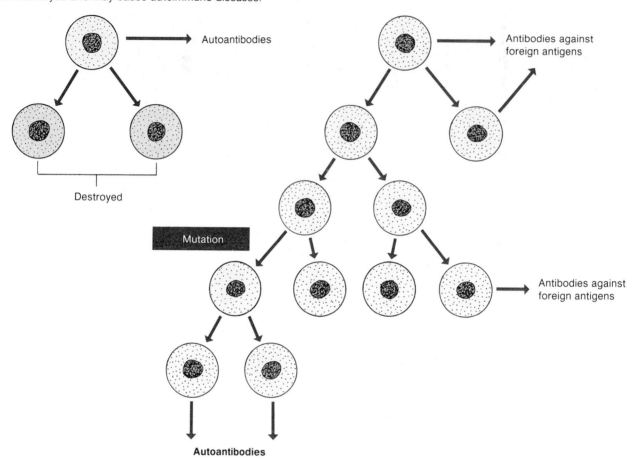

4. **Antibodies produced against foreign antigens may cross-react with self-antigens.** Antoimmune diseases of this sort can occur, for example, as a result of *streptococcus* bacterial infections. Antibodies produced in response to antigens in this bacterium may cross-react with self-antigens in the heart and kidneys. The inflammation induced by such autoantibodies can produce heart damage (including the valve defects characteristic of *rheumatic fever*) and damage to the glomerular capillaries in the kidneys *(glomerulonephritis [glo-mer''u-lo-ně-fri'tis])*.
5. **Genes that code for non-self-antibodies may mutate to produce autoantibodies** (fig. 22.26). This would not cause particular diseases but would lead to the increased frequency of autoimmune diseases in general. An increased frequency of autoimmune diseases does in fact occur with age, as predicted by this mutation theory. Table 22.11 provides some examples of autoimmune diseases.

Immune Complex Diseases

The term *immune complexes* refers to combinations of antibodies with antigens that are free, rather than attached to bacterial or other cells. Formation of such complexes activates complement proteins and promotes inflammation. This inflammation normally is self-limiting because the immune complexes are removed by phagocytic cells. When large numbers of immune complexes are continuously formed, however, the inflammation may be prolonged. Also, the dispersion of immune complexes to other sites can lead to widespread inflammations and organ damage. The damage produced by the inflammatory response to antigens is called **immune complex disease.**

Immune complex diseases can result from infections by bacteria, parasites, and viruses. In viral hepatitis B, for example, an immune complex that consists of viral antigens and antibodies can cause widespread inflammation of arteries *(periarteritis [per''e-ar''tě-ri'tis])*. Note that the arterial damage is not caused by the hepatitis virus itself but by the inflammatory process.

Table 22.11 Examples of autoimmune diseases

Disease	Antigen	Ig and/or T cell response
Postvaccinal and postinfectious encephalomyelitis	Myelin, cross-reactive	T cell
Aspermatogenesis	Sperm	T cell
Sympathetic ophthalmia	Uvea	T cell
Hashimoto's disease	Thyroglobulin	IgG and T cell
Graves' disease	Receptor proteins for TSH	Long-acting thyroid stimulator (LATS)
Autoimmune hemolytic disease	I, Rh, and others on surface of RBCs	IgM and IgG
Thrombocytopenic purpura	Hapten-platelet or hapten-adsorbed antigen complex	IgG
Myasthenia gravis	Myosin	IgG
Rheumatic fever	Streptococcal cross-reactive with heart	IgG and IgM
Glomerulonephritis	Streptococcal cross-reactive with kidney	IgG and IgM
Rheumatoid arthritis	IgG	IgM to Fc(γ)
Systemic lupus erythematosus	DNA, nucleoprotein, RNA, etc.	IgG

Source: From Barrett, James T.: Textbook of immunology, ed. 4, St. Louis, 1983, The C. V. Mosby Co.

Immune complex diseases can also result from the formation of complexes between self-antigens and autoantibodies. This is the case in rheumatoid arthritis, where the inflammation is produced by complexes of altered IgG antibodies (the antigens in this case) and IgM antibodies. Another immune complex disease that has an autoimmune basis is *systemic lupus erythematosus (SLE)*. People with SLE produce antibodies against their own DNA and nuclear proteins. This can result in the formation of immune complexes throughout the body, including the glomerular capillaries where glomerulonephritis may be produced.

Allergy

The term *allergy,* usually used synonymously with *hypersensitivity,* refers to particular types of abnormal immune responses to antigens, which are called *allergens (al'er-jen)* in these cases. There are two major forms of allergy: (1) **immediate hypersensitivity,** which is due to an abnormal B lymphocyte response to an allergen that produces symptoms within seconds or minutes; and (2) **delayed hypersensitivity,** which is an abnormal T cell response that produces symptoms within about forty-eight hours after exposure to an allergen. A comparison between these two types of hypersensitivity is provided in table 22.12.

Immediate Hypersensitivity. Immediate hypersensitivity can produce such symptoms as allergic rhinitis (chronic runny or stuffy nose), conjunctivitis (red eyes), allergic asthma, atopic dermatitis (urticaria or hives), and others. These symptoms result from the production of antibodies of the IgE subclass, instead of the normal IgG antibodies.

Unlike IgG antibodies, IgE antibodies do not circulate in the blood but instead attach to tissue mast cells (which have membrane receptors for these antibodies).

Table 22.12 Allergy: comparison between immediate and delayed hypersensitivity reaction

Characteristic	Immediate reaction	Delayed reaction
Time for onset of symptoms	Within several minutes	Within one to three days
Lymphocytes involved	B cells	T cells
Immune effector	IgE antibodies	Cell-mediated immunity
Allergies most commonly produced	Hay fever, asthma, and most other allergic conditions	Contact dermatitis (such as to poison ivy and poison oak)
Therapy	Antihistamines and adrenergic drugs	Corticosteroids (such as cortisone)

Source: Fox, Stuart Ira, *Human Physiology.* © 1984 Wm. C. Brown Publishers, Dubuque, Iowa. All Rights Reserved. Reprinted by permission.

When the person is again exposed to the same allergen, the allergen bonds to the antibodies attached to the mast cells. This stimulates the mast cells to secrete various chemicals, including **histamine** (fig. 22.27). During this process, leukocytes may also secrete **prostaglandins** and related molecules called **leukotrienes.** These chemicals (table 22.13) produce the symptoms of the allergic reactions.

The itching, sneezing, tearing, and runny nose of persons suffering from hay fever are produced largely by histamine and can be treated effectively by antihistamine drugs. Food allergies, causing diarrhea and colic, are mediated primarily by prostaglandins and can be treated with aspirin (these are the only allergies which respond positively to aspirin). Asthma, produced by smooth muscle constriction in the bronchioles in the lungs, is due to the release of leukotrienes. Since there are no antileukotriene drugs presently available, asthma is treated with epinephrinelike compounds (which cause bronchodilation) and corticosteroids.

Figure 22.27.   Allergy (immediate hypersensitivity) is produced when antibodies of the IgE subclass attach to tissue mast cells. The combination of these antibodies with allergens (antigens that provoke an allergic reaction) cause the mast cell to secrete histamine and other chemicals that produce the symptoms of allergy.

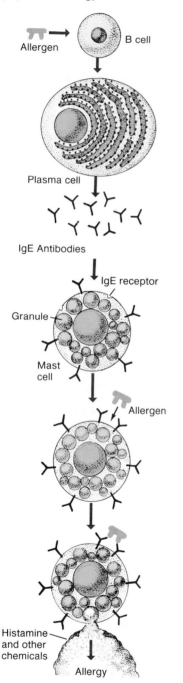

Allergen   B cell

Plasma cell

IgE Antibodies

IgE receptor

Granule

Mast cell

Allergen

Histamine and other chemicals

Allergy

Table 22.13 Some of the chemicals released by mast cells and leukocytes that are responsible for immediate hypersensitivity

Chemical	Derivation	Action
Histamine	From histidine	Contracts smooth muscles in bronchioles; dilates blood vessels; increases capillary permeability
Serotonin	From tryptophane	Contracts smooth muscles
Prostaglandins and leukotrienes	Arachidonic acid	Prolonged contraction of smooth muscle; increased capillary permeability
Eosinophil chemotactic factor	Polypeptides	Attracts eosinophils
Bradykinin and related compounds	Polypeptides	Slow smooth muscle contraction

Source: Fox, Stuart Ira, *Human Physiology.* © 1984 Wm. C. Brown Publishers, Dubuque, Iowa. All Rights Reserved. Reprinted by permission.

Immediate hypersensitivity to a particular antigen is commonly tested by injecting various antigens under the skin (fig. 22.28). Within a short time a *flare-and-wheal reaction* is produced if the person is allergic to that antigen. This reaction is due to the release of histamine and other chemical mediators: the flare is due to vasodilation, and the wheal results from local edema.

Allergens that provoke immediate hypersensitivity include various foods, bee stings, and pollen grains. The most common allergy of this type is seasonal hay fever, which may be provoked by ragweed *(Ambrosia)* pollen grains. People with chronic allergic rhinitis due to an allergy to "dust" or "feathers" are actually allergic to a tiny mite (fig. 22.29) that lives in dust and eats the scales of skin that are constantly shed from the body.

Delayed Hypersensitivity.   As the name implies, a longer time is required for the development of symptoms in delayed hypersensitivity (hours to days) than in immediate hypersensitivity. This may be due to the fact that immediate hypersensitivity is mediated by antibodies, whereas delayed hypersensitivity is a cell-mediated T lymphocyte response. Since the symptoms are caused by the secretion of lymphokines, rather than by the secretion of histamine, treatment with antihistamines provides little benefit. At present, corticosteroids are the only drugs that can effectively treat delayed hypersensitivity.

One of the best-known examples of delayed hypersensitivity is **contact dermatitis,** caused by poison ivy, poison oak, and poison sumac. The skin test for tuberculosis—the tine test and the Mantoux test—also rely on delayed hypersensitivity reactions. If a person has been exposed to the tubercle bacillus and has, as a result, developed T cell clones, skin reactions appear within a few days after the tubercle antigens are rubbed into the skin with small needles (tine test) or are injected under the skin (Mantoux test).

Figure 22.28. Skin test for allergy. If an allergen is injected into the skin of a sensitive individual, a typical flare-and-wheal response occurs within several minutes.

Figure 22.29. (*a*) a scanning electron micrograph of ragweed (*Ambrosia*), which is responsible for hay fever. (*b*) a scanning electron micrograph of the house dust mite (*Dermatophagoides farinae*), which lives in dust and is often responsible for yearlong allergic rhinitis and asthma.

(a)

(b)

Chapter Summary

I. Defense Mechanisms
   The immune system includes all of the structures and processes that provide a defense against potential pathogens. These defenses can be grouped into nonspecific and specific categories.
   A. Nonspecific defense mechanisms include barriers to penetration of the body and internal defenses.
      1. Phagocytic cells engulf invading pathogens.
      2. Interferons are polypeptides secreted by cells infected with viruses that help protect other cells from viral infections.
   B. Specific immune responses are directed against antigens.
      1. Antigens are molecules, or parts of molecules, that are usually large, complex, and foreign.

2. A given molecule can have a number of antigenic determinant sites that stimulate the production of different antibodies.
   C. Specific immunity is a function of lymphocytes.
      1. B lymphocytes secrete antibodies and provide humoral immunity.
      2. T lymphocytes provide cell-mediated immunity.
II. Humoral Immunity
   The structure of antibodies allows them to combine with antigens in a specific fashion and, by this means, to activate other elements of the immune system to protect against pathogens.
   A. There are five subclasses of antibodies, or immunoglobulins.
      1. These subclasses differ in regard to the polypeptides in the constant region of the heavy chains.

2. The five subclasses are IgG, IgA, IgM, IgD, and IgE.
   3. Each type of antibody has two variable regions that combine with specific antigens.
   4. The combination of antibodies with antigens promotes phagocytosis, a process called opsonization.
   B. Antigen-antibody complexes activate a system of proteins called the complement system.
      1. This results in complement fixation, where complement proteins attach to a cell membrane and promote the destruction of the cell.

2. Free complement proteins promote opsonization and chemotaxis and stimulate the release of histamine from tissue mast cells.
C. Specific and nonspecific immune mechanisms cooperate in the development of a local inflammation.

III. Cell-Mediated Immunity
Killer T lymphocytes, acting through the secretion of lymphokines, provide cell-mediated immunity. Subpopulations of helper and suppressor T lymphocytes regulate the immune response.
A. T lymphocytes are activated by antigens; this process is aided by
   1. Killer T lymphocytes secrete chemicals called lymphokines when they are in close proximity to the victim cells.
   2. Lymphokines include factors that directly destroy the victim cells, factors that attract phagocytes and stimulate their activity, and factors that stimulate the proliferation of T lymphocytes.
   3. Helper and suppressor T lymphocytes modify the immune function of killer T lymphocytes and regulate the secretion of antibodies by B lymphocytes.
B. The thymus processes T lymphocytes and secretes hormones that stimulate the development of T lymphocytes.

IV. Immunizations
Resistance to pathogens can be gained by both active and passive immunizations. The change from a primary to a secondary response in active immunity is best explained by the clonal selection theory.
A. A primary response is produced when a person is first exposed to a pathogen; a subsequent exposure results in a secondary response.
   1. During the primary response IgM antibodies are produced slowly, and the person is likely to get sick.
   2. During the secondary response IgG antibodies are produced quickly, and the person has resistance to the pathogen.
   3. In active immunizations, the person is exposed to pathogens of attenuated virulence that have the same antigenicity as the virulent pathogen.
   4. The secondary response is believed to be due to the development of lymphocyte clones after the first exposure as a result of the antigen-stimulated proliferation of appropriate lymphocytes.
B. Passive immunity is provided by antibodies made by a different organism.
   1. This occurs naturally from mother to fetus.
   2. Injections of antiserum provide passive immunity to some pathogenic organisms and toxins.
C. Monoclonal antibodies are made by hybridomas, which are formed artificially by the fusion of B lymphocytes with multiple myeloma cells.

V. Tolerance and Non-Self-Recognition
The immune system normally can tolerate self-antigens while attacking non-self-antigens. A system of histocompatibility cell antigens identifies the cells of each person.
A. Autoantibodies are not normally made against self-antigens, perhaps because lymphocytes that could make these antibodies are destroyed or because they are suppressed.
B. Each tissue cell has surface antigens called histocompatibility antigens, which help to provide self-recognition.
C. Red blood cells contain surface antigens of the ABO system and may have an Rh antigen.
   1. According to the antigens on the red blood cell surface, a person may be type O, type A, type B, or type AB.
   2. People who are one blood type have antibodies against the other blood types, so blood types must be matched in blood transfusions.
   3. Mothers who are Rh negative may become sensitized to the Rh antigen if they have a baby who is Rh positive.
D. Immunological surveillance against cancer is provided primarily by T cells and natural killer cells.
   1. Natural killer cells are nonspecific; T cells can recognize certain antigens on the tumor cell surface.
   2. Immunological surveillance against cancer is weakened by stress.

Review Activities

Objective Questions

1. Which of the following offers a nonspecific defense against viral infection?
   (a) antibodies
   (b) leukotrienes
   (c) interferon
   (d) histamine

Match the cell type with its secretion
2. killer T cells   (a) antibodies
3. mast cells   (b) lymphokines
4. plasma cells   (c) lysosomal enzymes
5. macrophages   (d) histamine

6. Which of the following statements about the F_{ab} portion of antibodies is true?
   (a) It bonds to antigens.
   (b) Its amino acid sequences are variable.
   (c) It consists of both H and L chains.
   (d) All of the above are true.
7. Which of the following statements about complement proteins C3a and C5a is *false*?
   (a) They are released during the complement fixation process.
   (b) They stimulate chemotaxis of phagocytic cells.
   (c) They promote the activity of phagocytic cells.
   (d) They produce pores in the victim cell membrane.

8. Mast cell secretion during an immediated hypersensitivity reaction is stimulated when antigens combine with
   (a) IgG antibodies
   (b) IgE antibodies
   (c) IgM antibodies
   (d) IgA antibodies
9. During a secondary response
   (a) antibodies are made quickly and in great amounts
   (b) antibody production lasts longer than in a primary response
   (c) antibodies of the IgG class are produced
   (d) lymphocyte clones are believed to develop
   (e) all of the above

10. Which of the following cells aids the activation of lymphocytes by antigens?
    (a) macrophages
    (b) neutrophils
    (c) mast cells
    (d) natural killer cells
11. Which of the following statements about T lymphocytes is *false*?
    (a) Some T cells promote the activity of B cells.
    (b) Some T cells suppress the activity of B cells.
    (c) Some T cells secrete interferon.
    (d) Some T cells produce antibodies.
12. Delayed hypersensitivity is mediated by
    (a) T cells
    (b) B cells
    (c) plasma cells
    (d) natural killer cells
13. Which of the following blood types does not have antibodies in the plasma against red blood cell antigens?
    (a) type A
    (b) type B
    (c) type AB
    (d) type O

Essay Questions

1. Explain how antibodies help destroy invading bacterial cells.
2. Explain how killer T lymphocytes help destroy cells infected with viruses.
3. Explain the possible roles of helper and suppressor T lymphocytes in (a) defense against infections and (b) tolerance to self-antigens.
4. Describe the clonal selection theory, and use this theory to explain how active immunity is produced.
5. Explain the physiological and clinical significance of histocompatibility and red blood cell antigens.

Selected Readings

Baglioni, C., and T. W. Nilsen. 1981. The action of interferon at the molecular level. *American Scientist* 69:392.

Barrett, James T. 1978. *Textbook of immunology.* 3rd ed. St. Louis: C. V. Mosby Company.

Biusseret, P. D. August 1982. Allergy. *Scientific American.*

Boxer, G. J., Curnette, J. T., and L. A. Boxer. 1985. Polymorphonuclear leukocyte function. *Hos. Prac.* 20:69.

Burnet, F. M. 1976. *Immunology: Readings from Scientific American.* San Francisco: W. H. Freeman and Co.

Capra, J. D., and A. B. Edmunson. January 1977. The antibody combining site. *Scientific American.*

Croce, C. M. 1985. Chromosomal translocations, oncogenes, and B-cell tumors. *Hosp. Prac.* 20:41.

Cunningham, B. A. October 1977. The structure and function of histocompatibility antigens. *Scientific American.*

Dannenberg, A. M. 1975. Macrophages in inflammation and infection. *New Engl. J. Med.* 293:489.

Dausset, J. 1981. The major histocompatibility complex in man: Past, present, and future concepts. *Science* 213:1469.

Edelson, R. L., and J. M. Fink. June 1985. The immunologic function of skin. *Scientific American.*

Hamburger, R. N. 1976. Allergy and the immune system. *American Scientist* 64:157.

Herberman, R. B., and J. R. Ortaldo. 1981. Natural killer cells: Their role in defense against disease. *Science* 214:24.

Kapp, J. A., C. W. Pierce, and C. M. Sorensen. August 1984. Antigen-specific suppressor T cell factors. *Hosp. Prac.* 19:85.

Koffler, D. July 1980. Systemic lupus erythematosus. *Scientific American.*

Leder, P. November 1982. The genetics of antibody diversity. *Scientific American.*

McDevitt, H. O. 1985. The HLa system and its relation to disease. *Hos. Prac.* 20:57.

Milstein, C. October 1980. Monoclonal antibodies. *Scientific American.*

Oettgen, H. F. 1981. Immunological aspects of cancer. *Hosp. Prac.* 16:93.

Old, L. J. May 1977. Cancer immunology. *Scientific American.*

Rose, N. R. February 1981. Autoimmune diseases. *Scientific American.*

Samuelsson, B. 1983. Leukotrienes: Mediators of immediate hypersensitivity and inflammation. *Science* 220:568.

Steinmetz, M., and L. Hood. 1983. Genes of the major histocompatibility complex in mice and men. *Science* 222:727.

Tannock, I. F. 1983. Biology of tumor growth. *Hosp. Prac.* 18:81.

Vaughan, J. A. 1984. Rheumatoid arthritis: Evidence of a defect in T-cell function. *Hosp. Prac.* 19:101.

Wolde-Mariam, W., and J. B. Peter. March 1984. Recent diagnostic advances in cellular immunology. *Diagnostic Med.*, p. 25.

Yelton, D. E., and M. D. Scharff. 1980. Monoclonal antibodies. *American Scientist* 63:510.

Zinkernagel, R. M. 1978. Major transplantation antigens in host responses to infection. *Hosp. Prac.* 13:83.

23

Respiratory System

Outline

Concepts

The respiratory tract can be divided into upper and lower divisions on the basis of function and embryological development.

The oral and nasal cavities, pharynx, trachea, and bronchial tree deliver warmed and moistened air to the respiratory zone of the lungs.

Alveoli are the functional units of the lungs where gas exchange occurs. Right and left lungs are separately contained in pleural membranes.

Movement of air into and out of the lungs occurs as a result of pressure differences induced by changes in lung volumes. Ventilation is thus influenced by the physical properties of the lungs.

Inspiration and expiration result from the contraction and relaxation of the diaphragm and thoracic muscles. Pulmonary function can be assessed by measuring the amounts of air inspired and expired under different conditions.

The lungs function to bring the blood into gaseous equilibrium with the alveolar air. The efficiency of this process is measured by the P_{O_2} and P_{CO_2} of arterial blood.

Breathing is directly controlled by centers in the brain stem. These centers are influenced by higher brain areas and by sensory input that makes breathing responsive to the needs of the body.

Oxygen is transported by hemoglobin from the lungs to the tissues. The unloading of oxygen by hemoglobin is influenced by a variety of factors.

The acid-base balance of the blood is affected by the concentrations of carbonic acid and bicarbonate, which are transport forms of carbon dioxide. These concentrations are regulated by ventilation.

Changes in respiratory function help to improve oxygen delivery during exercise and to compensate for reduced P_{O_2} at a high altitude.

Selected Terms

diverticulum   (di''ver-tik'u-lum)
pharynx   (far'inks)
bronchus   (brong'kus)
larynx   (lar'inks)
trachea   (tra'ke-ah)
alveolus   (al-ve'o-lus)
mediastinum   (me''de-as-ti'num)
pleura   (ploor'ah)
compliance   (kom-pli'ans)
pneumothorax   (nu''mo-tho'raks)
surfactant   (sur-fak'tant)
spirometry   (spi-rom'ē-tre)
dyspnea   (disp'ne-ah)
solubility   (sol''u-bil'ĭ-te)
apneustic   (ap-nu'stik)
chemoreceptors   (ke''mo-re-cep'tors)
hypoxemia   (hi''pok-se'me-ah)
methemoglobin   (met-he'mo-glo'bin)
myoglobin   (mi''o-glo'bin)
carbaminohemoglobin
    (kar-bam''i-no-he''mo-glo'bin)
erythropoietin   (ē-rith''ro-poi'ē-tin)
hyperpnea   (hi''perp-ne'ah)

Functions and Development of the Respiratory System

The respiratory tract can be divided into upper and lower divisions on the basis of function and embryological development.

Objective 1. Describe the functions included in the term *respiration.*

Objective 2. Identify the organs of the respiratory system, and describe their locations.

Objective 3. Compare the development of the upper respiratory tract with that of the lower.

The term *respiration* includes three separate but related functions: (1) **ventilation** (breathing); (2) **gas exchange** between the air and blood in the lungs and between the blood and tissues; and (3) **oxygen utilization** by the tissues in the energy-liberating reactions of cell respiration. The exchange of gases (oxygen and carbon dioxide) between the air and blood is called *external respiration.* Gas exchange between the blood and tissues is known as *internal respiration.*

It is estimated that a relaxed adult breathes about nine to twenty times a minute, ventilating about 5 to 6 liters of air during this period. Strenuous exercise increases the demand for oxygen and increases the respiratory rate fifteenfold to twentyfold, so that about 100 liters of air are breathed each minute. If breathing stops a person will lose consciousness after four or five minutes, may suffer brain damage after seven to eight minutes, and will die after ten minutes. A knowledge of the structure and function of the respiratory system, therefore, is of the greatest clinical importance.

The major passages and structures of the respiratory system (fig. 23.1) are the **nasal cavity, pharynx, larynx,** and **trachea,** and the **bronchi, bronchioles,** and **alveoli** within the **lungs.** The respiratory system is frequently divided into the *conducting zone* and the *respiratory zone.* The conducting zone includes all of the cavities and structures that transport gases to the respiratory zone. The respiratory zone, including the respiratory bronchioles and alveoli, is the region in which gas exchange with the blood occurs.

Embryological Development of the Respiratory System

The development of the respiratory system is initiated early in embryonic development and involves both ectoderm and endoderm. Although the structures of the upper respiratory system (nose and pharynx) develop simultaneously with those of the lower (trachea and lungs), it is best to discuss the development separately because of the different germ layers involved.

Development of the Upper Respiratory System. *Cephalization (sef''al-i-za'shun)* means that the cephalic, or head, end of an organism structurally and functionally differentiates from the rest of the body. In humans, cephalization is apparent early in development. One of the initial events is the formation of the nasal cavity at three and one-half to four weeks of embryonic life.

---

respiration: L. *re*, back; *spirare*, to breathe
cephalization: Gk. *kephale*, head

Figure 23.1.   The basic anatomy of the respiratory system.

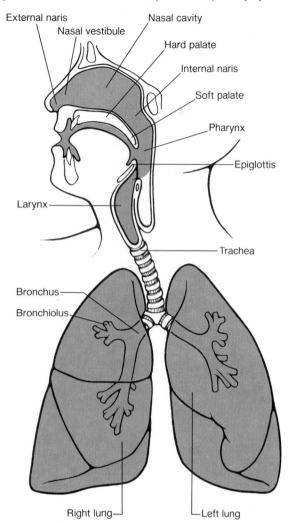

Figure 23.2.   The development of the upper respiratory system from the fourth to the twelfth week. (a) through (e) development in the fourth week, seen in midsagittal section.

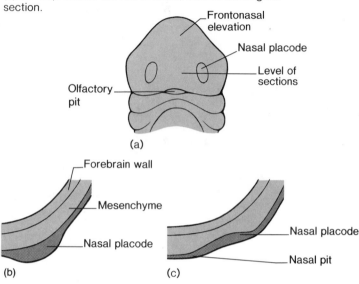

and cosmetically treated. The more immediate and serious problem facing an infant with a cleft palate is that it may be unable to create enough suction to nurse properly.

Development of the Lower Respiratory System.   The lower respiratory system begins as a diverticulum *(di″ver-tik′u-lum),* or outpouching, from the ventral surface of endoderm along the lower pharyngeal region (fig. 23.3). This diverticulum forms during the fourth week of development and is referred to as the **laryngotracheal** *(lah-ring″go-tra′ke-al)* **bud.** As the bud grows, the proximal portion forms the trachea and the distal portion bifurcates (splits) into a right and left bronchus.

The buds continue to elongate and split until all the tubular network within the lower respiratory tract is formed (fig. 23.3). As the terminal portion forms air sacs, called **alveoli,** at about eight weeks of development, the supporting lung tissue begins to form. The complete structure of the lungs, however, is not fully developed until about twenty-six weeks of age, so premature infants born prior to this time require special artificial respiratory equipment to live.

A region of thickened ectoderm called the **olfactory** (nasal) **placode** *(plak′ōd)* appears on the front and inferior part of the head (fig. 23.2). The placode invaginates to form the **olfactory pit,** which extends posteriorly to connect with the **foregut.** The foregut later develops into the pharynx.

The mouth, or oral cavity, develops at the same time as the nasal cavity, and for a short time there is a thin **oronasal** *(or″o-na′zal)* **membrane** separating the two cavities. This membrane ruptures during the seventh week, and a single, large **oronasal cavity** forms. Shortly after, tissue plates begin to grow horizontally across the cavity. At approximately the same time, a vertical plate develops inferiorly from the roof of the nasal cavity. These plates complete their formation by three months of development. The vertical plate forms the nasal septum, and the horizontal plates form the hard palate.

A *cleft palate* forms when the horizontal plates fail to meet in the midline. This condition can be surgically

1. *Define the terms* external respiration *and* internal respiration.
2. *Explain the role of ectoderm and endoderm in the embryological development of the respiratory system.*
3. *At what age might a cleft palate develop? At what age are the lungs sufficiently developed to sustain independent life?*
4. *Explain the difference between invagination and diverticulum in relation to the development of the respiratory system.*

---

alveolus: L. diminutive of *alveus,* cavity

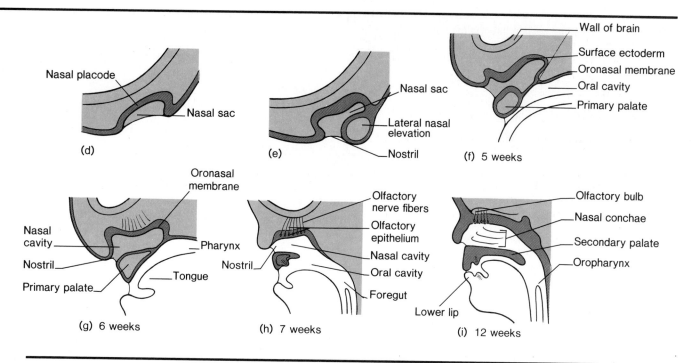

Figure 23.3. The development of the lower respiratory system. (a) through (h) a sequence of diverticula development: (a) through (d) occurs by the fourth week; (e) through (h) occurs by the eighth week.

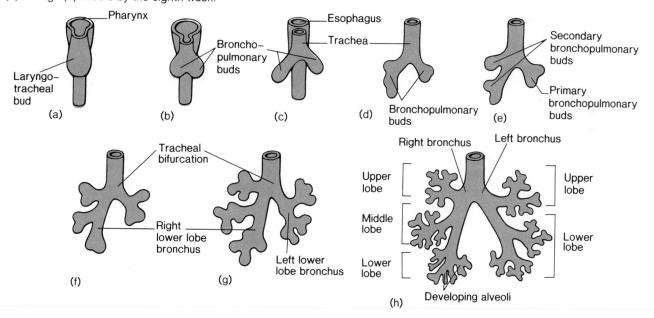

Conducting Passages

Air is conducted through the oral and nasal cavities to the pharynx and then through the larynx to the trachea and bronchial tree. These structures deliver warmed and humidified air to the respiratory zone in the lungs.

Objective 4. List the types of epithelial tissue present in each region of the respiratory tract, and discuss the significance of these differences.

Objective 5. Identify the boundaries of the nasal cavity, and discuss the relationship of the paranasal sinuses to the rest of the respiratory system.

Objective 6. Describe the three regions of the pharynx, and identify the structures located in each.

Objective 7. Discuss the role of the laryngeal region in digestion and respiration.

Objective 8. Identify the anatomical features of the larynx associated with sound production and respiration.

Figure 23.4. A sagittal section through the nose showing the parts of the nasal septum.

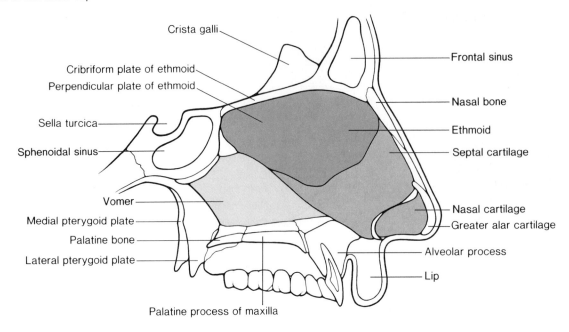

- Crista galli
- Cribriform plate of ethmoid
- Perpendicular plate of ethmoid
- Sella turcica
- Sphenoidal sinus
- Vomer
- Medial pterygoid plate
- Palatine bone
- Lateral pterygoid plate
- Palatine process of maxilla
- Frontal sinus
- Nasal bone
- Ethmoid
- Septal cartilage
- Nasal cartilage
- Greater alar cartilage
- Alveolar process
- Lip

The conducting passages serve to transport air to the respiratory structures of the lungs. The passageways are lined with various types of epithelia to prepare the air properly for utilization. The majority of the conducting passages are held permanently open by muscle or a bony or cartilagenous framework.

Nose

The **nose** includes an external portion that juts out from the face and an internal **nasal cavity** for the passage of air. The external portion of the nose is covered with skin and supported by paired **nasal bones,** forming the bridge, and pliable cartilage, forming the distal portions (figs. 23.4 and 23.5). The **septal cartilage** forms the anterior portion of the **nasal septum,** and the paired **lateral cartilages** and **alar cartilages** form the framework around the **nostrils.**

The perpendicular plate of the **ethmoid** bone and the **vomer** bone together with the septal cartilage constitute the supporting framework of the **nasal septum,** which divides the nasal cavity into two lateral halves, each referred to as a **nasal fossa.** The vestibule is the anterior expanded portion of the nasal fossa (fig. 23.6). Each nasal fossa opens anteriorly through the **nostrils,** or **anterior nares** *(na'rēz)* and communicates posteriorly with the **nasopharynx** through the **choanae** *(ko-a'ne),* or **posterior nares.** The roof of the nasal cavity is formed anteriorly by the frontal bone and paired nasal bones, medially by the cribriform plate of the ethmoid, and posteriorly by the sphenoid bone. The palatine and maxillary bones form the floor of the cavity. On the lateral walls are the superior,

Figure 23.5. The supporting framework of the external nose.

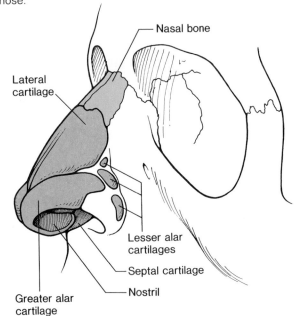

- Nasal bone
- Lateral cartilage
- Lesser alar cartilages
- Septal cartilage
- Nostril
- Greater alar cartilage

middle, and inferior **conchae** *(kong'ke)* (turbinate bones). Air passages between the conchae are referred to as **meatuses** (fig. 23.6). The anterior openings of the nasal cavity are lined with stratified squamous epithelium, whereas the conchae are lined with ciliated columnar epithelium (fig. 23.7). Mucus-secreting goblet cells are present in great abundance throughout both regions.

nose: O.E. *nosu,* nose
nostril: O.E. *nosu,* nose; *thyrel,* hole

choane: Gk. *choane,* funnel
meatus: L. *meatus,* path

Figure 23.6. The lateral wall of the nasal cavity. There are several openings into the nasal cavity, including the openings of the various paranasal sinuses and the nasolacrimal ducts that drain from the eyes.

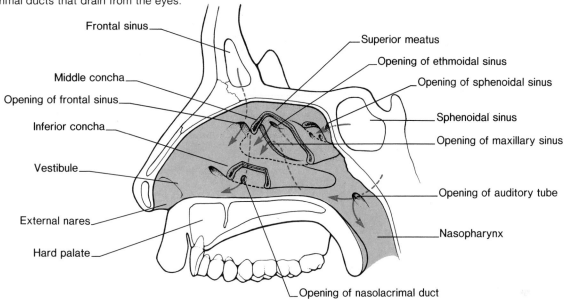

Frontal sinus

Middle concha

Opening of frontal sinus

Inferior concha

Vestibule

External nares

Hard palate

Superior meatus

Opening of ethmoidal sinus

Opening of sphenoidal sinus

Sphenoidal sinus

Opening of maxillary sinus

Opening of auditory tube

Nasopharynx

Opening of nasolacrimal duct

Figure 23.7. The various types of epithelial tissues present throughout the respiratory system.

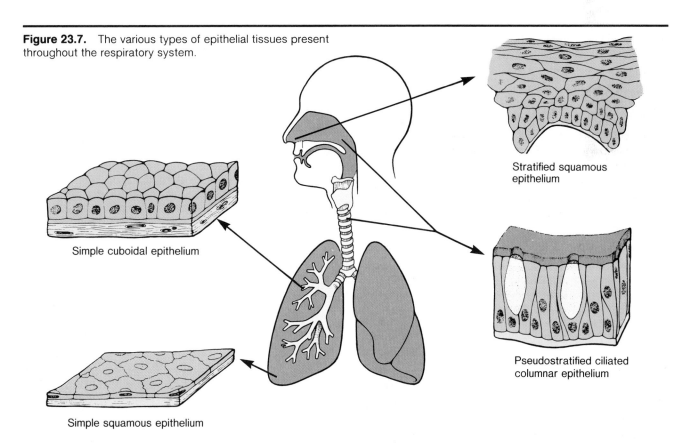

Simple cuboidal epithelium

Simple squamous epithelium

Stratified squamous epithelium

Pseudostratified ciliated columnar epithelium

Figure 23.8.   Paranasal sinuses.

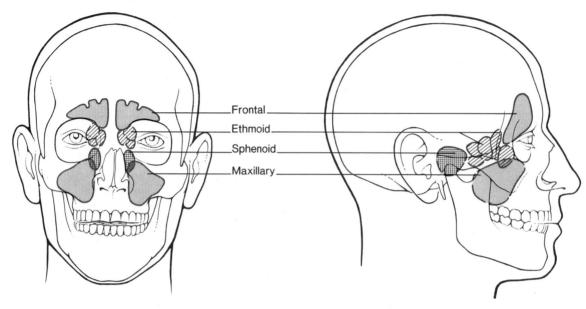

Three functions are associated with the nasal cavity and contents:

1. The nasal epithelium covering the conchae serves to warm, moisten, and cleanse the air. The nasal epithelium is highly vascular and covers an extensive surface area. This is important for warming the air but unfortunately also makes us susceptible to nosebleeds. Nasal hairs called **vibrissae** *(vi-bris'e),* which often extend from the nostrils, filter macroparticles that might otherwise be inhaled. Fine particles such as dust, pollen, or smoke are trapped along the moist mucous membrane lining the nasal cavity.
2. Olfactory epithelium in the upper medial portion of the nasal cavity is concerned with the sense of smell.
3. The nasal cavity is associated with voice phonetics by functioning as a resonating chamber.

There are several drainage openings into the nasal cavity. The paranasal ducts drain mucus from the paranasal sinuses, and the nasolacrimal ducts drain tears from the eyes. An excessive secretion of tears causes the nose to run as the tears drain into the nasal cavity. The auditory tube from the middle ear enters the upper respiratory tract posterior to the nasal cavity in the nasopharynx. With all these accessory connections, it is no wonder that infections can spread so easily from one chamber to another throughout the facial area. To avoid causing damage or spreading infections to other areas, you must be careful not to blow your nose too forcefully.

Paranasal Sinuses

Paired air spaces in certain bones of the facial region are called **paranasal sinuses.** These sinuses are named according to the bones in which they are found: thus there are the **maxillary, frontal, sphenoidal,** and **ethmoidal sinuses** (fig. 23.8). Each sinus communicates via drainage ducts within the nasal cavity on its own side. Paranasal sinuses may help to warm and moisten the inspired air. These sinuses are responsible for some sound resonance, but most importantly, they function to decrease the weight of the skull while giving structural strength.

You can observe your own paranasal sinuses. Face a mirror in a darkened room and shine a small flashlight into your face. The frontal sinuses will be illuminated by directing the light just below the eyebrow. The maxillary sinuses are illuminated by shining the light into the oral cavity and closing your mouth around the flashlight.

Pharynx

The **pharynx** *(far'inks)* is a funnel-shaped passageway approximately 13 cm (5 in.) in length that connects the nasal and oral cavities to the larynx at the base of the skull. The supporting walls of the pharynx are composed of skeletal muscle, and the lumen is lined with a mucous membrane. There are several paired lymphoid organs called **tonsils** in the pharynx (fig. 23.9). The pharynx, commonly referred to as the throat or gullet, has both respiratory and digestive functions. It also provides a resonating chamber for certain speech sounds. The pharynx is divided on the basis of location and function into three regions: nasal, oral, and laryngeal.

---

vibrissa: L. *vibrare,* to vibrate

---

sinus: L. *sinus,* bend or curve
pharynx: L. *pharynx,* throat
tonsil: L. *toles,* goiter or swelling

Figure 23.9. A sagittal section of the head showing the structures of the upper respiratory tract.

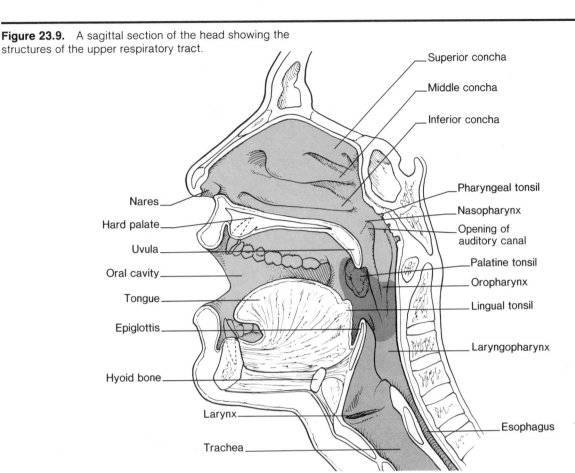

The **nasopharynx** *(na″zo-far′inks)* has only a respiratory function. It is the uppermost portion of the pharynx, directly posterior to the nasal cavity and above the soft palate. A pendulous **uvula** hangs from the middle, lower border of the soft palate. The paired **auditory,** or **eustachian, tubes** connect the nasopharynx with the middle ear cavities. **Pharyngeal tonsils,** or **adenoids,** are situated in the posterior wall of this cavity. During the act of swallowing, the soft palate and uvula are elevated to block the nasal cavity and prevent food from entering. Occasionally a person may suddenly exhale air (as with a laugh) while in the process of swallowing fluid. If this occurs before the uvula effectively blocks the nasopharynx, fluid will be discharged through the nasal cavity.

The **oropharynx** *(o″ro-far′inks)* is the middle portion of the pharynx between the soft palate and the level of the hyoid bone. The base of the tongue forms the anterior wall of the oropharynx. Paired **palatine tonsils** are located along the posterior lateral wall of the oropharynx, and the **lingual tonsils** are found on the base of the tongue. This portion of the pharynx has both a respiratory and digestive function.

The **laryngopharynx** *(lah-ring″go-far′inks)* is the lowermost portion of the pharynx. It extends posteriorly

from the level of the hyoid bone to the larynx and opens into the esophagus. It is at the lower laryngopharynx that the respiratory and digestive systems become distinct. Food is directed into the esophagus, whereas air is moved anteriorly into the larynx.

During a physical examination, a physician commonly depresses the patient's tongue and examines the condition of the palatine tonsils. Tonsils are pharyngeal lymphoid organs and tend to become swollen and inflamed after persistent infections. Tonsils have to be surgically removed when they become so overrun with pathogens after repeated infections that they themselves become the source of the infection. The removal of the palatine tonsils is called a **tonsillectomy,** whereas the removal of the pharyngeal tonsils is called an **adenoidectomy.**

Larynx

The **larynx** *(lar′inks),* or voice box, forms the entrance into the lower respiratory system as it connects the laryngopharynx with the trachea. It is positioned in the anterior midline of the neck at the level of the fourth through sixth

uvula: L. *uvula,* small grape
adenoid: Gk. *adenoeides,* glandlike

larynx: Gk. *larynx,* upper windpipe

Figure 23.10. The structure of the larynx: (*a*) an anterior view; (*b*) a lateral view.

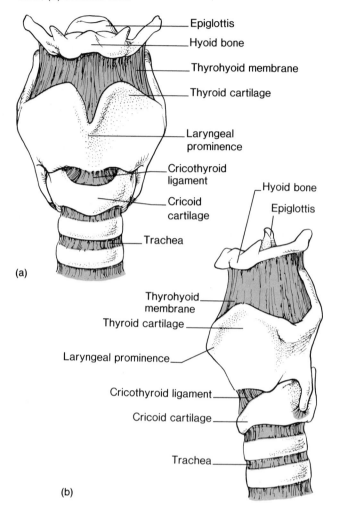

(a)

(b)

Figure 23.11. A superior view of the vocal cords as seen with a laryngoscope: (*a*) vocal cords are taut; (*b*) vocal cords are relaxed and the glottis is opened.

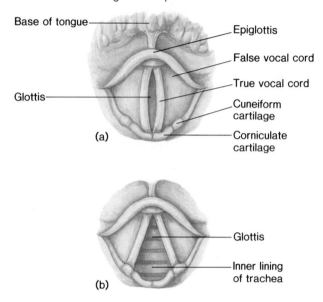

(a)

(b)

cervical vertebrae. The larynx has two functions. Its primary function is to prevent food or fluid from entering the trachea and lungs during swallowing and to permit passage of air while breathing. A secondary function is to produce sounds.

In structure, the larynx is shaped like a triangular box (fig. 23.10). It is composed of a framework involving nine cartilages: three are large, single structures and six are smaller, paired structures. The largest of the unpaired cartilages is the anterior **thyroid cartilage,** commonly called the "Adam's apple." The **laryngeal prominence** of the thyroid cartilage forms an anterior vertical ridge along the larynx that can be palpated on the midline of the neck. The thyroid cartilage is typically larger and more prominent in males than females because of the effect of male sex hormones on the development of the larynx during puberty.

The spoon-shaped **epiglottis** has a cartilaginous framework. It is behind the root of the tongue and aids in closing the **glottis,** or laryngeal opening, during swallowing.

thyroid: Gk. *thyreos,* shieldlike

The lower end of the larynx is formed by the ring-shaped **cricoid (*kri'koid*) cartilage.** This third unpaired cartilage connects the thyroid cartilage above and the trachea below. The paired **arytenoid *(ar''ē-te'noid)* cartilages,** located above the cricoid and behind the thyroid, furnish the attachment of the **vocal cords.** The other paired **cuneiform cartilages** and **corniculate *(kor-nik'u-lāt)* cartilages** are small accessory cartilages that are closely associated with the arytenoid cartilages (fig. 23.11).

Two pairs of strong connective tissue bands are stretched across the upper opening of the larynx from anterior to posterior. These are the **true vocal cords** and the **false vocal cords.** The false vocal cords, or **ventricular folds,** support the true vocal cords and, as their name implies, are not used in sound production. The true vocal cords vibrate in the production of sound.

The **laryngeal muscles** are extremely important in closing the glottis during swallowing and in speech. There are two groups of laryngeal muscles: **extrinsic muscles,** responsible for elevating the larynx during swallowing, and **intrinsic muscles** that, when contracted, change the length, position, and tension of the vocal cords. Various pitches are produced as air passes over the altered vocal cords. If the vocal cords are taut, vibration is more rapid and causes a higher pitch. Less tension on the cords produces lower sounds. Mature males generally have thicker and longer vocal cords than females; therefore, the vocal cords of males vibrate more slowly and produce lower pitches. The loudness of vocal sound is determined by the force of the air passed over the vocal cords and the amount of vibration. The vocal cords do not vibrate when whispering.

cricoid: Gk. *krikos,* ring; *eidos,* form
arytenoid: Gk. *arytaina,* ladle- or cup-shaped
cuneiform: L. *cuneus,* wedge-shaped
corniculate: L. *corniculum,* diminutive of *cornu,* horn

Figure 23.12.   An anterior view of the trachea and esophagus.

Thyroid cartilage

Cricoid cartilage

Trachea

Carina

Bronchi

Esophagus

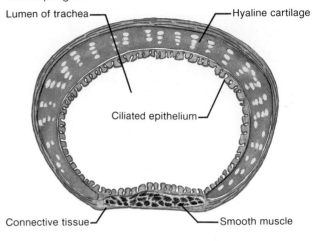

Figure 23.13.   A transverse section of the trachea. The connective tissue portion faces posteriorly in contact with the esophagus.

Lumen of trachea

Hyaline cartilage

Ciliated epithelium

Connective tissue

Smooth muscle

Sounds originate in the larynx, but other structures are necessary to convert sound into recognizable speech. Vowel sounds, for example, are produced by constriction of the walls of the pharynx. The pharynx, paranasal sinuses, and oral and nasal cavities act as resonating chambers. The final enunciation of words is accomplished through movements of the lips and tongue.

Swallowing is a complex action involving several neurological pathways and various groups of muscles. One of the swallowing responses is the elevation of the larynx to close the glottis against the epiglottis. You can note this movement on yourself by cupping your fingers lightly over your own larynx and then swallowing. Food may become lodged within the glottis if it is not closed as it should be. In this case, the abdominal-thrust (Heimlich) maneuver can be used to prevent suffocation.

Trachea

The **trachea** *(tra'ke-ah)*, commonly called the windpipe, is a rigid tube, approximately 12 cm (4½ in.) long and 2.5 cm (1 in.) in diameter, connecting the larynx to the bronchi (fig. 23.12). It is positioned anterior to the esophagus as it extends into the thoracic cavity. A series of sixteen to twenty C-shaped rings of hyaline cartilage form the walls of the trachea (fig. 23.13). The open part of the C is positioned posteriorly and is covered by fibrous connective tissue and smooth muscle. This soft tissue allows the adjacent esophagus to expand as swallowed food is transported to the stomach. These cartilages provide a rigid but flexible tube in which the lumen, or airway, is permanently open. The mucosa (surface lining of the lumen) consists of pseudostratified ciliated columnar epithelium containing many mucus-secreting **goblet cells** (see fig. 23.7). Inhaled dust particles stick to the mucus, and the

Figure 23.14.   The site for a surgical tracheostomy.

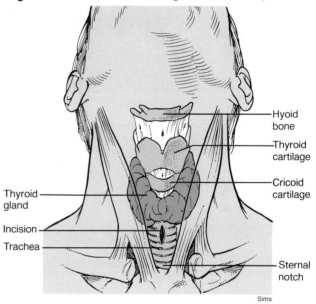

Hyoid bone

Thyroid cartilage

Cricoid cartilage

Thyroid gland

Incision

Trachea

Sternal notch

Sims

cilia sweep it upward to the pharynx, where it is removed through a cough reflex. The tracheal bifurcation forming the right and left **bronchi** is reinforced by the **carina** *(kah-ri'nah)*, a keel-like cartilage plate.

If the trachea becomes occluded through inflammation, excessive secretion, trauma, or aspiration of a foreign object, it may be necessary to create an emergency opening into this tube so that ventilation can still occur. A **tracheotomy** is the process of surgically opening the trachea, and a **tracheostomy** is the procedure of inserting a tube into the trachea to permit breathing and to keep the passageway open (fig. 23.14). A tracheotomy should be performed only by a competent physician, however, as there is great risk of cutting a recurrent laryngeal nerve or carotid artery, which are also in this area.

---

trachea: L. *trachia,* rough air vessel

---

bronchus: L. *bronchus,* windpipe
carina: L. *carina,* keel

Figure 23.15.   A photograph of a plastic cast of the conducting airways from the trachea to the terminal bronchioles.

Bronchial Tree

The **bronchial tree** is so named because it is composed of a series of respiratory tubes that branch into progressively narrower tubes as they extend into the lung (fig. 23.15). The trachea bifurcates into a right and left **primary bronchus** *(brong'kus)* at the level of the sternal angle behind the manubrium. Each bronchus has hyaline cartilage rings surrounding its lumen to keep it open as it extends into the lung. Because of the more vertical position of the right bronchus, foreign particles are more likely to lodge here than in the left bronchus.

The bronchus divides deeper in the lungs to form **secondary bronchi** and **segmental (tertiary) bronchi** (fig. 23.16). The bronchial tree continues to branch into yet smaller tubules called **bronchioles** *(brong'ke-ōl)* (fig. 23.17). There is little cartilage in the bronchioles, which contain thick smooth muscle that can constrict or dilate these airways. Bronchioles provide the greatest resistance to air flow in the conducting passages and thus are analogous to the function of arterioles in the vascular system. A simple cuboidal epithelium lines the bronchioles rather than a pseudostratified columnar epithelium as in the bronchi (fig. 23.7). Numerous **terminal bronchioles** mark the end of the air-conducting pathway.

Figure 23.16.   An anterior view of the lower respiratory system showing the bronchial tree, alveoli, and lungs.

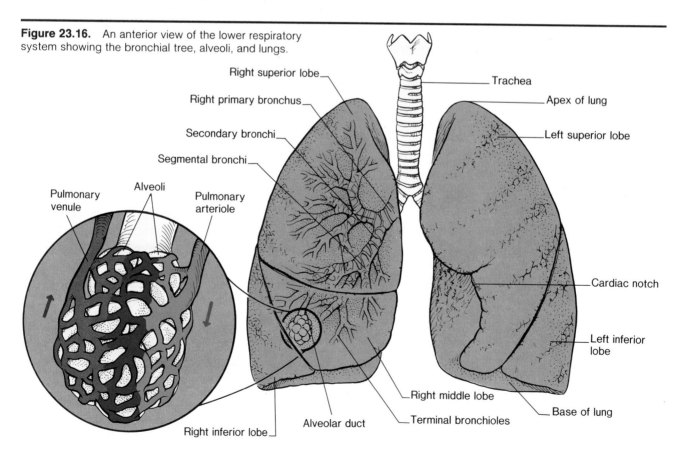

1. What are the functions of the nasal cavity?
2. Identify the structures that make up the nasal septum.
3. Identify the location of the nasopharynx, and list the structures found in this cavity.
4. List the paired and unpaired cartilages of the larynx, and describe the functions of the larynx.
5. Describe the structure of the conducting airways from the trachea to the terminal bronchioles.

Alveoli, Lungs, and Pleura

Alveoli are the functional units of the lungs where gas exchange occurs. Right and left lungs are separately contained in pleural membranes.

Objective 9. Describe the structure and function of the alveoli.

Objective 10. Describe the surface anatomy of the lungs in relation to the thorax.

Objective 11. Discuss the structural arrangement of the thoracic serous membranes, and explain their functions.

Figure 23.17. An anteroposterior bronchogram of the lungs.

Trachea

Left bronchus

Right bronchus

Bronchioles

Respiratory Sacs

The terminal bronchioles branch into many **alveolar ducts,** which lead directly into **alveolar sacs** and **alveoli.** The latter three structures comprise the respiratory zone of the lungs. Gas exchange occurs across the walls of the tiny (0.25–0.50 mm in diameter) alveoli, which are therefore the functional units of the respiratory system. The enormous number of these structures (about 350 million per lung) provides a very high surface area (60–80 square meters, or 760 square feet) for the diffusion of gases. The diffusion rate is further increased by the fact that each alveolus *(al-ve'o-lus)* is only one cell layer thick, so that the total "air-blood barrier" is only one alveolar cell and one blood capillary cell across, or about 2 μm. This is an average distance because the type II alveolar cells are thicker than the type I alveolar cells (fig. 23.18).

Alveoli are polyhedral in shape and are usually clustered together, like the units of a honeycomb, in groups called alveolar sacs (fig. 23.19). These clusters usually occur at the ends of alveolar ducts. Individual alveoli also occur as separate outpouchings along the length of the alveolar ducts. Although the distance between each alveolar duct and its terminal alveoli is only about 0.5 mm, these units together comprise most of the mass of the lungs.

Figure 23.18. A diagram showing the relationship between lung alveoli and pulmonary capillaries.

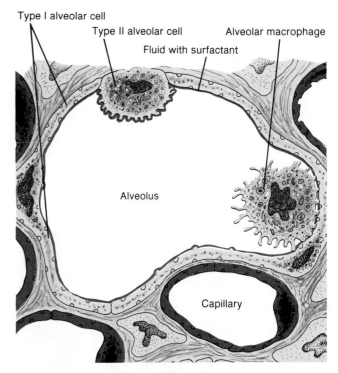

Type I alveolar cell

Type II alveolar cell

Alveolar macrophage

Fluid with surfactant

Alveolus

Capillary

Figure 23.19. (*a*) a scanning electron micrograph showing lung alveoli and a small bronchiole. (*b*) the alveoli under higher power (the arrow points to an alveolar pore through which air can pass from one alveolus to another).

(a)

(b)

Figure 23.20. A cross section of the thoracic cavity showing the mediastinum and pleural membranes.

Sternum

Anterior mediastinum

Pericardial cavity

Heart (in **middle mediastinum**)

Lung

Thoracic wall

Visceral pleura of mediastinum

Parietal pleura of mediastinum

Parietal pleura

Visceral pleura

Parietal pericardium

Visceral pericardium

Pleural cavity

Bronchus

Esophagus

Posterior mediastinum

Thoracic vertebra

Figure 23.21. Lobes, lobules, and bronchopulmonary segments of the lungs.

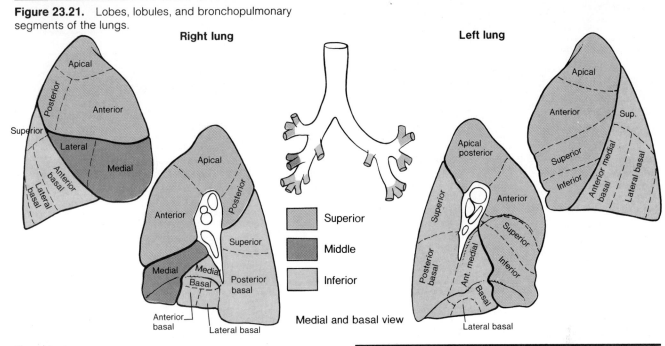

Right lung

Left lung

Medial and basal view

Superior
Middle
Inferior

Lungs

The *lungs* are large, spongy, paired organs within the thoracic cavity. They lie against the rib cage anteriorly and posteriorly and extend from the diaphragm to a point just above the clavicles. The lungs are separated from one another by the heart and other structures of the **mediastinum** *(me″de-as-ti′num),* as shown in figure 23.20. All structures of the respiratory system beyond the bronchi, including the bronchial tree and alveoli, are contained in the lungs.

There are some basic similarities and distinct differences between the right and left lung. Each lung presents four borders or surfaces that match the contour of the thoracic cavity. The **mediastinal (medial) surface** of each lung is slightly concave and contains a vertical slit, the **hilum** *(hi′lum),* through which pulmonary vessels, nerves, and bronchi traverse. The inferior border of the lung, called the **base,** is concave as it fits over the convex dome of the diaphragm. Anteriorly, the portion of the lungs that extends above the level of the clavicle is called the **apex,** or **cupola.** Finally, the broad, rounded surface in contact with the membranes covering the ribs is called the **costal surface.**

The left lung is somewhat smaller than the right and has a **cardiac notch** on its medial surface to accommodate the heart. The left lung is subdivided into a **superior lobe** and an **inferior lobe** by a single fissure. The right lung is subdivided by two fissures into three lobes: **superior, middle,** and **inferior** (fig. 23.16). Each lobe of the lung is divided into many small lobules, which in turn contain the alveoli. Lobular divisions of the lungs comprise specific bronchial segments. The right lung contains ten bronchial segments, and the left lung contains eight (fig. 23.21).

mediastinum: L. *mediastinus,* intermediate
hilum: L. *hilum,* a trifle (little significance)
cupola: L. *cupula,* diminutive of *cupa,* dome, tub

Figure 23.22. The position of the lungs and associated pleurae.

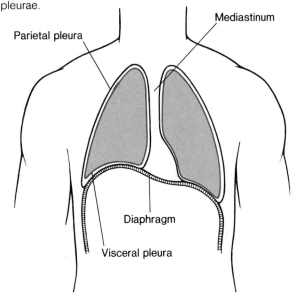

Parietal pleura
Mediastinum
Diaphragm
Visceral pleura

Pleura

Pleurae are serous membranes investing the lungs (fig. 23.20 and fig. 23.22). The pleura of each lung is composed of both a visceral and parietal portion. The **visceral pleura** *(ploo′rah)* adheres to the outer surface of the lung and extends into each of the interlobular fissures. The **parietal pleura** lines the thoracic walls and the thoracic surface of the diaphragm. A continuation of the parietal pleura around the heart and between the lungs forms the boundary of the mediastinum. Between the visceral and parietal pleurae is a moistened space called the **pleural**

pleura: Gk. *pleura,* side or rib

Table 23.1 Major structures of the respiratory system

Structure	Description	Function
Nose	Jutting external portion that is part of the face, plus internal nasal cavity	Warms, moistens, and filters inhaled air as it is conducted to the pharynx
Paranasal sinuses	Air spaces in certain facial bones	Produce mucus, provide sound resonance, lighten the skull
Pharynx	Chamber connecting oral and nasal cavities to larynx	Passageway for air into larynx and for food into esophagus
Larynx	Voice box; short passageway that connects the pharynx to the trachea	Passageway for air; sound production; prevents foreign materials from entering trachea
Trachea	Flexible, tubular connection between larynx and bronchial tree	Passageway for air; pseudostratified ciliated columnar epithelium filters the air
Bronchial tree	Bronchi and branching bronchioles in the lung; tubular connection between trachea and alveoli	Passageway for air; continued filtering of air
Alveoli	Microscopic, membranous air sacs within the lungs	Functional units of respiration: site of gaseous exchange between the respiratory and circulatory systems
Lungs	Major organs of respiratory system; located in pleural cavities in the thorax	Contain bronchial trees, alveoli, and associated pulmonary vessels
Pleurae	Serous membranes covering the lungs and lining the thoracic cavity	Compartmentalizes, protects, and lubricates the lungs

Source: Van De Graaff, Kent M., *Human Anatomy.* © 1984 Wm. C. Brown Publishers, Dubuque, Iowa. All Rights Reserved. Reprinted by permission.

cavity. An inferiorly extending reflection of the pleural layers around the roots of each lung is called the **pulmonary ligament.** The pulmonary ligaments help to support the lungs.

The arrangement of the pleurae serve three functions. First, a small amount of fluid within the pleural cavities acts as a lubricant to allow the lungs to slide along the chest wall. Second, the pressure in the pleural cavity is lower than the pressure in the lungs, which is needed for ventilation. The third function of the pleurae is the effective separation of the thoracic organs. Although the serous membrane is composed of simple squamous epithelium and fibrous connective tissue, it provides tight compartments for each major thoracic organ.

There are four distinct compartments in the thoracic cavity. A pleural cavity surrounds each lung, the heart is situated in the pericardial cavity, and the esophagus, thoracic duct, major vessels, various nerves, and portions of the respiratory tract are located in the mediastinum. This *compartmentalization* has a protective value because infections are usually confined to one compartment and damage to one organ will not usually involve another. *Pleurisy,* for example, which is an inflamed pleura, is generally confined to one side. A penetrating injury to one side, like a knife wound, might cause one lung to collapse but not the other.

The lungs of a newborn are pink in color but may become discolored in an adult as a result of smoking or air pollution. Smoking not only discolors the lungs, it may also cause deterioration of the alveoli. **Emphysema** *(em″fĭ-se′mah)* and **lung cancer** are diseases that have been linked to heavy smoking. If a person moves to a less polluted environment or gives up smoking, the lungs will become more pinkish and healthy.

pulmonary: Gk. *pleumon,* lung

The functions of the major organs of the respiratory system along with a description of each is presented in table 23.1.

1. Describe the structure of the respiratory zone of the lungs, and explain how this structure aids in gas exchange.
2. Compare the structure of the right and left lung.
3. Describe the structure and location of the mediastinum, and list the organs it contains.
4. Describe the location of the visceral pleura, parietal pleura, and pleural cavity. Explain the functional significance of the pleural membranes.

Physical Aspects of Ventilation

The movement of air into and out of the lungs occurs as a result of pressure differences induced by changes in lung volumes. Ventilation is thus influenced by the physical properties of the lungs, including their compliance, elasticity, and surface tension.

Objective 12. Define intrapulmonary and intrapleural pressures, and describe how they change during breathing.

Objective 13. Define the terms *compliance, elasticity,* and *surface tension,* and describe how they affect lung function.

Objective 14. Describe Boyle's law and explain how it relates to lung function.

The wet, serous membranes of the visceral and parietal pleurae are normally against each other, so that the lungs are stuck to the chest wall in the same manner that two wet pieces of glass stick to each other. The *intrapleural space* between the two wet membranes contains only a thin

Table **23.2** Intrapulmonary and intrapleural pressures in normal, quiet breathing, and the transpulmonary pressure (intrapulmonary minus intrapleural pressure) acting to expand the lungs		
.	**Inspiration**	**Expiration**
Intrapulmonary pressure (mm Hg)	−3 (to zero)	+3 (to zero)
Intrapleural pressure (mm Hg)	−6	−3
Transpulmonary pressure (mm Hg)	+3	+6

Note: Pressures indicate mm Hg below or above atmospheric pressure. Intrapleural pressure is normally always negative (subatmospheric).
Source: Fox, Stuart Ira, *Human Physiology.* © 1984 Wm. C. Brown Publishers, Dubuque, Iowa. All Rights Reserved. Reprinted by permission.

Figure 23.23. The volume of air inspired as a function of the transpulmonary pressure. The slope of each line (change in volume per unit change in pressure) in the linear regions is called the *compliance* and is a measure of the distensibility of the lungs.

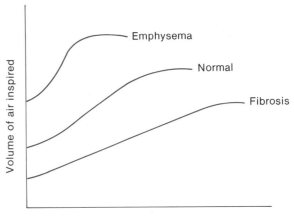

layer of fluid secreted by the pleural membranes. The pleural cavity in a healthy, living organism is thus potential rather than real; it can become real only in abnormal situations when air enters the intrapleural space. Since the lungs normally remain against the chest wall, they get larger and smaller together with the thoracic cavity during respiratory movements.

Intrapulmonary and Intrapleural Pressures

Air enters the lungs during inspiration because the atmospheric pressure is greater than the **intrapulmonary,** or **intra-alveolar pressure.** Since the atmospheric pressure does not usually change, the intrapulmonary pressure must fall below atmospheric pressure to cause inspiration. A pressure below that of the atmosphere is called a *subatmospheric pressure,* or *negative pressure.* During quiet inspiration, for example, the intrapulmonary pressure may become 3 mm Hg less than the pressure of the atmosphere. This subatmospheric pressure is commonly shown as −3 mm Hg. Expiration, conversely, occurs when the intrapulmonary pressure is greater than the atmospheric pressure. During quiet expiration, for example, the intrapulmonary may rise to at least +3 mm Hg over the atmospheric pressure.

The lack of air in the intrapleural space produces a subatmospheric **intrapleural pressure,** which is lower than the intrapulmonary pressure (table 23.2). There is thus a pressure difference across the wall of the lung—called the **transpulmonary pressure**—which is the difference between the intrapulmonary pressure and the intrapleural pressure. Since the pressure within the lungs (intrapulmonary pressure) is greater than outside the lungs (intrapleural pressure), the difference in pressure (transpulmonary pressure) acts to expand the lungs as the thoracic volume expands during inspiration.

Boyle's Law. Changes in intrapulmonary pressure occur as a result of changes in lung volume. This follows from **Boyle's law,** which states that *the pressure of a gas is inversely proportional to its volume.* An increase in lung

volume during inspiration decreases intrapulmonary pressure to subatmospheric levels; air therefore goes in. A decrease in lung volume raises the intrapulmonary pressure above that of the atmosphere, thus pushing air out. These changes in lung volume occur as a consequence of changes in thoracic volume produced by muscle contraction.

Physical Properties of the Lungs

In order for inspiration to occur the lungs must be able to expand when stretched; they must have high *compliance.* In order for expiration to occur the lungs must get smaller when this stretching force is released; they must have *elasticity.* The tendency to get smaller is also aided by *surface tension* forces within the alveoli.

Compliance. The lungs are very distensible—they are, in fact, about one hundred times more distensible than a toy balloon. Another term for distensibility is **compliance** *(kom-pli'ans),* which is defined as *the change in lung volume per change in transpulmonary pressure.* A given transpulmonary pressure, in other words, will cause greater or lesser expansion, depending on the compliance of the lungs.

The compliance of the lungs is reduced whenever there is more resistance to distension. If the lungs were filled with concrete (as an extreme example), a given transpulmonary pressure would produce no increase in lung volume and no air would enter; the compliance would be zero. The infiltration of lung tissue with connective tissue proteins, a condition called *pulmonary fibrosis,* similarly decreases lung compliance (fig. 23.23). In *emphysema,* in which alveolar tissue is destroyed, the lungs are less resistant to distension and have a greater compliance (if no fibrosis is present).

Boyle's law: from Robert Boyle, British physicist, 1627–1691

Figure 23.24.   A pneumothorax of the right lung. The right side of the thorax appears uniformly dark because it is filled with air; the spaces between the ribs is also greater than on the left due to release from the elastic tension of the lungs. The left lung appears denser (less dark) because of shunting of blood from the right to the left lung.

Figure 23.25.   Water molecules at the surface have a greater attraction for other water molecules than for air. The surface molecules are thus attracted to each other and pulled tightly together by the attractive forces of water underneath. This produces surface tension.

Air

Water surface

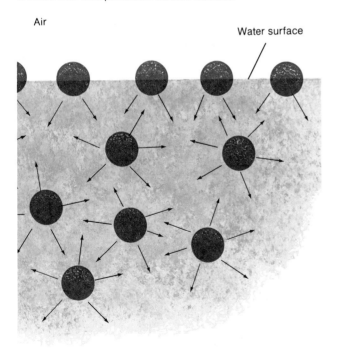

Figure 23.26.   According to physical laws, the surface tension that tends to collapse the alveoli should be greater in the smaller alveolus (*right*) than in the larger alveolus (*left*). This implies that (without surfactant) smaller alveoli would collapse and empty their air into larger alveoli.

Law of LaPlace

$$P = \frac{4 \times T}{r}$$

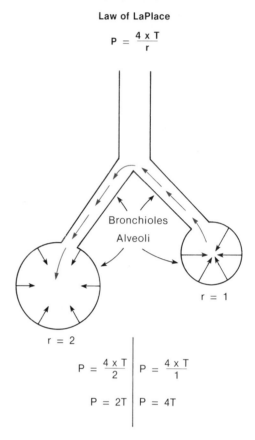

Bronchioles

Alveoli

r = 1

r = 2

$$P = \frac{4 \times T}{2} \quad P = \frac{4 \times T}{1}$$

$$P = 2T \quad P = 4T$$

Elasticity.   The term **elasticity** refers to the tendency of a structure to return to its initial size after being distended. The lungs are very elastic, due to a high content of elastin proteins, and resist distension. Since the lungs are normally stuck to the chest wall, they are always in a state of elastic tension. This tension increases during inspiration when the lungs are stretched and is reduced by elastic recoil during expiration. The elasticity of the lungs and of other thoracic structures thus aids in pushing the air out during expiration.

The elastic nature of lung tissue is revealed when air enters the intrapleural space (as a result of an open chest wound, for example). This condition is called a *pneumothorax (nu"mo-tho'raks)*, which is shown in figure 23.24. As air enters the intrapleural space, the intrapleural pressure rises until it is equal to the atmospheric pressure. When the intrapleural pressure is the same as the intrapulmonary pressure the lung can no longer expand. Not only does the lung not expand during inspiration, it actually collapses away from the chest wall as a result of

pneumothorax: Gk. *pneumon*, spirit (breath); L. *thorax*, thorax, chest

Figure 23.27. The production of pulmonary surfactant by type II alveolar cells. Surfactant appears to be composed of a derivative of lecithin combined with protein.

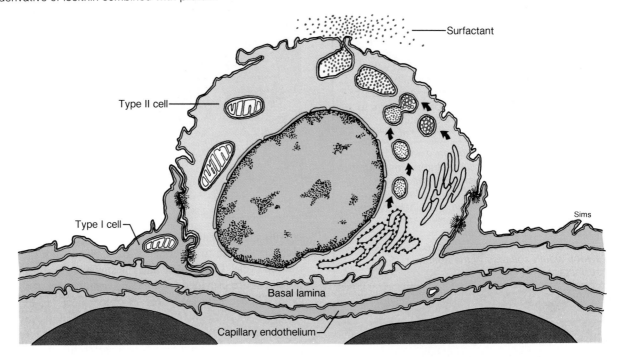

Surfactant

Type II cell

Type I cell

Sims

Basal lamina

Capillary endothelium

elastic recoil. Fortunately, a pneumothorax usually occurs only in one lung since each lung is in a separate pleural compartment.

Surface Tension. The forces that act to resist distension include elastic resistance and **surface tension** forces exerted by fluid in the alveoli. Although the alveoli are relatively dry, they do contain a very thin film of fluid, much like soap bubbles. Surface tension is created by the fact that water molecules at the surface are attracted more to other water molecules than to air. As a result, the surface water molecules are pulled tightly together by attractive forces from underneath (fig. 23.25).

When water molecules are arranged in a bubble, the force generated by their surface tension is directed inward to the center (similar to the effects of tightening a slipknot). According to physical laws, this surface tension force is inversely proportional to the radius of the bubble. The surface tension is thus greater in smaller bubbles than in larger ones.

Like tightening a slipknot, the surface tension force should make soap bubbles and alveoli collapse to a smaller radius. Accordingly, as the alveoli become smaller their surface tension force should increase, which acts to make them become smaller yet. A positive feedback loop is thus created. Since some alveoli are smaller than others and since alveoli decrease in size during expiration, one might expect that smaller alveoli would collapse and empty their air into larger alveoli (fig. 23.26). The reasons why this does not normally occur are discussed in the next section.

Surfactant and the Respiratory Distress Syndrome
Alveolar fluid contains a phospholipid, known as dipalmitoyl lecithin, probably attached to a protein, that functions to lower surface tension. This compound is called **lung surfactant** *(sur-fak'tant),* which is a contraction of the term *surface active agent.* Because of the presence of surfactant, the surface tension in the alveoli is lower at any given lung volume than would be predicted if surfactant were absent. Further, the ability of surfactant to lower surface tension improves as the alveoli get smaller during expiration. Surfactant thus helps to prevent the alveoli from collapsing as a result of surface tension. Even after a forceful expiration, the alveoli remain open and a *residual volume* of air remains in the lungs. Since the alveoli do not collapse, less surface tension has to be overcome to inflate them at the next inspiration.

Surfactant is produced by type II alveolar cells (fig. 23.27) in late fetal life. Since surfactant does not start to be produced until about the eighth month, premature babies are sometimes born with lungs that lack sufficient surfactant, and their alveoli are collapsed as a result. This condition is called **respiratory distress syndrome.** It is also called **hyaline** *(hi'ah-lin)* **membrane disease** because the high surface tension causes plasma fluid to leak into the alveoli, producing a glistening "membrane" appearance (and pulmonary edema). This condition does not occur in all premature babies; the rate of lung development depends on hormonal conditions (thyroxine and hydrocortisone primarily) and on genetic factors.

Even under normal conditions, the first breath of life is a difficult one because the newborn must overcome large surface tension forces in order to inflate its partially collapsed alveoli. The transpulmonary pressure required for the first breath is fifteen to twenty times that required for subsequent breaths, and an infant with respiratory distress syndrome must duplicate this effort with every breath. Fortunately, many babies with this condition can be saved by mechanical ventilators that keep them alive long enough for their lungs to mature and manufacture sufficient surfactant.

1. Describe how the intrapulmonary and intrapleural pressures change during inspiration and expiration, and explain the reasons for these changes in terms of Boyle's law.
2. Define the terms compliance and elasticity, and explain how these lung properties affect inspiration and expiration.
3. Describe lung surfactant, and explain why the alveoli would collapse in the absence of surfactant.

Mechanics of Breathing

Normal, quiet inspiration results from muscle contraction, expiration from muscle relaxation and elastic recoil. These actions can be forced by contractions of the accessory respiratory muscles. The amount of air inspired and expired can be measured by different procedures and used as tests of pulmonary function.

Objective 15. Identify and describe the actions of the muscles involved in both quiet and forced inspiration.

Objective 16. Describe how quiet expiration is produced, and identify and describe the actions of the muscles involved in forced expiration.

Objective 17. Define the concepts of restrictive and obstructive pulmonary disease, and give examples of each type.

Objective 18. Identify different lung volumes and capacities, and describe how pulmonary function tests help diagnose lung disorders.

Breathing, or **pulmonary ventilation,** refers to the movement of air into and out of the respiratory system. This movement of air occurs as a result of differences between the atmospheric and the intrapulmonary pressures; air goes in when the intrapulmonary pressure is subatmospheric, and air goes out when the intrapulmonary pressure rises above that of the atmosphere. As previously discussed, these pressure changes within the lungs occur as a consequence of changes in thoracic volume.

The thorax must be semirigid yet flexible. It must be sufficiently rigid so that it can protect vital organs and

Figure 23.28.   A change in lung volume, as shown by radiographs, during expiration (a) and inspiration (b). The increase in lung volume during full inspiration is shown by comparison with the lung volume in full expiration (dashed lines).

(a)   (b)

provide attachments for many short, powerful muscles. It must be flexible to function as a bellows during the ventilation cycle. The rigidity and the surfaces for muscle attachment are provided by the bony composition of the rib cage. The rib cage is pliable, however, because the ribs are separate from one another and because most ribs (the upper ten of the twelve pairs) are attached to the sternum by resilient costal cartilages. The vertebral attachments likewise provide considerable mobility. The structure of the rib cage and associated cartilages provide continuous elastic tension, so that when stretched by muscle contraction during inspiration, the rib cage can return passively to its resting dimensions when the muscles relax. This elastic recoil is greatly aided by the elasticity of the lungs.

Pulmonary ventilation consists of two phases, inspiration and expiration. Inspiration (inhalation) and expiration (exhalation) are accomplished by alternately increasing and decreasing the volumes of the thorax and lungs (fig. 23.28).

Inspiration and Expiration

The thoracic cavity increases in size during inspiration in three directions: anteroposteriorly, laterally, and vertically. This is accomplished by contractions of the **diaphragm** and the **external intercostal** muscles (fig. 23.29). A contraction of the dome-shaped diaphragm downward increases the thoracic volume vertically. A simultaneous contraction of the external intercostals increases the lateral and anteroposterior dimensions of the thorax.

Other thoracic muscles become involved in forced (deep) inspiration. The most important of these is the scalenus, followed by the pectoralis minor, and in extreme cases the sternocleidomastoid. A contraction of these muscles elevates the ribs in an anteroposterior direction (fig. 23.30), while at the same time the upper rib cage is stabilized so that the intercostals become more effective.

Quiet expiration is a passive process. After becoming stretched by contractions of the diaphragm and thoracic muscles, the thorax and lungs recoil as a result of their

diaphragm: Gk. dia, across; phragma, fence

Figure 23.29.   The position of the principal muscles of inspiration.

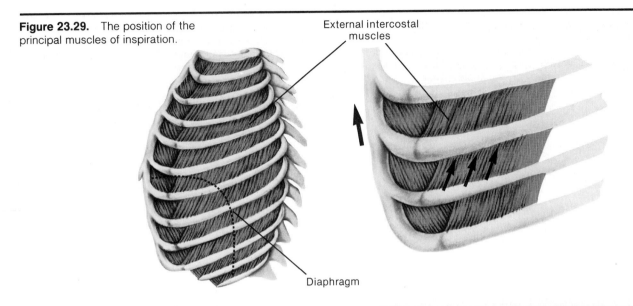

External intercostal muscles

Diaphragm

Figure 23.30.   The movement of the third and ninth rib as viewed anteroposteriorly and superiorly during inspiration and expiration.

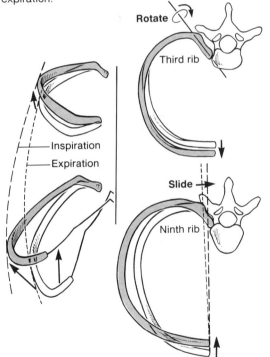

Rotate

Third rib

Inspiration

Expiration

Slide

Ninth rib

Table **23.3** Events of inspiration and expiration during a pulmonary ventilatory cycle*	
Nerve stimulus	**Event**
Inspiration Phrenic nerves	The diaphragm contracts inferiorly, increasing the volume of the thorax. The diaphragm is the principal muscle involved in quiet inspiration.
Intercostal nerves	Contraction of the external intercostal muscles elevates the ribs, thus increasing the capacity of the thoracic cavity.
Accessory, cervical, and thoracic nerves	Forced inspiration is accomplished through contraction of the sternocleidomastoid, scalenus, and pectoralis minor muscles, which increases the dimension of the thoracic cavity anteroposteriorly. Pulmonary ventilation during forced inspiration usually occurs through the mouth rather than the nose.
	As the dimension of the thoracic cavity is increased, the pressure within the pleural cavities decreases and the lungs inflate because of the greater atmospheric pressure.
Expiration	Nerve stimuli to the inspiratory muscles ceases and the muscles relax.
	The rib cage and lungs recoil as air is forced out of the lungs due to the increased pressure.
Intercostal and lower spinal nerves	Forced expiration occurs when the internal intercostal and abdominal muscles are contracted.

*Some of the events during inspiration and expiration may occur simultaneously.

Source: Van De Graaff, Kent M., *Human Anatomy*. © 1984 Wm. C. Brown Publishers, Dubuque, Iowa. All Rights Reserved. Reprinted by permission.

elastic tension when the respiratory muscles relax. The decrease in lung volume raises the pressure within the alveoli above the atmospheric pressure and pushes the air out.

During forced expiration the **internal intercostal** muscles contract and depress the rib cage. The abdominal muscles may also aid expiration, because when they contract they force abdominal organs up against the diaphragm and further decrease the volume of the thorax. By this means the intrapulmonary pressure can rise 20 or 30 mm Hg above the atmospheric pressure. The events that occur during inspiration and expiration are summarized in table 23.3 and shown in figure 23.31.

Figure 23.31. The mechanics of pulmonary ventilation. (a) the lungs and pleural cavities prior to inspiration. (b) during inspiration, the diaphragm is contracted and the rib cage elevated; intrapleural pressure is reduced and air inflates the lungs. (c) inspiration is completed as intrapulmonary pressure is equal to atmospheric pressure. (d) as the muscles of inspiration are relaxed, the rib cage recoils and the intrapleural and intrapulmonary pressures are raised until the intrapulmonary pressure equals the atmospheric pressure (a).

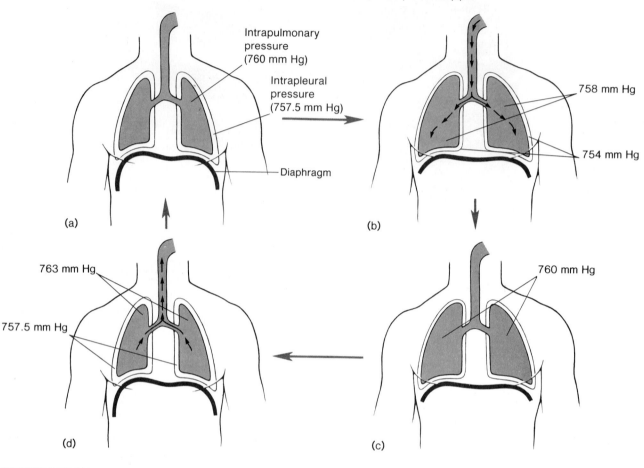

Intrapulmonary pressure (760 mm Hg)

Intrapleural pressure (757.5 mm Hg)

Diaphragm

(a)

758 mm Hg

754 mm Hg

(b)

763 mm Hg

757.5 mm Hg

(d)

760 mm Hg

(c)

Figure 23.32. A spirometer (Collins 9L respirometer) used to measure lung volumes and capacities.

Oxygen bell

Kymograph

Ventilometer pen
Respiration pen
Paper roll

Thermometer

O$_2$ inlet

Off, slow, medium, fast switch

Figure 23.33. A spirogram showing lung volumes and capacities at rest.

Table **23.4** Definitions of terms used to describe lung volumes and capacities	
Term	**Definition**
Lung volumes	The four non-overlapping components of the total lung capacity
Tidal volume	The volume of gas inspired or expired in an unforced respiratory cycle
Inspiratory reserve volume	The maximum volume of gas that can be inspired from the end of a tidal inspiration
Expiratory reserve volume	The maximum volume of gas that can be expired from the end of a tidal expiration
Residual volume	The volume of gas remaining in the lungs after a maximum expiration
Lung capacities	Measurements that are the sum of two or more lung volumes
Total lung capacity	The total amount of gas in the lungs at the end of a maximum inspiration
Vital capacity	The maximum amount of gas that can be expired after a maximum inspiration
Inspiratory capacity	The maximum amount of gas that can be inspired at the end of a tidal expiration
Functional residual capacity	The amount of gas remaining in the lungs at the end of a tidal expiration

Pulmonary Function Tests

Pulmonary function may be assessed clinically by means of a technique known as *spirometry (spi-rom'ĕ-tre)*. In this procedure, a subject breathes in a closed system, in which air is trapped within a light plastic bell floating in water. The bell moves up when the subject exhales and down when the subject inhales. The movements of the bell cause corresponding movements of a pen, which traces a record of the breathing on a rotating drum recorder (fig. 23.32).

The appearance of a spirogram taken during quiet, resting breathing is shown in figure 23.33, and the various lung volumes and capacities are defined in table 23.4.

spirometry: L. *spiro*, to breathe; Gk. *metron*, measure

During quiet breathing, the amount of air expired in each breath (the **tidal volume**) is equal to about 500 ml. This is about 12% of the **vital capacity** (the maximum amount of air that can be expired after a maximum inspiration). Multiplying the tidal volume at rest times the number of breaths per minute yields a **total minute volume** of about 6 L per minute. During exercise, the tidal volume can increase to as much as 50% of the vital capacity, and the number of breaths per minute likewise increases to produce a total minute volume as high as 100–200 L per minute.

Spirometry is useful in the diagnosis of lung diseases. In purely **restrictive disorders,** such as pulmonary fibrosis, the vital capacity is reduced below normal. In disorders that are only obstructive, such as asthma, the vital capacity is normal because lung tissue is not damaged. Bronchoconstriction and the increased airway resistance that results, however, makes expiration more difficult and takes a longer time. **Obstructive disorders** are thus diagnosed by tests that measure the rate of expiration. One such test is the **forced expiratory volume (FEV),** in which the percent of the vital capacity that can be exhaled in the first second ($FEV_{1.0}$), second ($FEV_{2.0}$), and third ($FEV_{3.0}$) seconds is measured (fig. 23.34). An $FEV_{1.0}$ that is significantly less than 75% to 85% suggests the presence of obstructive pulmonary disease.

Bronchoconstriction often occurs in response to the inhalation of noxious agents in the air, such as from smoke or smog. The $FEV_{1.0}$ has, therefore, been used by researchers to determine the effects of various components of smog and of passive cigarette smoke inhalation on pulmonary function. These studies have shown that it is unhealthy to exercise on very smoggy days and that inhalation of smoke from other people's cigarettes in a closed environment can measurably affect pulmonary function.

Figure 23.34. An illustration of the one-second forced expiratory volume ($FEV_{1.0}$) spirometry test for detecting obstructive pulmonary disorders.

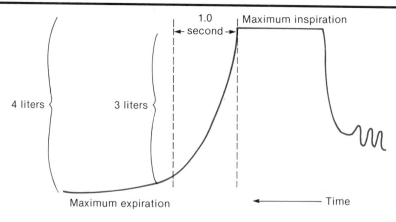

$$FEV_{1.0} = \frac{3L}{4L} \times 100\% = 75\%$$

Table 23.5 Definitions of some terms used to describe ventilation

Term	Definition
Air spaces	Alveolar ducts, alveolar sacs, and alveoli
Airways	Structures that conduct air from the mouth and nose to the respiratory bronchioles
Alveolar ventilation	Removal and replacement of gas in alveoli; equal to the tidal volume minus the volume of dead space
Anatomical dead space	Volume of the conducting airways to the zone where gas exchange occurs
Apnea	Cessation of breathing
Dyspnea	Unpleasant subjective feeling of difficult or labored breathing
Eupnea	Normal, comfortable breathing at rest
Hyperventilation	An alveolar ventilation that is excessive in relation to metabolic rate; results in abnormally low alveolar CO_2
Hypoventilation	An alveolar ventilation that is low in relation to metabolic rate; results in abnormally high alveolar CO_2
Physiological dead space	Combination of anatomical dead space and underventilated alveoli that do not contribute normally to blood-gas exchange
Pneumothorax	Presence of gas in the pleural space (the space between the visceral and parietal pleural membranes) causing lung collapse
Torr	Synonymous with millimeters of mercury (760 mm Hg = 760 torr)

Source: Fox, Stuart Ira, *Human Physiology.* © 1984 Wm. C. Brown Publishers, Dubuque, Iowa. All Rights Reserved. Reprinted by permission.

1. Describe the actions of the diaphragm and external intercostal muscles during inspiration, and explain how quiet expiration is produced.
2. Describe how forced inspiration and forced expiration are produced.
3. Define the terms *tidal volume* and *vital capacity*. Describe how the total minute volume is calculated, and explain how this value changes during exercise.
4. Describe how the vital capacity and the forced expiratory volume measurements are affected by asthma and pulmonary fibrosis, and explain why these measurements are affected.

Gas Exchange in the Lungs

The lungs function to bring the blood into gaseous equilibrium with the alveolar air. The efficiency of this process is measured by the P_{O_2} and P_{CO_2} of arterial blood.

Objective 19. Describe how the partial pressures in a gas mixture are calculated, and explain how these values are affected by water vapor pressure and altitude.

Objective 20. Describe how the P_{O_2} and P_{CO_2} of blood is measured, and list the factors that affect these measurements.

Objective 21. Explain the significance of the blood P_{O_2} measurement in terms of the oxygen content of blood and the delivery of oxygen to the tissues.

People with pulmonary disorders frequently complain of *dyspnea (disp'ne-ah),* which is a subjective feeling of "shortness of breath." Dyspnea may occur even when ventilation is normal, however, and may not occur even when total minute volume is very high as in exercise. Some of the terms used to define ventilation are defined in table 23.5.

The atmosphere is an ocean of gas that exerts pressure on all objects within it. The amount of this pressure can be measured with a glass U-tube filled with fluid. One end of the U-tube is exposed to the atmosphere, while the other end is evacuated of air and closed. Since the atmosphere presses on the open end but not on the end exposed to a vacuum, atmospheric pressure pushes fluid up the closed end to a height determined by the atmospheric pressure

dyspnea: Gk. *dys*, bad; *pnoe*, breathing

Figure 23.35. Atmospheric pressure at sea level can push a column of mercury to a height of 760 millimeters. This is also described as 760 torr, or one atmospheric pressure.

Table 23.6 The effect of altitude on P_{O_2}

Changes in P_{O_2} at various altitudes

Altitude (feet above sea level)	Atmospheric pressure (mm Hg)	P_{O_2} in air (mm Hg)	P_{O_2} in alveoli (mm Hg)	P_{O_2} in arterial blood (mm Hg)
0	760	159	105	100
2,000	707	148	97	92
4,000	656	137	90	85
6,000	609	127	84	79
8,000	564	118	79	74
10,000	523	109	74	69
20,000	349	73	40	35
30,000	226	47	21	19

Source: Fox, Stuart Ira, *Human Physiology.* © 1984 Wm. C. Brown Publishers, Dubuque, Iowa. All Rights Reserved. Reprinted by permission.

and the density of the fluid (fig. 23.35). Water, for example, would be pushed up to a height of 10,332 mm (33.9 feet) at sea level, whereas mercury (Hg)—which is more dense—is raised to a height of 760 mm. As a matter of convenience, therefore, devices used to measure atmospheric pressure (barometers) use mercury rather than water. The atmospheric pressure at sea level is thus said to be equal to *760 mm Hg (760 torr),* which is also described as *one atmosphere.*

According to **Dalton's law,** the total pressure of a gas mixture (such as air) is equal to the sum of the pressures that each gas in the mixture would exert independently. The pressure that each gas would exert independently is equal to the product of the total pressure times the fractional composition of the gas in the mixture. The total pressure of the gas mixture is thus equal to the sum of the **partial pressures** of the constituent gases. Since oxygen comprises about 21% of the atmosphere, for example, its partial pressure (abbreviated P_{O_2}) is 21% of 760, or about 159 mm Hg. Since nitrogen comprises about 79% of the

Figure 23.36. Partial pressures of gases in the inspired air and the alveolar air.

atmosphere, its partial pressure is equal to $0.79 \times 760 = 601$ mm Hg. These two gases thus contribute about 99% of the total pressure:

$$P_{dry\ atmospheric} = P_{N_2} + P_{O_2} + P_{CO_2} = 760\ mm\ Hg$$

Effects of Altitude and Diving. As you go to a higher altitude, the total atmospheric pressure and the partial pressure of the constituent gases decrease. At Denver, for example (5,000 feet above sea level), the atmospheric pressure is decreased to 619 mm Hg, and the P_{O_2} is therefore reduced to $619 \times 0.21 = 130$ mm Hg (table 23.6). At the peak of Mount Everest (29,000 feet), the P_{O_2} is only about 49 mm Hg. During ocean diving the pressure increases by one atmosphere for every 33 feet. At 33 feet, therefore, the pressure equals $2 \times 760 = 1,520$ mm Hg. At 66 feet, the pressure equals three atmospheres.

Effects of Water Vapor Pressure. Inspired air contains variable amounts of moisture. By the time the air has passed into the respiratory zone of the lungs, however, it is normally saturated with water vapor (has a relative humidity of 100%). The capacity of air to contain water vapor depends on its temperature; since the temperature of the respiratory zone is constant at 37°C, its water vapor pressure is also constant (47 mm Hg).

Water vapor, like the other constituent gases, contributes a partial pressure to the total atmospheric pressure. Since the total atmospheric pressure is constant (depending only on the height of the air mass), the water vapor "dilutes" the contribution of the other gases to the total pressure:

$$P_{dry\ atmosphere} = P_{N_2} + P_{O_2} + P_{CO_2} + P_{H_2O}$$

When the effect of water vapor pressure is considered, the partial pressure of oxygen in the inspired air is decreased at sea level to

$$P_{O_2}\ (sea\ level) = 0.21 \times (760 - 47) = 150\ mm\ Hg$$

As a result of gas exchange in the alveoli, the P_{O_2} of the *alveolar air* is further reduced to about 105 mm Hg. A comparison of the partial pressures of the inspired air with the partial pressures of the alveolar air is shown in figure 23.36.

Dalton's law: from John Dalton, English chemist, 1766–1844

Figure 23.37. Scanning electron micrographs of plastic casts of alveolar capillaries. In the left photograph only the capillaries are visible. In the right photograph, plastic also entered the alveoli so that the relationship between alveoli and capillaries can be seen. (From: *Tissues and Organs: A Text Atlas of Scanning Electron Microscopy* by R. G. Kessel and R. Kardon. W. H. Freeman and Company. © 1979.)

Partial Pressures of Gases in Blood

The enormous surface area of alveoli and the short diffusion distance between the alveolar air and the capillary blood quickly help to bring the blood into gaseous equilibrium with the alveolar air. This function is further aided by the fact that each alveolus is surrounded by so many capillaries that they form an almost continuous sheet of blood around the alveoli (fig. 23.37).

When a fluid and a gas, such as blood and alveolar air, are at equilibrium, the amount of gas dissolved in the fluid reaches a maximum value. This value depends, according to **Henry's law** (table 23.7), on (1) the solubility *(sol″u-bil′i-te)* of the gas in the fluid, which is a physical constant; (2) the temperature of the fluid—more gas can be dissolved in cold water than warm water; and (3) the partial pressure of the gas. Since the temperature of the blood does not vary significantly, the amount of dissolved gas depends directly on its partial pressure. When water— or plasma—is brought into equilibrium with air at a P_{O_2} of 100 mm Hg, for example, the fluid will contain 0.3 ml O_2 per 100 ml of fluid at 37°C. If the P_{O_2} of the gas were reduced by half, the amount of dissolved oxygen would also be reduced by half.

Blood-Gas Measurements. Measurement of the oxygen content of blood (in ml of O_2 per 100 ml blood) is a laborious procedure. Fortunately, an **oxygen electrode** that produces an electric current in proportion to the amount of *dissolved oxygen* has been developed. If this electrode is placed in a fluid while oxygen is artificially bubbled into it, the current produced by the oxygen electrode will increase up to a maximum value. At this maximum value the fluid is *saturated* with oxygen—that is, all of the oxygen that can be dissolved at that temperature and P_{O_2}

Henry's law: from William Henry, English chemist, 1775–1837

Table 23.7 Some of the physical laws that are important in ventilation and gas exchange

Physical law	Description
Boyle's law	The volume of a gas is inversely proportional to its pressure when the temperature and mass are constant (PV = constant).
Dalton's law	The total pressure of a gas mixture is equal to the sum of the partial pressures of its constituent gases. The partial pressure of each gas is the pressure it would exert if it alone occupied the total volume of the mixture.
Graham's law	The rate of diffusion of a gas through a liquid is directly proportional to its solubility and inversely proportional to its density (or gram molecular weight).
Henry's law	The weight of a gas dissolved in a liquid at a given temperature is proportional to the partial pressure of the gas.

Source: Fox, Stuart Ira, *Human Physiology.* © 1984 Wm. C. Brown Publishers, Dubuque, Iowa. All Rights Reserved. Reprinted by permission.

is dissolved. At a constant temperature, the amount dissolved and, thus, the electric current depend only on the P_{O_2} of the gas.

As a matter of convenience, it can now be said that the *fluid has the same P_{O_2} as the gas.* If it is known that the gas has a P_{O_2} of 150 mm Hg, for example, the deflection of the needle produced on a scale by the oxygen electrode can be calibrated at 150 mm Hg (fig. 23.38). The actual amount of dissolved oxygen under these circumstances is not known or very important (it can be looked up in solubility tables, if desired), since this is simply a linear function of the P_{O_2}. A lower P_{O_2} indicates that less oxygen is dissolved, and a higher P_{O_2} indicates that more oxygen is dissolved, in comparison to the calibration sample.

If the oxygen electrode is next inserted into an unknown sample of blood, the P_{O_2} of that sample can be read directly from the previously calibrated scale. Suppose, as

Figure 23.38.   Blood-gas measurements using a P_{O_2} electrode. (a) the electrical current generated by the oxygen electrode is calibrated so that the needle of the blood-gas machine points to the P_{O_2} of the gas with which the fluid is in equilibrium. (b) once standardized in this way, the electrode can be inserted in a fluid, such as blood, and the P_{O_2} of this solution can be measured.

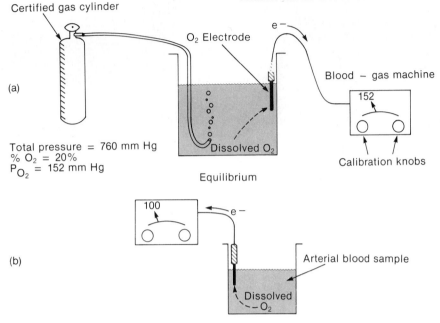

Certified gas cylinder

O_2 Electrode

(a)

Total pressure = 760 mm Hg
% O_2 = 20%
P_{O_2} = 152 mm Hg

Dissolved O_2

Equilibrium

Blood – gas machine

152

Calibration knobs

(b)

100

Arterial blood sample

Dissolved O_2

illustrated in figure 23.38, the blood sample has a P_{O_2} of 100 mm Hg. Since alveolar air has a P_{O_2} of about 105 mm Hg, this reading indicates that the blood is almost in complete equilibrium with the alveolar air.

The oxygen electrode responds only to the oxygen dissolved in water or plasma; it cannot respond to the oxygen that is hidden in red blood cells. Most of the oxygen in blood, however, is located in the red blood cells attached to hemoglobin. The oxygen content of whole blood thus depends on both its P_{O_2} and its red blood cell and hemoglobin content. At a P_{O_2} of about 100 mm Hg, whole blood normally contains 20 ml O_2 per 100 ml blood; of this amount, only 0.3 ml O_2 is dissolved in the plasma and 19.7 ml O_2 is within the red blood cells. Since only the 0.3 ml O_2 per 100 ml blood affects the P_{O_2} measurement, this measurement would be unchanged if the red blood cells were removed from the sample.

Significance of the Blood P_{O_2} and P_{CO_2} Measurements

Since blood P_{O_2} measurements are not directly affected by the oxygen in red blood cells, the P_{O_2} does not provide a measurement of the oxygen content of whole blood. It does, however, provide a good index of *lung function.* If the inspired air has a normal P_{O_2} but the arterial P_{O_2} is below normal, for example, gas exchange in the lungs must be impaired. Measurements of arterial P_{O_2} thus provide valuable information in the treatment of people with pulmonary diseases, during surgery when breathing may be depressed by anesthesia and in the care of premature babies with respiratory distress syndrome.

When the lungs are functioning properly, the P_{O_2} of systemic arterial blood is only 5 mm Hg less than the P_{O_2}

of alveolar air. Hyperventilation, therefore, cannot significantly increase the blood P_{O_2}. At a normal P_{O_2} of about 100 mm Hg, hemoglobin is almost completely loaded with oxygen. An increase in blood P_{O_2}—produced, for example, by breathing 100% oxygen from a gas tank—thus cannot significantly increase the amount of oxygen contained in the blood. This can, however, significantly increase the amount of oxygen dissolved in the plasma (because the amount dissolved is directly determined by the P_{O_2}). If the P_{O_2} doubles, the amount of oxygen dissolved in the plasma also doubles, but the total oxygen content of whole blood increases only slightly since most of the oxygen by far is not in plasma but in the red blood cells.

Since the oxygen carried by red blood cells must first dissolve in plasma before it can diffuse to the tissue cells, however, a doubling of the blood P_{O_2} means that the *rate of oxygen delivery* to the tissues would double under these conditions. For this reason, breathing from a tank of 100% oxygen (with a P_{O_2} of 760 mm Hg) would significantly increase oxygen delivery to the tissues, though it would have little effect on the total oxygen content of blood.

An electrode that produces a current in response to dissolved carbon dioxide has also been developed, so that the P_{CO_2} of blood can be measured together with its P_{O_2}. Blood in the systemic veins, which is delivered to the lungs by the pulmonary arteries, usually has a P_{O_2} of 40 mm Hg and a P_{CO_2} of 46 mm Hg. After gas exchange in the alveoli of the lungs, blood in the pulmonary veins and systemic arteries has a P_{O_2} of about 100 mm Hg and a P_{CO_2} of 40 mm Hg (fig. 23.39). The values in arterial blood are relatively constant and clinically significant, because they

Figure 23.39. The P_{O_2} and P_{CO_2} of blood as a result of gas exchange in the lung alveoli and gas exchange between systemic capillaries and body cells.

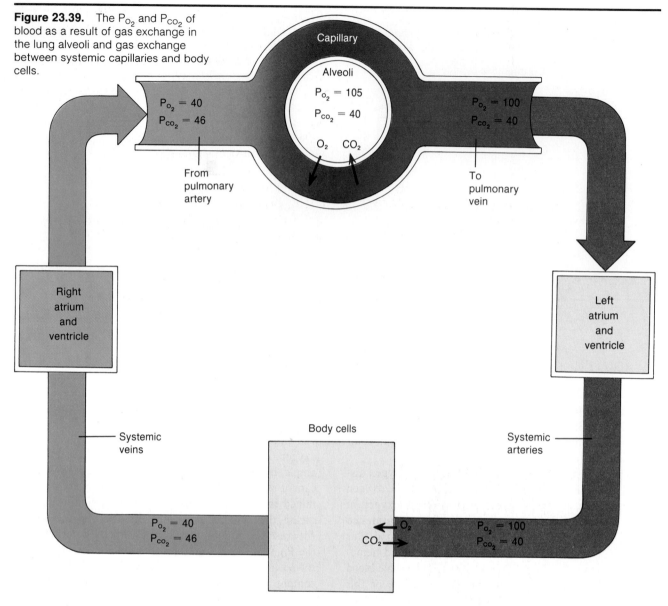

reflect lung function. Blood-gas measurements of venous blood are not performed clinically because these values are far more variable; venous P_{O_2} is much lower and P_{CO_2} much higher after exercise, for example, than at rest, whereas arterial values are not significantly affected by usual changes in physical activity.

1. Describe how the P_{O_2} of air is calculated and how this value is affected by altitude, diving, and water vapor pressure.
2. Describe how blood P_{O_2} measurements are taken, and explain the physiological and clinical significance of these measurements.
3. Explain how the arterial P_{O_2} and the oxygen content of whole blood is affected by (1) hyperventilation; (2) breathing from a tank containing 100% oxygen; (3) anemia (low red blood cell count and hemoglobin concentration); and (4) high altitude.

Regulation of Breathing

The rhythm of breathing is controlled by centers in the brain stem. These centers are influenced by higher brain areas and are regulated by sensory information that makes breathing responsive to the changing respiratory needs of the body.

Objective 22. Describe the functions of the pneumotaxic, apneustic, and rhythmicity centers in the brain stem.

Objective 23. Identify the chemoreceptors, and describe how they help to regulate ventilation.

Objective 24. Explain how blood P_{O_2}, P_{CO_2}, and pH affect ventilation.

Inspiration and expiration are produced by the contractions and relaxations of skeletal muscles in response to activity in somatic motor neurons. The activity of these motor

Figure 23.40.   Approximate locations of the brain stem respiratory centers.

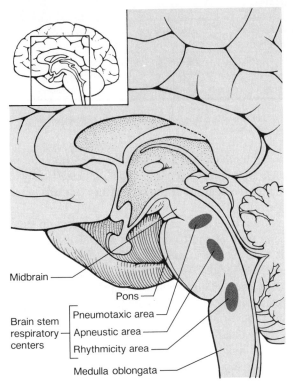

Figure 23.41.   The peripheral chemoreceptors (aortic and carotid bodies) regulate the brain stem respiratory centers by means of sensory nerve stimulation.

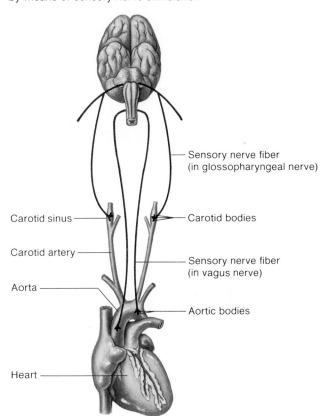

neurons, in turn, is directly controlled by neurons in the respiratory centers within the medulla oblongata and pons areas of the brain stem.

Respiratory Centers

A loose aggregation of neurons in the reticular formation of the *medulla oblongata* forms the **rhythmicity *(rith-mis'i-te)* center** that directly controls breathing. The discharges of some of these neurons produce inspiration, whereas the discharges of other neurons produce expiration. The activity of inspiratory and expiratory neurons varies in a reciprocal way, so that a rhythmic pattern of breathing is produced. The cycle of inspiration and expiration is thus intrinsic, or built-in, to the neural activity of the medulla.

The activity of the medullary rhythmicity center, however, is influenced by centers in the *pons.* As a result of research in which different areas of the pons are selectively destroyed, two respiratory centers have been identified in this brain stem region. One area—the **pneumotaxic center**—apparently promotes the activity of the inspiratory neurons of the medulla. The other pontine area—the **apneustic *(ap-nu'stik)* center**—seems to inhibit inspiration (fig. 23.40).

The brain stem respiratory centers produce rhythmic breathing even in the absence of other neural input. This intrinsic respiratory pattern, however, is modified by input from higher brain centers and by input from receptors sensitive to the chemical composition of the blood. The influence of higher brain centers is evidenced by the fact

that you can voluntarily hyperventilate and hypoventilate (as in breath-holding). The inability to hold your breath longer than a short period is due to reflex breathing in response to input from the chemoreceptors *(ke"mo-re-cep'tors).*

There are two groups of chemoreceptors that respond to changes in blood P_{CO_2}, pH, and P_{O_2}. These are the *central chemoreceptors* in the medulla oblongata and the *peripheral chemoreceptors.* The peripheral chemoreceptors include the **aortic bodies,** located in the aortic arch, and the **carotid bodies** (fig. 23.41). These peripheral chemoreceptors control breathing indirectly via sensory nerve fibers to the medulla. The aortic bodies send sensory information to the medulla in the vagus (tenth) cranial nerve; the carotid bodies stimulate sensory fibers in the glossopharyngeal (ninth) cranial nerve. The neural and sensory control of ventilation is summarized in figure 23.42.

Effects of Blood P_{CO_2} and pH on Ventilation

Chemoreceptor input to the brain stem modifies the rate and depth of breathing so that, under normal conditions, arterial P_{CO_2}, pH, and P_{O_2} remain relatively constant. If hypoventilation (inadequate ventilation) occurs, P_{CO_2} quickly rises and pH falls. The fall in pH is due to the fact

Figure 23.42. A schematic illustration of the control of ventilation by the central nervous system. The feedback effects of pulmonary stretch receptors and "irritant" receptors are not shown.

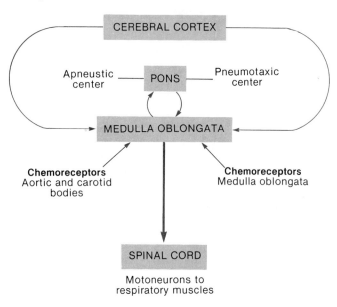

Table 23.8 The effect of ventilation, as measured by total minute volume (breathing rate × tidal volume), on the P_{CO_2} of arterial blood

Total minute volume	Arterial P_{CO_2}	Type of Ventilation
2 L/min	80 mm Hg	Hypoventilation
4–5 L/min	40 mm Hg	Normal ventilation
8 L/min	20–25 mm Hg	Hyperventilation

Source: Fox, Stuart Ira, *Human Physiology.* © 1984 Wm. C. Brown Publishers, Dubuque, Iowa. All Rights Reserved. Reprinted by permission.

that carbon dioxide can combine with water to form carbonic acid. The oxygen content of the blood decreases much more slowly, because there is a large "reservoir" of oxygen attached to hemoglobin. During hyperventilation, conversely, blood P_{CO_2} quickly falls and pH rises due to the excessive elimination of carbonic acid. The oxygen content of blood, on the other hand, is not significantly increased by hyperventilation (hemoglobin in arterial blood is 97% saturated during normal ventilation).

The blood P_{CO_2} and pH are, therefore, more immediately affected by changes in ventilation than is the oxygen content. Indeed, changes in P_{CO_2} provide a sensitive index of ventilation, as shown in table 23.8. In view of these facts, it is not surprising that changes in P_{CO_2} provide the most potent stimulus for the reflex control of ventilation. Ventilation, in other words, is adjusted to maintain a constant P_{CO_2}; proper oxygenation of the blood occurs naturally as a side product of this reflex control.

The rate and depth of ventilation is normally adjusted to maintain an arterial P_{CO_2} of 40 mm Hg. Hypoventilation causes a rise in P_{CO_2}—a condition called *hypercapnia (hi''per-kap'ne-ah)* (or hypercarbia). Hyperventilation, conversely, results in *hypocapnia*. Chemoreceptor regulation of breathing in response to changes in P_{CO_2} is illustrated in figure 23.43.

Chemoreceptors in the Medulla. The chemoreceptors most sensitive to changes in the arterial P_{CO_2} are located in the ventral area of the medulla, near the exit of the

ninth and tenth cranial nerves. These chemoreceptor neurons are anatomically separate from, but electrically connected to, the neurons of the respiratory control center in the medulla.

An increase in arterial P_{CO_2} causes a rise in the H^+ concentration of the blood as a result of increased carbonic acid concentrations. The H^+ in the blood, however, cannot cross the blood-brain barrier and, therefore, cannot influence the medullary chemoreceptors. Carbon dioxide in the arterial blood can cross the blood-brain barrier and, through the formation of carbonic acid, can lower the pH of cerebrospinal fluid. This fall in cerebrospinal fluid pH directly stimulates the medullary chemoreceptors when there is a rise in arterial P_{CO_2}.

Peripheral Chemoreceptors. The aortic and carotid bodies are stimulated directly by a rise in the H^+ concentration (fall in pH) of arterial blood. This is normally produced by an increase in P_{CO_2} and carbonic acid. The retention of CO_2 during hypoventilation thus stimulates the medullary chemoreceptors through a lowering of cerebrospinal fluid pH and stimulates peripheral chemoreceptors through a lowering of blood pH.

People who hyperventilate during psychological stress are sometimes told to breathe into a paper bag so that they re-breathe their expired air that is enriched in CO_2. This procedure helps to raise their blood P_{CO_2} back up to the normal range. This is needed because hypocapnia causes cerebral vasoconstriction. In addition to producing dizziness, the cerebral ischemia that results can lead to acidotic conditions in the brain that, through stimulation of medullary chemoreceptors, can cause further hyperventilation. Breathing into a paper bag can thus relieve the hypocapnia and stop the hyperventilation.

Effects of Blood P_{O_2} on Ventilation

Under normal conditions, blood P_{O_2} affects breathing only indirectly, by influencing the chemoreceptor sensitivity to changes in P_{CO_2}. Chemoreceptor sensitivity to P_{CO_2} is augmented by a low P_{O_2} (so ventilation is increased at a high altitude, for example) and is decreased by a high P_{O_2}. If

Figure 23.43. Negative feedback control of ventilation through changes in blood P_{CO_2} and pH.

<table>
<tr><td colspan="2">Table 23.9 Definitions of some terms used to describe blood oxygen and carbon dioxide levels</td></tr>
</table>

Term	Definition
Hypoxemia	The oxygen content or P_{O_2} is lower than normal in arterial blood.
Hypoxia	The oxygen content or P_{O_2} is lower than normal in the lungs, blood, or tissues. This is a more general term than hypoxemia. Tissues can be hypoxic, for example, even though there is no hypoxemia (as when the blood flow is occluded).
Hypercapnia, or hypercarbia	The P_{CO_2} of systemic arteries is higher than 40 mm Hg. Usually this occurs when the ventilation is inadequate for a given metabolic rate (hypoventilation). Antonyms are *hypocapnia* and *hypocarbia* (usually produced by hyperventilation).

Source: Fox, Stuart Ira, *Human Physiology.* © 1984 Wm. C. Brown Publishers, Dubuque, Iowa. All Rights Reserved. Reprinted by permission.

the blood P_{O_2} is raised by breathing 100% oxygen, therefore, the breath can be held longer because the response to increased P_{CO_2} is blunted.

When the blood P_{CO_2} is held constant by experimental techniques, the P_{O_2} of arterial blood must fall from 100 mm Hg to below 50 mm Hg before ventilation is significantly stimulated. This stimulation is apparently due to a direct effect of P_{O_2} on the carotid bodies. Since this degree of *hypoxemia (hi''pok-se'me-ah),* or low blood oxygen (table 23.9), does not normally occur even in breath-holding, P_{O_2} does not normally exert this direct effect on breathing.

In emphysema, when there is a chronic retention of carbon dioxide, the chemoreceptors eventually lose their ability to respond to increases in the arterial P_{CO_2}. The abnormally high P_{CO_2}, however, enhances the sensitivity of the carotid bodies to a fall in P_{O_2}. For people with emphysema, breathing may thus be stimulated by a *hypoxic drive* rather than by increases in blood P_{CO_2}.

<table>
<tr><td colspan="3">Table 23.10 The sensitivity of chemoreceptors to changes in blood gases and pH</td></tr>
</table>

Stimulus	Chemoreceptor(s)	Comments
↑ P_{CO_2}	Medulla oblongata	Medullary chemoreceptors are sensitive to the pH of cerebrospinal fluid (CSF). Diffusion of CO_2 from the blood into the CSF lowers the pH of CSF by forming carbonic acid.
↓ pH	Aortic and carotid bodies	Peripheral chemoreceptors are stimulated by decreased blood pH independent of the effect of blood CO_2. Chemoreceptors in the medulla are not affected by changes in blood pH because H^+ cannot cross the blood-brain barrier.
↓ P_{O_2}	Carotid bodies	Low blood P_{O_2} (hypoxemia) augments the chemoreceptor response to blood P_{CO_2}, and can stimulate ventilation directly when the P_{O_2} falls below 50 mm Hg.

Source: Fox, Stuart Ira, *Human Physiology.* © 1984 Wm. C. Brown Publishers, Dubuque, Iowa. All Rights Reserved. Reprinted by permission.

When people with emphysema are given oxygen therapy, the resulting rise in the arterial P_{O_2} may cause them to "forget to breathe" because their hypoxic drive is satisfied. Oxygen therapy in these patients must thus be performed carefully, with a gas mixture that is only slightly enriched with oxygen. After therapy, the patients must consciously force themselves to breathe at regular intervals for a period of time.

The effect of changes in the blood P_{CO_2}, pH, and P_{O_2} on chemoreceptors and the regulation of ventilation are summarized in table 23.10.

Table 23.11 Pulmonary receptors and reflexes served by sensory nerve fibers in the vagus (tenth cranial nerve)

Stimulus	Receptors	Comments
Stretch of lungs at inspiration	Stretch receptors	The Hering-Breuer reflex. This reflex is believed to be important in the control of breathing in the newborn but not in the adult at normal tidal volumes.
Stretch of lungs at expiration	Stretch receptors	Like the Hering-Breuer reflex, this is not believed to be significant in adults at normal tidal volumes.
Pulmonary congestion	Type J (juxta-capillary) receptors	This reflex may produce feelings of dyspnea at high altitudes and during severe exercise.
Irritation	"Irritant receptors"	This reflex causes reflex constriction of bronchioles in response to irritation from smoke, smog, and other noxious agents.

Source: Fox, Stuart Ira, *Human Physiology.* © 1984 Wm. C. Brown Publishers, Dubuque, Iowa. All Rights Reserved. Reprinted by permission.

Pulmonary Stretch and Irritant Reflexes

The lungs contain various types of receptors that influence the brain stem respiratory control centers via sensory fibers in the vagus (table 23.11). Irritant receptors in the lungs, for example, stimulate reflex constriction of the bronchioles in response to smoke and smog. Similarly, sneezing, sniffing, and coughing may be stimulated by irritant receptors in the nose, larynx, and trachea.

The **Hering-Breuer reflex** is stimulated by pulmonary stretch receptors. The activation of these receptors during inspiration inhibits the respiratory control centers, making further inspiration increasingly difficult. This helps to prevent undue distension of the lungs and may contribute to the smoothness of the ventilation cycles. A similar inhibitory reflex may occur during expiration. The Hering-Breuer reflex appears to be important in the control of normal ventilation in the newborn. Pulmonary stretch receptors in adults, however, are probably not active at normal resting tidal volumes (500 ml per breath) but may contribute to respiratory control at high tidal volumes.

1. Describe the effects of voluntary hyperventilation and breath-holding on arterial P_{CO_2}, pH, and oxygen content. Indicate the relative degree of changes in these values.
2. Using a flow chart to show a negative feedback loop, explain the relationship between ventilation and arterial P_{CO_2}.
3. Explain the effect of increased arterial P_{CO_2} on (1) chemoreceptors in the medulla oblongata and (2) chemoreceptors in the aortic and carotid bodies.
4. Explain the role of arterial P_{CO_2} in the regulation of breathing. Explain why ventilation increases when a person goes to a high altitude.

Hering-Breuer reflex: from Heinrich E. Hering, German physiologist, 1866–1948, and Josef Breuer, German physician, 1842–1925

Hemoglobin and Oxygen Transport

Hemoglobin loads with oxygen in the lungs and unloads oxygen in the tissue capillaries. The extent of this unloading is influenced by a variety of factors that act to insure an adequate delivery of oxygen to the tissues while maintaining an adequate reserve of oxygen in the blood.

Objective 25. Distinguish between the different forms of hemoglobin.

Objective 26. Describe the loading and unloading reactions, and explain the effects of P_{O_2} and hemoglobin affinity for oxygen on these reactions.

Objective 27. Describe the effects of changes in pH, temperature, and 2,3-DPG on the unloading reaction, and explain the significance of these effects.

If the lungs are functioning properly, blood leaving in the pulmonary veins and traveling in systemic arteries has a P_{O_2} of about 100 mm Hg, indicating a plasma oxygen concentration of about 0.3 ml O_2 per 100 ml blood. The total oxygen content of the blood, however, is not known; this depends not only on the P_{O_2} but also on the hemoglobin concentration. If the P_{O_2} and hemoglobin concentration are normal, arterial blood contains about 20 ml of O_2 per 100 ml of blood (fig. 23.44).

Hemoglobin

Most of the oxygen in the blood is contained within the red blood cells chemically bonded to **hemoglobin.** Each hemoglobin molecule consists of (1) a protein globin part composed of four polypeptide chains; and (2) four nitrogen-containing, disc-shaped organic pigment molecules, called *hemes* (fig. 23.45).

The protein part of hemoglobin is composed of two identical *alpha chains,* which are each 141 amino acids long, and two identical *beta chains,* which are each 146 amino acids long. Each of the four polypeptide chains is combined with one heme group. In the center of each heme

hemoglobin: Gk. *haima,* blood; L. *globus,* globe

Figure 23.44. Plasma and whole blood that are brought into equilibrium with the same gas mixture have the same P_{O_2} and thus the same amount of dissolved oxygen molecules (shown as black dots). The oxygen content of whole blood, however, is much higher than that of plasma because of the binding of oxygen to hemoglobin.

Figure 23.45. (a) an illustration of the three-dimensional structure of hemoglobin, in which the two alpha and two beta polypeptide chains are shown; the four heme groups are represented as flat structures with iron (*dark spheres*) in the centers. (b) the chemical structure of heme.

(a)

(b)

group is a single atom of iron, which can combine with one molecule of oxygen (O_2). One hemoglobin can thus combine with four molecules of oxygen; since there are about 280 million hemoglobin molecules per red blood cell, each red blood cell can carry over a billion molecules of oxygen.

Normal heme contains iron in the reduced form (Fe^{++}, or ferrous iron). In this form, the iron can share electrons and bond to oxygen to form **oxyhemoglobin.** When oxyhemoglobin dissociates to release oxygen to the tissues, the heme iron is still in the reduced (Fe^{++}) form and the hemoglobin is called **deoxyhemoglobin,** or **reduced hemoglobin.** The term *oxyhemoglobin* is thus not equivalent to *oxidized* hemoglobin; hemoglobin does not lose an electron (and become oxidized) when it combines with oxygen. Oxidized hemoglobin, or **methemoglobin** *(met-he'mo-glo'bin),* has iron in the oxidized (Fe^{+++}, or ferric) state. Methemoglobin thus lacks the electron it needs to form a bond with oxygen and cannot participate in oxygen transport. Blood normally contains only a small amount of methemoglobin but certain drugs can increase this amount.

Table 23.12 The relationship between percent oxyhemoglobin saturation and P_{O_2} (at pH = 7.40 and temperature = 37°C)

P_{O_2}(mm Hg)		100	80	61	45	40	36	30	26	23	21	19
Percent oxyhemoglobin		97	95	90	80	75	70	60	50	40	35	30
			Arterial blood				Venous blood					

Source: Fox, Stuart Ira, *Human Physiology.* © 1984 Wm. C. Brown Publishers, Dubuque, Iowa. All Rights Reserved. Reprinted by permission.

Carboxyhemoglobin is another abnormal form of hemoglobin in which the reduced heme is combined with *carbon monoxide* instead of oxygen. Since the bond with carbon monoxide is about 210 times stronger than the bond with oxygen, carbon monoxide tends to displace oxygen in hemoglobin and remains attached to hemoglobin as the blood passes through systemic capillaries. The transport of oxygen to the tissues is thus reduced.

According to federal standards, the percent carboxyhemoglobin in the blood of active nonsmokers should not be higher than 1.5%. Studies have shown, however, that these values can range as high as 3% or more in nonsmokers and as high as 10% or more in smokers in some cities. Although these high levels may not cause immediate problems in healthy people, long-term effects on health might occur. People with respiratory or cardiovascular diseases would be particularly vulnerable to the negative effects of carboxyhemoglobin on oxygen transport.

Hemoglobin Concentration. The *oxygen-carrying capacity* of whole blood is determined by its concentration of normal hemoglobin. If the hemoglobin concentration is below normal—a condition called **anemia**—the oxygen concentration of the blood is reduced below normal. Conversely, when the hemoglobin concentration is increased above the normal range—as occurs in **polycythemia** *(pol"e-si-the'me-ah)* (high red blood cell count)—the oxygen-carrying capacity of blood is increased accordingly. This can occur as an adaptation to life at a high altitude.

The production of hemoglobin and red blood cells in bone marrow is controlled by a hormone called **erythropoietin** *(e-rith"ro-poi'e-tin).* This hormone is produced by conversion in the blood of **renal erythropoietic factor** secreted by the kidneys. The production of erythropoietin—and thus the production of red blood cells—is stimulated when the delivery of oxygen to the kidneys and other organs is lower than normal. Red blood cell production is also promoted by androgens, which explains why the hemoglobin concentration in men averages 1–2 g per 100 ml higher than in women.

anemia: Gk. *a*, negative; *haima*, blood
polycythemia: Gk. *polys*, much; *kytos*, cell; *haima*, blood
erythropoietin: Gk. *erythros*, red; *poiesis*, a making

The Loading and Unloading Reactions. Reduced hemoglobin and oxygen combine to form oxyhemoglobin; this is called the **loading reaction.** Oxyhemoglobin, in turn, dissociates to yield deoxyhemoglobin and free oxygen molecules; this is the **unloading reaction.** The loading reaction occurs in the lungs and the unloading reaction occurs in the systemic capillaries.

Loading and unloading can thus be shown as a reversible reaction:

$$\text{Deoxyhemoglobin} + O_2 \underset{\text{(tissues)}}{\overset{\text{(lungs)}}{\rightleftharpoons}} \text{Oxyhemoglobin}$$

The extent that the reaction will go in each direction depends on two factors: (1) the P_{O_2} of the environment and (2) the *affinity,* or bond strength, between hemoglobin and oxygen. High P_{O_2} drives the loading reaction; at the high P_{O_2} of the pulmonary capillaries almost all the deoxyhemoglobin molecules combine with oxygen. Low P_{O_2} in the systemic capillaries drives the reaction in the opposite direction to promote unloading. The extent of this unloading depends on how low the P_{O_2} values are.

The affinity (bond strength) between hemoglobin and oxygen also influences the loading and unloading reactions. A very strong bond would favor loading but inhibit unloading; a weak bond would hinder loading but improve unloading. The bond strength between hemoglobin and oxygen is normally strong enough so that 97% of the hemoglobin leaving the lungs is in the form of oxyhemoglobin, yet the bond is sufficiently weak so that adequate amounts of oxygen are unloaded to sustain aerobic respiration in the tissues. Under normal resting conditions only about 22% of the oxygen is unloaded; this satisfies the tissue needs for oxygen yet maintains an oxygen reserve in the blood for emergency conditions.

The Oxyhemoglobin Dissociation Curve

Blood in the systemic arteries, at a P_{O_2} of 100 mm Hg, has a *percent oxyhemoglobin saturation* of 97% (which means that 97% of the hemoglobin is in the form of oxyhemoglobin). This blood is delivered to the systemic capillaries, where oxygen diffuses into the tissue cells and is consumed in aerobic respiration. Blood leaving in the systemic veins is thus reduced in oxygen; it has a P_{O_2} of about 40 mm Hg and a percent saturation of about 75% (table 23.12). Expressed another way, blood entering the tissues contains 20 ml O_2 per 100 ml blood, and blood leaving the tissues contains 15.5 ml O_2 per 100 ml blood (fig. 23.46).

Figure 23.46. The percent of oxyhemoglobin saturation and the blood oxygen content are shown at different values of P_{O_2}. Notice that there is about a 25% decrease in percent oxyhemoglobin as the blood passes through the tissue from arteries to veins, resulting in the unloading of approximately 5 ml O_2 per 100 ml to the tissues.

Thus, 22%, or 4.5 ml of O_2 out of 20 ml O_2 per 100 ml blood, is unloaded to the tissues.

A graphic illustration of the percent oxyhemoglobin saturation at different values of P_{O_2} is called an **oxyhemoglobin dissociation curve** (fig. 23.46). The values in this graph are obtained by subjecting samples of blood *in vitro* (outside the body) to different partial oxygen pressures. The percent oxyhemoglobin saturation obtained by this procedure, however, can be used to predict what the percent unloading would be *in vivo* (within the body) with a given difference in arterial and venous P_{O_2} values.

Figure 23.46 shows the difference between the arterial and venous P_{O_2} and the percent oxyhemoglobin saturation at rest. The relatively large amount of oxyhemoglobin remaining in the venous blood at rest functions as an oxygen reservoir. If a person stops breathing, there is a sufficient reserve of oxygen in the blood to keep the brain and heart alive for approximately four to five minutes in the absence of cardiopulmonary resuscitation (CPR) techniques. This reserve supply of oxygen can also be tapped when the tissue's requirements for oxygen are raised.

The oxyhemoglobin dissociation curve is S-shaped, or *sigmoidal*. The fact that it is relatively flat at high P_{O_2} values indicates that changes in P_{O_2} within this range have little effect on the loading reaction. One would have to ascend as high as 10,000 feet, for example, before the oxyhemoglobin saturation or arterial blood would decrease from 97% to 93%. At more common elevations the percent oxyhemoglobin saturation would not be significantly different from the 97% value at sea level.

At the steep part of the sigmoidal curve, however, small changes in P_{O_2} values produce large differences in percent saturation. A decrease in *venous* P_{O_2} from 40 mm Hg to 30 mm Hg, as might occur during exercise, corresponds to a change in percent saturation from 75% to 58%. Since the *arterial* percent saturation is usually still 97% during exercise, this change in venous percent saturation indicates that more oxygen has been unloaded to the tissues. The difference between the arterial and venous percent saturations indicates the percent unloading: in these examples, 97% minus 75% = 22% unloading at rest, and 97% minus 58% = 39% unloading during exercise.

Effects of pH and Temperature on Oxygen Unloading

In addition to changes in P_{O_2}, changes in the affinity of hemoglobin for oxygen can affect the unloading reaction. The strength of the bond between hemoglobin and oxygen is weakened by a decrease in pH. This is called the **Bohr effect.** A weakening of the bond between hemoglobin and oxygen has less effect at higher P_{O_2} values than at lower P_{O_2} values. A decrease in pH thus results in slightly less oxygen loading in the lungs, and significantly more oxygen unloading in the tissues. The net effect is that the tissues receive more oxygen. These facts are shown in table 23.13. When percent saturation is graphed as a function of P_{O_2}, the oxyhemoglobin dissociation curve is seen to be shifted to the right by a lowering of pH (fig. 23.47).

Bohr effect: from Christian Bohr, Scandinavian physiologist, 1855–1911

Table 23.13 Effect of pH on hemoglobin affinity for oxygen and oxyhemoglobin "unloading" (the difference between arterial and venous blood oxygen content)

pH	Affinity	Arterial O$_2$ content per 100 ml	Venous O$_2$ content per 100 ml	O$_2$ unloaded to tissues per 100 ml
7.40	Normal	19.8 ml O$_2$	14.8 ml O$_2$	5.0 ml O$_2$
7.60	Increased	20.0 ml O$_2$	17.0 ml O$_2$	3.0 ml O$_2$
7.20	Decreased	19.2 ml O$_2$	12.6 ml O$_2$	6.6 ml O$_2$

Source: Fox, Stuart Ira, *Human Physiology.* © 1984 Wm. C. Brown Publishers, Dubuque, Iowa. All Rights Reserved. Reprinted by permission.

Figure 23.47. A decrease in blood pH (an increase in H$^+$ concentration) decreases the affinity of hemoglobin for oxygen at each P$_{O_2}$ value, resulting in a "shift to the right" of the oxyhemoglobin dissociation curve. A curve that is shifted to the right has a lower percent oxyhemoglobin saturation at each P$_{O_2}$, but the effect is more marked at lower P$_{O_2}$ values. This is called the *Bohr effect.*

Figure 23.48. The oxyhemoglobin dissociation curve is shifted to the right as the temperature increases, indicating a lowered affinity of hemoglobin for oxygen at each P$_{O_2}$. This effect, like the Bohr effect (see figure 23.47), is more marked at lower P$_{O_2}$ values.

Table 23.14 Factors that affect the affinity of hemoglobin for oxygen and the position of the oxyhemoglobin dissociation curve

Factor	Affinity	Position of curve	Comments
↓ pH	Decreased	Shift to the right	Called the Bohr effect; increases oxygen delivery during hypercapnia
↑ Temperature	Decreased	Shift to the right	Increases oxygen unloading during exercise and fever
↑ 2,3-DPG	Decreased	Shift to the right	Increases oxygen unloading when there is a decrease in total hemoglobin or total oxygen content; an adaptation to anemia and high-altitude living

Source: Fox, Stuart Ira, *Human Physiology.* © 1984 Wm. C. Brown Publishers, Dubuque, Iowa. All Rights Reserved. Reprinted by permission.

When oxyhemoglobin dissociation curves are constructed at constant pH values but at different temperatures, it can be seen that the affinity of hemoglobin for oxygen is decreased by a rise in temperature. An increase in temperature weakens the bond between hemoglobin and oxygen and thus has the same effect as a fall in pH; the oxyhemoglobin dissociation curve is shifted to the right (fig. 23.48). At higher temperatures, therefore, more oxygen is unloaded to the tissues than would be the case if the bond strength were constant. This effect can significantly increase the delivery of oxygen to muscles that are warmed during exercise.

Effect of 2,3-DPG on Oxygen Unloading

Mature red blood cells lack both nuclei and mitochondria. Without mitochondria they cannot respire aerobically and thus cannot use the oxygen they carry. Red blood cells, therefore, obtain energy through the anaerobic respiration of glucose. At a certain point in the glycolytic pathway there is a "side reaction" in red blood cells that results in a unique product—**2,3-diphosphoglyceric acid (2,3-DPG).**

The enzyme that produces 2,3-DPG is inhibited by oxyhemoglobin. When the oxyhemoglobin concentration is decreased, therefore, the production of 2,3-DPG is increased. This increase in 2,3-DPG production can occur when the total hemoglobin concentration is low (in anemia) or when the P_{O_2} is low (at a high altitude, for example). 2,3-DPG combines with deoxyhemoglobin and makes it more stable. At the P_{O_2} values in the tissue capillaries, therefore, a higher proportion of the oxyhemoglobin will be converted to deoxyhemoglobin by unloading its oxygen. Increased concentrations of 2,3-DPG in red blood cells thus increase oxygen unloading (table 23.14) and shift the oxyhemoglobin dissociation curve to the right.

The importance of 2,3-DPG within red blood cells is now recognized in blood banking. Old, stored red blood cells can lose their ability to produce 2,3-DPG as they lose their ability to metabolize glucose. Modern techniques for blood storage, therefore, include the addition of energy substrates for respiration and phosphate sources needed for the production of 2,3-DPG.

Anemia. When the total blood hemoglobin concentration is reduced below normal in anemia, each red blood cell produces increased amounts of 2,3-DPG. A normal hemoglobin concentration of 15 g per 100 ml unloads about 4.5 ml O_2 per 100 ml at rest. If the hemoglobin concentration were reduced by half, you might expect that the tissues would receive only half the normal amount of oxygen (2.25 ml O_2 per 100 ml). It has been shown, however, that as much as 3.3 ml O_2 per 100 ml are unloaded to the tissues under these conditions. This occurs as a result of a rise in 2,3-DPG production that produces a decrease in hemoglobin affinity for oxygen.

Hemoglobin F. The effects of 2,3-DPG are also important in the transfer of oxygen from maternal to fetal blood. Normal adult hemoglobin in the mother (*hemoglobin A*) is able to bond to 2,3-DPG. Fetal hemoglobin (*hemoglobin F*), in contrast, cannot bond to 2,3-DPG, and thus has a higher affinity for oxygen at a given P_{O_2} than does hemoglobin A. Since hemoglobin F can have a higher percent oxyhemoglobin saturation than hemoglobin A at a given P_{O_2}, oxygen is transferred from the maternal to the fetal blood as these two come into close proximity in the placenta.

Muscle Myoglobin

Myoglobin (*mi″o-glo′bin*) is a red pigment found exclusively in striated muscle fibers. In particular, slow-twitch, aerobically respiring skeletal fibers and cardiac muscle fibers are rich in myoglobin. Myoglobin is similar to hemoglobin, but it has one rather than four hemes and, therefore, can combine with only one molecule of oxygen.

As a result of the difference in structure between myoglobin and hemoglobin, myoglobin has a higher affinity for oxygen than does hemoglobin. The dissociation curve for myoglobin is, therefore, to the left of the oxyhemoglobin dissociation curve (fig. 23.49). The curve for myoglobin is also shaped differently than the oxyhemoglobin dissociation curve; this is a consequence of the fact that myoglobin has only one heme group, compared to four hemes in a hemoglobin molecule. The more rectangular shape of the myoglobin curve indicates that oxygen will be released only when the P_{O_2} gets very low.

Figure 23.49.   A comparison of the dissociation curves for hemoglobin and for myoglobin. At the P_{O_2} of venous blood, the myoglobin retains almost all of its oxygen, indicating a higher affinity than hemoglobin for oxygen. The myoglobin does, however, release its oxygen at the very low P_{O_2} values found inside the mitochondria.

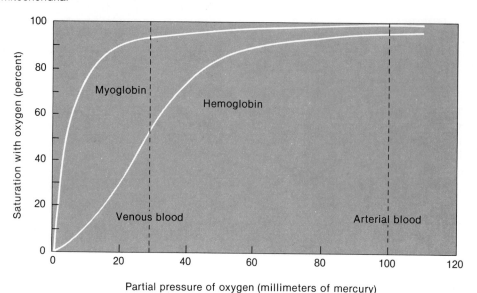

Partial pressure of oxygen (millimeters of mercury)

Since the P_{O_2} in mitochondria is very low (because oxygen is incorporated into water here), myoglobin may act as a "middleman" in the transfer of oxygen from blood to the mitochondria within muscle cells. Myoglobin may also have an oxygen storage function, which is of particular importance in the heart. During diastole, when the coronary blood flow is greatest, the myoglobin can load up with oxygen. This stored oxygen can then be released during systole, when the coronary arteries are squeezed closed by the contracting myocardium.

---

1. *Describe the effect of P_{O_2} on the loading and unloading reactions. Illustrate this with a graph, and label the regions of the graph that show when loading and unloading occur.*
2. *Draw an oxyhemoglobin dissociation curve, and label the values found in the arterial blood and venous blood under nonexercise conditions. Use this graph to show an increase in the percent unloading that can occur during exercise.*
3. *Describe how changes in pH and temperature affect the hemoglobin affinity for oxygen and the position of the oxyhemoglobin dissociation curve. Explain the effect of these changes on oxygen transport.*
4. *Explain how a person who is anemic or a person at high altitude could have an increase in the percent unloading of oxygen by hemoglobin.*

Carbon Dioxide Transport and Acid-Base Balance

Carbonic acid and bicarbonate are the major transport forms of carbon dioxide and the major contributors to the acid-base balance of the blood. Since the concentration of these compounds is partly controlled by ventilation, pulmonary function is an important regulator of the acid-base balance in the body.

Objective 28.   Describe how carbon dioxide is transported in the blood, and describe the chloride shift.

Objective 29.   Define the terms respiratory acidosis, respiratory alkalosis, metabolic acidosis, and metabolic alkalosis.

Objective 30.   Explain how ventilation helps to regulate the acid-base balance of the blood.

Carbon dioxide is carried by the blood in three forms: (1) as *dissolved CO_2*—since carbon dioxide is about twenty-one times more soluble than oxygen in water, a substantial portion (about one-tenth) of the total blood CO_2 is dissolved in plasma; (2) as *carbaminohemoglobin (kar-bam″i-no-he″mo-glo′bin)*—about one-fifth of the total blood CO_2 is carried attached to an amino acid in hemoglobin (carbaminohemoglobin should not be confused with carboxyhemoglobin, which is a combination of hemoglobin with carbon monoxide); and (3) as *carbonic acid* and *bicarbonate,* which account for most of the CO_2 carried by the blood.

Figure 23.50. An illustration of carbon dioxide transport by the blood and the "chloride shift." Carbon dioxide is transported in three forms: as dissolved CO_2 gas, attached to hemoglobin as carbaminohemoglobin, and as carbonic acid and bicarbonate.

Carbon dioxide is able to combine with water to form carbonic acid. This reaction occurs spontaneously in the plasma at a slow rate but occurs much more rapidly within the red blood cells due to the catalytic action of the enzyme **carbonic anhydrase.** Since this enzyme is confined to the red blood cells, most of the carbonic acid is produced there rather than in the plasma. The formation of carbonic acid from CO_2 and water is favored by the high P_{CO_2} found in tissue capillaries.

$$CO_2 + H_2O \xrightarrow[\text{high } P_{CO_2}]{\text{carbonic anhydrase}} H_2CO_3$$

The Chloride Shift

As a result of catalysis by carbonic anhydrase within the red blood cells, large amounts of carbonic acid are produced as blood passes through systemic capillaries. The buildup of carbonic acid concentrations within red blood cells favors the dissociation of these molecules into H^+ (protons, which contribute to the acidity of a solution) and HCO_3^- (bicarbonate):

$$H_2CO_3 \longrightarrow H^+ + HCO_3^-$$

The H^+ released by dissociation of carbonic acid are largely buffered by their combination with deoxyhemoglobin within the red blood cells. Although the unbuffered H^+ are free to diffuse out of the red blood cells, more bicarbonate diffuses outward into the plasma than does H^+. As a result of the "trapping" of H^+ within the red blood cells by their attachment to hemoglobin and the outward diffusion of bicarbonate, the inside of the red blood cell

gains a net positive charge. This attracts chloride ions (Cl^-), which move into the red blood cells as HCO_3^- moves out. This exchange of anions as blood moves through the tissue capillaries is called the *chloride shift* (fig. 23.50).

When blood reaches the pulmonary capillaries, deoxyhemoglobin is converted to oxyhemoglobin. Since oxyhemoglobin has a lower affinity for H^+ than does deoxyhemoglobin, H^+ are released within the red blood cells. This attracts HCO_3^- from the plasma, which combines with H^+ to form carbonic acid:

$$H^+ + HCO_3^- \longrightarrow H_2CO_3$$

Under conditions of low P_{CO_2}, as occurs in the pulmonary capillaries, carbonic anhydrase catalyzes the conversion of carbonic acid to carbon dioxide and water:

$$H_2CO_3 \xrightarrow[\text{low } P_{CO_2}]{\text{carbonic anhydrase}} CO_2 + H_2O$$

In summary, the carbon dioxide produced by the tissue cells is converted within the systemic capillaries, mostly through the action of carbonic anhydrase in the red blood cells, to carbonic acid. With the buildup of carbonic acid concentrations in the RBCs, the carbonic acid dissociates into bicarbonate and H^+, which results in the chloride shift. A reverse chloride shift operates in the pulmonary capillaries to convert carbonic acid and bicarbonate to CO_2 gas, which is eliminated in the expired breath (fig. 23.51). The P_{CO_2}, carbonic acid, H^+, and bicarbonate concentrations in the systemic arteries are thus maintained relatively constant by normal ventilation.

Figure 23.51. Carbon dioxide is released from the blood as it travels through the pulmonary capillaries. During this time a "reverse chloride shift" occurs and carbonic acid is transformed into CO_2 and H_2O.

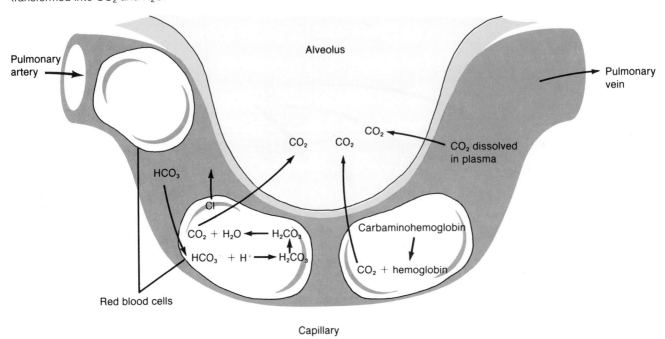

Table 23.15	Definition of terms used to describe acid-base balance
Terms	**Definitions**
Acidosis, respiratory	Increased carbon dioxide retention (due to hypoventilation), which can result in the accumulation of carbonic acid and thus a fall in blood pH below normal
Acidosis, metabolic	Increased production of "nonvolatile" acids such as lactic acid, fatty acids, and ketone bodies, or loss of blood bicarbonate (such as by diarrhea) resulting in a fall in blood pH below normal
Alkalosis, respiratory	A rise in blood pH due to loss of CO_2 and carbonic acid (through hyperventilation)
Alkalosis, metabolic	A rise in blood pH produced by loss of nonvolatile acids (as in excessive vomiting) or by excessive accumulation of bicarbonate base
Compensated acidosis or alkalosis	Metabolic acidosis or alkalosis is partially compensated by opposite changes in blood carbonic acid levels (through changes in ventilation). Respiratory acidosis or alkalosis is partially compensated by increased retention or excretion of bicarbonate in the urine.

Source: Fox, Stuart Ira, *Human Physiology.* © 1984 Wm. C. Brown Publishers, Dubuque, Iowa. All Rights Reserved. Reprinted by permission.

Ventilation and Acid-Base Balance

Normal systemic arterial blood has a pH of 7.35 to 7.45. Expressed another way, arterial blood has a H$^+$ concentration of about $10^{-7.4}$ molar. Some of these H$^+$ are derived from carbonic acid and some are derived from nonvolatile *metabolic acids* (fatty acids, ketone bodies, lactic acid, and others) that cannot be eliminated in the expired breath.

Under normal conditions the H$^+$ released by metabolic acids do not affect blood pH because these H$^+$ combine with HCO_3^- and are thereby removed from solution. *Bicarbonate is the major buffer in the plasma* and acts to maintain a blood pH of 7.4 despite the constant production of nonvolatile metabolic acids by the tissues. In this buffering process, some of the HCO_3^- released from the red blood cells during the chloride shift is converted into H_2CO_3 in the plasma. Normally, however, there is still a buffer reserve of free bicarbonate that can help protect against unusually large additions of metabolic acids to the blood. These processes are illustrated in figure 23.52.

Normal plasma, therefore, contains free bicarbonate, carbonic acid, and H$^+$ concentrations indicated by a pH of 7.4. If the H$^+$ concentration of the blood should fall, the carbonic acid produced by the buffering reaction can dissociate and serve as a source of additional H$^+$. If the H$^+$ concentration should rise, bicarbonate can remove this excess H$^+$ from solution. Carbonic acid and bicarbonate are thus said to function as a *buffer pair*.

Acidosis and Alkalosis.   A fall in blood pH below 7.35 is called **acidosis** *(as"i-do'sis)*, because the pH is to the acid-side of normal. Acidosis does not mean acidic (pH

Figure 23.52. Bicarbonate released into the plasma from red blood cells functions to buffer H^+ produced by the ionization of metabolic acids (lactic acid, fatty acids, ketone bodies, and others).

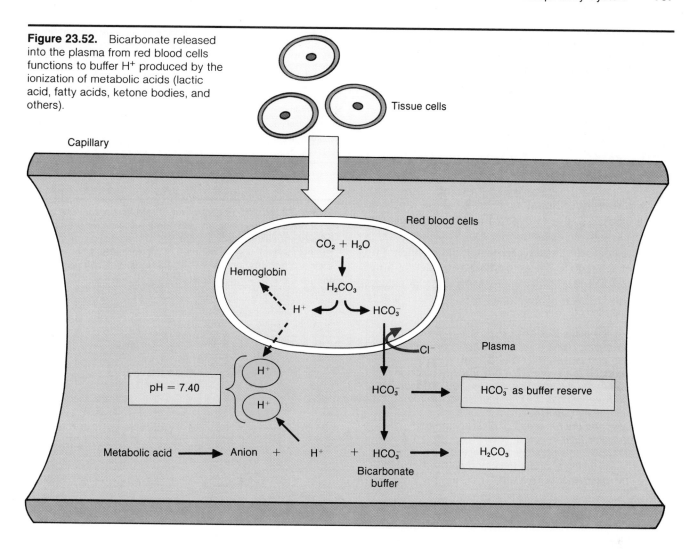

less than 7); blood pH of 7.2, for example, represents serious acidosis. Similarly, a rise in blood pH above 7.45 is known as **alkalosis** *(al''kah-lo'sis)*. There are two components to acid-base balance: *respiratory* and *metabolic*. Respiratory acidosis and alkalosis are due to abnormal concentrations of carbonic acid, as a result of abnormal ventilation; metabolic acidosis and alkalosis result from abnormal amounts of H^+ derived from nonvolatile metabolic acids (table 23.15).

Ventilation is normally adjusted to keep pace with the metabolic rate. **Hypoventilation** occurs when the rate of CO_2 production exceeds the rate at which it is "blown off" in ventilation. Under these conditions, carbonic acid production is excessively high compared to ventilation, and **respiratory acidosis** occurs. In **hyperventilation,** the depletion of carbonic acid raises the pH, and **respiratory alkalosis** occurs.

Metabolic acidosis can occur when the production of nonvolatile acids is abnormally increased. In uncontrolled diabetes mellitus, for example, ketone bodies (derived from fatty acids) may accumulate and produce *ketoacidosis (ke''to-ah''si-do'sis)*. In order for metabolic acidosis to occur, however, the buffer reserve of bicarbonate must first

be depleted (this is why metabolic acidosis can also be produced by the excess loss of bicarbonate, as in diarrhea). Until the buffer reserve is depleted the pH remains normal—ketosis can occur, for example, without ketoacidosis. **Metabolic alkalosis,** less common than metabolic acidosis, can result when there is a loss of acidic gastric juice by vomiting or when there is an excessive intake of bicarbonate (from stomach antacids or from an intravenous solution).

Compensations for Acidosis and Alkalosis.   A change in blood pH, produced by alterations in either the respiratory or metabolic component of acid-base balance, can be compensated for by a change in the other component. Metabolic acidosis, for example, is partially compensated for by hyperventilation (the aortic and carotid bodies are directly stimulated by blood H^+), which causes a decrease in carbonic acid. People with metabolic acidosis would thus have a low pH accompanied by a low blood P_{CO_2} as a result of the hyperventilation. Metabolic alkalosis, similarly, is partially compensated for by the retention of carbonic acid due to hypoventilation (table 23.16).

Table 23.16 The effect of lung function on blood acid-base balance

Condition	pH	P_{CO_2}	Ventilation	Cause or compensation
Normal	7.35–7.45	39–41 mm Hg	Normal	Not applicable
Respiratory acidosis	Low	High	Hypoventilation	Cause of the acidosis
Respiratory alkalosis	High	Low	Hyperventilation	Cause of the alkalosis
Metabolic acidosis	Low	Low	Hyperventilation	Compensation for acidosis
Metabolic alkalosis	High	High	Hypoventilation	Compensation for alkalosis

Source: Fox, Stuart Ira, *Human Physiology.* © 1984 Wm. C. Brown Publishers, Dubuque, Iowa. All Rights Reserved. Reprinted by permission.

A person with respiratory acidosis has a low pH and a high blood P_{CO_2} due to hypoventilation. This condition can be partially compensated for by the kidneys, which help to regulate the blood bicarbonate concentration. Two organs thus regulate blood acid-base balance: the lungs (regulating the respiratory component) and the kidneys (regulating the metabolic component).

1. List the ways that carbon dioxide is carried by the blood. Using equations, show how carbonic acid and bicarbonate are formed.
2. Describe the events that occur in the chloride shift in the systemic capillaries, and describe the reverse chloride shift that occurs in the pulmonary capillaries.
3. Describe the functions of bicarbonate and carbonic acid in blood.
4. Describe how hyperventilation and hypoventilation affect the blood pH, and explain the mechanisms involved.

Effect of Exercise and High Altitude on Respiratory Function

Changes in ventilation and oxygen delivery occur during exercise and during acclimatization to a high altitude. These changes help to compensate for the increased metabolic rate during exercise and for the decreased arterial P_{O_2} at a high altitude.

Objective 31. Describe the effect of exercise on blood P_{O_2}, P_{CO_2}, and pH, and describe the theories that have been proposed to explain the increase in ventilation that occurs during exercise.

Objective 32. Explain the effect of endurance training on the anaerobic threshold.

Objective 33. Explain the respiratory adaptations that occur during acclimatization to life at a high altitude.

Immediately upon exercise, the rate and depth of breathing increase to produce a total minute volume that is many times the resting value. This increased ventilation, particularly in well-trained athletes, is exquisitely matched to the simultaneous increase in oxygen consumption and

Figure 23.53. The effect of moderate and heavy exercise on arterial blood gases and pH. Notice that there are no consistent and significant changes in these measurements during the first several minutes of moderate and heavy exercise and that only the P_{CO_2} changes (actually decreases) during more prolonged exercise.

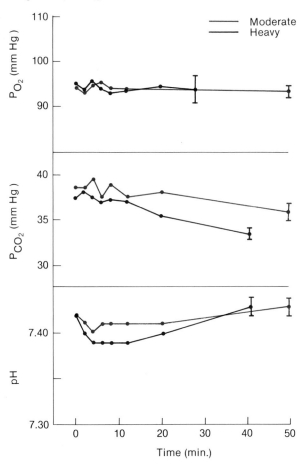

carbon dioxide production by the exercising muscles. The arterial blood P_{O_2}, P_{CO_2}, and pH, thus, remain surprisingly constant during exercise (fig. 23.53).

It is tempting to suppose that ventilation increases during exercise as a result of the increased CO_2 production by the exercising muscles. Ventilation increases together with increased CO_2 production, however, so that blood measurements of P_{CO_2} during exercise are not significantly higher than at rest. The mechanisms responsible for the increased ventilation during exercise must therefore be more complex.

Table 23.17	Changes in respiratory function during exercise	
Variable	**Change**	**Comments**
Ventilation	Increased	This is not hyperventilation because ventilation is matched to increased metabolic rate. Mechanisms responsible for increased ventilation are not well understood.
Blood gases	No change	Blood-gas measurements during light, moderate, and heavy exercise show little change because ventilation is increased to match increased muscle oxygen consumption and carbon dioxide production.
Oxygen delivery to muscles	Increased	Although the total oxygen content and P_{O_2} do not increase during exercise, there is an increased rate of blood flow to the exercising muscles.
Oxygen extraction by muscles	Increased	Increased oxygen consumption lowers the tissue P_{O_2} and lowers the affinity of hemoglobin for oxygen (due to the effect of increased temperature). More oxygen, as a result, is unloaded so that venous blood contains a lower oxyhemoglobin saturation than at rest. This effect is enhanced by endurance training.

Source: Fox, Stuart Ira, *Human Physiology.* © 1984 Wm. C. Brown Publishers, Dubuque, Iowa. All Rights Reserved. Reprinted by permission.

Two groups of mechanisms—*neurogenic (nu''ro-jen'ik)* and *humoral*—have been proposed to explain the increased ventilation that occurs during exercise. Possible neurogenic mechanisms include the following: (1) sensory nerve activity from the exercising limbs may stimulate the respiratory muscles, either directly or via the brain stem respiratory centers; and/or (2) input from the cerebral cortex may stimulate the brain stem centers to modify ventilation. These neurogenic theories help to explain the immediate increase in ventilation that occurs at the beginning of exercise.

Rapid and deep ventilation continues after exercise has stopped, suggesting that humoral (chemical) factors in the blood may also stimulate ventilation during exercise. Since the P_{O_2}, P_{CO_2}, and pH of the blood samples from exercising subjects are within the resting range, these humoral theories propose that (1) the P_{CO_2} and pH in the region of the chemoreceptors may be different from these values "downstream" where blood samples are taken; and/or (2) there may be cyclic variations in these values that stimulate the chemoreceptors but cannot be detected by blood samples. The evidence suggests that both neurogenic and humoral mechanisms are involved in the **hyperpnea** *(hi''perp-ne'ah)*, or increased ventilation, of exercise.

Anaerobic Threshold and Endurance Training

The ability of the cardiopulmonary system to deliver adequate amounts of oxygen to the exercising muscles at the beginning of exercise may be insufficient, as a result of the time lag required to make proper cardiovascular adjustments. During this time, therefore, the muscles respire anaerobically and a "stitch in the side"—probably due to hypoxia of the diaphragm—may develop. After the cardiovascular adjustments have been made, a person may experience a "second wind" when the muscles receive sufficient oxygen for their needs.

Continued heavy exercise can cause a person to reach the **anaerobic threshold,** which is the maximum amount of oxygen consumption that can be attained before blood lactic acid levels rise as a result of anaerobic respiration. This rise in lactic acid levels is not due to a malfunction of the cardiopulmonary system; indeed, the arterial oxygen

hemoglobin saturation remains at 97% and venous blood draining the muscles contains unused oxygen. The anaerobic threshold, however, is higher in endurance-trained athletes than it is in other people.

The rise in blood lactic acid that occurs when the anaerobic threshold is exceeded is due to the inability of the exercising muscles to increase their oxygen consumption any further. Endurance training increases the skeletal muscle content of myoglobin, mitochondria, and Krebs cycle enzymes. These muscles, therefore, are able to utilize more of the oxygen delivered to them by the arterial blood. At a given level of exercise, in other words, the venous blood draining muscles in endurance-trained people contain a lower percent oxyhemoglobin than in other people. The effects of exercise and endurance training on respiratory function are summarized in table 23.17.

Acclimatization to High Altitude

When a person who is from a region near sea level moves to a significantly higher altitude, several adjustments in the respiratory system must be made to compensate for the decreased atmospheric pressure and P_{O_2} at the higher elevation. These adjustments include changes in ventilation, in the hemoglobin affinity for oxygen, and in the total hemoglobin concentration.

Changes in Ventilation. A decrease in arterial P_{O_2} makes the chemoreceptors more sensitive to increases in P_{CO_2}; a decrease in P_{O_2} below 50 mm Hg directly stimulates the carotid bodies to increase ventilation. These changes cause *hyperventilation,* which becomes stabilized after a few days at about 2–3 L per minute more than the total minute volume at sea level. As a result of this hyperventilation, the arterial P_{CO_2} decreases from 40 mm Hg (its value at sea level) to about 29 mm Hg.

Hyperventilation cannot, of course, increase blood P_{O_2} above that of the inspired air. The P_{O_2} of arterial blood is therefore low at high altitudes. In the Peruvian Andes, for example, the normal arterial P_{O_2} is reduced from 100 mm Hg (at sea level) to 45 mm Hg. The loading of hemoglobin with oxygen is therefore incomplete, producing an oxyhemoglobin saturation that is decreased from 97% (at sea level) to 81%.

Table **23.18**	Changes in respiratory function during acclimatization to high altitude	
Variable	Change	Comments
Partial oxygen pressure	Decreased	Due to decreased total atmospheric pressure
Percent oxyhemoglobin saturation	Decreased	Due to lower P_{O_2} in pulmonary capillaries
Ventilation	Increased	Due at first to low arterial P_{O_2}; later to changes in blood P_{CO_2} as chemoreceptor sensitivity increases
Total hemoglobin	Increased	Due to stimulation by erythropoietin; raises oxygen capacity of blood to partially or completely compensate for the reduced partial pressure
Oxyhemoglobin affinity	Decreased	Due to increased DPG within the red blood cells; results in a higher percent unloading of oxygen to the tissues, which may partially or completely compensate for the reduced arterial oxyhemoglobin saturation

Source: Fox, Stuart Ira, *Human Physiology.* © 1984 Wm. C. Brown Publishers, Dubuque, Iowa. All Rights Reserved. Reprinted by permission.

Decreased Hemoglobin Affinity. Normal arterial blood at sea level only unloads about 22% of its oxygen to the tissues at rest; the percent saturation is reduced from 97% in arterial blood to 75% in venous blood. As a partial compensation for the decrease in oxygen content at high altitude, the affinity of hemoglobin for oxygen is reduced, so that a higher proportion of oxygen is unloaded. This occurs because the low oxyhemoglobin content of red blood cells stimulates the production of 2,3-DPG, which in turn decreases the hemoglobin affinity for oxygen.

Increased Hemoglobin and Red Blood Cell Production. In response to tissue hypoxia, the kidneys secrete *renal erythropoietic (ĕ-rith″ro-poi-et′ik) factor,* which reacts with a plasma protein to produce the hormone **erythropoietin** *(ĕ-rith″ro-poi′ĕ-tin).* Erythropoietin stimulates the bone marrow to increase its production of hemoglobin and red blood cells. In the Peruvian Andes, for example, people have a total hemoglobin concentration that is increased from 15 g per 100 ml (at sea level) to 19.8 g per 100 ml. Although the percent oxyhemoglobin saturation is still lower than at sea level, the total oxygen content of the blood is actually greater—22.4 ml O_2 per 100 ml compared to a sea level value of about 20 ml O_2 per 100 ml. These compensations of the respiratory system to high altitude are summarized in table 23.18.

---

1. Describe the effect of exercise on the blood values of P_{O_2}, P_{CO_2}, and pH, and explain how ventilation might be increased during exercise.
2. Explain why endurance-trained athletes have a higher anaerobic threshold than other people.
3. Describe the changes that occur in the respiratory system during acclimatization to life at a high altitude.

Clinical Considerations

The respiratory system is particularly vulnerable to inflammatory and infectious diseases simply because many pathogens are airborne, humans are highly social, and the warm, moistened environment along the respiratory tract

is very susceptible. Injury and trauma are likewise frequent problems. Protruding noses may become broken, the large, spongy lungs are easily penetrated by broken ribs, and portions of the respiratory tract are liable to become occluded because they also have a digestive function.

Developmental Problems

Birth defects, inherited disorders, and premature births commonly cause problems in the respiratory system of infants. A **cleft palate** is a developmental deformity of the hard palate of the mouth in which an opening persists between the oral and nasal cavities, making it difficult, if not impossible, for an infant to nurse. A cleft palate may be hereditary or a complication of some disease (e.g., German measles) contracted by the mother during pregnancy. A **cleft lip** is a genetically based developmental disorder in which the two sides of the upper lip fail to fuse. Cleft palates and cleft lips can be treated very effectively with cosmetic surgery.

As mentioned earlier in this chapter, **hyaline membrane disease** is a fairly common respiratory disorder that accounts for about one-third of neonatal deaths. This condition results from the deficient production of surfactant. **Cystic fibrosis** *(sis′tik fĭ-bro′sis)* is a genetic disorder that affects the respiratory system, as well as other systems of the body, and accounts for approximately one childhood death in twenty. The effect of this disease on the respiratory system is usually a persistent inflammation and infection of the respiratory tract.

There are a number of hemoglobin diseases that are produced by inherited defects in the protein part of hemoglobin. **Sickle-cell anemia**—a disease that affects 8%–11% of the black population of the United States—is caused by an abnormal form of hemoglobin called *hemoglobin S.* This abnormal hemoglobin becomes gel-like when the P_{O_2} of the blood decreases; this causes a characteristic sickling of the red blood cells and organ damage as a result of ischemia and the release of hemoglobin from the damaged erythrocytes. A related hemoglobin disorder is **thalassemia** *(thal″ah-se′me-ah),* found predominantly among people of Mediterranean ancestry.

Trauma or Injury

Humans are especially susceptible to **epistaxis** *(ep''i-stak'sis)* (nosebleeds) because the nose is located where it can be easily bumped and because the nasal mucosa has extensive vascularity for warming the inspired air. Certain conditions or diseases such as high blood pressure or leukemia may cause epistaxis.

When air enters the pleural cavity surrounding either lung and causes the lung to collapse, it is referred to as a **pneumothorax.** A pneumothorax can result from an external injury, such as a stabbing, bullet wound, or penetrating fractured rib, or it can occur internally. A severely diseased lung, as in emphysema, can create a pneumothorax as the wall of the lung deteriorates and permits air to enter the pleural cavity.

Choking on a foreign object such as aspirated food is a common serious trauma to the respiratory system. More than eight Americans choke to death each day on food lodged in their trachea. A simple process termed the **abdominal thrust (Heimlich) maneuver** can save the life of a person who is choking. The abdominal thrust maneuver is performed as follows:

If the victim is standing or sitting:

1. Stand behind the victim or the victim's chair, and wrap your arms around his or her waist.
2. Grasp your fist with your other hand, and place the fist against the victim's abdomen, slightly above the navel and below the rib cage.
3. Press your fist into the victim's abdomen with a quick upward thrust.
4. Repeat several times if necessary.

If the victim is lying down:

1. Position the victim on his or her back.
2. Face the victim, and kneel on his or her hips.
3. With one of your hands on top of the other, place the heel of your bottom hand on the abdomen, slightly above the navel and below the rib cage.
4. Press into the victim's abdomen with a quick upward thrust.
5. Repeat several times if necessary.

If you are alone and choking: Use anything that applies force just below your diaphragm. Press into a table or a sink, or use your own fist.

Persons saved from drowning and frequently shock victims experience apnea (cessation of breathing) and will soon die if not revived by someone performing artificial respiration. Figure 23.54 illustrates the accepted treatment for reviving a person who has stopped breathing.

---

epistaxis: Gk. *epi,* upon; *stazein,* spill in drops
pneumothorax: Gk. *pneumon,* spirit (breath); L. *thorax,* thorax, chest

Common Respiratory Disorders

A cough is the most common symptom of respiratory disorders. Acute problems may be accompanied by dyspnea or wheezing. Respiratory or circulatory problems may cause **cyanosis,** which is a blue discoloration of the skin caused by blood with a low oxygen content.

Pulmonary disorders are classified as **obstructive,** when there is increased resistance to air flow in the bronchioles, and as **restrictive,** when alveolar tissue is damaged. Asthma and acute bronchitis are usually just obstructive; chronic bronchitis and emphysema are both obstructive and restrictive. Pulmonary fibrosis, in contrast, is a purely restrictive disorder.

Asthma.   The obstruction of air flow through the bronchioles may occur as a result of excessive mucus secretion, inflammation, and/or contraction of the smooth muscles in the bronchioles. *Asthma* results from bronchiolar constriction, which increases airway resistance and makes breathing difficult. Constriction of the bronchiolar smooth muscles is stimulated by leukotrienes and to a lesser degree by histamine, released by leukocytes and mast cells. This can be provoked by an allergic reaction or by the release of acetylcholine from parasympathetic nerve endings. There is evidence that some people with asthma may have a decreased sensitivity to the beta-adrenergic effects of epinephrine, which stimulate bronchodilation; the effects of chemicals that promote bronchoconstriction may, therefore, be enhanced.

> Epinephrine and related compounds stimulate beta-adrenergic receptors in the bronchioles and by this means promote bronchodilation. Asthmatics can, therefore, take epinephrine, as an inhaled spray to relieve the symptoms of an asthma attack. It has been learned that there are two subtypes of beta receptors for epinephrine and that the subtype in the heart (called beta–1) is different from the one in the bronchioles (beta–2). By utilizing these differences, compounds have been developed that can more selectively stimulate the beta–2 adrenergic receptors and cause bronchodilation, without having as great an effect on the heart as epinephrine does.

Emphysema.   Alveolar tissue is destroyed in emphysema, resulting in fewer but larger alveoli (fig. 23.55). This produces a decreased surface area for gas exchange and a decreased ability of the bronchioles to remain open during expiration. Collapse of the bronchioles as a result of the compression of the lungs during expiration produces *air trapping,* which further decreases the efficiency of gas exchange in the alveoli.

---

cyanosis: Gk. *kyanosis,* dark-blue color
asthma: Gk. *asthma,* panting
emphysema: Gk. *emphysan,* blow up, inflate

Figure 23.54. The mouth-to-mouth method of artificial respiration. (American Red Cross Poster)

WHEN BREATHING STOPS

IF A VICTIM APPEARS TO BE UNCONSCIOUS TAP VICTIM ON THE SHOULDER AND SHOUT, "ARE YOU OKAY?"

IF THERE IS NO RESPONSE TILT THE VICTIM'S HEAD, CHIN POINTING UP. Place one hand under the victim's neck and gently lift. At the same time, push with the other hand on the victim's forehead. This will move the tongue away from the back of the throat to open the airway.

IMMEDIATELY LOOK, LISTEN, AND FEEL FOR AIR.
While maintaining the backward head tilt position, place your cheek and ear close to the victim's mouth and nose. Look for the chest to rise and fall while you listen and feel for the return of air. Check for about 5 seconds.

IF THE VICTIM IS NOT BREATHING GIVE FOUR QUICK BREATHS.
Maintain the backward head tilt, pinch the victim's nose with the hand that is on the victim's forehead to prevent leakage of air, open your mouth wide, take a deep breath, seal your mouth around the victim's mouth, and blow into the victim's mouth with four quick but full breaths just as fast as you can. When blowing, use only enough time between breaths to lift your head slightly for better inhalation. **For an infant,** give gentle puffs and blow through the mouth *and* nose and do not tilt the head back as far as for an adult.

If you do not get an air exchange when you blow, it may help to reposition the head and try again.

AGAIN, LOOK, LISTEN, AND FEEL FOR AIR EXCHANGE.

IF THERE IS STILL NO BREATHING CHANGE RATE TO ONE BREATH EVERY 5 SECONDS **FOR AN ADULT.**

FOR AN INFANT, GIVE ONE GENTLE PUFF EVERY 3 SECONDS.

MOUTH-TO-NOSE METHOD

The mouth-to-nose method can be used with the sequence described above instead of the mouth-to-mouth method. Maintain the backward head-tilt position with the hand on the victim's forehead. Remove the hand from under the neck and close the victim's mouth. Blow into the victim's nose. Open the victim's mouth for the look, listen, and feel step.

For more information about these and other lifesaving techniques, contact your Red Cross chapter for training.

Figure 23.55. Photomicrographs of tissue from a normal lung (*a*) and from the lung of a person with emphysema (*b*). In emphysema, lung tissue is destroyed, resulting in the presence of fewer and larger alveoli.

(a)

(b)

Table 23.19 Pulmonary disorders caused by the inhalation of noxious agents containing particles less than 6 μm in size, which can lodge in the alveoli

Condition	Agent	Pulmonary disorders
Silicosis	Silica	Fibrosis, obstructive emphysema, tuberculosis, cor pulmonale
Anthracosis (black lung)	Coal dust	Similar to silicosis (most coal dust contains silica)
Asbestosis	Asbestos fibers	Diffuse fibrosis, cor pulmonale, pulmonary malignancy

Source: Fox, Stuart Ira, *Human Physiology.* © 1984 Wm. C. Brown Publishers, Dubuque, Iowa. All Rights Reserved. Reprinted by permission.

There are different types of emphysema. The most common type occurs almost exclusively in people who have smoked cigarettes heavily over a period of years. A component of cigarette smoke apparently stimulates the macrophages and leukocytes to secrete proteolytic (protein-digesting) enzymes that destroy lung tissues. A less common type of emphysema results from the genetic inability to produce a plasma protein called alpha-1-antitrypsin. This protein normally inhibits proteolytic enzymes such as trypsin and thus normally protects the lungs against the effects of enzymes that are released from alveolar macrophages.

Chronic bronchitis and emphysema, the most common causes of respiratory failure, together are called **chronic obstructive pulmonary disease (COPD).** In addition to the more direct obstructive and restrictive aspects of these conditions, other pathological changes may occur. These include edema, inflammation, hyperplasia (increased cell number), zones of pulmonary fibrosis, pneumonia, pulmonary emboli (traveling blood clots), and heart failure. Patients with severe emphysema may eventually develop *cor pulmonale*—pulmonary hypertension with hypertrophy and the eventual failure of the right ventricle.

Pulmonary Fibrosis. Under certain conditions, for reasons that are poorly understood, lung damage leads to pulmonary fibrosis instead of emphysema. In this condition the normal structure of the lungs is disrupted by the accumulation of fibrous connective tissue proteins. Fibrosis can result, for example, from the inhalation of particles less than 6 μm in size, which are able to accumulate in the respiratory zone of the lungs. Included in this category is *anthracosis,* or black lung, which is produced by the inhalation of carbon particles from coal dust (table 23.19).

Other Respiratory Disorders. The **common cold** is the most widespread of all respiratory diseases. No cure is presently available, only symptomatic relief. Colds occur repeatedly because acquired immunity to one virus does not protect against other viruses that cause colds. Cold viruses cause acute inflammation of the respiratory mucosa, resulting in a flow of mucus, a fever, and often a headache.

Nearly all the structure and regions of the respiratory tract can become infected and inflamed. **Influenza** is a viral disease that causes inflammation of the upper respiratory tract. Influenza can become epidemic, but fortunately vaccines are available. **Sinusitis** *(si″nū-si′tis)* is an inflammation of the paranasal sinuses. Sinusitis can be quite painful if the drainage ducts from the sinuses into the nasal cavity become blocked. **Tonsillitis** may involve one or all of the tonsils and frequently follows other lingering diseases of the oral or pharyngeal region. Chronic tonsillitis often requires a tonsillectomy. **Laryngitis** is inflammation of the larynx, which often produces a hoarse voice and limits the ability to talk. **Tracheobronchitis** and **bronchitis** are infections of the regions that give them their names. Severe inflammation of the bronchioles can cause significant airway resistance and dyspnea.

influenza: L. *influentia,* a flowing in

Diseases of the lungs are common and may be serious. **Pneumonia** is an acute infection and inflammation of lung tissue accompanied by exudation (the accumulation of fluid). It can have many causes, but the most common is a bacterial pneumonia. **Tuberculosis (TB)** is an inflammatory disease of the lungs caused by the presence of *tubercle bacilli*. Tuberculosis softens and leads to the ulceration of lung tissue. **Pleurisy** is an inflammation of the pleura and is usually secondary to some other respiratory disease. Inspiration may become painful, and fluid may collect within the pleural space.

Cancer in the respiratory system is often caused by repeated inhalation of irritating substances such as cigarette smoke. Cancers of the lips, larynx, and lungs are especially common in smokers over the age of fifty.

Disorders of Respiratory Control

There are a variety of disease processes that can produce cessation of breathing during sleep, or *sleep apnea*. **Sudden infant death syndrome** is an especially tragic form of sleep apnea that claims the lives of about ten thousand babies annually in the United States. Victims of this condition are apparently healthy two-to-five-month old babies who die in their sleep without apparent reason—hence, the layman's term of *crib death*. These deaths seem to be caused by failure of the respiratory control mechanisms in the brain stem and/or by failure of the carotid bodies to be stimulated by reduced arterial oxygen.

Abnormal breathing patterns often appear prior to death by brain damage or heart disease. The most common of these abnormal patterns is **Cheyne-Stokes breathing**, in which the depth of breathing progressively increases and then progressively decreases. These cycles of increasing and decreasing tidal volumes may be followed by periods of apnea of varying durations. Cheyne-Stokes breathing may be caused by neurological damage or by insufficient oxygen delivery to the brain. The latter may result from heart disease or from a brain tumor that diverts a large part of the vascular supply from the respiratory centers.

Disorders Caused by High Partial Pressure of Gases

The total atmospheric pressure increases by one atmosphere (760 mm Hg) for every 10 m (33 feet) below sea level. If one dives 10 m below sea level, therefore, the partial pressures and amounts of dissolved gases in the plasma are twice that at sea level. At 66 feet they are three times, and at 100 feet they are four times the values at sea level. The increased amounts of nitrogen and oxygen dissolved in the blood plasma under these conditions can have serious effects on the body.

Oxygen Toxicity. Although breathing 100% oxygen at one or two atmospheres pressure can be safely tolerated for a few hours, higher partial oxygen pressures can be very dangerous. *Oxygen toxicity* develops rapidly when the P_{O_2} rises above about 2.5 atmospheres. This is apparently caused by the oxidation of enzymes and other destructive changes that can damage the nervous system and lead to coma and death. For these reasons, deep sea divers commonly use gas mixtures, in which oxygen is diluted with inert gases such as nitrogen (as in ordinary air) or helium.

Hyperbaric oxygen—oxygen at greater than one atmosphere pressure—is often used to treat conditions such as carbon monoxide poisoning, circulatory shock, and gas gangrene. Before the dangers of oxygen toxicity were realized these hyperbaric oxygen treatments sometimes resulted in tragedy. Particularly tragic were the cases of **retrolental fibroplasia** *(re''tro-len'tal fi''bro-pla'se-ah)*, in which damage to the retina and blindness resulted from the hyperbaric oxygen treatment of premature babies with hyaline membrane disease.

Nitrogen Narcosis. Although at sea level nitrogen is physiologically inert, larger amounts of dissolved nitrogen under hyperbaric conditions have deleterious effects. Since it takes time for the nitrogen to dissolve, these effects usually don't appear until the person has remained submerged over an hour. *Nitrogen narcosis* resembles alcohol intoxication; depending on the depth of the dive, the diver may experience "rapture of the deep" or may become so drowsy that he or she is totally incapacitated.

Decompression Sickness. The amount of nitrogen dissolved in the plasma decreases as the diver ascends to sea level, due to a progressive decrease in the P_{N_2}. If the diver surfaces slowly, a large amount of nitrogen can diffuse through the alveoli and be eliminated in the expired breath. If decompression occurs too rapidly, however, bubbles of nitrogen gas (N_2) can form in the blood and block small blood channels, producing muscle and joint pain as well as more serious damage. These effects are known as *decompression sickness*, or the bends.

The cabins of airplanes that fly long distances at high altitudes (30,000 to 40,000 feet) are pressurized so that the passengers and crew are not exposed to the very low atmospheric pressures of these altitudes. If a cabin were to become rapidly depressurized at high altitude, much less nitrogen could remain dissolved at the greatly lowered pressure. People in this situation, like the divers that ascend too rapidly, would thus experience decompression sickness.

---

tuberculosis: L. *tuberculum,* diminutive of tuber, swelling
Cheyne-Stokes breathing: from John Cheyne, Scottish
physician, 1777–1836 and William Stokes, Irish physician,
1804–1878

Chapter Summary

I. Functions and Development of the Respiratory System
The respiratory tract can be divided into upper and lower divisions on the basis of function and embryological development. The upper respiratory system develops from ectoderm lining the oronasal cavity, whereas the lower respiratory system develops as an endodermal outpouching from the foregut.

II. Conducting Passages
The oral and nasal cavities, pharynx, trachea, and bronchial tree deliver warmed and moistened air to the respiratory zone of the lungs.
A. The nose is supported by nasal bones and cartilages.
   1. The nasal epithelium warms, moistens, and cleanses the inspired air.
   2. Olfactory epithelium provides a sense of smell, and the nasal cavity acts as a resonating chamber for voice phonetics.
B. The paranasal sinuses are found in the maxillary, frontal, sphenoid, and ethmoid bones.
C. The pharynx is a funnel-shaped passageway that connects the oral and nasal cavities with the larynx.
   1. The nasopharynx is connected by the auditory tubes to the middle ear cavities and contains the pharyngeal tonsils or adenoids.
   2. The oropharynx is the middle portion, extending from the soft palate, which contains the palatine and lingual tonsils.
   3. The laryngopharynx extends from the hyoid bone to the larynx.
D. The larynx is composed of a number of cartilages that admit air to the trachea during breathing and that close the respiratory passages during swallowing.
   1. The epiglottis is a spoon-shaped structure that aids in closing the laryngeal opening, or glottis, during swallowing.
   2. The larynx contains vocal cords that are controlled by intrinsic muscles and used in sound production.
E. The trachea is a rigid tube, supported by rings of cartilage, that leads from the larynx to the bronchial tree.
F. The bronchial tree includes a right and left primary bronchus, that divides to produce secondary bronchi, tertiary bronchi, and bronchioles; the conducting zone ends with the terminal bronchioles.

III. Alveoli, Lungs, and Pleura
Alveoli are the functional units of the lungs where gas exchange occurs. The right and left lungs are separately packaged in pleural membranes.
A. Alveoli are small, numerous, thin-walled air sacs that provide an enormous surface area for gas diffusion.
B. The right and left lungs are separated by the mediastinum; each lung is divided into lobes and lobules.
C. The lungs are covered by a visceral pleura, and the thoracic cavity is lined by a parietal pleural membrane.
   1. There is a potential space between these two pleural membranes called the pleural cavity.
   2. The pleural membranes package each lung separately and exclude the structures located in the mediastinum.

IV. Physical Aspects of Ventilation
Movement of air into and out of the lungs occurs as a result of pressure differences induced by changes in lung volumes. Ventilation is thus influenced by the physical properties of the lungs.
A. An increase in lung volume during inspiration creates a subatmospheric intrapulmonary pressure.
   1. During expiration the lung volume decreases, and the intrapulmonary pressure rises above the pressure of the atmosphere.
   2. The intrapleural pressure is normally always subatmospheric and less than the intrapulmonary pressure.
B. The lungs have great compliance, which allows them to expand, and great elasticity, which allows them to recoil after being expanded.
   1. The elastic recoil of the lungs causes the lungs to collapse when air enters the intrapleural space in a pneumothorax.
   2. The surface tension of fluid in the alveoli also aids the return of the lungs to their pre-expansion volume.
C. Collapse of the lungs due to excess surface tension is normally prevented by lung surfactant; the absence of surfactant in some premature babies causes respiratory distress syndrome (hyaline membrane disease).

V. Mechanics of Breathing
Inspiration and expiration result from the contraction and relaxation of the diaphragm and thoracic muscles. Pulmonary function can be assessed by measuring the amounts of air inspired and expired under different conditions.
A. Quiet (unforced) inspiration is due to contraction of the diaphragm and the external intercostal muscles. Forced inspiration is aided by the scalenus, pectoralis minor, and sternocleidomastoid.
B. Quiet expiration is produced by relaxation of the respiratory muscles and elastic recoil of the lungs and thorax; forced expiration is aided by abdominal muscles.

VI. Gas Exchange in the Lungs
The lungs function to bring the blood into gaseous equilibrium with the alveolar air. The efficiency of this process is measured by the P_{O_2} and P_{CO_2}.
A. The partial pressure of a gas in a gas mixture is equal to the total pressure times the percent contribution of that gas to the mixture.
B. The concentration of dissolved oxygen and carbon dioxide in plasma at saturation is proportional to the partial pressure of these gases.
   1. The concentrations of these gases in plasma are measured by special electrodes and are indicated in partial pressure units.
   2. Systemic arterial blood normally has a P_{O_2} of about 100 mm Hg; at this P_{O_2}, whole blood contains about 0.3 ml O_2 in the plasma and 19.7 ml O_2 in the red blood cells per 100 ml of blood.
C. Measurments of the P_{O_2} and P_{CO_2} of arterial blood are used as indices of lung function.

VII. Regulation of Breathing
Breathing is directly controlled by centers in the brain stem. These centers are influenced by higher brain areas and by sensory input that makes breathing responsive to the needs of the body.
A. Ventilation is directly controlled by the rhythmicity center in the medulla oblongata, which in turn is influenced by the pneumotaxic and apneustic centers in the pons.
   1. These brain stem areas are affected by higher brain function and by sensory input from chemoreceptors.
   2. There are central chemoreceptors in the medulla oblongata and peripheral chemoreceptors in the aortic and carotid bodies.

B. P_{CO_2} in arterial blood stimulates the chemoreceptors in the medulla by affecting the pH of cerebrospinal fluid; the peripheral chemoreceptors are stimulated by increases in blood P_{CO_2} directly and by decreases in pH caused by the accumulation of carbonic acid.
   1. The P_{CO_2} of blood is the most important stimulus for regulating breathing.
   2. Blood P_{CO_2} is increased in hypoventilation and decreased by hyperventilation.
C. A fall in blood P_{O_2} makes the peripheral chemoreceptors more sensitive to a rise in blood P_{CO_2}; a decrease in blood P_{O_2} below half its normal value stimulates the carotid bodies directly.
D. Pulmonary stretch receptors act to inhibit further inspiration when the lungs are stretched; this is part of the Hering-Breuer reflex.

VIII. Hemoglobin and Oxygen Transport
Oxygen is transported by hemoglobin from the lungs to the tissues. The unloading of oxygen by hemoglobin is influenced by a variety of factors.
A. Each hemoglobin molecule consists of four polypeptide chains and four heme groups that contain iron.
   1. Hemoglobin with ferric iron is called deoxyhemoglobin or when it bonds oxygen, oxyhemoglobin.
   2. Abnormal forms of hemoglobin include carboxyhemoglobin and methemoglobin.
   3. The amount of oxygen carried by the blood at a given P_{O_2} depends directly on the hemoglobin concentration.

4. The formation of oxyhemoglobin, called the loading reaction, depends on the P_{O_2} and on the affinity (bond strength) of hemoglobin for oxygen.
B. A graphic display of the relationship between P_{O_2} and percent oxyhemoglobin saturation is called an oxyhemoglobin dissociation curve.
C. A fall in pH and a rise in temperature weaken the affinity of hemoglobin for oxygen; this promotes the unloading of oxygen to the tissues.
D. A rise in the red blood cell content of 2,3-DPG, as may occur in anemia or hypoxemia, also decreases the hemoglobin affinity for oxygen and promotes oxygen unloading.
E. Striated muscles contain myoglobin, which is related to hemoglobin and acts as an oxygen storage molecule.

IX. Carbon Dioxide Transport and Acid-Base Balance
The acid-base balance of the blood is affected by the concentrations of carbonic acid and bicarbonate, which are transport forms of carbon dioxide. These concentrations are regulated by ventilation.
A. The combination of carbon dioxide and water to form carbonic acid is catalyzed within red blood cells by carbonic anhydrase.
   1. Carbonic acid formed within red blood cells ionizes to H^+ and HCO_3^- (bicarbonate); the H^+ combine with hemoglobin, and the bicarbonate diffuses out of the red blood cell.
   2. When bicarbonate diffuses out, chloride diffuses into the red blood cell; this is called the chloride shift.

3. A reverse chloride shift occurs in the pulmonary capillaries.
B. Hypoventilation causes respiratory acidosis, and hyperventilation causes respiratory alkalosis.
   1. Bicarbonate buffers nonvolatile metabolic acids.
   2. Changes in ventilation can partially compensate for metabolic acidosis and metabolic alkalosis.

X. Effect of Exercise and High Altitude on Respiratory Function
Changes in respiratory function help to improve oxygen delivery during exercise and to compensate for reduced P_{O_2} at a high altitude.
A. The measurement of arterial blood gases and pH do not change appreciably during exercise, because neurogenic and humoral mechanisms match ventilation to the metabolic rate.
B. Endurance-trained athletes extract a higher percentage of the oxygen delivered to their exercising muscles than other people and, thus, have a higher anaerobic threshold.
C. A person who goes to a high altitude hyperventilates as a result of the decrease in P_{O_2}.
   1. There is an increase in 2,3-DPG production in the red blood cells that promotes an increase in oxygen unloading to the tissues.
   2. After a period of time, erythropoietin stimulates the increased production of red blood cells and hemoglobin, so that the blood can carry more oxygen despite the lower P_{O_2}.

Review Activities

Objective Questions

1. Which of the following statements about intrapulmonary and intrapleural pressure is true?
   (a) The intrapulmonary pressure is always subatmospheric.
   (b) The intrapleural pressure is always greater than the intrapulmonary pressure.
   (c) The intrapulmonary pressure is greater than the intrapleural pressure.
   (d) The intrapleural pressure equals the atmospheric pressure.
2. If the transpulmonary pressure equals zero,
   (a) a pneumothorax has probably occurred
   (b) the lungs cannot inflate
   (c) elastic recoil causes the lungs to collapse
   (d) all of the above

3. The maximum amount of air that can be expired after a maximum inspiration is the
   (a) tidal volume
   (b) forced expiratory volume
   (c) vital capacity
   (d) maximum expiratory flow rate
4. If the blood lacked red blood cells but the lungs were functioning normally,
   (a) the arterial P_{O_2} would be normal
   (b) the oxygen content of arterial blood would be normal
   (c) both a and b
   (d) neither a nor b
5. If a person were to dive with scuba equipment to a depth of sixty-six feet, which of the following statements would be *false*?
   (a) The arterial P_{O_2} would be three times normal.
   (b) The oxygen content of plasma would be three times normal.
   (c) The oxygen content of whole blood would be three times normal.

6. Which of the following would be most affected by a decrease in the affinity of hemoglobin for oxygen? The
   (a) arterial P_{O_2}
   (b) arterial percent oxyhemoglobin saturation
   (c) venous oxyhemoglobin saturation
   (d) arterial P_{CO_2}
7. If a person with normal lung function were to hyperventilate for several seconds, there would be a significant
   (a) increase in the arterial P_{O_2}
   (b) decrease in the arterial P_{CO_2}
   (c) increase in the arterial percent oxyhemoglobin saturation
   (d) decrease in the arterial pH
8. Erythropoietic factor is produced by the
   (a) kidneys
   (b) liver
   (c) lungs
   (d) bone marrow

9. The affinity of hemoglobin for oxygen is decreased under conditions of
   (a) acidosis
   (b) fever
   (c) anemia
   (d) acclimatization to a high altitude
   (e) all of the above

10. Most of the carbon dioxide in the blood is carried in the form of
    (a) dissolved CO_2
    (b) carbaminohemoglobin
    (c) carbonic acid and bicarbonate
    (d) carboxyhemoglobin

11. The bicarbonate concentration of the blood would be decreased during
    (a) metabolic acidosis
    (b) respiratory acidosis
    (c) metabolic alkalosis
    (d) respiratory alkalosis

12. The chemoreceptors in the medulla are directly stimulated by
    (a) CO_2 from the blood
    (b) H^+ from the blood
    (c) H^+ in cerebrospinal fluid that is derived from blood CO_2
    (d) decreased arterial P_{O_2}

13. The rhythmic control of breathing is produced by the activity of inspiratory and expiratory neurons in the
    (a) medulla oblongata
    (b) apneustic center of the pons
    (c) pneumotaxic center of the pons
    (d) cerebral cortex

14. Which of the following occurs during hypoxemia?
    (a) increased ventilation
    (b) increased production of 2,3-DPG
    (c) increased production of erythropoietin
    (d) all of the above

15. During exercise, which of the following statements is true?
    (a) The arterial percent oxyhemoglobin saturation is decreased.
    (b) The venous percent oxyhemoglobin saturation is decreased.
    (c) The arterial P_{CO_2} is measurably increased.
    (d) The arterial pH is measurably decreased.

Essay Questions

1. Using a flow diagram to show cause and effect, explain how contraction of the diaphragm produces inspiration.

2. Radiographic (X-ray) pictures show that the ribs of a person with a pneumothorax are expanded and farther apart. Explain why this occurs.

3. Explain, using a flow chart, how a rise in blood P_{CO_2} stimulates breathing. Include both the central and peripheral chemoreceptors in your answer.

4. A person with ketoacidosis may hyperventilate. Explain why this occurs, and explain why this hyperventilation can be stopped by an intravenous fluid containing bicarbonate.

5. What blood measurements can be performed to detect (a) anemia, (b) carbon monoxide poisoning, and (c) poor lung function.

6. Explain how measurements of blood P_{CO_2}, bicarbonate, and pH are affected by hypoventilation and hyperventilation.

7. Explain how blood pH and bicarbonate concentrations are affected by respiratory and metabolic acidosis.

8. How would an increase in the red blood cell content of 2,3-DPG affect the P_{O_2} of venous blood? Explain your answer.

Selected Readings

Avery, M. E., N. S. Wang, and H. W. Taeusch, Jr. March 1975. The lung of the newborn infant. *Scientific American.*

Berger, A. J., R. A. Mitchel, and J. W. Severinghaus. 1977. Regulation of respiration. *New Engl. J. Med.* 297: first part, p. 92; second part, p. 138; third part, p. 194.

Brample, D. M., and D. R. Carrier. 1983. Running and breathing in mammals. *Science* 219:251.

Browning, R. J. 1982. Pulmonary disease: Back to basics (part 1); Putting blood gases to work (part 2). *Diagnostic Med.* First part: Jan/Feb, p. 39; second part: March/April, p. 59.

Finch, C. A., and C. Lenfant. 1972. Oxygen transport in man. *New Engl. J. Med.* 286:407.

Flenley, D. C., and P. M. Warren. 1983. Ventilatory response to O_2 and CO_2 during exercise. *Ann. Rev. Physiol.* 45:415.

Fraser, R. G., and J. A. P. Pare. 1977. *Structure and function of the lung.* 2d ed. Philadelphia: W. B. Saunders Co.

Guz, A. 1975. Regulation of respiration in man. *Ann. Rev. Physiol.* 37:303.

Haddad, G. G., and R. B. Mellins. 1984. Hypoxia and respiratory control in early life. *Ann. Rev. Physiol.* 46:629.

Irsigler, G. B., and J. W. Severinghaus. 1980. Clinical problems of ventilatory control. *Ann. Rev. Med.* 31:109.

Kassirer, J. P., and N. E. Madias. 1980. Respiratory acid-base disorders. *Hosp. Prac.* 15:57.

Naeye, R. L. April 1980. Sudden infant death. *Scientific American.*

Perutz, M. F. December 1978. Hemoglobin structure and respiratory transport. *Scientific American.*

Rigatto, H. 1984. Control of ventilation in the newborn. *Ann. Rev. Physiol.* 46:661.

Roussos, C., and P. T. Macklem. 1982. The respiratory muscles. *New Engl. J. Med.* 307:786.

Shannon, D. C., and D. H. Kelly. 1982. SIDS and near-SIDS. *New Engl. J. Med.* 306: first part, p. 959; second part, p. 1022.

Walker, D. W. 1984. Peripheral and central chemoreceptors in the fetus and newborn. *Ann. Rev. Physiol.* 46:687.

West, J. B. 1984. Human physiology at extreme altitudes on Mount Everest. *Science* 223:784.

Whipp, B. J. 1983. Ventilatory control during exercise in humans. *Ann. Rev. Physiol.* 45:393.

24

Digestive System

Outline

Concepts

The digestive system is derived
    primarily from endoderm. It consists
    of the gastrointestinal tract and
    accessory digestive organs.

Ingested food is changed by the
    mechanical action of teeth and by the
    chemical activity of saliva.

Food is passed from the esophagus to
    the stomach, where it is mixed with
    HCl and pepsin. The resulting chyme
    is sent past the pyloric sphincter to
    the duodenum.

The villi and microvilli in the small
    intestine provide a very high surface
    area for digestion by brush border
    enzymes and for absorption of the
    products of digestion.

The large intestine absorbs water and
    electrolytes from the chyme and
    passes waste products out of the body
    through the rectum and anal canal.

Blood from the intestine flows to the
    liver. The liver produces bile and
    secretes it into the small intestine
    and gallbladder. Both the common
    bile duct and the pancreatic duct,
    carrying pancreatic juice, empty into
    the duodenum.

Polysaccharides and polypeptides are
    digested into amino acids and
    monosaccharides, respectively, which
    are absorbed and secreted into the
    blood. Triglycerides are secreted into
    lymphatic vessels as chylomicrons.

Neural and hormonal mechanisms
    coordinate the activities of different
    regions of the GI tract and the
    actions of the liver, gallbladder, and
    pancreas.

Selected Terms

cloaca (klo-a′kah)
peritoneum (per″i-to-ne′um)
stomodeum (sto″mo-de′um)
viscera (vis′er-ah)
buccal (buk′al)
dentin (den′tin)
gingiva (jin-ji′vah)
rugae (roo′je)
deglutition (deg″loo-tish′un)
esophagus (ē-sof′ah-gus)
hiatus (hi-a′tus)
duodenum
    (du″o-de′num or du-od′ē-num)
ileum (il′e-um)
jejunum (je-joo′num)
mesenteric (mes″en-ter′ik)
cecum (se′kum)
haustrum (hows′trum)
taenia coli (te′ne-ah co′li)
bilirubin (bil″i-roo′bin)
micelle (mi-sel′)
chylomicrons (ki″lo-mi′krons)
chymotrypsin (ki″mo-trip′sin)
ptyalin (ti′ah-lin)
cholecystokinin (ko″le-sis″to-kin′in)

Introduction to the Digestive System

The digestive system is derived primarily from embryonic endoderm that develops into regions and organs specialized for the digestion and absorption of food. The digestive system consists of a tubular gastrointestinal tract and accessory digestive organs.

Objective 1. List the activities of the digestive system, and distinguish between digestion and absorption.

Objective 2. Identify the major structures and regions of the digestive system.

Objective 3. Describe the embryological development of the gastrointestinal tract and accessory digestive organs.

Objective 4. Describe the serous membranes of the gastrointestinal tract.

Unlike plants, which can form organic molecules using inorganic compounds such as carbon dioxide, water, and ammonia, humans and other animals must obtain their basic organic molecules from food. Some of the ingested food molecules are needed for their energy (caloric) value—obtained by the reactions of cell respiration and used in the production of ATP—and the balance is used to make additional tissue.

Most of the organic molecules that are ingested are similar to the molecules that form the structure of human tissues. These are generally large molecules *(polymers),* which are composed of subunits *(monomers).* Within the gastrointestinal tract, these large molecules are **digested** into their monomers by *hydrolysis* reactions (reviewed in fig. 24.1). The monomers thus formed are transported across the wall of the intestine into the blood and lymph; this process is called **absorption.**

Since the composition of food is similar to the composition of body tissues, enzymes that digest food could also digest the person's own tissues. This does not normally occur because there is a variety of protective devices that inactivate digestive enzymes in the body and keep these enzymes separated from the cytoplasm of tissue cells. The fully active digestive enzymes are normally limited in their location to the lumen (cavity) of the gastrointestinal tract.

The lumen of the gastrointestinal tract is continuous with the environment because it is open at both ends (mouth and anus). Indigestible material, such as cellulose from plant walls, pass from one end to the other without crossing the epithelial lining of the digestive tract (that is, without being absorbed). In this sense, these indigestible materials never enter the body, and the harsh conditions required for digestion thus occur *outside* the body.

Figure 24.1.   The digestion of food molecules occurs by means of hydrolysis reactions.

$$\text{Maltose} + \text{Water} \xrightarrow{\text{Carbohydrate digesting enzymes}} \text{Glucose} + \text{Glucose}$$

Disaccharide → **Monosaccharides**

$$\text{Peptide (portion of protein molecule)} + \text{Water} \xrightarrow{\text{Protein digesting enzyme}} \text{Amino acid} + \text{Amino acid}$$

$$\text{Fat} + \text{Water} \xrightarrow{\text{Fat digesting enzyme}} \text{Fatty acids} + \text{Glycerol}$$

In *planaria* (a type of flatworm) the gastrointestinal tract has only one opening—the mouth is also the anus. Each cell that lines the gastrointestinal tract is thus exposed to food, absorbable digestion products, and waste products. The two-ended digestive tract of higher organisms, in contrast, allows one-way transport, which is insured by wavelike muscle contractions and by the action of sphincter muscles. This one-way transport allows different regions of the gastrointestinal tract to be specialized for different functions, as a "dis-assembly line." There are seven functions:

1. **Ingestion.** The taking of food into the digestive system by way of the mouth.
2. **Mastication.** Chewing movements to pulverize food and mix it with saliva.
3. **Deglutition.** The swallowing of food to move it from the mouth to the stomach.
4. **Digestion.** Mechanical and chemical breakdown of food into absorbable molecules.
5. **Absorption.** The passage of monomers of food molecules through the mucous membrane of the intestine into blood or lymph.
6. **Peristalsis.** Rhythmic, wavelike contractions that move food through the digestive tract.
7. **Defecation** *(def''e-ka'shun).* The discharge of indigestible wastes, called feces, from the body.

Anatomically and functionally the digestive system can be divided into a tubular **gastrointestinal (GI) tract,** or an **alimentary canal,** and **accessory organs.** The GI tract is approximately 9 m (30 ft) long and extends from the mouth to the anus. It traverses the thoracic cavity and enters the abdominal cavity at the level of the diaphragm. The anus is located at the inferior portion of the pelvic cavity. The organs of the GI tract include the **oral (buccal) cavity, pharynx, esophagus, stomach, small intestine,** and

ingestion: L. *ingerere,* carry in
mastication: Gk. *mastichan,* gnash the teeth
deglutition: L. *deglutire,* swallow down

peristalsis: Gk. *peri,* around; *stellein,* compress
defecation: L. *de,* from, away; *faecare,* cleanse

Figure 24.2. The digestive system, including the alimentary canal and the accessory digestive organs.

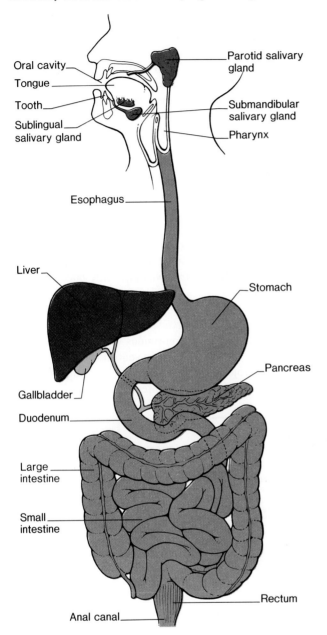

Table **24.1**	Regions of the GI tract and basic functions
Region	**Function**
Oral cavity	Ingests food; receives saliva; mastication; initiates digestion of carbohydrates; forms bolus; deglutition
Pharynx	Receives bolus from oral cavity; autonomically continues deglutition of bolus to esophagus
Esophagus	Transports bolus to stomach by peristalsis; esophageal sphincter restricts backflow of food
Stomach	Receives bolus from esophagus; churns bolus with gastric juice; initiates digestion of proteins; limited absorption; moves chyme into duodenum and prohibits backflow of chyme; regurgitates when necessary
Small intestine	Receives chyme from stomach and secretions from liver and pancreas; chemically and mechanically breaks down chyme; absorbs nutrients; transports wastes through peristalsis to large intestine; prohibits backflow of intestinal wastes from large intestine
Large intestine	Receives undigested wastes from small intestine; absorbs water and electrolytes; forms, stores, and expels feces through defecation reflex

Source: Van De Graaff, Kent M., *Human Anatomy.* © 1984 Wm. C. Brown Publishers, Dubuque, Iowa. All Rights Reserved. Reprinted by permission.

Although there is an abundance of food in the United States, too many people eat the wrong kinds of foods, eat irregularly, or overeat, to the point that eating patterns have become a critical public health concern. Obesity is a major health problem; obese persons are at greater risk for cardiovascular disease, hypertension, osteoarthritis, and diabetes mellitus. People with good nutritional habits are better able to withstand trauma, are less likely to get sick, and are usually less seriously ill when they do become sick.

Embryological Development of the Digestive System

The entire digestive system develops from modifications of an elongated tubular structure called the **primitive gut.** These modifications are initiated during the fourth week of embryonic development. The primitive gut is composed solely of endoderm and for descriptive purposes can be divided into three regions: foregut, midgut, and hindgut (fig. 24.3).

Foregut.   The **stomodeum** *(sto"moh-de'um), or **oral pit,** is not part of the foregut but an invagination of ectoderm that breaks through a thin **oral membrane** to become continuous with the foregut and form the oral cavity or mouth. Structures in the mouth, therefore, are ectodermal in origin. The esophagus, stomach, a portion of the duodenum, the pancreas, liver, and gallbladder are the organs that develop from the foregut (fig. 24.4). Along the alimentary

large intestine (fig. 24.2). The accessory digestive organs include the **teeth, tongue, salivary glands, liver,** and **pancreas.** The term *viscera (vis'er-ah)* is frequently used to refer to the abdominal organs of digestion, but this term can actually be used to indicate any organ in the thoracic and abdominal cavities. **Gut** is an anatomical term that generally refers to the developing stomach and intestine in the embryo.

The regions of the GI tract and their functions are summarized in table 24.1.

---

stomodeum: Gk. *stoma*, mouth; *hodaios*, on the way to

Figure 24.3. A midsagittal section of a four-week-old embryo showing the regions of the primitive gut and the associated blood supply.

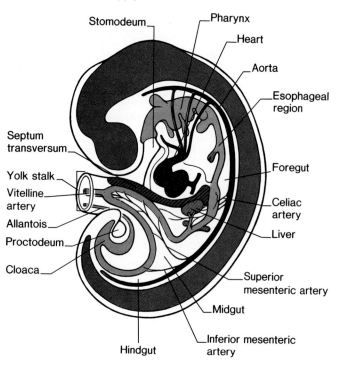

canal, only the inside epithelial lining of the lumen is derived from endoderm. The vascular portion and smooth muscle layers are formed from mesoderm that develops from the surrounding splanchnic mesenchyme. The stomach first appears as an elongated dilation of the foregut. The dorsal border of the stomach undergoes more rapid growth than the ventral border, forming a distinct curvature. The caudal portion of the foregut and the cranial portion of the midgut form the duodenum. The liver and pancreas arise as small **hepatic and pancreatic buds,** respectively, from the wall of the duodenum. The hepatic bud experiences incredible growth to form the gallbladder, associated ducts, and the various lobes of the liver (fig. 24.4). By the sixth week, the liver is carrying out hemopoiesis (the formation of blood cells); and by the ninth week, the liver represents 10% of the total weight of the fetus.

The pancreas develops from dorsal and ventral pancreatic buds of endodermal cells. As the duodenum grows, it rotates clockwise and the two pancreatic buds fuse (fig. 24.4).

Midgut. During the fourth week of the embryonic stage (fig. 24.3), the midgut is continuous with the yolk sac. By the fifth week, the midgut forms a ventral U-shaped **midgut loop,** which projects into the umbilical cord (fig. 24.5). As development continues, the anterior limb of the midgut

Figure 24.4. Progressive stages of development of the foregut to form the stomach, duodenum, liver, gallbladder, and pancreas. (a) four weeks, (b, c) five weeks, and (d) six weeks.

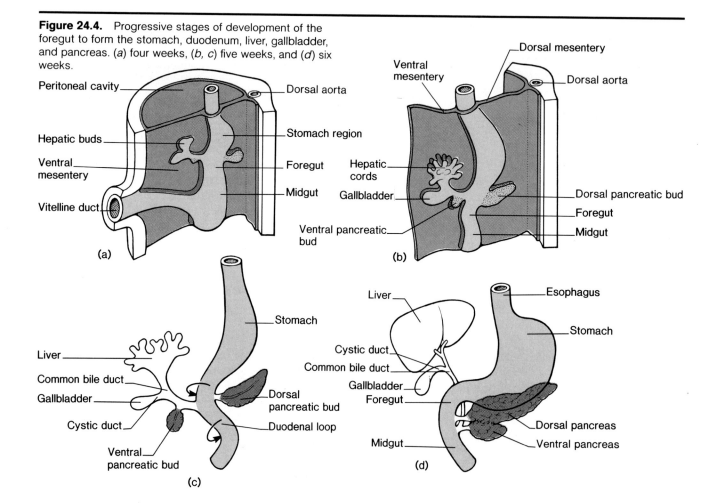

loop coils to form most of the small intestine. The posterior limb of the midgut loop expands to form a portion of the small and large intestines. A **cecal diverticulum** *(di″ver-tik′u-lum)* appears during the fifth week.

During the tenth week, the intestines are withdrawn to the abdominal cavity and further differentiation and rotation occur. The cecal diverticulum continues to develop, forming the cecum and appendix. The remainder of the midgut gives rise to the ascending colon and hepatic flexure (fig. 24.5).

Hindgut.   The hindgut extends from the midgut to the **cloacal membrane** (fig. 24.3). The **proctodeum** *(prok″to-de′um)*, or **anal pit,** is a depression in the anal region formed from an invagination of ectoderm that contributes

---

proctodeum: Gk. *proktos*, anus; *hodaios*, on the way to

to the cloacal membrane. The *allantois (ah-lan′to-is),* which receives urinary wastes from the fetus, connects to the hindgut at a region called the **cloaca** *(klo-a′kah)*, as seen in figure 24.6. A band of mesenchymal cells called the **urorectal septum** grows caudally during the fourth through seventh week until a complete partition separates the cloaca into a dorsal **anal canal** and a ventral **urogenital sinus.** With the completion of the urorectal septum, the cloacal membrane is divided into an anterior **urogenital membrane** and a posterior **anal membrane.** Toward the end of the seventh week, the anal membrane perforates and forms the anal opening, which is lined with ectodermal cells. About this time, the urogenital membrane ruptures to provide further development of the urinary and reproductive systems.

---

cloaca: L. *cloaca*, sewer

Figure 24.5.   Progressive stages of the development of the midgut to form the distal portion of the small intestine and the proximal portion of the large intestine.

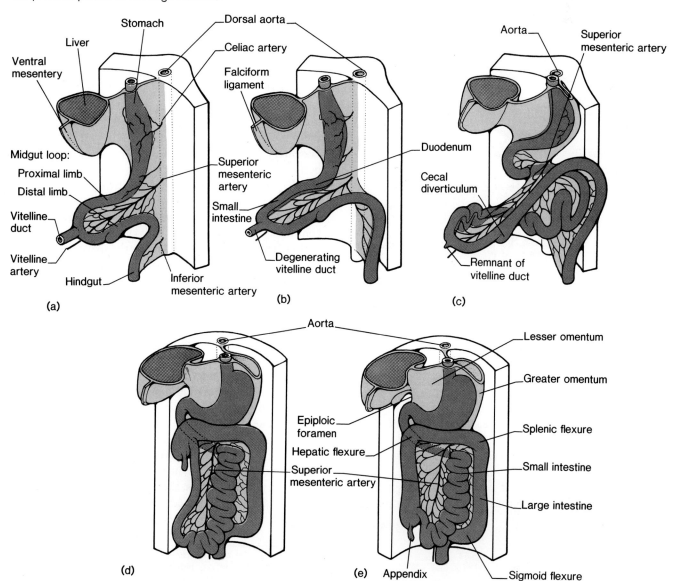

Figure 24.6. The progressive development of the hindgut, illustrating the developmental separation of the digestive system from the urogenital system. (*a, b, c*) four, six, and seven weeks, respectively. (*a₁, b₁, c₁*) transverse sections through the cloaca.

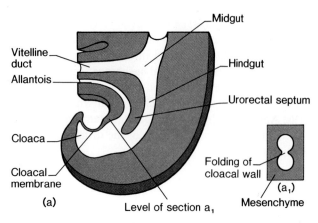

(a)   Level of section a₁

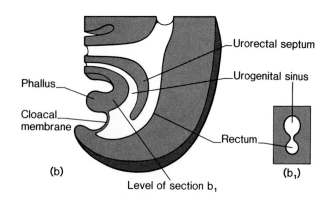

(b)   Level of section b₁

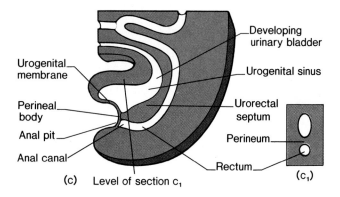

(c)   Level of section c₁

Figure 24.7. A diagrammatic representation of the serous membranes.

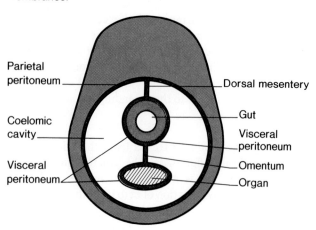

Serous Membranes

Most of the GI tract and abdominal accessory digestive organs are positioned within the **abdominal cavity.** These organs are not firmly embedded in solid tissue but are supported and covered by **serous membranes.** A serous membrane is an epithelial membrane that lines the thoracic and abdominal cavities and covers the organs that lie within these cavities. A serous membrane has a parietal portion lining the body wall and a visceral portion covering the internal organs. The serous membranes associated with the lungs are called pleurae. The serous membranes of the abdominal cavity are called **peritoneal membranes,** or **peritoneum** *(per″i-to-ne′um).* The peritoneum is the largest serous membrane of the body. It is composed of simple squamous epithelium with portions reinforced with connective tissue.

The **parietal peritoneum** lines the wall of the abdominal cavity (fig. 24.7). Along the dorsal, or posterior, aspect of the abdominal cavity the parietal peritoneum comes together to form a double-layered peritoneal fold called the **mesentery,** which supports the GI tract. The **dorsal mesentery** gives the pendulous small intestine freedom for peristaltic movement and provides a structure through which intestinal nerves and vessels traverse. The **mesocolon** is a specific portion of the mesentery that supports the large intestine (fig. 24.8). The peritoneal covering continues around the intestinal viscera as the **visceral peritoneum.** The **peritoneal cavity** is the space between the parietal and visceral portions of the peritoneum.

Extensions of the parietal peritoneum, located in the peritoneal cavity, serve specific functions (fig. 24.8). The **falciform** *(fal′si-form)* **ligament,** a serous membrane reinforced with connective tissue, attaches the liver to the diaphragm and anterior abdominal wall. The **lesser omentum** *(o-men′tum)* passes from the lesser curvature of the stomach and the upper duodenum to the inferior surface of the liver. The **greater omentum** extends from the greater curvature of the stomach to the transverse colon, forming an apronlike structure over the small intestine. Functions of the omentum include storing fat, cushioning visceral organs, supporting lymph nodes, and protecting against the spread of infections. In cases of localized inflammation, such as appendicitis, the greater omentum may compartmentalize the inflamed area, sealing it off from the rest of the peritoneal cavity.

---

peritoneum: Gk. *peritonaion,* stretched over
mesentery: Gk. *mesos,* middle; *enteron,* intestine
omentum: L. *omentum,* apron

Figure 24.8.   The structural arrangement of the abdominal organs and peritoneal membranes. (*a*) the greater omentum; (*b*) the lesser omentum with the liver lifted; (*c*) the mesentery with the greater omentum lifted; (*d*) the relationship of the peritoneal membranes to the visceral organs as shown in a sagittal view.

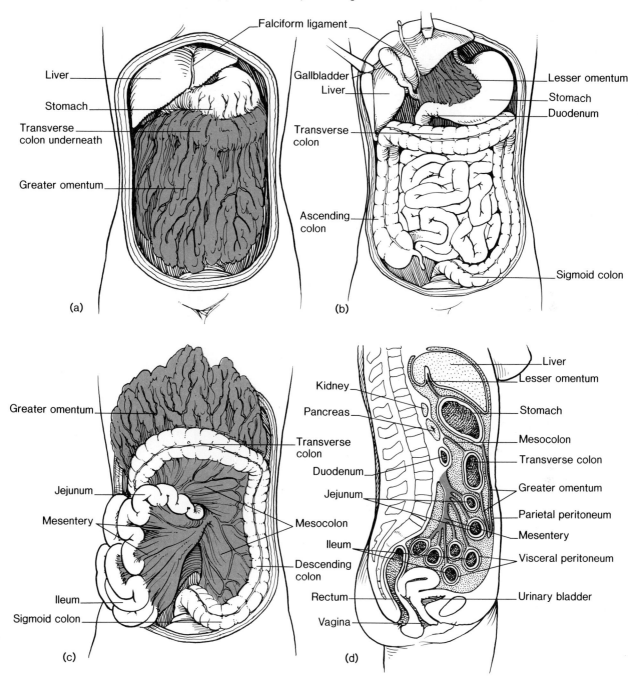

Certain abdominal organs are not within the peritoneal cavity and are therefore not covered by peritoneum. These organs are said to be *retroperitoneal* and include most of the pancreas, kidneys, a portion of the duodenum, and the abdominal aorta. The urinary bladder is situated in front of the peritoneum and is considered *antiperitoneal.*

Peritonitis is a bacterial inflammation of the peritoneum. Peritonitis may be caused by trauma, rupture of a visceral organ, ectopic pregnancy, or postoperative complications. Peritonitis is usually extremely painful and serious. Treatment usually involves the injection of massive doses of antibiotics and perhaps peritoneal intubation to permit drainage.

Figure 24.9.   (a) an illustration of the major layers of the intestine. The insert shows how folds of mucosa form projections called *villi.* (b) an illustration of a cross section of the intestine showing layers and glands.

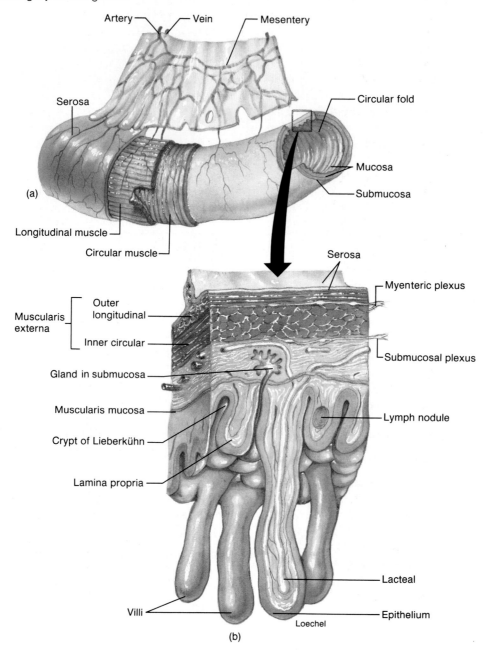

Layers of the Gastrointestinal Tract

The GI tract from the esophagus to the anal canal is composed of four layers, or **tunics.** Each tunic contains a dominant tissue type that performs specific functions in the digestive process. The four tunics of the GI tract, from the inside out, are the **mucosa, submucosa, muscularis,** and **serosa** (fig. 24.9a).

tunica: L. *tunica,* covering or coat

Mucosa.   The mucosa surrounds the lumen of the GI tract and is the absorptive and major secretory layer. It consists of a simple columnar epithelium supported by the **lamina *(lam'i-nah)* propria,** which is a thin layer of connective tissue. The lamina propria contains numerous lymph nodules that are important in protecting against disease (fig. 24.9b). External to the lamina propria are thin layers of smooth muscle called the **muscularis mucosa.** This is the muscle layer that causes portions of the

GI tract to have numerous small folds, thus greatly increasing the absorptive surface area. Specialized **goblet cells** in the mucosa secrete mucus throughout most of the GI tract.

Submucosa.   The relatively thick submucosa is a highly vascular layer of connective tissue serving the mucosa. Absorbed molecules that pass through the columnar epithelial cells of the mucosa enter into blood vessels of the submucosa. In addition to blood vessels, the submucosa contains glands and nerve plexuses. The *submucosal plexus (Meissner's plexus)* (fig. 24.9b) provides an autonomic nerve supply to the muscularis mucosa.

Muscularis.   The muscularis (also called the muscularis externa) is responsible for segmental contractions and peristaltic movement through the GI tract. The muscularis has an inner circular and an outer longitudinal layer of smooth muscle. Contractions of these layers move the food peristaltically through the tract and physically pulverize and churn the food with digestive enzymes. The *myenteric plexus (Auerbach's plexus)* located between the two muscle layers provides the major nerve supply to the GI tract and includes fibers and ganglia from both the sympathetic and parasympathetic divisions of the autonomic system.

Serosa.   The outer serosa layer completes the wall of the GI tract. It is a binding and protective layer consisting of areolar connective tissue covered with a layer of simple squamous epithelium and subjacent connective tissue. The simple squamous epithelium is actually the visceral peritoneum.

The body has several defense mechanisms to protect against ingested material that may be harmful if absorbed. The acidic environment of the stomach and the lymphatic system kill many harmful bacteria. A mucus lining throughout the GI tract serves as a protective layer. Vomiting and in certain cases diarrhea are reactions to substances that irritate the GI tract. Vomiting is a reflexive response to many toxic chemicals and as such can be beneficial even though unpleasant.

Innervation.   The GI tract is innervated by the sympathetic and parasympathetic divisions of the autonomic system. The vagus nerve is the source of parasympathetic activity in the esophagus, stomach, pancreas, gallbladder, small intestine, and upper portion of the large intestine. The lower portion of the large intestine receives parasympathetic innervation from spinal nerves in the sacral region. The submucosal plexus and myenteric plexus are the

---

Meissner's plexus: from Georg Meissner, German histologist, 1829–1905
Auerbach's plexus: from Leopold Auerback, German anatomist, 1828–97

---

sites where preganglionic fibers synapse with postganglionic fibers that innervate the smooth muscle of the GI tract. Stimulation of the parasympathetic fibers increases peristalsis and the secretions of the GI tract.

Postganglionic sympathetic fibers pass through the submucosal and myenteric plexuses and innervate the GI tract. Sympathetic nerve stimulation acts antagonistically to the effects of parasympathetic nerves by reducing peristalsis and secretions and stimulating the contraction of sphincter muscles along the GI tract.

---

1. *Define the terms* digestion *and* absorption. *Give examples of molecules that are absorbed.*
2. *Describe the embryonic derivation of the mouth and anus. Identify the portion of the gut that gives rise to the pancreas and liver.*
3. *Describe the functions of the dorsal mesentery and greater omentum. Identify the location of the peritoneum, and list the organs that are retroperitoneal.*
4. *List the four tunics of the GI tract, and identify their major tissue types. Explain the functions of these four tunics.*

Mouth, Pharynx, and Associated Structures

Ingested food is changed by the mechanical action of teeth and by the chemical activity of saliva.

Objective 5.   Describe the anatomy of the oral cavity.
Objective 6.   Contrast deciduous and permanent dentitions, and describe the structure of a typical tooth.
Objective 7.   Describe the location and histological structures of the salivary glands, and list the functions of saliva.

The functions of the **mouth** and associated structures are to form a receptacle for food, to initiate digestion through mastication, to swallow food, and to form words in speech. The mouth can also assist the respiratory system in the passage of air. The **pharynx,** which is posterior to the mouth, serves as a common passageway for both the respiratory and digestive systems. Both the mouth and pharynx are lined with nonkeratinized, stratified squamous epithelium, which is constantly moistened by the secretion of saliva. The mouth is referred to as the **oral,** or **buccal** *(buk′al)* cavity (fig. 24.10). It is formed by the **cheeks, lips, hard** and **soft palates,** and **tongue.** The **vestibule** of the oral cavity is the depression between the cheeks and lips externally and the gums and teeth internally (fig. 24.11). The opening of the oral cavity is referred to as the **oral orifice,** and the opening between the oral cavity and the pharynx is called the **fauces.**

---

pharynx: L. *pharynx,* throat
buccal: L. *bucca,* cheek
fauces: L. *fauces,* throat

Figure 24.10.   The structures of the oral cavity as seen through the oral orifice.

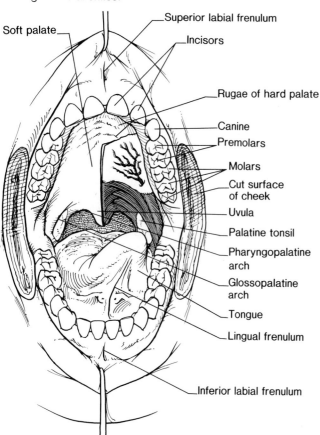

Soft palate

Superior labial frenulum

Incisors

Rugae of hard palate

Canine

Premolars

Molars

Cut surface of cheek

Uvula

Palatine tonsil

Pharyngopalatine arch

Glossopalatine arch

Tongue

Lingual frenulum

Inferior labial frenulum

Cheeks and Lips

The cheeks form the lateral walls of the oral cavity and consist of outer layers of skin, subcutaneous fat, facial muscles that assist in manipulating food in the oral cavity, and inner linings of moistened, stratified squamous epithelium. The anterior portion of the cheeks terminates in the superior and inferior lips that surround the oral orifice.

The lips are fleshy, highly mobile organs whose principal function in humans is associated with speech. Lips also serve for suckling, manipulating food, and keeping food between the upper and lower teeth. Each lip is attached from its inner surface to the gum by a midline fold of mucous membrane called the **labial frenulum** *(fren'u-lum)* (fig. 24.10). The lips are formed from the orbicularis oris muscle and associated connective tissue and covered with soft, pliable skin. Between the outer skin and the mucous membrane of the oral cavity is a transition zone called the **vermilion.** Lips are red to reddish-brown in color because of blood vessels close to the surface. The many sensory receptors in the lips aid in determining the temperature and texture of food.

Suckling is an innate characteristic of newborn infants. The lips are well formed for this activity and even contain blisterlike "milk pads" that aid in suckling. The wide nostrils and receding lower jaw of infants also facilitates suckling.

vermilion: O.E. *vermeylion,* red colored

Figure 24.11.   A midsagittal section of the head, showing the oral and pharyngeal regions.

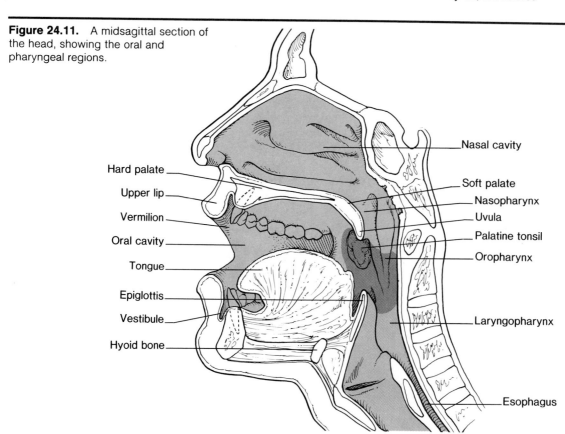

Hard palate

Upper lip

Vermilion

Oral cavity

Tongue

Epiglottis

Vestibule

Hyoid bone

Nasal cavity

Soft palate

Nasopharynx

Uvula

Palatine tonsil

Oropharynx

Laryngopharynx

Esophagus

Tongue

As a digestive organ, the tongue functions to move food around in the mouth during mastication and to assist in swallowing food. It contains **taste buds,** that sense various food tastes. The tongue is also essential in producing articulated speech. The tongue is a mass of striated muscle covered with a mucous membrane. Extrinsic tongue muscles (those that insert upon the tongue) move the tongue from side to side and in and out. Only the anterior two-thirds of the tongue lies in the oral cavity; the remaining one-third lies in the pharynx (fig. 24.11) and is attached to the hyoid bone. Rounded masses of **lingual tonsils** are located on the dorsal surface of the base of the tongue (fig. 24.12). The undersurface of the tongue is connected along the midline anteriorly to the floor of the mouth by the vertically positioned **lingual frenulum** (fig. 24.10).

When a short lingual frenulum restricts tongue movements, the person is said to be "tongue-tied." If this developmental problem is severe, the infant may have difficulty suckling. Older children with this problem may have faulty speech. These functional problems can be easily corrected through surgery.

The dorsal surface of the tongue has numerous small elevations, called **papillae,** that give the tongue a distinct roughened surface, which aids the handling of food. These papillae also contain taste buds that can distinguish sweet, salty, sour, and bitter sensations. Three types of papillae are present on the dorsum of the tongue: **filiform, fungiform,** and **vallate (circumvallate)** (fig. 24.12). Filiform papillae are sensitive to touch, have tapered tips, and are by far the most numerous. The larger and rounded fungiform papillae are scattered among the filiform type. The few vallate papillae are arranged in a V-shape on the posterior surface of the tongue.

Palate

The palate is the roof of the oral cavity and consists of the bony hard palate anteriorly and the soft palate posteriorly (figs. 24.10 and 24.11). The hard palate is formed by the palatine processes of the maxillae and the horizontal plates of the palatine bones and is covered with a mucous membrane. Transverse ridges called **palatal rugae** *(roo'je)* are located along the mucous membrane of the hard palate. These structures serve as friction ridges against which the tongue is placed during swallowing. The soft palate is a muscular arch covered with mucous membrane and is continuous anteriorly with the hard palate. Suspended

tonsil: L. *toles,* swelling
frenulum: L. diminutive of *frenum,* bridle
papilla: L. *papula,* swelling or pimple
filiform: L. *filum,* thread; *forma,* form
fungiform: L. *fungus,* fungus; *forma,* form
vallate: L. *vallatus,* surrounded with a rampart

Figure 24.12.   The dorsum of the tongue.

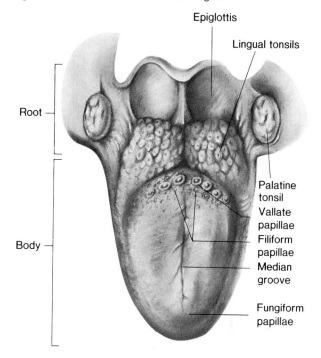

from the middle lower border of the soft palate is a cone-shaped projection called the **uvula.** During swallowing, the soft palate and uvula are drawn upward, closing the nasopharynx and preventing food and fluid from entering the nasal cavity.

Two muscular folds extend downward from both sides of the base of the uvula (fig. 24.10). The anterior fold is called the **glossopalatine arch,** and the posterior fold is the **pharyngopalatine** *(fah-ring''go-pal'ah-tīn)* **arch.** Between these two arches, toward the posterior lateral portion of the oral cavity, is the **palatine tonsil.**

Teeth

Humans, being mammals, have **heterodont dentition,** which means that we have teeth, or **dentes** *(den'tēz),* that differ in structure and are adapted to handle food in different ways (fig. 24.13). The four pairs (upper and lower jaws) of anteriormost teeth are the **incisors** *(in-si'zer).* The chisel-shaped incisors are adapted for cutting and shearing food. The two pairs of cone-shaped **canines,** or **cuspids,** are located at the anterior corners of the mouth and are adapted for holding and tearing. Incisors and canines are further characterized by a single root on each tooth. Located behind the canines are the **premolars,** or **bicuspids,** and **molars.** These teeth have two or three roots and have somewhat rounded, irregular surfaces called *cusps* for crushing and grinding food.

uvula: L. *uvula,* small grapes
incisor: L. *incidere,* to cut
canine: L. *canis,* dog
molar: L. *mola,* millstone

Figure 24.13. Dentitions and the sequence of eruptions. (a) deciduous teeth; (b) permanent teeth; (c) deciduous and permanent teeth viewed in a panoramic X-ray of a female age ten years and four months.

Eruption (months)

Central incisor — 6–8
Lateral incisor — 7–12
Canine — 16–20
First molar — 12–16
Second molar — 20–30

18–22
16–24
14–18
7–9
6–8

(a)

Eruption (years)

Central incisor — 7–8
Lateral incisor — 7–10
Canine — 9–14
First premolar — 9–13
Second premolar — 10–14
First molar — 5–8
Second molar — 10–14
Third molar (wisdom tooth) — 17–24

(b)

(c)

Humans are **diphyodont** *(dif″ĭ-o-dont″);* that is, normally two sets of teeth develop in a person's lifetime. There are twenty **deciduous (milk) teeth,** which begin to erupt at about six months of age (fig. 24.14 and tables 24.2, 24.3), beginning with the incisors. All of the deciduous teeth have erupted by the age of two and a half. There are thirty-two **permanent teeth,** which replace the deciduous teeth in a predictable sequence. This process begins at about age six and continues until about age seventeen. The **third molars,** or **wisdom teeth,** are the last to erupt. Wisdom teeth are less predictable, and if they do erupt, it is between the ages of seventeen and twenty-five. Because the jaws are formed by this time and other teeth are in place, the eruption of wisdom teeth may cause serious problems of crowding or impaction.

deciduous: L. *deciduus,* to fall away

Figure 24.14. The skull of a youth (nine to twelve years) showing the eruption of teeth.

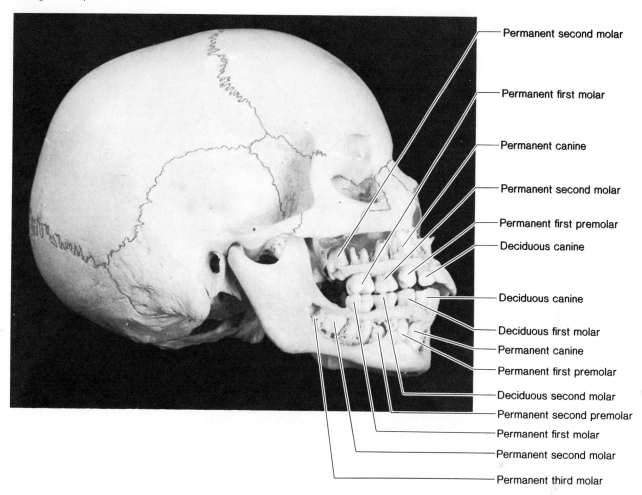

- Permanent second molar
- Permanent first molar
- Permanent canine
- Permanent second molar
- Permanent first premolar
- Deciduous canine
- Deciduous canine
- Deciduous first molar
- Permanent canine
- Permanent first premolar
- Deciduous second molar
- Permanent second premolar
- Permanent first molar
- Permanent second molar
- Permanent third molar

Table 24.2 Eruption sequence and loss of deciduous teeth

Type of tooth	Average age of eruption		Average age of loss
	Lower	Upper	
Central incisors	6 mos	7½ mos	7 yrs
Lateral incisors	7 mos	9 mos	8 yrs
First molars	12 mos	14 mos	10 yrs
Canines (cuspids)	16 mos	18 mos	10 yrs
Second molars	20 mos	24 mos	11–12 yrs

Source: Van De Graaff, Kent M., *Human Anatomy.* © 1984 Wm. C. Brown Publishers, Dubuque, Iowa. All Rights Reserved. Reprinted by permission.

Table 24.3 Eruption sequence of permanent teeth

Type of tooth	Average age of eruption	
	Lower	Upper
First molars	6–7 yrs	6–7 yrs
Central incisors	6–7 yrs	7–8 yrs
Lateral incisors	7–8 yrs	8–9 yrs
Canines (cuspids)	9–10 yrs	11–12 yrs
First bicuspids (premolars)	10–12 yrs	10–11 yrs
Second bicuspids (premolars)	11–12 yrs	10–12 yrs
Second molars	11–13 yrs	12–13 yrs
Third molars (wisdom)	17–25 yrs	17–25 yrs

Source: Van De Graaff, Kent M., *Human Anatomy.* © 1984 Wm. C. Brown Publishers, Dubuque, Iowa. All Rights Reserved. Reprinted by permission.

A **dental formula** is a graphic representation of the types, number, and position of teeth in the oral cavity. All mammals are heterodont and have a constant tooth count, so a dental formula can be written for each species of mammal. Following are the deciduous and permanent dental formulae for humans:

Formula for deciduous dentition:
I2/2, C1/1, DM2/2 = 10 × 2 = 20 teeth

Formula for permanent dentition:
I2/2, C1/1, P2/2, M3/3 = 16 × 2 = 32 teeth

(*I* = incisor; *C* = canine; *P* = premolar; *DM* = deciduous molar; *M* = molar)

The cusps of the upper and lower premolar and molar teeth occlude for chewing food in a process called *mastication,* whereas the upper incisors normally form an overbite with the incisors of the lower jaw. An overbite of the upper incisors creates a shearing action as these teeth slide past one another. Masticated food is mixed with saliva, which initiates chemical digestion and facilitates swallowing. The soft, flexible mass of food that is swallowed is called a **bolus.**

A tooth consists of an exposed **crown,** supported by a **neck** anchored firmly into the jaw by one or more **roots** (fig. 24.15). The roots of teeth fit into sockets called **alveoli** in the alveolar processes of the mandible and maxillae. Each socket is lined with a connective tissue periosteum, specifically called the **periodontal membrane.** The root of a tooth is covered with a bonelike material called the **cementum;** fibers in the periodontal membrane insert into the cementum and fasten the tooth in its socket. The **gingiva** *(jin-ji'vah),* or **gum,** is the mucous membrane surrounding the alveolar processes in the oral cavity.

The bulk of a tooth consists of **dentin** *(den'tin),* a substance similar to bone but harder. Covering the dentin on the outside and forming the crown is a tough, durable layer of **enamel.** Enamel is composed primarily of calcium phosphate and is the hardest substance in the body. The central region of the tooth contains the **pulp cavity.** The pulp cavity contains the **pulp,** which is composed of connective tissue with blood vessels, lymph vessels, and nerves. A **root canal** is continuous with the pulp cavity through the root to an opening at the base called the **apical foramen.** The tooth receives nourishment through vessels traversing the apical foramen. Proper nourishment is particularly important during embryonic development. The diet of the mother should contain an abundance of calcium and vitamin D during pregnancy to insure the proper development of the baby's teeth.

Enamel is the hardest substance in the body, but it can be dissolved by bacterial activity that results in dental caries. These caries must be artificially filled because new enamel is not produced after a tooth erupts. The rate of tooth decay decreases after age thirty-five, but then periodontal diseases may develop. Periodontal diseases result from plaque or tartar buildup at the gum line, which wedges the gum away from the teeth, allowing bacterial infections to develop.

Salivary Glands

Salivary glands are accessory digestive glands that produce a fluid secretion called saliva. Saliva functions as a solvent in cleansing the teeth and dissolving food chemicals so they can be tasted. Saliva also contains enzymes,

---

bolus: Gk. *bolos,* lump
gingiva: L. *gingiva,* gum
dentin: L. *dens,* tooth

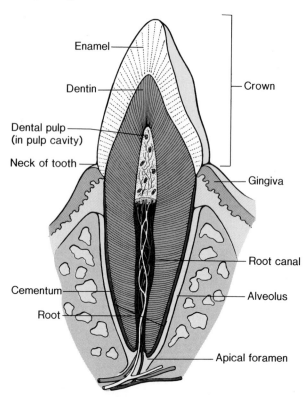

Figure 24.15. The structure of a tooth shown in a vertical section through a canine tooth.

which digest starch and fat, and mucus, which lubricates the pharynx to facilitate swallowing. Saliva is secreted continuously but usually only in sufficient amounts to keep the mucous membranes of the oral cavity moist. Numerous minor salivary glands, called **buccal glands,** are located in the mucous membranes of the palatal region of the oral cavity. But most of the saliva is produced by three pairs of salivary glands outside of the oral cavity and is transported to the mouth via **salivary ducts.** The three major pairs of salivary glands are the parotid, submandibular, and sublingual (fig. 24.16).

The **parotid** *(pah-rot'id)* **gland** is the largest and is positioned below and in front of the ear, between the skin and the masseter muscle. The **parotid (Stensen's) duct** parallels the zygomatic arch across the masseter muscle, pierces the buccinator muscle, and drains into the oral cavity opposite the second upper molar. It is the parotid gland that becomes infected and swollen with the mumps.

The **submandibular** *(sub"man-dib'u-lar)* **(submaxillary) gland** is positioned inferiorly to the body of the mandible about midway along the inner side of the jaw. This gland is covered by the more superficial mylohyoid muscle. The **submandibular (Wharton's) duct** empties into the floor of the mouth on either side of the lingual frenulum.

---

parotid: Gk. *para,* beside; *otos,* ear
Stensen's duct: from Nicholaus Stensen, Danish anatomist, 1638–86
Wharton's duct: from Thomas Wharton, English anatomist, 1614–73

Figure 24.16.   The salivary glands.

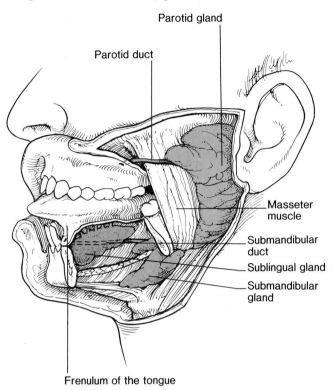

Parotid gland

Parotid duct

Masseter muscle

Submandibular duct

Sublingual gland

Submandibular gland

Frenulum of the tongue

Figure 24.17.   The histology of the salivary glands. (*a*) the parotid gland; (*b*) the submandibular gland; and (*c*) the sublingual gland.

Interlobular parotid duct

Seromucous acini

(a)

Mucous cells

Lumen of submandibular intralobular duct

Serous cells

(b)

Mucous cells

Serous cells

Intralobular sublingual duct

(c)

Two types of secretory cells, called **serous** and **mucous** cells, are found in all salivary glands in various proportions (fig. 24.17). Serous cells produce a watery fluid containing digestive enzymes; mucous cells secrete a thick, stringy, mucus. Cuboidal epithelial cells line the lumina of the salivary ducts.

The salivary glands are innervated by both divisions of the autonomic nervous system. Sympathetic impulses stimulate the secretion of small amounts of viscous saliva. Parasympathetic stimulation causes the secretion of large volumes of watery saliva. Physiological responses of this type occur whenever a person sees, smells, tastes, or even thinks about desirable food. The amount of saliva secreted daily ranges from 1,000 to 1,500 ml. Information about the salivary glands is summarized in table 24.4.

1. *Define the terms* diphyodont *and* heterodont. *Which of the four kinds of teeth is absent in deciduous dentition?*
2. *Identify the locations of the enamel, dentin, cementum, and pulp of a tooth. Explain how a tooth is anchored into its socket.*
3. *Identify the location of the parotid and the submandibular ducts, and describe where in the oral cavity they secrete.*
4. *Describe the effects of autonomic stimulation on salivary secretion.*

Table 24.4 Major salivary glands

Gland	Location	Duct	Entry into oral cavity	Type of secretion
Parotid	Anterior and inferior to auricle; subcutaneous over masseter muscle	Parotid (Stensen's duct)	Lateral to upper second molar	Watery serous fluid, salts, and enzyme
Submandibular	Inferior to the base of the tongue	Submandibular (Wharton's duct)	Papilla lateral to lingual frenulum	Watery serous fluid with some mucus
Sublingual	Anterior to submandibular; under tongue	Several small ducts (Rivinus's ducts)	With submandibular duct	Mostly thick, stringy mucus, salts, and enzyme

Source: Van De Graaff, Kent M., *Human Anatomy.* © 1984 Wm. C. Brown Publishers, Dubuque, Iowa. All Rights Reserved. Reprinted by permission.

Esophagus and Stomach

A bolus of food is passed from the esophagus to the stomach, where it is churned and mixed with hydrochloric acid and pepsin. The chyme thus produced is sent past the pyloric sphincter to the duodenum.

Objective 8. Describe the histological structure of the esophagus and stomach. List the cell types in the gastric mucosa and their secretory products.

Objective 9. Describe the steps in deglutition.

Objective 10. Describe the functions of hydrochloric acid and pepsin.

Objective 11. Explain how gastric and duodenal peptic ulcers might be produced.

Esophagus

The **esophagus** *(ĕ-sof'ah-gus)* is that portion of the GI tract that connects the pharynx to the stomach. It is a collapsible muscular tube approximately 25 cm (10 in.) long, originating at the larynx and located posterior to the trachea.

The esophagus is located within the mediastinum of the thorax and passes through the diaphragm just above the opening into the stomach. The opening through the diaphragm is called the **esophageal hiatus** *(ĕ-sof''ah-je'al hi-a'tus).* The esophagus is lined with a nonkeratinized stratified squamous epithelium; its walls contain either striated or smooth muscle, depending on the location. The upper third of the esophagus contains striated muscle; the middle third contains a mixture of striated and smooth muscle, and the terminal portion contains only smooth muscle (fig. 24.18).

The lumen of the terminal portion of the esophagus is slightly narrowed because of a thickening of the circular muscle fibers in its wall. This portion is referred to as the **lower esophageal (gastroesophageal) sphincter.** The muscle fibers of this region constrict after food passes into the stomach to help prevent the stomach contents from regurgitating into the esophagus. Regurgitation would

esophagus: Gk. *oisein,* to carry; *phagema,* food

occur because the pressure in the abdominal cavity is greater than the pressure in the thoracic cavity as a result of respiratory movements. The lower esophageal sphincter must remain closed, therefore, until food is pushed through it by peristalsis into the stomach.

The lower esophageal sphincter is not a true sphincter muscle that can be identified histologically, and it does at times permit the acidic contents of the stomach to enter the esophagus. This can create a burning sensation commonly called **heartburn,** although the heart is not involved. Infants under one year old have difficulty controlling their lower esophageal sphincter and thus often "spit up" following meals. Certain mammals, such as rodents, have a true gastroesophageal sphincter and cannot regurgitate, which is why poison grains can kill mice and rats effectively.

Mechanism of Swallowing

Swallowing, or **deglutition** *(deg''loo-tish'un),* is the complex mechanical and physiological act of moving food or fluid from the oral cavity to the stomach. For descriptive purposes, deglutition is divided into three stages (fig. 24.19).

The first stage is voluntary and follows mastication, if food is involved. During this stage, the mouth is closed and breathing is temporarily interrupted. A bolus is formed as the tongue is elevated against the palate through contraction of the mylohyoid and styloglossus muscles and the intrinsic muscles of the tongue.

The second stage of deglutition is the passage of the bolus through the pharynx. The events of this stage are involuntary and are elicited by stimulation of sensory receptors located at the opening of the oral pharynx. Pressure of the tongue against the rugae of the hard palate seals the pharynx from the oral cavity, creates a pressure, and forces the bolus into the oral pharynx. The soft palate and pendulant uvula are elevated to close the nasal pharynx as the bolus passes. The hyoid bone and the larynx are also elevated. Elevation of the larynx against the epiglottis seals the glottis so that food or fluid is less likely to enter the trachea. Sequential contraction of the constrictor muscles of the pharynx moves the bolus through the pharynx to the esophagus. This stage is completed in one second or less.

Figure 24.18. The microscopic appearance of the esophagus.

Figure 24.19. The esophagus. (*a-d*) the deglutition reflex and peristalsis of the esophagus, and (*e*) an anteroposterior X-ray. (*a*) a bolus is formed as the tongue is elevated against the palate. (*b*) the bolus passes through the oropharynx as the nasopharynx is sealed by the elevated uvula. (*c*) the bolus enters the esophagus as the glottis is sealed against the epiglottis. (*d*) a peristaltic wave moves the bolus through the esophagus to the stomach.

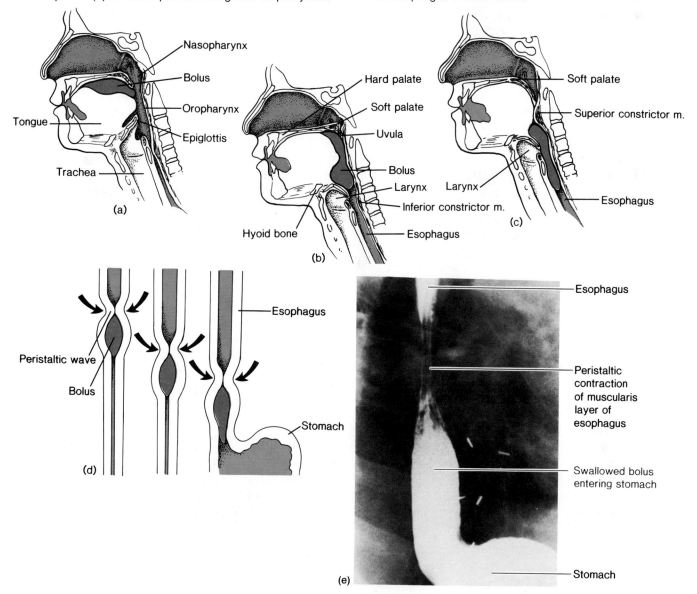

Figure 24.20. The major regions and structures of the stomach.

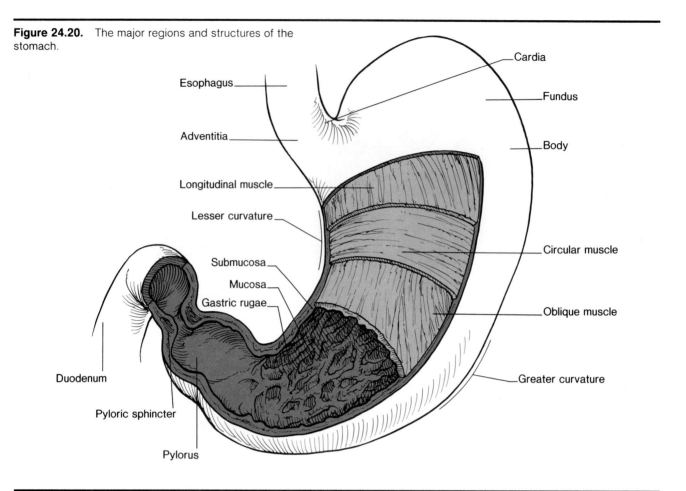

Esophagus

Adventitia

Longitudinal muscle

Lesser curvature

Submucosa

Mucosa

Gastric rugae

Duodenum

Pyloric sphincter

Pylorus

Cardia

Fundus

Body

Circular muscle

Oblique muscle

Greater curvature

Figure 24.21. An X-ray of a stomach.

Pyloric sphincter

Duodenum

Pylorus

Fundus

Rugae

Body

Figure 24.22. Microscopic structures of the mucosa of the stomach.

— Columnar epithelium of mucosal ridge

— Gastric pit

— Gastric glands with chief and parietal cells

— Lamina propria

Figure 24.23. Gastric pits and gastric glands of the mucosa. (*a*) gastric pits are the openings of the gastric glands. (*b*) gastric glands consist of mucous cells, chief cells, and parietal cells, each of which produces a specific secretion.

Gastric pits

(a)

Mucosa

Submucosa

(b)

— Gastric pit
— Mucous cell
— Gastric gland

— Parietal cell

— Chief cell
— Submucosa

The third stage, the entry and passage of food through the esophagus, is also involuntary. The bolus is moved through the esophagus by peristalsis. The entire time for deglutition varies, but it is slightly more than one second in the case of fluids and five to eight seconds with solid food material.

Stomach

The **stomach** is the most distensible part of the GI tract and is positioned in the upper left quadrant of the peritoneal cavity immediately below the diaphragm. It is a J-shaped pouch that is continuous with the esophagus superiorly and empties into the duodenal portion of the small intestine inferiorly. The functions of the stomach are to store food as it is mechanically churned with gastric secretions; to initiate the digestion of proteins; to carry on limited absorption; and to move food into the small intestine as a pasty material called **chyme** *(kīm)*.

The stomach, shown in figures 24.20 and 24.21, can be divided into four regions: the cardia, fundus, body, and pylorus. The **cardia** is the upper, narrow region immediately below the gastroesophageal sphincter. The **fundus** is the dome-shaped portion to the left and in direct contact with the diaphragm. The **body** is the large central portion, and the **pyloris** is the funnel-shaped terminal portion. The pyloris communicates with the duodenal portion of the small intestine through a **pyloric sphincter.** *Pyloris* is a Greek word meaning "gatekeeper," and this junction is just that, regulating the movement of chyme into the small intestine and prohibiting backflow.

The stomach has two surfaces and two borders. The broadly rounded surfaces are referred to as the **anterior** and **posterior surfaces.** The medial concave border is the **lesser curvature** (fig. 24.20), and the lateral convex border is the **greater curvature.** The lesser omentum extends between the lesser curvature and the liver, and the greater omentum is attached to the greater curvature.

The wall of the stomach consists of the same four layers found in other regions of the GI tract with certain modifications. The muscularis is composed of three layers of smooth muscle named according to the direction of fiber arrangement: an outer **longitudinal layer,** a middle **circular layer,** and an inner **oblique layer.** The circular muscle layer is further thickened at the gastroduodenal junction to form the pyloric sphincter.

The mucosa is shaped into numerous longitudinal folds, called **gastric rugae,** which permit stomach distension. The gastric rugae gradually smooth out as the stomach fills. The mucosal layer contains a simple columnar epithelium. The entire mucosal layer is folded to form **gastric pits** and **gastric glands** (figs. 24.22 and 24.23).

chyme: L. *chymus,* juice
cardia: Gk. *kardia,* heart (upper portion, nearer the heart)
fundus: L. *fundus,* bottom
pyloris: Gk. *pyloros,* gatekeeper

Table 24.5 Secretions of the fundus and pyloric regions of the stomach

Stomach region	Cell type	Secretions
Fundus	Parietal cells	Hydrochloric acid; intrinsic factor
	Chief cells	Pepsinogen
	Goblet cells	Mucus
	Argentaffen cells	Histamine, serotonin
Pyloric	G cells	Gastrin
	Chief cells	Pepsinogen
	Goblet cells	Mucus

Source: Fox, Stuart Ira, *Human Physiology*. © 1984 Wm. C. Brown Publishers, Dubuque, Iowa. All Rights Reserved. Reprinted by permission.

Figure 24.24.   The gastric mucosa secretes the inactive enzyme pepsinogen and hydrochloric acid (HCl). Through a process of autocatalysis, the active enzyme pepsin is produced. Pepsin digests proteins into shorter polypeptides.

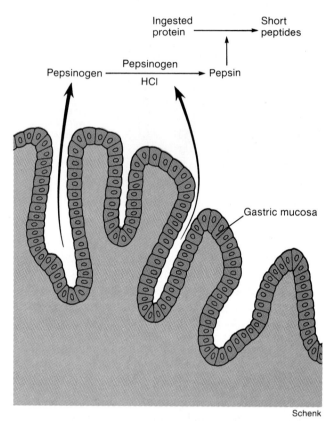

Gastric lumen

Ingested protein → Short peptides

Pepsinogen — (Pepsinogen HCl) → Pepsin

Gastric mucosa

Schenk

There are several different types of cells in the gastric glands that secrete different products: (1) **goblet cells,** which secrete mucus; (2) **parietal cells,** which secrete hydrochloric acid; (3) **chief cells,** which secrete pepsinogen, an inactive form of the protein-digesting enzyme pepsin; (4) **argentaffen** *(ar-jen'tah-fin)* **cells,** which secrete serotonin and histamine; and (5) **G cells,** which secrete the hormone gastrin. In addition to these products (table 24.5), the gastric mucosa (probably the parietal cells) secretes a polypeptide called *intrinsic factor,* which is required for absorption of vitamin B_{12} in the intestine.

Pepsin and Hydrochloric Acid.   The secretion of hydrochloric acid by parietal cells makes gastric juice very acidic, with a pH less than 2. This strong acidity serves three functions: (1) ingested proteins are denatured at low pH—that is, their tertiary structure is altered so that they become more digestible; (2) under acidic conditions, weak pepsinogen enzymes partially digest each other—this frees the active pepsin enzyme as small peptide fragments are removed (fig. 24.24); and (3) pepsin is more active under acidic conditions—it has a pH optimum of about 2.0. The peptide bonds of ingested protein are broken (through hydrolysis reactions) by pepsin under acidic conditions; the HCl itself does not directly digest proteins.

Digestion and Absorption in the Stomach.   Proteins are only partially digested in the stomach by the action of pepsin. Fats are also partially digested, through the action of a lipase secreted by the lingual salivary glands and activated by low pH in the stomach. Carbohydrates are not digested at all in the stomach. The complete digestion of food molecules occurs later, when chyme enters the small

intestine. Patients with partial gastric resections, therefore, and even those with complete gastrectomies (removal of the stomach), can still adequately digest and absorb their food.

Almost all of the products of digestion are absorbed through the wall of the intestine; the only commonly ingested substances that can be absorbed across the stomach wall are alcohol and aspirin. This occurs as a result of the lipid solubility of these molecules. The passage of aspirin through the gastric mucosa has been shown to cause bleeding, which may be significant if large amounts of aspirin are taken.

argentaffen cells: L. *argentum,* silver; *affinis,* attraction (become colored with silver stain)

The only function of the stomach that appears to be essential for life is the secretion of intrinsic factor. This polypeptide is needed for the intestinal absorption of vitamin B_{12}, required for maturation of red blood cells in the bone marrow. A patient with a gastrectomy has to receive B_{12} orally (together with intrinsic factor) or through injections so that he or she will not develop **pernicious anemia.**

Table 24.6 Gastric secretion of hydrochloric acid, measured in milliequivalents (mEq) (one milliequivalent is equal to one millimole of H^+)

Condition	Day (mEq/hr)	Night (mEq/12 hrs)	Maximal (mEq/hr)
Normal	2–3	18	16–20
Gastric ulcer	2–4	8	16–20
Duodenal ulcer	4–10	60	25–40
Zollinger-Ellison syndrome (G cell tumor)	30	120	45

Source: Fox, Stuart Ira, *Human Physiology.* © 1984 Wm. C. Brown Publishers, Dubuque, Iowa. All Rights Reserved. Reprinted by permission.

Gastritis and Peptic Ulcers. The gastric mucosa is normally resistant to the harsh conditions of low pH and pepsin activity in gastric juice. The reasons for its resistance to pepsin digestion are not well understood. Its resistance to hydrochloric acid appears to be due to three interrelated mechanisms: (1) the stomach lining is covered with a thin layer of alkaline mucus, which may offer some protection; (2) the epithelial cells of the mucosa are joined together by tight junctions, which prevent acid from leaking into the submucosa; and (3) the epithelial cells that are damaged are exfoliated (shed) and replaced by new cells. This latter process results in the loss of about one-half million cells a minute, so that the entire epithelial lining is replaced every three days.

These three mechanisms provide a barrier that prevents self-digestion of the stomach. Breakdown of the gastric mucosa barrier may occur at times, however, perhaps as a result of the detergent action of bile salts that are regurgitated from the small intestine through the pyloric sphincter. When this occurs, acid can leak through the mucosal barrier to the submucosa, which causes direct damage and stimulates inflammation. The histamine released from mast cells during inflammation may stimulate further acid secretion and result in further damage to the mucosa. The inflammation that occurs during these events is called **acute gastritis.**

Once the mucosal barrier is broken, the acid can produce craterlike holes and even complete perforations of the stomach wall. These are called gastric ulcers. More commonly encountered, however, are ulcers of the first 3–4 cm of the duodenum. Ulcers of the stomach and duodenum are known as **peptic ulcers.** It has been estimated that about 10% of American men and 4% of American women will develop peptic ulcers.

The duodenum is normally protected against gastric acid by the buffering action of bicarbonate in alkaline pancreatic juice. People who develop duodenal ulcers, however, produce excessive amounts of gastric acid (table 24.6) that cannot be neutralized by pancreatic juice. This excessive gastric acid secretion is apparently due to excessive parasympathetic stimulation of the stomach. People with gastritis and peptic ulcers must avoid substances that stimulate acid secretion, including caffeine and alcohol, and often must take antacids.

1. Describe the structure and function of the lower esophageal sphincter.
2. Describe the three stages in deglutition, including the structures involved.
3. List the secretory cells of the gastric mucosa and the products they secrete.
4. Explain the roles of hydrochloric acid in the stomach and how pepsinogen is activated.
5. Explain the cause of peptic ulcers and why they are more likely to be duodenal than gastric in location.

Small Intestine

The structure of the small intestine is adapted to provide a very high surface area for the exposure of chyme to brush border enzymes and for the absorption of the products of digestion. The spread of action potentials through intestinal smooth muscle produces movements that mix the chyme and propel it by peristalsis to the large intestine.

Objective 12. Identify the regions of the small intestine, and describe how bile and pancreatic juice are delivered to the small intestine.

Objective 13. Describe the structure and function of the intestinal villi, the microvilli, and the crypts of Lieberkühn.

Objective 14. Identify the location and action of brush border enzymes.

Objective 15. Describe the types of intestinal motility, and explain how intestinal smooth muscle contraction is produced.

The **small intestine** is that portion of the GI tract between the pyloric sphincter of the stomach and the ileocecal valve opening into the large intestine. It is positioned in the central and lower portions of the abdominal cavity and is supported, except for the first portion, by mesentery (figs. 24.25 and 24.33). The fan-shaped attachment of mesentery to the small intestine maintains mobility while allowing little chance of the intestine becoming twisted or kinked. Enclosed within the mesentery are blood vessels, nerves, and lymphatic vessels that supply the intestinal wall.

Figure 24.25.   The mesenteric attachment of the small intestine.

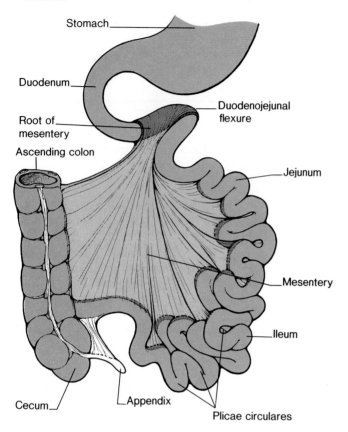

The small intestine is approximately 3 m (12 ft) long and 2.5 cm (1 in.) wide in a living person, but it will measure nearly twice this length in a cadaver when the muscle wall is relaxed. It is called the "small" intestine because of its relatively small diameter compared to that of the large intestine. The small intestine serves as the major site of digestion and absorption in the GI tract.

The small intestine is innervated by the **superior mesenteric** *(mes''en-ter'ik)* **plexus.** The branches of the plexus contain afferent fibers, postganglionic sympathetic fibers, and preganglionic parasympathetic fibers. The small intestine's arterial blood supply comes through the superior mesenteric artery and small branches from the celiac and inferior mesenteric arteries. Venous drainage is through the superior mesenteric vein, which unites with the splenic vein to form the hepatic portal vein carrying nutrient-rich blood to the liver.

Regions of the Small Intestine

The small intestine is divided into three regions on the basis of function and histological structure. These three regions are the duodenum, jejunum, and ileum.

The **duodenum** *(du''o-de'num* or *du-od'ĕ-num)* is a relatively fixed, C-shaped tube measuring approximately

25 cm (10 in.) from the pyloric sphincter of the stomach to the **duodenojejunal** *(du-od''ĕ-no''jĕ-joo'nal)* **flexure.** The concave surface of the duodenum faces to the left, where it receives bile secretions through the **common bile duct** from the liver and gallbladder and pancreatic secretions through the **pancreatic duct** of the pancreas (fig. 24.26). These two ducts unite to form a common entry into the duodenum called the **hepatopancreatic ampulla** *(hep''ah-to-pan''kre-at'ik am-pul'lah)* (or **ampulla of Vater**), which pierces the duodenal wall and drains into the duodenum from an elevation called the **duodenal papilla.** The duodenum is retroperitoneal except for a short portion near the stomach. The duodenum differs histologically from the rest of the small intestine by the presence of **duodenal (Brunner's) glands** in the submucosa (fig. 24.27). These compound tubuloalveolar glands secrete mucus and are most numerous near the superior end of the duodenum.

The **jejunum** *(je-joo'num)* is approximately 1 m (3 ft) long and extends from the duodenum to the ileum. The jejunum has a slightly larger lumen and more internal folds than does the ileum but lacks unique histological structures.

The **ileum** *(il'e-um)* makes up the remaining 2 m (6–7 ft) of the small intestine. The terminal portion of the ileum empties into the medial side of the cecum through the **ileocecal** *(il''e-o-se'kal)* **valve.** The walls of the ileum have an abundance of lymphatic tissue. Aggregates of lymph nodules, called **mesenteric (Peyer's) patches,** are characteristic of the ileum.

Structural Modifications of the Small Intestinal Wall

The products of digestion are absorbed across the epithelial lining of the intestinal mucosa. Absorption occurs primarily in the jejunum, although some also occurs in the duodenum and ileum. Absorption occurs at a rapid rate as a result of the high mucosal surface area in the small intestine, provided by folds. The mucosa and submucosa form large folds called the **plicae** *(pli'se)* **circularis,** which can be observed with the unaided eye. The surface area is further increased by the microscopic folds of mucosa called villi and by the foldings of the apical cell membrane of epithelial cells (which can only be seen with an electron microscope), called microvilli.

Each **villus** is a fingerlike fold of mucosa that projects into the intestinal lumen. The villi are covered with columnar epithelial cells and interspersed among these are the mucus-secreting goblet cells. The lamina propria forms a connective core of each villus and contains numerous lymphocytes, blood capillaries, and a lymphatic vessel

---

duodenum: L. *duodeni,* twelve each (length of twelve fingers' breadth)

ampulla of Vater: from Abraham Vater, German anatomist, 1684–1751
Brunner's glands: from Johann C. Brunner, Swiss anatomist, 1653–1727
Peyer's patches: from Johann K. Peyer, Swiss anatomist, 1653–1712
plica: L. *plicatus,* folded
villus: L. *villosus,* shaggy

Figure 24.26.   The duodenum and associated structures.

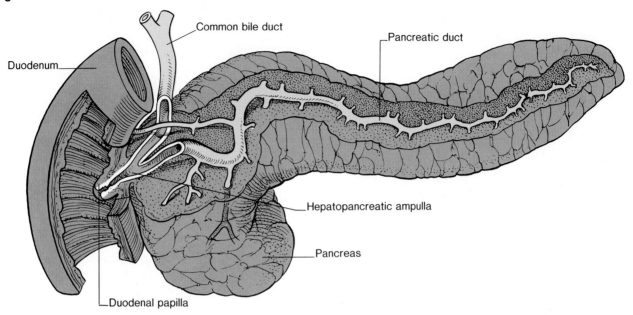

Figure 24.27.   The microscopic structure of the duodenum.

Figure 24.28.   A diagram of the structure of an intestinal villus.

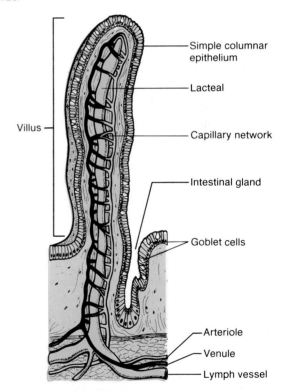

called the **central lacteal** (fig. 24.28). Absorbed monosaccharides and amino acids are secreted into the blood capillaries; absorbed fat enters the central lacteals.

Epithelial cells at the tips of the villi are continuously exfoliated and are replaced by cells that are pushed up from the bases of the villi. The epithelium at the base of the villi invaginates downward at various points to form

narrow pouches that open through pores to the intestinal lumen. These structures are called the **crypts** of **Lieberkühn** (fig. 24.29). Microvilli are fingerlike projections formed by foldings of the cell membrane, which can only

lacteal: L. *lacteus,* milk

crypts of Lieberkühn: from Johann N. Lieberkühn, German anatomist, 1711–56

Figure 24.29. Intestinal villi and crypts of Lieberkühn. The crypts serve as sites for production of new epithelial cells. The time required for migration of these new cells to the tip of the villi is shown in (b). Epithelial cells are exfoliated from the tips of the villi.

(a)

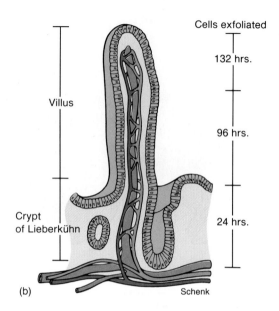

(b)

be clearly seen in an electron microscope. In a light microscope, the microvilli produce a somewhat vague **brush border** on the edges of the columnar epithelial cells. The term *brush border* is thus often used synonymously with microvilli in descriptions of the intestine (fig. 24.30).

Intestinal Enzymes

In addition to providing a large surface area for absorption, the cell membranes of the microvilli contain digestive enzymes. These enzymes are not secreted into the lumen, but instead remain attached to the cell membrane with their active sites exposed to the chyme. These **brush border enzymes** hydrolyze disaccharides, polypeptides, and other substrates (table 24.7). One brush border enzyme, *enterokinase (en"ter-o-ki'nās),* is required for activation of the protein-digesting enzyme *trypsin,* which enters the intestine in pancreatic juice.

The ability to digest milk sugar, or lactose, depends on the presence of a brush border enzyme called lactase. This enzyme is present in all children under the age of four but becomes inactive in most adults. This can result in **lactose intolerance.** The presence of large amounts of undigested lactose in the intestine causes diarrhea, gas, cramps, and other unpleasant symptoms. Yogurt is better tolerated than milk because it contains lactase produced by the yogurt bacteria, which becomes activated in the duodenum and digests lactose.

Intestinal Contractions and Motility

Like cardiac muscle, intestinal smooth muscle is capable of spontaneous electrical activity and automatic, rhythmic contractions. Spontaneous depolarization begins in the longitudinal smooth muscle and is conducted to the circular smooth muscle layer across *nexuses.* The term *nexus* is used here to indicate an electrical synapse between smooth muscle cells. The spontaneous depolarizations, called **pacesetter potentials,** decrease in amplitude as they are conducted from one muscle cell to another, much like excitatory postsynaptic potentials (EPSPs). Also like EPSPs, pacesetter potentials stimulate the production of action potentials in the smooth muscle cells through which they are conducted (fig. 24.31).

The nexuses conduct the pacesetter potentials, not the action potentials. Action potentials are, therefore, limited to those smooth muscle cells that are depolarized to threshold by the spreading pacesetter potentials. When action potentials are produced, they stimulate smooth muscle contraction. The rate at which this automatic activity occurs is influenced by autonomic nerves. Contraction is stimulated by parasympathetic (vagus nerve) innervation and is reduced by sympathetic nerve activity.

The small intestine has two major types of contractions: peristalsis and segmentation. Peristalsis is much weaker in the small intestine than in the esophagus and stomach. **Intestinal motility**—the movement of chyme through the intestine—is relatively slow and is due primarily to the fact that the pressure at the pyloric end of the small intestine is greater than at the distal end.

---

nexus: L. *nexus,* interconnection

Figure 24.30. (*a*) the cell membrane of epithelial cells that line the small intestine are folded into microvilli. (*b*) glycoproteins within the membranes of the microvilli contain polysaccharides that extend out into the lumen and form the glycocalyx, which covers the brush border epithelium. (*c*) a diagram of an intestinal epithelial cell. (*d*) the microvilli contain actin filaments attached to a terminal web of myosin filaments. This allows the microvilli to shorten.

(a)

(b)

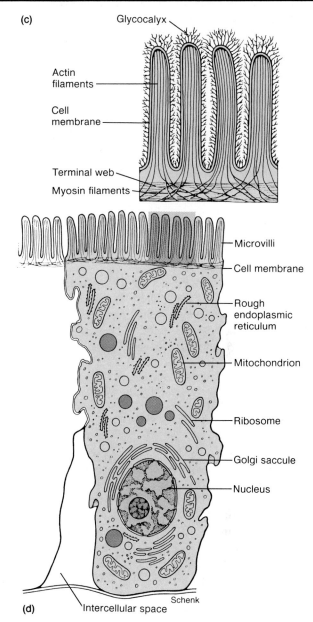

(c)

(d)

Table **24.7**	Brush border enzymes attached to the cell membrane of microvilli in the small intestine	
Category	**Enzyme**	**Comments**
Disaccharidase	Sucrase	Digests sucrose to glucose and fructose; deficiency produces gastrointestinal disturbances
	Maltase	Digests maltose to glucose
	Lactose	Digests lactose to glucose and galactose; deficiency produces gastrointestinal disturbances (lactose intolerance)
Peptidase	Aminopeptidase	Produces free amino acids, dipeptides, and tripeptides
	Enterokinase	Activates trypsin (and indirectly other pancreatic juice enzymes); deficiency results in protein malnutrition
Phosphatase	Ca⁺⁺, Mg⁺⁺—ATPase	Needed for absorption of dietary calcium; enzyme activity regulated by vitamin D
	Alkaline phosphatase	Removes phosphate groups from organic molecules; enzyme activity may be regulated by vitamin D

Figure 24.31. The smooth muscle of the gastrointestinal tract produces and conducts spontaneous pacesetter potentials. As these potential changes reach a threshold level of depolarization, they stimulate the production of action potentials, which in turn stimulate smooth muscle contraction.

Figure 24.32. Segmentation of the small intestine. Simultaneous contractions of many segments of the intestine help to mix the chyme with digestive enzymes and mucus.

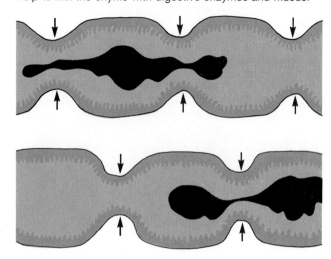

The major contractile activity of the small intestine is **segmentation.** This term refers to muscular constrictions of the lumen, which occur simultaneously at different intestinal segments (fig. 24.32). This action serves to mix the chyme more thoroughly.

---

1. *Describe the structures involved in the delivery of pancreatic juice and bile to the small intestine.*
2. *Describe the intestinal structures that increase surface area, and explain the function of the crypts of Lieberkühn.*
3. *Define the term brush border enzymes, and list examples. Explain why many adults cannot tolerate milk.*
4. *Describe how smooth muscle contraction in the small intestine is regulated, and explain the function of segmentation.*

Large Intestine

The large intestine absorbs water and electrolytes from the chyme it receives from the small intestine and, through the action of sphincter muscles, passes waste products out of the body through the rectum and anal canal.

Objective 16. Identify the regions of the large intestine, and describe its gross and histological structure.

Objective 17. Describe the functions of the large intestine, and explain how defecation is accomplished.

The **large intestine** is about 1.5 m (5 ft) long and 6.5 cm (2.5 in.) in diameter. It is named the "large" intestine because its diameter is larger than that of the small intestine. The large intestine begins at the terminal end of the ileum in the lower right quadrant of the abdominal cavity.

From there, the large intestine leads superiorly on the right side to just below the liver where it crosses to the left and descends into the pelvis and terminates at the anus. A specialized portion of the dorsal mesentery, called the **mesocolon,** supports the large intestine along the posterior abdominal wall.

The large intestine has little or no digestive function, but it does absorb water and electrolytes from the remaining chyme. In addition, the large intestine functions to form, store, and expel feces from the body.

Regions and Structures of the Large Intestine

The large intestine is structurally divided into the cecum, colon, rectum, and anal canal (fig. 24.33).

The **cecum** *(se'kum)* is a dilated pouch that hangs inferiorly, slightly below the ileocecal valve. The ileocecal valve is a fold of mucous membrane at the junction of the small and large intestine, which prohibits the backflow of chyme. A fingerlike projection called the **vermiform appendix** is attached to the inferior medial margin of the cecum. The 8 cm (3 in.) appendix has an abundance of lymphatic tissue (fig. 24.34), which may serve to resist infection.

The open, superior portion of the cecum is continuous with the colon. The colon consists of ascending, transverse, descending, and sigmoid portions (fig. 24.33). The **ascending colon** extends superiorly from the cecum along the right abdominal wall to the inferior surface of the liver. Here the colon bends sharply to the left at the **hepatic flexure** and transversely crosses the upper abdominal cavity as the **transverse colon.** At the left abdominal wall, another right angle bend called the **splenic** *(splen'ik)* **flexure**

---

cecum: L. *caecum,* blind pouch
vermiform appendix: L. *vermiformis,* wormlike; *appendix,* attachment
colon: Gk. *kolon,* member of the whole

Figure 24.33. The large intestine.

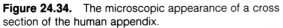

Figure 24.34. The microscopic appearance of a cross section of the human appendix.

Figure 24.35.   The anal canal.

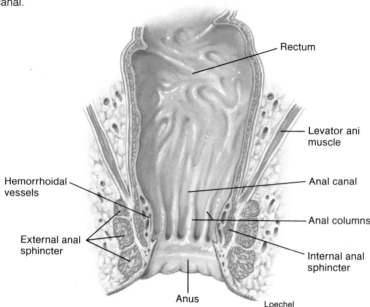

Rectum

Levator ani
muscle

Anal canal

Hemorrhoidal
vessels

Anal columns

External anal
sphincter

Internal anal
sphincter

Anus

Loechel

denotes the beginning of the **descending colon.** The descending colon traverses inferiorly along the left abdominal wall to the pelvic region. The colon then angles medially from the brim of the pelvis to form an S-shaped bend known as the **sigmoid colon.**

The terminal 20 cm (7.5 in.) of the GI tract is the **rectum,** and the last 2 to 3 cm of the rectum is referred to as the **anal canal** (fig. 24.35). The rectum lies anterior to the sacrum where it is firmly attached by peritoneum. The **anus** is the external opening of the anal canal. Two sphincter muscles guard the anal opening: the **internal anal sphincter,** which is composed of smooth muscle fibers, and the **external anal sphincter,** consisting of striated muscle. The mucous membrane of the anal canal is arranged into highly vascular, longitudinal folds called **anal columns.** Varicose veins in the anal area are known as *hemorrhoids (hem'o-roid),* or piles.

Although the large intestine consists of the same tunicas as the small intestine, there are some structural differences. The large intestine lacks villi but does have numerous goblet cells in the mucosal layer. The longitudinal muscle layer of the muscularis externa forms three distinct muscle bands, called **taeniae coli** *(te'ne-ah co'li),* which run the length of the large intestine. A series of bulges in the walls of the large intestine form sacculations, or **haustra** *(hows'tra),* along its entire length (fig. 24.36). Finally, the large intestine has small but numerous fat-filled pouches called **epiploic appendages** (fig. 24.33), which are attached superficially to the taeniae coli in the serous layer.

sigmoid: Gk. *sigmoeides,* shaped like the letter C
rectum: L. *rectum,* straight tube
anus: L. *anus,* ring
taenia: Gk. *teinein,* to stretch
haustrum: L. *haustrum,* bucket or scoop
epiploic: Gk. *epiplein,* to float on

Figure 24.36.   A radiograph after a barium enema showing the haustra of the large intestine.

The large intestine has both types of autonomic innervation. The sympathetic innervation arises from superior and inferior mesenteric plexuses as well as from the celiac plexus. The parasympathetic innervation arises from the paired pelvic splanchnic and vagus nerves. Sensory fibers from the large intestine respond to bowel pressure and signal the need to defecate.

Inflammation of the pancreas may result when the various safeguards against self-digestion are insufficient. **Acute pancreatitis** is believed to be caused by the reflux of pancreatic juice and bile from the duodenum into the pancreatic duct. The leakage of trypsin into the blood also occurs, but trypsin is inactive in the blood because of the inhibitory action of two plasma proteins, alpha–1-antitrypsin and alpha–2-macroglobulin. Pancreatic amylase may also leak into the blood, but it is not active because its substrate (glycogen) is not present in blood. Pancreatic amylase activity can be measured *in vitro*, however, and these measurements are commonly performed to assess the health of the pancreas.

1. *Describe the structure of liver lobules, and trace the pathways for the flow of blood and bile in the lobules.*
2. *Describe the composition and function of bile, and trace the flow of bile from the liver and gallbladder to the duodenum.*
3. *Explain how the liver inactivates and excretes compounds such as hormones and drugs.*
4. *Describe the enterohepatic circulation of bilirubin and urobilinogen.*
5. *Explain how the liver helps maintain a constant blood glucose concentration and how the pattern of venous blood flow permits this task to be performed.*
6. *Describe the endocrine and exocrine structures and functions of the pancreas, and explain why the pancreas does not normally digest itself.*

Digestion and Absorption of Carbohydrates, Lipids, and Proteins

Polysaccharides and polypeptides are digested into their subunits, which are absorbed and secreted into the blood capillaries. Emulsified fat is digested, absorbed into the intestinal cells, and then resynthesized into triglycerides, which are secreted as small particles in the lymph.

Objective 23. Describe the regions of the digestive system and the enzymes involved with the digestion and absorption of carbohydrates.

Objective 24. Describe the digestion and absorption of proteins, indicating the enzymes involved and their location in the GI tract.

Objective 25. Describe the actions of bile and enzymes in lipid digestion, and explain how lipids are absorbed.

The caloric (energy) value of food is found predominantly in its content of carbohydrates, lipids, and proteins. In the average American diet, carbohydrates account for approximately 50% of the total calories, protein accounts for 11% to 14%, and lipids make up the balance. These food molecules consist primarily of long combinations of subunits (monomers), which must be digested by hydrolysis reactions into the free monomers before absorption can occur. The characteristics of the major digestive enzymes are summarized in table 24.13.

Digestion and Absorption of Carbohydrates

Most of the ingested carbohydrates are in the form of starch, which is a long polysaccharide of glucose in the form of straight chains with occasional branchings. The most commonly ingested sugars are the disaccharides sucrose (table sugar, consisting of glucose and fructose) and lactose (milk sugar, consisting of glucose and galactose). The digestion of starch begins in the mouth with the action of **salivary amylase,** or **ptyalin** *(ti'ah-lin).* This enzyme cleaves some of the bonds between adjacent glucose molecules, but most people don't chew their food long enough for sufficient digestion to occur in the mouth. The digestive action of salivary amylase stops when the bolus enters the stomach because this enzyme is inactivated at the low pH of gastric juice.

Table 24.13 Summary of the sources and activities of the major digestive enzymes

Region or source					
Organ	**Source**	**Substrate**	**Enzymes**	**Optimum pH**	**Products**
Mouth	Saliva	Starch	Salivary amylase	6.7	Maltose
Stomach	Gastric glands	Protein	Pepsin	1.6–2.4	Shorter polypeptides
Duodenum	Pancreatic juice	Starch	Pancreatic amylase	6.7–7.0	Maltose, maltriose, and oligosaccharides
		Polypeptides	Trypsin, chymotrypsin, carboxypeptidase	8.0	Amino acids, dipeptides, and tripeptides
		Triglycerides	Pancreatic lipase	8.0	Fatty acids and monoglycerides
	Epithelial membranes	Maltose	Maltase	5.0–7.0	Glucose
		Sucrose	Sucrase	5.0–7.0	Glucose + fructose
		Lactose	Lactase	5.8–6.2	Glucose + galactose
		Polypeptides	Aminopeptidase	8.0	Amino acids, dipeptides, tripeptides

Source: Fox, Stuart Ira, *Human Physiology.* © 1984 Wm. C. Brown Publishers, Dubuque, Iowa. All Rights Reserved. Reprinted by permission.

Figure 24.48. Pancreatic amylase digests starch into maltrose, maltriose, and short oligosaccharides containing branch points in the chain of glucose molecules.

The digestion of starch, therefore, occurs mainly in the duodenum as a result of the action of **pancreatic amylase.** This enzyme cleaves the straight chains of starch to produce the disaccharide *maltose* and the trisaccharide *maltriose.* Pancreatic amylase, however, cannot hydrolyze the bond between glucose molecules at the branch points in the starch. As a result, short, branched chains of glucose molecules, called *oligosaccharides,* are released together with maltose and maltriose by the activity of this enzyme (fig. 24.48).

Maltose, maltriose, and oligosaccharides of glucose released from partially digested starch, together with the disaccharides sucrose and lactose, are hydrolyzed to their monosaccharides by brush border enzymes located on the microvilli of the epithelial cells in the small intestine. The absorption of these monosaccharides across the membrane of the microvilli occurs by means of coupled transport. In this process, glucose binds to the same carrier as Na^+ and enters the epithelial cell as Na^+ diffuses down its electrochemical gradient. This is a type of active transport, because energy from ATP is needed to maintain the Na^+ gradient. Glucose is then secreted from the epithelial cells into capillaries within the villi.

Digestion and Absorption of Proteins

Protein digestion begins in the stomach with the action of pepsin. Pepsin results in the liberation of some amino acids, but the major products of pepsin digestion are short-chain polypeptides. This activity helps to produce a more homogenous chyme, but it is not essential for the complete digestion of protein that occurs—even in people with total gastrectomies—in the small intestine.

Most protein digestion occurs in the duodenum and jejunum. The pancreatic juice enzymes **trypsin, chymotrypsin** *(ki''mo-trip'sin),* and **elastase** cleave peptide bonds within the interior of the polypeptide chains. These enzymes are thus grouped together as *endopeptidases (en''do-pep'ti-dās).* Enzymes that remove amino acids from the ends of polypeptide chains, in contrast, are *exopeptidases.* These include the pancreatic juice enzyme **carboxypeptidase,** which removes amino acids from the carboxyl-terminal end of polypeptide chains, and the brush border enzyme **aminopeptidase.** Aminopeptidase cleaves amino acids from the amino-terminal end of polypeptide chains.

As a result of the action of these enzymes, polypeptide chains are digested into free amino acids, dipeptides, and tripeptides. The free amino acids are absorbed through the epithelial cells of the intestinal mucosa and secreted into blood capillaries. The dipeptides and tripeptides may enter epithelial cells, but they are then digested within these cells into amino acids, which are secreted into the blood (fig. 24.49). Newborn babies appear to be capable of absorbing a substantial amount of undigested proteins; in adults, however, only the free amino acids enter the portal vein. Foreign food protein, which would be very antigenic, does not normally enter the blood.

---

oligosaccharide: Gk. *oligos,* little; *sakcharon,* sugar

Figure 24.49. Polypeptide chains are digested into free amino acids, dipeptides, and tripeptides by the action of pancreatic juice enzymes and brush border enzymes. The amino acids, dipeptides, and tripeptides enter duodenal epithelial cells. Dipeptides and tripeptides are hydrolyzed into free amino acids within the epithelial cells, and these products are secreted into capillaries that carry them to the hepatic portal vein.

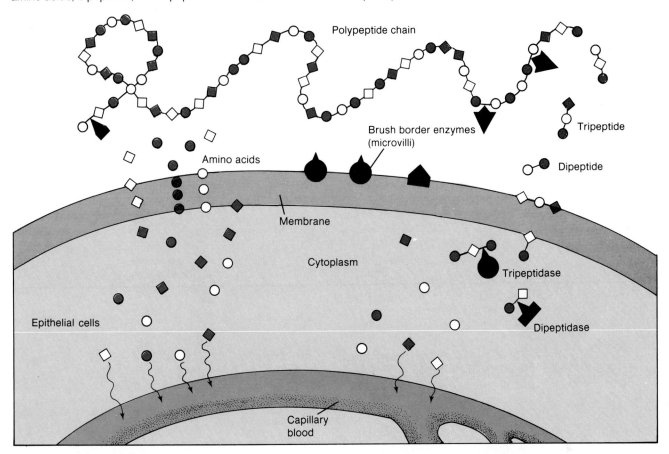

Digestion and Absorption of Lipids

Fat digestion begins in the stomach through the action of a **lingual lipase** produced in the lingual salivary glands and activated by the low pH of gastric juice. Most of the ingested fat, however, enters the duodenum undigested in the form of fat globules. Bile salts released into the duodenum function as detergents and break up these droplets into much finer droplets; this process is called **emulsification.** Emulsification is not digestion—the bonds joining the subunits together are not hydrolyzed—but the total surface area is increased by the emulsification process. An increased surface area aids digestion because the fat-digesting action of **pancreatic lipase,** which is a water-soluble enzyme, can only occur at the surface of the droplets.

Almost all fat digestion occurs in the small intestine as a result of the action of pancreatic lipase and phospholipase A. Lipase removes two of the three fatty acids from each triglyceride molecule and thus liberates free fatty acids and monoglycerides (fig. 24.50). Phospholipase A likewise digests phospholipids such as lecithin into fatty acids and lysolecithin (the remainder of the lecithin molecule after two fatty acids are removed).

Free fatty acids, monoglycerides, and lysolecithin are more polar than the undigested lipids and are able to be moved more easily into the micelles of bile salts, cholesterol, and lecithin that are released in the bile. The entry of the products of lipid digestion into the micelles produces "mixed micelles" (fig. 24.51) which move to the brush border of the intestinal epithelium.

Absorption of Lipids. Free fatty acids, monoglycerides, and lysolecithin can leave the micelles and pass through the membrane of the microvilli to enter the intestinal epithelial cells. There is also some evidence that the micelles may be transported intact into the epithelial cells and that the lipid digestion products may be removed intracellularly from the micelles. In either event, these products are used to *resynthesize* triglycerides and phospholipids within the epithelial cells. This is different from the absorption of amino acids and monosaccharides, which pass through the epithelial cells without being altered.

Triglycerides, phospholipids, and cholesterol are then combined with protein inside the epithelial cells to form small particles called **chylomicrons** *(ki″lo-mi′krons).*

Figure 24.50. Pancreatic lipase digests fat (triglycerides) by cleaving off the first and third fatty acids. This produces free fatty acids and monoglycerides. Sawtooth structures indicate hydrocarbon chains in the fatty acids.

Figure 24.51. Steps in the digestion of fat (triglycerides) and the entry of fat digestion products (fatty acids and monoglycerides) into micelles of bile salts secreted by the liver.

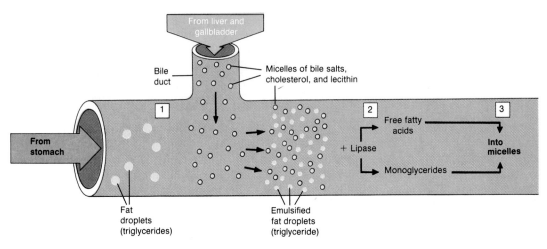

Step 1 Emulsification of fat droplets by bile salts

Step 2 Hydrolysis of triglycerides in emulsified fat droplets into fatty acid and monoglycerides

Step 3 Dissolving of fatty acids and monoglycerides into micelles to produce "mixed micelles"

These tiny combinations of lipid and protein are secreted into the lymphatic capillaries of the intestinal villi (fig. 24.52). Absorbed lipids thus pass through the lymphatic system, eventually entering the venous blood by way of the thoracic duct. The absorption of lipids is thus significantly different from that of amino acids and monosaccharides, which enter the portal blood.

Once the chylomicrons are in the blood, their triglyceride content is removed by the enzyme *lipoprotein lipase,* found in plasma. This enzyme hydrolyzes triglycerides and thus provides free fatty acids and glycerol for use by the tissue cells. The remaining "remnant particle" is combined with proteins made by the liver to produce lipoproteins that are important in the transport of cholesterol.

Figure 24.52. Fatty acids and monoglycerides from the micelles within the small intestine are absorbed by epithelial cells and converted intracellularly into triglycerides. These are then combined with protein to form chylomicrons, which enter the lymphatic vessels (lacteals) of the villi. These lymphatic vessels transport the chylomicrons to the thoracic duct, which empties them into the venous blood.

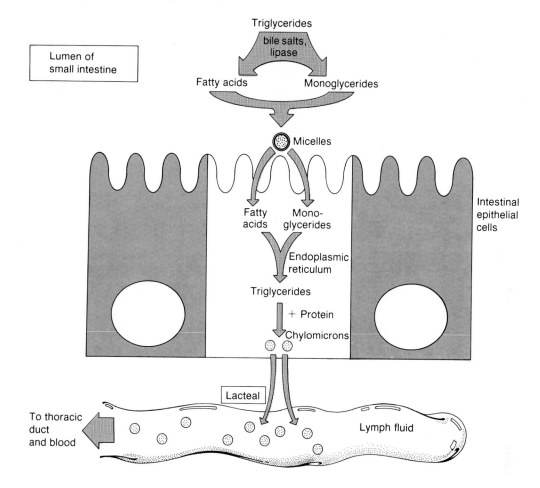

1. List the enzymes involved in carbohydrate digestion, indicating their origin, sites of action, substrates, and products.
2. List the enzymes involved in protein digestion, indicating their origins, sites of action, and mode of action (endopeptidase or exopeptidase). Compare the characteristics of pepsin and trypsin.
3. Describe how bile aids both the digestion and absorption of fats. Explain how the absorption of fat differs from the absorption of amino acids and monosaccharides.

Neural and Endocrine Regulation of the Digestive System

Neural and hormonal mechanisms coordinate the activities of different regions of the GI tract and the actions of the liver, gallbladder, and pancreas.

Objective 26. Describe the mechanisms that regulate gastric juice secretion, and indicate those that belong to the cephalic, gastric, and intestinal phases of regulation.

Objective 27. Describe the neural and hormonal mechanisms that regulate the secretion of pancreatic juice and bile.

Objective 28. Explain the significance of the trophic effects of gastrointestinal hormones.

The motility and glandular secretions of the GI tract are, to a large degree, automatic. Neural and endocrine control mechanisms, however, can stimulate or inhibit these automatic functions to help coordinate the different stages of digestion. The sight, smell, or taste of food, for example, can stimulate salivary and gastric secretions via activation of the vagus nerve, which helps to "prime" the digestive system in preparation for a meal. Stimulation of the vagus, in this case, originates in the brain and is a conditioned reflex (as Pavlov demonstrated by training dogs to salivate in response to a bell). The vagus nerve is also involved in the reflex control of one part of the digestive system by another—these are "short reflexes," which don't involve the brain.

Pavlov: Ivan Petrovich Pavlov, Russian physiologist, 1849–1936

Table 24.14 Summary of the physiological effects of gastrointestinal hormones

Secreted by	Hormone	Effects
Stomach	Gastrin	Stimulates parietal cells to secrete HCl Stimulates chief cells to secrete pepsinogen Maintains structure of gastric mucosa
Small intestine	Secretin	Stimulates water and bicarbonate secretion in pancreatic juice Potentiates actions of cholecystokinin on pancreas
Small intestine	Cholecystokinin (CCK)	Stimulates contraction of the gallbladder Stimulates secretion of pancreatic juice enzymes Potentiates action of secretin on pancreas Maintains structure of exocrine pancreas (acini)
Small intestine	Gastric inhibitory peptide (GIP)	Inhibits gastric emptying Inhibits gastric acid secretion Stimulates secretion of insulin from endocrine pancreas (islets of Langerhans)

Source: Fox, Stuart Ira, *Human Physiology.* © 1984 Wm. C. Brown Publishers, Dubuque, Iowa. All Rights Reserved. Reprinted by permission.

The GI tract is both an endocrine gland and a target for the action of various hormones. Indeed, the first hormones to be discovered were gastrointestinal hormones. In 1902, Bayliss and Starling discovered that the duodenum produced a chemical regulator, which they named **secretin** *(se-kre′tin);* in 1905, these scientists proposed that secretin was but one of many yet undiscovered chemical regulators produced by the body. They coined the term *hormones* for this new class of regulators. Other investigators in 1905 discovered that an extract from the stomach antrum (pyloric region) stimulated gastric secretion. The hormone **gastrin** was thus the second hormone to be discovered.

The chemical structures of gastrin, secretin, and the duodenal hormone **cholecystokinin** *(ko″le-sis″to-kin′in),* CCK, were determined in the 1960s. More recently, a fourth hormone produced by the small intestine, **gastric inhibitory peptide** *(GIP),* has been added to the list of proven GI tract hormones. The effects of these hormones are summarized in table 24.14.

Regulation of Gastric Function

Gastric motility and secretion are, to some extent, automatic. Waves of contraction that serve to push chyme through the pyloric sphincter, for example, are initiated spontaneously by pacesetter cells in the greater curvature of the stomach. The secretion of hydrochloric acid (HCl) and pepsinogen, likewise, can be stimulated in the absence of neural and hormonal influences by the presence of cooked or partially digested protein in the stomach. The effects of autonomic nerves and hormones are superimposed on this automatic activity. The extrinsic control of gastric function is conveniently divided into three phases: (1) the cephalic phase; (2) the gastric phase; and (3) the intestinal phase. These are summarized in table 24.15.

Cephalic Phase. The cephalic phase of gastric regulation refers to control by the brain via the vagus nerve. As previously discussed, various conditioned stimuli can evoke gastric secretion. Activation of the vagus nerve can stimulate HCl and pepsinogen secretion by three mechanisms:

Starling: Ernest Henry Starling, English physiologist, 1866–1927

Table 24.15 The cephalic, gastric, and intestinal phases in the regulation of gastric acid secretion

Phase of Regulation	Description

Cephalic phase	1. Sight, smell, and taste of food cause stimulation of vagus nuclei in brain
	2. Vagus stimulates acid secretion
	  a) Direct stimulation of parietal cells (major effect)
	  b) Stimulation of gastrin secretion; gastrin stimulates acid secretion (lesser effect)

Gastric phase	1. Distension of stomach stimulates vagus nerve; vagus stimulates acid secretion
	2. Amino acids and peptides in stomach lumen stimulate acid secretion
	  a) Direct stimulation of parietal cells (lesser effect)
	  b) Stimulation of gastrin secretion; gastrin stimulates acid secretion (major effect)
	3. Gastrin secretion inhibited when pH of gastric juice falls below 2.5

Intestinal phase	1. Neural inhibition of gastric emptying and acid secretion
	  a) Arrival of chyme in duodenum causes distension, increase in osmotic pressure
	  b) These stimuli activate a neural reflex that inhibits gastric activity
	2. Gastric inhibitory peptide (GIP) secreted by duodenum in response to fat in chyme; GIP inhibits gastric acid secretion

Source: Fox, Stuart Ira, *Human Physiology.* © 1984 Wm. C. Brown Publishers, Dubuque, Iowa. All Rights Reserved. Reprinted by permission.

(1) direct vagal stimulation of the gastric parietal and chief cells (the primary mechanism); (2) vagal stimulation of gastrin secretion by the G cells, which in turn stimulates the parietal and chief cells to secrete HCl and pepsinogen, respectively; and (3) stimulation of secretion by increases in gastric blood flow.

Gastric Phase. The presence of short polypeptides and amino acids in the stomach stimulates the G cells to secrete gastrin and the parietal and chief cells to secrete HCl and pepsinogen; this is called the gastric phase of regulation. Since gastrin also stimulates HCl and pepsinogen secretion, a *positive feedback mechanism* develops: as more HCl and pepsinogen are secreted, more short polypeptides and amino acids are released from the ingested

Figure 24.53. The stimulation of gastric acid (HCl) secretion by the presence of proteins in the stomach lumen and by the hormone gastrin. The secretion of gastrin is inhibited by gastric acidity. This forms a negative feedback loop.

protein, thus stimulating more secretion of gastrin and, therefore, more secretion of HCl and pepsinogen (fig. 24.53).

Secretion of HCl during the gastric phase is also regulated by a *negative feedback mechanism*. As the pH of gastric juice drops, so does the secretion of gastrin—at a pH of 2.5 gastrin secretion is reduced, and at a pH of 1.0 gastrin secretion is totally abolished. The secretion of HCl thus declines accordingly. The presence of proteins and polypeptides in the stomach help to buffer the acid and thus to prevent a rapid fall in gastric pH; more acid can thus be secreted when proteins are present than when they are absent. The arrival of protein into the stomach thus stimulates acid secretion two ways—by the positive feedback mechanism previously discussed and by lessening of the negative feedback control of acid secretion. The amount of acid secreted is, by means of these effects, closely matched to the amount of protein ingested.

Intestinal Phase.   The intestinal phase of gastric regulation refers to the inhibition of gastric activity when chyme enters the small intestine. Investigators in 1886 demonstrated that the addition of olive oil to a meal inhibits gastric emptying, and in 1929 it was shown that the presence of fat inhibits gastric juice secretion. This inhibitory intestinal phase of gastric regulation is due to both a neural reflex originating from the duodenum and to a chemical hormone secreted by the duodenum.

The arrival of chyme into the duodenum increases its osmolality. This stimulus, together with stretch of the duodenum and possibly other stimuli, produces a neural reflex that results in the inhibition of gastric motility and secretion. The presence of fat in the chyme also stimulates the duodenum to secrete a hormone that inhibits gastric function. This inhibitory hormone is believed to be gastric inhibitory peptide (GIP).

The inhibitory neural and endocrine mechanisms during the intestinal phase prevent the further passage of chyme from the stomach to the duodenum. This gives the duodenum time to process the load of chyme that it has previously received. Since GIP is stimulated by fat in the chyme, a breakfast of bacon and eggs takes longer to pass through the stomach—and makes one feel ''fuller'' for a longer time—than does a breakfast of pancakes and syrup.

Regulation of Pancreatic Juice and Bile Secretion

The arrival of chyme into the duodenum stimulates the intestinal phase of gastric regulation and, at the same time, stimulates reflex secretion of pancreatic juice and bile. The entry of new chyme is thus retarded as the previous load is digested. The secretion of pancreatic juice and bile is stimulated by both neural reflexes initiated in the duodenum and by secretion of the duodenal hormones cholecystokinin (CCK) and secretin.

Pancreatic Juice.   The secretion of pancreatic juice is stimulated by both secretin and CCK. The release of secretin occurs in response to a fall in duodenal pH below 4.5; this pH fall occurs for only a short time, however,

Table 24.16 Regulation of pancreatic juice and bile secretion by the hormones *secretin* and *cholecystokinin* (CCK) and by the neurotransmitter *acetylcholine,* released from parasympathetic nerve endings

Description	Secretin	CCK	Acetylcholine (vagus nerve)
Stimulus for release	Acidity of chyme decreases duodenal pH below 4.5	Fat and protein in chyme	Sight, smell of food; distension of stomach
Second messenger	Cyclic AMP	Ca^{++}	Ca^{++}
Effect on pancreatic juice	Stimulates water and bicarbonate secretion; potentiates action of CCK	Stimulates enzyme secretion; potentiates action of secretin	Stimulates enzyme secretion
Effect on bile	Stimulates secretion	Potentiates action of secretin; stimulates contraction of gallbladder	Stimulates contraction of gallbladder

Source: Fox, Stuart Ira, *Human Physiology.* © 1984 Wm. C. Brown Publishers, Dubuque, Iowa. All Rights Reserved. Reprinted by permission.

because the acidic chyme is rapidly neutralized by alkaline pancreatic juice. The secretion of CCK occurs in response to the fat content of chyme in the duodenum.

Secretin stimulates the production of bicarbonate by the pancreas. Since bicarbonate neutralizes the acidic chyme and since secretin is released in response to the low pH of chyme, this completes a negative feedback loop where the effects of secretin inhibit its secretion. Cholecystokinin, in contrast, stimulates the production of pancreatic enzymes such as trypsin, lipase, and amylase. Secretin and CCK can have different effects on the same cells (the pancreatic acinar cells) because their actions are mediated by different intracellular compounds that act as second messengers. The second messenger of secretin action is cyclic AMP, whereas the second messenger for CCK is Ca^{++} (table 24.16).

Secretion of Bile. The liver secretes bile continuously, but this secretion is greatly augmented following a meal. This increased secretion is due to the release of secretin and CCK from the duodenum. Secretin is the major stimulator of bile secretion by the liver, and CCK enhances this effect. The arrival of chyme in the duodenum also causes the gallbladder to contract and eject bile. Contraction of the gallbladder occurs in response to neural reflexes from the duodenum and in response to stimulation by CCK.

Trophic Effects of Gastrointestinal Hormones

Patients with tumors of the stomach pylorus have high acid secretion and hyperplasia (growth) of the gastric mucosa. Surgical removal of the pylorus reduces gastric secretion and prevents growth of the gastric mucosa. Patients with peptic ulcers are sometimes treated by vagotomy—cutting of the vagus nerve. Vagotomy also reduces acid secretion but has no effect on the gastric mucosa. These observations suggest that the hormone gastrin, secreted by the pyloric mucosa, may exert stimulatory, or *trophic* effects on the gastric mucosa. The structure of the gastric mucosa, in other words, is dependent on the effects of gastrin.

In the same way, the structure of the acinar (exocrine) cells of the pancreas is dependent upon the trophic effects of CCK. Perhaps this explains why the pancreas, as well as the stomach, atrophies during starvation. Since neural reflexes appear to be capable of regulating digestion, perhaps the primary function of the GI hormone is trophic—that is, maintenance of the structure of their target organs.

---

1. Describe the positive and negative feedback mechanisms that operate during the gastric phase of HCl and pepsinogen secretion.
2. Describe the mechanisms involved in the intestinal phase of gastric regulation, and explain why a fatty meal takes longer to leave the stomach than a meal low in fat.
3. Explain the hormonal mechanisms involved in the production and release of pancreatic juice and bile.

Clinical Considerations

Developmental Disorders

Most of the congenital disorders of the digestive system have unknown causes and occur during the fourth or fifth week of embryonic development. A **cleft palate** is a congenital opening between the oral and nasal cavities and therefore involves both the digestive and respiratory systems. It occurs in about one in 1,000 live births and has immediate importance because it interferes with suckling and swallowing mechanisms. **Esophageal atresia,** or failure to develop the normal structure of the esophageal-stomach area, is another disorder of the upper digestive tract that requires surgery to correct. **Pyloric stenosis** is a common abnormality in which the pyloric sphincter muscle is hypertrophied, reducing the size of the lumen. This condition affects one in about 200 newborn males and one in 1,000 newborn females. Stenoses, atresias, and rotations of various portions of the GI tract may occur as the gut develops. Umbilical problems involving the GI tract are fairly common, as is some form of *imperforate anus,* which occurs in about one in 5,000 births.

---

trophic: Gk. *trophikos,* pertaining to nutrition

Figure 24.54. Clinical problems of the teeth. (*a*) trench mouth and dental caries. (*b*) severe alveolar bone destruction from periodontitis. (*c*) pyogenic granuloma and dental caries. (*d*) malposition of teeth.

(a)

(b)

(c)

(d)

Pathogens and Poisons

The GI tract is a suitable environment for an array of microorganisms. Many of these are beneficial, but some bacteria and protozoa can cause diseases. Only a few examples of these will be discussed.

Dysentery *(dis'en-ter''e)* is an inflammation of the intestinal mucosa, characterized by the discharge of loose stools that contain mucus, pus, and blood. The most common dysentery is **amoebic dysentery,** which is caused by the protozoan *Entamoeba histolytica.* Cysts from this organism are ingested in contaminated food, and after the protective coat is removed by HCl in the stomach, the vegetative form invades the mucosal walls of the ileum and colon.

Food poisoning comes from ingesting pathogenic bacteria or their toxins. *Salmonella* is a bacterium that commonly causes food poisoning. **Botulism** is the most serious type of food poisoning and is caused by ingesting food contaminated with the bacterium *Clostridium botulinum.* This organism is widely distributed in nature, and the spores from it are frequently on food being processed by canning. For this reason food must be heated to 120°C (248°F) before it is canned. It is the toxins produced by the bacterium growing in the food that are pathogenic rather than the organisms themselves. The poison is a neurotoxin that is readily absorbed in the blood where it affects the nervous system.

Clinical Problems of the Teeth and Salivary Glands

Dental caries, or tooth decay, is the gradual decalcification of tooth enamel (fig. 24.54) and underlying dentin, produced by the acid products of bacteria. These bacteria thrive between teeth where food particles accumulate and form part of the thin layer of bacteria, proteins, and other debris called *plaque* that covers teeth. The development of dental caries can thus be reduced by brushing at least once a day and by flossing between teeth at regular intervals.

People over the age of thirty-five are particularly susceptible to **periodontal diseases.** There are many types of periodontal disease, but basically they all cause inflammation and deterioration of the gingiva, alveolar sockets, periodontal membranes, and teeth cementum. Some of the symptoms are loosening of the teeth, bad breath, bleeding gums when brushing, and some edema. Periodontal diseases are caused by many things, including impacted plaque, cigarette smoking, malocclusion, and poor diet.

Mumps is a viral disease of the parotid salivary glands and in advanced stages may involve the pancreas and testes. It is generally not serious in children but can be very serious in adults where it may cause deafness and destroy the pancreatic islet tissue or testicular cells.

Disorders of the Liver

The liver is a remarkable organ that has the ability to regenerate itself even if up to 80 percent has been removed. The most serious diseases of the liver (hepatitis, cirrhosis, and hepatomas) affect the liver throughout so that it cannot repair itself. **Hepatitis** is inflammation of the liver. Certain chemicals may cause hepatitis, but generally it is caused by infectious viral agents. **Infectious hepatitis** is a viral disease transmitted through contaminated foods and liquids. **Serum hepatitis** is also caused by a virus and is transmitted in serum or plasma during transfusions or by improperly sterilized needles and syringes.

---

dysentery: Gk. *dys*, bad; *entera*, intestine

In **cirrhosis,** large numbers of liver lobules are destroyed and replaced with permanent connective tissue and "regenerative nodules" of hepatocytes. These nodules do not have the platelike structure of normal liver tissue and are therefore less functional. Cirrhosis may be caused by chronic alcohol abuse, viral hepatitis, and other agents that attack liver cells.

Jaundice is a yellow staining of the tissue produced by high blood concentrations of either free or conjugated bilirubin. Since free bilirubin is derived from heme, abnormally high concentrations of this pigment may result from an unusually high rate of red blood cell destruction. This can occur, for example, as a result of Rh disease (erythroblastosis fetalis) in an Rh positive baby born to a sensitized Rh negative mother. Jaundice may also occur in otherwise healthy infants because of the fact that red blood cells are normally destroyed at about the time of birth (hemoglobin concentrations decrease from 19 g per 100 ml to 14 g per 100 ml near the time of birth). This condition is called *physiological jaundice of the newborn* and is not indicative of disease. Premature infants may also develop jaundice due to the fact that the hepatic enzymes that conjugate bilirubin (a reaction needed for the excretion of bilirubin in the bile) don't normally mature until about the tenth month of fetal life. Jaundice due to high levels of conjugated bilirubin in the blood is commonly produced in adults when the excretion of bile is blocked by gallstones.

Hepatomas *(hep''ah-to'mah)* are malignant tumors that originate in or secondarily invade the liver. Those that originate in the liver (primary hepatomas) are relatively rare, but those that secondarily metastasize to the liver from other organs (secondary hepatomas) are common. Carcinoma of the liver is usually fatal.

Disorders of the GI Tract

Peptic ulcers are erosions of the mucous membranes of the stomach or duodenum produced by the action of HCl. Agents that weaken the mucosal lining of the stomach, such as alcohol and aspirin, and abnormally high secretions of HCl thus increase the likelihood of developing peptic ulcers. Many people subject to chronic stress produce too much gastric acid and develop a peptic ulcer as a result.

Enteritis is inflammation of the intestinal mucosa and is frequently referred to as intestinal flu. Causes of enteritis include bacterial or viral infections, irritating foods or fluids (including alcohol), and emotional stress. The symptoms are abdominal pain, nausea, and diarrhea. **Diarrhea** is the passage of watery, unformed stools. This condition is symptomatic of inflammation, stress, and other body dysfunctions.

cirrhosis: Gk. *kirrhos,* yellow-orange
jaundice: L. *galbus,* yellow

A **hernia** is a protrusion of a portion of a visceral organ, usually the small intestine, through a weakened portion of the abdominal wall. Inguinal, femoral, umbilical, and hiatal hernias are the most common. A **hiatal hernia** is when a portion of the stomach pushes superiorly through the esophageal hiatus in the diaphragm and into the thorax. The potential dangers of a hernia are strangulation of the blood supply followed by gangrene, blockage of chyme, or rupture—each of which can threaten life.

Diverticulosis is a condition where the intestinal wall weakens and an outpouching occurs. Recent studies suggest that suppressing the passage of flatus (intestinal gas) may contribute to diverticulosis, especially in the sigmoid colon. **Diverticulitis,** or inflammation of a diverticulum, can develop if fecal material becomes impacted in these pockets.

Peritonitis is inflammation of the peritoneum lining the abdominal cavity and covering the viscera. The causes of peritonitis include bacterial contamination of the abdominal cavity through accidental or surgical wounds in the abdominal wall or perforation of the intestinal wall (as with ulcers or a ruptured appendix).

Other Important Clinical Terminology

chilitis   Inflammation of the lips.

colitis   Inflammation of the colon and rectum.

colostomy   The formation of an abdominal exit from the GI tract by bringing a loop of the colon to the surface of the abdomen. If the rectum is removed because of cancer, the colostomy provides a permanent outlet for the feces.

cystic fibrosis   An inherited disease of the exocrine glands, particularly the pancreas. Pancreatic secretions are too thick to drain easily, causing the ducts to become inflamed and promoting connective tissue formation that occludes drainage from the ducts still further.

gingivitis   Inflammation of the gums. It may result from improper hygiene, poorly fitted dentures, improper diet, or certain infections.

halitosis   Offensive breath odor. It may result from dental caries, certain diseases, or eating particular foods.

heartburn   A burning sensation of the esophagus and stomach. It may result from the regurgitation of gastric juice into the lower portion of the esophagus.

hemorrhoids   Varicose veins of the rectum and anus.

jejunoileal bypass   A surgical procedure for creating a bypass of a considerable portion of the small intestine. It reduces the absorptive capacity of the small intestine and is thus used to control extreme obesity.

nausea   Gastric discomfort and sensations of illness with a tendency to vomit. This feeling is symptomatic of many conditions (e.g., motion sickness, foul odors or sights, pregnancy, diseases, etc.).

pyorrhea   The discharge of pus at the base of the teeth at the gum line.

regurgitation (vomiting)   The forceful expulsion of gastric contents into the mouth. Nausea and vomiting are common symptoms of almost any dysfunction of the digestive system.

trench mouth   A contagious bacterial infection that causes inflammation, ulceration, and painful swelling of the floor of the mouth. Generally it is contracted through direct contact by kissing an infected person. Trench mouth can be treated by penicillin and other medications.

vagotomy   The surgical removal of a section of the vagus nerve where it enters the stomach to eliminate nerve impulses that stimulate gastric secretion. This procedure is used to treat ulcers.

Chapter Summary

I. Introduction to the Digestive System
The digestive system is derived primarily from endoderm. It consists of the gastrointestinal tract and accessory digestive organs.
  A. The digestion of food molecules involves hydrolysis reactions, which occur in the lumen of the GI tract.
  B. The digestive tract is derived embryologically from the primitive gut, which is composed of endoderm.
      1. The structures of the mouth are derived from stomodeal ectoderm, and the epithelium of the anal canal is derived from ectoderm of the proctodeum; these structures become continuous with the gut endoderm.
      2. The liver and pancreas develop from buds that arise from the wall of the small intestine.
  C. Peritoneal membranes line the abdominal wall and cover the visceral organs; the GI tract is supported by a double layer of peritoneum called the mesentery.
      1. The lesser omentum and greater omentum are folds of peritoneum that extend from the stomach.
      2. Visceral organs that are not in the peritoneal cavity and not covered with peritoneum are said to be retroperitoneal in location.
  D. The layers (tunics) of the GI tract are, from the inside outward, mucosa, submucosa, muscularis, and serosa.
      1. The mucosa consists of a simple columnar epithelium, a thin layer of connective tissue called the lamina propria, and a thin layer of smooth muscle called muscularis mucosa.
      2. The submucosa is composed of connective tissue, the muscularis consists of layers of smooth muscles, and the serosa is connective tissue covered with the visceral peritoneum.
      3. The submucosa contains the submucosal plexus, and the muscularis contains the myenteric plexus of autonomic nerves.

II. Mouth, Pharynx, and Associated Structures
Ingested food is changed by the mechanical action of teeth and by the chemical activity of saliva.
  A. The cheeks form the lateral walls of the buccal (oral) cavity.
  B. The tongue contains lingual tonsils and papillae with taste buds.
  C. The palate contains rugae, a cone-shaped projection called the uvula, and palatine tonsils.
  D. The four pairs of anteriormost teeth are the incisors and canines, which have one root each; the bicuspids and molars have two or three roots.
      1. Humans are diphyodont; they have deciduous and permanent sets of teeth.
      2. The tooth is anchored to its bony alveolar socket by a periodontal membrane, which contains fibers that insert into the cementum of the tooth root.
      3. Enamel forms the outer layer of the tooth crown; beneath the enamel is dentin.
      4. The interior of a tooth contains a pulp cavity, which is continuous through the apical foramen of the root with the connective tissue around the tooth.
  E. The major salivary glands are the parotid glands and the submandibular glands.

III. Esophagus and Stomach
Food is passed from the esophagus to the stomach, where it is mixed with HCl and pepsin. The resulting chyme is sent past the pyloric sphincter to the duodenum.
  A. Peristaltic waves of contraction push food through the lower esophageal sphincter into the stomach.
  B. Swallowing, or deglutition, occurs in three phases, and involves structures of the buccal cavity, pharynx, and esophagus.
  C. The stomach consists of a cardia, fundus, body, and pyloris, which ends with the pyloric sphincter.
      1. The lining of the stomach is thrown into folds, or rugae, and the mucosa is formed into gastric pits and gastric glands.
      2. The parietal cells of the gastric glands secrete HCl, and the chief cells secrete pepsinogen.
      3. In the acidic environment of gastric juice, pepsinogen is converted into the active protein-digesting enzyme called pepsin.
      4. Some digestion of protein and fat occurs in the stomach; the most important function of the stomach is the secretion of intrinsic factor, which is needed for the absorption of vitamin B_{12} in the intestine.

IV. Small Intestine
The villi and microvilli in the small intestine provide a very high surface area for digestion by brush border enzymes and for absorption of the products of digestion.
  A. Regions of the small intestine include the duodenum, jejunum, and ileum; the common bile duct and pancreatic duct empty into the duodenum.
  B. Fingerlike extensions of mucosa called villi project into the lumen, and at the bases of the villi the mucosa forms narrow pouches called the crypts of Lieberkühn.
      1. New epithelial cells are formed in the crypts.
      2. The membrane of intestinal epithelial cells is folded to form microvilli; this is called the brush border of the mucosa and serves to increase surface area.
  C. Digestive enzymes, called brush border enzymes, are located in the membranes of the microvilli.
  D. The small intestine exhibits two major types of movements—peristalsis and segmentation.

V. Large Intestine

The large intestine absorbs water and electrolytes from the chyme and passes waste products out of the body through the rectum and anal canal.

A. The large intestine is divided into the cecum, colon, rectum, and anal canal.
1. The vermiform appendix is attached to the inferior medial margin of the cecum.
2. The colon consists of ascending, transverse, descending, and sigmoid portions.
3. Bulges in the walls of the large intestine are called haustra.

B. Three types of movements occur in the large intestine: peristalsis, haustral churning, and mass movement.

VI. Liver, Gallbladder, and Pancreas

Blood from the intestine flows to the liver. The liver produces bile and secretes it into the large intestine and gallbladder. Both the common bile duct and the pancreatic duct, carrying pancreatic juice, empty into the duodenum.

A. The liver, the largest internal organ, is composed of functional units called lobules.
1. Liver lobules consist of plates of hepatic cells separated by capillary sinusoids.
2. Blood flows from the periphery of each lobule, where the hepatic artery and portal vein empty, through the sinusoids, and out the central vein.
3. Bile flows within the hepatocyte plates, in canaliculi, to the bile ducts.
4. Substances excreted in the bile can be returned to the liver in the hepatic portal blood; this is called an enterohepatic circulation.
5. Bile consists of a pigment called bilirubin, bile salts, cholesterol, and other molecules.
6. The liver detoxifies the blood by excreting substances in the bile, by phagocytosis, and by chemical inactivation.

7. The liver modifies the plasma concentrations of proteins, glucose, triglycerides, and ketone bodies.

B. The gallbladder serves to store and concentrate the bile, and it releases bile through the cystic duct and common bile duct to the duodenum.

C. The pancreas is both an exocrine and an endocrine gland.
1. The endocrine portion is known as the islets of Langerhans, and secretes the hormones insulin and glucagon.
2. The exocrine acini of the pancreas produce pancreatic juice, which contains various digestive enzymes and bicarbonate.

VII. Digestion and Absorption of Carbohydrates, Lipids, and Proteins

Polysaccharides and polypeptides are digested into monosaccharides and amino acids, respectively, and are secreted into the blood. Triglycerides are secreted into lymphatic vessels as chylomicrons.

A. The digestion of starch begins in the mouth through the action of salivary amylase.
1. Pancreatic amylase digests starch into disaccharides and short-chain oligosaccharides.
2. Complete digestion into monosaccharides is accomplished by brush border enzymes.

B. Protein digestion begins in the stomach by the action of pepsin.
1. Pancreatic juice contains protein-digesting enzymes, including trypsin, chymotrypsin, and others.
2. The brush border contains digestive enzymes that help to complete the digestion of proteins into amino acids.
3. Amino acids, like monosaccharides, are absorbed and secreted into capillary blood entering the portal vein.

C. Lipids are digested in the small intestine after being emulsified by bile salts.
1. Free fatty acids and monoglycerides enter particles called micelles, formed in large part by bile salts, and in this form, or as free molecules, they are absorbed.

2. Once inside the mucosal epithelial cells, these subunits are used to resynthesize triglycerides.
3. Triglycerides in the epithelial cells, together with proteins, form chylomicrons, which are secreted into the central lacteals of the villi.
4. Chylomicrons are transported by lymph to the thoracic duct and there enter the blood.

VIII. Neural and Endocrine Regulation of the Digestive System

Neural and hormonal mechanisms coordinate the activities of different regions of the GI tract and the actions of the liver, gallbladder, and pancreas.

A. The regulation of gastric function occurs in three phases.
1. In the cephalic phase, the activity of higher brain centers, acting via the vagus nerve, stimulate gastric juice secretion.
2. In the gastric phase, the secretion of HCl and pepsin is controlled by the gastric contents and by the hormone gastrin, secreted by the gastric mucosa.
3. In the intestinal phase, the activity of the stomach is inhibited by neural reflexes from the duodenum and by gastric inhibitory peptide (GIP), secreted by the duodenum.

B. The secretion of the hormones secretin and cholecystokinin (CCK) regulate pancreatic juice and bile secretion.
1. Secretin secretion is stimulated by the arrival of acidic chyme into the duodenum.
2. CCK secretion is stimulated by the presence of fat in the chyme arriving in the duodenum.
3. Contraction of the gallbladder occurs in response to a neural reflex and to the secretion of CCK by the duodenum.

C. Gastrointestinal hormones may be needed for the maintenance of the GI tract and accessory digestive organs.

Review Activities

Objective Questions

1. Which of the following types of teeth are found in the permanent but not in the deciduous dentition?
   (a) incisors
   (b) canines
   (c) premolars
   (d) molars
2. The epithelium of the mouth is derived from the
   (a) ectoderm of the stomodeum
   (b) ectoderm of the proctodeum
   (c) endoderm of the foregut
   (d) all of the above
3. A double layer of peritoneum that supports the GI tract is called the
   (a) visceral peritoneum
   (b) dorsal mesentery
   (c) greater omentum
   (d) lesser omentum
4. Which of the following tissue layers in the small intestine contains the central lacteals?
   (a) submucosa
   (b) muscularis mucosa
   (c) lamina propria
   (d) muscularis externa
5. Which of the following statements about gastric secretion of HCl is *false?*
   (a) HCl is secreted by parietal cells.
   (b) HCl hydrolyzes peptide bonds.
   (c) HCl is needed for the conversion of pepsinogen to pepsin.
   (d) HCl is needed for maximum activity of pepsin.
6. Intestinal enzymes such as lactase are
   (a) secreted by the intestine into the chyme
   (b) produced by the crypts of Lieberkühn
   (c) produced by the pancreas
   (d) attached to the cell membrane of the microvilli
7. Most digestion occurs in the
   (a) mouth
   (b) stomach
   (c) small intestine
   (d) large intestine
8. Which of the following statements about trypsin is true?
   (a) Trypsin is derived from trypsinogen by the digestive action of pepsin.
   (b) Active trypsin is secreted into the pancreatic acini.
   (c) Trypsin is produced in the crypts of Lieberkühn.
   (d) Trypsinogen is converted to trypsin by the brush border enzyme enterokinase.
9. During the gastric phase, the secretion of HCl and pepsinogen is stimulated by
   (a) vagus nerve stimulation that originates in the brain
   (b) polypeptides in the gastric lumen and by gastrin secretion
   (c) secretin and cholecystokinin from the duodenum
   (d) all of the above
10. The secretion of HCl by the stomach mucosa is inhibited by
   (a) neural reflexes from the duodenum
   (b) the secretion of gastric inhibitory peptide from the duodenum
   (c) the lowering of gastric pH
   (d) all of the above
11. The first organ to receive the blood-borne products of digestion is the
   (a) liver
   (b) pancreas
   (c) heart
   (d) brain
12. Which of the following statements about hepatic portal blood is true?
   (a) It contains absorbed fat.
   (b) It contains ingested proteins.
   (c) It is mixed with bile in the liver.
   (d) It is mixed with blood from the hepatic artery in the liver.

Essay Questions

1. Explain the embryological and anatomical relationships that the liver and pancreas have with the intestine.
2. Explain how the gastric secretion of HCl and pepsin is regulated during the cephalic, gastric, and intestinal phases.
3. Describe how pancreatic enzymes become activated in the lumen of the intestine, and explain the need for these mechanisms.
4. What is the function of bicarbonate in pancreatic juice? Explain why stress ulcers are more likely to be located in the duodenum than in the stomach.
5. Explain why the pancreas is considered to be both an exocrine and an endocrine gland. Given this information, predict what effects tying of the pancreatic duct would have on pancreatic structure and function.

Selected Readings

Binder, H. J. 1984. The pathophysiology of diarrhea. *Hos. Prac.* 19:107.
Bleich, H. L., and E. S. Boro. 1979. Protein digestion and absorption. *New Engl. J. Med.* 300:659.
Bortoff, A. 1972. Digestion. *Ann. Rev. Physiol.* 28:201.
Carey, M. C., D. M. Small, and C. M. Bliss. 1983. Lipid digestion and absorption. *Ann. Rev. Physiol.* 45:651.
Chou, C. C. 1982. Relationship between intestinal blood flow and motility. *Ann. Rev. Physiol.* 44:29.
Christensen, R. R. 1971. The controls of gastrointestinal movements; Some old and new views. *New Engl. J. Med.* 285:483.
Cohen, S. 1983. Neuromuscular disorders of the gastrointestinal tract. *Hosp. Prac.* 18:121.
Davenport, H. W. 1982. *Physiology of the digestive tract.* 5th ed. Chicago: Year Book Medical Publishers.
Dockray, G. J. 1979. Comparative biochemistry and physiology of gut hormones. *Ann. Rev. Physiol.* 41:83.

Freeman, H. J., and Y. S. Kim. 1978. Digestion and absorption of proteins. *Ann. Rev. Med.* 29:99.
Gray, G. M. 1975. Carbohydrate digestion and absorption: Role of the small intestine. *New Engl. J. Med.* 292:1225.
Grossman, M. I. 1979. Neural and hormonal regulation of gastrointestinal function: An overview. *Ann. Rev. Physiol.* 41:27.
Guth, P. H. 1982. Stomach blood flow and acid secretion. *Ann. Rev. Physiol.* 44:3.
Hersey, S. J., S. H. Norris, and A. J. Gilbert. Cellular control of pepsinogen secretion. *Ann. Rev. Physiol.* 46:393.
Holt, K. M., and J. I. Isenberg. 1985. Peptic ulcer disease: Physiology and pathophysiology *Hos. Prac.* 20:89.
Kappas, A. and A. P. Alvarez. June 1975. How the liver metabolizes foreign substances. *Scientific American.*
McGuigan, J. E. 1978. Gastrointestinal hormones. *Ann. Rev. Physiol.* 29:99.
Moog, F. November 1981. The lining of the small intestine. *Scientific American.*
Richardson, I. and P. G. Withrington. 1982. Physiological regulation of the hepatic circulation. *Ann. Rev. Physiol.* 44:57.

Salen, G. and S. Shefer. 1983. Bile acid synthesis. *Ann. Rev. Physiol.* 45:679.
Soll, A. and J. H. Walsh. 1979. Regulation of gastric acid secretion. *Ann. Rev. Physiol.* 41:35.
Walsh, J. H. and M. I. Grossman. 1975. Gastrin. *New Engl. J. Med.* 292: first part, p. 1324; second part, p. 1377.
Weisbrodt, N. W. 1981. Patterns of intestinal motility. *Ann. Rev. Physiol.* 43:21.
Williams, J. A. 1984. Regulatory mechanisms in pancreas and salivary acini. *Ann. Rev. Physiol.* 46:361.
Wood, J. D. 1981. Intrinsic neural control of intestinal motility. *Ann. Rev. Physiol.* 43:33.

25

Regulation of Metabolism

Outline

Nutritional Requirements
Vitamins and Minerals
Caloric Requirements
Regulation of Energy Metabolism
Eating
Hormonal Regulation of Metabolism
Energy Regulation by the Islets of Langerhans
Regulation of Insulin and Glucagon Secretion
Insulin and Glucagon: Period of Absorption
Insulin and Glucagon: Period of Fasting
Metabolic Regulation by the Adrenal Hormones, Thyroxine, and Growth Hormone
Adrenal Hormones
Thyroxine
Growth Hormone
Regulation of Calcium and Phosphate Balance
Parathyroid Hormone and 1,25-Dihydroxy Vitamin D_3
Negative Feedback Control of Calcium and Phosphate Balance
Calcitonin
Clinical Considerations
Diabetes Mellitus
Hypoglycemia
Hypothyroidism and Hyperthyroidism
Gigantism, Acromegaly, and Dwarfism
Chapter Summary
Review Activities
Selected Readings

Concepts

Vitamins, minerals, essential amino acids, and essential fatty acids are required in the diet for metabolism. The energy needs of the body are met by the caloric value of food.

Food and stored energy reserves can provide circulating energy substrates for the tissue cells. The utilization and storage of energy is regulated by a variety of hormones.

The secretion of insulin is increased and that of glucagon is decreased, during the intestinal absorption of a meal. These two hormones have antagonistic effects in the body.

Adrenal hormones, thyroxine, and growth hormone stimulate the catabolism of carbohydrates and lipids. In addition, thyroxine and growth hormone stimulate protein synthesis and are needed for proper body growth and development.

The calcium and phosphate concentration in the blood is affected by the actions of parathyroid hormone, 1,25-dihydroxy vitamin D_3, and calcitonin on bone, intestinal absorption, and urinary excretion of these ions.

Nutritional Requirements

Vitamins and minerals are needed in the diet for enzymatic reactions to proceed, and the essential amino acids and fatty acids are required for the metabolism of protein and fat. The energy needs of the body are met by the caloric value of food.

Objective 1. Distinguish between the water-soluble and fat-soluble vitamins, and describe some of the functions that vitamins serve in the body.

Objective 2. Explain the importance of the essential amino acids and essential fatty acids.

Objective 3. Explain the caloric needs of the body and some of the dangers of excessive caloric intake.

Living tissue is maintained by the constant expenditure of energy. This energy is obtained directly from ATP and indirectly from the cell respiration of glucose, fatty acids, ketone bodies, amino acids, and other organic molecules. These molecules are ultimately obtained from food, but they can also be obtained from the glycogen, fat, and protein stored in the body. The energy value of food is commonly measured in *kilocalories,* which are also called "big calories" and spelled with a capital letter (*C*alories).

In addition to its caloric value, food also supplies the essential amino acids and fatty acids. The nine **essential amino acids** are lysine, methionine, valine, leucine, isoleucine, tryptophane, phenylalanine, threonine, and histidine. The **essential fatty acids** are linoleic acid and linolenic acid. These molecules are termed essential because the body cannot make them and thus is forced to obtain them in the diet for proper protein and fat synthesis.

Vitamins and Minerals

Vitamins are small organic molecules that cannot be made by the body and that serve as coenzymes in metabolic reactions. There are two groups of vitamins—the fat-soluble vitamins (A, D, E, and K) and the water-soluble vitamins. Water-soluble vitamins include thiamine *(thi'ah-min),* B_1, riboflavin (B_2), niacin (B_3), pyridoxine (B_6), pantothenic acid, biotin, folic acid, B_{12}, and vitamin C (ascorbic acid). Recommended daily allowances for these vitamins are shown in table 25.1.

Many of the water-soluble vitamins serve as coenzymes in the metabolism of carbohydrates, lipids, and proteins. Thiamine, for example, is needed for the activity of the enzyme that converts pyruvic acid to acetyl coenzyme A. Riboflavin and niacin are needed for the production of FAD and NAD, respectively; these latter compounds serve as coenzymes that transfer hydrogens during cell respiration. Pyridoxine is a cofactor for the enzymes involved in amino acid metabolism. Deficiencies of the water-soluble vitamins can, for obvious reasons, have widespread effects in the body.

851

Table 25.1 Recommended daily allowances for vitamins and minerals

	Infants 0–6 mos	Children 4–6 yrs	Adolescents (15–18 yrs) Males	Adolescents (15–18 yrs) Females	Adults (23–50 yrs) Males	Adults (23–50 yrs) Females
Weight, kg (lb)	6 (13)	20 (44)	66 (145)	55 (120)	70 (154)	55 (120)
Height, cm (in.)	60 (24)	112 (44)	176 (69)	163 (64)	178 (70)	163 (64)
Protein, g	kg × 2.2	30	56	46	56	44
Fat-soluble vitamins†						
Vitamin A, µg	420	500	1000	800	1000	800
Vitamin D, µg	10	10	10	10	5	5
Vitamin E activity, mg	3	6	10	8	10	8
Water-soluble vitamins						
Ascorbic acid, mg	35	45	60	60	60	60
Folic acid, µg	30	200	400	400	400	400
Niacin, mg	6	11	18	14	18	13
Riboflavin, mg	0.4	1.0	1.7	1.3	1.6	1.2
Thiamine, mg	0.3	0.9	1.4	1.1	1.4	1.0
Vitamin B_6, mg	0.3	1.3	2.0	2.0	2.2	2.0
Vitamin B_{12}, µg	0.5	2.5	3.0	3.0	3.0	3.0
Minerals						
Calcium, mg	360	800	1200	1200	800	800
Phosphorus, mg	240	800	1200	1200	800	800
Iodine, µg	40	90	150	150	150	150
Iron, mg	10	10	18	18	10	18
Magnesium, mg	50	200	400	300	350	300
Zinc, mg	3	10	15	15	15	15

†Microgram

Source: Food and Nutrition Board, *Recommended Dietary Allowances*, 9th ed., National Academy of Sciences—National Research Council, Washington, D.C., 1980.

Many fat-soluble vitamins have highly specialized functions. Vitamin K, for example, is required for the production of prothrombin and for clotting factors VII, IX, and X. Vitamin D is converted into a hormone that participates in the regulation of calcium balance. The visual pigments in the rods and cones of the retina are derived from vitamin A. Vitamin A and related compounds, called *retinoids (ret'ĭ-noid),* also have effects on genetic expression in epithelial cells; these compounds are now used clinically in the treatment of some skin conditions, and researchers are attempting to derive related compounds that may aid the treatment of some cancers.

Minerals are needed as cofactors for specific enzymes and for a wide variety of other critical functions. Minerals that are required in relatively large amounts per day include sodium, potassium, magnesium, calcium, phosphorous, and chlorine (table 25.1). In addition, the following **trace elements** are recognized as essential: iron, zinc, manganese, fluorine, copper, molybdenum *(mo-lib'dē-num),* chromium, selenium, and iron. These must be ingested in amounts ranging from fifty micrograms to eighteen milligrams per day (table 25.2).

Caloric Requirements

There are tremendous variations in the energy requirements of people. These are partly due to differences in "fuel efficiency," so that some people consume more calories than others do at comparable levels of physical activity. This "fuel efficiency" is regulated in part by the thyroid gland.

Table 25.2 Recommended daily intake of trace elements

Element	Safe and Adequate Intake (mg/day)
Iron (males)	10
Iron (females)	18
Zinc	15
Manganese	2.5 to 5.0
Fluorine	1.5 to 4.0
Copper	2.0 to 3.0
Molybdenum	0.15 to 0.5
Chromium	0.05 to 0.2
Selenium	0.05 to 0.2
Iodine	0.15

Source: From the Food and Nutrition Board of the National Academy of Science, 1980.

The differences in daily energy requirements among people, however, are largely due to differences in physical activity.

Average daily energy expenditures may range from about 1,300 to 5,000 kilocalories per day. The average values for people not engaged in heavy manual labor but who are active during their leisure time is about 2,900 kilocalories per day for a man and 2,100 kilocalories per day for a woman. People engaged in office work, the professions, sales, and comparable occupations consume approximately 5 kilocalories per minute during work. More physically demanding occupations may require energy expenditures of 7.5 to 10 kilocalories per minute.

Table 25.3 Desirable body weights, according to sex, age, height, and body frame

	Height Feet	Height Inches	Small Frame	Medium Frame	Large Frame		Height Feet	Height Inches	Small Frame	Medium Frame	Large Frame
Men	5	2	128–134	131–141	138–150	**Women**	4	10	102–111	109–121	118–131
	5	3	130–136	133–143	140–153		4	11	103–113	111–123	120–134
	5	4	132–138	135–145	142–156		5	0	104–115	113–126	122–137
	5	5	134–140	137–148	144–160		5	1	106–118	115–129	125–140
	5	6	136–142	139–151	146–164		5	2	108–121	118–132	128–143
	5	7	138–145	142–154	149–168		5	3	111–124	121–135	131–147
	5	8	140–148	145–157	152–172		5	4	114–127	124–138	134–151
	5	9	142–151	148–160	155–176		5	5	117–130	127–141	137–155
	5	10	144–154	151–163	158–180		5	6	120–133	130–144	140–159
	5	11	146–157	154–166	161–184		5	7	123–136	133–147	143–163
	6	0	149–160	157–170	164–188		5	8	126–139	136–150	146–167
	6	1	152–164	160–174	168–192		5	9	129–142	139–153	149–170
	6	2	155–168	164–178	172–197		5	10	132–145	142–156	152–173
	6	3	158–172	167–182	176–202		5	11	135–148	145–159	155–176
	6	4	162–176	171–187	181–207		6	0	138–151	148–162	158–179

Weights at ages 25–59 based on lowest mortality. Weight in pounds according to frame (in indoor clothing weighing 5 lbs., shoes with 1″ heels).

Weights at ages 25–59 based on lowest mortality. Weight in pounds according to frame (in indoor clothing weighing 3 lbs., shoes with 1″ heels).

Reprinted with permission of the Metropolitan Insurance Companies.

When the caloric intake is greater than the energy expenditures, excess calories are stored primarily as fat. This is true regardless of the source of the calories—carbohydrates, protein, or fat—because these molecules can be converted to fat by the metabolic pathways described in chapter 4. Appropriate body weights are indicated in table 25.3.

Obesity is a risk factor for cardiovascular diseases, renal disease, diabetes mellitus, gallbladder disease, the development of kidney stones, and some malignancies (particularly endometrial and breast cancer). Obesity in childhood is due to an increase in both the size and number of adipose cells; weight gain in adulthood is due mainly to an increase in adipose cell size, although the number of these cells may also increase in extreme weight gains. When weight is lost, the size of the adipose cells decreases, but the number of adipose cells does not decrease. It is thus important to prevent further increases in weight in all overweight people but particularly so in children.

The degree of obesity can be most conveniently determined by the **body mass index.** This is equal to the nude body weight (in kilograms) divided by the square of the barefoot height (in meters). For example, a person who weighs 170 pounds (77.1 kg) and is five-feet-nine-inches tall (1.75 m), has a body mass index of $77.1 \div 1.75^2 = 25.1$. The normal body mass index is 20–25 for males and 19–24 for females. Statistically, the morbidity and mortality rates are significantly increased when the body mass index is greater than 30. When the body mass index is over 40, the risk factor for cardiovascular disease due to obesity is comparable to risk factors due to smoking, hypertension, or hyperlipidemia.

1. List the fat-soluble vitamins, and describe some of their functions.
2. Explain the roles of vitamins B_1, B_2, and B_3 in energy metabolism.
3. List the essential amino acids and fatty acids, and explain their significance.
4. In a single sentence, completely explain how a person can gain fat. Describe the effects of weight gain and loss on adipose cells.

Regulation of Energy Metabolism

Food and stored energy reserves in the body can provide circulating energy substrates for the tissue cells. The utilization and storage of energy is regulated by a variety of hormones.

Objective 4. List molecules that function as energy reserves and as circulating energy substrates, and explain why these terms are used to describe these molecules.

Objective 5. Identify the parts of the brain that appear to be involved in eating behavior, and describe how some chemicals affect hunger and satiety.

Objective 6. Describe the types of reactions that occur during anabolism and catabolism, and list the hormones that promote these processes.

The term **metabolism** refers to all of the chemical changes that occur within the cells of the body. On the basis of energy flow, these reactions comprise two categories: **anabolism** *(ah-nab'o-lizm)* and **catabolism** *(kah-tab'o-lizm)*. Anabolic reactions require the input of energy (obtained from the hydrolysis of ATP) and result in the formation of large, energy-rich molecules such as triglycerides, glycogen, and protein. Catabolism refers to the hydrolysis of these molecules into their subunits and to the use of these subunits in cell respiration for the generation of energy used to make ATP.

The molecules that can be oxidized for energy by the processes of cell respiration may be derived from the **energy reserves** of glycogen, fat, or protein in the body, or they can be derived from the products of digestion that are absorbed through the small intestine. Since these molecules—glucose, fatty acids, amino acids, and others—are carried by the blood to the tissue cells for use in cell respiration, they can be called **circulating energy substrates** (fig. 25.1).

Because of differences in cellular enzyme content, different organs have different *preferred energy sources*. The brain has an almost absolute requirement for blood glucose as its energy source, for example. A fall in the plasma concentration of glucose below about 50 mg per 100 ml can thus "starve" the brain and have disastrous consequences. Resting skeletal muscles, in contrast, use fatty acids as their preferred energy source. Similarly, ketone bodies, lactic acid, and amino acids can be used to different degrees as energy sources by various organs. The plasma normally contains adequate concentrations of all of these circulating energy substrates to meet the energy needs of the body.

Eating

Ideally, one should eat the kinds and amounts of foods that provide adequate vitamins, minerals, essential amino acids and fatty acids, and calories. Proper caloric intake is that which maintains energy reserves (primarily fat and glycogen) and results in a body weight within an optimum range for health.

There is a tendency for body weight to be stable despite short-term changes in caloric intake. It has thus been proposed that there may be some mechanism that is sensitive to the amount of body fat. Although this mechanism is not known, it is clear that there is a relationship between body fat and endocrine function. The secretion of anterior pituitary hormones is affected in a variety of ways. Obese women, for example, may experience menstrual cycle abnormalities and hirsutism (hairiness), whereas the sudden weight loss seen in women with *anorexia nervosa* may produce amenorrhea (cessation of the menstrual cycle). Abnormalities in growth hormone, ACTH, and prolactin secretion have also been observed in obese people.

Eating behavior appears to be at least partially controlled by areas of the hypothalamus. Lesions (destruction) in the ventromedial area of the hypothalamus produce *hyperphagia (hi''per-fa'je-ah)*, or overeating, and obesity in experimental animals. Lesions of the lateral hypothalamus, in contrast, produce *hypophagia* and weight loss.

The chemical neurotransmitters that may be involved in neural pathways mediating eating behavior are being investigated. There is evidence, for example, that endorphins may be involved, because injections of noloxone (a morphine-blocking drug) suppresses overeating in rats.

Figure 25.1. A schematic flow chart of energy pathways in the body.

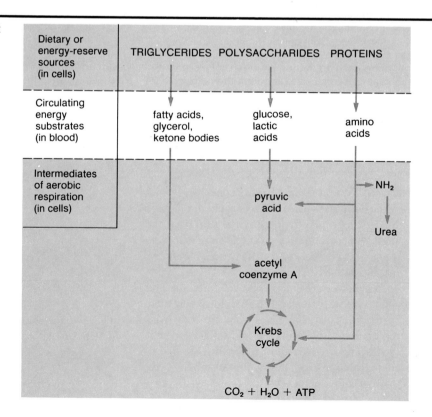

There is also evidence that the neurotransmitters norepinephrine and serotonin may be involved; injections of norepinephrine into the brain of rats cause overeating, whereas injections of serotonin have the opposite effect. Interestingly, the intestinal hormone cholecystokinin (CCK) also appears to function as a neurotransmitter in the brain, and it has been shown that injections of CCK cause experimental animals to stop eating. More research is needed to elucidate the structure and processes involved in the regulation of eating.

Hormonal Regulation of Metabolism

The absorption of energy carriers from the intestine is not continuous; it rises to high levels following meals and tapers toward zero between meals. Despite this, the plasma concentration of glucose and other energy substrates does not remain high during periods of absorption and does not normally fall below a certain level during periods of fasting. During the absorption of digestion products from the intestine, energy substrates are removed from the blood and deposited as energy reserves from which withdrawals can be made during times of fasting (fig. 25.2). This assures that there will be an adequate plasma concentration of energy substrates to sustain tissue metabolism at all times.

The rate of deposit and withdrawal of energy substrates into and from the energy reserves and the conversion of one type of energy substrate into another are regulated by the actions of hormones. The balance between anabolism and catabolism is determined by the antagonistic effects of hormones such as insulin, glucagon, growth hormone, thyroxine, and others (fig. 25.2). The specific metabolic effects of these hormones are summarized in table 25.4.

Figure 25.2. The balance of metabolism can be tilted toward anabolism (synthesis of energy reserves) or catabolism (utilization of energy reserves) by the combined actions of various hormones. Growth hormone and thyroxine have both anabolic and catabolic effects.

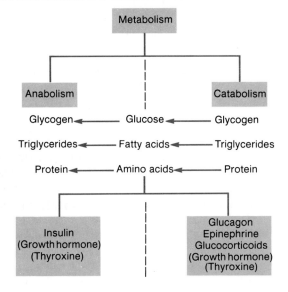

1. Define the terms energy reserves and circulating energy carriers, and give examples.
2. Describe the structures and neurotransmitters that may be involved in the regulation of eating.
3. Which hormones promote an increase in blood glucose? Which promote a decrease? List the hormones that stimulate fat synthesis (lipogenesis) and fat breakdown (lipolysis).

Energy Regulation by the Islets of Langerhans

The secretion of insulin is increased and that of glucagon is decreased during intestinal absorption of a meal. These two hormones have antagonistic effects on energy balance in the body.

Objective 7. Describe how the secretion of insulin and glucagon is affected by periods of absorption and fasting, and explain how the

Table 25.4 Summary of the endocrine regulation of metabolism

Hormone	Blood glucose	Carbohydrate metabolism	Protein metabolism	Lipid metabolism
Insulin	Decreased	↑Glycogen formation ↓Glycogenolysis ↓Gluconeogenesis	↑Amino acid transport	↑Lipogenesis ↓Lipolysis ↓Ketogenesis
Glucagon	Increased	↓Glycogen formation ↑Glycogenolysis ↑Gluconeogenesis	No direct effect	↑Lipolysis ↑Ketogenesis
Growth hormone	Increased	↑Glycogen formation ↑Gluconeogenesis ↓Glucose utilization	↑Protein synthesis	↓Lipogenesis ↑Lipolysis ↑Ketogenesis
Glucocorticoids	Increased	↑Glycogen formation ↑Gluconeogenesis	↓Protein synthesis	↓Lipogenesis ↑Lipolysis ↑Ketogenesis
Epinephrine	Increased	↓Glycogen formation ↑Glycogenolysis ↑Gluconeogenesis	No direct effect	↑Lipolysis ↑Ketogenesis
Thyroxine	No effect	↑Glucose utilization	↑Protein synthesis	No direct effect

secretion of these hormones is regulated.

Objective 8. Describe the effects of insulin and glucagon on the metabolism of glycogen, fat, and protein. Explain the physiological significance of the effects of these two islet hormones.

Scattered within a "sea" of pancreatic exocrine tissue (the acini) are islands of hormone-secreting cells. These **islets of Langerhans** contain three distinct cell types that secrete different hormones (fig. 25.3). The most numerous are the **beta cells;** these comprise 60% of each islet and are the cells that secrete the hormone **insulin.** The **alpha cells** comprise about 25% of each islet and secrete the hormone **glucagon.** The least numerous cell type, the **delta cells,** produce **somatostatin.** The latter hormone is identical to the somatostatin produced by the hypothalamus, which acts to inhibit growth-hormone secretion from the pituitary.

All three hormones are polypeptides. Insulin consists of two polypeptide chains—one twenty-one amino acids long and another that is thirty amino acids long—joined together by disulfide bonds. Glucagon is a twenty-one-amino-acid polypeptide, and somatostatin contains fourteen amino acids. Insulin was the first of these hormones to be discovered (in 1921). The importance of insulin in diabetes mellitus was immediately appreciated, and clinical use of insulin in the treatment of this disease began almost immediately after its discovery. The physiological role of glucagon was discovered later, and the importance of glucagon in the development of diabetes has only recently been suspected. The physiological significance of islet-secreted somatostatin is not currently known.

Regulation of Insulin and Glucagon Secretion

Insulin and glucagon secretion is largely regulated by the plasma concentrations of glucose and, to a lesser degree, of amino acids. The alpha and beta cells, therefore, act as both the sensors and effectors in this control system. Since the plasma concentration of glucose and amino acids rises during the absorption of a meal and falls during fasting, the secretion of insulin and glucagon likewise changes between periods of absorption and periods of fasting. These changes in insulin and glucagon secretion, in turn, cause changes in plasma glucose and amino acid concentrations and thus help to maintain homeostasis via negative feedback loops (fig. 25.4).

Effect of Glucose. During the absorption of a carbohydrate meal, the plasma glucose concentration rises. This rise in plasma glucose (1) stimulates the beta cells to secrete insulin and (2) inhibits the secretion of glucagon from the alpha cells. Insulin acts to *stimulate the cellular uptake of plasma glucose.* A rise in insulin secretion therefore lowers the plasma glucose concentration. Since glucagon has the antagonistic effect of raising the plasma glucose concentration, the inhibition of glucagon secretion complements the effect of increased insulin during the

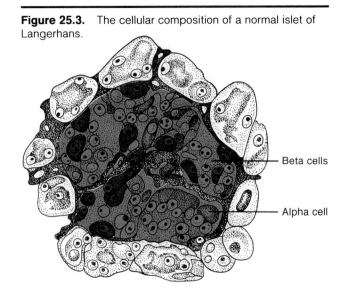

Figure 25.3. The cellular composition of a normal islet of Langerhans.

Beta cells

Alpha cell

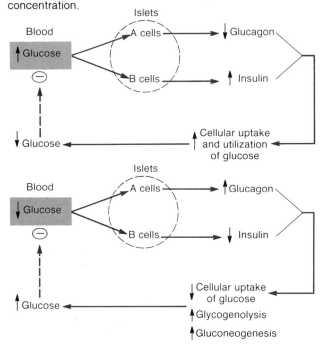

Figure 25.4. The secretion from the B (beta) cells and A (alpha) cells of the islets of Langerhans is regulated to a large degree by the blood glucose concentration. A high blood glucose concentration stimulates insulin and inhibits glucagon secretion. A low blood glucose concentration, conversely, stimulates glucagon and inhibits insulin secretion. The actions of insulin and glucagon provide negative feedback control of the blood glucose concentration.

absorption of a carbohydrate meal. A rise in insulin and a fall in glucagon secretion thus help to lower the high plasma glucose concentration that occurs during periods of absorption.

During fasting, the plasma glucose concentration falls. During fasting, therefore, (1) insulin secretion decreases and (2) glucagon secretion increases. These changes in hormone secretion prevent the cellular uptake of blood glucose into organs such as the muscles, liver, and adipose tissue and promote the release of glucose from the liver

Figure 25.5. Changes in blood glucose and plasma insulin concentrations after the ingestion of 100 grams of glucose in an oral glucose tolerance test.

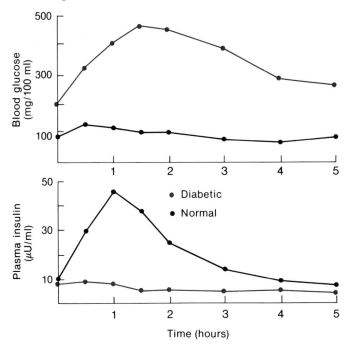

Table **25.5** Regulation of insulin and glucagon secretion		
Regulator	Effect on insulin secretion	Effect on glucagon secretion
Hyperglycemia	Stimulates	Inhibits
Hypoglycemia	Inhibits	Stimulates
Gastric inhibitory peptide	Stimulates	Stimulates (?)
Sympathetic nerve impulses	Inhibits	Stimulates
Parasympathetic nerve impulses	Stimulates	Inhibits
Amino acids	Stimulates	Stimulates
Somatostatin	Inhibits*	Inhibits*

*Inhibitory effects of somatostatin on insulin and glucagon secretion may not occur under normal conditions.
Source: Fox, Stuart Ira, *Human Physiology.* © 1984 Wm. C. Brown Publishers, Dubuque, Iowa. All Rights Reserved. Reprinted by permission.

(through the actions of glucagon). A negative feedback loop is therefore completed (fig. 25.4), which helps to retard the fall in plasma glucose concentration that occurs during fasting.

The ability of the beta cells to secrete insulin, as well as the ability of insulin to lower blood glucose, is measured clinically by the **oral glucose tolerance test** (fig. 25.5). In this procedure, a person drinks a glucose solution and blood samples are taken periodically for plasma glucose measurements. In a normal person the rise in blood glucose produced by drinking this solution is reversed to normal levels within two hours following glucose ingestion.

> Pople with **diabetes mellitus** (di''ah-be'tez mel'li-tus)—due to the inadequate secretion or action of insulin—maintain a state of high plasma glucose concentration (hyperglycemia) during the oral glucose tolerance test (fig. 25.5). People who have **reactive hypoglycemia** (low plasma glucose concentration due to excessive insulin secretion) have lower-than-normal blood glucose concentrations five hours following glucose ingestion.

Effects of Ingested Protein. Although the ingestion of carbohydrates causes a rise in insulin and a fall in glucagon secretion, the ingestion of proteins stimulates the secretion of both hormones. This rise in insulin secretion helps to complete a negative feedback loop because insulin promotes the uptake of amino acids into the tissues and the incorporation of these amino acids into proteins. The importance of a simultaneous rise in glucagon secretion can be understood in terms of the need to maintain a

constant blood glucose concentration. A rise in insulin secretion, if not accompanied by a simultaneous rise in glucagon secretion, would cause hypoglycemia in the absence of ingested carbohydrates. A rise in both insulin and glucagon secretions thus promotes lowering of the amino acid concentration of the blood, while the blood glucose concentration remains relatively constant due to the antagonistic effects of insulin and glucagon.

Since most meals contain both carbohydrates and proteins, insulin secretion is normally always raised following meals, but the effects on glucagon secretion are variable. In an average meal that is rich in carbohydrates, the suppressive effects of high plasma glucose are more potent than the stimulatory effects of amino acids on glucagon secretion. In general, therefore, the insulin secretion is increased and the glucagon secretion is decreased during the period of absorption.

Effects of Autonomic Nerves. The islets of Langerhans receive both parasympathetic and sympathetic innervation. The activation of the parasympathetic system during meals stimulates insulin secretion at the same time that gastrointestinal function is stimulated. The activation of the sympathetic system, in contrast, stimulates glucagon secretion and inhibits insulin secretion. The effects of glucagon, together with those of epinephrine, produce a "stress hyperglycemia" when the sympathoadrenal system is activated.

Effects of Gastric Inhibitory Peptide (GIP). Surprisingly, insulin secretion increases more rapidly following glucose ingestion than it does following an intravenous injection of glucose. This is due to the fact that the intestine, in response to glucose ingestion, secretes a hormone that stimulates insulin secretion before the glucose is absorbed. Insulin secretion thus begins to rise in anticipation of a rise in blood glucose. The intestinal hormone that mediates this effect is believed to be gastric inhibitory peptide (GIP).

Table 25.5 summarizes the effects of various factors that regulate insulin and glucagon secretion.

The mechanisms that regulate insulin and glucagon secretion and the actions of these hormones normally prevent the plasma glucose concentration from rising above 170 mg per 100 ml after a meal or from falling below about 50 mg per 100 ml between meals. This regulation is important because abnormally high blood glucose can damage tissue cells (as may occur in diabetes mellitus), and abnormally low blood glucose can damage the brain. The later effect results from the fact that glucose enters the brain by facilitated diffusion; when the rate of this diffusion is too low, due to low plasma glucose concentrations, the supply of metabolic energy for the brain may become inadequate. This can result in weakness, dizziness, personality changes, and ultimately in coma and death.

Insulin and Glucagon: Period of Absorption

The lowering of plasma glucose by insulin is, in a sense, a side effect of the primary action of this hormone. Insulin is the major hormone that promotes anabolism in the body. During absorption of the products of digestion and the subsequent rise in the plasma concentrations of circulating energy substrates, insulin promotes the cellular uptake of plasma glucose and its incorporation into glycogen and fat. As previously described, insulin also promotes the cellular uptake of amino acids and their incorporation into proteins. The stores of large, energy reserve molecules are thus increased while the plasma concentrations of glucose and amino acids are decreased.

The synthesis of triglycerides (fat) within adipose cells depends upon insulin-stimulated glucose uptake from plasma. Once inside the adipose cells, glucose can be converted into α-glycerol phosphate and acetyl coenzyme A (acetyl CoA). The formation of fatty acids from acetyl CoA and condensation of three fatty acids with glycerol yields triglycerides. Although the adipose cells can also utilize free fatty acids from the blood, they cannot incorporate glycerol from the blood into triglycerides. This is because adipose cells lack the enzymes needed to convert glycerol to α-glycerol phosphate, which is the precursor needed in triglyceride synthesis (fig. 25.6). Entry of blood glucose into adipose cells—which is directly dependent on insulin—thus determines the rate at which fat is produced.

A nonobese 70 kg man has approximately 10 kg of stored fat. Since 250 g of fat can supply the energy requirements for one day, this reserve fuel is sufficient for about forty days. Glycogen is less efficient as an energy reserve, and less is stored in the body; there is about 100 g of glycogen stored in the liver and about 200 g in skeletal muscles. Insulin promotes the cellular uptake of glucose into the liver and muscles and the conversion of glucose into glucose–6-phosphate. In the liver and muscles, this can be changed into glucose–1-phosphate, which is used as the precursor of glycogen. Once the stores of glycogen are filled, the continued ingestion of excess calories results in the continued production of fat rather than of glycogen.

Figure 25.6.   The synthesis of triglycerides (fat) within adipose cells. Notice that fat can be produced from glucose and from the fatty acids released by the hydrolysis of plasma lipoproteins.

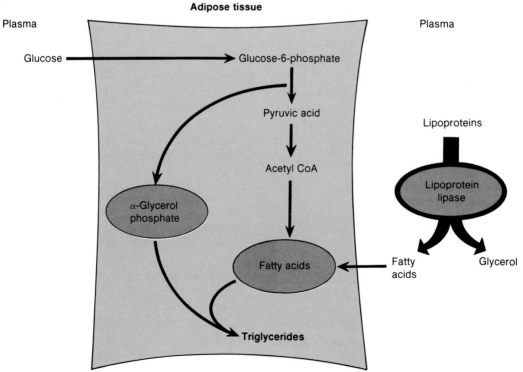

Insulin and Glucagon: Period of Fasting

Glucagon stimulates and insulin suppresses the hydrolysis of liver glycogen, or **glycogenolysis** *(gli″ko-jĕ-nol′ĭ-sis).* During times of fasting, when glucagon secretion is high and insulin secretion is low, therefore, liver glycogen is used as a source of additional blood glucose. This process is essentially the reverse of that which formed glycogen and results in the liberation of free glucose from glucose–6-phosphate by the action of an enzyme called *glucose–6-phosphatase.* Only the liver has this enzyme and therefore only the liver can use its stored glycogen as a source of additional blood glucose. Since muscles lack glucose–6-phosphatase, the glucose–6-phosphate produced from muscle glycogen can only be used for glycolysis by the muscle cells themselves.

Since there are only about 100 g of stored glycogen in the liver, adequate blood glucose levels could not be maintained for very long during fasting using this source alone. The low levels of insulin secretion during fasting, together with elevated glucagon secretion, however, promote **gluconeogenesis:** the formation of glucose from non-carbohydrate molecules. Low insulin allows the release of amino acids from skeletal muscles, while glucagon stimulates the production of enzymes in the liver that convert amino acids to pyruvic acid and pyruvic acid to glucose.

During prolonged fasting and exercise, gluconeogenesis in the liver using amino acids from muscles may be the only source of blood glucose.

The secretion of glucose from the liver during fasting thus compensates for the low plasma glucose concentrations and helps to provide the glucose needed by the brain. In the presence of low insulin secretion, however, other organs cannot utilize blood glucose as an energy source. This helps to spare glucose for the brain, but alternative energy sources are needed for these other organs. The major alternative energy sources that are used by the liver and skeletal muscles are free fatty acids and ketone bodies.

Glucagon, in the presence of low insulin levels, stimulates in adipose cells an enzyme called hormone-sensitive lipase. This enzyme catalyzes the hydrolysis of stored triglycerides and the release of free fatty acids and glycerol into blood. Glucagon, in the presence of low insulin levels, also stimulates enzymes in the liver that convert some of these fatty acids into ketone bodies. These derivatives of fatty acids—including acetoacetic acid, β-hydroxybutyric acid, and acetone—can also be released into the blood (fig. 25.7). Several organs in the body can use ketone bodies, like fatty acids, as a source of acetyl CoA in aerobic respiration.

Through the stimulation of **lipolysis** *(lĭ-pol′ĭ-sis),* the breakdown of fat, and **ketogenesis,** the formation of ketone bodies, the high glucagon and low insulin levels that

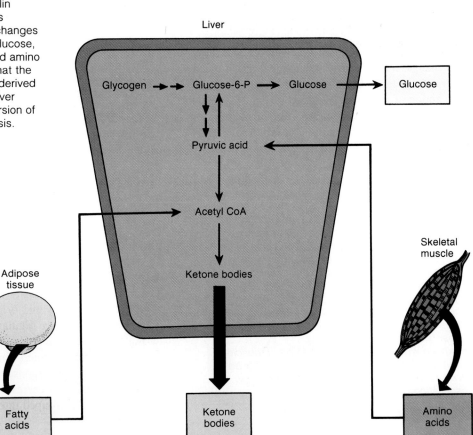

Figure 25.7. Increased glucagon secretion and decreased insulin secretion during fasting favors catabolism. These hormonal changes result in elevated release of glucose, fatty acids, ketone bodies, and amino acids into the blood. Notice that the liver secretes glucose that is derived both from the breakdown of liver glycogen and from the conversion of amino acids in gluconeogenesis.

Figure 25.8. The inverse relationship between insulin and glucagon secretion during the absorption of a meal and during fasting. Changes in the insulin : glucagon ratio tilts metabolism toward anabolism during the absorption of food and toward catabolism during fasting.

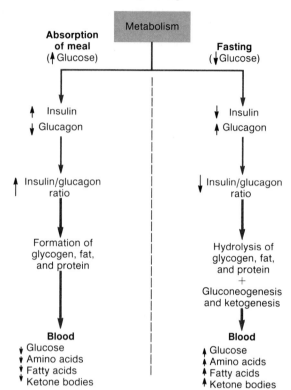

Table **25.6** Comparison of the metabolic effects of insulin and glucagon		
Actions	**Insulin**	**Glucagon**
Cellular glucose transport	↑	0
Glycogen synthesis	↑	↓
Glycogenolysis in liver	↓	↑
Gluconeogenesis	↓	↑
Amino acid uptake; protein synthesis	↑	0
Inhibition of amino acid release; protein degradation	↓	0
Lipogenesis	↑	0
Lipolysis	↓	↑
Ketogenesis	↓	↑

Code: ↑ = Increased
 ↓ = Decreased
 0 = No direct effect
Source: Fox, Stuart Ira, *Human Physiology*. © 1984 Wm. C. Brown Publishers, Dubuque, Iowa. All Rights Reserved. Reprinted by permission.

are found during fasting provide circulating energy substrates for use by the muscles, liver, and other organs. Through liver glycogenolysis and gluconeogenesis, these hormonal changes help to provide adequate levels of blood glucose to sustain the metabolism of the brain. The antagonistic action of insulin and glucagon (fig. 25.8) thus promotes appropriate metabolic responses during periods of fasting and periods of absorption. The actions of insulin and glucagon are summarized in table 25.6.

1. *Describe how the secretions of insulin and glucagon change during periods of absorption and periods of fasting. Explain how these changes in hormone secretion are produced.*
2. *Describe how the synthesis of fat in adipose cells is regulated by insulin. Explain how fat metabolism is regulated by insulin and glucagon during periods of absorption and fasting.*
3. *Define the following terms: glycogenolysis, gluconeogenesis, and ketogenesis. Explain how insulin and glucagon affect the processes these terms describe during periods of absorption and fasting.*
4. *Explain why the liver, but not skeletal muscles, can secrete glucose into the blood. Describe two pathways that contribute to the hepatic secretion of glucose.*

Metabolic Regulation by the Adrenal Hormones, Thyroxine, and Growth Hormone

Epinephrine, glucocorticoids, thyroxine, and growth hormone stimulate the catabolism of carbohydrates and lipids. In addition, thyroxine and growth hormone stimulate protein synthesis and are needed for proper body growth and development.

Objective 9. Describe the metabolic effects of epinephrine and the glucocorticoids.

Objective 10. Describe the metabolic effects of thyroxine, and explain how these effects may be produced.

Objective 11. Describe the metabolic effects of growth hormone, and explain the role of somatomedins in these effects.

The anabolic effects of insulin are antagonized by glucagon, as previously described, and by the actions of a variety of other hormones. The hormones of the adrenals, thyroid, and anterior pituitary (specifically growth hormone) antagonize the action of insulin on carbohydrate and lipid metabolism. The actions of insulin, thyroxine, and growth hormone, however, can act synergistically in the stimulation of protein synthesis.

Adrenal Hormones

The adrenal gland consists of two different parts that have different embryonic origins, that secrete different hormones, and that are regulated by different control systems. The **adrenal** *(ah-dre'nal)* **medulla** secretes catecholamine hormones—epinephrine and lesser amounts of norepinephrine—in response to sympathetic nerve stimulation. The **adrenal cortex** secretes corticosteroid hormones. These are grouped into two functional categories:

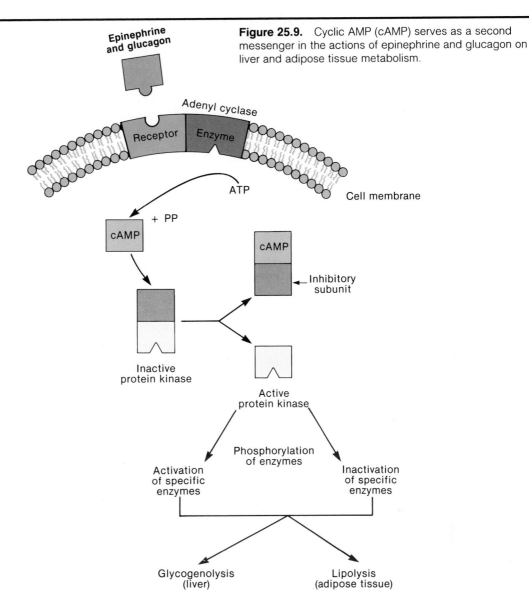

Figure 25.9. Cyclic AMP (cAMP) serves as a second messenger in the actions of epinephrine and glucagon on liver and adipose tissue metabolism.

mineralocorticoids *(min″er-al-o-kor′ti-koid),* such as aldosterone, which regulate Na+ and K+ balance and **glucocorticoids** *(gloo″ko-kor′ti-koids),* such as hydrocortisone (cortisol), which participate in metabolic regulation. The secretion of the glucocorticoids is stimulated by the anterior pituitary hormone ACTH.

Metabolic Effects of Epinephrine. The metabolic effects of epinephrine are similar to those of glucagon. Both stimulate glycogenolysis and the release of glucose from the liver and lipolysis and the release of fatty acids from adipose tissue. These actions occur in response to glucagon during fasting, when low blood glucose stimulates glucagon secretion, and in response to epinephrine during the "fight or flight" reaction to stress. The latter effect provides circulating energy substrates in anticipation of the need for intense physical activity. Glucagon and epinephrine have similar mechanisms of action; the actions of both are mediated by cyclic AMP (figure 25.9).

Metabolic Effects of Glucocorticoids. Hydrocortisone and other glucocorticoids are secreted by the adrenal cortex in response to ACTH stimulation. The secretion of ACTH from the anterior pituitary occurs as part of the general adaptation syndrome in response to stress. Since prolonged fasting or prolonged exercise certainly qualify as stressors, ACTH—and thus glucocorticoid secretion—is stimulated under these conditions. The increased secretion of glucocorticoids during prolonged fasting or exercise supports the effects of increased glucagon and decreased insulin secretion from the islets.

Like glucagon, hydrocortisone promotes lipolysis and ketogenesis; it also stimulates the synthesis of hepatic enzymes that promote gluconeogenesis. Although it stimulates enzyme (protein) synthesis in liver, hydrocortisone promotes protein breakdown in the muscles. This latter effect increases the blood levels of amino acids and thus provides the substrates needed by the liver for gluconeogenesis. The release of circulating energy substrates—amino acids, glucose, fatty acids, and ketone bodies—into

Figure 25.10. The catabolic actions of glucocorticoids help raise the blood concentration of glucose and other energy-carrier molecules.

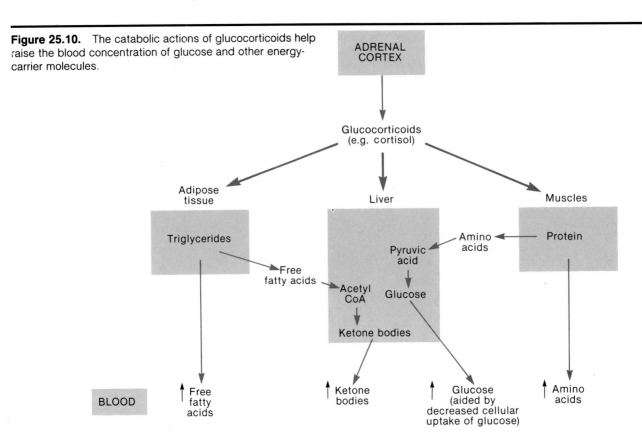

the blood in response to hydrocortisone (fig. 25.10) helps to compensate for a state of prolonged fasting or exercise. Whether these metabolic responses are beneficial in other stressful states is open to question.

Thyroxine

The thyroid follicles secrete thyroxine also called tetraiodothyronine (T_4), in response to stimulation by TSH from the anterior pituitary. Almost all organs in the body are targets of thyroxine action. Thyroxine *(thi-rok'sin)* itself, however, is not the active form of the hormone within the target cells; thyroxine is a prehormone that must first be converted to triiodothyronine (T_3) within the target cells to be active. Acting via its conversion to T_3, thyroxine (1) regulates the rate of cell respiration and (2) contributes to proper growth and development, particularly during early childhood.

Thyroxine and Cell Respiration. Thyroxine stimulates the rate of cell respiration in almost all cells in the body. This is believed to be due to thyroxine-induced lowering of cellular ATP concentrations. ATP exerts an end-product inhibition of cell respiration so that when ATP concentrations increase, the rate of cell respiration decreases. Conversely, a lowering of ATP concentrations, as may occur in response to thyroxine, stimulates cell respiration.

The mechanisms of thyroxine action are not entirely understood. One effect of thyroxine, however—the stimulation of Na^+/K^+ pump activity in cell membranes—could account for the thyroxine-induced lowering of cell

ATP concentrations. The active transport of Na^+ and K^+ represents a significant energy "sink" in the cell, accounting for about 12% of the calories consumed at rest. Through stimulation of Na^+/K^+ pumps, therefore, thyroxine could significantly decrease ATP concentrations and thus stimulate the rate of cell respiration (fig. 25.11).

When a person wakes up in the morning the total energy consumption of the body is at its lowest or basal level, known as the **basal metabolic rate (BMR)**. Most commonly measured by the rate of oxygen consumption, the BMR indicates the "idling speed" of the body. Since the activity of the Na^+/K^+ pumps contributes significantly to the energy consumed in the basal state and since the activity of these pumps is set by thyroxine secretion, the BMR can be used as an index of thyroid function. Indeed, such measurements were used clinically to evaluate thyroid function prior to the development of direct chemical determinations of T_4 and T_3 in the blood.

The coupling of energy-releasing reactions to energy-requiring reactions is never 100% efficient; a proportion of the energy is always lost as heat. Much of the energy liberated during cell respiration and much of the energy released by the hydrolysis of ATP escapes as heat. Since thyroxine stimulates both ATP consumption and cell respiration, the actions of thyroxine result in the production of metabolic heat.

The heat-producing, or *calorigenic (kah-lor"i-jen'ik)* (calor = heat), *effects* of thyroxine are required for cold adaptation. This does not mean that people who are cold-adapted have high levels of thyroxine secretion. Rather,

Figure 25.11. A mechanism proposed to explain the effects of thyroid hormones on basal metabolic rate. Through the activation of genes and the stimulation of protein synthesis, thyroid hormones increase the activity of the Na^+/K^+ pump. This active transport carrier accounts for a large percentage of the energy expenditures in the cell. The concentration of ATP, therefore, declines as a result of this increased energy usage, and the decreased ATP concentrations stimulate increased cellular respiration.

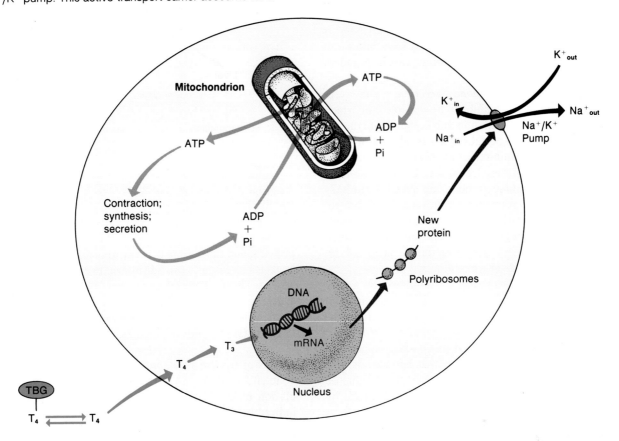

thyroxine levels in the normal range coupled with the increased activity of the sympathoadrenal system are responsible for cold adaptation. Thyroxine exerts a permissive effect on the ability of the sympathoadrenal system to increase heat production in response to cold stress.

Thyroxine in Growth and Development. Through its stimulation of cell respiration, thyroxine stimulates the increased consumption of circulating energy substrates such as glucose, fatty acids, and other molecules. These effects, however, are mediated at least in part by the activation of genes; thyroxine thus stimulates both RNA and protein synthesis. As a result of its stimulation of protein synthesis throughout the body, thyroxine is considered to be an anabolic hormone like insulin and growth hormone.

Because of its stimulation of protein synthesis, thyroxine is needed for growth of the skeleton and, most importantly, for the proper development of the central nervous system. This latter effect is particularly significant during prenatal development and the first two years after birth. Hypothyroidism during this time may result in **cretinism** *(kre'tin-izm)* (fig. 25.12). Unlike dwarfs, who have normal thyroxine secretion but a low secretion of growth hormone, cretins suffer from severe mental retardation.

Figure 25.12. Cretinism is a disease of infancy caused by an underactive thyroid gland.

Growth Hormone

The anterior pituitary secretes growth hormone, also called somatotropic hormone, in larger amounts than any other of its hormones. As its name implies, growth hormone stimulates growth in children and adolescents. The continued high secretion of growth hormone in adults, particularly under the conditions of fasting and other forms of stress, implies that this hormone can have important metabolic effects even after the growing years have ended.

Regulation of Growth Hormone Secretion. The secretion of growth hormone is inhibited by somatostatin, which is produced by the hypothalamus and secreted into the hypothalamo-hypophyseal portal system. In addition, a newly discovered hypothalamic-releasing hormone appears to stimulate growth hormone secretion. Growth hormone thus appears to be unique among the anterior pituitary hormones in that its secretion is controlled by both a releasing and an inhibiting hormone from the hypothalamus. The secretion of growth hormone follows a circadian ("about a day") pattern, increasing during sleep and decreasing during periods of wakefulness.

Growth hormone secretion is stimulated by an increase in the plasma concentrations of amino acids and by a decrease in the plasma glucose concentration. The secretion of growth hormone is, therefore, stimulated during absorption of a high protein meal, when amino acids are absorbed. The secretion of growth hormone is also stimulated during prolonged fasting when plasma glucose is low and plasma amino acid concentration is raised by the breakdown of muscle protein.

Effects of Growth Hormone on Metabolism. The fact that growth hormone secretion is increased during fasting and also during absorption of a protein meal reflects the complex nature of this hormone's action. Growth hormone has both anabolic and catabolic effects; it promotes protein synthesis (anabolism) and it also stimulates the catabolism of fat and release of fatty acids from adipose tissue. Growth hormone is similar to insulin in its former effect and similar to glucagon in its latter effect.

In terms of its action on lipid and carbohydrate metabolism, growth hormone is said to have an anti-insulin effect. A rise in the plasma fatty acid concentration induced by growth hormone results in decreased rates of glycolysis in many organs. This inhibition of glycolysis by fatty acids, perhaps together with a more direct action of growth hormone, results in decreased glucose utilization by the tissues. As a consequence, experimental animals that have had their pituitary removed (and thus lack growth hormone) experience hypoglycemia and are more sensitive to the effects of insulin than intact animals.

Growth hormone stimulates the cellular uptake of amino acids and protein synthesis in many organs of the body. These actions are useful during a protein-rich meal; amino acids are removed from the blood and used to form

Figure 25.13. The effects of growth hormone. The growth promoting, or anabolic, effects of growth hormone are mediated indirectly via stimulation of somatomedin production by the liver.

proteins, and the plasma concentration of glucose and fatty acids is increased to provide alternate energy sources (fig. 25.13). The anabolic effect of growth hormone on protein synthesis is particularly important during the growing years, when it contributes to increases in bone length and in the mass of many soft tissues.

Effects of Growth Hormone on Body Growth. The anabolic effects of growth hormone and their resultant stimulation of skeletal growth are believed to be produced indirectly. This was demonstrated by the observation that growth hormone by itself may not stimulate bone growth *in vitro* (outside the body); the plasma from growth hormone-stimulated animals must be present. This is because the growth-promoting effects are stimulated indirectly by polypeptides called **somatomedins** *(so"mah-to-me'din)*, which are produced by the liver and secreted into the blood in response to growth hormone stimulation. The somatomedins are therefore responsible for, or mediate, the stimulation of skeletal growth induced by growth hormone (fig. 25.14). Skeletal growth is produced by the stimulation of chondrocytes in the epiphyseal disc cartilages of bones in growing children and adolescents. This skeletal growth stops when the epiphyseal discs are converted to bone after the growth spurt during puberty, despite the fact that growth hormone secretion remains as high in adults as in children.

An adequate diet, particularly of proteins, is required for the production of somatomedins. This helps to explain the common observation that many children are significantly taller than their immigrant parents, who may not have had an adequate diet in their youth. Children with protein malnutrition **(kwashiorkor** *[kwash-e-or'kor]*) have low growth rates and low somatomedin levels, despite the fact that their growth hormone levels are abnormally elevated. When these children eat an adequate diet, somatomedin levels and growth rates increase.

Figure 25.14.   The hepatic production of polypeptides called *somatomedins* is stimulated by growth hormone. These compounds in turn produce the anabolic effects that are characteristic of the actions of growth hormone in the body.

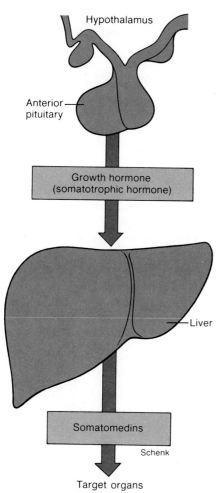

Figure 25.15.   (*a*) the resorption of bone by osteoclasts and (*b*) the formation of new bone by osteoblasts. Both resorption and deposition (formation) occur simultaneously throughout the body.

1. *Describe the effects of epinephrine and of the glucocorticoids on the metabolism of carbohydrates and lipids, and explain the significance of these effects as a response to stress.*
2. *Explain the actions of thyroxine on the basal metabolic rate, and explain why people who are hypothyroid have a tendency to gain weight and are less resistant to cold stress.*
3. *Describe the effects of growth hormone on the metabolism of lipids, glucose, and amino acids.*
4. *Explain how growth hormone stimulates skeletal growth.*

Objective 12. Describe the relationships between bone and plasma calcium levels and the importance of maintaining normal blood calcium concentrations.

Objective 13. Describe how parathyroid hormone and calcitonin secretions are regulated and how these hormones regulate the plasma calcium and phosphate concentrations.

Objective 14. Explain how vitamin D functions as a prehormone, and describe its effects on calcium and phosphate balance.

Regulation of Calcium and Phosphate Balance

The calcium and phosphate concentrations of plasma are affected by bone formation and resorption, intestinal absorption, and urinary excretion of these ions. These processes are regulated by parathyroid hormone, 1,25-dihydroxy vitamin D_3, and calcitonin.

The skeleton, in addition to providing support for the body, serves as a large store of calcium and phosphate in the form of *hydroxyapatite (hi-drok″se-ap′ah-tīt)* crystals. The calcium phosphate in these hydroxyapatite crystals is derived from the blood by the action of bone-forming cells, or **osteoblasts.** Bone resorption (dissolution), produced by the action of **osteoclasts** (fig. 25.15), results in the return of bone calcium and phosphate to the blood.

The formation and resorption of bone occur constantly at rates determined by the hormonal balance. Body growth during the first two decades of life occurs because bone formation proceeds at a faster rate than bone resorption. By age fifty or sixty, the rate of bone resorption often exceeds the rate of bone formation. The constant activity of osteoblasts and osteoclasts allows bone to be remodeled throughout life. The position of the teeth, for example, can be changed by orthodontic appliances (braces), which cause bone resorption on the pressure-bearing side and bone formation on the opposite side of the alveolar sockets.

The plasma concentrations of calcium and phosphate are maintained, despite the changing rates of bone formation and resorption, by hormonal control of the intestinal absorption and urinary excretion of these ions. These hormonal control mechanisms are very effective at maintaining the plasma calcium and phosphate concentrations within narrow limits. Plasma calcium, for example, is normally maintained at about 2.5 millimolar, or 5 milliequivalents per liter (a milliequivalent equals a millimole times the valence of the ion, in this case, times two).

The maintenance of normal plasma calcium concentrations is important because of the wide variety of effects that calcium has in the body. In addition to its role in bone

formation, excitation-contraction coupling in muscles, and as a second messenger in the action of some hormones, calcium is needed to maintain proper membrane permeability. An abnormally low plasma calcium concentration increases the permeability of the cell membranes to Na^+ and other ions. Hypocalcemia, therefore, enhances the excitability of nerves and muscles and can result in muscle spasm (tetany).

Parathyroid Hormone and 1,25-Dihydroxy Vitamin D₃

Parathyroid hormone, an eighty-four-amino-acid polypeptide secreted by the parathyroid glands, acts to help raise the plasma concentration of calcium. The effects of parathyroid hormone include (1) the stimulation of bone resorption; (2) the stimulation of calcium reabsorption in the kidneys, a process that returns calcium to the blood, so that less calcium is excreted in the urine; and (3) the inhibition of renal phosphate reabsorption, so that more phosphate is excreted in the urine. Parathyroid hormone also promotes the formation of 1,25-dihydroxy vitamin D₃, and so it indirectly helps to raise plasma calcium levels via the effects of this other hormone.

Figure 25.16. The pathway for the production of the hormone 1,25-dihydroxy vitamin D₃. This hormone is produced in the kidneys from the inactive precursor, 25-hydroxy vitamin D₃ (formed in the liver). This latter molecule is produced from vitamin D₃ secreted by the skin.

The production of **1,25-dihydroxy vitamin D₃** begins in the skin, where vitamin D₃ is produced from its precursor molecule (7-dehydrocholesterol) under the influence of sunlight. When the skin does not make sufficient vitamin D₃ because of insufficient exposure to sunlight, this compound must be ingested in the diet—that is why it is called a vitamin. Whether this compound is secreted into the blood from the skin or enters the blood after being absorbed from the intestine, vitamin D₃ functions as a *prehormone,* which must be chemically changed in order to be biologically active.

An enzyme in the liver adds a hydroxyl group (OH) to carbon 25, which converts vitamin D₃ into 25-hydroxy vitamin D₃. In order to be active, however, another hydroxyl group must be added to the first carbon. Hydroxylation of the first carbon is accomplished by an enzyme in the kidneys, which converts the molecule to 1,25-dihydroxy vitamin D₃ (fig. 25.16). The activity of this enzyme in the kidneys is stimulated by parathyroid hormone (fig. 25.17).

The hormone 1,25-dihydroxy vitamin D₃ helps to raise the plasma concentrations of calcium and phosphate by stimulating (1) the intestinal absorption of calcium and phosphate, (2) the resorption of bones, and (3) the renal reabsorption of phosphate so that less is excreted in the urine. Notice that 1,25-dihydroxy vitamin D₃, but not parathyroid hormone, directly stimulates intestinal absorption of calcium and phosphate and promotes the reabsorption of phosphate in the kidneys.

Negative Feedback Control of Calcium and Phosphate Balance

The secretion of parathyroid hormone is controlled by the plasma calcium concentrations. Its secretion is stimulated by low-calcium and inhibited by high-calcium concentrations. Since parathyroid hormone stimulates the final hydroxylation step in the formation of 1,25-dihydroxy vitamin D₃, a rise in parathyroid hormone results in an increase in production of 1,25-dihydroxy vitamin D₃. Low blood calcium can thus be corrected by the effects of increased parathyroid hormone and 1,25-dihydroxy vitamin D₃ (fig. 25.18).

Figure 25.17. A decrease in plasma Ca⁺⁺ directly stimulates the secretion of parathyroid hormone (*PTH*). The production of 1,25-dihydroxy vitamin D₃ also rises when Ca⁺⁺ is low because PTH stimulates the final hydroxylation step in the formation of this compound in the kidneys.

Figure 25.18. The negative feedback loop that returns low blood Ca⁺⁺ concentrations to normal without simultaneously raising blood phosphate levels above normal.

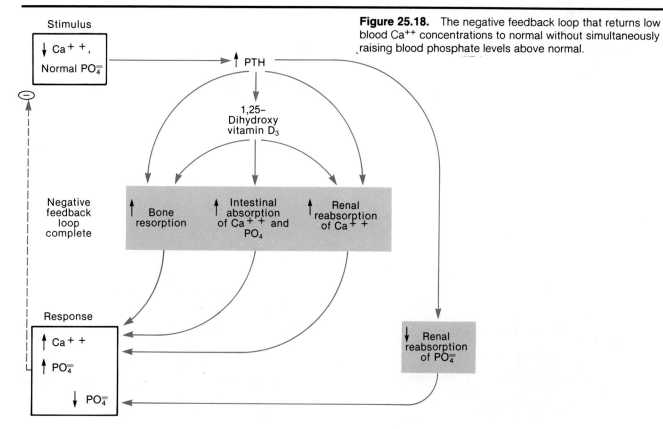

It is possible that plasma calcium levels might fall while phosphate levels remain normal. In this case, the increased secretion of parathyroid hormone and the production of 1,25-dihydroxy vitamin D_3 that result could abnormally raise phosphate levels while acting to restore normal calcium levels. This is prevented by the fact that parathyroid hormone inhibits renal reabsorption in the kidneys, so that more phosphate is excreted in the urine (fig. 25.18). In this way blood calcium levels can be raised to normal without excessively raising blood phosphate concentrations.

Calcitonin

Experiments in the 1960s revealed that high blood calcium in dogs may be lowered by a hormone secreted from the thyroid gland. This hormone thus has an effect opposite to that of parathyroid hormone and 1,25-dihydroxy vitamin D_3. The calcium-lowering hormone, called **calcitonin** *(kal''si-to'nin)*, was found to be a thirty-two-amino-acid polypeptide secreted by *parafollicular cells,* or *C cells,* in the thyroid, which are distinct from the follicular cells that secreted thyroxine.

The secretion of calcitonin is stimulated by high plasma calcium levels and acts to lower calcium levels by (1) inhibiting the activity of osteoclasts, thus reducing bone resorption and (2) stimulating the urinary excretion of calcium and phosphate by inhibiting their reabsorption in the kidneys (fig. 25.19).

Although it is attractive to think that calcium balance is regulated by the effects of antagonistic hormones, the significance of calcitonin in human physiology remains unclear. Patients with their thyroid gland surgically removed (as for thyroid cancer) are *not* hypercalcemic, as would have been expected if calcitonin were needed to lower blood calcium levels. Similarly, patients who receive injections of many times the normal amounts of calcitonin do not become hypocalcemic. The ability of very large, pharmacological doses of calcitonin to inhibit osteoclast activity and bone resorption, however, is clinically useful in the treatment of *Paget's disease,* in which osteoclast activity causes softening of bone.

Figure 25.19. Negative feedback control of calcitonin secretion.

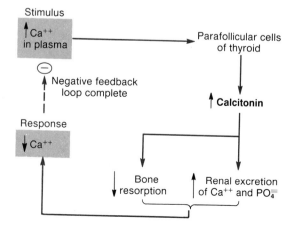

There are a variety of bone disorders associated with abnormal calcium and phosphate balance. In **rickets,** bone demineralization and softening occur as a result of inadequate vitamin D. Excessive parathyroid hormone secretion, resulting in excessive demineralization of bone and high levels of blood calcium, produces **osteitis fibrosa cystica.** The most common of these bone disorders is **osteoporosis** *(os''te-o-po-ro'sis),* in which decreased bone mass per volume results in fractures in response to ordinary bone stress. This condition occurs in both sexes but advances more rapidly in women after menopause. The causes of osteoporosis are not known, but it may be associated with age-related decreases in the intestinal absorption of calcium. The treatment of this condition includes estrogen therapy (for postmenopausal women), increased dietary calcium, and sometimes small amounts of vitamin D or its derivatives.

1. Describe how the secretion of parathyroid hormone and of calcitonin is regulated.
2. List the steps involved in the formation of 1,25-dihydroxy vitamin D_3, and indicate how this formation is influenced by parathyroid hormone.
3. Describe the actions of parathyroid hormone, 1,25-dihydroxy vitamin D_3, and calcitonin on the intestine, skeletal system, and kidneys, and explain how these actions affect the blood levels of calcium.

Clinical Considerations

Diabetes Mellitus

Chronic high blood glucose, or hyperglycemia, is the hallmark of the disease **diabetes mellitus.** The name of this disease is derived from the fact that glucose "spills over" into the urine when the blood glucose concentration is too

Table 25.7 Comparison of juvenile-onset and maturity-onset diabetes mellitus

Characteristics	Juvenile-onset (Type I)	Maturity-onset (Type II)
Usual age at onset	Under 20 years	Over 40 years
Development of symptoms	Rapid	Slow
Percent of diabetic	About 10%	About 90%
Development of ketoacidosis	Common	Rare
Association with obesity	Rare	Common
Beta cells of islets	Destroyed	Usually not destroyed
Insulin secretion	Decreased	Normal or increased
Autoantibodies to islet cells	Present	Absent
Associated with particular HLA antigens	Yes	No
Usual treatment	Insulin injections	Diet; oral stimulators of insulin secretion

Source: Fox, Stuart Ira, *Human Physiology.* © 1984 Wm. C. Brown Publishers, Dubuque, Iowa. All Rights Reserved. Reprinted by permission.

high. The hyperglycemia of diabetes mellitus results from either the insufficient secretion of insulin by the beta cells of the islets of Langerhans or the inability of secreted insulin to stimulate the cellular uptake of glucose from the blood. Diabetes mellitus, in short, results from the inadequate secretion or action of insulin.

There are two forms of diabetes mellitus. In **type I,** or **insulin-dependent** diabetes, the beta cells are destroyed and secrete little or no insulin. This form of the disease accounts for only about 10% of the cases of diabetes in the country. About 90% of the people who have diabetes have **type II,** or **non-insulin-dependent** diabetes mellitus. Type I diabetes is also called *juvenile-onset diabetes,* because this condition is usually diagnosed in people under the age of thirty. Type II, or *maturity-onset diabetes,* is usually diagnosed in people over the age of thirty. Some comparisons of these two forms of diabetes mellitus are shown in table 25.7.

Type I Diabetes Mellitus.   Type I diabetes mellitus results when the beta cells of the islets of Langerhans are destroyed by the effects of a virus or other environmental agent or by the action of autoantibodies. Removal of the insulin-secreting cells in this way causes hyperglycemia

and the appearance of glucose in the urine. Without insulin, glucose cannot enter the adipose cells; the rate of fat synthesis thus lags behind the rate of fat breakdown and large amounts of free fatty acids are released from the adipose cells.

In a person with uncontrolled type I diabetes, many of the fatty acids released from adipose cells are converted into ketone bodies in the liver. This may result in an elevated ketone body concentration in the blood (ketosis), and if the buffer reserve of bicarbonate is neutralized, it may also result in *ketoacidosis.* During this time the glucose and excess ketone bodies that are excreted in the urine act as osmotic diuretics, and cause the excessive excretion of water in the urine. This can produce severe dehydration, which together with ketoacidosis and associated disturbances in electrolyte balance, may lead to coma and death (fig. 25.20).

In addition to the lack of insulin, people with type I diabetes have an abnormally high secretion of glucagon from the alpha cells of the islets. Glucagon stimulates glycogenolysis in the liver and thus helps to raise the blood glucose concentration. Glucagon also stimulates the production of enzymes in the liver that convert fatty acids to ketone bodies. Some researchers believe that the full

Figure 25.20.   The sequence of events by which an insulin deficiency may lead to coma and death.

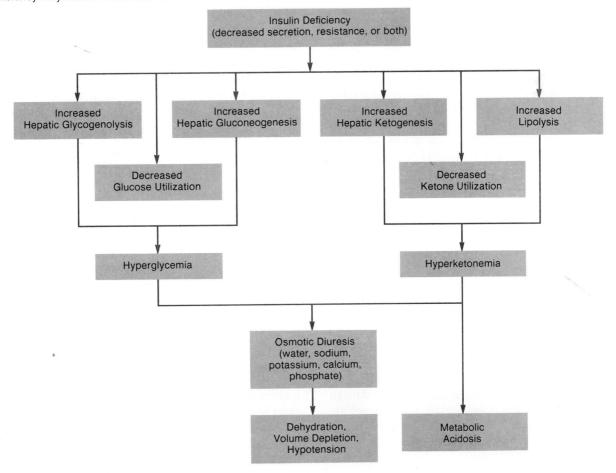

Table 25.8 Comparison of coma due to diabetic ketoacidosis and to hypoglycemia

	Diabetic ketoacidosis	Hypoglycemia
Onset	Hours to days	Minutes
Causes	Insufficient insulin; other diseases	Excess insulin; insufficient food; excessive exercise
Symptoms	Excessive urination and thirst; headache, nausea, and vomiting	Hunger, headache, confusion, stupor
Physical Findings	Deep, labored breathing; breath has acetone odor; blood pressure decreased, pulse weak; skin is dry	Pulse, blood pressure, and respiration are normal; skin is pale and moist
Laboratory Findings	Urine: glucose present, ketone bodies increased Plasma: glucose and ketone bodies increased, bicarbonate decreased	Urine: no glucose; ketone bodies at normal concentration Plasma: glucose concentration low, bicarbonate normal

Source: Fox, Stuart Ira, *Human Physiology.* © 1984 Wm. C. Brown Publishers, Dubuque, Iowa. All Rights Reserved. Reprinted by permission.

Figure 25.21. An idealized oral glucose tolerance test in a person with reactive hypoglycemia. The blood glucose concentration falls below the normal range within five hours of glucose ingestion as a result of excessive insulin secretion.

symptoms of diabetes result from high glucagon secretion as well as from the absence of insulin. The lack of insulin may be largely responsible for hyperglycemia and for the release of large amounts of fatty acids into the blood. The high glucagon secretion may contribute to the hyperglycemia and be largely responsible for the development of ketoacidosis.

Insulin Resistance: Obesity and Type II Diabetes. The effects produced by insulin, or any hormone, depend on the concentration of that hormone in the blood and on the sensitivity of the target tissue to given amounts of the hormone. Tissue responsiveness to insulin, for example, varies under normal conditions. For reasons that are incompletely understood, exercise increases insulin sensitivity and obesity decreases insulin sensitivity of the target tissues. The islets of a nondiabetic obese person, therefore, must secrete high amounts of insulin to maintain the blood glucose concentration in the normal range.

Type II diabetes is usually slow to develop, is hereditary, and occurs most often in people who are overweight. Unlike type I diabetes mellitus, people who have type II diabetes have normal or even elevated levels of insulin in their blood. Despite this, people with type II diabetes have hyperglycemia if untreated. This must mean that they have a decreased tissue sensitivity to the effects of insulin.

Since obesity decreases insulin sensitivity, people who are genetically predisposed to insulin resistance may develop symptoms of diabetes when they gain weight. Conversely, this type of diabetes mellitus can usually be controlled by increasing tissue sensitivity to insulin through diet and exercise. If this is not sufficient, oral drugs are available that increase insulin secretion and also stimulate tissue responsiveness to insulin.

People with type II diabetes don't usually develop ketoacidosis. The hyperglycemia itself, however, can be dangerous on a long-term basis. Diabetes is the second leading cause of blindness in the United States, and people with diabetes frequently have circulatory problems that increase the tendency to get gangrene and increase the risk of atherosclerosis. The causes of damage to the retina and lens of the eyes and to blood vessels are not well understood. It is believed, however, that these problems may result from a long-term exposure to high blood glucose.

Hypoglycemia

People with type I diabetes mellitus are dependent upon insulin injections to prevent hyperglycemia and ketoacidosis. If inadequate insulin is injected, the person may enter a coma as a result of the ketoacidosis, electrolyte imbalance, and dehydration that develop. An overdose of insulin, however, can also produce a coma as a result of the hypoglycemia (abnormally low blood glucose levels) produced. The physical signs and symptoms of diabetic and hypoglycemic coma are sufficiently different (table 25.8) to allow hospital personnel to distinguish between these two types.

Less severe symptoms of hypoglycemia are usually produced by an oversecretion of insulin from the islets of Langerhans after a carbohydrate meal. This **reactive hypoglycemia** is caused by an exaggerated response of the beta cells to a rise in blood glucose and is most commonly seen in adults who are genetically predisposed to type II diabetes. People with reactive hypoglycemia must, therefore, limit their intake of carbohydrates and eat small meals at frequent intervals, rather than two or three meals per day.

The symptoms of reactive hypoglycemia include tremor, hunger, weakness, hypoglycemia, blurred vision, and impaired mental ability. The appearance of some of these symptoms, however, does not necessarily indicate reactive hypoglycemia and a given level of blood glucose does not always produce these symptoms. For these reasons a number of tests must be performed, including the

Table **25.9** Comparison of hypothyroidism and hyperthyroidism		
	Hypothyroid	**Hyperthyroid**
Growth and Development	Impaired growth	Accelerated growth
Activity and Sleep	Decreased activity; increased sleep	Increased activity; decreased sleep
Temperature Tolerance	Intolerance to cold	Intolerance to heat
Skin Characteristics	Coarse, dry skin	Smooth skin
Perspiration	Absent	Excessive
Pulse	Slow	Rapid
Gastrointestinal Symptoms	Constipation; decreased appetite; increased weight	Frequent bowel movements; increased appetite; decreased weight
Reflexes	Slow	Rapid
Psychological Aspects	Depression and apathy	Nervous, "emotional"
Plasma T_4 Levels	Decreased	Increased

Source: Fox, Stuart Ira, *Human Physiology.* © 1984 Wm. C. Brown Publishers, Dubuque, Iowa. All Rights Reserved. Reprinted by permission.

oral glucose tolerance test, to confirm the diagnosis of reactive hypoglycemia. In the glucose tolerance test, reactive hypoglycemia is shown when the initial rise in blood glucose produced by the ingestion of a glucose solution triggers excessive insulin secretion, so that the blood glucose levels fall below normal within five hours (fig. 25.21).

Hypothyroidism and Hyperthyroidism

As might be predicted from the effects of thyroxine, people who are hypothyroid have an abnormally low basal metabolic rate (BMR) and experience weight gain and lethargy. There is also a decreased ability to adapt to cold stress when there is a thyroxine deficiency. Another symptom of hypothyroidism is **myxedema** *(mik″sĕ-de′mah)*—accumulation of mucoproteins in subcutaneous connective tissues. Hypothyroidism can be produced by a variety of causes, including insufficient TRH secretion from the hypothalamus, insufficient TSH secretion from the pituitary, or insufficient iodine in the diet. Hypothyroidism due to lack of iodine is accompanied by excessive TSH secretion, which stimulates growth of the thyroid and the production of a goiter.

A goiter can also be produced by another mechanism. In **Graves's disease,** autoantibodies are produced that have TSH-like effects on the thyroid. Since the production of these autoantibodies is not controlled by negative feedback, this results in the excessive stimulation of the thyroid. A goiter is thus produced that is associated with a hyperthyroid state. Hyperthyroidism produces a high BMR accompanied by weight loss, nervousness, irritability, and an intolerance to heat. The symptoms of hypothyroidism and hyperthyroidism are compared in table 25.9.

Grave's disease: from Robert James Graves, Irish physician, 1796–1853

Figure 25.22. The progression of acromegaly in one individual, from age nine (a), sixteen (b), thirty-three (c), and fifty-two (d) years. The coarsening of features and disfigurement are evident by age thirty-three and severe at age fifty-two.

(a)   (b)   (c)   (d)

Gigantism, Acromegaly, and Dwarfism

An excessive secretion of growth hormone in children can produce **gigantism** *(ji-gan′tizm)* in which people may grow up to eight feet tall. An excessive growth hormone secretion that occurs after the epiphyseal discs have sealed, however, cannot produce increases in height. The oversecretion of growth hormone in adults results in an elongation of the jaw and deformities in the bones of the face, hands, and feet. This condition, called **acromegaly** *(ak″ro-meg′ah-le),* is accompanied by the growth of soft tissues and coarsening of the skin (fig. 25.22).

An inadequate secretion of growth hormone during the growing years results in **dwarfism.** An interesting variant of this is *Laron dwarfism,* in which there is a genetic insensitivity to the effects of growth hormone. It is also believed that the short stature of African pygmies may be due to a genetically low sensitivity to the effects of growth hormone.

acromegaly: Gk. *akron,* extremity; *melos,* limb

Chapter Summary

I. Nutritional Requirements
Vitamins, minerals, essential amino acids, and essential fatty acids are required in the diet for metabolism. The energy needs of the body are met by the caloric value of food.
   A. Vitamins and minerals serve as cofactors and coenzymes.
      1. Vitamins are divided into those that are fat-soluble (A, D, E, and K) and those that are water-soluble.
      2. Many water-soluble vitamins are needed for the activity of the enzymes involved in cell respiration.
   B. The caloric intake must be sufficient to meet the energy expenditures of the body; an excessive caloric intake results in obesity.

II. Regulation of Energy Metabolism
Food and stored energy reserves can provide circulating energy substrates for the tissue cells. The utilization and storage of energy is regulated by a variety of hormones.
   A. The body tissues can use circulating energy substrates, including glucose, fatty acids, ketone bodies, lactic acid, amino acid, and others, for cell respiration.
      1. Different organs have different preferred energy sources.
      2. Circulating energy substrates can be obtained from food or from the energy reserves of glycogen, fat, and protein in the body.
   B. Eating behavior is regulated, at least in part, by the hypothalamus.
      1. Lesions of the ventromedial area of the hypothalamus produce hyperphagia, whereas lesions of the lateral hypothalamus produce hypophagia.
      2. A variety of neurotransmitters have been implicated in the control of eating behavior; these include the endorphins, norepinephrine, serotonin, and cholecystokinin.
   C. The control of energy balance in the body is regulated by the anabolic and catabolic effects of a variety of hormones.

III. Energy Regulation by the Islets of Langerhans
The secretion of insulin is increased, and that of glucagon is decreased, during the intestinal absorption of a meal. These two hormones have antagonistic effects in the body.
   A. A rise in plasma glucose concentration stimulates insulin and inhibits glucagon secretion.
      1. Amino acids stimulate the secretion of both insulin and glucagon.
      2. Insulin secretion is also stimulated by parasympathetic innervation of the islets and by

the action of gastric inhibitory peptide (GIP) secreted by the intestine.
   B. During the intestinal absorption of a meal, insulin promotes the uptake of blood glucose into tissue cells.
      1. This lowers the blood glucose concentration and increases the energy reserves of glycogen, fat, and protein.
      2. Insulin is required for the production of fat by adipose cells.
   C. During periods of fasting, insulin secretion decreases and glucagon secretion increases.
      1. Glucagon stimulates glycogenolysis in the liver, gluconeogenesis, lipolysis, and ketogenesis.
      2. These effects help maintain adequate levels of blood glucose for the brain and provide alternate energy sources for other organs.

IV. Metabolic Regulation by the Adrenal Hormones, Thyroxine, and Growth Hormone
The adrenal hormones, thyroxine, and growth hormone stimulate the catabolism of carbohydrates and lipids. In addition, thyroxine and growth hormone stimulate protein synthesis and are needed for proper body growth and development.
   A. The adrenal hormones involved in energy regulation include epinephrine from the adrenal medulla and the glucocorticoids (mainly hydrocortisone) from the adrenal cortex.
      1. The effects of epinephrine are similar to those of glucagon.
      2. Glucocorticoids promote the breakdown of muscle protein and the conversion of amino acids to glucose in the liver.
   B. Thyroxine stimulates the rate of cell respiration in almost all cells in the body.
      1. Thyroxine thus sets the basal metabolic rate (BMR), which is the rate at which energy is consumed by the body under resting conditions.
      2. Thyroxine also promotes protein synthesis and by this means is needed for proper body growth and development, particularly of the central nervous system.
   C. The secretion of growth hormone is regulated by releasing and inhibiting hormones from the hypothalamus.
      1. The secretion of growth hormone is stimulated by a protein meal and by a fall in glucose, as occurs during fasting.
      2. Growth hormone stimulates catabolism of lipids and inhibits glucose utilization.
      3. Growth hormone also stimulates protein synthesis and thus promotes body growth.

      4. The anabolic effects of growth hormone, including the stimulation of bone growth in childhood, is believed to be produced indirectly via polypeptides called somatomedins.

V. Regulation of Calcium and Phosphate Balance
The calcium and phosphate concentrations in the blood are affected by the actions of parathyroid hormone, 1,25-dihydroxy vitamin D_3, and calcitonin on bone, intestinal absorption, and urinary excretion of these ions.
   A. Bone contains calcium and phosphate in the form of hydroxyapatite crystals; this serves as a reserve supply of calcium and phosphate for the blood.
      1. The formation and resorption of bone is produced by the action of osteoblasts and osteoclasts, respectively.
      2. The plasma concentrations of calcium and phosphate are also affected by absorption from the intestine and by the urinary excretion of these ions.
   B. Parathyroid hormone stimulates bone resorption and calcium reabsorption in the kidneys; this hormone thus acts to raise the blood calcium concentration.
      1. The secretion of parathyroid hormone is stimulated by a fall in blood calcium levels.
      2. Parathyroid hormone also inhibits reabsorption of phosphate in the kidneys, so that more phosphate is excreted in the urine.
   C. 1,25-dihydroxy vitamin D_3 is derived from vitamin D by hydroxylation reactions in the liver and kidneys.
      1. The last hydroxylation step is stimulated by parathyroid hormone.
      2. 1,25-dihydroxy vitamin D_3 stimulates the intestinal absorption of calcium and phosphate, resorption of bone, and renal reabsorption of phosphate.
   D. A rise in parathyroid hormone, accompanied by the increased production of 1,25-dihydroxy vitamin D_3, helps to maintain proper blood levels of calcium and phosphate in response to a fall in calcium levels.
   E. Calcitonin is secreted by the parafollicular cells of the thyroid gland.
      1. Calcitonin secretion is stimulated by a rise in blood calcium levels.
      2. Calcitonin, at least at pharmacological levels, acts to lower blood calcium by inhibiting bone resorption and stimulating the urinary excretion of calcium and phosphate.

Review Activities

Objective Questions:

Match the following:

1. Absorption of carbohydrate meal
2. Absorption of protein meal
3. Fasting

    (a) rise in insulin, rise in glucagon
    (b) fall in insulin, rise in glucagon
    (c) rise in insulin, fall in glucagon
    (d) fall in insulin, fall in glucagon

Match the following:

4. Growth hormone
5. Thyroxine
6. Hydrocortisone

    (a) increased protein synthesis, increased cell respiration
    (b) protein catabolism in muscles; gluconeogenesis in liver
    (c) protein synthesis in muscles, decreased glucose utilization
    (d) fall in blood glucose, increased fat synthesis

7. A lowering of blood glucose concentration promotes
    (a) decreased lipogenesis
    (b) increased lipolysis
    (c) glycogenolysis
    (d) all of the above

8. Glucose can be secreted into the blood by the
    (a) liver
    (b) muscles
    (c) liver and muscles
    (d) liver, muscles, and brain

9. The basal metabolic rate is determined primarily by
    (a) hydrocortisone
    (b) insulin
    (c) growth hormone
    (d) thyroxine

10. Somatomedins are required for the anabolic effects of
    (a) hydrocortisone
    (b) insulin
    (c) growth hormone
    (d) thyroxine

11. The increased intestinal absorption of calcium is stimulated directly by
    (a) parathyroid hormone
    (b) 1,25-dihydroxy vitamin D_3
    (c) calcitonin
    (d) all of the above

12. A rise in blood calcium levels directly stimulates
    (a) parathyroid hormone secretion
    (b) calcitonin secretion
    (c) 1,25-dihydroxy vitamin D_3 formation
    (d) all of the above

Essay Questions

1. Compare the metabolic effects of fasting to the state of uncontrolled type I diabetes. Explain the hormonal similarities of these conditions.
2. Glucocorticoids stimulate the breakdown of protein in muscles but the synthesis of protein in the liver. Explain the significance of these differences.
3. Describe how thyroxine affects cell respiration, and explain why a hypothyroid person has a tendency to gain weight and has a reduced tolerance to cold.
4. Compare and contrast the metabolic effects of thyroxine and growth hormone.
5. Why is vitamin D considered to be both a vitamin and a prehormone? Explain why people with osteoporosis might be helped by taking controlled amounts of vitamin D.

Selected Readings

Austin, L. A., and H. Heath, III. 1981. Calcitonin: Physiology and pathophysiology. *New Engl. J. Med.* 304:269.

Barret, E. J., and R. A. DeFronzo. 1984. Diabetic ketoacidosis: Diagnosis and treatment. *Hosp. Prac.* 19:89.

Cahill, G. F., and H. O. McDevitt. 1981. Insulin-dependent diabetes mellitus: The initial lesion. *New Engl. J. Med.* 304:454.

DeLuca, H. F. 1980. The vitamin D hormonal system: Implications for bone disease. *Hosp. Prac.* 15:57.

Gardner, L. I. July 1972. Deprivation dwarfism. *Scientific American.*

Goodman, D. S. 1984. Vitamin A and retinoids in health and disease. *New Engl. J. Med.* 310:1023.

Habener, J. F., and J. E. Mahaffey. 1978. Osteomalacia and disorders of vitamin D metabolism. *Ann. Rev. Med.* 29:327.

Haussler, M. R., and T. A. McCain. 1977. Basic and clinical concepts related to vitamin D metabolism and action. *New Engl. J. Med.* 297: first part, p. 974; second part, p. 1041.

Hirsch, J. 1984. Hypothalamic control of appetite. *Hosp. Prac.* 19:131.

Notkins, A. L. November 1979. The cause of diabetes. *Scientific American.*

Oppenheimer, J. H. 1979. Thyroid hormone action at the cellular level. *Science* 203:971.

Phillips, L. S., and R. Vassilopoulou-Sellin. 1980. Somatomedins. *New Engl. J. Med.* 302: first part, p. 371; second part, p. 438.

Raisz, L. G., and B. E. Kream. 1981. Hormonal control of skeletal growth. *Ann. Rev. Physiol.* 43:225.

Sterling, S. 1979. Thyroid hormone action at the cellular level. *New Engl. J. Med.* 300: first part, p. 117; second part, p. 173.

Tepperman, J. 1980. *Metabolic and Endocrine Physiology.* 4th ed. Chicago: Year Book Medical Publishers.

Unger, R. H., R. E. Dobbs, and L. Orci. 1979. Insulin, glucagon, and somatostatin secretion in the regulation of metabolism. *Ann. Rev. Physiol.* 40:307.

Unger, R. H., and L. Orci. 1981. Glucagon and the A cell: Physiology and pathophysiology. *New Engl. J. Med.* 304: first part, p. 1518; second part, p. 1575.

Van Wyk, J., and L. E. Underwood. 1978. Growth hormone, somatomedins, and growth failure. *Hosp. Prac.* 13:57.

Wynder, E. L., and D. P. Rose. 1984. Diet and breast cancer. *Hosp. Prac.* 19:73.

26

Urinary System

Concepts

Humans have metanephric kidneys,
   which develop after pronephric and
   mesonephric kidneys degenerate
   during embryonic development.

A blood filtrate is produced, which
   enters the nephron tubules in the
   cortex of the kidney. This filtrate
   becomes urine as it is passed through
   different tubular segments and is
   drained into the calyces.

Large volumes of a plasma filtrate low
   in proteins are produced by the
   glomeruli. The glomerular filtration
   rate is controlled by the sympathetic
   nervous system and by
   autoregulatory mechanisms.

Most of the filtered salt and water is
   reabsorbed in the proximal tubule.
   Reabsorption of most of the
   remaining water occurs as a result of
   the presence of a hypertonic renal
   medulla and the action of ADH.

Compounds in the blood are cleared by
   the kidneys through filtration and
   secretion into the tubules.
   Measurements of renal plasma
   clearance rates are of clinical
   importance.

Sodium reabsorption and K^+ or H^+
   secretion in the distal convoluted
   tubules are regulated by aldosterone.
   The secretion of aldosterone is
   regulated by blood potassium and by
   the secretion of renin from the
   juxtaglomerular apparatus.

The kidneys are responsible for the
   metabolic component of acid-base
   regulation through their excretion of
   H^+ and their reabsorption of HCO_3^-.

Urine is channeled from the kidneys to
   the urinary bladder by the ureters
   and expelled from the body through
   the urethra. Muscles of the urinary
   bladder and urethra are used in the
   control of micturition.

Development of the Urinary System

Humans have metanephric kidneys, which develop after pronephric and mesonephric kidneys degenerate during embryonic development.

Objective 1. List the parts of the urinary system.
Objective 2. Describe the development of the pronephric, mesonephric, and metanephric kidneys.

The urinary system consists of two kidneys, two ureters, the urinary bladder, and the urethra (fig. 26.1). The urinary and reproductive systems both originate from a specialized elevation of mesodermal tissue called the **urogenital** *(u''ro-jen'i-tal)* **ridge.** The two systems share common structures during a portion of development, but by the time of birth two separate systems have formed. The separation in the male is not totally complete, however, since the urethra serves to transport both urine and semen. The development of both systems is initiated during the embryonic stage, but the development of the urinary system starts and finishes sooner than the reproductive system.

Three successive types of kidneys develop in the human embryo: the pronephros, mesonephros, and metanephros (fig. 26.2). The third type, or metanephric kidney, remains as the permanent kidney.

Figure 26.1. The anatomical locations of the kidneys, ureters, and urinary bladder.

Figure 26.2. The sequential embryonic and fetal development of the kidney. During the fifth week (a), all three kidney types are represented. Note that the metanephros forms, in part, from the mesonephric duct (b) and then migrates from the pelvic (c) to the abdominal region (d). Sexual differences in development are indicated in (e) and (f).

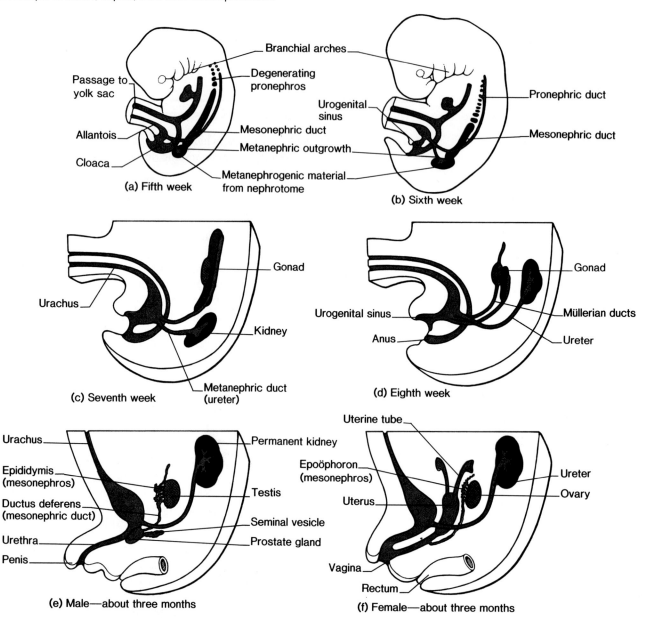

(a) Fifth week

Passage to yolk sac
Allantois
Cloaca
Branchial arches
Degenerating pronephros
Urogenital sinus
Mesonephric duct
Metanephric outgrowth
Metanephrogenic material from nephrotome

(b) Sixth week

Pronephric duct
Mesonephric duct

(c) Seventh week

Urachus
Gonad
Kidney
Metanephric duct (ureter)

(d) Eighth week

Urogenital sinus
Anus
Gonad
Müllerian ducts
Ureter

(e) Male—about three months

Urachus
Epididymis (mesonephros)
Ductus deferens (mesonephric duct)
Urethra
Penis
Permanent kidney
Testis
Seminal vesicle
Prostate gland

(f) Female—about three months

Uterine tube
Epoöphoron (mesonephros)
Uterus
Vagina
Rectum
Ureter
Ovary

The **pronephros** *(pro-nef′ros)* develops during the fourth week after conception and persists only through the sixth week. It is the most superior in position of the three kidneys and is connected to the embryonic **cloaca** by the **pronephric duct.** Although the pronephros is nonfunctional and degenerates in humans, most of its duct is used by the mesonephric kidney (fig. 26.2), and a portion of it is important in the formation of the metanephros.

pronephros: Gk. *pro,* before; *nephros,* kidney
cloaca: L. *cloaca,* sewer

The **mesonephros** *(mes″o-nef′ros)* develops toward the end of the fourth week as the pronephros becomes vestigial. The mesonephros forms from an intermediate portion of the urogenital ridge and functions throughout the embryonic period of development.

Although the **metanephros** *(met″ah-nef′ros)* begins its formation during the fifth week, it does not become functional until immediately before the start of the fetal stage of development at the end of the eighth week. The paired metanephric kidneys continue to form urine throughout fetal development. The urine is expelled through the urinary system into the amniotic fluid.

Figure 26.3. The development of the metanephric kidney from the ureteric bud and metanephrogenic mass. At the same time the metanephros is forming, the urinary bladder is developing from the urogenital sinus (see fig. 16.2).

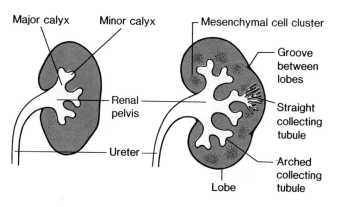

The metanephros develops from two mesodermal sources (fig. 26.3). The meaty part of the kidney forms from a specialized caudal portion of the urogenital ridge called the **metanephrogenic mass.** The tubular drainage portion of the kidneys forms as a diverticulum emerges from the wall of the mesonephric duct near the cloaca. This diverticulum, known as the **ureteric** *(u"rĕ-ter'ik)* **bud,** expands into the metanephrogenic mass to form the drainage pathway for urine. The stalk of the ureteric bud develops into the ureter, whereas the expanded terminal portion forms the renal pelvis, calyces, and collecting tubules. A combination of the ureteric bud and metanephrogenic mass forms the other tubular channels within the kidney.

Once the metanephric kidneys are formed, they begin to migrate from the pelvis to the upper, posterior portion of the abdomen. The renal blood supply develops as the kidneys become positioned in the posterior body wall.

The development of the kidneys illustrates the concept that ontogeny (embryonic development) recapitulates phylogeny (evolution). This means that the development of the three kidney types follows an evolutionary pattern. The larvae of a few of the "lower" vertebrates have functional pronephric kidneys. Adult fishes and amphibians have mesonephric kidneys, whereas adult reptiles, birds, and mammals have metanephric kidneys.

The urinary bladder develops from the urogenital sinus, which is connected to the embryonic allantois through the umbilical cord (fig. 26.2). By the twelfth week, the two ureters are emptying into the urinary bladder, the urethra is draining, and the connection of the urinary bladder to the allantois has been reduced to a supporting structure called the **urachus** *(u'rah-kus).*

Occasionally a **patent urachus** is present in a newborn and is discovered when urine is passed through the umbilicus, especially if there is a urethral obstruction. Usually, however, it remains undiscovered until old age, when an enlarged prostate obstructs the urethra and forces urine through the patent urachus and out the umbilicus (belly button).

1. *Which type of kidney is functional during the embryonic stage of development? When do the metanephric kidneys form and become functional?*
2. *Explain how kidney development illustrates the concept that ontogeny recapitulates phylogeny.*

Kidney Structure

A blood filtrate is produced, which enters the nephron tubules in the cortex of the kidney. This filtrate becomes urine as it is passed through different tubular segments in the cortex and medulla and is drained into the calyces.

Objective 3. Describe the gross structure of the kidney.

Objective 4. Describe the tubular and vascular components of a nephron.

Objective 5. Locate the position of cortical and juxtamedullary nephrons in the gross structure of the kidney.

The reddish brown kidneys lie on each side of the vertebral column high in the abdominal cavity from the level of the twelfth thoracic to the third lumbar vertebra (fig. 26.4). The right kidney is usually 1.5–2.0 cm lower than the left because of the large area occupied by the liver.

The kidneys are retroperitoneal, which means that they are positioned behind the peritoneum and are thus not, in a strict sense, within the peritoneal cavity. Each adult kidney is an anteroposteriorly compressed, lima bean-shaped organ, which measures about 11.25 cm (4 in.) long, 5.5–7.7 cm (2–3 in.) wide, and 2.5 cm (1 in.) thick.

The lateral surface of each kidney is convex, whereas the medial surface is strongly concave (fig. 26.5). Toward the center of the medial surface, within the kidney, is a deep depression called the **renal sinus.** The entrance to the

urachus: Gk. *ouron,* urine; *echein,* to hold

Figure 26.4. An X-ray of the calyces and renal pelvises of the kidneys, the ureters, and the urinary bladder. Notice the position of the kidneys relative to the vertebral column and ribs.

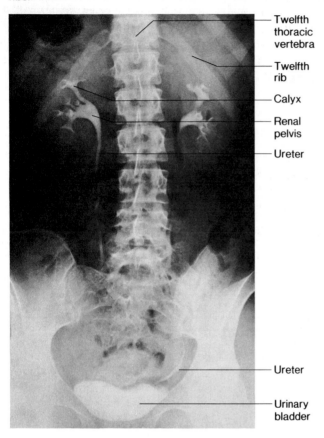

- Twelfth thoracic vertebra
- Twelfth rib
- Calyx
- Renal pelvis
- Ureter
- Ureter
- Urinary bladder

sinus is called the **hilum,** through which the **renal artery** enters and the **renal vein** and **ureter** *(u-re′ter)* exit. Innervation to the kidney is also at the hilum.

Each kidney is embedded in a fatty, fibrous pouch consisting of three layers. The **renal capsule** is the innermost layer that forms a strong, transparent fibrous attachment to the kidney. The renal capsule protects the kidney from trauma and the spread of infections. The second layer is formed by a firm, protective layer of adipose tissue called the **adipose capsule.** The outermost layer, called the **renal fascia,** is a supportive layer that anchors the kidney to the peritoneum and the abdominal wall.

Although the kidneys are firmly supported by the adipose capsule, renal fascia, and even the renal vessels, under certain conditions these structures may give in to the force of gravity and the kidneys may drop a bit in position. This condition is called **renal ptosis** *(to′sis)* and generally occurs in elderly persons who are extremely thin and have insufficient amounts of supportive fat in the adipose capsular layer. It may also affect victims of anorexia nervosa who suffer from extreme weight loss. The potential danger of renal ptosis is that the ureter may kink, blocking the flow of urine from the affected kidney.

Gross Structure

A coronal section of the kidney shows two distinct regions and a major cavity (fig. 26.6). The outer **cortex,** in contact with the capsule, is reddish brown and granular in appearance because of its many capillaries. The deeper

Figure 26.5. An anterior view of the kidneys, adrenal glands, and associated vessels. The right kidney is longitudinally dissected to show the internal structures.

- Cortex
- Renal pyramid
- Medulla
- Calyx
- Renal column
- Papilla
- Renal pelvis
- Adrenal gland
- Renal capsule
- Renal artery
- Hilum
- Renal vein
- Ureter

Figure 26.6. The internal structures of a kidney. (*a*) a coronal section showing the structure of the cortex, medulla, and renal sinus. (*b*) a diagrammatic magnification of a renal pyramid and cortex to depict the tubules. (*c*) a diagrammatic view of a single nephron.

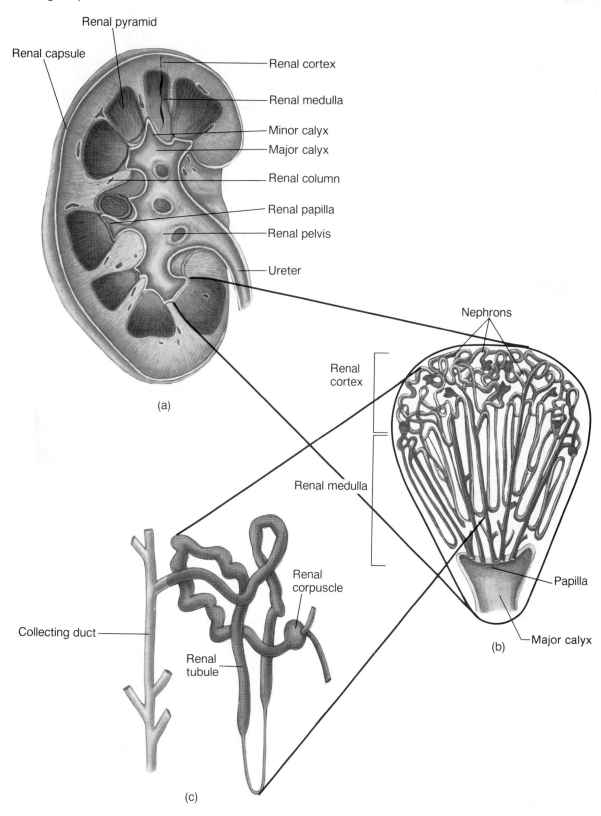

Figure 26.7. The vascular structure of the kidneys. (*a*) an illustration of the major arterial supply and (*b*) a scanning electron micrograph of the glomeruli.

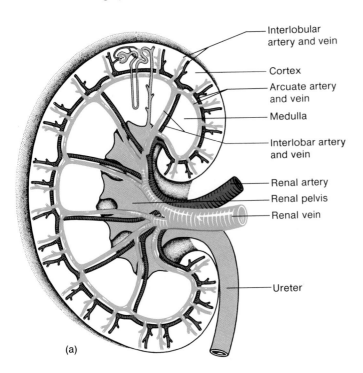

Interlobular artery and vein
Cortex
Arcuate artery and vein
Medulla
Interlobar artery and vein
Renal artery
Renal pelvis
Renal vein
Ureter

(a)

(b)

Kessel and Kardon

region, or **medulla,** is lighter in color and striped in appearance because of the presence of microscopic tubules and blood vessels. The medulla is composed of eight to fifteen conical **renal pyramids** separated by **renal columns.** The apexes of the renal pyramids are all directed into the renal sinus and are known as the renal papillae.

The cavity of the kidney collects and transports urine from the kidney to the ureter. It is divided into several portions. Each papilla of a pyramid projects into a small depression called the **minor calyx** *(ka'liks)*—the plural form is *calyces*. Several minor calyces unite to form a **major calyx.** In turn, the major calyces join to form the funnel-shaped **renal pelvis** within the renal sinus. The renal pelvis is actually an expanded portion of the ureter in the kidney and serves to collect urine from the calyces and transport it to the ureter.

Microscopic Structure

The **nephron** *(nef'ron)* is the functional unit of the kidney that is responsible for the formation of urine. Each kidney contains more than a million nephrons. A nephron consists of **urinary tubules** and associated small blood vessels. Fluid formed by capillary filtration enters the tubules and is subsequently modified by transport processes; the resulting fluid that leaves the tubules is urine.

Renal Blood Vessels. Arterial blood enters the kidney at the hilum through the **renal artery,** which divides into **interlobar** *(in''ter-lo'bar)* **arteries** (fig. 26.7), which pass between the pyramids through the renal columns. **Arcuate** *(ar'ku-āt)* **arteries** branch from the interlobar arteries at the boundary of the cortex and medulla. Many **interlobular arteries** radiate from the arcuate arteries and subdivide into numerous **afferent arterioles,** which are microscopic in size. The afferent arterioles deliver blood into capillary networks, called **glomeruli,** which produce a blood filtrate that enters the urinary tubules. The blood remaining in the glomerulus leaves through an **efferent arteriole,** which delivers the blood into another capillary network, the **peritubular capillaries,** that surround the tubules (fig. 26.8). This blood is drained into veins that parallel the course of the arteries in the kidney and are named the **interlobular veins, arcuate veins,** and **interlobar veins.** The interlobar veins descend between the pyramids, converge, and leave the kidney as a single **renal vein** that empties into the inferior vena cava.

Although the kidneys are generally well protected by being encapsulated retroperitoneally, they may be injured by a hard blow to the lumbar region. The immense vascularity of the kidney makes it highly susceptible to hemorrhage, and as a result such kidney damage can produce blood in the urine.

arcuate: L. *arcuare,* to bend
glomerulus: L. diminutive of *glomus,* ball

Figure 26.8. A simplified illustration of blood flow from a glomerulus to an efferent arteriole, to the peritubular capillaries, to the venous drainage of the kidneys.

Urinary Tubules.   The tubular portion of a nephron consists of a glomerular capsule, proximal convoluted tubule, descending limb of the loop of Henle, ascending limb of the loop of Henle, and distal convoluted tubule (fig. 26.8).

The **glomerular (Bowman's) capsule** surrounds the glomerulus. The glomerular capsule and its associated glomerulus is located in the cortex of the kidney and together constitute the **renal corpuscle.** The glomerular capsule contains an inner visceral layer of epithelium around the glomerular capillaries and an outer parietal layer. The space between these two layers, called the **capsular space,** receives the glomerular filtrate.

Filtrate in the glomerular capsule passes into the lumen of the **proximal convoluted tubule.** The wall of the proximal convoluted tubule consists of a single layer of cuboidal cells containing millions of microvilli; these serve to increase the surface area for reabsorption. In the process of reabsorption, salt, water, and other molecules needed by the body are transported from the lumen, through the tubular cells, and into the surrounding peritubular capillaries.

The glomerulus, glomerular capsule, and proximal convoluted tubule are located in the renal cortex. Fluid

passes from the proximal convoluted tubule to the **loop of Henle.** This fluid is carried into the medulla in the **descending limb** of the loop and returns to the cortex in the **ascending limb** of the loop. Back in the cortex, the tubule becomes coiled again and is called the **distal convoluted tubule.** In contrast to the proximal tubule, the distal convoluted tubule is shorter and has relatively few microvilli. The distal convoluted tubule is the last segment of the nephron and terminates as it empties into a collecting duct.

There are two types of nephrons, which are classified according to their position in the kidney and the lengths of their loops of Henle. Nephrons that originate in the inner one-third of the cortex—called **juxtamedullary nephrons**—have longer loops of Henle than the **cortical nephrons** that originate in the outer two-thirds of the cortex. In addition, the thin segments of the loops (fig. 26.9) are longer in the juxtamedullary nephrons.

The distal convoluted tubules of several nephrons drain into a **collecting duct.** Fluid is then drained by the collecting duct from the cortex into the medulla as the collecting duct passes through a renal pyramid. This fluid, now called urine, passes out of a renal papilla into a minor calyx. Urine is then funneled through the renal pelvis and out of the kidney into the ureter.

Bowman's capsule: from Sir William Bowman, English anatomist, 1816–92

loop of Henle: from Friedrich G. J. Henle, German anatomist, 1809–85

Figure 26.9.   The two types of nephrons are names according to their position in the kidney.

Cortical nephron   Juxtamedullary nephron

Renal cortex

Renal medulla

Collecting tubule

1. *Describe the gross appearance of the renal cortex and medulla, and indicate the microscopic structures found in each layer.*
2. *Trace the course of blood flow through the kidney from the renal artery to the renal vein.*
3. *Trace the course of tubular fluid from the glomerular capsules to the ureter.*
4. *Draw a diagram of the tubular component of a nephron. Label the segments, and indicate which parts are in the cortex and which are in the medulla.*

Glomerular Filtration

The structures of the glomerulus and glomerular capsule allow large volumes of plasma filtrate with greatly reduced protein concentration to be produced. The glomerular filtration rate is controlled by the sympathetic nervous system and by autoregulatory mechanisms.

Figure 26.10.   (*a*) the inner (visceral) layer of Bowman's capsule is composed of podocytes, as shown in this scanning electron micrograph. Very fine extensions of these podocytes form foot processes, or pedicels, that interdigitate around the glomerular capillaries. Spaces between adjacent

Podocyte

(a)

Objective 6. Describe the fine structure of the glomerular capillaries and of the podocytes.
Objective 7. Describe the composition of glomerular ultrafiltrate, and explain why it contains a low concentration of proteins.
Objective 8. Define the glomerular filtration rate and explain how it is regulated.

The glomerular capillaries have extremely large pores (200–500 Å in diameter) and are thus said to be *fenestrated*. As a result of these large pores (fenestra), glomerular capillaries are one hundred to four hundred times more permeable to plasma water and dissolved solutes than are the capillaries of skeletal muscles. Though the pores of glomerular capillaries are large, they are still small enough to prevent the passage of red blood cells, white blood cells, and platelets into the filtrate.

Before the filtrate can enter the capsular space of the glomerular capsule it must pass through the capillary pores, the basement membrane (a thin layer of glycoproteins immediately outside the endothelial cells), and the inner, visceral layer of the glomerular capsule (fig. 26.11). The inner layer of the glomerular capsule is composed of unique cells called **podocytes** *(pod'o-sīt)* with numerous cytoplasmic extensions known as **pedicels** (fig. 26.10).

podocyte: Gk. *pous*, foot; *kytos*, cell
pedicel: L. *pedicellus*, footplate

pedicels form the "filtration slits." (*b*) an illustration of the relationship between glomerular capillaries and the inner layer of Bowman's capsule.

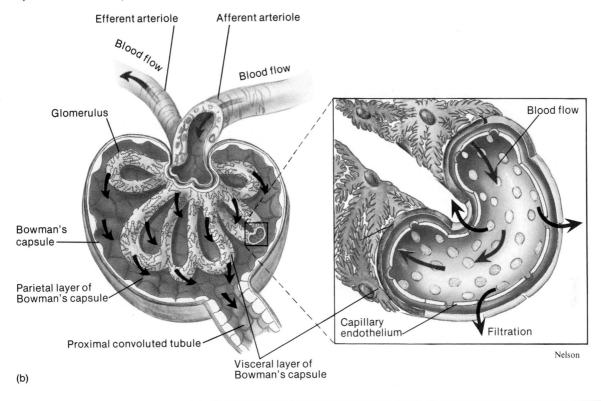

(b)

Figure 26.11.   (*a*) an electron micrograph and (*b*) an illustration of the "filtration barrier" between the capillary lumen and the cavity of Bowman's capsule.

(a)

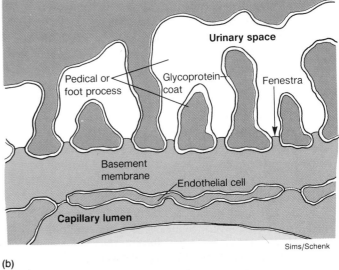

(b)

These pedicels—to mix analogies—interdigitate like fingers wrapped around the glomerular capillaries. The narrow slits between adjacent pedicels provide the passageways through which filtered molecules must pass to enter the interior of the glomerular capsule (fig. 26.11).

Although the glomerular capillary pores are apparently large enough to permit the passage of proteins, the fluid that enters the capsular space is almost completely free of plasma proteins. This exclusion of plasma proteins

Figure 26.12.   The formation of glomerular ultrafiltrate. Proteins (*large circles*) are not filtered, but smaller plasma solutes (*dots*) easily enter the glomerular ultrafiltrate.

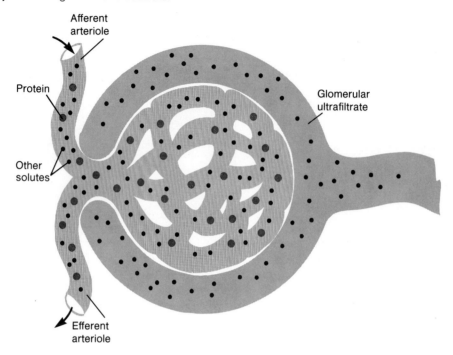

from the filtrate is partially a result of their negative charges, which hinder their passage through the negatively charged glycoproteins in the basement membrane of the capillaries. The large size and negative charges of plasma proteins may also restrict their movement through the filtration slits between pedicels.

Glomerular Ultrafiltrate

The fluid that enters the glomerular capsule is called *ultrafiltrate* (fig. 26.12) because it is formed under pressure (the hydrostatic pressure of the blood). This is similar to the formation of tissue fluid by other capillary beds in the body. The force favoring filtration is opposed by a counter force developed by the hydrostatic pressure of fluid in the glomerular capsule. Also, since the protein concentration of the tubular fluid is low (less than 2–5 mg per 100 ml) compared to that of plasma (6–8 g per 100 ml), the greater colloid osmotic pressure of plasma promotes the osmotic return of filtered water. When opposing forces are subtracted from the hydrostatic pressure of the glomerular capillaries, a *net filtration pressure* of approximately 10 mm Hg is obtained.

Because glomerular capillaries are extremely permeable and have a high surface area, this modest net filtration pressure produces an extraordinarily large volume of filtrate. The **glomerular filtration rate (GFR)** averages 115 ml per minute in women and 125 ml per minute in men. This is equivalent to 7.5 L per hour or 180 L per day (about 45 gallons)! Since the total blood volume averages about

5 L, this means that the total blood volume is filtered into the urinary tubules every forty minutes. Most of the filtered water must obviously be returned immediately to the vascular system or people would literally urinate to death within several minutes.

Regulation of Glomerular Filtration Rate

Vasoconstriction or dilation of afferent arterioles affects the rate of blood flow to the glomerulus and thus affects the glomerular filtration rate. Changes in the diameter of the afferent arterioles results from both extrinsic (sympathetic innervation) and intrinsic regulatory mechanisms.

Sympathetic Nerve Effects.   An increase in sympathetic nerve activity, as occurs during the fight or flight reaction and exercise, stimulates constriction of afferent arterioles. This is an alpha-adrenergic effect, which helps to preserve blood volume and to divert blood to the muscles and heart. A similar effect occurs during cardiovascular shock in which sympathetic nerve activity stimulates vasoconstriction. The decreased GFR and the resulting decreased rate of urine formation help to compensate for the rapid drop of blood pressure under these circumstances (fig. 26.13).

Renal Autoregulation.   When the direct effect of sympathetic stimulation is experimentally removed, the effect of systemic blood pressure on GFR can be observed. Under these conditions, surprisingly, the GFR remains relatively constant despite changes in mean arterial pressure within

Figure 26.13. The effect of increased sympathetic nerve activity on kidney function and other physiological processes.

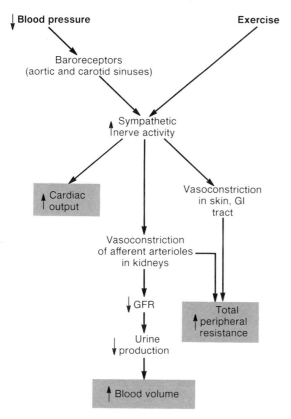

Table 26.1	Regulation of glomerular filtration rate (GFR)		
Regulation	**Stimulus**	**Afferent arteriole**	**GFR**
Sympathetic nerves	Activation by aortic and carotid baroreceptors or by higher brain centers	Constricts	Decreases
Autoregulation	Decreased blood pressure	Dilates	No change
Autoregulation	Increased blood pressure	Constricts	No change

a range of 70–180 mm Hg (normal mean arterial pressure is 100 mm Hg). The ability of the kidneys to maintain a relatively constant GFR in the face of fluctuating blood pressures is called **renal autoregulation.**

Renal autoregulation results from the effects of locally produced chemicals on the afferent arterioles (effects on the efferent arterioles are believed to be of secondary importance). When systemic arterial pressure falls toward a mean of 70 mm Hg, the afferent arterioles dilate, and when the pressure rises, the afferent arterioles constrict. Blood flow to the glomeruli and GFR can thus remain within the autoregulatory range of blood pressure values. The effects of different regulatory mechanisms on GFR are summarized in table 26.1.

1. Describe the structures that plasma fluid must pass through to enter the capsular space. Explain how proteins are excluded from the filtrate.
2. Describe the forces that affect the formation of glomerular ultrafiltrate.
3. Describe the effect of sympathetic innervation on the glomerular filtrate rate, and explain the meaning of the term "renal autoregulation."

Reabsorption of Salt and Water

The proximal tubule reabsorbs most of the filtered salt and water, and the transport of salt and water in the loop of Henle creates a hypertonic renal medulla. This hypertonic medulla, as regulated by the action of ADH, permits most of the remaining water to be reabsorbed.

Objective 9. Describe how salt and water is reabsorbed in the proximal tubule, and explain the significance of this process.

Objective 10. Describe the transport processes that occur in the descending and ascending limbs of the loop of Henle, and explain how the countercurrent multiplier effect is produced.

Objective 11. Describe the structure and function of the vasa recta.

Objective 12. Explain the significance of a hypertonic renal medulla, and describe the mechanism of ADH action.

Although about 180 L per day of glomerular ultrafiltrate are produced, the kidneys normally excrete only 1–2 L per day of urine. Approximately 99% of the filtrate must thus be returned to the vascular system, while 1% is excreted in the urine. The urine volume, however, varies according to the needs of the body. When a well-hydrated person drinks a liter or more of water, urine volume increases to 16 ml per minute (the equivalent of 23 L per day if this were to continue for twenty-four hours). In severe dehydration, when the body needs to conserve water, only 0.3 ml per minute, or 400 ml per day, of urine are produced. A volume of 400 ml per day of urine is needed to excrete the amount of metabolic wastes produced by the body; this is called the *obligatory water loss.* When water in excess of this amount is excreted, the urine volume is increased and its concentration is decreased.

Regardless of the body's state of hydration, it is clear that most of the filtered water must be returned to the vascular system to maintain blood volume and pressure.

Figure 26.14. Plasma water and its dissolved solutes (except proteins) enter the glomerular ultrafiltrate, but most of these filtered molecules are reabsorbed. The term *reabsorption* refers to the transport of molecules out of the tubular filtrate back into the blood.

Figure 26.15. An illustration of the appearance of tubule cells in the electron microscope. Molecules that are reabsorbed pass through the tubule cells from the apical membrane (facing the filtrate) to the basolateral membrane (facing the blood).

Figure 26.16. Mechanisms of salt and water reabsorption in the proximal tubule. Sodium is actively transported out of the filtrate, and chloride follows passively by electrical attraction. Water follows the salt out of the tubular filtrate into the peritubular capillaries by osmosis.

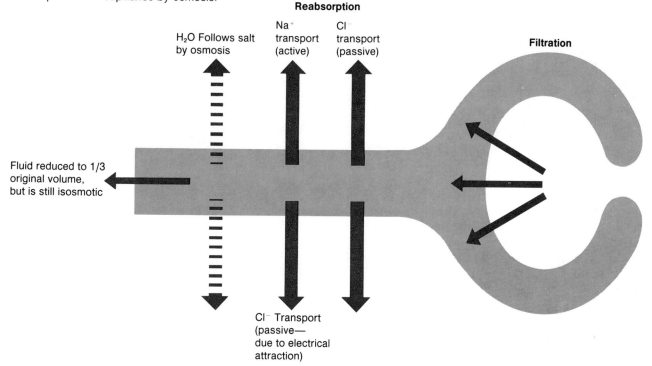

The return of filtered molecules from the tubules to the blood is called **reabsorption** *(re″ab-sorp′shun)*—see figure 26.14. It is important to realize that the transport of water always occurs passively by *osmosis;* there is no such thing as active transport of water. A concentration gradient must thus be created between tubular fluid and blood that favors the osmotic return of water to the vascular system.

Reabsorption in the Proximal Tubule

Since all plasma solutes, with the exception of proteins, are able to freely enter the glomerular ultrafiltrate, the osmolality of the filtrate is essentially the same as that of the plasma (300 milliosmoles per liter, or 300 mOsm). The filtrate is thus said to be isosmotic with the plasma. Osmosis thus cannot occur unless the concentration of plasma in the peritubular capillaries and the concentration of filtrate are altered by active transport processes. This is achieved by the active transport of Na^+ from the filtrate to the peritubular blood.

Active and Passive Transport.   The epithelial cells that compose the wall of the proximal tubule have two exposed sides. The apical sides of these cells face the tubular fluid in the lumen, and contain numerous microvilli. The basolateral sides face the outside of the tubule, and thus are near the peritubular capillaries (fig. 26.15).

The concentration of Na^+ in the glomerular ultrafiltrate—and thus in the fluid entering the proximal tubule—is the same as in plasma and the same as in the tissue fluid outside of the proximal convoluted tubules. The tubule cells that separate filtrate from tissue fluid, however, have a lower Na^+ concentration. This lower Na^+ concentration is partially due to low permeabilities of cell membranes to Na^+ and partially due to the action of active transport Na^+ pumps, which extrude Na^+ from the cells.

In the cells of the proximal convoluted tubule, the active carriers of Na^+ are located only in the basolateral membrane. As a result, Na^+ that diffuses across the apical membrane from the tubular fluid is extruded across the basolateral membrane into the surrounding tissue fluid. This raises the Na^+ concentration of tissue fluid and results in the diffusion of Na^+ into adjacent peritubular capillaries.

The reabsorption of Na^+ in the proximal tubule is thus dependent on *active transport*. As a result of the active movement of Na^+ from the tubular fluid to tissue fluid, an electrical gradient is created that allows Cl^- to follow the Na^+. The movement of Cl^-, in this case, occurs by *passive transport*. These transport processes create an osmotic gradient because they raise the osmolality and osmotic pressure of the surrounding tissue fluid above that of the tubular fluid. Since the membranes of the proximal tubule cells are permeable to water, movement of water occurs by osmosis from the tubular fluid to the surrounding tissue fluid (fig. 26.16) and from there to the peritubular capillaries.

Significance of Proximal Tubule Reabsorption. Approximately 65% of the salt and water in the original glomerular ultrafiltrate is reabsorbed across the proximal tubule and returned to the vascular system. The volume of tubular fluid remaining is reduced accordingly, but this fluid is still isosmotic with the blood (has a concentration of 300 mOsm). This results from the fact that the cell membranes in the proximal tubule are freely permeable to water so that water and salt are removed in proportionate amounts.

An additional smaller amount of salt and water is returned to the vascular system by reabsorption in the loop of Henle. This reabsorption, like that in the proximal tubule, occurs constantly regardless of the person's state of hydration. Unlike reabsorption in later regions of the nephron, it is not subject to hormonal regulation. Approximately 85% of the filtered salt and water is, therefore, reabsorbed in a constant, unregulated fashion in the early regions of the nephron (proximal tubule and loop of Henle). This reabsorption is very costly in terms of energy expenditures, accounting for as much as 6% of the calories consumed by the body at rest.

Since 85% of the original glomerular ultrafiltrate is immediately reabsorbed in the early region of the nephron, only 15% of the initial filtrate remains to enter the distal convoluted tubule and collecting duct. This is still a large volume of fluid—15% × GFR (180 L per day) = 27 L per day—that must be reabsorbed to varying degrees in accordance with the body's state of hydration. This "fine tuning" of the percent reabsorption and urine volume is accomplished by the action of hormones on the later regions of the nephron.

The Countercurrent Multiplier System

Water cannot be actively transported across the tubule wall, and osmosis of water cannot occur if the tubular fluid and surrounding tissue fluid are isotonic to each other. In order for water to be reabsorbed by osmosis, the surrounding tissue fluid must be hypertonic. The osmotic pressure of the tissue fluid in the renal medulla is, in fact, raised to over four times that of plasma. This results partly from the fact that the tubule bends; the geometry of the loop of Henle allows interaction to occur between the descending and ascending limbs. Since the ascending limb is the active partner in this interaction, its properties will be described before those of the descending limb.

Figure 26.17. The countercurrent multiplier system. The active extrusion of Cl^- followed by Na^+ from the ascending limb makes the surrounding tissue fluid more concentrated. This concentration is multiplied by the fact that the descending limb is passively permeable so that its fluid increases in concentration as the surrounding tissue fluid becomes more concentrated. The transport properties of the loop and their effect on tubular fluid concentration is shown in (*a*). The values of these changes in osmolality, together with the effect on surrounding tissue fluid concentration, are shown in (*b*).

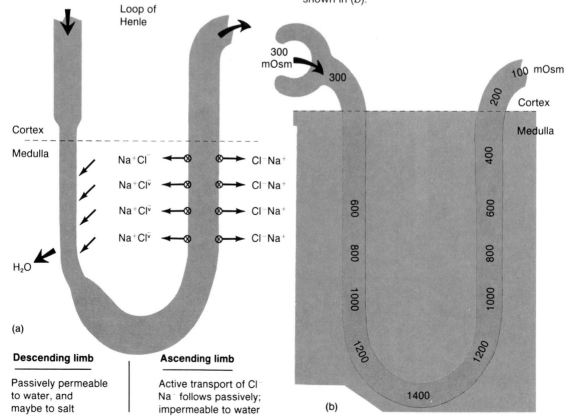

Ascending Limb of the Loop of Henle.

Chloride ion (Cl^-) is actively extruded from the ascending limb of the loop, and Na^+ follows the chloride passively. This is the reverse of salt transport in the proximal tubule, but the effects are the same—salt is pumped out of the tubule and enters the surrounding tissue fluid. Unlike the proximal tubule, however, the walls of the ascending limb are *not permeable to water.* The tubular fluid thus becomes increasingly dilute as it ascends toward the cortex, while the tissue fluid around the loops of Henle in the medulla become increasingly more concentrated. By means of these processes, the tubular fluid that enters the distal tubule in the cortex is hypotonic (with a concentration of 100 mOsm), whereas the tissue fluid in the medulla is hypertonic.

Descending Limb of the Loop of Henle.

The deeper regions of the medulla, around the tips of the loops of juxtamedullary nephrons, reach a concentration of 1200–1400 mOsm. In order to reach this high a concentration, the salt pumped out of the ascending limb must accumulate in the tissue fluid. This occurs as a result of the properties of the descending limb and as a result of the fact that blood vessels around the loop do not carry back all of the extruded salt to the general circulation.

The descending limb does not actively transport salt. Instead, it is *passively permeable* to water and perhaps also to salt (this is currently controversial). The wall of the descending limb is like a porous plastic membrane; water and perhaps salt are free to diffuse according to their concentration gradients. Since the renal medulla is hypertonic to the fluid entering the descending limb, water moves by osmosis out of the descending limb and is removed by peritubular capillaries. At the same time, salt may diffuse into the descending tubule since it is present at a higher concentration in the surrounding tissue fluid. The concentration of tubular fluid is thus increased, and its volume is decreased, as it descends toward the tips of the loops.

As a result of these passive transport processes in the descending limb, the fluid that "rounds the bend" at the tip of the loop has the same osmolality as the surrounding tissue fluid (1200–1400 mOsm). There is, therefore, a higher salt concentration arriving in the ascending limb than there would be if the descending limb delivered isotonic fluid. Salt transport by the ascending limb is increased accordingly, so that the "saltiness" of the tissue fluid is multiplied (fig. 26.17).

Countercurrent Multiplication.

Countercurrent flow (flow in opposite directions) in the ascending and descending limbs and the close proximity of the two limbs allow interaction to occur. Since the concentration of the tubular fluid in the descending limb reflects the concentration of surrounding tissue fluid, and since the concentration of this tissue fluid is raised by the active extrusion of salt from the ascending limb, a *positive feedback* mechanism is created. The more salt the ascending limb extrudes, the more concentrated will be the fluid that returns to it from the descending limb. This positive feedback mechanism multiplies the concentration of tissue fluid and descending limb fluid and is thus called the **countercurrent multiplier system.**

The countercurrent multiplier system recirculates salt and thus traps some of the salt that enters the loop of Henle in the tissue fluid of the renal medulla. This system results in a gradually increasing concentration of renal tissue fluid from the cortex to the inner medulla; the osmolality of tissue fluid increases from 300 mOsm (isotonic) in the cortex to 1200–1400 mOsm in the deepest part of the medulla.

Vasa Recta.

In order for the countercurrent multiplier system to be effective, only a small amount of salt that is extruded from the ascending limbs must be carried away by the vascular system. Hypertonicity of the renal medulla is maintained by the **countercurrent exchange** of salt and water in vessels known as the **vasa recta.** These vessels form long loops that parallel the long loops of Henle of juxtamedullary nephrons. Salt that diffuses into the blood as it descends these loops subsequently diffuses out of the blood in the ascending segments of these vessels (fig. 26.18). This countercurrent exchange helps to maintain the hypertonicity of the medullary tissue fluid.

Figure 26.18. Countercurrent exchange in the vasa recta. The diffusion of salt and water first into and then out of these blood vessels helps to maintain the "saltiness" (hypertonicity) of the interstitial fluid in the renal medulla (numbers indicate osmolality).

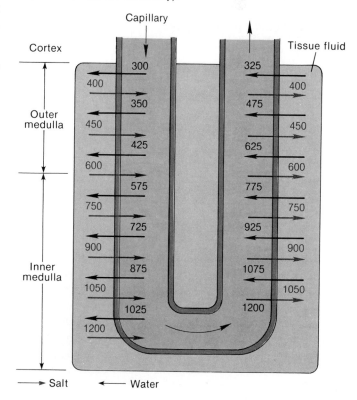

Figure 26.19. According to some authorities, urea diffuses out of the collecting duct and contributes significantly to the concentration of the interstitial fluid in the renal medulla. The active transport of Cl^- out of the thick segments of the ascending limbs also contributes to the hypertonicity of the medulla so that water is reabsorbed by osmosis from the collecting ducts.

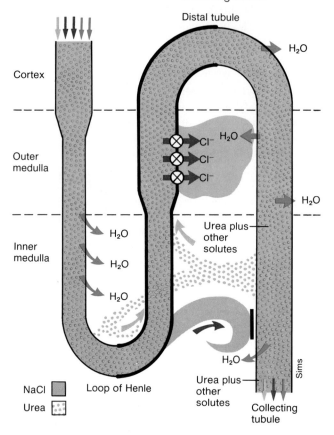

Table 26.2 Properties of different nephron segments in regard to the concentrating-diluting mechanisms of the kidney

Nephron segment	Active transport	Passive transport		
		Salt	Water	Urea
Proximal tubule	Na^+	Cl^-	Yes	Yes
Descending limb of Henle's loop	None	Maybe	Yes	No
Thin segment of ascending limb*	Cl^- or None	Na^+ or NaCl	No	Yes
Thick segment of ascending limb	Cl^-	Na^+	No	No
Distal tubule	Na^+	No	No	No
Collecting duct**	Slight Na^+	No	Yes (ADH) or Slight (no ADH)	Yes

*According to one theory, the thin segment of the ascending limb actively transports Cl^-, and Na^+ follows passively; according to another theory, the thin segment does not actively extrude salt. Both theories are presented in this table.
**The permeability of the collecting duct to water depends on the presence of ADH.

Source: Fox, Stuart Ira, *Human Physiology.* © 1984 Wm. C. Brown Publishers, Dubuque, Iowa. All Rights Reserved. Reprinted by permission.

Possible Effects of Urea. Experimental evidence suggests that the active transport of Cl^- may only occur in the thick segments of the ascending limbs. The thin segments of the ascending limbs, which are located in the deeper regions of the medulla, may not be able to extrude salt actively. Since salt does leave the thin segments, it must leave by diffusion even though the surrounding tissue fluid is at least as concentrated as the tubular fluid. Some investigators therefore conclude that molecules other than salt—specifically urea—contribute to the hypertonicity of the tissue fluid. Urea may accumulate in the medullary tissue fluid as a result of the recycling of urea between the collecting duct and the loop of Henle (fig. 26.19). The transport properties of different tubule segments are summarized in table 26.2.

Figure 26.20.   The countercurrent multiplier system in the loop of Henle and the countercurrent exchange in the vasa recta help create a hypertonic renal medulla. Under the influence of antidiuretic hormone (ADH), the collecting duct is more permeable to water so that water is drawn by osmosis out into the hypertonic renal medulla and into the peritubular capillaries.

Collecting Duct: Effect of Antidiuretic Hormone (ADH)

As a result of the recycling of salt between the ascending and descending limbs and possibly of the recycling of urea between the collecting duct and the loop of Henle, the medullary tissue fluid is made very hypertonic. The collecting ducts must transport their fluid through this hypertonic environment in order to empty their contents of urine into the calyces. Since the distal convoluted tubules are impermeable to water, the fluid that enters the collecting ducts in the cortex is hypotonic as a result of the active extrusion of salt by the ascending limbs of the loops.

The walls of the collecting ducts are *permeable to water but not to salt*. Since the surrounding tissue fluid in the renal medulla is very hypertonic, as a result of the countercurrent multiplier system, water is drawn out of the collecting ducts by osmosis. This water does not dilute the surrounding tissue fluid because it is transported by capillaries to the general circulation. In this way, most of the water remaining in the tubules after reabsorption in the proximal tubules is returned to the vascular system (fig. 26.20).

Table **26.3** Antidiuretic hormone secretion and action

Stimulus	Receptors	Secretion of ADH	Effects on	
			Urine volume	Blood
↑Osmolality (dehydration)	Osmoreceptors in hypothalamus	Increased	Decreased	Increased water retention; decreased blood osmolality
↓Osmolality	Osmoreceptors in hypothalamus	Decreased	Increased	Water loss increases blood osmolality
↑Blood volume	Stretch receptors in left atrium	Decreased	Increased	Decreased blood volume
↓Blood volume	Stretch receptors in left atrium	Increased	Decreased	Increased blood volume

Source: Fox, Stuart Ira, *Human Physiology.* © 1984 Wm. C. Brown Publishers, Dubuque, Iowa. All Rights Reserved. Reprinted by permission.

The osmotic gradient created by the countercurrent multiplier system provides the force for water reabsorption through the collecting ducts. The rate of this reabsorption, however, is determined by the permeability of the collecting duct cell membranes to water. The permeability of the collecting duct to water, in turn, is determined by the concentration of **antidiuretic** *(an''ti-di''u-ret'ik)* **hormone (ADH)** in the blood. When the concentration of ADH is increased, the collecting ducts become more permeable to water, and more water is reabsorbed. A decrease in ADH, conversely, results in less reabsorption of water and thus in the excretion of a larger volume of more dilute urine.

ADH is produced by neurons in the hypothalamus and is secreted from the posterior pituitary gland. The secretion of ADH is stimulated when osmoreceptors in the hypothalamus respond to an increase in blood osmotic pressure. During dehydration, therefore, when the plasma becomes more concentrated, increased secretion of ADH promotes increased permeability of the collecting ducts to water. In severe dehydration only the minimal amount of water needed to eliminate the body's wastes is excreted. This minimum, about 400 ml per day, is limited by the fact that urine cannot become more concentrated than the medullary tissue fluid surrounding the collecting ducts. Under these conditions about 99.8% of the initial glomerular ultrafiltrate is reabsorbed.

A person in a state of normal hydration excretes about 1.5 L per day of urine, indicating that 99.2% of the glomerular ultrafiltrate volume is reabsorbed. Notice that small changes in percent reabsorption translate into large changes in urine volume. Increasing water ingestion—and thus decreasing ADH secretion (table 26.3)—results in correspondingly larger volumes of urine excretion. It should be noted that even in the complete absence of ADH some water is still reabsorbed through the collecting ducts.

Diabetes insipidus is a disease associated with the inadequate secretion or action of ADH. The collecting ducts are thus not very permeable to water and, therefore, a large volume (5–10 L per day) of dilute urine is produced. The dehydration that results causes intense thirst, but a person with this condition has difficulty drinking enough to compensate for the large volumes of water lost in the urine.

1. Describe the mechanisms for salt and water reabsorption in the proximal tubule.
2. Compare the transport of Na^+, Cl^-, and water across the walls of the proximal tubule, ascending and descending limbs of the loop of Henle, and collecting duct.
3. Explain the interaction of the ascending and descending limbs of the loop and how this interaction results in a hypertonic renal medulla.
4. Explain how ADH helps the body to conserve water, and describe how variations in ADH secretion affect the volume and concentration of urine.

Renal Clearance Rates

Solutes in the plasma that are filtered or are secreted into the tubules may be cleared from the blood if they are not reabsorbed. Measurements of the renal plasma clearance rates of different compounds indicate how the kidneys "handle" these substances and provide information that is clinically important.

Objective 13. Define the terms *reabsorption* and *secretion,* and explain how these processes affect renal clearance rates.

Objective 14. Describe how the glomerular filtration rate is measured by the clearance rate of inulin.

Objective 15. Describe how the renal blood flow is measured by the clearance rate of PAH.

Objective 16. Describe how the kidneys reabsorb glucose and amino acids, and explain how glycosuria is produced.

One of the major functions of the kidneys is the excretion of waste products such as urea, creatinine, and other molecules. These molecules are filtered through the glomerulus into the glomerular capsule along with water, salt, and other plasma solutes. In addition, some waste products can gain access to the urine by a process called **secretion** (fig. 26.21). Secretion is the opposite of reabsorption by active transport. Molecules that are secreted are transported actively from the peritubular capillaries,

Figure 26.21.   Secretion refers to the active transport of substances from the peritubular capillaries into the tubular fluid. This transport is in a direction opposite to that of reabsorption.

Secretion

Reabsorption

Filtration

Excretion

Table 26.4   The effects of filtration, reabsorption, and secretion on renal clearance rates

Term	Means	Effect on renal clearance
Filtered	A substance enters the glomerular ultrafiltrate	Some or all of a filtered substance may enter the urine and be "cleared" from the blood.
Reabsorbed	The transport of a substance from the filtrate, through tubular cells, and into the blood	Reabsorption decreases the rate at which a substance is cleared; clearance rate is less than the glomerular filtration rate (GFR).
Secreted	The transport of a substance from peritubular blood, through tubular cells, and into the filtrate	When a substance is secreted by the nephrons, its clearance rate is greater than the GFR.

across the tubule wall, and into the tubular fluid. In this way, molecules that were not filtered from the blood in the glomerulus can still be excreted in the urine.

Although most (about 99%) of the filtered water is returned to the vascular system by reabsorption, most of the wastes that are filtered or secreted are eliminated in the urine. The concentration of these substances in the renal vein leaving the kidneys is therefore lower than their concentrations in the blood entering the kidneys in the renal artery. Some of the blood that passes through the kidneys, in other words, is "cleared" of these waste products; this is known as the **renal plasma clearance.**

The quantity of a substance excreted in the urine within a given period of time depends on the (1) quantity *filtered* through the glomeruli into the tubular fluid; (2) quantity *secreted* from the unfiltered blood in peritubular capillaries by active transport into the tubules; and the (3) quantity *reabsorbed* by transport from the tubules into the peritubular blood. Filtration and secretion increase renal plasma clearance; reabsorption decreases the amount excreted and thus decreases the clearance rate (table 26.4).

Renal Clearance of Inulin: Measurement of GFR

If a substance is neither reabsorbed nor secreted by the tubules, the amount excreted per minute in the urine will be equal to the amount that is filtered out of the glomeruli. There does not seem to be a single substance produced by the body, however, that is not reabsorbed or secreted to some degree. Plants such as artichokes, dahlias, onions, and garlic, fortunately, do produce such a compound. This compound, a polymer of the monosaccharide fructose, is **inulin** *(in'u-lin).* Once injected into the blood, inulin is filtered by the glomeruli, and the amount of inulin excreted per minute is exactly equal to the amount that was filtered per minute (fig. 26.22).

If the concentration of inulin in urine is measured and the rate of urine formation is determined, the rate of inulin excretion can easily be calculated:

$$\underset{\text{(mg/min)}}{\textit{Quantity excreted per minute}} = \underset{\left(\frac{\text{ml}}{\text{min}}\right)}{V} \times \underset{\left(\frac{\text{mg}}{\text{ml}}\right)}{U}$$

Where V = rate of urine formation
U = inulin concentration in urine

Figure 26.22. The renal clearance of inulin. (a) inulin is present in the blood entering the glomeruli, and (b) some of this blood, together with its dissolved inulin, is filtered. All of this filtered inulin enters the urine, whereas most of the filtered water is returned to the vascular system (is reabsorbed). (c) the blood leaving the kidneys in the renal vein, therefore contains less inulin than the blood that entered the kidneys in the renal artery. Since inulin is filtered but neither reabsorbed nor secreted, the inulin clearance rate equals the glomerular filtration rate (GFR).

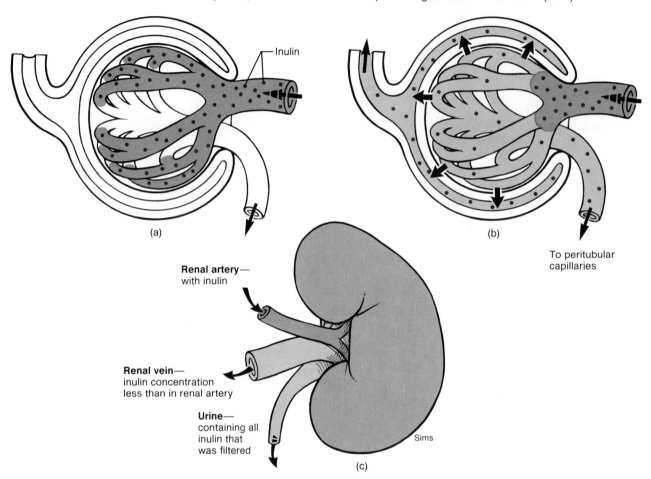

Inulin

(a)

(b)

To peritubular capillaries

Renal artery—
with inulin

Renal vein—
inulin concentration
less than in renal artery

Urine—
containing all
inulin that
was filtered

Sims

(c)

In order to calculate the rate at which a substance is filtered by the glomeruli (in mg per minute), multiply the ml per minute of plasma that is filtered (the **glomerular filtration rate,** or **GFR**) by the concentration of that substance in the plasma. This is shown in the following equation:

$$Quantity\ filtered = GFR \times P$$
$$(mg/min) \qquad \left(\frac{ml}{min}\right) \quad \left(\frac{mg}{ml}\right)$$

Where P = inulin concentration in plasma

Since inulin is neither reabsorbed nor secreted, the amount filtered equals the amount excreted:

$$GFR \times P \quad = \quad V \times U$$
$$\text{(amount filtered)} \quad \text{(amount excreted)}$$

If the last equation is now solved for the glomerular filtration rate,

$$GFR_{(ml/min)} = \frac{V_{(ml/min)} \times U_{(mg/ml)}}{P_{(mg/ml)}}$$

Suppose, for example, that inulin is infused into a vein and its concentration in the urine and plasma are found to be 30 mg per ml and 0.5 mg per ml, respectively. If the rate of urine formation is 2 ml per minute, the GFR can be calculated as follows:

$$GFR = \frac{2\ ml/min \times 30\ mg/ml}{0.5\ mg/ml} = 120\ ml/min$$

This equation states that at a plasma inulin concentration of 0.5 mg per ml, 120 ml of plasma must have been filtered in order to excrete the measured amount of 60 mg that appear in the urine per minute. The glomerular filtration rate is thus 120 ml per minute in this example.

Measurements of the plasma concentration of **creatinine** *(kre-at'i-nin)* are often used clinically as an index of kidney function. Creatinine, produced as a waste product of muscle creatine, is secreted to a slight degree by the renal tubules so that its excretion rate is a little above that of inulin. Since it is released into the blood at a constant rate and since its excretion is closely matched to the GFR, an abnormal decrease in GFR causes the plasma creatinine concentration to rise. A simple measurement of blood creatinine concentration can thus provide information about the health of the kidneys.

Table 26.5	Renal "handling" of different plasma molecules			
If substance is	**Example**	**Concentration in renal vein**	**Renal clearance rate**	
Not filtered	Proteins	Same as in renal artery	Zero	
Filtered, not reabsorbed nor secreted	Inulin	Less than in renal artery	Equal to GFR (115–125 ml/min)	
Filtered, partially reabsorbed	Urea	Less than in renal artery	Less than GFR	
Filtered, completely reabsorbed	Glucose	Same as in renal artery	Zero	
Filtered and secreted	PAH	Less than in renal artery; approaches zero	Greater than GFR; equal to total plasma flow rate (~625 ml/min)	
Filtered, reabsorbed, and secreted	K⁺	Variable	Variable	

Source: Fox, Stuart Ira, *Human Physiology.* © 1984 Wm. C. Brown Publishers, Dubuque, Iowa. All Rights Reserved. Reprinted by permission.

Clearance Rates. If one knows the amount of inulin excreted per minute and the plasma inulin concentration, one can calculate the volume of plasma that was filtered per minute (the GFR). Since 120 ml per minute are filtered (enter the glomerular capsules) in the previous example, the amount of inulin contained in 120 ml of plasma is excreted per minute. Since creatinine is filtered like inulin but is also secreted to a slight degree, the excretion rate of creatinine is slightly greater than the filtration rate. The amount of creatinine excreted per minute, in other words, is greater than the amount contained in the 120 ml of plasma that was filtered.

If a substance is reabsorbed to some degree, conversely, the excretion rate will be less than the amount contained in the 120 ml of plasma that was filtered. In order to compare the renal "handling" of various substances in terms of their reabsorption or secretion, the **renal plasma clearance rate** of these substances can be calculated using the same formula used for the determination of the GFR:

$$Clearance\ Rate = \frac{V \times U}{P}$$

Where V = urine volume per minute
U = concentration of substance in urine
P = concentration of substance in plasma

In the case of inulin, the clearance rate is equal to the glomerular filtration rate. If a substance is secreted by the tubules, its clearance rate is greater than the GFR; if a substance is reabsorbed, its clearance rate is less than the GFR (table 26.5). In each of these cases, the clearance rate, in ml per minute, indicates the amount of plasma that originally contained the amount of the substance excreted in a minute's time.

Clearance of Urea

Urea *(u-re'ah)* is a waste product of amino acid metabolism that is secreted by the liver into the blood. Despite the fact that it is a waste product, a significant proportion of the filtered urea (40%–60%) is reabsorbed passively, as a result of the high permeability of the tubule membranes to this compound. Not all of the urea contained in the filtered plasma, therefore, is cleared. Using the formula for renal clearance rate previously described,

If $V = 2$ ml/min
$U = 7.5$ mg/ml of urea
$P = 0.2$ mg/ml of urea

$$urea\ clearance\ rate = \frac{2\ ml/min \times 7.5\ mg/ml}{0.2\ mg/ml} = 75\ ml/min$$

The amount of urea excreted per minute, in other words, was originally contained in 75 ml of plasma. Since 120 ml of plasma were filtered, as determined by the inulin clearance rate, this indicates that approximately 40% of the filtered urea must have been reabsorbed in this example.

Clearance of PAH: Measurement of Renal Blood Flow

Not all of the blood delivered to the glomeruli is filtered into the glomerular capsules; most of the glomerular blood passes through to the efferent arterioles and peritubular capillaries. The inulin and urea in this unfiltered blood is not excreted but instead returns to the general circulation. Blood must thus make many passes through the kidneys before it can be completely cleared of a given amount of inulin or urea.

In order for compounds in the unfiltered renal blood to be cleared, they must be *secreted* into the tubules by active transport from the peritubular capillaries. In this way all of the blood going to the kidneys can potentially be cleared of a secreted compound in a single pass. This is the case for a molecule called **para-aminohippuric acid, or PAH,** (fig. 26.23). The clearance rate (in ml/min) of PAH can be used to measure the *total renal blood flow* (in ml/min). The normal PAH clearance rate has been found to average 625 ml/min. Since the glomerular filtration rate averages about 120 ml/min, this indicates that only about 120/625, or roughly 20%, of the renal blood flow is filtered. The remaining 80% passes on to the efferent arterioles.

Since filtration and secretion clear only the molecules dissolved in plasma, the PAH clearance rate measures the renal plasma flow rate. In order to convert this to the total renal blood flow, the volume of blood occupied by erythrocytes must be taken into account. If the hematocrit is 45, for example, erythrocytes occupy 45% of the blood volume and plasma accounts for the remaining 55% of the blood volume. The **total renal blood flow** is calculated by

Figure 26.23.   Some of the para-aminohippuric acid (PAH) in glomerular blood (a) is filtered into Bowman's capsules (b). The PAH present in the unfiltered blood is secreted from peritubular capillaries into the nephron (c), so that all of the blood leaving the kidneys is free of PAH (d). The clearance rate of PAH therefore equals the total plasma flow to the glomeruli.

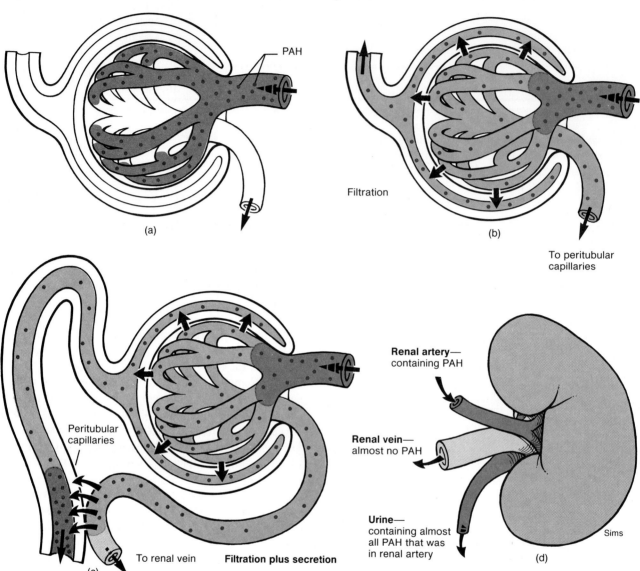

(a)

Filtration

(b)

To peritubular capillaries

Peritubular capillaries

To renal vein   **Filtration plus secretion**

(c)

Renal artery— containing PAH

Renal vein— almost no PAH

Urine— containing almost all PAH that was in renal artery

Sims

(d)

dividing the PAH clearance rate by the fractional blood volume occupied by plasma (0.55, in this example). The total renal blood flow in this example is thus 625 ml/min divided by 0.55, or 1.1 L/min.

Many antibiotics, like penicillin, are secreted by the renal tubules, and thus have clearance rates greater than the glomerular filtration rate. Penicillin is thus rapidly removed from the blood by renal clearance, and consequently large amounts must be administered to be effective. The ability of the kidneys to be visualized in radiographs is improved by injection of Diodrast, a material that absorbs X rays and is secreted into the tubules. Many drugs and some hormones are inactivated in the liver by chemical transformations and are rapidly cleared from the blood by active secretion in the nephrons.

Reabsorption of Glucose and Amino Acids

Glucose and amino acids in the blood are easily filtered by the glomeruli into the renal tubules. These molecules, however, are usually not present in the urine. It can be concluded, therefore, that filtered glucose and amino acids are normally completely reabsorbed by the nephrons.

The reabsorption of glucose and amino acids is an energy-requiring process, which occurs primarily in the proximal convoluted tubules. The energy required for movement of these compounds from the filtrate into the tubule cells is provided by cotransport with Na^+, which diffuses down its electrochemical gradient when it enters the cells. The glucose and amino acids appear to share carriers with Na^+ in the apical membrane. The extrusion

Figure 26.24.   The reabsorption of glucose in the proximal tubule. Glucose is reabsorbed by cotransport with Na^+ from the tubular fluid and then transported by Na^+-independent carriers through the basolateral membrane into the peritubular blood. By this means normally all of the filtered glucose is reabsorbed.

of glucose and amino acids from the other side of the cell (across the basolateral membranes) occurs by means of a Na^+-independent active transport carrier (fig. 26.24).

Carrier-mediated transport displays the property of *saturation.* This means that when the transported molecule (such as glucose) is present in sufficiently high concentrations, all of the carriers become "busy," and the transport rate reaches a maximal value. The concentration of transported molecules needed to just saturate the carriers and to just achieve the maximal transport rate is called the **transport maximum** (abbreviated T_m).

The carriers for glucose and amino acids in the renal tubules are not normally saturated and so are able to remove the filtered molecules completely. The T_m for glucose, for example, averages 375 mg per minute, which is well above the rate at which glucose is delivered to the tubules. The rate of glucose delivery can be calculated by multiplying the plasma glucose concentration (about 1 mg per ml) by the GFR (about 125 ml per minute). Approximately 125 mg per minute are thus delivered to the tubules, whereas a rate of 375 mg per minute are required to reach saturation.

Glycosuria.   Glucose appears in the urine—a condition called glycosuria *(gli″ko-su′re-ah)*—when more glucose passes through the tubules than can be reabsorbed. This occurs when the plasma glucose concentration reaches 180–200 mg per 100 ml. Since the rate of glucose delivery under these conditions is still below the average T_m for glucose, one must conclude that some nephrons have considerably lower T_m values than the average.

The **renal plasma threshold** is the minimum plasma concentration of a substance that results in the excretion of that substance in the urine. The renal plasma threshold for glucose, for example, is 180–200 mg per 100 ml. Glucose is normally absent from urine because plasma glucose concentrations normally remain below this threshold value. The appearance of glucose in the urine (glycosuria) thus occurs only when the plasma concentration of glucose is abnormally high (hyperglycemia).

Fasting hyperglycemia is caused by the inadequate secretion or action of insulin. When this hyperglycemia results in glycosuria, the disease is called **diabetes mellitus.** A person with uncontrolled diabetes mellitus also excretes a large volume of urine, because the excreted glucose carries water with it as a result of the osmotic pressure it generates in the tubules. This condition should not be confused with diabetes insipidus, in which a large volume of dilute urine is excreted as a result of inadequate ADH secretion.

The excretion of amino acids in the urine is not usually due to the presence of high–amino acid concentration in the blood. Rather, specific amino acids are excreted when their carriers are missing or defective due to a genetic disease. Since different classes of amino acids are reabsorbed by different carriers, the types of amino acids that "spill" into the urine are characteristic of the genetic defect (table 26.6).

Table 26.6 Inherited diseases associated with the presence of specific amino acids in the urine

Disease	Cause of disease	Effect of defect	Treatment
Cystinuria	Renal carriers for cystine and related amino acids are defective.	Kidney stones	Bicarbonate and diuretic administration
Hartnup disease	Renal carriers for tryptophane are defective.	Decreased NAD and NADP within body cells	Nicotinamide administration
Homocystinuria	Enzyme defect results in excessive blood levels of homocystine.	Speech defects, mental retardation	Diet low in methionine, high in cystine
Phenylketonuria	Enzyme defect results in excessive accumulation of phenylalanine and in urinary excretion of phenylpyruvic acid.	Severe mental retardation	Diet low in phenylalanine

Source: Fox, Stuart Ira, *Human Physiology.* © 1984 Wm. C. Brown Publishers, Dubuque, Iowa. All Rights Reserved. Reprinted by permission.

1. Define clearance rate, and describe how it is measured. Explain why the glomerular filtration rate is equal to the clearance rate of inulin.
2. Define the terms reabsorption *and* secretion. Describe how the clearance rate is affected by the processes of reabsorption and secretion, and give examples.
3. Explain why the total renal blood flow can be measured by the clearance rate of PAH.
4. Define transport maximum and renal plasma threshold. Explain why people with diabetes mellitus have glycosuria.

Renal Control of Electrolyte Balance

The reabsorption of Na^+ in exchange for the secretion of K^+ or H^+ in the distal convoluted tubules is regulated by aldosterone. Aldosterone secretion is controlled in a negative feedback fashion by the secretion of renin from the juxtaglomerular apparatus of the kidneys and by a direct effect of blood potassium on the adrenal cortex.

Objective 17. Describe how aldosterone regulates the Na^+/K^+ balance of the blood.
Objective 18. Describe how the secretion of renin and of aldosterone is regulated.
Objective 19. Explain how changes in blood pH affect the plasma concentration of potassium.

The kidneys help regulate the concentrations of plasma electrolytes—sodium, potassium, chloride, bicarbonate, and phosphate—by matching the urinary excretion of these compounds to the amounts ingested. The control of plasma Na^+ is important in the regulation of blood volume and pressure; the control of plasma K^+ is required to maintain proper function of cardiac and skeletal muscles. The regulation of Na^+/K^+ balance is also intimately related to renal control of acid-base balance.

Role of Aldosterone in Na^+/K^+ Balance

Approximately 90% of the filtered Na^+ and K^+ is reabsorbed in the early part of the nephron before the filtrate reaches the distal tubule. This reabsorption occurs at a constant rate and is not subject to hormonal regulation. The final concentration of Na^+ and K^+ in the urine is varied according to the needs of the body by processes that occur in the distal tubule. These processes are regulated by aldosterone *(al″do-ster′ōn),* a steroid hormone secreted from the adrenal cortex.

Sodium Reabsorption. Although 90% of the filtered sodium is reabsorbed in the early region of the nephron, the amount left in the filtrate delivered to the distal convoluted tubule is still quite large. In the absence of aldosterone, 80% of this amount is automatically reabsorbed through the wall of the distal tubule into the peritubular blood; this is 8% of the amount filtered. The amount of sodium excreted without aldosterone is thus 2% of the amount filtered. Although this percentage seems small, the actual amount of sodium this represents is an impressive 30 g per day excreted in the urine. When aldosterone is secreted in maximal amounts, in contrast, all of the sodium delivered to the distal tubule is reabsorbed. Under these conditions urine contains no Na^+ at all.

Potassium Secretion. About 90% of the filtered K^+ is reabsorbed in the early regions of the nephron (mainly in the proximal tubule). In the absence of aldosterone, all of the filtered K^+ that remains is reabsorbed in the distal tubule. In the absence of aldosterone, therefore, no K^+ is excreted in the urine. The presence of aldosterone stimulates the *secretion of K^+* from the peritubular blood into the distal tubule (fig. 26.25). This secretion is the only means by which K^+ can be eliminated in the urine. When aldosterone secretion is maximal, as much as fifty times more K^+ is excreted in the urine, because of secretion into the distal tubule, than was originally filtered through the glomeruli.

In summary, aldosterone promotes sodium retention and potassium loss from the blood by stimulating the

Figure 26.25. Potassium is almost completely reabsorbed in the proximal tubule, but under aldosterone stimulation, it is secreted into the distal tubule. All of the K⁺ in urine is derived from secretion rather than from filtration.

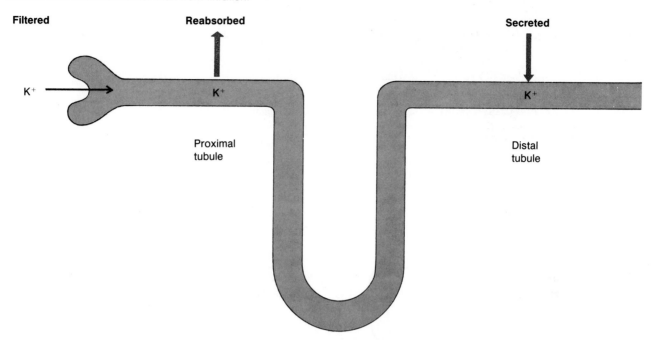

reabsorption of Na⁺ and the secretion of K⁺ across the wall of the distal convoluted tubules. Since aldosterone promotes the retention of Na⁺, it contributes to an increased blood volume and pressure.

> The body cannot get rid of excess K⁺ in the absence of aldosterone-stimulated secretion of K⁺ into the distal tubules. Indeed, when both adrenal glands are removed from an experimental animal, the **hyperkalemia** (hi''per-kah-le'me-ah) (high blood K⁺) that results can produce fatal cardiac arrhythmias. Abnormally low plasma K⁺ concentrations, as might result from excess aldosterone secretion, can produce arrhythmias, muscle weakness, and cramps.

Control of Aldosterone Secretion
Since aldosterone promotes Na⁺ retention and K⁺ loss, one might predict (on the basis of negative feedback) that aldosterone secretion will be increased when there is a low Na⁺ or a high K⁺ concentration in the blood. This indeed is the case. A rise in blood K⁺ *directly* stimulates the secretion of aldosterone from the adrenal cortex. Decreases in plasma Na⁺ concentrations also promote aldosterone secretion, but they do so indirectly.

Juxtaglomerular Apparatus. The juxtaglomerular (juks''tah-glo-mer'u-lar) apparatus is the region in each nephron where the afferent arteriole and distal tubule come into contact (fig. 26.26). The microscopic appearance of the afferent arteriole and distal tubule in this small region differs from the appearance in other regions. **Granular cells** within the afferent arteriole secrete the enzyme **renin** (re'nin) into the blood; this enzyme catalyzes the conversion of angiotensinogen (a protein) into angiotensin I (a ten-amino-acid polypeptide).

Secretion of renin into the blood thus results in the formation of angiotensin I, which is then converted to **angiotensin** (an''je-o-ten'sin) **II** by a *converting enzyme* as blood passes through the lungs and other organs. Angiotensin II, in addition to other effects, stimulates the adrenal cortex to secrete aldosterone. Secretion of renin from the granular cells of the juxtaglomerular apparatus is thus said to initiate the **renin-angiotensin-aldosterone system.** Conditions that result in renin secretion thus cause increased aldosterone secretion and, by this means, promote the reabsorption of Na⁺ in the distal convoluted tubules.

Regulation of Renin Secretion. A fall in plasma Na⁺ concentration is always accompanied by a fall in blood volume. This is because ADH secretion is inhibited by the decreased plasma concentration (osmolality); with less ADH, less water is reabsorbed through the collecting ducts and more is excreted in the urine. The fall in blood volume and the fall in renal blood flow that results causes increased renin secretion. Renin secretion is believed to be due in part to a direct effect of blood flow on the granular cells, which may function as stretch receptors in the afferent arterioles. Renin secretion is also stimulated by sympathetic nerve activity, which is increased when the blood volume and pressure fall.

Figure 26.26. The juxtaglomerular apparatus (*a*) includes the region of contact of the afferent arteriole with the distal tubule. The afferent arterioles in this region contain granular cells with renin, and the distal tubule cells in contact with the granular cells form an area called the macula densa (*b*). The granular cells of the afferent arteriole are innervated by renal sympathetic nerve fibers.

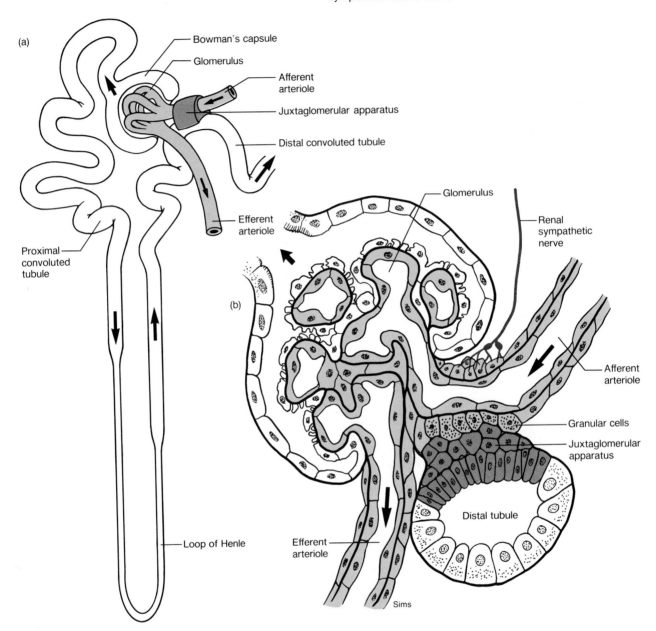

If there is inadequate sodium intake, therefore, the fall in plasma volume that results acts to increase renin secretion. The increased renin secretion acts, via the increased production of angiotensin II, to stimulate aldosterone secretion. In consequence, there is less Na⁺ excreted in the urine. This negative feedback system is illustrated in figure 26.27.

Role of the Macula Densa. The region of the distal tubule in contact with the granular cells of the afferent arteriole is called the **macula densa** (fig. 26.26). There is

evidence that this region helps to inhibit renin secretion when the blood Na^+ concentration is raised.

According to the proposed mechanism, the cells of the macula densa respond to Na^+ within the filtrate delivered to the distal tubule. When the plasma Na^+ concentration is raised, the rate of Na^+ delivered to the distal tubule is also increased. Through an effect on the macula densa, this increase in filtered Na^+ may inhibit the granular cells from secreting renin. Aldosterone secretion thus decreases, and since less Na^+ is reabsorbed in the distal tubule, more Na^+ is excreted in the urine. The regulation of renin and aldosterone secretion is summarized in table 26.7.

macula densa: L. *macula*, a spot; *densitas*, thick

Figure 26.27. The sequence of events by which a low sodium (salt) intake leads to increased sodium reabsorption by the kidneys. The dotted arrow and negative sign show the completion of the negative feedback loop.

Table **26.7**	Regulation of renin and aldosterone secretion			
Stimulus	**Effect on renin secretion**	**Mechanisms**	**Angiotensin II production**	**Aldosterone secretion**
↓Na⁺	Increased	Low blood volume stimulates renal baroreceptors; granular cells release renin.	Increased	Increased
↑Na⁺	Decreased	Increased blood volume inhibits baroreceptors; increased Na⁺ in distal tubule acts via macula densa to inhibit release of renin from granular cells.	Decreased	Decreased
↑K⁺	None	Not applicable	Not changed	Increased
↑Sympathetic nerve activity	Increased	α-adrenergic effect stimulates constriction of afferent arterioles; β-adrenergic effect stimulates renin secretion directly.	Increased	Increased

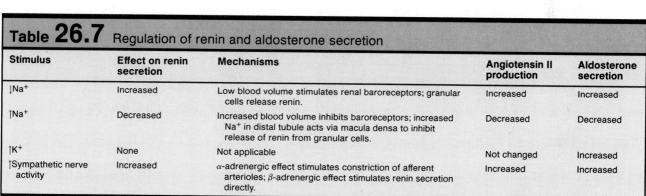

Source: Fox, Stuart Ira, *Human Physiology.* © 1984 Wm. C. Brown Publishers, Dubuque, Iowa. All Rights Reserved. Reprinted by permission.

Natriuretic Hormone. Expansion of the blood volume causes increased salt and water excretion in the urine. This is due in part to an inhibition of aldosterone secretion, as previously described. There is much experimental evidence, however, that the increased salt excretion that occurs under these conditions is due not only to the inhibition of aldosterone secretion, but also to the increased secretion of another substance with hormone properties. This other substance is called **natriuretic** *(na″tre-u-ret′ik)* **hormone** and is so named because it stimulates salt excretion (the opposite of aldosterone's action). The source and chemical nature of natriuretic hormone are not currently known.

---

natriuretic: L. *natrium,* sodium; *urina,* urine

Relationship between Na+, K+, and H+

Hormones, as a general rule, alter the rate of already existing processes. In the absence of aldosterone, the distal tubule reabsorbs 80% of the Na^+ delivered to it; aldosterone can increase this reabsorption to 100%. At a given level of aldosterone, the amount of Na^+ reabsorbed in the distal tubule is a given proportion of the Na^+ delivered to it. Some diuretic drugs inhibit Na^+ reabsorption in the loop of Henle and, therefore, increase the delivery of Na^+ to the distal tubule. As a result, there is an increased reabsorption of Na^+ in the distal convoluted tubule when a person takes these types of diuretics.

The reabsorption of Na^+ in the distal convoluted tubules occurs together with K^+ secretion. These two ions may share the same carrier, but if they do, this carrier does not appear to be the same as the Na^+/K^+ pump in cell membranes and the ratio of Na^+ to K^+ transported by this carrier does not appear to be constant. Because of the Na^+/K^+ exchange in the distal tubule, an increase in Na^+ reabsorption is coupled to an increase in K^+ secretion. People who take diuretics that inhibit Na^+ reabsorption in the loop of Henle, for these reasons, tend to have excessive K^+ secretion into the distal tubules and, therefore, excessive K^+ loss in the urine.

The K^+ loss that occurs with many diuretics may present serious side effects to these medications. If K^+ secretion into the distal convoluted tubules is significantly increased, a condition of **hypokalemia** (low blood K^+) may be produced, which must be compensated for by the increased ingestion of potassium. People who take diuretics for the treatment of high blood pressure are usually on a low-sodium diet and often must supplement their meals with potassium chloride (KCl).

Relationship between K+ and H+.   The plasma K^+ concentration indirectly affects the plasma H^+ concentration (pH). Changes in plasma pH likewise affect the K^+ concentration of the blood. These effects serve to stabilize the *ratio* of K^+ to H^+. When the extracellular H^+ concentration increases, for example, some of the H^+ moves into the tissue cells and causes cellular K^+ to diffuse outward into the extracellular fluid. The plasma concentration of H^+ is thus decreased while the K^+ increases, helping to restabilize the ratio of these ions in the extracellular fluid.

This exchange of H^+ for K^+ also occurs in the cells of the distal tubule when the plasma pH is lowered (fig. 26.28). Under these conditions the tubule cells contain more H^+ and less K^+ than before. Whenever the tubule

Figure 26.28.   In the distal tubule, K^+ and H^+ are secreted in exchange for Na^+. High concentrations of H^+ may therefore decrease K^+ secretion, and vice versa.

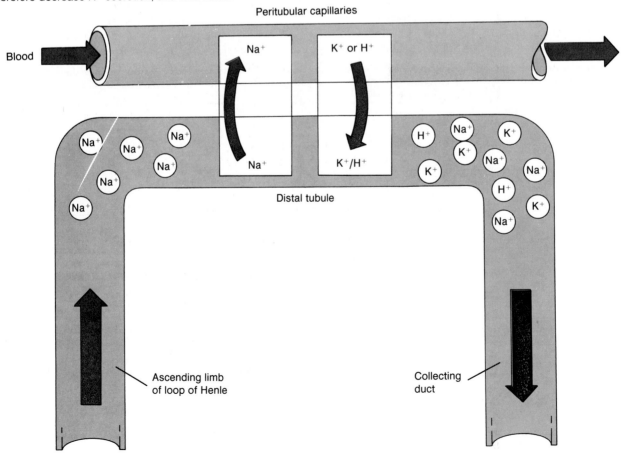

cells reabsorb Na$^+$ from the filtrate, they can replace it by secreting either K$^+$ or H$^+$ (they usually secrete both ions). When the tubule cells contain increased amounts of H$^+$ as a result of the fall in plasma pH, however, they secrete increased amounts of H$^+$ at the expense of K$^+$, which helps to lower the plasma H$^+$ concentration and make the urine more acidic. As a side effect, less K$^+$ is secreted; the plasma concentration of K$^+$ may thus be increased under these conditions.

Hyperkalemia (high blood K$^+$) may thus occur in metabolic acidosis. Metabolic alkalosis can have the opposite effect, resulting in lowered plasma K$^+$ concentrations. This lowering of plasma K$^+$ is sometimes induced intentionally by the infusion of bicarbonate in patients with heart or renal failure, who are susceptible to hyperkalemia as a result of decreased glomerular filtration.

Aldosterone appears to stimulate the secretion of H$^+$ as well as K$^+$ into the distal tubules. Abnormally high aldosterone secretion, in **primary aldosteronism,** or **Conn's syndrome,** therefore results in both hypokalemia and metabolic alkalosis. Conversely, abnormally low aldosterone secretion, as occurs in **Addison's disease,** can produce hyperkalemia and metabolic acidosis.

---

1. *Describe the effects of aldosterone on the renal nephrons, and explain how aldosterone secretion is regulated.*
2. *Describe how changes in plasma Na$^+$ concentrations regulate renin secretion, and explain how the secretion of renin acts to help regulate the plasma Na$^+$ concentration.*
3. *Explain why people who take some diuretic drugs may suffer from hypokalemia and why they must supplement their diets with potassium.*
4. *Explain the mechanisms involved in the lowering of plasma K$^+$ concentrations by the intravenous infusion of bicarbonate.*

Renal Control of Acid-Base Balance

The kidneys help regulate the acid-base balance of the blood through their excretion of H$^+$ and their reabsorption of HCO$_3^-$. By means of these processes, the kidneys are responsible for the metabolic component of acid-base balance.

Objective 20. Describe the respiratory and metabolic components of acid-base balance, and explain how these relate to the bicarbonate and carbonic acid concentrations of the blood.

---

hypokalemia: Gk. *hypo,* under; L. *kalium,* potassium;
Gk. *haima,* blood
Addison's disease: from Christopher Addison, English
anatomist, 1869–1951

Objective 21. Describe how filtered bicarbonate is reabsorbed by the nephron.
Objective 22. Explain how the nephrons help compensate for acidosis and alkalosis, and describe how acid is excreted in the urine.

Normal arterial blood has a pH of 7.37–7.45. It should be recalled that the pH number is *inversely* related to the H$^+$ concentration. An increase in H$^+$, derived from carbonic acid or the nonvolatile metabolic acids, can lower blood pH. An increase in H$^+$ derived from metabolic acid, however, is normally prevented from changing the blood pH by **bicarbonate buffer,** which combines with and thus removes from solution the excess H$^+$.

Respiratory and Metabolic Components of Acid-Base Balance

Acid-base balance has both respiratory and metabolic components. The respiratory component describes the effect of ventilation on arterial P$_{CO_2}$ and thus on the production of carbonic acid (H$_2$CO$_3$). The metabolic component describes the effect of nonvolatile metabolic acids—lactic acid, fatty acids, and ketone bodies—on blood pH. Since these acids are normally buffered by bicarbonate (HCO$_3^-$), the metabolic component can be described in terms of the free HCO$_3^-$ concentration. An increase in metabolic acids "uses up" free bicarbonate, as HCO$_3^-$ is converted to H$_2$CO$_3$, and is thus associated with a fall in plasma HCO$_3^-$ concentrations. A decrease in metabolic acids, conversely, is associated with a rise in free HCO$_3^-$.

Since the respiratory component of acid-base balance is represented by the plasma P$_{CO_2}$ and carbonic acid concentrations and the metabolic component is represented by the free bicarbonate concentrations, the study of acid-base balance can be considerably simplified. A normal plasma pH of 7.40 is obtained when the ratio of bicarbonate to carbonic acid concentrations is twenty to one:

$$pH \propto \frac{HCO_3^- \text{ concentration}}{H_2CO_3 \text{ concentration}} = \frac{20}{1} = 7.40$$

($\propto$ means "is proportional to")

A change in this ratio results in an abnormal blood pH. Pure respiratory acidosis *(as''i-do'sis)* or alkalosis *(al''kah-lo'sis)* occurs when the HCO$_3^-$ concentration is normal but the P$_{CO_2}$ and H$_2$CO$_3$ concentration are altered. Pure metabolic acidosis or alkalosis occurs when the P$_{CO_2}$ and H$_2$CO$_3$ are normal but the HCO$_3^-$ concentration is abnormal. This classification is summarized in table 26.8.

A normal blood pH is produced when the P$_{CO_2}$ is 40 mm Hg and the bicarbonate concentration is 24 milliequivalents (mEq) per liter (one mEq is equal to one millimole times the valence of the ion; in this case, since the valence is 1, mEq = mmole). Table 26.9 shows how changes in P$_{CO_2}$ alter the ratio of HCO$_3^-$ to H$_2$CO$_3$ and thus affect blood pH. For the sake of simplicity, this table

Table 26.8 Classification of metabolic and respiratory components of acidosis and alkalosis

P_{CO_2}	HCO_3^-	Condition	Causes
Normal	Low	Metabolic acidosis	Increased production of "nonvolatile" acids (lactic acid, ketone bodies, and others), or loss of HCO_3^- in diarrhea
Normal	High	Metabolic alkalosis	Vomiting of gastric acid; hypokalemia; excessive steroid administration
Low	Normal	Respiratory alkalosis	Hyperventilation
High	Normal	Respiratory acidosis	Hypoventilation

Source: Fox, Stuart Ira, *Human Physiology.* © 1984 Wm. C. Brown Publishers, Dubuque, Iowa. All Rights Reserved. Reprinted by permission.

Table 26.9 The effect of changes in P_{CO_2} on the blood pH

P_{CO_2} (mm Hg)	H_2CO_3 (mEq/L)	HCO_3^- (mEq/L)	HCO_3^-/H_2CO_3 ratio	pH	Condition
20	0.6	24	40/1	7.70	Respiratory alkalosis
30	0.9	24	26.7/1	7.53	Respiratory alkalosis
40	1.2	24	20/1	7.40	Normal
50	1.5	24	16/1	7.30	Respiratory acidosis
60	1.8	24	13.3/1	7.22	Respiratory acidosis

Note: A blood pH of less than 7.35 due to high P_{CO_2} is called respiratory acidosis. A blood pH greater than 7.45 due to low P_{CO_2} is called respiratory alkalosis.
Source: Fox, Stuart Ira, *Human Physiology.* © 1984 Wm. C. Brown Publishers, Dubuque, Iowa. All Rights Reserved. Reprinted by permission.

shows the effect of changing P_{CO_2} when the bicarbonate concentration remains at a constant, normal value. Such changes in P_{CO_2} are produced by hyperventilation and hypoventilation and result in respiratory alkalosis and acidosis, respectively.

A normal plasma P_{CO_2} is maintained by proper lung function. A normal plasma bicarbonate concentration of 24 mEq per L is maintained, despite the continuous production of metabolic acids, by proper kidney function. The kidneys normally reabsorb all of the filtered bicarbonate. During alkalosis, however, the kidneys can reabsorb bicarbonate in excess of H^+, which is excreted in the urine. In summary, the plasma bicarbonate concentration is the responsibility of the kidneys, whereas the plasma carbonic acid concentration is regulated by the lungs:

$$pH \propto \frac{HCO_3^-}{H_2CO_3} \text{ or } \frac{\text{Kidney function}}{\text{Lung function}}$$

Mechanisms of Renal Acid-Base Regulation

The kidneys help regulate the blood pH by excreting H^+ in the urine and by the reabsorption of bicarbonate. These two mechanisms are interdependent—the reabsorption of bicarbonate occurs as a result of the filtration and secretion of H^+. The kidneys normally reabsorb all of the filtered bicarbonate and excrete about 40–80 mEq of H^+ per day. Normal urine, therefore, is free of bicarbonate and is slightly acidic (with a pH range between 5 and 7).

The apical membranes of the tubule cells are impermeable to bicarbonate. The reabsorption of bicarbonate must therefore occur indirectly. When the urine is acidic, HCO_3^- combines with H^+ to form carbonic acid. Carbonic acid in the filtrate is then converted to CO_2 and H_2O by the action of **carbonic anhydrase.** This enzyme is located in the apical cell membrane that faces the filtrate.

Notice that the reaction that occurs in the tubule is the same one that occurs within the red blood cells in pulmonary capillaries.

Carbon dioxide, unlike bicarbonate, can easily pass from the filtrate, through the tubule cells, and enter the blood. Once inside the red blood cells the CO_2 combines with water to form carbonic acid. This reaction is also catalyzed by carbonic anhydrase and results in the addition of equal amounts of H^+ and HCO_3^- to the blood (fig. 26.29).

Production of Bicarbonate in Tubules. The tubule cell cytoplasm also contains carbonic anhydrase. Some of the CO_2 that enters the tubule cells can thus be converted to carbonic acid, which can in turn dissociate to HCO_3^- and H^+ within the tubule cells. Under acidotic conditions, the H^+ produced in this way is secreted back into the filtrate, whereas the bicarbonate diffuses across the basolateral membranes and enters the blood. More bicarbonate than H^+ is thus returned to the blood as a partial compensation for the state of acidosis.

During alkalosis there is less H^+ secreted into the filtrate. Since the reabsorption of filtered bicarbonate requires the combination of HCO_3^- with H^+ to form carbonic acid, less bicarbonate is reabsorbed. This results in urinary excretion of bicarbonate, which helps to partially compensate for the alkalosis.

In this way, disturbances in acid-base balance caused by respiratory problems can be partially compensated for by changes in plasma bicarbonate concentrations. Metabolic acidosis or alkalosis—in which changes in bicarbonate concentrations occur as the primary disturbance—can be similarly compensated for in part by changes in ventilation. These interactions of the respiratory and metabolic components of acid-base balance are shown in table 26.10.

Figure 26.29. Mechanisms of bicarbonate reabsorption.

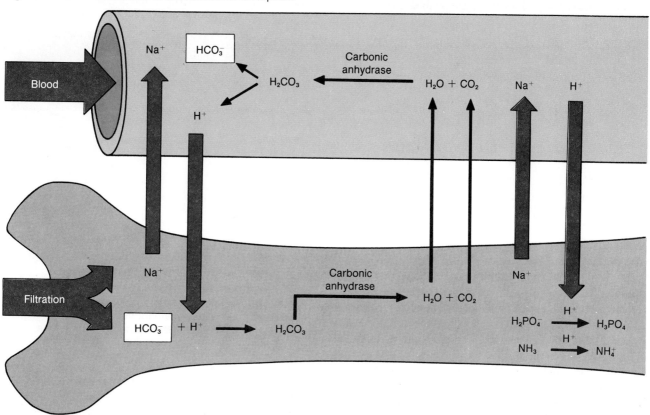

Table **26.10**	Categories of disturbances in acid-base balance, including those that involve mixtures of respiratory and metabolic components		

P_{CO_2} (mmHg)		Bicarbonate (mEq/L)	
	Less than 21	21–26	More than 26
More than 45	Combined metabolic and respiratory acidosis	Respiratory acidosis	Metabolic alkalosis and respiratory acidosis
35–45	Metabolic acidosis	Normal	Metabolic alkalosis
Less than 35	Metabolic acidosis and respiratory alkalosis	Respiratory alkalosis	Combined metabolic and respiratory alkalosis

Source: Fox, Stuart Ira, *Human Physiology.* © 1984 Wm. C. Brown Publishers, Dubuque, Iowa. All Rights Reserved. Reprinted by permission.

Urinary Buffers. In order for H^+ to be excreted in the urine without making the urine damagingly (and perhaps painfully) acidic, the acid must be buffered. Bicarbonate cannot serve this function because it is normally completely reabsorbed. Instead, the urine pH is usually prevented from falling below a value of about 4.6 by the buffering action of phosphates (mainly $H_2PO_4^-$) and ammonia (NH_3). Phosphate enters the urine by filtration. Ammonia (whose presence is strongly evident in a diaper pail or kitty litter box) is produced in the tubule cells by deamination of amino acids. These molecules buffer H^+ as described in the following equations:

$$NH_3 + H^+ \rightarrow NH_4^+ \text{ (ammonium ion)}$$
$$H_2PO_4^- + H^+ \rightarrow H_3PO_4$$

1. Describe the respiratory and metabolic components of acid-base balance, and explain how these components are represented by the carbonic acid and bicarbonate concentration of blood.
2. Explain how the kidneys reabsorb filtered bicarbonate and how this process is affected by acidosis and alkalosis.
3. Suppose a person with diabetes mellitus has an arterial pH of 7.30, an abnormally low arterial P_{CO_2}, and an abnormally low bicarbonate concentration. Identify the type of acid-base disturbance, and explain how these values might have been produced.

Ureters, Urinary Bladder, and Urethra

Urine is channeled from the kidneys to the urinary bladder by the ureters and expelled from the body through the urethra. Muscles of the urinary bladder and urethra are used in the control of micturition.

Objective 23. Describe the location, structure, and function of the ureters.

Objective 24. Discuss the gross and histological structure of the urinary bladder and its innervation.

Objective 25. Explain the process of micturition.

Objective 26. Compare and contrast the structure of the male and female urethra.

Ureters

The **ureters,** like the kidneys, are retroperitoneal in location. Each ureter is a tubular organ about 25 cm (10 in.) long, which begins at the renal pelvis and courses inferiorly to enter the urinary bladder at the superior lateral angle of its base. The thickest portion of the ureter is near where it enters the urinary bladder and is approximately 1.7 cm (0.5 in.) in diameter.

The wall of the ureter consists of three layers, or tunicas. The inner **mucosa** *(mu-ko′sah)* is continuous with the linings of the renal tubules and the urinary bladder. The mucosa consists of transitional epithelium (fig. 26.30). The cells of this layer secrete a mucus that lubricates the walls of the ureter with a protective film. The middle layer of the ureter is called the **muscularis.** It consists of an inner longitudinal and an outer circular layer of smooth muscle. In addition, the proximal one-third of the ureter contains another longitudinal layer to the outside of the circular layer. Muscular peristaltic waves move the urine through the ureter. The peristaltic waves are initiated by the presence of urine in the renal pelvis, and their frequency is determined by the volume of urine. The waves, which occur from every few seconds to every few minutes, force urine through the ureter and cause it to spurt into the urinary bladder. The outer layer of the ureter is called the **fibrous coat.** The fibrous coat is composed of areolar connective tissue that not only covers the ureter but has extensions that anchor the ureter in place.

The arterial supply of the ureter comes from several sources. Branches from the renal artery serve the superior portion. The testicular (or ovarian) artery supplies the middle portion, and the superior vesicular artery serves the pelvic region. The venous return is through corresponding veins.

Figure 26.30. A photomicrograph of a ureter in transverse section at a magnification of 63✕.

Lamina propria

Transitional epithelium

A **calculus,** or kidney stone, may obstruct the ureter and produce a tremendous amount of peristaltic waves in an attempt to pass the stone. The pain from a lodged calculus is extreme and extends throughout the pelvic area. A lodged calculus also causes a sympathetic ureterorenal reflex that results in constriction of renal arterioles, which reduces the production of urine in the kidney on the affected side.

Urinary Bladder

The **urinary bladder** is a storage sac for urine. It is located posterior to the symphysis pubis and anterior to the rectum. In females, the urinary bladder is in contact with the uterus and vagina. In males, the prostate gland is positioned below the urinary bladder (fig. 26.31).

The shape of the urinary bladder is determined by the volume of urine it contains. An empty urinary bladder is pyramidal in shape, having an anteroinferior apex, a superior surface, two inferolateral surfaces, a base (posterior surface), and a neck. The apex of the urinary bladder is superior to the symphysis pubis and is secured to the **median umbilical ligament** by a fibrous cord called the **urachus.** The base of the urinary bladder receives the ureters along the superolateral angles, and the urethra exits at the inferior angle. The urethra is a tubular continuation of the neck of the urinary bladder.

As the urinary bladder fills, it loses its pyramidal shape and becomes ovoid as the superior surface enlarges and bulges upward into the abdominal cavity (fig. 26.32). The urinary bladder is antiperitoneal when it is empty, but the peritoneal cover is retracted back along the superior surface when the bladder distends.

The wall of the urinary bladder consists of four layers: the mucosa, submucosa, muscularis, and serosa (adventitia). The **mucosa** is the innermost layer. It is composed

---

calculus: L. *calculus,* small stone

Figure 26.31.   The male urinary bladder and urethra. (*a*) a coronal view; (*b*) a posterior view.

(a)

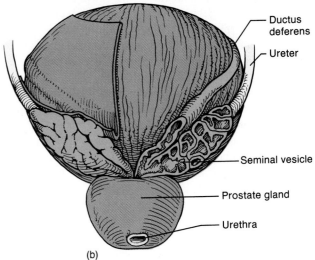

(b)

Figure 26.32.   The position of an empty and a distended urinary bladder.

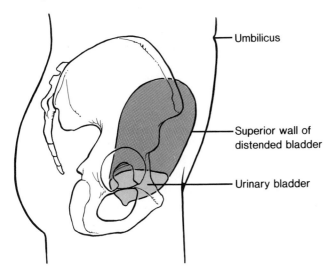

of transitional epithelium, which decreases in thickness as the urinary bladder distends and the cells are stretched. Further distension is permitted by folds of the mucosa, called **rugae,** which can be seen when the urinary bladder is empty (fig. 26.31). Fleshy flaps of mucosa located where the ureters pierce into the urinary bladder act as valves over the openings of the ureters to prevent a reverse flow of urine toward the kidneys as the bladder fills. A triangular area known as the **trigone** *(tri'gōn)* is formed on the mucosa between the two ureter openings and the single urethral opening. The internal trigone lacks rugae and is therefore smooth in appearance and remains relatively fixed in position as the urinary bladder changes shape during distension and contraction.

trigone: L. *trigonum,* triangle

The second layer of the urinary bladder is the **submucosa,** which functions to support the mucosa. The **muscularis** consists of three interlaced smooth muscle layers and is referred to as the **detrusor muscle.** At the neck of the urinary bladder, the detrusor muscle is modified to form the upper of two muscular sphincters surrounding the urethra. The outer covering of the urinary bladder is the **serosa.** It appears only on the superior surface of the bladder and is actually a continuation of the peritoneum.

The arterial supply to the urinary bladder comes from the **superior** and **inferior vesicular arteries,** which arise from the internal iliac arteries. Blood draining from the urinary bladder enters from a **vesicular venous plexus** and empties into the internal iliac veins.

The nerves associated with the urinary bladder are derived from pelvic plexuses. The sympathetic fibers arise from the last thoracic and the first and second lumbar spinal nerves. They innervate the trigone, ureteral openings, and blood vessels of the urinary bladder. The parasympathetic innervation arises from the second, third, and fourth sacral nerves. These fibers serve the detrusor muscle. The afferent sensory fibers are specialized stretch receptors that respond to distension and relay impulses to the central nervous system via the pelvic spinal nerves.

The urinary bladder becomes infected easily, particularly in women because of their short urethra, which increases the possibility of contamination. A urinary bladder infection, called **cystitis,** may easily ascend from the bladder to the ureters since the mucous linings are continuous. An infection that involves the renal pelvis is called **pyelitis;** if it continues into the nephrons, it is known as **nephritis.**

detrusor: L. *detrudere,* thrust or forced down

Urethra

The tubular **urethra** *(u-re'thrah)* conveys urine from the urinary bladder to the outside of the body. The urethral wall has an inside lining of mucous membrane surrounded by a relatively thick layer of smooth muscle whose fibers are directed longitudinally. Specialized **urethral glands** are embedded in the urethral wall and function to secrete mucus into the urethral canal.

Two muscular sphincters surround the urethra. The upper, involuntary smooth muscle sphincter is the **internal urethral sphincter** (sphincter vesicae), which is formed from the detrusor muscle of the urinary bladder. The lower sphincter is composed of voluntary, striated muscle fibers and is called the **external urethral sphincter** (sphincter urethrae) (fig. 26.31).

The urethra of the female is a simple tube about 4 cm (1.5 in.) long, which empties urine through the **urethral orifice** *(or'i-fis)* into the vestibule between the labia minora. The urethral orifice is positioned anterior to the vaginal orifice and about 2.5 cm posterior to the clitoris.

The urethra of the male serves both the urinary and reproductive systems. It is about 20 cm (8 in.) long and S-shaped because of the shape of the penis. Three regions can be identified in the male urethra: the prostatic urethra, membranous urethra, and penile urethra (fig. 26.33).

The **prostatic urethra** is the proximal portion, about 2.5 cm long, that passes through the **prostate gland** located near the neck of the urinary bladder. The prostatic urethra receives drainage from small ducts of the prostate gland and two **ejaculatory ducts** of the reproductive system.

The **membranous urethra** is the short (0.5 cm) portion of the urethra that passes through the urogenital diaphragm. The external urethral muscle is located in this region.

The **penile urethra** (cavernous urethra) is the longest portion (15 cm), extending from the outer edge of the urogenital diaphragm to the external urethral orifice on the glans penis. This portion is surrounded by erectile tissue as it passes through the corpus spongiosum of the penis. The paired ducts of the **bulbourethral glands** (Cowper's glands) of the reproductive system attach to the penile urethra near the urogenital diaphragm.

Micturition

Micturition *(mik"tu-rish'un)*, commonly called urination or voiding, is a reflex action that expels urine from the urinary bladder. It is a complex function that requires a stimulus from the urinary bladder and a combination of involuntary and voluntary nerve impulses to the appropriate muscular structures of the bladder and urethra.

In young children, micturition is a simple reflex action that occurs when the urinary bladder becomes sufficiently distended. Voluntary control of micturition is

micturition: L. *micturire*, to urinate

Figure 26.33.   A longitudinal section of a male urethra.

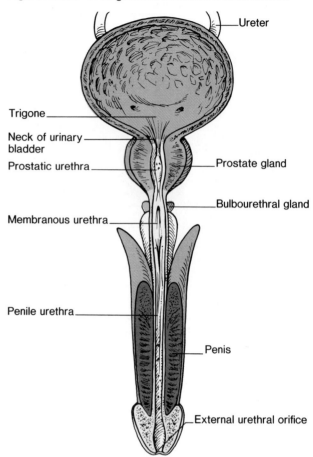

normally developed at the time a child is two or three years old. Voluntary control requires the development of an inhibitory ability by the cerebral cortex and a maturing of various portions of the spinal cord.

The volume of urine produced by an adult averages about 1,200 ml per day, but it can vary from 600–2,500 ml. The average capacity of the urinary bladder is 700–800 ml. A volume of 200–300 ml will distend the bladder enough to stimulate stretch receptors and trigger the micturition reflex, creating a desire to urinate.

The micturition reflex center is located in the second, third, and fouth sacral segments of the spinal cord. Following stimulation of this center by impulses arising from stretch receptors in the urinary bladder, parasympathetic nerves that stimulate the detrusor muscle and the internal urethral sphincter are activated. Stimulation of these muscles causes a rhythmic contraction of the bladder wall and a relaxation of the internal sphincter. At this point, a sensation of urgency is perceived in the brain, but there is still voluntary control over the external urethral sphincter. At the appropriate time, the conscious activity of the brain activates the motor nerve fibers (S4) to the external urethral sphincter via the pudendal nerve (S2, S3, and S4), causing the sphincter to relax and urination to occur.

Figure 26.34.   Innervation of the ureter, urinary bladder, and urethra.

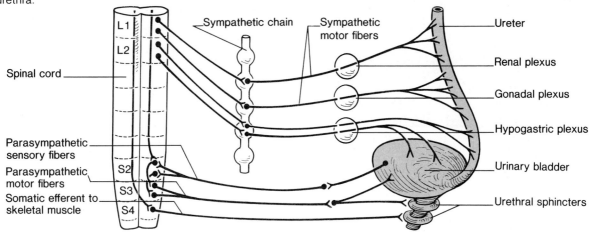

Table **26.11** Events of micturition	
1. The urinary bladder becomes distended as it fills with urine.	6. Urination is prevented by voluntary contraction of the external urethral sphincter and by inhibition of the micturition reflex by impulses from the midbrain and cerebral cortex.
2. Stretch receptors in the bladder wall are stimulated, and impulses are sent to the micturition center in the spinal cord.	7. Following the decision to urinate, the external urethral sphincter is relaxed, and the micturition reflex is facilitated by impulses from the pons and the hypothalamus.
3. Parasympathetic nerve impulses travel to the detrusor muscle and the internal urethral sphincter.	8. The detrusor muscle contracts, and urine is expelled through the urethra.
4. The detrusor muscle contracts rhythmically, and the internal urethral sphincter relaxes.	9. Neurons of the micturition reflex center are inactivated, the detrusor muscle relaxes, and the bladder begins to fill with urine.
5. The need to urinate is sensed as urgent.	

Source: Van De Graaff, Kent M., *Human Anatomy.* © 1984 Wm. C. Brown Publishers, Dubuque, Iowa. All Rights Reserved. Reprinted by permission.

The innervation of the ureter, urinary bladder, and urethra is shown in figure 26.34, and a summary of the process of micturition is presented in table 26.11.

Urinary retention, or the inability to void, may occur postoperatively, especially following surgery of the rectum, colon, or internal reproductive organs. The difficulty may be due to nervous tension, the effects of anesthetics, or pain and edema at the site of the operation. If urine is retained beyond six to eight hours, **catheterization** may become necessary. In this procedure, a tube or catheter is passed through the urethra into the urinary bladder so that urine can flow freely.

1. *Describe the structure of a ureter, and indicate the function of its muscularis layer.*
2. *Describe the structure of the urinary bladder, and indicate the structures that allow it to distend.*
3. *Compare the male urinary system with that of the female.*
4. *Explain the structures and processes involved in the control of micturition.*

catheterization: L. *catheter,* to let down

Clinical Considerations

The importance of kidney function in maintaining homeostasis and the ease with which urine can be collected and used as a mirror of the plasma's chemical composition make the clinical study of renal function and urine composition particularly significant. **Urology** is the medical specialty concerned with dysfunctions of the urinary system. Urinary dysfunctions can be congenital, acquired, due to physical trauma, or the result of conditions that secondarily involve the urinary organs.

Use of Diuretics
People who need to lower their blood volume because of hypertension, congestive heart failure, or edema take medications that increase the volume of urine excreted. Such medications are called **diuretics.** There are a large number of diuretic drugs in clinical use that act on the renal nephrons in different ways (table 26.12). Based on their chemical structure and characteristics, the commonly used clinical diuretics are categorized as carbonic anhydrase inhibitors, loop diuretics, thiazides, and potassium-sparing diuretics. The loop diuretics are the most

Table 26.12 Actions of different classes of diuretics			
Category of diuretic	**Example**	**Mechanism of action**	**Major site of action**
Carbonic anhydrase inhibitors	Acetazolamide	Inhibits reabsorption of bicarbonate	Proximal tubule
Loop diuretics	Furosemide	Inhibits chloride transport	Thick segments of ascending limbs
Thiazides	Hydrochlorothiazide	Inhibits sodium transport	Distal convoluted tubule
Potassium-sparing diuretics	Spironolactone	Inhibits action of aldosterone	Distal convoluted tubule
	Triamterene	Inhibits Na^+/K^+ exchange	Distal convoluted tubule

powerful, followed by the thiazides. Carbonic acid inhibitors promote water excretion by inhibiting the reabsorption of water that occurs when bicarbonate is reabsorbed. The other diuretics promote the excretion of Na^+. Except for the potassium-sparing diuretics, a side effect of diuretic use is the excessive loss of K^+ in the urine. The potassium-sparing diuretics act by inhibiting aldosterone action or by more directly preventing the secretion of K^+ into the tubular fluid.

Congenital Malformations of the Urinary Tract
Abnormalities of the organs of the urinary tract occur in about 12% of newborn babies. These deviations range from insignificant anomalies to those that are incompatible with life.

Kidneys. Common malformations of the kidneys are illustrated in figure 26.35. One common deformity is the unilateral absence of a kidney, called **renal agenesis** *(ah-jen'ē-sis),* caused when a uretic bulb fails to develop. **Renal ectopia** means that one or both kidneys are in an abnormal position. Generally, this condition occurs when a kidney remains in the pelvic area. **Horseshoe kidney** refers to a fusion of the kidneys across the midline. The incidence of this asymptomatic condition is about one in 600.

Ureters. **Duplication of the ureters** and the associated renal pelvis is a frequent anomaly of the urinary tract. Unilateral duplication occurs in one in 200 births, whereas bilateral duplication occurs in one in 1,200 births. Occasionally there is partial duplication, which may become very important because of the propensity for urinary infections. A completely duplicated ureter frequently opens into areas other than the urinary bladder and requires surgical correction.

Urinary Bladder. Protrusion of the posterior wall of the urinary bladder is called **exstrophy of the bladder** and occurs when the wall of the perineum fails to close. Associated with this condition, which occurs in about one in 40,000 births, are defective urethral sphincter muscles.

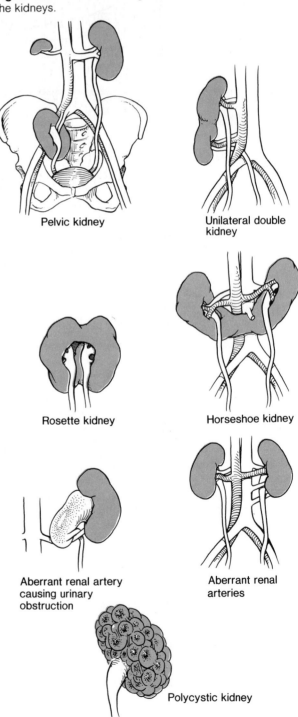

Figure 26.35. Various congenital anomalies involving the kidneys.

Pelvic kidney

Unilateral double kidney

Rosette kidney

Horseshoe kidney

Aberrant renal artery causing urinary obstruction

Aberrant renal arteries

Polycystic kidney

Figure 26.36. Hypospadias and epispadias of the male urethra.

Epispadias   Hypospadias

Figure 26.37. Cystoscopic examination of a male.

Urethra. The most common anomaly of the urethra is a condition called **hypospadias** *(hi″po-spa′de-ah),* where the urethra of the male opens on the ventral surface of the penis instead of at the tip of the glans (fig. 26.36); there is a similar defect in the female in which the urethra opens into the vagina. **Epispadias** is a failure of closure on the dorsal surface of the penis. Hypospadias and epispadias can be corrected surgically.

Symptoms and Diagnosis of Urinary Disorders

Normal micturition is painless. **Dysuria** *(dis-u′re-ah),* or painful urination, is a sign of a urinary tract infection or obstruction of the urethra—as in an enlarged prostate gland in a male. **Hematuria** means blood in the urine and is usually associated with trauma. **Bacteriuria** means bacteria in the urine, and **pyuria** is the term for pus in the urine, which may result from a prolonged infection. **Oliguria** is a scanty output of urine, whereas **polyuria** is an excessive output. Low blood pressure and kidney failure are two causes of oliguria. **Uremia** is a condition in which substances ordinarily excreted in the urine accumulate in the blood. **Enuresis** *(en″u-re′sis),* or **incontinence,** is the inability to control micturition. Its causes range from psychosomatic sources to actual physical impairment.

The palpation and inspection of urinary organs is an important aspect of physical assessment. The right kidney is palpable in the supine position; the left kidney usually is not. The distended urinary bladder is palpable along the superior pelvic rim.

The urinary system may be examined using X-ray techniques. An **intravenous pyelogram** (IVP) permits X-ray examination of the kidneys following the injection of radiopaque dye. In this procedure, the dye that has been injected intravenously is excreted by the kidneys so that the renal pelvises and the outlines of the ureters and urinary bladder can be observed in an X-ray.

Cystoscopy *(sis-tos′ko-pe)* is the inspection of the inside of the urinary bladder by means of an instrument called a cystoscope (fig. 26.37). By using this technique, tissue samples can be obtained as well as urine samples from each kidney prior to mixing in the bladder. Once the cystoscope is in the bladder, the ureters and pelvis can be viewed through urethral catheterization and inspection for obstructions.

A **renal biopsy** is a diagnostic test for evaluating certain types and stages of kidney diseases. The biopsy is performed either through a skin puncture (closed biopsy) or through a surgical incision (open biopsy).

Urinalysis is a simple and important laboratory aspect of a physical examination. The voided urine specimen is tested for color, specific gravity, chemical composition, and for the presence of microscopic bacteria, crystals, and *casts.* Casts are accumulations of proteins that have leaked through the glomeruli and have been pushed through the tubules like toothpaste through a tube.

Infections of Urinary Organs

Urinary tract infections are a significant cause of illness and are also a major factor in the development of chronic renal failure. Females are more predisposed to urinary tract infections than are males, and the incidence of infection increases directly with sexual activity and aging. The higher infection rate in females has been attributed to their shorter urethra, which has a close proximity to the rectum, and to the lack of protection offered by prostatic secretions in males. To reduce the risk of urinary infections, a young girl should be taught to wipe her anal

hypospadias: Gk. *hypo,* under; *spadix,* torn branch
epispadias: Gk. *epi,* upon; *spadix,* torn branch

region in a posterior direction, away from the urethral orifice, after a bowel movement.

Infections of the urinary tract are named according to the infected organ. An infection of the urethra is called **urethritis,** and involvement of the urinary bladder is **cystitis.** Cystitis is frequently a secondary infection from some other part of the urinary tract.

Nephritis means inflammation of the kidney tissue. **Glomerulonephritis** *(glo-mer″u-lo-nĕ-fri′tis)* is inflammation of the glomeruli. Glomerulonephritis frequently occurs following an upper respiratory tract infection, because antibodies produced against streptococci bacteria can produce an autoimmune inflammation in the glomeruli. This inflammation may permanently change the glomeruli and figure significantly in the development of chronic renal disease and renal failure.

Any interference with the normal flow of urine, such as from a kidney stone or an enlarged prostate gland in a male, causes stagnation of urine in the renal pelvis and the development of pyelitis. **Pyelitis** is an inflammation of the renal pelvis and its calyces. **Pyelonephritis** is inflammation involving the renal pelvis, the calyces, and the tubules of the nephron within one or both kidneys. Bacterial invasion from the blood or from the lower urinary tract is another cause of both pyelitis and pyelonephritis.

Trauma to Urinary Organs

A sharp blow to a lumbar region of the back may cause a contusion or rupture of a kidney. Symptoms of kidney trauma include hematuria and pain of the upper abdominal quadrant and flank on the injured side.

Pelvic fractures from accidents may result in perforation of the urinary bladder and urethral tearing. When driving an automobile, it is advisable to stop periodically to urinate because an attached seat belt over the region of a full urinary bladder can cause it to rupture in even a relatively minor accident. Urethral injuries are more common in men than women because of the position of the urethra in the penis. "Straddle" injuries are those where, for example, a man walking along a raised beam slips and compresses his urethra and penis between the hard surface and his pubic arch, rupturing the urethra.

Obstruction. The urinary system can become obstructed anywhere along the tract. Calculi (stones) are the most common cause, but blockage can also come from trauma, strictures, tumors or cysts, spasms or kinks of the ureters, or congenital anomaly. If not corrected, an obstruction causes urine to collect behind the blockage and generate pressure that may cause permanent functional and anatomic damage to one or both kidneys. Pressure buildup in a ureter causes a distended ureter to develop, called a **hydroureter.** Dilation in the renal pelvis is called **hydronephrosis.**

Calculi, or kidney stones, are generally the result of infections or metabolic disorders that cause the excretion of large amounts of organic and inorganic substances. As the urine becomes concentrated these substances may crystalize and form granules. The granule then serves as a core for further precipitation and development into a larger calculus. This becomes dangerous when the calculus is sufficiently large to cause an obstruction. It also causes intense pain when it passes through the urinary tract.

Renal Failure. An output of 50–60 cc of urine per hour is considered normal. If the output drops to less than 30 cc per hour, it may indicate renal failure. Renal failure is the loss of the kidney's ability to maintain fluid and electrolyte balance and to excrete waste products.

Renal failure can be either acute or chronic. **Acute renal failure** is the sudden loss of kidney function caused by shock and hemorrhage, thrombosis, or other physical trauma to the kidneys. The kidneys may sustain a 90% loss of their nephrons through tissue death and still not have an obvious loss of function. If a patient suffering acute renal failure is stabilized, the nephrons have an excellent capacity to regenerate.

A person with **chronic renal failure** cannot sustain life independently. Chronic renal failure is the end result of kidney disease in which the kidney tissue is progressively destroyed. As renal tissue continues to deteriorate, the options for sustaining life are hemodialysis or kidney transplantation.

Hemodialysis. Hemodialysis equipment is designed to filter the wastes from the blood of a patient who has chronic renal failure. During hemodialysis, the blood of a patient is pumped via a tube from the radial artery through the machine where it is cleansed and then returned to the body via a vein (fig. 26.38). The cleaning process involves pumping the blood past a semipermeable cellophane membrane, which separates the blood from an isotonic solution containing molecules needed by the body (such as glucose). In a process called **dialysis,** waste products diffuse out of the blood through the membrane while glucose and other molecules needed by the body remain in the blood.

Urinary Incontinence. The inability to voluntarily retain urine in the urinary bladder is known as urinary incontinence. It has a number of temporary or permanent causes. Emotional stress is a cause of temporary incontinence in adults. Causes of permanent incontinence include neurological trauma, various urinary diseases, and tissue damage within the urinary bladder or urethra. Remarkable advances have been made in treating permanent urinary incontinence through the implantation of an artificial urethral sphincter (fig. 26.39).

Figure 26.38. A diagram of a hemodialysis machine.

Blood outlet

Blood inlet

Vein

Bubble trap

Artery

Rotary pump

O_2 and CO_2

Wash solution

Drain

Coils

Circulating pump

Figure 26.39. An artificial urethral sphincter. The entire device is implanted internally and consists of an inflation bulb that is manually pumped to inflate the urethral cuff from the fluid stored in a reservoir. Manual pumping of the deflation bulb releases the pressure of the urethral cuff and allows urine to flow from the bladder. The device for a female is similar except that the inflation and deflation bulbs are implanted in the tissue of the labia minora.

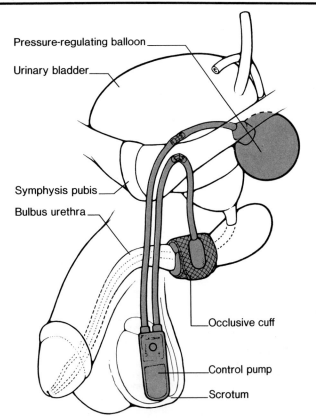

Pressure-regulating balloon

Urinary bladder

Symphysis pubis

Bulbus urethra

Occlusive cuff

Control pump

Scrotum

Chapter Summary

I. Development of the Urinary System
Humans have metanephric kidneys, which develop after pronephric and mesonephric kidneys degenerate during embryonic development.
   A. Three successive types of kidneys develop in the human embryo.
      1. The pronephric kidney persists only through the sixth week.
      2. The mesonephric kidney functions throughout the remainder of embryonic development.
      3. The metanephric kidneys function during fetal development and after birth.
   B. The urinary and reproductive systems both originate from mesodermal tissue called the urogenital ridge.

II. Kidney Structure
A blood filtrate is formed, which enters the nephron tubules in the cortex of the kidney. This filtrate becomes urine as it is passed through different tubular segments and is drained into the calyces.
   A. The kidney is divided into an outer cortex and inner medulla.
      1. The medulla is composed of renal pyramids separated by renal columns.
      2. The renal papillae empty urine into the calyces that drain into the renal pelvis, and from there to the ureter.
   B. Each kidney contains more than a million microscopic functional units called nephrons, which consist of vascular and tubular components.
      1. Filtration occurs in the glomerulus, which receives blood from an afferent arteriole.
      2. Glomerular blood is drained by an efferent arteriole, which delivers blood to peritubular capillaries that surround the nephron tubules.
      3. The glomerular capsule and the proximal and distal convoluted tubules are located in the cortex.
      4. The loop of Henle is located in the medulla.
      5. Filtrate from the distal convoluted tubule is drained into collecting ducts, which plunge through the medulla to empty urine into the calyces.

III. Glomerular Filtration
Large volumes of plasma filtrate low in proteins are produced by the glomeruli. The glomerular filtration rate is controlled by the sympathetic nervous system and by autoregulatory mechanisms.
   A. A filtrate derived from plasma in the glomerulus must pass through a basement membrane of the glomerular capillaries and through slits in the processes of the podocytes, which comprise the inner layer of the glomerular capsule.
      1. The glomerular ultrafiltrate is formed under the force of blood pressure and has a low protein concentration.
      2. The glomerular filtration rate is 115–125 ml/min.
   B. The glomerular filtration rate can be regulated by constriction or dilation of the afferent arterioles.
      1. Sympathetic innervation causes constriction of the afferent arterioles.
      2. Intrinsic mechanisms help to autoregulate the rate of renal blood flow and the glomerular filtration rate.

IV. Reabsorption of Salt and Water
Most of the filtered salt and water is reabsorbed in the proximal tubule. The reabsorption of most of the remaining water occurs as a result of the presence of a hypertonic renal medulla and the action of ADH.
   A. Approximately 65% of the filtered salt and water is reabsorbed across the proximal convoluted tubules.
      1. Sodium is actively transported, chloride follows passively by electrical attraction, and water follows the salt out of the proximal tubule.
      2. Salt transport in the proximal tubules is not under hormonal regulation.
   B. The reabsorption of most of the remaining water occurs as a result of the action of the countercurrent multiplier system.
      1. Chloride is actively extruded from the ascending limb, followed passively by sodium.
      2. Since the ascending limb is impermeable to water, the remaining filtrate becomes hypotonic.
      3. Because of this salt transport and because of countercurrent exchange in the vasa recta, the tissue fluid of the medulla becomes hypertonic.
      4. The hypertonicity of the medulla is multiplied by a positive feedback mechanism involving the descending limb, which is passively permeable to water and perhaps to salt.
   C. The collecting duct is permeable to water but not to salt.
      1. As the collecting ducts pass through the hypertonic renal medulla, water leaves by osmosis and is carried away in surrounding capillaries.
      2. The permeability of the collecting ducts to water is stimulated by antidiuretic hormone (ADH).

V. Renal Clearance Rates
Compounds in the blood are cleared by the kidneys through filtration and secretion into the tubules. Measurements of renal plasma clearance rates are of clinical importance.
   A. Inulin is filtered but neither reabsorbed nor secreted; its clearance rate is thus equal to the glomerular filtration rate.
   B. Some of the filtered urea is reabsorbed; its clearance rate is therefore less than the glomerular filtration rate.
   C. Since almost all the PAH in blood going through the kidneys is cleared by filtration and secretion, the PAH clearance rate is a measure of the total renal blood flow.
   D. Normally all of the filtered glucose and amino acids are reabsorbed; glycosuria occurs when the transport carriers for glucose become saturated due to hyperglycemia.

VI. Renal Control of Electrolyte Balance
Sodium reabsorption and K^+ or H^+ secretion in the distal convoluted tubules are regulated by aldosterone. The secretion of aldosterone is regulated by blood potassium and by the secretion of renin from the juxtaglomerular apparatus.
   A. Aldosterone stimulates sodium reabsorption and potassium secretion in the distal convoluted tubule.
   B. Aldosterone secretion is stimulated directly by a rise in blood potassium and indirectly by a fall in blood sodium.
      1. Decreased blood flow through the kidneys stimulates the secretion of the enzyme renin from the juxtaglomerular apparatus.
      2. Renin catalyzes the formation of angiotensin I, which is then converted to angiotensin II.
      3. Angiotensin II stimulates the adrenal cortex to secrete aldosterone.
   C. Aldosterone stimulates the secretion of H^+ as well as potassium into the filtrate in exchange for sodium.

VII. Renal Control of Acid-Base Balance
The kidneys are responsible for the metabolic component of acid-base regulation through their excretion of H^+ and their reabsorption of HCO_3^-.
   A. The lungs regulate the P_{CO_2} and carbonic acid concentration of the blood, whereas the kidneys regulate the bicarbonate concentration.
   B. Filtered bicarbonate combines with H^+ to form carbonic acid in the filtrate.
      1. Carbonic anhydrase in the membranes of microvilli in the tubules catalyzes the conversion of carbonic acid to carbon dioxide and water.
      2. Carbon dioxide is reabsorbed and converted in either the tubule cells or the red blood

cells to carbonic acid, which dissociates to bicarbonate and H+.

   3. In addition to reabsorbing bicarbonate, the kidneys excrete H+, which is buffered by ammonium and phosphate buffers.

VIII. Ureters, Urinary Bladder, and Urethra
Urine is channeled from the kidneys to the urinary bladder by the ureters and expelled from the body through the urethra. Muscles of the urinary bladder and urethra are used in the control of micturition.

   A. The ureters contain three layers: the mucosa, muscularis, and fibrous coats.

   B. The urinary bladder is lined by a transitional epithelium that is folded into rugae; these structures permit great distension of the bladder.

   C. The urethra has an internal sphincter of smooth muscle and an external sphincter is striated muscle.

   D. Micturition, or urination, is controlled by reflex centers in the second, third, and fourth sacral segments of the spinal cord.

Review Activities

Objective Questions

1. Which of the following statements about metanephric kidneys is *true?*
   (a) They become functional at the end of the eighth week.
   (b) They are active throughout fetal development.
   (c) They are the third pair of kidneys to develop.
   (d) All of the above are true.

2. Which of the following statements about the renal pyramids is *false?*
   (a) They are located in the medulla.
   (b) They contain glomeruli.
   (c) They contain collecting ducts.
   (d) They open by renal papillae into the renal sinus.

Match the following:

3. Active transport of sodium; water follows passively
4. Active transport of chloride; impermeable to water
5. Passively permeable to salt and water
6. Passively permeable to water only

   (a) proximal tubule
   (b) descending limb
   (c) ascending limb
   (d) distal tubule
   (e) collecting duct

7. Antidiuretic hormone promotes the retention of water by stimulating the
   (a) active transport of water
   (b) active transport of chloride
   (c) active transport of sodium
   (d) permeability of the collecting duct to water

8. Aldosterone stimulates sodium reabsorption and potassium secretion in the
   (a) proximal convoluted tubule
   (b) descending limb of loop
   (c) ascending limb of loop
   (d) distal convoluted tubule
   (e) collecting duct

9. Substance X has a clearance rate greater than zero but less than that of inulin. What can you conclude about substance X?
   (a) It is not filtered.
   (b) It is filtered, but neither reabsorbed nor secreted.
   (c) It is filtered and partially reabsorbed.
   (d) It is filtered and secreted.

10. Substance Y has a clearance rate greater than that of inulin. What can you conclude about Y?
    (a) It is not filtered.
    (b) It is filtered, but neither reabsorbed nor secreted.
    (c) It is filtered and partially reabsorbed.
    (d) It is filtered and secreted.

11. About 65% of the glomerular ultrafiltrate is reabsorbed in the
    (a) proximal tubule
    (b) distal tubule
    (c) loop of Henle
    (d) collecting duct

12. Diuretic drugs that act in the loop of Henle
    (a) inhibit active chloride transport
    (b) result in the increased flow of filtrate to the distal convoluted tubule
    (c) cause the increased secretion of potassium into the tubule
    (d) promote the excretion of salt and water
    (e) all of the above

13. The appearance of glucose in the urine
    (a) occurs normally
    (b) indicates the presence of kidney disease
    (c) occurs only when the transport carriers for glucose become saturated
    (d) is a result of hypoglycemia

14. The detrusor muscle is located in the
    (a) kidneys
    (b) ureters
    (c) urinary bladder
    (d) urethra

15. The internal urethral sphincter is innervated by
    (a) sympathetic nerve fibers
    (b) parasympathetic nerve fibers
    (c) somatic motor nerve fibers
    (d) all of the above

Essay Questions

1. Explain how glomerular ultrafiltrate is produced and why it has a low protein concentration.
2. Explain how the countercurrent multiplier system works, and describe its functional significance.
3. Explain how countercurrent exchange occurs in the vasa recta, and describe its functional significance.
4. Explain the mechanisms whereby diuretic drugs may cause an excessive loss of potassium; also explain how the potassium-sparing diuretics work.
5. Describe the micturition reflex, and explain the structures involved in the voluntary control of micturition.

Selected Readings

Anderson, B. 1977. Regulation of body fluids. *Ann. Rev. Physiol.* 39:185.

Bauman, J. W., and F. P. Chinard. 1975. *Renal Function: Physiological and Medical Aspects.* St. Louis: C. V. Mosby Co.

Beeuwkes, R., III. 1980. The vascular organization of the kidney. *Ann. Rev. Physiol.* 42:531.

Brenner, B. M., and R. Beeuwkes, III. 1978. The renal circulation. *Hosp. Prac.* 13:35.

Brenner, B. M., T. H. Hostetter, and H. D. Humes. 1978. Molecular basis of proteinuria of glomerular origin. *New Engl. J. Med.* 298:826.

Buckalew, V. M., Jr., and K. A. Gruber. 1984. Natriuretic hormone. *Ann. Rev. Physiol.* 46:343.

Burg, H., and L. Stoner. 1976. Renal tubular chloride transport and the mode of action of some diuretics. *Hosp. Prac.* 38:37.

Giebisch, G. H., and B. Stanton. 1979. Potassium transport in the nephron. *Ann. Rev. Physiol.* 41:241.

Hays, R. M. 1978. Principles of ion and water transport in the kidneys. *Hosp. Prac.* 13:79.

Kokko, J. S. 1979. Renal concentrating and diluting mechanisms. *Hosp. Prac.* 14:110.

Peart, W. S. 1975. Renin-angiotensin system. *New Engl. J. Med.* 292:302.

Rector, F. C., Jr., and M. G. Cogan. 1980. The renal acidoses. *Hosp. Prac.* 15:99.

Reid, I. A., B. J. Morris, and W. F. Ganong. 1978. The renin-angiotensin system. *Ann. Rev. Physiol.* 40:377.

Renkin, E. M., and R. R. Robinson. 1974. Glomerular filtration. *New Engl. J. Med.* 290:79.

Steinmetz, P. R., and B. M. Koeppen. 1984. Cellular mechanisms of diuretic action along the nephron. *Hosp. Prac.* 19:125.

Tanner, R. L. 1980. Control of acid excretion by the kidney. *Ann. Rev. Med.* 31:35.

Vander, A. J. 1980. *Renal Physiology.* 2d ed. New York: McGraw-Hill Book Company.

Walker, L. A., and H. Valtin. 1982. Biological importance of nephron heterogeneity. *Ann. Rev. Physiol.* 44:203.

Warnock, D. G., and F. C. Rector, Jr. 1979. Proton secretion by the kidney. *Ann. Rev. Physiol.* 41:197.

Unit 5

Continuance of the

The chapters of Unit V are concerned with the reproduction, genetics, development, growth, and aging of humans. They describe the anatomy and physiology of the male and female reproductive systems and how these systems produce a new and genetically unique individual. The fascinating process of prenatal development is described. The final chapter discusses the anatomical and physiological changes that occur in each body system as a person ages.

The chapters and principal concepts of Unit V follow:

Chapter 27—Male Reproductive System
The organs of the male reproductive system produce and transport sperm to the female reproductive system. It also produces the male sex hormones, which have various regulatory functions within the body.

Human Species

27

Male Reproductive System

Concepts

The organs of the male and female reproductive system produce and unite gametes that contain specific genes. A random combination of the genes during sexual reproduction permits the propagation of individuals with genetic differences.

Sexual distinction is determined upon fertilization of an ovum by a sperm containing either an X or a Y chromosome. Sexual differentiation under hormonal control is a progressive process through embryonic and early fetal development.

Gonadotrophic hormones from the anterior pituitary and maturational changes in the central nervous system result in an increased secretion of sex steroids from the testes, which in turn stimulates a growth spurt and sexual maturation during puberty.

The testes within the scrotum produce spermatozoa and androgens. Androgens regulate spermatogenesis and the development and functioning of the secondary sex organs.

The spermatic ducts store spermatozoa and transport them from the testes to the urethra. The accessory reproductive glands provide additives to the spermatozoa in the formation of semen, which is discharged from the erect penis during ejaculation.

Erection of the penis results from parasympathetic induced vasodilation of arteries within the penis. Emission and ejaculation are stimulated by sympathetic impulses resulting in the forceful expulsion of semen from the penis.

Introduction to the Reproductive System

The organs of the male and female reproductive system are adapted to produce and unite gametes that contain specific genes. A random combination of the genes during sexual reproduction permits the propagation of individuals with genetic differences.

Objective 1. Explain why sexual reproduction is biologically advantageous.

Objective 2. List the functions of the male reproductive system, and compare them with those of the female.

Objective 3. Distinguish between primary and secondary sex organs.

Sexual reproduction is common in organisms. Through this process, individuals of a species are propagated and have a genetic diversity inherited from both parents. Sexual reproduction provides the important advantage of genetic recombination, whereby a genetic change becomes a part of the gene pool. Genetic differences in the offspring help some of them adapt to environmental changes that might otherwise cause the extinction of a population.

Because sexual reproduction requires the production of two types of **gametes** *(gam'ēts),* or sex cells, the species has a male and female, each with its own unique reproductive system. The functions of the female reproductive system are more complex than those of the male. The functions of the male reproductive system are to produce the male gametes, **spermatozoa** (sperm), and to transfer them to the female through the process of *sexual intercourse (coitus),* or *copulation.* The female not only produces her own gametes, called **ova,** and receives the sperm from the male, but her reproductive organs are specialized to provide a site for fertilization, implantation of the blastocyst, pregnancy, and delivery of a baby. The reproductive system of the female also provides a means of nourishing the baby through the secretions of the mammary glands in the breast.

The reproductive system of the male and female is a unique body system in three respects. First, whereas all the other body systems function to sustain the individual, the reproductive system is specialized to perpetuate the species and pass genetic material from generation to generation. Secondly, the anatomy and physiology of the reproductive organs are the major differences between the male and female. Other systems, such as the integumentary, skeletal, and urinary, have minor sexual differences, but none to the extent of the reproductive system. The third uniqueness of the reproductive system is its latent development under hormonal control. The other body systems are totally functional at birth or shortly thereafter, whereas the reproductive system does not become functional until it is acted on by hormones during puberty. *Puberty (pu'ber-te)* is the period of human development during which the reproductive organs become functional.

---

gamete: Gk. *gameta,* husband or wife
spermatozoa: Gk. *sperma,* seed; *zoon,* animal
puberty: L. *puberty,* grown up

Figure 27.1. Organs of the male reproductive system seen in parasagittal view.

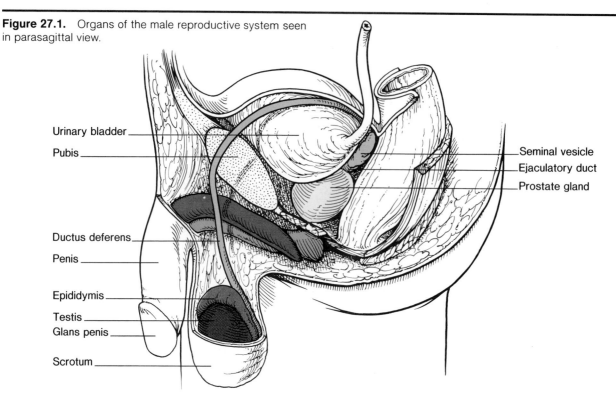

Urinary bladder

Pubis

Ductus deferens

Penis

Epididymis

Testis

Glans penis

Scrotum

Seminal vesicle

Ejaculatory duct

Prostate gland

The structures of the male reproductive system can be divided into three categories based on function:

1. **Primary sex organs.** The primary sex organs are called **gonads** (go'nad), specifically the **testes** in the male. Gonads produce the gametes, or sperm, and produce and secrete sex hormones. The appropriate amounts and timing of the production and secretion of male sex hormones cause the development of secondary sex organs and the expression of secondary sex characteristics.

2. **Secondary sex organs.** Secondary sex organs are those structures that are essential in caring for and transporting sperm. The three categories of secondary sex organs are the sperm-transporting ducts, the accessory glands, and the copulatory organ. The ducts that transport sperm include the **epididymides, ductus deferentia, ejaculatory ducts,** and **urethra.** The accessory glands are the **seminal vesicles,** the **prostate gland,** and the **bulbourethral** (bul''bo-u-re'thral) **glands.** The **penis,** which contains erectile tissue, is the copulatory organ. The **scrotum** is a pouch of skin that encloses the testes.

3. **Secondary sex characteristics.** Secondary sex characteristics are features that are not essential for the reproductive process but are generally considered sexually attractant features. Body physique, body hair, and voice pitch are examples.

Figure 27.1 depicts the external and internal organs of the male reproductive system, and table 27.1 summarizes their functions.

Table 27.1 Functions of the organs of the male reproductive system

Organ	Function
Testes	
Seminiferous tubules	Produce spermatozoa
Interstitial cells	Produce and secrete male sex hormones
Epididymides	Storage and maturation of spermatozoa; convey spermatozoa to ductus deferentia
Ductus deferentia	Store spermatozoa; convey spermatozoa to ejaculatory ducts
Ejaculatory ducts	Receive spermatozoa and additives to produce seminal fluid
Seminal vesicles	Secrete alkaline fluid containing nutrients and prostaglandins
Prostate gland	Secretes alkaline fluid that helps neutralize acidic seminal fluid and enhances motility of spermatozoa
Bulbourethral glands	Secrete fluid that lubricates urethra and end of penis
Scrotum	Encloses and protects testes
Penis	Conveys urine and seminal fluid to outside of body; organ of copulation

Source: Van De Graaff, Kent M., *Human Anatomy.* © 1984 Wm. C. Brown Publishers, Dubuque, Iowa. All Rights Reserved. Reprinted by permission.

1. *What is the principal value of sexual reproduction?*
2. *What are the functions of the male reproductive system?*
3. *List the organs of the male reproductive system, and indicate whether they are primary or secondary sex organs.*
4. *Define latent development with respect to the reproductive system.*

Figure 27.2.   The development of the gonadal ridge and the primary sex cords: (*a*) a midsagittal view of a five-week embryo; (*b*) a three-dimensional view through the pelvic region of a five-week embryo showing the gonadal ridges; (*c*) a transverse section through a six-week embryo showing the differentiation of the gonadal ridge and the formation of the primary sex cords.

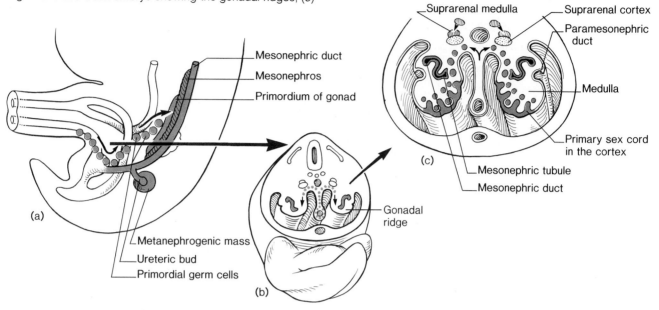

Development of the Male Reproductive System

Sexual distinction is determined upon fertilization of an ovum by a sperm containing either an X or a Y chromosome. Sexual differentiation under hormonal control is a progressive process through embryonic and early fetal development.

>   *Objective 4.*   Discuss how the increased production of testosterone affects the development of the secondary sex organs and secondary sex characteristics.
>
>   *Objective 5.*   Explain how the genetic sex of a person is determined by the action of an XX or XY chromosome combination.
>
>   *Objective 6.*   Trace the development of the reproductive organs to the indifferent stage.
>
>   *Objective 7.*   Describe the sequence of events in the development of the male reproductive organs during the fetal period—including the descent of the testes into the scrotum.

Sex Determination

Sexual identity is initiated at the moment of conception when the *genetic sex* of the zygote is determined. The ovum is fertilized by a sperm containing either an X or a Y chromosome. If the sperm contains an X chromosome, it will pair with the X chromosome of the ovum and a female child will develop. A sperm carrying a Y chromosome results in an XY combination, and a male child will develop.

Genetic sex determines whether the gonads will be testes or ovaries. If testes develop, they will produce and secrete male sex hormones during late embryonic and early fetal development and cause the secondary sex organs of the male to develop.

Embryonic Development

The first sign of development of either the male or the female reproductive organs occurs during the fifth week as the medial aspect of each mesonephros enlarges to form the **gonadal ridge.** The gonadal ridge continues to grow behind the developing peritoneal membrane. By the sixth week, stringlike masses called **primary sex cords** form within the enlarging gonadal ridge (fig. 27.2). Externally, a swelling called the **genital tubercle** appears cephalad to the cloacal membrane (see fig. 27.6).

Reproductive development is well progressed by the eighth week, but it is still in what is known as the *indifferent stage,* because although the sex organs are apparent, external distinction is not (see fig. 27.6). The gonads at this stage are relatively large and have a distinct outer cortex composed of primary sex cords and an inner medulla (fig. 27.3). During this time, cells that will give rise to sperm (called *spermatogonia*) and cells that will give rise to ova (called *oogonia [o"o-go'ne-ah]*) migrate from the yolk sac to the embryonic gonads. At this stage in development the gonads have the potential to become either testes or ovaries (fig. 27.4). The significance of the indifferent stage is that the sexual organs for both male and female are derived from the same embryonic tissues and are considered *homologous* structures.

It is currently believed that the Y chromosome in XY embryos masculinize the gonads (cause them to develop

---

spermatogonia: Gk. *sperma*, seed; *gone*, generation
oogonia: Gk. *oon*, egg; *gone*, generation
homologous: Gk. *homos*, the same

Figure 27.3. Differentiation of the male and female gonads: (*a*) the indifferent stage of gonads at 6 weeks; (*b*) the testes at 7 weeks; (*c*) the ovaries at 7 weeks; (*d*) the testes at 20 weeks showing the seminiferous tubules and rete testis; (*e*) the ovaries at 20 weeks showing primordial follicles.

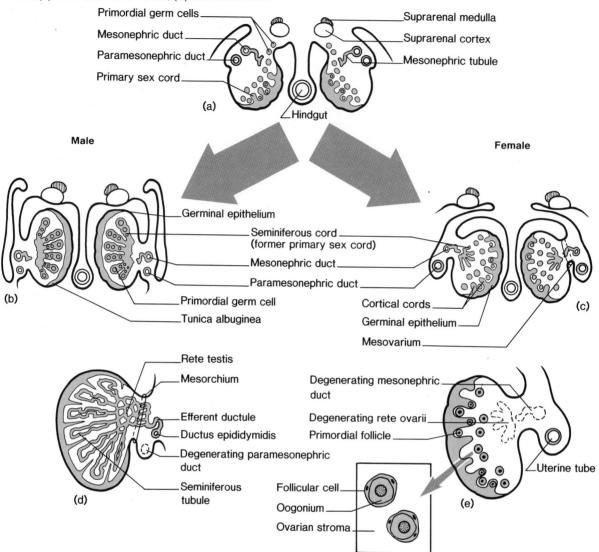

into testes) through the production of cell-surface proteins coded by genes in the Y chromosome. These male "markers" are a type of histocompatibility antigen and are therefore known as **H-Y antigens.** Embryos that are chromosomal females (XX) lack Y chromosomes and therefore lack H-Y antigens. In the absence of these antigens the gonads develop into ovaries.

Notice that it is the presence or absence of the Y chromosome that determines whether the embryo will have testes or ovaries. This point is well illustrated by two genetic abnormalities. In **Klinefelter's syndrome** the affected person has forty-seven instead of forty-six chromosomes because of the presence of an extra X chromosome. These people, with XXY genotypes, develop testes despite the presence of two X chromosomes. Patients with **Turner's syndrome,** who have the genotype XO (and therefore have only forty-five chromosomes) develop ovaries.

Once the testes have differentiated, large amounts of male sex hormones, called *androgens (an'dro-jens),* are secreted from **interstitial cells (cells of Leydig).** The major androgen secreted by these cells is *testosterone (tes-tos'tĕ-rōn).* Testosterone secretion begins as early as eight to ten weeks after conception, reaches a peak at twelve to fourteen weeks, and thereafter declines to very low levels by the end of the second trimester (about twenty-one weeks). High levels of testosterone will not appear again in the life of the individual until the time of puberty.

The **seminiferous** *(se''mĭ-nif'er-us)* **tubules,** which will eventually produce sperm within the testes, appear about forty-five to fifty days following conception. Although spermatogenesis begins during embryonic life, it is arrested until the onset of puberty.

In addition to gonads, various internal accessory sexual organs are needed for reproductive functioning. Most of these sex accessory organs are derived from two systems of embryonic ducts. Male accessory organs are

Figure 27.4.   The formation of the chromosomal sex of the embryo and the development of the gonads. The very early embryo has "indifferent gonads" that can develop into either testes or ovaries. If the embryonic cells have Y chromosomes, which code for H-Y antigens, the gonads become testes. If no Y chromosomes and therefore no H-Y antigens are present, the gonads become ovaries. Embryonic testes develop quickly, forming seminiferous tubules (which will produce sperm later in life) and interstitial cells (which secrete testosterone during embryonic development).

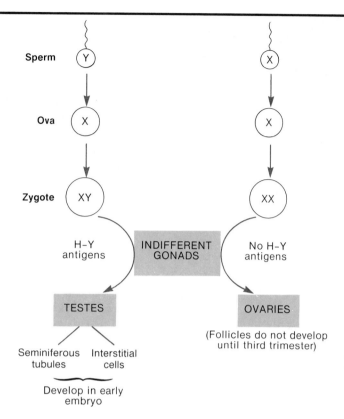

Figure 27.5.   The embryonic development of male and female accessory organs and external genitalia. In the presence of testosterone and müllerian inhibition factor (MIF) secreted by the testes, male structures develop. In the absence of these secretions, female structures develop.

derived from mesonephric (wolffian) ducts, and female accessory organs are derived from paramesonephric (müllerian) ducts (fig. 27.5). Interestingly, both male and female embryos between day twenty-five and day fifty have both duct systems and, therefore, have the potential to form the accessory organs characteristic of either sex.

The experimental removal of the testes (castration) from male embryonic animals results in the regression of the mesonephric ducts and the development of the paramesonephric ducts into female accessory organs: the uterus and uterine tubes. *Female sex accessory organs, therefore, develop as a result of the absence of testes rather than as a result of the presence of ovaries.*

Figure 27.6. Differentiation of the external genitalia in the male and female. (a) the genital tubercle, urogenital fold, and labioscrotal swelling have differentiated from the genital tubercle by week 6. (b) a distinct phallus is present during the indifferent stage at week 8. (c, e, g) sequential stages of differentiation of the male external genitalia through weeks 9, 11, and 12 respectively. (d, f, h) sequential stages of differentiation of the female external genitalia through weeks 9, 11, and 12 respectively. (i and j) photographs at week 10 of male and female genetalia, respectively.

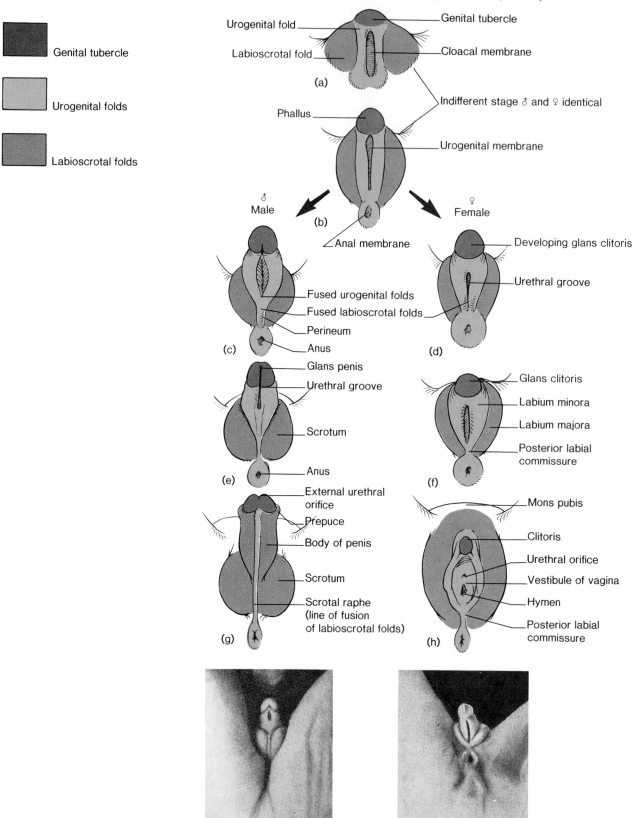

Figure 27.7. The conversion of testosterone, secreted by the interstitial cells of the testes, into dihydrotestosterone (DHT) within the target cells. This reaction involves the addition of a hydrogen (and the removal of the double carbon bond) in the first (*A*) ring of the steroid.

Figure 27.8. The descent of the testes. During the twenty-eighth week of development each testis descends through an inguinal canal in front of the symphysis pubis and enters the scrotum.

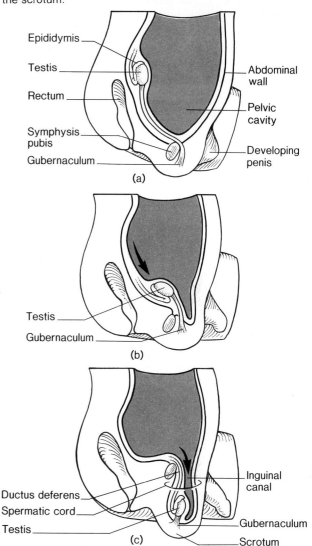

The developing seminiferous tubules within the testes secrete a polypeptide called *Müllerian inhibition factor* (MIF), which causes the regression of the paramesonephric ducts beginning about day sixty. The secretion of testosterone by the interstitial cells of the testes subsequently causes the growth and development of the mesonephric ducts into male secondary sex organs.

Other embryonic reproductive structures (**urogenital sinus, genital tubercle, urogenital folds,** and **labioscrotal folds**) are also masculinized by secretions of the testes. The prostate gland derives from the urogenital sinus, and the other embryonic structures differentiate into the external genitalia (fig. 27.6). In the absence of testicular secretions, the female genitalia are formed.

The masculinization of the embryonic reproductive structures occurs as a result of testosterone secreted by the embryonic testes. Testosterone itself, however, is not the active agent within these target organs. Once inside the target cells, testosterone is converted by means of an enzyme called *5α-reductase* into the active hormone known as **dihydrotestosterone** *(di-hi''dro-tes-tos'ter-ōn)* **(DHT),** which directly mediates the androgen effect in these organs (fig. 27.7).

In summary, the genetic sex is determined by whether a Y-bearing or an X-bearing sperm fertilizes the ovum; the presence or absence of a Y chromosome in turn determines whether the gonads of the embryo will be testes or ovaries; and the presence or absence of testes, finally, determines whether the sex accessory organs and external genitalia will be male or female. This regulatory pattern of sex determination makes sense in light of the fact that both male and female embryos develop within an environment high in estrogen, which is secreted by the mother's ovaries and the placenta. If estrogen determined the sex, all embryos would become feminized.

Fetal Development and Descent of the Testes

By the beginning of the fetal period at nine weeks, male differentiation of the gonads into testes is well under way. Internal changes include the formation of the tubular **seminiferous tubules** and **rete** *(re'te)* **testis** from the primary sex cord (see fig. 27.3). Developing on the outside surface of each testis is a fibromuscular cord called the **gubernaculum** *(gu''ber-nak'u-lum)* (fig. 27.8), which attaches to the inferior pole and extends to the labioscrotal

---

rete: L. *rete,* a net
gubernaculum: L. *gubernaculum,* helm

fold of the same side. At the same time, a portion of the embryonic mesonephric duct adjacent to the testis becomes attached and convoluted and forms the **epididymis.** Another portion of the mesonephric duct becomes the **ductus deferens.**

The accessory reproductive glands in a male develop from two different sources. The **seminal vesicles** form from lateral outgrowths of the caudal ends of each mesonephric duct. The **prostate gland** arises from endodermal outgrowths of the urogenital sinus. The **bulbourethral glands** develop from outgrowths in the membranous portion of the urethra.

The external genitalia of the male become distinct from those of a female by the end of the ninth week (see fig. 27.6). Prior to that, the genital tubercle in both sexes elongates and is called a **phallus** *(fal'us).* A **urethral groove** forms on the ventral surface of the phallus.

The differentiation of the external genitalia of a male from the indifferent stage to discernible organs is caused by the androgens produced and secreted by the testes. These changes include the elongation and differentiation of the phallus into a **penis,** a fusion of the **urogenital folds** surrounding the urethral groove along the ventral surface of the penis, and a midventral fusion of the labioscrotal folds to form the wall of the **scrotum.** The external genitalia of a male are completely formed by the end of the twelfth week.

The descent of the testes from the site of development begins between the sixth and tenth week. Descent into the scrotal sac, however, does not occur until about the twenty-eighth week, when paired inguinal canals form in the abdominal wall to provide openings from the pelvic cavity to the scrotal sac. The process by which a testis descends is not well understood, but it seems to be associated with the shortening of the **gubernaculum,** which is attached to the testis and extends through the inguinal canal to the wall of the scrotum (fig. 27.8). As the testis descends, it passes to the side of the urinary bladder and anterior to the symphysis pubis. It carries with it the ductus deferens, the testicular vessels and nerve, a portion of the abdominal muscle, and lymph vessels. All of these structures remain attached to the testis and form what is known as the **spermatic cord.** By the time the testis is in the scrotal sac, the gubernaculum is no more than a remnant of scarlike tissue.

During the physical examination of a neonatal male child, a physician will palpate the scrotum to determine if the testes are in position. If one or both are not in the scrotal sac, it may be possible to induce descent by administering certain hormones. If this procedure does not work, surgery is necessary. The surgery is generally performed before the age of five. Failure to correct the situation may result in sterility and possibly the development of a tumorous testicle.

phallus: Gk. *phallus,* penis

1. Describe the hormonal process of genital masculinization.
2. Explain how the chromosomal sex determines whether testes or ovaries will be formed.
3. What early embryonic structures give rise to the gonads, and what structures give rise to the external genitalia?
4. List the homologous structures of the male and female reproductive systems.
5. Describe the reproductive changes that occur during the fetal stage of development. How and why do the testes descend into the scrotum?

Puberty

Gonadotrophic hormones from the anterior pituitary and maturational changes in the central nervous system result in an increased secretion of sex steroids from the testes, which in turn stimulate a growth spurt and sexual maturation during puberty.

Objective 8. Describe the hormonal changes and feedback mechanisms that result in puberty in males.

Objective 9. List the physical changes that occur in the male secondary sex organs during puberty.

The testes are formed and receive spermatogonia (sperm-forming cells) early in embryonic development. During the first trimester of pregnancy, the embryonic testes are active endocrine glands, secreting the high amounts of testosterone needed to masculinize the male embryo's secondary sex organs. During the second trimester of pregnancy, testosterone secretion in a male fetus declines so that at birth the testes are relatively inactive.

There is no difference in the blood concentrations of *sex steroids (ste'roids)*—androgens and estrogens—in prepubertal boys and girls. This does not appear to be due to deficiencies in the ability of the gonads to produce these hormones, but rather to lack of sufficient stimulation. During *puberty,* the gonads secrete increased amounts of sex steroid hormones as a result of increased stimulation by **gonadotrophic hormones** from the anterior pituitary gland.

Interactions between the Hypothalamus, Pituitary, and Gonads

The anterior pituitary gland produces and secretes **FSH (follicle-stimulating hormone)** and **LH (luteinizing hormone).** Although these two hormones are named according to their actions in the female, the same hormones are secreted by the male's pituitary. The gonadotrophic hormones of both sexes have three primary effects on the gonads: (1) stimulation of spermatogenesis or oogenesis (formation of sperm or ova); (2) stimulation of gonadal hormone secretion; and (3) maintenance of the structure of the gonads (gonads atrophy if the pituitary is removed).

Figure 27.9. Interactions between the hypothalamus, anterior pituitary, and gonads. Sex steroids secreted by the gonads have a negative feedback effect on the secretion of GnRH (gonadotrophin-releasing hormone) and on the secretion of gonadotrophins. The gonads may also secrete a polypeptide hormone called *inhibin* that exerts negative feedback control of FSH secretion.

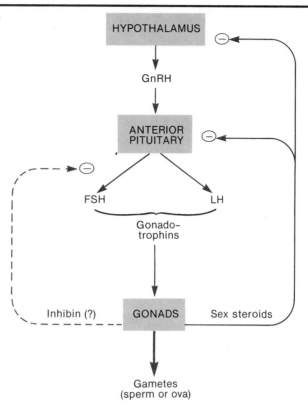

The secretion of both LH and FSH from the anterior pituitary is stimulated by a hormone produced by the hypothalamus and secreted into the hypothalamo-hypophyseal portal vessels. This releasing hormone is called *LHRH (luteinizing hormone–releasing hormone)*. Since attempts to find a separate FSH-releasing hormone have thus far failed and since LHRH stimulates FSH as well as LH secretion, LHRH is often referred to as **gonadotrophin-releasing hormone** (and accordingly abbreviated **GnRH**).

If a male or female animal is castrated (has had its gonads surgically removed), the secretion of FSH and LH increases to much higher levels than those measured in the intact animal. This demonstrates that the gonads secrete products that have a **negative feedback inhibition** of gonadotrophin secretion. This negative feedback is exerted in a large part by sex steroids: testosterone in the male and estrogen and progesterone in the female.

The negative feedback effects of steroid hormones are believed to occur by means of two mechanisms: (1) inhibition of GnRH secretion, and (2) inhibition of the pituitary's response to a given amount of GnRH secreted from the hypothalamus. In addition to steroid hormones, there is evidence that the testes (and perhaps the ovaries as well) may secrete a polypeptide hormone, called **inhibin** *(in-hib'in),* that specifically inhibits FSH secretion in response to GnRH, without affecting the secretion of LH.

Figure 27.9 illustrates the similarities in the gonadal regulation of males and females. Important differences in hypothalamus-pituitary-gonad interactions exist, however, between males and females. The secretion of gonad-

otrophins and sex steroids are more or less constant in adult males. The secretion of gonadotrophins and sex steroids in adult females, in contrast, shows cyclic variations (during the menstrual cycle). Also, during the normal female cycle, estrogen exerts a positive effect on LH secretion.

The Onset of Puberty

The secretion of FSH and LH is high in the newborn but falls to very low levels a few weeks after birth. Gonadotrophin secretion remains low until the beginning of puberty, which is marked by rising levels of FSH followed by LH secretion. Experimental evidence suggests that this rise in gonadotrophin secretion is the result of two processes: (1) maturational changes in the brain that result in increased GnRH secretion by the hypothalamus and (2) a decreased sensitivity of gonadotrophin secretion to the negative feedback effects of sex steroid hormones.

The maturation of the hypothalamus or other regions of the brain that lead to increased GnRH secretion at the time of puberty appear to be programmed—children without gonads show increased FSH secretion at the normal time. Also during this period of time, a given amount of sex steroids has less of a suppressive effect on gonadotrophin secretion than the same dose would have if administered prior to puberty. This suggests that the sensitivity of the hypothalamus and the pituitary to negative feedback effects decreases at puberty, which would also help account for the rising gonadotrophin secretion at this time.

Table 27.2 Development of secondary sexual characteristics and other changes that occur during puberty in males

Characteristic	Age of first appearance	Hormonal stimulation
Growth of testes	10–14	Testosterone, FSH, growth hormone
Pubic hair	10–15	Testosterone
Body growth	11–16	Testosterone, growth hormone
Growth of penis	11–15	Testosterone
Growth of larynx (voice lowers)	Same time as growth of penis	Testosterone
Facial and axillary (underarm) hair	About two years after appearance of pubic hair	Testosterone
Eccrine sweat glands and sebaceous glands; acne (from blocked sebaceous glands)	About same time as facial and axillary hair growth	Testosterone

Source: Fox, Stuart Ira, *Human Physiology.* © 1984 Wm. C. Brown Publishers, Dubuque, Iowa. All Rights Reserved. Reprinted by permission.

During late puberty there is a "pulsatile" secretion of gonadotrophins—FSH and LH secretion *increases during periods of sleep* and decreases during periods of wakefulness. These pulses of increased gonadotrophin secretion during puberty stimulate a rise in sex steroid secretion from the gonads. The increased secretion of testosterone from the testes during puberty causes the male **secondary sexual characteristics** to be manifested (table 27.2).

---

1. *Using a flow diagram, show the negative feedback control that the gonads exert on GnRH and gonadotrophin secretion. Explain the effects of castration on FSH and LH secretion and the effects of the removal of the pituitary on the structure of the gonads and accessory sex organs.*
2. *Describe the two mechanisms that have been proposed to explain the rise in sex steroid secretion that occurs at puberty.*
3. *Contrast the relative levels of testosterone during early fetal development, at birth, and during puberty.*
4. *Describe the physical changes that occur in a male during puberty.*

Structure and Function of the Testes

The testes within the scrotum produce spermatozoa and androgens. Androgens regulate spermatogenesis and the development and functioning of the secondary sex organs.

Objective 10. Describe the location, structure, and functions of the testes.

Objective 11. Explain the hormonal actions of testosterone, and specifically describe the interactions between the interstitial cells and the seminiferous tubules in each testis.

Objective 12. List the events of spermatogenesis, and distinguish between spermatogenesis and spermiogenesis.

Objective 13. Diagram the structure of a sperm, and explain the function of each of its parts.

Scrotum

The saclike scrotum *(skro'tum)* is suspended immediately behind the base of the penis, anterior to the anal opening, in a region known as the **perineum** *(per''ĭ-ne'um)*. The functions of the scrotum are to support and protect the testes and to regulate their position relative to the pelvic region of the body. The soft textured skin of the scrotum is covered with sparse hair in mature males and is darker in color than most of the other skin of the body. It also contains numerous sebaceous glands. The external appearance of the scrotum varies at different times in the same individual depending on environmental conditions and the contraction of the dartos and cremaster muscles. The **dartos** *(dar'tos)* is a layer of smooth muscle fibers in the subcutaneous tissue of the scrotum, and the **cremaster** *(kre-mas'ter)* is a thin strand of skeletal muscle associated with the spermatic cord. Both muscles involuntarily contract to cold temperatures to move the testes closer to the heat of the body in the pelvic region. The cremaster muscle is a continuation of the internal oblique muscle of the abdominal wall, which is derived as the testes descend into the scrotum. Because it is a skeletal muscle, it can contract voluntarily as well. When these muscles are contracted, the scrotum appears tightly wrinkled. Warm temperatures cause the dartos and cremaster muscles to relax and the testes to become pendent in the flaccid scrotum. The temperature of the testes is maintained at about 35°C (95°F or about 3.6°F below normal body temperature) by the contraction or relaxation of the scrotal muscles. This temperature has been determined to be an optimum temperature for the production and storage of sperm.

The scrotum is subdivided into two longitudinal compartments by a fibrous **median septum.** The purpose of the median septum is to compartmentalize each testis so that the trauma or infections of one will generally not affect the other. Another protective feature is that the left testis is generally suspended lower in the scrotum than the right so that the two will not as likely be compressed forcefully together. The site of the median septum is apparent on the surface of the scrotum along a median longitudinal ridge

---

dartos: Gk. *dartos*, skinned or flayed
cremaster: Gk. *cremaster*, a suspender, to hang
septum: L. *septum*, a partition

Figure 27.10. Structural features of the testis and epididymis: (*a*) a longitudinal view; (*b*) a cross-sectional view.

(a)

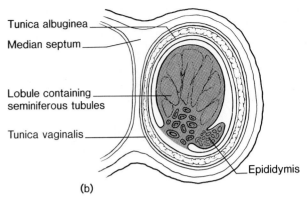

(b)

called the **perineal raphe** *(ra'fe)*. The perineal raphe extends forward to the undersurface of the penis and backward to the anal opening. The blood supply and innervation of the scrotum are extensive. The arteries that serve the scrotum are the internal pudendal branch of the internal iliac artery, the external pudendal branch from the femoral artery, and the cremasteric branch of the inferior epigastric artery. The venous drainage follows a pattern similar to the arteries. The scrotal nerves are primarily sensory and include the pudendal nerves, ilioinguinal nerves, and posterior cutaneous nerves of the thigh.

> Although uncommon, male infertility may result from an excessively high temperature of the testes over an extended period of time. Tight clothing that keeps the testes close to the body or frequent hot baths or saunas may destroy sperm to the extent that the sperm count of discharged semen is below that necessary to cause fertilization.

Testes

Structure of the Testis. The testes are paired, whitish, ovoid organs, which measure about 4 cm (1.5 in.) in length and 2.5 cm (1 in.) in diameter. Each testis weighs between 10 and 14 g.

Two tissue layers, or tunicas, cover the testis. The outer **tunica vaginalis** is a thin, serous sac derived from the peritoneum during the descent of the testis. The **tunica albuginea** *(al"bu-jin'e-ah)* is a tough, fibrous membrane that directly encapsules the testis (fig. 27.10). Fibrous, inward extensions of the tunica albuginea partition the testis into 250 to 300 wedge-shaped **lobules.**

Each lobule of the testis contains one to three tightly convoluted **seminiferous tubules,** which may exceed 70 cm (28 in.) in length if uncoiled. The seminiferous tubules are the functional units of the testis because it is here that

spermatogenesis occurs. Sperm are produced at the rate of thousands per second throughout the life of a healthy, sexually mature male.

Various stages of meiosis can be observed in a histological section of seminiferous tubules (fig. 27.11). The germinal cells, called **spermatogonia** *(sper"mah-to-go'ne-ah),* are in contact with the basement membrane. Spermatogonia are in a constant state of mitotic division to produce, in order of advancing maturity, the primary spermatocytes, secondary spermatocytes, and spermatids (see fig. 27.14). Forming the walls of the seminiferous tubules are **nurse cells,** or **Sertoli cells,** which produce and secrete nutrients to the developing spermatozoa embedded between them. The spermatozoa are formed, but not fully matured, by the time they reach the lumina of the seminiferous tubules.

Between the seminiferous tubules are specialized endocrine cells called **interstitial cells (cells of Leydig).** The function of these cells is to produce and secrete the male sex hormones. The testes are considered mixed exocrine and endocrine glands because they produce both sperm and androgens.

Once the sperm are produced, they move through the seminiferous tubules and enter the **rete testis** for further maturation. Cilia are located on some of the cells of the rete testis, presumably for moving sperm. The sperm are transported out of the testis and into the epididymis through a series of **efferent ductules.**

The testes receive blood through the **testicular arteries,** which arise from the abdominal aorta immediately below the origin of the renal arteries. The **testicular veins** drain the testes. The testicular vein of the right side enters directly into the inferior vena cava, whereas the testicular vein of the left side drains into the left renal vein.

Testicular nerves innervate the testes with both efferent and afferent fibers arising from the tenth thoracic segment of the spinal cord. Innervation is primarily through sympathetic fibers, but there is limited parasympathetic stimulation as well.

---

raphe: Gk. *raphe*, a seam
pudendal: L. *pudeo*, to feel ashamed
tunica: L. *tunica*, a coat
vaginalis: L. *vagina*, a sheath
albuginea: L. *albus*, white

Sertoli cells: from Enrico Sertoli, Italian histologist, 1842–1910
efferent ductules: L. *efferre*, to bring out; *ducere*, to lead

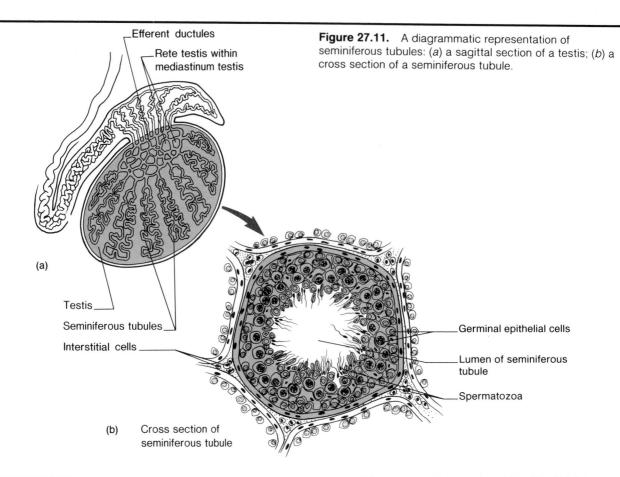

Efferent ductules

Rete testis within mediastinum testis

(a)

Testis

Seminiferous tubules

Interstitial cells

Germinal epithelial cells

Lumen of seminiferous tubule

Spermatozoa

(b) Cross section of seminiferous tubule

Figure 27.11. A diagrammatic representation of seminiferous tubules: (*a*) a sagittal section of a testis; (*b*) a cross section of a seminiferous tubule.

The primary cause of male infertility is a condition called **varicocele** (var′ĭ-ko-sēl″). Varicocele occurs when one or both of the testicular veins draining from the testes becomes swollen, resulting in poor vascular circulation in the testes. A varicocele generally occurs on the left side, because the left spermatic vein drains into the renal vein where the blood pressure is higher than it is in the inferior vena cava, into which the right testicular vein empties.

Table 27.3 Summary of some of the actions of androgens in the male

Category	Action
Sex determination	Growth and development of mesonephric ducts into epididymides, ductus deferentia, seminal vesicles, and ejaculatory ducts Development of the urogenital sinus and tubercle into prostate gland Development of male external genitalia
Spermatogenesis	At puberty: completion of meiotic division and early maturation of spermatids After puberty: maintenance of spermatogenesis
Secondary sex characteristics	Growth and maintenance of accessory sex organs Growth of the penis Growth of facial and axillary hair Body growth
Anabolic effects	Protein synthesis and muscle growth Growth of bones Growth of other organs (including the larynx) Erythropoiesis (red blood cell formation)

Source: Fox, Stuart Ira, *Human Physiology.* © 1984 Wm. C. Brown Publishers, Dubuque, Iowa. All Rights Reserved. Reprinted by permission.

Endocrine Functions of the Testis. Testosterone is by far the major androgen secreted by the adult testis. This hormone (or derivatives of it—the 5 α-reduced androgens and estradiol) is responsible for the initiation and maintenance of the body changes of puberty in males. Androgens are sometimes called *anabolic steroids* because they stimulate the growth of muscles and other structures (table 27.3). Increased testosterone secretion during puberty is also required for the growth of the sex accessory organs, primarily the seminal vesicles and prostate. The removal of androgens by castration results in atrophy of these organs.

Androgens stimulate the growth of the larynx (causing lowering of the voice), increased hemoglobin synthesis (males have higher hemoglobin levels than females), and

bone growth. The effect of androgens on bone growth is self-limiting, however, because androgens ultimately cause the conversion of cartilage to bone in the epiphyseal discs, thus "sealing" the discs and preventing further lengthening of the bones.

It was once thought that the interstitial cells were the only source of testicular androgens. Although the interstitial cells are the major source of testosterone, it is now

---

varicocele: L. *varico*, a dilated vein; Gk. *kele*, tumor or hernia

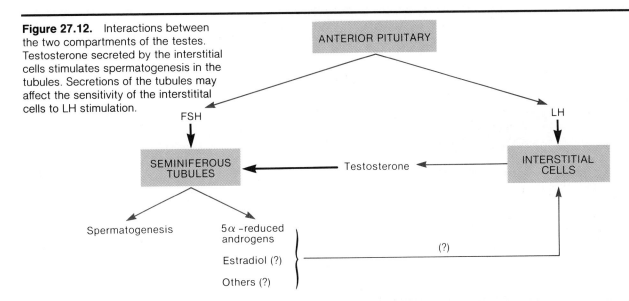

Figure 27.12. Interactions between the two compartments of the testes. Testosterone secreted by the interstitial cells stimulates spermatogenesis in the tubules. Secretions of the tubules may affect the sensitivity of the interstitial cells to LH stimulation.

known that the seminiferous tubules are also able to produce small amounts of testosterone if they are supplied with appropriate precursors (such as progesterone). Probably of more importance is the fact that the tubules in adult testes are able to convert testosterone, supplied to them by the interstitial cells, into 5α-reduced androgens (DHT and 3α-diol). The physiological significance of the endocrine function of the tubules is not currently understood.

Although androgens are by far the major secretory product of the testes, the testes do produce and secrete small amounts of estradiol (fig. 27.12). The source of these estrogens is not known—some evidence suggests that the tubules produce estradiol as a result of FSH stimulation, whereas other evidence suggests that estradiol comes from the interstitial cells as a result of LH stimulation. The physiological significance of estrogens secreted into the blood by the testes is unclear.

The testes contain estrogen receptors, which are confined to the interstitial tissue. Receptors for androgens, in contrast, are confined to the seminiferous tubules. This suggests the possibility that androgens and estrogens might have regulatory functions within the testes themselves. This possibility has proven correct for androgens (they are needed for spermatogenesis). The physiological role of estrogens in the testes remains to be demonstrated.

The two compartments of the testes interact with each other. Testosterone from the interstitial cells is metabolized by the tubules into other active androgens and serves to regulate tubular functions. The tubules also secrete products—5α-reduced androgens and possibly other hormones—which could conceivably influence interstitial cell function. Such interactions are suggested by evidence that in the pubertal male rat, exposure to FSH augments the responsiveness of the interstitial cells to LH. Since FSH can only directly affect the tubules, the FSH-induced enhancement of LH responsiveness must be mediated by products secreted from the tubules. The nature of these regulators is not currently known.

Table **27.4** Stages of meiosis	
Stage	**Events**
First meiotic division	
Prophase I	Chromosomes appear double-stranded. Each strand, called a chromatid, contains duplicate DNA joined together by a structure known as a centromere. Homologous chromosomes pair up side by side.
Metaphase I	Homologous chromosome pairs line up at equator. Spindle apparatus is completed.
Anaphase I	Homologous chromosomes are separated; each member of a homologous pair moves to opposite poles.
Telophase I	Cytoplasm divides to produce two haploid cells.
Second meiotic division	
Prophase II	Chromosomes appear, each containing two chromatids.
Metaphase II	Chromosomes line up single file along equator as spindle formation is completed.
Anaphase II	Centromeres split and chromatids move to opposite poles.
Telophase II	Cytoplasm divides to produce two haploid cells from each of the haploid cells formed at telophase I.

Source: Fox, Stuart Ira, *Human Physiology.* © 1984 Wm. C. Brown Publishers, Dubuque, Iowa. All Rights Reserved. Reprinted by permission.

Spermatogenesis. The germ cells that migrate from the yolk sac to the testes during early embryonic development become "stem cells" called **spermatogonia** within the outer region of the seminiferous tubules. Spermatogonia are diploid cells (with forty-six chromosomes) that ultimately give rise to mature haploid gametes by a process of cell division called *meiosis (mi-o'sis).*

Meiosis occurs in two parts as summarized in table 27.4. During the first part of this process, the DNA duplicates (prophase I), and homologous chromosomes are

Figure 27.13. Meiosis, or reduction division. In the first meiotic division the homologous chromosomes of a diploid parent cell are separated into two haploid daughter cells. Each of these chromosomes contain duplicate strands, or chromatids. In the second meiotic division these chromatids are distributed to two new haploid daughter cells.

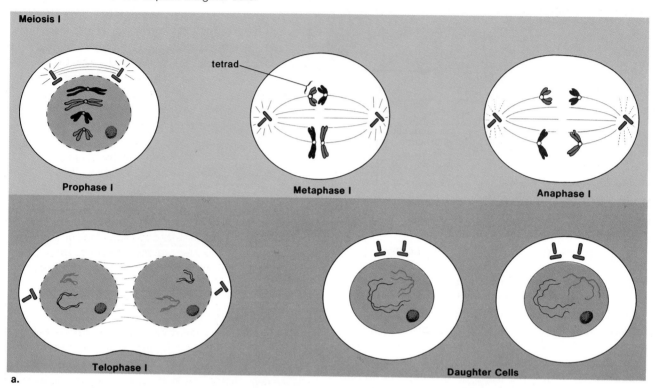

Meiosis I

tetrad

Prophase I   Metaphase I   Anaphase I

Telophase I   Daughter Cells

a.

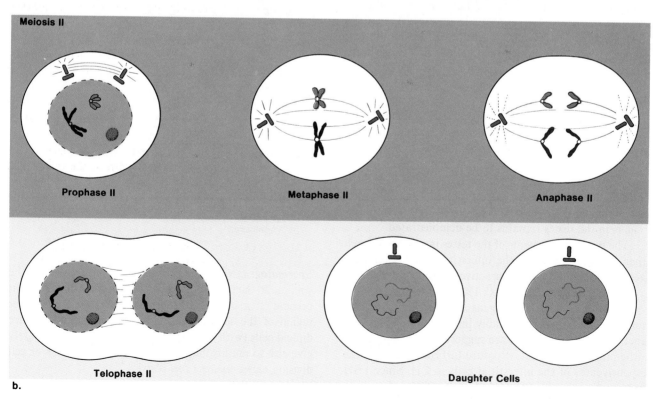

Meiosis II

Prophase II   Metaphase II   Anaphase II

Telophase II   Daughter Cells

b.

Figure 27.14. Spermatogonia undergo mitotic division to replace themselves and produce a daughter cell that will undergo meiotic division. This cell is called a primary spermatocyte. Upon completion of the first meiotic division, the daughter cells are called secondary spermatocytes. Each of these completes a second meiotic division to form spermatids. Notice that the four spermatids produced by the meiosis of a primary spermatocyte are interconnected. Each spermatid forms a mature spermatozoan.

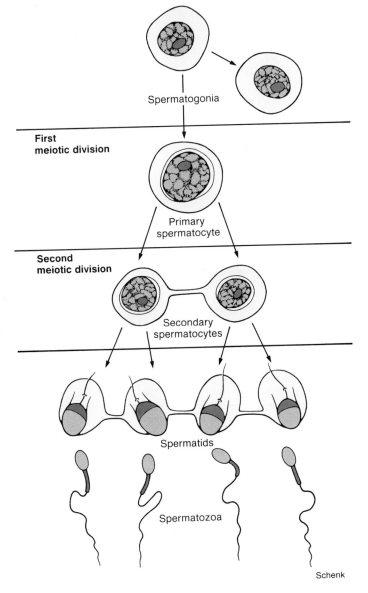

Spermatogonia

First meiotic division

Primary spermatocyte

Second meiotic division

Secondary spermatocytes

Spermatids

Spermatozoa

Schenk

Actually, only about 1,000–2,000 stem cells migrate from the yolk sac into the embryonic testes. In order to produce many millions of sperm throughout adult life, these spermatogonia cells duplicate themselves by mitotic division, and only one of the two cells—now called a **primary spermatocyte**—undergoes meiotic division (fig. 27.14). In this way, spermatogenesis can occur continuously without exhausting the number of spermatogonia.

When a diploid primary spermatocyte completes the first meiotic division (at telophase I), the two haploid daughter cells thus produced are called **secondary spermatocytes.** At the end of the second meiotic division, each of the two secondary spermatocytes produce two haploid **spermatids.** One primary spermatocyte therefore produces four spermatids.

The stages of spermatogenesis are arranged sequentially in the wall of the seminiferous tubule. The epithelial wall of the tubule, called the **germinal epithelium,** is indeed composed of germ cells in different stages of spermatogenesis. The spermatogonia and primary spermatocytes are located toward the outer side of the tubule, whereas the spermatids and mature spermatozoa are located on the side of the tubule facing the lumen (fig. 27.15).

At the end of the second meiotic division, the four spermatids produced by meiosis of one primary spermatocyte are interconnected with each other—their cytoplasm does not completely pinch off at the end of each division. The development of these interconnected spermatids into separate, mature **spermatozoa** (a process called **spermiogenesis**) requires the participation of another type of cell in the tubules, the nurse cells.

Nurse Cells. The nurse (Sertoli) cells are the only nongerminal cell type in the tubules. They form a continuous layer, connected by tight junctions, around the circumference of each tubule (fig. 27.15). In this way the nurse cells comprise a *blood-testis barrier,* because molecules from the blood must pass through the cytoplasm of nurse cells before entering germinal cells. The cytoplasm of the nurse cells extends through the width of the tubule and envelops the developing germ cells, so that it is often difficult to tell where the cytoplasm of nurse cells and germ cells are separated.

In the process of *spermiogenesis* (the conversion of spermatids to spermatozoa), most of the spermatid cytoplasm is eliminated (fig. 27.15). This occurs through phagocytosis by the nurse cells of the "residual bodies" of cytoplasm from the spermatids. Many believe that phagocytosis of the residual bodies may transmit informational molecules from the germ cells to the nurse cells. The nurse cells, in turn, may provide many molecules needed by the germ cells. It is known, for example, that the X chromosome of the germ cells is inactive during meiosis. Since this chromosome contains the genes needed to produce many essential molecules, it is believed that these molecules are provided by the nurse cells during this time.

separated (during anaphase I) into two daughter cells (telophase I). Since each daughter cell contains only one of each homologous pair of chromosomes, the cells formed at the end of this *first meiotic division* contain twenty-three chromosomes each and are haploid (fig. 27.13). Each of the twenty-three chromosomes at this stage, however, consists of two strands (called *chromatids*) of identical DNA. During the *second meiotic division,* these duplicate chromatids are separated (at anaphase II) into daughter cells at telophase II. Meiosis of one diploid spermatogonia cell therefore produces four haploid daughter cells.

Figure 27.15. Seminiferous tubules. (*a*) a cross section with surrounding interstitial tissue; (*b*) the stages of spermatogenesis within the germinal epithelium of a seminiferous tubule, showing the relationship between nurse cells and developing spermatozoa.

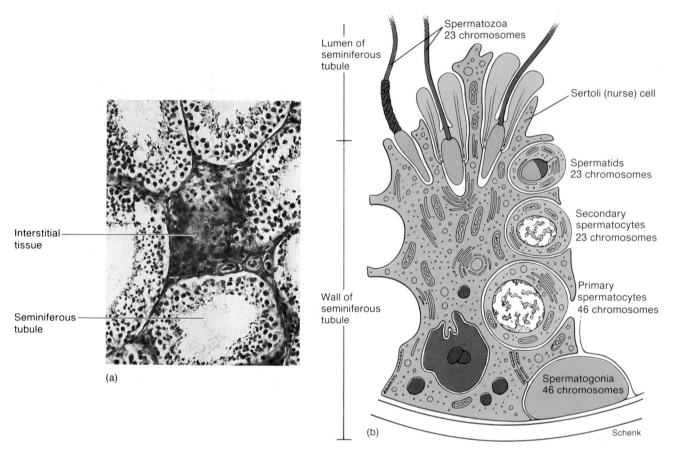

Figure 27.16. The endocrine control of spermatogenesis. During puberty both testosterone and FSH are required to initiate spermatogenesis. In the adult, however, testosterone alone can maintain spermatogenesis.

Figure 27.17. A human spermatozoan: (*a*) a diagrammatic representation; (*b*) a scanning electron micrograph. (From: *Tissues and Organs: A Text Atlas of Scanning Electron Microscopy* by R. G. Kessel and R. Kardon. W. H. Freeman and Company. © 1979.)

The importance of the nurse cells in tubular function is further evidenced by the fact that *FSH receptors are confined to the nurse cells.* Any effect of FSH on the tubules, therefore, must be mediated through the action of the nurse cells. The mechanisms by which these effects are accomplished and, indeed, the nature of all chemical interactions between the nurse and germ cells are currently unknown.

Hormonal Control of Spermatogenesis. The very beginning of spermatogenesis—the formation of primary spermatocytes and entry into early prophase I—is apparently somewhat independent of hormonal control and, in fact, starts during embryonic development. Spermatogenesis is arrested at this stage, however, until puberty, when testosterone secretion rises (fig. 27.16). Testosterone is required for the completion of meiotic division and for the early stages of spermatid maturation. This effect is probably not mediated directly by testosterone but, rather, by some of the molecules produced from testosterone by the tubules.

The later stages of spermatid maturation during puberty appear to require stimulation by FSH. This FSH effect is, as previously mentioned, mediated by the nurse cells, which contain FSH receptors and surround and interact with the spermatids. During puberty, therefore, both FSH and androgens are needed for the initiation of spermatogenesis.

Experiments in rats, and more recently in humans, have revealed that spermatogenesis within the adult testis can be maintained by androgens alone, in the absence of FSH. It appears, in other words, that FSH is needed to initiate spermatogenesis at puberty, but that once spermatogenesis has begun, FSH may no longer be required. The physiological importance of FSH in the adult male is, as a result of these experiments, currently controversial.

Structure of Spermatozoa

A mature sperm cell, or spermatozoa *(sper''mah-to-zo'ah),* is a microscopic, tadpole-shaped structure about 0.06 mm long (fig. 27.17). It consists of an oval-shaped **head,** a cylindrical **body,** and an elongated **tail.** The head of a sperm contains a nucleus with twenty-three chromosomes. The tip of the head, called the *acrosome,* contains enzymes that help the sperm penetrate into the ovum. The body of the sperm contains numerous mitochondria spiraled around a filamentous core. The mitochondria provide the energy necessary for locomotion. The tail of the sperm is a flagellum that propels the sperm with a lashing movement. The maximum unassisted rate of sperm movement is about 3 mm per hour.

> The life expectancy of ejaculated sperm is between forty-eight and seventy-two hours at body temperature. Many of the ejaculated sperm, however, are defective and are of no value. It is not uncommon for sperm to have enlarged heads, dwarfed and misshapen heads, two flagella, or a flagellum that is bent. Sperm such as these are unable to propel themselves adequately.

---

acrosome: Gk. *akros,* extremity; *soma* body

1. Describe the location and structure of the testes.
2. Describe the effects that castration has in the male on FSH and LH secretion, and explain the experimental evidence suggesting that the testes produce a polypeptide that specifically inhibits FSH secretion.
3. Describe the two compartments of the testes with respect to their structure, function, and response to gonadotrophin stimulation. Describe two ways that these compartments appear to interact.
4. Using a diagram, describe the stages of spermatogenesis. Explain why spermatogenesis can continue throughout life without using up all of the spermatogonia.
5. Diagram a sperm and its adjacent nurse cell, and explain the functions of nurse cells in the seminiferous tubules.
6. Explain how FSH and androgens synergize to stimulate sperm production at puberty. Describe the hormonal requirements for spermatogenesis after puberty.

Spermatic Ducts, Accessory Glands, and the Penis

The spermatic ducts store spermatozoa and transport them from the testes to the urethra. The accessory reproductive glands provide additives to the spermatozoa in the formation of semen, which is discharged from the erect penis during ejaculation.

Objective 14. List the various spermatic ducts and describe the location and structure of each segment.

Objective 15. Describe the structure and contents of the spermatic cord.

Objective 16. Describe the location, structure, and function of the ejaculatory ducts, seminal vesicles, prostate gland, and bulbourethral gland.

Objective 17. Describe the structure and function of the penis.

Spermatic Ducts

The duct system, which stores and transports spermatozoa from the testes to the urethra, includes the epididymides, the ductus deferentia, and the ejaculatory ducts.

Epididymis. The **epididymis** *(ep''ĭ-did'ĭ-mis)*—pl., *epididymides*—is a long, flattened organ, attached to the posterior surface of the testis (see fig. 27.10). The tubular portion of the epididymis is highly coiled and contains millions of sperm in their final stages of maturation. It is estimated that if the epididymis were uncoiled, it would measure 5–6 m (about 17 ft). The upper, expanded portion of the epididymis is the **head,** the tapering middle section is the **body,** and the lower, tubular portion is the **tail.**

Figure 27.18. A photomicrograph of the epididymis showing sperm in the lumen (50×).

— Pseudostratified columnar epithelium with stereocilia

— Smooth muscle

— Sperm in lumen

— Connective tissue

Figure 27.19. A photomicrograph of the ductus deferens (250×).

— Sperm in lumen

— Pseudostratified columnar epithelium with stereocilia

— Inner longitudinal smooth muscle layer

The tail of the epididymis is continuous with the ductus deferens. Both the tail of the epididymis and the ductus deferens store the sperm that is to be discharged during ejaculation (fig. 27.18). The time required to produce mature sperm—from meiosis in the seminiferous tubules to storage in the ductus deferens—is approximately two months.

Ductus Deferens. The **ductus deferens** *(duk'tus def'er-enz)*—pl., *ductus deferentia*—also called *vas deferens,* is a fibromuscular tube about 45 cm (18 in.) long and 2.5 mm thick (see fig. 27.1), which conveys sperm from the epididymis to the ejaculatory duct. The ductus deferens originates where the tail of the epididymis becomes less convoluted and is no longer attached to the testis. This first portion of the ductus deferens is important for the

deferens: L. *deferens,* conducting away

Figure 27.20. The male accessory glands and ducts associated with the urethra and urinary bladder.

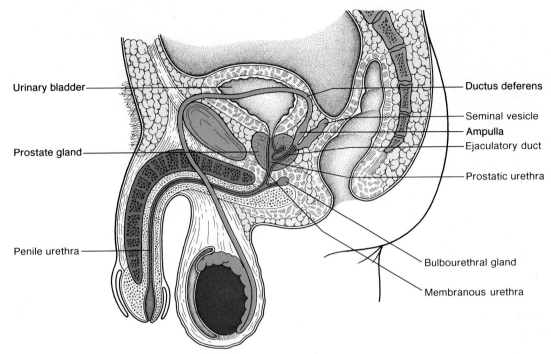

Urinary bladder

Prostate gland

Penile urethra

Ductus deferens

Seminal vesicle

Ampulla

Ejaculatory duct

Prostatic urethra

Bulbourethral gland

Membranous urethra

storage of sperm. The ductus deferens exits the scrotum as it ascends along the posterior border of the testis. From here, it penetrates the inguinal canal, enters the pelvic cavity, and passes to the side of the urinary bladder medial to the ureter. The **ampulla** of the ductus deferens is the terminal portion that joins the ejaculatory duct.

The histological structure of the ductus deferens includes a layer of pseudostratified epithelium in contact with the tubular lumen and surrounded by three layers of tightly packed smooth muscle (fig. 27.19). Sympathetic nerves from the pelvic plexus serve the ductus deferens. Stimulation through these nerves causes peristaltic contractions of the muscular layer, which forcefully ejects the stored sperm toward the ejaculatory duct.

Much of the ductus deferens is located within a structure known as the **spermatic cord.** The spermatic cord extends from the testis to the inguinal ring and consists of the ductus deferens, spermatic vessels, nerves, cremaster muscle, lymph vessels, and connective tissue. The portion of the spermatic cord positioned anterior to the pubic bone can be palpated on a male as it is compressed between the skin and the bone. The **inguinal** *(ing'gwi-nal)* **canal** is an opening for the spermatic cord to traverse the inguinal ligament and is a potentially weak area and common site for a hernia to develop.

Ejaculatory Duct. The ejaculatory *(e-jak'u-lah-to"re)* duct is 2 cm (1 in.) long and is formed by the union of the ampulla of the ductus deferens and the duct of the seminal vesicle. The ejaculatory duct then pierces the capsule

of the prostate gland on its posterior surface and continues through this gland (fig. 27.20). Both ejaculatory ducts receive secretions from the seminal vesicles and prostate gland and then eject the sperm with its additives into the prostatic urethra.

Accessory Glands

Accessory reproductive glands include the seminal vesicles, the prostate gland, and the bulbourethral glands (fig. 27.20). The contents of the seminal vesicles and the prostate gland are mixed with the sperm during ejaculation to form *semen (se'men) (seminal fluid).* The fluid from the bulbourethral glands is released in response to sexual stimulation prior to ejaculation. Both the seminal vesicles and prostate are androgen-dependent organs. They atrophy if androgen is deprived by castration.

Seminal Vesicles. The seminal vesicles are convoluted, club-shaped glands about 5 cm (2 in.) long and are positioned immediately posterior to and at the base of the urinary bladder. They secrete a sticky, slightly alkaline, yellowish substance, which serves as a fluid medium to sperm movement and longevity. The secretion from the seminal vesicles contains a variety of nutrients, including fructose, a monosaccharide that provides sperm with an energy source. It also contains citric acid, coagulation proteins, and prostaglandins. The discharge from the seminal vesicles makes up about 60% of the volume of semen.

---

prostate: Gk. *prostate*, one standing before

---

ampulla: L. *ampulla*, a two-handled bottle
vesicle: L. *vesicula*, a blister; diminutive of *vesica*, bladder

Figure 27.21. A photomicrograph of the seminal vesicle.

Seminal crypt

Smooth muscle

Glandular epithelium

Figure 27.22. A photomicrograph of the prostate gland.

Glandular alveoli

Smooth muscle

Urethra

Ejaculatory duct

Histologically, the seminal vesicle appears as a mass of cuts embedded in connective tissue (fig. 27.21). The extensively coiled mucosal layer breaks the lumen into numerous intercommunicating spaces that are lined by pseudostratified columnar and cuboidal secretory epithelia (referred to as glandular epithelium).

Blood is supplied to the seminal vesicles by the branches from the middle rectal arteries. The seminal vesicles are innervated by both sympathetic and parasympathetic fibers. Sympathetic stimulation causes the contents of the seminal vesicles to empty into the ejaculatory ducts of their respective sides.

Prostate Gland. The firm prostate *(pros'tāt)* gland is the size and shape of a horse chestnut. It is about 4 cm (1.6 in.) across and 3 cm (1.2 in.) thick and positioned immediately below the urinary bladder surrounding the beginning of the urethra (see figs. 27.1, 27.20). It is enclosed by a fibrous capsule and divided into five distinct lobules. The lobules are formed by the urethra and the ejaculatory ducts that extend through the gland. The ducts from the lobules open into the urethra. Extensive bands of smooth muscular tissue course throughout the prostate to form a meshwork that supports the glandular tissue (fig. 27.22). Contraction of the smooth muscle within the prostate empties the contents from the gland and provides part of the propulsive force needed to ejaculate the semen. The thin, milky-colored, prostatic secretion assists sperm motility as a liquefying agent, and its alkalinity protects sperm

in their passage through the acidic environment of the female vagina. The prostate also secretes the enzyme *acid phosphatase,* which is often measured clinically to assess prostate function.

Blood is supplied to the prostate from branches of the middle rectal and inferior vesical arteries. The venous return forms the prostatic venous plexus along with blood draining from the penis. The prostatic venous plexus drains into the internal iliac veins. The prostate gland has both sympathetic and parasympathetic innervation arising from the pelvic plexuses.

A routine physical examination of the male includes rectal palpation of the prostate gland. Enlargement or overgrowth of the glandular substance of the prostate, called benign prostatic hypertrophy, is relatively common in older men. This may constrict the urethra and cause difficult micturition. An enlarged prostate gland usually requires surgery. If the obstruction is slight, the surgery may be accomplished through the urethral canal using a technique called a transurethral prostatic resection, in which excessive tissue is cut and cauterized.

Bulbourethral Glands. The paired, pea-sized bulbourethral (Cowper's) glands are located inferior to the prostate gland. Each bulbourethral gland is brownish in color, about 1 cm in diameter, and drains by a 2.5 cm (1 in.) duct into the urethra (see fig. 27.20). Upon sexual excitement and prior to ejaculation, the bulbourethral glands are stimulated to secrete a mucoid substance, which coats the lining of the urethra to neutralize the pH of the urine residue and lubricates the tip of the penis in preparation for sexual intercourse.

cauterize: Gk. *kauterion,* a branding iron
Cowper's gland: from William Cowper, English anatomist, 1666–1709

Figure 27.23. A photomicrograph of the urethra (10×).

—Paraurethral glands

—Lumen of urethra

—Transitional epithelium

—Corpus cavernosum urethra

Urethra

The urethra of the male serves as a common tube for both the urinary and reproductive systems. However, urine and semen cannot simultaneously pass through the urethra because the nervous reflex during ejaculation automatically inhibits micturition. The urethra of the male is about 20 cm (8 in.) long and S-shaped due to the shape of the penis. Three regions can be identified—the prostatic urethra, the membranous urethra, and the penile urethra (see fig. 27.20).

The **prostatic urethra** is the proximal (2.5 cm) portion of the urethra that passes through the prostate gland. The prostatic urethra receives drainage from small ducts of the prostate gland and the two ejaculatory ducts.

The **membranous** *(mem'brah-nus)* **urethra** is the short (0.5 cm) portion of the urethra that passes through the urogenital diaphragm. The external urethral sphincter muscle is located in this region.

The **penile urethra** (cavernous urethra) is the longest portion (15 cm), extending from the outer edge of the urogenital diaphragm to the external urethral orifice on the glans penis. This portion is surrounded by erectile tissue as it passes through the corpus spongiosum of the penis. The paired ducts from the bulbourethral glands attach to the penile urethra near the urogenital diaphragm.

The wall of the urethra has an inside lining of mucous membrane surrounded by a relatively thick layer of smooth muscle tissue (fig. 27.23). Specialized **urethral glands** are embedded in the urethral wall and function to secrete mucus into the urethral canal.

Penis

The penis is composed mainly of erectile tissue and when distended, serves as the intromittent, or copulatory, organ of the male reproductive system. **Erectile tissue** contains numerous vascular spaces that become engorged with blood. The penis is a pendent structure, positioned anterior to the scrotum and attached to the pubic arch. It is divided into a proximally attached root, an elongated tubular shaft, and a distal cone-shaped glans (fig. 27.24).

The **root of the penis** expands posteriorly to form the **bulb of the penis** and the **crus** *(krus)* **of the penis.** The bulb is positioned in the urogenital triangle of the perineum where it is attached to the undersurface of the perineal membrane and enveloped by the bulbospongiosus muscle. The crus, in turn, attaches the root of the penis to the pubic arch (ischiopubic ramus) and to the perineal membrane. The crus is positioned superior to the bulb and is enveloped by the ischiocavernosus muscle.

The **shaft,** or body, **of the penis** is composed of three cylindrical columns of erectile tissue that are bound together by fibrous tissue and covered with skin (fig. 27.24). The paired dorsally positioned masses are named the **corpora cavernosa penis.** The fibrous tissue between the two corpora forms a **median septum.** The **corpus spongiosum penis (corpus cavernosum urethra)** lies ventral to the other two and surrounds the penile urethra. The penis is flaccid and relaxed when the spongelike tissue is not engorged with blood but becomes firm and erect when the spaces are filled.

The **glans penis** is the cone-shaped terminal portion of the penis that is formed from the expanded corpus spongiosum. The opening of the urethra at the tip of the glans is called the **external urinary meatus.** The **corona glandis** is the prominent posterior ridge of the glans. On the undersurface of the glans, a vertical fold of tissue, called the **frenulum** *(fren'u-lum),* attaches the skin covering the penis to the glans.

The skin covering the penis is hairless, contains no fat cells and generally has a darker pigment than the other body skin. The skin of the shaft is loosely attached and is continuous over the glans as a retractable sheath called the **prepuce** *(pre'pūs),* or **foreskin.** The prepuce is commonly removed in a newborn by a surgical procedure called *circumcision.*

A circumcision is generally performed for hygienic purposes because the glans is easier to clean if exposed. A sebaceous secretion from the glans, called **smegma,** will accumulate along the border of the corona glandis if good hygiene is not practiced. Smegma can foster bacteria that may cause infections and therefore should be removed through washing. Cleaning the glans of an uncircumcised male requires retraction of the prepuce. Occasionally, a child is born with a prepuce that is too tight to permit retraction. This condition is called **phimosis** and necessitates circumcision.

---

crus: L. *crus,* leg; resembling a leg
cavernous: L. *cavus,* hollow
glans: L. *glans,* acorn
corona: L. *corona,* garland, crown
frenulum: L. diminutive of *frenum,* a bridle
prepuce: L. *prae,* before; *putium,* penis
phimosis: Gk. *phimosis,* a muzzling

Figure 27.24. The structure of the penis showing the attachment, blood and nerve supply, and the arrangement of the erectile tissue.

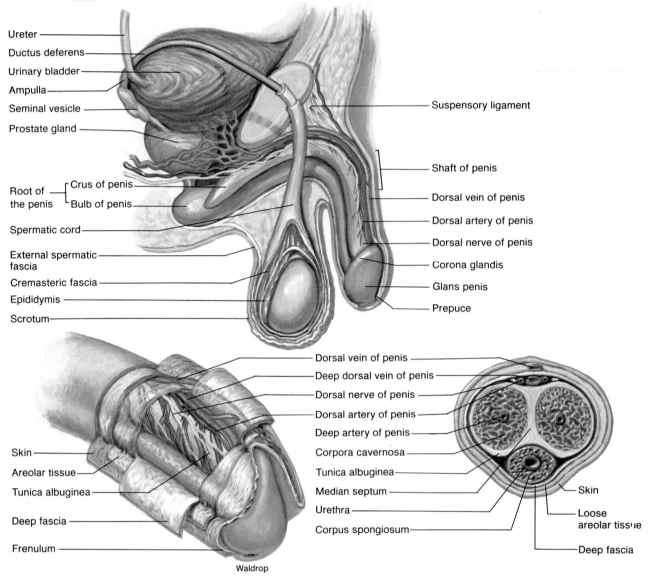

Ureter
Ductus deferens
Urinary bladder
Ampulla
Seminal vesicle
Prostate gland

Root of the penis — Crus of penis / Bulb of penis
Spermatic cord
External spermatic fascia
Cremasteric fascia
Epididymis
Scrotum

Suspensory ligament
Shaft of penis
Dorsal vein of penis
Dorsal artery of penis
Dorsal nerve of penis
Corona glandis
Glans penis
Prepuce

Skin
Areolar tissue
Tunica albuginea
Deep fascia
Frenulum

Dorsal vein of penis
Deep dorsal vein of penis
Dorsal nerve of penis
Dorsal artery of penis
Deep artery of penis
Corpora cavernosa
Tunica albuginea
Median septum
Urethra
Corpus spongiosum

Skin
Loose areolar tissue
Deep fascia

Waldrop

The penis is supplied with blood primarily through the superficial external pudendal branch of the femoral artery. The venous return is through a superficial median dorsal vein that drains into the great saphenous vein in the thigh and through the deep median vein that drains into the prostatic plexus.

The penis has many afferent tactile receptors, especially in the glans, that make it a highly sensitive organ. In addition, the penis has extensive efferent innervation from both parasympathetic and sympathetic fibers.

1. What are the functions of each of the regions of the spermatic duct?
2. Differentiate between the ductus deferens and the spermatic cord.
3. Describe the structure and location of each of the accessory glands.
4. Why is the position of the prostate gland of clinical importance?
5. What do each of the accessory glands secrete?
6. Describe the external structure of the penis and the internal arrangement of the erectile tissue within the penis.

Figure 27.25. The mechanism of erection of the penis.

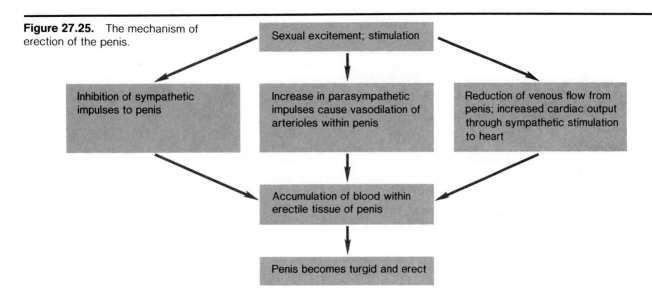

Sexual excitement; stimulation

Inhibition of sympathetic impulses to penis

Increase in parasympathetic impulses cause vasodilation of arterioles within penis

Reduction of venous flow from penis; increased cardiac output through sympathetic stimulation to heart

Accumulation of blood within erectile tissue of penis

Penis becomes turgid and erect

Mechanisms of Erection, Emission, and Ejaculation

Erection of the penis results from parasympathetic induced vasodilation of arteries within the penis. Emission and ejaculation are stimulated by sympathetic impulses, resulting in the forceful expulsion of semen from the penis.

> *Objective 18.* Distinguish between erection, emission, and ejaculation.
>
> *Objective 19.* Describe the events that result in erection of the penis.
>
> *Objective 20.* Explain the physiological process of ejaculation.
>
> *Objective 21.* Describe the characteristic properties of semen.

Erection, emission, and ejaculation are interrelated events that are necessary for the deposition of semen into the vagina by the natural means of sexual intercourse (coitus), or copulation. *Erection* usually occurs as a male becomes sexually aroused; the erectile tissue of the penis becomes engorged with blood, causing the penis to become wider, longer, and firmer. *Emission (e-mish'un)* is the movement of spermatozoa from the epididymides to the ejaculatory ducts. *Ejaculation (e-jak''u-la'shun)* is the forceful expulsion of the ejaculate, or *semen* (seminal fluid), from the ejaculatory ducts and urethra of the penis. Emission and ejaculation do not have to follow erection of the penis and only occur if there is sufficient stimulation of the sensory receptors in the penis to elicit the ejaculatory response.

Erection of the Penis

Erection of the penis depends on the volume of blood that enters the arteries of the penis as compared to the volume that exits through venous drainage. Normally, constant sympathetic stimuli to the arterioles of the penis maintain a partial constriction of smooth muscles within the arteriole walls so that there is an even flow of blood throughout the penis. During sexual excitement, however, parasympathetic impulses cause marked vasodilation within the arterioles of the penis, resulting in more blood entering than venous blood draining. At the same time, there may be slight vasoconstriction of the dorsal vein of the penis and an increase in cardiac output. These combined events cause the spongy tissue of the corpora cavernosa and the corpus spongiosum to become distended with blood and the penis to become turgid. In this condition, the penis can be inserted into the vagina of the female and function as an intromittent, or copulatory, organ to discharge semen.

Erection is controlled by two portions of the central nervous system—the hypothalamus in the brain and the sacral portion of the spinal cord. The hypothalamus controls conscious sexual thoughts that originate in the cerebral cortex. Nerve impulses from the hypothalamus elicit parasympathetic responses from the sacral region, which cause vasodilation of the arterioles within the penis. Conscious thought is not required for an erection, however, and stimulation of the penis can cause an erection because of a reflex response in the spinal cord. This reflexive action makes possible an erection in a sleeping male or in an infant—perhaps from the stimulus of a diaper.

The mechanism of erection of the penis is summarized in figure 27.25.

Ejaculation is the expulsion of semen through the urethra of the penis. In contrast to erection, ejaculation is a response involving the sympathetic innervation of the accessory reproductive organs. Ejaculation is preceded by continued sexual stimulation, usually through activated tactile receptors in the glans penis and the skin of the shaft.

turgid: L. *turgeo*, to swell

emission: L. *emittere*, expel or eject

Rhythmic friction of these structures during sexual intercourse causes afferent impulses to be transmitted to the thalamus and cerebral cortex. The first sympathetic response, which occurs prior to ejaculation, is the discharge of the contents from the bulbourethral glands. The fluids from these glands are usually discharged before penetration into the vagina and serve to lubricate the urethra and the glans penis.

Emission and Ejaculation of Semen

Emission. Continued sexual stimulation following erection of the penis causes emission. Emission is the movement of sperm from the epididymides to the ejaculatory ducts and the secretions of the accessory glands into the ejaculatory ducts and urethra in the formation of semen. Emission occurs as sympathetic impulses from the pelvic plexus cause a rhythmic contraction of the smooth muscle layers of the testes, epididymides, ductus deferentia, ejaculatory ducts, seminal vesicles, and prostate gland.

Ejaculation. Ejaculation immediately follows emission and is accompanied by *orgasm,* which is considered the climax of the sex act. Ejaculation occurs in a series of spurts of semen from the urethra. This takes place as parasympathetic impulses traveling through the pudendal nerves stimulate the bulbocavernosus muscles at the base of the penis and cause them to contract rhythmically. There is also sympathetic stimulation of the smooth muscles in the urethral wall that peristaltically contract to help eject the semen.

Sexual function in the male thus requires the synergistic action (rather than antagonistic action) of the parasympathetic and sympathetic systems. The mechanism of emission and ejaculation is summarized in figure 27.26.

Immediately following ejaculation or a cessation of sexual stimulus, sympathetic impulses cause vasoconstriction of the arterioles within the penis, reducing the inflow of blood. At the same time, cardiac output returns to normal as does venous return of blood from the penis. With the normal flow of blood through the penis, it returns to its flaccid condition. If an ejaculation of semen occurred while the penis was erect, another erection and ejaculation cannot be triggered for ten to thirty minutes or longer.

> Adolescent males may experience erection of the penis and spontaneous emission and ejaculation of semen during sleep. These **nocturnal emissions** are thought to be caused by changes in hormonal concentrations that accompany adolescent development.

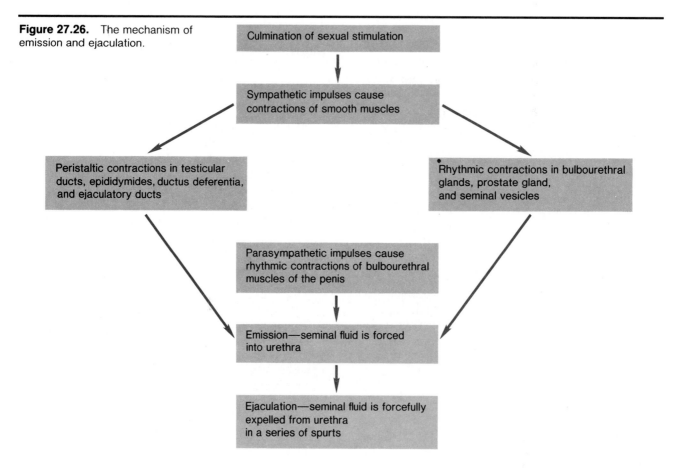

Figure 27.26. The mechanism of emission and ejaculation.

Semen. Semen, also called seminal *(sem'ĭ-nal)* fluid, which consists of spermatozoa plus the additives from the accessory reproductive glands, is the substance discharged during ejaculation (table 27.5). Generally, between 1.5 and 5.0 ml of semen are ejected during ejaculation. The bulk of the fluid (about 60%) is produced by the seminal vesicles, and the remaining (about 40%) is contributed by the prostate gland. There are usually between 60 and 150 million sperm per milliliter of ejaculate. In the condition of *oligospermia,* the male ejaculates fewer than 10 million sperm per milliliter and is likely to have fertility problems.

Human semen can be frozen and stored in sperm banks for future artificial insemination. In this procedure, the semen is diluted with 10% glycerol, monosaccharide, and distilled water buffer and frozen in liquid nitrogen. The freezing process destroys defective and abnormal sperm. For some unknown reason, however, not all human sperm is suitable for freezing.

---

1. *Define the terms* erection, emission *and* ejaculation.
2. *Explain the statement that male sexual function is an autonomic synergistic action.*
3. *Compose a flow chart to explain the physiological and physical events of erection, emission, and ejaculation.*
4. *Describe the components of a "normal" ejaculate.*

Clinical Considerations

Sexual dysfunction is a broad area of medical concern that includes developmental and psychogenic problems as well as conditions resulting from various diseases. Psychogenic problems of the reproductive system are extremely complex, poorly understood, and beyond the scope of this book.

Table 27.5 Some characteristics and reference values used in clinical examination of semen

Characteristic	Reference value
Volume of ejaculate	1.5–5.0 ml
Sperm count	40–250 million/ml
Sperm motility	
Percent of motile forms:	
1 hour after ejaculation	70% or more
3 hours after ejaculation	60% or more
Leukocyte count	0–2000/ml
pH	7.2–7.8
Fructose concentration	150–600 mg/100 ml

Source: Modified from Glasser, L. Seminal fluid and subfertility. *Diagnostic Medicine,* July/August 1981, p. 28. By permission.

fertility: L. *fere,* to bear
oligospermia: Gk. *oligos,* few; *sperma,* seed

Only a few of the principal developmental conditions, reproductive treatments, and diseases that affect the physical structure and function of the male reproductive system will be discussed.

Developmental Abnormalities

The reproductive organs of both sexes develop from similar embryonic tissue that follows a consistent pattern of formation well into the fetal period. Because an embryo has the potential to differentiate into a male or a female, developmental errors can result in various degrees of intermediate sex, or **hermaphroditism** *(her-maf'ro-di-tizm'').* A person with undifferentiated or ambiguous external genitalia is called a **hermaphrodite.**

True hermaphroditism—in which both male and female gonadal tissues are present, in either the same or opposite gonads—is a rare anomaly. True hermaphrodites usually have a 46, XX chromosome constitution. Male **pseudohermaphroditism** occurs more commonly and generally results from hormonal influences during early fetal development. This condition is caused either by inadequate amounts of androgenic hormones being secreted or by the delayed development of the reproductive organs after the period of tissue sensitivity has passed. These persons have a 46, XY chromosome constitution and male gonads, but the genitalia are intersexual and variable. The treatment of hermaphroditism varies, depending on the extent of ambiguity of the reproductive organs. These persons are sterile but may marry and live a normal life following hormonal therapy and plastic surgery.

Chromosomal anomalies result from the improper separation of the chromosomes during meiosis and are usually expressed in deviations of the reproductive organs. The two most frequent chromosomal anomalies cause Turner's syndrome and Klinefelter's syndrome. **Turner's syndrome** occurs when only one X chromosome is present. About 97% of embryos lacking an X chromosome die; the remaining 3% survive and appear to be females, but their gonads are rudimentary or absent, and they do not mature at puberty. A person with **Klinefelter's syndrome** has an XXY chromosome constitution, develops breasts and male genitalia, but has underdeveloped seminiferous tubules and is generally retarded.

A more common developmental problem than genetic abnormalities, and fortunately less serious, is cryptorchidism. **Cryptorchidism** *(krip-tor'kĭ-dizm)* means "hidden testis" and is characterized by the failure of one or both testes to descend into the scrotum. A cryptorchid testis is

hermaphrodite: Gk. (mythology) *Hermaphroditos,* son of Hermes (Mercury)
Turner's syndrome: from Henry H. Turner, American endocrinologist, 1892–1970
Klinefelter's syndrome: from Harry F. Klinefelter, Jr., American physician, b. 1912
cryptorchidism: Gk. *crypto,* hidden; *orchis,* testis

Figure 27.27. Cryptorchidism. (a) incomplete descent of a testis may involve four separate regions: (1) in the pelvic cavity, (2) in the inguinal canal, (3) at the superficial inguinal ring, (4) in the upper scrotum. (b) an ectopic testis may be (1) in the superficial fascia of the anterior pelvic wall, (2) at the root of the penis, (3) in the perineum, in the thigh alongside the femoral vessels.

(a)

(b)

usually located along the path of descent but can be anywhere in the pelvic cavity (fig. 27.27). It occurs in about 3% of male infants and should be treated before the infant is five years old to reduce the chance of infertility or other complications.

Functional Considerations

Functional disorders of the male reproductive system include impotence, infertility, and sterility. **Impotence** *(im'po-tens)* is the inability of a sexually mature male to maintain erection long enough to have an ejaculation. The causes of impotence may be physical, such as abnormalities of the penis, vascular irregularities, neurological disorders, or the result of diseases. Generally, however, the cause of impotence is psychological, and the patient requires skilled counseling by a sex therapist.

Infertility is the inability of the sperm to fertilize the ovum and may be the fault of the male or female, or both. The term *impotence* should not be used when referring to infertility. Infertility in males may be caused by a number of things, the most common of which is the inadequate production of viable sperm. Some of the causes of infertility in males are alcoholism, dietary deficiencies, local injury, varicocele, excessive heat, or exposure to X rays. A hormonal imbalance may also contribute to infertility. Many of the causes of infertility can be treated through proper nutrition, gonadotrophic hormone treatment, or microsurgery. If corrective treatment is not possible, however, it may be possible to concentrate the sperm obtained through *masturbation* (self-stimulation to the point of ejaculation) and use this concentrate to inseminate the woman artificially.

Sterility is similar to infertility except that it is a permanent condition. Sterility may be genetically caused, or it may be the result of degenerative changes in the seminiferous tubules (for example, mumps in a mature male may secondarily infect the testes and cause irreversible tissue damage).

Voluntary sterilization of the male in a procedure called a **vasectomy** is a common technique of birth control. In this procedure, a small section of each ductus deferens near the epididymis is surgically removed, and the cut ends of the ducts are tied (fig. 27.28). Occasionally a valve or similar device is inserted. This procedure interferes with sperm transport but does not directly affect the secretion of androgens from interstitial cells in the interstitial tissue. Since spermatogenesis continues, the sperm cannot be drained from the testes and instead accumulate in the "crypts" that form in the seminiferous tubules and ductus deferens. These crypts present sites of inflammatory reactions in which spermatozoa are phagocytosed and destroyed by the immune system.

impotence: L. *im*, not; *potens*, potent
sterility: L. *sterilis*, barren
vasectomy: L. *vas*, vessel; Gk. *ektome*, excision

Figure 27.28.   A simplified illustration of a vasectomy, in which segments of the ductus deferentia are removed through incisions in the scrotum.

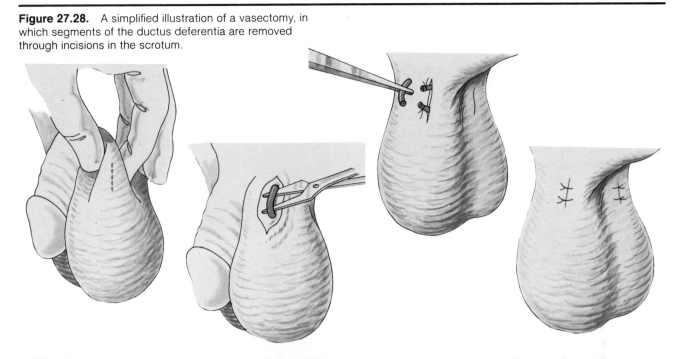

Table **27.6** Kinds of venereal diseases			
Name	**Organism**	**Resulting condition**	**Treatment**
Gonorrhea*	*Gonococcus* (bacterium)	Adult: sterility due to scarring of epididymides and tubes; rarely: septicemia; newborn: blindness	Penicillin injections; tetracycline tablets; eye drops (silver nitrate or penicillin)
Syphilis*	*Treponema pallidum* (bacterium)	Adult: gummas, cardiovascular neurosyphilis; newborn: congenital syphilis (abnormalities, blindness)	Penicillin injections; tetracycline tablets
Chancroid** (soft chancre)	*Hemophilus ducreyi* (bacterium)	Chancres, buboes	Tetracycline; sulfa drugs
Urethritis in men*	Various microorganisms	Clear discharge	Tetracycline
Vaginitis*	*Trichomonas* (protozoan)	Frothy white or yellow discharge	Metronidazole
	Candida albicans (yeast)	Thick, white, curdy discharge (moniliasis)	Nystatin
Lymphogranuloma venereum (LGV)**	Microorganism	Ulcerating buboes; rectal stricture	Tetracycline; sulfa drugs
Granuloma venereum (inguinale)**	*Donovania granulomatis*	Raw, open, extended sore	Tetracycline
Venereal warts	Virus	Warts	Podophyllin
Genital herpes*	Herpes simplex virus	Sores	Palliative treatment
Crabs*	Arthropod	Itching	Gamma benzene hexachloride

*Fairly common   **More common in warmer climates

Source: Van De Graaff, Kent M., *Human Anatomy.* © 1984 Wm. C. Brown Publishers, Dubuque, Iowa. All Rights Reserved. Reprinted by permission.

Although androgen secretion from the interstitial cells is not directly affected by a vasectomy, it is clear that tubular function may be affected by the inflammatory reactions that result. Since the tubules have been shown to secrete 5α-reduced androgens and perhaps other hormones and since the physiological importance of this endocrine function is not presently understood, it is clear that more research is needed to establish the long-term safety of this contraceptive procedure.

Diseases of the Male Reproductive System

Venereal Diseases.   Venereal diseases (VD) are contagious diseases that affect the reproductive systems of both the male and the female and are transmitted during sexual activity. The frequency of venereal diseases in the United States is regarded by health authorities as epidemic; venereal diseases are second to the common cold as a communicable disease. Although there are many kinds of venereal diseases (table 27.6), gonorrhea and syphilis are

---

venereal: L. (mythology) from *Venus,* the goddess of love the most common. These two diseases have not been eradicated mainly because humans cannot develop immunity to them and increased sexual promiscuity increases the chances of reinfection.

Gonorrhea *(gon''o-re'ah),* commonly called "clap," is caused by the bacterium gonococcus, or *Neisseria gonorrhoeae.* Males with this disease suffer inflammation of the urethra, accompanied by painful urination and frequently the discharge of pus. In females, the condition is usually asymptomatic, and therefore many women may be unsuspecting carriers of the disease. Advanced stages of gonorrhea in females may infect the uterus and the uterine tubes. A pregnant woman with gonorrhea that is not treated may transmit the disease to the eyes of her newborn, causing blindness.

Syphilis *(sif'ĭ-lis)* is caused by the bacterium *Treponema pallidum.* Syphilis is less common than gonorrhea but is the more serious of the two diseases. During the *primary stage* of syphilis, a lesion called a *chancre* develops at the point where contact was made with a similar sore from an infected person. The chancre is an ulcerated sore that has hard edges and endures for ten days to three months. It is only during the primary stage that syphilis can be spread to another sexual partner. The chancre will heal with time, but if not treated, it will be followed by secondary and tertiary stages of syphilis. During the initial contact, the bacteria enter the bloodstream and spread throughout the body. The *secondary stage* of syphilis is expressed by lesions or a rash of the skin and mucous membranes, accompanied by fever (fig. 27.29). This stage lasts from two weeks to six months, and the symptoms disappear of their own accord. The *tertiary stage* occurs ten to twenty years following the primary infection. The circulatory, integumentary, skeletal, and nervous systems are particularly vulnerable to the degenerative changes caused by this disease. The end result of untreated syphilis is frequently blindness and insanity.

Disorders of the Prostate Gland.   The prostate gland is subject to several disorders, most of which are common in older men. The four most frequent prostatic problems are acute prostatitis, chronic prostatitis, benign prostatic hyperplasia, and carcinoma of the prostate.

Acute prostatitis is common in sexually active young men through infections acquired from a gonococcus bacterium. The symptoms of acute prostatitis are a swollen and tender prostate gland, painful urination, and in extreme conditions, pus dripping from the penis. It is treated with penicillin, bed rest, and increased fluid intake.

Chronic prostatitis is one of the most common afflictions of middle-aged and elderly men. The symptoms of this condition vary considerably from irritation and slight

gonorrhea: L. *gonos,* seed; *rhoia,* a flow
chancre: Fr. *chancre,* indirectly from L. *cancer,* a crab

Figure 27.29.   The secondary stages of syphilis as expressed by lesions of the skin of this young woman.

difficulty in urination to extreme pain and urine blockage, which commonly causes secondary renal infections. In this disease, several kinds of infectious microorganisms are believed to be harbored in the prostate gland and are responsible for inflammations elsewhere in the body, such as in the nerves (neuritis), the joints (arthritis), the muscles (myositis), and the iris (iritis).

Prostatic hyperplasia, or an enlarged prostate gland, occurs in approximately one-third of all males over the age of sixty. In this condition, an overgrowth of granular material compresses the prostatic urethra. The cause of prostatic hyperplasia is not known. As the prostate enlarges, urination becomes painful and difficult. If the urinary bladder is not emptied completely, cystitis eventually occurs. Persons with cystitis may become incontinent and dribble urine continuously. Prostatic hypertrophy is usually treated by the surgical removal of portions of the gland through transurethral curetting (cutting and removal of a small section) or the removal of the entire prostate gland, called **prostatectomy.**

Prostatic carcinoma, or cancer of the prostate gland, is the second leading cause of death from cancer in males in the United States. It is common in males over sixty and accounts for 19,000 deaths annually. When prostatic cancer is confined to the prostate gland, it is generally small and asymptomatic. But as the cancer grows and invades surrounding nerve plexuses, it becomes extremely painful and easily detected. The metastases of this cancer to the spinal column and brain are generally what kills the patient.

As prostatic carcinoma develops, it has symptoms nearly identical to prostatic hyperplasia—painful urination and cystitis. When examined by rectal palpation with a gloved finger, however, a hard cancerous mass can be detected in contrast to the enlarged, soft, and tender prostate diagnostic of prostatic hypertrophy. Prostatic carcinoma is treated by prostatectomy and frequently by the removal of the testes (called **orchiectomy**) as well. An orchiectomy inhibits metastases by eliminating testosterone secretion.

Disorders of the Testes and Scrotum. A **hydrocele** *(hi'dro-sēl)* is a benign fluid mass within the tunica vaginalis that causes swelling of the scrotum. It is a frequent, minor disorder in infant boys as well as in adults. The cause is unknown.

An infection in the testes is called **orchitis.** Orchitis may develop from a primary infection from a tubercle bacterium or as a secondary complication of mumps contracted after puberty. If orchitis from mumps involves both testes, it usually causes sterility.

Trauma to the testes and scrotum is common because of their pendent position. The testes are extremely sensitive to pain, and a male responds reflexively to protect the groin area.

Chapter Summary

I. Introduction to the Reproductive System
The organs of the male and female reproductive system produce and unite gametes that contain specific genes. A random combination of the genes during sexual reproduction permits the propagation of individuals with genetic differences.
  A. The functions of the male reproductive system are to produce sperm, secrete testosterone, and transfer sperm to the reproductive system of the female.
  B. The male reproductive system is divided into primary sex organs (the testes), secondary sex organs (those that are essential for sexual reproduction), and secondary sex characteristics (features that are sexual attractants, which are expressed after puberty).

II. Development of the Male Reproductive System
Sexual distinction is determined upon fertilization of an ovum by a sperm containing either an X or a Y chromosome. Sexual differentiation under hormonal control is a progressive process through embryonic and early fetal development.
  A. An XY-chromosome combination produces a male, and an XX-chromosome combination produces a female.
  B. The appearance of the gonadal ridge during the fifth week is the first indication of sex organ formation.
    1. Primary sex cords and a genital tubercle develop during the sixth week.
    2. Reproductive development is well progressed by the eighth week but is in the indifferent stage because sexual distinction is not apparent.

  C. Testes secrete testosterone, which stimulates the development of male accessory sex organs and male genitalia, and Müllerian inhibition factor, which prevents the development of female accessory sex organs.
  D. The penis and scrotum are formed by the end of the twelfth week, and the descent of the testes occurs during the twenty-eighth week.

III. Puberty
Gonadotrophic hormones from the anterior pituitary and maturational changes in the central nervous system result in the increased secretion of sex steroids from the testes, which in turn stimulates a growth spurt and sexual maturation during puberty.
  A. Androgens stimulate the development of secondary sex characteristics.
  B. The sensitivity of the hypothalamus and pituitary to negative feedback may decrease at puberty and thus contribute to the rise in gonadotrophin secretion.

IV. Structure and Function of the Testes
The testes within the scrotum produce spermatozoa and androgens. Androgens regulate spermatogenesis and the development and functioning of the secondary sex organs.
  A. The saclike scrotum supports and protects the testes and regulates their position relative to the pelvic region of the body.
  B. The testes are separated into lobules, composed of seminiferous tubules that produce sperm and interstitial tissue that produces androgens.
    1. LH stimulates the interstitial (Leydig) cells to secrete androgens; FSH stimulates the seminiferous tubules to produce sperm.

    2. Testosterone from the interstitial cells exerts negative feedback inhibition on LH secretion; inhibin, secreted from the tubules, is believed to exert negative feedback inhibition on FSH secretion.
  C. Spermatogenesis occurs by meiotic division of the cells that line the seminiferous tubules.
    1. At the end of the first meiotic division, two secondary spermatocytes are produced.
    2. At the end of the second meiotic division, four haploid spermatids are produced.
  D. The conversion of spermatids to spermatozoa is called spermiogenesis.
    1. Nurse cells contain FSH receptors, so that the stimulation of spermatogenesis by FSH must act via the nurse cells.
    2. The initiation of sperm production at puberty requires stimulation by both FSH and androgens.
  E. A sperm consists of a head, body, and tail and matures in the epididymides prior to ejaculation.

V. Spermatic Ducts, Accessory Glands, and the Penis
The spermatic ducts store spermatozoa and transport them from the testes to the urethra. The accessory reproductive glands provide additives to the spermatozoa in the formation of semen, which is discharged from the erect penis during ejaculation.
  A. The epididymides, ductus (vas) deferentia, and ejaculatory ducts are the components of the spermatic ducts.

1. The highly coiled epididymides are the tubular structures on the testes where sperm mature and are stored.
2. The ductus deferentia convey sperm from the epididymides to the ejaculatory ducts during emission. Each ductus deferens forms a component of a spermatic cord.

B. The seminal vesicles and prostate gland provide additives to the sperm in the formation of semen.
1. The seminal vesicles are positioned posterior to the base of the urinary bladder and secrete about 60% of the additive fluid of semen.
2. The prostate gland surrounds the urethra just below the urinary bladder.

3. The small bulbourethral glands secrete fluid that serves as a lubricant for the erect penis in preparation for copulation.

C. The male urethra, which serves both the urinary and reproductive systems, is divided into the prostatic, membranous, and penile portions.

D. The penis is specialized to become erect for insertion into the vagina during copulation.
1. The body of the penis is composed of three columns of erectile tissue, the penile urethra, and associated vessels and nerves.
2. The root of the penis is attached to the pubic arch and perineal membrane.
3. The glans penis is the terminal end, which is covered with the prepuce in an uncircumsized male.

VI. Mechanisms of Erection, Emission, and Ejaculation

Erection of the penis results from parasympathetic induced vasodilation of arteries within the penis. Emission and ejaculation are stimulated by sympathetic impulses, resulting in the forceful expulsion of semen from the penis.

A. Erection of the penis occurs as the erectile tissue becomes engorged with blood; emission is the movement of the spermatozoa from the epididymides to the ejaculatory ducts; and ejaculation is the forceful expulsion of semen from the ejaculatory ducts and urethra of the penis.

B. Parasympathetic stimuli to arteries in the penis cause the erectile tissue to engorge with blood as arteriole flow increases and venous drainage decreases.

C. Ejaculation is the result of sympathetic reflexes to the smooth muscles in the reproductive organs.

Review Activities

Objective Questions

1. An embryo with the genotype XY develops male accessory sex organs because of
   (a) androgens
   (b) estrogens
   (c) the absence of androgens
   (d) the absence of estrogens
2. Which of the following does *not* arise from the embryonic mesonephric duct? The
   (a) epididymis
   (b) ductus deferens
   (c) seminal vesicle
   (d) prostate gland
3. The external genitalia of a male are completely formed by the end of the
   (a) embryonic period
   (b) ninth week
   (c) tenth week
   (d) twelfth week
4. In the male, FSH
   (a) is not secreted by the pituitary
   (b) receptors are located in the interstitial cells
   (c) receptors are located in the spermatogonia
   (d) receptors are located in the nurse cells
5. The secretion of FSH in a male is inhibited by negative feedback effects of
   (a) inhibin secreted from the tubules
   (b) inhibin secreted from the interstitial cells
   (c) testosterone secreted from the tubules
   (d) testosterone secreted from the interstitial cells

6. Which of the following is *not* a spermatic duct? The
   (a) epididymis
   (b) spermatic cord
   (c) ejaculatory duct
   (d) ductus deferens
7. Spermatozoa are stored prior to emission and ejaculation in the
   (a) epididymides
   (b) seminal vesicles
   (c) penile urethra
   (d) prostate gland
8. Urethral glands function to
   (a) secrete mucus
   (b) produce nutrients
   (c) secrete hormones
   (d) regulate sperm production
9. Which statement is *false* regarding erection of the penis?
   (a) It is a parasympathetic response.
   (b) It may be both a voluntary and involuntary response.
   (c) It has to be followed by emission and ejaculation.
   (d) It is controlled by the hypothalamus of the brain and sacral portion of the spinal cord.
10. The condition where one or both testes fail to descend into the scrotum is
   (a) cryptorchidism
   (b) Turner's syndrome
   (c) hermaphroditism
   (d) Klinefelter's syndrome

Essay Questions

1. Explain what is meant by the latent development of the reproductive organs during puberty, and discuss the physiological mechanisms that cause puberty in a male.

2. Discuss the significance of the indifferent stage in reproductive development. When is the indifferent stage, and what are its implications in the potential development of abnormalities of the reproductive organs?
3. Describe the location and structure of the scrotum. Explain how the scrotal muscles regulate the position of the testes in the scrotum. Why is this important?
4. Describe the internal structure of a testis. Discuss the function of the nurse cells, interstitial cells, seminiferous tubules, rete testis, and efferent ductules.
5. List the structures that constitute the spermatic cord. Where is the inguinal canal? Why are the inguinal canal and inguinal ring clinically important?
6. Define the term *semen*. What amount of semen is ejected during ejaculation, and what are the properties and substances of semen?
7. Compare the seminal vesicles and the prostate gland in terms of location, structure, and function.
8. Describe the structure of the penis, and explain the mechanisms that result in erection, emission, and ejaculation.
9. Distinguish between impotence, infertility, and sterility.
10. Distinguish between gonorrhea and syphilis. Describe the stages through which syphilis will progress if untreated.

Selected Readings

Bardin, C. W. 1979. The neuroendocrinology of male reproduction. *Hosp. Prac.* 14:65.

Bartke, A., A. A. Hafiez, F. J. Bex, and S. Dalterio. 1978. Hormonal interaction in the regulation of androgen secretion. *Biology of Reproduction* 18:44.

Chan, L., and B. W. O'Malley. 1976. Recent studies on the mechanisms of action of the steroid hormones. *New Engl. J. Med.* 294:1322.

Davidson, J. M. September 1975. Hormones and sexual behavior in the male. *Hosp. Prac.*

Epel, D. November 1977. The program of fertilization. *Scientific American.*

Friedman, T. November 1971. Prenatal diagnosis of genetic disease. *Scientific American.*

Goldstein, B. 1976. *Human sexuality.* New York: McGraw-Hill.

Jones, K. L. et al. 1973. *Sex.* 2d ed. New York: Harper and Row Publishers.

Katchadourian, H. A., and D. T. Lunde. 1980. *Fundamentals of human sexuality.* 3d ed. New York: Holt, Rinehart and Winston.

Mader, S. S. 1980. *Human reproductive biology.* Dubuque, Iowa: Wm. C. Brown Co.

Marx, J. L. 1973. Birth control: current technology, future prospects. *Science* 179:1222.

Oaks, W. W. et al. 1976. *Sex and the life cycle.* New York: Grune and Stratton.

Odell, W. D., and D. L. Moyer. 1971. *Physiology of reproduction.* St. Louis: C. V. Mosby Co.

Page, E. W. et al. 1976. *Human reproduction: The core content of obstetrics, gynecology, and perinatal medicine.* 2d ed. Philadelphia: W. B. Saunders Co.

Parkes, A. 1976. *Patterns of sexuality and reproduction.* New York: Oxford Univ. Press.

Pengelley, E. T. 1978. *Sex and human life.* 2d ed. Reading, Mass.: Addison-Wesley.

Reiter, E. O., and M. M. Grumback. 1982. Neuroendocrine control mechanisms and the onset of puberty. *Ann. Rev. of Physiol.* 44:595.

Segal, S. J. September 1974. The physiology of human reproduction. *Scientific American.*

Swanson, H. D. 1974. *Human reproduction: Biology and social change.* New York: Oxford Univ. Press.

Witters, W. L., and P. Jones-Witters. 1975. *Drugs and sex.* New York: Macmillan.

28

Female Reproductive

Outline

Structure and Development of the Female Reproductive System
Secondary Sex Organs and Characteristics
Development of the Female Reproductive System
Ovaries
Position and Structure of the Ovaries
Ovarian Cycle
Secondary Sex Organs
Uterine Tubes
Uterus
Vagina
Vulva
Mechanism of Erection and Orgasm
Mammary Glands and Lactation
Structure of Breasts and Mammary Glands
Lactation
Menstrual Cycle
Phases of the Menstrual Cycle
Cyclic Changes in the Endometrium
Clinical Considerations
Diagnostic Procedures
Developmental Problems
Problems Involving the Ovaries and Uterine Tubes
Problems Involving the Uterus
Diseases of the Vagina and Vulva
Diseases of the Breasts and Mammary Glands
Methods of Contraception
Chapter Summary
Review Activities
Selected Readings

Concepts

In the course of embryonic development, follicles containing oogonia appear in the ovaries, and the sex accessory organs are formed. The female external genitalia develop from the same indifferent structures that give rise to the external genitalia of a male.

During an ovarian cycle primary follicles become secondary follicles, and one of these becomes a graafian follicle. When the graafian follicle ruptures at ovulation, the empty follicle becomes a corpus luteum.

The uterine tubes conduct the zygote to the uterus, where implantation into the endometrial layer may occur. Sperm enters the female genital duct through the vagina.

The structure and function of mammary glands is dependent on the effects of a number of hormones. Prolactin and oxytocin are required for the production and delivery of milk to a suckling infant.

During the course of the ovarian cycle, there are cyclic changes in the secretion of ovarian hormones, which in turn, cause cyclic changes in the endometrium of the uterus.

System

Structure and Development of the Female Reproductive System

In the course of embryonic development, follicles containing oogonia appear in the ovaries. The female secondary sex organs, including the vagina, uterine tubes, uterus, and mammary glands, appear, and the external genitalia are derived from the same indifferent structures that give rise to the external genitalia of a male.

Objective 1. Identify the primary and secondary sex organs in a female, and describe the secondary sex characteristics.

Objective 2. Describe the embryonic origin of the female genital ducts.

Objective 3. Explain why the glans penis and clitoris are considered homologous structures, and list other structures that are homologous in the two sexes.

The primary sex organs are called **gonads;** in the female they are more specifically known as the **ovaries.** Ovaries produce gametes—the ova, or eggs—and secrete sex steroid hormones. Secretion of the female sex hormones at puberty contributes to the development of secondary sexual characteristics and causes cyclic changes in the secondary sex organs that are required for reproductive function.

Secondary Sex Organs and Characteristics

Secondary sex organs (fig. 28.1 and table 28.1) are those structures that are essential for successful fertilization, implantation of the embryo, and development of the embryo and fetus. The secondary sex organs include the **vagina** *(vah-ji′nah),* which receives the penis and ejaculated semen during copulation and through which the baby passes during delivery; the **uterine** *(u′ter-īn)* **(fallopian) tubes,** through which an ovum is transported toward the uterus after ovulation, and in which fertilization normally occurs; and the **uterus** (womb), where implantation and development occur. The muscular walls of the uterus play an active role in *parturition* (labor and delivery). **Mammary glands** are also considered secondary sex organs because the milk secreted after parturition provides nourishment to the child.

 Secondary sex characteristics are features that are not essential for the reproductive process but that are considered to be sexual attractants. Body physique, contributed largely by the female distribution of subcutaneous fat, the pattern of body hair, and breast development are examples. Although the breasts contain the mammary glands, large breasts are not essential for nursing the young. In fact, all female mammals have mammary glands, but only human females have protruding breasts that function as a sexual attractant.

---

vagina: L. *vagina,* sheath or scabbard

Figure 28.1.   Organs of the female reproductive system seen in sagittal section.

Uterine tube
Fimbriae
Ovary
Uterus
Urinary bladder
Symphysis pubis
Urethra
Clitoris
Labium minus
Labium majus
Vaginal orifice

Posterior fornix
Cervix
Rectum
Vagina
Anus
Perineal body

Table 28.1   Functions of the organs of the female reproductive system

Organ(s)	Function(s)
Ovaries	Produce ova and female sex hormones
Uterine tubes	Convey ovum toward uterus; site of fertilization; convey developing embryo to uterus
Uterus	Site of implantation; protects and sustains life of embryo and fetus during pregnancy; active role in parturition
Vagina	Conveys uterine secretions to outside of body; receives erect penis and semen during copulation and ejaculation; passage for fetus during parturition
Labia majora	Form margins of pudendal cleft; enclose and protect other external reproductive organs
Labia minora	Form margins of vestibule; protect openings of vagina and urethra
Clitoris	Glans of the clitoris is richly supplied with sensory nerve endings associated with feeling of pleasure during sexual stimulation
Vestibule	Cleft between labia minora that includes vaginal and urethral openings
Vestibular glands	Secrete fluid that moistens and lubricates the vestibule and vaginal opening during intercourse
Mammary glands	Produce and secrete milk for nourishment of an infant

Source: Van De Graaff, Kent M., *Human Anatomy.* © 1984 Wm. C. Brown Publishers, Dubuque, Iowa. All Rights Reserved. Reprinted by permission.

Puberty occurs at age twelve to fourteen, varying with the nutritional condition, genetic background, and even sexual exposure of the individual. Generally, girls attain puberty six months to one year earlier than boys, and the transition is more abrupt because of the onset of menstruation, or *menarche (mĕ-nar′ke).* Puberty results from the increased secretion of gonadotrophins from the anterior pituitary, which stimulates the ovaries to begin their cycles of ova development and sex steroid secretion.

Development of the Female Reproductive System

Although the genetic sex is determined at fertilization, both sexes develop similarly through the indifferent stage of the eighth week. The gonads of both sexes develop from the **gonadal ridges** medial to the mesonephros. **Primary sex cords** form within the gonadal ridges by the sixth week. The **genital tubercle** also develops during the sixth week as an external swelling cephalic to the cloacal membranes.

The ovaries develop more slowly than do the testes. Ovarian development begins at about the tenth week when **primordial follicles** begin to form within the medulla of the gonads. Each of the primordial follicles consists of an **oogonium** *(o″o-go′ne-um)* surrounded by a layer of **follicular cells.** Mitosis of the oogonia occurs during fetal development, so that thousands of germ cells are formed. Unlike the male, in which spermatogonia are formed by mitosis throughout life, all oogonia are formed prenatally and their number continuously decreases after birth.

The **genital ducts** include the uterus and the uterine tubes. These organs develop from a pair of embryonic tubes called the **paramesonephric (Müllerian) ducts.** The paramesonephric ducts form to the sides of the mesonephric ducts that develop into the kidneys. As the mesonephric ducts regress, the paramesonephric ducts develop into the female genital tract (fig. 28.2). The lower portion of both ducts fuse to form the uterus. The upper portions remain unfused and give rise to the uterine tubes.

The epithelial lining of the vagina develops from the endoderm of the urogenital sinus. A thin membrane called the **hymen** forms to separate the lumen of the vagina from the urethral sinus. The hymen usually is perforated during later fetal development.

primordial: L. *prima,* first; *ordior,* to begin
follicle: L. diminutive of *follis,* bag
hymen: Gk. (mythology) Hymen was god of marriage; *hymen,* thin skin or membrane

Figure 28.2. The development of the female genital tract: (*a*) a lateral view; (*b*) an anterior view.

(a)

(b)

Table **28.2** Homologous reproductive organs in the male and female and the undifferentiated structures from which they develop		
Indifferent stage	**Male**	**Female**
Gonads	Testes	Ovaries
Urogenital groove	Membranous urethra	Vestibule
Genital tubercle	Glans penis	Clitoris
Urogenital folds	Penile urethra	Labia minora
Genital swelling	Scrotum	Labia majora
	Bulbourethral glands	Vestibular glands

1. Distinguish between the primary sex organs, secondary sex organs, and secondary sexual characteristics.
2. Describe the location and fate of the paramesonephric (müllerian) ducts in females.
3. Explain why the external genitalia of males and females are considered to be homologous, and explain what may happen if a female embryo is exposed to androgens at a critical stage of development.

The external genitalia of both sexes appear the same during the indifferent stage of the eighth week. A prominent **phallus** *(fal'us)* forms from the genital tubercle, and a **urethral groove** forms on the ventral side of the phallus. Paired **urogenital folds** surround the urethral groove on the lateral sides. In the male embryo these indifferent structures become masculinized by testosterone secreted by the testes. In the female embryo, in the absence of testosterone, feminization occurs.

These feminizing changes include an inhibition in the size of the phallus to form the relatively small **clitoris** *(kli'tor-is);* the urogenital folds remain unfused to form an inner **labia minora;** and the labioscrotal folds remain unfused to form the prominent **labia majora** (table 28.2). The external genitalia of a female are completely formed by the end of the twelfth week.

There is tremendous individual variation in the structure of the hymen. The hymen of a baby girl may be absent, or it may partially, or occasionally completely, cover the vaginal orifice, in which case it is called an **imperforate hymen.** An imperforate hymen is usually not detected until the first menstruation (menarche) when the discharge cannot be expelled. If the hymen is present, it may be ruptured during childhood in the course of normal exercise. On the other hand, a hymen may be so elastic that it persists even after sexual intercourse. The presence of a hymen is not, therefore, a reliable sign of virginity.

Ovaries

The ovary contains numerous primary follicles with primary oocytes. During an ovarian cycle some of these become secondary follicles, and one becomes a graafian follicle, which ruptures and releases its oocyte in ovulation. The ruptured follicle becomes a corpus luteum and regresses to become a corpus albicans. These cyclic changes in structure are associated with cyclic changes in hormone secretion.

Objective 4. Describe the position of the ovaries and the ligaments associated with the ovaries and genital ducts.

Objective 5. Describe the structural changes that occur in the ovaries leading to and following ovulation.

Objective 6. Describe oogenesis, and explain why meiosis of one primary oocyte results in the formation of only one mature ovum.

Objective 7. Describe the hormonal secretions of the ovaries during an ovarian cycle.

The ovaries of sexually mature females are solid, ovoid structures that are about 3.5 cm (1.4 in.) long, 2 cm (0.8 in.) wide, and 1 cm (0.4 in.) thick. On the medial portion of each ovary is a **hilum,** which is the point of entrance for ovarian blood vessels and nerves. The lateral portion of the ovary is in contact with the open ends of the uterine tube (fig. 28.3).

ovaries: L. *ovaries,* egg holders; ovum, egg
hilum: L. *hilum,* a trifle (little significance)

Figure 28.3. A dorsal view of the female reproductive organs showing the relationship of the ovaries, uterine tubes, uterus, cervix, and vagina.

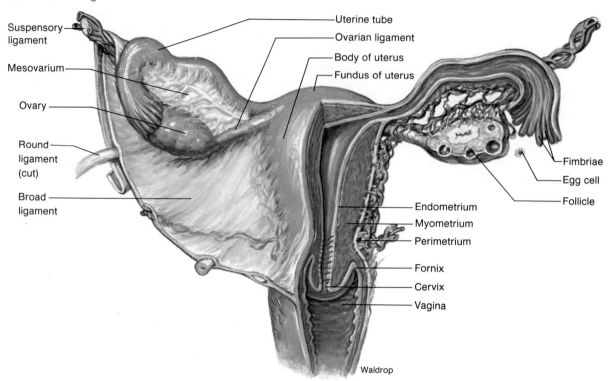

Waldrop

Position and Structure of the Ovaries

The ovaries are positioned in the upper pelvic cavity on both sides of the uterus. Each ovary is situated in a shallow depression of the posterior body wall, called the **ovarian fossa,** and is secured by several membranous attachments. The principal supporting membrane of the female reproductive tract is the **broad ligament.** The broad ligament is the parietal peritoneum that supports the uterine tubes and uterus. The **mesovarium** *(mes"o-va're-um)* is a specialized posterior extension of the broad ligament that attaches to an ovary. Each ovary is additionally supported by an **ovarian ligament** anchored to the uterus and a **suspensory ligament** attached to the pelvic wall (fig. 28.3).

Each ovary consists of four layers. The **germinal epithelium** is the thin, outermost layer composed of cuboidal epithelial cells. A collagenous connective tissue layer called the **tunica albuginea** *(al"bu-jin'e-ah)* is immediately below the germinal epithelium. The principal substance of the ovary is divided into an outer **cortex** and an inner, vascular **medulla,** although the boundary between these layers is not distinct. The **stroma** is the material of the ovary in which follicles and blood vessels are embedded and lies in both cortical and medullary layers.

Blood Supply and Innervation. Blood is supplied by ovarian arteries that arise from the lateral sides of the abdominal aorta just below the origin of the renal arteries.

An additional supply comes from the ovarian branches of the uterine arteries. Venous return is through the ovarian veins. The right ovarian vein empties into the inferior vena cava, whereas the left ovarian vein drains into the left renal vein.

The ovaries are innervated by sympathetic and parasympathetic fibers of the ovarian plexus. Nerve fibers to the ovaries innervate only the vascular networks and not the follicular substance within the stroma. All the vessels and nerves to an ovary enter by way of the hilum, which is supported by the ovarian ligament.

Normal, healthy ovaries usually cannot be palpated either by vaginal or abdominal examination. If the ovaries become swollen or displaced, however, they are palpable through the vagina. There is a great variety of nonmalignant tumors of the ovaries, most of which cause swelling and some localized tenderness. The ovaries atrophy during menopause, and ovarian enlargement in postmenopausal women is usually serious.

Ovarian Cycle

The germ cells that migrate into the ovaries during early embryonic development multiply, so that by about five months of gestation the ovaries contain approximately 6 to 7 million oogonia. Production of new oogonia stops at this point and never resumes again. Toward the end of gestation the oogonia begin meiosis, at which time they are called **primary oocytes** *(o'o-sīts).* Like spermatogenesis in the male, oogenesis is arrested at prophase I of the

---

stroma: Gk. *stroma,* a couch or bed
oogonium: Gk. *oion,* egg; *gonos,* generation

Figure 28.4. (*a*) a photomicrograph of primordial and secondary follicles. (*b*) a diagram of the parts of a secondary follicle.

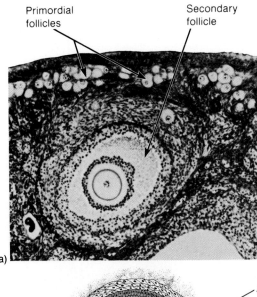

Primordial follicles

Secondary follicle

(a)

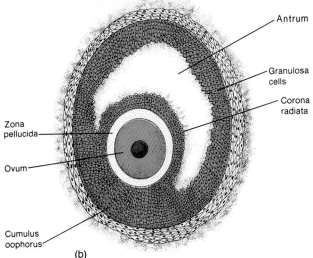

Antrum

Granulosa cells

Corona radiata

Zona pellucida

Ovum

Cumulus oophorus

(b)

Figure 28.5. (*a*) a primary oocyte at metaphase I of meiosis. Note the alignment of chromosomes (arrow). (*b*) a secondary oocyte formed at the end of the first meiotic division and the first polar body (arrow).

(a)

(b)

first meiotic division. The number of primary oocytes decreases throughout a woman's reproductive years. The ovaries of a newborn girl contain about 2 million oocytes, but this number is reduced to about 300,000–400,000 by the time the girl enters puberty. Oogenesis ceases entirely at menopause (the time menstruation stops).

Primary oocytes that are not stimulated to complete the first meiotic division are contained within tiny follicles called **primordial follicles.** In response to gonadotrophin stimulation, some of these oocytes and follicles get larger, and the follicular cells divide to produce numerous small **granulosa cells** that surround the oocyte and fill the follicle. A follicle at this stage in development is called a **primary follicle.**

Some primary follicles will be stimulated to grow still bigger and develop a fluid-filled cavity, called an *antrum,* at which time they are called **secondary follicles** (fig. 28.4). The granulosa cells of secondary follicles form a ring around the circumference of the follicle and form a mound

that supports the ovum. This mound is called the *cumulus oophorous (o-of'o-rus).* Some granulosa cells also encircle the oocyte, forming a *corona radiata.* Between the oocyte and the corona radiata is a thin gel-like layer of proteins and polysaccharides called the *zona pellucida (pel-lu'si-dah).* Under stimulation of FSH from the anterior pituitary, the granulosa cells secrete increasing amounts of estrogen as the follicles grow. Interestingly, the granulosa cells produce estrogen from its precursor testosterone, which is supplied by cells of the *theca interna* layer immediately outside the follicle.

As the follicle develops, the primary oocyte completes its first meiotic division. This does not form two complete cells, however, because only one cell—the **secondary oocyte**—gets all the cytoplasm. The other cell formed at this time becomes a small *polar body* (fig. 28.5) which eventually fragments and disappears. The secondary oocyte enters the second meiotic division, but meiosis is arrested at metaphase II and is never completed unless fertilization occurs.

cumulus oophorous: L. *cumulus,* a mound; G. *oophoros,* egg-bearing
corona radiata: G. *korone,* crown; L. *radiata,* radiate
zona pellucida: G. *zone,* girdle; L. *pellis,* skin
theca interna: G. *theke,* a box; L. *internus,* interior

Figure 28.6.   A graafian follicle within the ovary of a monkey.

Germinal epithelium   Primordial follicle   Graafian follicle   Atretic follicle   Blood vessel

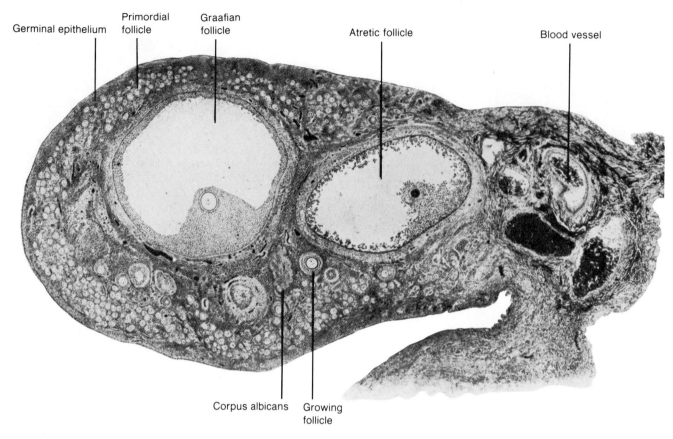

Corpus albicans   Growing follicle

Figure 28.7.   Ovulation from a rabbit ovary.

Oocyte

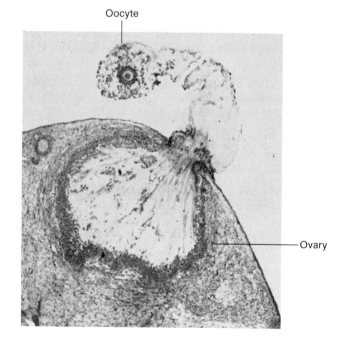

Ovary

Figure 28.8.   A schematic diagram of the process of oogenesis. During meiosis, each primary oocyte produces a single haploid gamete. If the secondary oocyte is fertilized, it forms a secondary polar body and becomes a zygote.

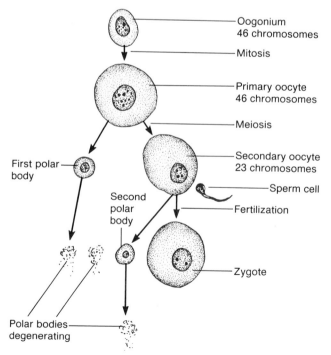

Oogonium 46 chromosomes

Mitosis

Primary oocyte 46 chromosomes

Meiosis

First polar body

Secondary oocyte 23 chromosomes

Sperm cell

Second polar body

Fertilization

Zygote

Polar bodies degenerating

Figure 28.9. A schematic diagram of an ovary showing the various stages of ovum and follicle development.

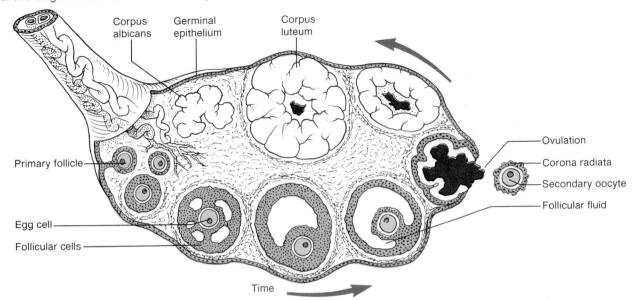

Ovulation. Usually, by about ten to fourteen days after the first day of menstruation, only one follicle has continued its growth to become a mature **graafian follicle** (fig. 28.6); other secondary follicles during that cycle regress and become *atretic (ah-tret'ik)*. The graafian follicle is so large that it forms a bulge on the surface of the ovary. Under proper hormonal stimulation this follicle will rupture—much like the popping of a blister—and extrude its oocyte into the uterine tube in the process of **ovulation** (fig. 28.7).

The released cell is a secondary oocyte, surrounded by the zona pellucida and corona radiata. If it is not fertilized it disintegrates in a couple of days. If a sperm passes through the corona radiata and zona pellucida and enters the cytoplasm of the secondary oocyte, the oocyte completes the second meiotic division. In this process the cytoplasm is again not divided equally; most of the cytoplasm remains in the zygote (fertilized egg), leaving another polar body which, like the first, disintegrates (fig. 28.8).

Changes continue in the ovary following ovulation. The empty follicle, under the influence of luteinizing hormone from the anterior pituitary, undergoes structural and biochemical changes to become a **corpus luteum.** Unlike the ovarian follicles, which secrete only estrogen, the corpus luteum secretes two sex steroid hormones: estrogen and progesterone. Toward the end of a nonfertile cycle the corpus luteum regresses and is changed into a nonfunctional **corpus albicans.** These cyclic changes in the ovary are summarized in figure 28.9.

---

graafian follicle: from Regnier de Graaf, Dutch anatomist and physician, 1641–73
atretic: Gk. *atretos,* not perforated
corpus luteum: L. *corpus,* body; *luteum,* yellow
albicans: L. *albicare,* to whiten

1. *Describe the position of the ovaries relative to the uterine tubes, and describe the position and functions of the broad ligament and mesovarium.*
2. *Compare the structure and contents of a primordial follicle, primary follicle, secondary follicle, and graafian follicle.*
3. *Define ovulation, and describe the changes that occur in the ovary following ovulation in a nonfertile cycle.*
4. *Describe oogenesis, and explain why only one mature ovum is produced by this process.*
5. *Compare the hormonal secretions of the ovarian follicles with those of a corpus luteum.*

Secondary Sex Organs

The uterine tubes conduct the zygote to the uterus, where implantation into the endometrial layer may occur. The muscular layer of the uterus, or myometrium, is functional in labor and delivery. Sperm enters the female genital duct through the vagina, which also serves as the birth canal during parturition.

Objective 8. Describe how an ovum is moved through a uterine tube to the uterus.

Objective 9. Describe the structure and function of each of the three layers of the uterus.

Objective 10. Describe the structure and functions of the vulva and vagina.

Objective 11. Describe the changes that occur in the female reproductive system during sexual excitement and intercourse.

The secondary sex organs described in this section include the internal organs of the uterine tubes, uterus, and vagina, as well as the external genitalia (vulva). The mammary glands, which are also considered to be secondary sex organs, are described in a separate section.

Uterine Tubes

The paired uterine tubes, also known as the **fallopian** *(fal-lo'pe-an)* **tubes** or **oviducts,** transport ova from the ovaries to the uterus. Fertilization normally occurs within the uterine tube. Each uterine tube is approximately 10 cm (4 in.) long and 0.7 cm (0.3 in.) in diameter and is positioned between the folds of the broad ligament of the uterus.

> The term **salpinx** is occasionally used to refer to the uterine tubes. It is a Greek word meaning "trumpet" or "tube" and is the root of such clinical terms as **salpingitis** *(sal''pin-ji'tis),* or inflammation of the uterine tubes, **salpingography** (radiography of the uterine tubes), and **salpingolysis** (the breaking up of adhesions of the uterine tube to correct female infertility).

The funnel-shaped, open-ended portion of the uterine tube is called the **infundibulum.** Although the infundibulum is close to the ovary, it is not attached. A number of fringed, fingerlike processes, called **fimbria,** project from the margins of the infundibulum over the lateral surface of the ovary. The function of the fimbria is to direct the ovum into the lumen of the uterine tube. From the infundibulum, the uterine tube extends medially and inferiorly to open into the superior-lateral cavity of the uterus at the **uterine opening.** The **ampulla** *(am-pul'ah)* is the longest and widest portion of the uterine tube.

The wall of the uterine tube consists of three histological layers (fig. 28.10). The internal **mucosa** lines the lumen and is composed of ciliated columnar cells that are drawn into numerous folds. The **muscularis** is the middle layer, composed of a thick, circular layer of smooth muscle and a thin, outer layer of smooth muscle. Peristaltic contractions of the muscularis and ciliary action of the mucosa move the ovum through the lumen of the uterine tube. The outer, lubricative **serous layer** of the uterine tube is part of the visceral peritoneum.

The oocyte takes four to five days to move through the uterine tube. If enough viable sperm are ejaculated into the vagina during sexual intercourse and an oocyte is in the uterine tube, fertilization will occur within hours after discharge of the semen. The zygote will move toward the uterus where implantation occurs. If the embryo implants instead into the uterine tube, an *ectopic pregnancy* will be produced. An ectopic pregnancy is an implantation of the embryo (called a blastocyst) in a site other than the uterus.

---

fallopian tubes: from Gabriele Fallopius, Italian anatomist, 1523–62
oviduct: L. *ovum,* egg; *ductus,* a leading
fimbria: L. *fimbria,* fringe
ectopic: Gk. *ex,* out; *topos,* place

Figure 28.10.   The histological structure of the uterine tube.

- Connective tissue layer
- Basement membrane
- Nucleus
- Cytoplasm
- Cilia
- Plasma membrane

> Since the infundibulum of the uterine tube is unattached, it provides a pathway for pathogens to enter the abdominopelvic cavity. The mucosa of the uterine tube is continuous with that of the uterus and vagina, and it is possible for infectious agents to enter the vagina and cause infections that may ultimately spread to the peritoneal linings resulting in pelvic inflammatory disease (PID). There is no opening into the abdominopelvic cavity other than through the uterine tubes. The abdominopelvic cavity of a male is totally sealed from external contamination.

The uterine tubes are supplied with blood through the ovarian and uterine arteries. Venous drainage is through uterine veins that parallel the arteries. Both the uterine artery and vein can be observed in the broad ligament that supports the uterine tube.

The uterine tubes have both sympathetic and parasympathetic innervation from the hypogastric plexus and pelvic splanchnic nerves. The nerve supply to the uterine tubes regulates the activity of the smooth muscles and blood vessels.

Uterus

The uterus receives the blastocyst that develops from a fertilized ovum and provides a site for implantation. Prenatal development continues within the uterus until gestation is completed, at which time the uterus plays an active role in the delivery of the baby.

Structure.   The uterus is a hollow, thick-walled, muscular organ that is shaped like an inverted pear. Although the shape and position of the uterus changes immensely during pregnancy (fig. 28.11), in its nonpregnant state it is about 7 cm (2.8 in.) long, 5 cm (2 in.) wide (through its broadest region), and 2.5 cm (1 in.) in diameter. The anatomical regions of the uterus include the uppermost,

Figure 28.11. The structure and position of the uterus of a pregnant woman in sagittal section.

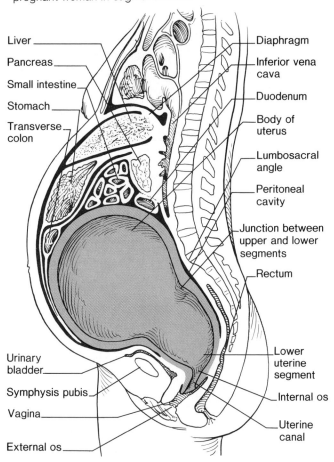

Figure 28.12. Layers of the uterine wall.

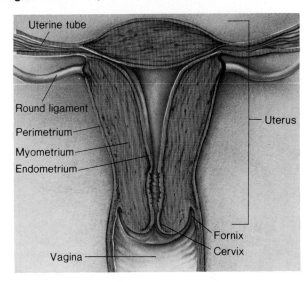

dome-shaped portion above the entrance of the uterine tubes called the **fundus,** the enlarged main portion called the **body,** and the inferior constricted portion opening into the vagina called the **cervix** (fig. 28.12). The uterus is located between the urinary bladder anteriorly and the rectum and sigmoid colon posteriorly. The fundus projects anteriorly and slightly superiorly over the urinary bladder. The cervix projects posteriorly and inferiorly, joining the vagina at nearly a right angle.

The **uterine cavity** is the space within the regions of the fundus and body. The lumina of the uterine tubes open into the uterine cavity on the superior-lateral portions. The uterine cavity is continuous inferiorly with the **cervical canal,** which extends through the cervix and opens into the lumen of the vagina. The junction of the uterine cavity with the cervical canal is called the **internal os,** whereas the opening of the uterine cavity into the cavity of the vagina is called the **external os.**

Support of the Uterus. The uterus is maintained in position by muscular support and ligaments that extend from the pelvic girdle or body wall to the uterus. Muscles of the perineum, especially the levator ani muscle, provide the principal muscular support. The ligaments that support

the uterus undergo marked hypertrophy during pregnancy, regress in size after parturition, and atrophy after menopause.

Four paired ligaments support the uterus in position within the pelvic cavity. The paired **broad ligaments** are folds of the peritoneum that extend from the pelvic walls and floor to the lateral walls of the uterus. The ovaries and uterine tubes are also supported by the broad ligaments. The paired **uterosacral ligaments** are also folds of peritoneum that curve along the lateral pelvic wall on either side of the rectum to connect the uterus to the sacrum. The **cardinal (lateral cervical) ligaments** are fibrous sheets of peritoneum that extend laterally from the cervix and vagina across the pelvic floor where they attach to the wall of the pelvis. The cardinal ligaments contain some smooth muscle as well as vessels and nerves to the cervix and vagina. The fourth paired ligaments are the **round ligaments.** Each round ligament extends from the lateral border of the uterus just below the position where the uterine tube attaches to the lateral pelvic wall. Similar to the course taken by the ductus deferens in the male, the round ligaments continue through the inguinal canal of the abdominal wall where they attach to the deep tissues of the labia majora.

Although the uterus has extensive support, considerable movement is possible. The uterus tilts slightly posteriorly as the urinary bladder fills and moves anteriorly during defecation. In some women, the uterus may go out of position and interfere with the normal progress of pregnancy. A posterior tilting of the uterus is called retroflexion, whereas an anterior tilting is called anteflexion.

Uterine Wall. The wall of the uterus is composed of three layers: the perimetrium, myometrium, and endometrium (fig. 28.12).

---

fundus: L. *fundus,* bottom
cervix: L. *cervix,* neck

Figure 28.13.   Vascular supply to the uterus.

The **perimetrium** is the thin, outer serosal covering and a part of the peritoneum. The lateral portion of the perimetrium merges with the broad ligament. A shallow pouch called the **vesicouterine** *(ves″i-ko-u′-ter-in)* **pouch** is formed as the peritoneum is reflected over the urinary bladder. The **rectouterine** *(rek″to-u′ter-in)* **pouch** (pouch of Douglas) is formed as the peritoneum is reflected onto the rectum. The rectouterine pouch is the lowest point in the pelvic cavity and provides a site for surgical entry into the peritoneal cavity.

The thick **myometrium** is composed of three poorly defined layers of smooth muscle arranged in longitudinal, circular, and spiral patterns. The myometrium is thickest in the fundus and thinnest in the cervix. During parturition, the muscles of this layer are stimulated to contract forcefully.

The **endometrium** *(en-do-me′tre-um)* is the inner mucosal lining of the uterus. The endometrium has two distinct layers. The superficial **stratum functionale** layer, composed of columnar epithelium and containing secretory glands, is shed as *menses* during menstruation and built up again under the stimulation of ovarian steroid hormones. The deeper **stratum basale** layer is highly vascular and serves to regenerate the stratum functionale after each menstruation.

---

os: L. *os*, mouth
perimetrium: Gk. *peri*, around; *metra*, uterus
vesicouterine: L. *vesico*, bladder; *uterus*, womb
pouch of Douglas: from James Douglas, English anatomist and physician, 1675–1742
myometrium: Gk. *mys*, muscle; *metra*, uterus
endometrium: Gk. *endon*, within; *metra*, uterus
menses: L. *menses*, plural of *mensis*, monthly

The size of the uterus changes tremendously during pregnancy. Its weight increases more than sixteen times (from about 60 g to about 1,000 g), and its capacity increases from about 2.5 ml to cover 5,000 ml. The principal changes in the myometrium is a marked hypertrophy, or elongation, of the individual muscle cells to as much as ten times their original length. There is some atrophy of the muscle cells after parturition, but the uterus never returns to its original size.

Uterine Blood Supply and Innervation.   The uterus is supplied with blood through the **uterine arteries** that arise from the internal iliac arteries and by the **uterine branches** of the **ovarian arteries** (fig. 28.13). Each pair of these two vessels anastomose on the upper lateral margin of the uterus. Branches from the uterine arteries penetrate the perimetrium to form a vascular arch, called the **arcuate artery,** along the outside of the myometrium. Numerous **radial arteries** arise from the arcuate artery to penetrate the myometrium and supply it with blood. **Spiral arterioles** arise from the radial arteries to serve the endometrium. The blood from the uterus returns through uterine veins that parallel the pattern of the arteries.

The uterus is innervated by sympathetic and parasympathetic fibers that arise from the pelvic and hypogastric plexuses. Both autonomic innervations serve the arteries of the uterus, whereas the smooth muscle of the myometrium receives only sympathetic innervation.

Vagina

The vagina is the organ that receives sperm through the urethra of the erect penis during sexual intercourse. It also serves as the birth canal during parturition and provides

Figure 28.14.   The histological structure of a vaginal ruga

Stratified
squamous
epithelium

Germinal
layer

Lamina
propria

Figure 28.15.   The external female genitalia.

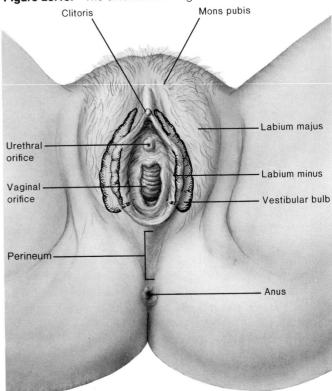

Clitoris

Mons pubis

Urethral
orifice

Vaginal
orifice

Perineum

Labium majus

Labium minus

Vestibular bulb

Anus

for the passage of menses to the outside. The musculo-membranous vagina is a tubular organ about 9 cm (3.6 in.) in length, passing from the cervix of the uterus to the vestibule. The vagina is positioned between the urinary bladder and urethra anteriorly and the rectum posteriorly, where it is continuous with the cervical canal of the uterus. The cervix attaches to the vagina at a nearly 90 degree angle. The deep recess behind the protrusion of the cervix into the vagina is called the **posterior fornix.** The smaller recesses in front and to the sides are called the **anterior** and **lateral fornices.**

> The fornices are of clinical importance since they permit the cervix to be palpated during a gynecological examination. Occasionally, the deep posterior fornix provides surgical access to the pelvic cavity through the vagina. In addition, the fornices are important in the placement of two forms of birth-control devices—the cervical cap and the diaphragm.

The exterior opening of the vagina, at its lower end, is called the **vaginal orifice.** A thin fold of mucous membrane, called the **hymen,** may partially cover the vaginal orifice.

The vaginal wall is composed of three layers: an inner mucosa, a middle muscularis, and an outer fibrous layer. The **mucosal layer** consists of stratified squamous epithelium, which forms a series of transverse folds called **vaginal rugae** (fig. 28.14). The vaginal rugae provide friction ridges for stimulation of the erect penis during sexual intercourse. They also permit considerable distension of the vagina to facilitate copulation. The mucosal layer contains few glands; the acidic mucus that is present in the vagina comes primarily from glands within the uterus. The acidic environment of the vagina retards microbial growth. The additives within semen, however, temporarily neutralize the acidity of the vagina to insure the survival of the ejaculated sperm deposited within the vagina.

The **muscularis layer** consists of longitudinal and circular bands of smooth muscle interlaced with distensible connective tissue. The distension of this layer is especially important during parturition. Striated muscle strands, including the levator ani muscle, near the vaginal orifice partially constrict this opening.

The **fibrous layer** covers the vagina and attaches it to surrounding pelvic organs. This layer consists of dense fibrous connective tissue interlaced with strands of elastic fibers.

The blood supply to the highly vascular vagina is primarily from the vaginal branches of the internal iliac artery. Blood is also supplied to the vagina through branches of the uterine, middle rectal, and internal pudendal arteries. All of these vessels form an extensive plexus surrounding the vagina called the **vaginal azygos** *(az'i-gos)* **arteries.** Blood draining from the vagina returns through vaginal veins that parallel the course of the arteries.

Innervation of the vagina is supplied by sympathetic fibers from the hypogastric plexuses and parasympathetic fibers from the second and third sacral nerves. Afferent innervation is through the pudendal plexus and is especially well developed near the vaginal orifice.

Vulva

The external genitalia of the female are referred to as the **vulva,** or **pudendum** *(pu-den'dum),* which is shown in figure 28.15. The structures of the vulva surround the vaginal orifice and include the mons pubis, labia majora, labia minora, clitoris, vestibule, and vestibular glands.

vulva: L. *volvere,* to roll, wrapper
pudendum: L. *pudere,* to be ashamed
mons pubis: L. *mons,* mountain; *pubis,* genital area

The **mons pubis** is the subcutaneous pad of adipose connective tissue covering the symphysis pubis. At puberty, the mons pubis becomes covered with a pattern of coarse pubic hair that is somewhat triangular and usually has a horizontal upper border. The elevated and padded mons pubis cushions the symphysis pubis and vulva during copulation.

The **labia majora** are two thickened longitudinal folds of skin that contain areolar and adipose tissue as well as some smooth muscle. The labia majora are continuous anteriorly with the mons pubis, are separated longitudinally by the **pudendal cleft,** and converge again posteriorly on the **perineum** *(per''i-ne'um).* They are also covered with hair and contain numerous sebaceous and sweat glands. The labia majora are homologous to the scrotum of the male and function to enclose and protect the other organs of the vulva.

The **labia minora** are two smaller longitudinal folds positioned close together between the labia majora. The labia minora are hairless but do contain sebaceous glands. On the anterior side, the labia minora split to form the **prepuce** *(pre'pūs)* of the clitoris. These inner folds of skin further protect the vaginal and urethral openings.

The **clitoris** is a small rounded projection at the upper portion of the pudendal cleft. The clitoris corresponds in structure and origin to the penis in the male; it is, however, smaller and has no urethra. Although most of the clitoris is embedded, it does have an exposed **glans** of erectile tissue that is richly innervated with afferent endings. The clitoris is about 2 cm (0.8 in.) long and 0.5 cm (0.2 in.) in diameter. The unexposed portion of the clitoris is composed of two columns of erectile tissue called the **corpora cavernosa,** which diverge posteriorly to form the **crura** and attach to the sides of the pubic arch.

The **vestibule** is the longitudinal cleft enclosed by the labia minora. The openings for the urethra and vagina are located in the vestibule. The external opening of the urethra is about 2.5 cm (1 in.) behind the glans of the clitoris and immediately in front of the vaginal orifice. The vaginal orifice is lubricated during sexual excitement by secretions from a pair of **vestibular glands** (Bartholin's glands) located within the wall of the region immediately inside the vaginal orifice. The ducts from these glands open into the vestibule near the lateral margins of the vaginal orifice. Bodies of vascular erectile tissue, called **vestibular bulbs,** are located immediately below the skin forming the lateral walls of the vestibule. The vestibular bulbs are separated from each other by the vagina and urethra and extend from the level of the vaginal orifice to the clitoris.

The vulva is highly vascular and is supplied with arterial blood from internal pudendal branches of the internal iliac arteries and external pudendal branches from the femoral arteries. Extensive vascular networks exist within most of the organs of the vulva. The venous return is through vessels that correspond in name and position to the arteries.

The vulva has both sympathetic and parasympathetic innervation, as well as extensive somatic fibers that respond to sensory stimulation. Parasympathetic stimulation causes a response similar to that of the male: dilation of the arterioles of the genital erectile tissue and compression of the venous return.

Mechanism of Erection and Orgasm

The homologous structures of the male and female reproductive systems respond to sexual stimulation in a similar fashion. The erectile tissues of a female, like those of a male, become engorged with blood and swollen during sexual arousal. During sexual excitement, the hypothalamus of the brain sends parasympathetic nerve impulses through the sacral segments of the spinal cord, which cause dilation of arteries serving the clitoris and vestibular bulbs. This increased blood flow causes the erectile tissues to swell. In addition, the erectile tissues in the areola of the breasts become engorged.

Simultaneous with the erection of the clitoris and vestibular bulbs, the vagina expands and elongates to accommodate the erect penis of the male, and parasympathetic impulses cause the vestibular glands to secrete mucus near the vaginal orifice. The vestibular secretion moistens and lubricates the tissues of the vestibule, thus facilitating the penetration of the erect penis into the vagina during copulation. Mucus continues to be secreted during copulation so that the male and female genitalia do not become irritated as they would if the vagina became dry.

The position of the sensitive clitoris usually permits its being stimulated during sexual intercourse. If stimulation of the clitoris is of sufficient intensity and duration, a woman will experience a culmination of pleasurable psychological and physiological release called *orgasm.*

If orgasm occurs, there is an associated rhythmic contraction of the muscles of the perineum and the muscular walls of the uterus and uterine tubes. These reflexive muscular actions are thought to aid the movement of sperm through the female reproductive tract toward the upper end of a uterine tube where an ovum might be located.

Following an orgasm or completion of the sexual act, sympathetic impulses cause a reduction in arterial flow to the erectile tissues, and their size diminishes to that prior to sexual stimulation.

1. Describe the structure and position of the uterine tubes, and explain how an ovum is moved through these tubes to the uterus.
2. Describe the structure and function of the three layers of the uterus, and explain why the endometrium is subdivided into a stratum functionale and a stratum basale.
3. Describe the structures of the vagina and vulva, and explain how these structures change during sexual excitement and intercourse.

vestibule: L. *vestibule,* an entrance, court
Bartholin's glands: from Casper Bartholin, Jr., Danish anatomist, 1655–1738

orgasm: Gk. *orgasmos,* to swell; to become excited

Mammary Glands and Lactation

The structure and function of the mammary glands is dependent on the action of a number of hormones. The secretion of prolactin and oxytocin is directly required for the production and delivery of milk to a suckling infant.

Objective 12. Describe the structure of breasts and of the mammary glands.

Objective 13. Describe the interaction of hormones required for breast and mammary gland development.

Objective 14. Explain the actions of prolactin and oxytocin on milk production and delivery, and explain how the secretion of these hormones is regulated during lactation.

In structure, the mammary glands, located in the **breasts,** are modified sweat glands and part of the integumentary system. In function, however, these glands are associated with the reproductive system since they secrete milk for the nourishment of the young. The size and shape of the breasts vary considerably from person to person because of genetic differences, age, percentage of body fat, or

pregnancy. At puberty, estrogen from the ovaries stimulates growth of the mammary glands and the deposition of adipose tissue within the breasts. Mammary glands hypertrophy in pregnant and lactating women and usually atrophy somewhat after menopause.

Structure of Breasts and Mammary Glands

Each breast is positioned over the second to the sixth rib and overlies the pectoralis major muscle and portions of the serratus anterior and external oblique muscles (fig. 28.16). The medial boundary of the breast is over the lateral margin of the sternum, and the lateral margin of the breast is along the anterior border of the axilla. The **axillary tail** of the breast extends upward and laterally toward the axilla where it comes into close relationship with the axillary vessels. This region of the breast is clinically significant because of the high incidence of breast cancer within the lymphatic drainage of the axillary tail.

Each mammary gland is composed of fifteen to twenty **lobes,** divided by adipose tissue and each with its own drainage pathway to the outside. The amount of adipose tissue determines the size and shape of the breast but has nothing to do with the ability of a woman to nurse. Each

Figure 28.16. The structure of the breast and mammary glands: (*a*) a sagittal section; (*b*) an anterior view partially sectioned.

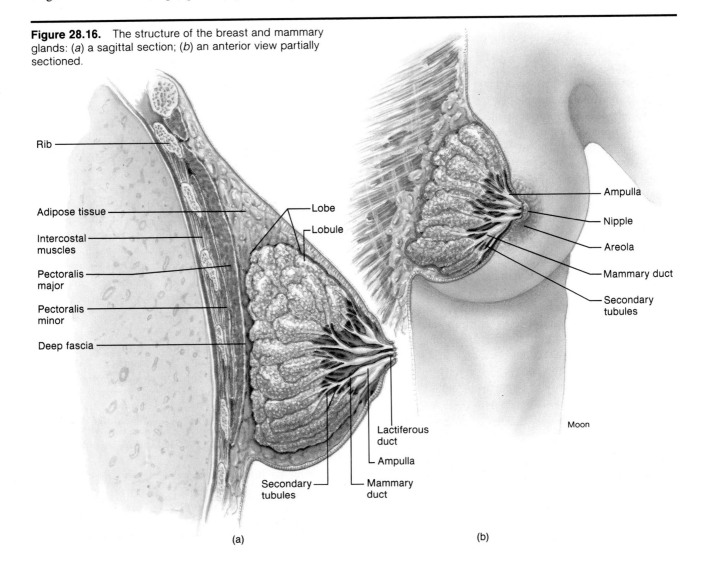

(a)   (b)

Figure 28.17. The histology of the mammary gland: (*a*) nonlactating (63×); (*b*) lactating (63×).

(a)

- Blood vessel
- Intralobular loose connective tissue
- Secretory tubules
- Interlobular dense connective tissue

(b)

- Large irregular alveoli with secretions
- Interlobular dense connective tissue
- Alveoli with secretion
- Adipose tissue

lobe is subdivided into **lobules,** which contain the glandular **alveoli** (fig. 28.17) that secrete the milk of a lactating female. **Suspensory ligaments of Cooper** between the lobules extend from the skin to the deep fascia overlying the pectoralis muscle and support the breasts. The clustered alveoli secrete milk into a series of **secondary tubules.** The secondary tubules in each lobe converge to form a **lactiferous** *(lak-tif'er-us)* **duct.** The lumen of each lactiferous duct expands near the nipple to form an **ampulla** *(am-pul'lah),* where milk may be stored before it drains at the tip of the nipple.

The **nipple** is a cylindrical projection containing some erectile tissue. A circular pigmented **areola** *(ah-re'o-lah)* surrounds the nipple. The surface of the areola may appear rough because of the presence of sebaceous **areolar glands** close to the surface. The secretions of the areolar glands keep the nipple pliable. The color of the areola and nipple varies with the complexion of the woman and whether or not she is pregnant. During pregnancy, the areola becomes darker and enlarges somewhat, presumably to become more conspicuous to a nursing infant.

Blood is supplied to the mammary gland through the perforating branches of the internal thoracic artery, which enter the breast through the second, third, and fourth intercostal spaces just lateral to the sternum, and through the more superficial mammary artery, branching from the lateral thoracic artery. Venous return is through a series of veins that parallel the pattern of the arteries. A superficial venous plexus may be apparent through the skin of the breast, especially during pregnancy and lactation.

The breast is innervated primarily through afferent somatic fibers that are derived from the anterior and lateral cutaneous branches of the fourth, fifth, and sixth thoracic nerves. Sensory nerve endings in the nipple and areola are especially important in stimulating the release of milk from the mammary glands to a suckling infant.

ligaments of Cooper: from Sir Astley P. Cooper, English anatomist and surgeon, 1768–1841

Lymphatic drainage and the location of lymph nodes within the breast are of considerable clinical importance because of the frequency of breast cancer and the high incidence of metastases. About 75% of the lymph drains through the axillary tail of the breast into the axillary lymph nodes. Some 20% of the lymph passes toward the sternum to the internal thoracic lymph nodes. The remaining 5% of the lymph is subcutaneous and follows the lymph drainage pathway in the skin toward the back where it reaches the intercostal nodes near the neck of the ribs.

Lactation

The changes that occur in the mammary glands during pregnancy and the regulation of lactation provide excellent examples of hormonal interactions and neuroendocrine regulation. Growth and development of the mammary glands during pregnancy requires the permissive actions of insulin, cortisol, and thyroid hormones; in the presence of adequate amounts of these hormones, high levels of estrogen stimulate the development of the mammary alveoli and progesterone stimulates proliferation of the tubules and ducts (fig. 28.18).

The production of milk proteins, including casein and lactalbumin, is stimulated after parturition by **prolactin,** a hormone secreted by the anterior pituitary gland. The secretion of prolactin is controlled primarily by *prolactin-inhibiting hormone (PIH),* which is produced by the hypothalamus and secreted into the portal blood vessels. The secretion of PIH is stimulated by high levels of estrogen. During pregnancy, consequently, the high levels of estrogen prepare the breasts for lactation but prevent prolactin secretion.

After parturition, when the placenta is eliminated, declining levels of estrogen are accompanied by an increase in the secretion of prolactin. Lactation, therefore, commences. If a woman does not wish to breast-feed her

Figure 28.18. The hormonal control of mammary gland development during pregnancy and lactation. Note that milk production is prevented during pregnancy by estrogen inhibition of prolactin secretion. This inhibition is accomplished by the stimulation of PIH (prolactin-inhibiting hormone) secretion from the hypothalamus.

Figure 28.19. Lactation occurs in two stages: milk production (stimulated by prolactin) and milk ejection (stimulated by oxytocin). The stimulus of suckling triggers a neuroendocrine reflex that results in increased secretion of oxytocin and prolactin.

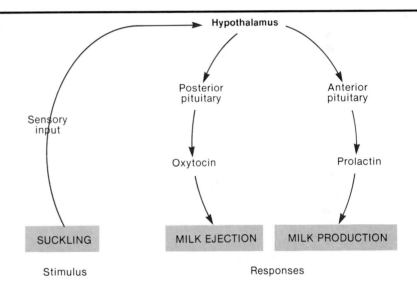

baby, she may be injected with a powerful synthetic estrogen (diethylstilbestrol, or DES), which inhibits further prolactin secretion.

The act of nursing helps to maintain high levels of prolactin secretion via a *neuroendocrine reflex* (fig. 28.19). Sensory endings in the breast, activated by the stimulus of suckling, relay impulses to the hypothalamus and inhibit PIH secretion. The continued secretion of high levels of prolactin results in the secretion of milk from the alveoli into the ducts. In order for the baby to get the milk, however, the action of another hormone is needed.

The stimulus of suckling also results in the reflex secretion of **oxytocin** from the posterior pituitary. This hormone is produced in the hypothalamus and stored in the posterior pituitary; its secretion results in the **milk-ejection reflex,** or milk let-down. Oxytocin stimulates contraction of the lactiferous ducts, as well as of the uterus (this is why women who breast-feed regain uterine muscle tone faster than women who bottle-feed).

For reasons that are poorly understood, breast-feeding often suppresses the release of FSH and LH from the hypothalamus. This inhibits ovulation for a period of months to a year or so producing a natural spacing of births. In many women, however, suckling does not suppress FSH and LH secretion, and ovulation commences soon after parturition. For this reason, lactation is not regarded as a reliable means of birth control.

1. Describe the structure of the breasts and mammary glands, and explain why variations in breast size do not affect the ability to lactate.
2. Describe the structural changes in the mammary glands that occur during pregnancy, and explain how these changes are produced.
3. Describe the roles of prolactin and oxytocin in lactation, and explain how the secretion of these hormones is regulated.

Menstrual Cycle

During the course of the ovarian cycle, there are cyclic changes in the secretion of ovarian sex steroid hormones. These cycles of hormonal secretion cause characteristic changes in the endometrium of the uterus, including shedding of the stratum functionale in the menstrual phase of the cycle.

Objective 15. Describe the hormonal changes that occur during the follicular phase of the cycle, and explain how these changes are regulated.

Objective 16. Describe the hormonal control of ovulation, and explain how the time for ovulation is regulated.

Objective 17. Describe the formation, function, and fate of the corpus luteum.

Objective 18. Describe the structural changes that occur in the endometrium during the cycle, and explain the hormonal control of these changes.

Humans, apes, and old-world monkeys have cycles of ovarian activity that repeat at approximately one-month intervals; hence the name *menstrual cycle.* The term *menstruation (men''stroo-a'shun)* is used to indicate the periodic shedding of the stratum functionale of the endometrium, which becomes thickened prior to menstruation under stimulation by ovarian steroid hormones. In primates (other than new-world monkeys) this shedding of the endometrium is accompanied by bleeding. There is no bleeding, in contrast, when other mammals shed the endometrium; their cycles, therefore, are not called menstrual cycles (even though the cycle of a cow is also about a month long).

Humans and other primates that have menstrual cycles may permit copulation at any time of the cycle. Nonprimate mammals, in contrast, are sexually receptive only at a particular time in their cycles (shortly before or shortly after ovulation). These animals are therefore said to have *estrus cycles.* Bleeding occurs in some animals (such as dogs and cats) that have estrus cycles shortly before they permit copulation. This bleeding is a result of high estrogen secretion and is not associated with shedding of the endometrium.

The bleeding that accompanies menstruation, in contrast, can be experimentally induced by removing the ovaries. This demonstrates that menstrual bleeding is caused by the withdrawal of ovarian hormones. During the normal menstrual cyle, the secretion of ovarian hormones rises and falls in a regular fashion, causing cyclic changes in the endometrium and other sex-steroid-dependent tissues.

Phases of the Menstrual Cycle

The average menstrual cycle has a duration of about twenty-eight days. Since it is a cycle there is no beginning or end, and the changes that occur are generally gradual. It is convenient, however, to call the first day of menstrua-

Figure 28.20. Average values for LH, FSH, progesterone, and estradiol during the menstrual cycle. The midcycle peak of LH is used as a reference day.

tion "day one" of the cycle, because menstrual flow is the most apparent of the changes that occur. It is also convenient to divide the cycle into four phases based on changes that occur in the ovary. These phases are: (1) follicular; (2) ovulatory; (3) luteal; and (4) menstrual.

Menstrual and Follicular Phases. The **menstrual phase** lasts from day one to day four or five of the average cycle. During this time, the secretion of ovarian steroid hormones are at their lowest ebb and the ovaries contain only primordial and primary follicles. During the **follicular phase,** which lasts from day five or six to day thirteen of the cycle, some of the primary follicles grow, form an antrum, and become secondary follicles. Toward the end of the follicular phase one follicle in one ovary reaches maturity and becomes a graafian follicle. As follicles grow, the granulosa cells secrete an increasing amount of **estradiol** *(es''trah-di'ol)* (the principal estrogen), which reaches its highest concentration in the blood at about day twelve of the cycle.

The growth and development of the follicles and the secretion of estradiol are stimulated by **FSH (follicle-stimulating hormone).** Although it has often been stated that FSH levels rise during the follicular phase, most

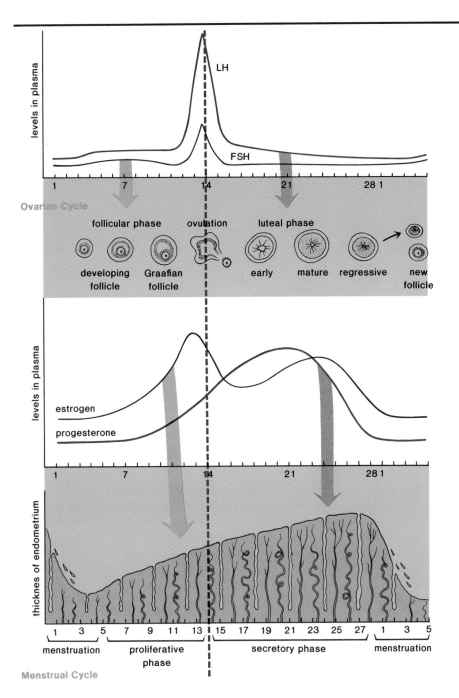

Figure 28.21. Phases of the menstrual cycle in relation to ovarian changes and hormone secretion, and the relationship between changes in the ovaries and the endometrium of the uterus during different phases of the menstrual cycle.

studies fail to show significant changes in FSH concentration during this time of the cycle (fig. 28.20). There are, however, increasing numbers of FSH receptor proteins in the granulosa cells at this time, so perhaps the increased growth and secretion of the follicles is due to the increased sensitivity of the follicles to FSH.

The rapid rise in estradiol secretion from the follicles during the follicular phase stimulates a "surge" of **LH (luteinizing hormone)** secretion from the anterior pituitary. This *LH surge* occurs at the last day of the follicular phase as a result of the **positive feedback** of estradiol on the pituitary and hypothalamus.

The LH surge begins about twenty-four hours before ovulation and reaches its peak about sixteen hours before ovulation. Since GnRH (gonadotrophin-releasing hormone) from the hypothalamus stimulates the pituitary to

secrete both FSH and LH, there is a simultaneous, though smaller, surge in FSH secretion. Since this is the only time of the cycle that FSH secretion significantly increases, many investigators believe that this midcycle peak in FSH acts as a stimulus for the development of new follicles for the next month's cycle.

Ovulation. Under the influence of FSH stimulation, the graafian follicle grows so large that it becomes a thin-walled "blister" on the surface of the ovary. The growth of the follicle is accompanied by a rapid rate of increase in estradiol secretion. This rapid increase in estradiol, in turn, triggers the LH surge. Finally, the surge in LH secretion causes the wall of the graafian follicle to rupture (fig. 28.21). Ovulation occurs, therefore, as a result of the

the sequential effects of FSH followed by LH on the ovarian follicles.

By means of the positive feedback effect of estradiol on LH secretion, the follicle in a sense sets the time for its own ovulation. This is because ovulation is triggered by an LH surge, and the LH surge is triggered by increased estradiol secretion, which occurs while the follicle grows. In this way the graafian follicle does not normally ovulate until it has reached the proper size and degree of maturation.

In ovulation, a secondary oocyte, arrested at metaphase II of meiosis, is released into a uterine tube. This oocyte is still surrounded by a zona pellucida and corona radiata as it begins its journey to the uterus. Normally only one ovary ovulates per cycle, with the left and right ovary alternating in successive cycles. Interestingly, if one ovary is removed the remaining ovary does not skip cycles, but ovulates every month. The mechanisms by which this regulation is achieved are not understood.

Luteal Phase. After ovulation, the empty follicle is stimulated by LH to become a new structure, the **corpus luteum** (fig. 28.22). This change in structure is accompanied by a change in function. Whereas the developing follicles secrete only estradiol, the corpus luteum secretes both estradiol and **progesterone.** Progesterone levels in the blood are negligible before ovulation but rise rapidly to reach a peak during the luteal *(lu'te-al)* phase, approximately one week after ovulation.

The combined high levels of estradiol and progesterone during the luteal phase exert a **negative feedback inhibition** of FSH and LH secretion. This serves to retard development of new follicles, so that further ovulation does not normally occur during that cycle. Although this might seem like locking the barn door after the horse (ovum) has escaped, it does prevent more horses (ova) from escaping. In this way multiple ovulations (and possible pregnancies) on succeeding days of the cycle are prevented.

High levels of estrogen and progesterone during the nonfertile cycle do not persist for very long, however, and new follicles do start to develop towards the end of one cycle, in preparation for the next cycle. Estrogen and progesterone levels fall during the late luteal phase (starting about day twenty-two), because the corpus luteum regresses and stops functioning. In lower mammals, the decline in corpus luteum function is caused by a hormone secreted by the uterus called *luteolysin.* A similar hormone has not yet been identified in humans, and the cause of corpus luteum regression in humans is not well understood. Luteolysis (breakdown of the corpus luteum) can be prevented by high LH secretion, but LH levels remain low during the luteal phase as a result of negative feedback inhibition by ovarian steroids.

With the declining function of the corpus luteum, estrogen and progesterone fall to very low levels by day

Figure 28.22. A corpus luteum in a human ovary.

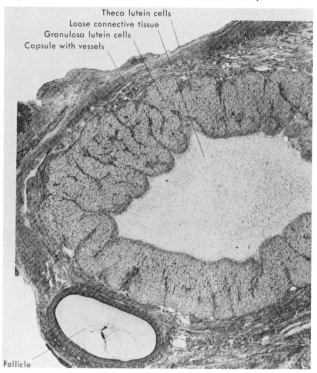

Theca lutein cells
Loose connective tissue
Granulosa lutein cells
Capsule with vessels

Follicle

twenty-eight of the cycle. The withdrawal of ovarian steroids causes menstruation and permits a new cycle of ovarian follicle development to progress.

Cyclic Changes in the Endometrium

In addition to a description of the female cycle in terms of the phases of ovarian function, the cycle can also be described in terms of the changes that occur in the endometrium. Three phases can be identified on this basis (fig. 28.21b): (1) the proliferative phase; (2) the secretory phase; and (3) the menstrual phase.

The **proliferative** *(pro-lif'er-a-tiv)* **phase** of the endometrium occurs while the ovary is in its follicular phase. The increasing amounts of estradiol secreted by the developing follicles stimulates growth (proliferation) of the stratum functionale of the endometrium. In humans and other primates, spiral arteries develop in the endometrium during this phase (fig. 28.21b). Estradiol may also stimulate the production of receptor proteins for progesterone at this time, in preparation for the next phase of the cycle.

The **secretory phase** of the endometrium occurs when the ovary is in its luteal phase. In this phase, increased progesterone secretion stimulates the development of mucus glands. As a result of the combined actions of estradiol and progesterone, the endometrium becomes thick, vascular, and "spongy" in appearance during the time of the cycle following ovulation. It is therefore well prepared to accept and nourish an embryo if fertilization occurs.

The **menstrual phase** occurs as a result of the fall in ovarian hormone secretion during the late luteal phase.

Table 28.3 Phases of the menstrual cycle

Phase of cycle		Hormonal changes		Tissue changes	
Ovarian	Endometrial	Pituitary	Ovary	Ovarian	Endometrial
Follicular (days 5–13)	Proliferative	No apparent change	Estradiol secretion rises (due to FSH stimulation of follicles)	Follicles grow; graafian follicle develops (due to FSH stimulation)	Mitotic division increases thickness of endometrium; spiral arteries develop (due to estradiol stimulation)
Ovulatory (day 14)	Proliferative	LH surge (and increased FSH) stimulated by positive feedback from estradiol	Slight fall in estradiol secretion	Graafian follicle is ruptured and ovum is extruded into fallopian tube	No change
Luteal (days 15–28)	Secretory	LH and FSH decrease (due to negative feedback of steroids)	Progesterone and estrogen secretion increase	Development of corpus luteum (due to LH stimulation)	Glandular development in endometrium (due to progesterone stimulation)
Menstrual (days 1–4)	Menstrual	FSH and LH remain low	Estradiol and progesterone fall	Corpus luteum regresses; primary follicles grow	Inner two-thirds of endometrium is shed with accompanying bleeding

Source: Fox, Stuart Ira, *Human Physiology.* © 1984 Wm. C. Brown Publishers, Dubuque, Iowa. All Rights Reserved. Reprinted by permission.

Necrosis (cellular death) and sloughing of the stratum functionale of the endometrium may be produced by constriction of the spiral arteries. The spiral arteries appear to be responsible for bleeding during menstruation because lower animals that lack spiral arteries don't bleed when they shed their endometrium. The phases of the menstrual cycle are summarized in table 28.3.

The cyclic changes in ovarian secretion also cause other cyclic changes in the female genital ducts. High levels of estradiol secretion, for example, cause cornification of the vaginal epithelium (the upper cells die and become filled with keratin). During the luteal phase of the cycle, the high levels of progesterone cause a change in the consistency of the cervical mucus. This allows sperm to penetrate the cervix and enter the uterus more easily.

---

1. *Describe the changes that occur in the ovary and endometrium during the follicular phase, and explain the hormonal control of these changes.*
2. *Describe the hormonal regulation of ovulation.*
3. *Describe the formation, function, and fate of the corpus luteum, and describe the changes that occur in the endometrium during the luteal phase.*
4. *Explain the significance of negative feedback inhibition during the luteal phase, and explain the hormonal control of menstruation.*

Clinical Considerations

Females are more prone to dysfunctions and diseases of the reproductive organs than are males because of cyclic changes in reproductive events, problems associated with pregnancy, and the susceptibility of the breasts to infections and neoplasms. The termination of reproductive capabilities at menopause can also cause complications due to hormonal alterations. *Gynecology* is the specialty of medicine concerned with dysfunction and diseases of the female reproductive system, whereas *obstetrics* is the specialty dealing with pregnancy and childbirth. Frequently a physician will specialize in both obstetrics and gynecology (OBGYN).

There are numerous clinical aspects of the female reproductive system, but only the most important conditions will be discussed. This portion of the chapter is divided into diagnostic procedures, developmental abnormalities, problems involving the ovaries and uterine tubes, problems involving the uterus, diseases of the vagina and vulva, and diseases of the breasts and mammary glands.

Diagnostic Procedures

A gynecological, or pelvic, examination is generally given in a thorough physical examination, especially prior to marriage, during pregnancy, or if problems involving the reproductive organs are suspected. In a gynecological examination, the physician inspects the vulva for irritations, lesions, or abnormal vaginal discharge, and palpates the vulva and internal organs. Most of the internal organs can be palpated through the vagina, especially if they are enlarged or tender. Inserting a lubricated *speculum* into the vagina allows visual examination of the cervix and vaginal walls. A speculum is an instrument for opening or distending a body opening to permit visual inspection.

In special cases, it may be necessary to examine the cavities of the uterus and uterine tubes by *hysterosalpingography (his″ter-o-sal″ping-gog′rah-fe)* (fig. 28.23). This technique involves injecting a radiopaque dye into the reproductive tract. The patency of the uterine tubes, irregular pregnancies, and various types of tumors may be

---

hysterosalpingogram: Gk. *hystera,* uterus; *salpinx,* trumpet (uterine tube); *graphein,* to record

Figure 28.23.   A hysterosalpingogram showing the cavities of the uterus and uterine tubes.

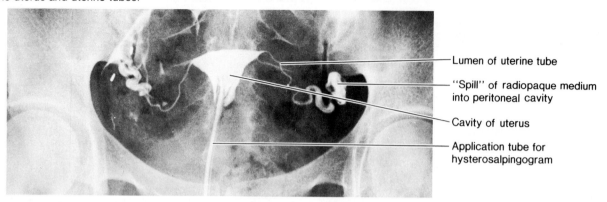

- Lumen of uterine tube
- "Spill" of radiopaque medium into peritoneal cavity
- Cavity of uterus
- Application tube for hysterosalpingogram

detected using this technique. A *laparoscopy (lap″ah-ros′ko-pe)* permits *in vivo* visualization of the internal reproductive organs. The entrance for the laparoscope may be the umbilicus, a small incision in the lower abdominal wall, or through the posterior fornix of the vagina into the rectouterine pouch. Although a laparoscope is used primarily in diagnosis, it can be used when performing a **tubal ligation** (fig. 28.24), which is a method of sterilizing a female.

One diagnostic procedure that should be routinely performed by a woman is a *breast self-examination* (BSE). The importance of a BSE is not to prevent diseases of the breast but to detect any problems before they become serious. A BSE should be performed monthly one week after the cessation of menstruation so that the breast will not be swollen or especially tender. The procedure for a BSE is presented in figure 28.25.

Another important diagnostic procedure is a *Papanicolaou (Pap) smear.* The Pap smear permits microscopic examination of cells obtained from near the external os of the cervix. Samples of cells are obtained by using a vaginal pipette or a specially designed wooded spatula. Women should have routine Pap smears for the early detection of cervical cancer.

Developmental Problems

Many of the developmental abnormalities of the female reproductive system also occur in the reproductive system of the male and are discussed in the previous chapter. Hermaphroditism and irregularities of the sex chromosomes, for example, are developmental conditions that cause a person to develop both male and female characteristics.

Other developmental abnormalities of the female reproductive system may occur during the formation of the uterus and vagina. Failure of the paramesonephric ducts to fuse normally can result in a **double uterus,** a **bicornuate** *(bi-kor′nu-āt)* **uterus,** or a **unicornuate uterus.** These types of uterine abnormalities and others are diagrammed in figure 28.26. In about one in 4,000 females, the vagina is absent. This is usually accompanied by an absence of the uterus as well.

Figure 28.24.   Tubal ligation.

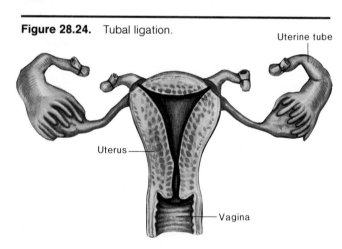

Uterine tube

Uterus

Vagina

Problems Involving the Ovaries and Uterine Tubes

Most **ovarian neoplasms** are nonmalignant ovarian cysts lined by cuboidal epithelium and filled with a serous albuminous fluid. These tumors may frequently be palpated during a gynecological examination and may require surgical removal if they exceed about 4 cm in diameter. They are generally removed as a precaution because it is impossible to determine by palpation whether the mass is malignant or benign.

Malignant ovarian tumors, which generally occur in women over the age of sixty, may reach massive size. Ovarian tumors of 5 kg (14 lbs) are not uncommon and ovarian tumors of 110 kg (300 lbs) have been reported. Some ovarian tumors produce estrogen and thus cause feminization in elderly women, including recommencement of menstrual periods. The prognosis for women with ovarian tumors varies depending on the type of tumor and whether or not metastasis has occurred.

Two frequent problems involving the uterine tubes are salpingitis and ectopic pregnancies. **Salpingitis** is an inflammation of one or both uterine tubes. Infection of the uterine tubes is generally caused by venereal disease, although secondary bacterial infections from the vagina may also cause salpingitis. Salpingitis may cause sterility if the uterine tubes become occluded.

---

bicornuate: L. *bi,* two; *cornu,* horn
neoplasm: Gk. *neos,* new; *plasma,* something formed

Figure 28.25. The procedures for a breast self-examination (BSE).

How to examine your breasts

1

In the shower:

Examine your breasts during bath or shower; hands glide easier over wet skin. Fingers flat, move gently over every part of each breast. Use right hand to examine left breast, left hand for right breast. Check for any lump, hard knot or thickening.

© 1975 AMERICAN CANCER SOCIETY INC

2

Before a mirror:

Inspect your breasts with arms at your sides. Next, raise your arms high overhead. Look for any changes in contour of each breast, a swelling, dimpling of skin or changes in the nipple.

Then, rest palms on hips and press down firmly to flex your chest muscles. Left and right breast will not exactly match—few women's breasts do.

Regular inspection shows what is normal for you and will give you confidence in your examination.

3

Lying down:

To examine your right breast, put a pillow or folded towel under your right shoulder. Place right hand behind your head—this distributes breast tissue more evenly on the chest. With left hand, fingers flat, press gently in small circular motions around an imaginary clock face. Begin at outermost top of your right breast for 12 o'clock, then move to 1 o'clock, and so on around the circle back to 12. A ridge of firm tissue in the lower curve of each breast is normal. Then move in an inch, toward the nipple, keep circling to examine *every part of your breast*, including nipple. This requires at least three more circles. Now slowly repeat procedure on your left breast with a pillow under your left shoulder and left hand behind head. Notice how your breast structure feels.

Finally, squeeze the nipple of each breast gently between thumb and index finger. Any discharge, clear or bloody, should be reported to your doctor immediately.

Figure 28.26. Congenital uterine abnormalities. (*a*) a double uterus and double vagina; (*b*) a double uterus with single vagina; (*c*) bicornuate uterus; (*d*) a bicornuate uterus with a rudimentary left horn; (*e*) a septate uterus; (*f*) a unicornuate uterus.

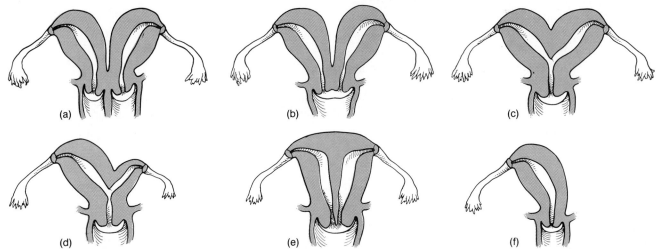

(a)   (b)   (c)

(d)   (e)   (f)

   Ectopic pregnancy results from implantation of the blastocyst in a location other than the body or fundus of the uterus. The most frequent ectopic site is in the uterine tube, where an implanted blastocyst causes what is commonly called a **tubular pregnancy.** One danger of a tubular pregnancy is the enlargement, rupture, and subsequent hemorrhage of the uterine tube where implantation has occurred. A tubular pregnancy is frequently treated by removing the affected tube.

   Infertility, or the inability to conceive, is a clinical problem that involves the male or female reproductive system and affects about 10% of couples. Generally, when a male is infertile, it is because of inadequate sperm counts. Female infertility is frequently caused by an obstruction of the uterine tubes or abnormal ovulation.

Problems Involving the Uterus

Abnormal menstruations are among the most common disorders of the female reproductive system. Abnormal menstruations may be directly related to problems of the reproductive organs and pituitary gland or associated with emotional and psychological stress.

Amenorrhea *(ah-men''o-re'ah)* is the absence of menstruation and can be categorized as normal, primary, or secondary. **Normal amenorrhea** follows menopause, occurs during pregnancy, and in some women may occur during lactation. **Primary amenorrhea** is when a woman has never menstruated although she has passed the age when menstruation normally begins. Primary amenorrhea is generally accompanied by failure of the secondary sexual characteristics to develop. Endocrine disorders may cause primary amenorrhea and abnormal development of the ovaries or uterus.

Secondary amenorrhea is the cessation of menstruation in women who previously have had normal menstrual periods and are not pregnant and have not gone through menopause. Various endocrine disturbances as well as psychological factors cause secondary amenorrhea. It is not uncommon, for example, for young women who are in the process of making major changes or adjustments in their lives to miss menstrual periods. Secondary amenorrhea is also frequent in women athletes during periods of intense training. A low percentage of body fat may be a contributing factor. Sickness, fatigue, poor nutrition, or emotional stress also cause secondary amenorrhea.

Dysmenorrhea is painful or difficult menstruation accompanied by severe menstrual cramps. The causes of dysmenorrhea are not totally understood but may include endocrine disturbances (inadequate progesterone levels), a faulty position of the uterus, emotional stress, or some type of obstruction that prohibits menstrual discharge.

Abnormal uterine bleeding includes **menorrhagia** *(men''o-ra'je-ah)*, or excessive bleeding during the menstrual period, and **metrorrhagia**, or spotting between menstrual periods. Other types of abnormal uterine bleeding are menstruations of excessive duration, too frequent menstruations, and postmenopausal bleeding. These abnormalities may be caused by hormonal irregularities, emotional factors, or various diseases and physical conditions.

Uterine neoplasms are an extremely common problem of the female reproductive tract. Most of the neoplasms are benign and include cysts, polyps, and smooth muscle tumors (leiomyoma). Any of these conditions may provoke irregular menstruations and may cause infertility if the neoplasms are massive.

Cancer of the uterus is the most common malignancy of the female reproductive tract. The most common site of uterine cancer is in the cervix (fig. 28.27). Cervical cancer is second only to cancer of the breast in frequency of occurrence and is a disease of young women (ages thirty through fifty), especially those who have had frequent intercourse with multiple partners during their teens and onward. If detected early through regular Pap smears, the disease can be cured before it metastasizes. The treatment of cervical cancer depends on the stage of the malignancy and the age and health of the woman. In the case of women who are through having children, a **hysterectomy** *(his''tĕ-rek'to-me)* (surgical removal of the uterus) is usually performed.

Figure 28.27. Sites of various conditions and diseases of the female reproductive tract, each of which could cause an abnormal discharge of blood.

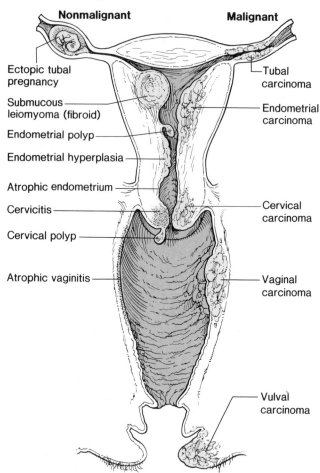

Endometriosis is a condition characterized by the presence of endometrial tissues at sites other than the inner lining of the uterus. Frequent sites of ectopic endometrial cells are on the ovaries, outer layer of the uterus, abdominal wall, and urinary bladder. Although it is not certain how endometrial cells become established outside the uterus, it is speculated that some discharged endometrial tissue might be flushed backward from the uterus and through the uterine tubes during menstruation. Women with endometriosis will bleed internally with each menstrual period because the ectopic endometrial cells are stimulated along with the normal endometrium by ovarian hormones. The most common symptoms of endometriosis are extreme dysmenorrhea and a feeling of fullness during each menstruation period. Endometriosis can cause infertility. It is treated by suppressing the endometrial tissues with oral contraceptive pills or by surgery. An oophorectomy, or removal of the ovaries, may be necessary in extreme cases.

Uterine displacements are relatively common in elderly women. When uterine displacements occur in younger women, they are important because of dysmenorrhea, fertility, or parturition problems. **Retroversion**, or **retroflexion**, is a displacement backward; **anteversion**, or

anteflexion, is displacement forward. **Prolapse of the uterus** is a marked downward displacement into the vagina.

An **abortion** is defined as the termination of a pregnancy before the twenty-eighth week of gestation. A **spontaneous abortion,** or **miscarriage,** occurs without mechanical aid or medicinal intervention and may happen in as many as 10% of all pregnancies. Spontaneous abortions usually occur when there is abnormal development of the fetus or disease of the maternal reproductive system. An **induced abortion** is removal of the fetus from the uterus by mechanical means or drugs. Induced abortions are a major issue of controversy because of questions regarding individual rights—those of the mother and those of the fetus—the definition of life, and moral concerns.

Diseases of the Vagina and Vulva

Pelvic inflammatory disease (PID) is a general term for inflammation of the female reproductive organs within the pelvis. The infection may be confined to a single organ, or it may involve all the internal reproductive organs. The pathogens generally enter through the vagina during copulation, induced abortion, childbirth, or postpartum. Inflammation of the ovaries is called **oophoritis,** and inflammation of the uterine tube is **salpingitis** *(sal''pin-ji'tis).*

The vagina and vulva are generally resistant to infection because of the acidity of the vaginal secretions. Occasionally, however, localized infections and inflammations do occur; these are termed *vaginitis,* if confined to the vagina, or **vulvovaginitis,** if both the vagina and external genitalia are affected. The symptoms of vaginitis are a discharge of pus *(leukorrhea)* and itching *(pruritus).* The two most common organisms that cause vaginitis are the protozoan *Trichomonas vaginalis* and the fungus *Candida albicans.*

Diseases of the Breasts and Mammary Glands

The breasts and mammary glands of females are highly susceptible to infections, cysts, and tumors. Infections involving the mammary glands usually follow the development of a dry and cracked nipple during lactation. Bacteria enter the wound and establish an infection within the lobules of the gland. During an infection of the mammary gland, a blocked duct frequently causes a lobe to become engorged with milk. This localized swelling is usually accompanied by redness, pain, and an elevation of temperature. Administering specific antibiotics and applying heat are the usual treatments.

Nonmalignant cysts are the most frequent diseases of the breast. These masses are generally of two types, neither of which is life threatening.

Dysplasia (fibrocystic disease) is a broad condition involving several nonmalignant diseases of the breast. All dysplasias are benign neoplasms of various sizes that may become painful during or prior to menstruation. Most of the masses are small and remain undetected. Dysplasia affects nearly one-half of women over thirty years of age prior to menopause.

Figure 28.28. *(a)* a mammogram of a patient with carcinoma of the upper breast. Note the sunburst or stellate pattern of the neoplasm indicated with an arrow. *(b)* in mammography, the breasts are placed alternately on a metal plate and X-rayed from the side and from above.

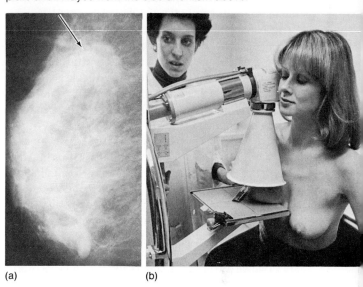

(a)   (b)

Fibroadenoma *(fi''bro-ad''ē-no'mah)* is a benign tumor of the breast that frequently occurs in women under the age of thirty-five. Fibroadenomas are nontender, rubbery masses, which are easily moved about in the mammary tissue. A fibroadenoma can be excised in a physician's office under local anesthetics.

Carcinoma of the breast is the most common malignancy in women. One in thirteen women, or about 7%, will develop breast cancer and one-third of these will die from the disease. Breast cancer is the leading cause of death in women between forty and fifty years of age. Men are also susceptible to breast cancer, but it is one hundred times more frequent in women. Breast cancer in men is usually fatal.

The causes of breast cancer are not known, but women who are most susceptible are those who are over age thirty-five, who have a family history of breast cancer, and who are nulliparus (no children). The early detection of breast cancer is important because the progressed state of the disease will determine the treatment and prognosis.

Confirming suspected breast cancer generally requires *mammography* (fig. 28.28). If the mammograph indicates breast cancer, surgery is performed so that a biopsy can be obtained and the tumor assessed. If the tumor is found to be malignant, extensive surgery is performed depending on the size of the tumor and whether metastasis has occurred. The surgical treatment for breast cancer is generally some degree of *mastectomy*. A *simple mastectomy* is removal of the entire breast but not the underlying lymph nodes. A *modified radical mastectomy* is the complete removal of the breast, the lymphatic drainage, and perhaps the pectoralis major muscle. A *radical mastectomy* is similar to a modified except that the pectoralis major muscle is always removed as well as the axillary lymph nodes and adjacent connective tissue.

Methods of Contraception

The *rhythm method* of birth control is used by many people, but the popularity of this technique has declined as more successful methods of contraception have been introduced. In the rhythm method of birth control, the woman attempts to predict the day of her ovulation and restricts intercourse to "safe" times of the cycle when fertilization does not occur. The day of ovulation can be determined by a drop in basal body temperature or by a change in mucus discharge from the vagina. This technique is subject to great error, however, because most women don't ovulate on precisely the same day of the cycle in different months.

More popular methods of birth control include (1) sterilization, (2) oral contraceptives, (3) intrauterine devices (IUDs), and (3) barrier methods—including condoms for the male and diaphragms for the female. All of these techniques are effective, but the safety, side effects, and efficacy of each technique are different.

Sterilization techniques include *vasectomy* for the male and *tubal ligation* for the female. In the latter technique (which currently accounts for over 60% of sterilization procedures performed in the United States), the uterine tubes are cut and tied. This is analogous to the procedure performed on the ductus (vas) deferens in a vasectomy and prevents fertilization of the ovulated ovum. The long-term safety of vasectomy, as previously discussed, has not been well established, and evidence in experiments with rhesus monkeys suggests that an increased tendency toward atherosclerosis may be a side effect of this procedure. Symptoms of pelvic pain and other problems are found in about 15% of women who have tubal ligations.

About 10 million women in the United States and 60 million women in the world are currently using **oral steroid contraceptives.** These contraceptives usually consist of a synthetic estrogen combined with a synthetic progesterone in the form of pills that are taken once each day for three weeks after the last day of a menstrual period. This procedure causes an immediate increase in blood levels of ovarian steroids (from the pill), which is maintained for the normal duration of a monthly cycle. As a result of *negative feedback inhibition* of gonadotrophin secretion, *ovulation never occurs.* The entire cycle is like a false luteal phase, with high levels of progesterone and estrogen and low levels of gonadotrophins.

Since the contraceptive pills contain ovarian steroid hormones, the endometrium proliferates and becomes secretory just as it does during a normal cycle. In order to prevent an abnormal growth of the endometrium, women stop taking the pill after three weeks. This causes estrogen and progesterone levels to fall, and permits menstruation to occur. The contraceptive pill is an extremely effective method of birth control, but it does have potentially serious side effects—including an increased incidence of thromboembolism, cardiovascular disorders, and endometrial and breast cancer. It has been pointed out, however, that the mortality risk of contraceptive pills is still much lower than the risk of death from the complications of pregnancy—or from automobile accidents.

Intrauterine *(in''trah-u'ter-in)* **devices (IUDs)** don't prevent ovulation, but instead prevent implantation of the embryo into the uterus in the event fertilization occurs. The mechanisms by which these contraceptive effects are produced are not well understood, but the efficiency of different IUDs appears to be related to their ability to cause inflammatory reactions in the uterus. Uterine perforations are the foremost complication of the use of IUDs.

Barrier contraceptives—condoms and diaphragms—are slightly less effective than the other methods of contraception, but they do not cause serious side effects. The failure rate of condoms—one of the oldest methods of contraception—is 12 to 20 pregnancies per 100 woman years of use, whereas the failure rate of the diaphragm is 12 to 18 pregnancies per 100 woman years.

Chapter Summary

I. Structure and Development of the Female Reproductive System
In the course of embryonic development, follicles containing oogonia appear in the ovaries, and the sex accessory organs are formed. The female external genitalia develop from the same indifferent structures that give rise to the external genitalia of a male.
  A. The secondary sex organs include the uterine tubes, uterus, vagina, and mammary glands.
  B. Oogonia contained within follicles appear in the ovaries, and the uterus and uterine tubes develop from paramesonephric (Müllerian) ducts.
II. Ovaries
During an ovarian cycle primary follicles become secondary follicles, and one of these becomes a graafian follicle. When the graafian follicle ruptures at ovulation, the empty follicle becomes a corpus luteum.

  A. The ovaries are supported by the mesovarium from the broad ligament, the ovarian ligament, and the suspensory ligament.
  B. The ovarian follicles undergo cyclic changes.
    1. Primary oocytes, arrested at prophase I of meiosis, are contained within primordial follicles.
    2. Some of these follicles become larger and gain granulosa cells during a cycle; these are primary follicles.
    3. When a follicle develops a fluid-filled antrum, it is called a secondary follicle.
    4. One follicle continues to grow and becomes a graafian follicle.
    5. The graafian follicle contains a secondary oocyte arrested at metaphase II of meiosis.

    6. The graafian follicle ruptures and releases its ovum in the process of ovulation.
    7. The empty follicle after ovulation becomes a corpus luteum.
III. Secondary Sex Organs
The uterine tubes conduct the zygote to the uterus, where implantation into the endometrial layer may occur. Sperm enters the female genital duct through the vagina.
  A. Ciliary action and peristalsis help move the ovum from the fimbria into the infundibulum of the uterine tube and toward the uterus.
  B. The uterus is supported by the broad ligaments, uterosacral ligaments, cardinal ligaments, and round ligaments.
    1. The endometrium consists of a stratum basale and a stratum functionale, which is shed in menstruation.

2. The myometrium produces the muscular contractions needed for labor and delivery (parturition).

C. The vagina serves to deliver sperm to the cervix of the uterus and also functions as the birth canal.

D. The external genitalia—mons pubis, labia majora and minora, clitoris, vestibule, and vestibular glands—are also called the vulva or pudendum.

E. Parasympathetic nerves stimulate erectile tissues in the clitoris and vestibular bulbs; in orgasm, muscular contraction occurs in the perineum, uterus, and uterine tubes.

IV. Mammary Glands and Lactation
The structure and function of mammary glands is dependent on the effects of a number of hormones. Prolactin and oxytocin are required for the production and delivery of milk to a suckling infant.

A. Milk is produced by glandular alveoli, which are associated into lobules and lobes; these secrete into secondary tubules that drain into lactiferous ducts.

B. Breast development and growth of the mammary glands and ducts are stimulated by high levels of estrogen and progesterone.

1. These effects require the permissive action of insulin, thyroid hormones, and cortisol.

2. Prolactin stimulates the production of milk proteins; oxytocin stimulates the milk ejection reflex.

3. The secretion of oxytocin and prolactin is stimulated by suckling; this is a neuroendocrine reflex.

V. Menstrual Cycle
During the course of the ovarian cycle, there are cyclic changes in the secretion of ovarian hormones, which in turn cause cyclic changes in the endometrium of the uterus.

A. Structural changes in the ovaries are associated with changes in hormone secretion.

1. During the follicular phase, FSH stimulates the follicles to grow and secrete increasing amounts of estradiol.

2. The rising levels of estradiol stimulate the hypothalamus to secrete GnRH, which in turn causes the anterior pituitary to secrete a surge of LH; this is positive feedback.

3. The LH surge causes the graafian follicle to break and ovulation to occur.

4. LH stimulates the ruptured follicle to become a corpus luteum; the ovary at this point is in the luteal phase.

5. The corpus luteum secretes both estradiol and progesterone, which exert negative feedback inhibition of FSH and LH during the luteal phase of the cycle.

6. Toward the end of the luteal phase the corpus luteum dies, estradiol and progesterone levels fall, and menstruation occurs.

B. High levels of estradiol during the follicular phase causes growth of the endometrium; this is known as the proliferative phase.

1. High levels of estradiol and progesterone during the luteal phase makes the endometrium more glandular and vascular; this is the secretory phase.

2. The fall in estradiol and progesterone that occurs at the end of the luteal phase causes the stratum functionale of the endometrium to be shed; this is the menstrual phase.

Review Activities

Objective Questions

Match the following:

1. Menstrual phase
2. Follicular phase
3. Luteal phase
4. Ovulation

(a) high estrogen and progesterone; low FSH and LH
(b) low estrogen and progesterone
(c) LH surge
(d) increasing estrogen; little or no progesterone

5. The secretory phase of the endometrium corresponds to which of the following ovarian phases?
   (a) follicular phase
   (b) ovulation
   (c) luteal phase
   (d) menstrual phase

6. Which of the following statements about oogenesis is *true*?
   (a) Oogonia, like spermatogonia, form continuously during postnatal life.
   (b) Primary oocytes are haploid.
   (c) Meiosis is completed prior to ovulation.
   (d) In ovulation, secondary oocytes are released from graafian follicles.

7. The paramesonephric (müllerian) ducts give rise to the
   (a) uterine tubes   (d) both (a) and (b)
   (b) uterus   (e) both (b) and (c)
   (c) pudendum

8. In a female, the homologue of the male scrotum is the
   (a) labia majora   (c) clitoris
   (b) labia minora   (d) vestibule

9. The cervix is a portion of the
   (a) vulva   (c) uterus
   (b) vagina   (d) uterine tubes

10. Fertilization normally occurs in the
    (a) ovaries   (c) uterus
    (b) uterine tubes (d) vagina

11. Contraction of the mammary ducts during the milk ejection reflex is stimulated by
    (a) prolactin   (c) estrogen
    (b) oxytocin   (d) luteinizing hormone

Essay Questions

1. Explain how the genital ducts develop and why the external genitalia of males and females are considered homologous.

2. Describe the gross and histologic structure of the uterus, and explain the significance of the strata functionale and basale of the endometrium.

3. Explain the hormonal interactions that control ovulation and make it occur at the proper time.

4. Compare menstrual bleeding and bleeding that occurs during the estrus cycle of a dog, in terms of hormonal control mechanisms and the ovarian cycle.

5. "The [contraceptive] pill tricks the brain into thinking you're pregnant." Interpret this popularized explanation in terms of physiological mechanisms.

6. Why does menstruation normally occur? Under what conditions does menstruation not occur? Explain.

Selected Readings

Boyar, R. M. 1978. Control of the onset of puberty. *Annual Review of Medicine* 31:329.

Fink, G. 1979. Feedback action of target hormones on hypothalamus and pituitary with special reference to gonadal steroids. *Ann. Rev. Physiol.* 41:571.

Frantz, A. G. 1978. Prolactin. *New Engl. J. Med.* 298:112.

Grumbach, M. M. 1979. The neuroendocrinology of puberty. *Hosp. Prac.* 14:65.

Leong, D. A., L. S. Frawley, and J. D. Neill. 1983. Neuroendocrine control of prolactin secretion. *Ann. Rev. Physiol.* 45:109.

Mader, S. S. 1980. *Human Reproductive Biology.* Dubuque: Wm. C. Brown Co.

Naftolin, F. March 1981. Understanding the basis of sex differences. *Science* 211 (No. 4488):1263.

Odell, W. D., and D. L. Moyer. 1971. *Physiology of Reproduction.* St. Louis: C. V. Mosby Co.

Segal, S. J. September 1974. The physiology of human reproduction. *Scientific American.*

Tyson, J. E. April, 1984. Reproductive endocrinology: New problems call for new solutions. *Diagnostic Med.* p. 25.

Warkentin, D. L. April, 1984. From A to hCG in pregnancy testing. *Diagnostic Med.* p. 35.

Wilson, J. D. 1978. Sexual differentiation. *Ann. Rev. Physiol.* 40:279.

Wilson, J. D., F. W. George, and J. E. Griffin. March 1981. The hormonal control of sexual development. *Science* 211 (No. 4488):1278.

Yen, S. S. C. 1977. Regulation of the hypothalamic-pituitary-ovarian axis in women. *Journal of Reproduction and Fertility* 51:181.

Yen, S. S. C. 1979. Neuroendocrine regulation of the menstrual cycle. *Hosp. Pract.* 14:83.

29 Prenatal Development and

Outline

Fertilization
Pre-Embryonic Stage
Cleavage and Formation of the Blastocyst
Implantation
Formation of Germ Layers
Embryonic Stage
Extraembryonic Membranes
Placenta
Umbilical Cord
Structural Changes of the Embryo by Weeks
Fetal Stage
Labor and Parturition
Inheritance
Clinical Considerations
Abnormal Implantation Sites
In Vitro Fertilization and Artificial
    Implantation
Multiple Pregnancy
Fetal Monitoring
Congenital Defects
Some Genetic Disorders of Clinical
    Importance
Chapter Summary
Review Activities
Selected Readings

Concepts

The fertilization of an ovum by a sperm in the uterine tube promotes the completion of meiotic development and the formation of a diploid zygote.

The events of the two-week pre-embryonic stage include fertilization, transportation of the fertilized ovum through the uterine tube, mitotic divisions, implantation, and the formation of primordial embryonic tissue.

The events of the six-week embryonic stage include the differentiation of the germ layers into specific body organs and the formation of the placenta, the umbilical cord, and the extraembryonic membranes that provide sustenance and protection to the embryo.

The fetal stage lasts from nine weeks until birth and is characterized by tremendous growth and the specialization of body structures.

Labor and parturition are the culmination of gestation and require the action of oxytocin from the posterior pituitary and prostaglandins, produced in the uterus.

Inheritance is the passage of hereditary traits carried on the genes of chromosomes from one generation to another.

Inheritance

Fertilization

The fertilization of an ovum by a sperm in the uterine tube promotes the completion of meiotic development and the formation of a diploid zygote.

Objective 1. Define capacitation, fertilization, and morphogenesis.
Objective 2. Describe the changes that occur in the sperm and ovum during fertilization.

The structure of the human body forms before birth, or *prenatally,* through a process called **morphogenesis** *(mor″fo-jen′ĕ-sis).* Through morphogenic events the organs and systems of the body are established in a functional relationship. Understanding morphogenesis is important for a complete perspective of the structure and function of each body system. There are "sensitive" stages of morphogenesis for each organ and system, during which genetic or environmental conditions (nutrition, smoking, or drugs taken by the mother) may affect the normal development of the baby. Many clinical problems are congenital in nature and originate during morphogenic development.

The prenatal development of a human is a fascinating and awesome event. It begins with a single fertilized egg and culminates some thirty-eight weeks later with a complex organization of over 200 million cells. Prenatal development can be divided into a *pre-embryonic stage,* which is initiated by the fertilization of a secondary oocyte; an *embryonic stage,* during which the body's organ systems are formed; and a *fetal stage,* which culminates in parturition, or the birth of the baby.

During coitus (sexual intercourse), a male ejaculates between 100 and 500 million sperm into the vagina. This tremendous number is needed because of the high rate of sperm fatality—only about 100 survive to contact the secondary oocyte in the uterine tube. During passage through the acidic female tract, sperm gain the ability to fertilize an ovum through a process called **capacitation.** The changes that occur in capacitation are not known. Experiments have shown, however, that freshly ejaculated sperm are infertile and must be in the female tract for at least seven hours before they can fertilize an ovum.

A woman usually ovulates one ovum a month, totaling approximately four hundred ova during her reproductive years. Each ovulation releases a secondary oocyte arrested at metaphase of the second meiotic division. As the secondary oocyte enters the uterine tube, it is surrounded by a thin transparent layer of protein and polysaccharides, called the **zona pellucida,** and a layer of granulosa cells, called the **corona radiata** (fig. 29.1).

---

morphogenesis: Gk. *morphe,* form; *genesis,* beginning
capacitation: L. *capacitas,* capable of
zona pellucida: L. *zone,* a girdle; *pellis,* skin
corona radiata: Gk. *korone,* crown; *radiata,* radiate

Figure 29.1.   The process of fertilization. (a, b) diagrammatic representations. (c) a scanning electron micrograph of sperm bound to the egg surface. As the head of the sperm encounters the gelatinous corona radiata of the egg (b2), the acrosomal vesicle ruptures and the sperm digests a path for itself by the action of enzymes released from the acrosome (b3, b4). When the cell membrane of the sperm contacts the cell membrane of the egg (b5), they become continuous, and the sperm nucleus and other contents move into the egg cytoplasm.

First polar body

Perivitelline space

Cytoplasm of ovum

(a)

Corona radiata

Second meiotic spindle

Zona pellucida

Cell membrane of ovum

(c)

Nucleus containing chromosomes

Acrosome containing enzymes

Perforations in acrosome wall

1   2   3   4   5

(b)

The head of each sperm is capped by an organelle called an **acrosome** (figs. 29.1 and 29.2). The acrosome contains a trypsinlike protein-digesting enzyme and *hyaluronidase (hi''ah-lu-ron'ĭ-dās),* which digests hyaluronic acid, an important constituent of connective tissue. When a sperm meets a secondary oocyte in the uterine tube, an *acrosomal reaction* occurs that exposes the acrosome's digestive enzymes and allows a sperm to penetrate through the corona radiata and the zona pellucida.

As the first sperm penetrates the zona pellucida, a chemical change in the zona occurs that prevents other sperm from entering. Only one sperm, therefore, is allowed to fertilize a secondary oocyte. As fertilization occurs, the secondary oocyte is stimulated to complete its second meiotic division (fig. 29.3). Like the first meiotic division, the second produces one cell that contains all of the cytoplasm and one polar body. The healthy cell is the mature ovum, and the second polar body, like the first, ultimately fragments and disintegrates.

At fertilization, the entire sperm enters the cytoplasm of the much larger egg cell (fig. 29.4). Within twelve hours the nuclear membrane in the ovum disappears, and the *haploid number* of chromosomes (twenty-three) in the

acrosome: Gk. *akron,* extremity; *soma,* body
haploid: Gk. *haplous,* single; L. *ploideus,* multiple in form

Figure 29.2.   An electron micrograph showing the head of a human sperm with its nucleus and acrosomal cap.

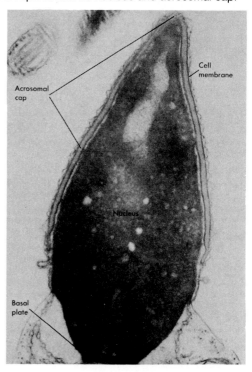

Acrosomal cap

Cell membrane

Nucleus

Basal plate

Figure 29.3. A secondary oocyte, arrested at metaphase II of meiosis, is released at ovulation. If this cell is fertilized, it completes its second meiotic division and produces a second polar body.

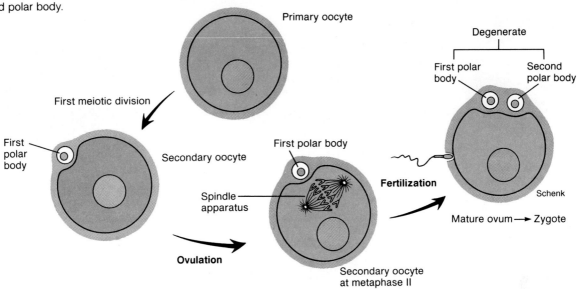

Figure 29.4. Fertilization and the union of chromosomes from the sperm and ovum to form the zygote.

ovum is joined by the haploid number of chromosomes from the sperm. A fertilized egg, or zygote *(zi′gōt)*, containing the *diploid number* of chromosomes (forty-six) is thus formed.

A secondary oocyte that is ovulated but not fertilized does not complete its second meiotic division, but instead it disintegrates twelve to twenty-four hours after ovulation. Fertilization therefore cannot occur if intercourse takes place beyond one day following ovulation. Sperm, in contrast, can survive up to three days in the female reproductive tract. Fertilization therefore can occur if coitus is performed within three days prior to the day of ovulation.

1. *Explain why capacitation and the acrosomal reaction of sperm are necessary to accomplish fertilization of a secondary oocyte.*
2. *Define the term* morphogenesis, *and discuss its importance in the development of a healthy baby.*
3. *Discuss the changes that occur in a sperm from the time of ejaculation to the time of fertilization. What changes occur in a secondary oocyte following ovulation to the time of fertilization?*

diploid: Gk. *diplous,* double; L. *ploideus,* multiple in form

Figure 29.5. Sequential illustrations from the first cleavage of the zygote to the formation of the blastocyst. Note the deterioration of the zona pellucida in the early blastocyst.

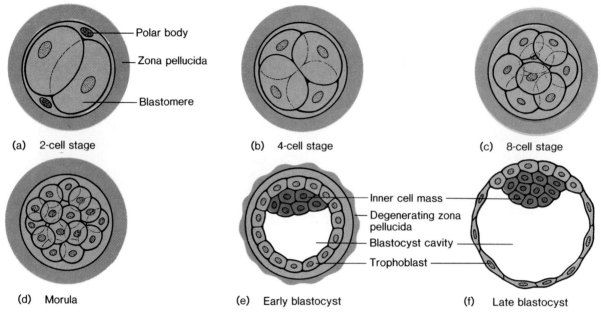

(a)   2-cell stage

(b)   4-cell stage

(c)   8-cell stage

(d)   Morula

(e)   Early blastocyst

(f)   Late blastocyst

Figure 29.6. Stages of pre-embryonic development of a human ovum fertilized in a laboratory (*in vitro*) as seen in scanning electron micrographs. (*a*) 4-cell stage; (*b*) cleavage at the 16-cell stage; (*c*) a morula; and (*d*) a blastocyst.

(a)

(b)

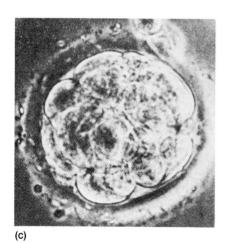

(c)

(d)

Figure 29.7. A diagrammatic representation of the ovarian cycle, fertilization, and the morphogenic events of the first week. The numbers indicate the days after fertilization.

Implantation of the blastocyst begins between the fifth and seventh day and is generally completed by the tenth day.

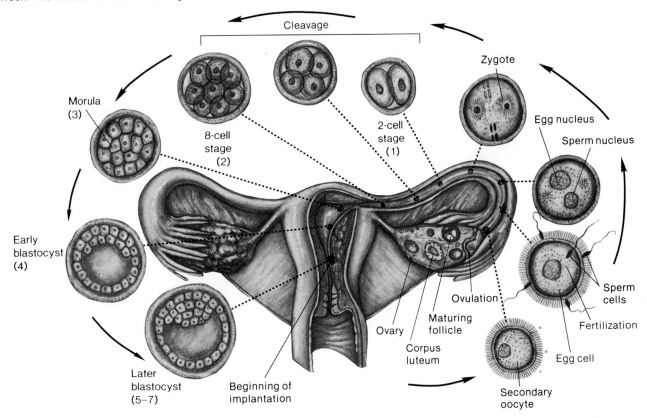

Pre-Embryonic Stage

The events of the two-week pre-embryonic stage include fertilization, transportation of the fertilized ovum through the uterine tube, mitotic divisions, implantation, and the formation of primordial embryonic tissue.

Objective 3. Describe the events of pre-embryonic development that result in the formation of the blastocyst.

Objective 4. Discuss the role of the trophoblast in the implantation and development of the placenta.

Objective 5. Explain how the primary germ layers develop, and list the structures produced by each layer.

Objective 6. Define *gestation,* and explain how the parturition date is determined.

Cleavage and Formation of the Blastocyst

Fertilization occurs within the uterine tube, usually about twelve to twenty-four hours following ovulation. The fertilized egg is referred to as a **zygote.** Within thirty hours the **cleavage** *(klēv'ij)* process begins with a mitotic division that results in the formation of two identical daughter cells called blastomeres (figs. 29.5 and 29.6). Several more

cleavages occur as the structure passes down the uterine tube and enters the uterus on about the third day. It is now composed of a ball of sixteen or more cells called a **morula** *(mor'u-lah).* Although the morula has undergone several mitotic divisions, it is not much larger than the zygote because there have not been additional nutrients entering the cells necessary for growth.

The developing structure remains unattached in the uterine cavity for about three days. During this time, the center of the morula fills with fluid passing in from the uterine cavity. As the fluid-filled space develops inside the morula, two distinct groups of cells form and the structure becomes known as a **blastocyst** *(blas'to-sist).* The single outer layer of cells forming the wall of the blastocyst is composed of **trophoblast** *(trof'o-blast)* **cells,** whereas the small, inner aggregation of cells becomes the *inner cell mass,* or **embryoblast.** With further development the trophoblast cells differentiate into a structure called the **chorion** *(ko're-on)* which will become a portion of the placenta, and the inner cell mass will become the embryo. The hollow, fluid-filled center of the blastocyst is called the **blastocyst cavity.** A diagrammatic summary of the ovarian cycle, fertilization, and the morphogenic events of the first week is presented in figure 29.7.

---

zygote: Gk. *zygotos,* yolked, joined
blastomere: Gk. *blastos,* germ; *meros,* a part

morula: L. *morum,* mulberry
blastocyst: Gk. *blastos,* germ; *kystis,* bag
trophoblast: Gk. *trophe,* nourishment; *blastos,* germ

Figure 29.8. The blastocyst adheres to the endometrium on about the sixth day as seen in (a) an electron micrograph and (b) a diagram of the specific cells of the blastocyst. By the seventh day (c), specialized syncytiotrophoblasts from the trophoblast invade the endometrium. The syncytiotrophoblasts secrete hCG to sustain pregnancy and will eventually participate in the formation of the placenta for embryonic and fetal sustenance.

(a)

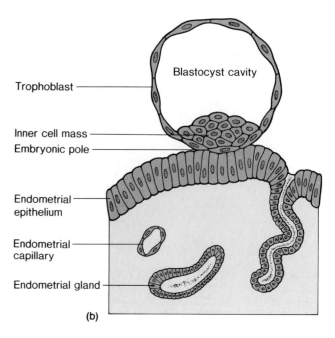

(b)

Implantation

The process of *implantation (im''plan-ta'shun),* or *nidation,* begins between the fifth and seventh day. Attachment is usually upon the posterior wall of the body of the uterus, with the side containing the inner cell mass against the endometrium (fig. 29.8). Implantation is made possible by the secretion of *proteolytic enzymes* by the trophoblast, which digest a portion of the endometrium. The blastula sinks into the depression, and endometrial cells move back to cover the defect in the wall. At the same time, the part of the uterine wall below the implanting blastocyst thickens, and specialized cells of the trophoblast produce fingerlike projections, called **syncytiotrophoblasts** *(sin-sit''e-o-trof'o-blast),* into the thickened area. The syncytiotrophoblasts arise from a specific portion of the trophoblast at the embryonic pole called the **cytotrophoblast.**

The blastocyst saves itself from being eliminated with the endometrium during a menstrual discharge by secreting a hormone that indirectly prevents menstruation. Even before the sixth day when implantation begins, the trophoblast cells secrete **chorionic gonadotrophin** *(ko''re-on'ik gon''ah-do-tro'fin),* or **hCG** (the h stands for *human*). This hormone is identical to LH in its effects and therefore is able to maintain the corpus luteum past the time when it would otherwise regress. The secretion of estrogen and progesterone is thus maintained and menstruation is normally prevented (fig. 29.9).

The secretion of hCG declines by the tenth week of pregnancy. Actually, this hormone is only required for the first five to six weeks of pregnancy, because the placenta itself becomes an active steroid-secreting gland. At the fifth to sixth week, the mother's corpus luteum begins to regress (even in the presence of hCG), but the placenta secretes more than sufficient amounts of steroids to maintain the endometrium and prevent menstruation.

All pregnancy tests assay for the presence of hCG in the blood or urine, because this is the only hormone that is secreted by the blastocyst but not by the mother's endocrine glands. Modern pregnancy tests detect the presence of hCG by use of antibodies against hCG or by the use of cellular receptor proteins for hCG.

Formation of Germ Layers

As the blastocyst completes implantation during the second week of development, the inner cell mass undergoes marked differentiation. A slitlike space, called the **amniotic cavity,** forms between the inner cell mass and the invading trophoblast (fig. 29.10). The inner cell mass flattens into the **embryonic disc** (see fig. 29.12), which consists of two layers: an upper **ectoderm,** which is closer to the amniotic cavity, and a lower **endoderm,** which borders the blastocyst cavity. A short time later, a third layer called the **mesoderm** forms between the endoderm and ectoderm. These three layers constitute the **primary germ**

implantation: L. *im,* in; *planto,* to plant
nidation: L. *nidus,* nest

ectoderm: Gk. *ecto,* outside; *derm,* skin
endoderm: Gk. *endo,* within; *derm,* skin
mesoderm: Gk. *meso,* middle; *derm,* skin

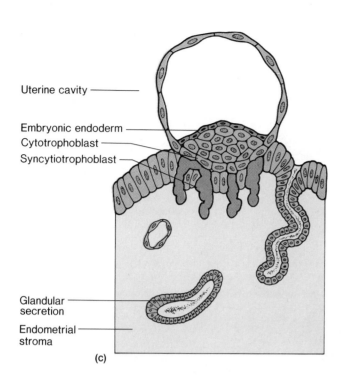

Uterine cavity
Embryonic endoderm
Cytotrophoblast
Syncytiotrophoblast

Glandular secretion
Endometrial stroma

(c)

Figure 29.9. Human chorionic gonadotrophin (hCG) is secreted by trophoblast cells during the first trimester of pregnancy. This hormone maintains the mother's corpus luteum for the first five and a half weeks. After that time the placenta becomes the major sex-hormone-producing gland, secreting increasing amounts of estrogen and progesterone throughout pregnancy.

Human chorionic gonadotropin

Estrogen

Progesterone

Increasing hormone concentration

Months of pregnancy

(a)

Figure 29.10. The completion of implantation. (a) an electron micrograph of an implanted blastocyst in the uterine wall. (b) the formation of the primary germ layers at the end of two weeks demarks the end of the pre-embryonic stage and the beginning of the embryonic stage. Note the formation of extraembryonic membranes at this early period of development.

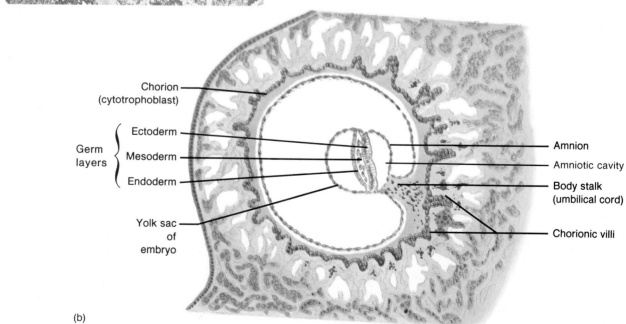

Chorion (cytotrophoblast)

Germ layers {
Ectoderm
Mesoderm
Endoderm
}

Yolk sac of embryo

Amnion
Amniotic cavity
Body stalk (umbilical cord)
Chorionic villi

(b)

Figure 29.11. The body systems and the primary germ layers from which they develop.

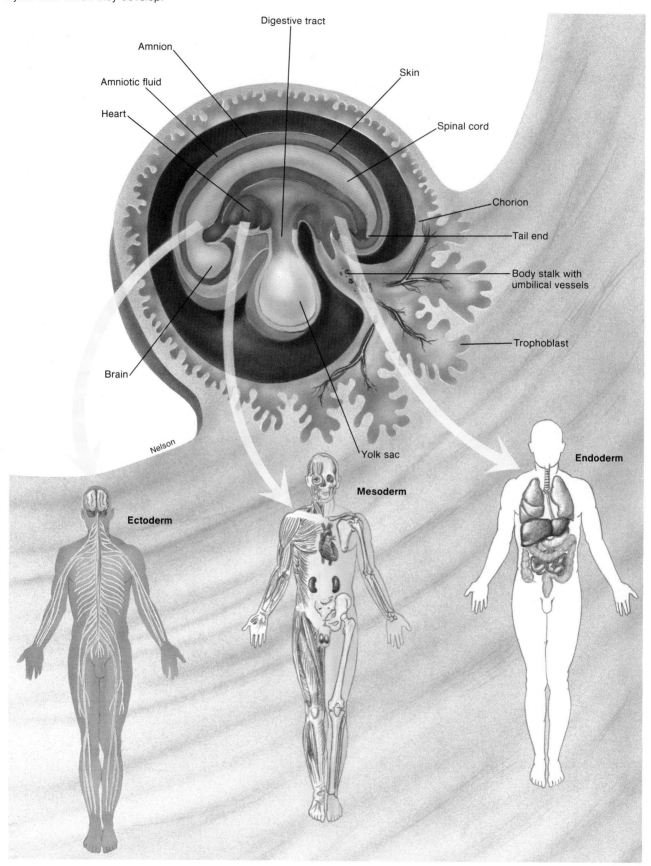

Table 29.1 Morphogenic stages and principal events during pre-embryonic development

Stage	Time period	Principal events
Zygote	Twelve to twenty-four hours following ovulation	Egg is fertilized; zygote has twenty-three pairs of chromosomes (diploid) from haploid sperm and haploid egg; genetically unique
Cleavage	Thirty hours to third day	Mitotic divisions produce increased number of cells
Morula	Third to fourth day	Hollow ball-like structure forms; single layer thick
Blastocyst	Fifth day to end of second week	Inner cell mass and trophoblast form; implantation occurs; embryonic disc forms followed by primary germ layers

Source: Van De Graaff, Kent M., *Human Anatomy.* © 1984 Wm. C. Brown Publishers, Dubuque, Iowa. All Rights Reserved. Reprinted by permission.

Table 29.2 Derivatives of germ layers

Ectoderm	Mesoderm	Endoderm
Epidermis of skin and epidermal derivatives: hair, nails, glands of the skin; linings of oral, nasal, anal, and vaginal cavities	Muscle: smooth, cardiac, and skeletal	Epithelium of pharynx, auditory canal, tonsils, thyroid, parathyroid, thymus, larynx, trachea, lungs, digestive tract, urinary bladder and urethra, and vagina
Nervous tissue; sense organs	Connective tissue: embryonic, connective tissue proper, cartilage, bone, blood	
Lens of eye; enamel of teeth	Dermis of skin; dentin of teeth	Liver and pancreas
Pituitary gland	Epithelium of blood vessels, lymphatic vessels, body cavities, joint cavities	
Adrenal medulla	Internal reproductive organs	
	Kidneys and ureters	
	Adrenal cortex	

Source: Van De Graaff, Kent M., *Human Anatomy.* © 1984 Wm. C. Brown Publishers, Dubuque, Iowa. All Rights Reserved. Reprinted by permission.

layers. Once they are formed, at the end of the second week, the pre-embryonic stage is completed, and the embryonic stage begins.

The primary germ layers are important because various cells and tissues of the body are derived from them. Ectodermal cells form the nervous system; the outer layer of skin (epidermis), including hair, nails, and skin glands; and portions of the sensory organs. Mesodermal cells form the skeleton, muscles, blood, reproductive organs, dermis of the skin, and connective tissue. Endodermal cells produce the lining of the digestive tract, the digestive organs, the respiratory tract and lungs, and the urinary bladder and urethra.

Figure 29.11 illustrates the organs and body systems that derive from each of the three primary germ layers. Table 29.1 summarizes the events of the pre-embryonic stage, and table 29.2 lists the derivatives of the primary germ layers.

The period of prenatal development is referred to as gestation. Normal gestation for humans is nine months. Knowing this and the pattern of menstruation make it possible to determine the delivery date of a baby. In a typical reproductive cycle, a woman ovulates fourteen days prior to the onset of the next menstruation and is fertile for approximately twenty to twenty-four hours following ovulation. Adding nine months, or thirty-eight weeks, to the time of ovulation gives one the estimated delivery date.

1. List the structural characteristics of a zygote, morula, and blastocyst. What is the general time of each of these stages of the pre-embryonic period of development?
2. Discuss the process of implantation, and describe the physiological events that ensure pregnancy.
3. Describe the development of the placenta.
4. List the major structures or organs that derive from each germ layer.
5. What is the length of time of each of the three principal stages of gestation? How is the parturition date determined?

Embryonic Stage

The events of the six-week embryonic stage include the differentiation of the germ layers into specific body organs and the formation of the placenta, the umbilical cord, and the extraembryonic membranes that provide sustenance and protection to the embryo.

Objective 7. Define the term *embryo,* give the time duration of the embryonic stage, and describe the major events of this period of development.

Objective 8. List the embryonic needs that must be met to avoid a spontaneous abortion.

Objective 9. Describe the structure and function of each of the extraembryonic membranes.

Objective 10. Describe the development and function of the placenta and the umbilical cord.

The embryonic stage lasts from the beginning of the third week to the end of the eighth week. At this stage, the developing organism can correctly be called an **embryo** *(em'bre-o)*. The science of *embryology* is concerned with development from conception through the eighth week. During the embryonic stage all of the body organs form, as well as the placenta, umbilical cord, and extraembryonic membranes. The term **conceptus** refers to the embryo, or fetus, and all of the extraembryonic structures—the products of conception. During **parturition** *(par"tu-rish'un)*, childbirth, the baby and the remaining portion of the conceptus are expelled from the uterus.

> Embryology is the study of the sequential changes in an organism as the various tissues, organs, and systems develop. Chick embryos are frequently studied because of the easy access through the shell and their rapid development. Mice and pig embryos are also extensively studied as mammalian models. Genetic manipulation, induction of drugs, exposure to disease, radioactive tagging or dyeing of developing tissues, and X-ray treatments are some of the commonly conducted experiments that provide information that can be applied to human development and birth defects.

During the pre-embryonic stage of cell divisions and differentiation, the developing structure is self-sustaining. The embryo, however, is not self-supporting and must derive sustenance from the mother. For morphogenesis to continue, certain immediate needs must be met, which include the (1) formation of a vascular connection between the uterus of the mother and the embryo so that nutrients and respiratory gases can be provided and metabolic wastes removed; (2) establishment of a constant, protective environment around the embryo, which is conducive to development; (3) establishment of a structural organization for embryonic morphogenesis along a longitudinal axis; (4) ensurance of structural support for the embryo both internally and externally; and (5) coordination of the morphogenic events through genetic expression. If these needs are not met, a spontaneous abortion will generally occur.

The first two of the aforementioned needs are provided by extraembryonic structures, whereas the other three are provided intraembryonically. The extraembryonic membranes, the placenta, and the umbilical cord will be discussed separately from the development of the embryo.

> Most serious developmental defects cause the embryo to be naturally aborted. About 25% of early aborted embryos have chromosomal abnormalities. Other abortions may be caused by environmental factors (infectious agents and teratogenic drugs). For example, an implanting, developing embryo is regarded as foreign tissue by the mother and is rejected and aborted unless maternal immune responses are suppressed.

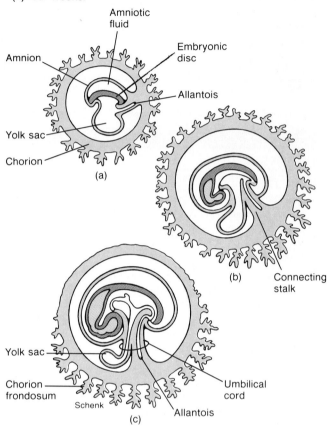

Figure 29.12. The formation of the extraembryonic membranes during a single week of rapid embryonic development: (*a*) three weeks; (*b*) three and one-half weeks; (*c*) four weeks.

Extraembryonic Membranes

While the many intraembryonic events are forming the body organs, a complex system of extraembryonic membranes is developing as well (fig. 29.12). The extraembryonic membranes are the amnion, yolk sac, allantois, and chorion. These membranes are responsible for the protection, respiration, excretion, and nutrition of the embryo and subsequent fetus. At parturition, the placenta, umbilical cord, and extraembryonic membranes separate from the fetus and are expelled from the uterus as the *afterbirth*.

Amnion. The amnion *(am'ne-on)* is a thin extraembryonic membrane, derived from ectoderm, which loosely envelops the embryo, forming an **amniotic sac**, filled with *amniotic fluid* (fig. 29.13). In later stages of fetal development, the amnion expands to come in contact with the chorion. Amniotic development is initiated during early embryonic development, at which time its margin is attached around the free edge of the embryonic disc (fig. 29.12). As the amniotic sac enlarges during the late embryonic period (about eight weeks), the amnion gradually sheaths the developing umbilical cord with an epithelial covering (see fig. 29.14).

As a buoyant medium, amniotic fluid performs four functions for the embryo and subsequent fetus: (1) it permits symmetrical structural development and growth;

Figure 29.13. An implanted embryo at approximately four and one-half weeks. (*a*) the interior of a uterus showing the implantation site and elevated *decidua capsularis* caused by the expanded chorion. (*b*) the developing embryo, extraembryonic membranes, and the formation of the placenta.

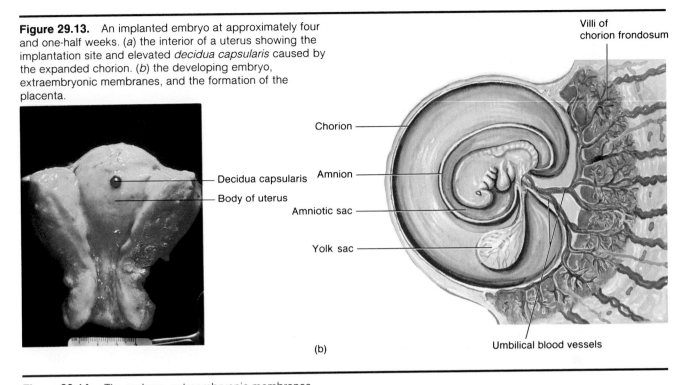

Villi of chorion frondosum

Chorion

Amnion

Amniotic sac

Yolk sac

Decidua capsularis

Body of uterus

Umbilical blood vessels

(b)

Figure 29.14. The embryo, extraembryonic membranes, and placenta at approximately seven weeks of development. Note that at this time, the amnion and chorion are adherent and are frequently referred to as the *amniochorionic membrane*. Blood from the embryo is carried to and from the chorion frondosum by umbilical arteries and vein. The maternal tissue between the chorionic villi is known as the decidua basalis, and this tissue, together with the villi, form the functioning placenta.

Decidua basalis

Chorion frondosum

Umbilical cord

Chorion

Amnion

Amniotic sac containing amniotic fluid

Placenta

Beck

Figure 29.15. Amniocentesis. In this procedure amniotic fluid, together with suspended cells, is withdrawn for examination. Various genetic diseases can be detected prenatally by this means.

Uterus

Amniotic sac

Placenta

Beck

(2) it cushions and protects by absorbing jolts that the mother may receive; (3) it helps to maintain consistent pressure and temperature; and (4) it enables freedom of fetal development, which is important for musculoskeletal development and blood flow.

Amniotic fluid is formed initially as an isotonic fluid absorbed from the maternal blood in the endometrium surrounding the developing embryo. Later, the volume is increased and the concentration changed by urine excreted from the fetus into the amniotic sac. Amniotic fluid also contains cells that are sloughed off from the fetus, placenta, and amniotic sac. Since all of these cells are derived from the same fertilized egg, all have the same genetic composition. Many genetic abnormalities can be detected by aspirating this fluid and examining the cells thus obtained in a procedure called *amniocentesis (am''ne-o-sen-te'sis)*.

Amniotic fluid is normally swallowed by the fetus and absorbed in the gastrointestinal tract. Prior to delivery,

the amnion is naturally or surgically ruptured, and the amniotic fluid ("bag of waters") is released.

Amniocentesis (fig. 29.15) is usually performed at the fourteenth or fifteenth weeks of pregnancy, when the amniotic sac contains 175–225 ml of fluid. Genetic diseases such as Down's syndrome (where there are three instead of two number–21 chromosomes) can be detected by examining chromosomes; diseases such as Tay-Sachs disease, in which there is a defective enzyme involved in formation of myelin sheaths, can be detected by biochemical techniques.

Yolk Sac. The yolk sac is established during the end of the second week as cells from the trophoblast form a thin *exocoelomic (ek''so-se-lo'mik) membrane.* Unlike many vertebrates, the human yolk sac contains no nutritive yolk

Down's syndrome: from John L. H. Down, English physician, 1828–96

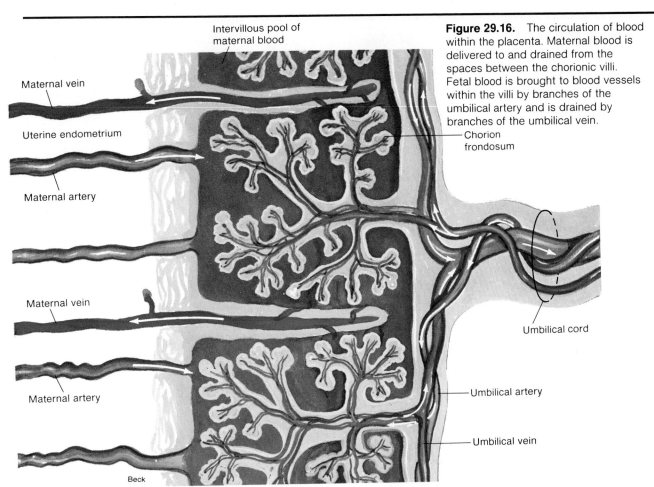

Intervillous pool of
maternal blood

Maternal vein

Uterine endometrium

Maternal artery

Maternal vein

Maternal artery

Beck

Figure 29.16. The circulation of blood within the placenta. Maternal blood is delivered to and drained from the spaces between the chorionic villi. Fetal blood is brought to blood vessels within the villi by branches of the umbilical artery and is drained by branches of the umbilical vein.

Chorion
frondosum

Umbilical cord

Umbilical artery

Umbilical vein

but is an essential structure during early embryonic development. It is attached to the underside of the embryonic disc (fig. 29.13), where it produces blood for the embryo until the liver forms during the sixth week. The dorsal portion of the yolk sac is involved in the formation of the primitive gut. In addition, primordial germ cells form in the wall of the yolk sac and migrate during the fourth week to the developing gonads where they become the primitive germ cells (spermatogonia or oogonia).

The stalk of the yolk sac usually detaches from the gut by the sixth week. Following this, the yolk sac gradually shrinks as pregnancy advances. Eventually it becomes very small and serves no additional developmental functions.

Allantois. The allantois forms during the third week as a small outpouching, or diverticulum, from the caudal wall of the yolk sac (see fig. 29.12). The allantois remains small but is involved in the formation of blood cells and gives rise to the fetal umbilical arteries and vein. It also contributes to the development of the urinary bladder.

The extraembryonic portion of the allantois degenerates during the second month. The intraembryonic portion involutes to form a thick urinary tube called the **urachus.** After birth, the urachus becomes a fibrous cord, called the median umbilical ligament, which supports the urinary bladder.

Chorion. The chorion is a highly specialized extraembryonic membrane that participates in the formation of the placenta (see fig. 29.12). It is the outermost membrane and originates from the trophoblast of the blastocyst. Numerous, small, fingerlike extensions, called **villi** (see fig. 29.13), form from the chorion and penetrate deeply into the uterine tissue. Initially, the entire surface of the chorion is covered with villi. But those villi on the surface toward the uterine cavity gradually degenerate and produce a smooth, bare area known as the **smooth chorion.** As this occurs, the villi associated with the uterine wall rapidly increase in number and branch out. This portion of the chorion is known as the **villous chorion.** The villous chorion becomes highly vascular, and as the embryonic heart begins to function, blood is pumped in close proximity to vascular sinuses of the uterine wall.

Placenta

The **placenta** *(plah-sen'tah)* is a vascular structure by which an unborn child is attached to its mother's uterine wall and through which metabolic exchange occurs (fig. 29.16). The placenta is formed in part from maternal tissue and in part from embryonic tissue. The embryonic portion

allantois: Gk. *allanto,* sausage; *iodos,* resemblance
chorion: Gk. *chorion,* external fetal membrane
placenta: L. *placenta,* a flat cake
villous: L. *villus,* tuft of hair

Table 29.3 Hormones secreted by placenta	
Hormones	**Effects**
Pituitary-like hormones	
Chorionic gonadotrophin (hCG)	Similar to LH; maintains mother's corpus luteum for first 5½ weeks of pregnancy; may be involved in suppressing immunological rejection of embryo
Placental lactogen (hPL)	Similar to prolactin; synergizes with growth hormone to promote, in the mother, increased fat breakdown and fatty acid release from adipose tissue, and to promote the sparing of glucose use by maternal tissues ("diabetic-like" effects)
Chorionic thyrotrophin (hCT)	Similar to TSH; physiological significance uncertain
Sex steroids	
Progesterone	Helps maintain endometrium during pregnancy; helps suppress gonadotrophin secretion; promotes uterine sensitivity to oxytocin; helps stimulate mammary gland development
Estrogens	Helps maintain endometrium during pregnancy; helps suppress gonadotrophin secretion; helps stimulate mammary gland development; inhibits prolactin secretion

Source: Fox, Stuart Ira, *Human Physiology.* © 1984 Wm. C. Brown Publishers, Dubuque, Iowa. All Rights Reserved. Reprinted by permission.

of the placenta consists of the villi of the **chorion frondosum,** whereas the maternal portion is composed of the area of the uterine wall called the **decidua basalis** (see fig. 29.14), to which the villi penetrate. Blood does not flow directly between these two portions, but because of the close membranous proximity, certain substances diffuse readily.

When fully formed, the placenta is a reddish brown oval disc with a diameter of 15–20 cm (8 in.) and a thickness of 2.5 cm (1 in.). It weighs 500–600 gm, being about one-sixth the weight of the fetus.

Exchange of Molecules across the Placenta. The **umbilical artery** delivers fetal blood to vessels within the villi of the chorion frondosum of the placenta. This blood circulates within the villi and returns to the fetus via the **umbilical vein.** Maternal blood is delivered to and drained from the cavities within the decidua basalis, which are located between the chorionic villi. In this way, maternal and fetal blood are brought close together but never mix within the placenta.

The placenta serves as a site for the exchange of gases and other molecules between the maternal and fetal blood. Oxygen diffuses from mother to fetus, and carbon dioxide diffuses in the opposite direction. Nutrient molecules and waste products likewise pass between maternal and fetal blood.

The placenta is not merely a passive conduit for exchange between maternal and fetal blood, however. It has a very high metabolic rate, utilizing about a third of all the oxygen and glucose supplied by the maternal blood. The rate of protein synthesis is, in fact, higher in the placenta than it is in the liver. Like the liver, the placenta produces a great variety of enzymes capable of converting biologically active molecules (such as hormones and drugs) into less active, more water-soluble forms. In this way, potentially dangerous molecules in the maternal blood are often prevented from harming the fetus.

Most drugs ingested by a pregnant woman can readily pass through the placenta to the developing baby. Many drugs that are not particularly harmful to the mother may be deleterious to her child. For example, the nicotine taken in by a chain-smoking pregnant woman will stunt the growth of her fetus. Hard drugs, such as heroin, can lead to fetal drug addiction. Depressant drugs given to a mother during labor can readily cross the placenta and, if given in high dosages, can cause respiratory depression in the newborn infant.

Although the placenta is an effective barrier against diseases of bacterial origin, viruses such as rubella and certain blood-borne diseases such as syphilis can diffuse through the placenta and affect the fetus. During parturition (childbirth), small amounts of fetal blood may pass across the placenta to the mother. If the fetus is Rh positive and the mother Rh negative, the antigens of the fetal red blood cells elicit an antibody response in the mother. In a subsequent pregnancy, the maternal antibodies then cross the placenta and cause a breakdown of fetal red blood cells, a condition called **erythroblastosis fetalis.**

Endocrine Functions of the Placenta. The placenta secretes both steroid hormones and glycoprotein hormones that have actions similar to those of some anterior pituitary hormones. This latter category of hormones includes chorionic gonadotrophin (hCG), placental lactogen (hPL), and chorionic thyrotrophin (hCT).

The importance of **chorionic gonadotrophin** in maintaining the mother's corpus luteum for the first five and a half weeks of pregnancy has been previously discussed (table 29.3). There is also some evidence that hCG may in some way help to prevent immunological rejection of the implanting embryo. **Placental lactogen** acts together (synergizes with) growth hormone from the mother's pituitary to produce a "diabetic-like" effect in the pregnant woman. The effects of these two hormones stimulate (1) lipolysis and increased plasma fatty acid concentration; (2) decreased maternal utilization of glucose and, therefore, increased blood glucose concentrations; and

Figure 29.17. The secretion of progesterone and estrogen from the placenta requires a supply of cholesterol from the mother's blood and the cooperation of fetal enzymes that convert progesterone to androgens.

(3) polyuria (excretion of large volumes of urine), thereby producing a degree of dehydration and thirst. This "diabetic-like" effect in the mother helps to spare glucose for the placenta and fetus that, like the brain, use glucose as their primary energy source.

After the first five and a half weeks of pregnancy, when the corpus luteum regresses, the placenta becomes the major sex-steroid-producing gland. The blood concentration of estrogens, as a result of placental secretion, rises to levels more than one hundred times greater than those existing at the beginning of pregnancy. The placenta also secretes large amounts of progesterone, changing the estrogen-progesterone ratio in the blood from 1:100 at the beginning of pregnancy to a ratio of close to 1:1 towards full-term.

The placenta, however, is an "incomplete endocrine gland" because it cannot produce estrogen and progesterone without the aid of precursors supplied to it by both the mother and the fetus (fig. 29.17). The placenta, for example, cannot produce cholesterol from acetate and so must be supplied with cholesterol from the mother's circulation. Cholesterol, which is a steroid containing twenty-seven carbons, can then be converted by enzymes in the placenta into steroids that contain twenty-one carbons—such as progesterone. The placenta, however, lacks the enzymes needed to convert progesterone into androgens (which have nineteen carbons) and estrogens (which have eighteen carbons).

In order for the placenta to produce estrogens it needs to cooperate with steroid-producing tissues in the fetus. The fetus and the placenta, therefore, form a single functioning system in terms of steroid hormone production. This system has been called the *fetal-placental unit* (fig. 29.17).

The ability of the placenta to convert androgens into estrogen helps to protect the female embryo from becoming masculinized by the androgens secreted from the mother's adrenal glands. In addition to producing estradiol, the placenta secretes large amounts of a weak estrogen called *estriol*. The production of estriol increases

tenfold during pregnancy, so that by the third trimester estriol accounts for about 90% of the estrogens excreted in the mother's urine. Since almost all of this estriol comes from the placenta (rather than from maternal tissues), measurements of urinary estriol can be used clinically to assess the health of the placenta.

Umbilical Cord

The **umbilical** *(um-bil′ĭ-kal)* **cord** forms as the yolk sac shrinks and the amnion expands to envelop the tissues on the underside of the embryo. Figure 29.18 illustrates the formation of the umbilical cord.

The umbilical cord usually attaches near the center of the placenta. When fully formed, the umbilical cord is about 1–2 cm (0.5–1 in.) in diameter and approximately 55 cm (2 ft) in length. The umbilical cord contains two **umbilical arteries,** which carry deoxygenated blood toward the placenta and one **umbilical vein,** which carries oxygenated blood from the placenta to the embryo. These vessels are surrounded by embryonic connective tissue called *Wharton's jelly.*

> The umbilical cord has natural twists because the umbilical vein is longer than the arteries. In about one-fifth of all deliveries, the cord is looped once around the baby's neck. If drawn tightly, the cord may cause death or serious perinatal problems.

Structural Changes of the Embryo by Weeks

Third Week. Early in the third week a thick linear band, known as the **primitive streak,** appears along the dorsal midline of the embryonic disc. The primitive streak is derived from the mesodermal cells, which are specialized from the ectodermal layer of the embryonic mass. The primitive streak establishes a structural foundation for embryonic morphogenesis along a longitudinal axis. As the primitive streak elongates, a prominent thickening

---

Wharton's jelly: from Thomas Wharton, English anatomist, 1614–73

Figure 29.18. The formation of the umbilical cord and other extraembryonic structures as seen in sagittal sections of the gravid uterus from the fourth to the twenty-second week. (a) a connecting stalk forms as the developing amnion expands around the embryo and finally meets ventrally. (b) the umbilical cord begins to take form as the amnion ensheathes the yolk sac. (c) a cross section of the umbilical cord showing the embryonic vessels, Wharton's jelly, and the tubular connection to the yolk sac. (d) by the twenty-second week, the amnion and chorion are fused, and the umbilical cord and placenta are well-developed structures. (e) 16-week-old fetus.

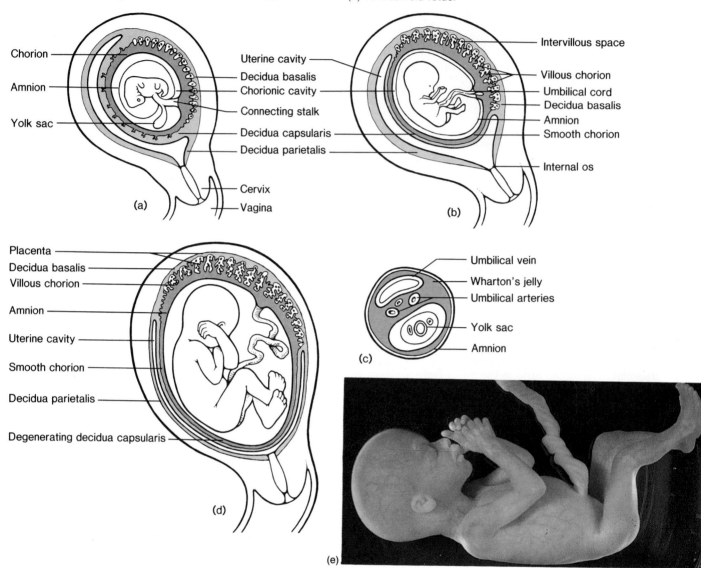

(a)
Chorion
Amnion
Yolk sac
Uterine cavity
Decidua basalis
Chorionic cavity
Connecting stalk
Decidua capsularis
Decidua parietalis
Cervix
Vagina

(b)
Intervillous space
Villous chorion
Umbilical cord
Decidua basalis
Amnion
Smooth chorion
Internal os

(c)
Umbilical vein
Wharton's jelly
Umbilical arteries
Yolk sac
Amnion

(d)
Placenta
Decidua basalis
Villous chorion
Amnion
Uterine cavity
Smooth chorion
Decidua parietalis
Degenerating decidua capsularis

(e)

known as the **primitive knot** appears as its cranial end (fig. 29.19). The primitive knot later forms the mesodermal structures of the head and the **notochord.** To give support to the embryo the notochord forms a midline axis which is the basis of the human skeleton. The primitive streak also gives rise to loose embryonic connective tissue called **mesenchyme,** which differentiates into all the various kinds of connective tissue found in the adult. One of the earliest formed organs is the skin, which develops to support and maintain homeostasis within the embryo.

A tremendous amount of change and specialization occurs during the embryonic stage. The factors that cause precise, sequential change from one cell or tissue type to another are not fully understood. We do know, however,

that the potential for change is programmed into the genetics of each cell and that under conducive environmental conditions these characteristics become expressed. The process of developmental change is referred to as **induction.** Induction occurs when one tissue, called the **inductor tissue,** has a marked effect on an adjacent **induced tissue** and stimulates it to differentiate.

Fourth Week. During the fourth week of development, the embryo is about 5 mm (0.2 in.) in length. A **connecting stalk,** which is later involved in the formation of the umbilical cord, is established from the body of the embryo to the developing placenta (fig 29.20). By this time,

induction: L. *inductus,* to lead in

Figure 29.19. The appearance of the primitive streak and primitive knot along the embryonic disc. These progressive changes occur through the process of induction.

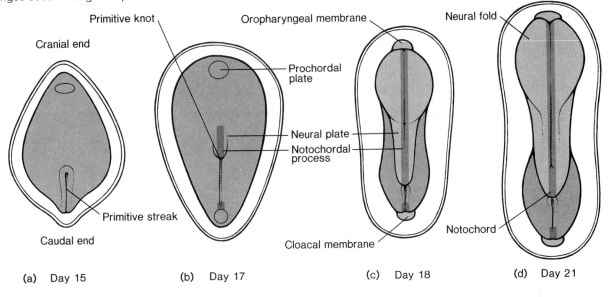

(a)   Day 15

(b)   Day 17

(c)   Day 18

(d)   Day 21

Figure 29.20. Photograph of four-week-old embryo; progressive illustrations of four-week-old embryos.

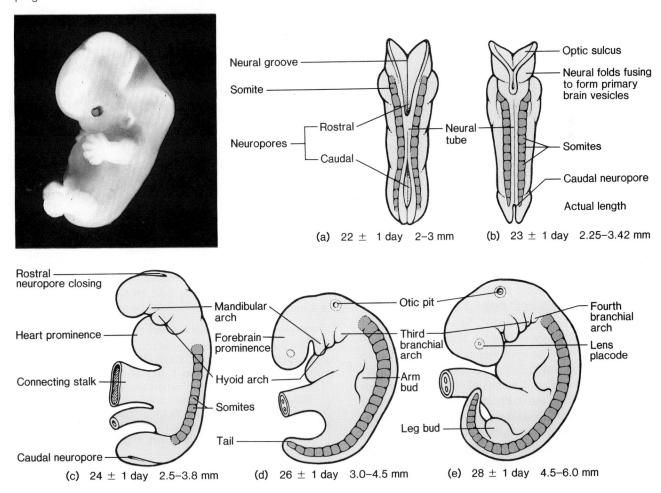

(a) 22 ± 1 day   2–3 mm

(b) 23 ± 1 day   2.25–3.42 mm

(c) 24 ± 1 day   2.5–3.8 mm

(d) 26 ± 1 day   3.0–4.5 mm

(e) 28 ± 1 day   4.5–6.0 mm

Figure 29.21. A diagram of a five-week-old embryo. Note the digital rays and the developing eyes and ears.

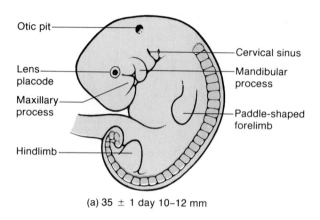

(a) 35 ± 1 day 10–12 mm

Otic pit
Lens placode
Maxillary process
Hindlimb
Cervical sinus
Mandibular process
Paddle-shaped forelimb

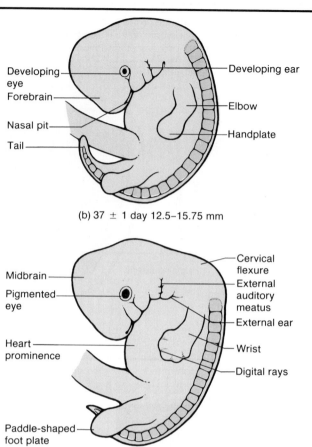

Developing eye
Forebrain
Nasal pit
Tail
Developing ear
Elbow
Handplate

(b) 37 ± 1 day 12.5–15.75 mm

Midbrain
Pigmented eye
Heart prominence
Paddle-shaped foot plate
Cervical flexure
External auditory meatus
External ear
Wrist
Digital rays

(c) 40 ± 1 day 16.0–21.0 mm

the heart is beating firmly to pump blood to all parts of the embryo. The head and jaws are apparent, and the primordial tissue that will form the eyes, brain, spinal cord, lungs, and digestive organs has developed. The **arm** and **leg buds** are recognizable as small swellings on the lateral body walls.

Fifth Week. Embryonic changes during the fifth week are not as extensive as those during the fourth week. The head enlarges, and the developing eyes, ears, and nasal pit are obvious (fig. 29.21). The appendages have formed from the limb buds, and paddle-shaped hand plates develop digital ridges called **finger** or **digital rays.**

Sixth Week. During the sixth week, the embryo is 14–24 mm (0.75 in.) long. The head is much larger relative to the trunk, and the brain has undergone marked differentiation. This is the most vulnerable period of development for many organs. An interruption at this critical time can easily cause congenital damage. The limbs undergo considerable change during this week. The forelimbs are lengthened, slightly flexed, and notches appear between the rays in the hand and foot plates.

Seventh and Eighth Weeks. During the last two weeks of the embryonic stage, the embryo, which is now 25–30 mm (1–1.25 in.) long, has distinct human characteristics (fig. 29.22). The body organs are formed, and the nervous system begins coordinating body activity. The neck region is apparent, and the abdomen is less protuberant. The eyes are well developed, but the lids are stuck together to protect against probing fingers during muscular movement. The nostrils are developed but plugged with mucus. The external genitalia are forming but are still undifferentiated. The body systems are developed by the end of the eighth week, and from this time on the embryo is called a **fetus** *(fe'tus).*

The most precarious time of prenatal development is during the embryonic stage when there is much tissue differentiation and organ formation. Frequently, however, a woman does not even realize that she is pregnant until she is well into this period of development. For this reason, a woman should consistently take good care of herself if there is even a chance that she might become pregnant.

Figure 29.22. Diagrams of six- and seven-week-old embryos.

(a) 45 ± 1 day 22–24 mm

Cervical flexure
External ear
Notches between digital rays
Toe rays

Ear
Eyelid
Webbed fingers
Early umbilical herniation
Notches between toe rays

(b) 49 ± 1 day 28–30 mm

Fingers separated
Fan-shaped webbed toes

(c) 52 ± 1 day 32–34 mm

Toes separated

(d) 56 ± 1 day 34–40 mm

1. *Distinguish between the terms* embryo *and* fetus. *Briefly summarize the structural changes that an embryo undergoes between the fourth and eighth weeks of development.*
2. *Describe the origin of hCG, and explain why it is needed to maintain pregnancy for the first ten weeks.*
3. *What are the five embryonic needs that must be met in order to avoid spontaneous abortion? Which of these needs are met by the extraembryonic membranes?*
4. *Explain the formation and function of each of the extraembryonic membranes.*
5. *Name the fetal and maternal components of the placenta, and describe the circulation in these two components. Explain how fetal and maternal gas exchange occurs.*
6. *List the names and major functions of the glycoprotein hormones and sex steroids secreted by the placenta.*

Fetal Stage

The fetal stage lasts from nine weeks until birth and is characterized by tremendous growth and the specialization of body structures.

> *Objective 11.* Define the term *fetus,* and discuss the major events of the fetal stage of development.
>
> *Objective 12.* Describe the various techniques available for fetal examination or monitoring fetal activity.

Since most of the tissues and organs of the body appear during the embryonic period, the **fetus** is recognizable as a human being at nine weeks and is far less vulnerable than the embryo to deformation from viruses, drugs, and radiation. A small amount of tissue differentiation and organ development still occurs during the fetal stage, but for the most part fetal development is primarily limited to body growth. Figure 29.23 summarizes development

Figure 29.23. Major changes in fetal development from the third week following conception through the tenth week.

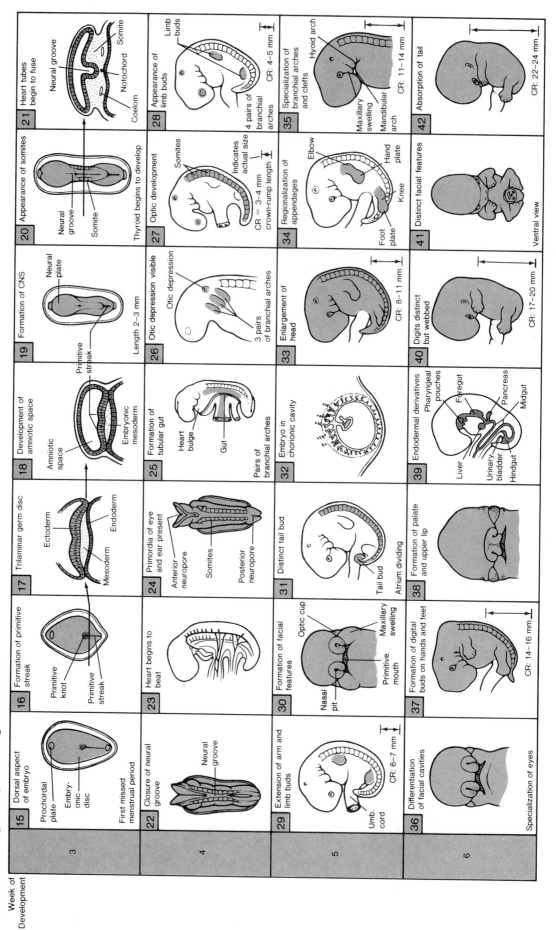

Week of Development						
7	**43** Formation of extraembryonic membranes Decidua capsularis — Uterus — Lumen of uterus — Placenta — Yolk sac	**44** Formation of eyelids	**45** Distinction of fingers CR = 23–25 mm	**46** Ossification centers begin to form Appearance of nipples	**47** Distinction of toes Toes CR = 26–27 mm	**48** Formation of undifferentiated external genitalia Genital tubercle — Urogenital membrane — Anal membrane ♀ or ♂	**49** Well established placenta Placenta — Amniotic cavity
8	**50** Trunk elongating and straightening CR = 28–30 mm	**51** Marked development of brain	**52** Anal membrane perforated	**53** Well-developed facial features	**54** Distinction between anus and genitalia Genital tubercle — Urethral groove — Anus ♀ or ♂ Testes and ovaries distinguishable	**55** Formation of dental buds for teeth All principal body organs are formed	**56** Delineation of appendicular muscles
9	**57** Beginning of fetal stage CR = 34–40 mm	**58** Position of fetus within uterus	**59** Distinct nails on digits	**60** Differentiation of female genitalia Phallus — Urogenital fold — Labioscrotal fold — Perineum	**61** Differentiation of male genitalia Phallus — Urogenital fold — Labioscrotal fold — Perineum	**62** Development of accessory reproductive organs within male	**63** Development of the cerebellum
10	**64** Refined facial profile	**65** Differentiation of inner and outer layers of the eye	**66** Enlargement of cerebrum	**67** Further differentiation of female genitalia Clitoris — Labium minus — Urogenital groove — Labium majus	**68** Further differentiation of male genitalia Glans penis — Urethral groove — Scrotum Fusion of urethral groove	**69** Developing ear bones	**70** Accelerated growth in body length CR = 45–50 mm

from the third week following conception through the tenth week, and figure 29.24 depicts changes in external appearance of the fetus from the eleventh through the thirty-eighth week. A discussion of the structural changes of the fetus by weeks follows.

Nine to Twelve Weeks.   At the beginning of the ninth week, the head is as large as the rest of the body. The eyes are widely spaced, and the ears are set low. Head growth slows during the next three weeks, whereas growth in body length accelerates. Ossification centers appear in most bones during the ninth week. Differentiation of the external genitalia becomes apparent at the end of the ninth week, but the genitalia are not developed to the point of sex determination until the twelfth week. By the end of the twelfth week, the fetus is 87 mm (3.5 in.) long and weighs about 45 g (1.6 oz). It can swallow, digest the fluid that passes through its system, and defecate and urinate into the amniotic fluid. The nervous system and muscle coordination are developed enough so that the fetus will withdraw its leg if tickled. The fetus begins inhaling through its nose but can take in only amniotic fluid.

> Major structural abnormalities, which may not be predictable from genetic analysis, can often be detected by **ultrasound** (see fig. 29.35). Sound wave vibrations are reflected from the interface of tissues with different densities—such as the interface between the fetus and amniotic fluid—and used to produce an image. This technique is so sensitive that it can be used to detect a fetal heartbeat several weeks before it can be detected by a stethoscope.

Thirteen to Sixteen Weeks.   During the period from thirteen through sixteen weeks, the facial features are well formed, and epidermal structures such as eyelashes, eyebrows, hair on the head, fingernails, and nipples begin to develop. The appendages lengthen, and by the sixteenth week the skeleton is sufficiently developed so that it shows clearly on X-ray films. During the sixteenth week, the fetal heartbeat can be heard by applying a stethoscope to the mother's abdomen. By the end of the sixteenth week, the fetus is 140 mm in length (5.5 in.) and weighs about 200 g (7 oz).

> After the sixteenth week, fetal length can be determined using X-ray films. The reported length of a fetus is generally derived from a straight line measurement from the crown of the head to the developing ischium (crown-rump length). Measurements made on an embryo prior to the fetal stage, however, are not reported as crown-rump measurements but as total length.

Seventeen to Twenty Weeks.   During the period from seventeen to twenty weeks, the legs achieve their final relative proportions, and fetal movements, known as **quickening,** are commonly felt by the mother. The skin is covered with a white, cheeselike material known as **vernix caseosa.** It consists of fatty secretions from the sebaceous glands and dead epidermal cells. The function of vernix caseosa is to protect the fetus while it is bathed in amniotic fluid. Twenty-week-old fetuses usually have fine, silklike fetal hair, called **lanugo** *(lah-nu'go),* covering the skin. Lanugo is thought to hold the vernix caseosa on the skin and produce a ciliarylike motion that moves amniotic fluid. The length of a twenty-week-old fetus is about 190 mm (7.5 in.) and the weight is 460 gm (16 oz). Because of cramped space the fetus develops a marked spinal flexure and is in what is commonly called the "fetal posture" with the head bent down in contact with the flexed knees.

Twenty-one to Twenty-five Weeks.   During the period from twenty-one to twenty-five weeks, the fetus increases its weight substantially to about 900 gm (32 oz). Body length increases only moderately (240 mm), however, so the weight is evenly proportioned. The skin is quite wrinkled and is translucent pinkish in color because the blood flowing in the capillaries is now visible.

Twenty-six to Twenty-nine Weeks.   Toward the end of the period from twenty-six to twenty-nine weeks, the fetus will be 275 mm (about 11 in.) in length and will weigh 1,300 gm (46 oz). A fetus might now survive if born prematurely, but the mortality rate is high. Its body metabolism cannot yet maintain a constant temperature, and the respiratory muscles have not matured enough to provide a regular respiratory rate. If, however, the premature infant is put in an incubator and a respirator is used to maintain its breathing, it may survive. The eyes open during this period, and the body is well covered with lanugo. If the fetus is a male, the testes should have begun descent into the scrotum. As the time of birth approaches, the fetus rotates to a **vertex position.** The head repositions toward the cervix because of the shape of the uterus and because the head is the heaviest part of the body.

Thirty to Thirty-eight Weeks.   By the end of thirty-eight weeks, the fetus is considered "full-term." It has reached a crown-rump length of 360 mm (14 in.) and weighs 3,400 gm (7.5 lbs). The average total length from crown to heel is 530 mm (21 in.). Most fetuses are plump with smooth skin because of the accumulation of subcutaneous fat. The skin is pinkish blue in color even on fetuses of dark-skinned parents, because melanocytes do not produce melanin until the skin is exposed to sunlight. Lanugo hair is sparse and is generally found on the head and back. The chest is prominent, and the mammary area protrudes in both sexes. The external genitalia are somewhat swollen.

---

vernix caseosa: L. *varnish;* L. *caseus,* cheese
lanugo: L. *lana,* wool
vertex: L. *vertex,* summit

Figure 29.24.   Changes in the external appearance of the fetus from the eleventh through the thirty-eighth week.

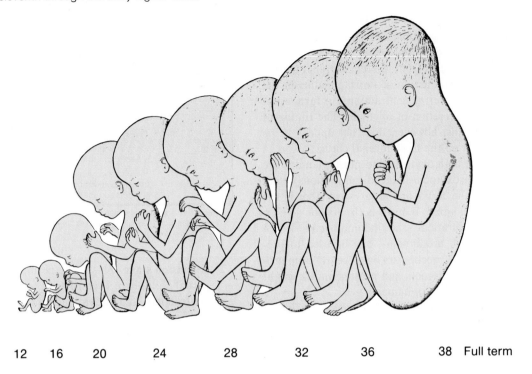

| 11 | 12 | 16 | 20 | 24 | 28 | 32 | 36 | 38 | Full term |

---

1. *Explain why the ninth week is designated as the beginning of the fetal stage of development.*
2. *List the approximate fetal age that each of the following occur: first detection of fetal heartbeat; presence of vernix caseosa and lanugo; fetal rotation into vertex position.*
3. *Compare the techniques of ultrasound and X rays in examining fetal development and structure.*

Labor and Parturition

Labor and parturition are the culmination of gestation and require the action of oxytocin from the posterior pituitary and prostaglandins, produced in the uterus.

> *Objective 13.* Describe the hormonal action that controls labor and parturition.
>
> *Objective 14.* Describe the three stages of labor.

The time of prenatal development, or the time of pregnancy, is called **gestation** *(jes-ta'shun).* The human gestational period is usually 266 days or about 280 days from the beginning of the last menstrual period to **parturition,** or birth. Most fetuses are born within 10 to 15 days before or after this time. Parturition is accompanied by a sequence of physiological and physical events called **labor.**

The *onset of labor* is denoted by rhythmic and forceful contractions of the myometrium layer of the uterus. In *true labor,* the pains from uterine contractions occur at regular intervals and intensify as the time between contractions shorten. A reliable indication of true labor is dilation of the cervix and a "show," or discharge of blood-containing mucus in the cervical canal and vagina. In *false labor,* abdominal pain is felt at irregular intervals, and there is a lack of cervical dilations and cervical "show."

The uterine contractions of labor are stimulated by two agents: (1) **oxytocin** *(ok"si-to'sin),* a polypeptide hormone produced in the hypothalamus and secreted by the posterior pituitary, and (2) **prostaglandins** *(pros"tah-glan'dins),* a class of fatty acids produced within the uterus itself. Labor can indeed be induced artificially by injections of oxytocin or by insertion of prostaglandins into the vagina as a suppository.

The hormone **relaxin,** produced by the corpus luteum, may also be involved in labor and parturition. Relaxin is known to soften the symphysis pubis in preparation for parturition and is thought to also soften the cervix in preparation for dilation. It may be that relaxin does not affect the uterus, but rather this function may in fact be carried out by progesterone and estradiol. Further research is necessary to understand the total physiological effect of these hormones.

---

gestation: L. *gestatus,* to bear

Although the mechanisms operating during labor are well understood, the factors that lead to the initiation of parturition in humans remain unclear. In some lower mammals, the concentrations of estrogen and progesterone drop prior to parturition; this does not appear to be true in humans. In sheep and goats, labor seems to be initiated by corticosteroids from the fetal adrenal cortex, which stimulate the production of prostaglandins in the uterus. The resulting uterine contractions, in turn, stimulate the increased secretion of oxytocin. The timing of parturition in humans, however, does not appear to be similarly dependent on the fetal adrenal cortex.

It appears that both oxytocin and prostaglandins are required in humans; oxytocin-induced contractions do not lead to dilation of the cervix and progressive labor in the absence of prostaglandins. Indeed, both alcohol—which inhibits oxytocin secretion—and indomethacin—which inhibits prostaglandin production—can be used to prevent pre-term births. A mechanism has recently been proposed to explain how oxytocin and prostaglandins might cooperate to initiate labor in humans.

It has been demonstrated that the concentration of oxytocin receptors increases dramatically in the uterus during gestation. Although the blood concentrations of oxytocin remain constant, the sensitivity of the uterus to oxytocin increases as a result of the increased receptors. The threshold amount of oxytocin required to induce uterine contractions in a woman at term is almost one-hundredth the amount required in a nonpregnant woman. The increase in oxytocin receptors and sensitivity may be one of the effects of the rising levels of estrogens that occur during gestation.

In addition to finding an increased concentration of oxytocin receptors in the myometrium (muscular layer of the uterus), an increased concentration of oxytocin receptors were also found in the non-muscular decidua. It has been proposed that oxytocin (possibly coming from the fetus as well as from the mother) may stimulate prostaglandin production in the decidua. Prostaglandins may then diffuse into the myometrium from the decidua and act together with the increased sensitivity of the myometrium to oxytocin to stimulate the onset of labor. These events are summarized in table 29.4.

Labor is divided into three stages (fig. 29.25).

1. **Dilation stage.** In this period the cervix dilates to a diameter of approximately 10 cm. There are regular contractions during this stage and usually a rupturing of the amniotic sac ("bag of waters"). If the amniotic sac does not rupture spontaneously, it is done surgically. The dilation stage may last eight to twenty-four hours depending on whether it is occurring in the first or subsequent pregnancies.

Table 29.4 Possible sequence of events leading to the onset of labor in humans

Step	Event
1	High estrogen secretion from the placenta stimulates production of oxytocin receptors in the uterus
2	Uterine muscle (myometrium) becomes increasingly sensitive to effects of oxytocin during pregnancy
3	Oxytocin may stimulate production of prostaglandins in the uterus
4	Prostaglandins may stimulate uterine contractions
5	Contractions of the uterus stimulate oxytocin secretion from the posterior pituitary
6	Increased oxytocin secretion stimulates increased uterine contractions, creating a positive feedback loop and resulting in labor

Source: Fox, Stuart Ira, *Human Physiology.* © 1984 Wm. C. Brown Publishers, Dubuque, Iowa. All Rights Reserved. Reprinted by permission.

2. **Expulsion stage.** This is the period of parturition, or actual childbirth. It consists of forceful uterine contractions and abdominal compressions to expel the fetus from the uterus and through the vagina. This stage may require thirty minutes in a first pregnancy, but only a few minutes in subsequent pregnancies.

3. **Placental stage.** Generally within ten to fifteen minutes after parturition, the placenta is separated from the uterine wall and expelled as the *"afterbirth."* Forceful uterine contractions characterize this stage, constricting uterine blood vessels to prevent hemorrhage. In a normal delivery, blood loss does not exceed 350 milliliters.

A *pudendal nerve block* may be administered during the early part of the expulsion stage to ease the trauma of delivery for the mother and make the performance of an episiotomy possible.

Five percent of newborns are born **breech.** In a breech birth, the fetus has not rotated and the buttocks are the presenting part. The principal concern of a breech birth is the increased time and difficulty of the expulsion stage of parturition. Attempts to rotate the fetus through the use of forceps may injure the infant. If an infant cannot be delivered breech, a **cesarean** (*se-sa're-an*) **section** must be performed. A cesarean section is delivery of the fetus through an incision made into the abdominal wall and the uterus.

1. Distinguish between labor and parturition.
2. Describe the hormonal mechanisms responsible for labor.
3. Explain two techniques for inducing labor.
4. Describe the three stages of labor, and list the relative time duration of each.

Figure 29.25. The stages of labor and parturition. (*a*) the position of the fetus prior to labor. (*b*) the ruptured amniotic sac and early dilation of the cervix. (*c*) expulsion stage, or the period of parturition. (*d*) the placental stage.

(a)

Placenta

Symphysis pubis

Urinary bladder

Urethra

Vagina

Cervix

Rectum

(b)

Ruptured amniotic sac

(c)

Placenta

(d)

Uterus

Umbilical cord

Placenta

Schenk

Inheritance

Inheritance is the passage of hereditary traits carried on the genes of chromosomes from one generation to another.

Objective 15. Define the term *genetics.*

Objective 16. Discuss the variables that account for a person's phenotype.

Objective 17. Explain how probability is involved in predicting inheritance, and use a Punnett square to illustrate selected probabilities.

Genetics is the branch of biology that deals with inheritance. Genetics and inheritance is important in anatomy and physiology because of the numerous developmental and functional disorders that have a genetic basis. Genetic counseling is the practical application of knowing which disorders and diseases are inherited. The genetic inheritance of an individual begins with conception.

Each zygote inherits twenty-three chromosomes from its mother and twenty-three chromosomes from its father. This does not produce forty-six different chromosomes, but rather, twenty-three pairs of *homologous chromosomes.* Each member of a homologous pair, with the important exception of the sex chromosomes, looks like the other and contains similar genes (such as those coding for eye color, height, and so on). These homologous pairs of chromosomes can be **karyotyped** (photographed or illustrated) and identified (as shown in figure 29.26). Each cell that contains forty-six chromosomes (that is *diploid*) has two chromosomes number 1, two chromosomes number 2, and so on through chromosomes number 22. The first twenty-two pairs of chromosomes are called **autosomal** *(aw″to-so′-mal)* **chromosomes.** The twenty-third pair of chromosomes are the **sex chromosomes,** which may look different and may carry different genes. In a female these consist of two X chromosomes, whereas in a male there is only one X chromosome and one Y chromosome.

Genes and Alleles. A **gene** is a portion of the DNA of a chromosome that contains the information needed to synthesize a particular protein molecule. Although each diploid cell has a pair of gene locations for each characteristic, there may be a number of alternate forms of each gene.

Figure 29.26. A karyotype of homologous pairs of chromosomes obtained from a human diploid cell. The first twenty-two pairs of chromosomes are called the autosomal chromosomes. The sex chromosomes are (a) XY for a male and (b) XX for a female.

Those forms of a gene that affect the same characteristic but produce different expressions of that characteristic are called **alleles** *(ah-lēls')*. One allele of each pair originates from the female parent and the other from the male. The shape of a person's ears, for example, is determined by the kind of allele received from each parent and how the alleles interact with one another. Alleles are always located on the same spot (**locus**) on homologous chromosomes (fig. 29.27).

For any particular pair of alleles in a person, the two alleles are either identical or not identical. If the alleles are identical, the person is said to be **homozygous** *(ho''mo-zi'gus)* for that particular characteristic. But if the two alleles are different, the person is **heterozygous** *(het''er-o-zi'gus)* for that particular trait.

Genotype and Phenotype. A person's DNA contains a catalog of genes known as the **genotype** *(je'no-tīp)* of that person. The expression of those genes results in certain observable characteristics referred to as the **phenotype** *(fe'no-tīp).*

If the alleles for a particular trait are homozygous, the characteristic expresses itself in a specific manner (two alleles for attached earlobes, for example, results in a person with attached earlobes). If the alleles for a particular trait are heterozygous, however, which of the alleles

Figure 29.27. A pair of homologous chromosomes. Homologous chromosomes contain genes for the same characteristic at the same locus.

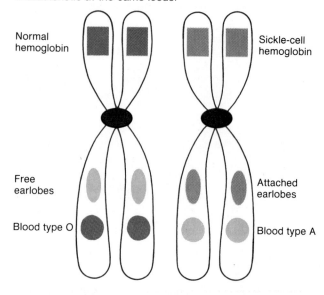

Table **29.5** Some hereditary traits in humans determined by single pairs of dominant and recessive alleles			
Dominant	**Recessive**	**Dominant**	**Recessive**
Free earlobes	Attached earlobes	Color vision	Color blindness
Dark brown hair	All other colors	Broad lips	Thin lips
Curly hair	Straight hair	Ability to roll tongue	Lack of this ability
Pattern baldness (♂♂)	Baldness (♀♀)	Arched feet	Flatfeet
Pigmented skin	Albinism	A or B blood factor	O blood factor
Brown eyes	Blue or green eyes	Rh blood factor	No Rh blood factor

expresses itself and how the genes for that trait interact will determine the phenotype. Often one of the alleles expresses itself as the **dominant allele,** whereas the other does not and is the **recessive allele.** The combinations of dominant and recessive alleles are responsible for a person's hereditary traits (table 29.5).

In describing genotypes, it is traditional to use letter symbols to refer to the alleles of an organism. The dominant alleles are symbolized with upper case letters and the recessive alleles are symbolized with lower case. Thus, the *genotype* of a person who is homozygous for free earlobes due to a dominant allele is symbolized *EE;* a heterozygous pair is symbolized *Ee.* In both of these instances, the *phenotypes* of the individuals would be free earlobes, since a dominant allele is present in each genotype. A person who inherited two recessive alleles for earlobes has the genotype *ee* and will have attached earlobes.

Thus, three genotypes are possible when gene pairing involves dominant and recessive alleles. They are *homozygous dominant (EE)*, *heterozygous (Ee)*, and *homozy-*

Figure 29.28. Inheritance of ear shape. Two heterozygous parents with free earlobes can have a child with attached earlobes.

Figure 29.29. Inheritance of the shape of earlobes.

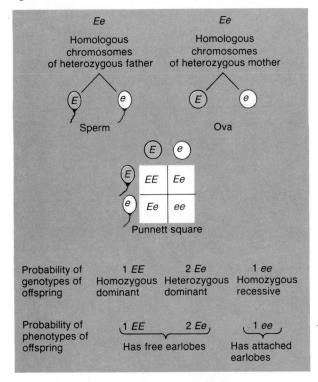

Figure 29.30. A dihybrid cross studies the probability of two characteristics at the same time. Any of the combinations of genes that have a *D* and an *E* (nine possibilities) will have free earlobes and dark hair. These are indicated with an asterisk (*). Three of the possible combinations have two alleles for attached earlobes (*ee*) and at least one allele for dark hair. They are indicated with a dot (•). Three of the combinations have free earlobes and light hair. These are indicated with a square (■). The remaining possibility has the genotype (*eedd*) for attached earlobes and light hair.

gous recessive (ee). Only two phenotypes are possible, however, since the dominant allele is expressed in both the homozygous dominant (*EE*) and the heterozygous (*Ee*) individuals. The recessive allele is expressed only in the homozygous recessive (*ee*) condition. Figure 29.28 illustrates how a homozygous recessive trait may be expressed in a child of parents who are heterozygous dominant.

Probability. A **Punnett square** is a convenient way to express the probabilities of allele combinations for a particular inheritable trait. In constructing a Punnett square,

the male gametes (spermatozoa) carrying a particular trait are placed at the side of the chart, and the female gametes (ova) at the top (as in figure 29.29). The four spaces on the chart represent the possible combinations of male and female gametes that could form zygotes. The probability of an offspring having a particular genotype is one in four (.25) for homozygous dominant and homozygous recessive, and one in two (.50) for heterozygous dominant.

A genetic study in which a single characteristic (e.g., ear shape) is followed from parents to offspring is referred to as a **monohybrid cross.** A genetic study in which two characteristics are followed from parents to offspring is referred to as a **dihybrid cross** (fig. 29.30). The term **hybrid** refers to an offspring descended from parents that have different genotypes.

Sex-linked Inheritance. Certain inherited traits are located on a sex-determining chromosome and are called **sex-linked** characteristics. The allele for red-green *color blindness,* for example, is determined by a recessive allele (designated c) found on the X chromosome but not the Y chromosome. Normal color vision (designated C) dominates. The ability to discern red-green colors, therefore, depends entirely on the X chromosomes. The genotype possibilities are

X^CY Normal male
X^cY Color-blind male
X^CX^C Normal female
X^CX^c Normal female carrying the recessive allele
X^cX^c Color-blind female

In order for a female to be red-green color-blind, she must have the recessive allele on both of her X chromosomes. A male with only one such allele on his X chromosome, however, will show the characteristic.

> **Hemophilia** is a sex-linked condition caused by a recessive allele. The blood in a person with hemophilia fails to clot or clots very slowly after an injury. If H represents normal clotting and h represents abnormal clotting, then males with X^HY will be normal and males with X^hY will be hemophiliac. Females with X^hX^h will have the disorder.

> 1. *Define the following terms:* genetics, genotype, phenotype, allele, dominant, recessive, homozygous, *and* heterozygous.
> 2. *List several dominant and recessive traits inherited in humans. What are some variables that determine a person's phenotype?*
> 3. *Construct a Punnett square to show the possible genotypes for color blindness of an X^cY male and an X^CX^c female.*

Clinical Considerations

Pregnancy and childbirth are natural events in human biology and generally progress smoothly without complications. Prenatal development is amazingly precise, and although traumatic, childbirth for most women in the world takes place without the aid of a physician. Occasionally, however, serious complications arise, and the knowledge of an obstetrician is required. The physician's knowledge of normal development and the causes of congenital malformations ensures the embryo every possible chance to develop normally. Many of the clinical aspects of prenatal development involve what might be referred to as applied developmental biology.

In clinical terms, gestation is frequently divided into three phases, or **trimesters,** each lasting three calendar months. By the end of the **first trimester** all of the major body systems are formed, the fetal heart can be detected, the external genitalia are developed, and the fetus is about

Figure 29.31. The various sites of ectopic pregnancies. The normal site is indicated by an X, and the abnormal sites are indicated by letters in order of frequency of occurrence.

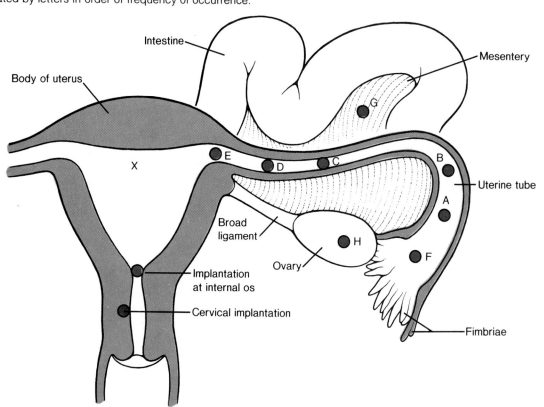

the width of the palm of an adult's hand. During the **second trimester,** fetal quickening can be detected, epidermal features are formed, and the vital body systems are functioning. The fetus, however, is still unlikely to survive if birth were to occur. At the end of the second trimester, fetal length is about equal to the length of an adult's hand. The fetus experiences a tremendous amount of growth and refinement in system functioning during the **third trimester.** A fetus of this age may survive if born prematurely, and of course, the chances of survival improve as the length of pregnancy approaches the natural delivery date.

There are many clinical considerations associated with prenatal development, some of which relate directly to the female reproductive system. Other developmental problems are genetically related and will be mentioned only briefly. Of clinical concern for developmental anatomy are topics such as implantation sites, test-tube development, multiple pregnancy, fetal monitoring, and regeneration.

Abnormal Implantation Sites

In an **ectopic pregnancy** the blastocyst implants outside the uterus or in an abnormal site within the uterus (fig. 29.31). The most common ectopic location (about 95%) is within the uterine tube and is referred to as a **tubal pregnancy.** Occasionally, implantation occurs near the cervix where development of the placenta blocks the cervical opening. This condition, called **placenta previa,** causes serious bleeding. Ectopic pregnancies will not develop

---

previa: L. *previa,* appearing before or in front of

Figure 29.32. A laparoscope, used for various abdominal operations including the extraction of a preovulatory ovum.

normally in unfavorable locations, and the fetus seldom survives beyond the first trimester. Tubular pregnancies are terminated through medical intervention. If a tubular pregnancy is permitted to progress, however, the uterine tube generally ruptures, followed by hemorrhaging. Depending on the location and the stage of development (hence vascularity) of tubal pregnancies, they may not be serious or the hemorrhaging and shock may cause the death of the woman.

In Vitro Fertilization and Artificial Implantation

Reproductive biologists have been able to fertilize a human ovum *in vitro* (outside the body), culture it to the blastocyst stage, and then perform artificial implantation, leading to a full-term development and delivery. This is the so-called test-tube baby. To obtain the ovum, a specialized laparoscope (fig. 29.32) is used to aspirate the preovulatory egg from a mature graafian follicle. The ovum is then placed in a suitable culture medium where it is fertilized with sperm. After the zygote forms, the sequential pre-embryonic development continues until the blastocyst stage, at which time implantation is performed. *In vitro* fertilization with artificial implantation is a means of overcoming infertility problems due to blocked uterine tubes in females or low sperm counts in males.

Multiple Pregnancy

Twins occur about once in eighty-five pregnancies. They can develop in two ways. **Dizygotic** (fraternal) **twins** develop from two zygotes resulting from two spermatozoa fertilizing two ova in the same ovulatory cycle (fig. 29.33). **Monozygotic** (identical) **twins** form from a single zygote (fig. 29.34). Approximately two-thirds of twins are dizygotic.

Dizygotic twins may be of the same sex or different sexes and are not any more alike than brothers or sisters born at different times. Dizygotic twins always have two chorions and two amnions, but the chorions and the placentas may be fused.

Monozygotic twins are of the same sex and are genetically identical. Any physical differences in monozygotic twins is caused by environmental factors during morphogenic development (e.g., there might be a differential vascular supply that causes slight differences to be expressed). Monozygotic twinning is usually initiated toward the end of the first week when the inner cell mass divides to form two embryonic primordia. Monozygotic twins have two amnions but only one chorion and a common placenta. If the inner mass fails to completely divide, **conjoined twins** (Siamese twins) may form.

Triplets occur about once in 7,600 pregnancies and may be (1) all from the same ovum and identical, (2) two identical and the third from another ovum, or (3) three zygotes from three different ova. Similar combinations occur in quadruplets, quintuplets, and so on.

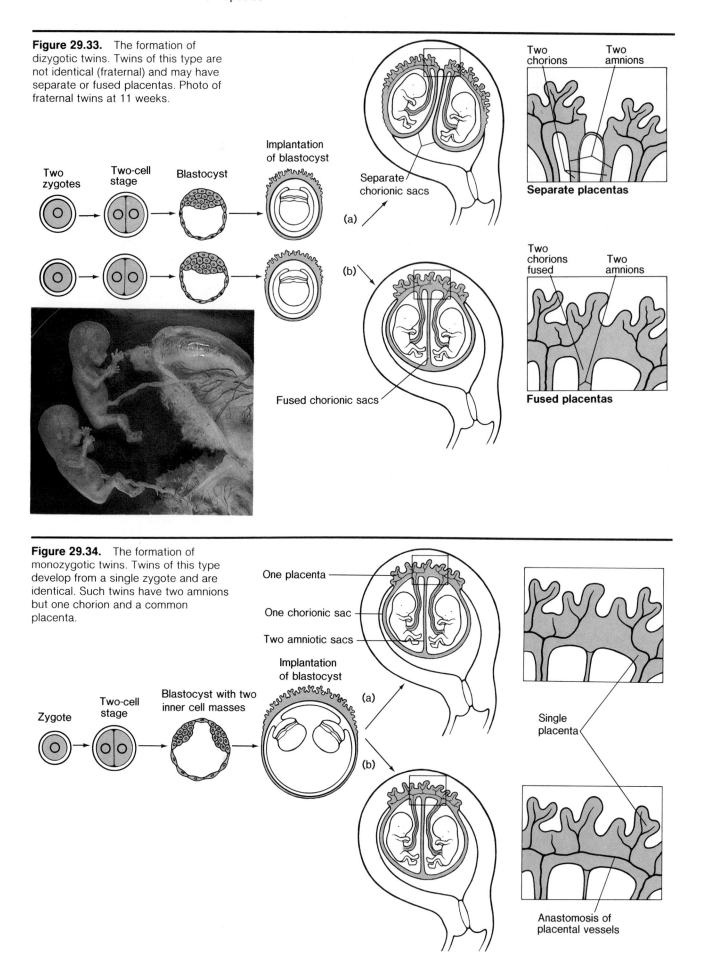

Figure 29.33. The formation of dizygotic twins. Twins of this type are not identical (fraternal) and may have separate or fused placentas. Photo of fraternal twins at 11 weeks.

Two zygotes   Two-cell stage   Blastocyst   Implantation of blastocyst

Separate chorionic sacs

(a)

Two chorions   Two amnions

Separate placentas

(b)

Fused chorionic sacs

Two chorions fused   Two amnions

Fused placentas

Figure 29.34. The formation of monozygotic twins. Twins of this type develop from a single zygote and are identical. Such twins have two amnions but one chorion and a common placenta.

One placenta

One chorionic sac

Two amniotic sacs

Zygote   Two-cell stage   Blastocyst with two inner cell masses   Implantation of blastocyst

(a)

(b)

Single placenta

Anastomosis of placental vessels

Figure 29.35. Ultrasonogram of a fetus during the third trimester. The left hand is raised, as if waving to the viewer.

— Amniotic fluid

— Placenta

— Left cerebral hemisphere

— Orbit of eye

— Left hand

— Uterine wall

— Thorax

Figure 29.36. Fetoscopy.

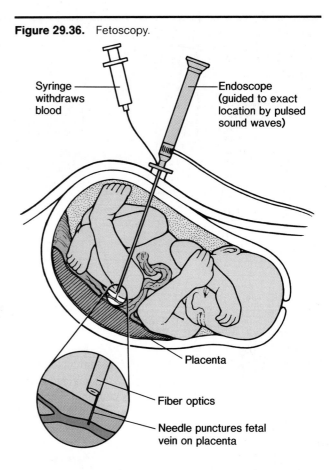

Syringe withdraws blood

Endoscope (guided to exact location by pulsed sound waves)

Placenta

Fiber optics

Needle punctures fetal vein on placenta

Fetal Monitoring

Obstetrics has benefited greatly from the advancements made in fetal monitoring in the last two decades. Before these techniques became available, physicians could determine the welfare of the unborn child only by auscultation of the fetal heart and palpation of the fetus. Currently, there are several tests that provide much information about the fetus during any stage of development. Fetal conditions that can now be diagnosed and evaluated include genetic disorders, hypoxia, blood disorders, growth retardation, placental functioning, prematurity, postmaturity, and intrauterine infections. These tests also help determine the advisability of an abortion.

X rays of the fetus were once commonly performed but were found harmful and have been replaced by other methods of evaluation that are safer and more informative. **Ultrasonography,** produced by a mechanical vibration of high frequency, produces a safe, high resolution (sharp image) of fetal structure (fig. 29.35). Ultrasonic imaging is a reliable way to determine pregnancy as early as six weeks after ovulation. It can also be used to determine fetal weight, length, and position, as well as to diagnose multiple fetuses.

Amniocentesis is a technique used to obtain a small sample (5–10 ml) of amniotic fluid with a syringe so that the fluid can be assessed. Amniocentesis is most often performed to determine fetal maturity, but it can also help predict serious disorders like Down's syndrome (mongolism) and Gaucher's disease (a metabolic disorder).

A new technique called **fetoscopy** (fig. 29.36) goes beyond amniocentesis by allowing direct examination of the fetus. Using fetoscopy, physicians scan the uterus with pulsed sound waves to locate fetal structures, the umbilical cord, and the placenta. Skin samples are taken from

the head of the fetus and blood samples extracted from the placenta. The principal advantage of fetoscopy is that external features of the fetus (such as fingers, eyes, ears, mouth, and genitals) can be carefully observed. Fetoscopy is also used to determine several diseases, including hemophilia, thalassemia, and the 40% of sickle-cell anemia cases missed by amniocentesis.

Most hospitals are now equipped with instruments that monitor fetal heart rate and uterine contractions during labor and can detect any complication that arises during the delivery. This procedure is called Electronic Monitoring of Fetal Heart Rate and Uterine Contractions (FHR-UC Monitoring). The stress to the fetus from uterine contractions can be determined through monitoring (fig. 29.37). Long, arduous deliveries can be taxing to both the mother and fetus. If the baby's health and vitality are diagnosed to be in danger because of a difficult delivery, the physician may decide to perform a cesarean section.

Congenital Defects

Major developmental problems called **congenital malformations** occur in approximately 2% of all newborn infants. The causes of congenital conditions include genetic inheritance, mutation (genetic change), and environment. About 15% of neonatal deaths are attributed to congenital

amniocentesis: Gk. *amnion,* lamb (fetal membrane); *kentesis,* puncture
fetoscopy: L. *fetus,* offspring, *skopein,* to view

congenital: L. *congenitus,* born with

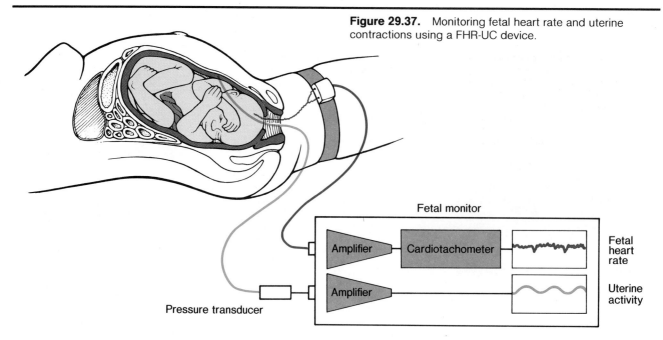

Figure 29.37. Monitoring fetal heart rate and uterine contractions using a FHR-UC device.

malformations. The branch of developmental biology concerned with abnormal development and congenital malformations is called *teratology.* Many congenital problems have been discussed in previous chapters with the body system in which they occur.

Some Genetic Disorders of Clinical Importance

cystic fibrosis An autosomal recessive disorder that is characterized by the formation of thick mucus in the lungs and pancreas, which interferes with normal breathing and digestion.

familial cretinism An autosomal recessive disorder that is characterized by a lack of thyroid secretion, due to a defect in the iodine transport mechanism. Untreated children are dwarfed, sterile, and may be mentally retarded.

galactosemia An autosomal recessive disorder that is characterized by an inability to metabolize galactose, a component of milk sugar. Patients with this disorder have cataracts, damaged livers, and mental retardation.

gout An autosomal dominant disorder that is characterized by an accumulation of uric acid in the blood and tissue due to an abnormal metabolism of purines.

hepatic porphyria An autosomal dominant disorder that is characterized by painful gastrointestinal disorders and neurologic disturbances due to an abnormal metabolism of porphyrins.

hereditary hemochromatosis A sex-influenced, autosomal dominant disorder that is characterized by an accumulation of iron in the pancreas, liver, and heart, resulting in diabetes, cirrhosis, and heart failure.

hereditary leukomelanopathy An autosomal recessive disorder that is characterized by decreased pigmentation in the skin, hair, and eyes, and abnormal white blood cells. Patients with this condition are generally susceptible to infections and early deaths.

Huntington's chorea An autosomal dominant disorder that is characterized by uncontrolled twitching of skeletal muscles and deterioration of mental capacities. A latent expression of this disorder allows the mutant gene to be passed to children before the symptoms develop.

Marfan's syndrome An autosomal dominant disorder that is characterized by tremendous growth of the extremities, dislocation of the lenses, and congenital cardiovascular defects.

phenylketonuria *(fen"il-ke"to-nu're-ah)* **(PKU)** An autosomal recessive disorder that is characterized by an inability to metabolize the amino acid phenylalanine. This is accompanied by brain and nerve damage and mental retardation.

pseudohypertrophic muscular dystrophy A sex-linked recessive disorder that is characterized by progressive muscle atrophy. It usually begins during childhood and causes death in adolescence.

retinitis pigmentosa A sex-linked recessive disorder that is characterized by progressive atrophy of the retina and eventual blindness.

Tay-Sachs disease An autosomal recessive disorder that is characterized by a deterioration of physical and mental abilities, early blindness, and death.

teratology: Gk. *teras,* monster; *logos,* study of

Huntington's chorea: from George Huntington, U.S. physician, 1850–1916
Marfan's syndrome: from Antoine Bernard-Jean Marfan, French physician, 1858–1942
Tay-Sachs disease: from Warren Tay, English physician, 1843–1927, and Bernard Sachs, U.S. neurologist, 1858–1944

Chapter Summary

I. Fertilization

The fertilization of an ovum by a sperm in the uterine tube promotes the completion of meiotic development and the formation of a diploid zygote.

  A. Morphogenesis is the sequential formation of body structures during the prenatal period of human life. The prenatal period lasts thirty-eight weeks and is divided into a pre-embryonic, an embryonic, and a fetal stage.

  B. A capacitated sperm digests its way through the zona pellucida and corona radiata layers of the ovum to complete the fertilization process and formation of a zygote.

II. Pre-Embryonic Stage

The events of the two-week pre-embryonic stage include fertilization, transportation of the fertilized ovum through the uterine tube, mitotic divisions, implantation, and the formation of primordial embryonic tissue.

  A. Cleavage of the zygote is initiated within thirty hours and continues until a morula forms, which enters the uterine cavity on about the third day.

  B. A hollow, fluid-filled space forms within the morula, and it is then called a blastocyst.

  C. Implantation begins between the fifth and seventh day and is made possible by the secretion of proteolytic enzymes that digest a portion of the endometrium.

    1. During implantation, the trophoblast cells secrete human chorionic gonadotrophin (hCG) to prevent menstruation.

    2. The secretion of hCG declines by the tenth week as the developed placenta secretes steroids to maintain the endometrium.

  D. The inner cell mass of the implanted blastocyst flattens into the embryonic disc, from which the primary germ layers of the embryo develop.

    1. Ectoderm gives rise to the nervous system, the epidermis of the skin and epidermal derivatives, and portions of sensory organs.

    2. Mesoderm gives rise to bones, muscles, blood, reproductive organs, the dermis of the skin, and connective tissue.

    3. Endoderm gives rise to linings of the digestive tract, digestive organs, the respiratory tract and lungs, and the urinary bladder and urethra.

III. Embryonic Stage

The events of the six-week embryonic stage include the differentiation of the germ layers into specific body organs and the formation of the placenta, the umbilical cord, and the extraembryonic membranes that provide sustenance and protection to the embryo.

  A. The extraembryonic membranes include the amnion, yolk sac, allantois, and chorion.

    1. The amnion is a thin membrane surrounding the embryo and contains amniotic fluid, which cushions and protects the embryo.

    2. The yolk sac produces blood for the embryo.

    3. The allantois also produces blood for the embryo and gives rise to the umbilical arteries and veins.

    4. The chorion participates in the formation of the placenta.

  B. The placenta, formed from both maternal and embryonic tissue, has a transportatory role in providing for metabolic needs of the fetus and in removing its wastes.

    1. The placenta produces steroids and hormones.

    2. Nicotine, drugs, alcohol, and viruses can cross the placenta to the fetus.

  C. The umbilical cord containing two umbilical arteries and one umbilical vein is formed as the amnion envelops the tissues on the underside of the embryo.

  D. During the third to the eighth week, the structure of all the body organs, except the genitalia, becomes apparent.

    1. During the third week, the primitive knot forms from the primitive streak, which later becomes the notochord and mesenchyme.

    2. By the end of the fourth week, the heart is beating, the primordial tissue of the eyes, brain, spinal cord, lungs, and digestive organs are in their proper place, and the arm and leg buds are recognizable.

    3. At the end of the fifth week, the sense organs are formed in the enlarged head and the appendages have developed with finger rays present.

    4. The brain is well developed by the end of the sixth week, and the digits are separate on elongated appendages.

    5. During the seventh and eighth weeks, the body organs, except the genitalia, are formed and the embryo appears distinctly human.

IV. Fetal Stage

The fetal stage lasts from nine weeks until birth and is characterized by tremendous growth and the specialization of body structures.

  A. A small amount of tissue differentiation and organ development occurs during the fetal stage, but for the most part fetal development is primarily limited to body growth.

    1. Between weeks nine and twelve, ossification centers appear, the genitalia are formed, and the digestive, urinary, respiratory, and the muscle systems show functional activity.

    2. Between weeks thirteen and sixteen, facial features are formed and the fetal heart beat can be detected with a stethoscope.

    3. During the period from seventeen to twenty weeks, quickening can be felt by the mother, and vernix caseosa and lanugo cover the skin.

    4. During the period from twenty-one to twenty-five weeks, substantial weight gain occurs and the fetal skin becomes wrinkled and pinkish.

    5. Toward the end of the period from twenty-six to twenty-nine weeks, the eyes have opened, the gonads have descended in a male, and the weight and development of the fetus is such that it may survive if born prematurely.

    6. By thirty-eight weeks, the fetus is full-term; the normal gestation is 266 days.

V. Labor and Parturition

Labor and parturition are the culmination of gestation and require the action of oxytocin from the posterior pituitary and prostaglandins, produced in the uterus.

  A. During pregnancy, the uterus becomes more sensitive to oxytocin, which may stimulate the production of prostaglandins.

  B. Labor is divided into the dilation, expulsion, and placental stages.

VI. Inheritance

Inheritance is the passage of hereditary traits carried on the genes of chromosomes from one generation to another.

  A. Each zygote contains twenty-two pairs of autosomal chromosomes and one pair of sex chromosomes; XX in a female and XY in a male.

B. A gene is a portion of a DNA molecule that contains information for the production of one kind of protein molecule; alleles are different forms of genes that occupy corresponding positions on homologous chromosomes.

C. The combination of genes present in a person's cells constitutes a genotype; the appearance of a person is a phenotype.

1. Dominant alleles are symbolized by uppercase letters and recessive alleles are symbolized by lowercase letters.

2. The three possible genotypes are homozygous dominant, heterozygous, and homozygous recessive.

D. A Punnett square is a convenient way to express probability.

1. The probability of a particular genotype is one in four (.25) for homozygous dominant and homozygous recessive, and one in two (.50) for heterozygous dominant.

2. A single trait is studied in a monohybrid cross, and two traits are studied in a dihybrid cross.

E. Sex-linked traits, such as color blindness or hemophilia, are carried on the sex-determining chromosome.

Review Activities

Objective Questions

1. The pre-embryonic stage is completed when the
   (a) blastocyst implants
   (b) placenta forms
   (c) blastocyst reaches the uterus
   (d) primary germ layers form

2. The yolk sac produces blood for the embryo until the
   (a) heart is functional
   (b) kidneys are functional
   (c) liver is functional
   (d) baby is delivered

3. Which of the following is *not* a function of the placenta?
   (a) production of steroids and hormones
   (b) diffusion of nutrients and oxygen
   (c) removal of metabolic wastes
   (d) production of enzymes

4. The decidua basalis is
   (a) a component of the umbilical cord
   (b) the embryonic portion of the villous chorion
   (c) the maternal portion of the placenta
   (d) a vascular membrane derived from the trophoblast

5. Which of the following could diffuse across the placenta?
   (a) nicotine
   (b) alcohol
   (c) heroin
   (d) all of the above

6. During which week following conception, does the embryonic heart begin pumping blood?
   (a) fourth
   (b) fifth
   (c) sixth
   (d) eighth

7. Twins that develop from two zygotes resulting from the fertilization of two ova by two sperm in the same ovulatory cycle are referred to as
   (a) monozygotic
   (b) conjoined
   (c) dizygotic
   (d) identical

8. Match the genotype descriptions with the correct symbols:
   homozygous recessive   Bb
   heterozygous   bb
   homozygous dominant   BB

9. An allele that is *not* expressed in a heterozygous genotype is called
   (a) recessive
   (b) dominant
   (c) genotypic
   (d) phenotypic

10. If the genotypes of both parents are Aa and Aa, the offspring probably will be
    (a) ½ AA and ½ aa
    (b) all Aa
    (c) ¼ AA, ½ Aa, ¼ aa
    (d) ¾ AA and ¼ aa

Essay Questions

1. Discuss the implantation of the trophoblast into the uterine wall and its involvement in the formation of the placenta.

2. Explain how the primary germ layers form. What major structures does each germ layer give rise to?

3. Explain why development during the embryonic stage is so critical, and list the embryonic needs that must be met during the embryonic stage for morphogenesis to continue.

4. Identify the approximate time period (in weeks) for the following occurrences:
   (a) The arm and leg buds appear.
   (b) The external genitalia differentiate.
   (c) Quickening is perceived by the mother.
   (d) The embryonic heart is functioning.
   (e) The ossification of bone is initiated.
   (f) Lanugo and vernix caseosa appear.
   (g) The fetus has a chance of survival if born prematurely.
   (h) All major body organs are formed.

5. State the features of a genetic disorder that would lead you to believe that it is a form of sex-linked inheritance.

Selected Readings

Beer, A. E., and R. E. Billingham. April 1974. The embryo as a transplant. *Scientific American.*

Birnholz, J. C., and E. E. Farrell. 1984. Ultrasound images of human fetal development. *American Scientist,* 72:608.

Danforth, D. N., ed. 1982. *Obstetrics and gynecology.* 4th ed. New York: Harper and Row Publishers.

England, M. A. 1983. *Color atlas of life before birth: Normal fetal development.* Chicago: Year Book Medical Publishers, Inc.

Gellis, S. S., ed. 1973. *Year book of pediatrics.* Chicago: Year Book Medical Publishers, Inc.

Grobstein, C. November 1977. External human fertilization. *Scientific American.*

Hole, J. W. 1981. *Human anatomy and physiology.* Dubuque: Wm. C. Brown Co.

Jackson, L. G. 1985. First trimester diagnosis of fetal genetic disorders. *Hosp. Prac.,* 20:39.

Moore, K. L. 1977. *Before we are born: Basic embryology and birth defects.* Rev. ed. Philadelphia: W. B. Saunders Co.

Nilsson, L., A. Ingelman-Sundbert, and C. Wirsen. 1977. *A child is born.* Rev. ed. New York: Dell Publishing Co.

Oppenheimer, S. B. 1980. *Introduction to embryonic development.* Boston: Allyn and Bacon.

Patten, B. M. 1968. *Human embryology.* New York: McGraw-Hill.

Rugh, R., and L. B. Shettles. 1971. *From conception to birth.* New York: Harper and Row Publishers.

Schaffer, A. J., and M. E. Avery. 1977. *Diseases of the newborn.* 4th ed. Philadelphia: W. B. Saunders Co.

Volpe, E. P. 1983. *Biology and human concerns.* 3d ed. Dubuque: Wm. C. Brown Co.

Warkentin, D. L. April 1984. From A to hCG in pregnancy testing. *Diagnostic Medicine,* p. 35.

30

Postnatal Growth,

Concepts

Mitotic potential is an important aspect of cellular specialization and body growth. Growth is not a linear process, and growth rates vary for different organs and structures of the body.

The human life following birth is divided on the basis of physical and physiological changes and maturity into the neonatal, infancy, childhood, adolescence, and adulthood periods.

Every living organism experiences chronological and biological aging. Senescence accompanies biological aging as each body system experiences a decrease in viability and an increase in susceptibility to injury and disease.

Biological aging and senescence is just as universal as chronological aging, but individuals, within limits, age biologically at different rates. Regardless of the rate of biological aging, each person has a life expectancy, and the individuals within the human population have a life span.

The many theories of aging are evidence that it is a complex, multifaceted process, which is not well understood. Evidence suggests that there is some genetic control of the aging process.

Development, and Aging

Selected Terms

neonatal   (ne″o-na′tal)
infancy (in′fan-se)
adolescence   (ad″o-les′ens)
puberty (pu′ber-te)
pubescence (pu-bes′ens)
sexually dimorphic (di-mor′fik)
menarche (mē-nar′ke)
senescence (se-nes′ens)
atrophy (at′ro-fe)
menopause (men′o-pawz)
gerontology (jer″on-tol′o-je)
mutation (mu-ta′shun)
autoimmunity (aw″to-im-mu′ni-te)

Introduction to Postnatal Growth, Development, and Aging

Mitotic potential is an important aspect of cellular specialization and body growth. Growth is not a linear process, and growth rates vary for different organs and structures of the body.

Objective 1. List the factors that presumably influence mitosis.
Objective 2. Discuss how an organism increases in body size.

Although the body organs are formed prenatally, the body undergoes continual growth and development into postnatal life. Body growth, developmental changes, and death are fundamental principles of life. Understanding the biological importance of these principles will make the material in this chapter more meaningful. Growth and physical development are necessary for an organism to sustain itself and contribute to its own species population. The principal contribution of an individual is reproduction, which ensures the continuance of the species. In a biological sense, once an individual has completed its reproductive potential, it is a liability to the species population because of competition for sustaining resources. Death of the aged is beneficial to a population because it eliminates those members who can no longer reproduce and provides room for offspring with genetic differences to add to the gene pool of the population.

Although we humans have the capability and opportunity long beyond our natural reproductive years to contribute far more to our population than simply producing and protecting offspring, we are still subject to the biological phenomena of growth, development, aging, and death. Clinical aspects of anatomy and physiology have been stressed throughout this book, and many of these conditions are, in part, the result of changes in the body during postnatal growth, development, and aging.

Cell Division

A zygote is a single fertilized egg cell, which has the potential to divide repeatedly, differentiate, and grow into a complex individual with a body that contains trillions of cells. As specialized cell types are formed, further mitotic divisions occur until distinct organ systems become functional. Embryonic and early fetal development are characterized by phenomenal cellular mitotic activity, differentiation and specialization, and growth.

Apparently included within cellular specialization of structure and function is mitotic potential. Certain cells do not require further division once the organ to which they contribute becomes functional. Others, as part of their specialization, require continuous mitosis to keep an organ healthy. Thus, in the adult, we find some cells that divide continually, some occasionally, and some not at all. For example, epidermal cells, hemopoietic cells, and cells that line the lumen of the gastrointestinal tract divide continually throughout life. Cells within specialized organs, such as the liver or kidneys, divide as the need becomes apparent. Disease or trauma from surgery or injury may necessitate mitosis in these organs. Still other cells, such as muscle or nerve cells, lose their mitotic ability as they become differentiated. Trauma to these cells frequently causes a permanent loss of function.

The factors regulating mitosis are unclear. Evidence suggests that mitotic ability is genetically controlled and, of those cells that do divide, even the number of divisions is predetermined. If this is true, it would certainly account for the aging process. Physical stress, nutrition, and hormones definitely have an effect on mitotic activity. It is thought that the reproductive activity of cells might be controlled through a feedback mechanism involving the release of a *growth-inhibiting substance*. Such a substance may slow or inhibit the divisions and growth of particular organs once the organ has a certain number of cells or reaches a certain size.

Growth

Growth is a normal process by which an organism increases in size as a result of the accretion of tissues similar to those already present. Growth is an integral part of de-

velopment that continues until adulthood. There are three aspects to the growth process: (1) an increase in the number of constituent cells; (2) an increase in the size of the existing cells through the addition of protoplasmic substance; and (3) an increase in the amount of intercellular substance.

Human growth is not a steady, linear process. It varies with age and the individual and is somewhat sex-dependent (fig. 30.1). At birth, an average full-term baby has a weight of 3,405 g (7.5 lb) and a length of about 50 cm (20 in.). The head circumference of newborns averages 35.5 cm (14 in.). In the first days of life, most newborns lose nearly 100 g (about 4 oz) before their bodies adapt to ingested food. The head of the newborn is disproportionately large compared to the trunk and appendages.

Growth rates vary for different organs and structures of the body. For example, the brain at birth is about 24% of its adult weight, whereas the neonatal body is only 6% of its adult weight. Phenomenally rapid growth of the brain continues, so that by the time a child is four years old the brain has reached 90% of its adult weight, whereas its body

Figure 30.1. A chronological sequence of heights, growth rates, and physical changes from infancy through adolescence. Each chart (females above and males below) displays a growth curve of total average heights. The sharp

peaks in the growth-rate curves are the adolescent growth spurts. Note that the adolescent growth spurt in males follows that of females by about two years. (Tanner, 1973)

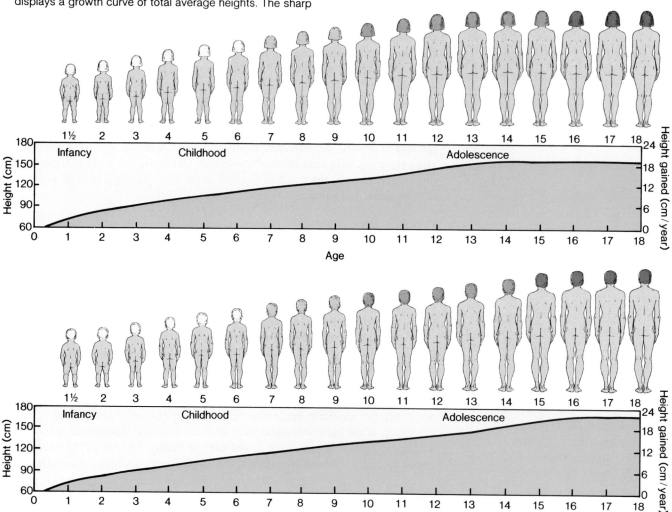

weight is only 25% of its eventual adult weight. The reproductive organs experience latent development because they are under hormonal control. These organs remain at less than 10% of their final weight until the onset of puberty.

Normal growth depends not only on proper nutrition but on the concerted effect of several hormones, including insulin, growth hormone, thyroid hormone, and (during adolescence) the androgens. The hypothalamus, through the action of hormone-releasing factors, seems to play the governing role in growth, including the initiation of puberty and the adolescent growth spurt. Both hyper- and hypoactivity of the hypothalamus may have profound effects on the growth process.

---

1. *List examples of cells that divide continuously, those that divide occasionally, and those that never divide in an adult person.*
2. *Diagram a cell, and construct a model that depicts the factors that may influence mitosis. Which factor is intrinsic to the cell, and which are extrinsic?*
3. *List the three aspects of the growth process that occur at the cellular level.*

Stages of Postnatal Growth

The human life following birth is divided on the basis of physical and physiological changes and maturity into the neonatal, infancy, childhood, adolescence, and adulthood periods.

Objective 3. Describe the growth and developmental events of the neonatal, infancy, childhood, and adolescent periods.

Objective 4. Define puberty, and discuss what determines its onset in males and females.

Objective 5. Define the term *adulthood,* and discuss sexual dimorphism in adult humans.

Neonatal Period

The **neonatal** *(ne″o-na′tal)* period extends from birth to the end of four weeks. Although growth is rapid during this period, the most drastic changes are physiological. The body of a newborn must immediately adapt to major environmental changes, including thermal stress; rapid bacterial colonization of the skin, oral cavity, and gastrointestinal tract; a barrage of sensory stimuli; and sudden demands on cardiorespiratory, gastrointestinal, and renal functions.

The most critical need of the newborn is the establishment of an adequate respiratory rate to ensure sufficient amounts of oxygen. The normal respiratory rate of a newborn is 30 to 40 times per minute. An adequate heart rate is also imperative. The heart of a newborn seems enlarged in respect to the thoracic cavity (compared to the heart of an adult) and has a rapid rate that ranges from 120 to 160 beats per minute.

Most full-term newborn babies appear chubby because of the deposition of fat within adipose tissue during the last trimester of pregnancy. Dehydration is a serious threat because of the inability of the kidneys to excrete concentrated urine; large volumes of dilute urine are eliminated. Immunity is not well developed and consists only of that obtained from the placental transfer of the mother. For this reason, newborns need to be guarded against exposure to infected persons.

Although virtually all the neurons of the nervous system are present in a newborn, they are immature and the newborn has little coordination. Most behavior such as sleep, hunger, and discomfort appears to be governed by lower cerebral centers and the spinal cord.

A newborn has many reflexes, some indicative of neuromuscular maturity and others essential for life itself. Four reflexes critical to survival are (1) the *suckling reflex,* which causes a newborn to suck anything that touches the lips; (2) the *rooting reflex,* which helps a baby find a nipple by causing it to turn its head and start suckling whenever something brushes its cheek; (3) the *crying reflex,* when its stomach is empty or it is experiencing other discomforts; and (4) the *breathing reflex,* which is apparent in a normal newborn even before the umbilical cord, with its supply of oxygen, is cut.

Babies born more than three weeks before the due date are generally considered **premature,** but because errors are commonly made in calculating the conception date, prematurity is defined by neonatal body weight rather than due date. Newborns weighing less than 2,500 grams (5.5 lbs) are considered premature. By this definition, approximately 8% of newborns in the United States are premature.

Postmature babies are those born two or more weeks after the due date. They frequently weigh less than they would have if they had been born at term because the placenta often becomes less efficient after a full-term pregnancy. Approximately 10% of newborns in the United States are postmature.

Infancy

The period of **infancy** *(in′fan-se)* extends from the end of the neonatal period at four weeks until two years of age. Infancy is characterized by tremendous growth, coordination, and mental development.

A full-term child will generally double its birth weight by five months and triple it in a year. The formation of subcutaneous adipose tissue reaches its peak at about nine months, causing the infant to appear chubby. Growth decelerates during the second year, during which time the infant gains only about 2.5 kg (5–6 lbs). During the second year, the infant develops locomotor and manipulative control and gradually becomes more lean and muscular.

---

neonatal: Gk. *neos,* new; *natus,* born
premature: L. *prae,* before; *maturus,* ripe
infancy: L. *in,* not; *fans,* speaking

Figure 30.2. The relative proportions of the body from embryo to adulthood. The head of a newborn accounts for a quarter of the total body length, and the lower appendages make up about one-third. In an adult, the head accounts for about 13% of the total body length, whereas the length of the lower appendages constitutes approximately one-half.

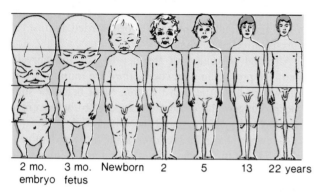

2 mo. embryo 3 mo. fetus Newborn 2 5 13 22 years

Body length increases during the first year by 25 to 30 cm (10–12 in.). There is an additional 12 cm (5 in.) of growth during the second year. The brain and circumference of the head also grow rapidly during the first year and only moderately so during the second. Head circumference increases approximately 12 cm (5 in.) during the first year and only an additional 2 cm during the second. The anterior fontanel gradually diminishes in size after six months and becomes effectively closed at any time from nine to eighteen months. It is the last of the fontanels to close. The brain is two-thirds of its adult size at the end of the first year and four-fifths of its adult size by the end of the second year.

By two years, most infants weigh approximately four times their birth weight and average between 81 and 91 cm (32–36 in.) in length. The body proportions of a two-year-old are certainly not the same as an adult (fig. 30.2). Growth is a differential process, resulting in gradual changes in body proportions.

Deciduous teeth begin to erupt in most infants between five and nine months. By one year of age, most infants have six to eight teeth. Eight more teeth erupt during the second year, making a total of fourteen to sixteen, including the first deciduous molars and canine teeth.

The growth rates of children vary tremendously. Body lengths and weights are not always reliable indicators of normal growth and development. A more objective evaluation of a child's physical developmental progress is determined through X-ray analysis of skeletal ossification in the carpal region (fig. 30.3).

Childhood

Childhood is the period of growth and development extending from infancy to adolescence, at which time puberty begins. The chronological duration of childhood varies because puberty begins at different ages for different people.

Figure 30.3. X rays of the right hand of (a) a child, (b) an adolescent, and (c) an adult.

(a)

Figure 30.4. Growth of the skull. The height of the cranial vault (distance between planes a and b) is drawn the same in both the infant and adult skulls. Growth of the skull occurs almost exclusively within the bones of the facial region.

Childhood years are a period of relatively steady growth until preadolescence, when there is a growth spurt. The average weight gain during childhood is about 3 to 3.5 kg (7 lb) per year. There is an average increase in height of 6 cm (2.5 in.) per year. The circumference of the head increases by only about 3 to 4 cm (1.5 in.) during childhood, and by adolescence the head and brain are virtually adult size.

The facial bones continue to develop during childhood (fig. 30.4). Especially significant is the enlargement of the sinuses. The first permanent teeth generally erupt during the seventh year and then the deciduous teeth are shed approximately in the same sequence as they were acquired. Deciduous teeth are replaced at a rate of about four per year over the next seven years.

(b)

(c)

Although there is an average rate of growth during childhood due to genetics, there is a tremendous range in what is considered normal growth. If, for example, the eight-year-olds who are within the tallest and heaviest 10% of their age group were to stop growing for a year while their classmates grew normally, they would still be taller than half of their contemporaries and heavier than three-quarters of them.

During childhood, the average child becomes thinner and stronger each year as he or she grows taller. The average ten-year-old, for example, can throw a ball twice as far as the average six-year-old. Visceral organs, particularly the heart and lungs, develop tremendously during this period, enabling a child to run faster and exercise longer.

Lymphatic tissue is at its peak of development during midchildhood and generally exceeds the amount of such tissue in the normal adult. Children need the extra lymphatic tissue to combat childhood diseases, which take a tremendous toll, particularly in countries where nutrition is poor and health care is minimal.

Childhood obesity can become a serious physical and psychological problem if not corrected. Overweight children usually exercise less and run a greater risk of serious illnesses. Frequently, they are teased and rejected by classmates, which causes psychological stress and learning impairments. At least 5% of children in the United States can be classified as obese. There is generally a correlation between childhood obesity and adult obesity. Obesity in adults is a major health problem, in light of the fact that one of five adults are 30% or more over their ideal, healthy weight. The obvious help for obese children and adults is a controlled diet and regular exercise.

Adolescence

Adolescence *(ad″o-les′ens)* is the period of growth and development between childhood and adulthood. It begins around the age of ten years in girls and the age of twelve years in boys. The end of adolescence is frequently said to be the age of twenty years, but it is not clearly delineated and varies with the developmental, physical, emotional, mental, or cultural criteria that define an adult.

Puberty *(pu′ber-te)* is the stage of early adolescence when the secondary sexual characteristics become expressed and the sexual organs become functional. **Pubescence** *(pu-bes′ens)* refers to the continuum of physical changes during puberty, particularly in regard to body hair.

For both sexes there is wide individual variation in the onset and duration of puberty. Although puberty is under hormonal control, a complex interaction of other factors, including nutrition and socioeconomic forces, has a decisive influence on the onset and duration of puberty. The end result of puberty is that the sexes are **sexually dimorphic** *(di-mor′fik);* that is, they are structurally different. The average adult male, for example, has a deeper voice and more body hair, and is taller than the average adult female. Prior to puberty, male and female children have few major structural differences aside from the general appearance of the external genitalia.

Puberty actually begins before it is physically expressed. In most instances, significant amounts of sex hormones appear in the blood of females by the age of ten years and in males by the age of eleven years. Sexual changes are usually thought of as the only features of puberty, but major musculoskeletal changes take place as

adolescence: L. *adolescere*, to grow up
puberty: L. *pubertas*, adult form
dimorphic: Gk. *di*, two; *morphe*, form

Table **30.1** Sequence of adolescent physical development			
Females	**Age span**		**Males**
Growth spurt begins; breast buds appear; sparse pubic hair	10–11 yrs	11.5–13 yrs	Growth of testes and scrotum; sparse pubic hair; growth spurt begins; penis growth begins
Appearance of straight, pigmented pubic hair; some deepening of voice; rapid growth of ovaries, uterus, and vagina; acidic vaginal secretion; menarche; further enlargement of breasts; kinky pubic hair; age of maximum growth	11–14 yrs	13–16 yrs	Appearance of straight, pigmented pubic hair; deepening of voice; maturation of penis, testes, scrotum, and accessory reproductive glands; ejaculation of semen; axillary hair; kinky pubic hair; sparce facial hair; age of maximum growth
Appearance of axillary hair; breasts are adult size and shape; culmination of physical growth	14–16 yrs	16–18 yrs	Increased body hair; marked vocal change; culmination of physical growth

Source: Van De Graaff, Kent M., *Human Anatomy.* © 1984 Wm. C. Brown Publishers, Dubuque, Iowa. All Rights Reserved. Reprinted by permission.

well. During late childhood, the body proportions of the sexes are similar, males being slightly taller. Under the influence of hormones, females experience a growth spurt in early adolescence that precedes that of males by nearly two years. During this time, females are temporarily taller. Once puberty begins in males, the heights of both sexes are soon the same, and at the culmination of puberty, males average about 10 cm (4 in.) taller than females. By the time growth is completed at the end of adolescence, males average 13 cm (5 in.) taller than females.

Other dimorphic differences involving skeletal structures include a broadening of the pelvic girdle in females. The muscles of males become more massive and stronger than those of females. Females acquire a thicker subcutaneous layer of the skin during adolescence, which gives them a softer appearance.

Sexual maturation during adolescence includes not only the development of the reproductive organs but the appearance of secondary sexual characteristics. The sequence of average sexual maturation and the expression of secondary sexual characteristics for both males and females is presented in table 30.1.

In females, the first physical indication of puberty is the appearance of **breast buds,** which are swellings of the breasts and slight enlargement and pigmentation of the areolar areas. The average age for breast buds to appear in healthy girls is about eleven years, but it ranges from nine to thirteen years. Approximately three years are required after the appearance of breast buds for the maturation of the breasts. Usually pubic hair begins to appear shortly after the breast buds, but in about one-third of all girls, sparse pubic hair appears before the breast buds. Axillary hair appears a year or two after pubic hair.

The first menstrual period, referred to as **menarche** *(mē-nar′ke),* is generally at the age of thirteen years but may range as much as from nine to seventeen years of age. During puberty, the vaginal secretions change from alkaline to acidic.

The onset of puberty in males varies just as much as in females but generally lags behind by about one and a half years. The first indication of puberty in boys is growth of the testes and the appearance of sparse pubic hair at the age of twelve years on the average. This is followed by growth of the penis, which continues for about two years, and the appearance of axillary hair. Vocal changes

generally begin during early puberty but are not completed until midpuberty. Facial hair and chest hair (which may or may not be present) come toward the end of puberty. The mean age when semen can be ejaculated is 13.7 years, but sufficient mature sperm for fertility are generally not produced until fourteen to sixteen years of age.

Acne is an inflammatory disease of the integument, which is common during adolescence. The increase in hormonal activity that is responsible for the physical changes taking place during puberty also affects the activity of the sebaceous glands and promotes the formation of inflamed superficial pustules and comedones. Tension and emotional stress may also promote acne. Acne in teenagers may cause serious psychological problems related to a concern about and need for peer acceptance.

Adulthood

Adulthood is the final stage of human physical change. It is the period of life beyond adolescence. An adult has reached maximum physical stature as determined by genetic, nutritional, and environmental factors. Although skeletal maturity is reached in early adulthood, anatomical and physiological changes continue throughout adulthood and are part of the aging process.

Sexual dimorphism exists in adults apart from the obvious primary and secondary sex differences. The sexes differ anatomically, physiologically, metabolically, and behaviorally (psychologically or socially). Some of these differences manifest themselves prenatally and during childhood. Others are characteristics of adolescence and adulthood. It is uncertain to what extent specific dimorphisms of the sexes are genetically determined through hormonal action or influenced by environmental (including cultural) factors. It is also unclear how these governing factors are expressed in observed physical characteristics.

The shape of the adult body is determined primarily by the skeleton and attached muscles as well as the subcutaneous connective tissue (especially adipose tissue) and extracellular body fluids. Although there is considerable variation in body proportions between adult males and females (fig. 30.5), in general, adult males have relatively longer appendages than females, their shoulders are relatively broader, and their pelvises are narrower. A male also has a relatively longer neck.

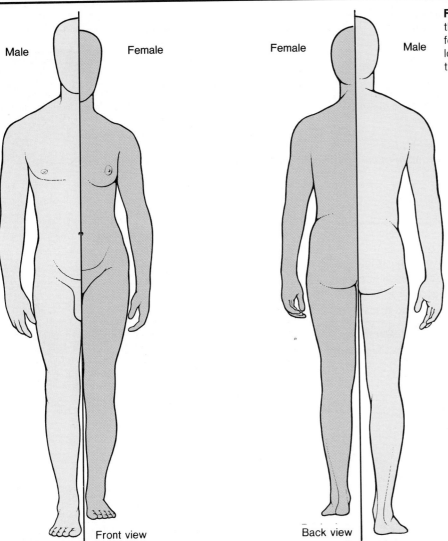

Male Female Female Male

Front view Back view

Figure 30.5. Relative differences in the physiques of an adult male and female. A male has proportionately longer appendages and a longer neck than a female.

Table **30.2** Body composition in average adult males and females at age 25 and 65

	Males				Females			
	Absolute		Relative (% body weight)		Absolute		Relative (% body weight)	
	Age 25	Age 65	Age 25	Age 65	Age 25	Age 65	Age 25	Age 65
Body weight	70.0 kg	70.0 kg			69.0 kg	60.0 kg		
Body fluids	41.0 liters	37.0 liters	58.9%	52.9%	30.8 liters	28.0 liters	51.3%	46.7%
Intracellular	24.0 liters	19.2 liters	34.3%	27.4%	16.6 liters	14.3 liters	27.7%	23.8%
Extracellular	17.2 liters	17.8 liters	24.6%	25.5%	14.2 liters	13.7 liters	23.6%	22.9%
Plasma volume	3,302 ml	2,940 ml	4.7%	4.2%	2,760 ml	2,462 ml	4.6%	4.1%
Lean body weight	56.3 kg	50.5 kg	80.4%	72.1%	42.0 kg	38.2 kg	70.2%	63.7%
Body fat	13.7 kg	19.5 kg	19.6%	27.9%	17.9 kg	21.8 kg	29.8%	36.3%
Skeletal weight	5.8 kg	5.7 kg	8.3%	8.1%	4.4 kg	4.2 kg	7.3%	7.0%

Source: From Oleson, K. H., "Body composition in normal adults," in *Human body composition*, J. Brozek, (ed.), 1965. Symposium of the Society for the Study of Human Body Composition, vol. 7 © Pergamon Press, Oxford, England. Reprinted by permission.

The general body composition of males and females can also be compared numerically. Mean data of body composition are summarized in table 30.2. These data show that total body fluid and skeletal weight is lower in adult females than in adult males. Females, however, have a greater percentage of body fat.

Other differences not shown in table 30.2 are that adult females have lower blood pressures, erythrocyte counts (hematocrits), basal metabolic rates, and respiratory rates than adult males. Females, however, have higher heart rates and oral temperatures.

1. *Construct a table that lists the stages from infancy through adolescence of postnatal growth, and indicate the events or characteristics of each.*
2. *List four reflexes in a newborn that are critical for survival.*
3. *Define the terms* puberty, pubescence, sexual dimorphism, *and* menarche. *What is the mean age of puberty, and what causes its occurrence?*
4. *Describe the physical characteristics of adulthood. Compare the body structure of an adult female to an adult male.*

Aging and Senescence

Every living organism experiences chronological and biological aging. Senescence accompanies biological aging as each body system experiences a decrease in viability and an increase in susceptibility to injury and disease.

Objective 6. Define the terms *aging* and *senescence.*
Objective 7. Discuss the senescent changes that occur in each body system.

Aging

Aging is a term that has a broad definition. *Chronological aging* is an expression of time that can apply to prenatal or postnatal life. The age of an embryo is given chronologically in days or weeks from the date of conception, whereas the age of the fetus is expressed in weeks or months from the date of birth. The age during the first year is given in weeks and months and thereafter is indicated in years. *Biological aging,* which begins with conception and ends at death, is a process characterized by changes in the structure and function of organ systems. These changes are generally predictable and, at least during certain age periods, not always detrimental. The changes in biological aging that take place during adolescence, for example, permit humans to be at peak performance as they enter early adulthood. Aging during midadulthood, on the other hand, results in a gradual reduction of capabilities.

Senescence of Body Systems

Senescence *(se-nes'ens)* refers to the biological aging process, characterized by a gradual deterioration of body structure and function. As senescence progresses through adulthood and old age, the viability of the body decreases and vulnerability increases. The body becomes more susceptible to injury and disease.

The following paragraphs of this section briefly discuss senescence of the body systems in the same sequence the systems were presented in the previous chapters of the text.

senescence: L. *senis,* old

Figure 30.6. Senescence of the skin results in a loss of elasticity and the appearance of wrinkles.

Integumentary System. With senescence, the skin becomes thin, dry, and inelastic. Collagen fibers in the dermis become thicker and less elastic, and the amount of adipose tissue in the hypodermis diminishes, making it thinner. Skinfold measurements indicate that the diminution of the hypodermis begins with regularity at the age of forty-five years. With a loss of elasticity and a reduction in the thickness of the hypodermis, wrinkling, or permanent infolding of the skin, becomes apparent (fig. 30.6).

During senescence of the integument, the number of hair follicles, sweat glands, and sebaceous glands also diminishes, as well as their activity. Consequently, there is a marked thinning of scalp hair and hair on the extremities, reduced sweating, and decreased sebum production. Since elderly people cannot perspire as freely, they are more likely to complain of heat and are more subject to heat exhaustion. They also become more sensitive to cold because of the loss of insulating adipose and diminished circulation. A decrease in the production of sebum causes the skin to dry and crack frequently.

The integument is not as well protected from the sun because of thinning, and melanocytes that produce the brown pigment melanin gradually atrophy. The loss of melanocytes accounts for graying of the hair and pallor of the skin. After the age of fifty, brown, plaquelike growths, called *seborrheic (seb''o-re'ik) hyperkeratoses,*

Figure 30.7.   The geriatric skull. Note the loss of teeth and the degeneration of bone, particularly in the facial region.

Figure 30.8.   A gradual diminishing of muscle strength occurs after the age of thirty-five, as shown with a graph of hand-grip strength.

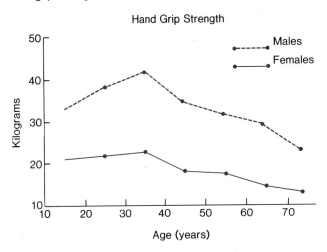

appear within the skin, particularly on exposed portions. Skin that has been exposed to excessive sunlight tends to develop more cutaneous carcinomas than less exposed skin. Caucasians are more susceptible to skin cancer than more darkly pigmented races.

Whereas a general loss of hair is characteristic of aging, females who have gone through menopause may develop more facial hair, particularly around the lips and on the chin. Changes in the androgen/estrogen ratio are responsible for the growth of this coarse, darkly pigmented hair. It can be removed through a procedure referred to as *depilation.*

Skeletal System.   Senescence affects the skeletal system by decreasing skeletal mass and density and increasing porosity and erosion (fig. 30.7). Bones become more brittle and susceptible to fracture. Articulating surfaces also deteriorate, contributing to arthritic conditions. Arthritic diseases are second to heart disease as the most common debilitation in elderly persons. *Osteoporosis (os"te-o-po-ro'sis)* is the most prevalent metabolic disorder of bone and develops in the aged. It is characterized by a decrease in skeletal mass and density. Postmenopausal women are most susceptible to this condition and because of it frequently sustain fractures of the hip, vertebrae, or wrist. Osteoporosis is attributed to immobilization, decreased estrogen levels, high steroid levels, and environmental factors.

Distinct losses in height occur during middle and old age. Between the ages of fifty and fifty-five years, there is a decrease of 0.5–2 cm (0.25–0.75 in.) because of compression and shrinkage of the intervertebral discs. Elderly persons may suffer a further major loss of height because of osteoporosis.

Muscular System.   Although the aged experience a general decrease in the strength, endurance, and agility of skeletal muscle (fig. 30.8), the extent of senescent changes varies considerably among individuals. Apparently the muscular system is one of the body systems in which a person may actively slow senescent changes. A decrease in muscle mass, in part, is due to changes in connective and circulatory tissues. Atrophy *(at'ro-fe)* of the muscles of the appendages causes the arms and legs to appear thin and bony. Degenerative changes in the nervous system decrease the effectiveness of motor activity. Muscle reflexes become less efficient, causing a marked reduction in physical capabilities.

Diminished muscular capabilities may affect the functioning of other body systems. A decrease in the strength of the respiratory muscles may limit the ability of the lungs to ventilate. Reduced muscularity of the urinary bladder causes difficult micturition and may cause urinary infections.

Exercise is important at all stages of life but is especially beneficial as one approaches old age (fig. 30.9). Exercise not only strengthens muscles, but it also contributes to a healthy circulatory system and thus ensures an adequate blood supply to all body tissues. If an elderly person does not maintain muscular strength through exercise, he or she will be more prone to debility. Learning to use a cane, crutches, or a walker would be extremely difficult for such a person.

Figure 30.9. Competitive runners in late adulthood, recognizing the benefits of exercise.

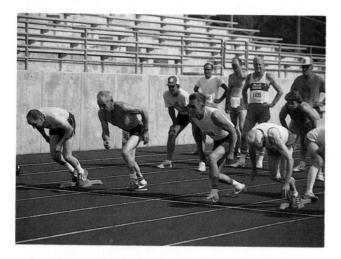

Figure 30.10. The decline in visual acuity in males and females due to senescence.

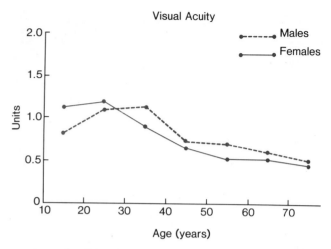

Nervous System. The extent of senescent changes within the brain is not known. Previous textbooks have reported that perhaps 100,000 neurons die each day of our adult life. Recent studies, however, show that such claims are unfounded. It is believed that relatively few neural cells are lost during the normal aging process. Neurons are, however, extremely sensitive and susceptible to various drugs or interruptions of vascular supply such as caused by a stroke or other cardiovascular diseases.

There is evidence that senescence alters neurotransmitters. Age-related conditions such as depression or specific diseases such as Parkinson's disease may be caused by an imbalance of neurotransmitter chemicals. Changes in sleeping patterns in aged persons also probably result from neurotransmitter problems.

The slowing of the nervous system with age is most apparent in tests of reaction time. It is not certain whether this is a result of a slowing in the transmission of impulses along neurons or in neurotransmitter relays at the synapses.

Although the nervous system does deteriorate with age (in ways that are not well understood), in most people it functions effectively throughout life. Brain dysfunction is not a common characteristic of senescence.

Sensory Organs. Senescence is undoubtedly characterized by sensory impairments. An elderly person may require more intense levels of stimulation in vision, hearing, taste, and smell in order for the sensory receptors to perform with precise acuity. Such declines in sensory capabilities, especially in vision and hearing, are important not only because they affect a person's ability to function, but because they may socially isolate the person and cause serious psychological problems.

Many conditions of the eye are related to age, and some cause visual impairments. Most elderly persons develop lipid infiltrates of the cornea in a condition called

arcus senilis. Visual acuity remains fairly constant until the ages of thirty-five to forty and then gradually declines (fig. 30.10). By the age of sixty-five, 40% of males and 60% of females have vision poorer than 20/70. The pupil of an elderly person's eye cannot dilate fully, and the amount of light reaching the retina may be only 50% of the amount that reaches the retina of a youth. In addition, the depth of the anterior chamber diminishes, creating poor drainage of aqueous humor, which results in increased intraocular pressure and a greater likelihood of glaucoma. Ptosis, or drooping, of the lids and diminished lacrimal secretion, contributing to dryness and irritation of the eyes, is typical in elderly persons.

The lens gradually loses its elasticity during senescence, and its ability to accommodate for viewing near objects diminishes. This change, called *presbyopia (pres''be-o'pe-ah),* makes it necessary to hold reading material farther from the eye for the print to be focused on the retina. A person with presbyopia may require bifocals to restore visual abilities.

Cataract is the major cause of visual disability in elderly persons. In this condition, the transparent lens becomes opaque (cataractal), restricting the passage of light waves.

Retinal blindness is caused by a slow degeneration of the macula of the retina. Focusing for reading is gradually impaired, whereas perimeter retinal vision remains unaffected.

Actually, the number of legally blind persons in the United States is relatively low (about 225 per 100,000 population), but 60% of blind people are older than seventy years. Most cases of senescent blindness are preventable with early detection and treatment.

Senescence affects hearing more frequently than sight. About 5% of the population have hearing impairments by the age of fifty; after the age of sixty-five, 30% have hearing difficulties. Senescent hearing loss is referred to as *presbycusis (pres''be-ku-sis)* and is generally a loss of the ability to perceive high-frequency sounds. Changes that

Figure 30.11. There is a marked increase in systolic blood pressure after the age of thirty-five as the heart and blood vessels undergo senescent changes.

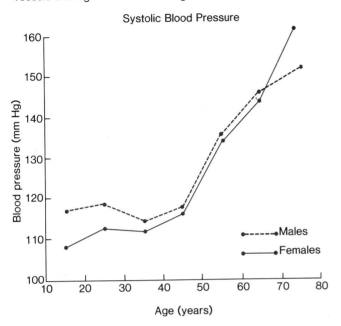

Systolic Blood Pressure

Figure 30.12. Senescence of the lungs and muscles of respiration reduces vital capacity. The graph represents data accumulated from healthy nonsmokers.

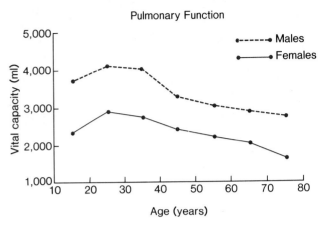

Pulmonary Function

accompany presbycusis are so gradual that many elderly people fail to recognize them until the disability is extreme.

The discernment of taste and smell declines with age. About 30% of people over the age of eighty have difficulty identifying common substances by smell. The ability to recognize common foods by taste likewise diminishes with age. The loss in the sensory perception of taste and smell may perhaps be responsible for the clinically noted frequency of complaints about food among elderly people.

Endocrine System. Although the organs of the endocrine system do undergo senescence, the consequences on overall body function are not well understood. The alteration in responsiveness to hormones is likely a major factor in the deterioration of biochemical activity in aging. The ability to secrete insulin and glucocorticoids is impaired in aged persons, and there seem to be changes in hormone transport, action, and feedback systems.

The organs of the endocrine system are susceptible to disorders of the circulatory system and to the diseases of other body organs that have feedback mechanisms with specific endocrine glands.

Circulatory System. Cardiovascular senescence is the leading cause of death in the aged. As the heart and blood vessels age, a person develops or becomes more susceptible to life-threatening diseases. The pulse and electrocardiogram give some insight into general circulatory changes in aging. Changes in blood vessels include an increased stiffness of the arterial walls (*arteriosclerosis*) and the deposition of fatty, plaquelike material (*atherosclerosis*) along the lumina of arteries.

The senescent heart may exhibit three principal characteristics: (1) structural changes in the heart and valves; (2) a diminished blood flow through the heart; and (3) an altered conduction system. The heart generally shrinks with age and accumulates adipose tissue along the coronary vessels. The endocardium increases 25% in thickness from the age of thirty to the age of eighty. After the age of thirty, the valves gradually become more rigid, thickened, and distorted because of sclerosis. The heart in an aged person reacts poorly to stress and cannot utilize oxygen as well. The maximum blood flow through the coronary arteries at the age of sixty is about 35% lower than in a person at the age of thirty. Blood pressure in elderly people may vary from 120 to 160 systolic (fig. 30.11) and from 70 to 90 diastolic. Electrocardiograms indicated slower and more erratic depolarization within the conduction system of the senescent heart.

Respiratory System. Senescence significantly changes the structure of the lung and affects the capacity for ventilation. After the age of thirty, there is a gradual loss of elasticity within the alveoli and in the pulmonary capillaries, so that by the age of seventy vital capacity has decreased by about 40% (fig. 30.12). Elderly persons thus require increased muscular exertion to maintain an adequate level of ventilation. Vital lung capacity decreases at a much faster rate if the person is a cigarette smoker.

In addition to a loss of elasticity, pulmonary aging causes a change in gaseous exchange across pulmonary membranes. By the age of sixty-five, there is about a 48% decrease in diffusion capacity from that at the age of thirty.

Digestive System. It is estimated that 60% of persons over fifty years of age have senescent problems of the digestive system that are generally more of an irritation than an impairment of health. Common problems include constipation, difficulties in swallowing, increased amounts of intestinal gas, heartburn, and the formation of diverticula along the large intestine. In addition, persons over the age of thirty-five are more susceptible to periodontal disease than those under the age of thirty-five.

Although gastrointestinal problems such as those listed above are normal in elderly persons, they may be symptomatic of more serious conditions or diseases such as hypertension, ulcers, or cancer.

The *basal metabolism rate (BMR)* is the minimum amount of energy a person utilizes in a state of rest and is an index of caloric usage and requirements. The BMR rates are usually higher for males than for females (fig. 30.13) and decline proportionally with age for both sexes. Young adults with a high BMR can eat almost anything and not get fat, whereas middle-aged people with a lower BMR must constantly control food intake to avoid obesity.

Urinary System. A progressive decline in kidney function accompanies senescence, due to general atrophy of the kidney and sclerosis of glomeruli. At the age of seventy the rate of glomerular filtration is only about 50% of the rate at the age of thirty. Renal blood flow diminishes from approximately 1,100 ml per minute at the age of thirty to only about 475 ml per minute at the age of eighty.

Figure 30.13. The basal metabolism rate.

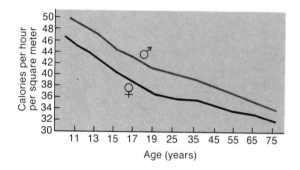

The kidneys are able to regulate acid-base balance during senescence, although they respond more slowly to a sudden, large acid load.

Reproductive System. The senescent changes in the reproductive system in both males and females are generally more important psychologically than pathologically. A decrease in sex hormones in elderly persons may decrease sexual desires or abilities. There is a decreased sensitivity of the genitalia, which may also affect sexual responsiveness.

A number of physical changes occur in aging males. The testes atrophy and the seminiferous tubules thicken and decrease in diameter. The prostate gland hypertrophies, which may cause pathological conditions. Although healthy males maintain the ability to ejaculate throughout old age, there is a reduction in the force of ejaculate, the volume, and the general quality of the seminal fluid.

Females experience major physiological changes at menopause *(men'o-pawz),* which generally occurs near the age of fifty. Hormonal changes bring about a cessation of menstruation and ovulation. Physical changes following menopause include atrophy of the uterus (fig. 30.14), with a reduction in the size of the cervix and a thinning of the vaginal wall and walls of the uterine tubes. The ovaries become irregular in shape. The vulva develops less pronounced folds as the skin becomes thinner. Vascularity and elasticity decrease, causing the vulva to become more susceptible to tissue trauma and pruritus (itching).

1. Using the term senescence, *distinguish between chronological and biological aging.*
2. *Construct a table that lists the senescence of the principal organs of each body system. Which body systems seem to age most rapidly? In which body systems does aging present a greater threat to life?*
3. *List three sensory impairments that result from senescence.*

Figure 30.14. Changes in the body and cervix of the uterus as a female ages. The uterus of a postmenopausal woman atrophies and becomes fibrous.

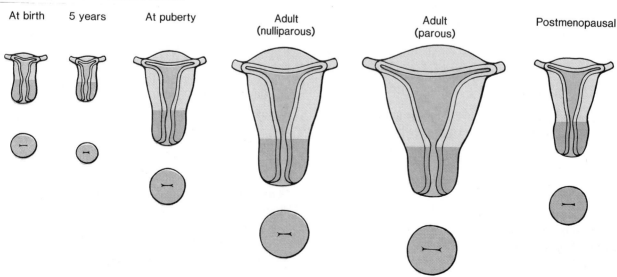

At birth   5 years   At puberty   Adult (nulliparous)   Adult (parous)   Postmenopausal

Life Expectancy and Life Span

Biological aging and senescence is just as universal as chronological aging, but individuals, within limits, age biologically at different rates. Regardless of the rate of biological aging, each person has a life expectancy, and the individuals within the human population have a life span.

> *Objective 8.* Define the terms *life expectancy, life span,* and *gerontology.*

Life expectancy is best defined as the average number of years lived by persons born in a particular location and period of time. A person born today in the United States has a greater life expectancy than a person born a century ago or a person born today in a Third World country. Life expectancy is relative to age. For example, life expectancy at birth is greater than it is at adolescence.

Life span is the maximum length of life possible in a particular culture. The average life span in the United States has increased twenty-six years in this century, from forty-seven to seventy-three years. Owing to cultural and environmental factors, the current average life span of males is about six years less (seventy years) than females (seventy-six years). The potential human life span is between 120 and 150 years if unaffected by accidents or diseases.

> Gerontology *(jer''on-tol'o-je)* is the study of senescence. This science is concerned primarily with the physical and psychological changes that occur between middle and old age until death. Gerontological research is concentrated in two principal areas: (1) studying the process of aging to determine if and how rates of senescence can be altered; and (2) examining and improving the quality of old age. Because of the increasing percentage of elderly people in our population, gerontology is becoming a more popular and vital science.

> 1. Discuss what determines life expectancy and what determines life span. Can either life expectancy or life span be increased? If so, how?
> 2. Define gerontology. List the kinds of classes a person would probably take who is studying to become a gerontologist.

Theories of Aging

The many theories of aging are evidence that it is a complex, multifaceted process, which is not well understood. Evidence suggests that there is some genetic control of the aging process.

> *Objective 9.* Discuss the five principal theories of aging.

Current literature in gerontology presents nearly two hundred theories of the mechanism of aging. These theories can be synthesized, however, into several principal types. Most of these theories continue to be scientifically scrutinized and are subject to continued modification.

Genetic Mutation Theory. The foundation for the mutation *(mu-ta'shun)* theory is that cells with unusual or different characteristics are more frequently noted as a person ages. It is thought that senescence is related to the accumulation of mutational damage within the genetic mechanism of somatic cells. The processes that DNA undergoes during mitosis and protein synthesis increase the chance of damage or defect to the copying mechanism. As DNA becomes a modified or inaccurate blueprint, defective proteins are synthesized. Some of these proteins are essential enzymes whose loss will cause the death of a cell. Senescence occurs as cells mutate. When a critical number of cells are altered or killed, the organism dies.

Autoimmunity Theory. The autoimmunity *(au''to-im-mu'ni-te)* theory suggests that senescence results when various body systems begin to reject their own tissues. As a person ages, the production of autoantibodies in the body increases. The antibodies are produced in response to somatic mutations or post-DNA errors. This causes organs to behave self-destructively, and the immune mechanism operates against its own body cells as if they were foreign bodies. Autoimmune reactions are known to occur in various cardiovascular diseases (e.g., giant cell arteries and amyloidosis), diabetes, rheumatoid arthritis, and certain types of cancer. It is also suggested that immunological surveillance becomes impaired as a person ages, so that cells that might give rise to neoplasm are no longer suppressed.

Cross-Linking of Molecules Theory. Homogeneity is a characteristic of some organic molecules. That is, different molecules tend to attract each other and become uniform. Chemical substances adhere to, or cross-link with, adjacent protein molecules so that each loses its functional identity. During senescence, tendons, skin, and even blood vessels lose elasticity. As the functional characteristics of collagen and elastic fibers change, the skin wrinkles, muscles sag, and wounds in the skin heal more slowly.

Cellular Aging Theory. The cellular aging theory suggests that there is a genetically determined limit to the doubling potential of each normal mitotically active cell. This means that cells have an intrinsic finite life span. As cells approach their final divisions, the time interval for mitosis progressively increases, causing a gradual cessation of activity and an accumulation of cellular debris. Eventually there is a total loss of vigor, and cellular death occurs. Degenerative phenomena are manifest after about fifty cell population doublings.

Programmed Aging Theory. The preceding theories fail to explain why the specific life span for each species (even species that are closely related biologically) is so consistently fixed. Also to be explained is the observation that the offspring of long-lived parents tend to live longer than the offspring of short-lived parents. Furthermore, the various cells constituting an organism have different fixed life spans. Epidermal cells, for example, live only a relatively short time before they are replaced, whereas muscle and nerve cells endure the lifetime of the body.

Clearly, there must be some genetic control of the aging process that is not only species and individual specific but also cell specific. If there is an aging gene responsible for senescence, then someday it should be possible to identify and perhaps eventually manipulate that gene.

Senescence is a complex process, and all of the theories are probably valid to some extent. The process almost certainly has numerous factors. A major challenge for gerontology is to explain how events that occur at the cellular level are transcribed into events at the organismic level.

---

1. *Using hypothetical situations, describe how laboratory experiments could be conducted to test each of the five current theories of aging.*
2. *Which of the five theories of aging accounts for closely related species having markedly different life spans?*

Chapter Summary

I. Introduction to Postnatal Growth, Development, and Aging
Mitotic potential is an important aspect of cellular specialization and body growth. Growth is not a linear process, and growth rates vary for different organs and structures of the body.
  A. Growth and development are necessary so that an individual organism can sustain itself and contribute to its own species population.
  B. Mitotic potential is apparently just as much of a cellular specialization as is any structural or functional feature of a cell.
  C. A growth-inhibiting substance released via a feedback mechanism may influence cellular reproductive activities to limit the number of cells or the size of specific organs.
  D. Growth is a normal process by which an organism increases in size as a result of the accretion of tissues similar to those already present.
II. Stages of Postnatal Growth
The human life following birth is divided on the basis of physical and physiological changes and maturity into the neonatal, infancy, childhood, adolescence, and adulthood periods.
  A. The neonatal period, extending from birth to the end of the fourth week, is characterized by major physiological changes.
    1. The most critical need of the newborn is to establish adequate respiratory and heart rates. A normal respiratory rate is 30 to 40 times per minute, and a normal heart rate ranges from 120 to 160 beats per minute.
    2. The four reflexes in the newborn critical to survival are the suckling reflex, the rooting reflex, the crying reflex, and the breathing reflex.
  B. Infancy, extending from four weeks until the age of two years, is characterized by tremendous growth, coordination, and mental development.
    1. By two years, most infants weigh about four times their birth weight and average between 32 and 36 inches in length.

  2. Growth is a differential process resulting in gradual changes from infant to adult body proportions.
  C. Childhood, extending from infancy to adolescence, is characterized by steady growth, until preadolescence when there is a marked growth spurt.
    1. During childhood, the average child becomes thinner and stronger each year as he or she grows taller.
    2. The fact that disease and death are more rare during childhood may be because lymphatic tissue is then at its peak of development and is present in greater amounts in children than in adults.
  D. Adolescence is the period of growth and development between childhood and adulthood.
    1. Puberty is the stage of early adolescence when the secondary sexual characteristics become expressed and the sexual organs become functional.
    2. The end result of puberty is the structural expression of gender, or sexual dimorphism.
    3. The onset of menarche is generally at the age of thirteen years but ranges from nine to seventeen years. At this time, vaginal secretions change from alkaline to acid.
    4. The first physical indications of puberty are the appearance of breast buds in females and the growth of the testes and the appearance of sparse pubic hair in males.
    5. Although semen may be ejaculated at the ages of thirteen or fourteen, sufficient mature sperm for fertility are not produced until fourteen to sixteen years of age.
  E. Adulthood, the final stage of human physical change, is characterized by gradual senescence as a person ages.
    1. Although skeletal maturity is reached in early adulthood, anatomical and physiological changes continue throughout adulthood and are part of the aging process.

  2. Sexual dimorphism exists anatomically, physiologically, metabolically, and behaviorally between males and females.
  3. Differences in body statures, proportions, and compositions between males and females may become more expressed with age.
III. Aging and Senescence
Every living organism experiences chronological and biological aging. Senescence accompanies biological aging as each body system experiences a decrease in viability and an increase in susceptibility to injury and disease.
  A. Aging can be expressed as chronological or as biological events. Senescence refers to the biological aging process, characterized by a gradual deterioration of body structure and function.
  B. Senescence affects each body system in a predictable but somewhat individual sequence.
    1. In the integumentary system, the skin thins, becomes dry, is less elastic, and is more fragile. There is an increased sensitivity to temperature extremes and an increased occurrence of cutaneous carcinomas in less pigmented races.
    2. In the skeletal system, the skeletal mass decreases in density and increases in porosity and erosion, predisposing humans to osteoporosis and arthritic conditions. A decrease in height during middle and old age results from the compression or shrinkage of intervertebral discs.
    3. In the muscular system, there is a general decrease in strength, endurance, and agility that varies considerably among individuals. Senescent changes may be slowed through regular exercise.
    4. In the nervous system, there is evidence that senescence causes an alteration in neurotransmitters, which may result in depression or conditions such as Parkinson's disease. The nervous system slows with age but functions effectively in most people throughout life.

5. Sense organs are impaired and become less acute in response to stimuli with aging. Vision may be decreased by corneal lipid infiltrates (arcus senilis), glaucoma, ptosis of the eyelids, presbyopia, cataracts, or retinal blindness. Presbycusis refers to the loss of high-frequency sound perception. The discernment of taste and smell also decline with age.

6. In the endocrine system, the consequences of senescence on overall body functions are not well understood. Organs of this system are susceptible to circulatory disorders and to the diseases of other body organs through feedback mechanisms.

7. Cardiovascular senescence is the leading cause of death in the aged. Blood vessels may develop arteriosclerosis or atherosclerosis, and the heart may develop structural changes in the valves, a decreased blood flow, and an altered conduction system.

8. In the respiratory system, the structure of the lung changes significantly with senescence to decrease its capacity for ventilation. Elasticity decreases as does the effectiveness of gaseous diffusion in lung tissue.

9. In the digestive system, senescence irritates rather than impairs health in most cases. Common problems include constipation, dysphagia, increased flatulence, heartburn, diverticuli formation along the colon, and an increased susceptibility to periodontal disease. These problems may be symptomatic of more serious diseases such as hypertension, ulcers, or cancer.

10. In the urinary system, kidney function progressively declines. The kidneys also respond slower in regulating acid-base balance.

11. In the reproductive system, senescent changes are more important psychologically in regard to maintaining sexuality than they are pathological. Although able to ejaculate, males experience a decrease in the force, volume, and general quality of the ejaculate. Females experience major physiological changes at menopause.

IV. Life Expectancy and Life Span
Biological aging and senescence is just as universal as chronological aging, but individuals, within limits, age biologically at different rates. Regardless of the rate of biological aging, each person has a life expectancy, and the individuals within the human population have a life span.

A. Life expectancy is the average number of years lived by persons born during a particular time period in a certain location.

B. Life span is the maximum length of life attainable in a culture.

V. Theories of Aging
The many theories of aging are evidence that it is a complex, multifaceted process, which is not well understood. Evidence suggests that there is some genetic control of the aging process. There are five principal theories of aging that are probably not mutually exclusive.

A. Genetic mutation theory—with age, somatic cells with mutational damage in the genetic mechanism accumulate.

B. Autoimmunity theory—senescence results when various body systems reject their own tissue.

C. Cross-linking of molecules theory—chemical substances adhere or cross-link adjacent protein molecules so that each loses its functional identity.

D. Cellular aging theory—cells have an intrinsic, finite life span due to the genetically determined, limited doubling potential characteristic of each normal mitotically active cell.

E. Programmed aging theory—evidence indicates that the aging process is genetically species specific, individual specific, and cell specific.

Review Activities

Objective Questions

1. Which of the following is the period of growth from birth to the end of the fourth week? The
   (a) neonatal
   (b) fetal
   (c) infant
   (d) suckling

2. The most drastic changes in the first four weeks of life are
   (a) anatomical
   (b) environmental
   (c) psychological
   (d) physiological

3. The normal newborn heart rate is
   (a) 70–80 beats/min
   (b) 120–160 beats/min
   (c) 100–120 beats/min
   (d) 180–200 beats/min

4. The continuum of physical change occurring in adolescence that regulates the growth of body hair is known as
   (a) puberty
   (b) pubal progression
   (c) pubescence
   (d) dimorphism

5. At birth, a baby is regarded as premature if it weighs less than
   (a) 2,500 grams (5.5 lbs)
   (b) 1,350 grams (3.0 lbs)
   (c) 1,050 grams (2.5 lbs)
   (d) 3,200 grams (7.0 lbs)

6. Which is not characteristic of the female when compared to the male?
   (a) lower blood pressure
   (b) higher basal metabolic rate
   (c) lower red blood cell count
   (d) faster heart rate

7. The first physical indication of puberty in females is generally
   (a) alkaline vaginal secretions
   (b) a widening pelvis
   (c) breast buds
   (d) axillary hair

8. Which statement is not true of a senescent heart of an elderly person?
   (a) The heart enlarges to pump additional blood.
   (b) The endocardium increases in thickness.
   (c) Adipose tissue accumulates along the coronary vessels.
   (d) Slower and more erratic depolarization within the conduction system.

9. Which statement is not true of life expectancy?
   (a) It can be changed with improved diet and health care.
   (b) It has an inverse ratio to chronological age.
   (c) It is the maximum length of life possible in a particular culture.
   (d) It varies from country to country.

10. The trend that the offspring of long-lived parents are often long-lived themselves supports which theory of aging? The
    (a) cross-linked
    (b) autoimmunity
    (c) programmed aging
    (d) genetic mutation

Essay Questions

1. Explain the biological importance of growth, development, aging, and death.

2. Discuss the growth and developmental events characteristic of the neonatal period.

3. Define the term infancy, and discuss the growth and developmental events that are characteristic of this stage of life.

4. Define the term adolescence, and discuss the role that puberty has in this stage of life for both males and females.

5. Distinguish between puberty and pubescence.

6. List the sexual dimorphic structural features of adult males and females.

7. For each body system, list two senescent changes.

8. Distinguish between chronological aging, biological aging, and senescence.

9. In which body systems do senescent changes threaten life?

10. Discuss gerontology as a contemporary science.

Selected Readings

Carter, N., ed. 1980. *Development, growth and aging.* London: Croon Helm.

Comfort, A. 1979. *The biology of senescence.* 3d ed. London: Churchill Livingstone.

Gersh, E. S., and I. Gersh. 1981. *Biology of women.* Baltimore: Univ. Park Press.

Lazarus, R. S., and A. DeLongis. 1983. Psychological stress and coping in aging. *American Psychologist* 38:245.

Lehman, H. C. 1978. *Age and achievement.* Princeton, N.J.: Princeton Univ. Press.

Miller, C. A. July 1985. Infant mortality in the U.S. *Scientific American.*

Neuhaus, R. H., and R. H. Neuhaus. 1982. *Successful aging.* New York: John Wiley and Sons.

Oleson, K. H. 1965. Body composition in normal adults. In *Human body composition,* edited by J. Brozek. Symposia of the Society for the Study of Human Body Composition 7:177. New York: Pergamon Press.

Palmore, E. 1970. *Normal aging.* Durham, N.C.: Duke Univ. Press.

Palmore, E., and E. Jeffers, eds. 1971. *Predictions of life span.* Lexington, Ky.: Heath Lexington Books.

Riley, M. W. et al. 1968. *Aging and society.* St. Louis: C. V. Mosby.

Rockstein, M. 1968. The biological aspects of aging. *The Gerontologist* 8:124.

Santrock, J. W. 1985. *Adult development and aging.* Dubuque: Wm. C. Brown.

Selye, H. 1970. Stress and aging. *J. Am. Geriatr. Soc.* 18:669–80.

Tanner, J. M. September 1973. Growing up. *Scientific American.*

Tanner, J. M. 1978. *Foetus into man.* Cambridge, Mass.: Harvard Univ. Press.

Wantz, M. S., and J. E. Gay. 1981. *The aging process: A health perspective.* Cambridge, Mass.: Winthrop.

Answers to Objective Questions

Chapter 1
1. (c) 5. (b) 8. (c)
2. (a) 6. (c) 9. (c)
3. (c) 7. (b) 10. (d)
4. (b)

Chapter 2
1. (c) 5. (c) 9. (d)
2. (b) 6. (b) 10. (b)
3. (a) 7. (c) 11. (a)
4. (d) 8. (d) 12. (d)

Chapter 3
1. (d) 5. (d) 8. (c)
2. (b) 6. (b) 9. (a)
3. (a) 7. (a) 10. (b)
4. (c)

Chapter 4
1. (d) 6. (d) 11. (a)
2. (a) 7. (d) 12. (d)
3. (d) 8. (c) 13. (b)
4. (e) 9. (a) 14. (d)
5. (e) 10. (c)

Chapter 5
1. (c) 5. (b) 8. (a)
2. (b) 6. (d) 9. (b)
3. (b) 7. (a) 10. (d)
4. (c)

Chapter 6
1. (b) 5. (d) 8. (a)
2. (c) 6. (a) 9. (b)
3. (a) 7. (c) 10. (d)
4. (b)

Chapter 7
1. (b) 3. (a) 5. (b)
2. (b) 4. (c) 6. (d)

Chapter 8
1. (a) 5. (d) 8. (b)
2. (b) 6. (a) 9. (b)
3. (b) 7. (c) 10. (d)
4. (a)

Chapter 9
1. (c) 5. (a) 8. (b)
2. (a) 6. (d) 9. (d)
3. (b) 7. (a) 10. (c)
4. (b)

Chapter 10
1. (b) 5. (a) 8. (b)
2. (a) 6. (d) 9. (c)
3. (d) 7. (c) 10. (d)
4. (d)

Chapter 11
1. (e) 5. (c) 8. (c)
2. (b) 6. (d) 9. (b)
3. (b) 7. (a) 10. (b)
4. (a)

Chapter 12
1. (e) 5. (d) 8. (b)
2. (b) 6. (a) 9. (c)
3. (d) 7. (a) 10. (a)
4. (d)

Chapter 13
1. (d) 6. (a) 11. (a)
2. (a) 7. (a) 12. (c)
3. (e) 8. (c) 13. (c)
4. (c) 9. (d) 14. (b)
5. (d) 10. (d)

Chapter 14
1. (b) 5. (c) 8. (a)
2. (a) 6. (d) 9. (b)
3. (d) 7. (c) 10. (d)
4. (b)

Chapter 15
1. (a) 5. (a) 8. (d)
2. (b) 6. (b) 9. (a)
3. (d) 7. (c) 10. (c)
4. (c)

Chapter 16
1. (d) 5. (a) 8. (b)
2. (d) 6. (c) 9. (e)
3. (c) 7. (b) 10. (c)
4. (c)

Chapter 17
1. (d) 5. (d) 8. (b)
2. (b) 6. (b) 9. (d)
3. (a) 7. (b) 10. (b)
4. (c)

Chapter 18
1. (e) 6. (d) 11. (a)
2. (b) 7. (d) 12. (b)
3. (d) 8. (e) 13. (e)
4. (c) 9. (e) 14. (d)
5. (a) 10. (d) 15. (a)

Chapter 19
1. (b) 5. (c) 9. (a)
2. (c) 6. (c) 10. (a)
3. (f) 7. (a) 11. (c)
4. (a) 8. (b) 12. (c)

Chapter 20
1. (a) 6. (c) 10. (b)
2. (d) 7. (a) 11. (c)
3. (c) 8. (c) 12. (d)
4. (e) 9. (d) 13. (c)
5. (b)

Chapter 21
1. (c) 5. (d) 9. (d)
2. (b) 6. (c) 10. (b)
3. (a) 7. (d) 11. (e)
4. (b) 8. (d) 12. (a)

Chapter 22
1. (c) 6. (d) 10. (a)
2. (b) 7. (d) 11. (d)
3. (d) 8. (b) 12. (a)
4. (a) 9. (e) 13. (c)
5. (c)

Chapter 23
1. (c) 6. (c) 11. (a)
2. (d) 7. (b) 12. (c)
3. (c) 8. (a) 13. (a)
4. (a) 9. (e) 14. (d)
5. (c) 10. (c) 15. (b)

Chapter 24
1. (c) 5. (b) 9. (b)
2. (a) 6. (d) 10. (d)
3. (b) 7. (c) 11. (a)
4. (c) 8. (d) 12. (d)

Chapter 25
1. (c) 5. (a) 9. (d)
2. (a) 6. (b) 10. (c)
3. (b) 7. (d) 11. (b)
4. (c) 8. (a) 12. (b)

Chapter 26
1. (d) 6. (e) 11. (a)
2. (b) 7. (d) 12. (e)
3. (a) 8. (d) 13. (c)
4. (c) 9. (c) 14. (c)
5. (b) 10. (d) 15. (b)

Chapter 27
1. (a) 5. (a) 8. (a)
2. (d) 6. (b) 9. (c)
3. (d) 7. (a) 10. (a)
4. (d)

Chapter 28
1. (b) 5. (c) 9. (c)
2. (d) 6. (d) 10. (a)
3. (a) 7. (d) 11. (b)
4. (c) 8. (a) 12. (b)

Chapter 29
1. (d) 8. homozygous
2. (c)   recessive (bb);
3. (a)   heterozygous
4. (c)   (Bb);
5. (d)   homozygous
6. (a)   dominant (BB)
7. (c) 9. (a)
        10. (c)

Chapter 30
1. (a) 5. (a) 8. (a)
2. (c) 6. (b) 9. (b)
3. (b) 7. (c) 10. (c)
4. (c)

Appendix B

Scientific Journals of Anatomy and Physiology

Acta Anatomica (Basel)
Acta Biologica et Medica Germanica (Berlin)
Acta Cytologica (St. Louis)
Acta Embryologiae Experimentalis (Palermo)
Acta Morphologica Academiae Scientiarum
  Hungaricae (Budapest)
Acta Morphologica Academiae Scientiarum
  Hungaricae Supplementum (Budapest)
Acta Morphologica Neerlando-Scandinavica
  (Utrecht)
Acta Physiologica et Pharmacologica
  Bulgarica (Sofia)
Acta Physiologica Latino Americana (Buenos
  Aires)
Acta Physiologica Scandinavica (Stockholm)
Activitas Nervosa Superior (Prague)
Age and Ageing (London)
Agressologie (Paris)
American Journal of Anatomy (New York)
American Journal of Physiology (Bethesda)
Anatomical Record (New York)
Anatomischer Anzeiger (Jena)
Anatomy and Embryology (Berlin)
Andrologia (Berlin)
Archives d'Anatomie, d'Histologie et
  d'Embryologie (Strasbourg)
Archives d'Anatomie et de Cytologie
  Pathologiques (Paris)
Archives d'Anatomie Microscopique et de
  Morphologie Experimentale (Paris)
Archives Internationales de Physiologie et de
  Biochimie (Liege)
Archivio di Fisiologia (Florence)
Archivio Italiano di Anatomia e di
  Embriologia (Florence)
Archivum Histologicum Japonicum. Nihon
  Soshikigaku Kiroku (Niigata)
Arkhiv Anatomii, Gistologii i Embriologii
  (Moscow)
Biology of the Neonate (Basel)
Brain, Behavior, and Evolution (Basel)
Calcified Tissue International (New York)
Canadian Journal of Genetics and Cytology
  (Ottawa)
Canadian Journal of Physiology and
  Pharmacology (Ottawa)
Cell (Cambridge MA)
Cell Biology International Reports (London)
Cell Differentiation (Limerick)
Cell and Tissue Kinetics (Oxford)
Cell and Tissue Research (Berlin)

Cellule (Brussels)
Ceskoslovenska Fysiologie (Prague)
Chronobiologia (Milan)
Clinical and Experimental Pharmacology and
  Physiology (Carlton)
Comparative Biochemistry and Physiology.
  A. Comparative Physiology (Oxford)
Computers and Biomedical Research (New
  York)
Connective Tissue Research (London)
Contraception (Los Altos CA)
Cytobiologie (Stuttgart)
Cytobios (Cambridge ENG)
Cytogenetics and Cell Genetics (Basel)
Cytologia (Tokyo)
Developmental Biology (New York)
Differentiation (New York)
EEG/EMG (Stuttgart)
Electroencephalography and Clinical
  Neurophysiology (Limerick)
European Journal of Applied Physiology and
  Occupational Physiology (Berlin)
Experimental Cell Research (New York)
Fiziologicheskii Zhurnal (Leningrad)
Folia Morphologica (Prague)
Growth (Lakeland FL)
Human Development (Basel)
Human Physiology (New York)
Indian Journal of Physiology and
  Pharmacology (New Delhi)
International Journal of Aging and Human
  Development (Farmingdale NY)
In Vitro (Gaithersburg MD)
Japanese Journal of Physiology (Tokyo)
Journal de Physiologie (Paris)
Journal of Anatomy (Cambridge ENG)
Journal of Applied Physiology: Respiratory,
  Environmental and Exercise Physiology
  (Bethesda)
Journal of Cell Biology (New York)
Journal of Cell Science (London)
Journal of Cellular Physiology (New York)
Journal of Comparative and Physiological
  Psychology (Washington)
Journal of Electron Microscopy (Tokyo)
Journal of Embryology and Experimental
  Morphology (Colchester)
Journal of General Physiology (New York)
Journal of Human Ergology (Tokyo)
Journal of Insect Physiology (Elmsford)
Journal of Membrane Biology (New York)

Journal of Microscopy (Oxford)
Journal of Molecular Evolution (Berlin)
Journal of Morphology (New York)
Journal of Neurocytology (London)
Journal of Neurophysiology (Bethesda)
Journal of Physiology (London)
Journal of Ultrastructure Research (New
  York)
Kaibogaku Zasshi. Journal of Anatomy
  (Tokyo)
Life Sciences (Oxford)
Mechanisms of Ageing and Development
  (Limerick)
Okajima's Folia Anatomica Japonica (Tokyo)
Patologicheskaya Fiziologiya i
  Eksperimental'naya Terapiya (Moscow)
Pfluegers Archiv. European Journal of
  Physiology (Berlin)
Physiologia Bohemoslovaca (Prague)
Physiological Reviews (Bethesda)
Physiologist (Bethesda)
Physiology and Behavior (Elmsford NY)
Prostaglandins, Leukotrienes, and Medicine
  (Edinburgh)
Psychophysiology (Baltimore)
Quarterly Journal of Experimental
  Physiology and Cognate Medical Sciences
  (Edinburgh)
Respiration Physiology (Amsterdam)
Revista Espanola De Fisiologia (Barcelona)
Scanning Electron Microscopy (Chicago)
Stain Technology (Baltimore)
Teratology (New York)
Tissue and Cell (Harlow)
Tsitologiya (Leningrad)
Tsitologiya i Genetika (Kiev)
Ultramicroscopy (Amsterdam)
Undersea Biomedical Research (Bethesda)
Uspekhi Fizicheskikh Nauk (Moscow)
Virchows Archiv. A. Pathological Anatomy
  and Histology (New York)
Zeitschrift fuer Mikroskopisch-Anatomische
  Forschung (Leipzig)
Zeitschrift fuer Morphologie und
  Anthropologie (Stuttgart)
Zeitschrift fuer Tierphysiologie
  Tierernaehrung und Futtermittelkunde
  (Hamburg)
Zhurnal Evolyutsionnoi Biokhimii i Fiziologii
  (Leningrad)

Medical and Pharmacological Abbreviations

aa	of each	feb	fever	Part. aeq.	equal parts
a.c.	before meals			PBI	protein-bound iodine
A/G	albumin globulin ratio	G.I.	gastrointestinal	p.c.	after meals
ANS	autonomic nervous system	gm.	gram	pCO₂	partial pressure of carbon dioxide
		gr.	grain		
Bib	drink	Grad.	gradually	PNS	peripheral nervous system
b.i.d.	twice a day	gtt.	drop(s)	P.O.	by mouth
bihor	during two hours			pO₂	partial pressure of oxygen
B.M.R.	basal metabolic rate	h.	hour	p.p.a.	having first shaken the bottle
B.P.	blood pressure	HCT	hematocrit		
BUN	blood urea nitrogen	Hg.	mercury	p.r.n.	as needed
b.v.	vapor bath	Hgb	hemoglobin	pro. us. ext.	for external use
		h.s.	at bedtime	pt.	let it be continued
c̄	with				
caps.	capsule	ibid	in the same place	q.	each; every
c.b.c.	complete blood count	I.M.	intramuscular	q.d.	every day
cc.	cubic centimeter(s)	incid.	cut	q.h.	every hour
cm.	centimeter(s)	in d.	in a day	q.____ h.	every ____ hours
CNS	central nervous system	inj.	an injection	q.i.d.	four times a day
Co., Comp.	compound	int. cib.	between meals	q. noct.	every night
cr	tomorrow	int. noct.	during the night	q.o.d.	every other day
C.S.F.	cerebrospinal fluid	IPPB	intermittent positive pressure breathing	q.q.	also
CVP	central venous pressure			q.s.	sufficient quantity
		I.V.	intravenous		
d.	a day			RBC	red blood cell
D. & C.	dilatation and curettage	kg.	kilogram		
D.C.	discontinue			s̄	without
D, Det	give	Lat. dol.	to the painful side	Semih.	half an hour
de d. in d.	from day to day			Sig.	write, label
Dieb, secund	every second day	M.	mix	S.O.S.	if needed
Dieb, tert.	every third day	man.	in the morning	sp. gr.	specific gravity
dim.	one-half	mEq.	milliequivalent	ss., s̄s̄	one-half
d. in dup.	give twice as much	mg.	milligram	s.s.s.	layer on layer
D. in p. aeq.	divide into equal parts	ml.	milliliter	stat.	immediately
dr.	dram			sum.	take
D.T.D.	give of such doses	Noct.	at night	s.v.r.	alcohol
Dur. dolor.	while pain lasts	Noct. maneq.	night and morning		
		N.P.O.	nothing by mouth	t.	three times
e	out of, with			tab.	tablet
ECG, EKG	electrocardiogram	O.D.	in the right eye	t.i.d.	three times a day
EEG	electroencephalogram	o.d.	every day		
e.m.p.	in the manner prescribed	Omn. hor.	every hour	ung.	ointment
		Omn. man.	every morning	Ut. dict.	as directed
		Omn. noct.	every night		
		O.S.	in the left eye	vic.	times
		O.U.	in each eye		
		oz.	ounce	WBC	white blood cell

Appendix D

Some Laboratory Tests

Test	Normal values (adult)	Clinical significance
Acetone and acetoacetate (serum)	0.3–2.0 mg/100 ml	Values increase in diabetic acidosis, toxemia of pregnancy, fasting, and high-fat diet.
Albumin-globulin ratio or A/G ratio (serum)	1.5:1 to 2.5:1	Ratio of albumin to globulin is lowered in kidney diseases and malnutrition.
Albumin (serum)	3.2–5.5 gm/100 ml	Values increase in multiple myeloma and decrease with proteinuria and as a result of severe burns.
Ammonia (plasma)	50–170 µg/100 ml	Values increase in severe liver disease, pneumonia, shock, and congestive heart failure.
Amylase (serum)	80–160 Somogyi units/100 ml	Values increase in acute pancreatitis, intestinal obstructions, and mumps. They decrease in chronic pancreatitis, cirrhosis of the liver, and toxemia of pregnancy.
Bilirubin, total (serum)	0.3–1.1 mg/100 ml	Values increase in conditions causing red blood cell destruction or biliary obstruction.
Blood urea nitrogen or BUN (plasma or serum)	10–20 mg/100 ml	Values increase in various kidney disorders and decrease in liver failure and during pregnancy.
Calcium (serum)	9.0–11.0 mg/100 ml	Values increase in hyperparathyroidism, hypervitaminosis D, and respiratory conditions that cause a rise in CO_2 concentration. They decrease in hypoparathyroidism, malnutrition, and severe diarrhea.
Carbon dioxide (serum)	24–30 mEq/l	Values increase in respiratory diseases, intestinal obstruction, and vomiting. They decrease in acidosis, nephritis, and diarrhea.
Chloride (serum)	96–106 mEq/l	Values increase in nephritis, Cushing's syndrome, and hyperventilation. They decrease in diabetic acidosis, Addison's disease, diarrhea, and following severe burns.
Cholesterol, total (serum)	150–250 mg/100 ml	Values increase in diabetes mellitus and hypothyroidism. They decrease in pernicious anemia, hyperthyroidism, and acute infections.
Creatine phosphokinase or CPK (serum)	Men: 0–20 IU/l Women: 0–14 IU/l	Values increase in myocardial infarction and skeletal muscle diseases such as muscular dystrophy.
Creatine (serum)	0.2–0.8 mg/100 ml	Values increase in muscular dystrophy, nephritis, severe damage to muscle tissue, and during pregnancy.
Creatinine (serum)	0.7–1.5 mg/100 ml	Values increase in various kidney diseases.
Erythrocyte count or red cell count (whole blood)	Men: 4,600,000–6,200,000/cu mm Women: 4,200,000–5,400,000/cu mm Children: 4,500,000–5,100,000/cu mm (varies with age)	Values increase as a result of severe dehydration or diarrhea and decrease in anemia, leukemia, and following severe hemorrhage.
Fatty acids, total (serum)	190–420 mg/100 ml	Values increase in diabetes mellitus, anemia, kidney disease, and hypothyroidism. They decrease in hyperthyroidism.
Globulin (serum)	2.5–3.5 gm/100 ml	Values increase as a result of chronic infections.

# of Clinical Importance

Common tests performed on blood (cont.)		
Test	**Normal values (adult)**	**Clinical significance**
Glucose (plasma)	70–115 mg/100 ml	Values increase in diabetes mellitus, liver diseases, nephritis, hyperthyroidism, and pregnancy. They decrease in hyperinsulinism, hypothyroidism, and Addison's disease.
Hematocrit (whole blood)	Men: 40–54 ml/100 ml Women: 37–47 ml/100 ml Children: 35–49 ml/100 ml (varies with age)	Values increase in polycythemia due to dehydration or shock. They decrease in anemia and following severe hemorrhage.
Hemoglobin (whole blood)	Men: 14–18 gm/100 ml Women: 12–16 gm/100 ml Children: 11.2–16.5 gm/100 ml (varies with age)	Values increase in polycythemia, obstructive pulmonary diseases, congestive heart failure, and at high altitudes. They decrease in anemia, pregnancy, and as a result of severe hemorrhage or excessive fluid intake.
Iron (serum)	75–175 µg/100 ml	Values increase in various anemias and liver disease. They decrease in iron deficiency anemia.
Iron-binding capacity (serum)	250–410 µg/100 ml	Values increase in iron deficiency anemia and pregnancy. They decrease in pernicious anemia, liver disease, and chronic infections.
Lactic acid (whole blood)	6–16 mg/100 ml	Values increase with muscular activity and in congestive heart failure, severe hemorrhage, and shock.
Lactic dehydrogenase or LDH (serum)	90–200 milliunits/ml	Values increase in pernicious anemia, myocardial infarction, liver diseases, acute leukemia, and widespread carcinoma.
Lipids, total (serum)	450–850 mg/100 ml	Values increase in hypothyroidism, diabetes mellitus, and nephritis. They decrease in hyperthyroidism.
Oxygen saturation (whole blood)	Arterial: 94–100% Venous: 60–85%	Values increase in polycythemia and decrease in anemia and obstructive pulmonary diseases.
ph (whole blood)	7.35–7.45	Values increase due to vomiting. Cushing's syndrome, and hyperventilation. They decrease as a result of hypoventilation, severe diarrhea, Addison's disease, and diabetic acidosis.
Phosphatase, acid (serum)	1.0–5.0 King-Armstrong units/ml	Values increase in cancer of the prostate gland, hyperparathyroidism, certain liver diseases, myocardial infarction, and pulmonary embolism.
Phosphatase, alkaline (serum)	5–13 King-Armstrong units/ml	Values increase in hyperparathyroidism (and in other conditions that promote resorption of bone), liver diseases, and pregnancy.
Phospholipids (serum)	6–12 mg/100 ml as lipid phosphorus	Values increase in diabetes mellitus and nephritis.
Phosphorus (serum)	3.0–4.5 mg/100 ml	Values increase in kidney diseases, hypoparathyroidism, acromegaly, and hypervitaminosis D. They decrease in hyperparathyroidism.
Platelet count (whole blood)	150,000–350,000/cu mm	Values increase in polycythemia and certain anemias. They decrease in acute leukemia and aplastic anemia.
Potassium (serum)	3.5–5.0 mEq/l	Values increase in Addison's disease, hypoventilation, and conditions that cause severe cellular destruction. They decrease in diarrhea, vomiting, diabetic acidosis, and chronic kidney disease.
Protein, total (serum)	6.0–8.0 gm/100 ml	Values increase in severe dehydration and shock. They decrease in severe malnutrition and hemorrhage.
Protein-bound iodine or PBI (serum)	3.5–8.0 µg/100 ml	Values increase in hyperthyroidism and liver disease. They decrease in hypothyroidism.
Prothrombin time (serum)	12–14 sec (one stage)	Values increase in certain hemorrhagic diseases, liver disease, vitamin K deficiency, and following the use of various drugs.

Common tests performed on blood (cont.)

Test	Normal values (adult)	Clinical significance
Sedimentation rate, Westergren (whole blood)	Men: 0–15 mm/hr Women: 0–20 mm/hr	Values increase in infectious diseases, menstruation, pregnancy, and as a result of severe tissue damage.
Sodium (serum)	136–145 mEq/l	Values increase in nephritis and severe dehydration. They decrease in Addison's disease, myxedema, kidney disease, and diarrhea.
Thyroxine or T$_4$ (serum)	2.9–6.4 μg/100 ml	Values increase in hyperthyroidism and pregnancy. They decrease in hypothyroidism.
Thromboplastin time, partial (plasma)	35–45 sec	Values increase in deficiencies of blood factors VIII, IX, and X.
Transaminases or SGOT (serum)	5–40 units/ml	Values increase in myocardial infarction, liver disease, and diseases of skeletal muscles.
Uric acid (serum)	Men: 2.5–8.0 mg/100 ml Women: 1.5–6.0 mg/100 ml	Values increase in gout, leukemia, pneumonia, toxemia of pregnancy, and as a result of severe tissue damage.
White blood cell count, differential (whole blood)	Neutrophils 54–62% Eosinophils 1–3% Basophils 0–1% Lymphocytes 25–33% Monocytes 3–7%	Neutrophils increase in bacterial diseases; lymphocytes and monocytes increase in viral diseases; eosinophils increase in collagen diseases, allergies, and in the presence of intestinal parasites.
White blood cell count, total (whole blood)	5,000–10,000/cu mm	Values increase in acute infections, acute leukemia, and following menstruation. They decrease in aplastic anemia and as a result of drug toxicity.

Common tests performed on urine

Test	Normal values	Clinical significance
Acetone and acetoacetate	0	Values increase in diabetic acidosis.
Albumin, qualitative	0 to trace	Values increase in kidney disease, hypertension, and heart failure.
Ammonia	20–70 mEq/l	Values increase in diabetes mellitus and liver diseases.
Bacterial count	Under 10,000/ml	Values increase in urinary tract infection.
Bile and bilirubin	0	Values increase in melanoma and biliary tract obstruction.
Calcium	Under 250 mg/24 hr	Values increase in hyperparathyroidism and decrease in hypoparathyroidism.
Creatinine clearance	100–140 ml/min	Values increase in renal diseases.
Creatinine	1–2 gm/24 hr	Values increase in infections and decrease in muscular atrophy, anemia, leukemia, and kidney diseases.
Glucose	0	Values increase in diabetes mellitus and various pituitary gland disorders.
17-hydroxycorticosteroids	2–10 mg/24 hr	Values increase in Cushing's syndrome and decrease in Addison's disease.
Phenylpyruvic acid	0	Values increase in phenylketonuria.
Urea clearance	Over 40 ml blood cleared of urea/min	Values increase in renal diseases.
Urobilinogen	0–4 mg/24 hr	Values increase in liver diseases and hemolytic anemia. They decrease in complete biliary obstruction and severe diarrhea.
Urea	25–35 gm/24 hr	Values increase as a result of excessive protein breakdown. They decrease as a result of impaired renal function.
Uric acid	0.6–1.0 gm/24 hr. as urate	Values increase in gout and decrease in various kidney diseases.

Glossary

The words in this glossary are followed by a phonetic guide to pronunciation. This is a simplified system that is standard in medical usage and terminology.

Any unmarked vowel that ends a syllable or stands alone as a syllable is long. Any unmarked vowel that is followed by a consonant has the short sound.

If a long vowel appears in the middle of a syllable (followed by a consonant), it is marked with a macron (-). Similarly, if a vowel stands alone or ends a syllable, but should have a short sound, it is marked with a breve (˘).

a

abdomen (ab-do′men) A region of the body between the diaphragm and pelvis.

abduction (ab-duk′shun) The movement of a body part away from the axis or midline of the body.

ABO system The most common system of classification for red blood cell antigens. On the basis of antigens on the red blood cell surface, individuals can be type A, type B, type AB, or type O.

absorption (ab-sorp′shun) The transport of molecules across epithelial membranes into the body fluids.

accessory organs (ak-ses′o-re) Organs that assist the functioning of other organs within a system.

accommodation (ah-kom″o-da′shun) To fit or adjust; specifically, the ability of the eyes to adjust their curvature so that an image of an object is focused on the retina at different distances.

acetabulum (as″ĕ-tab′u-lum) A socket in the lateral surface of the hipbone (os coxae) into which the head of the femur articulates.

acetone (as′e-tōn) A ketone body produced as a result of the oxidation of fats.

acetylcholine (ACh) (as″ĕ-til-ko′lēn) A molecule—which is an acetic acid ester of choline—that functions as a neurotransmitter chemical in somatic motor nerve and parasympathetic nerve fibers.

acetylcholinesterase (as″ ĕ-til-ko″lin-es′ter-ās) An enzyme in the membrane of postsynaptic cells that catalyzes the conversion of ACh into choline and acetic acid. This enzymatic reaction inactivates the neurotransmitter.

Achilles tendon (ah-kil′ēz) Tendon that attaches the calf muscles to the calcaneus bone.

acid (as′id) A substance that releases hydrogen ions when ionized in water.

acidosis (as″i-do′sis) An abnormal increase in the H^+ concentration of the blood that lowers arterial pH below 7.35.

acromegaly (ak″ro-meg′ah-le) A condition caused by the hypersecretion of growth hormone from the pituitary after maturity and characterized by enlargement of the extremities, such as the nose, jaws, fingers, and toes.

actin (ak′tin) A protein in muscle fibers that together with myosin is responsible for contraction.

action potential An all-or-none electrical event in an axon or muscle fiber, in which the polarity of the membrane potential is rapidly reversed and reestablished.

active immunity (ĭ-mu′nĭ-te) Immunity involving sensitization, in which antibody production is stimulated by prior exposure to an antigen.

active transport The movement of molecules or ions across the cell membranes of epithelial cells by membrane carriers; an expenditure of cellular energy (ATP) is required.

adduction (ah′duk′shun) The movement of a body part toward the axis or midline of the body.

adenoids (ad′ĕ-noids) The tonsils located in the nasopharynx; pharyngeal tonsils.

adenyl cyclase (ad″ĕ-nil si′klās) An enzyme, found in cell membranes, that catalyzes the conversion of ATP to cyclic AMP and pyrophosphate (PP_1). This enzyme is activated by an interaction between a specific hormone and its membrane receptor protein.

ADH (antidiuretic hormone) Also known as *vasopressin*. A hormone produced by the hypothalamus and secreted by the posterior pituitary gland; it acts on the kidneys to promote water reabsorption.

ADP Adenosine diphosphate; a molecule that, together with inorganic phosphate, is used to make ATP (adenosine triphosphate).

adrenal cortex (ah-dre′nal kor′teks) The outer part of the adrenal gland. Derived from embryonic mesoderm, the adrenal cortex secretes corticosteroid hormones (such as aldosterone and hydrocortisone).

adrenal medulla (mĕ-dul′ah) The inner part of the adrenal gland. Derived from embryonic postganglionic sympathetic neurons, the adrenal medulla secretes catecholamine hormones—epinephrine and (to a lesser degree) norepinephrine.

adrenergic (ad″ren-er′jik) An adjective describing the actions of epinephrine, norepinephrine, or other molecules with similar activity (as in *adrenergic receptor* and *adrenergic stimulation*).

aerobic capacity (a-er-o′bik) The ability of an organ to utilize oxygen and respire aerobically to meet its energy needs.

afferent (af′er-ent) Conveying or transmitting to.

afferent arteriole (ar-te′re-ōl) A blood vessel within the kidney that supplies blood to the glomerulus.

afferent neuron (nu′ron) A sensory nerve cell that transmits an impulse toward the central nervous system.

agglutinate (ah-gloo′tĭ-nāt) A clump of cells (usually erythrocytes) due to specific chemical interaction between surface antigens and antibodies.

agonist (ag′o-nist) The "prime mover" muscle, which is directly engaged in the contraction that produces the desired movement.

agranular leukocytes (ah-gran′u-ler loo′ko-sĭts) White blood cells (leukocytes) that do not contain cytoplasmic granules; specifically, lymphocytes and monocytes.

albumin (al-bu′min) A water-soluble protein, produced in the liver, that is the major component of the plasma proteins.

aldosterone (al-dos′ter-ōn) The principal corticosteroid hormone involved in the regulation of electrolyte balance (mineralocorticoid).

alimentary canal (al″ĕ-men′tar-e) The tubular portion of the digestive tract.

allergens (al′er-jens) Antigens that evoke an allergic response rather than a normal immune response.

allergy (al′er-je) A state of hypersensitivity caused by exposure to allergens; it results in the liberation of histamine and other molecules with histamine-like effects.

all-or-none principle The statement of the fact that muscle fibers of a motor unit contract to their maximum extent when exposed to a stimulus of threshold strength.

allosteric (al″o-ster′ik) Denoting the alteration of an enzyme's activity by its combination with a regulator molecule; allosteric inhibition by an end product represents negative feedback control of an enzyme's activity.

alveolar sacs (al-ve′o-lar) A cluster of alveoli that share a common chamber or central atrium.

alveolus (al-ve′o-lus) An individual air capsule within the lung. The alveoli are the basic functional units of respiration.

amniocentesis (am″ne-o-sen-te′sis) A procedure to obtain amniotic fluid and fetal cells in this fluid through transabdominal perforation of the uterus.

amnion (am′ne-on) A developmental membrane that surrounds the fetus and contains amniotic fluid.

amphiarthrosis (am″fe-ar-thro′sis) A slightly movable articulation.

amphoteric (am-fo-ter′ik) Pertaining to having opposite characteristics; denoting a molecule that can be positively or negatively charged, depending on the pH of its environment.

ampulla (am-pul′lah) A saclike enlargement of a duct or tube.

ampulla of Vater (fah′ter) A small, elevated area within the duodenum where the combined pancreatic and common bile duct empties.

anabolic steroids (an″ah-bol′ik ste′roids) Steroids with androgenlike stimulatory effects on protein synthesis.

anabolism (ah′nab′o-lizm) Chemical reactions within cells that result in the production of larger molecules from smaller ones; specifically, the synthesis of protein, glycogen, and fat.

anaerobic respiration (an″a-er-o′bik res″pi-ra′shun) A form of cell respiration, involving the conversion of glucose to lactic acid, in which energy is obtained without the use of molecular oxygen.

anal canal (a′nal) The terminal tubular portion of the rectum that opens through the anus of the alimentary canal.

anal glands Enlarged and modified sweat glands that empty into the anal opening.

anaphylaxis (an″ah-fi-lak′sis) An unusually severe allergic reaction, which can result in cardiovascular shock and death.

anastomosis (ah-nas″to-mo′sis) An interconnecting aggregation of blood vessels or nerves that form a network plexus.

anatomical position (an″ah-tom′e-kal) An erect body stance with the eyes directed forward, the arms at the sides, and the palms of the hands facing forward.

anatomy (ah-nat′o-me) The branch of science concerned with the structure of the body and the relationship of its organs.

androgens (an′dro-jens) Steroids, containing eighteen carbons, that have masculinizing effects; primarily those hormones (such as testosterone) secreted by the testes, although weaker androgens are also secreted by the adrenal cortex.

anemia (ah-ne′me-ah) An abnormal reduction in the red blood cell count, hemoglobin concentration, or hematocrit, or any combination of these measurements. This condition is associated with a decreased ability of the blood to carry oxygen.

angina pectoris (an-ji′nah pek′to-ris) A thoracic pain, often referred to the left pectoral and arm area, caused by myocardial ischemia.

angiotensin II (an″je-o-ten′sin) An eight-amino-acid polypeptide formed from angiotensin I (a ten-amino-acid precursor), which in turn is formed from cleavage of a protein (angiotensinogen) by the action of renin (which is an enzyme secreted by the kidneys). Angiotensin II is a powerful vasoconstrictor and a stimulator of aldosterone secretion from the adrenal cortex.

anions (an′i-ons) Ions that are negatively charged, such as chloride, bicarbonate, and phosphate.

antagonist (an-tag′-o-nist) A muscle that acts in opposition to a "prime mover," or an agonist.

antebrachium (an″te-bra′ke-um) The forearm.

anterior (ventral) (an-te′re-or) Toward the front; the opposite of *posterior,* or *dorsal.*

anterior pituitary (pi-tu′i-tar″e) Also called the *adenohypophysis;* this part of the pituitary gland secretes FSH (follicle-stimulating hormone), LH (luteinizing hormone), ACTH (adrenocorticotrophic hormone), TSH (thyroid-stimulating hormone), GH (growth hormone), and prolactin. Secretions of the anterior pituitary are controlled by hormones secreted by the hypothalamus.

anterior root The anterior projection of the spinal cord, which is composed of axons of motor or efferent fibers.

antibodies (an″ti-bod′es) Immunoglobin proteins secreted by B lymphocytes that have transformed into plasma cells. Antibodies are responsible for humoral immunity. Their synthesis is induced by specific antigens, and they combine with these specific antigens but not with unrelated antigens.

anticodon (an″ti-ko′don) A base triplet provided by three nucleotides within a loop of transfer RNA, which is complementary in its base pairing properties to a triplet (the codon) in mRNA; the matching of codon to anticodon provides the mechanism for translating of genetic code into a specific sequence of amino acids.

antigen (an′ti-jen) A molecule able to induce the production of antibodies and to react in a specific manner with antibodies.

antigenic determinant site (an-ti-jen′ik) The region of an antigen molecule that specifically reacts with particular antibodies. A large antigen molecule may have a number of such sites.

antiserum (an″ti-se′rum) A serum that contains specific antibodies.

anus (a-nus) The terminal portion and outlet of the alimentary canal.

aorta (a-or′tah) The major systemic vessel of the arterial system of the body, emerging from the left ventricle.

apex (a′peks) The tip or pointed end of a conical structure.

apneustic center (ap-nu′stik) A collection of neurons in the brain stem that participates in the rhythmic control of breathing.

apocrine gland (ap′o-krin) A type of sweat gland that functions in evaporative cooling. It may respond during periods of emotional stress.

aponeurosis (ap″o-nu-ro′sis) A fibrous or membranous sheetlike tendon.

aqueous humor (a′-kwe-us hum′or) The watery fluid that fills the anterior and posterior chambers of the eye.

arachnoid (ah-rak′noid) The weblike middle covering (one of the three meninges) of the central nervous system.

arbor vitae (ar′bor vi′tah) The branching arrangement of white matter within the cerebellum.

arch of aorta The superior left bend of the aorta between the ascending and descending portions.

arm (brachium) The portion of the upper extremity from the shoulder to the elbow.

arrector pili (ah-rek′tor pih′le) The smooth muscle attached to a hair follicle, which upon contraction, pulls the hair vertical, resulting in "goose bumps."

arteriole (ar-te′re-ōl) A minute arterial branch.

arteriosclerosis (ar-te″re-o-skle-ro′sis) A group of diseases characterized by thickening and hardening of the artery wall and in the narrowing of its lumen.

arteriovenous anastomoses (ar-te″re-o-ve′nus ah-nas″to-mo′sis) Direct connections between arteries and veins that bypass capillary beds.

artery (ar′ter-e) A blood vessel that carries blood away from the heart.

arthrology (ar-throl′o-je) The scientific study of the structure and function of joints.

articular cartilage (ar-tik′u-lar kar′ti-lij) A hyaline cartilaginous covering over the articulating surface of the bones of synovial joints.

articulation (ar-tik″u-la′shun) A joint.

arytenoid cartilages (ar″ĕ-te′noid) A pair of small cartilages located on the superior aspect of the larynx.

ascending colon (ko′lon) The portion of the large intestine between the cecum and the hepatic flexure.

association neuron (nu′ron) A nerve cell located completely within the central nervous system. It conveys impulses in an arc from afferent to efferent neurons.

astigmatism (ah-stig′mah-tizm) Unequal curvature of the refractive surfaces of the eye (cornea and/or lens), so that light that enters the eye along certain meridians does not focus on the retina.

atherosclerosis (ath″er-o″skle-ro′sis) A common type of arteriosclerosis found in medium and large arteries in which raised areas, or plaque, within the tunica intima are formed from smooth muscle cells, cholesterol, and other lipids. These plaques occlude arteries and serve as sites for the formation of thrombi.

atomic number (ah-tom′ik) A whole number representing the number of positively charged protons in the nucleus of an atom.

atopic dermatitis (ah-top′ik der″mah-ti′tis) An allergic skin reaction to agents such as poison ivy and poison oak; a type of delayed hypersensitivity.

ATP Adenosine triphosphate; the universal energy donor of the cell.

atretic (ah-tret′ik) Without an opening; atretic ovarian follicles are those that fail to ovulate.

atrioventricular bundle (a″tre-o-ven-trik′u-lar) A group of specialized cardiac fibers that conduct impulses from the atrioventricular node to the ventricular muscles or the heart; also called the *bundle of His* or *AV bundle.*

atrioventricular node A microscopic aggregation of specialized cardiac fibers located in the interatrial septum of the heart that are a part of the conduction system of the heart; *AV node.*

atrioventricular valve A cardiac valve located between an atrium and a ventricle of the heart; *AV valve.*

atrium (a′tre-um) Either of the two superior chambers of the heart that receive venous blood.

atrophy (at′ro-fe) A gradual wasting away or decrease in the size of a tissue or an organ.

atropine (at′ro-pēn) An alkaloid drug obtained from a plant of the species *Belladonna* that acts as an anticholinergic agent. Used medically to inhibit parasympathetic nerve effects, dilate the pupils of the eye, increase the heart rate, and inhibit movements of the intestine.

auditory (aw′dĭ-to″re) Pertaining to the structures of the ear that are associated with hearing.

auricle (aw′rĕ-kl) 1. The fleshy pinna of the ear. 2. An ear-shaped appendage of each atrium of the heart.

autoantibodies (aw″to-an′tĭ-bod″e) Antibodies that are formed in response to, and which react with, molecules that are part of one's own body.

autonomic nervous system (aw″to-nom′ik) The sympathetic and parasympathetic portions of the nervous system that function to control the actions of the visceral organs and skin.

autosomal chromosomes (aw″to-so′mal kro′mo-sōms) The paired chromosomes; those other than the sex chromosomes.

axilla (ak-sil′ah) Pertaining to the depressed hollow commonly called the armpit.

axon (ak′son) The elongated process of a nerve cell that transmits an impulse away from the cell body.

b

ball-and-socket joint The most freely movable type of diarthrosis (e.g., the shoulder or hip joint).

Barr body (bahr) A microscopic structure in the cell nucleus produced from an inactive X chromosome in females.

basal ganglion (ba′sal gang′gle-on) A mass of nerve cell bodies located deep within a cerebral hemisphere of the brain.

basal metabolic rate (BMR) (ba′sal met″ah-bol′ik) The rate of metabolism (expressed as oxygen consumption or heat production) under resting or basal conditions (fourteen to eighteen hours after eating).

base (bās) A chemical substance that ionizes in water to release hydroxyl ions (OH^-) or other ions that combine with hydrogen ions.

basement membrane A thin sheet of extracellular substance to which the basal surfaces of epithelial cells are attached; also called the *basal lamina.*

basophil (ba′so-fil) A granular leukocyte that readily stains with basophilic dye.

B cell lymphocytes Lymphocytes that can be transformed by antigens into plasma cells that secrete antibodies (and are thus responsible for humoral immunity). The *B* stands for *bursa equivalent.*

belly The thickest circumference of a skeletal muscle.

benign (be-nīn′) Not malignant.

bifurcation (bi″fur-ka′shun) Forked; divided into two branches.

bile (bīl) Fluid produced by the liver and stored in the gallbladder that contains bile salts, bile pigments, cholesterol, and other molecules. The bile is secreted into the small intestine.

bilirubin (bil″i-roo′bin) Bile pigment derived from the breakdown of the heme portion of hemoglobin.

bipennate (bi-pen′āt) Denoting muscles that have a fiber architecture coursing obliquely on both sides of a tendon.

blocking antibodies An antibody that is specific for antigens attacked by other antibodies or by T cells and, therefore, may block this attack.

blood The fluid connective tissue that circulates through the cardiovascular system to transport substances throughout the body.

blood-brain barrier A specialized mechanism that inhibits the passage of certain materials from the blood into brain tissue and cerebrospinal fluid.

bone A solid, rigid, ossified connective tissue forming the skeletal system.

bony labyrinth (lab′i-rinth) A series of chambers within the petrous portion of the temporal bone associated with the vestibular organs and the cochlea.

Bowman's capsule (bo′manz kap′sūl) The double-walled proximal portion of a renal tubule that encloses the glomerulus of a nephron.

brachial plexus (bra′ke-al plek′sus) A network of nerve fibers that arise from C5–C8 and T1. Nerves arising from the brachial plexuses supply the upper extremities.

bradycardia (brad″e-kar′de-ah) A slow cardiac rate; less than sixty beats per minute.

bradykinins (brad″e-ki′nins) Short polypeptides that stimulate vasodilation and other cardiovascular changes.

brain The enlarged superior portion of the central nervous system located in the cranial cavity of the skull.

brain stem The portion of the brain consisting of the medulla oblongata, pons, and midbrain.

bronchial tree (brong′ke-al) The bronchi and their branching bronchioles.

bronchiole (brong′ke-ōl) A small division of a bronchus within the lung.

bronchus (brong′kus) A branch of the trachea that leads to a lung.

buccal cavity (buk′al) The mouth, or oral cavity.

buffer (buf′er) A molecule that serves to prevent large changes in pH by either combining with H^+ or by releasing H^+ into solution.

bulbourethral glands (bul″bo-u-re′thral) A pair of glands that secrete a viscous fluid into the male urethra during sexual excitement; also called *Cowper's glands.*

bundle of His A band of rapidly conducting cardiac fibers that originate in the AV node and extend down the atrioventricular septum to the apex of the heart. This tissue conducts action potentials from the atria into the ventricles.

bursa (ber′sah) A saclike structure filled with synovial fluid, which occurs around joints and over which tendons can slide without contacting bone.

buttocks (but′oks) The rump or fleshy masses on the posterior aspect of the lower trunk, formed primarily by the gluteal muscles.

c

calcitonin (kal″si-to′nin) Also called *thyrocalcitonin.* A polypeptide hormone produced by the parafollicular cells of the thyroid and secreted in response to hypercalcemia. It acts to lower blood calcium and phosphate concentrations and may serve as an antagonist of parathyroid hormones.

calmodulin (kal′mod-u′lin) A receptor protein for Ca^{++} located within the cytoplasm of target cells. It appears to mediate the effects of this ion on cellular activities.

calorie (kal′o-re) A unit of heat equal to the amount of heat needed to raise the temperature of one gram of water by 1° C.

calyx (ka′liks)   A cup-shaped portion of the renal pelvis that encircles renal papillae.

cAMP   Cyclic adenosine monophosphate; a second messenger in the action of many hormones, such as catecholamines, polypeptides, and glycoproteins. It serves to mediate the effects of these hormones on their target cells.

canaliculus (kan″ah-lik′u-lus)   A microscopic channel in bone tissue that connects lacunae.

canal of Schlemm (schlem)   A circular venous drainage for the aqueous humor from the anterior chambers; located at the junction of the sclera and the cornea.

cancellous bone (kan′se-lus)   Spongy bone; bone tissue with a latticelike structure.

cancer   A tumor characterized by abnormally rapid cell division and the loss of specialized tissue characteristics. This term usually refers to malignant tumors.

capacitation (kah-pas″ĭ-ta′shun)   Changes that occur within spermatozoa in the female reproductive tract that enable the sperm to fertilize ova; sperm that have not been capacitated in the female tract cannot fertilize ova.

capillary (cap′ĭ-lar″e)   A microscopic blood vessel that connects an arteriole and a venule; the functional unit of the circulatory system.

carbonic anhydrase (kar-bon′ik an-hi′drās)   An enzyme that catalyzes the formation or breakdown of carbonic acid. When carbon dioxide concentrations are relatively high this enzyme catalyzes the formation of carbonic acid from CO_2 and H_2O. When carbon dioxide concentrations are low, the breakdown of carbonic acid to CO_2 and H_2O is catalyzed. These reactions aid the transport of carbon dioxide from tissues to alveolar air.

cardiac muscle (kar′de-ak)   Muscle of the heart, consisting of striated muscle cells. These cells are interconnected into a mass called the myocardium.

cardiac output   The volume of blood pumped by either the right or left ventricle per minute.

cardiogenic shock (kar″de-o-jen′ik)   Shock that results from low cardiac output in heart disease.

carpus (kar′pus)   Pertaining to the wrist; collectively, the eight wrist bones.

carrier-mediated transport   The transport of molecules or ions across a cell membrane by means of specific protein carriers. It includes both facilitated diffusion and active transport.

cartilage (kar′ti-lij)   A type of connective tissue with a solid elastic matrix.

cartilaginous joint (kar″ti-laj′ĭ-nus)   A joint that lacks a joint cavity and permits little movement between the bones held together by cartilage.

cast   An accumulation of proteins that produces molds of kidney tubules and that appears in urine sediment.

catabolism (kah-tab′o-lizm)   Chemical reactions in a cell whereby larger, more complex molecules are converted into smaller molecules.

catecholamines (kat″ĕ-kol′ah-mēns)   A group of molecules including epinephrine, norepinephrine, L-dopa, and related molecules with effects that are similar to those produced by activation of the sympathetic nervous system.

cations (kat′i-ons)   Positively charged ions, such as sodium, potassium, calcium, and magnesium.

cauda equina (kaw′dah e-kwi′nah)   The lower end of the spinal cord where the roots of spinal nerves have a tail-like appearance.

caudal (kaw′dal)   Referring to a position more toward the tail.

cecum (se′kum)   The pouchlike portion of the large intestine to which the ileum of the small intestine is attached.

cell   The structural and functional unit of an organism; the smallest structure capable of performing all the functions necessary for life.

cell-mediated immunity (ĭ-mu′nĭ-te)   Immunological defense provided by T cell lymphocytes, which come into close proximity to their victim cells (as opposed to humoral immunity provided by the secretion of antibodies by plasma cells).

cellular respiration (sel′u-lar res″pĭ-ra′shun)   The energy-releasing metabolic pathways in a cell that oxidize organic molecules such as glucose, fatty acids, and others.

cementum (se-men′tum)   Bonelike material that binds the root of a tooth into its bony socket.

central canal   A small tube that runs the length of the spinal cord in the gray commissure and contains cerebrospinal fluid.

central nervous system   The brain and the spinal cord; *CNS*.

centrioles (sen′tri-ōl)   Cell organelles that form the spindle apparatus during cell division.

centromere (sen′tro-mēr)   The central region of a chromosome to which the chromosomal arms are attached.

centrosome (sen′tro-sōm)   A dense body near the nucleus of a cell that contains a pair of centrioles.

cerebellar peduncle (ser″ĕ-bel′ar pe-dung′-k′l)   An aggregation of nerve fibers that connect the cerebellum with the brain stem.

cerebellum (ser″ĕ-bel′um)   The portion of the brain concerned with the coordination of movements. Part of the metencephalon, it consists of two hemispheres and a central vermis.

cerebral aqueduct (ser′ĕ-bral ak′we-dukt″)   The channel that connects the third and fourth ventricles of the brain; also called the *aqueduct of Sylvius*.

cerebral peduncles   A paired bundle of nerve fibers along the ventral surface of the midbrain, conducting impulses between the pons and the cerebral hemispheres.

cerebrospinal fluid (ser′ĕ-bro-spi′nal)   A fluid produced by the choroid plexus of the ventricles of the brain. It fills the ventricles and surrounds the central nervous system in association with the meninges.

cerebrum (ser′ĕ-brum)   The largest portion of the brain, composed of the right and left hemispheres.

cervical (ser′vĭ-kal)   Pertaining to the neck or a necklike portion of an organ.

cervical ganglion (gang′gle-on)   A cluster of postganglionic sympathetic nerve cell bodies located in the neck, near the cervical vertebrae.

cervical plexus (plek′sus)   A network of spinal nerves formed by the anterior branches of the first four cervical nerves.

cervix (ser′viks)   1. The narrow necklike portion of an organ. 2. The inferior end of the uterus that adjoins the vagina.

chemoreceptor (ke″mo-re-sep′tor)   A neuroreceptor that is stimulated by the presence of chemical molecules.

chemotaxis (ke″mo-tak′sis)   The movement of an organism or a cell, such as a leukocyte, toward a chemical stimulus.

Cheyne-Stokes respiration (chān′stōkes res″pĭ-ra′shun)   Breathing characterized by rhythmic waxing and waning of the depth of respiration, with regularly occurring periods of apnea (lack of breathing).

cholesterol (ko-les′ter-ol)   A twenty-seven-carbon steroid that serves as the precursor of steroid hormones.

cholinergic (ko″lin-er′jik)   Denoting nerve endings that liberate acetylcholine as a neurotransmitter, such as those of the parasympathetic system.

chondrocranium (kon″dro-kra′ne-um)   The portion of the skull that supports the brain and is derived from endochondral bone.

chondrocytes (kon′dro-sīts)   Cartilage-forming cells.

chordae tendineae (kor′de ten-din′e-e)   Chordlike tendinous bands that connect papillary muscles to the atrioventricular valves within the ventricles of the heart.

chorea (ko-re′ah)   The occurrence of a wide variety of rapid, complex, jerky movements that appear to be well coordinated but are performed involuntarily.

choroid (ko′roid)   The vascular, pigmented middle layer of the wall of the eye.

choroid plexus (ko′roid plek′sus)   A mass of vascular capillaries from which cerebrospinal fluid is secreted into the ventricles of the brain.

chromatids (kro′mah-tids)   Duplicated chromosomes that are joined together at the centromere and separate during cell division.

chromatin (kro′mah-tin)   Threadlike structures in the cell nucleus consisting primarily of DNA and protein. They represent the extended form of chromosomes during interphase.

chromosome (kro'mo-sōm)   Structures in the nucleus that contain the genes for genetic expression.

chyme (kīm)   The mixture of partially digested food and digestive juices within the stomach and small intestine.

cilia (sil'e-ah)   Plural of cilium; tiny hairlike processes that extend from the cell surface and beat in a coordinated fashion.

ciliary body (sil'e-er''e)   A portion of the choroid layer of the eye that secretes aqueous humor and contains the ciliary muscle.

circadian rhythms (ser''kah-de'an)   Physiological changes that repeat at about a twenty-four-hour period. These are often synchronized to changes in the external environment, such as the day-night cycles.

circle of Willis   An arterial vessel located on the ventral surface of the brain around the pituitary gland.

circumduction (ser''kum-duk'shun)   A conelike movement of a body part, such that the distal end moves in a circle while the proximal portion remains relatively stable.

circumvallate papilla (sir''kum-val'āt pah-pil'ah)   The largest papillae on the dorsal surface of the tongue. They are arranged in an inverted V-shaped pattern at the posterior portion of the tongue.

cirrhosis (sir-ro'sis)   Liver disease characterized by loss of normal microscopic structure, which is replaced by fibrosis and nodular regeneration.

clitoris (kli'to-ris)   A small, erectile structure in the vulva of the female, homologous to the glans penis in the male.

clone (klōn)   1. A group of cells derived from a single parent cell by mitotic cell division; since reproduction is asexual, the descendants of the parent cell are genetically identical. 2. A term used when cells are separate individuals (as in white blood cells) rather than part of a growing organ.

CNS   Central nervous system; part of the nervous system consisting of the brain and spinal cord.

coccygeal (kok-sij'e-al)   Pertaining to the region of the coccyx; the caudal termination of the vertebral column.

cochlea (kok'le-ah)   The organ of hearing in the inner ear where nerve impulses are generated in response to sound waves.

codon (ko'don)   The sequence of three nucleotide bases in mRNA that specifies a given amino acid and determines the position of that amino acid in a polypeptide chain through complementary base pairing with an anticodon in RNA.

coelom (se'lom)   The abdominal cavity.

coenzyme (ko-en'zīm)   An organic molecule, usually derived from a water-soluble vitamin, that combines with and activates specific enzyme proteins.

cofactor (ko'fak-tor)   A substance needed for the catalytic action of an enzyme; usually refers to inorganic ions such as Ca^{++} and Mg^{++}.

collateral (kŏ-lat'er-al)   A small branch of a blood vessel or nerve fiber.

colloid osmotic pressure (kol'oid oz-mot'ik)   Osmotic pressure exerted by plasma proteins, which are present as a colloidal suspension. Also called oncotic pressure.

colon (ko'lon)   The large intestine.

common bile duct   A tube that is formed by the union of the hepatic duct and cystic duct and transports bile to the duodenum.

compact (dense) bone   Tightly packed bone that is superficial to spongy bone and covered by the periosteum.

compliance (kom-pli'ans)   A measure of the ease with which a structure such as the lungs expands under pressure; a measure of the change in volume as a function of pressure changes.

condyle (kon'dīl)   A rounded process at the end of a long bone that forms an articulation.

cone (kōn)   A color receptor cell in the retina of the eye.

congenital (kon-jen'i-tal)   Present at the time of birth.

congestive heart failure (kon-jes'tiv)   The inability of the heart to deliver an adequate blood flow, due to heart disease or hypertension; it is associated with breathlessness, salt and water retention, and edema.

conjunctiva (kon''junk-ti'vah)   The thin membranous covering on the anterior surface of the eyeball and lining the eyelids.

conjunctivitis (kon''junk''ti-vi'tis)   Inflammation of the conjunctiva of the eye which sometimes is called "pink eye."

connective tissue   One of the four basic tissue types within the body. It is a binding and supportive tissue with abundant matrix.

Conn's syndrome (konz)   Primary hyperaldosteronism; excessive secretion of aldosterone produces electrolyte imbalances.

contralateral (kon''trah-lat'er-al)   Affecting the opposite side of the body.

conus medullaris (ko'nus med''u-lār'is)   The caudal, tapering portion of the spinal cord.

convolution (kon-vo-lu'shun)   An elevation on the surface of a structure and an infolding of the tissue upon itself.

cornea (kor'ne-ah)   The transparent convex, anterior portion of the outer layer of the eyeball.

coronary circulation (kor'ŏ-na-re)   The arterial and venous blood circulation to the wall of the heart.

coronary sinus   A large venous channel on the posterior surface of the heart into which the cardiac veins drain.

corpora quadrigemina (kor'po-rah kwod''ri-jem'i-nah)   Four dorsal lobes of the midbrain concerned with visual and auditory functions.

corpus callosum (kor'pus kah-lo'sum)   A major tract within the brain that is composed of white matter and connects the right and left cerebral hemispheres.

cortex (kor'teks)   The outer layer of an organ such as the convoluted cerebrum, adrenal gland, or kidney.

corticosteroids (kor''ti-ko-ste'roids)   Steroid hormones of the adrenal cortex, consisting of glucocorticoids (such as hydrocortisone) and mineralocorticoids (such as aldosterone).

costal cartilage (kos'tal)   The cartilage that connects the ribs to the sternum.

cranial (kra'ne-al)   Pertaining to the cranium.

cranial nerve   One of twelve pairs of nerves that arise from the ventral surface of the brain.

cranium (kra'ne-um)   The endochondral bones of the skull that enclose or support the brain and the organs of sight, hearing, and balance.

creatine phosphate (kre'ah-tin fos'fāt)   An organic phosphate molecule in muscle cells that serves as a source of high-energy phosphate for the synthesis of ATP. Also called *phosphocreatine*.

crenation (kre-na'shun)   A notched or scalloped appearance of the red blood cell membrane caused by the osmotic loss of water from these cells.

crest   A thickened ridge of bone for the attachment of muscle.

cretinism (kre'tin-izm)   A condition caused by insufficient thyroid secretion during prenatal development or the years of early childhood; results in stunted growth and inadequate mental development.

cricoid cartilage (kri'koid)   A ring-shaped cartilage that forms the inferior end of the larynx.

crista (kris'tä)   A crest, such as the crista galli extending superiorly from the cribriform plate.

cryptorchidism (krip-tor'ki-dizm)   A developmental defect in which the testes fail to descend into the scrotum and, instead, remain in the body cavity.

cubital (ku'bi-tal)   Pertaining to the forearm. The cubital fossa is the anterior aspect of the elbow joint.

curare (koo-rah're)   A chemical derived from plant sources that causes flaccid paralysis by blocking ACh receptor proteins in muscle cell membranes.

Cushing's syndrome (koosh'ingz)   Symptoms caused by the hypersecretion of adrenal steroid hormones, due to tumors of the adrenal cortex or to ACTH-secreting tumors of the anterior pituitary.

cyanosis (si''ah-no'sis)   A blue color given to the skin or mucous membranes by deoxyhemoglobin; indicates inadequate oxygen concentration in the blood.

cystic duct (sis'tik dukt)   The tube that transports bile from the gallbladder to the common bile duct.

cytochrome (si'to-krōm)   A pigment in mitochondria that transports electrons in the process of aerobic respiration.

cytokinesis (si″to-ki-ne′sis) The division of the cytoplasm that occurs in mitosis and meiosis, when a parent cell divides to produce two daughter cells.

cytology (si-tol′o-je) The science dealing with the study of cells.

cytoplasm (si′to-plazm″) In a cell, the protoplasm located outside of the nucleus.

cytoskeleton (si″to-skel′ĕ-ton) A latticework of structural proteins in the cytoplasm arranged in the form of microfilaments and microtubules.

d

deciduous (de-sid′u-us) Not permanent. Deciduous teeth are shed and replaced by permanent teeth during development.

defecation (def″ĕ-ka′shun) The elimination of feces from the rectum through the anus.

deglutition (deg″loo-tish′un) The process of swallowing.

delayed hypersensitivity An allergic response in which the onset of symptoms takes as long as two to three days after exposure to an antigen. Produced by T cells, it is a type of cell-mediated immunity.

dendrite (den′drīt) A nerve cell process that transmits impulses toward a neuron cell body.

dentin (den′tīn) The main substance of a tooth, covered by enamel over the crown of the tooth and by cementum on the root.

dentition (den-tish′un) The number, arrangement, and shape of teeth.

depolarization (de-po″lar-i-za′shun) The loss of membrane polarity in which the inside of the cell membrane becomes less negative in comparison to the outside of the membrane. The term is also used to indicate the reversal of membrane polarity that occurs during the production of action potentials in nerve and muscle cells.

dermis (der′mis) The second, or deep, layer of skin beneath the epidermis.

descending colon The segment of the large intestine that descends on the left side from the level of the spleen to the level of the left iliac crest.

diabetes insipidus (di″ah-be′tēz in-sip′id-es) A condition in which inadequate amounts of antidiuretic hormone (ADH) are secreted by the posterior pituitary. It results in the inadequate reabsorption of water by the kidney tubules and, thus, in the excretion of a large volume of dilute urine.

diabetes mellitus (di″ah-be′tēz mel′li-tus) The appearance of glucose in the urine due to the presence of high plasma glucose concentrations, even in the fasting state. This disease is caused by either lack of sufficient insulin secretion or inadequate responsiveness of the target tissues to the effects of insulin.

diapedesis (di″ah-pĕ′de′sis) The migration of white blood cells through the endothelial walls of blood capillaries into the surrounding connective tissues.

diaphragm (di′ah-fram) A sheetlike dome of muscle and connective tissue that separates the thoracic and abdominal cavities.

diaphysis (di-af′i-sis) The shaft of a long bone.

diarrhea (di″ah-re′ah) Abnormal frequence of defecation accompanied by abnormal liquidity of the feces.

diarthrosis (di″ar-thro′sis) A type of joint in which the articulating bones are freely movable; also called a *synovial joint.*

diastole (di-as′to-le) The sequence of the cardiac cycle during which a heart chamber wall is relaxed.

diencephalon (di″en-sef′ah-lon) A major region of the brain that includes the third ventricle, thalamus, hypothalamus, and pituitary.

diffusion (di-fu′zhun) The net movement of molecules or ions from regions of higher to regions of lower concentration.

digestion The process by which larger molecules of food substance are broken down mechanically and chemically into smaller molecules that can be absorbed.

diploid (dip′loid) Denoting cells having two of each chromosome, or twice the number of chromosomes that are present in sperm or ova.

disaccharide (di-sak′ah-rīd) A class of double sugars; carbohydrates that yield two simple sugars, or monosaccharides, upon hydrolysis.

distal (dis′tal) Away from the midline or origin; the opposite of *proximal.*

DNA Deoxyribonucleic acid; composed of nucleotide bases and deoxyribose sugar; contains the genetic code.

dopa (do′pah) A derivative of the amino acid tyrosine, L-dopa serves as the precursor for the neurotransmitter molecule dopamine. L-dopa is given to patients with Parkinson's disease to stimulate dopamine production.

dopamine (do′pah-mēn) A type of neurotransmitter in the central nervous system; also is the precursor of norepinephrine, another neurotransmitter molecule.

dorsal (dor′sal) Pertaining to the back or posterior portion of a body part; the opposite of *ventral.*

dorsal root ganglion Collections of cell bodies of sensory neurons that form swellings in the dorsal roots of spinal nerves.

ductus arteriosus (duk′tus ar-te″re-o′sus) The blood vessel that connects the pulmonary trunk and the aorta in a fetus.

ductus deferens (def′er-enz) A tube that carries spermatozoa from the epididymis to the ejaculatory duct; also called the *vas deferens* or *seminal duct.*

ductus venosus (ven-o′sus) A fetal blood vessel that connects the umbilical vein and the inferior vena cava.

duodenum (du″o-de′num) The first portion of the small intestine that leads from the pyloric sphincter of the stomach to the jejunum.

dura mater (du′rah ma′ter) The outermost meninx.

dwarfism A condition in which a person is undersized due to inadequate secretion of growth hormone.

dyspnea (disp′ne-ah) Subjective difficulty in breathing.

e

eccrine gland (ek′rin) A sweat gland that functions in thermoregulation.

ECG Electrocardiogram; EKG.

E. coli A species of bacteria normally found in the human intestine; full name is *Escherichia coli.*

ectoderm (ek′to-derm) The outermost of the three primary germ layers of an embryo.

ectopic (ek-top′ik) Foreign, out of place.

ectopic focus An area of the heart other than the SA node that assumes pacemaker activity.

ectopic pregnancy Embryonic development that occurs anywhere other than in the uterus (as in the uterine tubes or body cavity).

edema (ĕ-de′mah) An excessive accumulation of fluid in the body tissues.

EEG Electroencephalogram; a recording of the electrical activity of the brain from electrodes placed on the scalp.

effector (ef-fek′tor) An organ such as a gland or muscle that responds to a motor stimulation.

efferent (ef′er-ent) Conveying away from the center of an organ or structure.

efferent arteriole (ar-te′re-ōl) An arteriole of the renal vascular system that conducts blood away from the glomerulus of a nephron.

efferent ductules (dukt′ūls) A series of coiled tubules that convey spermatozoa from the rete testis to the epididymis.

efferent neuron (nu′ron) A motor nerve cell that conducts impulses from the central nervous system to effector organs such as muscles or glands.

ejaculation (e-jak″u-la′shun) The discharge of semen from the male urethra during climax.

ejaculatory duct (e-jak′u-lah-to″re) A tube that transports spermatozoa from the ductus deferens to the prostatic urethra.

elastic fibers (e-las′tik) Protein strands that are found in certain connective tissue and have contractile properties.

elbow The diarthrotic joint between the brachium and the forearm.

electrocardiogram (e-lek″tro-kar′de-o-gram″) A recording of the electrical activity that accompanies the cardiac cycle; ECG or EKG.

electroencephalogram (e-lek″tro-en-sef′ah-lo-gram″) A recording of the brain wave patterns or electrical impulses of the brain; EEG.

electrolytes (e-lek′tro-līts)   Ions and molecules that are able to ionize and thus carry an electric current. The most common electrolytes in the plasma are Na$^+$, HCO$_3^-$, and K$^+$.

electromyogram (e-lek″tro-mi′o-gram″)   A recording of the electrical impulses or activity of a muscle; EMG.

electrophoresis (e-lek″tro-fo-re′sis)   A biochemical technique in which different molecules can be separated and identified by their rate of movement in an electric field.

elephantiasis (el″ĕ-fan-ti′ah-sis)   A disease caused by infection with a nematode worm, in which the larvae block lymphatic drainage and produce edema; the lower areas of the body can become enormously swollen as a result.

embryology (em′bre-ol′o-je)   The study of prenatal development from conception through the eighth week in utero.

EMG   Electromyogram; electrical recordings of the activity of skeletal muscles using surface electrodes.

emmetropia (em-ĕ-tro′pe-ah)   A condition of normal vision in which the image of objects is focused on the retina, as opposed to nearsightedness (myopia) or farsightedness (hypermetropia).

emphysema (em″fĭ-se′mah, em″fĭ-ze′mah)   A lung disease in which the alveoli are destroyed and the remaining alveoli become larger; results in decreased vital capacity and increased airways resistance.

emulsification (e-mul″si-fĭ-kā′shun)   The process of producing an emulsion or fine suspension; in the intestine, fat globules are emulsified by the detergent action of bile.

enamel (en-am′el)   The outer, dense substance covering the crown of a tooth.

endergonic (end″er-gon′ik)   A chemical reaction that requires the input of energy from an external source in order to proceed.

endocardium (en″do-kar′de-um)   The endothelial lining of the heart chambers and valves.

endocrine gland (en′do-krīn)   A ductless gland that secretes hormones directly into the bloodstream or body fluids.

endocytosis (en″do-si-to′sis)   The cellular uptake of particles that are too large to cross the cell membrane; occurs by invagination of the cell membrane until a membrane-enclosed vesicle is inched off within the cytoplasm.

endoderm (en′do-derm)   The innermost of the three primary germ layers of an embryo.

endogenous (en-doj′ĕ-nus)   A product or process arising from within the body; as opposed to exogenous products, or influences from external sources.

endolymph (en′do-limf)   The fluid within the membranous labyrinth of the inner ear.

endometrium (en″do-me′tre-um)   The inner lining of the uterus.

endomysium (en″do-mis′e-um)   The connective tissue sheath surrounding each skeletal muscle fiber, separating the muscle cells from one another.

endoneurium (en″do-nu′re-um)   The connective tissue sheath surrounding each nerve fiber, separating the nerve fibers one from another within a nerve.

endoplasmic reticulum (en-do-plas′mik re-tik′u-lum)   A cytoplasmic organelle composed of a network of canals running through the cytoplasm of a cell.

endorphins (en-dor′fins, en′dor-fins)   A group of endogenous opiate molecules that may act as a natural analgesic.

endothelium (en″do-the′le-um)   The layer of epithelial tissue that forms the thin inner lining of blood vessels and heart chambers.

endotoxin (en″do-tok′sin)   A toxin contained within certain types of bacteria that is able to stimulate the release of endogenous pyrogen and produce a fever.

enkephalins (en-kef′ah-lins)   Short polypeptides, containing five amino acids, that have analgesic effects; may function as neurotransmitters in the brain. The two known enkephalins (which differ in only one amino acid) are endorphins.

enteric (en-ter′ik)   A term referring to the intestine.

entropy (en′tro-pe)   The energy of a system that is not available to perform work; a measure of the degree of disorder in a system, entropy increases whenever energy is transformed.

enzyme (en′zīm)   A protein catalyst that increases the rate of specific chemical reactions.

eosinophil (e″o-sin′o-fil)   A type of white blood cell characterized by the presence of cytoplasmic granules that become stained by acidic eosin dye; eosinophils normally constitute about 2%–4% of the white blood cells.

epicardium (ep″i-kar′de-um)   A thin, outer layer of the heart; also called the *visceral pericardium.*

epicondyle (ep″i-kon′dīl)   A projection of bone above a condyle.

epidermis (ep″i-der′mis)   The outermost layer of the skin, composed of several stratified squamous epithelial layers.

epididymis (ep″i-did′i-mis)   A highly coiled tube located along the posterior border of the testis. It stores spermatozoa and transports them from the seminiferous tubules of the testis to the vas deferens.

epidural space (ep″i-du′ral)   A space between the spinal dura mater and the bone of the vertebral canal.

epiglottis (ep″i-glot′is)   A cartilaginous leaflike structure positioned on top of the larynx that covers the glottis during swallowing.

epimysium (ep″i-mis′e-um)   A fibrous, outer sheath of connective tissue surrounding a skeletal muscle.

epinephrine (ep″i-nef′rin)   Also known as adrenalin; a catecholamine hormone secreted by the adrenal medulla in response to sympathetic nerve stimulation; acts together with norepinephrine released from sympathetic nerve endings to prepare the organism for "fight or flight."

epineurium (ep″i-nu′re-um)   A fibrous, outer sheath of connective tissue surrounding a nerve.

epiphyseal plate (ep″i-fiz′e-al)   A cartilaginous layer that is located between the epiphysis and diaphysis of a long bone and functions as a longitudinal growing region.

epiphysis (ĕ-pif′i-sis)   The end segment of a long bone, separated from the diaphysis early in life by an epiphyseal plate but later becoming part of the larger bone.

episiotomy (ĕ-piz″e-ot′o-me)   An incision of the perineum at the end of the second stage of labor to facilitate delivery and avoid tearing the perineum.

epithelial tissue (ep″i-the′le-al)   One of the four basic tissue types; the type of tissue that covers or lines all exposed body surfaces.

eponychium (ep″o-nik′e-um)   The thin layer of stratum corneum of the epidermis of the skin, which overlaps and protects the lunula of the nail.

EPSP   Excitatory postsynaptic potential; a graded depolarization of a postsynaptic membrane in response to stimulation by a neutrotransmitter chemical. EPSPs can be summated but can only be transmitted short distances; EPSPs can stimulate the production of action potentials when a threshold level of depolarization is attained.

erythroblastosis fetalis (ĕ-rith″ro-blas-to′sis fi-tal′is)   Hemolytic anemia in a newborn Rh positive baby caused by maternal antibodies against the Rh factor that have crossed the placenta.

erythrocyte (ĕ-rith′ro-sīt)   A corpuscle or red blood cell.

esophagus (ĕ-sof′ah-gus)   A tubular portion of the digestive tract that leads from the pharynx to the stomach as it passes through the thoracic cavity.

essential amino acids   Those eight amino acids in adults or nine amino acids in children that cannot be made by the human body and therefore must be obtained in the diet.

estrus cycle (es′trus)   Cyclic changes in the structure and function of the ovaries and female reproductive tract, accompanied by periods of "heat" (estrus) or sexual receptivity; the lower mammalian equivalent of the menstrual cycle, but differing from the menstrual cycle in that the endometrium is not shed with accompanying bleeding.

etiology (e″te-ol′o-je)   The study of cause, especially of disease, including the origin and what pathogens, if any, are involved.

eustachian canal (u-sta′ke-an)   A narrow tube that connects the middle ear chamber to the pharynx; also called the *auditory tube.*

eversion (e-ver′zhun)   A movement of the foot in which the sole is turned outward.

exergonic (ek″ser-gon′ik)   Chemical reactions that liberate energy.

exocrine gland (ek′so-krin)   A gland that secretes its product to an epithelial surface, directly or through ducts.

exocytosis (eks″o-si-to′sis) The process of cellular secretion in which the secretory products are contained within a membrane-enclosed vesicle; the vesicle fuses with the cell membrane so that the lumen of the vesicle is open to the extracellular environment.

expiration (ek″spi-ra′shun) The process of expelling air from the lungs through breathing out; also called *exhalation.*

extension (ek-sten′shun) A movement that increases the angle between parts of a joint.

extensor (eks-ten′sor) A muscle that upon contraction increases the angle of a joint.

external (superficial) Located on or toward the surface.

external auditory meatus (aw′di-to″re me-a′tus) An opening through the temporal bone that connects with the tympanum and the middle ear chamber and through which sound vibrations pass.

external ear The outer portion of the ear, consisting of the pinna, external auditory meatus, and tympanum.

external nares (na′rez) The opening into the nasal cavity; also called the *nostrils.*

exteroreceptors (eks″ter-o-sep′tors) Sensory receptors that are sensitive to changes in the external environment (as opposed to interoreceptors).

extraocular muscles (eks″trah-ok′u-lar) The muscles that insert into the sclera of the eye and act to change the position of the eye in its orbit, as opposed to the intraocular muscles such as those of the iris and ciliary body within the eye.

extrinsic (eks-trin′sik) Pertaining to an outside or external origin.

f

face 1. The anterior aspect of the head not supporting or covering the brain. 2. The exposed surface of a structure.

facet (fas′et) A small, smooth surface of a bone where articulation occurs.

facilitated diffusion (fah-sil′i-ta″tid) The carrier-mediated transport of molecules through the cell membrane along the direction of their concentration gradients; it does not require the expenditure of metabolic energy.

FAD Flavin adenine dinucleotide; a coenzyme derived from riboflavin that participates in electron transport within the mitochondria.

falciform ligament (fal′si-form lig′ah-ment) An extension of parietal peritoneum that separates the two major lobes of the liver.

fallopian tube (fal-lo′pe-an) The tube through which the ovum is transported to the uterus and the site of fertilization; also called the *oviduct* or *uterine tube.*

false vocal cords The supporting folds of tissue for the true vocal cords within the larynx.

fascia (fash′e-ah) A tough sheet of fibrous tissue binding the skin to underlying muscles or supporting and separating muscles.

fasciculus (fah-sik′u-lus) A small bundle of muscle or nerve fibers.

fauces (faw′sez) The passageway between the mouth and the pharynx.

feces (fe′sez) Material expelled from the digestive tract during defecation, composed of food residue, bacteria, and secretions; also called *stool.*

fertilization The fusion of an ovum and sperm.

fetus (fe′tus) A prenatal human after eight weeks of development.

fibrillation (fi-bri-la′shun) A condition of cardiac muscle characterized electrically by random and continuously changing patterns of electrical activity and resulting in the inability of the myocardium to contract as a unit and pump blood. It can be fatal if it occurs in the ventricles.

fibrin (fi′brin) The insoluble protein formed from fibrinogen by the enzymatic action of thrombin during the process of blood clot formation.

fibrinogen (fi-brin′o-jen) Also called *factor I;* a soluble plasma protein that serves as the precursor of fibrin.

fibrous joint (fi′brus) A type of articulation that allows little or no movement (e.g., syndesmosis).

filiform papillae (fil′i-form pah-pil′e) The numerous small projections over the entire dorsal surface of the tongue that contain no taste buds.

filum terminale (fi′lum ter-mi-nal′e) A fibrous, threadlike continuation of the pia mater, extending inferiorly from the terminal end of the spinal cord to the coccyx.

fimbriae (fim′bre-e) Fringelike extensions from the open end of the uterine tube.

fissure (fish′ur) A groove or narrow cleft that separates two parts such as the cerebral hemispheres of the brain.

flagellum (flah-jel′um) A whiplike structure that provides motility for sperm.

flare-and-wheal reaction (hwel, wel) A cutaneous reaction to skin injury or the administration of antigens, produced by release of histamine and related molecules and characterized by local edema and a red flare.

flavoprotein (fla″vo-pro′te-in) A conjugated protein that contains a flavin pigment and involved in electron transport within the mitochondria.

flexion (flek′shun) A movement that decreases the angle between parts of a joint.

flexor (flek′sor) A muscle that decreases the angle of a joint when it contracts.

fontanel (fon″tah-nel) A membranous-covered region on the skull of a fetus or baby where ossification has not yet occurred; commonly called a *soft spot.*

foot The terminal portion of the lower extremity, consisting of the tarsus, metatarsus, and phalanges.

foramen (fo-ra′men), pl. *foramina* An opening, usually in a bone, for the passage of a blood vessel or a nerve.

foramen ovale (o-val′e) An opening through the interatrial septum of the fetal heart.

forearm (for′arm) The portion of the upper extremity between the elbow and the wrist; also called the *antebrachium.*

fornix (for′niks) 1. A recess around the cervix of the uterus where it protrudes into the vagina. 2. A tract within the brain connecting the hippocampus with the mammillary bodies.

fossa (fos′ah) A depressed area, usually on a bone.

fourth ventricle (ven′tri-k′l) A cavity within the brain, between the cerebellum and the medulla and the pons, containing cerebrospinal fluid.

fovea centralis (fo′ve-ah sen-tra′lis) A depression on the macula lutea of the eye, where only cones are located, which is the area of keenest vision.

frenulum (fren′u-lum) A membranous tissue that serves to anchor and limit the movement of a body part.

frontal 1. Pertaining to the region of the forehead. 2. A plane through the body, dividing the body into anterior and posterior portions; also called the *coronal plane.*

FSH Follicle-stimulating hormone; one of the two gonadotrophic hormones secreted from the anterior pituitary. In females FSH stimulates the development of the ovarian follicles, whereas in males it stimulates the production of sperm in the seminiferous tubules.

fungiform papillae (fun′ji-form pah-pil′e) Flattened, mushroom-shaped projections that are interspersed over the dorsal surface of the tongue and contain taste buds.

g

GABA Gamma-aminobutyric acid; believed to function as an inhibitory neurotransmitter in the central nervous system.

gallbladder A pouchlike organ that is attached to the underside of the liver and stores and concentrates bile.

gamete (gam′et) A haploid sex cell; either an egg cell or a sperm cell.

ganglion (gang′gle-on) An aggregation of nerve cell bodies occurring outside the central nervous system.

gastric intrinsic factor (gas′trik) A glycoprotein secreted by the stomach and needed for the absorption of vitamin B_{12}.

gastrin (gas′trin) A hormone secreted by the stomach that stimulates the gastric secretion of hydrochloric acid and pepsin.

gastrointestinal tract (GI tract) (gas″tro-in-tes′ti-nal) The portion of the digestive tract that includes the stomach and the small and large intestines.

gates A term used to describe structures within the cell membrane that regulate the passage of ions through membrane channels. Such gates may be chemically regulated (by neurotransmitters) or voltage regulated (in which case, they open in response to a threshold level of depolarization).

genetic recombination (jĕ-net'ik re''kom-bĭ-na'shun)   The formation of new combinations of genes, as by crossing-over between homologous chromosomes.

genetic transcription (trans-krip'shun)   The process by which RNA is produced with a sequence of nucleotide bases that is complementary to a region of DNA.

genetic translation (trans-la'shun)   The process by which proteins are produced with amino acid sequences specified by the sequence of codons in messenger RNA.

gigantism (ji-gan'tism; ji'gan-tizm)   Abnormal body growth due to the excessive secretion of growth hormone.

gingiva (jin-ji'vah)   The fleshy covering over the mandible and maxilla through which the teeth protrude within the mouth; also called the *gum*.

gland   An organ that produces a specific substance or secretion.

glans penis (glanz pe'nis)   The enlarged, sensitive, distal end of the penis.

gliding joint   A type of diarthrotic joint in which the articular surfaces are flat, permitting only side-to-side and back-and-forth movements.

glomerular ultrafiltrate (glo-mer'u-lar ul''trah-fil'trāt)   Fluid filtered through the glomerular capillaries into Bowman's capsule of the kidney tubules.

glomerulonephritis (glo-mer''u-lo-nĕ-fri'tis)   Inflammation of the renal glomeruli, associated with fluid retention, edema, hypertension, and the appearance of protein in the uterine.

glomerulus (glo-mer'u-lus)   A coiled tuft of capillaries that is surrounded by Bowman's capsule and filtrates urine from the blood.

glottis (glot'is)   A slitlike opening into the larynx, positioned between the true vocal cords.

glucagon (gloo'kah-gon)   A polypeptide hormone that is secreted by the alpha cells of the islets of Langerhans in the pancreas and acts to promote glycogenolysis and raise the blood glucose levels.

glucocorticoids (gloo''ko-kor'ti-koids)   Steroid hormones secreted by the adrenal cortex (corticosteroids) that affect the metabolism of glucose, protein, and fat. These hormones also have anti-inflammatory and immunosuppressive effects; the major glucocorticoid in humans is hydrocortisone (cortisol).

gluconeogenesis (gloo''ko-ne''o-jen'ĕ-sis)   The formation of glucose from non-carbohydrate molecules such as amino acids and lactic acid.

glycogen (gli'ko-jen)   A polysaccharide of glucose—also called *animal starch*—produced primarily in the liver and skeletal muscles. Similar to plant starch in composition, glycogen contains more highly branched chains of glucose subunits than does plant starch.

glycogenesis (gli''ko-jen'ĕ-sis)   The formation of glycogen from glucose.

glycogenolysis (gli''ko-jĕ-nol'ĭ-sis)   The hydrolysis of glycogen to glucose-1-phosphate, which can be converted to glucose-6-phosphate, which then may be oxidized via glycolysis or (in the liver) converted to free glucose.

glycolysis (gli''kol'ĭ-sis)   The metabolic pathway that converts glucose to pyruvic acid and yields a net production of two ATP molecules and two molecules of reduced NAD. In anaerobic respiration the reduced NAD is oxidized by the conversion of pyruvic acid to lactic acid. In aerobic respiration pyruvic acid enters the Krebs cycle in mitochondria, and reduced NAD is ultimately oxidized by oxygen to yield water.

glycosuria (gli''ko-su're-ah)   The excretion of an abnormal amount of glucose in the urine (urine normally only contains trace amounts of glucose).

goblet cell   A unicellular gland that secretes mucus and is associated with columnar epithelia.

Golgi apparatus   Stacks of flattened membranous sacs within the cytoplasm of cells that are believed to bud off vesicles containing secretory proteins.

Golgi tendon organ   An afferent receptor found near the junction of tendons and muscles.

gonad (go'nad)   A reproductive organ, testis or ovary, that produces gametes and sex hormones.

gonadotrophin hormones (gon''ah-do-tro'fin)   Hormones of the anterior pituitary gland that stimulate gonadal function—the formation of gametes and secretion of sex steroids. The two gonadotrophins are FSH (follicle-stimulating hormone) and LH (luteinizing hormone), which are essentially the same in males and females.

graafian follicle (graf'e-an)   A mature ovarian follicle, containing a single fluid-filled cavity, with the ovum located toward one side of the follicle and perched on top of a hill of granulosa cells.

granular leukocytes (gran'u-lar loo'ko-sīts)   Leukocytes with granules in the cytoplasm; on the basis of the staining properties of the granules, these cells are of three types: neutrophils, eosinophils, and basophils.

Graves' disease   A hyperthyroid condition believed to be caused by excessive stimulation of the thyroid gland by autoantibodies; it is associated with exophthalmos (bulging eyes), high pulse rate, high metabolic rate, and other symptoms of hyperthyroidism.

gray matter   The region of the central nervous system that is composed of nonmyelinated nerve tissue.

greater omentum (o-men'tum)   A double-layered serosa membrane that originates on the greater curvature of the stomach and hangs inferiorly like an apron over the contents of the abdominal cavity.

gross anatomy   A branch of anatomy concerned with structures of the body that can be studied without a microscope.

growth hormone   A hormone secreted by the anterior pituitary that stimulates growth of the skeleton and soft tissues during the growing years and that influences the metabolism of protein, carbohydrate, and fat throughout life.

gustatory (gus'tah-to''re)   Pertaining to the sense of taste.

gut (gŭt)   Pertaining to the intestines; generally a developmental term.

gyrus (ji'rus)   A convoluted elevation or ridge.

h

hair   A threadlike appendage of the epidermis, consisting of keratinized dead cells that have been pushed up from a dividing germinativum layer.

hair cells   Specialized receptor nerve endings for detecting sensations, such as in the organ of Corti.

hair follicle (fol'li-k'l)   A tubular depression in the dermis of the skin in which a hair develops.

hand   The terminal portion of the upper extremity, consisting of the carpus, metacarpus, and phalanges.

haploid (hap'loid)   A cell that has one of each chromosome type and therefore half the number of chromosomes present in most other body cells; only the gametes (sperm and ova) are haploid.

haptens (hap'tens)   Small molecules that are not antigenic by themselves, but which—when combined with proteins—become antigenic and thus able to stimulate the production of specific antibodies.

hard palate (pal'at)   The bony partition between the oral and nasal cavities, formed by the maxillae and palatine bones and lined by mucous membrane.

haustra (haws'trah)   Sacculations or pouches of the colon.

haversian canal (ha-ver'shan)   An elongated, longitudinal channel in the center of a haversian system in bone, containing nutrient vessels and nerves.

haversian system   A group of osteocytes and concentric lamellae surrounding a haversian canal, constituting the basic unit of structure in osseous tissue; also called an *osteon*.

hay fever   A seasonal type of allergic rhinitis caused by pollen; it is characterized by itching and tearing of the eyes, swelling of the nasal mucosa, attacks of sneezing, and often by asthma.

head   The superior portion of a human that contains the brain and major sense organs.

heart   A four-chambered, muscular, pumping organ positioned in the thoracic cavity slightly to the left of midline.

heart murmur   Abnormal heart sounds caused by an abnormal flow of blood in the heart due to structural defects, usually of the valves or septum.

helper T cells A subpopulation of T cells (lymphocytes), which help stimulate the antibody production of B lymphocytes by antigens.

hematocrit (he-mat′o-krit) The ratio of packed red blood cells to total blood volume in a centrifuged sample of blood, expressed as a percentage.

heme (hēm) The iron-containing red pigment that, together with the protein globin, forms hemoglobin.

hemoglobin (he′′mo-glo′bin) The pigment of red blood cells that constitutes about 33% of the cell volume and transports oxygen and carbon dioxide.

hemopoiesis (he′′mo-poi-e′sis) The production of red blood cells.

heparin (hep′ah-rin) A mucopolysaccharide, found in many tissues but most abundantly in the lungs and liver, that is used medically as an anticoagulant.

hepatic duct (he-pat′ik) Tubules that drain bile from the liver and merge with the cystic duct from the gallbladder to form the common bile duct.

hepatic portal circulation (por′tal) The return of venous blood from the digestive organs through a capillary network within the liver before draining into the heart.

hepatitis (hep′′ah-ti′tis) Inflammation of the liver.

Hering-Breuer reflex A reflex in which distension of the lungs stimulates stretch receptors, which in turn act to inhibit further distension of the lungs.

hermaphrodite (her-maf′ro-dīt) An organism with both testes and ovaries.

heterochromatin (het′′er-o-kro′mah-tin) A condensed, inactive form of chromatin.

hiatal hernia (hi-a′tal her′ne-ah) A protrusion of an abdominal structure through the esophageal hiatus of the diaphragm into the thoracic cavity.

high-density lipoproteins (lip′′o-pro′te-ins) Combinations of lipids and proteins that migrate rapidly to the bottom of a test tube during centrifugation; carrier proteins for lipids, such as cholesterol, which appear to offer some protection from atherosclerosis.

hilus (hi′lus) A concave or depressed area where vessels or nerves enter or exit an organ.

hinge joint A type of diarthrotic articulation characterized by a convex surface of one bone fitting into a concave surface of another so that movement is confined to one plane, such as in the knee or interphalangeal joint.

histamine (his′′tah-mēn) A compound secreted by tissue mast cells and other connective tissue cells that stimulates vasodilation and increases capillary permeability; it is responsible for many of the symptoms of inflammation and allergy.

histology (his-tol′o-je) Microscopic anatomy of the structure and function of tissues.

histone (his′tōn) A basic protein associated with DNA that is believed to repress genetic expression.

homeostasis (ho′′me-o-sta′sis) The dynamic constancy of the internal environment, which serves as the principal function of physiological regulatory mechanisms. The concept of homeostasis provides a framework for the understanding of most physiological processes.

homologous chromosomes (ho-mol′o-gus) The matching pairs of chromosomes in a diploid cell.

horizontal (transverse) A directional plane that divides the body, organ, or appendage into superior and inferior or proximal and distal portions.

hormone (hor′mōn) A chemical substance that is produced in an endocrine gland and secreted into the bloodstream to cause an effect in a specific target organ.

humoral immunity (hu′mor-al i-mu′ni-te) The form of acquired immunity in which antibody molecules are secreted in response to antigenic stimulation (as opposed to cell-mediated immunity).

hyaline cartilage (hi′ah-līn) A cartilage with a homogeneous matrix. It is the most common type, occurring at the articular ends of bones, in the trachea, and within the nose, and forms the precursor to most of the bones in the body.

hyaline membrane disease A disease of some premature infants who lack pulmonary surfactant, it is characterized by collapse of the alveoli (atelectasis) and pulmonary edema. Also called *respiratory distress syndrome.*

hydrocortisone (hi′′dro-kor′ti-son) Also called *cortisol;* the principal corticosteroid hormone secreted by the adrenal cortex, with glucocorticoid action.

hydrophilic (hy′′dro-fil′ik) A substance that readily absorbs water, literally "water loving."

hydrophobic (hi′′dro-fo′bik) A substance that repels, and that is repelled by, water; "water fearing."

hymen (hi′men) A developmental remnant of membranous tissue that partially covers the vaginal opening.

hyperbaric oxygen (hi′′per-bār′ik) Oxygen gas present at greater than atmospheric pressure.

hypercapnia (hi′′per-kap′ne-ah) Excessive concentration of carbon dioxide in the blood.

hyperextension (hi′′per-ek-sten′shun) Extension beyond the normal anatomical position or 180°.

hyperglycemia (hi′′per-gli-se′me-ah) Abnormally increased concentration of glucose in the blood.

hyperkalemia (hi′′per-kal-le′me-ah) Abnormally high concentration of potassium in the blood.

hyperopia (hi′′per-o′pe-ah) Also called *farsightedness:* a refractive disorder in which rays of light are brought to a focus behind the retina as a result of the eyeball being too short.

hyperplasia (hi′′per-pla′ze-ah) An increase in organ size due to an increase in cell numbers as a result of mitotic cell division.

hyperpolarization (hi′′per-po′′lar-i-za′shun) An increase in the negativity of the inside of a cell membrane with respect to the resting membrane potential.

hypersensitivity (hi′′per-sen′si-tiv′′i-te) Another name for *allergy;* abnormal immune response that may be immediate (due to antibodies of the IgE class) or delayed (due to cell-mediated immunity).

hypertension (hi′′per-ten′shun) Elevated or excessive blood pressure.

hypertonic (hi′′per-ton′ik) A solution with a greater solute concentration and thus a greater osmotic pressure than plasma.

hypertrophy (hi-per′tro-fe) Growth of an organ due to an increase in the size of its cells.

hyperventilation (hi-per-ven′′ti-la′shun) A high rate and depth of breathing that results in a decrease in the blood carbon dioxide concentration below normal.

hypodermis (hi′′po-der′mis) A layer of fat beneath the dermis of the skin.

hyponychium (hi′′po-nik′e-um) A thickened, supportive layer of stratum corneum at the distal end of a digit under the free edge of the nail.

hypothalamic hormones (hi′′po-thah-lam′ik) Hormones produced by the hypothalamus; these include antidiuretic hormone and oxytocin, which are secreted by the posterior pituitary gland; and both releasing and inhibiting hormones that regulate the secretion of the anterior pituitary.

hypothalamo-hypophyseal portal system (hi′′po-fiz′e-al) A vascular system that transports releasing and inhibiting hormones from the hypothalamus to the anterior pituitary.

hypothalamo-hypophyseal tract The tract of nerve fibers (axons) that transports antidiuretic hormone and oxytocin from the hypothalamus to the posterior pituitary gland.

hypothalamus (hi′′po-thal′ah-mus) A portion of the forebrain within the diencephalonic region that is positioned below the thalamus and functions as an autonomic nerve center and regulates the pituitary gland.

hypovolemic shock (hi′′po-vo-le′mik) A rapid fall in blood pressure due to diminished blood volume.

hypoxemia (hi′′pok-se′me-ah) A low oxygen concentration of the arterial blood.

i

ileocecal valve (il′′e-o-se′kal) A modification of the mucosa at the junction of the small and large intestine that forms a one-way passage and prevents the backflow of food materials.

ileum (il′e-um) The terminal portion of the small intestine between the jejunum and cecum.

immediate hypersensitivity (hi′′per-sen′si-tiv′′i-te) Hypersensitivity (allergy) that is mediated by antibodies of the IgE class and results in the release of histamine and related compounds from tissue cells.

immunization (im″u-ni-za′shun) The process of increasing one's resistance to pathogens. In active immunity a person is injected with antigens that stimulate the development of clones of specific B or T lymphocytes; in passive immunity a person is injected with the antibodies made by another organism.

immunoassay (im″u-no-as′sa) The detection or measurement of a molecule that acts as an antigen by its reaction with specific antibodies. The antigen-antibody reaction may be followed in a variety of ways, such as agglutination (if the antigen is attached to visible cells or particles), fluorescence, or radioactivity (a radioimmunoassay, or RIA).

immunoglobins (im″u-no-glob′u-lins) Subclasses of the gamma globulin fraction of plasma proteins that have antibody functions, providing humoral immunity.

immunosurveillance (im″u-no-ser-va′lens) The function of the immune system to recognize and attack malignant cells that produce antigens not recognized as "self." This function is believed to be cell mediated rather than humoral.

implantation (im″plan-ta′shun) The process by which a blastocyst attaches itself to and penetrates into the endometrium of the uterus.

incus (ing′kus) The middle of three ear ossicles within the middle ear chamber; commonly called the *anvil.*

infundibulum (in″fun-dib′u-lum) The flesh stalk that attaches the pituitary gland to the hypothalamus of the brain.

ingestion (in-jes′chun) The process of taking food or liquid into the body by way of the oral cavity.

inguinal (ing′gwĭ-nal) Pertaining to the groin region.

inguinal canal The circular passage through which a testis descends into the scrotum.

inhibin (in-hib′in) Believed to be a water-soluble hormone secreted by the seminiferous tubules of the testes that specifically exerts negative feedback inhibition of FSH secretion from the anterior pituitary gland.

insertion The more movable attachment of a muscle, usually more distal.

inspiration (in″spi-ra′shun) The act of breathing air into the alveoli of the lungs; also called *inhalation.*

insulin (in′su-lin) A polypeptide hormone, secreted by the beta cells of the islets of Langerhans in the pancreas, that promotes the anabolism of carbohydrates, fat, and protein. Insulin acts to promote the cellular uptake of blood glucose and, therefore, to lower the blood glucose concentration; insulin deficiency produces hyperglycemia and diabetes mellitus.

integument (in-teg′u-ment) Pertaining to the skin.

intercalated disc (in-ter′kah-lāt-ed) A thickened portion of the sarcolemma that extends across a cardiac muscle fiber and indicates the boundary between cells.

intercellular substance (in″ter-sel′u-lar) The matrix or material between cells that largely determines tissue types.

interferons (in″ter-fēr′ons) A group of small proteins that inhibit the multiplication of viruses inside host cells and also have antitumor properties.

internal (deep) Toward the center, away from the surface of the body.

internal ear The innermost portion or chamber of the ear, containing the cochlea and the vestibular organs.

internal nares (na′rēz) The two posterior openings from the nasal cavity into the nasopharynx; also called the *choanae.*

interneurons (in″ter-nu′rons) Also called *association neurons;* those neurons within the central nervous system that do not extend into the peripheral nervous system; they are interposed between sensory (afferent) and motor (efferent) neurons.

interoreceptors (in″ter-o-re-sep′tors) Sensory receptors that respond to changes in the internal environment (as opposed to exteroreceptors).

interphase The interval between successive cell divisions; during this time the chromosomes are in an extended state and are active in directing RNA synthesis.

intervertebral disc (in″ter-ver′tĕ-bral) A pad of fibrocartilage located between the bodies of adjacent vertebrae.

intrafusal fibers (in″trah-fu′zal) Modified muscle fibers that are encapsulated to form muscle spindle organs, which are muscle stretch receptors.

intramembranous ossification (in″trah-mem′brah-nus os″i-fi-ka′shun) A type of bone formation in which there is a membranous, rather than cartilaginous, precursor.

intrapleural space (in″trah-ploor′al) An actual or potential space between the visceral pleural membrane covering the lungs and the somatic pleural membrane lining the thoracic wall.

intrapulmonary space (in″trah-pul′mo-ner″e) The space within the air sacs and airways of the lungs.

intrinsic (in-trin′sik) Situated in or pertaining to internal origin.

inulin (in′u-lin) A polysaccharide of fructose, produced by certain plants, that is filtered by the human kidneys but neither reabsorbed nor secreted. The clearance rate of injected inulin is thus used to measure the glomerular filtration rate.

inversion (in-ver′zhun) A movement of the foot in which the sole is turned inward.

in vitro (in ve′tro) Occurring outside the body, in a test tube or other artificial environment.

in vivo (in ve′vo) Occurring within the body.

ion (i′on) An atom or group of atoms that has either lost or gained electrons and thus has a net positive or a net negative charge.

ionization (i″on-i-za′shun) The dissociation of a solute to form ions.

ipsilateral (ip″si-lat′er-al) On the same side (as opposed to contralateral).

IPSP Inhibitory postsynaptic potential; hyperpolarization of the postsynaptic membrane in response to a particular neurotransmitter chemical, which makes it more difficult for the postsynaptic cell to attain a threshold level of depolarization required to produce action potentials; it is responsible for postsynaptic inhibition.

iris (i′ris) The pigmented muscular portion of the eye that surrounds the pupil and regulates its diameter.

ischemia (is-ke′me-ah) A rate of blood flow to an organ that is inadequate to supply sufficient oxygen and maintain aerobic respiration in that organ.

islets of Langerhans (i′lets of lahng′er-hanz) A cluster of cells within the pancreas that forms the endocrine portion and secretes insulin and glucagon.

isoenzymes (i″so-en′zīms) Enzymes, usually produced by different organs, that catalyze the same reaction but that differ from each other in amino acid composition.

isometric contraction (i″so-met′rik) Muscle contraction in which there is no appreciable shortening of the muscle.

isotonic contraction (i″so-ton′ik) Muscle contraction in which the muscle shortens in length and maintains approximately the same amount of tension throughout the shortening process.

isotonic solution A solution having the same total solute concentration, osmolality, and osmotic pressure as the solution with which it is compared; a solution with the same solute concentration and osmotic pressure as plasma.

isthmus (is′mus) A narrow neck or portion of tissue connecting two structures.

j

jaundice (jawn′dis) A condition characterized by high blood bilirubin levels and staining of the tissues with bilirubin, which gives skin and mucous membranes a yellow color.

jejunum (je-joo′num) The middle portion of the small intestine, located between the duodenum and the ileum.

joint capsule (kap′sūl) A fibrous tissue cuff surrounding a diarthrotic joint.

junctional complexes (junk′shun-al) The structures that join adjacent epithelial cells together, including the zonula occludens, zonula adherens, and macula adherens (desmosome).

k

keratin (ker′ah-tin) An insoluble protein present in the epidermis and in epidermal derivatives such as hair and nails.

ketoacidosis (ke″to-ah″si-do′sis) A type of metabolic acidosis produced by the excessive production of ketone bodies, as in diabetes mellitus.

ketogenesis (ke″to-jen′ĕ-sis) The production of ketone bodies.

ketone bodies (ke′tōn) The substances acetone, acetoacetic acid, and β hydroxybutyric acid, which are derived from fatty acids via acetyl coenzyme A in the liver. Ketone bodies are oxidized by skeletal muscles for energy.

ketosis (ke-to′sis) An abnormal elevation in the blood concentration of ketone bodies that does not necessarily produce acidosis.

kidney (kid′ne) One of the paired organs of the urinary system that contains nephrons and filters urine from the blood.

kilocalorie (kil′o-kal″o-re) A unit of measurement equal to 1,000 calories, which are units of heat (a kilocalorie is the amount of heat required to raise the temperature of 1 kilogram of water by 1° C). In nutrition the kilocalorie is called a big calorie (Calorie).

Klinefelter's syndrome (klīn′fel-terz sin′drōm) The syndrome produced in a male by the presence of an extra X chromosome (genotype XXY).

knee A region in the lower extremity, between the thigh and the ankle, containing a diarthrotic hinge joint.

Krebs cycle (krebz) A cyclic metabolic pathway in the matrix of mitochondria by which the acetic acid part of acetyl CoA is oxidized and substrates are provided for reactions that are coupled to the formation of ATP.

Kupffer cells (koop′fer) Phagocytic cells that line the sinusoids of the liver, and which are part of the reticuloendothelial system.

l

labial frenulum (la′be-al fren′u-lum) A longitudinal fold of mucous membrane that attaches the lips to the gum along the midline of both the upper and lower lip.

labia majora (la′be-ah ma-jor′ah) A portion of the external genitalia of a female, consisting of two longitudinal folds of skin extending downward and backward from the mons pubis.

labia minora (ma-nor′-ah) Two small folds of skin devoid of hair and sweat glands, lying between the labia majora of the external genitalia of a female.

labyrinth (lab′ĭ-rinth) The complex system of interconnecting tubes within the inner ear, which includes the semicircular canals, cochlea, and vestibule.

lacrimal canal (lak′rĭ-mal) A drainage duct for tears, located at the medial corner of an eyelid and conveying the tears medially into the nasolacrimal sac.

lacrimal gland A tear-secreting gland, located on the superior lateral portion of the eyeball underneath the upper eyelid.

lactation (lak-ta′shun) The production and secretion of milk by the mammary glands.

lacteal (lak′te-al) A small lymphatic duct associated with a villus of the small intestine.

lactose (lak′tōs) Milk sugar; a disaccharide of glucose and galactose.

lactose intolerance The inability of many adults to digest lactose due to a loss of the ability of the intestine to produce lactase enzyme.

lacuna (lah-ku′nah) A small, hollow chamber that houses an osteocyte in mature bone tissue or a chondrocyte in cartilage tissue.

lamella (lah-mel′ah) A concentric ring of matrix surrounding the haversian canal in a haversian system of mature bone tissue.

lamina (lam′ĭ-nah) A thin plate of bone that extends superiorly from the body of a vertebra to form either side of the arch of a vertebra.

lanugo (lah-nu′go) Short, silky fetal hair, which may be present for a short time on a premature infant.

large intestine The last major portion of the alimentary canal, consisting of the cecum, colon, rectum, and anal canal.

laryngopharynx (lah-ring″go-far′inks) The inferior or lower portion of the pharynx in contact with the larynx.

larynx (lar′inks) The structure located between the pharynx and trachea that houses the vocal cords; commonly called the *voice box.*

lateral (lat′er-al) Pertaining to the side; farther from the midline.

lateral ventricle (ven′tri-k′l) A cavity located in the cerebral hemisphere of the brain and filled with cerebrospinal fluid.

leg The portion of the lower extremity between the knee and ankle.

lens (lenz) A transparent refractive organ of the eye, derived from ectoderm and positioned posterior to the pupil and iris.

lesion (le′zhun) A wounded or damaged area.

lesser omentum (o-men′tum) A peritoneal fold of tissue extending from the lesser curvature of the stomach to the liver.

leukocyte (loo′ko-sīt) A white blood cell; also spelled leucocyte.

ligament (lig′ah-ment) A tough chord or fibrous band of connective tissue that binds bone to bone to strengthen and provide flexibility to a joint; it also may support viscera.

limbic system (lim′bik) A portion of the brain concerned with emotions and autonomic activity.

linea alba (lin′e-ah al′bah) A vertical fibrous band extending down the anterior medial portion of the abdominal wall.

lingual frenulum (ling′gwal fren′u-lum) A longitudinal fold of mucous membrane that attaches the tongue to the floor of the mouth.

lipogenesis (lip″o-jen′ĕ-sis) The formation of fat or triglycerides.

lipolysis (li-pol′ĭ-sis) The hydrolysis of triglycerides into free fatty acids and glycerol.

low-density lipoproteins (lip″o-pro′te-ins) Plasma proteins that transport triglycerides and cholesterol; they are believed to contribute to arteriosclerosis.

lower extremity A lower appendage, including the thigh, leg, and foot.

lumbar (lum′ber) Pertaining to the region of the loins.

lumbar plexus (plek′sus) A network of nerves formed by the anterior branches of spinal nerves L1 through L4.

lumen (lu′men) The space within a tubular structure through which a substance passes.

lung One of the two major organs of respiration positioned within the thoracic cavity on either side of the mediastinum.

lung surfactant (sur-fak′tant) A mixture of lipoproteins (containing phospholipids) secreted by type II alveolar cells into the alveoli of the lungs; it lowers surface tension and prevents collapse of the lungs as occurs in hyaline membrane disease, in which surfactant is absent.

lunula (lu′nu-lah) The half-moon-shaped whitish area at the proximal portion of a nail.

luteinizing hormone (LH) (loo″te-in″i-zing) A gonadotrophic hormone secreted by the anterior pituitary, which in a female, stimulates ovulation and the development of a corpus luteum. In a male, LH stimulates the Leydig cells to secrete androgens.

lymph (limf) A clear, plasmalike fluid that flows through lymphatic vessels.

lymphatic system (lim-fat′ik) The lymphatic vessels and lymph nodes.

lymph node A small, oval mass of reticular tissue located along the course of lymph vessels.

lymphocyte (lim′fo-sīt) A type of white blood cell characterized by agranular cytoplasm. Lymphocytes usually constitute about 20%–25% of the white cell count.

lymphokines (lim′fo-kīns) A group of chemicals released from T cells that contribute to cell-mediated immunity.

lysosome (li′so-sōm) Organelles containing digestive enzymes and responsible for intracellular digestion.

m

macromolecules (mak″ro-mol′ĕ-kūl) Large molecules; a term that usually refers to protein, RNA, and DNA.

macrophage (mak′ro-fāj) A wandering phagocytic cell.

macula lutea (mak′u-lah lu′te-ah) A yellowish depression in the retina of the eye that contains the fovea centralis, the area of keenest vision.

malignant (mah-lig′nant) Tending to become worse and end in death.

malleus (mal′e-us) The first of three ear ossicles attached to the tympanum; commonly called the *hammer.*

mammary gland (mam′er-e) The gland of the female breast responsible for lactation and nourishment of the young.

marrow (mar′o) The soft connective tissue that occupies the inner cavity of certain bones and produces red blood cells.

mast cells A type of connective tissue cells that produce and secrete histamine and heparin.

mastication (mas″ti-ka′shun) Pertaining to the chewing of food.

matrix (ma′triks) The intercellular substance of a tissue.

maximal oxygen uptake The maximum amount of oxygen that can be consumed by the body per unit time during heavy exercise.

meatus (me-a′tus) A passageway or opening into a structure.

mechanoreceptor (mek″ah-no-re-sep′tor) A sensory receptor that responds to a mechanical stimulus.

medial (me′de-al) Toward or nearer the midline of the body.

mediastinum (me″de-as-ti′num) The space in the center of the thorax between the two pleural cavities.

medulla (mĕ-dul′ah) The center portion of an organ.

medulla oblongata (ob″long-ga′tah) A portion of the brain stem located between the spinal cord and the pons.

medullary (marrow) cavity (med′u-lar″e) The hollow core of the diaphysis of a long bone, occupied by marrow.

megakaryocyte (meg″ah-kar′e-o-sit) A bone marrow cell that gives rise to blood platelets.

meiosis (mi-o′sis) A specialized type of cell division by which gametes or haploid sex cells are formed.

Meissner's corpuscle (mis′nerz kor′pus′l) A touch sensory receptor found in the papillary layer of the dermis of the skin.

melanin (mel′ah-nin) A dark pigment found within the epidermis or epidermal derivatives of the skin.

melanocyte (mel′ah-no-sit) A specialized melanin-producing cell found in the deepest layer of epidermis.

melatonin (mel″ah-to′nin) A hormone secreted by the pineal gland that produces lightening of the skin in lower animals and may contribute to the regulation of gonadal function in mammals.

membrane potential The potential difference or voltage that exists between the inner and outer sides of a cell membrane; it exists in all cells but is capable of being changed by excitable cells (neurons and muscle cells).

membranous bone (mem′brah-nus) Bone that forms from membranous connective tissue rather than from cartilage; also called *intramembranous bone.*

membranous labyrinth (lab′i-rinth) A system of communicating sacs and ducts within the bony labyrinth of the inner ear.

menarche (mĕ-nar′ke) The first menstrual discharge.

Meniere's disease (men″e-arz′) Deafness, tinnitus, and vertigo resulting from a disease of the labyrinth.

meninges (mĕ-nin′jez) A group of three fibrous membranes that cover the central nervous system, composed of the dura, arachnoid, and pia maters.

menisci (mĕn-is′si) Wedge-shaped fibrocartilages in certain movable joints; also called *semilunar cartilages.*

menopause (men′o-pawz) The cessation of menstrual periods in the human female.

menstrual cycle (men′stroo-al) The rhythmic female reproductive cycle, which is characterized by changes in hormone levels and physical changes in the uterine lining.

menstruation (men″stroo-a′shun) The discharge of blood and tissue from the uterus at the end of the female reproductive cycle.

mesentery (mes′en-ter″e) A fold of peritoneal membrane that attaches an abdominal organ to the abdominal wall.

mesoderm (mes′o-derm) The middle of the three germ layers.

mesovarium (mes″o-va′re-um) The peritoneal fold that attaches an ovary to the broad ligament of the uterus.

messenger RNA (mRNA) A type of RNA that contains a base sequence complementary to a part of the DNA that specifies the synthesis of a particular protein.

metabolism (mĕ-tab′o-lizm) The sum total of the chemical changes that occur within a cell.

metacarpus (met″ah-kar′pus) The region of the hand between the wrist and the phalanges, including the five bones that compose the palm of the hand.

metastasis (mĕ-tas′tah-sis) The spread of a disease from one organ or body part to another.

metatarsus (met″ah-tar′sus) The region of the foot between the ankle and the phalanges, comprising five bones.

micelles (mi-sels′) Colloidal particles formed by the aggregation of many molecules.

microglia (mi-krog′le-ah) Small phagocytic cells found in the central nervous system.

microvilli (mi″kro-vil′i) Microscopic, hairlike projections of cell membranes on certain epithelial cells.

micturition (mik″tu-rish′un) The process of voiding urine; also called *urination.*

midbrain The portion of the brain between the pons and the forebrain.

middle ear The middle of the three ear chambers, which contains the three ear ossicles.

midsagittal (mid-saj′i-tal) A plane that divides the body or an organ into right and left halves.

mineralocorticoids (min″er-al-o-kor′ti-koids) Steroid hormones of the adrenal cortex (corticosteroids) that regulate electrolyte balance.

mitosis (mi-to′sis) The process of cell division, in which the two daughter cells are identical and contain the same number of chromosomes.

mitral valve (mi′tral) The left atrioventricular heart valve; also called the *bicuspid valve.*

mixed nerve A nerve that contains both motor and sensory nerve fibers.

molal (mo′lal) Pertaining to the number of moles of solute per kilogram of solvent.

molar (mo′lar) Pertaining to the number of moles of solute per liter of solution.

mole (mol) The number of grams of a chemical that is equal to its formula weight (atomic weight for an element or molecular weight for a compound).

monoclonal antibodies (mon″o-klon′al an″ti-bod″es) Identical antibodies derived from a clone of genetically identical plasma cells.

monocyte (mon′o-sit) A phagocytic type of white blood cell, normally constituting about 3%–8% of the white blood cell count.

monomer (mon′o-mer) A single molecular unit of a longer, more complex molecule; monomers are joined together to form dimers, trimers, and polymers; the hydrolysis of polymers eventually yields separate monomers.

monosaccharide (mon″o-sak′ah-rid) Also called a *simple sugar;* the monomers of the more complex carbohydrates. Examples include glucose, fructose, and galactose.

mons pubis (monz pu′bis) A fatty tissue pad covering the symphysis pubis, which is covered by pubic hair in the female.

morula (mor′u-lah) An early stage of embryonic development characterized by a solid ball of cells.

motile (mo′til; mo′t′l, mo′til) Capable of self-propelled movement.

motor area A region of the cerebral cortex from which originate motor impulses to muscles or glands.

motor nerve A nerve composed of motor nerve fibers.

motor neuron (nu′ron) An efferent neuron that conducts action potentials away from the central nervous system and innervates effector organs (muscles and glands). It forms the ventral roots of the spinal nerves.

motor unit A single motor neuron and the muscle fibers it innervates.

mucosa (mu-ko′sah) A mucous membrane that lines cavities and tracts opening to the exterior.

mucous cell (mu′kus) A specialized unicellular gland that produces and secretes mucus; also called a *goblet cell.*

mucous membrane The layers of visceral organs that include the lining epithelium, submucosal connective tissue, and (in some cases) a thin layer of smooth muscle (the muscularis mucosa).

multipolar neuron (mul″ti-po′lar nu′ron) A nerve cell with many processes originating from the cell body.

muscle (mus′el) A major type of tissue that is adapted to contract. The three kinds of muscle are cardiac, smooth, and skeletal.

muscle spindles Sensory organs within skeletal muscles that are composed of intrafusal fibers and are sensitive to muscle stretch; they provide a length detector within muscles.

muscularis (mus″ku-la′ris) A muscular layer or tunic of an organ, composed of smooth muscle tissue.

myelin (mi′ĕ-lin) A lipoprotein material that forms a sheathlike covering around nerve fibers.

myelin sheath (shēth) A sheath surrounding axons, which is formed by successive wrappings of a neuroglial cell membrane. Myelin sheaths are formed by Schwann cells in the peripheral nervous system and by oligodendrocytes within the central nervous system.

myocardial infarction (mi′′o-kar′de-al in-fark′shun) An area of necrotic tissue in the myocardium, which is filled in by scar (connective) tissue.

myocardium (mi′′o-kar′de-um) The cardiac muscle layer of the heart.

myofibril (mi′′o-fi′bril) A bundle of contractile fibers within muscle cells.

myogenic (mi′′o-jen′ik) Originating within muscle cells; used to describe self-excitation by cardiac and smooth muscle cells.

myoglobin (mi′′o-glo′bin) A molecule composed of globin protein and heme pigment; it is related to hemoglobin but contains only one subunit (instead of the four in hemoglobin) and is found in striated muscles. Myoglobin serves to store oxygen in skeletal and cardiac muscle cells.

myogram (mi′o-gram) A recording of electrical activity within a muscle.

myology (mi-ol′o-je) The science or study of muscle structure function.

myometrium (mi′′o-me′tre-um) The layer or tunic of smooth muscle within the uterine wall.

myoneural junction (mi′′o-nu′ral) The site of contact between an axon of a motor neuron and a muscle fiber.

myosin (mi′o-sin) A thick filament protein that together with actin causes muscle contraction.

myxedema (mik′′sĕ-de′mah) A type of edema associated with hypothyroidism; it is characterized by the accumulation of mucoproteins in tissue fluid.

n

NAD Nicotinamide adenine dinucleotide; a coenzyme derived from niacin, which functions to transport electrons in oxidation-reduction reactions; it helps to transport electrons to the electron transport chain within mitochondria.

nail A hardened, keratinized plate that develops from the epidermis and forms a protective covering on the dorsal surface of the distal phalanges of fingers and toes.

naloxone (nal-oks′ōn) A drug that antagonizes the effects of morphine and endorphins.

nasal cavity (na′zal) A mucosa-lined space above the oral cavity that is divided by a nasal septum and is the first chamber of the respiratory system.

nasal concha (kong′kah) A scroll-like bone extending medially from the lateral wall of the nasal cavity; also called a *turbinate bone.*

nasal septum (sep′tum) A bony and cartilaginous partition that separates the nasal cavity into two portions.

nasopharynx (na′′zo-far′inks) The first or uppermost chamber of the pharynx, positioned posterior to the nasal cavity and extending down to the soft palate.

neck 1. Any constricted portion, such as the neck of an organ. 2. The cervical region of the body between the head and thorax.

necrosis (nĕ-kro′sis) Cellular death within tissues and organs.

negative feedback Mechanisms in the body that act to maintain a state of internal constancy, or homeostasis; effectors are activated by changes in the internal environment, and the actions of the effectors serve to counteract these changes and maintain a state of balance.

neonatal (ne′′o-na′tal) The stage of life from birth to the end of four weeks.

neoplasm (ne′o-plazm) A new, abnormal growth of tissue, as in a tumor.

nephron (nef′ron) The functional unit of the kidney, consisting of a renal corpuscle, Bowman's capsule, convoluted tubules, and a loop of Henle.

nerve A bundle of nerve fibers outside the central nervous system.

neurilemma (nu′′ri-lem′ah) A thin, membranous covering surrounding the myelin sheath of a nerve fiber.

neurofibril (nu′′ro-fi′bril) One of many delicate threadlike structures within the cytoplasm of a cell body and the axon hillock of a neuron.

neuroglia (nu-rog′le-ah) Specialized supportive cells of the central nervous system.

neurohypophysis (nu′′ro-hi-pof′ĭ-sis) The posterior lobe of the pituitary gland.

neuron (nu′ron) The structural and functional unit of the nervous system, composed of a cell body, dendrites, and an axon; also called a *nerve cell.*

neurotransmitter (nu′′ro-trans′mit-er) A chemical contained in synaptic vesicles in nerve endings, which is released into the synaptic cleft and stimulates the production of either excitatory or inhibitory postsynaptic potentials.

neutrons (nu′trons) Electrically neutral particles that exist together with positively charged protons in the nucleus of atoms.

neutrophil (nu′tro-fil) A type of phagocytic white blood cell, normally constituting about 60%–70% of the white blood cell count.

nexus (nek′sus) A bond between members of a group; the type of bonds present in single-unit smooth muscles.

nidation (ni-da′shun) Implantation of the blastocyst into the endometrium of the uterus.

nipple A dark pigmented, rounded projection at the tip of the breast.

Nissl bodies (nis′zl) Clumps of rough endoplasmic reticulum in the cell bodies of neurons.

node of Ranvier (rah-ve-a′) A gap in the myelin sheath of a nerve fiber.

norepinephrine (nor′′ep-i-ne′frin) A catecholamine released as a neurotransmitter from postganglionic sympathetic nerve endings and as a hormone (together with epinephrine) from the adrenal medulla.

nucleolus (nu-kle′o-lus) A dark-staining area within a cell nucleus; the site where ribosomal RNA is produced.

nucleoplasm (nu′kle-o-plazm′′) The protoplasmic contents of the nucleus of a cell.

nucleotide (nu′kle-o-tīd) The subunit of DNA and RNA macromolecules; each nucleotide is composed of a nitrogenous base (adenine, guanine, cytosine, and thymine or uracil), a sugar (deoxyribose or ribose), and a phosphate group.

nucleus (nu′kle-us) A spheroid body within a cell that contains the genetic factors of the cell.

nucleus pulposus (pul-po′sus) The soft, pulpy core of an intervertebral disc; a remnant of the notochord.

nystagmus (nis-tag′mus) Involuntary, oscillatory movements of the eye.

o

obese (o-bēs′) Excessively fat.

olfactory (ol-fak′to-re) Pertaining to the sense of smell.

olfactory bulb An aggregation of sensory neurons of an olfactory nerve, lying inferior to the frontal lobe of the cerebrum on either side of the crista galli of the ethmoid bone.

olfactory tract The olfactory sensory tract of axons, which conveys impulses from the olfactory bulb to the olfactory portion of the cerebral cortex.

oligodendrocyte (ol′′ĭ-go-den′dro-sīt) A type of neuroglial cell concerned with the formation of myelin of nerve fibers within the central nervous system.

oncology (ong-kol′o-je) The study of tumors.

oncotic pressure (ong-kot′ik) The colloid osmotic pressure of solutions produced by proteins; in plasma, it serves to counterbalance the outward filtration of fluid from capillaries due to hydrostatic pressure.

oocyte (o′o-sīt) A developing egg cell.

oogenesis (o′′o-jen′ĕ-sis) The process of female gamete formation.

oogenesis (o′′o-jen′ĕ-sis) The formation of ova in the ovaries.

opsonization (op′′so-ni-za′shun) The role of antibodies in enhancing the ability of phagocytic cells to attack bacteria.

optic (op′tik) Pertaining to the eye.

optic chiasma (ki-as′mah) An X-shaped structure on the inferior aspect of the brain, anterior to the pituitary gland, where there is a partial crossing over of fibers in the optic nerves.

optic disc A small region of the retina where the fibers of the ganglion neurons exit from the eyeball to form the optic nerve; also called the *blind spot.*

optic tract A bundle of sensory axons located between the optic chiasma and the thalamus that functions to convey visual impulses from the photoreceptors within the eye.

oral Pertaining to the mouth; also called *buccal.*

organ A structure consisting of two or more tissues, which performs a specific function.

organ of Corti (kor'te) The spiral organ, or functional unit of hearing, consisting of a basilar membrane supporting receptor hair cells and a tectorial membrane within the endolymph of the cochlear duct.

organelle (or''gan-el') A minute living structure of a cell with a specific function.

organism (or'gah-nizm) An individual living creature.

orifice (or'i-fis) An opening into a body cavity or tube.

origin (or'i-jin) The place of muscle attachment—usually the more stationary point or the proximal bone; opposite the insertion.

oropharynx (o''ro-far'inks) The second portion of the pharynx, located in a position posterior to the oral cavity and extending from the soft palate to the hyoid bone.

osmolality (oz''mo-lal'i-te) A measure of the total concentration of a solution; the number of moles of solute per kilogram of solvent.

osmoreceptors (oz''mo-re-cep'tors) Sensory neurons that respond to changes in the osmotic pressure of the surrounding fluid.

osmosis (oz-mo'sis) The passage of solvent (water) from a more dilute to a more concentrated solution through a membrane that is more permeable to water than to the solute.

osmotic pressure (oz-mot'ik) A measure of the tendency for a solution to gain water by osmosis when separated by a membrane from pure water; directly related to the osmolality of the solution, it is the pressure required to just prevent osmosis.

osseous tissue (os'e-us) Bone tissue.

ossification (os''i-fi-ka'shun) The process of bone tissue formation.

osteoblast (os'te-o-blast'') A bone-forming cell.

osteoclast (os'te-o-klast'') A cell that causes erosion and resorption of bone tissue.

osteocyte (os''te-o-sīt'') A mature bone cell.

osteology (os''te-ol'o-je) The study of the structure and function of bone and the entire skeleton.

osteomalacia (os''te-o-mah-la'she-ah) Softening of bones due to a deficiency of vitamin D and calcium.

osteoporosis (os''te-o-po-ro'sis) Demineralization of bone, seen most commonly in the elderly. It may be accompanied by pain, loss of stature, and other deformities and fractures.

otoliths (o'to-liths) Small, hardened particles of calcium carbonate in the saccule and utricle of the inner ear, associated with the receptors of equilibrium.

oval window An oval opening in the bony wall between the middle and inner ear, into which the footplate of the stapes fits.

ovarian follicle (o-va're-an fol'li-k'l) A developing ovum and its surrounding epithelial cells.

ovarian ligament (lig'ah-ment) A cordlike connective tissue that attaches the ovary to the uterus.

ovary (o'vah-re) The female gonad in which ova and certain sexual hormones are produced.

oviduct (o'vi-dukt) The tube that transports ova from the ovary to the uterus; also called the *uterine tube* or *fallopian tube*.

ovulation (o''vu-la'shun) The rupture of an ovarian follicle with the release of an ovum.

ovum (o'vum) A mature egg cell.

oxidative phosphorylation (ok''si-da'tiv fos''fōr-i-la'shun) The formation of ATP using energy derived from electron transport to oxygen; it occurs in the mitochondria.

oxidizing agent (ok''si-dīz-ing) An atom that accepts electrons in an oxidation-reduction reaction.

oxyhemoglobin (ok''se-he''mo-glo'bin) A compound formed by the bonding of molecular oxygen to hemoglobin.

oxyhemoglobin saturation The ratio, expressed as a percentage, of the amount of oxyhemoglobin compared to the total amount of hemoglobin in blood.

oxytocin (ok''si-to'sin) One of the two hormones produced in the hypothalamus and secreted by the posterior pituitary (the other hormone is vasopressin); oxytocin stimulates the contraction of uterine smooth muscles and promotes milk ejection in females.

p

pacemaker (pās'māk-er) A group of cells that has the fastest spontaneous rate of depolarization and contraction in a mass of electrically coupled cells; in the heart, this is the sinoatrial, or SA, node.

pacinian corpuscle (pah-sin'e-an kor'pusl) A sensory receptor for pressure, found in tendons, around joints, and in visceral tissues.

PAH Para-aminohippuric acid; a substance used to measure total renal plasma flow because its clearance rate is equal to the total rate of plasma flow to the kidneys; PAH is filtered and secreted but not reabsorbed by the renal nephrons.

palate (pal'at) The roof of the mouth or oral cavity.

palatine (pal'ah-tīn) Pertaining to the palate.

palmar (pal'mar) Pertaining to the palm of the hand.

palpebra (pal'pĕ-brah) An eyelid.

pancreas (pan'kre-as) A mixed organ in the abdominal cavity that secretes gastric juices into the digestive tract and insulin and glucagon into the blood.

pancreatic duct (pan''kre-at'ik) A drainage tube that carries pancreatic juice from the pancreas into the duodenum of the ampulla of Vater.

papillae (pah-pil'e) Small, nipplelike projections.

papillary muscle (pap'i-ler''e) Muscular projections from the ventricular walls of the heart to which the chordae tendineae are attached.

paranasal sinus (par''ah-na'zal si'nus) A mucous-lined air chamber that communicates with the nasal cavity.

parasympathetic (par''ah-sim''pah-thet'ik) Pertaining to the division of the autonomic nervous system concerned with activities that restore and conserve metabolic energy.

parathyroid hormone (PTH) A polypeptide hormone secreted by the parathyroid glands, PTH acts to raise the blood Ca^{++} levels primarily by stimulating reabsorption of bone.

parathyroids (par''ah-thi'roid) Small endocrine glands that are embedded on the posterior surface of the thyroid glands and are concerned with calcium metabolism.

parietal (pah-ri'e-tal) Pertaining to a wall of an organ or cavity.

parietal pleura (ploor'ah) The thin serous membrane attached to the thoracic walls of the pleural cavity.

Parkinson's disease (par'kin-sunz) A tremor of the resting muscles and other symptoms caused by inadequate dopamine-producing neurons in the basal ganglia of the cerebrum. Also called *paralysis agitans*.

parotid gland (pah-rot'id) One of the paired salivary glands located on the sides of the face over the masseter muscle just anterior to the ear and connected to the oral cavity through a salivary duct.

parturition (par''tu-rish'un) The process of childbirth.

passive immunity (i-mu'ni-te) Specific immunity granted by the administration of antibodies made by another organism.

Pasteur effect (pas-ter) A decrease in the rate of glucose utilization and lactic acid production in tissues or microorganisms by their exposure to oxygen.

pathogen (path'o-jen) Any disease-producing microorganism or substance.

pectoral (pek'to-ral) Pertaining to the chest region.

pectoral girdle (ger'd'l) The portion of the skeleton that supports the upper extremities.

pedicle (ped'i-k'l) The portion of a vertebra that connects and attaches the lamina to the body.

pelvic (pel'vik) Pertaining to the pelvis.

pelvic girdle (ger'd'l) The portion of the skeleton to which the lower extremities are attached.

pelvis (pel'vis) A basinlike bony structure formed by the sacrum and os coxae.

penis (pe'nis) The external male genital organ, through which urine passes during urination and which transports spermatozoa to the female during sexual intercourse.

pepsin The protein-digesting enzyme secreted in gastric juice.

peptic ulcer (pep′tik ul′ser)   An injury to the mucosa of the esophagus, stomach, or small intestine, caused by acidic gastric juice.

pericardium (per″ĭ-kar′de-um)   A protective serous membrane that surrounds the heart.

perichondrium (per″ĭ-kon′dre-um)   A toughened connective sheet that covers cartilage.

perikaryon (per″ĭ-kar′e-on)   The cell body of a neuron.

perilymph (per′ĭ-limf)   The fluid between the membranous and bony labyrinth of the inner ear.

perimysium (per″ĭ-mis′e-um)   Fascia or connective tissue surrounding a bundle (fascicle) of muscle fibers.

perineum (per″ĭ-ne′um)   The floor of the pelvis, which is the region between the anus and the scrotum in the male and between the anus and the vulva in the female.

perineurium (per″ĭ-nu′re-um)   Connective tissue surrounding a bundle (fascicle) of nerve fibers.

periodontal membrane (per″e-o-don′tal)   A fibrous connective tissue lining the sockets of teeth.

periosteum (per″e-os′te-um)   A fibrous connective tissue covering the surface of bone.

peripheral nervous system (pĕ-rif′er-al)   The nerves and ganglia of the nervous system that lie outside of the brain and spinal cord.

peristalsis (per″ĭ-stal′sis)   Rhythmic contractions of smooth muscle in the walls of various tubular organs, which move the contents along.

peritoneum (per″ĭ-to-ne′um)   The serous membrane that lines the abdominal cavity and covers the abdominal visceral organs.

permease (per′me-ās)   A term used to indicate membrane transport carriers and to emphasize the similarity of specificity and other properties that transport carriers have with enzymes.

Peyer's patches (pi′erz)   Clusters of lymph nodes on the walls of the small intestine.

pH   The pH of a solution is equal to 1 over the logarithm of the hydrogen ion concentration. The pH scale goes from zero to 14; a pH of 7.0 is neutral, whereas solutions with lower pH are acidic and solutions with higher pH are basic.

phagocytosis (fag″o-si-to′sis)   Cellular eating; the ability of some cells (such as white blood cells) to engulf large particles such as bacteria and digest these particles by merging the food vacuole containing these particles with a lysosome containing digestive enzymes.

phalanx (fa′lanks), pl. *phalanges*   A bone of a finger or toe.

pharynx (far′inks)   The region of the digestive system and respiratory system located at the back of the oral and nasal cavities and extending to the larynx anteriorly and the esophagus posteriorly; also called the *throat*.

phonocardiogram (fo″no-kar′de-o-gram)   A visual display of the heart sounds.

photoreceptor (fo″to-re-sep′tor)   A sensory nerve ending that responds to the stimulation light.

physiology (fiz″e-ol′o-je)   The science that deals with the study of body functions.

pia mater (pi′ah ma′ter)   The innermost meninx that is in direct contact with the brain and spinal cord.

pineal gland (pin′e-al)   A small cone-shaped gland located in the root of the third ventricle.

pinna (pin′nah)   The outer, fleshy portion of the external ear; also called the *auricle*.

pinocytosis (pi″no-, pin″-o-si-to′sis)   Cell drinking; invagination of the cell membrane forming narrow channels that pinch off into vacuoles; this provides cellular intake of extracellular fluid and dissolved molecules.

pituitary gland (pĭ-tu′ĭ-tār′e)   A small, pea-shaped endocrine gland situated on the interior surface of the diencephalonic region of the brain, consisting of anterior and posterior lobes; also called the *hypophysis* and commonly called the "master gland."

pivot joint (piv′ut)   A diarthrotic joint in which a rounded head of one bone articulates with a depressed cup of another to permit a rotational type of movement.

placenta (plah-sen′tah)   The organ of metabolic exchange between the mother and the fetus.

plantar (plan′tar)   Pertaining to the sole of the foot.

plasma (plaz′mah)   The fluid, extracellular portion of circulating blood.

plasma cells   Cells derived from B lymphocytes that produce and secrete large amounts of antibodies; they are responsible for humoral immunity.

platelets (plāt′lets)   Small fragments of specific bone marrow cells that function in blood coagulation; also called *thrombocytes*.

pleural (ploor′al)   Pertaining to the serous membranes associated with the lungs.

pleural cavity   The potential space between the visceral pleural and parietal pleural membranes.

pleural membranes   Serous membranes that surround the lungs and provide protection and compartmentalization.

plexus (plek′sus)   A network of interlaced nerves or vessels.

plexus of Auerbach (ow′er-bahk)   A network of sympathetic and parasympathetic nerve fibers located in the muscularis tunic of the small intestine.

plexus of Meissner (mīs′ner)   A network of autonomic nerve fibers located in the submucosa of the small intestine.

plica circularis (pli′kah ser-ku-lar′is)   A deep fold within the wall of the small intestine, which increases the absorptive surface area.

pluripotent (ploo-rip′o-tent)   A property of early embryonic cells; the ability to specialize in a number of ways to produce tissues characteristic of different organs.

pneumotaxic area (nu″mo-tak′sik)   The region of the respiratory control center, located in the pons of the brain.

polar body   A small daughter cell formed by meiosis that degenerates in the process of oocyte production.

polar molecule   A molecule in which the shared electrons are not evenly distributed, so that one side of the molecule is relatively negatively (or positively) charged in comparison with the other side; polar molecules are soluble in polar solvents such as water.

polydipsia (pol′e-dip′se-ah)   Excessive thirst.

polymer (pol′i-mer)   A large molecule formed by the combination of smaller subunits, or monomers.

polymorphonuclear leukocyte (pol″e-mor″fo-nu′kle-ar loo′ko-sīt)   A granular leukocyte containing a nucleus with a number of lobes connected by thin, cytoplasmic strands; this type includes neutrophils, eosinophils, and basophils.

polypeptide (pol″e-pep′tĭd)   A chain of amino acids connected by covalent bonds called peptide bonds. A very large polypeptide is called a protein.

polyphagia (pol′e-fa′je-ah)   Excessive eating.

polysaccharide (pol″e-sak′ah-rīd)   A carbohydrate formed by covalent bonding of numerous monosaccharides; examples include glycogen and starch.

polyuria (pol″e-u′re-ah)   Excretion of an excessively large volume of urine in a given period.

pons (ponz)   The portion of the brain stem just above the medulla oblongata and anterior to the cerebellum.

popliteal (pop-lit′e-al, pop″li-te′al)   Pertaining to the concave region on the posterior aspect of the knee joint.

posterior (pos-tēr′e-or)   Toward the back; also called *dorsal*.

posterior pituitary (pĭ-tu′ĭ-tār″e)   The part of the pituitary gland that is derived from the brain; it secretes vasopressin (ADH) and oxytocin, produced in the hypothalamus. Also called the *neurohypophysis*.

posterior root   An aggregation of sensory neuron fibers lying between a spinal nerve and the dorsolateral aspect of the spinal cord; also called the *dorsal root* or *sensory root*.

posterior root ganglion (gang′gle-on)   A cluster of cell bodies of sensory neurons located along the posterior root of a spinal nerve.

postganglionic neuron (pōst″gang-gle-on′ik)   The second neuron in an autonomic efferent pathway. Its cell body is outside the central nervous system, and it terminates at an effector organ.

postnatal (pōst-na′tal)   After birth.

postsynaptic inhibition (pōst″sĭ-nap′tik)   The inhibition of a postsynaptic neuron by axon endings that release a neurotransmitter that induces hyperpolarization (inhibitory postsynaptic potentials).

preganglionic neuron (pre''gang-gle-on'ik) The first neuron in an autonomic efferent pathway. Its cell body is inside the central nervous system, and it terminates on a postganglionic neuron.

pregnancy A condition where a female has a developing offspring in the uterus.

prenatal (pre-na'tal) The period of offspring development during pregnancy; before birth.

prepuce (pre'pūs) A fold of loose, retractable skin covering the glans of the penis or clitoris; also called the *foreskin*.

presynaptic inhibition (pre''si-nap'tik) Neural inhibition in which axoaxonic synapses inhibit the release of neurotransmitter chemicals from the presynaptic axon.

prime mover The muscle most directly responsible for a particular movement; an agonist.

prolactin (pro-lak'tin) A hormone secreted by the anterior pituitary that stimulates lactation (acting together with other hormones) in the postpartum female. It may also participate (along with the gonadotrophins) in regulating gonadal function in some mammals.

pronation (pro-na'shun) A rotational movement of the forearm in which the palm of the hand is turned posteriorly.

prophylaxis (pro''fi-lak'sis) Prevention or protection.

proprioceptor (pro''pre-o-sep'tor) A sensory nerve ending that responds to changes in tension in a muscle or tendon.

prostate gland (pros'tāt) A walnut-shaped gland that surrounds the male urethra just below the urinary bladder and secretes an additive to seminal fluid during ejaculation.

proton A unit of positive charge in the nucleus of atoms.

protoplasm (pro'to-plazm) A general term that includes cytoplasm and nucleoplasm.

protraction (pro-trak'shun) The movement of a body part, such as the mandible, forward on a plane parallel with the ground; the opposite of *retraction*.

proximal (prok'si-mal) Closer to the midline of the body or origin of an appendage; the opposite of *distal*.

pseudohermaphrodite (soo''do-her-maf'ro-dīt) An individual with some of the physical characteristics of both sexes, but who lacks functioning gonads of both sexes; a true hermaphrodite has both testes and ovaries.

pseudopods (soo'do-pods) Footlike extensions of the cytoplasm that enable some cells (with amoeboid motion) to move across a substrate; pseudopods also are used to surround food particles in the process of phagocytosis.

ptyalin (ti'ah-lin) Also called *salivary amylase;* an enzyme in saliva that catalyzes the hydrolysis of starch into smaller molecules.

puberty (pu'ber-te) The period of development in which the reproductive organs become functional.

pulmonary (pul'mo-ner''e) Pertaining to the lungs.

pulmonary circulation The system of blood vessels from the right ventricle of the heart to the lungs, transporting deoxygenated blood and returning oxygenated blood from the lungs to the left atrium of the heart.

pulp cavity A cavity within the center of a tooth, containing blood vessels, nerves, and lymphatics.

pupil The opening through the iris that permits light to enter the posterior cavity of the eyeball and be refracted by the lens.

Purkinje fibers (pur-kin'je) Specialized cardiac muscle fibers that conduct electrical impulses from the AV bundle into the ventricular walls.

pyloric sphincter (pi-lor'ik sfingk'ter) A modification of the muscularis tunica between the stomach and the duodenum, which functions to regulate the food material leaving the stomach.

pyramid (pir'ah-mid) Any of several structures that have a pyramidal shape, including the renal pyramids in the kidney and the medullary pyramids on the ventral surface of the brain.

pyrogen (pi'ro-jen) A fever-producing substance.

q

QRS complex The part of an electrocardiogram that is produced by depolarization of the ventricles.

r

ramus (ra'mus) A branch of a bone, artery, or nerve.

raphe (ra'fe) A ridge or a seamlike structure.

receptor (re-sep'tor) A sense organ or a specialized distal end of a sensory neuron that receives stimuli from the environment.

rectouterine pouch (rek''to-u'ter-in) A pocket of parietal peritoneal membrane between the uterus and the rectum, forming the lowest point in the pelvic cavity; also called the *pouch of Douglas*.

rectum (rek'tum) The terminal portion of the gastrointestinal tract, from the sigmoid colon to the anus.

red marrow (mar'o) A tissue that forms blood cells, located in the medullary cavity of certain bones.

red nucleus (nu'kle-us) An aggregation of gray matter of a reddish color that is located in the upper portion of the midbrain and sends fibers to certain brain tracts.

reduced hemoglobin (he''mo-glo'bin) Hemoglobin with iron in the reduced ferrous state that is able to bond with oxygen but is not combined with oxygen. Also called *deoxyhemoglobin*.

reducing agent An electron donor in a coupled oxidation-reduction reaction.

reflex (re'fleks) A rapid involuntary response to a stimulus.

reflex arc The basic conduction pathway through the nervous system, consisting of a sensory neuron, interneuron, and a motor neuron.

regional anatomy The division of anatomy concerned with structural arrangement in specific areas of the body such as the head, neck, thorax, or abdomen.

REM sleep The stage of sleep in which dreaming occurs; associated with rapid eye movements (REM). REM sleep occurs three to four times each night and lasts from a few minutes to over an hour.

renal (re'nal) Pertaining to the kidney.

renal corpuscle (kor'pus'l) The portion of the nephron consisting of the glomerulus and a Bowman's capsule; also called the *Malpighian corpuscle*.

renal cortex The outer portion of the kidney, primarily vascular.

renal medulla (mĕ-dul'ah) The inner portion of the kidney, including the renal pyramids and renal columns.

renal pelvis The inner cavity of the kidney formed by the expanded ureter and into which the calyces open.

renal plasma clearance rate The milliliters of plasma that are cleared of a particular solute per minute by the excretion of that solute in the urine; if there is no reabsorption or secretion of that solute by the nephron tubules, this is equal to the glomerular filtration rate.

renal pyramid A triangular structure within the renal medulla, composed of the loops of Henle and the collecting ducts.

repolarization (re-po''lar-i-za'shun) The reestablishment of the resting membrane potential after depolarization has occurred.

respiration (res''pi-ra'shun) The exchange of gases between the external environment and the cells of an organism.

respiratory acidosis (re-spi'rah-to''re as''i-do'sis) A lowering of the blood pH below 7.35 due to accumulation of CO_2 as a result of hypoventilation.

respiratory alkalosis (al''kah-lo'sis) A rise in blood pH above 7.45 due to excessive elimination of blood CO_2 as a result of hyperventilation.

respiratory center The structure or portion of the brain stem that regulates the depth and rate of breathing.

respiratory distress syndrome Also called *hyaline membrane disease;* most frequently occurring in premature infants, this syndrome is caused by abnormally high alveolar surface tension as a result of a deficiency in lung surfactant.

respiratory membrane A thin, moistened membrane within the lungs, composed of an alveolar portion and a capillary portion, through which gaseous exchange occurs.

rete testis (re'te tes'tis) A network of ducts in the center of the testis, associated with the production of spermatozoa.

reticular formation (rē-tik′u-lar) A network of nervous tissue fibers in the brain stem that arouses the higher brain centers.

retina (ret′ĭ-nah) The inner layer of the eyeball that contains the photoreceptors.

retraction (re-trak′shun) The movement of a body part, such as the mandible, backward on a plane parallel with the ground; the opposite of *protraction*.

retroperitoneal (ret″ro-per″ĭ-to-ne′al) Positioned behind the parietal peritoneum.

rhodopsin (ro-dop′sin) Visual purple; a pigment in rod cells that undergoes a photochemical dissociation in response to light and in so doing, stimulates electrical activity in the photoreceptors.

rhythmicity area (rith-mis′ĭ-te) A portion of the respiratory control center that is located in the medulla and controls inspiratory and expiratory phases.

ribosome (ri′bo-sōm) A cytoplasmic organelle composed of protein and RNA, which is the site of protein synthesis.

rickets (rik′ets) A condition caused by a deficiency of vitamin D and associated with an interference of the normal ossification of bone.

right lymphatic duct (lim-fat′ik) A major vessel of the lymphatic system that drains lymph from the upper right portion of the body into the right subclavian vein.

rigor mortis The stiffening of a dead body, due to the depletion of ATP and the production of rigor complexes between actin and myosin in muscles.

RNA Ribonucleic acid; a nucleic acid consisting of the nitrogenous bases adenine, guanine, cytosine, and uracil; the sugar ribose; and phosphate groups. There are three types of RNA found in cytoplasm: messenger RNA (mRNA), transfer RNA (tRNA), and ribosomal RNA (rRNA).

rod A photoreceptor in the retina of the eye that is specialized for colorless, dim light vision.

root canal The hollow, tubular extension of the pulp cavity into the root of the tooth, containing vessels and nerves.

rotation (ro-ta′shun) The movement of a bone around its own longitudinal axis.

round window A round, membrane-covered opening between the middle and inner ear, directly below the oval window.

rugae (ru′je) The folds or ridges of the mucosa of an organ.

s

saccadic eye movements (sah-kad′ik) Very rapid eye movements that occur constantly and that change the focus on the retina from one point to another.

saccule (sak′ūl) A saclike cavity in the membranous labyrinth inside the vestibule of the inner ear, containing a vestibular organ for equilibrium.

sacral (sa′kral) Pertaining to the sacrum.

sacral plexus (plek′sus) A network of nerve fibers that arise from spinal nerves L4 through S3. Nerves arising from the sacral plexus merge with those from the lumbar plexus to form the lumbosacral plexus and supply the lower extremity.

saddle joint A diarthrotic joint in which the articular surfaces of both bones are concave in one plane and convex, or saddle shaped, in the other plane, such as in the distal carpal and proximal metacarpal joint of the thumb.

sagittal (saj′ĭ-tal) A vertical plane through the body that divides it into right and left portions.

salivary gland (sal′ĭ-ver-e) An accessory digestive gland that secretes saliva into the oral cavity.

saltatory conduction (sal′tah-to″re) The rapid passage of action potentials from one node of Ranvier to another in myelinated axons.

sarcolemma (sar″ko-lem′ah) The cell membrane of a muscle fiber.

sarcomere (sar′ko-mēr) The portion of a striated muscle fiber between the two adjacent Z lines that is considered the functional unit of a myofibril.

sarcoplasm (sar′ko-plazm) The cytoplasm within a muscle fiber.

sarcoplasmic reticulum (sar′ko-plaz′mik rē-tik′u-lum) The smooth or agranular endoplasmic reticulum of skeletal muscle cells; it surrounds each myofibril and serves to store Ca^{++} when the muscle is at rest.

scala tympani (ska′lah tim′pah-ne) The lower channel of a cochlea that is filled with perilymph.

scala vestibuli (ves-tib′u-le) The upper channel of the cochlea that is filled with perilymph.

Schwann cell (shwahn) A specialized neuroglia cell that surrounds an axon fiber of a peripheral nerve and forms the neurilemmal sheath.

sclera (skle′rah) The outer white layer of fibrous connective tissue that forms the protective covering of the eyeball.

scrotum (skro′tum) A pouch of skin that contains the testes and their accessory organs.

sebaceous gland (se-ba′shus) An exocrine gland of the skin that secretes sebum.

sebum (se′bum) An oily, waterproofing secretion of the sebaceous glands.

second messenger A molecule or ion whose concentration within a target cell is increased by the action of a regulator compound (e.g., a hormone or neurotransmitter) and which stimulates the metabolism of that target cell in a way characteristic of the actions of that regulator molecule—that is, in a way that mediates the intracellular effects of that regulatory compound.

secretin (se-kre′tin) A polypeptide hormone secreted by the small intestine in response to acidity of the intestinal lumen; along with cholecystokinin, secretin stimulates the secretion of pancreatic juice into the small intestine.

semen (se′men) The thick, whitish fluid secretion of the reproductive organs of the male, consisting of spermatozoa and additives from several other accessory organs.

semicircular canals Tubular channels within the inner ear that contain receptors for equilibrium.

semilunar valve (sem″e-lu′nar) Crescent or half-moon-shaped heart valves positioned at the entrances to the aorta and the pulmonary trunk.

seminal vesicles (sem′ĭ-nal ves′ĭ-k′lz) A pair of accessory male reproductive organs lying posterior and inferior to the urinary bladder, which secrete additives to spermatozoa into the ejaculatory ducts.

seminiferous tubule (se″mi-nif′er-us too′būl) Numerous small ducts in the testis where spermatozoa are produced.

semipermeable membrane (sem″e-per′me-ah-b′l) A membrane with pores of a size that permits the passage of solvent and some solute molecules but restricts the passage of other solute molecules.

senescence (sē-nes′ens) The process of aging.

sensory area A region of the cerebral cortex that receives and interprets sensory nerve impulses.

sensory neuron (nu′ron) A nerve cell that conducts an impulse from a receptor organ to the central nervous system.

septum (sep′tum) A membranous or fleshy wall dividing two cavities.

serosa (se-ro′sah, se-ro′zah) An outer epithelial membrane that covers the surface of a visceral organ.

serous membrane (se′rus) An epithelial and connective tissue membrane that lines body cavities and covers visceral organs within these cavities; also called *serosa*.

Sertoli cells (ser-to′le) Specialized cells within the testes that supply nutrients to developing spermatozoa; also called *nurse cells*.

serum (se′rum) Blood plasma with the clotting elements removed.

sesamoid bone (ses′ah-moid) A membranous bone formed in a tendon in response to joint stress (e.g., the patella).

sex chromosomes The X and Y chromosomes; the unequal pairs of chromosomes involved in sex determination (which is due to the presence or absence of a Y chromosome). Females lack a Y chromosome and normally have the genotype XX; males have a Y chromosome and normally have the genotype XY.

shock As it relates to the cardiovascular system, this term refers to a rapid, uncontrolled fall in blood pressure, which in some cases becomes irreversible and leads to death.

shoulder The region of the body where the humerus articulates with the scapula; also called the *omos*.

sickle-cell anemia A hereditary, autosomal recessive trait that occurs primarily in people of African ancestry, in which it evolved apparently as a protection (in the carrier state) against malaria. In the homozygous state, hemoglobin S is made instead of hemoglobin A; this leads to the characteristic sickling of red blood cells, hemolytic anemia, and organ damage.

sigmoid colon (sig'moid ko'lon) The S-shaped portion of the large intestine between the descending colon and the rectum.

sinoatrial node (sin''o-a'tre-al) A mass of specialized cardiac tissue in the wall of the right atrium that initiates the cardiac cycle; the SA node; also called the *pacemaker.*

sinus (si'nus) A cavity or hollow space within a body organ such as a bone.

sinusoid (si'nŭ-soid) A small, blood-filled space in certain organs such as the spleen or liver.

sleep apnea (ap-ne'ah) A temporary cessation of breathing during sleep, usually lasting for several seconds.

sliding filament theory The theory that the thick and thin filaments of a myofibril slide past each other, while maintaining their initial length, during muscle contraction.

small intestine The portion of the alimentary canal between the stomach and the cecum whose function is the absorption of food nutrients.

smooth muscle A specialized type of muscle tissue that is nonstriated, composed of fusiform, single-nucleated fibers, and contracts in an involuntary, rhythmic fashion with the walls of visceral organs.

sodium/potassium pump (so'de-um po-tas'e-um) An active transport carrier, with ATPase enzymatic activity, that acts to accumulate K⁺ within cells and extrude Na⁺ from cells, thus maintaining gradients for these ions across the cell membrane.

soft palate (pal'at) The fleshy, posterior portion of the roof of the mouth from the palatine bones to the uvula.

somatic (so-mat'ik) Pertaining to the nonvisceral parts of the body.

somatomedins (so''mah-to-me'dins) A group of small polypeptides that are believed to be produced in the liver in response to growth hormone stimulation and to mediate the actions of growth hormone on the skeleton and other tissues.

somatostatin (so''mah-to-stat'in) A polypeptide produced in the hypothalamus that acts to inhibit the secretion of growth hormone from the anterior pituitary; somatostatin is also produced in the islets of Langerhans of the pancreas, but its function there has not been established.

somatotrophic hormone (so''mah-to-trōf'ik) Growth hormone; an anabolic hormone secreted by the anterior pituitary that stimulates skeletal growth and protein synthesis in many organs.

sounds of Korotkoff (ko-rot'kof) The sounds heard when blood pressure measurements are taken. These sounds are produced by the turbulent flow of blood through an artery that has been partially constricted by a pressure cuff.

spermatic cord (sper-mat'ik) The structure of the male reproductive system composed of the ductus deferens, spermatic vessels, nerves, cremasteric muscle, and connective tissue. The spermatic cord extends from a testis to the inguinal ring.

spermatogenesis (sper''mah-to-jen'ĕ-sis) The production of male sex gametes, or spermatozoa.

spermatozoon (sper''mah-to-zo'on) A mature male sperm cell, or gamete.

spermiogenesis (sper''me-o-jen'ĕ-sis) The maturational changes that transform spermatids into spermatozoa.

sphincter (sfingk'ter) A circular muscle that functions to constrict a body opening or the lumen of a tubular structure.

sphincter of Oddi (o'de) The muscular constriction at the opening of the common bile and pancreatic ducts.

sphygmomanometer (sfig''mo-mah-nom'ĕ-ter) A manometer (pressure transducer) used to measure the blood pressure.

spinal cord (spi'nal) The portion of the central nervous system that extends downward from the brain stem through the vertebral canal.

spinal nerve One of the thirty-one pairs of nerves that arise from the spinal cord.

spindle fibers Filaments that extend from the poles of a cell to its equator and attach to the chromosomes during the metaphase stage of cell division. Contraction of the spindle fibers pulls the chromosomes to opposite poles of the cell.

spinous process (spi'nus) A sharp projection of bone or a ridge of bone, such as on the scapula.

spironolactones (spi-ro''no-lak'tōns) Diuretic drugs that act as an aldosterone antagonist.

spleen (splēn) A large, blood-filled, glandular organ located in the upper left of the abdomen and attached by mesenteries to the stomach.

spongy bone (spun'je) A type of bone that contains many porous spaces; also called *cancellous bone.*

squamous (skwa'mus) Flat or scalelike.

stapes (sta'pēz) The innermost of the ossicles of the ear, which fits against the oval window of the inner ear; also called the *stirrup.*

steroid (ste'roid) A lipid, derived from cholesterol, that has three six-sided carbon rings and one five-sided carbon ring. These form the steroid hormones of the adrenal cortex and gonads.

stomach A pouchlike digestive organ located between the esophagus and the duodenum.

stratified (strat'ĭ-fīd) Arranged in layers, or strata.

stratum corneum (stra'tum kor'ne-um) The outer, cornified layer of the epidermis of the skin.

stratum germinativum (stra'tum jer'mĭ-na''tiv-um) The deepest epidermal layer, where mitotic activity occurs.

striated muscle (stri'āt-ed) A specialized type of muscle tissue that is multinucleated, occurs in bundles, has crossbands of proteins, and contracts in either a voluntary or involuntary fashion.

stroke volume The amount of blood ejected from each ventricle at each heartbeat.

stroma (stro'mah) A connective tissue framework in an organ, gland, or other tissue.

subarachnoid space (sub''ah-rak'noid) The space within the meninges, between the arachnoid and pia maters, where cerebrospinal fluid flows.

subdural space (sub-du'ral) A space within the meninges between the dura and arachnoid of the brain and spinal cord.

sublingual gland (sub-ling'gwal) One of the three pairs of salivary glands; it is located below the tongue and secretes to the side of the lingual frenulum.

submandibular gland (sub-man-dib'u-lar) One of the three pairs of salivary glands; it is located below the mandible and secretes to the side of the lingual frenulum.

submucosa (sub''mu-ko'sah) A layer of supportive connective tissue that underlies a mucous membrane.

substrate (sub'strat) In enzymatic reactions, the molecules that combine with the active sites of an enzyme and are converted to products by catalysis of the enzyme.

sulcus (sul'kus) A shallow impression or groove.

superficial (su''per-fish'al) Toward or near the surface.

superficial fascia (fash'e-ah) A binding layer of connective tissue between the dermis of the skin and the underlying muscle.

superior Toward the upper part of a structure or toward the head; also called *cephalic.*

supination (su''pi-na'shun) Rotation of the arm so that the palm is directed forward or anteriorly; the opposite of *pronation.*

suppressor T cells A subpopulation of T lymphocytes that acts to inhibit the production of antibodies against specific antigens by B lymphocytes.

surface anatomy The division of anatomy concerned with the structures that can be identified from the outside of the body.

surfactant (sur-fak'tant) A substance produced by the lungs that decreases the surface tension within the alveoli.

suspensory ligament (sus-pen'so-re) A portion of the peritoneum that extends laterally from the surface of the ovary to the wall of the pelvic cavity.

suture (su'chur) A type of immovable joint articulation found between bones of the skull.

sweat gland A skin gland that secretes a fluid substance for evaporative cooling.

sympathetic (sim''pah-thet'ik) Pertaining to that part of the autonomic nervous system concerned with processes involving the utilization of energy; also called *thoracolumbar division.*

symphysis (sim'fĭ-sis) A type of articulation characterized by a fibrocartilaginous pad between the articulating bones, which provides slight movement.

synapse (sin'aps) A region where a nerve fiber comes into close or actual contact with another cell and across which nerve impulses are transmitted either directly or indirectly (via release of chemical neurotransmitters).

synarthrosis (sin″ar-thro′sis) An immovable joint, such as a synchondrosis or a suture.

synchondrosis (sin″kon-dro′sis) An immovable cartilaginous joint in which the articulating bones are separated by hyaline cartilage.

syndesmosis (sin″des-mo′sis) A type of slightly movable, fibrous joint in which two bones are united by an interosseous ligament.

synergist (sin′er-jist) A muscle that assists the action of the prime mover.

synergistic (sin″er-jis′tik) Pertaining to regulatory processes or molecules (such as hormones) that have complementary or additive effects.

synovial cavity (si-no′ve-al) A space between the two bones of a diarthrotic joint, filled with synovial fluid.

synovial joint A freely movable joint in which there is a synovial cavity between the articulating bones; also called a *diarthrotic joint*.

synovial membrane The inner membrane of a synovial capsule, which secretes synovial fluid into the joint cavity.

system A group of body organs that function together.

systemic (sis-tem′ik) Relating to the entire organism rather than individual parts.

systemic anatomy The division of anatomy concerned with the structure and function of the various systems.

systemic circulation The portion of the circulatory system concerned with blood flow from the left ventricle of the heart to the entire body and back to the heart via the right atrium; as opposed to the pulmonary system, which involves the lungs.

systole (sis′to-le) The muscular contraction of a heart chamber during the cardiac cycle.

systolic pressure (sis′tol′ik) Arterial blood pressure during the ventricular systolic phase of the cardiac cycle.

t

tachycardia (tak″e-kar′de-ah) Excessively rapid heart rate, usually applied to rates in excess of 100 beats per minute. In contrast to an excessively slow heart rate (below 60 beats per minute), which is termed bradycardia.

tactile (tak′til) Pertaining to the sense of touch.

taeniae coli (te′ne-e ko′li) The three longitudinal bands of muscle in the wall of the large intestine.

target organ The specific body organ that a particular hormone affects.

tarsus (tahr′sus) The seven bones that form the ankle.

taste bud An organ containing the chemoreceptors associated with the sense of taste.

T cell A type of lymphocyte that provides cell-mediated immunity, in contrast to B lymphocytes, which provide humoral immunity through the secretion of antibodies. There are three subpopulations of T cells: cytotoxic, helper, and suppressor.

tectorial membrane (tek-to′re-al) A gelatinous membrane positioned over the hair cells of the organ of Corti in the cochlea.

tendon (ten′dun) A band of dense fibrous connective tissue that attaches muscle to bone.

tendon sheath A covering of synovial membrane surrounding certain tendons.

tentorium cerebelli (ten-to′re-um ser″e-bel′e) An extension of dura mater that forms a partition between the cerebral hemispheres and the cerebellum and covers the cerebellum.

testis (tes′tis) The primary reproductive organ of a male, which produces spermatozoa and male sex hormones.

testosterone (tes-tos′te-ron) The major androgenic steroid secreted by the Leydig cells of the testes after puberty.

tetanus (tet′a-nus) A term used to mean either a smooth contraction of a muscle (as opposed to muscle twitching), or a state of maintained contracture of high tension.

thalamus (thal′ah-mus) An oval mass of gray matter within the diencephalon, which serves as a sensory relay area.

thalassemia (thal″ah-se′me-ah) A group of hemolytic anemias caused by the hereditary inability to produce either the alpha or beta chain of hemoglobin. It is found primarily among Mediterranean people.

thigh The proximal portion of the lower extremity between the hip and the knee, containing the femur bone.

third ventricle (ven′tri-k′l) A narrow cavity between the right and left halves of the thalamus and between the lateral ventricles, containing cerebrospinal fluid.

thoracic (tho-ras′ik) Pertaining to the chest region.

thoracic duct The major lymphatic vessel of the body, which drains lymph from the entire body except the upper right quadrant and returns it to the left subclavian vein.

thorax (tho′raks) The chest.

threshold The minimum stimulus that just produces a response.

thrombocyte (throm′bo-sit) A blood platelet formed from a fragmented megakaryocyte.

thrombus A blood clot produced by the formation of fibrin threads around a platelet plug.

thymus gland (thi′mus) A bilobed lymphoid organ positioned in the upper mediastinum, posterior to the sternum and between the lungs.

thyroid cartilage (thi′roid kar′ti-lij) The largest cartilage in the larynx, which supports and protects the vocal cords; also called the *Adam's apple*.

thyroxine (thi-rok′sin) Also called tetraiodothyronine, or T_4. The major hormone secreted by the thyroid gland, which regulates the basal metabolic rate and stimulates protein synthesis in many organs; a deficiency of this hormone in early childhood produces cretinism.

tinnitus (ti-ni′tus) A ringing sound or other noise that is heard but is not related to external sounds.

tissue An aggregation of similar cells and their binding intercellular substance, joined to perform a specific function.

tongue A protrusible muscular organ on the floor of the oral cavity.

tonsil (ton′sil) A node of lymphoid tissue located in the mucous membrane of the pharynx.

toxin (tok′sin) A poison.

toxoid (tok′soid) A modified bacterial endotoxin that has lost toxicity but retains its ability to act as an antigen and stimulate antibody production.

trabeculae (trah-bek′u-le) A supporting framework of fibers crossing the substance of a structure, as in the lamellae of spongy bone.

trachea (tra′ke-ah) The airway leading from the larynx to the bronchi, composed of cartilaginous rings and a ciliated mucosal lining of the lumen; also called the *windpipe*.

tract A bundle of nerve fibers within the central nervous system.

transamination (trans″am-i-na′shun) The transfer of an amino group from an amino acid to an alpha-keto acid, forming a new keto acid and a new amino acid, without the appearance of free ammonia.

transection (tran-sek′shun) A cross-sectional cut.

transpulmonary pressure (trans″pul′mo-ner″e) The pressure difference across the wall of the lung, equal to the difference between intrapulmonary pressure and intrapleural pressure.

transverse colon (trans-vers′ ko′lon) A portion of the large intestine that extends from right to left across the abdomen between the hepatic and splenic flexures.

transverse fissure (fish′ur) The prominent cleft that horizontally separates the cerebrum from the cerebellum.

tricuspid valve (tri-kus′pid) The heart valve located between the right atrium and the right ventricle.

trigone (tri′gon) A triangular area in the urinary bladder between the openings of the ureters and the urethra.

triiodothyronine (tri″i-o″do-thi′ro-nen) Abbreviated T_3; a hormone secreted in small amounts by the thyroid; the active hormone in target cells, formed from thyroxine.

trochanter (tro-kan′ter) A broad, prominent process on a bone, specifically on the femur.

trochlea (trok′le-ah) A pulley-shaped structure.

tropomyosin (tro″po-mi′o-sin) A filamentous protein that attaches to actin in the thin filaments and that acts, together with another protein called troponin, to inhibit and regulate the attachment of myosin cross-bridges to actin.

true vocal cords Folds of the mucous membrane in the larynx that produce sound as they are pulled taut and vibrated.

trunk The thorax and abdomen together.

trypsin (trip′sin) A protein-digesting enzyme in pancreatic juice that is released into the small intestine.

tubercle (tu′ber-k′l) A small, elevated process on a bone.

tuberosity (tu″bĕ-ros′i-te) An elevation or protuberance on a bone.

tunica albuginea (tu′ni-kah al″bu-jin′e-ah) A tough, fibrous tissue surrounding the testis.

tympanic membrane (tim-pan′ik) The membranous eardrum positioned between the external and middle ear.

u

umbilical cord (um-bil′i-kal) A cordlike structure containing the umbilical arteries and vein, which connects the fetus with the placenta.

umbilicus (um-bil′i-kus) The site where the umbilical cord was attached to the fetus; commonly called the *navel*.

unipolar neuron (u″ni-po′lar nu′ron) A nerve cell that has a single nerve fiber extending from its cell body.

universal donor A person with blood type O, who is able to donate blood to people with other blood types in emergency blood transfusions.

universal recipient A person with blood type AB, who can receive blood of any type in emergency transfusions.

upper extremity The appendage attached to the pectoral girdle, consisting of the brachium, forearm, and hand.

urea (u-re′ah) The chief nitrogenous waste product of protein catabolism in the urine, formed in the liver from amino acids.

uremia (u-re′me-ah) The retention of urea and other products of protein catabolism due to inadequate kidney function.

ureter (u-re′ter) A tube that transports urine from the kidney to the urinary bladder.

urethra (u-re′thrah) A tube that transports urine from the urinary bladder to the outside of the body.

urinary bladder (u′ri-ner″e) A distensible sac situated in the pelvic cavity posterior to the symphysis pubis, which stores urine.

urobilinogen (u″ro-bi-lin′o-jen) A compound formed from bilirubin in the intestine; some is excreted in the feces, and some is absorbed and enters the enterohepatic circulation, where it may be excreted either in the bile or in the urine.

urogenital triangle (u″ro-jen′i-tal) The region of the pelvic floor containing the external genitalia.

uterine tube (u′ter-in) A tube that leads from the ovary to the uterus; also called the *oviduct* or *fallopian tube*.

uterus (u″ter-us) A hollow, muscular organ in which a fetus develops. It is located within the female pelvis between the urinary bladder and the rectum; commonly called the *womb*.

utricle (u′tre-k′l) An enlarged portion of the membranous labyrinth, located within the vestibule of the inner ear.

uvula (u′vu-lah) A fleshy, pendulous portion of the soft palate that blocks the nasopharynx during swallowing.

v

vacuole (vak′u-ōl) Small spaces or cavities within the cytoplasm of a cell.

vagina (vah-ji′nah) A tubular organ that leads from the uterus to the vestibule of the female reproductive tract and receives the male penis during intercourse.

vasa vasorum (va′sah va-so′rum) Tiny blood vessels that form a network through the tunica of larger blood vessels to supply them with nutrients.

vasectomy (vah-sek′to-me) Surgical removal of a portion of the vas (ductus) deferens to induce infertility.

vasoconstriction (vas″o-kon-strik′shun) Narrowing of the lumen of blood vessels due to contraction of the smooth muscles in their walls.

vasodilation (vas″o-di-la′shun) Widening of the lumen of blood vessels due to relaxation of the smooth muscles in their walls.

vein A blood vessel that conveys blood toward the heart.

vena cava (ve′nah ka′vah) One of two large vessels that return deoxygenated blood to the right atrium of the heart.

ventilation (ven″ti-la′shun) Breathing; the process of moving air into and out of the lungs.

ventral (ven′tral) Toward the front or belly surface; the opposite of *dorsal*.

ventricle (ven′tri-k′l) A cavity within an organ; especially those in the brain that contain cerebrospinal fluid and those in the heart that contain blood to be pumped from the heart.

venule (ven′ūl) A small vessel that carries venous blood from capillaries to a vein.

vermiform appendix (ver′mi-form ah-pen′diks) A short, blind pouch that attaches to the cecum; commonly called the *appendix*.

vermis (ver′mis) The coiled, middle lobular structure that separates the two cerebellar hemispheres.

vertebral canal (ver′te-bral) The tubelike cavity that extends through the vertebral column and contains the spinal cord; also called the *spinal canal*.

vertigo (ver′ti-go, ver-ti′go) A feeling of movement or loss of equilibrium.

vestibule (ves′ti-būl) A space or cavity at the entrance to a canal, especially that of the nose, inner ear, and vagina.

villus (vil′lus) A minute projection that extends outward into the lumen from the mucosal layer of the small intestine.

virulent (vir′u-lent) Pathogenic, or able to cause disease.

viscera (vis′er-ah) The organs within the abdominal or thoracic cavities.

visceral (vis′er-al) Pertaining to the membranous covering of the viscera.

visceral peritoneum (per″i-to-ne′um) A serous membrane that covers the surfaces of abdominal viscera.

visceral pleura (ploor′ah) A serous membrane that covers the surfaces of the lungs.

visceroceptor (vis″er-o-sep′tor) A sensory receptor that is located within body organs and responds to information concerning the internal environment.

vitreous humor (vit′re-us hu′mor) The transparent gel that occupies the space between the lens and retina of the eyeball.

Volkmann's canal (fōlk′mahnz) Minute ducts through compact bone by which blood vessels and nerves penetrate.

vulva (vul′vah) The external genitalia of the female that surround the opening of the vagina; also called the *pudendum*.

w

white matter Bundles of myelinated axons located in the central nervous system.

wormian bone (wer′me-an) A small bone positioned within a suture of certain cranial bones; also called a *sutural bone*.

y

yellow marrow (mar′o) Specialized lipid storage tissue within bone cavities.

z

zygote (zi′gōt) A fertilized egg cell formed by the union of a sperm and an egg.

zymogens (zi′mo-jens) Inactive enzymes that become active when part of their structure is removed by another enzyme or by some other means.

Credits

Kage/Peter Arnold, Inc.; **Fig. 24.30a:** © Keith Porter; **Fig. 24.30b:** Alpers, D. H. and Seetharam, D. *New England Journal of Medicine*, 296(1977)1047; **Fig. 24.34:** After Sobotta, from Bloom, William and Fawcett, Don W., A TEXTBOOK OF HISTOLOGY, 10th ed., W. B. Saunders Company, 1975; **Fig. 24.36:** Sodeman, W. A. and Watson, T. M. *Pathologic Physiology*, 6th ed., W. B. Saunders Company, 1969; **Fig. 24.37b:** Courtesy of Utah Valley Hospital, Dept. of Radiology; **Fig. 24.38a:** From: TISSUES AND ORGANS: A TEXT ATLAS OF SCANNING ELECTRON MICROSCOPY by R. G. Kessel and R. Kardon. W. H. Freeman and Company © 1979; **Fig. 24.39c:** © Nels H. Granholm; **Fig. 24.45a:** © Carroll H. Weiss, RBP; **Fig. 24.45b:** © Dr. Sheril D. Burton; **Fig. 24.54a–d:** John Van De Graaff, DDS and Larry S. Pierce, DDS both at Northwestern University, School of Dentistry, Chicago.

Chapter 25

Fig. 25.12: © Lester V. Bergmann and Associates; **Fig. 25.15a,b:** Bhasker, S. N., editor. *Orban's Oral Histology and Embryology,* 9th ed., C. V. Mosby Company, 1980; **Fig. 25.22a–d:** *American Journal of Medicine*, 20(1956)133.

Chapter 26

Fig. 26.4: Courtesy of Utah Valley Hospital, Dept. of Radiology; **Fig. 26.7:** © R. B. Wilson, University of Nebraska; **Fig. 26.10:** © Gordon F. Leedale/ Biophoto Associates; **Fig. 26.11a:** Daniel Friend, from Bloom, William and Fawcett, Don W., A TEXTBOOK OF HISTOLOGY, 10th ed., W. B. Saunders Company, 1975; **Fig. 26.30:** © Dr. Kerry L. Openshaw

Chapter 27

Fig. 27.6i,j: © Dr. Landrum Shettles; **Fig. 27.15a:** © Biophoto Associates, Science Source/Photo Researchers; **Fig. 27.17b:** From: TISSUES AND ORGANS: A TEXT ATLAS OF SCANNING ELECTRON MICROSCOPY by R. G. Kessel and R. Kardon. W. H. Freeman and Company © 1979; **Figs. 27.18, 27.19:** © Edwin Reschke; **Figs. 27.21, 27.22:** © Manfred Kage/Peter Arnold, Inc.; **Fig. 27.23:** © Edwin Reschke; **Fig. 27.29:** Dept. of Health, Education and Welfare

Chapter 28

Fig. 28.4: Bloom, William and Fawcett, Don W., A TEXTBOOK OF HISTOLOGY, 10th ed., W. B. Saunders Company, 1975; **Figs. 28.5a,b, 28.6:** R. J. Blandau: A TEXTBOOK OF HISTOLOGY, 10th ed., W. B. Saunders and Company, 1975; **Fig. 28.7:** Hafez, E. S. E. and Blandau, R. J., THE MAMMALIAN OVIDUCT, 1969, University of Chicago Press; **Figs. 28.10, 28.14:** © Edwin Reschke; **Fig. 28.17a,b:** © Dr. Kerry L. Openshaw; **Fig. 28.22:** Bloom, William and Fawcett, Don W., A TEXTBOOK OF HISTOLOGY, 10th ed., W. B. Saunders Company, 1975; **Figs. 28.23, 28.28a:** Courtesy of Utah Valley Hospital, Dept. of Radiology; **Fig. 28.28b:** American Cancer Association

Chapter 29

Fig. 29.1c: "Sea Urchin Sperm-Egg Interactions Studied with the Scanning Electron Microscope," Tegner, M. J., and Epel, D., *Science*, vol. 179, pp. 685–688, 16 February 1973; **Fig. 29.2:** Lucian Zamboni from Greep, Roy and Weiss, Leon, HISTOLOGY, 3d ed., McGraw-Hill Book Company, 1973; **Fig. 29.6a–d:** © R. G. Edwards; **Fig. 29.8:** © Ronan O'Rahilly/Carnegie Laboratories of Embryology, University of California; **Fig. 29.10a:** Carnegie Institute of Washington, Dept. of Embryology, Davis Division; **Fig. 29.13a:** © Donald Yeager/Carroll Weiss; **Figs. 29.18e, 29.20 (top left):** © Landrum Shettles; **Fig. 29.26a,b:** March of Dimes; **Fig. 29.33 (bottom left):** © Landrum Shettles; **Fig. 29.35:** © Gregory DeVore, M.D., and Steven L. Clark, M.D.

Chapter 30

Fig. 30.3a–c: Courtesy of Utah Valley Hospital, Dept. of Radiology; **Fig. 30.6:** From *Science Year, The World Book Science Annual.* © 1973 Field Enterprises Educational Corporation. Reproduced by permission of World Book, Inc.; **Fig. 30.7:** Carolina Biological Supply; **Fig. 30.9:** © John Bird

Index

Prefixes and Suffixes of Anatomical and Medical Terminology

Element	Definition and Example	Element	Definition and Example
a-	absent, deficient, without: *atrophy*	dis-	apart, away from: *disarticulate*
ab-	off, away from: *abduct*	duct-	lead, conduct: *ductus deferens*
abdomin-	relating to the abdomen: *abdominal*	dur-	hard: *dura mater*
-able	capable of: *viable*	dys-	bad, difficult, painful: *dysentery*
ac-	toward, to: *actin*	e-	out, from: *eccrine*
acou-	hear: *acoustic*	ecto-	outside, outer, external: *ectoderm*
ad-	toward, to: *adduct*	-ectomy	surgical removal: *tonsillectomy*
af-	movement toward a central point: *afferent artery*	ede-	swelling: *edema*
		-emia	pertaining to a condition of the blood: *lipemia*
alba	pale or white: *linea alba*		
-alg	pain: *neuralgia*	end-	within: *endoderm*
ambi-	both: *ambidextrous*	entero-	intestine: *enteritis*
angi-	pertaining to the vessels: *angiology*	epi-	upon, after, in addition: *epigenesis*
ante-	before, in front of: *antebrachium*	erythro-	red: *erythrocyte*
anti-	against: *anticoagulant*	ex-	out of: *excise*
aqua-	water: *aqueous*	exo-	outside: *exocrine*
archi-	first: *archeteron*	extra-	outside of, beyond, in addition: *extracellular*
arthr-	joint: *arthritis*		
-asis	condition or state of: *homeostasis*	fasci-	band: *fascia*
aud-	pertaining to the ear: *auditory*	febr-	fever: *febrile*
auto-	self: *autolysis*	-ferent	bear, carry: *efferent*
bi-	two: *bipedal*	fiss-	split: *fissure*
bio-	life: *biopsy*	for-	opening: *foramen*
blast-	generative or germ bud: *osteoblast*	-form	shape: *fusiform*
brachi-	arm: *brachialis*	gastro-	relating to the stomach: *gastrointestinal*
brachy-	short: *brachydont*	-gen	an agent that produces or originates: *pathogen*
brady-	slow: *bradycardia*		
bucc-	cheek: *buccal cavity*	-genic	produced from, producing: *carcinogenic*
cac-	bad, ill: *cachexia*	gloss-	tongue: *glassopharyngeal*
calc-	stone: *calculus*	glyco-	sugar: *glycosuria*
capit-	head: *capitis*	-gram	a record, recording: *electroencephalogram*
carcin-	cancer: *carcinogenic*		
cardi-	heart: *cardiac*	gran-	grain, particle: *agranulocyte*
cata-	lower, under, against: *catabolism*	-graph	instrument for recording: *electrocardiograph*
caud-	tail: *cauda equina*		
cephal-	head: *cephalic*	gravi-	heavy: *gravid*
cerebro-	brain: *cerebrospinal fluid*	gyn-	female sex: *gynecology*
chol-	bile: *cholic*	haplo-	simple or single: *haploid*
chondr-	cartilage: *chondrocyte*	hema(o)-	blood: *hematology*
chrom-	color: *chromocyte*	hemi-	half: *hemiplagia*
-cide	destroy: *germicide*	hepat-	liver: *hepatic portal*
circum-	around: *circumduct*	hetero-	other, different: *heterosexual*
co-	together: *copulation*	histo-	web, tissue: *histology*
coel-	hollow cavity: *coelom*	holo-	whole, entire: *holocrine*
-coele	swelling, an enlarged space or cavity: *blastocoele*	homo-	same, alike: *homologous*
		hydro-	water: *hydrocoel*
con-	with, together: *congenital*	hyper-	beyond, above, excessive: *hypertension*
contra-	against, opposite: *contraception*	hypo-	under, below: *hypoglycemia*
corn-	denoting hardness: *cornified*	-ia	state or condition: *hypoglycemia*
corp-	body: *corpus*	-iatrics	medical specialties: *pediatrics*
crypt-	hidden: *cryptorchidism*	idio-	self, separate, distinct: *idiopathic*
cyan-	blue color: *cyanosis*	ilio-	ilium: *iliosacral*
cysti-	sac or bladder: *cystoscope*	infra-	beneath: *infraspinatus*
cyto-	cell: *cytology*	inter-	among, between: *interosseous*
		intra-	inside, within: *intracellular*
de-	down, from: *descent*	-ion	process: *acromion*
derm-	skin: *dermatology*	-ism	condition or state: *rheumatism*
di-	two: *diarthrotic*	iso-	equal, like: *isotonic*
dipl-	double: *diploid*	-itis	inflammation: *meningitis*